formulas in geometry

rectangle

$Area = lw$

$Perimeter = 2l + 2w$

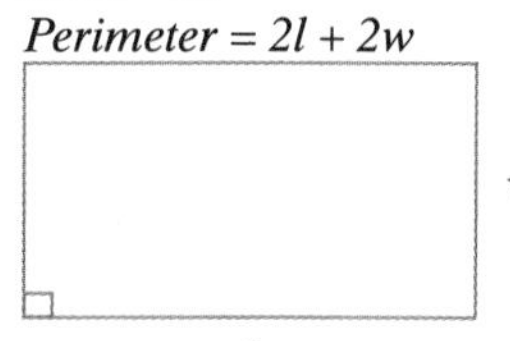

rectangular box

$Volume = lwh$

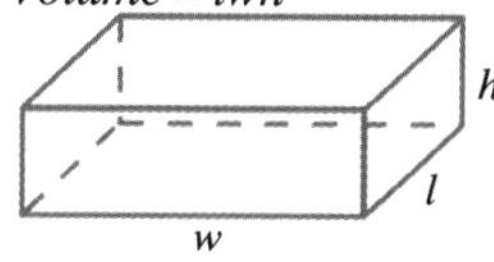

triangle

$Area = \frac{1}{2}bh$

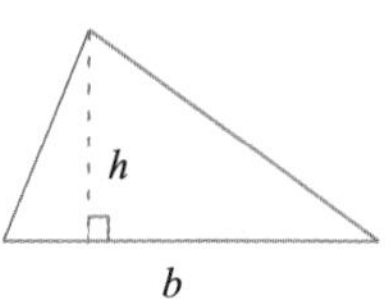

pyramid

$Volume = \frac{1}{3}lwh$

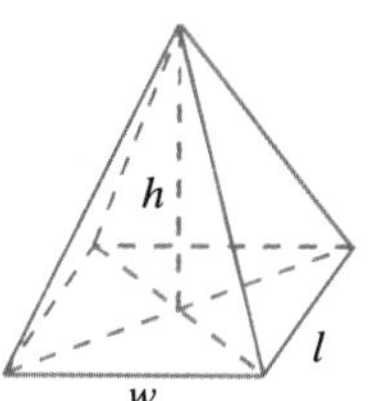

circle

$Area = \pi r^2$

$Circumference = 2\pi r$

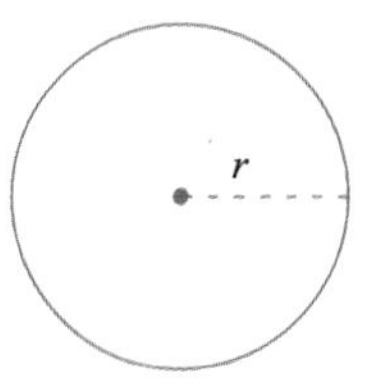

sphere

$Volume = \frac{4}{3}\pi r^3$

$Surface\ Area = 4\pi r^2$

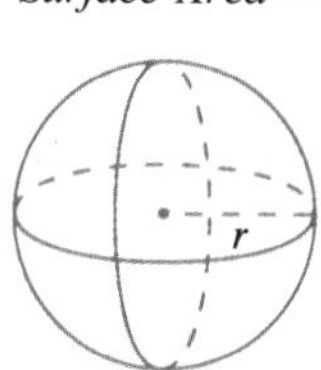

trapezoid

$Area = \frac{1}{2}h(b + c)$

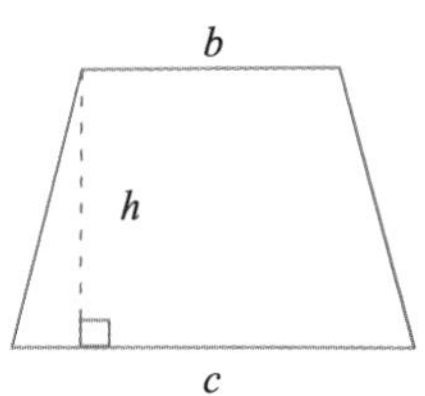

right circular cylinder

$Volume = \pi r^2 h$

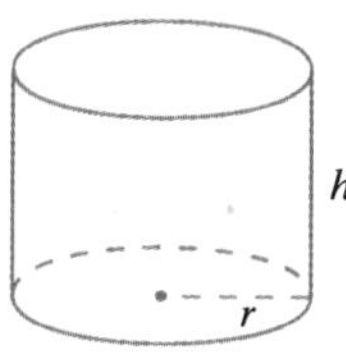

cone

$Volume = \frac{1}{3}\pi r^2 h$

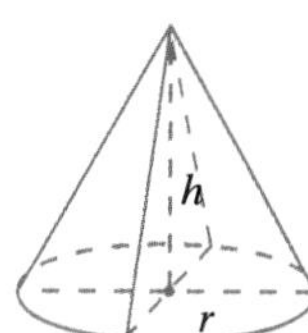

right cylinder

$Volume = (Area\ of\ Base)(Height,\ h)$

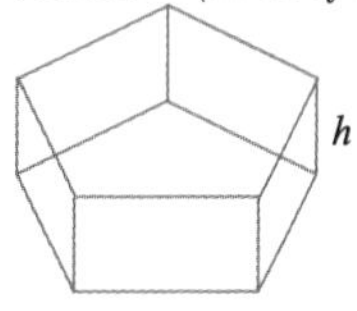

heron's formula

$A = \sqrt{s(s-a)(s-b)(s-c)}$

where $s = \frac{a+b+c}{2}$

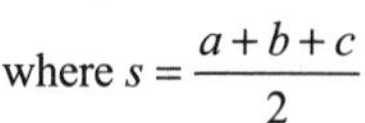

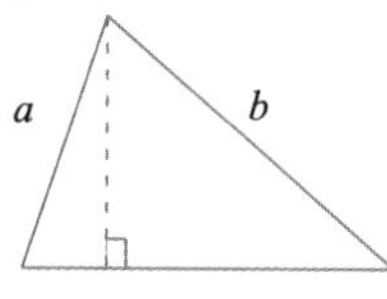

properties of exponents and radicals

$a^m \cdot a^n = a^{m+n}$ $\qquad (a^n)^m = a^{nm}$

$\frac{a^n}{a^m} = a^{n-m}$ $\qquad (ab)^n = a^n b^n$

$a^{-n} = \frac{1}{a^n}$ $\qquad \left(\frac{a}{b}\right)^n = \frac{a^n}{b^n}$

$(a)^{1/n} = \sqrt[n]{a}$ $\qquad (a)^{m/n} = \sqrt[n]{a^m} = \left(\sqrt[n]{a}\right)^m$

$\sqrt[n]{ab} = \sqrt[n]{a} \cdot \sqrt[n]{b}$ $\qquad \sqrt[n]{\frac{a}{b}} = \frac{\sqrt[n]{a}}{\sqrt[n]{b}}$

$\sqrt[m]{\sqrt[n]{a}} = \sqrt[n]{\sqrt[m]{a}} = \sqrt[mn]{a}$

special product formulas

$A^2 - B^2 = (A-B)(A+B)$

$A^3 - B^3 = (A-B)(A^2 + AB + B^2)$

$A^3 + B^3 = (A+B)(A^2 - AB + B^2)$

$A^2 + 2AB + B^2 = (A+B)^2$

$A^2 - 2AB + B^2 = (A-B)^2$

the quadratic formula

The solutions of the equation $ax^2 + bx + c = 0$ are:

$$x = \frac{-b \pm \sqrt{b^2 - 4ac}}{2a}$$

the pythagorean theorem

Given a right triangle with legs a and b and hypotenuse c:

$$a^2 + b^2 = c^2$$

distance formula

$$d = \sqrt{(x_2 - x_1)^2 + (y_2 - y_1)^2}$$

midpoint formula

$$\left(\frac{x_1 + x_2}{2}, \frac{y_1 + y_2}{2}\right)$$

slope of a line

Given two points (x_1, y_1) and (x_2, y_2) on a line, then the slope of the line is the ratio:

$$\frac{y_2 - y_1}{x_2 - x_1}$$

Horizontal lines have slope 0.
Vertical lines have an undefined slope.
Given a line with slope m:

slope of parallel line $= m$.

slope of perpendicular line $= -\frac{1}{m}$.

compound interest

An investment of P dollars at an annual interest rate of r, compounded n times per year for t years has an accumulated value of

$$A(t) = P\left(1+\frac{r}{n}\right)^{nt}$$

An investment compounded continuously has an accumulated value of $A(t) = Pe^{rt}$.

properties of logarithms

For $a > 0$, a is not equal to 1, $x, y > 0$ and r is a real number:

$\log_a(x) = y$ and $x = a^y$ are equivalent

$\log_a(1) = 0$

$\log_a(a) = 1$

$\log_a(a^x) = x$

$a^{\log_a(x)} = x$

$\log_a(xy) = \log_a(x) + \log_a(y)$

$\log_a\left(\frac{x}{y}\right) = \log_a(x) - \log_a(y)$

$\log_a(x^r) = r\log_a x$

change of base formula

$a, b, x > 0;\ a, b \neq 1;$

$$\log_b(x) = \frac{\log_a(x)}{\log_a(b)}$$

summation formulas

$$\sum_{i=1}^{n} 1 = n \qquad \sum_{i=1}^{n} i = \frac{n(n+1)}{2}$$

$$\sum_{i=1}^{n} i^2 = \frac{n(n+1)(2n+1)}{6} \qquad \sum_{i=1}^{n} i^3 = \frac{n^2(n+1)^2}{4}$$

sequences and series

arithmetic

General term: (where d is the common difference)

$$a_n = a_1 + (n-1)d$$

Partial sum: $S_n = na_1 + d\left(\frac{(n-1)n}{2}\right) = \left(\frac{n}{2}\right)(a_1 + a_n)$

geometric

General term: $a_n = a_1 r^{n-1}$ $\left(r = \frac{a_{n+1}}{a_n}\right)$

Partial sum: $S_n = \frac{a_1(1-r^n)}{1-r}$

Infinite sum: $S = \sum_{n=0}^{\infty} a_1 r^n = \frac{a_1}{1-r}$

permutation formula

$${}_nP_k = \frac{n!}{(n-k)!}$$

combination formula

$${}_nC_k = \binom{n}{k} = \frac{n!}{k!(n-k)!}$$

binomial theorem

$$(A+B)^n = \sum_{k=0}^{n} \binom{n}{k} A^{n-k} B^k$$

multinomial coefficients

$$\binom{n}{k_1, k_2, \ldots, k_r} = \frac{n!}{k_1!\,k_2!\ldots k_r!}$$

multinomial theorem

$$(A_1 + A_2 + \ldots + A_r)^n = \sum_{k_1+k_2+\ldots+k_r=n} \binom{n}{k_1, k_2, \ldots, k_r} A_1^{k_1} A_2^{k_2} \ldots A_r^{k_r}$$

classifying conics

Assuming the graph of the equation $Ax^2 + Bxy + Cy^2 + Dx + Ey + F = 0$ is a non-degenerate conic section, it is classified by its discriminant as follows:

1. **Ellipse** if $B^2 - 4AC < 0$
2. **Parabola** if $B^2 - 4AC = 0$
3. **Hyperbola** if $B^2 - 4AC > 0$

complex numbers & demoivre's theorem

$z = a + bi$ $\qquad$ $|z| = \sqrt{a^2 + b^2}$

$\tan\theta = \frac{b}{a}$ $\qquad$ $z = |z|(\cos\theta + i\sin\theta)$

$z^n = |z|^n(\cos n\theta + i\sin n\theta)$ $\qquad$ $z^n = |z|^n e^{in\theta}$

$$w_k = |z|^{\frac{1}{n}}\left[\cos\left(\frac{\theta + 2k\pi}{n}\right) + i\sin\left(\frac{\theta + 2k\pi}{n}\right)\right]$$

$$w_k = |z|^{\frac{1}{n}} e^{i\left(\frac{\theta + 2k\pi}{n}\right)}$$

where $k = 0, 1, \ldots, n-1$

PAUL SISSON

PRECALCULUS

Editor: Mandy Glover
Assistant Editor: Kimberly Scott
Development Director: Marcel Prevuznak
Contributors: Greg Hill, Nina Miller
Production Editors: Priyanka Bihani, Jennifer Butler, Mary Janelle Cady
Editorial Assistants: Kelly Epperson, Cynthia Fort, K.V. Jagannadham, Ashley Rankin
Layout: QSI (Pvt.) Ltd.: U. Nagesh, E. Jeevan Kumar
Art: Ayvin Samonte
Cover Art and Design: Johnson Design

A division of Quant Systems, Inc.

Library of Congress Control Number: 2004114372

Printed in the United States of America

ISBN (Student): 0-918091-89-6
Student Solutions Manual: 0-918091-99-3

contents

chapter three 187 Relations, Functions, and Their Graphs

chapter four 297 Polynomial Functions

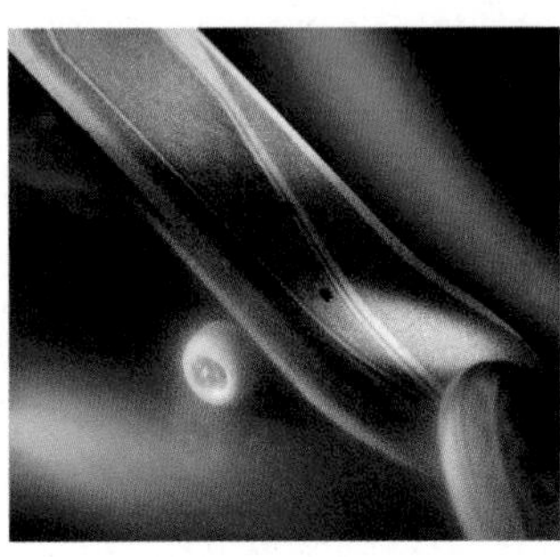

chapter five 375 Exponential and Logarithmic Functions

chapter six 453 Trigonometric Functions

chapter seven 543 Trigonometric Identities and Equations

chapter eight 597 Additional Topics in Trigonometry

preface

Introduction

Why do students study precalculus? Is it because of an innate love of math? For those with such a love, for those who have glorious visions of a world beyond algebra, precalculus unlocks the door to calculus and higher mathematics. For other students, those for whom calculus is a tool and not necessarily a thing of beauty, precalculus provides a solid foundation for successfully tackling the subject.

Whether you dream of integrals and derivatives or perceive yourself to be merely laying the groundwork for success in later math, this book prepares you for the next step. Its primary purpose is to provide you with the skills and concepts necessary to achieve a mastery of calculus. If you are coming from a course with a title such as "College Algebra" or "Algebra", you will encounter many familiar topics in this book. They are presented, however, more thoroughly and at a level intended for students who will soon be going on to calculus. You will also discover new concepts that are not usually included in College Algebra courses, such as the elements of trigonometry.

The book begins with two chapters that will ease the transition from your previous courses. While you may find a few topics that are unfamiliar, most of the material is included to help refresh your memory and for you to have as a reference throughout the course. The rest of the book is devoted to covering concepts from algebra in greater depth and introducing new topics which will prepare you for calculus. Whether calculus is near the beginning or end of your math experience, best wishes to you on this step in your journey.

Features

Topics:

Each section begins with a list of topics. These concise objectives are a helpful guide for both reference and class preparation.

section 1.1 Real Numbers and Algebraic Expressions

1.1 Real Numbers and Algebraic Expressions

TOPICS

1. Common subsets of real numbers
2. The real number line
3. Order on the real number line
4. Set-builder notation and interval notation
5. Basic set operations and Venn diagrams
6. Absolute value and distance
7. Components of algebraic expressions
8. The field properties and their use in algebra

Topic 1: Common Subsets of Real Numbers

Certain types of numbers occur so frequently in mathematics that they have been given special names and symbols. These names will be used throughout this book and in later ...lasses when referring to members of the following sets:

...mbers

Natural (or Counting) Numbers: This is the set of numbers $\mathbb{N} = \{1, 2, 3, 4, 5, \ldots\}$.

Whole Numbers: This is the set of natural numbers with 0 added: ...2, 3, 4, 5, ...}.

Integers: This is the set of natural numbers, their negatives, and 0. As a list, this ...set $\mathbb{Z} = \{\ldots, -4, -3, -2, -1, 0, 1, 2, 3, 4, \ldots\}$.

Rational Numbers: This is the set, with symbol $\mathbb{Q}$ for quotient, of *ratios* of ...ers. That is, any rational number can be written in the form $\frac{p}{q}$, where p and q ...oth integers and $q \neq 0$. When written in decimal form, rational numbers either ...inate or repeat a pattern of digits past some point.

Irrational Numbers: Every real number that is not rational is, by definition, ...onal. In decimal form, irrational numbers are non-terminating and non-repeating.

Real Numbers: Every set above is a subset of the set of real numbers, which is ...ted $\mathbb{R}$. Every real number is either rational or irrational, and no real number ...th.

3

chapter ONE

NUMBER SYSTEMS AND EQUATIONS AND INEQUALITIES OF ONE VARIABLE

OVERVIEW

1.1 Real Numbers and Algebraic Expressions
1.2 Properties of Exponents and Radicals
1.3 Polynomials and Factoring
1.4 The Complex Number System
1.5 Linear Equations in One Variable
1.6 Linear Inequalities in One Variable
1.7 Quadratic Equations
1.8 Rational and Radical Equations

By the end of this chapter you should be able to work with and manipulate expressions involving exponents and radicals. You will be able to restate radical equations in terms of variables which appear inaccessible – or "trapped". When solving this type of equation be sure to check your potential solutions in the original expression to see if they are true solutions. On page 103, you will find a problem on determining rate of work. You will master this type of problem, using tools such as those learned in *Solving Rational Equations.*

1

Chapter Openers:

Each chapter begins with a list of sections and an engaging preview of an application appearing later in the chapter.

chapter one Number Systems and Fundamental Concepts of Algebra

Definitions and Theorems:

All of the primary definitions and theorems are clearly identified and set off in yellow boxes that are highly visible and easily located. All key terms appear in bold print when first formally defined, and other useful terms appear in italic font when informally defined.

Sets that consist of all real numbers bounded by two endpoints, possibly including those endpoints, are called **intervals**. Intervals can also consist of a portion of the real line extending indefinitely (in either direction) from just one endpoint. We could describe such sets with set-builder notation, but intervals occur frequently enough that special notation has been devised to define them succinctly.

Interval Notation

Notation	Meaning	Graphical Representation
(a, b)	$\{x \mid a < x < b\}$, or all real numbers strictly between a and b.	a b
$[a, b]$	$\{x \mid a \le x \le b\}$, or all real numbers between a and b, including both a and b.	a b
$(a, b]$	$\{x \mid a < x \le b\}$, or all real numbers between a and b, including b but not a.	a b
$(-\infty, b)$	$\{x \mid x < b\}$, or all real numbers less than b.	b
$[a, \infty)$	$\{x \mid x \ge a\}$, or all real numbers greater than or equal to a.	a
$(-\infty, \infty)$	$\mathbb{R}$ set of all real numbers	

In interval notation, the left endpoint is always written first. Intervals of the form (a, b) are called **open** intervals, while those of the form $[a, b]$ are **closed** intervals. The interval $(a, b]$ is **half-open** (or **half-closed**). Of course, a half-open interval may be open at either endpoint, as long as it is closed at the other. The symbols $-\infty$ and ∞ indicate that the interval extends indefinitely in, respectively, the left and the right directions. Note that $(-\infty, b)$ excludes the endpoint b, while $[a, \infty)$ includes the endpoint a.

Cautions:

Common pitfalls are highlighted, and examples of common errors and their corrections abound.

caution!

The symbols $-\infty$ and ∞ are just that: symbols! They are not real numbers, and so they cannot, for instance, be solutions to a given equation. The fact that they are symbols, and not numbers, also means that as endpoints of intervals they can never be included. For this reason, a parenthesis always appears next to either $-\infty$ or ∞; a bracket should never appear next to either infinity symbol.

example 3

If $A = \{-9, 2, 3, 5, x, y\}$, $B = \{-9, 3, 4, 7, z\}$, $C = \{-8, 1, 6\}$

Find: **a.** $A \cup B$ **b.** $A \cap B$ **c.** $B \cap C$

Solutions:

a. $A \cup B = \{-9, 2, 3, 4, 5, x, y, z\}$ Since this is a union of two sets, the new set will consist of all the values from the individual

b. $A \cap B = \{-9, 3\}$

c. $B \cap C = \varnothing$

Numerous Examples:

Each section contains many examples that illustrate the concepts presented and the skills to be mastered. The examples are clearly set off from the accompanying text, and the exercises in each section refer the student to the relevant examples to study.

Historical Contexts:

Each chapter begins with a brief introduction setting the historical context of the math that follows. Mathematics is a human endeavor, and knowledge of how and why a particular idea developed is of great help in understanding it. Too often, math is presented in cold, abstract chunks completely divorced from the rest of reality. While a (very) few students may be able to master material this way, most benefit from an explanation of how math ties into the rest of what people were doing at the time it was created.

chapter one Number Systems and Fundamental Concepts of Algebra

Introduction

In this chapter, we review the terminology of the real number system, the notation and properties frequently encountered in algebra, the extension of the real number system to the larger set of complex numbers, and the basic algebraic methods we use to solve equations and inequalities.

We begin with a discussion of common subsets of the set of real numbers. Certain types of numbers are important from both a historical and a mathematical perspective. There is archeological evidence that people used the simplest sort of numbers, the *counting* or *natural numbers*, as far back as 50,000 years ago. Over time, many cultures discovered needs for various refinements to the number system, resulting in the development of such classes of numbers as the *integers*, the *rational numbers*, the *irrational numbers*, and ultimately, the *complex numbers*, a number system which contains the real numbers.

Many of the ideas in this chapter have a history dating as far back as Egyptian and Babylonian civilizations of around 3000 BC, with later developments and additions due to Greek, Hindu, and Arabic mathematicians. Much of the material was also developed independently by Chinese mathematicians. It is a tribute to the necessity, utility, and objectivity of mathematics that so many civilizations adopted so much mathematics from abroad, and that different cultures operating independently developed identical mathematical concepts so frequently.

Pythagoras

As an example of the historical development of just one concept, consider the notion of an irrational number. The very idea that a real number could be irrational (which simply means not rational, or not the ratio of two integers) is fairly sophisticated, and it took some time for mathematicians to come to this realization. The Pythagoreans, members of a school founded by the Greek philosopher Pythagoras in southern Italy around 540 BC, discovered that the square root of 2 was such a number, and there is evidence that for a long time $\sqrt{2}$ was the only known irrational number. A member
us of Cyrene, later showed (c. 425 BC) that
$\sqrt{13}$, $\sqrt{14}$, $\sqrt{15}$ and $\sqrt{17}$ also are irrational. It
thematician, Johann Lambert, showed that the
modern rigorous mathematical description of
ichard Dedekind in 1872.

1, keep the larger picture firmly in mind. All of
as developed over long periods of time by many
g problems important to them.

chapter one Number Systems and Fundamental Concepts of Algebra

technology note

Computer algebra systems, as well as some calculators, are programmed to apply the properties of exponents and can demonstrate the result of simplifying expressions. In *Mathematica*, the command to do this is **Simplify**, illustrated with the first expression below. In many cases, *Mathematica* will simplify a given expression automatically upon evaluation, making the **Simplify** command unnecessary. This is illustrated with the second and third expressions below. *(See Appendix A for guidance on using computer algebra systems like Mathematica.)*

Section 1-3 Technology Note.nb

```
In[1]:= Simplify[x^3*(-2y)^5/(8*x^7*y^3)]
Out[1]= -4y^2/x^4
In[2]:= (3x^5*y^2)^0
Out[2]= 1
In[3]:= 1/(-1/z^(-5))
Out[3]= -1/z^5
```

Technology Notes and Exercises:

Where appropriate, the use of technology (in the form of graphing calculators or computer software) is illustrated. Such notes are then followed by specially identified problems in the Exercises for which technology is helpful. *See the Technology Index on page 1066 for a complete listing of these notes and exercises.*

chapter one project

Polynomials

A chemistry professor calculates final grades for her class using the polynomial

$$A = 0.3f + 0.15h + 0.4t + .15p,$$

where A is the final grade, f is the final exam, h is the homework average, t is the chapter test average, and p is the semester project.

The following is a table containing the grades for various students in the class:

Name	Final Exam	Homework Avg.	Test Avg.	Project
Alex	77	95	79	85
Ashley	91	95	88	90
Darren	82	85	81	75
Elizabeth	75	100	84	80
Gabe	94	90	90	85
Lynn	88	85	80	75

1. Find the course average for each student, rounded to the nearest tenth.
2. Who has the highest total score?
3. Why is the total grade raised more with a grade of 100 on the final exam than with a grade of 100 on the semester project?
4. Assume you are a student in this class. With 1 week until the final exam, you have a homework average of 85, a test average of 85, and a 95 on the semester project. What score must you make on the final exam to achieve at least a 90.0 overall? (Round

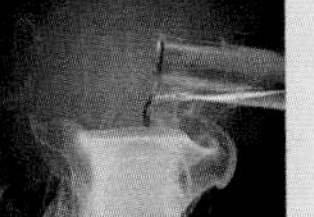

Chapter Projects:

Each project describes a plausible scenario and related constraints, and is suitable for individual or group assignments.

example 5

The manufacturer of a certain brand of 16-foot ladder recommends that, when in use, the angle between the ground and the ladder should equal 75°. What distance should the foot of the ladder be from the base of the wall it is leaning against?

Solution:

Since we are given information about an angle, its adjacent side, and the hypotenuse of a right triangle, cosine is the logical trigonometric function to use in solving this problem (equivalently, secant could be used, but calculators are equipped with a "cos" and not a "sec" button so our current technology tends to lead to the use of cosine).

We want to determine the length of the adjacent side when the ladder is resting against the wall with its recommended angle of 75°, and we note that

$$\cos 75^\circ = \frac{\text{adj}}{16}.$$

This gives us $\text{adj} = 16\cos 75^\circ \approx 4.14$ feet, or a bit less than 4 feet, 2 inches.

In many surveying problems, it is frequently necessary to determine the height of some distant object when it is impossible or impractical to measure how far away the object is. One way to determine the height anyway begins with the diagram in Figure 2. In the diagram, assume that distance d and angles α and β can be measured, but that distance x is unknown. How can we determine height h?

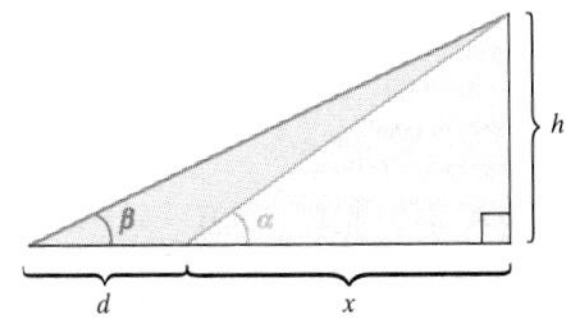

Figure 2: Determining h from Two Angles

Graphing Calculators:

Throughout the text, graphing calculators (with an emphasis on the TI-83 Plus) are integrated into the text to demonstrate usage, with helpful operating hints where appropriate.

Applications:

Many exercises and examples illustrate practical applications, keeping students engaged. *See the Application Index on page 1062 for a complete list of application problems and examples.*

section 1.2 Properties of Exponents and Radicals

Topic 4: Interlude: Working with Geometric Formulas

Exponents occur in a very natural way when geometric formulas are considered. Some of the problems that you encounter in this book (and elsewhere) will require nothing more than using one of the basic geometric formulas found on the inside front cover of the text, but others will require a bit more work. Often, the exact geometric formula that you need to solve a given problem can be derived from simpler formulas.

We will look at several examples of how a new geometric formula is built up from known formulas. The general rule of thumb in each case is to break down the task at hand into smaller pieces that can be easily handled.

example 6

Find formulas for each of the following:

a. The surface area of a box.
b. The surface area of a soup can.
c. The volume of a birdbath in the shape of half of a sphere.
d. The volume of a gold ingot whose shape is a right trapezoidal cylinder.

Solutions:

a. A box whose six faces are all rectangular is characterized by its length l, its width w, and its height h. The area of a rectangle is one of the basic formulas that you should be familiar with, and can be found on the inside front cover of th text. The formula for the surface area of a bo then, is just the sum of the areas of the six sides. we let S stand for the total surface area, we obtai the formula $S = lw + lw + lh + lh + hw + hw$, $S = 2lw + 2lh + 2hw$.

h

w

b. A soup can, an example of a right circular cylinde is characterized by its height h and the radius r the circle that makes up the base (or the top).

Mathematical Interludes:

Many sections end with an Interlude which ties together some of the ideas just presented, extends the math to a higher level, or showcases some concept in an important application. Depending on the class in which this book is used, Interludes may be integral to the course or may serve as nothing more than extracurricular reading for interested students.

section 1.6 Linear Inequalities in One Variable

exercises

Solve the following linear inequalities. Describe the solution set using interval notation and by graphing. See Example 1.

1. $4+3t \le t-2$
2. $-\frac{v+2}{3} > \frac{5-v}{2}$
3. $4.2x-5.6 < 1.6+x$
4. $8.5y-3.5 \ge 2.5(3-y)$
5. $-2(3-x) < -2x$
6. $\frac{1-x}{5} > \frac{-x}{10}$
7. $4w+7 \le -7w+4$
8. $-5(p-3) > 19.8-p$
9. $\frac{6f-2}{5} < \frac{5f-3}{4}$
10. $\frac{u-6}{7} \ge \frac{2u-1}{3}$
11. $.04n+1.7 < 0.13n-1.45$
12. $2k+\frac{3}{2} < 5k-\frac{7}{3}$

Solve the following compound inequalities. Describe the solution set using interval notation and by graphing. See Example 2.

13. $-4 < 3x-7 \le 8$
14. $-10 < -2(4+y) \le 9$
15. $-8 \le \frac{z}{2}-4 < -5$
16. $3 < \frac{w+3}{8} \le 9$
17. $5 \le 2m-3 \le 13$
18. $4 \le \frac{p+7}{-2} < 9$
19. $\frac{1}{3} < \frac{7}{6}(l-3) < \frac{2}{3}$
20. $\frac{1}{4} \le \frac{g}{2}-3 < 5$
21. $0.08 < 0.03c+0.13 \le 0.16$

Solve the following absolute value inequalities. Describe the solution set using interval notation and by graphing. See Example 3.

22. $4+|3-2y| \le 6$
23. $4+|3-2y| > 6$
24. $|x-2| \ge 5$
25. $2|z+5| < 12$
26. $7-\left|\frac{q}{2}+3\right| \ge 12$
27. $|4-2x| > 11$
28. $5.5+|x-7.2| \le 3.5$
29. $6-5|x+2| \ge -4$
30. $|2x-1| < x+4$
31. $-3|4-t| < -6$
32. $-3|4-t| > -6$
33. $3|4-t| < -6$
34. $|3t+4| > -8$
35. $2 < |6w-2|+7$

79

Exercises:

Each section concludes with a selection of exercises designed to allow the student to practice skills and master concepts. References to appropriate chapter examples are clearly labeled for those who desire assistance. Many levels of difficulty exist within each exercise set, allowing teachers to adapt the exercises as necessary and offering students the opportunity to practice elementary skills or stretch themselves, as appropriate.

chapter one Number Systems and Equations and Inequalities of One Variable

CHAPTER REVIEW

A summary of concepts and skills follows each chapter. Refer to these summaries to make sure you feel comfortable with the material in the chapter. The concepts and skills are organized according to the Section title and Topic title in which the material is first discussed.

1.1 Real Numbers and Algebraic Expressions

topics	pages	test exercises
Common subsets of real numbers • The sets ℕ, ℤ, ℚ, and ℝ, as well as *whole* numbers and *irrational* numbers • Identifying numbers as elements of one or more of the common sets	p. 3 – 4	1 – 2
The real number line • Plotting numbers on the real number line • The *origin* of the real number line, and its r… and positive numbers	p. 4 – 5	3 – 4
Order on the real number line • The four inequality symbols <, ≤, >, and ≥ • Distinguishing between strict and non-strict		
Set-builder notation and interval notation • Using set-builder notation to define sets • The empty set and its two common symbols		
Basic Set Operations and Venn Diagrams • The meaning and use of Venn diagrams • The definition of the set operations *union* a… their application to intervals • Interval notation and its relation to inequal…		
Absolute value and distance • The definition of absolute value on the real… $\lvert a\rvert = \begin{cases} a & \text{if } a \ge 0 \\ -a & \text{if } a < 0 \end{cases}$ • The relationship between absolute value an… • Distance between two real numbers		

114

Chapter Review:

Each chapter ends with a concise summary of the concepts and skills to be mastered, arranged by section and topic.

chapter one Number Systems and Fundamental Concepts of Algebra

chapter test

Section 1.1

Which elements of the following sets are ***(a)*** *natural numbers,* ***(b)*** *whole numbers,* ***(c)*** *integers,* ***(d)*** *rational numbers,* ***(e)*** *irrational numbers,* ***(f)*** *real numbers,* ***(g)*** *undefined.*

1. $\left\{-15.75,\ 7, \frac{-8}{0}, \frac{0}{5}, 3^3, -1, -\sqrt{4}\right\}$ **2.** $\left\{2\sqrt{3},\ 5\pi,\ \sqrt{1},\ 7.\overline{6},\ -1,\ \frac{2}{9},\ \lvert -21\rvert\right\}$

Plot the following real numbers on a number line. Choose the unit length appropriately for each set.

3. $\{-3, 0, 3, 5\}$ **4.** $\{-4.5, -4.2, -4.0\}$

Select all of the symbols from the set $\{<, \le, >, \ge\}$ *that can be placed in the blank to make true statements.*

5. 6.1 ____ 8.3 **6.** −12 ____ −11

Describe each of the following sets using set-builder notation. There may be more than one correct way to do this.

7. $\{-2, -1, 0, 1, 2, 3\}$ **8.** $\{0, 2, 4, 6, \ldots\}$

Simplify the following unions and intersections of intervals.

9. $[4, 8] \cup (8, 11]$ **10.** $[4, 8] \cap (8, 11]$ **11.** $(-2, 3] \cap [0, 3)$
12. $(-\infty, 4] \cap (2, \infty)$

Write each set as an interval using interval notation.

13. $\{x \mid -7 < x \le 9\}$ **14.** $\{x \mid x \ge 14\}$

Write the following rational numbers as ratios of integers.

15. $7.\overline{6}$ **16.** $-2.0\overline{42}$

Evaluate the absolute value expressions.

17. $-\lvert 11-2\rvert$ **18.** $-\lvert -4-3\rvert$ **19.** $\lvert \sqrt{5}-\sqrt{11}\rvert$

122

Chapter Test:

Immediately following each Chapter Review is a Chapter Test that presents several more problems pertaining to the major ideas of the chapter.

A Note on Technology

The purpose of this book is to help the reader acquire a solid understanding of precalculus. To the extent that technology (in whatever form) can aid in pursuit of that goal, it should be enthusiastically embraced by both student and teacher. Aside from masochism, there is no excuse for not using a calculator or computer program when faced with tedious, pedagogically unenlightening tasks. However, technology can be terribly abused as well, and the risks of relying too heavily on technological aids when attempting to understand a concept cannot be stressed too much. There is great temptation, on the part of students and teachers alike, to let calculators and computers turn every problem into a mindless "plug and chug" operation.

Experienced educators have long realized the dangers of teaching students rote algorithmic procedures for solving problems; chief among them is the fact that if a problem is not presented in "just the right way" a student is unable to solve it. A textbook that takes a technology-based cookbook approach to mathematics has exactly the same problem, and far too many students suffer later in life because they were taught how to use a calculator, but not mathematics.

The above should by no means be taken as a dismissal of technology, however. Once the mathematics has been learned, calculators and (especially) computers are essential everyday tools for many people. The important thing is to make the distinction between learning mathematics and using mathematics. In general, technology is much more useful in using mathematics than in learning it.

I have tried to take great care in this book to note the exceptions. For example, the software package *Mathematica®* is featured in many of the Technology Notes and Exercises, and an Appendix on its elementary use is included. For many of the Technology Exercises a graphing calculator can also be substituted. An additional and highly useful example of how technology can be used as an aid to learning mathematics is seen in the multimedia courseware that has been designed as a companion to this text. See the section entitled *Hawkes Learning Systems: Precalculus* for more details on installing and using this tutorial and testing package.

Supplements

The **Instructor's Annotated Edition** of this text contains, in the margins, solutions to all of the exercises.

Hawkes Learning Systems: Precalculus is a comprehensive tutoring and testing software package that has been specifically designed to complement this book. While the book stands alone and can be used independently of the software, the computer component provides an effective second approach to learning and to developing problem-solving skills.

The **Students' Solution Manual** contains the complete "worked out" solutions to all the odd numbered section exercises and all the Chapter Test exercises.

To the Student: Learning Mathematics

There is a saying among math teachers that you might have heard: "Math is not a spectator sport." While trite, it is undeniably true (and while trite, at least it's concise). Mathematics is not something you can learn by watching someone else do it. You have probably already had the experience of watching a teacher solve a problem and marveling at how easy it seems, only to discover that a nearly identical problem is much harder to solve at home or (far worse) on a test.

The lesson to be learned is that you have to practice mathematics in order to learn mathematics. Make sure you do the homework problems your teacher assigns you, not only because that's how mathematics is learned but also because it gives you insight into the types of problems your teacher thinks are important.

Beyond that, there are some other key ideas to keep in mind. One is that very few people will fully grasp a mathematical concept or master a skill the first time they are presented with it. If you find yourself lost after the first reading of a section of this book (or any math book), don't despair. Just remember what is puzzling you, take a break, and try it again when you're fresh. Most math is learned in a cyclic process of plowing ahead until lost, backing up and re-reading, and then plowing ahead a bit further. A math book is not like a novel—you shouldn't expect to be able to read it quickly in a purely linear fashion.

A third point related to the first two is that math must be read with a pencil (or, the writing instrument of your choice). Mathematical reading is slow, and must be broken up with your own writing as you make sure you understand the worked examples and maybe do a few homework problems. Take your time as you read, and work carefully and deliberately.

Finally, make sure you take advantage of your teacher and peers. Go to class and pay attention to what your teacher emphasizes and learn from his or her unique insight into the material. Work with friends when possible, and ask others for help if they understand something you haven't gotten yet. When you have the opportunity to explain math to someone else, take advantage of that, too. Teaching mathematics to others is an amazingly effective way to improve your own understanding.

Acknowledgements

I am very grateful to all the people at Hawkes Learning Systems for their support, their dedication, and their indefatigable cheerfulness. In particular, many thanks to Marcel Prevuznak (development director), Mandy Glover (editor), Kim Scott (assistant editor), Emily Omlor (marketing), and James Hawkes.

I am also very grateful to the following for their many insightful comments and reviews:

Russ Baker, *Howard Community College*
Brenda Cates, *Mount Olive College*
Deanna Caveny, *College of Charleston*
Brenda Chapman, *Trident Technical College*
Michelle DeDeo, *University of North Florida*
Hamidullah Farhat, *Hampton University*
Terry Fung, *Kean University*
Mark Goldstein, *West Virginia Northern Community College-New Martinsville*
Bill Lepowsky, *Laney College*
Harriette Roadman, *Community College of Allegheny County*
Joan Sallenger, *Midlands Technical College-Beltline*
Elizabeth Schubert, *Saddleback College*
Chris Schroeder, *Morehead State University*
Barbara Sehr, *Indiana University-Kokomo*
Vance Waggener, *Trident Technical College*
Elizabeth White, *Trident Technical College*

Thanks also to Drs. Richard Mabry, Carlos Spaht, and Al Vekovius of LSU-Shreveport, and to their precalculus students, for giving the text a trial run in the Fall 2003 semester. My thanks also to Drs. Larry Anderson and Mary Ellen Foley of LSUS for their reading of Chapter 1.

I would also like to thank John Scofield, Charlie Hunter, Miles Davis, Herbie Hancock, the Dave Matthews Band, Groove Collective, Carlos Santana, and Medeski Martin & Wood, among others, for their unwitting but much-appreciated assistance.

Finally, my very great thanks to my wife Cindy for her support and understanding at all times.

Hawkes Learning Systems: Precalculus

Overview:

This multimedia courseware allows students to become better problem-solvers by creating a mastery level of learning in the classroom. The software includes an "Instruct," "Practice," "Tutor," and "Certify" mode in each lesson, allowing students to learn through step-by-step interactions with the software. This automated homework system's tutorial and assessment modes extend instructional influence beyond the classroom. Artificial intelligence is what makes the tutorials so unique. By offering intelligent tutoring and mastery level testing to measure what has been learned, the software extends the instructor's ability to influence students to solve problems. This courseware can be ordered either separately or bundled together with this text.

Minimum Requirements:

In order to run ***HLS: Precalculus***, you will need:

Intel® Pentium® 1 GHz or faster processor
Windows® 2000 or later
256 MB RAM
200 MB hard drive space
256 color display (1024x768, 16-bit color recommended)
Internet Explorer 6.0 or later
CD-ROM drive

Getting Started:

Before you can run ***HLS: Precalculus***, you will need an access code. This 30 character code is <u>your</u> personal access code. To obtain an access code, go to http://www.hawkeslearning.com and follow the links to the access code request page (unless directed otherwise by your instructor).

Installation:

Insert the ***HLS: Precalculus*** installation CD-ROM into the CD-ROM drive. Select the Start/Run command, type in the CD-ROM drive letter followed by \setup.exe. (For example, d:\setup.exe where d is the CD-ROM drive letter.)

The complete installation will use over 140 MB of hard drive space and will install the entire product, except the multimedia files, on your hard drive.

After selecting the desired installation option, follow the on-screen instructions to complete your installation of ***HLS: Precalculus***.

Starting the Courseware:

After you have installed ***HLS: Precalculus*** on your computer, to run the courseware select Start/Programs/Hawkes Learning Systems/Precalculus.

You will be prompted to enter your access code with a message box similar to the following:

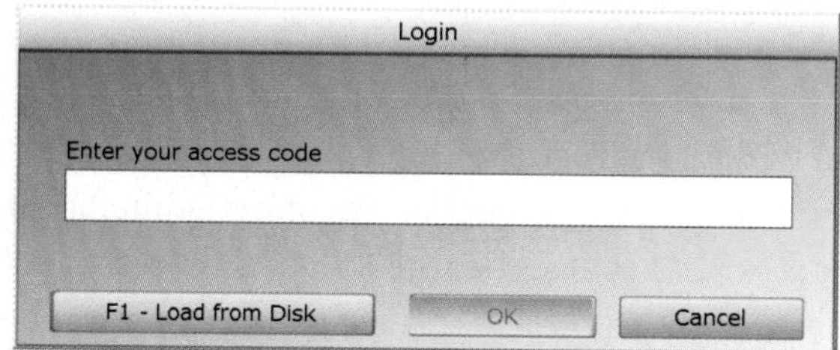

Type your entire access code in the box. No spaces or dashes are necessary; they will be supplied automatically. When you are finished, press OK.

If you typed in your access code correctly, you will be prompted to save the code to disk. If you choose to save your code to disk, typing in the authorization code each time you run ***HLS: Precalculus*** will not be necessary. Instead, select the F1 - Load from Disk button when prompted to enter your access code and choose the path to your saved access code.

Now that you have entered your authorization code and saved it to diskette, you are ready to run a lesson. From the table of contents screen, choose the appropriate chapter and then choose the lesson you wish to run.

Features:

Each lesson in ***HLS: Precalculus*** has four modes: Instruct, Practice, Tutor, and Certify.

Instruct: Instruct provides a multimedia presentation of the material covered in the lesson. Instruct offers example problems, animation, and helpful tips to enhance the learning experience.

Practice: Practice allows you to hone your problem-solving skills. It provides an unlimited number of randomly generated problems. Practice also provides access to the Tutor mode with the Tutor button at the bottom of the screen.

Tutor: The Interactive Tutor mode is broken up into several parts: Instruct, Explain Error, Step by Step, and Solution.

1. **Instruct**, which can also be selected directly from Practice mode, contains a multimedia lecture of the material covered in a lesson.

2. **Explain Error** is active whenever a problem is answered incorrectly. It attempts to explain the error that caused you to arrive at the incorrect solution to the problem.

3. **Step by Step** is an interactive explanation of the problem. It breaks each problem into several steps, explains to you each step in solving the problem, and asks you a question about the step. After you answer the last step correctly, you have solved the problem.

4. **Solution** provides a detailed "worked-out" solution to the problem.

Throughout the Tutor, you will see words or phrases that are underlined and colored green. These are called Hot Words. Clicking on a Hot Word will provide you with more information on these word(s) or phrases.

Certify: Certify is the testing mode. Each certification session consists of a finite number of problems and allows a certain number of strikes (incorrect answers). If you correctly answer the required number of questions, you will receive a certification code and a certificate. Write down your certification code and/or print out your certificate. The certification code will be used by your instructor to update your records. Note that the Tutor is not available in Certify.

Integration of Courseware and Textbook:

HLS: Precalculus courseware has been carefully designed to complement this textbook. Each section of this text has a corresponding lesson in the courseware which provides additional instruction, examples and problems.

Support:

If you have questions about ***HLS: Precalculus*** or are having technical difficulties, we can be contacted as follows:

Phone: (843) 571-2825
Email: techsupport@hawkeslearning.com
Web: www.hawkeslearning.com

Support hours are 8:30 am to 5:30 pm, Eastern Time, Monday through Friday.

chapter

ONE

NUMBER SYSTEMS AND EQUATIONS AND INEQUALITIES OF ONE VARIABLE

OVERVIEW

By the end of this chapter you should be able to work with and manipulate expressions involving exponents and radicals. You will be able to restate radical equations in terms of variables which appear inaccessible – or "trapped". When solving this type of equation be sure to check your potential solutions in the original expression to see if they are true solutions. On page 110, you will find a problem on determining rate of work. You will master this type of problem, using tools such as those learned in *Work-Rate Problems* on pages 103 – 105.

Introduction

In this chapter, we review the terminology of the real number system, the notation and properties frequently encountered in algebra, the extension of the real number system to the larger set of complex numbers, and the basic algebraic methods we use to solve equations and inequalities.

We begin with a discussion of common subsets of the set of real numbers. Certain types of numbers are important from both a historical and a mathematical perspective. There is archeological evidence that people used the simplest sort of numbers, the *counting* or *natural numbers*, as far back as 50,000 years ago. Over time, many cultures discovered needs for various refinements to the number system, resulting in the development of such classes of numbers as the *integers*, the *rational numbers*, the *irrational numbers*, and ultimately, the *complex numbers*, a number system which contains the real numbers.

Many of the ideas in this chapter have a history dating as far back as Egyptian and Babylonian civilizations of around 3000 BC, with later developments and additions due to Greek, Hindu, and Arabic mathematicians. Much of the material was also developed independently by Chinese mathematicians. It is a tribute to the necessity, utility, and objectivity of mathematics that so many civilizations adopted so much mathematics from abroad, and that different cultures operating independently developed identical mathematical concepts so frequently.

Pythagoras

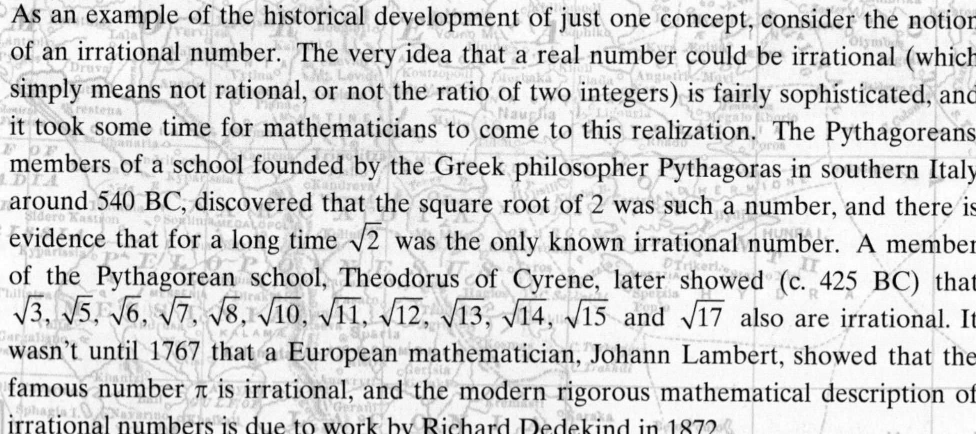

As an example of the historical development of just one concept, consider the notion of an irrational number. The very idea that a real number could be irrational (which simply means not rational, or not the ratio of two integers) is fairly sophisticated, and it took some time for mathematicians to come to this realization. The Pythagoreans, members of a school founded by the Greek philosopher Pythagoras in southern Italy around 540 BC, discovered that the square root of 2 was such a number, and there is evidence that for a long time $\sqrt{2}$ was the only known irrational number. A member of the Pythagorean school, Theodorus of Cyrene, later showed (c. 425 BC) that $\sqrt{3}$, $\sqrt{5}$, $\sqrt{6}$, $\sqrt{7}$, $\sqrt{8}$, $\sqrt{10}$, $\sqrt{11}$, $\sqrt{12}$, $\sqrt{13}$, $\sqrt{14}$, $\sqrt{15}$ and $\sqrt{17}$ also are irrational. It wasn't until 1767 that a European mathematician, Johann Lambert, showed that the famous number π is irrational, and the modern rigorous mathematical description of irrational numbers is due to work by Richard Dedekind in 1872.

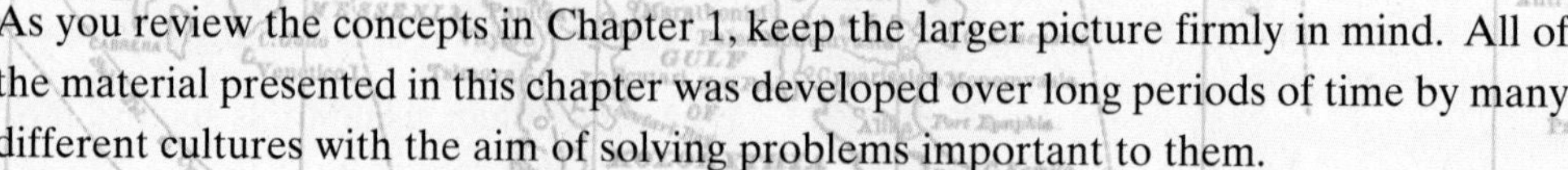

As you review the concepts in Chapter 1, keep the larger picture firmly in mind. All of the material presented in this chapter was developed over long periods of time by many different cultures with the aim of solving problems important to them.

1.1 Real Numbers and Algebraic Expressions

TOPICS

1. Common subsets of real numbers
2. The real number line
3. Order on the real number line
4. Set-builder notation and interval notation
5. Basic set operations and Venn diagrams
6. Absolute value and distance
7. Components of algebraic expressions
8. The field properties and their use in algebra

Topic 1: Common Subsets of Real Numbers

Certain types of numbers occur so frequently in mathematics that they have been given special names and symbols. These names will be used throughout this book and in later math classes when referring to members of the following sets:

Types of Real Numbers

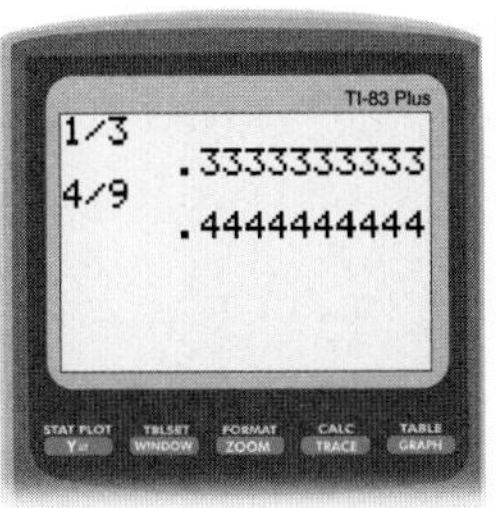

Examples of rational numbers

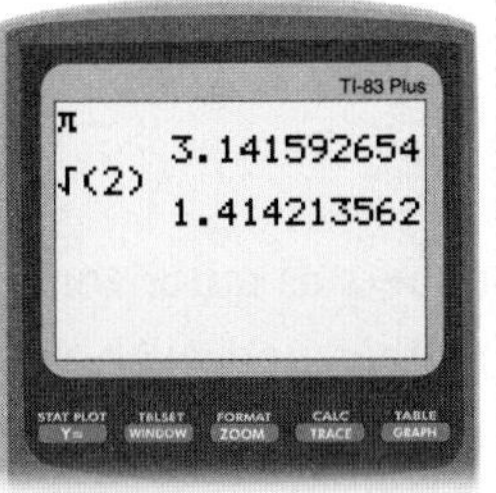

Examples of irrational numbers

The Natural (or Counting) Numbers: This is the set of numbers $\mathbb{N} = \{1, 2, 3, 4, 5, \ldots\}$.

The Whole Numbers: This is the set of natural numbers with 0 added: $\{0, 1, 2, 3, 4, 5, \ldots\}$.

The Integers: This is the set of natural numbers, their negatives, and 0. As a list, this is the set $\mathbb{Z} = \{\ldots, -4, -3, -2, -1, 0, 1, 2, 3, 4, \ldots\}$.

The Rational Numbers: This is the set, with symbol $\mathbb{Q}$ for quotient, of *ratios* of integers. That is, any rational number can be written in the form $\frac{p}{q}$, where p and q are both integers and $q \neq 0$. When written in decimal form, rational numbers either terminate or repeat a pattern of digits past some point.

The Irrational Numbers: Every real number that is not rational is, by definition, irrational. In decimal form, irrational numbers are non-terminating and non-repeating.

The Real Numbers: Every set above is a subset of the set of real numbers, which is denoted $\mathbb{R}$. Every real number is either rational or irrational, and no real number is both.

example 1

Consider the set $S = \left\{-15, -7.5, -\frac{7}{3}, 0, \sqrt{2}, 1.87\overline{35}, \sqrt{9}, \pi, 10^{17}\right\}$.

a. The natural numbers in S are $\sqrt{9}$ and 10^{17}. $\sqrt{9}$ is a natural number since $\sqrt{9} = 3$.

b. The whole numbers in S are 0, $\sqrt{9}$, and 10^{17}.

c. The integers in S are $-15, 0$, $\sqrt{9}$, and 10^{17}.

d. The rational numbers in S are $-15, -7.5, -\frac{7}{3}, 0, 1.87\overline{35}, \sqrt{9}$, and 10^{17}. The numbers -7.5 and $1.87\overline{35}$ qualify as rational numbers since $-7.5 = \frac{-15}{2}$ and $1.87\overline{35} = \frac{4637}{2475}$ (the bar over the last two digits indicates that these digits constitute a pattern that is repeated indefinitely). Any integer p automatically qualifies as a rational number since it can be written as $\frac{p}{1}$.

e. The only irrational numbers in S are $\sqrt{2}$ and π.

The following figure shows the relationships among the subsets of $\mathbb{R}$ defined above. This figure indicates, for example, that every natural number is automatically a whole number, and also an integer, and also a rational number.

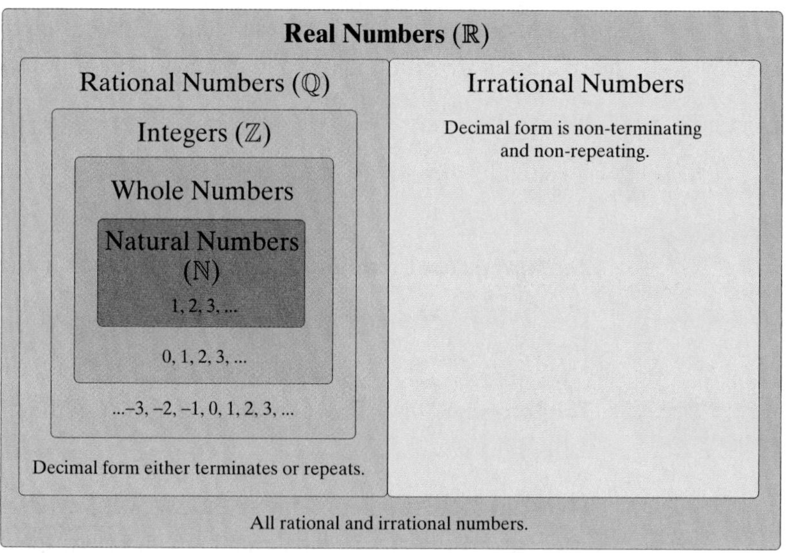

Figure 1: The Real Numbers

Topic 2: The Real Number Line

It is often convenient and instructive to depict the set of real numbers as a horizontal line, with each point on the line representing a unique real number and each real number associated with a unique point on the line. The real number corresponding to a given point is called the **coordinate** of that point. Thus, one (and only one) point on the line represents the number 0, and this point is called the **origin** of the real number

line. Points to the right of the origin represent positive real numbers, while points to the left of the origin represent negative real numbers. Figure 2 is an illustration of the real number line with several points plotted. Note that two irrational numbers are plotted, though their locations on the line are necessarily approximations.

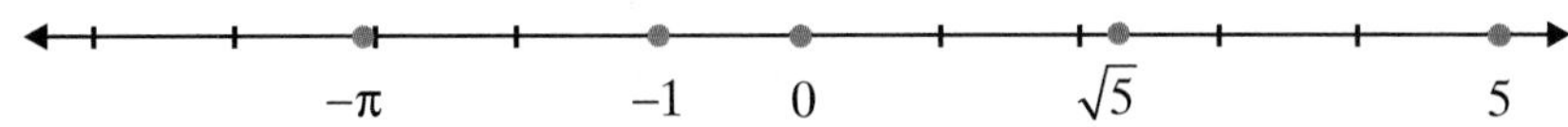

Figure 2: The Real Number Line

We choose which portion of the real number line to depict and the physical length that represents one unit based on the numbers that we wish to plot.

Topic 3: Order on the Real Number Line

The representation of the real numbers as a line leads naturally to the idea of *ordering* the real numbers.

Inequality Symbols (Order)

Symbol	Meaning
$a < b$ (read "a is less than b")	a lies to the left of b on the number line.
$a \leq b$ (read "a is less than or equal to b")	a lies to the left of b or is equal to b.
$b > a$ (read "b is greater than a")	b lies to the right of a on the number line.
$b \geq a$ (read "b is greater than or equal to a")	b lies to the right of a or is equal to a.

The two symbols $<$ and $>$ are called *strict* inequality signs, while the symbols $\leq$ and $\geq$ are *non-strict* inequality signs.

caution!

Remember that order is defined by placement of real numbers on the number line, *not* by magnitude. For instance, $-36 < 5$, because -36 lies to the left of 5 on the number line. Also, be aware that the negation of the statement $a \leq b$ is the statement $a > b$. Furthermore, if $a \leq b$ and $a \geq b$, then it must be the case that $a = b$.

Topic 4: Set-Builder Notation and Interval Notation

In describing the solutions to equations and inequalities, we will often need a precise way of defining sets of real numbers. **Set-builder notation** is a general method of describing the elements that belong to a given set. **Interval notation** is a way of describing certain subsets of the real number line.

Set-Builder Notation

The notation $\{x \mid x \text{ has property } P\}$ is used to describe a set of real numbers, all of which have the property P. This can be read "the set of all real numbers x having property P."

example 2

a. $\{x \mid x \text{ is an even integer}\}$ is a more precise way of describing the set $\{\ldots, -4, -2, 0, 2, 4, \ldots\}$. We could also describe this set as $\{2n \mid n \text{ is an integer}\}$, since every even integer is a multiple of 2.

b. $\{x \mid x \text{ is an integer such that } -3 \le x < 2\}$ describes the set $\{-3, -2, -1, 0, 1\}$.

c. $\{y \mid y > 1 \text{ and } y \le -4\}$ describes the **empty set**, which is denoted $\varnothing$ or $\{\ \}$, as no real number y satisfies the stated properties such that both $y > 1$ and $y \le -4$.

Topic 5: Basic Set Operations and Venn Diagrams

The sets that arise most frequently in algebra are sets of real numbers, and these sets are often the solutions of equations or inequalities. We will, on occasion, find it necessary to combine two or more such sets through the set operations of **union** and **intersection**. These operations are defined on sets in general, not just sets of real numbers, and are perhaps best illustrated by means of Venn diagrams.

A **Venn diagram** is a pictorial representation of a set or sets, and its aim is to indicate, through shading, the outcome of set operations such as union and intersection. In the definitions on the next page, these two operations are first defined with set-builder notation and then demonstrated with a Venn diagram. The symbol $\in$ is read "is an element of."

Union

In this definition, A and B denote two sets, and are represented in the Venn diagrams by circles. The operation of union is demonstrated in the diagram by means of shading.

The **union** of A and B, denoted $A \cup B$, is the set $\{x \mid x \in A \text{ or } x \in B\}$. That is, an element x is in $A \cup B$ if it is in the set A, the set B, or both. Note that the union of A and B contains both individual sets.

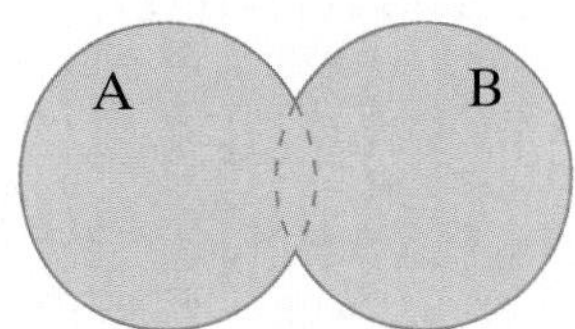

Intersection

In this definition, A and B denote two sets, and are represented in the Venn diagrams by circles. The operation of intersection is demonstrated in the diagram by means of shading.

The **intersection** of A and B, denoted $A \cap B$, is the set $\{x \mid x \in A \text{ and } x \in B\}$. That is, an element x is in $A \cap B$ if it is in both A and B. Note that the intersection of A and B is contained in each individual set.

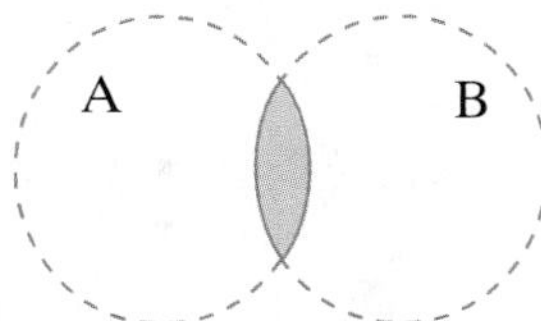

example 3

$A = \{-9, 2, 3, 5, x, y\}$, $B = \{-9, 3, 4, 7, z\}$, $C = \{-8, 1, 6\}$

Find: **a.** $A \cup B$ **b.** $A \cap B$ **c.** $B \cap C$

Solutions:

a. $A \cup B = \{-9, 2, 3, 4, 5, 7, x, y, z\}$

Since this is a union of two sets, the new set will consist of all the values from the individual sets.

b. $A \cap B = \{-9, 3\}$

The intersection of two sets only consists of those values which appear in both sets.

c. $B \cap C = \varnothing$

These two sets have no elements in common, and their intersection is the empty set.

Sets that consist of all real numbers bounded by two endpoints, possibly including those endpoints, are called **intervals**. Intervals can also consist of a portion of the real number line extending indefinitely (in either direction) from just one endpoint. We could describe such sets with set-builder notation, but intervals occur frequently enough that special notation has been devised to define them succinctly.

Interval Notation

Notation	Meaning	Graphical Representation
(a, b)	$\{x \mid a < x < b\}$, or all real numbers strictly between a and b	a b
$[a, b]$	$\{x \mid a \le x \le b\}$, or all real numbers between a and b, including both a and b	a b
$(a, b]$	$\{x \mid a < x \le b\}$, or all real numbers between a and b, including b but not a	a b
$(-\infty, b)$	$\{x \mid x < b\}$, or all real numbers less than b	b
$[a, \infty)$	$\{x \mid x \ge a\}$, or all real numbers greater than or equal to a	a
$(-\infty, \infty)$	$\mathbb{R}$, set of all real numbers	

In interval notation, the left endpoint is always written first. Intervals of the form (a, b) are called **open** intervals, while those of the form $[a, b]$ are **closed** intervals. The interval $(a, b]$ is **half-open** (or **half-closed**). Of course, a half-open interval may be open at either endpoint, as long as it is closed at the other. The symbols $-\infty$ and ∞ indicate that the interval extends indefinitely in, respectively, the left and the right directions. Note that $(-\infty, b)$ excludes the endpoint b, while $[a, \infty)$ includes the endpoint a.

caution!

The symbols $-\infty$ and ∞ are just that: symbols! They are not real numbers, and so they cannot, for instance, be solutions to a given equation. The fact that they are symbols, and not numbers, also means that as endpoints of intervals they can never be included. For this reason, a parenthesis always appears next to either $-\infty$ or ∞; a bracket should never appear next to either infinity symbol.

example 4

Simplify each of the following set expressions, if possible.

a. $(-2,4]\cup[0,9]$ **b.** $(-2,4]\cap[0,9]$ **c.** $(-\infty,4]\cup(-1,\infty)$

Solutions:

a. $(-2,4]\cup[0,9]=(-2,9]$

Since these two intervals overlap, their union is best described with a single interval.

b. $(-2,4]\cap[0,9]=[0,4]$

This intersection of two intervals can also be described with a single interval.

c. $(-\infty,4]\cup(-1,\infty)=(-\infty,\infty)$

The union of these two intervals constitutes the entire set of real numbers.

Topic 6: Absolute Value and Distance

In addition to order, the depiction of the set of real numbers as a line leads to the notion of *distance*. Physically, distance is a well-understood concept: the distance between two objects is a non-negative number, dependent on a choice of measuring system, indicating how close the objects are to one another. The mathematical idea of **absolute value** gives us a means of defining distance in a mathematical setting.

Absolute Value

The **absolute value** of a real number a, denoted as $|a|$, is defined by:

$$|a|=\begin{cases} a & \text{if } a\geq 0 \\ -a & \text{if } a<0 \end{cases}$$

The absolute value of a number is also referred to as its **magnitude**; it is the non-negative number corresponding to its distance from the origin. Note also that 0 is the only real number whose absolute value is 0.

Distance on the Real Number Line

Given two real numbers a and b, the distance between them is defined to be $|a-b|$. In particular, the distance between a and 0 is $|a-0|$, or just $|a|$.

Given two distinct real numbers a and b, exactly one of the two differences $a-b$ and $b-a$ will be negative, and since $b-a=-(a-b)$, the definition of absolute value makes it clear that these two differences have the same magnitude. That is, $|a-b|=|b-a|$.

A few examples should make the above points clear.

example 5

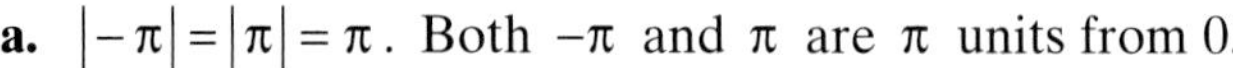

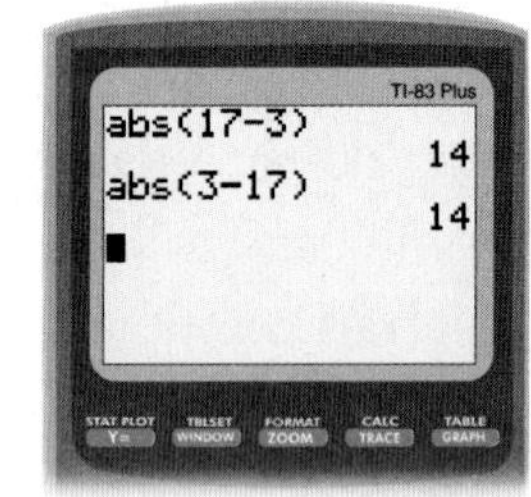

a. $|-\pi|=|\pi|=\pi$. Both $-\pi$ and π are π units from 0.

b. $|17-3|=|3-17|=14$. 17 and 3 are 14 units apart.

c. $-|-5|=-5$. Note that the negative sign outside the absolute value symbol is not affected by the absolute value. Compare this with the fact that $-(-5)=5$.

d. $|\sqrt{7}-19|=19-\sqrt{7}$. In contrast to the last example, we know $\sqrt{7}-19$ is negative, so its absolute value is $-(\sqrt{7}-19)=19-\sqrt{7}$.

The examples above illustrate some of the basic properties of absolute value. The list of properties below can all be derived from the definition of absolute value.

Properties of Absolute Value

Property	Example	Description
1. $\lvert a\rvert \geq 0$	$\lvert -5\rvert = 5 \geq 0$	The absolute value of a number is always positive or zero.
2. $\lvert -a\rvert = \lvert a\rvert$	$\lvert -2\rvert = \lvert 2\rvert$	A number and its negative have the same absolute value.
3. $a \leq \lvert a\rvert$	$-4 \leq \lvert -4\rvert$	A number is always less than or equal to its absolute value.
4. $\lvert ab\rvert = \lvert a\rvert \lvert b\rvert$	$\lvert 3\cdot 4\rvert = \lvert 3\rvert \lvert 4\rvert$	The absolute value of a product is the product of the absolute values.
5. $\left\lvert \frac{a}{b}\right\rvert = \frac{\lvert a\rvert}{\lvert b\rvert},\ b \neq 0$	$\left\lvert \frac{-10}{2}\right\rvert = \frac{\lvert -10\rvert}{\lvert 2\rvert}$	The absolute value of a quotient is the quotient of the absolute values.
6. $\lvert a+b\rvert \leq \lvert a\rvert + \lvert b\rvert$	$\lvert -1+7\rvert \leq \lvert -1\rvert + \lvert 7\rvert$	This is called the **triangle inequality**, as it is a reflection of the fact that one side of a triangle is never longer than the sum of the other two sides.

example 6

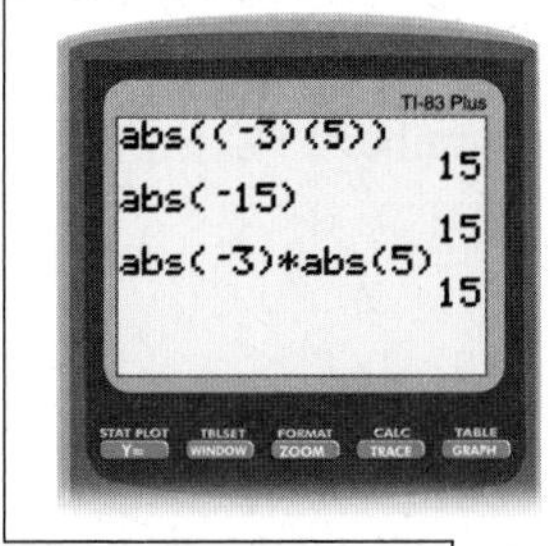

a. $|(-3)(5)| = |-15| = 15 = |-3||5|$ Use Property 4 and let $a = -3$ and $b = 5$.

b. $1 = |-3+4| \le |-3| + |4| = 7$

c. $7 = |-3-4| \le |-3| + |-4| = 7$

d. $\left|\frac{-3}{7}\right| = \frac{|-3|}{|7|} = \frac{3}{7}$

Topic 7: Components of Algebraic Expressions

Algebraic expressions are made up of constants and variables, combined by the operations of addition, subtraction, multiplication, division, exponentiation, and the taking of roots. The basic operations of addition, subtraction, multiplication, and division are familiar to you, and exponents and roots will be discussed in the following two sections. **Constants** are fixed numbers, while **variables** are letters that represent unspecified numbers. To *evaluate* a given expression means to replace the variables with specific numbers and then to perform the indicated mathematical operations and simplify the result.

The **terms** of an algebraic expression are those parts joined by addition (or subtraction), while the **factors** are the individual parts of the term that are joined by multiplication (or division). The **coefficient** is the constant factor of the term, while the remaining part of the term constitutes the **variable factor**.

example 7

Look at the algebraic expression $-17x^3(x^2+4y)+5\sqrt{x}-13(x-y)$.

The three terms are $-17x^3(x^2+4y)$, $5\sqrt{x}$, and $-13(x-y)$.
The factors of $-17x^3(x^2+4y)$ are -17, x^3, and (x^2+4y).
The coefficient of $-17x^3(x^2+4y)$ is -17 while the variable part is $x^3(x^2+4y)$.

Now evaluating at $x = 1$ and $y = -2$,
$-17(1)^3\left((1)^2+4(-2)\right)+5\sqrt{(1)}-13\left((1)-(-2)\right) = -17(1-8)+5-13(3) = 119+5-39 = 85.$

Topic 8: The Field Properties and Their Use in Algebra

The set of real numbers forms what is known, mathematically, as a *field*, and consequently the properties below are called *field properties*. These properties also apply to the set of complex numbers, which is a larger field containing the real numbers. Complex numbers will be discussed in Section 4 of this chapter.

Field Properties

In this table, *a*, *b*, and *c* represent arbitrary real numbers. The first five properties apply to addition and multiplication, while the last combines the two.

Name of Property	Additive Version	Multiplicative Version
Closure	$a + b$ is a real number	ab is a real number
Commutative	$a+b=b+a$ $5+4=4+5$	$ab=ba$ $2\cdot 3=3\cdot 2$
Associative	$a+(b+c)=(a+b)+c$ $4+(3+2)=(4+3)+2$	$a(bc)=(ab)c$ $7\cdot(8\cdot 3)=(7\cdot 8)\cdot 3$
Identity	$a+0=0+a=a$ $8+0=0+8=8$	$a\cdot 1=1\cdot a=a$ $5\cdot 1=1\cdot 5=5$
Inverse	$a+(-a)=0$ $4+(-4)=0$	$a\cdot\frac{1}{a}=1(\text{for } a\neq 0)$ $9\cdot\frac{1}{9}=1$
Distributive	$a(b+c)=ab+ac$ $3(6+7)=3\cdot 6+3\cdot 7$	

While the field properties are of fundamental importance to algebra, they imply further properties that are often of more immediate use.

Zero-Factor Property

Let A and B represent algebraic expressions. If the product of A and B is 0, then at least one of A and B is itself 0. Using the symbol $\Rightarrow$ for "implies," we write

$$AB=0 \Rightarrow A=0 \text{ or } B=0.$$

Cancellation Properties

Throughout this table, A, B, and C represent algebraic expressions. The symbol $\Leftrightarrow$ can be read as "if and only if" or "is equivalent to."

Property	Description
$A = B \Leftrightarrow A + C = B + C$	Adding the same quantity to both sides of an equation results in an equivalent equation.
For $C \neq 0$, $A = B \Leftrightarrow A \cdot C = B \cdot C$	Multiplying both sides of an equation by the same non-zero quantity results in an equivalent equation.

Properties of Negatives

Property	Example
1. $(-1)a = -a$	$(-1)7 = -7$
2. $-(-a) = a$	$-(-8) = 8$
3. $(-a)b = a(-b) = -(ab)$	$(-3)2 = 3(-2) = -(3 \cdot 2)$
4. $(-a)(-b) = ab$	$(-6)(-4) = (6 \cdot 4)$
5. $-(a + b) = -a - b$	$-(5 + 4) = -5 - 4$
6. $-(a - b) = b - a$	$-(3 - 7) = 7 - 3$

example 8

a. The equation $x^5 + y = 3xy^3 + y$ is equivalent to the equation $x^5 = 3xy^3$. Adding the quantity $-y$ to both sides of the first equation leads to the second equation.

b. The equation $-6\left(x - \frac{y}{2}\right)^2 = 18$ is equivalent to the equation $\left(x - \frac{y}{2}\right)^2 = -3$. We have multiplied both sides of the original equation by $-\frac{1}{6}$ to obtain the second.

c. The equation $(x - y)(x + y) = 0$ means that either $x - y = 0$ or $x + y = 0$.

exercises

*Which elements of the following sets are **(a)** natural numbers, **(b)** whole numbers, **(c)** integers, **(d)** rational numbers, **(e)** irrational numbers, **(f)** real numbers, **(g)** undefined. See Example 1.*

1. $\left\{19,\ -4.3,\ -\sqrt{3},\ \frac{15}{0},\ \frac{0}{15},\ 2^5,\ -33\right\}$ **2.** $\left\{5\sqrt{7},\ 4\pi,\ \sqrt{16},\ 3.\bar{3},\ -1,\ \frac{22}{7},\ \lvert -8 \rvert\right\}$

A rational number that appears as a decimal with a repeating pattern of digits can be written as a ratio of integers by following the procedure outlined here. As a sum, the decimal number $1.87\overline{35}$ stands for $1+\frac{87}{100}+0.00\overline{35}$. The first two terms pose no problem, but we will have to work to write the third term as a ratio of integers.

Let $x = 0.00\overline{35}$. Then $100x = 0.\overline{35}$, so $100x = 0.35 + 0.00\overline{35}$, or $100x = 0.35 + x$. Thus, $99x = 0.35$, or $99x = \frac{35}{100}$. Solving for x, we have $x = \frac{35}{9900}$. (We have just solved a linear equation; we will study these in detail in Section 1.5.) Altogether, we have $1.87\overline{35} = 1 + \frac{87}{100} + \frac{35}{9900}$ or, after adding the three terms, $x = \frac{18548}{9900}$, which is equal to $\frac{4637}{2475}$ when reduced.

Write the following rational numbers as ratios of integers by following the procedure above.

3. $2.\overline{3}$ **4.** $-5.0\overline{82}$ **5.** $0.\overline{41836}$

6. $0.\overline{9}$ **7.** $-1.\overline{01}$ **8.** $7.\overline{421}$

Plot the following real numbers on a number line. Choose the unit length appropriately for each set.

9. $\{-4.5, -1, 2.5\}$ **10.** $\{5.1, 5.2, 5.8\}$

11. $\{-24, 2, 15\}$ **12.** $\left\{0, \frac{1}{2}, \frac{5}{6}\right\}$

Select all of the symbols from the set $\{<, \leq, >, \geq\}$ that can be placed in the blank to make true statements.

13. $-3.4 \_\_\_ -3.5$ **14.** $-102 \_\_\_ 9$ **15.** $3 \_\_\_ 3$

16. $-50 \_\_\_ -45$ **17.** $-\frac{1}{4} \_\_\_ -\frac{1}{3}$ **18.** $.0087 \_\_\_ -42.9$

Describe each of the following sets using set-builder notation. There may be more than one correct way to do this. See Example 2.

19. $\{-6, -3, 0, 3, 6, 9\}$ **20.** $\{5, 6, 7, \ldots, 105\}$ **21.** $\{2, 3, 5, 7, 11, 13, 17, \ldots\}$

22. $\{1, 2, 4, 8, 16, 32, \ldots\}$ **23.** $\left\{\ldots, \frac{1}{3}, \frac{1}{5}, \frac{1}{7}, \frac{1}{9}, \ldots\right\}$ **24.** $\{0, 1, 2, 3, 4, 5, \ldots\}$

Simplify the following unions and intersections of sets. See Example 3 and 4.

25. $(-5,2]\cup(2,4]$ **26.** $(-5,2]\cap(2,4]$ **27.** $[3,5]\cap[2,4]$

28. $(-\infty,4]\cup(0,\infty)$ **29.** $\mathbb{Q}\cap\mathbb{Z}$ **30.** $\mathbb{N}\cup\mathbb{R}$

31. $(-\infty,\infty)\cap[-\pi,21)$ **32.** $[2,\infty)\cap(-4,7)\cap(-3,2]$ **33.** $(3,5]\cup[5,9]$

34. $[-\pi,2\pi)\cap[0,4\pi]$ **35.** $\mathbb{N}\cup\mathbb{Z}\cap\mathbb{Q}$ **36.** $(-4.8,-3.5)\cap\mathbb{Z}$

Write each set as an interval using interval notation.

37. $\{x \mid -3 \le x < 19\}$ **38.** $\{x \mid x < 4\}$ **39.** The positive real numbers

40. $\left\{x \,\middle|\, -\frac{1}{2} < x < \frac{2}{5}\right\}$ **41.** $\{x \mid 1 \le x \le 2\}$ **42.** All integers that are irrational

Graph the following intervals.

43. $[5, 14)$ **44.** $[-9, -1]$ **45.** $(0, 2)$

46. $(-3, 18]$ **47.** $(-\infty, 7]$ **48.** $(25, \infty)$

Evaluate the absolute value expressions. See Examples 5 and 6.

49. $-|-11|$ **50.** $|3-7|$ **51.** $-|4-9|$

52. $\left|\sqrt{3}-\sqrt{5}\right|$ **53.** $\sqrt{|-4|}$ **54.** $-\left|-4-|-11|\right|$

55. $\left|-\sqrt{2}\right|$ **56.** $\frac{|-x|}{|x|}\ (x \ne 0)$

technology exercises

A computer algebra system such as *Mathematica* can be used to check many of your answers. As an example, *Mathematica* can be used to verify the truth or falsehood of inequality statements. In the image below, the lines appearing in bold font have been typed and executed within *Mathematica*; the response (either True or False) appears in normal font. Although *Mathematica* has changed the appearance below, the relation $\leq$ is entered as <=. (Similarly, $\geq$ is entered as >=.)

If available, check your answers to the following problems with a computer algebra system. (See Appendix A for guidance on using computer algebra systems like Mathematica.)

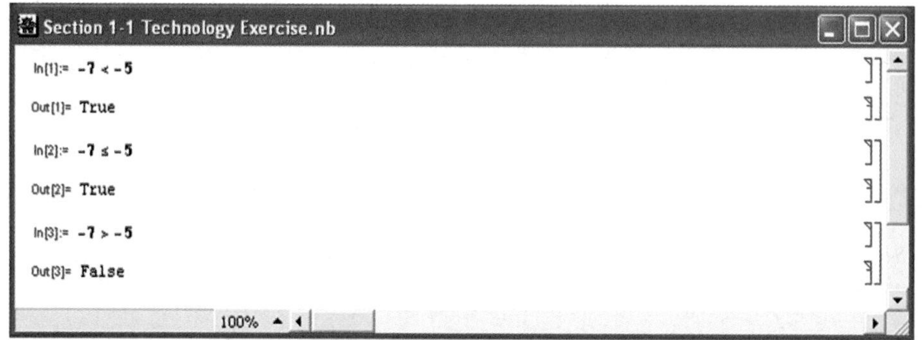

57. $-3.1 \_\_\_ -2.9$ **58.** $2.1 \_\_\_ -5.5$ **59.** $-4 \_\_\_ 100$

60. $.001 \_\_\_ -99.8$ **61.** $\frac{1}{3} \_\_\_ \frac{1}{4}$ **62.** $-\frac{3}{4} \_\_\_ -\frac{1}{5}$

Identify the components of the algebraic expressions, as indicated. See Example 7.

63. Identify the terms in the expression $3x^2y^3 - 2\sqrt{x+y} + 7z$.

64. Identify the coefficients in the expression $3x^2y^3 - 2\sqrt{x+y} + 7z$.

65. Identify the factors in the term $-2\sqrt{x+y}$.

66. Identify the terms in the expression $x^2 + 8.5x - 14y^3$.

67. Identify the coefficients in the expression $x^2 + 8.5x - 14y^3$.

68. Identify the factors in the term $8.5x$.

Evaluate each expression for the given values of the variables. See Example 7.

69. $3x^2y^3 - 2\sqrt{x+y} + 7z$ for $x = -1$, $y = 2$, and $z = -2$

70. $-3\pi y + 8x + y^3$ for $x = 2$ and $y = -2$

71. $\dfrac{|x|\sqrt{2}}{x^3y^2} - \dfrac{3y}{x}$ for $x = -3$ and $y = 2$

72. $y\sqrt{x^3 - 2} + \sqrt{x - 2y} - 3y$ for $x = 3$ and $y = -\dfrac{1}{2}$

73. $\left|-x^2 + 2xy - y^2\right|$ for $x = -3$ and $y = -5$

74. $\dfrac{1}{32}x^2y^3 + y\sqrt{x} - 7y$ for $x = 4$ and $y = 2$

Identify the property that justifies each of the following statements. See Example 8.

75. $(x-y)(z^2) = (z^2)(x-y)$

76. $3 - 7 = -7 + 3$

77. $(3x+2) + z = 3x + (2+z)$

78. $4(y-3) = 4y - 12$

79. $-3(4x^6z) = (-3)(4)(x^6z)$

80. $4 + (-3+x) = (4-3) + x$

81. $-2(4-x) = -8 + 2x$

82. $(x+y)\left(\dfrac{1}{x+y}\right) = 1$

Identify the property that justifies each of the following statements. If one of the cancellation properties is being used to transform an equation, identify the quantity that is being added to both sides or the quantity that both sides are being multiplied by. See Example 8.

83. $25x^3 = 10y \Leftrightarrow 5x^3 = 2y$

84. $-14y = 7 \Leftrightarrow y = -\dfrac{1}{2}$

85. $14 - x = 2x \Leftrightarrow 14 = 3x$

86. $5 + 3x - y = 2x - y \Leftrightarrow 5 + x = 0$

87. $x^2z = 0 \Rightarrow x^2 = 0$ or $z = 0$

88. $(a+b)(x) = 0 \Rightarrow a + b = 0$ or $x = 0$

89. $\dfrac{x}{6} + \dfrac{y}{3} - 2 = 0 \Leftrightarrow x + 2y - 12 = 0$

90. $(x-3)(x+2) = 0 \Rightarrow x - 3 = 0$ or $x + 2 = 0$

Translate the following directions into an arithmetic expression.

91. Begin with 3. Add 7, and multiply the result by 3. Subtract 5. Take the square root, raise the result to the 3rd power, and then multiply by $-\dfrac{1}{5}$.

technology exercises

Use a calculator or computer software to assign the indicated values to the variables and evaluate the corresponding expressions. In *Mathematica*, for example, the expression $\sqrt{x^3y^2}+5z-14$ can be evaluated for the values $x=2$, $y=-1$, and $z=-5$ by typing and executing the lines shown in bold below; the output appears in normal font. *(See Appendix A for guidance on using computer algebra systems like Mathematica.)*

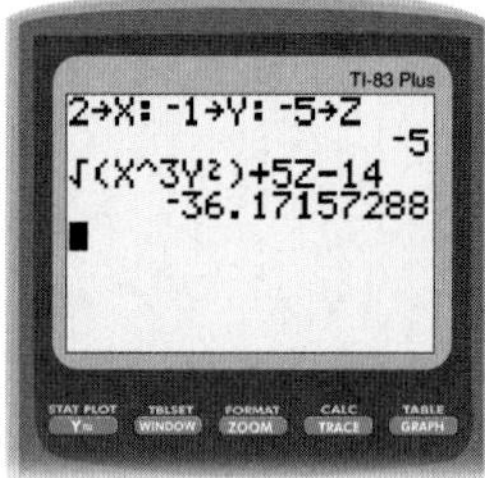

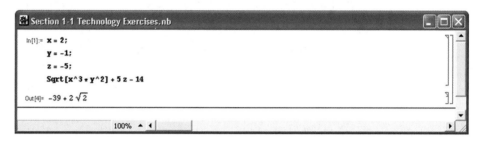

To define X, Y, and Z, enter the value of the variable, press **STO►**, and then press **ALPHA** followed by the desired variable name. Separate each variable definition with a colon (:) by pressing **ALPHA** and then : (colon). Once the variables have been defined enter the given equation. Your screen should appear as above.

92. $\sqrt{x^4y-z}+\dfrac{x-y^3}{z^2}$ for $x=-3, y=2,$ and $z=-2$

93. $\dfrac{\left(x-pq^2\right)^3}{2q^3}$ for $x=-5, p=2,$ and $q=-3$

94. $\dfrac{\left|x^2-y^3\right|-4x}{3y^5}$ for $x=2$ and $y=3$

95. $\sqrt{p^2q-q^3}-\left|p+q^2\right|$ for $p=-5$ and $q=2$

1.2 Properties of Exponents and Radicals

TOPICS

1. Natural number and integer exponents
2. Properties of exponents and their use
3. Scientific notation
4. Interlude: working with geometric formulas
5. Roots and radical notation
6. Simplifying and combining radical expressions
7. Rational number exponents

Topic 1: Natural Number and Integer Exponents

As we progress in Chapter 1, we will encounter a variety of algebraic expressions. As discussed in Section 1.1, algebraic expressions consist of constants and variables combined by the basic operations of addition, subtraction, multiplication, and division, along with exponentiation and the taking of roots. In this section, we will explore the meaning of exponentiation and the properties of exponents.

Natural Number Exponents

If a is any real number and if n is any natural number, then $a^n = \underbrace{a \cdot a \cdot \cdots \cdot a}_{n \text{ factors}}$.

In the expression, the number a is called the **base**, and n is the **exponent**. The process of multiplying n factors of a is called "raising a to the nth power," and the expression a^n may be referred to as "the nth power of a" or "a to the nth power." Note that $a^1 = a$.

Other phrases are also commonly used to denote the raising of something to a power, especially when the power is 2 or 3. For instance, a^2 is often referred to as "a squared" and a^3 is often referred to as "a cubed."

If the base and exponent are known constants, the expression a^n may be evaluated and written as a simple number. Even if an expression with one or more natural number exponents contains a variable, some simplification may be possible based only on the above definition.

example 1

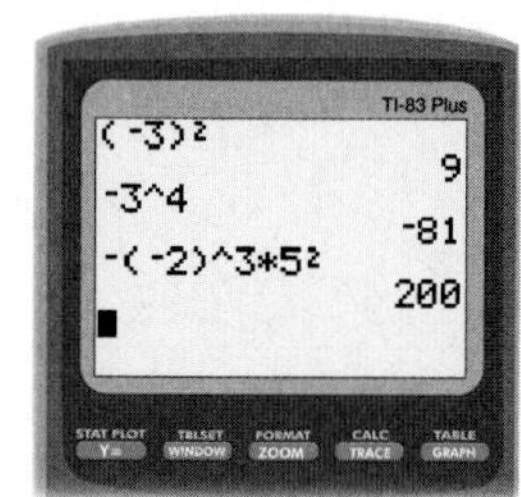

a. $4^3 = 4 \cdot 4 \cdot 4 = 64$. Thus, "four cubed is sixty-four."

b. $(-3)^2 = (-3)(-3) = 9$. Thus, "negative three, squared, is nine."

c. $-3^4 = -(3 \cdot 3 \cdot 3 \cdot 3) = -81$. Note that, by our order of operations, the exponent of 4 applies only to the number 3. After raising 3 to the 4th power, the result is multiplied by -1.

d. $-(-2)^3 \cdot 5^2 = -(-8)(25) = 200$.

e. $x^3 \cdot x^4 = (x \cdot x \cdot x)(x \cdot x \cdot x \cdot x) = x^7$.

The ultimate goal is to give meaning to the expression a^n for any real number n and to do so in such a way that certain properties of exponents hold consistently. For example, analysis of Example 1e above leads to the observation that if n and m are natural numbers, then

$$a^n \cdot a^m = \underbrace{a \cdot a \cdot \cdots \cdot a}_{n \text{ factors}} \cdot \underbrace{a \cdot a \cdot \cdots \cdot a}_{m \text{ factors}} = \underbrace{a \cdot a \cdot \cdots \cdot a}_{n+m \text{ factors}} = a^{n+m}.$$

$$a^n \cdot a^m = a^{n+m}$$

To extend the meaning of a^n to the case where $n = 0$, then, we might start by noting that we would like the following to hold:

$$a^0 \cdot a^m = a^{0+m} = a^m.$$

This suggests the following:

0 as an Exponent

For any real number $a \neq 0$, we define

$$a^0 = 1.$$

0^0 is undefined, just as division by 0 is undefined.

With this small extension of the meaning of exponents as inspiration, let us continue. Consider the following table:

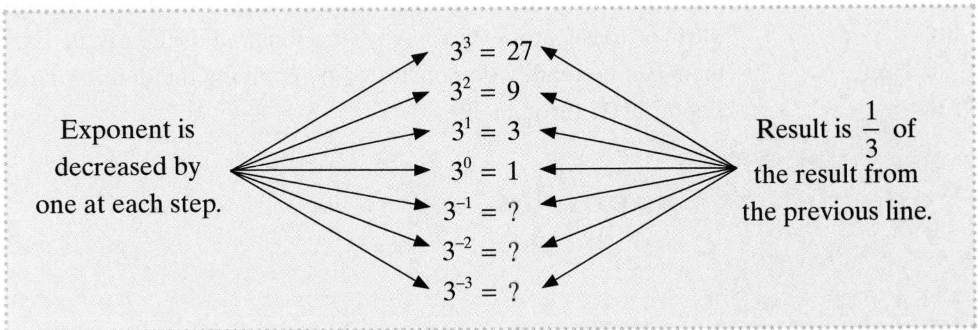

Exponent is decreased by one at each step.
$3^3 = 27$
$3^2 = 9$
$3^1 = 3$
$3^0 = 1$
$3^{-1} = ?$
$3^{-2} = ?$
$3^{-3} = ?$
Result is $\frac{1}{3}$ of the result from the previous line.

In order to maintain the pattern that has begun to emerge, we are led to complete the table with $3^{-1} = \frac{1}{3}$, $3^{-2} = \frac{1}{9}$ and $3^{-3} = \frac{1}{27}$. In general, we define negative integer exponents as follows:

Negative Integer Exponents

For any real number $a \neq 0$, and for any natural number n,

$$a^{-n} = \frac{1}{a^n}.$$

Since any negative integer is the negative of a natural number, this defines exponentiation by negative integers.

example 2

a. $\frac{y^2}{y^7} = \frac{y \cdot y}{y \cdot y \cdot y \cdot y \cdot y \cdot y \cdot y} = \frac{1}{y \cdot y \cdot y \cdot y \cdot y} = \frac{1}{y^5} = y^{-5}.$

b. $\frac{6x^2}{-3x^2} = \frac{6}{-3} = -2.$ Note that the variable x cancelled out entirely.

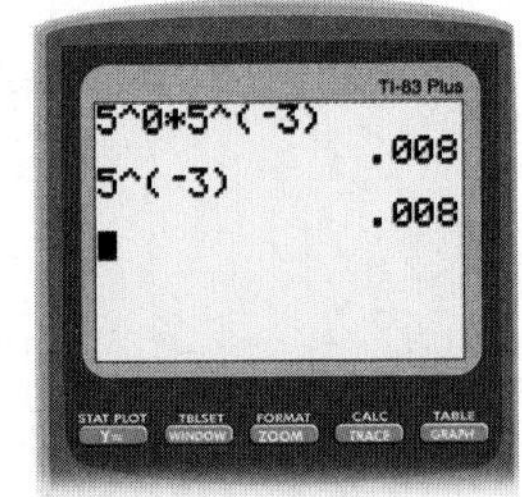

c. $5^0 5^{-3} = 5^{0-3} = 5^{-3} = \frac{1}{125}.$ Note that $5^0 = 1$, as does a^0 for any $a \neq 0$.

d. $\frac{1}{t^{-3}} = \frac{1}{\frac{1}{t^3}} = 1 \cdot \frac{t^3}{1} = t^3.$

e. $(x^2 y)^3 = (x^2 y)(x^2 y)(x^2 y) = x^2 \cdot x^2 \cdot x^2 \cdot y \cdot y \cdot y = x^{2+2+2} y^{1+1+1} = x^6 y^3.$

Topic 2: Properties of Exponents and Their Use

The table below lists the properties of exponents that are used frequently in algebra. Most of these properties have been illustrated already in Examples 1 and 2. All of them can be readily demonstrated by applying the definition of integer exponents (see Exercises 25 through 32).

Properties of Exponents

Throughout this table, *a* and *b* may be taken to represent constants, variables, or more complicated algebraic expressions. The letters *n* and *m* represent integers.

Property	Example	Description
$a^n \cdot a^m = a^{n+m}$	$3^3 \cdot 3^{-1} = 3^{3+(-1)} = 3^2 = 9$	Add exponents when multiplying two powers of the same number.
$\dfrac{a^n}{a^m} = a^{n-m}$	$\dfrac{7^9}{7^{10}} = 7^{9-10} = 7^{-1}$	Subtract exponents when dividing.
$a^{-n} = \dfrac{1}{a^n}$	$5^{-2} = \dfrac{1}{5^2} = \dfrac{1}{25}$ and $x^3 = \dfrac{1}{x^{-3}}$	Invert the fraction and change the sign of the exponent.
$\left(a^n\right)^m = a^{nm}$	$\left(2^3\right)^2 = 2^{3\cdot 2} = 2^6 = 64$	Multiply exponents when raising to a new power.
$(ab)^n = a^n b^n$	$(7x)^3 = 7^3 x^3 = 343x^3$ $\left(-2x^5\right)^2 = (-2)^2\left(x^5\right)^2 = 4x^{10}$	Raise each factor to the power.
$\left(\dfrac{a}{b}\right)^n = \dfrac{a^n}{b^n}$	$\left(\dfrac{3}{x}\right)^2 = \dfrac{3^2}{x^2} = \dfrac{9}{x^2}$ and $\left(\dfrac{1}{3z}\right)^2 = \dfrac{1^2}{(3z)^2} = \dfrac{1}{9z^2}$	Raise both numerator and denominator to the power.
$\left(\dfrac{a}{b}\right)^{-n} = \dfrac{b^n}{a^n}$	$\left(\dfrac{5}{4}\right)^{-3} = \dfrac{4^3}{5^3} = \dfrac{64}{125}$	Raise both numerator and denominator to the power. Then invert the fraction and change the sign of the exponents.
$\dfrac{a^{-m}}{b^{-n}} = \dfrac{b^n}{a^m}$	$\dfrac{3^{-2}}{2^{-4}} = \dfrac{2^4}{3^2} = \dfrac{16}{9}$	Invert the fraction and change the sign of the exponents.

In the above table, it is assumed that every expression is defined.

example 3

Simplify the following expressions by using the properties of exponents. Write the final answers with only positive exponents. (As in the table on the previous page, it is assumed that every expression is defined.)

a. $(17x^4 - 5x^2 + 2)^0$ **b.** $\dfrac{(x^2y^3)^{-1}z^{-2}}{x^3z^{-3}}$ **c.** $\dfrac{(-2x^3y^{-1})^{-3}}{(18x^{-3})^0(xy)^{-2}}$

d. $(7xz^{-2})^2(5x^2y)^{-1}$

Solutions:

a. $(17x^4 - 5x^2 + 2)^0 = 1$

Any non-zero expression with an exponent of 0 is 1.

b. $$\frac{(x^2y^3)^{-1}z^{-2}}{x^3z^{-3}} = \frac{x^{-2}y^{-3}z^{-2}}{x^3z^{-3}} = \frac{z^3}{x^3x^2y^3z^2} = \frac{z}{x^5y^3}$$

Note that we have used several properties in this example. We could have used the applicable properties in many different orders to achieve the same result. Also note that the final answer contains only positive exponents. If we had not been told to write the answer in this way, we could have written the result as $zx^{-5}y^{-3}$.

c. $$\frac{(-2x^3y^{-1})^{-3}}{(18x^{-3})^0(xy)^{-2}} = \frac{(-2)^{-3}x^{-9}y^3}{(1)x^{-2}y^{-2}} = (-2)^{-3}x^{-9-(-2)}y^{3-(-2)} = (-2)^{-3}x^{-7}y^5 = \frac{y^5}{-8x^7} = -\frac{y^5}{8x^7}$$

We have chosen to apply the appropriate properties in a slightly different order than in the previous example, just to illustrate an alternative way to go about the task of simplifying such an expression.

d. $$(7xz^{-2})^2(5x^2y)^{-1} = \frac{49x^2z^{-4}}{5x^2y} = \frac{49}{5yz^4}$$

Note that the variable x disappeared entirely from the expression. If we had simplified the expression in a slightly different order, we would have obtained a factor of x^{2-2}, which is 1.

technology note

Computer algebra systems, as well as some calculators, are programmed to apply the properties of exponents and can demonstrate the result of simplifying expressions. In *Mathematica*, the command to do this is **Simplify**, illustrated with the first expression below. In many cases, *Mathematica* will simplify a given expression automatically upon evaluation, making the **Simplify** command unnecessary. This is illustrated with the second and third expressions below. *(See Appendix A for guidance on using computer algebra systems like Mathematica.)*

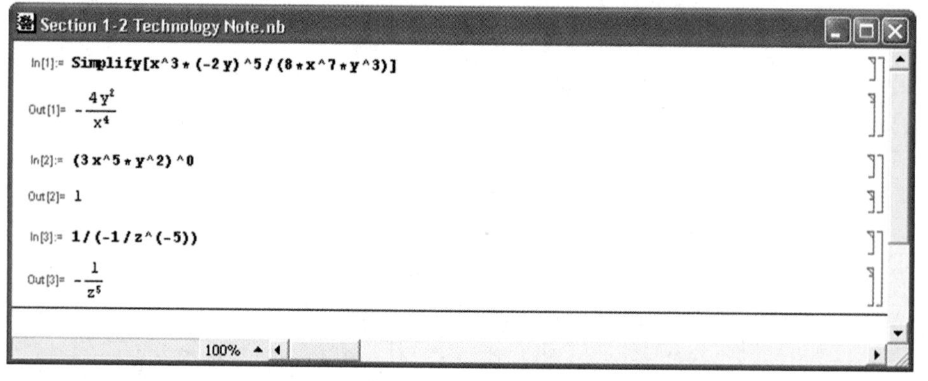

caution!

Many errors are commonly made in applying the properties of exponents and are a result of forgetting the exact form of the properties. The first column below contains examples of some common errors. The second column contains the corrected statements.

Incorrect Statements	Corrected Statements
$x^2x^5 = x^{10}$	$x^2x^5 = x^{2+5} = x^7$
$2^42^3 = 4^7$	$2^42^3 = 2^{4+3} = 2^7$
$\left(x^2+3y\right)^{-1} = \frac{1}{x^2} + \frac{1}{3y}$	$\left(x^2+3y\right)^{-1} = \frac{1}{x^2+3y}$
$(3x)^2 = 3x^2$	$(3x)^2 = 3^2x^2 = 9x^2$
$\frac{x^5}{x^{-2}} = x^3$	$\frac{x^5}{x^{-2}} = x^{5-(-2)} = x^7$

Topic 3: Scientific Notation

Scientific notation is an important application of exponents. Very large and very small numbers arise naturally in a variety of situations, and working with them without scientific notation is an unwieldy and error-prone process. Scientific notation uses the properties of exponents to rewrite very large and very small numbers in a less clumsy form. It applies to those numbers which contain a long string of 0's before the decimal point (very large numbers) or to those which contain a long string of 0's between the decimal point and the non-zero digits (very small numbers).

Scientific Notation

A number is in scientific notation when it is written in the form

$$a \times 10^n,$$

where $1 \le |a| < 10$ and n is an integer.
If n is a positive integer, the number is large in magnitude, and if n is a negative integer, the number is small in magnitude (close to 0). The number a itself can be either positive or negative, and the sign of a determines the sign of the number as a whole.

caution!

The sign of the exponent n in scientific notation does *not* determine the sign of the number as a whole. The sign of n only determines if the number is large (positive n), or small (negative n), in magnitude.

example 4

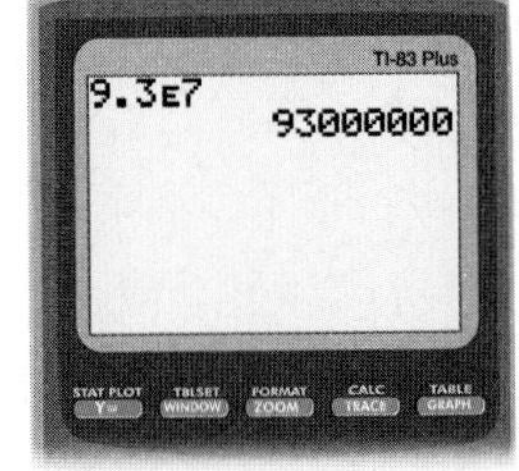

a. The distance from the earth to the sun is approximately 93,000,000 miles. Scientific notation takes advantage of the observation that multiplication of a number by 10 moves the decimal point one place to the right, and we can repeat this process as many times as necessary. Thus, 9.3×10^7 is equal to 93,000,000, as 93,000,000 is obtained from 9.3 by moving the decimal point 7 places to the right:

$$9.3 \times 10^7 = 9\underbrace{3000000}_{\text{7 places}}$$

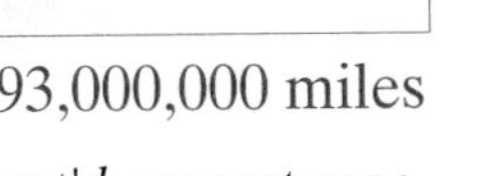

93,000,000 miles

cont'd. on next page ...

b. The mass of an electron, in kilograms, is approximately

$$0.000000000000000000000000000000911,$$

clearly not a convenient number to work with. Scientific notation takes advantage of the observation that multiplication of a number by 10^{-1} (which is equivalent to division by 10) moves the decimal point one place to the left. Again, we can repeat this process as many times as necessary. Thus, in scientific notation,

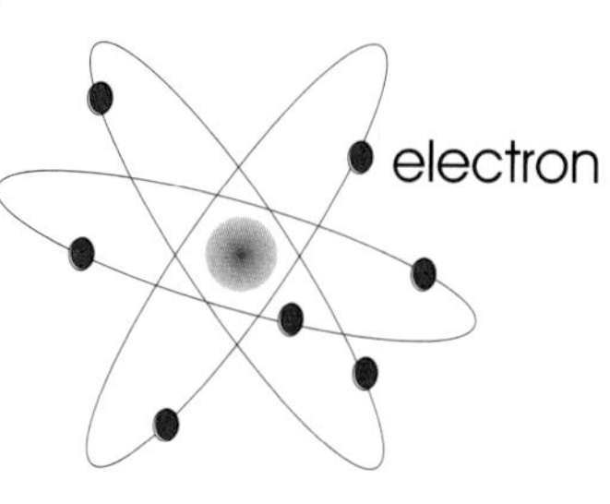

$$0.\underbrace{000000000000000000000000000000 9}_{31 \text{ places}}11 = 9.11 \times 10^{-31}.$$

c. The speed of light in a vacuum is approximately 3×10^{8} meters/second. In standard (non-scientific) notation, this number is written as 300,000,000.

We can also use the properties of exponents to simplify computations involving two or more numbers that are large or small in magnitude, as illustrated by the next set of examples.

example 5

Simplify the following expressions, writing your answer either in scientific or standard notation, as appropriate.

a. $\dfrac{\left(3.6 \times 10^{-12}\right)\left(-6 \times 10^{4}\right)}{1.8 \times 10^{-6}}$ **b.** $\dfrac{\left(7 \times 10^{34}\right)\left(3 \times 10^{-12}\right)}{6 \times 10^{-7}}$

Solutions:

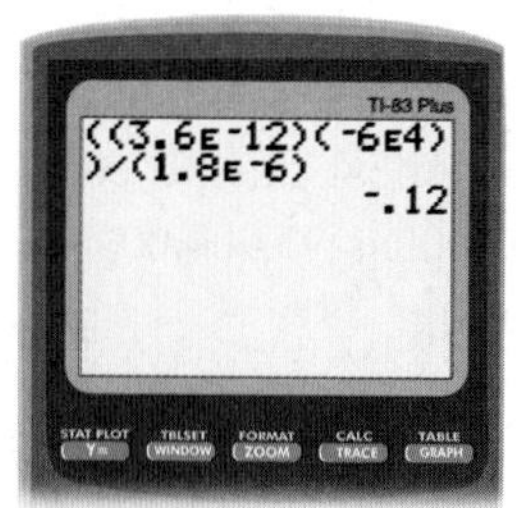

a. $$\frac{\left(3.6 \times 10^{-12}\right)\left(-6 \times 10^{4}\right)}{1.8 \times 10^{-6}} = \frac{(3.6)(-6)}{1.8} \times 10^{-12+4-(-6)}$$
$$= -12 \times 10^{-2}$$
$$= -0.12$$

We have written the final answer in standard notation, as it is not inconvenient to do so. Note that in scientific notation, it would be written as -1.2×10^{-1}.

b. $$\frac{\left(7 \times 10^{34}\right)\left(3 \times 10^{-12}\right)}{6 \times 10^{-7}} = \frac{(7)(3)}{6} \times 10^{34+(-12)-(-7)}$$
$$= 3.5 \times 10^{29}$$

This answer is best written in scientific notation.

Topic 4: Interlude: Working with Geometric Formulas

Exponents occur in a very natural way when geometric formulas are considered. Some of the problems that you encounter in this book (and elsewhere) will require nothing more than using one of the basic geometric formulas found on the inside front cover of the text, but others will require a bit more work. Often, the exact geometric formula that you need to solve a given problem can be derived from simpler formulas.

We will look at several examples of how a new geometric formula is built up from known formulas. The general rule of thumb in each case is to break down the task at hand into smaller pieces that can be easily handled.

example 6

Find formulas for each of the following:

a. The surface area of a box.

b. The surface area of a soup can.

c. The volume of a birdbath in the shape of half of a sphere.

d. The volume of a gold ingot whose shape is a right trapezoidal cylinder.

Solutions:

a. A box whose six faces are all rectangular is characterized by its length l, its width w, and its height h. The area of a rectangle is one of the basic formulas that you should be familiar with, and can be found on the inside front cover of this text. The formula for the surface area of a box, then, is just the sum of the areas of the six sides. If we let S stand for the total surface area, we obtain the formula $S = lw + lw + lh + lh + hw + hw$, or $S = 2lw + 2lh + 2hw$.

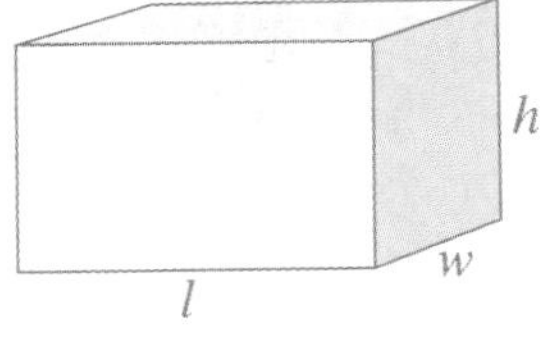

b. A soup can, an example of a right circular cylinder, is characterized by its height h and the radius r of the circle that makes up the base (or the top).

cont'd. on next page ...

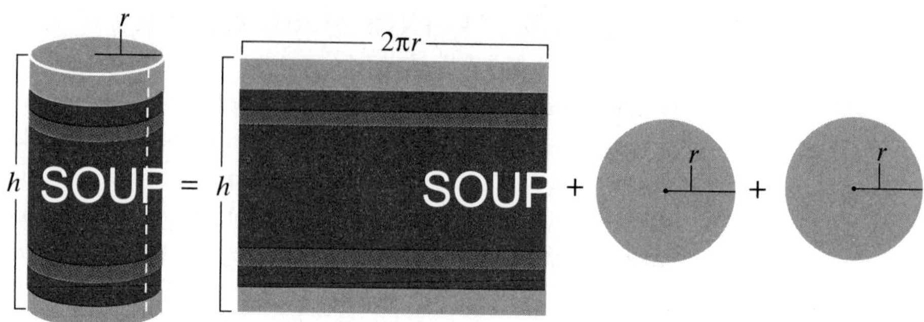

To determine the surface area of such a shape, imagine removing the top and bottom surfaces, cutting the soup can along the dotted line as shown, and flattening out the curved piece of metal making up the side. The flattened piece of metal would be a rectangle with height h and width $2\pi r$. Do you see why? The width of the rectangle would be the same as the circumference of the circular top and base, and the circumference of a circle is $2\pi r$. So the surface area of the curved side is $2\pi rh$. We also know that the area of a circle is πr^2, so if we let S stand for the surface area of the entire can, we have $S = \pi r^2 + \pi r^2 + 2\pi rh$, or $S = 2\pi r^2 + 2\pi rh$.

c. The volume of a sphere of radius r is $\frac{4}{3}\pi r^3$, and the birdbath we are to find the volume of has the shape of half a sphere. So if we let V stand for the birdbath's volume,

$$V = \left(\frac{1}{2}\right)\left(\frac{4}{3}\pi r^3\right), \text{ or } V = \frac{2}{3}\pi r^3.$$

d. A *right cylinder* is the three-dimensional object generated by extending a plane region along an axis perpendicular to itself for a certain distance. (Such objects are often called prisms when the plane region is a polygon.) As indicated on the inside front cover, the volume of any right cylinder is the product of the area of the plane region and the distance that region has been extended perpendicular to itself. The gold ingot under consideration in this example is a right cylinder based on a trapezoid, as shown. The area of the trapezoid is $\frac{1}{2}(B+b)h$, and the ingot has length l, so its volume is $V = \frac{1}{2}(B+b)hl$. This could also be written as

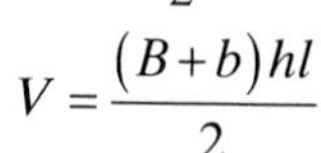

$$V = \frac{(B+b)hl}{2}.$$

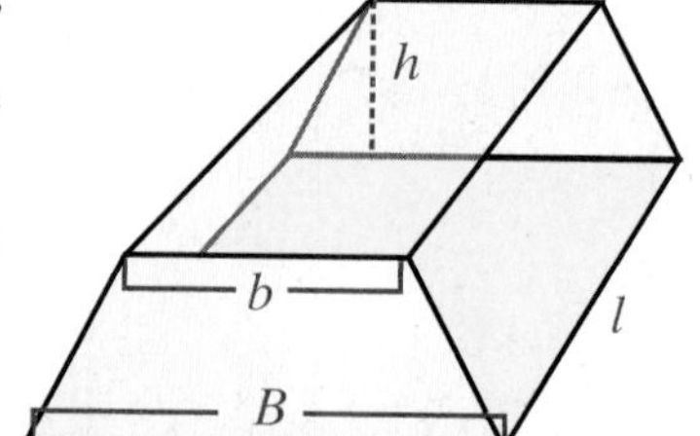

Topic 5: Roots and Radical Notation

The nth root of an expression, where n is a natural number, is related to exponentiation by n in the sense of being the opposite operation. For example, the statement $3^3 = 27$ (which can be read as "three cubed is twenty-seven") implies that the *cube root* of 27 is 3 (written $\sqrt[3]{27} = 3$). By analogy, the statement $2^4 = 16$ would lead us to write $\sqrt[4]{16} = 2$ (read "the *fourth root* of sixteen is two").

n^{th} Roots and Radical Notation

Case 1: *n* is an even natural number. If a is a non-negative real number and n is an even natural number, $\sqrt[n]{a}$ is the non-negative real number b with the property that $b^n = a$. That is,

$$\sqrt[n]{a} = b \Leftrightarrow a = b^n.$$

In this case, note that $\left(\sqrt[n]{a}\right)^n = a$ and $\sqrt[n]{a^n} = a$.

Case 2: *n* is an odd natural number. If a is any real number and n is an odd natural number, $\sqrt[n]{a}$ is the real number b (whose sign will be the same as the sign of a) with the property that $b^n = a$. Again, $\sqrt[n]{a} = b \Leftrightarrow a = b^n$, $\left(\sqrt[n]{a}\right)^n = a$ and $\sqrt[n]{a^n} = a$.

The expression $\sqrt[n]{a}$ expresses the nth root of a in **radical notation**. The natural number n is called the **index**, a is the **radicand**, and $\sqrt{}$ is called a **radical sign**. By convention, $\sqrt[2]{}$ is usually simply written as $\sqrt{}$.

The important distinction between the two cases is that when n is even, $\sqrt[n]{a}$ is defined only when a is non-negative, whereas if n is odd, $\sqrt[n]{a}$ is defined for all real numbers a. This asymmetrical situation will be remedied in Section 1.4, but to do so requires the introduction of *complex numbers*.

example 7

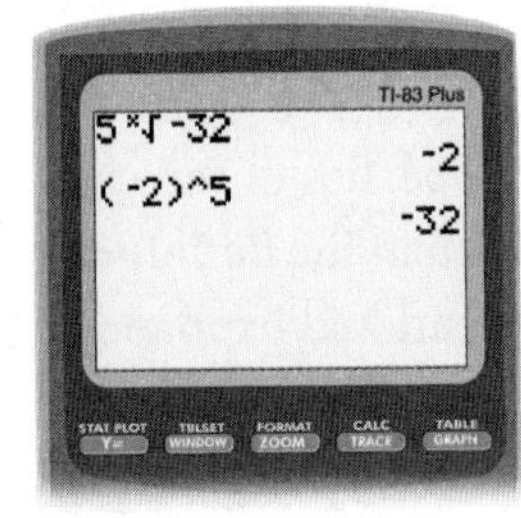

a. $\sqrt[5]{-32} = -2$ because $(-2)^5 = -32$.

b. $\sqrt[4]{-16}$ is not a real number, as no real number raised to the fourth power is -16.

c. $-\sqrt[4]{16} = -2$. Note that the fourth root of 16 is a real number, which in this expression is then multiplied by -1.

d. $\sqrt{0} = 0$. In fact, $\sqrt[n]{0} = 0$ for any natural number n.

cont'd. on next page ...

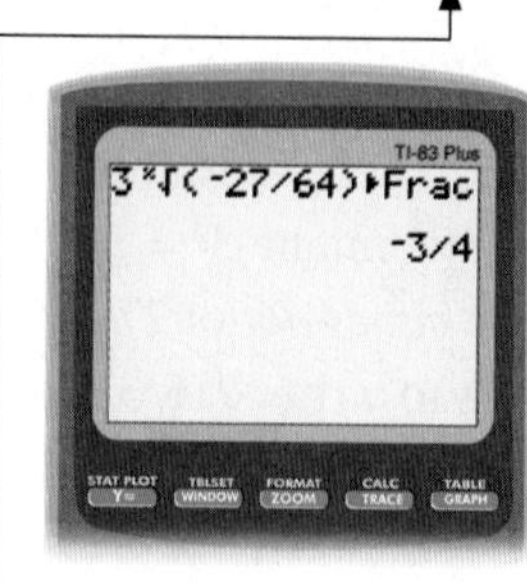

To force your answer to be in fraction form, go to the Math menu and select Frac.

e. $\sqrt[n]{1}=1$ for any natural number n. $\sqrt[n]{-1}=-1$ for any odd natural number n.

f. $\sqrt[3]{-\frac{27}{64}}=-\frac{3}{4}$ because $\left(-\frac{3}{4}\right)^3=-\frac{27}{64}$.

g. $\sqrt[5]{-\pi^5}=-\pi$ because $(-\pi)^5=(-1)^5\pi^5=-\pi^5$.

h. $\sqrt[4]{(-3)^4}=\sqrt[4]{81}=3$. In general, if n is an even natural number, $\sqrt[n]{a^n}=|a|$ for any real number a. Remember, though, that $\sqrt[n]{a^n}=a$ if n is an odd natural number.

Topic 6: Simplifying and Combining Radical Expressions

When solving equations that contain radical expressions, it is often helpful to simplify the expressions first. The word *simplify* is one which arises frequently in mathematics, and its meaning depends on the context.

Simplified Radical Form

A radical expression is in **simplified form** when:

1. The radicand contains no factor with an exponent greater than or equal to the index of the radical.
2. The radicand contains no fractions.
3. The denominator, if there is one, contains no radical.
4. The greatest common factor of the index and any exponent occurring in the radicand is 1. That is, the index and any exponent in the radicand have no common factor other than 1.

Before illustrating the process of simplifying radicals, a few useful properties of radicals will be listed. These properties can all be proved using nothing more than the definition of roots, but their validity will be more clear once we have discussed rational exponents later in this section.

Properties of Radicals

Throughout the following table, *a* and *b* may be taken to represent constants, variables, or more complicated algebraic expressions. The letters *n* and *m* represent natural numbers. It is assumed that all expressions are defined and are real numbers.

Property	Example
$\sqrt[n]{a^n} = a$ if n is odd	$\sqrt[3]{(-5)^3} = -5, \ \sqrt[7]{3^7} = 3$
$\sqrt[n]{a^n} = \lvert a \rvert$ if n is even	$\sqrt[4]{(-6)^4} = \lvert -6 \rvert = 6$
$\sqrt[n]{ab} = \sqrt[n]{a} \cdot \sqrt[n]{b}$	$\sqrt[3]{3x^6y^2} = \sqrt[3]{3} \cdot \sqrt[3]{x^3} \cdot \sqrt[3]{x^3} \cdot \sqrt[3]{y^2} = \sqrt[3]{3} \cdot x \cdot x \cdot \sqrt[3]{y^2} = x^2\sqrt[3]{3y^2}$
$\sqrt[n]{\frac{a}{b}} = \frac{\sqrt[n]{a}}{\sqrt[n]{b}}$	$\sqrt[4]{\frac{x^4}{16}} = \frac{\sqrt[4]{x^4}}{\sqrt[4]{16}} = \frac{\lvert x \rvert}{2}$
$\sqrt[m]{\sqrt[n]{a}} = \sqrt[mn]{a}$	$\sqrt[3]{\sqrt{64}} = \sqrt[3]{\sqrt[2]{64}} = \sqrt[6]{64} = 2$

example 8

Simplify the following radical expressions:

a. $\sqrt[3]{-16x^8y^4}$ **b.** $\sqrt{8z^6}$ **c.** $\sqrt[3]{\frac{72x^2}{y^3}}$

Solutions:

a. $\sqrt[3]{-16x^8y^4} = \sqrt[3]{(-2)^3 \cdot 2 \cdot x^3 \cdot x^3 \cdot x^2 \cdot y^3 \cdot y}$
$= -2x^2y\sqrt[3]{2x^2y}$

Note that since the index is 3, we look for all of the *perfect cubes* in the radicand.

b. $\sqrt{8z^6} = \sqrt{2^2 \cdot 2 \cdot (z^3)^2} = \lvert 2z^3 \rvert \sqrt{2} = 2\lvert z^3 \rvert \sqrt{2}$

Remember that $\sqrt[n]{a^n} = \lvert a \rvert$ if n is even, so absolute value signs are necessary.

c. $\sqrt[3]{\frac{72x^2}{y^3}} = \frac{\sqrt[3]{8 \cdot 9 \cdot x^2}}{\sqrt[3]{y^3}} = \frac{2\sqrt[3]{9x^2}}{y}$

All perfect cubes have been brought out from under the radical, and the denominator has been rationalized.

caution!

Notice that the expressions are not equal.

As with the properties of exponents, many mistakes arise from forgetting the properties of radicals. One common error, for instance, is to rewrite $\sqrt{a+b}$ as $\sqrt{a}+\sqrt{b}$. These two expressions are not equal! To convince yourself of this, evaluate the two expressions with actual constants in place of a and b. For example, note that $5=\sqrt{9+16}\neq\sqrt{9}+\sqrt{16}=7$.

Rationalizing denominators sometimes requires more effort than in Example 8c, and in fact is sometimes impossible. The following methods will, however, take care of two common cases.

Two Rationalization Methods

Case 1: Denominator is a single term containing a root.

If the denominator is a single term containing a factor of $\sqrt[n]{a^m}$, we will take advantage of the fact that

$$\sqrt[n]{a^m}\cdot\sqrt[n]{a^{n-m}}=\sqrt[n]{a^m\cdot a^{n-m}}=\sqrt[n]{a^n},$$

and this last expression is either a or $|a|$, depending on whether n is odd or even. Of course, we can't multiply the denominator by a factor of $\sqrt[n]{a^{n-m}}$ without multiplying the numerator by the same factor, as we otherwise change the expression. The method in this case is thus to multiply the fraction by $\frac{\sqrt[n]{a^{n-m}}}{\sqrt[n]{a^{n-m}}}$.

For example, $\frac{1}{\sqrt{a}}=\frac{1}{\sqrt{a}}\cdot\frac{1}{1}=\frac{1}{\sqrt{a}}\cdot\frac{\sqrt{a}}{\sqrt{a}}=\frac{\sqrt{a}}{a}$.

Case 2: Denominator consists of two terms, one or both of which are square roots.

To discuss the method used in this case, let $A+B$ represent the denominator of the fraction under consideration, where at least one of A and B stands for a square root term. We will take advantage of the fact that

$$(A+B)(A-B)=A(A-B)+B(A-B)=A^2-AB+BA-B^2=A^2-B^2,$$

and the exponents of 2 in the end result negate the square root (or roots) initially in the denominator. Just as in Case 1, though, we can't multiply the denominator by $A-B$ unless we multiply the numerator by this same factor. The method is thus

to multiply the fraction by $\frac{A-B}{A-B}$. The factor $A-B$ is called the **conjugate radical expression** of $A+B$.

For example, $\frac{1}{\sqrt{a}+\sqrt{b}}=\frac{1}{\sqrt{a}+\sqrt{b}}\cdot\frac{\sqrt{a}-\sqrt{b}}{\sqrt{a}-\sqrt{b}}=\frac{\sqrt{a}-\sqrt{b}}{a^2-b^2}$.

example 9

Simplify the following radical expressions:

a. $\sqrt[5]{\frac{-4x^6}{8y^2}}$ **b.** $\frac{4}{\sqrt{7}+\sqrt{3}}$ **c.** $\frac{-\sqrt{5x}}{5-\sqrt{x}}$

Solutions:

a. $\sqrt[5]{\frac{-4x^6}{8y^2}}=\frac{\sqrt[5]{-4x\cdot x^5}}{\sqrt[5]{2^3y^2}}=\frac{-x\sqrt[5]{4x}}{\sqrt[5]{2^3y^2}}\cdot\frac{\sqrt[5]{2^2y^3}}{\sqrt[5]{2^2y^3}}=\frac{-x\sqrt[5]{16xy^3}}{\sqrt[5]{2^5y^5}}=\frac{-x\sqrt[5]{16xy^3}}{2y}$

The first step is to simplify the numerator and denominator and determine what the denominator must be multiplied by in order to eliminate the radical. Remember to multiply the numerator by the same factor.

b. $\frac{4}{\sqrt{7}+\sqrt{3}}=\left(\frac{4}{\sqrt{7}+\sqrt{3}}\right)\left(\frac{\sqrt{7}-\sqrt{3}}{\sqrt{7}-\sqrt{3}}\right)=\frac{4\left(\sqrt{7}-\sqrt{3}\right)}{7-3}=\frac{4\left(\sqrt{7}-\sqrt{3}\right)}{4}=\sqrt{7}-\sqrt{3}$

Again, we have multiplied the fraction by 1, but this time we have had to multiply the numerator and denominator by the conjugate radical of the original denominator.

c. $\frac{-\sqrt{5x}}{5-\sqrt{x}}=\left(\frac{-\sqrt{5x}}{5-\sqrt{x}}\right)\left(\frac{5+\sqrt{x}}{5+\sqrt{x}}\right)=\frac{-5\sqrt{5x}-\sqrt{5x^2}}{25-x}=\frac{-5\sqrt{5x}-x\sqrt{5}}{25-x}$

In this example, the original denominator is the sum of 5 and $-\sqrt{x}$, so the conjugate radical of the denominator is $5+\sqrt{x}$. Note that in simplifying the term $-\sqrt{5x^2}$, we can simply write $-x\sqrt{5}$ instead of $-|x|\sqrt{5}$, since the original expression is not real if x is negative.

Often, a sum of two or more radical expressions can be combined into one. This can be done if the radical expressions are **like radicals**, meaning that they have the same index and the same radicand. It is frequently necessary to simplify the radical expressions before it can be determined if they are like or not.

example 10

Combine the radical expressions, if possible.

a. $-3\sqrt{8x^5}+\sqrt{18x}$ **b.** $\sqrt[3]{54x^3}+\sqrt{50x^2}$ **c.** $\sqrt{\frac{1}{12}}-\sqrt{\frac{25}{48}}$

Solutions:

a. $-3\sqrt{8x^5}+\sqrt{18x}=-3\sqrt{2^2\cdot 2\cdot x^4\cdot x}+\sqrt{2\cdot 3^2\cdot x}=-6|x^2|\sqrt{2x}+3\sqrt{2x}$
$$=-6x^2\sqrt{2x}+3\sqrt{2x}=\left(-6x^2+3\right)\sqrt{2x}$$

Start by simplifying the radicals. Upon simplification, the two radicals have the same index and radicand. The absolute value bars around the factor of x^2 are unnecessary, since x^2 is always non-negative.

b. $\sqrt[3]{54x^3}+\sqrt{50x^2}=\sqrt[3]{2\cdot 3^3\cdot x^3}+\sqrt{2\cdot 5^2\cdot x^2}=3x\sqrt[3]{2}+5|x|\sqrt{2}$

Upon simplification, the radicands are the same, but the indices are not. We have written the radicals in simplest form, but they cannot be combined.

c. $\sqrt{\frac{1}{12}}-\sqrt{\frac{25}{48}}=\frac{1}{\sqrt{2^2\cdot 3}}-\frac{\sqrt{5^2}}{\sqrt{4^2\cdot 3}}=\frac{1}{2\sqrt{3}}\cdot\frac{\sqrt{3}}{\sqrt{3}}-\frac{5}{4\sqrt{3}}\cdot\frac{\sqrt{3}}{\sqrt{3}}=\frac{2\sqrt{3}}{4\cdot 3}-\frac{5\cdot\sqrt{3}}{4\cdot 3}=-\frac{3\sqrt{3}}{12}=-\frac{\sqrt{3}}{4}$

First, simplify the radicals. Note that both denominators are being rationalized in this step. The first fraction must be multiplied by $\frac{2}{2}$ in order to get a common denominator. After subtraction, the fraction can be reduced.

Topic 7: Rational Number Exponents

We can now return to the task of defining exponentiation and give meaning to a^r when r is a rational number.

Rational Number Exponents

Meaning of $a^{1/n}$: If n is a natural number and if $\sqrt[n]{a}$ is a real number, then $a^{1/n}=\sqrt[n]{a}$.

Meaning of $a^{m/n}$: If m and n are natural numbers with $n\neq 0$, if m and n have no common factors greater than 1, and if $\sqrt[n]{a}$ is a real number, then
$$a^{m/n}=\sqrt[n]{a^m}=\left(\sqrt[n]{a}\right)^m.$$

Either $\sqrt[n]{a^m}$ or $\left(\sqrt[n]{a}\right)^m$ can be used to evaluate $a^{m/n}$. $a^{-m/n}$ is defined to be $\frac{1}{a^{m/n}}$.

In addition to giving meaning to rational exponentiation, the previous definition provides us with the means of converting between *radical notation* and *exponential notation*. In some cases, one notation or the other is more convenient, and the choice of notation should be based on which makes the task at hand easier.

Although originally stated only for integer exponents, the properties of exponents listed earlier also hold for rational exponents (and for real exponents as well).

example 11

Simplify each of the following expressions, writing your answer using the same notation as the original expression.

a. $27^{-2/3}$ **b.** $\sqrt[9]{-8x^6}$ **c.** $(5x^2+3)^{8/3}(5x^2+3)^{-2/3}$

d. $\sqrt[5]{\sqrt[3]{x^2}}$ **e.** $\dfrac{5x-y}{(5x-y)^{-1/3}}$

Solutions:

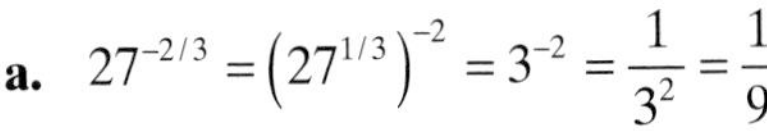

a. $27^{-2/3}=\left(27^{1/3}\right)^{-2}=3^{-2}=\dfrac{1}{3^2}=\dfrac{1}{9}$

The only task here is to evaluate the expression. We could also have begun by noting that $27^{-2/3}=\left(27^{-2}\right)^{1/3}$, but this would have made the calculations much more tedious.

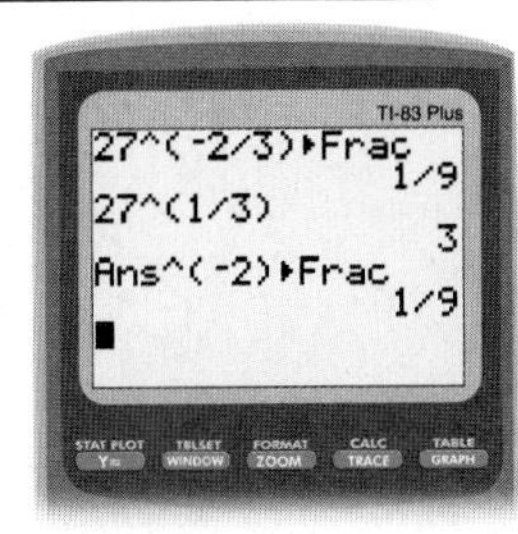

To force your answer to be in fraction form, go to the **Math** menu and select `Frac`.

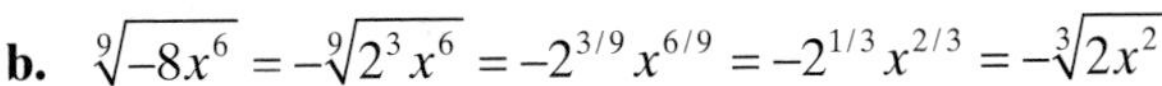

b. $\sqrt[9]{-8x^6}=-\sqrt[9]{2^3x^6}=-2^{3/9}x^{6/9}=-2^{1/3}x^{2/3}=-\sqrt[3]{2x^2}$

The only factor that can be brought out from under the radical is -1, but we can still simplify the expression by reducing the exponents and index. Note that we switched to exponential form temporarily in doing this.

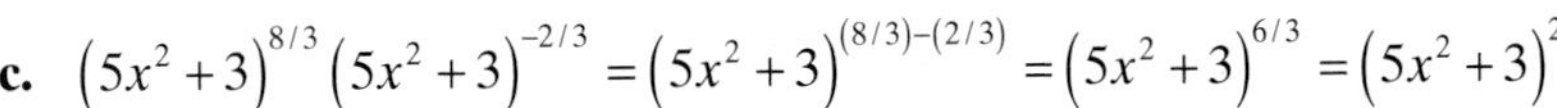

c. $(5x^2+3)^{8/3}(5x^2+3)^{-2/3}=(5x^2+3)^{(8/3)-(2/3)}=(5x^2+3)^{6/3}=(5x^2+3)^2$

Since the bases are the same, we only have to apply the property $a^n a^m=a^{n+m}$.

d. $\sqrt[5]{\sqrt[3]{x^2}}=\sqrt[15]{x^2}$ One property simplifies this expression.

e. $\dfrac{5x-y}{(5x-y)^{-1/3}}=(5x-y)^{1-(-1/3)}=(5x-y)^{4/3}$

We use the property $\dfrac{a^n}{a^m}=a^{n-m}$ to simplify this expression.

The following example requires a bit more work and/or caution.

example 12

a. Simplify the expression $\sqrt[4]{x^2}$.

b. Write $\sqrt[3]{2}\cdot\sqrt{3}$ as a single radical.

Solutions:

a. $\sqrt[4]{x^2} = \left(x^2\right)^{1/4} = |x|^{1/2} = \sqrt{|x|}$

We might be tempted to write $\sqrt[4]{x^2}$ as simply $\sqrt{x}$, but note that $\sqrt[4]{x^2}$ is defined for *all* real numbers x, while $\sqrt{x}$ is defined only for non-negative real numbers.

b. $\sqrt[3]{2}\cdot\sqrt{3} = 2^{1/3}\cdot 3^{1/2} = 2^{2/6}\cdot 3^{3/6} = \left(2^2\right)^{1/6}\left(3^3\right)^{1/6} = 4^{1/6}\cdot 27^{1/6} = 108^{1/6} = \sqrt[6]{108}$

We can make use of the property $a^n b^n = (ab)^n$ if we can first make the exponents equal. We do so by finding the least common denominator of $\frac{1}{3}$ and $\frac{1}{2}$, writing both fractions with this common denominator, and then making use of the property $a^{nm} = \left(a^n\right)^m$.

exercises

Simplify each of the following expressions, writing your answer with only positive exponents. See Examples 1 and 2.

1. $(-2)^4$ **2.** -2^4 **3.** $3^2\cdot 3^2$ **4.** $2^3\cdot 3^2$ **5.** $4\cdot 4^2$

6. $(-3)^3$ **7.** $\dfrac{8^2}{4^3}$ **8.** $2^2\cdot 2^3$ **9.** $\dfrac{7^4}{7^5}$ **10.** $n^2\cdot n^5$

11. $\dfrac{x^5}{x^2}$ **12.** $\dfrac{y^3\cdot y^8}{y^2}$

Simplify each of the following expressions, writing your answer with only positive exponents. See Example 2.

13. $\dfrac{3t^{-2}}{t^3}$ **14.** $-2y^0$ **15.** $\dfrac{1}{7x^{-5}}$ **16.** $9^0x^3y^0$ **17.** $\dfrac{2n^3}{n^{-5}}$

18. $\dfrac{11^{21}}{11^{19}x^{-7}}$ **19.** $\dfrac{x^7y^{-3}z^{12}}{x^{-1}z^9}$ **20.** $\dfrac{x^4\left(-x^{-3}\right)}{-y^0}$ **21.** $\dfrac{s^3}{s^{-2}}$ **22.** $\dfrac{x^{-1}}{x}$

23. $x^{\left(y^0\right)}\cdot x^9$ **24.** $\dfrac{x^2y^{-2}}{x^{-1}y^{-5}}$

Use the properties of exponents to simplify each of the following expressions, writing your answer with only positive exponents. See Example 3.

25. $\dfrac{-9^0\left(x^2y^{-2}\right)^{-3}}{3x^{-4}y}$ **26.** $\left[\left(2x^{-1}z^3\right)^{-2}\right]^{-1}$ **27.** $\dfrac{\left(3z^{-2}y\right)^0}{3zy^2}$

28. $\left(12a^2-3b^4\right)^0$ **29.** $\dfrac{3^{-1}}{\left(3^2xy^2\right)^{-2}}$ **30.** $\left[9m^2-\left(2n^2\right)^3\right]^{-1}$

31. $\left[\left(12x^{-6}y^4z^3\right)^5\right]^0$ **32.** $\dfrac{x\left(x^{-2}y^3\right)^3}{\left(2x^4\right)^{-2}y}$

Convert each number from scientific notation to standard notation, or vice versa, as indicated. See Example 4.

33. -1.76×10^{-5}; convert to standard.
34. $-912{,}000{,}000$; convert to scientific.
35. 0.00000021; convert to scientific.
36. 3.2×10^7; convert to standard.
37. 5100; convert to scientific.
38. -0.000187; convert to scientific.
39. 3.1212×10^2; convert to standard.
40. 1.934×10^{-4}; convert to standard.

Evaluate each expression, using the properties of exponents. Use a calculator only to check your final answer. See Example 5.

41. $\left(2.3\times10^{13}\right)\left(2\times10^{12}\right)$ **42.** $\dfrac{\left(8\times10^{-3}\right)\left(3\times10^{-2}\right)}{2\times10^5}$

43. $\left(2\times10^{-13}\right)\left(5.5\times10^{10}\right)\left(-1\times10^3\right)$ **44.** $\dfrac{\left(4\times10^{34}\right)\left(3\times10^{-32}\right)}{24}$

45. $\left(6\times10^{21}\right)\left(5\times10^{-19}\right)\left(5\times10^4\right)$ **46.** $\left(3.2\times10^7\right)\left(5\times10^{-4}\right)\left(2\times10^{-10}\right)$

47. $\dfrac{4\times10^{-6}}{\left(5\times10^4\right)\left(8\times10^{-3}\right)}$ **48.** $\dfrac{\left(4.6\times10^{12}\right)\left(9\times10^3\right)}{\left(1.5\times10^8\right)\left(2.3\times10^{-5}\right)}$

Apply the definition of integer exponents to demonstrate the following properties.

49. $a^n\cdot a^m=a^{n+m}$ **50.** $\left(a^n\right)^m=a^{nm}$ **51.** $(ab)^n=a^nb^n$

52. A farmer fences in three square garden plots that are situated along a road, as shown. Each square plot has a side-length of s, and he doesn't put fence along the road-side. Find an expression, in the variable s, for the amount of fencing used.

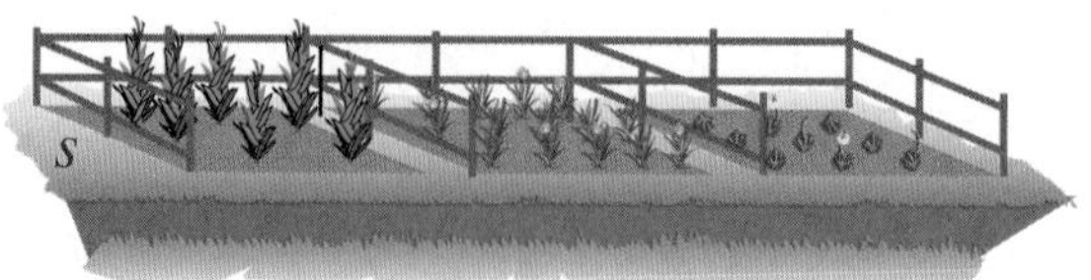

53. The prism shown below is a right triangular cylinder, where the triangular base is a right triangle. Find the volume of the prism in terms of $b, h,$ and l.

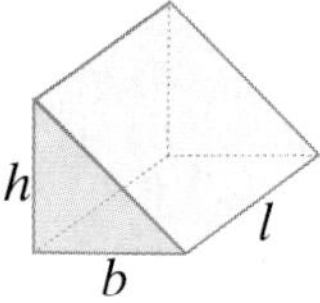

54. Determine the volume of the right circular cylinder shown, in terms of r and h.

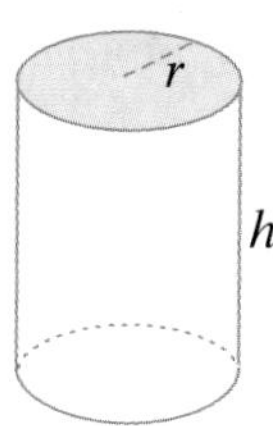

55. Matt wants to let people in the future know what life is like today, so he goes shopping for a time capsule. Capacity, along with price and quality, is an important consideration for him. One time capsule he looks at is a right circular cylinder with a hemisphere on each end. Find the volume of the time capsule, given that the height h is 16 inches and the radius r is 3 inches.

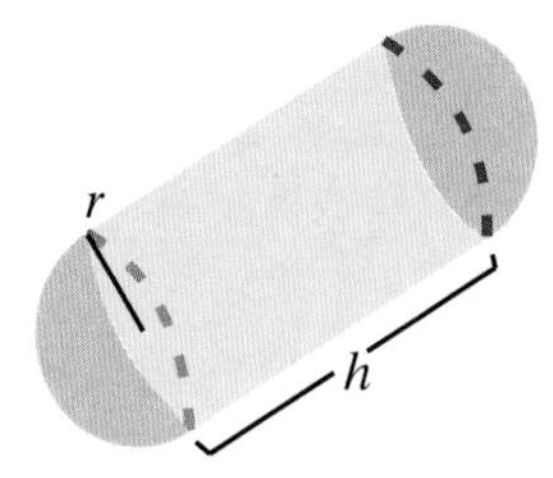

56. Bill and Dee are buying a new house. The house is a right cylinder based on a trapezoid atop a rectangular prism. The bases of the trapezoid are $B = 10$ m and $b = 8$ m, and the length of the house is $l = 15$ m. The height of the house up to the bottom of the roof is $H = 3$ m, and the height of the roof is $h = 1$ m. Find the volume of the house.

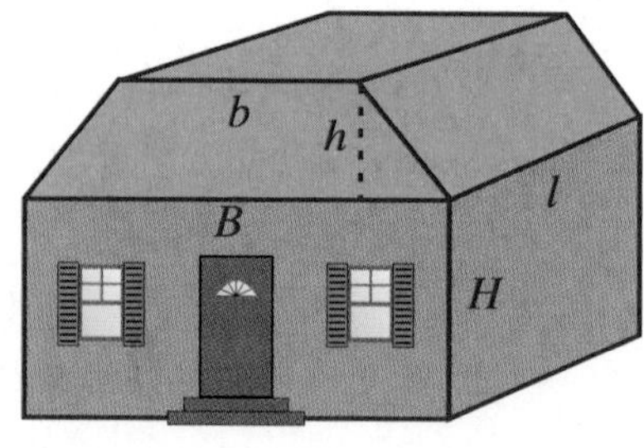

57. Construct the expression for the volume of water contained in a rectangular swimming pool of length l feet and width w feet, assuming the water has a uniform depth of 6 feet.

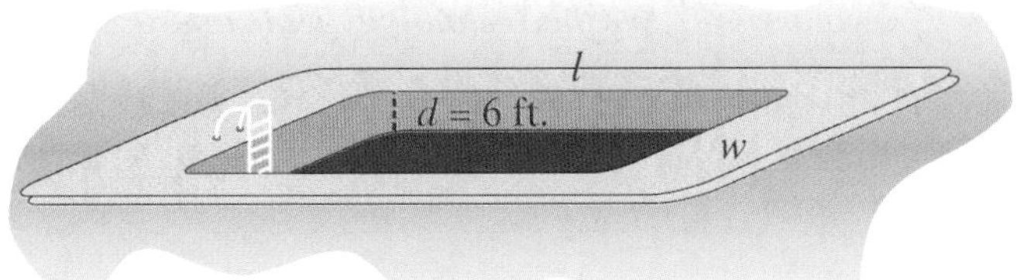

58. Construct the expression for the volume of water contained in an above-ground circular swimming pool that has a diameter of 18 feet, assuming the water has a uniform depth of d feet.

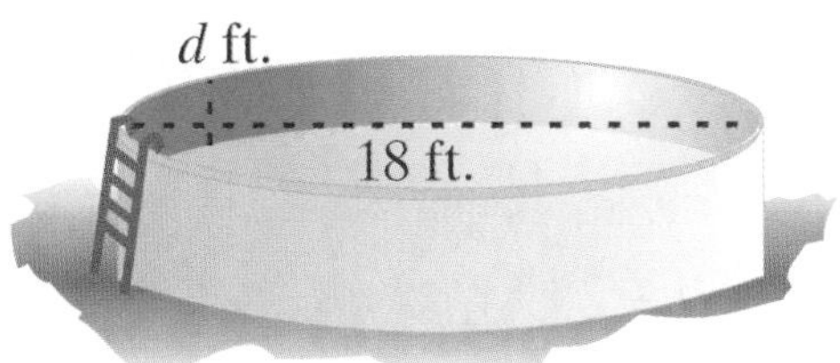

59. The floor of a rectangular bedroom measures N feet wide by M feet long. The height of the walls is 7 feet. Construct an expression for the number of square feet of wallpaper needed to cover all the walls. (Ignore the presence of doors and windows.)

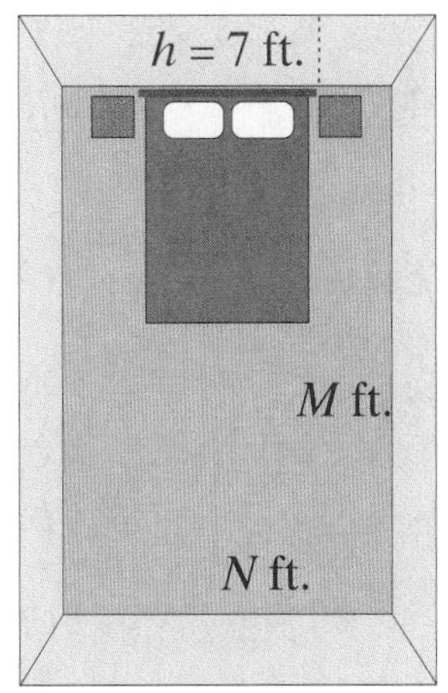

60. The interior surface of the birdbath in Example 6c needs to be painted with a waterproof (and nontoxic) coating. Construct the expression for the interior surface area.

technology exercises

Simplify each of the following expressions, writing your answer with only positive exponents. Then use a calculator (if possible) or a computer algebra system to check your work. See the Technology Note after Example 3 for guidance.

61. $\left[\left(5xy^0z^{-2}\right)^{-1}\left(25x^4z\right)^2\right]^{-2}$

62. $\dfrac{\left(3pq^{-1}r^{-2}\right)^{-3}}{\left(pqr\right)^3}$

63. $\left(\left(2x^{-2}yz^3\right)^{-1}\right)^2$

64. $\left(\dfrac{3x^{-2}}{(2y)^{-3}}\right)^{-1}$

Evaluate the following radical expressions. See Example 7.

65. $-\sqrt{9}$ **66.** $\sqrt[3]{-27}$ **67.** $\sqrt{-25}$ **68.** $\sqrt[6]{-64}$ **69.** $-\sqrt[6]{64}$

70. $\sqrt[3]{\dfrac{-8}{64}}$ **71.** $\sqrt{\dfrac{1}{4}}$ **72.** $-\sqrt[3]{-8}$ **73.** $\sqrt[4]{\sqrt{16}-\sqrt[3]{-27}+\sqrt{81}}$

74. $\sqrt{\dfrac{\sqrt[3]{-64}}{-\sqrt{144}-\sqrt{169}}}$

Simplify the following radical expressions. See Example 8.

75. $\sqrt{9x^2}$ **76.** $\sqrt[3]{-8x^6y^9}$ **77.** $\sqrt[4]{\dfrac{x^8z^4}{16}}$ **78.** $\sqrt{2x^6y}$ **79.** $\sqrt[7]{x^{14}y^{49}z^{21}}$

80. $\sqrt[3]{81m^4n^7}$ **81.** $\sqrt{\dfrac{x^2}{4x^4y^6}}$ **82.** $\sqrt[5]{32x^7y^{10}}$ **83.** $\sqrt[3]{\dfrac{a^3b^{12}}{27c^6}}$

Simplify the following radicals by rationalizing the denominators. See Example 9.

84. $\sqrt[3]{\dfrac{4x^2}{3y^4}}$ **85.** $\dfrac{-\sqrt{3a^3}}{\sqrt{6a}}$ **86.** $\dfrac{3}{\sqrt{2}-\sqrt{5}}$ **87.** $\dfrac{10}{\sqrt{7}-\sqrt{2}}$ **88.** $\dfrac{\sqrt{x}}{\sqrt{x}-\sqrt{2}}$

89. $\dfrac{x-y}{\sqrt{x}+\sqrt{y}}$ **90.** $\dfrac{\sqrt{x}+\sqrt{y}}{\sqrt{x}-\sqrt{y}}$ **91.** $\dfrac{1}{2-\sqrt{x}}$ **92.** $\dfrac{\sqrt{y}}{\sqrt{y}+2}$

Combine the radical expressions, if possible. See Example 10.

93. $\sqrt[3]{-16x^4}+5x\sqrt[3]{2x}$ **94.** $\sqrt{27xy^2}-4\sqrt{3xy^2}$ **95.** $\sqrt{7x}-\sqrt[3]{7x}$

96. $|x|\sqrt{8xy^2z^3} - |yz|\sqrt{18x^3z}$ **97.** $-x^2\sqrt[3]{54x} + 3\sqrt[3]{2x^7}$ **98.** $\sqrt[5]{32x^{13}} + 3x\sqrt[5]{x^8}$

Simplify the following expressions, writing your answer using the same notation as the original expression. See Examples 11 and 12.

99. $\sqrt[3]{\sqrt[4]{x^{36}}}$ **100.** $\left(3x^2-4\right)^{1/3}\left(3x^2-4\right)^{5/3}$ **101.** $32^{-3/5}$ **102.** $81^{3/4}$

103. $\dfrac{(x-z)^y}{(x-z)^4}$ **104.** $\sqrt[7]{n^9}\sqrt[7]{n^5}$ **105.** $(-8)^{2/3}$ **106.** $\dfrac{x^{1/5}y^{-2/3}}{x^{-3/5}y}$ **107.** $\dfrac{\sqrt[3]{a^2}}{\sqrt[3]{a^5}}$

Simplify the following expressions.

108. $\sqrt{5\sqrt[4]{5}}$ **109.** $\sqrt[16]{y^4}$ **110.** $\sqrt[4]{36}$ **111.** $\sqrt[3]{x^7}\sqrt[9]{x^6}$

Apply the definition of rational exponents to demonstrate the following properties.

112. $\sqrt[n]{ab} = \sqrt[n]{a}\cdot\sqrt[n]{b}$ **113.** $\sqrt[n]{\dfrac{a}{b}} = \dfrac{\sqrt[n]{a}}{\sqrt[n]{b}}$

114. The prism shown below is a triangular right cylinder, where the triangular base is a right triangle. Find the surface area of the prism in terms of b, h, and l.

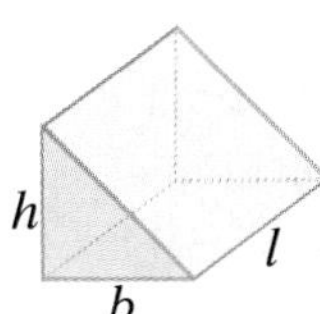

115. The pyramids in Egypt each consist of a square base and four triangular sides. For a class project, Karim constructs a model pyramid with equilateral triangles as sides. The side-length is $s = 43$ cm. Remember that the area of an equilateral triangle of side-length s is $A = \dfrac{s^2\sqrt{3}}{4}$. Find the total surface area of the pyramid.

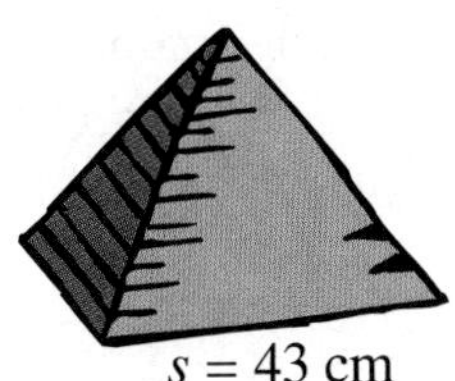

1.3 Polynomials and Factoring

TOPICS

1. The terminology of polynomial expressions
2. The algebra of polynomials
3. Common factoring methods:
   - greatest common factor
   - factoring by grouping
   - factoring special binomials
   - factoring trinomials
   - factoring expressions containing fractional exponents

Topic 1: The Terminology of Polynomial Expressions

Polynomials are algebraic expressions that meet further criteria. These criteria are that each term in a polynomial consists only of a number multiplied (if it is multiplied by anything at all) by variables raised to positive integer exponents. The number in any such term is called the **coefficient** of the term, and the sum of the exponents of the variables is the **degree of the term**. If a given term consists only of a non-zero number a (known as a **constant term**), its degree is defined to be 0, while the one specific term 0 is not assigned any degree at all. The polynomial as a whole is also given a degree: the **degree of a polynomial** is the largest of the degrees of the individual terms.

Every polynomial consists of some finite number of terms. Polynomials consisting of a single term are called **monomials**, those consisting of two terms are **binomials**, and those consisting of three terms are **trinomials**. The following example illustrates the use of the above terminology as applied to some specific polynomials.

example 1

a. The polynomial $-3x^4y^2+5.4x^3y^4$ is a binomial in the two variables x and y. The degree of the first term is 6, and the degree of the second term is 7, so the degree of the polynomial as a whole is 7. The coefficient of the first term is –3 and the coefficient of the second term is 5.4.

b. The single number 5 can be considered a polynomial. In particular, it is a monomial of degree 0. The rationale for assigning degree 0 to non-zero constants such as 5 is that 5 could be thought of as $5x^0$ (or 5 times *any* variable raised to the 0 power). The coefficient of this monomial is itself: 5.

c. The polynomial $\frac{2}{3}x^3y^5-z+y^{10}+3$ has four terms and is a polynomial in three variables. If one of the terms of a polynomial consists only of a number, it is referred to as the constant term. The degree of this polynomial is 10, and the degrees of the individual terms are, from left to right, 8, 1, 10, and 0. The coefficients are $\frac{2}{3}$, –1, 1, and 3.

The majority of the polynomials that occur in this book will be polynomials of a single variable, and can be described generically as follows.

Polynomials of a Single Variable

A polynomial in the variable x of degree n can be written in the form

$$a_nx^n+a_{n-1}x^{n-1}+\ldots+a_1x+a_0,$$

where $a_n, a_{n-1}, \ldots, a_1, a_0$ are numbers, $a_n \neq 0$, and n is a positive integer. This form is called **descending order**, because the powers descend from left to right. The **leading coefficient** of this polynomial is a_n.

Topic 2: The Algebra of Polynomials

The variables in polynomials, as in all algebraic expressions, represent numbers (albeit possibly unknown), and consequently polynomials can be added, subtracted, multiplied, and divided according to the field properties discussed in Section 1 of this chapter.

Addition, subtraction, and multiplication of polynomials will be discussed here, while division of polynomials will be taken up in Chapters 2 and 6.

To add two or more polynomials, we use the field properties to combine **like**, or **similar**, terms. These are the terms among all the polynomials being added that have the same variables raised to the same powers. Subtraction of a polynomial from another is accomplished by distributing the minus sign over the terms of the polynomial being subtracted, and then adding.

example 2

Add or subtract the polynomials as indicated.

a. $\left(2x^3y-3y+xz^2\right)+\left(3y+z^2-3xz^2+4\right)$ **b.** $\left(4ab^3-b^3c\right)-\left(4-b^3c\right)$

Solutions:

a.
$$\begin{aligned}&\left(2x^3y-3y+xz^2\right)+\left(3y+z^2-3xz^2+4\right)\\&=2x^3y+\left(-3y+3y\right)+\left(xz^2-3xz^2\right)+z^2+4\\&=2x^3y-2xz^2+z^2+4\end{aligned}$$

First, identify like terms and group them together. The like terms are then combined by using the distributive property.

b.
$$\begin{aligned}\left(4ab^3-b^3c\right)-\left(4-b^3c\right)&=4ab^3-b^3c-4+b^3c\\&=4ab^3-4+\left(-b^3c+b^3c\right)\\&=4ab^3-4\end{aligned}$$

Like terms are grouped together, after distributing the minus sign over all the terms in the second polynomial.

Polynomials are multiplied using the same properties. The distributive property is of particular importance, and its use is the key to multiplying polynomials correctly. When a binomial is multiplied by a binomial, as in Example 3b, the acronym FOIL is commonly used as a reminder of the four necessary products. For instance, in the product $\left(3ab-a^2\right)\left(ab+a^2\right)$, the product of the **F**irst terms is $3a^2b^2$, the product of the **O**uter terms is $3a^3b$, the product of the **I**nner terms is $-a^3b$, and the product of the **L**ast terms is $-a^4$. This acronym is, however, nothing more than a specialized application of the distributive property.

$$(a+b)(c+d)=ac+ad+bc+bd$$

example 3

Multiply the polynomials, as indicated.

a. $\left(2x^2y - z^3\right)\left(4 + 3z - 3xy\right)$ **b.** $\left(3ab - a^2\right)\left(ab + a^2\right)$

Solutions:

a.

$$\begin{aligned}&\left(2x^2y - z^3\right)\left(4 + 3z - 3xy\right)\\&= 2x^2y\left(4 + 3z - 3xy\right) - z^3\left(4 + 3z - 3xy\right)\\&= 8x^2y + 6x^2yz - 6x^3y^2 - 4z^3 - 3z^4 + 3xyz^3\end{aligned}$$

We use the distributive property first, and multiply each term of the first polynomial by each term of the second. None of the resulting terms are similar, so the final answer is a polynomial of 6 terms.

b.

$$\begin{aligned}&\left(3ab - a^2\right)\left(ab + a^2\right)\\&= 3a^2b^2 + 3a^3b - a^3b - a^4\\&= 3a^2b^2 + 2a^3b - a^4\end{aligned}$$

In this case, we will use the FOIL method to multiply the two binomials. The four terms that result from the initial multiplication contain two similar terms which are combined to obtain the final trinomial.

Special Product Formulas

1. $(A - B)(A + B) = A^2 - B^2$
2. $(A + B)^2 = A^2 + 2AB + B^2$
3. $(A - B)^2 = A^2 - 2AB + B^2$
4. $(A + B)^3 = A^3 + 3A^2B + 3AB^2 + B^3$
5. $(A - B)^3 = A^3 - 3A^2B + 3AB^2 - B^3$

Topic 3: Common Factoring Methods

Factoring, in general, means reversing the process of multiplication in order to find two or more expressions whose product is the original expression. For instance, the factored form of the polynomial $x^2 - 3x + 2$ is $(x-2)(x-1)$, since if the product $(x-2)(x-1)$ is expanded we obtain $x^2 - 3x + 2$ (as you should verify). Factoring is thus a logical topic to discuss immediately after multiplication, but the process of factoring is sometimes more complicated to carry out than multiplication.

The first four of the five methods presented here apply to polynomials. Unless otherwise indicated, we say that a polynomial with integer coefficients is **factorable** if it can be written as a product of two or more polynomials, all of which also have integer coefficients. If this cannot be done, we say the polynomial is **irreducible** (over the integers), or **prime**. The goal of factoring a polynomial is to **completely factor** it: to write it as a product of prime polynomials.

Method 1: Greatest Common Factor. Factoring out those factors common to all the terms in an expression is the least complex factoring method to apply, and should be done first if possible. The **Greatest Common Factor** (GCF) among all the terms is the product of all the factors common to each.

example 4

Use the Greatest Common Factor method to factor the following polynomials.

a. $12x^5 - 4x^2 + 8x^3z^3$ **b.** $(a^2 - b) - 3(a^2 - b)$

Solutions:

a. $12x^5 - 4x^2 + 8x^3z^3$
$= (4x^2)(3x^3) + (4x^2)(-1) + (4x^2)(2xz^3)$
$= (4x^2)(3x^3 - 1 + 2xz^3)$

As we noted above, $4x^2$ is the greatest common factor. Applying the distributive property in reverse leads to the factored form of this degree 6 trinomial.

b. $(a^2 - b) - 3(a^2 - b)$
$= (a^2 - b)(1) + (a^2 - b)(-3)$
$= (a^2 - b)(1 - 3)$
$= -2(a^2 - b)$

In factoring out the greatest common factor $a^2 - b$, remember that it is being multiplied first by 1 and then by –3. One common source of error in factoring is to forget factors of 1.

Method 2: Factoring by Grouping. If the terms of the polynomial are grouped in a suitable way, the GCF method may apply to each group, and a common factor might subsequently be found among the groups. Factoring by Grouping is the name given to this process, and it is important to realize that this is a trial and error process. Your first attempt at grouping and factoring may not succeed, and you may have to try several different ways of grouping the terms.

example 5

Use the Factoring by Grouping method to factor the following polynomials.

a. $6x^2 - y + 2x - 3xy$ **b.** $ax - ay - bx + by$

Solutions:

a.
$$\begin{aligned} &6x^2 - y + 2x - 3xy \\ &= \left(6x^2 + 2x\right) + \left(-y - 3xy\right) \\ &= 2x(3x+1) + y(-1-3x) \\ &= 2x(3x+1) - y(3x+1) \\ &= (3x+1)(2x-y) \end{aligned}$$

The first and third terms have a GCF of $2x$, while the second and fourth have a GCF of y, so we group accordingly. After factoring the two groups, we notice that $3x + 1$ and $-1 - 3x$ differ only by a minus sign (and the order). This means $3x + 1$ can be factored out.

b.
$$\begin{aligned} &ax - ay - bx + by \\ &= a(x-y) - b(x-y) \\ &= (x-y)(a-b) \end{aligned}$$

The first two terms have a common factor, as do the last two, so we proceed accordingly. In this problem, we could also have grouped the first and third terms, and the second and fourth terms, and obtained the same result.

caution!

One common error in factoring is to stop after groups within the original polynomial have been factored. For instance, while we have done some factoring to achieve the expression $2x(3x+1) + y(-1-3x)$ in Example 5a, this is *not* in factored form. An expression is only factored if it is written as a *product* of two or more factors. The expression $2x(3x+1) + y(-1-3x)$ is a sum of two smaller expressions.

Method 3: Factoring Special Binomials. Three types of binomials can always be factored by following the patterns outlined below. You should verify the truth of the patterns by multiplying out the products on the right-hand side of each one.

Factoring Special Binomials

In the following, A and B represent algebraic expressions.

Difference of Two Squares: $A^2 - B^2 = (A-B)(A+B)$

Difference of Two Cubes: $A^3 - B^3 = (A-B)(A^2 + AB + B^2)$

Sum of Two Cubes: $A^3 + B^3 = (A+B)(A^2 - AB + B^2)$

Note that $A^2 + B^2$ cannot be factored in the real sense.

example 6

Factor the following binomials.

a. $49x^2 - 9y^6$ **b.** $27a^6b^{12} + c^3$ **c.** $64 - (x+y)^3$

Solutions:

a.
$$49x^2 - 9y^6$$
$$= (7x)^2 - (3y^3)^2$$
$$= (7x - 3y^3)(7x + 3y^3)$$

The first step is to realize that the binomial is a difference of two squares, and to identify the two expressions that are being squared. Then follow the pattern to factor the binomial.

b.
$$27a^6b^{12} + c^3$$
$$= (3a^2b^4)^3 + (c)^3$$
$$= \left(\underbrace{3a^2b^4}_{A} + \underbrace{c}_{B}\right)\left(\underbrace{(3a^2b^4)^2}_{A^2} - \underbrace{(3a^2b^4)(c)}_{AB} + \underbrace{(c)^2}_{B^2}\right)$$
$$= (3a^2b^4 + c)(9a^4b^8 - 3a^2b^4c + c^2)$$

First recognize the binomial as a sum of two cubes, and then identify the two expressions being cubed. Remember that all of A and B must be squared when applying the pattern. Also, remember the minus sign in front of the product AB in the second factor.

c. $64-(x+y)^3$
$=4^3-(x+y)^3$
$=(4-(x+y))(4^2+4(x+y)+(x+y)^2)$
$=(4-x-y)(16+4x+4y+x^2+2xy+y^2)$

In this difference of two cubes, the second cube is itself a binomial. But the factoring pattern still applies, leading to the final factored form of the original binomial.

Method 4: Factoring Trinomials. In factoring a trinomial of the form ax^2+bx+c, the goal is to find two binomials $px+q$ and $rx+s$ such that

$$ax^2+bx+c=(px+q)(rx+s).$$

Since $(px+q)(rx+s)=prx^2+(ps+qr)x+qs$, we seek p, q, r, and s such that $a=pr$, $b=ps+qr$ and $c=qs$:

$$ax^2+bx+c=\underbrace{pr}_{a}x^2+\underbrace{(ps+qr)}_{b}x+\underbrace{qs}_{c}.$$

In general, this may require much trial and error, but the following guidelines will help.

Case 1: Leading Coefficient is 1. In this case, p and r will both be 1, so we only need q and s such that $x^2+bx+c=x^2+(q+s)x+qs$. That is, we need two integers whose sum is b, the coefficient of x, and whose product is c, the constant term.

example 7

To factor x^2+x-12, we can begin by writing $x^2+x-12=(x+\boxed{?})(x+\boxed{?})$ and then try to find two integers to replace the question marks. The two integers we seek must have a product of –12, and the fact that the product is negative means that one integer must be positive and one negative. The only possibilities are $\{1,-12\}$, $\{-1,12\}$, $\{2,-6\}$, $\{-2,6\}$, $\{3,-4\}$, and $\{-3,4\}$, and when we add the requirement that the sum must be 1, we are left with $\{-3,4\}$. Thus, $x^2+x-12=(x-3)(x+4)$.

Case 2: Leading Coefficient is not 1. In this case, trial and error may still be an effective way to factor the trinomial ax^2+bx+c, especially if a, b, and c are relatively small in magnitude and if you have lots of practice. If, however, trial and error seems to be taking too long, the following steps use factoring by grouping to minimize the amount of guessing required.

Factoring Trinomials by Grouping

For the trinomial $ax^2 + bx + c$,

Step 1: Multiply a and c.

Step 2: Factor ac into two integers whose sum is b. If no such factors exist, the trinomial is irreducible over the integers.

Step 3: Rewrite b in the trinomial with the sum found in Step 2, and distribute. The resulting polynomial of four terms may now be factored by grouping.

example 8

To factor the trinomial $6x^2 - x - 12$ by trial and error, we would begin by noting that if it can be factored, the factors must be of the form $\left(x+\boxed{?}\right)\left(6x+\boxed{?}\right)$ or $\left(2x+\boxed{?}\right)\left(3x+\boxed{?}\right)$. If we use the method outlined above, we form the product $(6)(-12) = -72$ and then factor -72 into two integers whose sum is -1. The two numbers -9 and 8 work, so we write $6x^2 - x - 12 = 6x^2 + (-9+8)x - 12 = 6x^2 - 9x + 8x - 12$. Now proceed by grouping:

$$\begin{aligned}&6x^2 - 9x + 8x - 12\\ &= 3x(2x-3) + 4(2x-3)\\ &= (2x-3)(3x+4).\end{aligned}$$

Method 5: Factoring Expressions Containing Fractional Exponents. This final method does not apply to polynomials, as polynomials cannot have fractional exponents. It will, however, be very useful in solving certain problems later in this book and possibly in other math classes. The method applies to negative fractional exponents as well as positive.

To factor an algebraic expression (in one variable) that has fractional exponents, identify the least exponent among the various terms, and factor the variable raised to that least exponent from each of the terms. Factor out any other common factors as well, and simplify if possible.

example 9

Factor each of the following algebraic expressions.

a. $3x^{-2/3} - 6x^{1/3} + 3x^{4/3}$

b. $(x-1)^{1/2} - (x-1)^{-1/2}$

Solutions:

a.

$$\begin{aligned} &3x^{-2/3} - 6x^{1/3} + 3x^{4/3} \\ &= 3x^{-2/3}\left(1 - 2x + x^2\right) \\ &= 3x^{-2/3}\left(x^2 - 2x + 1\right) \\ &= 3x^{-2/3}(x-1)(x-1) \\ &= 3x^{-2/3}(x-1)^2 \end{aligned}$$

Under the guidelines above, we factor out $3x^{-2/3}$. Note that we use the properties of exponents to obtain the terms in the second factor. After factoring out $3x^{-2/3}$, we notice that the second factor is a second-degree trinomial, and is itself factorable. In fact, it is an example of a perfect square trinomial (one that can be written as the square of a binomial).

b.

$$\begin{aligned} &(x-1)^{1/2} - (x-1)^{-1/2} \\ &= (x-1)^{-1/2}\left((x-1) - 1\right) \\ &= (x-1)^{-1/2}(x-2) \end{aligned}$$

In this example we factor out $(x-1)^{-1/2}$, again using the properties of exponents to obtain the terms in the second factor. The second factor in this example can then be simplified.

exercises

Classify each of the following expressions as either a polynomial or a non-polynomial. For those that are polynomials, identify the degree of the polynomial and the number of terms (using the words monomial, binomial and trinomial if applicable). See Example 1.

1. $3x^{3/2} - 2x$

2. $17x^2y^5 + 2z^3 - 4$

3. $5x^{10} + 3x^3 - 2y^3z^8 + 9$

4. πx^3

5. 8

6. 0

7. $7^3xy^2 + 4y^4$

8. abc^2d^3

9. $4x^2 + 7xy + 5y^2$

10. $3n^4m^{-3} + n^2m$

11. $\dfrac{y^2z}{4} + 2yz^4$

12. $6x^4y + 3x^2y^2 + xy^5$

*Write each of the following polynomials in descending order, and identify **(a)** the degree of the polynomial, and **(b)** the leading coefficient.*

13. $8z^2 + \pi z^5 - 2z + 1$

14. $9x^8 - 9x^{10}$

15. $4s^3 - 10s^5 + 2s^6$

16. $4 - 2x^5 + x^2$

17. $9y^6 - 2 + y - 3y^5$

18. $4n + 6n^2 - 3$

Add or subtract the polynomials, as indicated. See Example 2.

19. $\left(-4x^3y + 2zx - 3y\right) - \left(2xz + 3y + x^2z\right)$

20. $\left(4x^3 - 9x^2 + 1\right) + \left(-2x^3 - 8\right)$

21. $\left(x^2y - xy - 6y\right) + \left(y^2x + yx + 6x\right)$

22. $\left(5x^2 - 6x + 2\right) - \left(4 - 6x - 3x^2\right)$

23. $\left(a^2b + 2ab + ab^2\right) - \left(b^2a + 5ba + ba^2\right)$

24. $\left(x^4 + 2x^3 - x + 5\right) - \left(x^3 - x - x^4\right)$

Multiply the polynomials, as indicated. See Example 3.

25. $\left(3a^2b + 2a - 3b\right)\left(ab^2 + 7ab\right)$

26. $\left(x^2 - 2y\right)\left(x^2 + y\right)$

27. $(3a + 4b)(a - 2b)$

28. $(x + xy + y)(x - y)$

Factor each polynomial by factoring out the greatest common factor. See Example 4.

29. $3a^2b + 3a^3b - 9a^2b^2$

30. $\left(x^3 - y\right)^2 - \left(x^3 - y\right)$

31. $2x^6 - 14x^3 + 8x$

32. $27x^7y + 9x^6y - 9x^4yz$

Factor each polynomial by grouping. See Example 5.

33. $a^3 + ab - a^2b - b^2$

34. $ax - 2bx - 2ay + 4by$

35. $z + z^2 + z^3 + z^4$

36. $x^2 + 3xy + 3y + x$

37. $nx^2 - 2y - 2x^2 + ny$

38. $2ac - 3bd + bc - 6ad$

Factor the following special binomials, if possible, or state that it is prime. See Example 6.

39. $25x^4y^2 - 9$

40. $27a^9 + 8b^{12}$

41. $x^3 - 1000y^3$

42. $25x^2 + 16y^2$

Factor the following trinomials. See Example 7.

43. $x^2 + 2x - 15$

44. $x^2 + 6x + 9$

45. $x^2 - 2x + 1$

46. $x^2 - 5x + 6$

47. $x^2 - 4x + 4$

48. $x^2 + 5x + 4$

Factor the following trinomials. See Example 8.

49. $6x^2 + 5x - 6$

50. $5a^2 - 37a - 24$

51. $25y^2 + 10y + 1$

52. $5x^2 + 27x - 18$

53. $6y^2 - 13y - 8$

54. $16y^2 - 25y + 9$

Factor the following algebraic expressions. See Example 9.

55. $(2x-1)^{-3/2} + (2x-1)^{-1/2}$

56. $2x^{-2} + 3x^{-1}$

57. $7a^{-1} - 2a^{-3}b$

58. $(3z+2)^{5/3} - (3z+2)^{2/3}$

technology exercises

Computer algebra systems can be used to multiply polynomials, and are useful when the polynomials to be multiplied contain many terms or when many polynomials are to be multiplied. In *Mathematica*, the command is **`Expand`**, and its use is illustrated below.

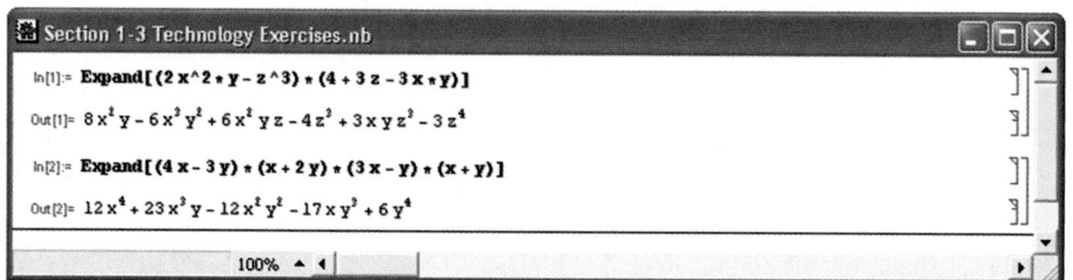

If available, use a computer algebra system to perform the polynomial multiplications that follow. (See Appendix A for guidance on using computer algebra systems like Mathematica.)

59. $(5x^2 - 2x + 3)(-x^2 + x - 3)(5x + 4)$

60. $(a + 2b - c)(3a - b + 4c)$

61. $(x^2 - 2xy + yz)(x - 3xy^2z + xz)$

62. $(pq - rs + t)(p^2 + r^2 + t^2)$

Computer algebra systems are also capable of reversing the result of polynomial multiplication, and such technology is useful when factoring by hand would be prohibitively tedious. In *Mathematica*, the command used to factor polynomials is, logically enough, **`Factor`**. The use of this command is illustrated below.

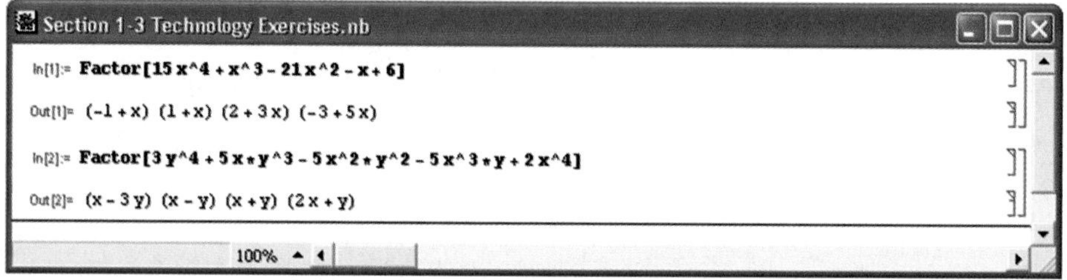

If available, use a computer algebra system to factor the following polynomials.

63. $3x^4 - 3x^2y + x^3y - xy^2 - 3x^3y^2 + 3xy^3 - x^2y^3 + y^4$

64. $16x^4 - 96x^3y + 216x^2y^2 - 216xy^3 + 81y^4$

65. $2p^4 - p^3q - 14p^2q^2 - 5pq^3 + 6q^4$

66. $ad + bd + cd + ae + be + ce + af + bf + cf$

1.4 The Complex Number System

TOPICS

1. The imaginary unit i and its properties
2. The algebra of complex numbers
3. Roots and complex numbers

Topic 1: The Imaginary Unit i and its Properties

Anything more than an elementary introduction to algebra soon reveals a troubling flaw with the real number system. The flaw manifests itself in a number of ways, one of which we have already alluded to: the lack of symmetry in the definition of roots of real numbers. Recall that we have thus far defined even roots only for non-negative numbers, but we have defined odd roots for both positive and negative numbers (as well as 0).

This asymmetry is a reflection of the fact that the real number system is not *algebraically complete*. The question of whether a field is algebraically complete is one that is dealt with in more advanced math classes (typically one called Abstract Algebra), but for the moment consider the following question:

For a given non-zero real number a, how many solutions does the equation $x^2 = a$ *have?*

As we have noted previously, the equation has two solutions ($x = \sqrt{a}$ and $x = -\sqrt{a}$) if a is positive, but no (real) solutions if a is negative. We improve this situation with the following definition.

The Imaginary Unit i

The imaginary unit i is defined as $i = \sqrt{-1}$. In other words, i has the property that its square is -1 : $i^2 = -1$.

This allows us to immediately define square roots of negative numbers in general, as follows.

Square Roots of Negative Numbers

If a is a positive real number, $\sqrt{-a} = i\sqrt{a}$. Note that by this definition, and by a logical extension of exponentiation, $\left(i\sqrt{a}\right)^2 = i^2\left(\sqrt{a}\right)^2 = -a$.

example 1

a. $\sqrt{-16} = i\sqrt{16} = i(4) = 4i$. As is customary, we write a constant such as 4 before letters in algebraic expressions, even if, as in this case, the letter is not a variable. Remember that i has a fixed meaning: i is the square root of -1.

b. $\sqrt{-8} = i\sqrt{8} = i\left(2\sqrt{2}\right) = 2i\sqrt{2}$. As is customary, again, we write the radical factor last. You should verify that $\left(2i\sqrt{2}\right)^2$ is indeed -8.

c. $i^3 = i^2 i = (-1)(i) = -i$, and $i^4 = i^2 i^2 = (-1)(-1) = 1$. The simple fact that $i^2 = -1$ allows us, by our extension of exponentiation, to determine i^n for any natural number n.

d. $(-i)^2 = (-1)^2 i^2 = i^2 = -1$. This observation shows that $-i$ also has the property that its square is -1.

Complex Numbers

For any two real numbers a and b, the sum $a + bi$ is a **complex number**. The collection $\mathbb{C} = \{a + bi \mid a \text{ and } b \text{ both real}\}$ is called the set of complex numbers and is another example of a field. The number a is called the **real part** of $a + bi$, and the number b is called the **imaginary part**. If the imaginary part of a given complex number is 0, the number is simply a real number. If the real part of a given complex number is 0, the number is a **pure imaginary number**.

Note: Two complex numbers are equal if and only if their real parts and their imaginary parts are equal.

The **powers of *i*** follow a repeating pattern that repeats every fourth power.

$$i = i \qquad i^{4n} = 1$$
$$i^2 = -1 \qquad i^{4n+1} = i$$
$$i^3 = -i \qquad i^{4n+2} = -1$$
$$i^4 = 1 \qquad i^{4n+3} = -i$$

Note that the set of real numbers is thus a subset of the complex numbers: every real number *is* a complex number with 0 as the imaginary part. The set of complex numbers is the largest set of numbers that will appear in this text.

Finally, a word on the nomenclature of complex numbers is in order. Don't be misled by the labels into thinking that complex numbers, with their possible imaginary parts, are somehow unimportant or physically meaningless. In many applications, complex numbers, even pure imaginary numbers, arise naturally and have important implications. As just one concrete example, the field of electrical engineering relies extensively on complex number arithmetic.

Topic 2: The Algebra of Complex Numbers

The set of complex numbers is a field, so the field properties discussed in Section 1.1 apply. In particular, every complex number has an additive inverse (its negative), and every non-zero complex number has a multiplicative inverse (its reciprocal). Further, sums and products (and hence differences and quotients) of complex numbers are complex numbers, and so can be written in the standard form $a + bi$. Given several complex numbers combined by the operations of addition, subtraction, multiplication, or division, the goal is to *simplify* the expression into the form $a + bi$.

Sums, differences, and products of complex numbers are easily simplified by remembering the definition of i and by thinking of every complex number $a + bi$ as a binomial. The simplification process is outlined below:

Simplifying Complex Expressions

Step 1: Add, subtract, or multiply the complex numbers, as required, by treating every complex number $a + bi$ as a polynomial expression. Remember, though, that i is not actually a variable. Treating $a + bi$ as a binomial in i is just a handy device.

Step 2: Complete the simplification by using the fact that $i^2 = -1$.

example 2

Simplify the following complex number expressions.

a. $(4+3i)+(-5+7i)$

b. $(-2+3i)-(-3+3i)$

c. $(3+2i)(-2+3i)$

d. $(2-3i)^2$

Solutions:

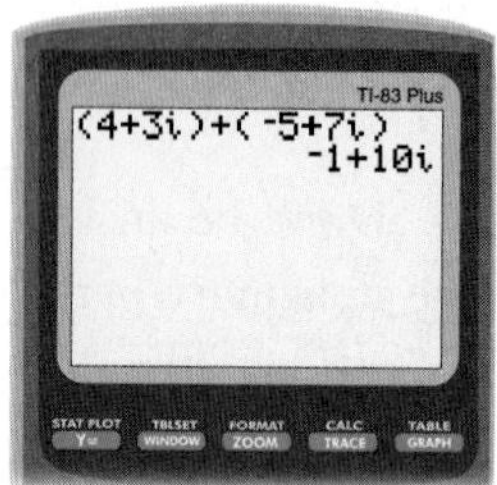

To enter complex numbers, you must first change the **Mode** from `Real` to `a+bi` by highlighting it with the arrow keys and pressing **Enter**. Remember to change it back when you are finished.

a. $(4+3i)+(-5+7i)=(4-5)+(3+7)i$
$=-1+10i$

Treating the two complex numbers as polynomials in i, we combine the real parts and then the imaginary parts.

b. $(-2+3i)-(-3+3i)=(-2+3i)+(-(-3)-3i)$
$=(-2+3)+(3-3)i$
$=1+0i$
$=1$

We begin by distributing the minus sign over the two terms of the second complex number, and then combine as in Example 2a.

c. $(3+2i)(-2+3i)=-6+9i-4i+6i^2$
$=-6+(9-4)i+6(-1)$
$=-6+5i-6$
$=-12+5i$

The product of two complex numbers leads to four products via the distributive property, as illustrated here. After multiplying, we combine the two terms containing i, and rewrite i^2 as -1.

d. $(2-3i)^2=(2-3i)(2-3i)$
$=4-6i-6i+9i^2$
$=4-12i+9(-1)$
$=-5-12i$

Squaring this complex number also leads to four products, which we simplify as in Example 2c. Remember that a complex number is not simplified until it has the form $a+bi$.

Division of one complex number by another is slightly more complicated. In order to rewrite a quotient in the standard form $a + bi$, we make use of the following observation:

$$(a+bi)(a-bi)=a^2-abi+abi-b^2i^2=a^2+b^2.$$

Given any complex number $a + bi$, the complex number $a - bi$ is called its **complex conjugate**. We simplify the quotient of two complex numbers by multiplying the numerator and denominator of the fraction by the complex conjugate of the denominator.

example 3

Simplify the quotients.

a. $\dfrac{2+3i}{3-i}$ **b.** $(4-3i)^{-1}$ **c.** $\dfrac{1}{i}$

Solutions:

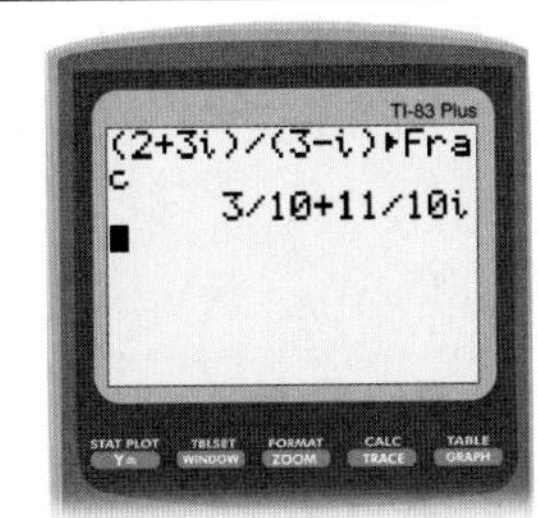

To force your answer to be in fraction form, go to the **Math** menu and select `Frac`. Enter `i` by selecting **2nd** then **.** (decimal point).

a.
$$\begin{aligned}\frac{2+3i}{3-i} &= \left(\frac{2+3i}{3-i}\right)\left(\frac{3+i}{3+i}\right)\\ &= \frac{(2+3i)(3+i)}{(3-i)(3+i)}\\ &= \frac{6+2i+9i+3i^2}{9+3i-3i-i^2}\\ &= \frac{3+11i}{10} = \frac{3}{10}+\frac{11}{10}i\end{aligned}$$

We multiply the top and bottom of the fraction by $3+i$, which is the complex conjugate of the denominator. The rest of the simplification involves multiplying complex numbers as in the last example.

We would often leave the answer in the form $\dfrac{3+11i}{10}$ unless it is necessary to identify the real and imaginary parts.

b.
$$\begin{aligned}(4-3i)^{-1} &= \frac{1}{4-3i}\\ &= \left(\frac{1}{4-3i}\right)\left(\frac{4+3i}{4+3i}\right)\\ &= \frac{4+3i}{(4-3i)(4+3i)}\\ &= \frac{4+3i}{16-9i^2}\\ &= \frac{4+3i}{25} = \frac{4}{25}+\frac{3}{25}i\end{aligned}$$

In this example, we simplify the reciprocal of the complex number $4-3i$. After writing the original expression as a fraction, we multiply the top and bottom by the complex conjugate of the denominator and proceed as in Example 3a.

c.
$$\begin{aligned}\frac{1}{i} &= \left(\frac{1}{i}\right)\left(\frac{-i}{-i}\right)\\ &= \frac{-i}{-i^2}\\ &= \frac{-i}{1} = -i\end{aligned}$$

This problem illustrates the process of writing the reciprocal of the imaginary unit as a complex number. Note that with this as a starting point, we could now calculate $i^{-2}, i^{-3}, \ldots$

Topic 3: Roots and Complex Numbers

We have now defined $\sqrt{a}$ for any real number a. We have done so in such a way that there is no ambiguity in the meaning of $\sqrt{a}$: given a positive real number a, $\sqrt{a}$ is defined to be the positive real number whose square is a, and $\sqrt{-a}$ is defined to be $i\sqrt{a}$. These are sometimes called the **principal square roots**, to distinguish them from $-\sqrt{a}$ and $-i\sqrt{a}$, respectively. (Remember, both $\sqrt{a}$ and $-\sqrt{a}$ are square roots of a.)

One more comment about roots should be made before we close this section. In simplifying radical expressions, we have made frequent use of the properties that if $\sqrt{a}$ and $\sqrt{b}$ are real numbers, then

$$\sqrt{a}\sqrt{b} = \sqrt{ab} \text{ and } \frac{\sqrt{a}}{\sqrt{b}} = \sqrt{\frac{a}{b}}.$$

There is a subtle but important condition in the above statement: $\sqrt{a}$ and $\sqrt{b}$ must both be real numbers. If this condition is not met, these properties of radicals do not necessarily hold. For instance,

$$\sqrt{(-9)(-4)} = \sqrt{36} = 6,$$

but

$$\sqrt{-9}\sqrt{-4} = (3i)(2i) = 6i^2 = -6.$$

In order to apply either of these two properties, then, first simplify any square roots of negative numbers by rewriting them as pure imaginary numbers.

example 4

Simplify the following expressions.

a. $\left(2-\sqrt{-3}\right)^2$

b. $\dfrac{\sqrt{4}}{\sqrt{-4}}$

Solutions:

a.
$$\begin{aligned}\left(2-\sqrt{-3}\right)^2 &= \left(2-\sqrt{-3}\right)\left(2-\sqrt{-3}\right)\\ &= 4-4\sqrt{-3}+\sqrt{-3}\sqrt{-3}\\ &= 4-4i\sqrt{3}+\left(i\sqrt{3}\right)^2\\ &= 4-4i\sqrt{3}-3\\ &= 1-4i\sqrt{3}\end{aligned}$$

Each $\sqrt{-3}$ is converted to $i\sqrt{3}$ before carrying out the associated multiplications. Note that incorrect use of one of the properties of radicals would have led to adding 3 instead of subtracting 3.

b. $\dfrac{\sqrt{4}}{\sqrt{-4}} = \dfrac{2}{2i}$
$= \dfrac{1}{i}$
$= -i$

We have already simplified $\frac{1}{i}$ in Example 3c, so we quickly obtain the correct answer of $-i$. If we had incorrectly rewritten the original fraction as $\sqrt{\frac{4}{-4}}$, we would have obtained $\sqrt{-1}$, or i, as the final answer.

exercises

Evaluate the following square root expressions. See Example 1.

1. $\sqrt{-25}$ **2.** $\sqrt{-12}$ **3.** $-\sqrt{-27}$ **4.** $-\sqrt{-100}$
5. $\sqrt{-32x}$ **6.** $\sqrt{-x^2}$ **7.** $\sqrt{-29}$ **8.** $(-i)^2\sqrt{-64}$

Simplify the following complex expressions by adding, subtracting, or multiplying as indicated. See Examples 1 and 2.

9. $(4-2i)-(3+i)$ **10.** $(1+i)+i$ **11.** $(4-i)(2+i)$
12. $i(5-i)$ **13.** $(3-i)^2$ **14.** i^7
15. $(3i)^2$ **16.** $(-5i)^3$ **17.** $(7i-2)+(3i^2-i)$
18. $(5-3i)^2$ **19.** $(3+i)(3-i)$ **20.** $(5+i)(2-9i)$
21. $(9-4i)(9+4i)$ **22.** i^{13} **23.** $i^{11}\left(\dfrac{6}{i^3}\right)$
24. $11i^{314}$

Simplify the following quotients. See Example 3.

25. $\dfrac{1+2i}{1-2i}$ **26.** $\dfrac{10}{3-i}$ **27.** $\dfrac{i}{2+i}$ **28.** $\dfrac{1}{i^9}$
29. $(2+5i)^{-1}$ **30.** i^{-25} **31.** $\dfrac{1}{i^{27}}$ **32.** $\dfrac{52}{5+i}$
33. $(2-3i)^{-1}$ **34.** $\dfrac{4i}{5+7i}$ **35.** i^{-4} **36.** $\dfrac{5+i}{4+i}$

Simplify the following square root expressions. See Example 4.

37. $\left(\sqrt{-9}\right)\left(\sqrt{-2}\right)$ **38.** $\left(1+\sqrt{-6}\right)^2$ **39.** $\dfrac{\sqrt{18}}{\sqrt{-2}}$

40. $\left(\sqrt{-32}\right)\left(-\sqrt{-2}\right)$ **41.** $\left(3+\sqrt{-2}\right)^2$ **42.** $\dfrac{\sqrt{-98}}{3i\sqrt{-2}}$

technology exercises

The advice given in this section for simplifying complex number expressions is to initially treat complex numbers as polynomials in the "variable" i, and the command used in *Mathematica* to simplify complex expressions reflects this approach. The *Mathematica* command for multiplying out polynomials is **Expand**, and this same command can be used to reduce complex number expressions to the form $a + bi$.

If available, use a computer algebra system to simplify the expressions that follow the illustration below. Note that the imaginary root i appears as the capital letter I in Mathematica. (See Appendix A for guidance on using computer algebra systems like Mathematica.)

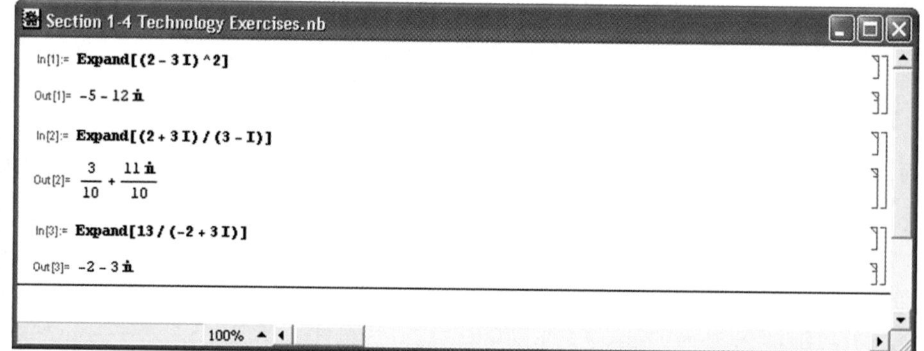

43. $\dfrac{3-2i}{1+i}$ **44.** $(3-2i)^4$ **45.** $\dfrac{2500}{(3+i)^4}$

46. $(2-5i)(3+7i)(1-4i)$ **47.** $(1+i)^5(3-i)^2$ **48.** $\dfrac{3+7i}{(2-5i)(1+3i)}$

1.5 Linear Equations in One Variable

TOPICS

1. Equivalent equations and the meaning of solutions
2. Solving linear equations
3. Solving absolute value equations
4. Solving linear equations for one variable
5. Interlude: distance and interest problems

Topic 1: Equivalent Equations and the Meaning of Solutions

In mathematics, an **equation** is a statement that two expressions are equal. If the statement is always true for any allowable value(s) of the variable(s), then the equation is an **identity**. If the statement is never true, it is a **contradiction**. The third possibility, when the equation is true for some values of the variable(s) and false for others, is called **conditional**, and the goal in solving these is to discover the set of values that the variable(s) can be replaced by to make the equation true. This set is called the **solution set** with any one element of the solution set being called a **solution**.

example 1

a. The equations $x^{1/2}(x+1)=x^{3/2}+x^{1/2}$ and $5+3=8$ are identities. Note that in the first equation the only allowable replacements for x are non-negative real numbers, but for all such numbers the equation is true. There are no variables in the second equation, so any value for any variable automatically satisfies this true statement.

b. The equation $t+3=t$ is an example of a contradiction. The solution set of this equation is the empty set, $\varnothing$, since no value for t satisfies the equation.

c. The equation $x^2=9$ is conditional. The solution set of the equation is $\{-3, 3\}$, as any other value for x results in a false statement.

The basic method for finding the solution set of an equation is to transform it into a simpler equation whose solution set is more clear. Two equations that have the same solution set are called **equivalent equations**. The field properties and the cancellation properties, discussed in Section 1.1, are the means by which equations are transformed into equivalent equations. Persistence and care are the key to successfully solving equations!

Topic 2: Solving Linear Equations

Linear equations in one variable are arguably the least complicated type of equation to solve.

Linear Equations

A **linear equation in one variable**, say the variable x, is an equation that can be transformed into the form $ax+b=0$, where a and b are real numbers and $a \neq 0$. Such equations are also called **first-degree** equations, as x appears to the first power.

A linear equation is an example of a polynomial equation, and the general method of solving polynomial equations will be discussed in Section 7 of this chapter.

Note that linear equations in one variable take very few steps to solve. Once the equation has been written in the form $ax+b=0$, the first cancellation property implies

$$ax+b=0 \Leftrightarrow ax=-b \qquad \text{(add } -b \text{ to both sides)}$$

and the second cancellation property implies

$$ax=-b \Leftrightarrow x=-\frac{b}{a} \qquad \text{(multiply both sides by } \frac{1}{a}\text{).}$$

example 2

Solve the following linear equations.

a. $3x-7=3(x-2)$

b. $3(x-2)+7x=1-2\left(x+\frac{1}{2}\right)$

c. $\frac{y}{6}+\frac{2y-1}{2}=\frac{y+1}{3}$

d. $-2(x-8)+x=16-x$

Solutions:

a.
$$\begin{aligned} 3x-7&=3(x-2)\\ 3x-7&=3x-6\\ 3x-3x&=-6+7\\ 0&=1 \quad \text{No Solution} \end{aligned}$$

First distribute, then add $-3x$ and 7 to both sides to combine like terms. The variable disappears and we are left with a false statement. Therefore, the solution set is $\varnothing$.

b. $3(x-2)+7x=1-2\left(x+\frac{1}{2}\right)$

$$3x-6+7x=1-2x-1$$
$$10x-6=-2x$$
$$12x-6=0$$
$$12x=6$$
$$x=\frac{1}{2}$$

As is typical, we must apply some of the field properties discussed in Section 1.1 to solve this equation. The distributive property leads to the second equation. Combining like terms leads to the third.

The cancellation properties then allow us to complete the process and solve the equation.

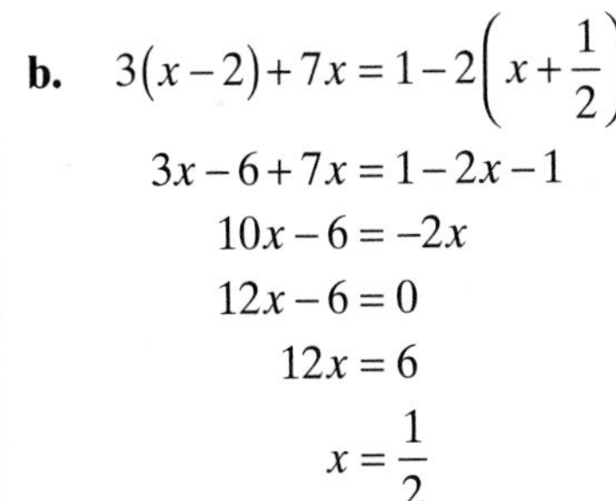

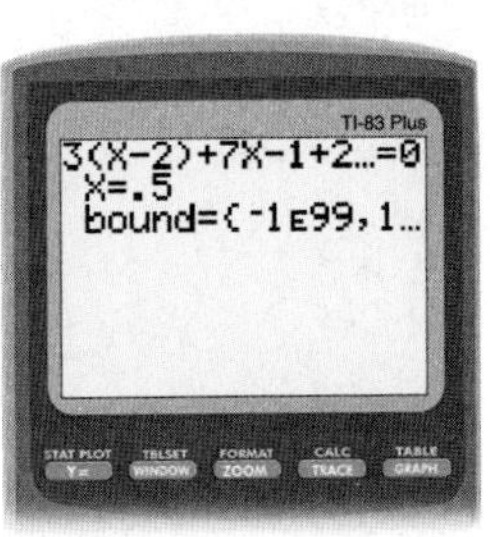

To get to equation solver, select **Math** and **0**. To solve, select **Alpha** and **Enter**.

c. $\frac{y}{6}+\frac{2y-1}{2}=\frac{y+1}{3}$

$$6\left(\frac{y}{6}+\frac{2y-1}{2}\right)=6\left(\frac{y+1}{3}\right)$$
$$6\cdot\frac{y}{6}+6\cdot\frac{2y-1}{2}=6\cdot\frac{y+1}{3}$$
$$y+3(2y-1)=2(y+1)$$
$$y+6y-3=2y+2$$
$$y+6y-2y=2+3$$
$$5y=5$$
$$y=1$$

Although it is not a necessary step, many people prefer to get rid of any fractions that might appear by multiplying both sides of the equation by the least common denominator (LCD). Remember to multiply every term by the LCD.

Note the cancellation that has occurred. We have replaced $\frac{6}{6}$ with 1, $\frac{6}{2}$ with 3, and $\frac{6}{3}$ with 2. Like terms are then combined, and multiplication by $\frac{1}{5}$ leads to the final answer.

d. $-2(x-8)+x=16-x$

$$-2x+16+x=16-x$$
$$16-x=16-x$$
$$0=0$$

First distribute, then combine like terms. Add -16 and x to both sides. Since the last equation, $0=0$, is always true, every real number is a solution.

Topic 3: Solving Absolute Value Equations

Linear absolute value equations in one variable are closely related to linear equations, and are solved in a similar fashion. The difference is that a linear absolute value equation contains at least one term inside the absolute value symbols; if these symbols were removed, the equation would be linear.

Such equations are solved by recognizing that the absolute value of any quantity is either the original quantity or its negative, depending on whether the quantity is positive or negative to start with. In general, every occurrence of an absolute value term in an equation leads to *two* equations with the absolute value signs removed. As an example,

$$|ax+b|=c \text{ means } ax+b=c \text{ or } -(ax+b)=c.$$

caution!

A word of warning: the apparent solutions obtained by the above method may not solve the original absolute value equation! Absolute value equations are one class of equations (there are others, as we shall see) in which it is very important to check your final answer in the original equation. An apparent solution that does not solve the original problem is called an **extraneous solution**.

example 3

Solve the absolute value equations.

a. $|3x-2|=1$ **b.** $|6x-7|+5=3$ **c.** $|x-4|=|2x+1|$

Solutions:

a. $|3x-2|=1$

$$3x-2=1 \quad\text{or}\quad -(3x-2)=1$$
$$3x-2=1 \quad\text{or}\quad 3x-2=-1$$
$$3x=3 \quad\text{or}\quad 3x=1$$
$$x=1 \quad\text{or}\quad x=\frac{1}{3}$$

The first step is to rewrite the absolute value equation without absolute values. The result is two linear equations that can be solved using the methods illustrated in this section. Note that this equation has two solutions, and they both solve the original equation.

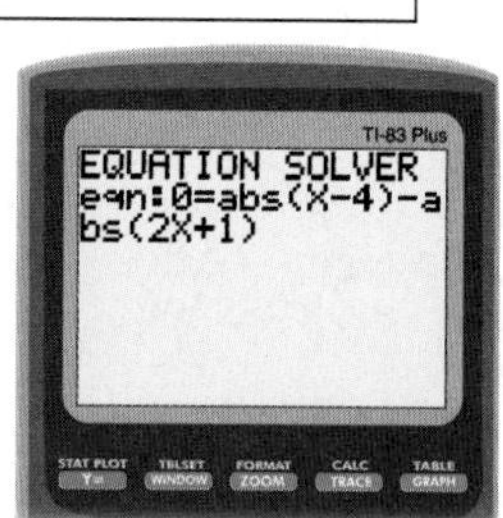

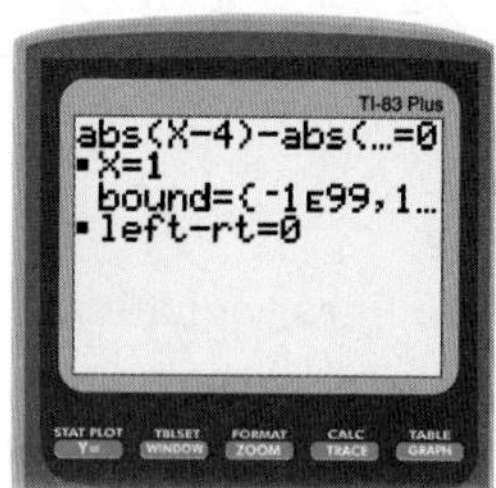

b. $|6x-7|+5=3$

$$|6x-7|=-2$$

solution is $\varnothing$

There is no solution because an absolute value can never be negative.

c. $|x-4|=|2x+1|$

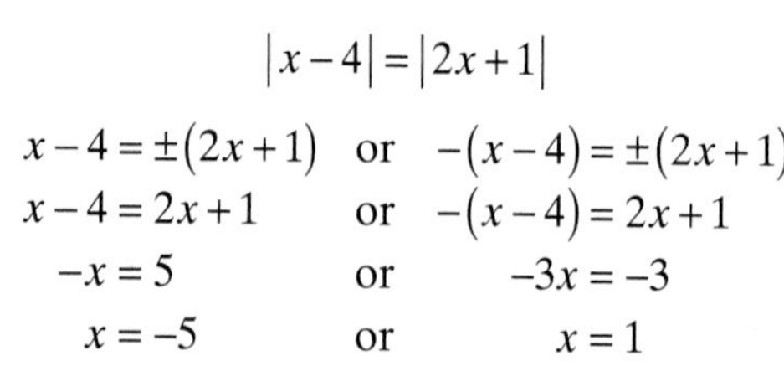

$$x-4=\pm(2x+1) \quad\text{or}\quad -(x-4)=\pm(2x+1)$$
$$x-4=2x+1 \quad\text{or}\quad -(x-4)=2x+1$$
$$-x=5 \quad\text{or}\quad -3x=-3$$
$$x=-5 \quad\text{or}\quad x=1$$

Since we have two absolute value quantities, the one original equation could potentially lead to four linear equations, with two of them being equivalent to the other two (as you should verify). We are thus left with two equations to solve.

Topic 4: Solving Linear Equations for One Variable

One common task in applied mathematics is to solve a given equation in two or more variables for one of the variables. **Solving for a variable** means to transform the equation into an equivalent one in which the specified variable is isolated on one side of the equation. For linear equations we accomplish this by the same methods we have used in the above examples.

example 4

Solve the following equations for the specified variable. All of the equations are formulas that arise in various applications, and they are linear in the specified variable.

a. $P = 2l + 2w$. Solve for w.

b. $A = P\left(1 + \frac{r}{n}\right)^{nt}$. Solve for P.

c. $S = 2\pi r^2 + 2\pi rh$. Solve for h.

Solutions:

a.

$$P = 2l + 2w$$
$$P - 2l = 2w$$
$$\frac{P - 2l}{2} = w$$
$$w = \frac{P - 2l}{2}$$

This is the formula for the perimeter, P, of a rectangle of length l and width w. First add $-2l$ to both sides, and then multiply by $\frac{1}{2}$.

The last equation is no different from the preceding one, but it is conventional to put the specified variable on the left side of the equation.

b.

$$A = P\left(1 + \frac{r}{n}\right)^{nt}$$
$$\frac{A}{\left(1 + \frac{r}{n}\right)^{nt}} = P$$
$$P = A\left(1 + \frac{r}{n}\right)^{-nt}$$

This is the equation for compound interest. If principal P is invested at an annual rate r for t years, compounded n times a year, the value of the investment at time t is A. This formula is linear in the variables P and A, though not in n, t, or r.

One step is all that is required to solve this equation for P, but the last equation uses one of the properties of exponents to make the result look neater.

c.

$$S = 2\pi r^2 + 2\pi rh$$
$$S - 2\pi r^2 = 2\pi rh$$
$$\frac{S - 2\pi r^2}{2\pi r} = h$$
$$h = \frac{S - 2\pi r^2}{2\pi r}$$

This is the formula for the surface area of a right circular cylinder of radius r and height h. It is linear in the variables S and h, but not in r.

Two steps are all that are necessary to solve this formula for h. Solving this formula for r requires a technique that will be discussed in Section 7 of this chapter.

Topic 5: Interlude: Distance and Interest Problems

Many applications lead to equations more complicated than those that we have studied so far, but good examples of linear equations arise from certain distance and simple interest problems. This is because the basic distance and simple interest formulas are linear in all of their variables:

$d = rt$, where d is the distance traveled at rate r for time t,

and

$I = Prt$, where I is the interest earned on principal P invested at rate r for time t.

example 5

The distance from Shreveport, LA to Austin, TX, by one route, is 325 miles. If Kevin made the trip in five and a half hours, what was his average speed?

Solution:

We know $d = 325$ miles and $t = 5.5$ hours, and need to solve for r.

$$d = rt$$

$$325 = 5.5r$$

$$\frac{325}{5.5} = r$$

$r = 59.1$ miles/hour (rounded to nearest tenth).

example 6

Julie invested \$1500 in a risky high-tech stock on January 1st. On July 1st, her stock is worth \$2100. She wants to determine her average annual rate of return at this point in the year. What effective annual rate of return has she earned so far?

Solution:

The interest that Julie has earned in half a year is $\$600\,(\text{or } \$2100 - \$1500)$. Replacing P with 1500, t with $\frac{1}{2}$, and I with 600 in the formula $I = Prt$, we have:

$$600 = (1500)\left(\frac{1}{2}\right)r$$

$$\frac{1200}{1500} = r$$

$r = .8$, or 80% average rate of return per year.

exercises

Solve the following linear equations. See Examples 1 and 2.

1. $-3(2t-4)=7(1-t)$
2. $5(2x-1)=3(1-x)+5x$
3. $\frac{y+5}{4}=\frac{1-5y}{6}$
4. $3x+5=3(x+3)-4$
5. $3w+5=2(w+3)-4$
6. $3x+5=3(x+3)-5$
7. $\frac{4s-3}{2}+\frac{7}{4}=\frac{8s+1}{4}$
8. $\frac{4x-3}{2}+\frac{3}{8}=\frac{7x+3}{4}$
9. $\frac{4z-3}{2}+\frac{3}{8}=\frac{8z+3}{4}$
10. $3(2w+13)=5w+w\left(7-\frac{3}{w}\right)$
11. $\frac{6}{7}(m-4)-\frac{11}{7}=1$
12. $0.08p+0.09=0.65$
13. $0.6x+0.08=2.3$
14. $0.9x+0.5=1.3x$
15. $0.73x+0.42(x-2)=0.35x$

Solve the following absolute value equations. See Example 3.

16. $|3x-2|=5$
17. $|3x-2|-1=|5-x|$
18. $|4x+3|+2=0$
19. $|6x-2|=0$
20. $|-8x+2|=14$
21. $|2x-109|=731$
22. $|4x-4|-40=0$
23. $|5x-3|=7$
24. $|4x+15|=3$
25. $-|6x+1|=11$
26. $|x+3|=|x-7|$
27. $|x-3|-|x-7|=0$
28. $|2-x|=|2+x|$
29. $|x|=|x+1|$
30. $|x+97|=|x+101|$
31. $\left|x+\frac{1}{4}\right|=\left|x-\frac{3}{4}\right|$

Solve the following equations for the indicated variable. See Example 4.

32. $C = 2\pi r$; solve for r.

33. $PV = nRT$; solve for T.

34. $v^2 = v_0^{\ 2} + 2ax$; solve for a.

35. $A = \frac{1}{2}(B+b)h$; solve for B.

36. $C = \frac{5}{9}(F-32)$; solve for F.

37. $V = \frac{1}{3}s^2h$; solve for h.

38. $A = 2lw + 2wh + 2hl$; solve for h.

39. $d = rt_1 + rt_2$; solve for r.

Solve the following application problems. See Examples 5 and 6.

40. Two trucks leave a warehouse at the same time. One travels due east at an average speed of 45 miles per hour, and the other travels due west at an average speed of 55 miles per hour. After how many hours will they be 450 miles apart?

41. A riverboat leaves port and proceeds to travel downstream at an average speed of 15 miles per hour. How long will it take for the boat to arrive at the next port, 95 miles downstream?

42. Two cars leave a rest stop at the same time, and proceed to travel down the highway. One travels at an average rate of 62 miles per hour, and the other at an average rate of 59 miles per hour. How far apart are the two cars after four and a half hours?

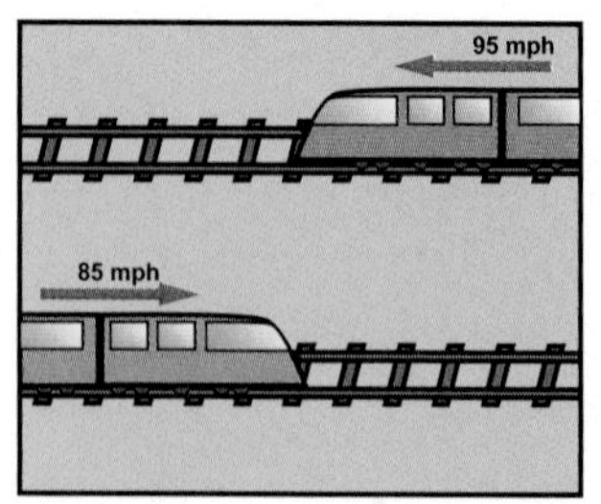

43. Two trains are 630 miles apart, heading directly toward each other. The first train is travelling at 95 mph, and the second train is travelling at 85 mph. How long will it be before the trains pass each other?

44. Two brothers, Rick and Tom, each inherit \$10,000. Rick invests his inheritance in a savings account with an annual return of 2.25%, while Tom invests his in a CD paying 6.15% annually. How much more money than Rick does Tom have after 1 year?

45. Sarah, sister to Rick and Tom in the previous problem, also inherits \$10,000, but she invests her inheritance in a global technology mutual fund. At the end of 1 year, her investment is worth \$12,800. What has her effective annual rate of return been?

46. Bob buys a large-screen digital TV priced at \$9,500, but pays \$10,212.50 with tax. What is the rate of tax where Bob lives?

47. Will and Matt are brothers. Will is 6 feet, 4 inches tall, and Matt is 6 feet, 7 inches tall. How tall is Will as a percentage of Matt's height? How tall is Matt as a percentage of Will's height?

48. A farmer wants to fence in three square garden plots situated along a road, as shown, and he decides not to install fencing along the road-edge. If he has 182 feet of fencing material total, what dimensions should he make each square plot?

49. Find three consecutive integers whose sum is 288. (Hint: if n represents the smallest of the three, then $n+1$ and $n+2$ represent the other two numbers.)

50. Find three consecutive odd integers whose sum is 165. (Hint: if n represents the smallest of the three, then $n+2$ and $n+4$ represent the other two numbers.)

51. Kathy buys last year's best-selling novel, in hardcover, for \$15.05. This is a 30% discount from the original price. What was the original price?

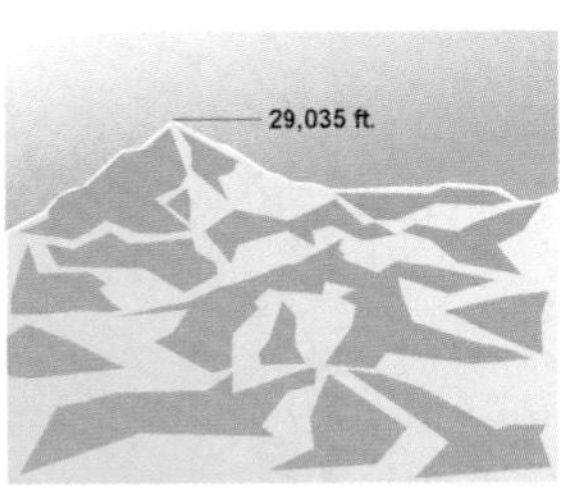

52. The highest point on Earth is the peak of Mount Everest. If you climbed to the top, you would be approximately 29,035 feet above sea level. Remembering that a mile is 5280 feet, what percentage of the height of the mountain would you have to climb to reach a point two miles above sea level?

technology exercises

Computer algebra systems are capable of solving simple equations for specific variables. In *Mathematica*, the command to do this is **Solve.** For instance, the equation relating the area A of a trapezoid to the lengths of the two bases B and b and the height h can be solved first for B and then for h as shown below. Notice the form that *Mathematica* uses to present the answer, and note particularly that *Mathematica*'s answer may not be in the simplest possible form.

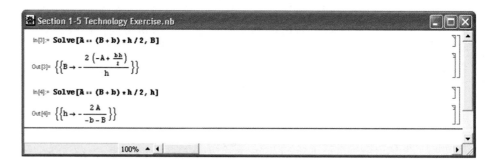

Section 1-5 Technology Exercise.nb
In[3]:= Solve[A == (B + b) * h / 2, B]
Out[3]= {{B → - 2 (-A + bh/2) / h}}
In[4]:= Solve[A == (B + b) * h / 2, h]
Out[4]= {{h → - 2 A / (-b - B)}}
100%

Use a computer algebra system, if available, to solve the following equations for the specified variable. (See Appendix A for guidance on using computer algebra systems like Mathematica.)

53. $S = 2\pi r^2 + 2\pi rh$; solve for h.

54. $3xyz - xy^3 = yz^2 + 5xz$; solve for x.

55. $ab - a(c-3) = ab + c$; solve for a.

56. $S = \dfrac{a}{1-r}$; solve for r.

1.6 Linear Inequalities in One Variable

TOPICS

1. Solving linear inequalities
2. Solving compound linear inequalities
3. Solving absolute value inequalities
4. Interlude: translating inequality phrases

Topic 1: Solving Linear Inequalities

If the equality symbol in a linear equation is replaced with $<$, $\leq$, $>$, or $\geq$, the result is a **linear inequality**. One difference between linear equations and linear inequalities is the way in which the solutions are described. Typically, the solution of a linear inequality consists of some interval of real numbers; such solutions can be described graphically or with interval notation. The process of obtaining the solution, however, is much the same as the process for solving linear equations, with the one important difference discussed below.

In solving linear inequalities, the field properties outlined in Section 1.1 are all still applicable, and it is often necessary to use, for example, the distributive and commutative properties in order to simplify one or both sides of an inequality. The additive version of the two cancellation laws is also used in the same way as in solving equations. The one difference lies in applying the multiplicative version of cancellation. Note that if both sides of the true statement $-3 < 2$ are multiplied by -1, and if no other changes are made, we obtain the false statement $3 < -2$. This is a basic illustration of the fact that the multiplicative cancellation law is slightly different for inequalities.

Cancellation Properties for Inequalities

Throughout this table, A, B, and C represent algebraic expressions. Each of the two properties is stated for the inequality symbol <, but they are also true for the other three symbols with the appropriate changes made.

Property	Description
$A < B \Leftrightarrow A + C < B + C$	Adding the same quantity to both sides of an inequality results in an equivalent inequality.
If $C > 0$, $A < B \Leftrightarrow A \cdot C < B \cdot C$ If $C < 0$, $A < B \Leftrightarrow A \cdot C > B \cdot C$	If both sides of an inequality are multiplied by a positive quantity, the sense of the inequality is unchanged. But if both sides are multiplied by a negative quantity, the sense of the inequality is reversed.

Keep in mind, then, that multiplying (or dividing) both sides of an inequality by a negative quantity necessitates reversing, or "flipping", the inequality symbol. This will be illustrated several times in the examples to follow.

example 1

Solve the following inequalities, using interval notation to describe the solution set.

a. $5-2(x-3) \le -(1-x)$ **b.** $\dfrac{3(a-2)}{2} < \dfrac{5a}{4}$

Solutions:

a.
$$5-2(x-3) \le -(1-x)$$
$$5-2x+6 \le -1+x$$
$$-2x+11 \le -1+x$$
$$-3x \le -12$$
$$x \ge 4$$
Solution is $[4, \infty)$.

We begin by using the distributive property, and then proceed to combine like terms.

At this point, division by –3 is all that is required to solve the inequality. Note the reversal of the inequality symbol.

b. $$\frac{3(a-2)}{2} < \frac{5a}{4}$$

$$4\left(\frac{3(a-2)}{2}\right) < 4\left(\frac{5a}{4}\right)$$

$$6(a-2) < 5a$$

$$6a-12 < 5a$$

$$a < 12$$

Solution is $(-\infty, 12)$.

Just as with equations, fractions in inequalities can be eliminated by multiplying both sides by the least common denominator.

The solution consists of all real numbers less than 12.

The solutions in Example 1 were described using interval notation. The solutions can also be described by graphing. Graphing a solution to an inequality can lead to a better understanding of which real numbers solve the inequality.

The symbols used in this text for graphing intervals are the same as the symbols in interval notation. Parentheses are used to indicate excluded endpoints of intervals and brackets are used when the endpoints are included in the interval. The portion of the number line that constitutes the interval is then shaded. (Other commonly used symbols in graphing are open circles for parentheses and filled-in circles for brackets.)

For example, the two solutions above are graphed as follows:

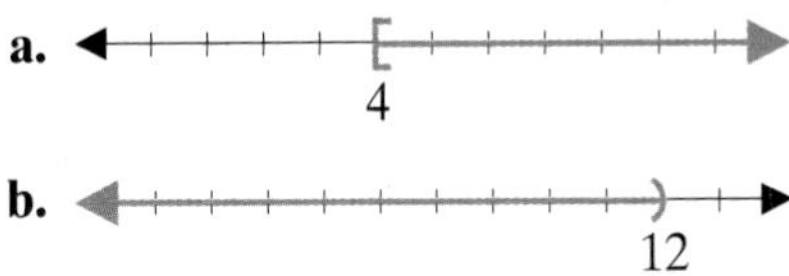

Topic 2: Solving Compound Linear Inequalities

A **compound inequality** is a statement containing two inequality symbols, and can be interpreted as two distinct inequalities joined by the word "and." Compound inequalities often arise in applications as a result of constraints placed upon some quantity.

example 2

Solve the compound inequalities.

a. $80 \le \dfrac{67 + 82 + 73 + 85 + x}{5} < 90$ **b.** $-1 < 3 - 2x \le 5$

Solutions:

a. $$80 \le \frac{67+82+73+85+x}{5} < 90$$
$$400 \le 307 + x < 450$$
$$93 \le x < 143$$
Solution is $[93, 143)$.

Given either of the two inequalities alone, multiplying through by 5 would probably be the first step. We do this with the compound inequality as well, and then subtract 307 from all three expressions. Of course, if this compound inequality relates to test scores on a scale of 0 to 100, the solution set is actually [93, 100].

b. $$-1 < 3 - 2x \le 5$$
$$-4 < -2x \le 2$$
$$2 > x \ge -1$$
$$-1 \le x < 2$$
Solution is $[-1, 2)$.

We begin by subtracting 3 from all three expressions, and then divide each expression by −2. Don't forget to reverse each inequality, since we are dividing by a negative number. The final compound inequality is identical to the one before it, but has been written so that the smaller number appears first.

Topic 3: Solving Absolute Value Inequalities

An **absolute value inequality** is an inequality in which some variable expression appears inside absolute value symbols. In the problems that we will study, the inequality would be linear if the absolute value symbols were not there. This means that absolute value inequalities can be written without absolute values as follows:

$$|x| < a \Leftrightarrow -a < x < a$$

and

$$|x| > a \Leftrightarrow x < -a \text{ or } x > a.$$

While most absolute value inequalities will be more complicated than the two above, they will serve as paradigms for the rewriting of such inequalities without the absolute values.

example 3

Solve the absolute value inequalities.

a. $|3y-2|+2 \le 6$ **b.** $|4-2x|>6$ **c.** $|5+2s| \le -3$ **d.** $|5+2s| \ge -3$

Solutions:

a.

$$|3y-2|+2 \le 6$$
$$|3y-2| \le 4$$
$$-4 \le 3y-2 \le 4$$
$$-2 \le 3y \le 6$$
$$-\frac{2}{3} \le y \le 2$$

Solution is $\left[-\frac{2}{3}, 2\right]$.

In order to remove the absolute value symbols, we first subtract 2 from both sides.

After rewriting the inequality as described above, we are faced with a compound inequality to solve.

The graph of the solution is:

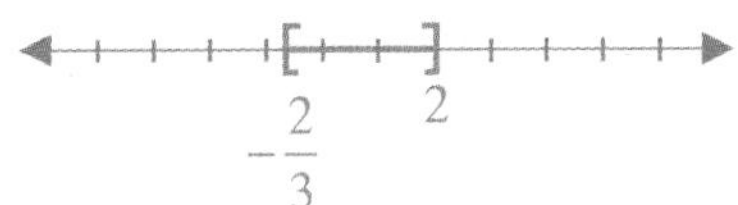

b.

$$|4-2x|>6$$
$$4-2x<-6 \quad \text{or} \quad 4-2x>6$$
$$-2x<-10 \quad \text{or} \quad -2x>2$$
$$x>5 \quad \text{or} \quad x<-1$$

Solution is $(-\infty,-1)\cup(5,\infty)$.

We can immediately rewrite the inequality without absolute values, and begin solving the two independent inequalities.

The graph of the solution is:

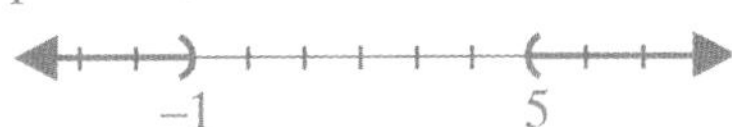

c. $|5+2s| \le -3$

Solution is $\varnothing$.

Just as in Example 3b of Section 1.5, we conclude that the solution set is the empty set, as it is impossible for the absolute value of any expression to be negative.

d. $|5+2s| \ge -3$

Solution is $\mathbb{R}$.

Since every absolute value is greater than or equal to 0, then the equation is true for all *s*.

Topic 4: Interlude: Translating Inequality Phrases

Many real-world applications leading to inequalities involve notions such as "is no greater than", "at least as large as", "does not exceed", and so on. Phrases such as these all have precise mathematical translations that use one of the four inequality symbols. The following example will illustrate how some sample phrases are translated.

example 4

Express each of the following problems as an inequality, and then solve the inequality.

a. The average daily high temperature in Santa Fe, NM over the course of three days exceeded 75. Given that the high on the first day was 72 and the high on the third day was 77, what was the minimum high temperature on the second day?

b. As a test for quality at a plant manufacturing silicon wafers for computer chips, a random sample of 10 batches of 1000 wafers each must not detect more than 5 defective wafers per batch on average. In the first 9 batches tested, the average number of defective wafers per batch is found to be 4.78 (to the nearest hundredth). What is the maximum number of defective wafers that can be found in the 10th batch for the plant to pass the quality test?

Solutions:

a.
$$\frac{72 + x + 77}{3} > 75$$
$$149 + x > 225$$
$$x > 76$$

Let x denote the high temperature on the second day. We obtain our inequality by setting the average of the three high temperatures greater than 75. We now know that the high temperature on the second day was greater than 76.

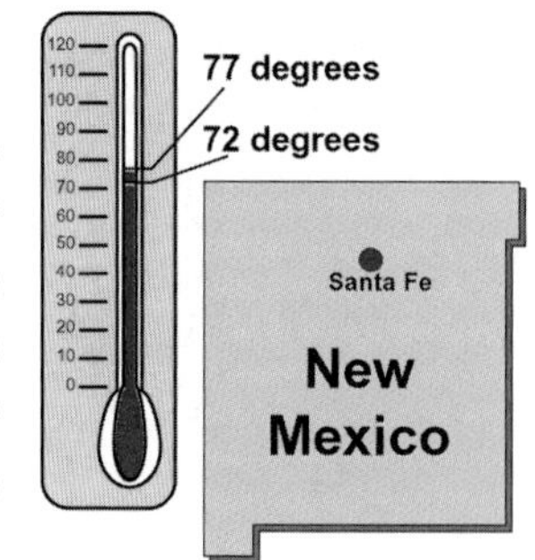

b.
$$\frac{(9)(4.78)+x}{10} \le 5$$
$$43.02 + x \le 50$$
$$x \le 6.98$$

Let x denote the maximum allowable number of defective wafers in the 10th batch. The number of defective wafers found in the first 9 batches is $(9)(4.78) = (43.02)$. When this number is added to x and divided by 10, the result must be less than or equal to 5. Of course, it is not possible to have a fractional number of wafers, so there must have actually been 43 defective wafers in the first 9 batches, and the maximum allowable number of defective wafers in the last batch is 7.

exercises

Solve the following linear inequalities. Describe the solution set using interval notation and by graphing. See Example 1.

1. $4+3t \le t-2$

2. $-\frac{v+2}{3} > \frac{5-v}{2}$

3. $4.2x-5.6 < 1.6+x$

4. $8.5y-3.5 \ge 2.5(3-y)$

5. $-2(3-x) < -2x$

6. $\frac{1-x}{5} > \frac{-x}{10}$

7. $4w+7 \le -7w+4$

8. $-5(p-3) > 19.8-p$

9. $\frac{6f-2}{5} < \frac{5f-3}{4}$

10. $\frac{u-6}{7} \ge \frac{2u-1}{3}$

11. $0.04n+1.7 < 0.13n-1.45$

12. $2k+\frac{3}{2} < 5k-\frac{7}{3}$

Solve the following compound inequalities. Describe the solution set using interval notation and by graphing. See Example 2.

13. $-4 < 3x-7 \le 8$

14. $-10 < -2(4+y) \le 9$

15. $-8 \le \frac{z}{2}-4 < -5$

16. $3 < \frac{w+3}{8} \le 9$

17. $5 \le 2m-3 \le 13$

18. $4 \le \frac{p+7}{-2} < 9$

19. $\frac{1}{3} < \frac{7}{6}(l-3) < \frac{2}{3}$

20. $\frac{1}{4} \le \frac{g}{2}-3 < 5$

21. $0.08 < 0.03c+0.13 \le 0.16$

Solve the following absolute value inequalities. Describe the solution set using interval notation and by graphing. See Example 3.

22. $4+|3-2y| \le 6$

23. $4+|3-2y| > 6$

24. $|x-2| \ge 5$

25. $2|z+5| < 12$

26. $7-\left|\frac{q}{2}+3\right| \ge 12$

27. $|4-2x| > 11$

28. $5.5+|x-7.2| \le 3.5$

29. $6-5|x+2| \ge -4$

30. $|2x-1| < x+4$

31. $-3|4-t| < -6$

32. $-3|4-t| > -6$

33. $3|4-t| < -6$

34. $|3t+4| > -8$

35. $2 < |6w-2|+7$

Solve the following application problems. See Example 4.

36. In a class in which the final course grade depends entirely on the average of four equally weighted 100-point tests, Cindy has scored 96, 94, and 97 on the first three. The professor has announced that there will be a fifteen-point bonus problem on the fourth test, and that anyone who finishes the semester with an average of more than 100 will receive an A+. What range of scores on the fourth test will give Cindy an A for the semester (an average between 90 and 100, inclusive), and what range will give Cindy an A+?

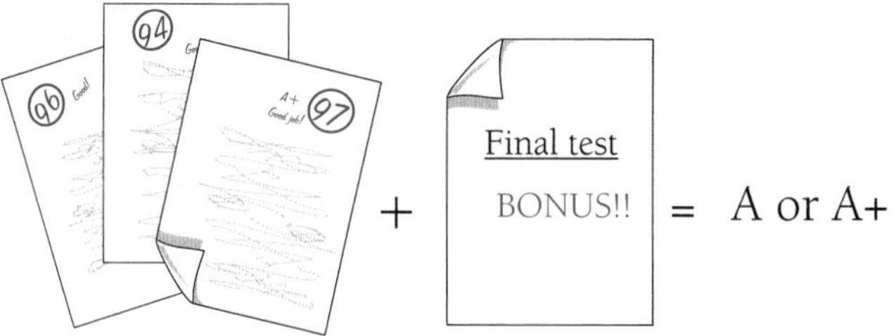

37. In a series of 30 racquetball games played to date, Larry has won 10, giving him a winning average so far of 33.3% (to the nearest tenth). If he continues to play, what interval describes the number of games he must now win in a row to have an overall winning average greater than 50%?

38. Assume that the national average SAT score for high school seniors is 1020 out of 1600. A group of 7 students receive their scores in the mail, and 6 of them look at their scores. Two students scored 1090, one got an 1120, two others each got a 910, and the sixth student received an 880. What range of scores can the seventh student receive to pull the group's average above the national average?

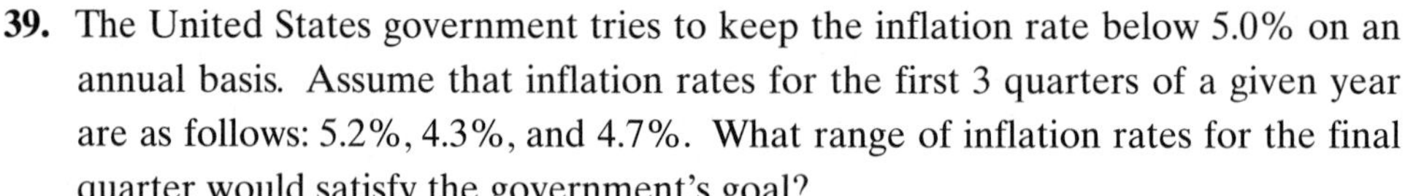

39. The United States government tries to keep the inflation rate below 5.0% on an annual basis. Assume that inflation rates for the first 3 quarters of a given year are as follows: 5.2%, 4.3%, and 4.7%. What range of inflation rates for the final quarter would satisfy the government's goal?

1.7 Quadratic Equations

TOPICS

1. Solving quadratic equations by factoring
2. Solving "perfect square" quadratic equations
3. Solving quadratic equations by completing the square
4. The quadratic formula
5. Interlude: gravity problems
6. Solving quadratic-like equations
7. Solving general polynomial equations by factoring
8. Solving polynomial-like equations by factoring

Topic 1: Solving Quadratic Equations by Factoring

In Section 1.5 of this chapter, we studied first-degree polynomial equations in one variable. We will now expand the class of one-variable polynomial equations that we can solve to include **quadratic** (or **second-degree polynomial**) equations.

Recall that the method of solving linear equations was particularly straightforward, and that the method always works for *any* such equation. We will, by the end of this section, develop a method for solving one-variable second-degree equations that is also guaranteed to work. This is in contrast to polynomial equations in general. In fact, it can be shown that for polynomial equations of degree five and higher there is *no* method that always works.

Our development will begin with a formal definition of quadratic equations, and we will then proceed to study those quadratic equations that can be solved by applying the factoring skills learned in Section 1.3.

Quadratic Equations

A **quadratic equation in one variable**, say the variable x, is an equation that can be transformed into the form $ax^2 + bx + c = 0$, where a, b, and c are real numbers and $a \neq 0$. Such equations are also called **second-degree** equations, as x appears to the second power. The name quadratic comes from the Latin word *quadrus*, meaning "square."

The key to using factoring to solve a quadratic equation, or indeed any polynomial equation, is to rewrite the equation so that 0 appears by itself on one side. This then allows us to use the Zero-Factor property discussed in Section 1.1. If the trinomial $ax^2 + bx + c$ can be factored, it can be written as a product of two linear factors A and B. The Zero-Factor property then implies that the only way for $ax^2 + bx + c$ to be 0 is if one (or both) of A and B is 0. This is all we need to solve the equation.

example 1

Solve the quadratic equations by factoring.

a. $x^2 + \frac{5x}{2} = \frac{3}{2}$ **b.** $s^2 + 9 = 6s$ **c.** $5x^2 + 10x = 0$

Solutions:

a.

$$x^2 + \frac{5x}{2} = \frac{3}{2}$$

To make the polynomial easier to factor, we multiply both sides by the LCD.

$$2x^2 + 5x = 3$$

Although we could factor $2x^2 + 5x$, this would not do us any good. We must have 0 on one side in order to apply the Zero-Factor property.

$$2x^2 + 5x - 3 = 0$$

$$(2x-1)(x+3) = 0$$

$$2x - 1 = 0 \quad \text{or} \quad x + 3 = 0$$

$$x = \frac{1}{2} \quad \text{or} \quad x = -3$$

After factoring, we have two linear equations to solve. The solution set is $\left\{\frac{1}{2}, -3\right\}$.

b.

$$s^2 + 9 = 6s$$

$$s^2 - 6s + 9 = 0$$

$$(s-3)^2 = 0$$

Again, we rewrite the equation with 0 on one side, and then factor the quadratic.

$$s - 3 = 0 \quad \text{or} \quad s - 3 = 0$$

$$s = 3$$

In this example, the two linear factors are the same. In such cases, the single solution is called a *double solution* or a *double root*.

c. $5x^2 + 10x = 0$

$5x(x+2) = 0$

$5x = 0$ or $x + 2 = 0$

$x = 0$ or $x = -2$

An alternative approach in this example would be to divide both sides by 5 at the very beginning. This would lead to the equation $x(x+2) = 0$, which gives us the same solution set of $\{0, -2\}$.

Topic 2: Solving "Perfect Square" Quadratic Equations

The factoring method is fine when it works, but there are two potential problems with the method: (1) the second-degree polynomial in question might not factor over the integers, and (2) even if the polynomial does factor, the factored form may not be obvious. In some cases where the factoring method is unsuitable, the solution can be obtained by using our knowledge of square roots. If A is an algebraic expression and if c is a constant, the equation $A^2 = c$ implies $A = \pm\sqrt{c}$.

If a given quadratic equation can be written in the form $A^2 = c$, we can use the above observation to obtain two linear equations that can be easily solved.

example 2

Solve the quadratic equations by taking square roots.

a. $(2x+3)^2 = 8$

b. $(x-5)^2 + 4 = 0$

Solutions:

a. $(2x+3)^2 = 8$

$2x + 3 = \pm\sqrt{8}$

$2x + 3 = \pm 2\sqrt{2}$

$2x = -3 \pm 2\sqrt{2}$

$x = \dfrac{-3 \pm 2\sqrt{2}}{2}$

We begin by taking the square root of each side, keeping in mind that there are two numbers whose square is 8.

We solve the two linear equations at once by subtracting 3 from both sides and then dividing both sides by 2. The solution set is $\left\{\dfrac{-3+2\sqrt{2}}{2}, \dfrac{-3-2\sqrt{2}}{2}\right\}$.

b. $(x-5)^2 + 4 = 0$

$(x-5)^2 = -4$

$x - 5 = \pm\sqrt{-4}$

$x - 5 = \pm 2i$

$x = 5 \pm 2i$

Before taking square roots, we isolate the perfect square algebraic expression on one side, and put the constant on the other.

In this example, taking square roots leads to two complex number solutions. (See Section 1.4 for a review of complex numbers.) The solution set is $\{5+2i, 5-2i\}$.

Topic 3: Solving Quadratic Equations by Completing the Square

There are potential pitfalls, once again, with the method just developed. If the quadratic equation under consideration appears in the form $A^2 = c$, the method works well (even if the ultimate solutions wind up being complex, as in Example 2b). But what if the equation doesn't have the form $A^2 = c$?

The method of **completing the square** allows us to write an arbitrary quadratic equation $ax^2 + bx + c = 0$ in the desired form. The method is outlined below:

Completing the Square

Step 1: Write the equation $ax^2 + bx + c = 0$ in the form $ax^2 + bx = -c$.

Step 2: Divide by a, if $a \neq 1$, so that the coefficient of x^2 is 1: $x^2 + \frac{b}{a}x = -\frac{c}{a}$.

Step 3: Divide the coefficient of x by 2, square the result, and add this to both sides: $x^2 + \frac{b}{a}x + \left(\frac{b}{2a}\right)^2 = -\frac{c}{a} + \left(\frac{b}{2a}\right)^2$.

Step 4: The trinomial on the left side will now be a perfect square. That is, it can be written as the square of an algebraic expression.

At this point, the equation can be solved by taking the square root of both sides.

example 3

Solve the quadratic equations by completing the square.

a. $x^2 - 2x - 6 = 0$ **b.** $9x^2 + 3x = 2$

Solutions:

a. $x^2 - 2x - 6 = 0$

$x^2 - 2x = 6$

$x^2 - 2x + 1 = 6 + 1$

$(x-1)^2 = 7$

After moving the constant term to the right-hand side, we divide -2 (the coefficient of x) by 2 to get -1, and add $(-1)^2$ to both sides. The trinomial on the left can now be factored.

$x - 1 = \pm\sqrt{7}$

$x = 1 \pm \sqrt{7}$

Taking square roots leads to two readily solved linear equations.

b.

$$9x^2 + 3x = 2$$

The constant term is already isolated on the right-hand side, so our first step is to divide by 9 (and simplify the resulting fractions, if possible).

$$x^2 + \frac{1}{3}x = \frac{2}{9}$$

$$x^2 + \frac{1}{3}x + \frac{1}{36} = \frac{2}{9} + \frac{1}{36}$$

Half of the coefficient of x is $\frac{1}{6}$, and the square of this is $\frac{1}{36}$.

$$\left(x + \frac{1}{6}\right)^2 = \frac{1}{4}$$

$$x + \frac{1}{6} = \pm\frac{1}{2}$$

After simplifying the sum of fractions on the right, we take the square root of each side.

$$x = -\frac{1}{6} \pm \frac{1}{2}$$

$$x = \frac{1}{3}, -\frac{2}{3}$$

Since the answer of $-\frac{1}{6} \pm \frac{1}{2}$ can be simplified, we do so to obtain the final answer.

If a quadratic can be factored as $(x-p)(x-q)$, then p and q solve the equation $(x-p)(x-q)=0$.

How does the quadratic $9x^2 + 3x - 2$ factor? This quadratic comes from Example 3b, so we might guess factors of $x - \frac{1}{3}$ and $x + \frac{2}{3}$. But,

$$\left(x - \frac{1}{3}\right)\left(x + \frac{2}{3}\right) = x^2 + \frac{1}{3}x - \frac{2}{9}.$$

It shouldn't be surprising that the product of these two factors has a leading coefficient of 1, since each of them individually has a leading coefficient of 1. To get the correct leading coefficient, we need to multiply by 9:

$$9\left(x - \frac{1}{3}\right)\left(x + \frac{2}{3}\right) = 9\left(x^2 + \frac{1}{3}x - \frac{2}{9}\right) = 9x^2 + 3x - 2.$$

And to make the factors look better, we can factor 9 into two factors of 3 and rearrange the products:

$$9\left(x - \frac{1}{3}\right)\left(x + \frac{2}{3}\right) = 3\left(x - \frac{1}{3}\right) \cdot 3\left(x + \frac{2}{3}\right) = (3x-1)(3x+2).$$

Topic 4: The Quadratic Formula

The method of completing the square will always serve to solve any equation of the form $ax^2+bx+c=0$. But this begs the question: why not just solve $ax^2+bx+c=0$ once and for all? Since a, b, and c represent arbitrary constants, the ideal situation would be to find a formula for the solutions of $ax^2+bx+c=0$ based on a, b, and c. That is exactly what the quadratic formula is: a formula that gives the solution to *any* equation of the form $ax^2+bx+c=0$. We will derive the formula now, using what we have learned.

$$ax^2+bx+c=0$$

$$x^2+\frac{b}{a}x=-\frac{c}{a}$$

We begin, as always in completing the square, by moving the constant to the right-hand side and dividing by a.

$$x^2+\frac{b}{a}x+\frac{b^2}{4a^2}=-\frac{c}{a}+\frac{b^2}{4a^2}$$

$$\left(x+\frac{b}{2a}\right)^2=-\frac{4ac}{4a^2}+\frac{b^2}{4a^2}$$

$$\left(x+\frac{b}{2a}\right)^2=\frac{b^2-4ac}{4a^2}$$

We next divide $\frac{b}{a}$ by 2 to get $\frac{b}{2a}$, and add $\left(\frac{b}{2a}\right)^2=\frac{b^2}{4a^2}$ to both sides. Note that to add the fractions on the right, we need a common denominator of $4a^2$.

$$x+\frac{b}{2a}=\pm\frac{\sqrt{b^2-4ac}}{2a}$$

Taking square roots leads to two linear equations, which we then solve for x.

$$x=\frac{-b}{2a}\pm\frac{\sqrt{b^2-4ac}}{2a}$$

$$x=\frac{-b\pm\sqrt{b^2-4ac}}{2a}$$

Since the fractions have the same denominator, they are easily added to obtain the final formula.

The Quadratic Formula

The solutions of the equation $ax^2+bx+c=0$ are

$$x=\frac{-b\pm\sqrt{b^2-4ac}}{2a}.$$

Note that the equation has a single solution if $b^2-4ac=0$, two real solutions if $b^2-4ac>0$, and two complex (and conjugates of one another) solutions if $b^2-4ac<0$.

The expression b^2-4ac is called the **discriminant**.

(It should be mentioned, for completeness' sake, that the quadratic formula also applies if a, b, and c are complex constants, not merely real constants.)

example 4

Solve the quadratic equations by using the quadratic formula.

```
(-B-√(B²-4AC))/(
2A)
        -.1830127019
```

To use the quadratic equation, first define the variables A, B, and C: enter the desired value, select **Sto**, then **Alpha Math (A)**. Select **Alpha** and . (decimal point) to enter a colon, and continue storing variables. Then enter the quadratic equation using **Alpha** and the defined variables (remember to use parentheses!) To obtain both answers, enter the formula with both the positive and the negative square root.

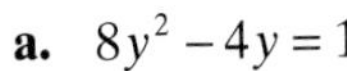

a. $8y^2 - 4y = 1$ **b.** $t^2 + 6t + 13 = 0$

Solutions:

a. $8y^2 - 4y = 1$

$$8y^2 - 4y - 1 = 0$$

Before applying the quadratic formula, move all the terms to one side so that $a, b,$ and c can be identified correctly.

$$y = \frac{-(-4) \pm \sqrt{(-4)^2 - (4)(8)(-1)}}{(2)(8)}$$

Apply the quadratic formula by making the appropriate replacements for $a, b,$ and c.

$$y = \frac{4 \pm \sqrt{16+32}}{16}$$

$$y = \frac{4 \pm \sqrt{48}}{16}$$

$$y = \frac{4 \pm 4\sqrt{3}}{16}$$

Note that there is a common factor of 4 that can be cancelled from the numerator and denominator in this case.

$$y = \frac{1 \pm \sqrt{3}}{4}$$

The two solutions are $\frac{1+\sqrt{3}}{4}$ and $\frac{1-\sqrt{3}}{4}$.

b. $t^2 + 6t + 13 = 0$

The equation is already in the proper form to apply the quadratic formula.

$$t = \frac{-6 \pm \sqrt{36-52}}{2}$$

$$t = \frac{-6 \pm \sqrt{-16}}{2}$$

Note that at this point we know the solutions will be complex numbers.

$$t = \frac{-6 \pm 4i}{2}$$

$$t = -3 \pm 2i$$

The two solutions are $-3+2i$ and $-3-2i$.

Topic 5: Interlude: Gravity Problems

When an object near the surface of the Earth is moving under the influence of gravity alone, its height above the surface is described by a quadratic polynomial in the variable t, where t stands for time and is usually measured in seconds.

The phrase "moving under the influence of gravity alone" means that all other forces that could potentially affect the object's motion, such as air resistance or mechanical lifting forces, are either negligible or absent. The phrase "near the surface of the earth" means that we are considering objects that travel short vertical distances relative to the earth's radius; the following formula doesn't apply, for instance, to rockets shot into orbit. As an example, think of someone throwing a baseball into the air on a windless day. After the ball is released, gravity is the only force acting on it.

If we let h stand for the height at time t of such an object,

$$h = -\frac{1}{2}gt^2 + v_0 t + h_0,$$

where g, v_0, and h_0 are all constants: g is the force due to gravity, v_0 is the initial velocity which the object has when $t = 0$, and h_0 is the height of the object when $t = 0$ (we normally say that ground-level corresponds to a height of 0). If t is measured in seconds and h in feet, g is 32 ft./s^2. If t is measured in seconds and h in meters, g is 9.8 m/s^2.

Many applications involving the above formula will result in a quadratic equation that must be solved for t. In some cases, one of the two solutions must be discarded as meaningless in the given problem.

example 5

Robert stands on the topmost tier of seats in a baseball stadium, and throws a ball out onto the field with a vertical upward velocity of 60 ft./s. The ball is 50 feet above the ground at the moment he releases the ball. When does the ball land?

Solution:

First, note that although the thrown ball has a horizontal velocity as well as a vertical velocity (otherwise it would go straight up and come straight back down), the horizontal velocity is irrelevant in such questions. All we are interested in is when the ball lands on the ground ($h = 0$). If we wanted to determine where in the field the ball lands, we would have to know the horizontal velocity as well.

We are given the following: $h_0 = 50$ ft. and $v_0 = 60$ ft./s. Since the units in the problem are feet and seconds, we know to use $g = 32$ ft./s^2. What we are interested in determining is the time t when the height h of the ball is 0. That is, we need to solve the equation $0 = -16t^2 + 60t + 50$ for t.

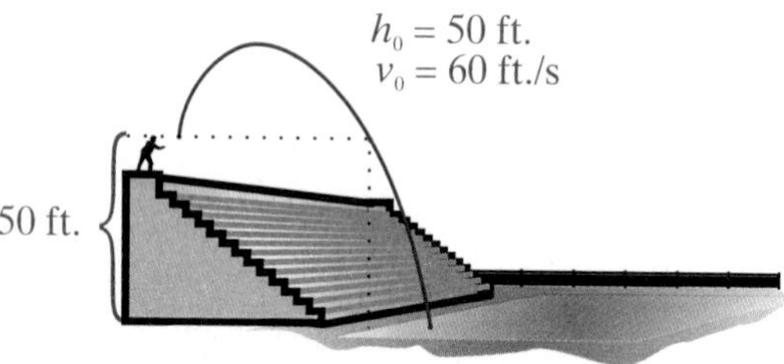

$0 = -16t^2 + 60t + 50$

$0 = 8t^2 - 30t - 25$

$t = \dfrac{30 \pm \sqrt{900 + 800}}{16}$

$t = \dfrac{30 \pm 10\sqrt{17}}{16}$

$t = \dfrac{15 \pm 5\sqrt{17}}{8}$

$t \approx \cancel{-0.70}, 4.45$

The quadratic polynomial in the equation doesn't factor over the integers, so the quadratic formula is a good method to use. To simplify the calculations, we can begin by dividing both sides of the equation by –2.

We then proceed to reduce the radical and simplify the resulting fraction (which actually describes two solutions).

These numbers are most meaningful in decimal form. The negative solution is immaterial in the problem, so it is discarded. The ball lands 4.45 seconds after being thrown.

Topic 6: Solving Quadratic-Like Equations

A polynomial equation of degree n in one variable, say x, is an equation that can be written in the form

$$a_n x^n + a_{n-1} x^{n-1} + \ldots + a_1 x + a_0 = 0,$$

where each a_i is a constant and $a_n \neq 0$. As we have seen, such equations can always be solved if $n = 1$ or $n = 2$, but in general there is no method for solving polynomial equations that is guaranteed to find all solutions. (There are formulas, called the *cubic* and *quartic* formulas, that solve third and fourth degree polynomial equations, but we will not use them in this text.) There are, moreover, many other types of equations for which no method is guaranteed to work. We are not entirely helpless, however. The Zero-Factor property applies whenever a product of any finite number of factors is equal to 0; if $A_1 \cdot A_2 \cdot \cdots \cdot A_n = 0$, then at least one of the A_i's must equal 0.

We can use the Zero-Factor property and our knowledge of quadratic equations to solve **quadratic-like equations**.

Quadratic-Like Equations

An equation is **quadratic-like**, or **quadratic in form**, if it can be written in the form

$$aA^2 + bA + c = 0,$$

where a, b, and c are constants, $a \neq 0$, and A is an algebraic expression. Such equations can be solved by first solving for A and then solving for the variable in the expression A.

example 6

Solve the quadratic-like equations.

a. $\left(x^2+2x\right)^2-7\left(x^2+2x\right)-8=0$

b. $y^{2/3}+4y^{1/3}-5=0$

Solutions:

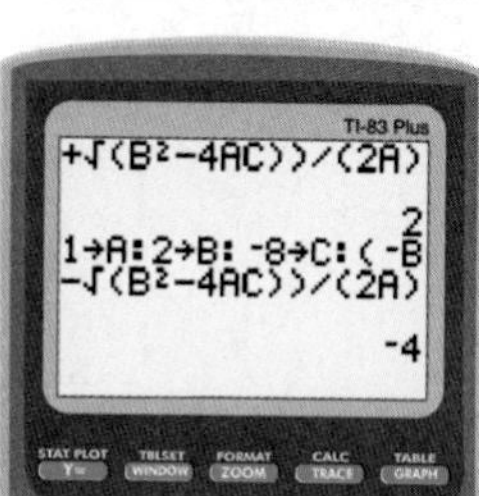

These screens are performed using the quadratic equation as described on page 87.

a.

$$\left(x^2+2x\right)^2-7\left(x^2+2x\right)-8=0$$

$$A^2-7A-8=0$$

$$(A-8)(A+1)=0$$

$$A=8 \quad \text{or} \quad A=-1$$

$$x^2+2x=8 \quad \text{or} \quad x^2+2x=-1$$

$$x^2+2x-8=0 \quad \text{or} \quad x^2+2x+1=0$$

$$(x+4)(x-2)=0 \quad \text{or} \quad (x+1)^2=0$$

$$x=-4 \text{ or } x=2 \text{ or } x=-1$$

By letting $A=x^2+2x$, the quadratic-like equation becomes a quadratic equation that can be solved by factoring.

Once the solutions for A have been obtained, we replace A with x^2+2x and solve for x. Note that –1 is a double root, while –4 and 2 are single roots.

b.

$$y^{2/3}+4y^{1/3}-5=0$$

$$\left(y^{1/3}\right)^2+4\left(y^{1/3}\right)-5=0$$

$$\left(\left(y^{1/3}\right)+5\right)\left(\left(y^{1/3}\right)-1\right)=0$$

$$y^{1/3}=-5 \quad \text{or} \quad y^{1/3}=1$$

$$y=(-5)^3 \quad \text{or} \quad y=1^3$$

$$y=-125 \quad \text{or} \quad y=1$$

This equation is not even a polynomial equation, but it is a polynomial equation in the expression $y^{1/3}$. In fact, it is another quadratic that can be solved for $y^{1/3}$, and then for y.

We could also use the substitution $A=y^{1/3}$ and get $A^2+4a-5=0$ which yields $A=-5$ or $A=1$.

Verify that these do indeed solve the equation $y^{2/3}+4y^{1/3}-5=0$.

Topic 7: Solving General Polynomial Equations by Factoring

As we have noted above, if an equation consists of a polynomial on one side and 0 on the other, and if the polynomial can be factored completely, then the equation can be solved by using the Zero-Factor property. If the coefficients in the polynomial are all real, the polynomial can, in principle, be factored into a product of first-degree and second-degree factors. In practice, however, this may be difficult to accomplish unless the degree of the polynomial is small or the polynomial is easily recognizable as a special product. (Several more factoring aids will be discussed in Chapter 4.)

example 7

Solve the equations by factoring.

a. $x^4 = 9$ **b.** $y^3 + y^2 - 4y - 4 = 0$ **c.** $8t^3 - 27 = 0$

Solutions:

a.

$$x^4 = 9$$
$$x^4 - 9 = 0$$
$$(x^2 - 3)(x^2 + 3) = 0$$
$$x^2 = 3 \text{ or } x^2 = -3$$
$$x = \sqrt{3} \text{ or } x = -\sqrt{3} \text{ or } x = i\sqrt{3} \text{ or } x = -i\sqrt{3}$$

After isolating 0 on one side, the polynomial is seen to be a difference of two squares, which can always be factored (see Section 1.3).

The Zero-Factor property gives us two equations, both of which can be solved by taking square roots.

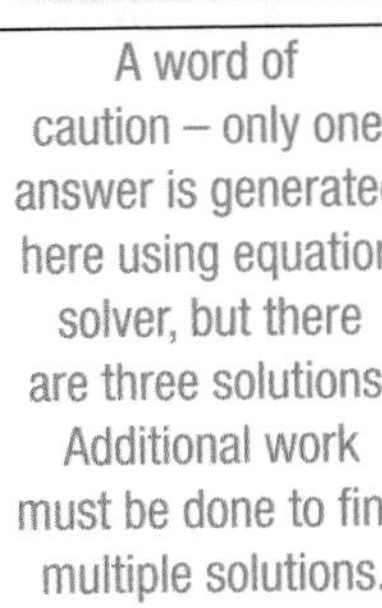

A word of caution – only one answer is generated here using equation solver, but there are three solutions. Additional work must be done to find multiple solutions.

b.

$$y^3 + y^2 - 4y - 4 = 0$$
$$y^2(y+1) - 4(y+1) = 0$$
$$(y+1)(y^2 - 4) = 0$$
$$(y+1)(y-2)(y+2) = 0$$
$$y = -1 \text{ or } y = 2 \text{ or } y = -2$$

The polynomial on the left can be factored by grouping.

After doing so, and after further factoring the difference of two squares, we obtain three solutions by the Zero-Factor property.

c.

$$8t^3 - 27 = 0$$
$$(2t)^3 - 3^3 = 0$$
$$(2t - 3)(4t^2 + 6t + 9) = 0$$
$$2t - 3 = 0 \quad \text{or} \quad 4t^2 + 6t + 9 = 0$$
$$2t = 3 \quad \text{or} \quad t = \frac{-6 \pm \sqrt{36 - 144}}{8}$$
$$t = \frac{3}{2} \quad \text{or} \quad t = \frac{-6 \pm \sqrt{-108}}{8}$$
$$t = \frac{3}{2} \quad \text{or} \quad t = \frac{-6 \pm 6i\sqrt{3}}{8}$$
$$t = \frac{3}{2} \quad \text{or} \quad t = \frac{-3 \pm 3i\sqrt{3}}{4}$$

The polynomial in this case is a difference of two cubes, which can always be factored. The Zero-Factor property gives us two equations, one of which is linear and straightforward to solve.

The second equation is a quadratic equation, and we can use the quadratic formula to solve it.

After simplifying the radical, and after further simplifying the fraction, we have three solutions of the original equation.

Topic 8: Solving Polynomial-Like Equations by Factoring

The last equations that we will consider in this section are equations that are not polynomial, but which can be solved using the methods we have developed above. We have already seen one such equation in Example 6b: the equation $y^{2/3}+4y^{1/3}-5=0$ is quadratic-like, and can be solved using polynomial methods. The general idea will be to rewrite the equation so that 0 appears on one side, and then to apply the Zero-Factor property. You may want to review some of the factoring techniques in Section 1.3 as you work through these problems.

example 8

Solve the following equations by factoring.

a. $x^{7/3}+x^{4/3}-2x^{1/3}=0$

b. $(x-1)^{1/2}-(x-1)^{-1/2}=0$

Solutions:

a.

$$x^{7/3}+x^{4/3}-2x^{1/3}=0$$
$$x^{1/3}\left(x^2+x-2\right)=0$$
$$x^{1/3}(x+2)(x-1)=0$$
$$x^{1/3}=0 \quad \text{or} \quad x+2=0 \quad \text{or} \quad x-1=0$$
$$x=0 \quad \text{or} \quad x=-2 \quad \text{or} \quad x=1$$

Recall that in cases like this, we factor out x raised to the lowest exponent. In this case, the remaining factor is a factorable trinomial.

The Zero-Factor property leads to three simple equations.

b.

$$(x-1)^{1/2}-(x-1)^{-1/2}=0$$
$$(x-1)^{-1/2}\left((x-1)-1\right)=0$$
$$(x-1)^{-1/2}(x-2)=0$$
$$(x-1)^{-1/2}=0 \quad \text{or} \quad x-2=0$$
$$\frac{1}{(x-1)^{1/2}}=0 \quad \text{or} \quad x=2$$
$$x=2$$

Again, we factor out the common algebraic expression raised to the lowest exponent.

This equation leads to two equations, only one of which has a solution. (Note that there is no value for x which would solve the first of the two equations.) The original equation has only one soluion.

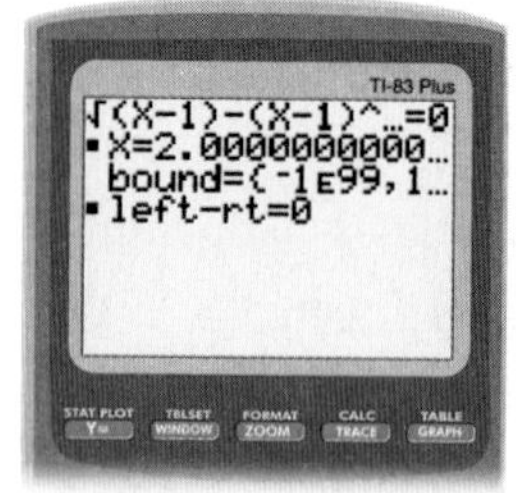

This screen was performed using the equation solver.

exercises

Solve the following quadratic equations by factoring. See Example 1.

1. $2x^2 - x = 3$

2. $x^2 - 14x + 49 = 0$

3. $9x - 5x^2 = -2$

4. $y(2y+9) = -9$

5. $2x^2 - 3x = x^2 + 18$

6. $(3x+2)(x-1) = 7 - 7x$

7. $15x^2 + x = 2$

8. $5x^2 + 2x + 3 = 4x^2 + 6x - 1$

9. $3x^2 + 33 = 2x^2 + 14x$

10. $(x-7)^2 = 16$

11. $3x^2 - 7x = 0$

12. $4x^2 - 9 = 0$

Solve the following quadratic equations by the square root method. See Example 2.

13. $(x-3)^2 = 9$

14. $(a-2)^2 = -5$

15. $x^2 - 6x + 9 = -16$

16. $(2x+1)^2 - 7 = 0$

17. $(8t-3)^2 = 0$

18. $9 = (3s+2)^2$

19. $(2x-1)^2 = 8$

20. $(y-18)^2 - 1 = 0$

21. $x^2 - 4x + 4 = 49$

22. $-3(n+7)^2 = -27$

23. $(3x-6)^2 = 4x^2$

24. $(2x+3)^2 + 9 = 0$

Solve the following quadratic equations by completing the square. See Example 3.

25. $y^2 + 11 = 12y$

26. $5y^2 + 6y - 3 = 0$

27. $-6z^2 + 4z = -1$

28. $2a^2 - 4a - 4 = 0$

29. $2b^2 + 10b + 5 = 0$

30. $x^2 - 4x = 16$

31. $x^2 + 8x + 7 = -8$

32. $2x^2 + 6x - 10 = 10$

33. $2x^2 + 7x - 15 = 0$

34. $4x^2 - 4x - 63 = 0$

35. $u^2 + 10u + 9 = 0$

36. $4x^2 - 56x + 195 = 0$

Solve the following quadratic equations by using the quadratic formula. See Example 4.

37. $4x^2 - 3x = -1$

38. $3x^2 - 4 = -x$

39. $2.1y^2 - 3.5y = 4$

40. $2.6z^2 - 0.9z + 2 = 0$

41. $a(a+2) = -1$

42. $3x^2 - 2x = 0$

43. $6x^2 + 5x - 4 = 3x - 2$

44. $7x^2 - 4x = 51$

45. $4x^2 - 14x - 27 = 3$

Solve the following application problems. See Example 5.

46. How long would it take for a ball dropped from the top of a 144-foot building to hit the ground?

47. Suppose that instead of being dropped, as in Exercise 46, a ball is thrown upward with a velocity of 40 feet per second from the top of a 144-foot building. Assuming it misses the building on the way back down, how long after being thrown will it hit the ground?

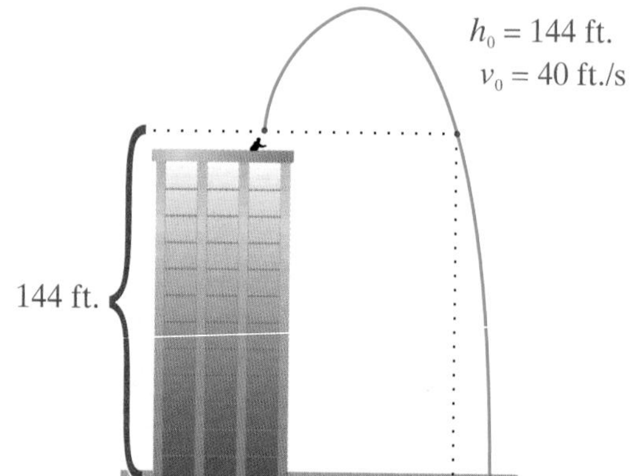

48. A slingshot is used to shoot a BB at a velocity of 96 feet per second straight up from ground level. When will the BB reach its maximum height of 144 feet?

49. A rock is thrown upward with a velocity of 20 meters per second from the top of a 24 meter high cliff, and it misses the cliff on the way back down. When will the rock be 7 meters from ground level? (Round your answer to the nearest tenth.)

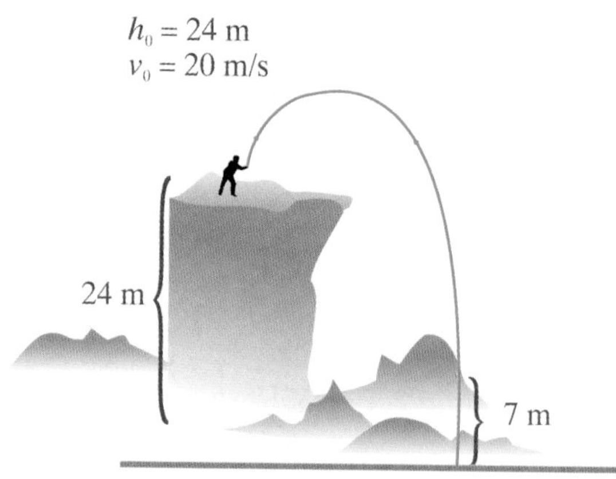

50. Luke, an experienced bungee jumper, leaps from a tall bridge and falls toward the river below. The bridge is 170 feet above the water and Luke's bungee cord is 110 feet long unstretched. When will Luke's cord begin to stretch? (Round your answer to the nearest tenth.)

Use the connection between solutions of quadratic equations and polynomial factoring to answer the following questions.

51. Factor the quadratic $x^2 - 6x + 13$.

52. Factor the quadratic $9x^2 - 6x - 4$.

53. Factor the quadratic $4x^2 + 12x + 1$.

54. Factor the quadratic $25x^2 - 10x + 2$.

55. Determine b and c so that the equation $x^2 + bx + c = 0$ has the solution set $\{-3, 8\}$.

Solve the following quadratic-like equations. See Example 6.

56. $(x-1)^2 + (x-1) - 12 = 0$

57. $(x^2-1)^2 + (x^2-1) - 12 = 0$

58. $(x^2+1)^2 + (x^2+1) - 12 = 0$

59. $(x^2-2x+1)^2 + (x^2-2x+1) - 12 = 0$

60. $2y^{2/3} + y^{1/3} - 1 = 0$

61. $2x^{2/3} - 7x^{1/3} + 3 = 0$

62. $(x^2-6x)^2 + 4(x^2-6x) - 5 = 0$

63. $(t^2-t)^2 - 8(t^2-t) + 12 = 0$

64. $2x^{1/2} - 5x^{1/4} + 2 = 0$

65. $3x^{2/3} - x^{1/3} - 2 = 0$

66. $(z-8)^2 - 7(z-8) + 12 = 0$

67. $(x^2-13)^2 + (x^2-13) - 12 = 0$

Solve the following polynomial equations by factoring. See Example 7.

68. $a^3 - 3a^2 = a - 3$

69. $2x^3 + x^2 + 2x + 1 = 0$

70. $2x^3 - x^2 = 15x$

71. $y^3 + 8 = 0$

72. $8a^3 - 27 = 0$

73. $16a^4 = 81$

74. $5s^3 + 6s^2 - 20s = 24$

75. $6x^3 + 8x^2 = 14x$

76. $14x^3 + 27x^2 - 20x = 0$

77. $27x^3 + 64 = 0$

78. $x^3 - 4x^2 + x = 4$

79. $x^4 + 5x^2 - 36 = 0$

Solve the following equations by factoring. See Example 8.

80. $3x^{11/3} + 2x^{8/3} - 5x^{5/3} = 0$

81. $(x-3)^{-1/2} + 2(x-3)^{1/2} = 0$

82. $2x^{13/5} - 5x^{8/5} + 2x^{3/5} = 0$

83. $(2x-5)^{1/3} - 3(2x-5)^{-2/3} = 0$

84. $(t+4)^{2/3} + 2(t+4)^{8/3} = 0$

85. $x^{11/2} - 6x^{9/2} + 9x^{7/2} = 0$

Use the connection between solutions of polynomial equations and polynomial factoring to answer the following questions.

86. Find b, c, and d so the equation $x^3+bx^2+cx+d=0$ has solutions of –3, –1, and 5.

87. Find b, c, and d so the equation $x^3+bx^2+cx+d=0$ has solutions of –2, 0, and 6.

88. Find a, b, and d so the equation $ax^3+bx^2+3x+d=0$ has solutions of –3, $-\frac{1}{2}$, and 0.

89. Find a, b, and c so the equation $ax^3+bx^2+cx+6=0$ has solutions of $-\frac{3}{5}, \frac{2}{3}$, and 1.

technology exercises

The Technology Exercises in Section 1.5 introduced *Mathematica*'s **`Solve`** command as used to solve a multi-variable formula for a specific variable. Not surprisingly, given the name of the command, **`Solve`** can also be used to solve many simple single-variable equations. The user must understand the math well enough to know the limitations of the computer algorithms used, to spot errors introduced by the user or the computer, and to correctly interpret the computer's output. For instance, when *Mathematica* is used to solve the problem in Example 5 of this section (as shown below), it is up to the user to realize that only the second answer makes physical sense.

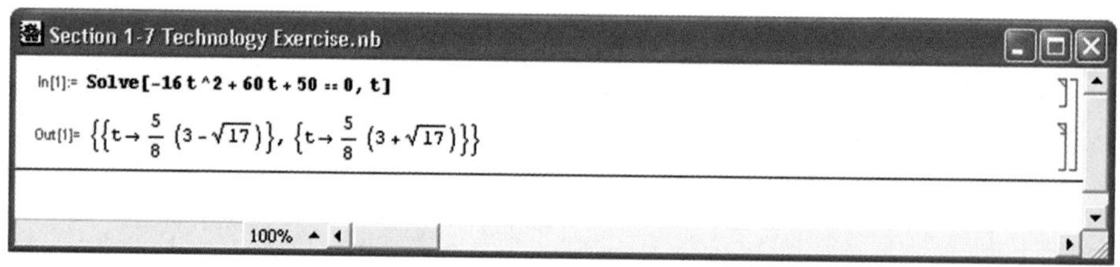

Section 1-7 Technology Exercise.nb
In[1]:= Solve[-16 t^2 + 60 t + 50 == 0, t]
Out[1]= {{t → 5/8 (3 - √17)}, {t → 5/8 (3 + √17)}}
100%

Use a computer algebra system, if available, to solve the following single-variable equations. (See Appendix A for guidance on using computer algebra systems like Mathematica.)

90. $5x^2-3x=17$

91. $5x^2-3x=-17$

92. $(a+4)(4a-3)=5$

93. $4.8x^2+3.5x-9.2=0$

94. $10\pi r+\pi r^2=107$

95. $(3x-1)(3-x)=2x+5$

1.8 Rational and Radical Equations

TOPICS

1. Simplifying rational expressions
2. Combining rational expressions
3. Simplifying complex rational expressions
4. Solving rational equations
5. Interlude: work-rate problems
6. Solving radical equations
7. Solving equations with positive rational exponents

Topic 1: Simplifying Rational Expressions

Many equations contain fractions in which the variable appears in the denominator, and the presence of such fractions can make the solution process challenging. We will learn how to work with a class of fractions called *rational expressions* and develop a general method for solving equations that contain such expressions.

Rational Expressions

A **rational expression** is an expression that can be written as a *ratio* of two polynomials $\frac{P}{Q}$. Of course, such a fraction is undefined for any value(s) of the variable(s) for which $Q = 0$. A given rational expression is *simplified* or *reduced* when P and Q contain no common factors (other than 1 and –1).

To simplify rational expressions, factor the polynomials in the numerator and denominator completely and then cancel any common factors. Remember that the simplified rational expression may be defined for values of the variable that the original expression is not, and the two versions are equal only where they are both defined. That is, if A, B, and C represent algebraic expressions,

$$\frac{AC}{BC} = \frac{A}{B} \text{ only where } B \neq 0 \text{ and } C \neq 0.$$

example 1

Simplify the following rational expressions, and indicate values of the variable that must be excluded.

a. $\dfrac{x^3-8}{x^2-2x}$ **b.** $\dfrac{x^2-x-6}{3-x}$

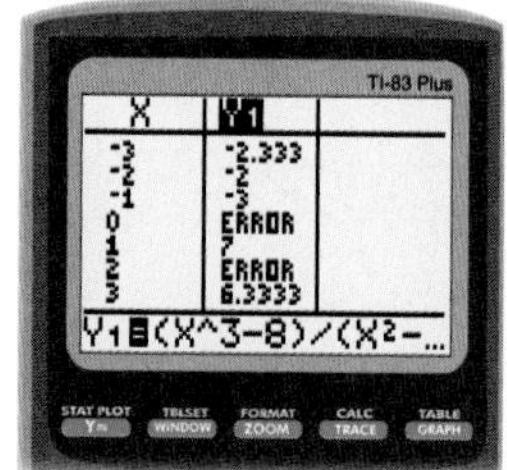

Notice that x-coordinates of 0 and 2 result in an error. To arrive at this screen, select **Y=** and enter the equation. Next, select **2nd** and **Graph** to view the table.

Solutions:

a. $\dfrac{x^3-8}{x^2-2x} = \dfrac{(x-2)(x^2+2x+4)}{x(x-2)}$

$= \dfrac{x^2+2x+4}{x} \qquad x \neq 0,2$

After factoring both polynomials, we cancel the common factor of $x-2$. Note that even though the final expression is defined when $x=2$, the first and last expressions are equal only where both are defined.

b. $\dfrac{x^2-x-6}{3-x} = \dfrac{(x+2)(x-3)}{-(x-3)}$

$= \dfrac{x+2}{-1}$

$= -x-2 \qquad x \neq 3$

In this rational expression, the denominator is already factored, but we bring out a factor of -1 from the denominator in order to cancel a common factor of $x-3$. Note that the original and simplified versions are only equal for values of x not equal to 3.

caution!

Remember that only common *factors* can be cancelled! A very common error is to think that common terms from the numerator and denominator can be cancelled. For instance, the statement $\dfrac{x+4}{x^2} = \dfrac{4}{x}$ is incorrect. It is not possible to factor $x+4$ at all, and the x that appears in the numerator is not a factor that can be cancelled with one of the x's in the denominator. The expression $\dfrac{x+4}{x^2}$ is already simplified as far as possible.

Topic 2: Combining Rational Expressions

Rational expressions are combined by the operations of addition, subtraction, multiplication, and division the same way that numerical fractions are. In order to add or subtract two rational expressions, a common denominator must first be found. In order to multiply two rational expressions, the numerators are multiplied and the denominators are multiplied. Finally, in order to divide one rational expression by another, the first is multiplied by the reciprocal of the second. It is generally best to factor all the numerators and denominators before combining rational expressions.

example 2

Add or subtract the rational expressions, as indicated.

a. $\dfrac{2x-1}{x^2+x-2}-\dfrac{2x}{x^2-4}$

b. $\dfrac{x+1}{x+3}+\dfrac{x^2+x-2}{x^2-x-6}-\dfrac{x^2-2x+9}{x^2-9}$

Solutions:

a.
$$\frac{2x-1}{x^2+x-2}-\frac{2x}{x^2-4}$$
$$=\frac{2x-1}{(x+2)(x-1)}-\frac{2x}{(x+2)(x-2)}$$
$$=\left[\frac{x-2}{x-2}\right]\cdot\frac{2x-1}{(x+2)(x-1)}-\left[\frac{x-1}{x-1}\right]\cdot\frac{2x}{(x+2)(x-2)}$$
$$=\frac{2x^2-5x+2}{(x-2)(x+2)(x-1)}-\frac{2x^2-2x}{(x-1)(x+2)(x-2)}$$
$$=\frac{-3x+2}{(x-2)(x+2)(x-1)}$$

We begin by factoring both denominators, and note that the LCD is $(x-2)(x+2)(x-1)$. In order to obtain this denominator in the first fraction, we multiply the top and bottom by $x-2$. In order to obtain the LCD in the second fraction, we multiply the top and bottom by $x-1$.

After subtracting the second numerator from the first, we are done. Note that there are no common factors to cancel.

b.
$$\frac{x+1}{x+3}+\frac{x^2+x-2}{x^2-x-6}-\frac{x^2-2x+9}{x^2-9}$$
$$=\frac{x+1}{x+3}+\frac{\cancel{(x+2)}(x-1)}{(x-3)\cancel{(x+2)}}-\frac{x^2-2x+9}{(x-3)(x+3)}$$
$$=\left[\frac{x-3}{x-3}\right]\cdot\frac{x+1}{x+3}+\left[\frac{x+3}{x+3}\right]\cdot\frac{x-1}{x-3}-\frac{x^2-2x+9}{(x-3)(x+3)}$$
$$=\frac{x^2-2x-3+x^2+2x-3-x^2+2x-9}{(x-3)(x+3)}$$
$$=\frac{x^2+2x-15}{(x-3)(x+3)}$$
$$=\frac{(x+5)\cancel{(x-3)}}{\cancel{(x-3)}(x+3)}$$
$$=\frac{x+5}{x+3}$$

We again factor all the polynomials, and note that the second rational expression can be reduced. We do this before determining that the LCD is $(x-3)(x+3)$.

After multiplying each numerator and denominator by the required factors in order to obtain a common denominator, we combine the numerators and simplify.

After factoring the resulting numerator, there is a common factor that can be cancelled.

example 3

Multiply or divide the rational expressions, as indicated.

a. $\frac{x^2+3x-10}{x+3}\cdot\frac{x-3}{x^2-x-2}$

b. $\frac{x^2+5x-14}{3x}\div\frac{x^2-4x+4}{9x^3}$

Solutions:

a. $\frac{x^2+3x-10}{x+3}\cdot\frac{x-3}{x^2-x-2}$

$=\frac{(x+5)(x-2)}{x+3}\cdot\frac{x-3}{(x-2)(x+1)}$

$=\frac{(x+5)\cancel{(x-2)}(x-3)}{(x+3)\cancel{(x-2)}(x+1)}$

$=\frac{(x+5)(x-3)}{(x+3)(x+1)}$

We begin by factoring both numerators and denominators, and then write the product of the two rational expressions as a single fraction.

Since we have already factored the polynomials completely, any common factors that can be cancelled are readily identified.

b. $\frac{x^2+5x-14}{3x}\div\frac{x^2-4x+4}{9x^3}$

$=\frac{(x+7)(x-2)}{3x}\cdot\frac{9x^3}{(x-2)^2}$

$=\frac{\overset{3}{\cancel{9}}x^{\overset{2}{\cancel{3}}}(x+7)\cancel{(x-2)}}{\cancel{3x}(x-2)^{\cancel{2}}}$

$=\frac{3x^2(x+7)}{x-2}$

We divide the first rational expression by the second by inverting the second fraction and multiplying. Note that we have factored all the polynomials and inverted the second fraction in one step.

Now we proceed to cancel common factors (including any common factors in the purely numerical factors) to obtain the final answer.

Topic 3: Simplifying Complex Rational Expressions

A **complex rational expression** is a fraction in which the numerator or denominator (or both) contains at least one rational expression. Complex rational expressions can always be rewritten as simple rational expressions. One way to do this is to simplify the numerator and denominator individually and then divide the numerator by the denominator as in Example 3b. Another way, which is frequently faster, is to multiply the numerator and denominator by the LCD of all the fractions that make up the complex rational expression. This method will be illustrated in the next example.

example 4

Simplify the complex rational expressions.

a. $\dfrac{\dfrac{1}{x+h}-\dfrac{1}{x}}{h}$

b. $\dfrac{x^{-1}-y^{-1}}{x^{-2}-y^{-2}}$

Solutions:

a. $$\frac{\frac{1}{x+h}-\frac{1}{x}}{h}=\frac{\frac{1}{x+h}-\frac{1}{x}}{\frac{h}{1}}\cdot\left[\frac{(x+h)(x)}{(x+h)(x)}\right]$$
$$=\frac{x-(x+h)}{(h)(x+h)(x)}$$
$$=\frac{-\cancel{h}}{(\cancel{h})(x+h)(x)}$$
$$=\frac{-1}{(x+h)(x)}$$

This complex rational expression contains two rational expressions in the numerator. It may be helpful to write the denominator as a fraction, as we have done here, in order to determine that the LCD of all the fractions making up the overall expression is $(x+h)(x)$. We multiply the numerator and denominator by the LCD (so we are multiplying the overall expression by 1), then cancel the common factor of h to get the final answer.

b. $$\frac{x^{-1}-y^{-1}}{x^{-2}-y^{-2}}=\frac{\frac{1}{x}-\frac{1}{y}}{\frac{1}{x^2}-\frac{1}{y^2}}$$

This expression is a complex rational expression, a fact that is more clear once we rewrite the terms that have negative exponents as fractions.

$$=\frac{\frac{1}{x}-\frac{1}{y}}{\frac{1}{x^2}-\frac{1}{y^2}}\cdot\left[\frac{x^2y^2}{x^2y^2}\right]$$

The LCD in this case is x^2y^2, so we multiply top and bottom by this and factor the resulting polynomials.

$$=\frac{xy^2-x^2y}{y^2-x^2}$$
$$=\frac{xy\cancel{(y-x)}}{\cancel{(y-x)}(y+x)}$$
$$=\frac{xy}{y+x}$$

There is a common factor that can be cancelled, so we do so to obtain the final simplified expression.

technology note

Mathematica, and other computer algebra systems, are capable of simplifying combinations of rational expressions. The command to do this in *Mathematica* is `Together`; when invoked, the command attempts to combine fractional expressions into one, eliminating common factors in the process. Of course, it is up to the user to realize that the resulting single fraction might not be defined for some values of the variable(s).

Output [1] below illustrates how *Mathematica* can be used to combine three rational expressions. Output [2] simply echoes Input [2], one of the complex rational expressions seen in Example 4. Output [3] then illustrates that `Together` can also be used to simplify this complex rational expression. (*See Appendix A for guidance on using computer algebra systems like Mathematica.*)

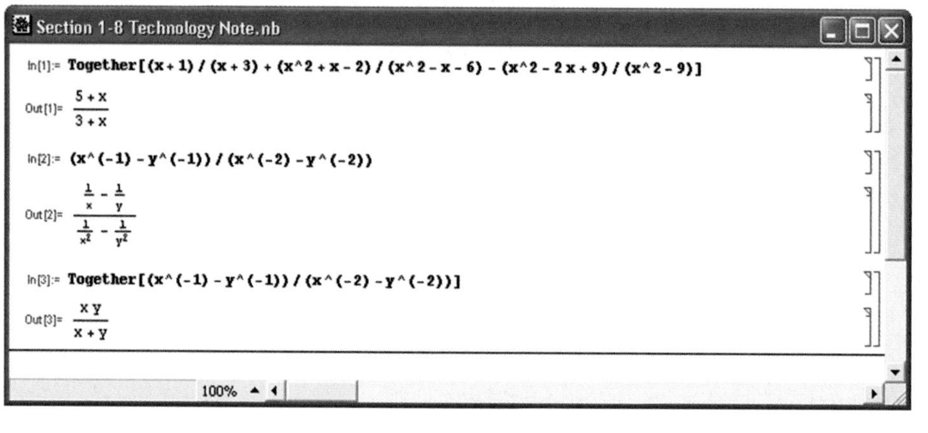

Topic 4: Solving Rational Equations

A **rational equation** is an equation that contains at least one rational expression, while any non-rational expressions are polynomials. Our general approach to solving such equations is to multiply each term in the equation by the LCD of all the rational expressions; this has the effect of converting rational equations into polynomial equations, which we have already learned how to solve. There is, however, one important difference between rational equations and polynomial equations: it is quite possible that one or more rational expressions in a rational equation are not defined for some values of the variable. Of course, these values cannot possibly be solutions of the equation, and must be excluded from the solution set. These excluded values may appear, however, as solutions of the polynomial equation that is derived from the original rational equation. Potential error can be avoided by keeping track of what values of the variable are disallowed and/or checking all solutions in the original equation.

example 5

Solve the following rational equations.

a. $\dfrac{x^3+3x^2}{x^2-2x-15}=\dfrac{4x+5}{x-5}$

b. $\dfrac{3x^2}{5x-1}-1=0$

Solutions:

a.

$$\frac{x^3+3x^2}{x^2-2x-15}=\frac{4x+5}{x-5}$$

$$\frac{x^2\cancel{(x+3)}}{(x-5)\cancel{(x+3)}}=\frac{4x+5}{x-5}$$

$$[x-5]\cdot\frac{x^2}{(x-5)}=[x-5]\cdot\frac{4x+5}{x-5}$$

$$x^2=4x+5$$

$$x^2-4x-5=0$$

$$(x-5)(x+1)=0$$

$$x=\cancel{5},-1$$

$$x=-1$$

In order to cancel any factors that might be cancellable, and in order to determine the LCD, we begin by factoring all the numerators and denominators. This also tells us the very important fact that 5 and −3 cannot be solutions of the equation.

After multiplying both sides of the equation by the LCD, we have a second-degree polynomial equation that can be solved by factoring. Note, however, that we have already determined that 5 cannot be a solution. It must be discarded.

b.

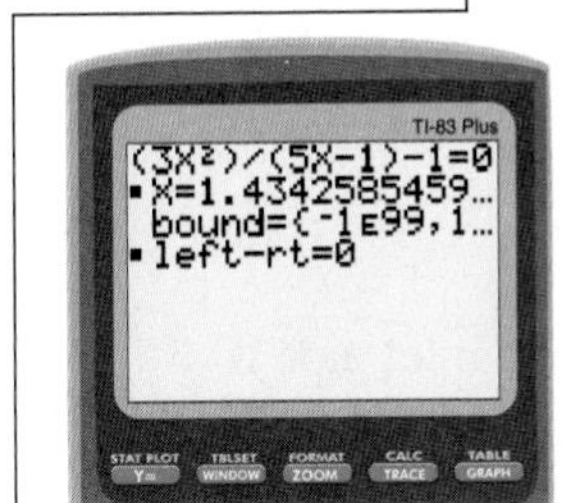

$$\frac{3x^2}{5x-1}-1=0$$

$$[5x-1]\cdot\frac{3x^2}{5x-1}-1\cdot[5x-1]=0$$

$$3x^2-5x+1=0$$

$$x=\frac{5\pm\sqrt{25-12}}{6}$$

$$x=\frac{5\pm\sqrt{13}}{6}$$

There is no factoring possible in this problem, so we merely note that $\frac{1}{5}$ is the one value for x that must be excluded. After multiplying through by the LCD, we have a second-degree polynomial equation that can be solved by the quadratic formula.

Since neither of the roots is $\frac{1}{5}$, both will solve the original equation.

Topic 5: Interlude: Work-Rate Problems

Many seemingly different applications fall into a class of problems known as work-rate problems. What these applications have in common is two or more "workers" acting in concert to complete a task. The workers can be, for example, employees on a job, machines manufacturing a part, or inlet and outlet pipes filling or draining a tank. Typically, each worker is capable of doing the task alone, but each worker works at his, her, or its own individual rate regardless of whether others are involved.

The goal in a work-rate problem is usually to determine how fast the task at hand can be completed, either by the workers together or by one of the workers individually. There are two keys to solving a work-rate problem:

1. **Rate of work is the reciprocal of the time needed to complete the task.** If a given job can be done by a worker in x units of time, the worker works at a rate of $\frac{1}{x}$ jobs per unit time. For instance, if Jane can overhaul an engine in 2 hours, her rate of work is $\frac{1}{2}$ of the job per hour. If a faucet can fill a sink in 5 minutes, its rate is $\frac{1}{5}$ of the job per minute. Of course, this also means that the time needed to complete a task is the reciprocal of the rate of work: if a faucet fills a sink at the rate of $\frac{1}{5}$ of the job per minute, it takes 5 minutes to fill the sink.
2. **Rates of work are "additive."** This means that, in the ideal situation, two workers working together on the same task will have a combined rate of work that is the sum of their individual rates. For instance, if Jane's rate in overhauling an engine is $\frac{1}{2}$ and Ted's rate is $\frac{1}{3}$, their combined rate is $\frac{1}{2}+\frac{1}{3}$, or $\frac{5}{6}$. That is, together they can overhaul an engine in $\frac{6}{5}$ of an hour, or one hour and twelve minutes.

example 6

One inlet pipe can fill a swimming pool in 12 hours. The owner installs an inlet pipe that can fill the pool at twice the rate of the first one. If both inlet pipes are used together, how long does it take to fill the pool?

Solution:

The rate of work of the first inlet pipe is $\frac{1}{12}$, so the rate of the second inlet pipe is $\frac{1}{6}$. If we let x denote the time needed to fill the pool when both inlet pipes are used together, the sum of the two individual rates must equal $\frac{1}{x}$. So we need to solve the equation $\frac{1}{12}+\frac{1}{6}=\frac{1}{x}$.

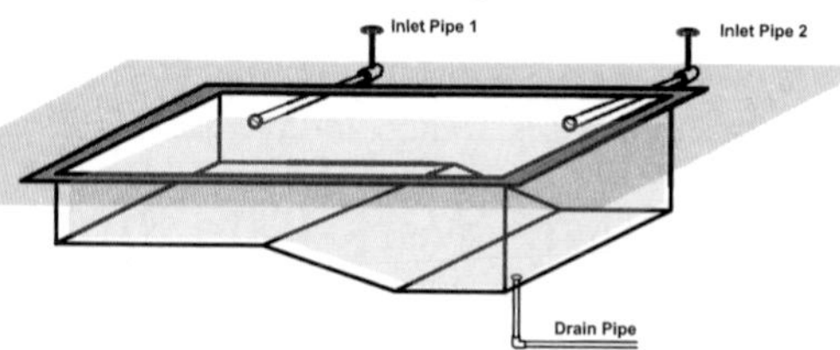

$$\frac{1}{12}+\frac{1}{6}=\frac{1}{x}$$
$$x+2x=12$$
$$3x=12$$
$$x=4$$

As is typical, this work-rate problem leads to a rational equation, and we can solve it as we solve any rational equation: multiply both sides by the LCD (which is $12x$ in this case) and proceed to solve the resulting polynomial equation.

Together, the two inlet pipes can fill the pool in 4 hours.

example 7

The pool-owner in Example 6 is a bit clumsy, and one day proceeds to fill his empty pool with the two inlet pipes but accidentally turns on the pump that drains the pool also. Fortunately, the pump rate is slower than the combined rate of the two inlet pipes, and the pool fills anyway, but it takes 10 hours to do so. At what rate can the pump empty the pool?

Solution:

If we let x denote the time it takes the pump to empty the pool, we can say that the pump has a *filling* rate of $-\frac{1}{x}$ (since emptying is the opposite of filling). The combined filling rate of the two inlet pipes is $\frac{1}{4}$, and the total rate at which the pool is filled on this unfortunate day is $\frac{1}{10}$.

$$\frac{1}{4}-\frac{1}{x}=\frac{1}{10}$$
$$5x-20=2x$$
$$3x=20$$
$$x=\frac{20}{3}$$

We again use the fact that the sum of the individual rates is equal to the combined rate, but in this case one of the individual rates is negative. We solve this rational equation as always, multiplying through by the LCD: $20x$.

The pump can empty the pool in $\frac{20}{3}$ hours, or 6 hours and 40 minutes.

Topic 6: Solving Radical Equations

The last one-variable equations that we will discuss are those that contain radical expressions. A **radical equation** is an equation that has at least one radical expression containing a variable, while any non-radical expressions are polynomial terms. As with the rational equations discussed in the last section, we will develop a general method of solution that converts a given radical equation into a polynomial equation. We will see, however, that just as with rational equations, we must check our potential solutions carefully to see if they actually solve the original radical equation.

Our method of solving rational equations involved multiplying both sides of the equation by an algebraic expression (the LCD of all the rational expressions), and in some cases potential solutions had to be discarded because they led to division by 0 in one or more of the rational expressions. Something similar can happen with radical equations. Since our goal is to convert a given radical equation into a polynomial equation, one reasonable approach is to raise both sides of the equation to whatever power is necessary to "undo" the radical (or radicals). The problem is that this does *not* result in an equivalent equation; the only means we have of transforming an equation into an equivalent equation is to add the same quantity to both sides or to multiply both sides by a non-zero quantity. We won't *lose* any solutions by raising both sides of an equation

to the same power, but we may *gain* some extraneous solutions. We identify these and discard them by simply checking all of our eventual solutions in the original equation.

A simple example will make this clear. Consider the equation

$$x = -3.$$

This equation is so basic that it is its own solution. But if, solely for the purposes of demonstration, we square both sides, we obtain the equation

$$x^2 = 9.$$

This second-degree equation can be solved by factoring the polynomial $x^2 - 9$ or by taking the square root of both sides, and in either case we obtain the solution set $\{-3, 3\}$. That is, by squaring both sides of the original equation, we gained a second (and false) solution.

The specifics of our method are as follows.

Method of Solving Radical Equations

Step 1: Begin by isolating the radical expression on one side of the equation. If there is more than one radical expression, choose one to isolate on one side.

Step 2: Raise both sides of the equation by the power necessary to "undo" the isolated radical. That is, if the radical is an nth root, raise both sides to the nth power.

Step 3: If any radical expressions remain, simplify the equation if possible and then repeat Steps 1 and 2 until the result is a polynomial equation. When a polynomial equation has been obtained, solve the equation using polynomial methods.

Step 4: Check your solutions in the original equation! Any extraneous solutions must be discarded.

(If the equation contains many radical expressions, and especially if they are of differing indices, eliminating all the radicals may be a long process! The equations that we will solve will not require more than several repetitions of Steps 1 and 2.)

example 8

Solve the radical equations.

a. $\sqrt{1-x}-1=x$ **b.** $\sqrt{x+1}+\sqrt{x+2}=1$ **c.** $\sqrt[4]{x^2+8x+7}-2=0$

Solutions:

a.
$$\sqrt{1-x}-1=x$$
$$\sqrt{1-x}=x+1$$
$$\left(\sqrt{1-x}\right)^2=(x+1)^2$$
$$1-x=x^2+2x+1$$
$$0=x^2+3x$$
$$0=x(x+3)$$
$$x=0, \cancel{-3}$$
$$x=0$$

There is only one radical expression, so we isolate it and proceed to square both sides. The result is a second-degree polynomial equation that can be solved by factoring.

Note, though, that $\sqrt{1-(-3)}\neq -3+1$, so -3 must be discarded. The solution is the single number 0.

b.
$$\sqrt{x+1}+\sqrt{x+2}=1$$
$$\sqrt{x+1}=1-\sqrt{x+2}$$
$$\left(\sqrt{x+1}\right)^2=\left(1-\sqrt{x+2}\right)^2$$
$$x+1=1-2\sqrt{x+2}+x+2$$
$$2\sqrt{x+2}=2$$
$$\sqrt{x+2}=1$$
$$x+2=1$$
$$x=-1$$

This equation has two radical expressions, so we isolate one of them initially.

We square both sides to eliminate the isolated radical, and then proceed to simplify and isolate the one remaining radical.

At this point, we square both sides again and solve the polynomial equation. Check that the one solution solves the original equation in this problem.

c.
$$\sqrt[4]{x^2+8x+7}-2=0$$
$$\sqrt[4]{x^2+8x+7}=2$$
$$\left(\sqrt[4]{x^2+8x+7}\right)^4=2^4$$
$$x^2+8x+7=16$$
$$x^2+8x-9=0$$
$$(x+9)(x-1)=0$$
$$x=-9, 1$$

We first isolate the radical (a fourth root in this case) and then raise both sides to the fourth power.

The resulting second-degree equation can again be solved by factoring.

Check that both roots satisfy the original equation.

Topic 7: Solving Equations with Positive Rational Exponents

We have already encountered several equations with rational exponents in Section 7 of this chapter, in which we solved such equations either by factoring or by quadratic methods. In the problems that we will consider here, equations containing terms with positive rational exponents can be viewed as radical equations. Rewriting each term that has a positive rational exponent as a radical will allow us to use the method developed previously to solve rational equations.

example 9

Solve the following equations with rational exponents.

a. $x^{2/3} - 9 = 0$

b. $\left(32x^2 - 32x + 17\right)^{1/4} = 3$

Solutions:

This screen was performed using the equation solver.

a.
$$x^{2/3} - 9 = 0$$
$$x^{2/3} = 9$$
$$\sqrt[3]{x^2} = 9$$
$$x^2 = 9^3$$
$$x = 9^{3/2}$$
$$x = \left(9^{1/2}\right)^3$$
$$x = \left(\pm 3\right)^3$$
$$x = \pm 27$$

Since the term containing the rational exponent can be rewritten as a radical expression, we will begin by isolating that term.

Cubing both sides eliminates the cube root.

Raising both sides to the $\frac{1}{2}$ power solves the equation for x, but we can evaluate the expression on the right-hand side. Note that both positive 3 and negative 3 must be considered. Verify that both numbers satisfy the original equation.

b.
$$\left(32x^2 - 32x + 17\right)^{1/4} = 3$$
$$\sqrt[4]{32x^2 - 32x + 17} = 3$$
$$32x^2 - 32x + 17 = 3^4$$
$$32x^2 - 32x + 17 = 81$$
$$32x^2 - 32x - 64 = 0$$
$$32\left(x^2 - x - 2\right) = 0$$
$$x^2 - x - 2 = 0$$
$$(x-2)(x+1) = 0$$
$$x = 2, -1$$

The exponent of $\frac{1}{4}$ indicates we should raise both sides to the fourth power in order to eliminate the radical.

We are left with a second-degree polynomial equation that can be solved by factoring. Check that both solutions again satisfy the original equation.

exercises

Simplify the following rational expressions, indicating which values of the variable must be excluded. See Example 1.

1. $\dfrac{2x^2+7x+3}{x^2-2x-15}$ **2.** $\dfrac{x^2+5x-6}{x^3+2x^2-3x}$ **3.** $\dfrac{x^3+2x^2-3x}{x+3}$ **4.** $\dfrac{x^2-4x+4}{x^2-4}$

5. $\dfrac{x^2+5x-6}{x^2+4x-5}$ **6.** $\dfrac{2x^2+7x-15}{x^2+3x-10}$ **7.** $\dfrac{x+1}{x^3+1}$ **8.** $\dfrac{x^3+x}{3x^2+3}$

9. $\dfrac{2x^2+11x+5}{x+5}$ **10.** $\dfrac{x^4-x^3}{x^2-3x+2}$ **11.** $\dfrac{2x^2+11x-21}{x+7}$ **12.** $\dfrac{8x^3-27}{2x-3}$

Add or subtract the rational expressions, as indicated, and simplify your answer. See Example 2.

13. $\dfrac{x-3}{x+5}+\dfrac{x^2+3x+2}{x-3}$ **14.** $\dfrac{x^2-1}{x-2}-\dfrac{x-1}{x+1}$

15. $\dfrac{x+2}{x-3}-\dfrac{x-3}{x+5}-\dfrac{1}{x^2+2x-15}$ **16.** $\dfrac{x+1}{x-3}+\dfrac{x^2+3x+2}{x^2-x-6}-\dfrac{x^2-2x-3}{x^2-6x+9}$

17. $\dfrac{x^2+1}{x-3}+\dfrac{x-5}{x+3}$ **18.** $\dfrac{x-37}{(x+3)(x-7)}+\dfrac{3x+6}{(x-7)(x+2)}-\dfrac{3}{x+3}$

Multiply or divide the rational expressions, as indicated, and simplify your answer. See Example 3.

19. $\dfrac{y-2}{y+1}\cdot\dfrac{y^2-1}{y-2}$ **20.** $\dfrac{a^2-3a-4}{a-2}\div\dfrac{a^2-2a-8}{a-2}$

21. $\dfrac{2x^2-5x-12}{x-3}\cdot\dfrac{x^2-x-6}{x-4}$ **22.** $\dfrac{z^2+2z+1}{2z^2+3z+1}\cdot\dfrac{2z^2-5z-3}{z+1}$

23. $\dfrac{3b^2+9b-84}{b^2-5b+4}\div\dfrac{5b^2+37b+14}{-10b^2+6b+4}$

24. $\dfrac{3x^2-x-10}{x-1}\cdot\dfrac{x^2-1}{6x^2+x-15}\div\dfrac{x^2-x-2}{2x^2+5x-12}$

Simplify the complex rational expressions. See Example 4.

25. $\dfrac{\frac{3}{x}+\frac{x}{3}}{2-\frac{1}{x}}$ **26.** $\dfrac{\frac{1}{x}-\frac{1}{y}}{\frac{1}{x}+\frac{1}{y}}$ **27.** $\dfrac{6x-6}{3-\frac{3}{x^2}}$ **28.** $\dfrac{x^{-2}-y^{-2}}{y-x}$

29. $\dfrac{\frac{1}{r}-\frac{1}{s}}{r+\frac{1}{r}}$ **30.** $\dfrac{\frac{1}{x^2}-\frac{1}{y^2}}{\frac{1}{y^3}-\frac{1}{xy^2}}$ **31.** $\dfrac{\frac{m}{n}-\frac{n}{m}}{m-n}$ **32.** $\dfrac{\frac{1}{y}-\frac{1}{x+3}}{\frac{1}{x}-\frac{y}{x^2+3x}}$

Solve the following rational equations. See Example 5.

33. $\dfrac{3}{x-2}+\dfrac{2}{x+1}=1$

34. $\dfrac{x}{x-1}+\dfrac{2}{x-3}=-\dfrac{2}{x^2-4x+3}$

35. $\dfrac{y}{y-1}+\dfrac{2}{y-3}=\dfrac{y^2}{y^2-4y+3}$

36. $\dfrac{2}{2x+1}-\dfrac{x}{x-4}=\dfrac{-3x^2+x-4}{2x^2-7x-4}$

37. $\dfrac{2}{2b+1}+\dfrac{2b^2-b+4}{2b^2-7b-4}=\dfrac{b}{b-4}$

38. $\dfrac{1}{x^3-27}=\dfrac{2}{x-3}+\dfrac{-2x}{x^2+3x+9}$

39. $\dfrac{1}{x-3}+\dfrac{1}{x+3}=\dfrac{2x}{x^2-9}$

40. $\dfrac{1}{t-3}+\dfrac{1}{t+2}=\dfrac{t}{t-3}$

41. $\dfrac{2}{n+3}+\dfrac{3}{n+2}=\dfrac{6}{n}$

42. $\dfrac{z}{6+z}+\dfrac{z-1}{6-z}=\dfrac{z}{6-z}$

Solve the following work-rate problems. See Examples 6 and 7.

43. If Joanne were to paint her living room alone, it would take 5 hours. Her sister Lisa could do the job in 7 hours. How long would it take them working together?

44. The hot water tap can fill a given sink in 4 minutes. If the cold water tap is turned on as well, the sink fills in 1 minute. How long would it take for the cold water tap to fill the sink alone?

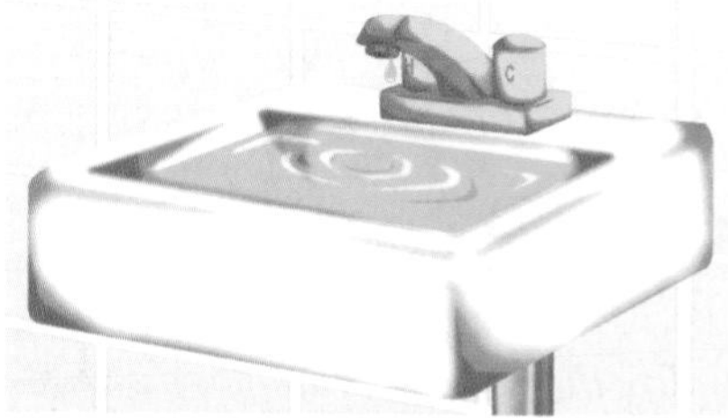

45. A farmer can plow a given field in 2 hours less time than it takes his son. If they acquire two tractors and work together, they can plow the field in 5 hours. How long does it take the father alone?

46. Two hoses, one of which has a flow-rate three times the other, can together fill a tank in 3 hours. How long does it take each of the hoses individually to fill the tank?

47. Officials begin to release water from a full man-made lake at a rate that would empty the lake in 12 weeks, but a river that can fill the lake in 30 weeks is replenishing the lake at the same time. How long does it take to empty the lake?

48. In order to flush deposits from a radiator, a drain that can empty the entire radiator in 45 minutes is left open at the same time it is being filled at a rate that would fill it in 30 minutes. How long does it take for the radiator to fill?

49. Jimmy and Janice are picking strawberries. Janice can fill a bucket in a half hour, but Jimmy continues to eat the strawberries that Janice has picked at a rate of one bucket per 1.5 hours. How long does it take Janice to fill her bucket?

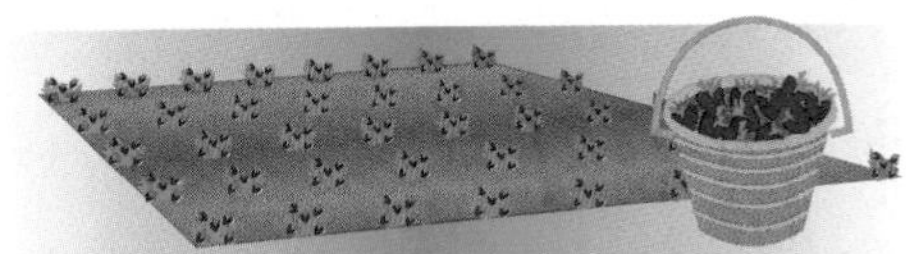

50. The hull of Jack's yacht needs to be cleaned. He can clean it by himself in 5 hours, but he asks his friend Thomas to help him. If it takes 3 hours for the two men to clean the hull of the boat, how long would it have taken Thomas alone?

technology exercises

Use a computer algebra system to combine the rational expressions or simplify the complex rational expressions, as indicated. See the Technology Note after Example 4 for guidance.

51. $\dfrac{3}{2x-10}-\dfrac{1}{2x+10}$

52. $\dfrac{5}{8x-40}+\dfrac{3}{8x+24}$

53. $\dfrac{1}{5}+\dfrac{6}{25x-5}$

54. $\dfrac{-3}{7x+14}+\dfrac{9}{21x-7}+\dfrac{2}{x+2}$

55. $\dfrac{x^{-3}-y^{-3}}{x^{-1}-y^{-1}}$

56. $\dfrac{\dfrac{1}{x}-\dfrac{y}{x^2}}{\dfrac{y}{x^2}+\dfrac{x}{y}}$

Solve the following radical equations. See Example 8.

57. $\sqrt{4-x}-x=2$

58. $\sqrt{x+1}+10=x-1$

59. $\sqrt{x+10}+1=x-1$

60. $\sqrt{x^2-4x+5}-x+2=0$

61. $\sqrt{x^2-4x+4}+2=3x$

62. $\sqrt{50+7s}-s=8$

63. $\sqrt[3]{3-2x}-\sqrt[3]{x+1}=0$

64. $\sqrt[4]{x^2-x}=\sqrt[4]{x-1}$

65. $\sqrt[4]{2x+3}=-1$

66. $\sqrt{11x+3}+4x=18$ **67.** $\sqrt{2b-1}+3=\sqrt{10b-6}$ **68.** $\sqrt{5x+5}=\sqrt{4x-7}+2$

69. $\sqrt{3-3x}-3=\sqrt{3x+2}$ **70.** $\sqrt{3y+4}+\sqrt{5y+6}=2$ **71.** $\sqrt{x^2-10}-1=x+1$

72. $\sqrt[3]{5x^2-14x}=-2$ **73.** $\sqrt[5]{7t^2+2t}=\sqrt[5]{5t^2+4}$ **74.** $\sqrt[4]{y^3-7y+2}=\sqrt[4]{2-3y}$

Solve the following equations. See Example 9.

75. $(x+3)^{1/4}+2=0$ **76.** $(2x-5)^{1/4}=(x-1)^{1/4}$ **77.** $(2x-1)^{2/3}=x^{1/3}$

78. $(3y^2+9y-5)^{1/2}=y+3$ **79.** $(3x-5)^{1/5}=(x+1)^{1/5}$ **80.** $w^{3/5}+8=0$

81. $(x^2+21)^{-3/2}=\frac{1}{125}$ **82.** $(x-2)^{2/3}=(14-x)^{1/3}$ **83.** $z^{4/3}-\frac{16}{81}=0$

Solve the following formulas for the indicated variable.

84. $T=2\pi\sqrt{\frac{l}{g}}$. This is the formula for the period T of a pendulum of length l. Solve this formula for l.

85. $c=\sqrt{a^2+b^2}$. This is the formula for the length of the hypotenuse c of a right triangle. Solve this formula for a.

86. $\omega=\sqrt{\frac{k}{m}}$. This is the formula for the angular frequency of a mass m suspended from a spring of spring constant k. Solve this formula for m.

87. Kepler's Third Law is $T^2=\frac{4\pi^2r^3}{GM}$. It relates the period T of a planet to the radius r of its orbit and the Sun's mass M. Solve this formula for r.

88. The total mechanical energy of an object with mass m at height h in a closed system can be written as $ME=\frac{1}{2}mv^2+mgh$. Solve for v, the velocity of the object, in terms of the given quantities.

89. In a circuit with an AC power source, the total impedance Z depends on the resistance R, the capacitance C, the inductance L, and the frequency of the current ω according to: $Z=\sqrt{R^2+\left(\omega L-\frac{1}{\omega C}\right)^2}$. Solve this equation for the inductance L.

chapter one project

Polynomials

A chemistry professor calculates final grades for her class using the polynomial

$$A = 0.3f + 0.15h + 0.4t + 0.15p,$$

where A is the final grade, f is the final exam, h is the homework average, t is the chapter test average, and p is the semester project.

The following is a table containing the grades for various students in the class:

Name	Final Exam	Homework Avg.	Test Avg.	Project
Alex	77	95	79	85
Ashley	91	95	88	90
Darren	82	85	81	75
Elizabeth	75	100	84	80
Gabe	94	90	90	85
Lynn	88	85	80	75

1. Find the course average for each student, rounded to the nearest tenth.
2. Who has the highest total score?
3. Why is the total grade raised more with a grade of 100 on the final exam than with a grade of 100 on the semester project?
4. Assume you are a student in this class. With 1 week until the final exam, you have a homework average of 85, a test average of 85, and a 95 on the semester project. What score must you make on the final exam to achieve at least a 90.0 overall? (Round to the nearest tenth.)

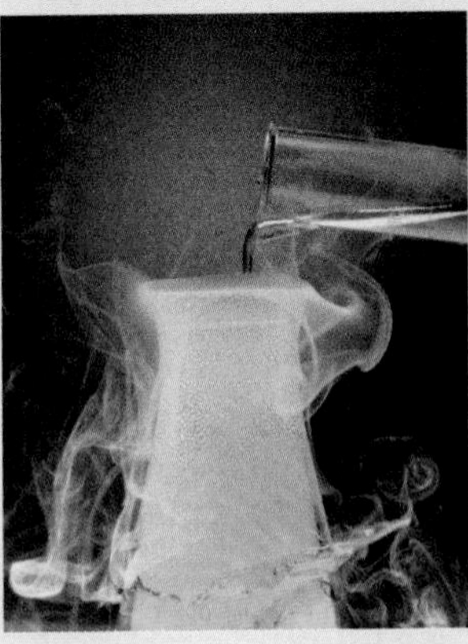

CHAPTER REVIEW

A summary of concepts and skills follows each chapter. Refer to these summaries to make sure you feel comfortable with the material in the chapter. The concepts and skills are organized according to the Section title and Topic title in which the material is first discussed.

1.1 Real Numbers and Algebraic Expressions

topics	pages	test exercises
Common subsets of real numbers • The sets $\mathbb{N}$, $\mathbb{Z}$, $\mathbb{Q}$, and $\mathbb{R}$, as well as *whole* numbers and *irrational* numbers • Identifying numbers as elements of one or more of the common sets	p. 3 – 4	1 – 2
The real number line • Plotting numbers on the real number line • The *origin* of the real number line, and its relation to negative and positive numbers	p. 4 – 5	3 – 4
Order on the real number line • The four inequality symbols $<, \leq, >,$ and $\geq$ • Distinguishing between strict and non-strict inequalities	p. 5	5 – 6
Set-builder notation and interval notation • Using set-builder notation to define sets • The empty set and its two common symbols: $\varnothing$ and $\{\ \}$	p. 6	7 – 8, 13 – 14
Basic Set Operations and Venn Diagrams • The meaning and use of Venn diagrams • The definition of the set operations *union* and *intersection* and their application to intervals • Interval notation and its relation to inequality statements	p. 6 – 9	9 – 12
Absolute value and distance • The definition of absolute value on the real line: $\lvert a\rvert = \begin{cases} a & \text{if } a \geq 0 \\ -a & \text{if } a < 0 \end{cases}$ • The relationship between absolute value and distance • Distance between two real numbers	p. 9 – 11	17 – 19

topics (continued)	pages	test exercises
• Properties of absolute values: 1. $\|a\| \geq 0$ 2. $\|-a\| = \|a\|$ 3. $a \leq \|a\|$ 4. $\|ab\| = \|a\|\|b\|$ 5. $\left\|\frac{a}{b}\right\| = \frac{\|a\|}{\|b\|},\ b \neq 0$ 6. $\|a+b\| \leq \|a\|+\|b\|$, the triangle inequality		
Components of algebraic expressions • Identifying *terms* of an expression, and identifying the *coefficient* of a term • Distinguishing between *constants* and *variables* in an expression • Identifying *factors* of a term	p. 11	20 – 21
The field properties and their use in algebra • The application of the *commutative, associative, identity, inverse*, and *distributive* properties of the real numbers • Using the *zero-factor* property: $AB = 0 \Rightarrow A = 0$ or $B = 0$ • Using the *cancellation* properties: $A = B \Leftrightarrow A+C = B+C$ and $C \neq 0$, $A = B \Leftrightarrow A \cdot C = B \cdot C$	p. 12 – 13	22 – 23

1.2 Properties of Exponents and Radicals

topics	pages	test exercises
Natural number and integer exponents • The meaning of exponential notation: $a^n = \underbrace{a \cdot a \cdot \cdots \cdot a}_{n \text{ factors}}$ and the distinction between *base* and *exponent* • The extension of natural number exponents to integer exponents • The definition of exponentiation by 0: $a^0 = 1$ $(a \neq 0)$ and 0^0 is undefined • The equivalence of a^{-n} and $\frac{1}{a^n}$ for $a \neq 0$	p. 19 – 21	24 – 25

topics (continued)	pages	test exercises
Properties of exponents and their use • The properties of exponents and their use in simplifying expressions: 1. $a^n \cdot a^m = a^{n+m}$ 2. $\frac{a^n}{a^m} = a^{n-m}$ 3. $a^{-n} = \frac{1}{a^n}$ 4. $\left(a^n\right)^m = a^{nm}$ 5. $(ab)^n = a^n b^n$ 6. $\left(\frac{a}{b}\right)^n = \frac{a^n}{b^n}$ 7. $\left(\frac{a}{b}\right)^{-n} = \frac{b^n}{a^n}$ 8. $\frac{a^{-m}}{b^{-n}} = \frac{b^n}{a^m}$ • Recognition of common errors made in trying to apply properties of exponents	p. 22 – 24	24 – 25
Scientific notation • The definition of the scientific notation form of a number: $a \times 10^n$ ($1 \le \lvert a \rvert < 10$ and n is an integer) • Using scientific notation in expressing numerical values	p. 25 – 26	26 – 27
Interlude: working with geometric formulas • How to solve word problems with geometric formulas	p. 27 – 28	
Roots and radical notation • The definition of nth roots, and the meaning of *index* and *radicand*	p. 29 – 30	28 – 33
Simplifying and combining radical expressions • The meaning of *simplification* as applied to radical expressions • The properties of radicals and their use in simplifying expressions: 1. $\sqrt[n]{a^n} = a$ if n is odd 2. $\sqrt[n]{a^n} = \lvert a \rvert$ if n is even 3. $\sqrt[n]{ab} = \sqrt[n]{a} \cdot \sqrt[n]{b}$ 4. $\sqrt[n]{\frac{a}{b}} = \frac{\sqrt[n]{a}}{\sqrt[n]{b}}$ 5. $\sqrt[m]{\sqrt[n]{a}} = \sqrt[mn]{a}$ • The use of *conjugate radical expressions* in rationalizing denominators (and numerators) • The meaning of *like radicals*	p. 30 – 34	28 – 33

topics (continued)	pages	test exercises
Rational number exponents • The extension of integer exponents to rational exponents: $a^{m/n} = \sqrt[n]{a^m} = \left(\sqrt[n]{a}\right)^m$ • Applying properties of exponents to simplify expressions containing rational exponents	p. 34 – 36	28 – 33

1.3 Polynomials and Factoring

topics	pages	test exercises
The terminology of polynomial expressions • The definition of *polynomial* • The meaning of *coefficient, leading coefficient, descending order, degree of a term,* and *degree of a polynomial* as applied to polynomials • The meaning of *monomial, binomial*, and *trinomial*	p. 42 – 43	34 – 36
The algebra of polynomials • The meaning of *like*, or *similar*, terms • Polynomial addition, subtraction, and multiplication	p. 43 – 45	37 – 40
Common factoring methods • The meaning of *factorable* and *irreducible* (or *prime*) as applied to polynomials • The mechanics of using the *greatest common factor, factoring by grouping, special binomials,* and *factoring trinomials* methods • The extension of these methods to expressions containing fractional exponents	p. 46 – 51	41 – 46

1.4 The Complex Number System

topics	pages	test exercises
The imaginary unit i and its properties • The definition of i: $i = \sqrt{-1}$ • Square roots of negative numbers • The definition of *complex numbers*, and the identification of their *real* and *imaginary* parts	p. 55 – 57	47 – 54
The algebra of complex numbers • Simplifying complex number expressions • The use of *complex conjugates*	p. 57 – 59	47 – 51
Roots and complex numbers • The meaning of *principal square root* • Understanding when properties of radicals apply and when they do not	p. 60 – 61	52 – 54

1.5 Linear Equations in One Variable

topics	pages	test exercises
Equivalent equations and the meaning of solutions • The categories of equations: *identities* (true for all allowable values), *contradictions* (not true for any allowable value), and *conditionals* (true for some allowable values) • The *solution set* of an equation (set of values that make a conditional equation true) and *equivalent equations* (two equations which have the same solution set)	p. 63	55 – 58
Solving linear equations • The definition of a *linear*, or *first-degree, equation in one variable* • General method of solving linear equations	p. 64 – 65	55 – 58

topics (continued)	pages	test exercises
Solving absolute value equations • Algebraic and geometric meaning of absolute value expressions in equations • *Extraneous solutions* to equations	p. 65 – 66	59 – 60
Solving linear equations for one variable • Solving an equation (or a formula) for a specified variable	p. 67	61 – 62
Interlude: distance and interest problems • Basic distance ($d = rt$) and interest ($I = Prt$) problems leading to linear equations	p. 68	63 – 64

1.6 Linear Inequalities in One Variable

topics	pages	test exercises
Solving linear inequalities • The definition of *linear inequality* • The use of *cancellation properties* in solving inequalities	p. 73 – 75	65 – 68
Solving compound linear inequalities • The definition of *compound inequality* (a statement containing two inequality symbols that can be interpreted as two distinct inequalities joined by the word "and")	p. 75 – 76	65 – 68
Solving absolute value inequalities • The definition of an *absolute value inequality* (an inequality in which some variable expression appears inside absolute value symbols) • The geometric meaning of inequalities containing an absolute value expression	p. 76 – 77	67
Interlude: translating inequality phrases • The meaning of commonly encountered inequality phrases	p. 78	69

1.7 Quadratic Equations

topics	pages	test exercises
Solving quadratic equations by factoring • The definition of a *quadratic,* or *second-degree, equation in one variable* (an equation that can be written in the form $ax^2 + bx + c = 0$) • Using the *zero-factor property* ($AB = 0 \Rightarrow A = 0$ or $B = 0$) to solve quadratic equations	p. 81– 83	70 – 73
Solving "perfect square" quadratic equations • Solving quadratic equations of the general form $A^2 = c$	p. 83	71
Solving quadratic equations by completing the square • Using the method of *completing the square* to solve arbitrary quadratic equations: 1. Write $ax^2 + bx + c = 0$ as $ax^2 + bx = -c$. 2. Divide by a so the coefficient of x^2 is 1. 3. Divide the coefficient of x by 2, square the result, and add this to both sides. 4. The trinomial on the left is a perfect square.	p. 84 – 85	
The quadratic formula • Apply the method of completing the square to a generic quadratic equation • The quadratic formula: $x = \dfrac{-b \pm \sqrt{b^2 - 4ac}}{2a}$	p. 86 – 87	72
Interlude: gravity problems • The general equation of motion of an object under the influence of gravity: $h = -\frac{1}{2}gt^2 + v_0t + h_0$	p. 87 – 89	74
Solving quadratic-like equations • The definition of a *degree n polynomial equation in one variable* • The definition of a quadratic-like equation (an equation that can be written in the form $aA^2 + bA + c = 0$ where A is an algebraic expression) • Solving quadratic-like equations	p. 89 – 90	75 – 76
Solving general polynomial equations by factoring • Extending the use of the zero-factor property to higher-degree polynomial equations	p. 90 – 91	77 – 78
Solving polynomial-like equations by factoring • Extending the use of the zero-factor property to selected non-polynomial equations	p. 92	79 – 80

1.8 Rational and Radical Equations

topics	pages	test exercises
Simplifying rational expressions • The definition of a *rational expression* (an expression that can be written as a ratio of two polynomials) • Simplifying rational expressions	p. 97 – 98	81 – 82
Combining rational expressions • Addition, subtraction, multiplication, and division of rational expressions	p. 98 – 100	83 – 86
Simplifying complex rational expressions • The definition of a *complex rational expression* (a fraction in which the denominator, the numerator, or both contain at least one rational expression) • Two methods for simplifying complex rational expressions	p. 100 – 102	87 – 88
Solving rational equations • The meaning of the phrase *rational equation* (an equation that contains at least one rational expression) • A general method of solving rational equations	p. 102 – 103	89 – 90
Interlude: work-rate problems • Types of problems that are classified as *work-rate problems* • The concept of rate of work as the reciprocal of time needed to complete the work • The concept of the *additivity* of rates of work	p. 103 – 105	91
Solving radical equations • The definition of a *radical equation* (an equation that has at least one radical expression containing a variable, while any non-radical expressions are polynomial terms) • A general method of solving radical equations: 1. Isolate one radical expression on one side of the equation. 2. Raise both sides of the equation by the power necessary to "undo" the isolated radical. 3. If any radical expressions remain, simplify the equation, and repeat steps 1 and 2 until the result is a polynomial equation; then solve. 4. Check your solutions in the original equation and discard any extraneous solutions.	p. 105 – 107	92 – 94
Solving equations with positive rational exponents • Using the method of solving radical equations to solve equations with positive rational exponents	p. 108	95 – 97

chapter test

Section 1.1

*Which elements of the following sets are **(a)** natural numbers, **(b)** whole numbers, **(c)** integers, **(d)** rational numbers, **(e)** irrational numbers, **(f)** real numbers, **(g)** undefined.*

1. $\left\{-15.75,\ 7, \frac{-8}{0}, \frac{0}{5}, 3^3, -1, -\sqrt{4}\right\}$

2. $\left\{2\sqrt{3},\ 5\pi,\ \sqrt{1},\ 7.\overline{6},\ -1,\ \frac{2}{9}, |-21|\right\}$

Plot the following real numbers on a number line. Choose the unit length appropriately for each set.

3. $\{-3, 0, 3, 5\}$

4. $\{-4.5, -4.2, -4.0\}$

Select all of the symbols from the set $\{<, \leq, >, \geq\}$ *that can be placed in the blank to make true statements.*

5. 6.1 ____ 8.3

6. −12 ____ −11

Describe each of the following sets using set-builder notation. There may be more than one correct way to do this.

7. $\{-2, -1, 0, 1, 2, 3\}$

8. $\{0, 2, 4, 6, \ldots\}$

Simplify the following unions and intersections of intervals.

9. $[4, 8] \cup (8, 11]$

10. $[4, 8] \cap (8, 11]$

11. $(-2, 3] \cap [0, 3)$

12. $(-\infty, 4] \cap (2, \infty)$

Write each set as an interval using interval notation.

13. $\{x \mid -7 < x \leq 9\}$

14. $\{x \mid x \geq 14\}$

Write the following rational numbers as ratios of integers.

15. $7.\overline{6}$

16. $-2.0\overline{42}$

Evaluate the absolute value expressions.

17. $-|11-2|$

18. $-|-4-3|$

19. $|\sqrt{5}-\sqrt{11}|$

Evaluate each expression for the given values of the variables.

20. $x^2z^3 + 5\sqrt{3x - 2y}$ for $x = 2, y = 1,$ and $z = -1$.

21. $7y^2 - \frac{1}{3}\pi xy + 8x^3$ for $x = -2$ and $y = 2$.

Identify the property that justifies each of the following statements. If one of the cancellation properties is being used to transform an equation, identify the quantity that is being added to both sides or the quantity that both sides are being multiplied by.

22. $12a^2 = 8b \Leftrightarrow 3a^2 = 2b$ **23.** $(x-3)(z-2) = 0 \Rightarrow x - 3 = 0$ or $z - 2 = 0$

Section 1.2

Use the properties of exponents to simplify each of the following expressions, writing your answer with only positive exponents.

24. $\dfrac{-4t^0\left(s^2t^{-2}\right)^{-3}}{2^3 st^{-3}}$

25. $\left[\left(3y^{-2}z\right)^{-1}\right]^{-3}$

Convert each number from scientific notation to standard notation, or vice versa, as indicated.

26. -2.004×10^{-4}; convert to standard.

27. 52,240,000; convert to scientific.

Simplify the following radical expressions. Rationalize all denominators and use only positive exponents.

28. $\sqrt[3]{\dfrac{8x^2}{3y^{-4}}}$

29. $\dfrac{\sqrt{3a^3}}{\sqrt{12a}}$

30. $\dfrac{4}{\sqrt{2} - \sqrt{6}}$

31. $\sqrt{16x^2}$

32. $\sqrt[3]{-64x^{-9}y^3}$

33. $\sqrt[4]{\dfrac{a^9b^{-4}}{81}}$

Section 1.3

Classify each of the following expressions as either a polynomial or a non-polynomial. For those that are polynomials, identify the degree of the polynomial and the number of terms (using the words monomial, binomial, and trinomial if applicable).

34. $4x^{-2} - 2x$

35. $5xyz + 7x^2 + y$

36. $4 + 8z^2y$

Add or subtract the polynomials, as indicated.

37. $(5x^2y+7xy-z)-(2x^2y+z-4xz)$

38. $(-2x^2y+xy+y^2)+(xy-y^2)$

Multiply the polynomials, as indicated.

39. $(5x^2y+2xy-3)(4x+2y)$

40. $(5x+2y)(5x-2y)$

Factor each of the following polynomials.

41. $nx+3mx-2ny-6my$

42. $36x^6-y^2$

43. $10a^3b-15a^3b^3+5a^2b^4$

44. $6a^2-7a-5$

45. x^2-x-12

46. $3x^4+3x^2y-x^2y^3-y^4$

Section 1.4

Simplify the following complex expressions by adding, subtracting, or multiplying.

47. $(8i+3i^3)-(4i^4-i^6)$

48. $\left(\frac{i^7}{10}\right)\left(\frac{5}{i^9}\right)$

Simplify the following quotients.

49. $\frac{3+4i}{3-4i}$

50. $\frac{17}{4-i}$

51. $\frac{2i}{3-5i}$

Simplify the following square root expressions.

52. $(\sqrt{-8})(\sqrt{-2})$

53. $(2-\sqrt{-4})^2$

54. $\frac{\sqrt{-24}}{\sqrt{6}}$

Section 1.5

Solve the following linear equations and identify whether each equation is an identity, conditional, or contradiction.

55. $5(3y-2)=(4y+4)+2y$

56. $7x-3x=3x+4$

57. $\frac{x-7}{2}=\frac{5-x}{4}$

58. $\frac{9a+4}{3}=\frac{6(2a+7)}{4}$

Solve the following absolute value equations.

59. $|2x-7|=1$

60. $|2y-5|-1=|3-y|$

Solve the following equations for the indicated variable.

61. $h = -16t^2 + v_0 t$; solve for v_0.

62. $A = 2lw + 2wh + 2hl$; solve for l.

Solve the following application problems.

63. Two trains leave the station at the same time in opposite directions. One travels at an average rate of 90 miles per hour, and the other at an average rate of 95 miles per hour. How far apart are the two trains after an hour and twenty minutes?

64. Two firefighters, Jake and Rose, each have $5,000 to invest. Jake invests his money in a money market account with an annual return of 3.25%, while Rose invests hers in a CD paying 4.95% annually. How much more money than Jake does Rose have after 1 year?

Section 1.6

Solve the following inequalities. Describe the solution set using interval notation and by graphing.

65. $-8 < 3x - 5 \le 16$

66. $-14 < -2(3+y) \le 8$

67. $-5|3+t| > -10$

68. $-3(x-1) < 12$ and $x - 4 \le 9$

69. Kim wants to keep her bills under $1800 a month. If each month her rent is $550, her utilities are $80, food is $420, entertainment is $250, and she has $80 in other expenses, how much can she afford to spend on car payments?

Section 1.7

Solve the following quadratic equations.

70. $2x^2 + 3x - 10 = 10$

71. $(x-3)^2 = 12$

72. $1.7z^2 - 3.8z - 2 = 0$

73. $2x^2 + 7x = x^2 + 2x - 6$

74. Ashley shoots an arrow into the air with an upward velocity of 83 ft./s. How long until the arrow hits the ground?

Solve the following quadratic-like equations.

75. $(x^2+2)^2 - 7(x^2+2) + 12 = 0$

76. $y^{2/3} + y^{1/3} - 6 = 0$

Solve the following polynomial equations by factoring.

77. $x^3 - 4x^2 - 2x + 8 = 0$

78. $2x^3 + 2x = 5x^2$

Solve the following polynomial-like equations by factoring.

79. $4x^{18/7} - 2x^{11/7} - 3x^{4/7} = 0$

80. $(2x+1)^{-1/2} - 3(2x+1)^{1/2} = 0$

Section 1.8

Simplify the following rational expressions, indicating which values of the variable must be excluded.

81. $\dfrac{2z^2 - 8z - 42}{z+3}$

82. $\dfrac{3t^3 + 23t^2 + 40t}{3t^2 + 27t + 60}$

Add or subtract the rational expressions as indicated and simplify.

83. $\dfrac{3a^3 + 5}{5a+1} + \dfrac{a-5}{5a+1}$

84. $\dfrac{4x+3}{x+1} - \dfrac{2x-3}{x-1}$

Multiply or divide the rational expressions as indicated and simplify.

85. $\dfrac{y^2 + 8y - 1}{7y+1} \cdot \dfrac{y-2}{y+2}$

86. $\dfrac{4x^2 - 30x - 16}{2x^3 - 19x^2 + 24x} \div \dfrac{4x^2 - 34x + 16}{2x^2 - 3x}$

Simplify the complex rational expressions.

87. $\dfrac{\frac{x}{3} - \frac{3}{x}}{-\frac{3}{x} + 1}$

88. $\dfrac{\frac{1}{2a} - \frac{1}{2b}}{\frac{2}{a} + \frac{2}{b}}$

Solve the following rational equations.

89. $\dfrac{y}{y-1} + \dfrac{1}{y-4} = \dfrac{y^2}{y^2 - 5y + 4}$

90. $\dfrac{2}{x+1} - \dfrac{x}{x-3} = \dfrac{3x-21}{x^2 - 2x - 3}$

91. If Mandy can paint a fence in 3 hours and Heather can paint it in 4 hours, how long would it take the two of them working together to paint the fence?

Solve the following equations.

92. $\sqrt{2x^2 + 8x + 1} - x - 3 = 0$

93. $\sqrt{5x-1} + \sqrt{x-1} = 4$

94. $\sqrt{2x-5} - 2 = \sqrt{x-2}$

95. $(2x-5)^{1/6} = (x-2)^{1/6}$

96. $(x^2 - 4x - 5)^{1/4} = 2$

97. $(2x^2 - 18x + 67)^{1/3} = 3$

chapter TWO

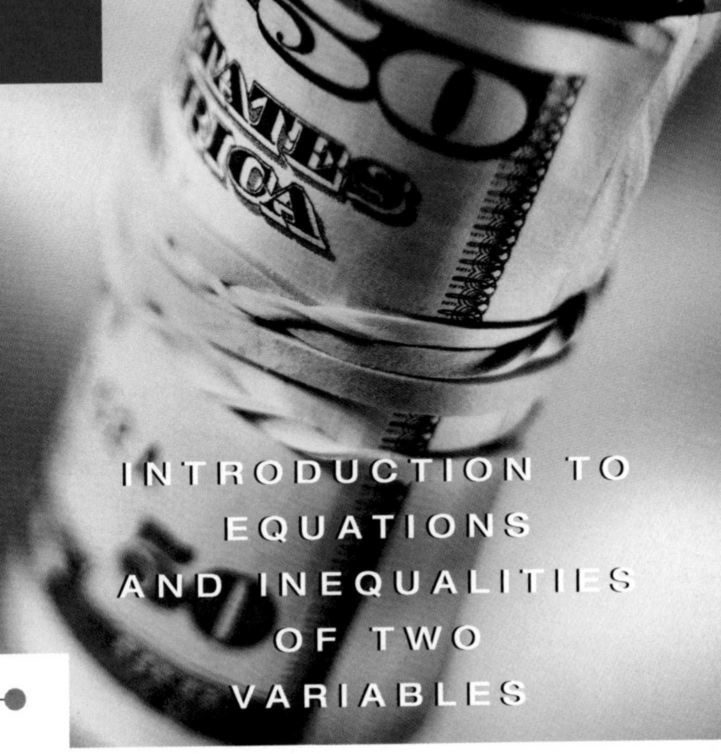

INTRODUCTION TO EQUATIONS AND INEQUALITIES OF TWO VARIABLES

OVERVIEW

By the end of the chapter, you should be able to use the skills you have learned regarding slopes and points in the Cartesian plane to estimate future outcomes given past data. On page 155, you will find a problem in which you are given a company's sales data for the past two years and asked to predict next year's sales if sales continue increasing at the current rate. You will master this type of problem using the *formula for the slope of a line* on page 146 and the equation of a line in two variables.

Introduction

This chapter introduces the Cartesian coordinate system, the two-dimensional framework that underlies most of the material throughout the remainder of this text.

Many problems that we seek to solve with mathematics are most naturally described with two variables. The existence of two variables in a problem leads naturally to the use of a two-dimensional system in which to work, especially when we attempt to depict the situation graphically, but it took many centuries for the ideas presented in this chapter to evolve. Some of the greatest accomplishments of the French mathematician and philosopher René Descartes (1596 - 1650) were his contributions to the then fledgling field of *analytic geometry,* the marriage of algebra and geometry. In *La géométrie,* an appendix to a volume of scientific philosophy, Descartes laid out the basic principles by which algebraic problems could be construed as geometric problems, and the methods by which solutions to the geometric problems could be interpreted algebraically. As many later mathematicians expanded upon Descartes' work, analytic geometry came to be an indispensable tool in understanding and solving problems of both an algebraic and a geometric nature.

Descartes

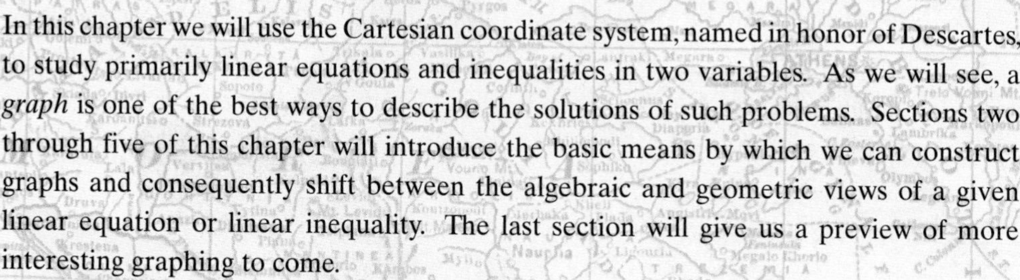

In this chapter we will use the Cartesian coordinate system, named in honor of Descartes, to study primarily linear equations and inequalities in two variables. As we will see, a *graph* is one of the best ways to describe the solutions of such problems. Sections two through five of this chapter will introduce the basic means by which we can construct graphs and consequently shift between the algebraic and geometric views of a given linear equation or linear inequality. The last section will give us a preview of more interesting graphing to come.

The Cartesian coordinate system will continue to play a prominent role as we proceed to other topics, such as relations and functions, in later chapters. Mastery of the foundational concepts in this chapter will be essential to understanding these related ideas.

2.1 The Cartesian Coordinate System

TOPICS

1. The components of the Cartesian coordinate system
2. The graph of an equation
3. The distance and midpoint formulas

Topic 1: The Components of the Cartesian Coordinate System

The last chapter dealt, essentially, with algebra of a single variable. Many of the methods learned in Chapter 1 revolve around solving equations and inequalities of one variable, and even in those equations with two or more variables we were interested in tasks such as solving the equation for one particular variable.

Our focus now shifts. While many important problems can be phrased, mathematically, as one-variable equations or inequalities, many more problems are most naturally expressed with two or more variables, and solving such a problem requires determining *all* of the values of *all* of the variables that make the equation or inequality true. This seemingly simple observation serves as the motivation for the material in this section. Consider, for example, an equation in the two variables x and y. A particular solution of the equation, if there is one, will consist of a value for x and a *corresponding* value for y; a solution of the equation cannot consist of a value for only one of the variables.

This leads us naturally to the concept of a two-dimensional coordinate system. Just as the real number line, a one-dimensional coordinate system, is the natural arena in which to depict solutions of one-variable equations and inequalities, a two-dimensional coordinate system is the natural place to graph solutions of two-variable equations and inequalities. The coordinate system we use is named after René Descartes, the 17th century French mathematician largely responsible for its development.

The Cartesian Coordinate System

The **Cartesian coordinate system**, also referred to as the **Cartesian plane** or the **rectangular coordinate system**, consists of two perpendicular real number lines (each of which is an **axis**) intersecting at the point 0 of each line (see Figure 1). The point of intersection is called the **origin** of the system, and the four quarters defined by the two lines are called the **quadrants** of the plane, numbered as indicated in Figure 1. Because the Cartesian plane consists of two crossed real lines, it is often given the symbol $\mathbb{R} \times \mathbb{R}$, or $\mathbb{R}^2$. Each point P in the plane is identified by a unique pair of numbers (a, b), called an **ordered pair**. In a given ordered pair (a, b) the number a (the **first coordinate**) indicates the horizontal displacement of the point from the origin, and the number b (the **second coordinate**) indicates the vertical displacement. Figure 1 is an example of a Cartesian coordinate system, and illustrates how several ordered pairs are **graphed**, or **plotted**.

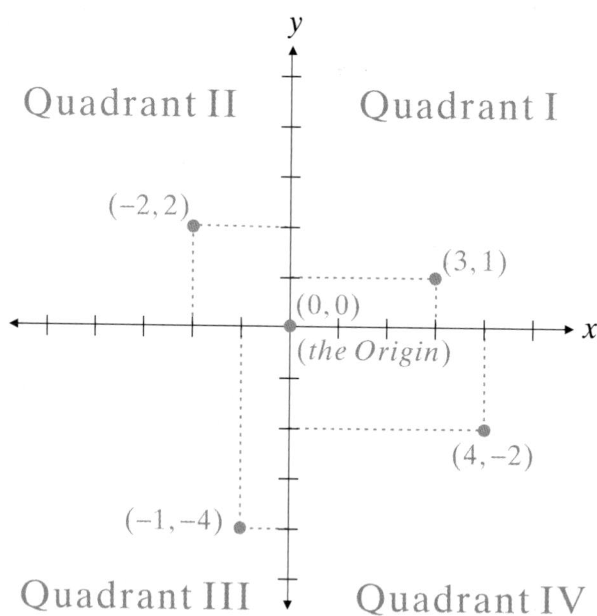

Figure 1: The Cartesian Plane $\mathbb{R}^2$

caution!

Although it is, unfortunately, a possible source of confusion, mathematics uses parentheses to denote ordered pairs as well as open intervals. You must rely on the context to determine the meaning of any parentheses you encounter. For instance, in the context of solving a one-variable inequality, the notation $(-2, 5)$ most likely refers to the open interval with endpoints at -2 and 5, whereas in the context of solving an equation in two variables, $(-2, 5)$ probably refers to a point in the Cartesian plane.

Topic 2: The Graph of an Equation

As mentioned above, a solution of an equation in x and y must consist of a value a for x and a corresponding value b for y. It is natural to present such a solution as an ordered pair (a, b), and equally natural to depict the solution, graphically, as a point in the plane whose coordinates are a and b. In this context, the horizontal number line would be referred to as the ***x*-axis**, the vertical number line as the ***y*-axis**, and the two coordinates of the ordered pair (a, b) as the ***x*-coordinate** and the ***y*-coordinate**.

As we shall see, a given equation in x and y will usually consist of far more than one ordered pair (a, b). The **graph of an equation** consists of a depiction in the Cartesian plane of *all* of those ordered pairs that make up the solution set of the equation.

At this point, we can make rough sketches of the graphs of many equations just by plotting enough ordered pairs that solve the equation to give us a sense of the entire solution set. We can find individual ordered pair solutions of a given equation by selecting numbers that seem appropriate for one of the variables and then solving the equation for the other variable. Doing so changes the task of solving a two-variable equation into that of solving a one-variable equation, and we have all the methods of Chapter 1 at our disposal to accomplish this. The process is illustrated in Example 1.

example 1

Sketch graphs of the following equations by plotting points.

a. $2x - 5y = 10$ **b.** $x^2 + y^2 - 6x = 0$ **c.** $y = x^2 - 2x$

Solutions:

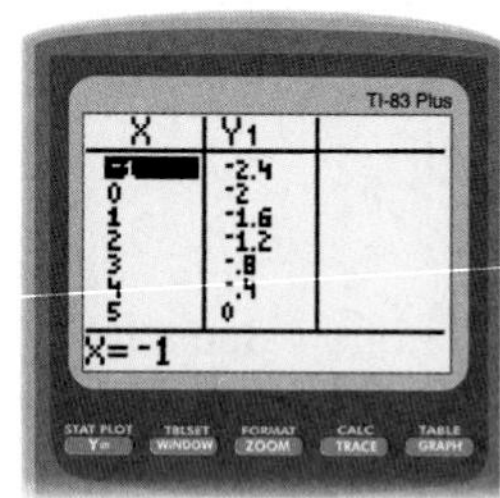

You can view a table of specific points on the graph of an equation by selecting Y= and entering the equation, and then selecting Table by pressing 2nd and then Graph.

a.

x	y
-3	?
0	?
?	0
?	5
1	?

$2x - 5y = 10$ →

x	y
-3	$-16/5$
0	-2
5	0
$35/2$	5
1	$-8/5$

In each row in the first table, values have been chosen essentially at random for one of the variables. Once these values have been chosen, the equation $2x - 5y = 10$ can be solved for the other variable. This gives us a list of 5 ordered pairs that can be plotted, though the ordered pair $\left(\frac{35}{2}, 5\right)$ is off the coordinate system drawn below.

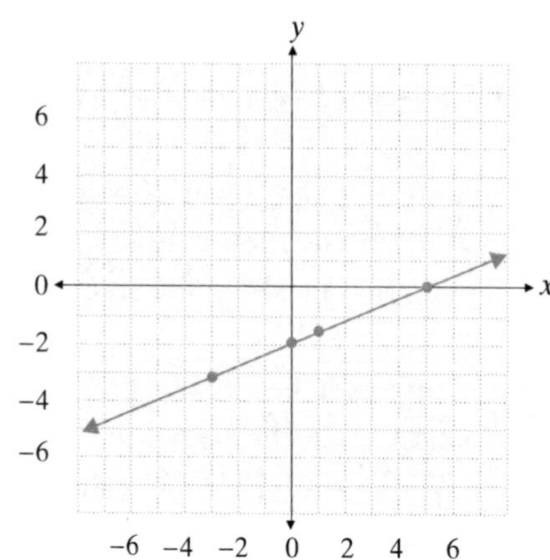

It appears that the four ordered pairs plotted lie on a straight line, and this is indeed the case. To gain more confidence in this fact, we could continue to plot more solutions of the equation, and we would find they all lie along the line that has been drawn through the four plotted ordered pairs. The infinite number of solutions of the equation are thus depicted by the blue line drawn as shown.

b.

x	y
0	?
1	?
2	?
3	?
4	?
5	?
6	?

$x^2 + y^2 - 6x = 0$ →

x	y
0	0
1	$\pm\sqrt{5} \approx \pm 2.2$
2	$\pm 2\sqrt{2} \approx \pm 2.8$
3	± 3
4	$\pm 2\sqrt{2} \approx \pm 2.8$
5	$\pm\sqrt{5} \approx \pm 2.2$
6	0

As in Example 1a, we determine enough ordered pairs that solve the equation $x^2 + y^2 - 6x = 0$ to feel confident in sketching the entire solution set. First, note that for any $x < 0$ and for any $x > 6$, the corresponding y would be imaginary, and thus irrelevant as far as graphing the equation is concerned. Similarly, for any $y < -3$ and for any $y > 3$, the corresponding x would be a complex number (use the quadratic formula to verify this fact).

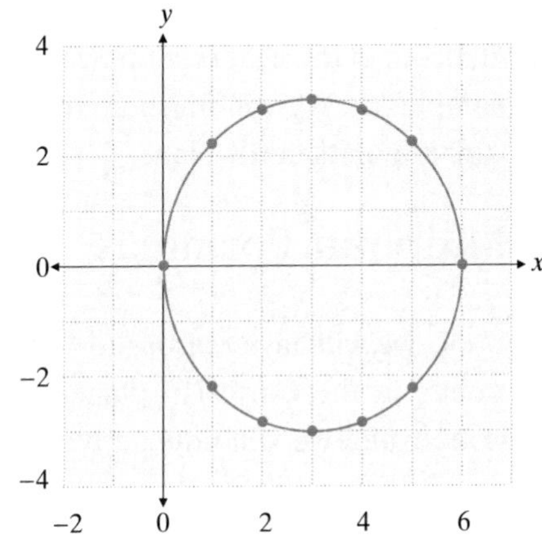

Once a sufficient number of ordered pairs have been calculated, the graph of the equation begins to take the shape of a circle. In fact, the graph is a circle, with center $(3,0)$ and a radius of 3, but we will not be able to prove this claim until Section 2.6.

c.

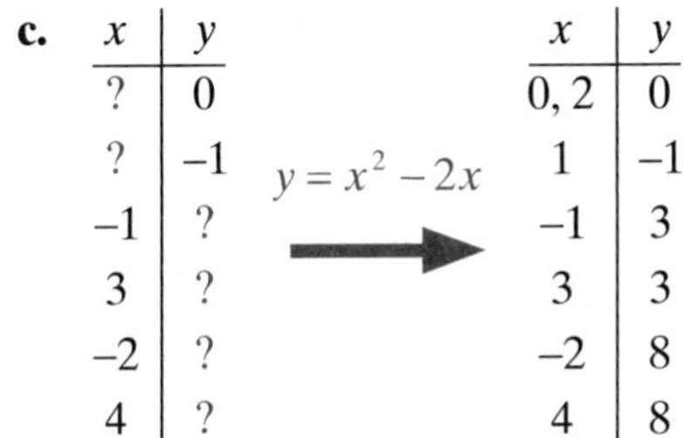

x	y
?	0
?	−1
−1	?
3	?
−2	?
4	?

$y = x^2 - 2x$ →

x	y
0, 2	0
1	−1
−1	3
3	3
−2	8
4	8

We again start with a table of values, with a number chosen for one of the two variables in the first table. Note that when y is 0, there are two corresponding values for x (this is determined by solving the resulting quadratic equation in x, which can be done by factoring). Again, enough points should be plotted to give some idea of the nature of the entire solution set of the equation.

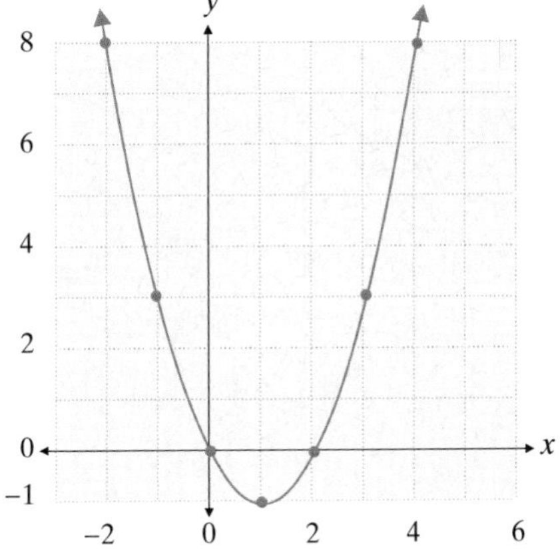

The graph of $y = x^2 - 2x$ is a shape known as a *parabola*. We will encounter these shapes again in Section 2 of Chapter 3, and we will be able, at that time, to prove that our rough sketch at the left is indeed the graph of the solution set of $y = x^2 - 2x$.

At this point, you may be uneasy about the graphs that were sketched in Example 1. You may be wondering if we have really plotted enough points to accurately "fill in" the gaps and sketch the entire solution set of each equation. Or you may be wondering if filling in the gaps is justified in the first place: what proof do we have that *all* the ordered pairs along our sketches actually solve the corresponding equation? Finally, you may be wondering if there isn't a faster and more sophisticated way to determine the graph of an equation. Plotting points is, for the most part, easily accomplished, but it is also a rather crude and tedious method.

These concerns are not trivial, and the resolution of them will constitute much of Chapters 2 through 9. We will begin in the next section by undertaking a careful study of equations like the one in Example 1a. We will finish up this section, however, with two useful formulas regarding pairs of points in the plane.

Topic 3: The Distance and Midpoint Formulas

Throughout the rest of this book, we will have reasons for wanting to know, on occasion, the *distance* between two points in the Cartesian plane. We already have the tools necessary to answer this question, and we will now derive a formula that we can apply whenever necessary.

Let (x_1, y_1) and (x_2, y_2) be the coordinates of two arbitrary points in the plane. By drawing the dotted lines parallel to the coordinate axes as shown in Figure 2, we can form a right triangle. Note that we are able to determine the coordinates of the vertex at the right angle from the two vertices (x_1, y_1) and (x_2, y_2).

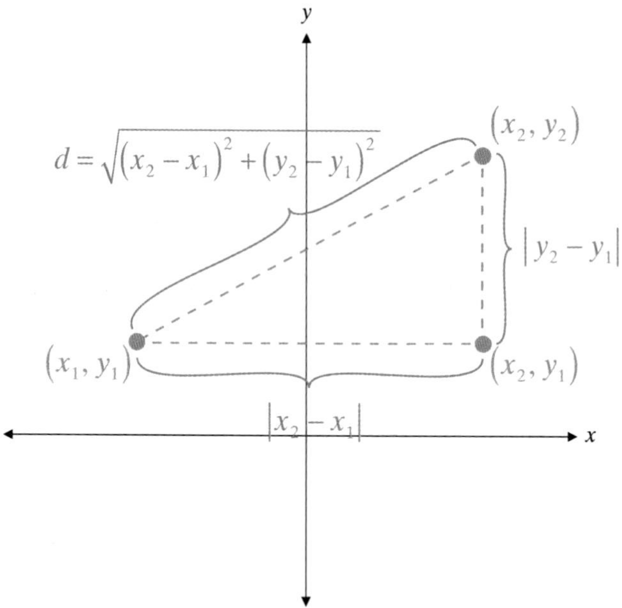

Figure 2: Distance Formula

The lengths of the two perpendicular sides of the triangle are easily calculated, as these lengths correspond to distances between numbers on real number lines. (The absolute value symbols are present as we don't necessarily know, in general, whether $x_2 - x_1$ or $x_1 - x_2$ is non-negative.) Now, we can apply the Pythagorean Theorem to determine the distance labeled d in Figure 2:

$$d^2 = (x_2 - x_1)^2 + (y_2 - y_1)^2 \text{, so}$$
$$d = \sqrt{(x_2 - x_1)^2 + (y_2 - y_1)^2}$$

(Notice that the absolute value symbols are not necessary in the formula, as any quantity squared is automatically non-negative.)

We will also, on occasion, want to be able to determine the midpoint of a line segment in the plane. That is, given two points (x_1, y_1) and (x_2, y_2), we will want to know the coordinates of the point exactly halfway between the two given points. Of course, such a formula will logically take the form of an ordered pair, as it must give us two numbers.

Consider the points as plotted in Figure 3. The x-coordinate of the midpoint should be the average of the two x-coordinates of the given points, and similarly for the y-coordinate.

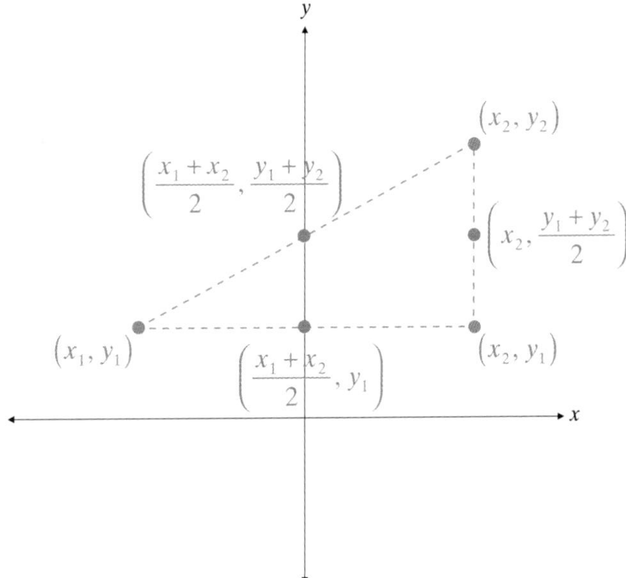

Figure 3: Midpoint Formula

Since x_1 and x_2 are numbers on a real number line, and y_1 and y_2 are numbers on a (different) real number line, determining the averages of these two pairs of numbers is straightforward: the average of the x-coordinates is $\frac{x_1 + x_2}{2}$ and the average of the y-coordinates is $\frac{y_1 + y_2}{2}$. Putting these two coordinates together gives us the desired formula: the midpoint is $\left(\frac{x_1 + x_2}{2}, \frac{y_1 + y_2}{2}\right)$.

We will end with an example of sample calculations.

example 2

Determine **(a)** the distance between $(5,1)$ and $(-1,3)$, and **(b)** the midpoint of the line segment joining $(5,1)$ and $(-1,3)$.

Solutions:

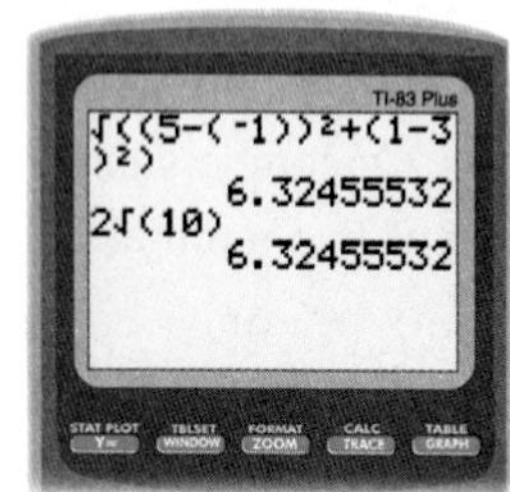

When entering the distance formula, be sure to include all the necessary parentheses.

a. $d=\sqrt{(5-(-1))^2+(1-3)^2}$

$=\sqrt{36+4}$

$=\sqrt{40}$

$=2\sqrt{10}$

All we have to do is apply the distance formula that we have derived, keeping in mind that it is the *differences* of the respective coordinates that get squared and added.

The two points $(5,1)$ and $(-1,3)$ are $2\sqrt{10}$ apart.

b. $\left(\dfrac{5+(-1)}{2},\dfrac{1+3}{2}\right)=(2,2)$

In determining the coordinates of the midpoint of the line segment, remember to add the respective coordinates (since we are finding an average), and then divide by 2.

exercises

Plot the following points in the Cartesian plane.

1. $\{(-3,2),(5,-1),(0,-2),(3,0)\}$ **2.** $\{(-4,0),(0,-4),(-3,-3),(3,-3)\}$

3. $\{(3,4),(-2,-1),(-1,-3),(-3,0)\}$ **4.** $\{(2,2),(0,3),(4,-5),(-1,3)\}$

Identify the quadrant in which each point lies, if possible. If a point lies on an axis, specify which part (positive or negative) of which axis (x or y).

5. $(-2,-4)$ **6.** $(0,-12)$ **7.** $(4,-7)$ **8.** $(-2,0)$ **9.** $(9,0)$ **10.** $(3,26)$

11. $(-4,-7)$ **12.** $(0,1)$ **13.** $(17,-2)$ **14.** $(-\sqrt{2},4)$ **15.** $(-1,1)$ **16.** $(-4,0)$

17. $(3,-9)$ **18.** $(0,0)$ **19.** $(4,3)$ **20.** $(-3,-11)$ **21.** $(0,-97)$ **22.** $\left(\frac{1}{3},0\right)$

For each of the following equations, determine the value of the missing entries in the accompanying table of ordered pairs. Then plot the ordered pairs and sketch your guess of the complete graph of the equation. See Example 1.

23. $6x - 4y = 12$

x	y
0	?
?	0
3	?
?	3

24. $y = x^2 + 2x + 1$

x	y
?	0
1	?
?	1
2	?
–3	?

25. $x = y^2$

x	y
0	?
1	?
4	?
9	?
?	$-\sqrt{2}$

26. $5x - 2 = -y$

x	y
?	0
0	?
1	?
?	7
–2	?

27. $x^2 + y^2 = 9$

x	y
0	?
?	0
–1	?
1	?
?	2

28. $y = -x^2$

x	y
0	?
–1	?
1	?
–2	?
2	?

*Determine **(a)** the distance between the following pairs of points, and **(b)** the midpoint of the line segment joining each pair of points. See Example 2.*

29. $(-2, 3)$ and $(-5, -2)$

30. $(-1, -2)$ and $(2, 2)$

31. $(0, 7)$ and $(3, 0)$

32. $\left(-\frac{1}{2}, 5\right)$ and $\left(\frac{9}{2}, -7\right)$

33. $(-2, 0)$ and $(0, -2)$

34. $(5, 6)$ and $(-3, -2)$

35. $(13, -14)$ and $(-7, -2)$

36. $(-8, 3)$ and $(2, 11)$

37. $(-3, -3)$ and $(5, -9)$

Find the perimeter of the triangle whose vertices are the specified points in the plane.

38. $(-2, 3), (-2, 1)$, and $(-5, -2)$

39. $(-1, -2), (2, -2)$, and $(2, 2)$

40. $(6, -1), (-6, 4)$, and $(9, 3)$

41. $(3, -4), (-7, 0)$, and $(-2, -5)$

42. $(-3, 7), (5, 1)$, and $(-3, -14)$

43. $(-12, -3), (-7, 9)$, and $(9, -3)$

44. Use the distance formula to prove that the triangle with vertices at the points $(1, 1)$, $(-2, -5)$, and $(3, 0)$ is a right triangle. Then determine the area of the triangle.

45. Use the distance formula to prove that the triangle with vertices at the points $(-2, 2)$, $(1, -2)$, and $(2, 5)$ is isosceles. Then determine the area of the triangle. (**Hint:** make use of the midpoint formula.)

Using the distance and midpoint formulas presented in this section, answer the following questions.

46. The navigator of a submarine keeps track of the position of the submarine and surrounding objects using a rectangular coordinate system.

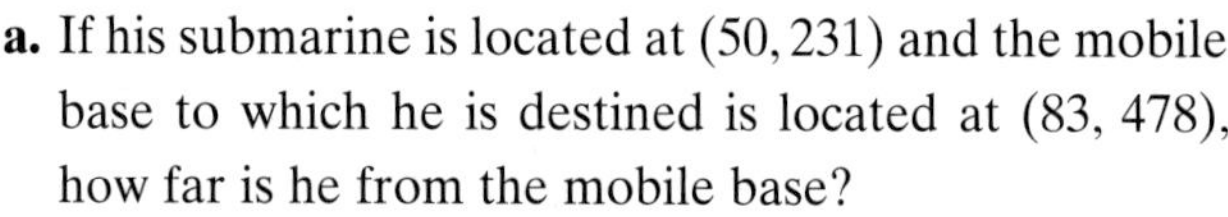

a. If his submarine is located at $(50, 231)$ and the mobile base to which he is destined is located at $(83, 478)$, how far is he from the mobile base?

b. Suppose there is another submarine located halfway between the first submarine and the mobile base. What is the position of the second sub?

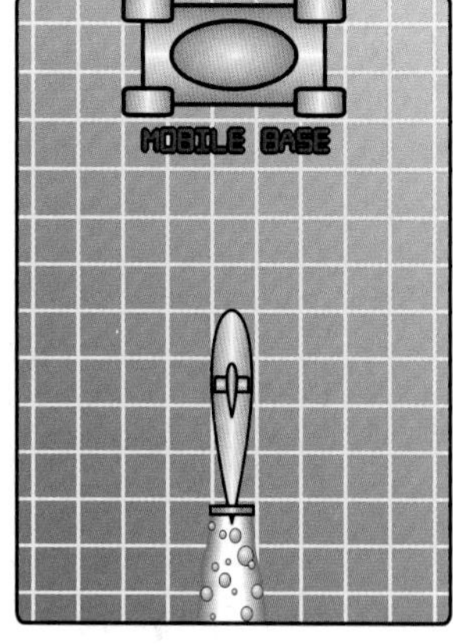

47. Two college friends are taking a weekend road trip. Friday they leave home and drive 87 miles north for a night of dinner and dancing in the city. The next morning they drive 116 miles east to spend a day at the beach. If they drive straight home from the beach the next day, how far do they have to travel on Sunday?

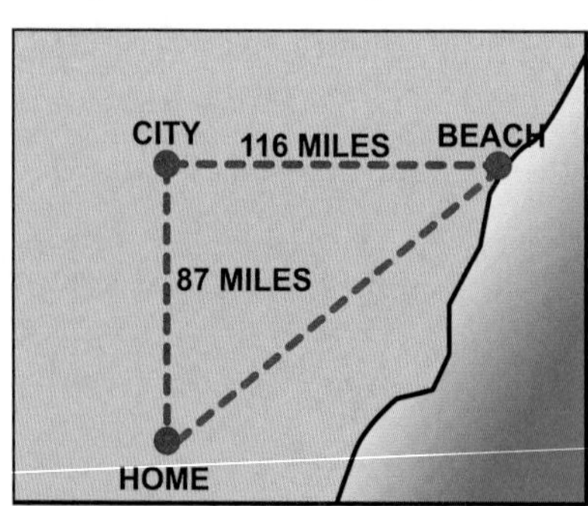

2.2 Linear Equations in Two Variables

TOPICS

1. Recognizing linear equations in two variables
2. Intercepts of the coordinate axes
3. Horizontal and vertical lines

Topic 1: Recognizing Linear Equations in Two Variables

We have encountered one linear equation in two variables already, in Example 1a of the last section. Our first goal in this section is to be able to recognize when an equation in two variables is linear; that is, we wish to know when the solution set of an equation consists of a straight line in the Cartesian plane.

Linear Equations in Two Variables

A **linear equation in two variables**, say the variables x and y, is an equation that can be written in the form $ax + by = c$, where a, b, and c are constants and a and b are not both zero. This form of such an equation is called the **standard form**.

Of course, an equation may be linear but not appear in standard form. Some algebraic manipulation is often necessary in order to determine if a given equation is linear.

example 1

Determine if the following equations are linear equations.

a. $3x-(2-4y)=x-y+1$

b. $3x+2(x+7)-2y=5x$

c. $\dfrac{x+2}{3}-y=\dfrac{y}{5}$

d. $7x-(4x-2)+y=y+3(x-1)$

e. $4x^3-2y=5x$

f. $x^2-(x-3)^2=3y$

cont'd. on next page ...

Solutions:

a.
$$3x-(2-4y)=x-y+1$$
$$3x-2+4y=x-y+1$$
$$3x-x+4y+y=1+2$$
$$2x+5y=3$$

As is typical, we need to use some algebra to rewrite the equation in an equivalent form which will be clearly either linear or not. After using the distributive and cancellation properties, we see that the equation is indeed linear.

b.
$$3x+2(x+7)-2y=5x$$
$$3x+2x+14-2y=5x$$
$$5x-5x-2y=-14$$
$$-2y=-14$$
$$y=7$$

We proceed as in the problem above, using familiar properties to rewrite the equation as an equivalent one in standard form. In this problem, however, one of the variables, namely x, is absent from the result. But since the coefficient of y is non-zero, this equation is also linear.

c.
$$\frac{x+2}{3}-y=\frac{y}{5}$$
$$\frac{1}{3}x+\frac{2}{3}-y=\frac{y}{5}$$
$$\frac{1}{3}x-y-\frac{y}{5}=-\frac{2}{3}$$
$$\frac{1}{3}x-\frac{6}{5}y=-\frac{2}{3}$$

Although this equation may appear at first to not fit the criteria to be linear, appearances are deceiving.

After splitting apart one of the fractions and moving all constant terms to the right-hand side and all variables to the left-hand side, we see that this is another example of a linear equation. Note that we could also have begun by clearing the equation of fractions first.

d.
$$7x-(4x-2)+y=y+3(x-1)$$
$$7x-4x+2+y=y+3x-3$$
$$3x-3x+y-y=-3-2$$
$$0=-5$$

In this equation, both variables are absent in the final (and equivalent) form. This tells us immediately that the equation is not linear, as the coefficients of x and y are both 0. Moreover, this equation has no solution: no values for x and y result in a true statement.

e.
$$4x^3-2y=5x$$

The presence of the cubed term in this already simplified equation makes it clearly not linear.

f.
$$x^2-(x-3)^2=3y$$
$$x^2-x^2+6x-9=3y$$
$$6x-3y=9$$

In contrast to the last equation, when we simplify this equation the result clearly is linear.

Topic 2: Intercepts of the Coordinate Axes

Quite often, the goal in working with a given linear equation is to graph its solution set. If the straight line whose points constitute the solution set crosses the horizontal and vertical axes in two distinct points, knowing the coordinates of these two points is sufficient to graph the complete solution; this follows from the fact that two points are enough to completely determine a line.

If the equation under consideration is in the two variables x and y, it is natural to call the point where the graph crosses the x-axis the ***x*-intercept**, and the point where it crosses the y-axis the ***y*-intercept**. If the line does indeed cross both axes, the two intercepts are easy to find: the y-coordinate of the x-intercept is 0, and the x-coordinate of the y-intercept is 0.

example 2

Find the x and y intercepts of the following equations, and then graph each equation.

a. $3x - 4y = 12$ **b.** $4x - (3 - x) + 2y = 7$

Solutions:

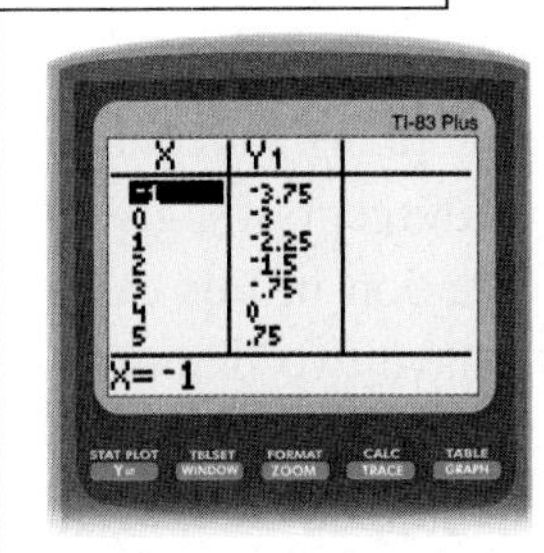

Notice that the y-value at x=4 is 0. To get to this screen, select **Y=** and enter the equation. Then select **2nd** and **Graph** to view the table.

a.

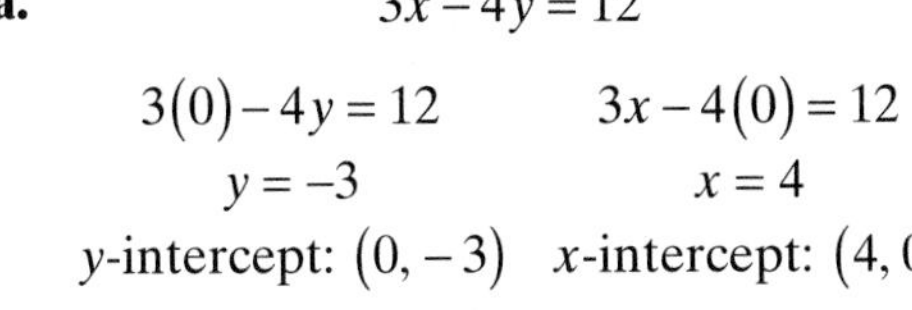

$$3x - 4y = 12$$

$$3(0) - 4y = 12 \qquad 3x - 4(0) = 12$$

$$y = -3 \qquad x = 4$$

$$y\text{-intercept: } (0, -3) \qquad x\text{-intercept: } (4, 0)$$

To find the two intercepts, we first set x equal to 0 and solve for y, then set y equal to 0 and solve for x. The coordinates of the two intercepts are shown to the left. Although it is slightly imprecise, it is common to say, for instance, that the y-intercept is –3 and the x-intercept is 4.

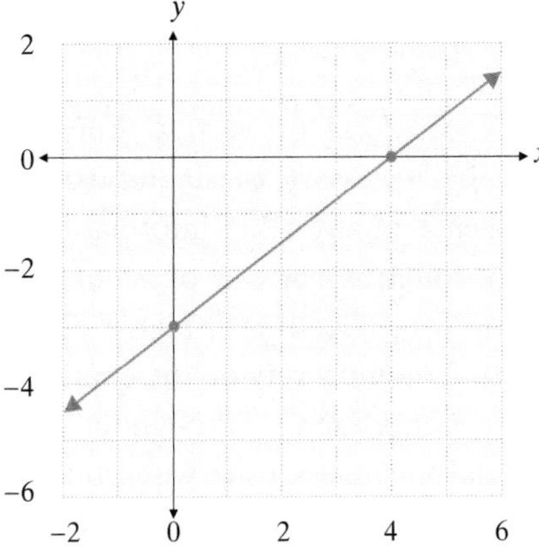

Once we have plotted the two intercepts, drawing a straight line through them gives us the graph of the equation.

cont'd. on next page ...

b.

$$4x-(3-x)+2y=7$$

$$5x+2y=10$$

$$5(0)+2y=10 \qquad 5x+2(0)=10$$

$$y=5 \qquad x=2$$

y-intercept: $(0,5)$ x-intercept: $(2,0)$

Again, find the two intercepts by setting the appropriate variables equal to 0. The y-intercept is 5 and the x-intercept is 2.

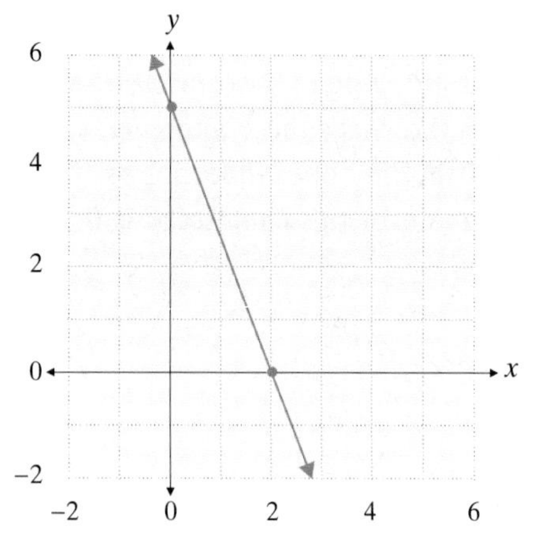

Note that in both graphs, the location of the origin has been chosen in order to conveniently plot the intercepts.

Topic 3: Horizontal and Vertical Lines

The above procedure for graphing linear equations may be undesirable or inadequate for several reasons. One such reason is that the x-intercept and y-intercept of a given line may be the same point: the origin. In order to graph the equation in this case, a second point distinct from the origin must be found in order to have two points to connect with a line. Another reason is that a given equation may not have one of the two types of intercepts. A moment's thought will reveal that this can happen only when the graph of the equation is a horizontal or vertical line.

Equations whose graphs are horizontal or vertical lines are missing one of the two variables (though it may be necessary to simplify the equation before this is apparent). However, in the absence of any other information, it is impossible to know if the solution of such a linear equation consists of a point on the real number line or a line in the Cartesian plane (or, indeed, of a higher-dimensional set in a higher-dimensional coordinate system). You must rely on the context of the problem to know how many variables should be considered. Throughout this chapter, all equations are assumed to be in two variables unless otherwise stated, so an equation of the form $ax=c$ or $by=c$ should be thought of as representing a line in $\mathbb{R}^2$.

Consider an equation of the form $ax = c$. The variable y is absent, so *any* value for this variable will suffice as long as we pair it with $x = \frac{c}{a}$. Thinking of the solution set as a set of ordered pairs, the solution consists of ordered pairs with a fixed first coordinate and arbitrary second coordinate. This describes, geometrically, a vertical line with an x-intercept of $\frac{c}{a}$. Similarly, the equation $by = c$ represents a horizontal line with y-intercept equal to $\frac{c}{b}$.

example 3

Graph the following equations.

a. $3x + 2(x+7) - 2y = 5x$ **b.** $5x = 0$ **c.** $2x - 2 = 3$

Solutions:

a. $3x + 2(x+7) - 2y = 5x$

$$y = 7$$

We encountered this equation in Example 1b, and have already written it in standard form as shown.

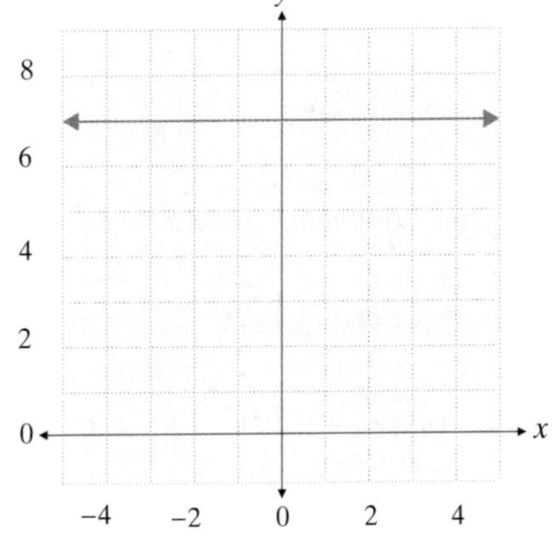

The graph of this equation is the horizontal line consisting of all those ordered pairs whose y-coordinate is 7.

To graph an equation, select **Y=** and enter the equation in terms of y. Then select **Graph** to view the graph.

b. $5x = 0$

$$x = 0$$

The first step in this example is to divide both sides by 5, leaving the simple equation $x = 0$.

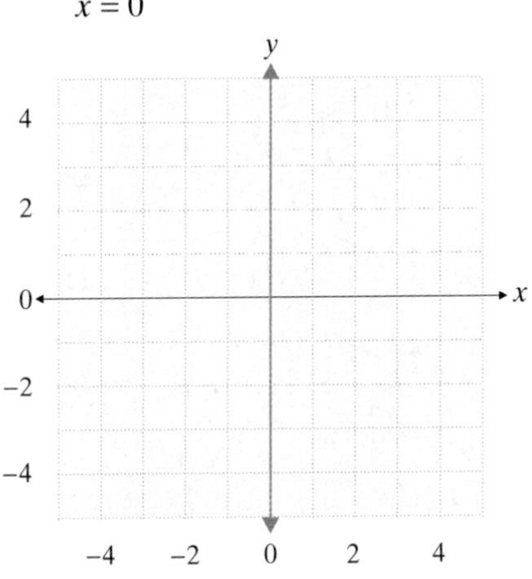

Note that the graph of this equation is the y-axis, as all ordered pairs on the y-axis have an x-coordinate of 0. This equation is unique in that it has an infinite number of y-intercepts (each point on the graph is also on the y-axis) and one x-intercept (the origin). Similarly, the equation $y = 0$ has an infinite number of x-intercepts and one y-intercept.

cont'd. on next page ...

c. 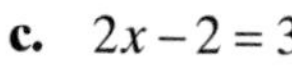

$2x - 2 = 3$

$x = \frac{5}{2}$

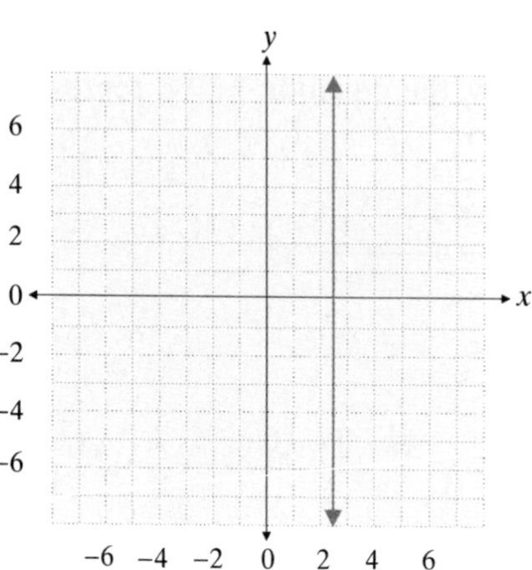

Upon simplifying, it is apparent that this equation also represents a vertical line, this time passing through $\frac{5}{2}$ on the x-axis.

exercises

Determine if the following equations are linear. See Example 1.

1. $3x + 2(x - 4y) = 2x - y$

2. $9x + 4(y - x) = 3$

3. $9x^2 - (x+1)^2 = y - 3$

4. $3x + xy = 2y$

5. $8 - 4xy = x - 2y$

6. $\frac{x-y}{2} + \frac{7y}{3} = 5$

7. $\frac{6}{x} - \frac{5}{y} = 2$

8. $3x - 3(x - 2y) = y + 1$

9. $2y - (x + y) = y + 1$

10. $(3 - y)^2 - y^2 = x + 2$

11. $x^2 - (x-1)^2 = y$

12. $(x+y)^2 - (x-y)^2 = 1$

13. $x(y+1) = 16 - y(1-x)$

14. $\frac{x-3}{2} = \frac{4+y}{5}$

15. $x - 2x^2 + 3 = \frac{x-7}{2}$

16. $x - 3 = \frac{4x+17}{5}$

17. $13x - 17y = y(7 - 2x)$

18. $y^2 - 3y = (1+y)^2 - 2x$

19. $x - 1 = \frac{2y}{x} - x$

20. $3x - 4 = 89(x - y) - y$

21. $x - x(1 + x) = y - 3x$

22. $x^2 - 2x = 3 - x^2 + y$

23. $\frac{2y-5}{14} = \frac{x-3}{9}$

24. $16x = y(4 + (x - 3)) - xy$

Determine the x- and y-intercepts of the following linear equations, if possible, and then graph the equations. See Examples 2 and 3.

25. $4x-3y=12$ **26.** $y-3x=9$ **27.** $5-y=10x$

28. $y-2x=y-4$ **29.** $3y=9$ **30.** $2x-(x+y)=x+1$

31. $x+2y=7$ **32.** $y-x=x-y$ **33.** $y=-x$

34. $2x-3=1-4y$ **35.** $3y+7x=7(3+x)$ **36.** $4-2y=-2-6x$

37. $x+y=1+2y$ **38.** $3y+x=2x+3y+4$ **39.** $3(x+y)+1=x-5$

2.3 Forms of Linear Equations

TOPICS

1. The slope of a line
2. Slope-intercept form of a line
3. Point-slope form of a line

Topic 1: The Slope of a Line

The methods we have used for graphing linear equations have been adequate for our purposes so far, but for many problems we will want more sophisticated ways to graph lines and to pick out the important characteristics of linear equations.

There are several ways to characterize a given line in the plane. We have already used one way repeatedly: two distinct points in the Cartesian plane determine a unique line. Another, and often more useful approach is to identify just one point on the line and to indicate how "steeply" the line is rising or falling as we scan the plane from left to right. It turns out that a single number is sufficient to convey this notion of "steepness."

Slope of a Line

Let L stand for a given line in the Cartesian plane, and let (x_1, y_1) and (x_2, y_2) be the coordinates of two distinct points on L. The **slope** of the line L is the ratio $\frac{y_2 - y_1}{x_2 - x_1}$, which can be described in words as "change in y over change in x" or "rise over run."

The phrase "rise over run" is motivated by the diagram in Figure 1.

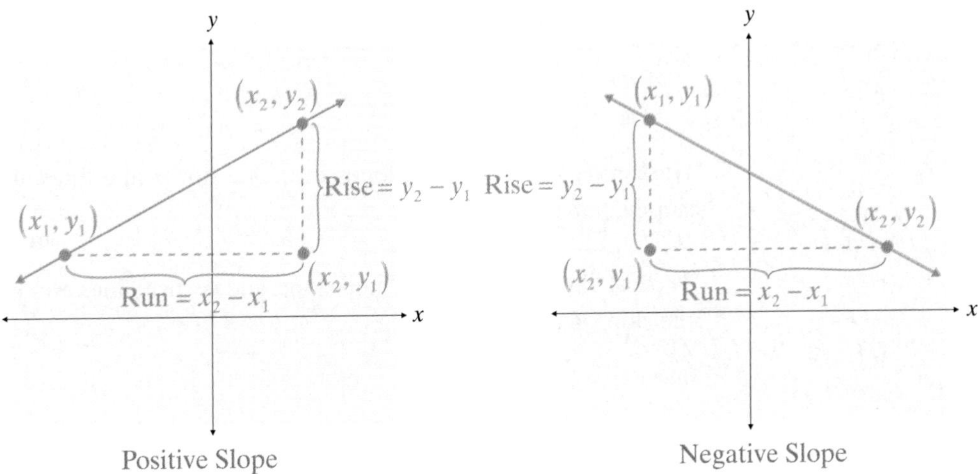

Figure 1: Rise and Run Between Two Points

As drawn in the graph on the left in Figure 1, the ratio $\frac{y_2 - y_1}{x_2 - x_1}$ is positive, and we say that the line has a positive slope. In the graph on the right in Figure 1, the rise and run have opposite signs, so the slope of the line is negative; the graph of the line falls from the upper left to the lower right.

caution!

It doesn't matter how you assign the labels (x_1, y_1) and (x_2, y_2) to the two points you are using to calculate slope, but it *is* important that you are consistent as you apply the formula. That is, don't change the order in which you are subtracting as you determine the numerator and denominator in the formula $\frac{y_2 - y_1}{x_2 - x_1}$.

The formula $\frac{y_2 - y_1}{x_2 - x_1}$ is only valid if $x_2 - x_1 \neq 0$, as division by zero is never defined. For what sorts of lines, then, are we unable to find the slope? If two points on a line have the same first coordinate, that is if $x_1 = x_2$, the line must be perfectly vertical. The other extreme is that the numerator is 0 (and the denominator is non-zero). In this case, the slope as a whole is 0. For what sorts of lines will this happen? If two points on a line have the same second coordinate, that is if $y_1 = y_2$, the line must be horizontal. These two cases are summarized on the following page.

Slopes of Horizontal and Vertical Lines

Horizontal lines all have slopes of 0, and horizontal lines are the *only* lines with slope equal to 0.

Vertical lines all have undefined slope, and vertical lines are the *only* lines for which the slope is undefined.

example 1

Determine the slopes of the lines passing through the following pairs of points in $\mathbb{R}^2$.

a. $(-4, -3)$ and $(2, -5)$ **b.** $\left(\frac{3}{2}, 1\right)$ and $\left(1, -\frac{4}{3}\right)$ **c.** $(-2, 7)$ and $(1, 7)$

Solutions:

a. $\frac{-3-(-5)}{-4-2} = \frac{2}{-6} = -\frac{1}{3}$

In the calculation to the left, the coordinates of the second point have been subtracted from the coordinates of the first. The result would be the same if we reversed the order and subtracted the coordinates of the first from those of the second.

b. $\frac{1-\left(-\frac{4}{3}\right)}{\frac{3}{2}-1} = \frac{\frac{7}{3}}{\frac{1}{2}} = \frac{7}{3}\cdot\frac{2}{1} = \frac{14}{3}$

The final answer tells us that the line through the two points rises 14 units for every run of 3 units horizontally.

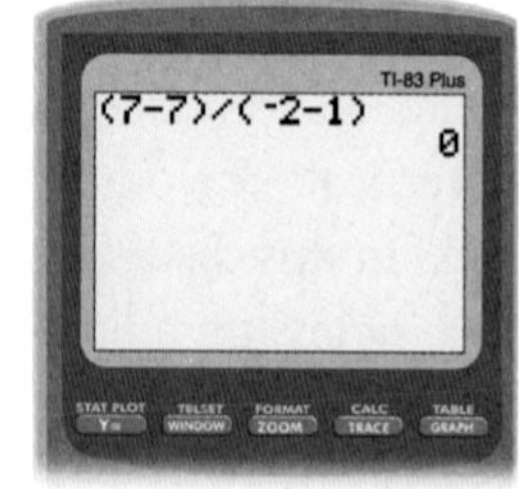

c. $\frac{7-7}{-2-1} = \frac{0}{-3} = 0$

The two points in part c have the same second coordinate, and thus lie on a horizontal line.

We already know how to identify any number of ordered pairs that lie on a line, given the equation for the line. Identifying just two such ordered pairs allows us to calculate the slope of a line defined by an equation.

example 2

Determine the slopes of the lines defined by the following equations.

a. $4x - 3y = 12$ **b.** $2x + 7y = 9$ **c.** $x = -\dfrac{3}{4}$

Solutions:

a. x-intercept: $(3, 0)$

y-intercept: $(0, -4)$

$$\text{slope} = \frac{-4-0}{0-3} = \frac{-4}{-3} = \frac{4}{3}$$

Determining the x- and y-intercepts of this line is fairly straightforward. Recall that the x-intercept is found by setting y equal to 0 and solving for x, and vice versa for the y-intercept. Once we have two points, we apply the slope formula.

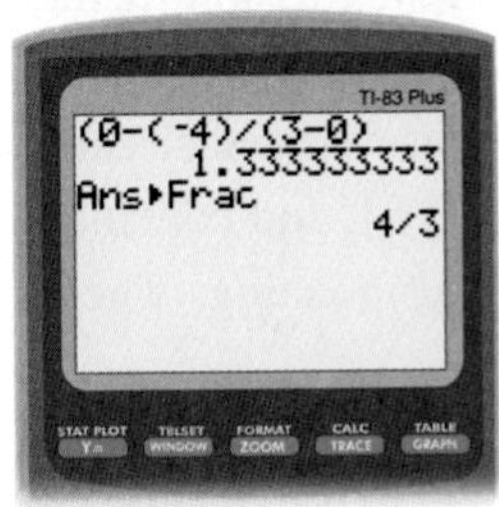

b. x-intercept: $\left(\dfrac{9}{2}, 0\right)$

second point on the line: $(1, 1)$

$$\text{slope} = \frac{1-0}{1-9/2} = \frac{1}{-7/2} = -\frac{2}{7}$$

In this example, we have found the x-intercept, but have used an ordered pair more easily determined than the y-intercept for the second point.

c. first point: $\left(-\dfrac{3}{4}, 2\right)$

second point: $\left(-\dfrac{3}{4}, 10\right)$

slope is undefined

As soon as we realize that the line defined by the equation is vertical, we can state that the slope is undefined. Two arbitrarily chosen points have been identified as lying on the line, but the critical observation is that *any* two points in the slope formula result in division by 0.

Note that the line in Example 2a has a positive slope and the line in Example 2b has a negative slope. So even without graphing these lines, we know that the first line will rise from the lower left to the upper right part of the plane, while the second line will fall from the upper left to the lower right. You should practice your graphing skills and verify that these observations are indeed correct.

Topic 2: Slope-Intercept Form of a Line

Example 2 illustrates the most elementary way of determining the slope of a line from an equation. With a small amount of work, we can develop a faster method for determining not only the slope of a line, but also the y-intercept (assuming the line *has* a y-intercept).

Consider a fixed non-vertical line in the plane. The variable y must appear in the linear equation that describes the line (otherwise the line would be vertical), so the equation can be solved for y. The result will be an equation of the form $y = mx + b$, where m and b (the letters used by convention in this context) are constants particular to the given line (note that the constant m will be 0 if the line is horizontal). Now suppose that (x_1, y_1) and (x_2, y_2) are two points that lie on the line $y = mx + b$. That is, suppose that (x_1, y_1) and (x_2, y_2) both solve the equation $y = mx + b$. Then it must be the case that $y_1 = mx_1 + b$ and $y_2 = mx_2 + b$. If we use these two points to determine the slope of the line, we obtain:

$$\text{slope} = \frac{y_2 - y_1}{x_2 - x_1} = \frac{(mx_2 + b) - (mx_1 + b)}{x_2 - x_1} = \frac{m\cancel{(x_2 - x_1)}}{\cancel{x_2 - x_1}} = m.$$

What we have just shown is that the slope of the line is automatically the coefficient of the variable x when the line's equation is solved for y.

Let us continue investigating the line in the form $y = mx + b$. Since we are assuming the line is non-vertical, it must have a y-intercept. We know how to find the y-intercept: set x equal to 0 and solve for y. This calculation can be performed directly, as we have already solved the equation for y. We find that the y-intercept is the point $(0, b)$, and conclude that the two constants in the equation $y = mx + b$ both have physical meaning. These observations are summarized below.

Slope-Intercept Form of a Line

If the equation of a non-vertical line in x and y is solved for y, the result is an equation of the form

$$y = mx + b.$$

The constant m is the slope of the line, and the line crosses the y-axis at b; that is, the y-intercept of the line is $(0, b)$. If the variable x does not appear in the equation, the slope is 0 and the equation is simply of the form $y = b$.

We can make use of the slope-intercept form of a line to graph the line, as illustrated in the next example.

example 3

Use the slope-intercept form of the line to graph the equation $4x-3y=6$.

Solution:

$$4x-3y=6$$
$$-3y=-4x+6$$
$$y=\frac{4}{3}x-2$$

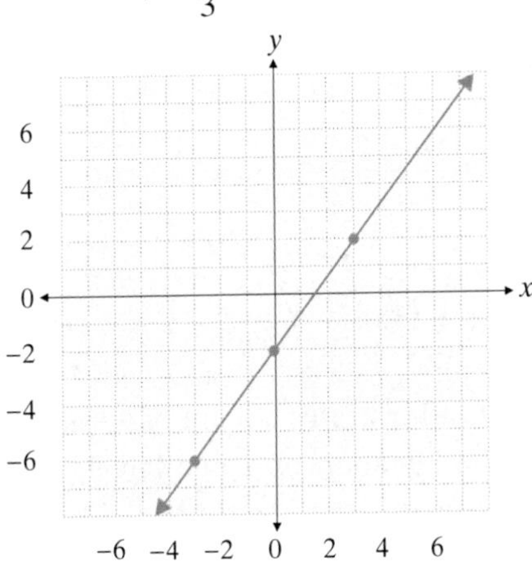

The first step is to solve the equation for y. Once we have done this, we know that the line has a slope of $\frac{4}{3}$ and crosses the y-axis at -2.

To use this information to sketch the graph of the line, we begin by plotting the y-intercept. A second point can now be found by using the fact that slope is "rise over run." This means a second point must lie, in this case, 4 units up and 3 units to the right, i.e. at $(3, 2)$. (Alternatively, we could locate a second point by moving down 4 units and then moving to the left 3 units.)

In some cases, we can also make use of the slope-intercept form to find the equation of a line that has certain properties.

example 4

Find the equation of the line that passes through the point $(0, 3)$ and has a slope of $-\frac{3}{5}$. Then graph the line.

Solution:

$$y=-\frac{3}{5}x+3$$
or
$$3x+5y=15$$

We can immediately write down the equation, as we have been given the y-intercept and the slope. If we wish, we can then write the equation in standard form.

cont'd. on next page ...

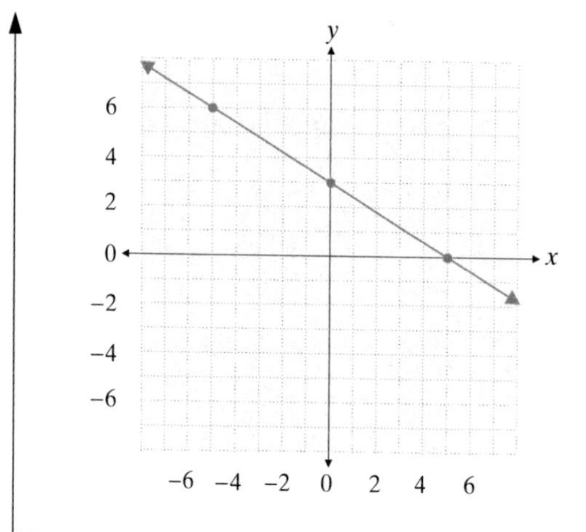

We can plot the line given only information about the slope and the y-intercept, just as we did in Example 3. We first plot the y-intercept, then move down 3 units and to the right 5 units to locate a second point. Or, we could have plotted a second point by moving up 3 units and to the left 5 units. Either method makes use of the fact that the slope is $-\frac{3}{5}$.

Topic 3: Point-Slope Form of a Line

A close examination of Example 4 illustrates a point of weakness in the slope-intercept form. We were able to find the equation of a line having certain properties in that example, but only because the point given to us happened to be the y-intercept. In the most general case, we need to be able to construct an equation of a line given the slope of the line and *any* point on the line (as opposed to the y-intercept). Or, we might need to be able to find the equation of a line given two points on the line. In order to meet these needs, we introduce here the point-slope form of a line, the final form of linear equations in two variables.

Suppose we know that a given line L has slope m and passes through the point (x_1, y_1). Given these three numbers (m, x_1, and y_1), we want to come up with an equation in the variables x and y whose solution set is the line L. We can achieve this goal by assuming that (x, y) represents an arbitrary point on the line. Then by the definition of slope, it must be the case that

$$\frac{y - y_1}{x - x_1} = m.$$

This is true because for any point (x, y) on the line, the slope formula applied to the two points (x, y) and (x_1, y_1) must yield the number m.

If we now multiply the above equation by $x - x_1$, we obtain the desired form, as summarized on the next page.

Point-Slope Form of a Line

Given an ordered pair (x_1, y_1) and a real number m, an equation for the line passing through the point (x_1, y_1) with slope m is

$$y - y_1 = m(x - x_1).$$

Note that m, x_1, and y_1 are all constants, and that x and y are variables. Note also that since the line, by definition, has slope m, vertical lines cannot be described in this form.

example 5

Find the equation, in slope-intercept form, of the line that passes through the point $(-2, 5)$ with slope 3.

Solution:

$$y - 5 = 3(x - (-2))$$
$$y - 5 = 3(x + 2)$$
$$y - 5 = 3x + 6$$
$$y = 3x + 11$$

A point and a slope are exactly the information we need to use the point-slope form. Replacing y_1 with 5 and x_1 with -2, we have an equation with the desired properties. The one remaining task is to write the equation in slope-intercept form, which we accomplish by solving for y.

We will conclude this section with one final example. We know that two distinct points in the plane are sufficient to determine a line. How do we use two such points to find the equation for a line?

example 6

Find the equation, in slope-intercept form, of the line that passes through the two points $(-3, -2)$ and $(1, 6)$.

Solution:

$$m = \frac{-2 - 6}{-3 - 1} = \frac{-8}{-4} = 2$$

We begin by using the slope-formula to determine the slope of the line from the two given points.

$$y - 6 = 2(x - 1)$$
$$y - 6 = 2x - 2$$
$$y = 2x + 4$$

Once we have found m, we can use this number and either of the two points that we know are on the line to find the equation in point-slope form. If we had used the point $(-3, -2)$ instead of the point $(1, 6)$, we would have obtained the same equation once it is written in slope-intercept form.

exercises

Determine the slopes of the lines passing through the specified points. See Example 1.

1. $(0,-3)$ and $(-2,5)$ **2.** $(-3,2)$ and $(7,-10)$ **3.** $(4,5)$ and $(-1,5)$

4. $(3,-1)$ and $(-7,-1)$ **5.** $(3,-5)$ and $(3,2)$ **6.** $(0,0)$ and $(-2,5)$

7. $(-2,1)$ and $(-5,-1)$ **8.** $(-2,4)$ and $(6,9)$ **9.** $(0,-21)$ and $(-3,0)$

10. $(-3,-5)$ and $(-2,8)$ **11.** $(29,-17)$ and $(31,-29)$ **12.** $(7,4)$ and $(-6,13)$

Determine the slopes of the lines defined by the following equations. See Example 2.

13. $8x-2y=11$ **14.** $2x+8y=11$ **15.** $4y=13$

16. $\dfrac{x-y}{3}+2=4$ **17.** $7x=2$ **18.** $3y-2=\dfrac{x}{5}$

19. $3(2y-1)=5(2-x)$ **20.** $12x-4y=-9$ **21.** $\dfrac{x+2}{3}+2(1-y)=-2x$

22. $2y-7x=4y+5x$ **23.** $x-7=\dfrac{2y-1}{-5}$ **24.** $3-y=2(5-x)$

Use the slope-intercept form of each line to graph the equations. See Example 3.

25. $6x-2y=4$ **26.** $3y+2x-9=0$ **27.** $5y-15=0$

28. $x+4y=20$ **29.** $\dfrac{x-y}{2}=-1$ **30.** $3x+7y=8y-x$

Find the equation, in slope-intercept form, of the line passing through the specified points with the specified slopes. See Example 4.

31. point $(0,-3)$; slope of $\dfrac{3}{4}$ **32.** point $(0,5)$; slope of -3

33. point $(0,-7)$; slope of $-\dfrac{5}{2}$ **34.** point $(0,6)$; slope of 4

Find the equation, in standard form, of the line passing through the specified points with the specified slopes. See Example 5.

35. point $(-1,-3)$; slope of $\dfrac{3}{2}$ **36.** point $(3,-1)$; slope of 10

37. point $(-3,5)$; slope of 0 **38.** point $(-2,-13)$; undefined slope

39. point $(6,0)$; slope of $\dfrac{5}{4}$ **40.** point $(-1,3)$; slope of $-\dfrac{2}{7}$

Find the equation, in standard form, of the line passing through the specified points. See Example 6.

41. $(-1, 3)$ and $(2, -1)$ **42.** $(1, 3)$ and $(-2, 3)$ **43.** $(2, -2)$ and $(2, 17)$

44. $(-9, 2)$ and $(1, 5)$ **45.** $(3, -1)$ and $(8, -1)$ **46.** $\left(\frac{4}{3}, 1\right)$ and $\left(\frac{2}{5}, \frac{3}{7}\right)$

47. A bottle manufacturer has determined that the total cost (C) in dollars of producing x bottles is $C = .25x + 2100$.

a. What is the cost of producing 500 units?

b. What are the fixed costs (costs incurred even when 0 units are produced)?

c. What is the increase in cost for each bottle produced?

48. Sales at Glover's Golf Emporium have been increasing at a steady rate for the past couple years. Last year sales were $163,000. This year sales were $215,000. If sales continue to increase at this rate, predict next year's sales.

49. For tax and accounting purposes, businesses often have to depreciate equipment values over time. One method of depreciation is the straight-line method. Three years ago Hilde Construction purchased a bulldozer for $51,500. Using the straight-line method, the bulldozer has now depreciated to a value of $43,200. If V equals the value at the end of year t, write a linear equation expressing the value of the bulldozer over time. How many years from the date of purchase will the value equal $0?

50. Amy owns stock in Trimetric Technologies. If the stock had a value of $2500 in 2003 when she purchased it, what has been the average change in value per year if in 2005 the stock was worth $3150?

technology exercises

The **Solve** command in *Mathematica* (and its equivalent in other computer algebra systems) is useful as an aid in a variety of contexts. When given a linear equation in two variables, the command can be used to quickly determine the slope and y-intercept of the line. In the first of the three commands illustrated below, **Solve** is used to write the equation in slope-intercept form (the result is an equation if we replace the arrow with the equality symbol; the **Simplify** command has been used to simplify the output of the **Solve** command so that the slope can be determined more easily). The result tells us that the slope of the line is $\frac{9}{2}$ and that the line has a y-intercept of $\left(0, \frac{7}{2}\right)$. In the second use of the command the output is the empty set, meaning the variable y disappears upon simplifying the equation in question. As we know, this indicates the line is vertical, so if we **Solve** for x (as illustrated by the third command) we can determine where it crosses the x-axis.

If available, use a computer algebra system to determine the slope and y-intercept of each of the following linear equations. (See Appendix A for guidance on using computer algebra systems like Mathematica.)

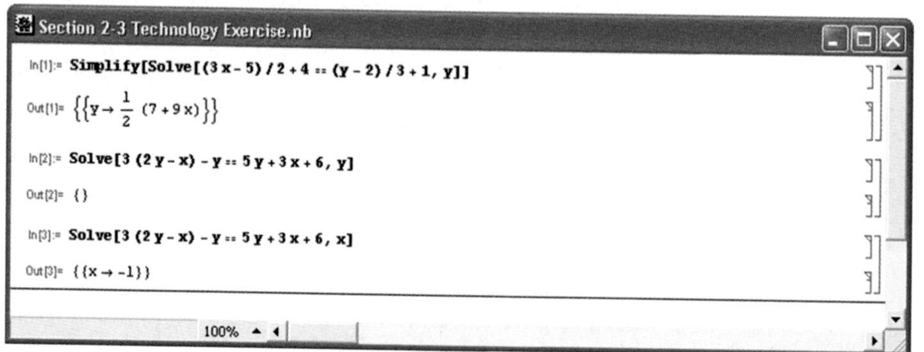

51. $\frac{3x-7y}{4}+5=y-\frac{x}{9}$

52. $4(2x-3y)+17y=3+5y$

53. $\frac{x-3y}{5}=\frac{1+2x}{7}$

54. $3-\frac{x-y}{7}=x+\frac{2y}{3}$

55. $\frac{x+y}{3}-\frac{x-y}{2}=\frac{y-x}{6}$

56. $\frac{7x-3y}{2}=\frac{2y-5x}{9}$

2.4 Parallel and Perpendicular Lines

TOPICS

1. Slopes of parallel lines
2. Slopes of perpendicular lines

Topic 1: Slopes of Parallel Lines

In this short section, we will explore the concept of slope in greater detail by studying the relationship between slope and the geometric concepts of parallel and perpendicular lines. We begin with a study of parallel lines.

As the slope of a line is a precise indication of its "steepness", it should come as no surprise that two lines are parallel if and only if they have the same slope (though in the case of vertical lines this means they both have undefined slope). This fact is also clear from the formula for slope: two lines are parallel if and only if they both "rise" vertically the same amount relative to the same horizontal "run." We can use this observation to derive equations for lines that are described in terms of other lines, as illustrated below.

example 1

Find the equation, in slope-intercept form, for the line which is parallel to the line $3x+5y=23$ and which passes through the point $(-2,1)$.

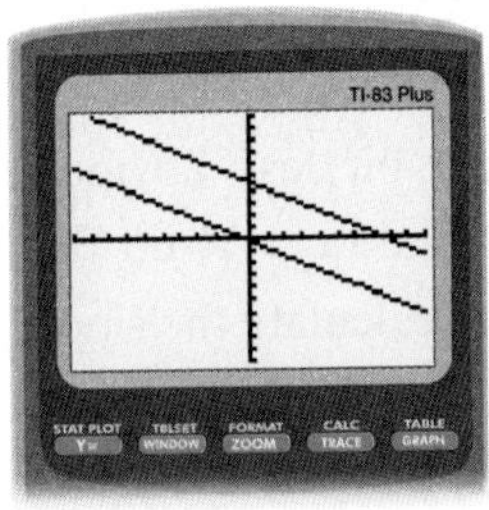

The screen above is the graphs of both $3x+5y=23$ and the parallel line passing through $(-2,1)$. To view multiple graphs, select Y= and enter the equations in terms of y. Then select Graph.

Solution:

We begin by writing the equation $3x+5y=23$ in slope-intercept form: $y=-\frac{3}{5}x+\frac{23}{5}$; this tells us that the slope of the line whose equation we seek is $-\frac{3}{5}$. With this slope and the knowledge that the line is to pass through $(-2,1)$, we use the point-slope form of a line to obtain the equation below.

$$y-1=-\frac{3}{5}(x-(-2))$$

$$y-1=-\frac{3}{5}(x+2)$$

$$y-1=-\frac{3}{5}x-\frac{6}{5}$$

$$y=-\frac{3}{5}x-\frac{1}{5}$$

The first equation to the left is the point-slope form of the line that passes through $(-2,1)$ with a slope of $-\frac{3}{5}$.

The instructions in this example asked for the equation in slope-intercept form, so we solve for y to obtain the final answer.

We can also use the knowledge that parallel lines have the same slope to answer questions that are more geometrical in nature.

example 2

Determine if the quadrilateral (four-sided figure) whose vertices are $(-4, 2)$, $(1, 1)$, $(2, 3)$, and $(-3, 4)$ is a parallelogram (a quadrilateral in which both pairs of opposite sides are parallel).

Solution:

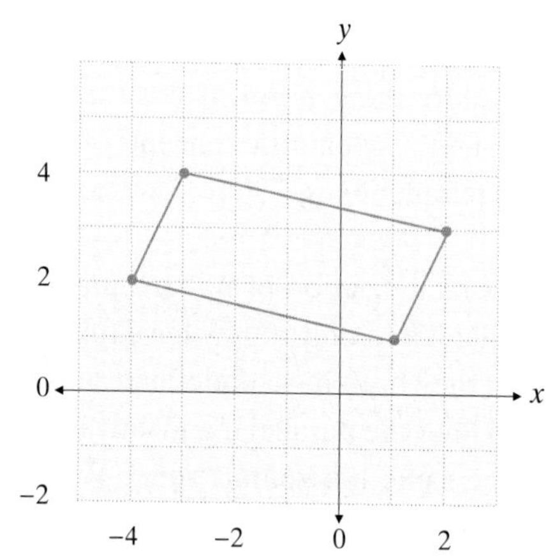

The four vertices are plotted in the picture to the left, and the sides of the quadrilateral drawn. The figure is a parallelogram if the left and right sides are parallel and the top and bottom sides are parallel. The slopes of the left and right sides are, respectively:

$$\frac{4-2}{-3-(-4)} = 2 \text{ and } \frac{3-1}{2-1} = 2,$$

and the slopes of the top and bottom sides are, respectively:

$$\frac{4-3}{-3-2} = -\frac{1}{5} \text{ and } \frac{2-1}{-4-1} = -\frac{1}{5}.$$

Thus, the figure is indeed a parallelogram.

Topic 2: Slopes of Perpendicular Lines

The relationship between the slopes of perpendicular lines is a bit less obvious. Consider a non-vertical line L_1, and consider two points (x_1, y_1) and (x_2, y_2) on the line, as shown in Figure 1. These two points can be used, of course, to calculate the slope m_1 of L_1, with the result that $m_1 = \frac{a}{b}$, where $a = y_2 - y_1$ and $b = x_2 - x_1$.

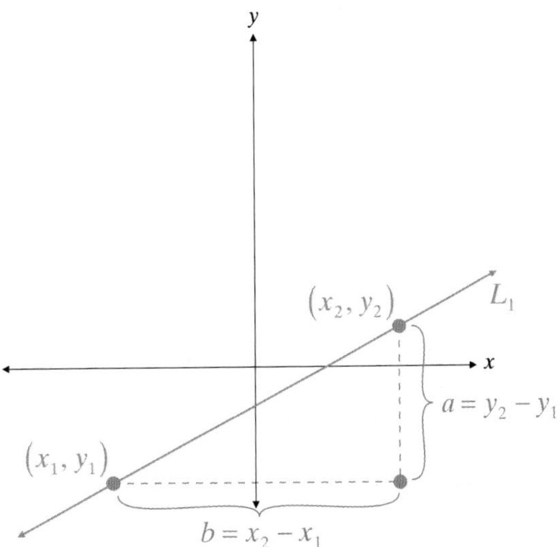

Figure 1: Definition of *a* and *b*

If we now draw a line L_2 perpendicular to L_1, we can use a and b to determine the slope m_2 of line L_2. There are an infinite number of lines that are perpendicular to L_1; one of them is drawn in Figure 2.

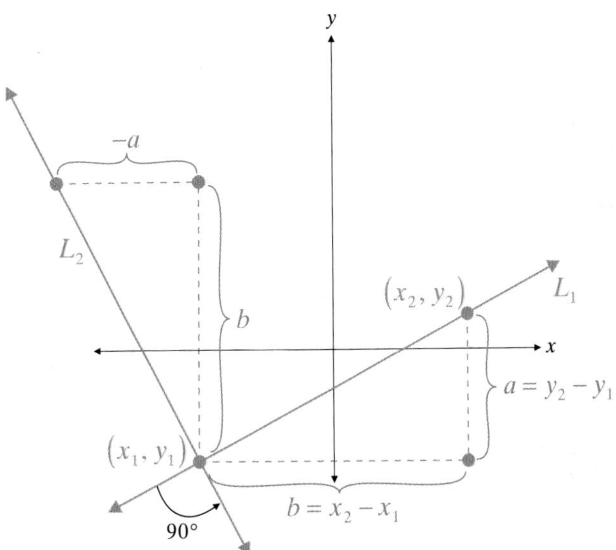

Figure 2: Perpendicular Lines

Note that in rotating the line L_1 by 90 degrees to obtain L_2, we have also rotated the right triangle drawn with dashed lines. But to travel along the line L_2 from the point (x_1, y_1) to the second point drawn requires a positive rise and a negative run, whereas the rise and run between (x_1, y_1) and (x_2, y_2) are both positive. In other words, $m_2 = -\frac{b}{a}$, the negative reciprocal of the slope m_1.

This relationship always exists between the slopes of two perpendicular lines, assuming neither one is vertical. Of course, if one line is vertical, any line perpendicular to it will be horizontal with a slope of zero, while if one line is horizontal, any line perpendicular to it will be vertical with undefined slope. This is summarized below.

Slopes of Perpendicular Lines

If m_1 and m_2 represent the slopes of two perpendicular lines, neither of which is vertical, then $m_1 = -\frac{1}{m_2}$ (equivalently, $m_2 = -\frac{1}{m_1}$). If one of two perpendicular lines is vertical, the other is horizontal, and the slopes are, respectively, undefined and zero.

The next two examples illustrate how we can use this knowledge of the relationship between slopes of perpendicular lines.

example 3

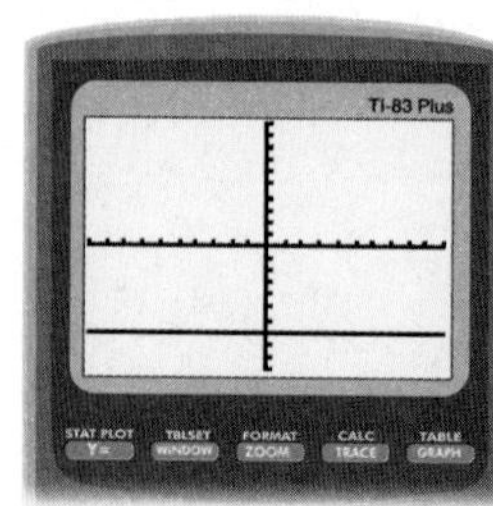

Find the equation, in standard form, of the line that passes through the point $(-3, 13)$ and that is perpendicular to the line $y = -7$.

Solution:

The line $y = -7$ is a horizontal line, and hence any line perpendicular to it must have undefined slope. In particular, the perpendicular line passing through $(-3, 13)$ must have the form $x = -3$, and this is the desired solution.

example 4

Prove that the two lines $3x - 7y = 12$ and $14x + 6y = -5$ are perpendicular to each other.

Solution:

This is most easily done by rewriting each equation in slope-intercept form. In this form, the first equation is $y = \frac{3}{7}x - \frac{12}{7}$ and the second equation is $y = -\frac{7}{3}x - \frac{5}{6}$. Hence, the slope of the first line is $\frac{3}{7}$, which is the negative reciprocal of the slope of the second line, $-\frac{7}{3}$.

exercises

Find the equation, in slope-intercept form, for the line parallel to the given line and passing through the indicated point. See Example 1.

1. Parallel to $y-4x=7$ and passing through $(-1,5)$.
2. Parallel to $3x+2y=3y-7$ and passing through $(3,-2)$.
3. Parallel to $2-\frac{y-3x}{3}=5$ and passing through $(0,-2)$.
4. Parallel to $y-4x=7-4x$ and passing through $(23,-9)$.
5. Parallel to $2(y-1)+\frac{x+3}{5}=-7$ and passing through $(-5,0)$.
6. Parallel to $6y-4=-3(1-2x)$ and passing through $(-2,-2)$.

Each set of four ordered pairs below defines the vertices, in counterclockwise order, of a quadrilateral. Determine if the quadrilateral is a parallelogram. See Example 2.

7. $\{(-2,2),(-5,-2),(2,-3),(5,1)\}$
8. $\{(-1,6),(-4,7),(-2,3),(1,1)\}$
9. $\{(-3,3),(-2,-2),(3,-1),(2,4)\}$
10. $\{(-2,-3),(-3,-6),(1,-2),(2,1)\}$

Determine if the two lines in each problem below are parallel.

11. $2x-3y=(x-1)-(y-x)$ and $-2y-x=9$
12. $\frac{2x-3y}{3}=\frac{x-1}{6}$ and $2y-x=3$
13. $x-5y=2$ and $5x-y=2$
14. $3-(2y+x)=7(x-y)$ and $\frac{5y+1}{4}=3+2x$
15. $\frac{x-y}{2}=\frac{x+y}{3}$ and $\frac{2x+3}{5}-4y=1+2y$
16. $6=-12(x-y)+y$ and $13y=-12x+3$

Find the equation, in slope-intercept form, for the line perpendicular to the given line and passing through the indicated point. See Example 3.

17. Perpendicular to $3x+2y=3y-7$ and passing through $(3,-2)$.
18. Perpendicular to $-y+3x=5-y$ and passing through $(-2,7)$.
19. Perpendicular to $x+y=5$ and passing through the origin.
20. Perpendicular to $x=\frac{1}{4}y-3$ and passing through $(1,-1)$.

21. Perpendicular to $6y+2x=1$ and passing through $(-4,-12)$.

22. Perpendicular to $2(y+x)-3(x-y)=-9$ and passing through $(2,5)$.

Determine if the two lines in each problem below are perpendicular. See Example 4.

23. $x-5y=2$ and $5x-y=2$

24. $3x+y=2$ and $x+3y=2$

25. $\frac{3x-y}{3}=x+2$ and $x=9$

26. $5x-6(x+1)=2y-x$ and $2y-(x+y)=4y+x$

27. $\frac{x-1}{2}+\frac{3y+2}{3}=-9$ and $3y-5x=x+5$

28. $-x=-\frac{2}{5}y+2$ and $5y=2x$

Each set of four ordered pairs below defines the vertices, in counterclockwise order, of a quadrilateral. Use the ideas from this section to determine if the quadrilateral is a rectangle.

29. $\{(-2,2),(-5,-2),(2,-3),(5,1)\}$

30. $\{(2,-1),(-2,1),(-3,-1),(1,-3)\}$

31. $\{(1,2),(3,-3),(9,-1),(7,4)\}$

32. $\{(5,-7),(1,-13),(28,-31),(32,-25)\}$

33. A construction company is building a new suspension bridge that has support cables attached to a center tower at various heights. One cable is attached at a height of 30 feet and connects to the ground 50 feet from the base of the tower. If the support cables should run parallel to each other, how far from the base should the company attach a cable whose other end is connected to the tower at a height of 25 feet?

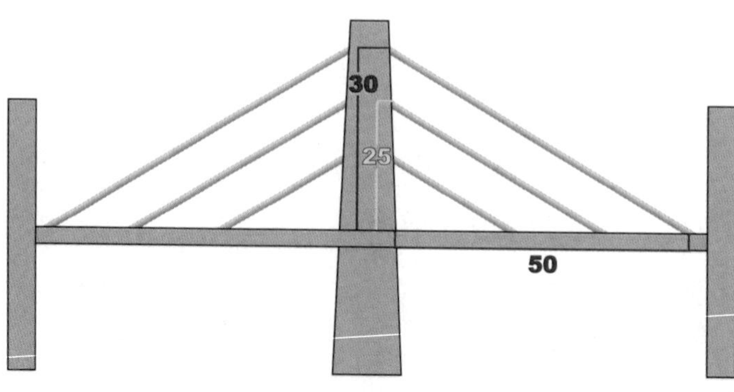

34. A light hits a mirror and is reflected off the mirror at a trajectory perpendicular to the original beam of light. If the line formed by the original beam of light can be described by the equation $y = -3.2x + b$, write a linear equation to describe the trajectory formed by the reflected beam.

technology exercises

Use a computer algebra system to determine the slopes of each of the following lines, as illustrated in the Technology Exercises of Section 2.3. Then use the results to classify the lines in each pair as parallel, perpendicular, or neither.

As an example, we know the two equations in the illustration below correspond to perpendicular lines, as the slopes are negative reciprocals of one another ($\frac{6}{5}$ and $-\frac{5}{6}$ respectively). *(See Appendix A for guidance on using computer algebra systems like Mathematica.)*

```
Section 2-4 Technology Exercise.nb

In[1]:= Solve[(2 x - y) / 2 + 4 == (y - 1) / 3, y]

Out[1]= {{y → -6/5 (-13/3 - x)}}

In[2]:= Solve[5 x + 6 y == -23, y]

Out[2]= {{y → 1/6 (-23 - 5 x)}}

100%
```

35. $3y = 2x - 7$ and $3x + 2y + 8 = 0$

36. $\frac{x-3y}{2}+\frac{y}{5}=2x-\frac{y}{3}$ and $29y + 45x = 17$

37. $\frac{2x-y}{3}+1=\frac{4y-x}{5}$ and $13y + 17x = 0$

38. $\frac{y-4x}{3}=x+2$ and $y-\frac{y-x}{5}=1-\frac{y}{2}$

39. $\frac{x-5}{3}+y=\frac{y-2x}{7}$ and $13y - 18x = 7$

40. $\frac{2y-x}{7}=\frac{x-2y}{3}+5$ and $2y = x - 5$

2.5 Linear Inequalities in Two Variables

TOPICS

1. Solving linear inequalities in two variables
2. Solving linear inequalities joined by "and" or "or"

Topic 1: Solving Linear Inequalities in Two Variables

Just as linear inequalities in one variable are similar to linear equations in one variable, linear inequalities in two variables have much in common with linear equations in two variables. The first similarity is in the definition: if the equality symbol in a linear equation in two variables is replaced with $<$, $>$, $\leq$, or $\geq$, the result is a **linear inequality in two variables.** In other words, a linear inequality in the two variables x and y is an inequality that can be written in the form

$$ax + by < c, \quad ax + by > c, \quad ax + by \leq c, \quad \text{or } ax + by \geq c,$$

where a, b, and c are constants and a and b are not both 0.

Another similarity lies in the solution process. The solution set of a linear inequality in two variables consists of all the ordered pairs in the Cartesian plane that lie on one side of a line in the plane, possibly including those points on the line. The first step in solving such an inequality then is to identify and graph this line. This is easily done, as the line is simply the graph of the equation that results from replacing the inequality symbol in the original problem with the equality symbol.

Any line in the Cartesian plane divides the plane into two **half-planes**, and in the context of linear inequalities all of the points in one of the two half-planes will solve the inequality. In addition, the points on the line (called the **boundary line** in this context) will also solve the inequality if the inequality symbol is $\leq$ or $\geq$, and this fact must somehow be denoted graphically. The entire process of solving linear inequalities in two variables, including the graphing conventions, is summarized on the next page.

Solving Linear Inequalities in Two Variables

Step 1: Graph the line in $\mathbb{R}^2$ that results from replacing the inequality symbol with =. Make the line solid if the inequality symbol is ≤ or ≥ (non-strict) and dashed if the symbol is < or > (strict). A solid line indicates that points on the line will be included in the eventual solution set, while a dashed line indicates that points on the line are to be excluded from the solution set.

Step 2: Determine which of the half-planes defined by the boundary line solve the inequality by substituting a **test point** from one of the two half-planes into the inequality. If the resulting numerical statement is true, all the points in the same half-plane as the test point solve the inequality. Otherwise, the points in the other half-plane solve the inequality. Shade in the half-plane that solves the inequality.

example 1

The line below was entered in the usual manner by selecting **Y=**. Before selecting **Graph**, choose the area to shade by pressing the left arrow until the cursor is to the left of Y. Then select **Enter** until the correct area is shaded.

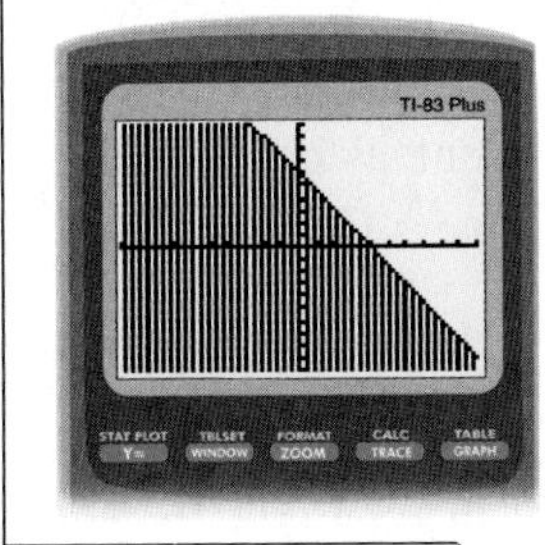

Solve the following linear inequalities by graphing their solution sets.

a. $3x+2y<12$ **b.** $x-y\le 0$ **c.** $x>3$

Solutions:

a. $3x+2y<12$

x-intercept of boundary: $(4,0)$

y-intercept of boundary: $(0,6)$

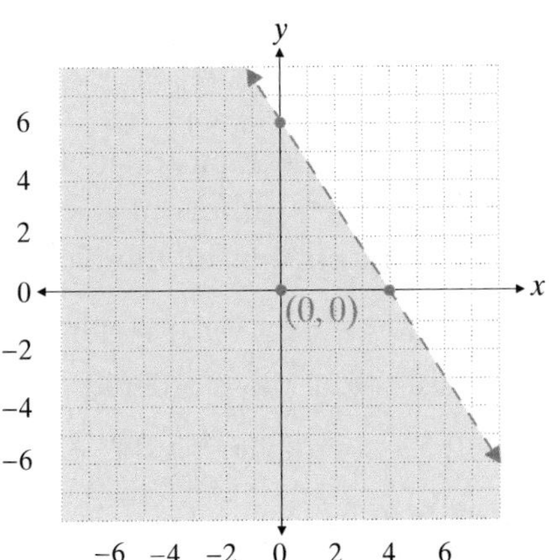

We begin by constructing the linear equation that defines the boundary line of the solution set. The equation $3x+2y=12$ is graphed by locating the x- and y-intercepts, as we have done at the left, or by determining the slope and y-intercept. The graph of the equation is then drawn with a dashed line since the inequality is strict.

A good test point in this problem is $(0,0)$, as it clearly lies on one side of the boundary line. When both x and y in the inequality are replaced with 0, we obtain the statement $0<12$, which is true. This tells us to shade the half-plane that contains $(0,0)$.

cont'd. on next page ...

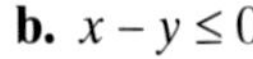

b. $x - y \leq 0$

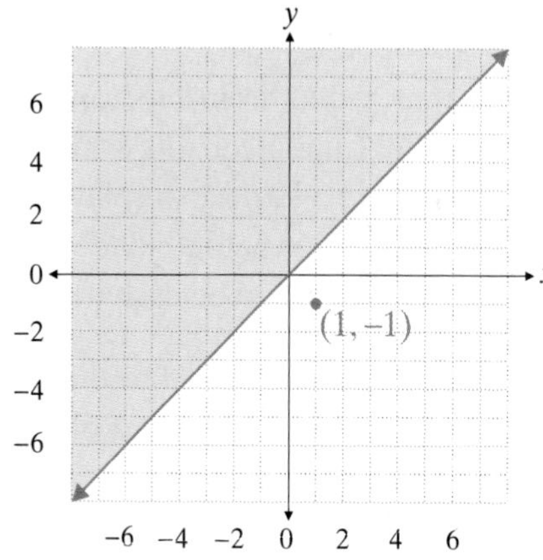

As in the first problem, we begin by graphing the boundary line. In slope-intercept form, the equation for the boundary is $y = x$, a line with a y-intercept of 0 and a slope of 1. The graph of the boundary has been drawn with a solid line since the inequality is not strict; that is, points on the boundary will solve the inequality.

This time, the origin cannot be used as a test point, as it lies directly on the boundary line. The point $(1, -1)$ clearly lies below the boundary, and if the values $x = 1$ and $y = -1$ are substituted into the inequality $x - y \leq 0$, we obtain the false statement $2 \leq 0$. Thus, the half-plane that does not contain $(1, -1)$ is the one we shade.

c. $x > 3$

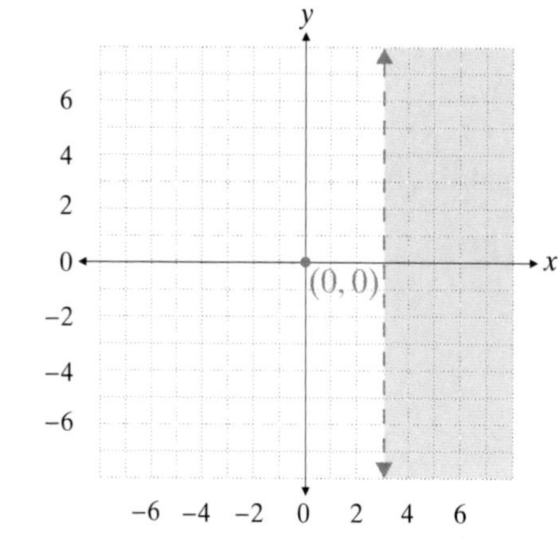

In this last example, the boundary line is a vertical line passing through 3 on the x-axis. It has been drawn with a dashed line since the inequality is strict.

The origin can be used as a convenient test point, and results in the false statement $0 > 3$, leading us to shade in the half-plane to the right of the boundary.

As an alternative to the test-point method demonstrated above, consider each of the inequalities in Example 1 after solving for y (or for x if y is not present, as in Example 1c). The inequality in Example 1a is satisfied by all those ordered pairs for which y is less than $-\frac{3}{2}x + 6$. That is, if the line $y = -\frac{3}{2}x + 6$ is graphed (with a dashed line), the solution set of the inequality consists of all those ordered pairs whose y-coordinate is *less than* (lies below) $-\frac{3}{2}$ times the x-coordinate, plus 6. Similarly, in Example 1b, we shade in those ordered pairs for which the y-coordinate is *greater than or equal to* the x-coordinate. Finally, in Example 1c, we shade in the ordered pairs for which the x-coordinate is *greater than* (lies to the right of) 3.

Topic 2: Solving Linear Inequalities Joined by "And" or "Or"

It is often necessary, in a variety of applications, to identify those points in the Cartesian plane that satisfy more than one condition. In the problems that we will examine, we will want to identify the portion of the plane that satisfies two or more linear inequalities joined by the word "and" or by the word "or." You may want to review the basic set operations of union and intersection that were discussed in Section 1.1. In that section, we defined the union of two sets A and B, denoted $A \cup B$, as the set containing all elements that are in set A **or** set B, and we defined the intersection of two sets A and B, denoted $A \cap B$, as the set containing all elements that are in both A **and** B. If we let A denote the portion of $\mathbb{R}^2$ that solves one inequality and B the portion of $\mathbb{R}^2$ that solves a second inequality, $A \cup B$ then represents the solution set of the two inequalities joined by the word "or" and $A \cap B$ represents the solution set of the two inequalities joined by the word "and." Keep in mind that $A \cup B$ *contains* both of the sets A and B (and so is at least as large as either one individually), while $A \cap B$ is *contained in* both A and B (and so is no larger than either individual set).

To find the solution sets in the following problems, we will solve each linear inequality individually and then form the union or the intersection of the individual solutions, as appropriate.

example 2

Choose the area to shade for each equation from the Y= screen by pressing the left arrow until the cursor is to the left of Y. Then select Enter until the correct area is shaded.

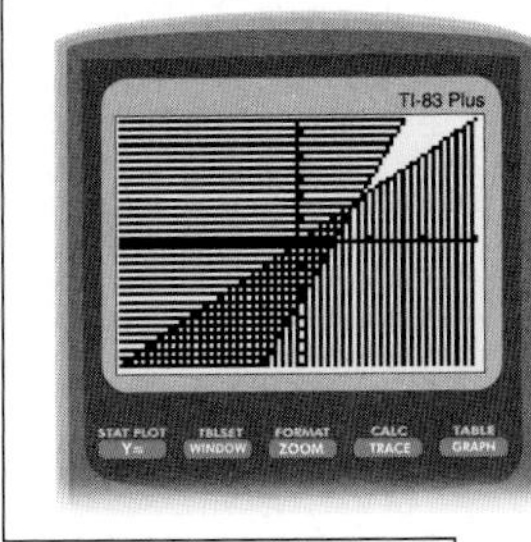

Graph the solution sets that satisfy the following inequalities.

a. $5x - 2y < 10$ and $y \le x$

b. $x + y < 4$ or $x \ge 4$

Solutions:

a. $5x - 2y < 10$ and $y \le x$

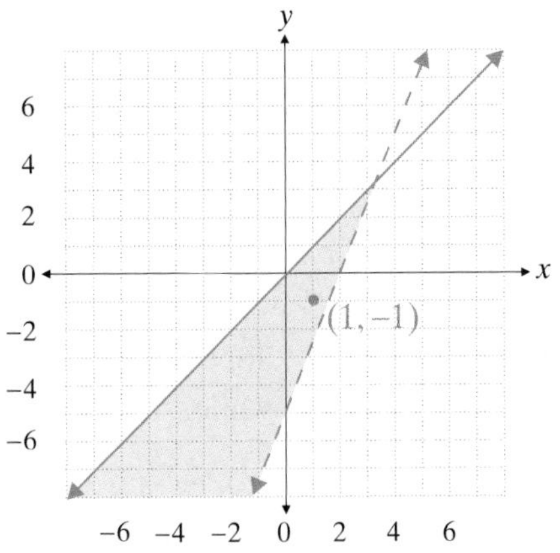

To solve the first inequality, we first graph the line $5x - 2y = 10$ with a dashed line and the line $y = x$ with a solid line. We then note that the half-plane lying above $5x - 2y = 10$ solves the first inequality (verify this by using the test point $(0, 0)$) and that the half-plane lying below $y = x$ solves the second inequality (this can be verified by using a test point such as $(1, -1)$).

Since the two inequalities are joined by the word "and", we shade in the *intersection* of the two half-planes, resulting in the picture at left.

cont'd. on next page ...

b. $x + y < 4$ or $x \geq 4$

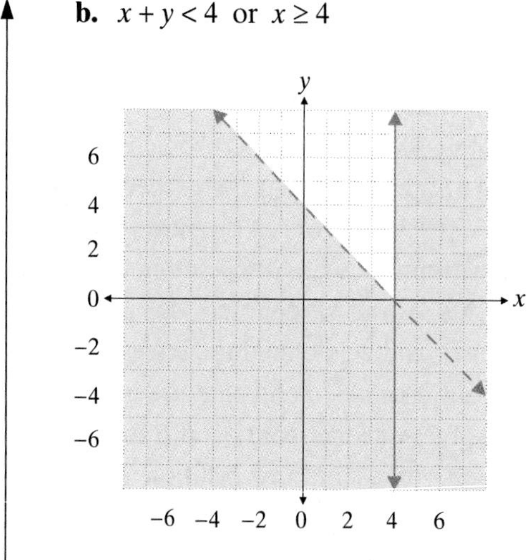

As always, we begin by solving each inequality separately. We graph the line $x + y = 4$ with a dashed line, and the line $x = 4$ with a solid line, and note that the half-plane below $x + y = 4$ solves the first inequality and that the half-plane to the right of $x = 4$ solves the second.

Since the two inequalities are joined by "or," we proceed to shade the *union* of the two half-planes. Note that this results in a larger solution set than either of the two individual solutions. Any ordered pair in the shaded region will satisfy one or the other (or both) of the inequalities.

Unions and intersections of regions of the plane can also arise when solving inequalities involving absolute values. In Section 1.6, we saw that an inequality of the form $|x| < a$ can be rewritten as the compound inequality $-a < x < a$. This in turn can be rewritten as the joint condition $x > -a$ and $x < a$, so an absolute value inequality of this form corresponds to the intersection of two sets. Similarly, an inequality of the form $|x| > a$ can be rewritten as $x < -a$ or $x > a$, so the solution of this form of absolute value inequality is a union of two sets.

If we are working in the Cartesian plane, solutions of such inequalities will be, respectively, intersections and unions of half-planes. The next example demonstrates how such solution sets can be found.

example 3

Graph the solution set in $\mathbb{R}^2$ that satisfies the joint conditions $|x-3| > 1$ and $|y-2| \leq 3$.

Solution:

We need to identify all ordered pairs (x, y) for which $x - 3 < -1$ or $x - 3 > 1$ while $-3 \leq y - 2 \leq 3$. That is, we need $x < 2$ or $x > 4$ while $-1 \leq y \leq 5$. The solution sets of the two conditions individually are:

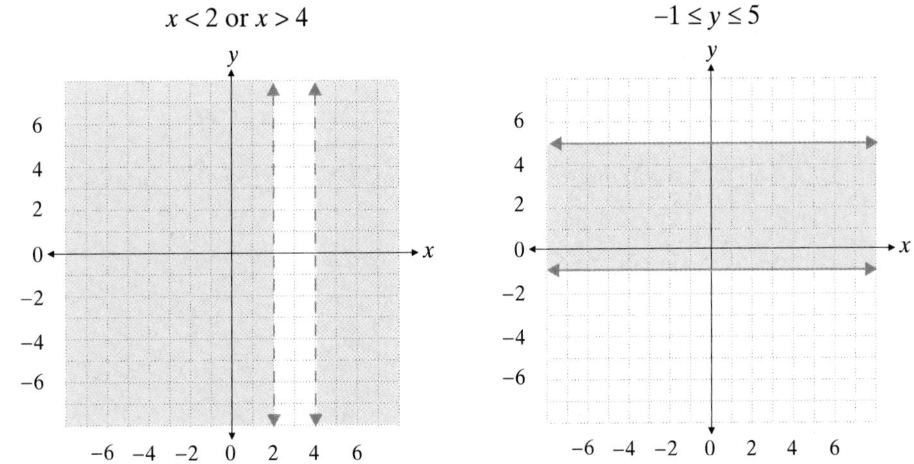

We now intersect the solution sets to obtain the final answer:

$$|x-3| > 1 \text{ and } |y-2| \le 3$$

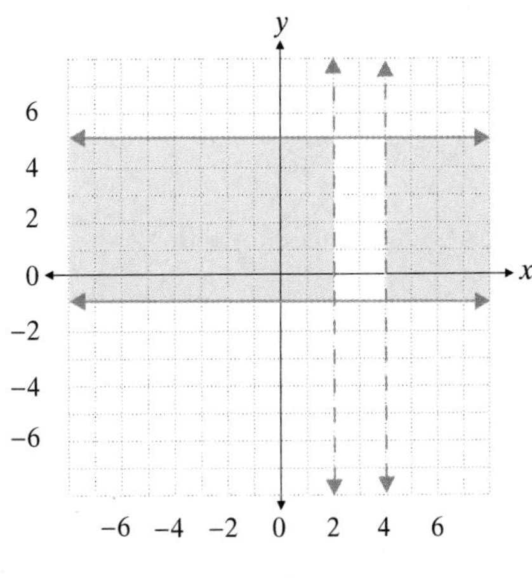

We will return to the notion of finding solutions to two or more inequalities when we study *linear programming* in Section 10.7.

exercises

Graph the solution sets of the following linear inequalities. See Example 1.

1. $x-3y<6$ **2.** $x-3y\geq 6$ **3.** $3x-y\leq 2$

4. $\frac{2x-y}{4}>1$ **5.** $y<-2$ **6.** $x+1\geq 0$

7. $x+y<0$ **8.** $x+y>0$ **9.** $y<2x-1$

10. $-(y-x)>-\frac{5}{2}-y$ **11.** $-2y\leq -x+4$ **12.** $x>\frac{3}{4}y$

13. $5(y+1)\geq -x$ **14.** $3x-7y\geq 7(1-y)+2$ **15.** $x-y<2y+3$

Graph the solution sets of the following linear inequalities. See Example 2.

16. $x-3y\geq 6$ and $y>-4$ **17.** $x-3y\geq 6$ or $y>-4$

18. $3x-y\leq 2$ and $x+y>0$ **19.** $x>1$ and $y>2$

20. $x>1$ or $y>2$ **21.** $x+y>-2$ and $x+y<2$

22. $y>-3x-6$ or $y\leq 2x-7$ **23.** $y\geq -2$ and $y>1$

24. $y>-2$ and $2y>-3x-4$ **25.** $y\geq -2x-5$ and $y\leq -6x-9$

26. $3y>x+2$ or $4y\leq -x-2$ **27.** $y\leq -x$ and $2y+3x>-4$

Graph the solution sets of the following absolute value inequalities. See Example 3.

28. $|x-3|<2$ **29.** $|x-3|>2$ **30.** $|3y-1|\leq 2$

31. $|2x-4|>2$ **32.** $1-|y+3|<-1$ **33.** $|x+1|<2$ and $|y-3|\leq 1$

34. $|x-3|\geq 1$ or $|y-2|\leq 1$ **35.** $|x-y|<1$ **36.** $|x+y|\geq 1$

37. $|4x-2y-3|\leq 5$ **38.** $|2x-3|\geq 1$ or $|2y+3|\geq 1$ **39.** $|y-3x|\leq 2$ and $|y|<2$

40. It costs Happy Land Toys \$5.50 in variable costs per doll produced. If fixed costs must remain less than \$200 dollars, write a linear inequality describing the relationship between cost and dolls produced.

41. Trish is having a garden party where she wants to have several arrangements of lilies and orchids for decoration. The lily arrangements cost \$12 apiece and the orchids cost \$22 each. If Trish wants to spend less than \$150 on flowers, write a linear inequality describing the number of each arrangement she can purchase.

42. Flowertown Canoes produces two types of canoes. The two person model costs \$73 to produce and the one person model costs \$46 to produce. Write a linear inequality describing the number of each canoe the company can produce and keep costs under \$1750. Graph the inequality.

43. Rob has 300 feet of fencing he can use to enclose a small rectangular area of his yard for a garden. Assuming Rob may or may not use all of the fence, write a linear inequality describing the possible dimensions of his garden. Graph the inequality.

2.6 Introduction to Circles

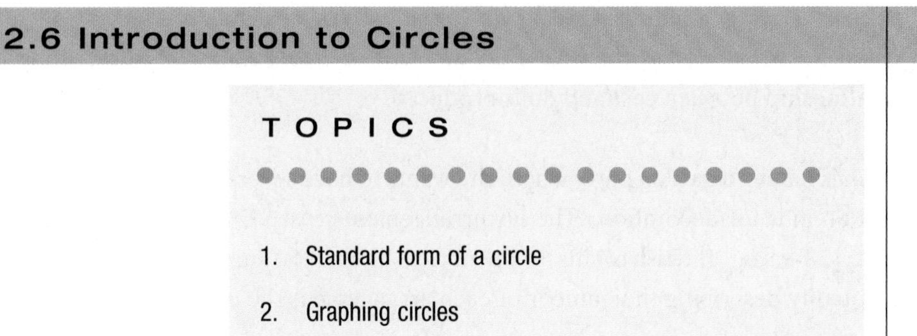

TOPICS
1. Standard form of a circle
2. Graphing circles

Topic 1: Standard Form of a Circle

The focus of this chapter so far has been on equations that define lines in $\mathbb{R}^2$. But as you no doubt suspect, linear equations are just the beginning. Much of the rest of this text will deal, directly or indirectly, with more interesting graphs in the plane. As a preview of the discussion to come, we will close out this chapter with a brief introduction to a particular class of nonlinear equations.

Circles in $\mathbb{R}^2$ can be visualized, graphed, and described mathematically. To begin with, note that two pieces of information are all we need to completely characterize a particular circle: the circle's center, and the circle's radius. To be specific, suppose (h, k) is the ordered pair corresponding to the circle's center, and suppose the radius is given by the positive real number r. Our goal is to develop an equation in the two variables x and y so that every solution (x, y) of the equation corresponds to a point on the circle.

The main tool that we need to achieve this goal is the distance formula derived in Section 2.1. Since every point (x, y) on the circle is a distance r from the circle's center (h, k), that formula tells us that

$$r = \sqrt{(x-h)^2 + (y-k)^2}.$$

This form of the equation for a circle is in many ways the most natural, but such equations are commonly presented in the radical-free form that results from squaring both sides:

$$r^2 = (x-h)^2 + (y-k)^2.$$

Standard Form of a Circle

The **standard form** of the equation for a circle of radius r and center (h, k) is

$$(x-h)^2+(y-k)^2=r^2.$$

example 1

Find the standard form of the equation for the circle with radius 3 and center $(-2, 7)$.

Solution:

We are given $h=-2$, $k=7$, and $r=3$, so the equation is determined to be

$$(x-(-2))^2+(y-7)^2=3^2.$$

This is better written as

$$(x+2)^2+(y-7)^2=9.$$

example 2

Find the standard form of the equation for each of the following circles:

a. A circle with a diameter whose endpoints are $(-4, -1)$ and $(2, 5)$.

b. A circle which is tangent to the line $x=-1$ and whose center is $(3, 5)$.

Solutions:

a. The midpoint of a diameter of a circle is the circle's center, so the first step is to use the Midpoint Formula as follows:

$$(h, k)=\left(\frac{-4+2}{2}, \frac{-1+5}{2}\right)=(-1, 2).$$

The distance from either diameter endpoint to the center defines the circle's radius. Since we ultimately will want r^2, we can use a slight variation of the Distance Formula to determine

$$r^2=(-4-(-1))^2+(-1-2)^2=9+9=18.$$

Thus, the equation we seek is

$$(x+1)^2+(y-2)^2=18.$$

Note that $(-4, -1)$ and $(2, 5)$, two points that by definition are on the circle, both satisfy this equation.

cont'd on next page ...

b. The word tangent in this context means that the circle just touches the line $x = -1$. In fact, by referring to a sketch, we can see that the described circle must touch the vertical line $x = -1$ at the point $(-1, 5)$. The distance between the center $(3, 5)$ and the point $(-1, 5)$ must then correspond to the radius of the circle, giving us $r = 4$. At this point, we are ready to construct the equation of the circle:

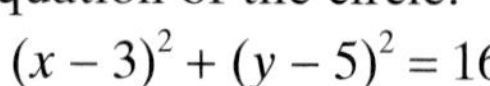

$$(x-3)^2 + (y-5)^2 = 16.$$

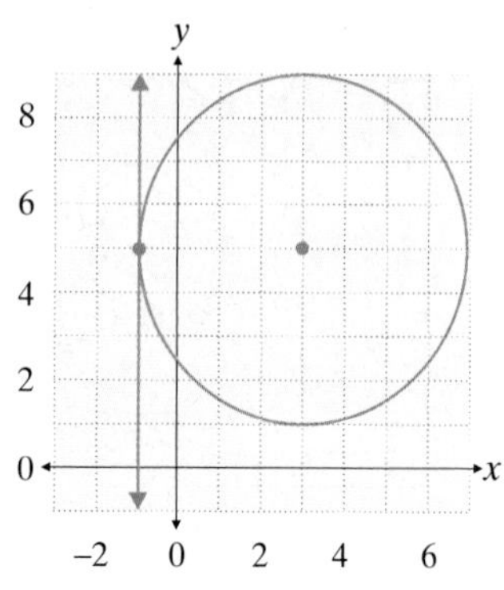

Topic 2: Graphing Circles

We will often need to reverse the process illustrated in the first two examples. That is, given an equation for a circle, we will need to determine the circle's center and radius and, possibly, graph the circle. If the equation is given in standard form, this is very easily accomplished.

example 3

Sketch the graph of the circle defined by

$$(x-2)^2 + (y+3)^2 = 4.$$

Solution:

The only preliminary step required is to slightly rewrite the equation in the form

$$(x-2)^2 + (y-(-3))^2 = 2^2.$$

From this, we see that $(h, k) = (2, -3)$ and that $r = 2$. The graph of the equation is thus

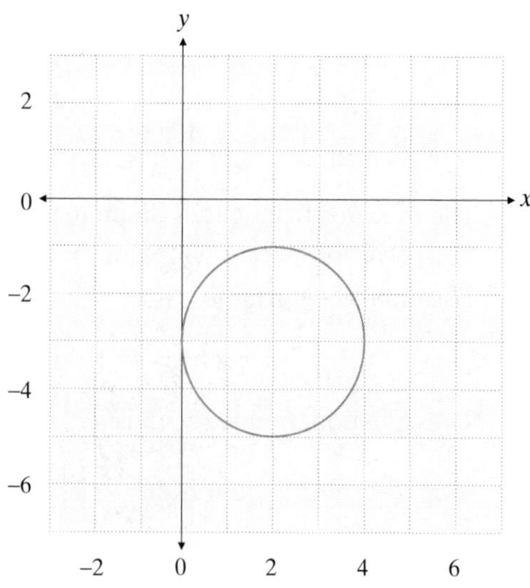

Typically, though, the equation for a circle will not be given to us in quite so neat a fashion. We may have to resort to a small amount of algebraic manipulation in order to determine that a given equation describes a circle and to determine the specifics of that circle. Fortunately, the algebraic technique of *completing the square* (Section 1.7) is usually all that is required.

example 4

Sketch the graph of the equation

$$x^2 + y^2 + 8x - 2y = -1.$$

Solution:

We need to complete the square in the variable x and the variable y, and we do so as follows:

$$(x^2 + 8x + 16) + (y^2 - 2y + 1) = -1 + 16 + 1 \qquad \text{Add 16 and 1 to both sides.}$$

$$(x+4)^2 + (y-1)^2 = 16.$$

We now see that the equation does indeed describe a circle, and that the center of the circle is (–4, 1) and the radius is 4. The graph appears below.

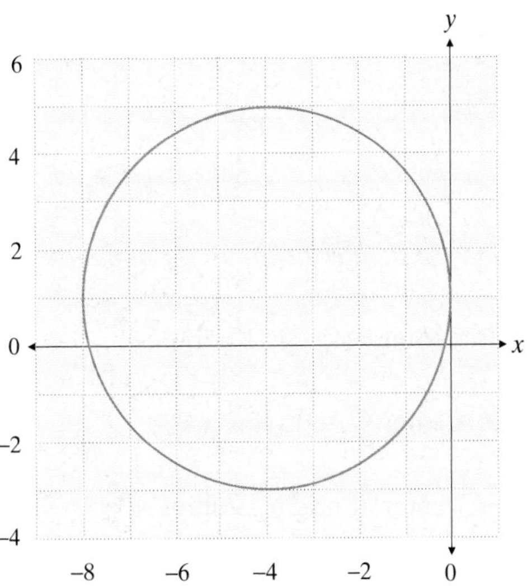

technology note

Graphing calculators and computer algebra systems, such as *Mathematica*, can be used to sketch accurate graphs of circles. To use *Mathematica* to do so, the command `<<Graphics`ImplicitPlot`` must be executed. This command allows *Mathematica* to graph equations, and needs to be executed just once per *Mathematica* session. The command used to then graph an equation is `ImplicitPlot`, and the details of its use are shown below.

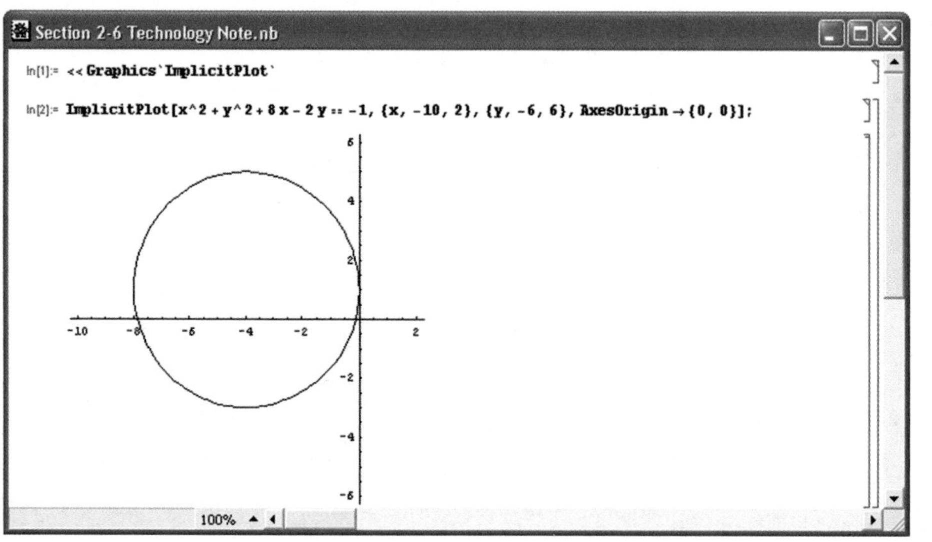

exercises

Find the standard form of the equation for the circle. See Examples 1 and 2.

1. Center $(-4, -3)$; Radius 5

2. Center $(0, 0)$; Radius $\sqrt{6}$

3. Center $(7, -9)$; Radius 3

4. Center $\left(-\sqrt{4}, 2\right)$; Radius 2

5. Center at origin; Radius $\sqrt{9}$

6. Center $(6, 3)$; Radius 8

7. Center $\left(\sqrt{5}, \sqrt{3}\right)$; Radius 4

8. Center $\left(\frac{5}{3}, \frac{8}{5}\right)$; Radius $\sqrt{8}$

9. Center $(7, 2)$; tangent to the x-axis

10. Center at $(3, 3)$; tangent to the x-axis

11. Center at $(-3, 8)$; passes through $(-4, 9)$

12. Center at $(0, 0)$; passes through $(2, 10)$

13. Center $(4, 8)$; passes through $(1, 9)$

14. Center $(12, -4)$; passes through $(-9, 5)$

15. Center at the origin; passes through $(6, -7)$

16. Center $(13, -2)$; passes through $(8, -3)$

17. Endpoints of a diameter are $(-8, 6)$ and $(1, 11)$

18. Endpoints of a diameter are $(5, 3)$ and $(8, -3)$

19. Endpoints of a diameter are $(-7, -4)$ and $(-5, 7)$

20. Endpoints of a diameter are $(2, 3)$ and $(7, 4)$

21. Endpoints of a diameter are $(0, 0)$ and $(-13, -14)$

22. Endpoints of a diameter are $(4, 10)$ and $(0, 3)$

23. Endpoints of a diameter are $(0, 6)$ and $(8, 0)$

24. Endpoints of a diameter are $(6, 9)$ and $(4, 9)$

25.

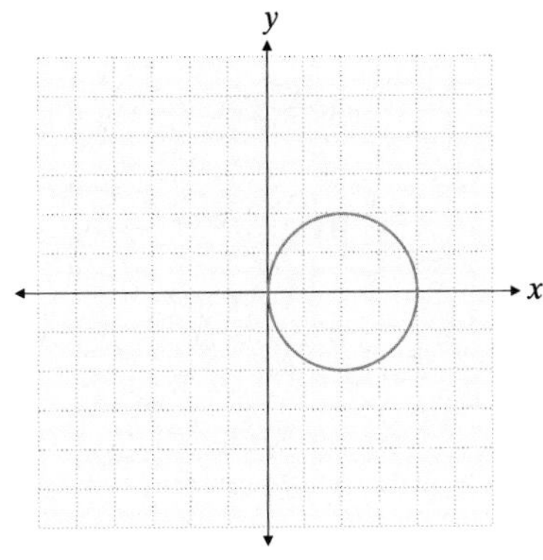

26.

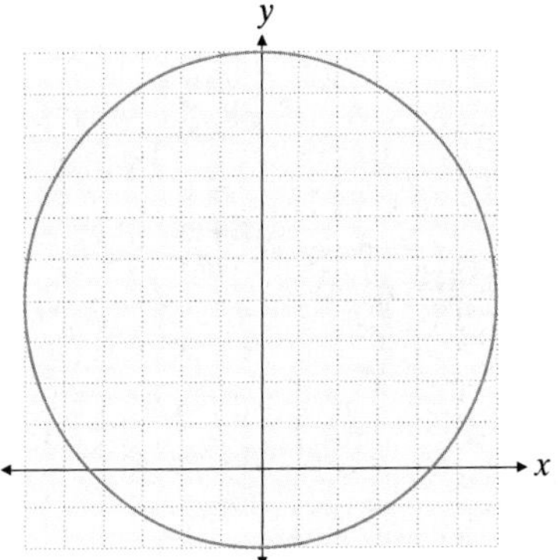

27.

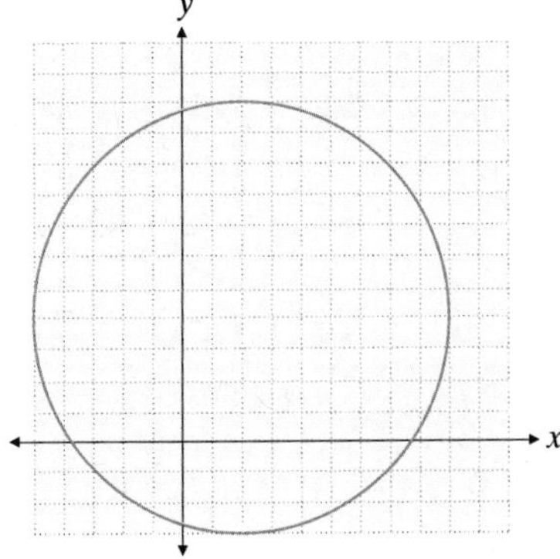

28.

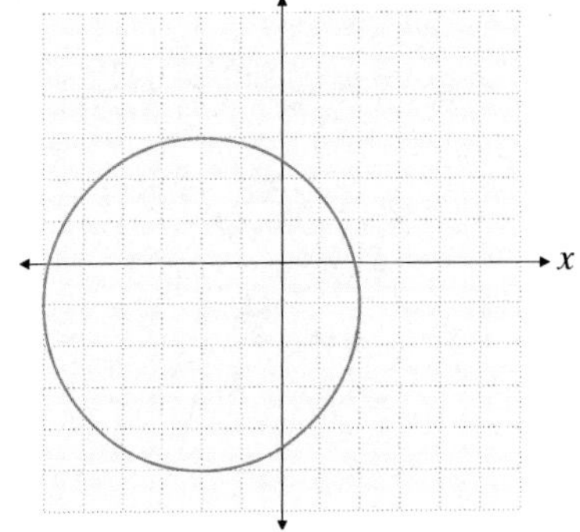

29.

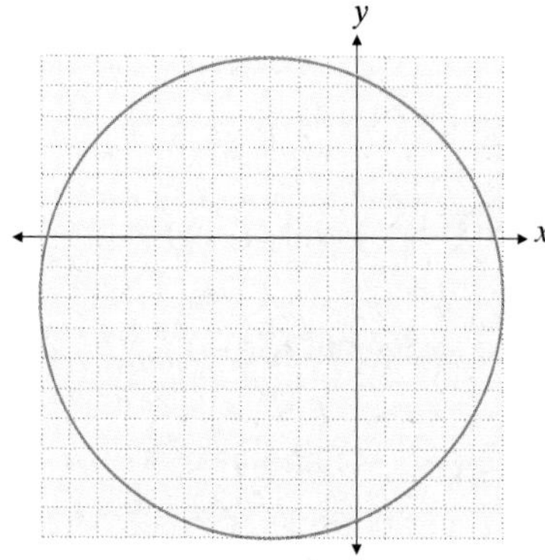

Sketch a graph of the equation. Then use a graphing calculator or computer to graph the circle. See Examples 3 and 4.

30. $x^2 + y^2 = 25$

31. $x^2 + y^2 = 36$

32. $x^2 + (y - 3)^2 = 16$

33. $x^2 + (y - 8)^2 = 9$

34. $(x + 2)^2 + y^2 = 169$

35. $(x - 8)^2 + y^2 = 8$

36. $(x - 9)^2 + (y - 4)^2 = 49$

37. $(x + 5)^2 + (y + 4)^2 = 4$

38. $(x + 2)^2 + (y - 7)^2 = 64$

39. $(x - 5)^2 + (y + 5)^2 = 5$

40. $x^2 + y^2 - 2x + 10y + 1 = 0$

41. $x^2 + y^2 - 4x + 4y - 8 = 0$

42. $x^2 + y^2 + 6x + 5 = 0$

43. $x^2 + y^2 + 10y + 9 = 0$

44. $x^2 + y^2 - x - y = 2$

45. $x^2 + y^2 + 6y - 2x = -2$

46. $(x - 5)^2 + y^2 = 225$

47. $4x^2 + 4y^2 = 256$

48. $(x - 3)^2 + (y + 2)^2 = 81$

49. $x^2 + y^2 - 6x + 4y - 3 = 0$

50. $(x + 2)^2 + (y - 1)^2 = 16$

51. $(x - 1)^2 + y^2 = 9$

52. $x^2 + (y + 2)^2 = 49$

53. $x^2 + y^2 - 4x + 8y - 16 = 0$

54. $x^2 + y^2 + 8x = 9$

55. $4x^2 + 4y^2 - 24x + 24y = 28$

chapter two project

Linear Equations and Inequalities

You have just been put in charge of production and sales of the newest pair of jeans at Strolling Along Jeans Company. In order to assure your product's success you need to examine what its demand may be so that you can determine how well it will sell. By researching sales of similar jeans produced by your competitors, you know that at a particular store if the price is \$30 dollars for a pair of jeans, then you will sell about 50 pairs. If the price is \$15 dollars then you will sell about 125 pairs of jeans.

1. Assuming the relationship between the price of the jeans and the number of sales is approximately linear, what is the slope of this function? What is the meaning of the slope in this context?

2. Determine the equation that describes this relationship and graph the first quadrant. (You can't have a negative amount of jeans or a negative price, so the other quadrants do not make sense.)

This graph is an example of a demand curve. The principles of supply and demand are essential to many areas of microeconomics. Economists use demand curves to determine total revenue. Total revenue (the total money made) is found by selecting a point (x, y) on the graph (where x represents price and y represents quantity sold) and multiplying x and y: TR = (price)(quantity).

3. At what two points on the demand curve will your total revenue be zero?

4. Discuss the advantages and disadvantages of charging a higher price for a pair of jeans.

5. It costs \$8.50 to produce a pair of jeans and you need to keep fixed costs under \$600 dollars. Write and graph the inequality describing the relationship between cost and pairs of jeans produced.

6. If you charge \$12 dollars for a pair of jeans what will be your total revenue according to the demand curve? What will be your total cost to produce those jeans assuming that your fixed costs ended up at \$525? What would be your profit, if Profit = Total Revenue – Total Cost? Do you think this is a good price to charge for the jeans, why or why not?

CHAPTER REVIEW

2.1 The Cartesian Coordinate System

topics	pages	test exercises
The components of the Cartesian coordinate system • The meaning of the *Cartesian plane*, as well as the terms *axes* (the two perpendicular real number lines), *origin* (the point of intersection of the two axes), *quadrant* (the four quarters defined by the axes), and *ordered pair* (pair of numbers identifying points in the plane) • The relation between ordered pairs and points in the plane	p. 129 – 131	1 – 4
The graph of an equation • The correspondence between an equation in two variables and its *graph* in the Cartesian plane	p. 131 – 134	5 – 7
The distance and midpoint formulas • The meaning of *distance* in the plane, and the use of the *distance formula* in determining distance: $d=\sqrt{(x_2-x_1)^2+(y_2-y_1)^2}$ • The use of the *midpoint formula* in finding the point midway between two given points in the plane: $\left(\frac{x_1+x_2}{2}, \frac{y_1+y_2}{2}\right)$	p. 134 – 136	8 – 12

2.2 Linear Equations in Two Variables

topics	pages	test exercises
Recognizing linear equations in two variables • The definition of a *linear equation in two variables* (an equation that can be written in the form $ax+by=c$) • The *standard form* of a linear equation: $ax+by=c$	p. 139 – 140	13 – 15
Intercepts of the coordinate axes • The geometric meaning of the *x-intercept* and the *y-intercept* of a line, and how to determine them	p. 141 – 142	16 – 18
Horizontal and vertical lines • The forms of equations that correspond to horizontal and vertical lines	p. 142 – 144	19 – 21

2.3 Forms of Linear Equations

topics	pages	test exercises
The slope of a line • The meaning of *slope* (the change in y over the change in x), and how to determine the slope of the line passing through two given points: $\frac{y_2 - y_1}{x_2 - x_1}$ • The meaning of a slope of zero and the meaning of an undefined slope	p. 146 – 149	22 – 24
Slope-intercept form of a line • The definition of *slope-intercept form* of a line: $y = mx + b$, where m is the slope and b is the y-intercept • Obtaining the slope-intercept form of a given line	p. 149 – 152	25 – 27
Point-slope form of a line • The definition of *point-slope form* of a line: $y - y_1 = m(x - x_1)$ • Using the point-slope form to obtain equations of lines with prescribed properties	p. 152 – 153	28 – 30

2.4 Parallel and Perpendicular Lines

topics	pages	test exercises
Slopes of parallel lines • The relationship between the slopes of parallel lines in the plane: $m_1 = m_2$	p. 157 – 158	31 – 36
Slopes of perpendicular lines • The relationship between the slopes of perpendicular lines in the plane: $m_1 = -\frac{1}{m_2}$ • The relationship between the slopes of perpendicular lines when one of the lines is vertical	p. 158 – 160	31 – 35

2.5 Linear Inequalities in Two Variables

topics	pages	test exercises
Solving linear inequalities in two variables • The meaning of a *linear inequality in two variables* and its solution set in the plane • The difference between *strict* and *non-strict* inequalities, and how the difference is indicated graphically • The use of *test points* in determining the solution set of an inequality	p. 164 – 166	37 – 40
Solving linear inequalities joined by "and" or "or" • The meaning of "and" and "or" when used to join two or more linear inequalities • The solution set of two or more joined inequalities	p. 167 – 169	37 – 39

2.6 Introduction to Circles

topics	pages	test exercises
Standard form of a circle • The definition of the standard form of the equation of a circle with radius r and center (h, k): $(x-h)^2+(y-k)^2=r^2$	p. 172 – 174	41
Graphing circles • Obtaining the center and radius of a circle given only the equation • Using method of *completing the square* to convert the equation of a circle to standard form	p. 174 – 176	42 – 43

chapter test

Section 2.1

Name the quadrant associated with each point.

1. $(-5, 6)$ **2.** $(0, 50)$ **3.** $(-2, -4)$ **4.** $(4, -3)$

Sketch graphs of the following equations by plotting points.

5. $5x + 2y = 20$ **6.** $y = x^2 + 2x + 3$ **7.** $x^2 + y^2 = 64$

*Determine **(a)** the distance between the following pairs of points, and **(b)** the midpoint of the line segment joining each pair of points.*

8. $(2, -6)$ and $(3, -7)$ **9.** $(-4, -3)$ and $(4, -9)$ **10.** $(-3, 6)$ and $(-7, 0)$

Find the perimeter of the triangle whose vertices are the specified points in the plane.

11. $(-3, 2)$, $(-3, 0)$, and $(-6, -3)$ **12.** $(8, -3)$, $(2, -3)$, and $(2, 5)$

Section 2.2

Determine if the following equations are linear.

13. $3x + y(4 - 2x) = 8$ **14.** $y - 3(y - x) = 8x$ **15.** $9x^2 - (3x + 1)^2 = y - 3$

Determine the x- and y-intercepts of the following linear equations.

16. $y = 3x$ **17.** $4x + 5y = 2 + 3y$ **18.** $5(x - y) - 3 = x + 17$

Graph the following.

19. $\frac{3}{4}y = 3$ **20.** $5(x - 3) - 2x = 7x + 1$ **21.** $5y + 3(x - 2y) - x = 2x + 10$

Section 2.3

Determine the slopes of the lines passing through the specified points.

22. $(-2, 5)$ and $(-3, -7)$ **23.** $(3, 6)$ and $(7, -10)$ **24.** $(3, 5)$ and $(3, -7)$

Use the slope-intercept form of each line to graph the equations.

25. $6x - 3y = 9$ **26.** $2y + 5x + 9 = 0$ **27.** $15y - 5x = 0$

Find the equation, in standard form, of the line passing through the specified points with the specified slopes.

28. point $(4, -1)$; slope of 1 **29.** point $(-2, 3)$; slope of $\frac{3}{2}$

Answer the following word problem.

30. The rental rate at Carrie's apartment home has increased at a linear rate since she moved in five years ago. If the rent per month when she moved in was \$480 and last year the rent was \$520/mo., how much can she expect to pay in rent per month this year?

Section 2.4

Determine if the two lines in each problem below are perpendicular, parallel, or neither.

31. $x - 4y = 3$ and $4x - y = 2$

32. $3x + y = 2$ and $x - 3y = 25$

33. $\frac{3x - y}{3} = x + 2$ and $\frac{y}{3} + x = 9$

Each set of four ordered pairs below defines the vertices, in counterclockwise order, of a quadrilateral. Determine if the quadrilateral is a rectangle.

34. $\{(-2, 1), (-1, -1), (3, 1), (2, 3)\}$

35. $\{(-2, 2), (-3, -1), (2, -3), (2, 1)\}$

Answer the following word problem.

36. A contractor is building a new 75-foot tall radio tower which needs several support cables to secure it in place. Suppose the support cables on one side run parallel to each other and are attached at the heights of 70 feet, 60 feet, and 50 feet. If the cable that is attached at a height of 50 feet is attached 20 feet from the base of the tower, how far from the base of the tower should the other two cables be attached?

Section 2.5

Graph the solution sets of the following linear inequalities.

37. $7x-2y\geq 8$ and $y<5$ **38.** $x-4y\geq 6$ or $y>-2$ **39.** $|2x+5|<3$

Answer the following word problem.

40. Tickets to a local play are \$10 for general admission and \$15 for premium seating. If the production company needs to make over \$1875 to make a profit, write an inequality expressing sales necessary to make a profit. Graph the inequality.

Section 2.6

41. Find the standard form of the equation for the circle with a diameter whose endpoints are $(1,2)$ and $(-5,8)$.

Sketch a graph of the circle defined by the given equation. Then state the radius and center of the circle.

42. $(x+5)^2+(y-2)^2=16$ **43.** $x^2+y^2+6x-10y=-5$

chapter THREE

RELATIONS, FUNCTIONS, AND THEIR GRAPHS

OVERVIEW

Many examples of variation exist in the natural and social realms. Distance fallen varies based on time traveled, demand varies based on supply, and the list goes on. By the end of this chapter you should be able to find the constant of proportionality given certain information about a variance, and formulate unknown information using this constant. On page 239 you will find a problem concerning a person's Body Mass Index (BMI) – a general indicator of acceptable weight based on height. Given the height, weight, and BMI of one person, you will be able to find the BMI of another, using tools such as the definition of *Direct and Inverse Variation* on pages 233 and 234.

Introduction

This chapter begins with a study of *relations*, which are generalizations of the equations in two variables discussed in Chapter 2, and then moves on to the more specialized topic of *functions*. As concepts, relations and functions are more abstract, but at the same time far more powerful and useful than the equations studied thus far in this text. Functions, in particular, lie at the heart of a great deal of the mathematics that you will encounter from this point on.

Leibniz

The history of the function concept serves as a good illustration of how mathematics develops. One of the first people to use the idea in a mathematical context was the German mathematician and philosopher Gottfried Leibniz (1646 - 1716), one of two people usually credited with the development of Calculus. Initially, Leibniz and other mathematicians tended to use the term to indicate that one quantity could be defined in terms of another by some sort of algebraic expression, and this (incomplete) definition of function is often encountered even today in elementary mathematics. As the problems that mathematicians were trying to solve increased in complexity, however, it became apparent that functional relations between quantities existed in situations where no algebraic expression defining the function was possible. One example came from the study of heat flow in materials, in which a description of the temperature at a given point at a given time was often given in terms of an infinite sum, not an algebraic expression.

The result of numerous refinements and revisions of the function concept is the definition that you will encounter in this chapter, and is essentially due to the German mathematician Lejeune Dirichlet (1805 - 1859). Dirichlet also refined our notion of what is meant by a *variable*, and gave us our modern understanding of *dependent* and *independent* variables, all of which you will encounter soon.

The proof of the power of functions lies in the multitude and diversity of their applications. And the subtle and easily overlooked advantage of functional notation deserves special mention; innovations in notation often go a long way toward solving difficult problems. In fact, if the notation is sufficiently advanced, the mere act of using the notation to state a problem does much of the work of actually solving it. As you work through Chapter 3, pay special attention to how functional notation works. A solid understanding of what functional notation means is essential to using functions.

3.1 Relations and Functions

TOPICS

1. Relations, domains, and ranges
2. Functions and the vertical line test
3. Functional notation and function evaluation
4. Implied domain of a function

Topic 1: Relations, Domains, and Ranges

In Chapter 2 we saw many examples of equations in two variables. Any such equation automatically defines a relation between the two variables present, in the sense that each ordered pair on the graph of the equation relates a value for one variable (namely, the first coordinate of the ordered pair) to a value for the second variable (the second coordinate). Many applications of mathematics involve relating one variable to another, and this notion merits much further study.

We begin by generalizing the above observation and define a relation to be simply a collection of ordered pairs; any collection of ordered pairs constitutes a relation, whether the ordered pairs correspond to the graph of an equation or not. This definition is repeated below, along with two related ideas.

Relations, Domains, and Ranges

A **relation** is a set of ordered pairs. Any set of ordered pairs automatically relates the set of first coordinates to the set of second coordinates, and these sets have special names. The **domain** of a relation is the set of all the first coordinates, and the **range** of a relation is the set of all second coordinates.

It is important to understand that relations can be described in many different ways. We have already noted that an equation in two variables describes a relation, as the solution set of the equation is a collection of ordered pairs. Relations can also be described with a simple list of ordered pairs (if the list is not too long), with a picture in the Cartesian plane, and by many other means.

The following example demonstrates some of the common ways of describing relations, and identifies the domain and range of each relation.

example 1

a. The set $R = \{(-4, 2), (6, -1), (0, 0), (-4, 0), (\pi, \pi^2)\}$ is a relation consisting of five ordered pairs. The domain of R is the set $\{-4, 6, 0, \pi\}$, as these four numbers appear as first coordinates in the relation. Note that it is not necessary to list the number -4 twice in the domain, even though it appears twice as a first coordinate in the relation. The range of R is the set $\{2, -1, 0, \pi^2\}$, as these are the numbers that appear as second coordinates. Again, it is not necessary to list 0 twice in the range, even though it is used twice as a second coordinate in the relation. The *graph* of this relation is simply a picture of the five ordered pairs plotted in the Cartesian plane, as shown below.

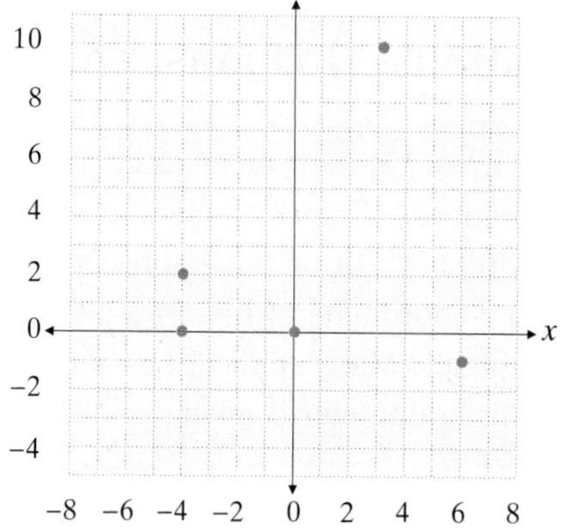

b. The equation $-3x + 7y = 13$ describes a relation. In contrast to the last example, this relation consists of an infinite number of ordered pairs, so it is not possible to list them all as a set. As an example, one of the ordered pairs in the relation is $(-2, 1)$, since $-3(-2) + 7(1) = 13$. Even though it is not possible to list all the elements of this relation, it is possible to graph it: it is the graph of the solution set of the equation, as drawn below. The domain and range of this relation are both the set of real numbers, as every real number appears as both a first coordinate and a second coordinate in the relation.

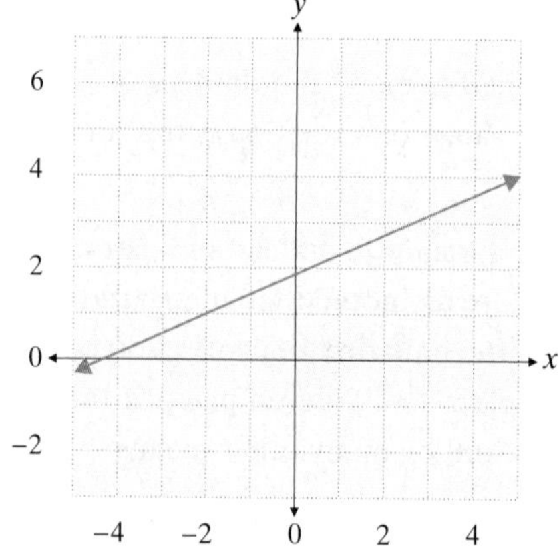

c. The picture below describes a relation. Some of the elements of the relation are $(-1, 1)$, $(-1, -2)$, $(-0.3, 2)$, $(0, -2)$, and $(1, -0.758)$, but this is another example of a relation with an infinite number of elements so we cannot list all of them. Using the interval notation defined in Section 1.1, the domain of this relation is the closed interval $[-1, 1]$ and the range is the closed interval $[-2, 2]$.

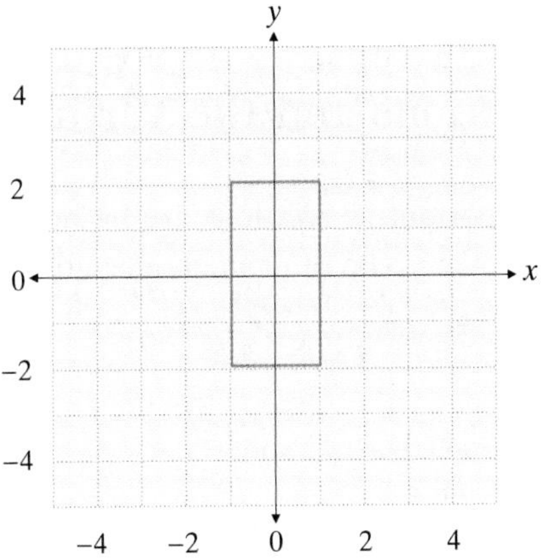

d. The picture below describes another relation, different from the last though with some similarities. The shading in the picture indicates that all ordered pairs lying inside the rectangle drawn in blue, as well as those actually *on* the rectangle, are elements of the relation. The domain is again the closed interval $[-1, 1]$ and the range is again the closed interval $[-2, 2]$, but this relation is not identical to the last example. For instance, the ordered pairs $(0, 0)$ and $(0.2, 1.5)$ are elements of this relation, but are not elements of the relation in Example 1c.

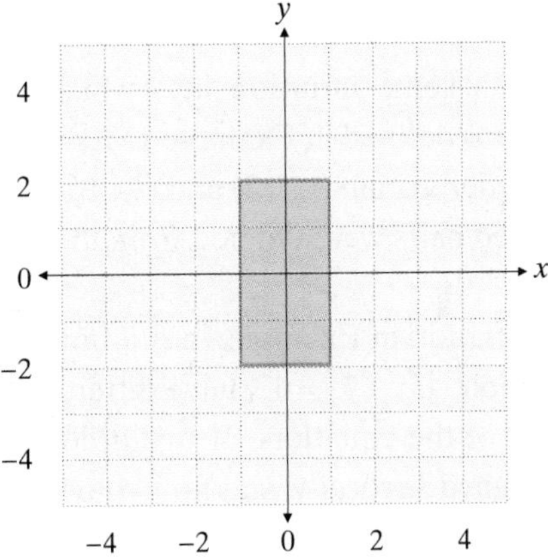

e. Although we will almost never encounter relations in this text that do not consist of ordered pairs of real numbers, there is nothing in our definition to prevent us from considering more exotic relations.

cont'd. on next page ...

For example, the set $S = \{(x, y) | x \text{ is the mother of } y\}$ is a relation among people. The domain of S is the set of all mothers, and the range of S is the set of all people. (Although advances in cloning are occurring rapidly, as of the writing of this text no one has yet been born without a mother!)

Topic 2: Functions and the Vertical Line Test

As important as relations are in mathematics, a subclass of relations, called functions, has proven to be of even greater utility.

Functions

A **function** is a relation in which every element of the domain is paired with *exactly one* element of the range. Equivalently, a function is a relation in which no two distinct ordered pairs have the same first coordinate.

Note that there is an asymmetry in the way domains and ranges are treated in the definition of function: the definition allows for the possibility that two distinct ordered pairs may have the same second coordinate, as long as their first coordinates differ.

To gain some intuition into which relations qualify as functions, consider the five relations defined in Example 1. Example 1a is *not* a function for the sole reason that the two ordered pairs $(-4, 2)$ and $(-4, 0)$ have the same first coordinate. If either one of these ordered pairs were deleted from the relation, the relation would be a function.

The relation in Example 1b *is* a function. Any two distinct ordered pairs that solve the equation $-3x + 7y = 13$ have different first coordinates. This can also be seen from the graph of the equation. If two ordered pairs have the same first coordinate, they must be aligned vertically, and no two ordered pairs on the graph of $-3x + 7y = 13$ have this property.

The relations in Examples 1c and 1d are also not functions. To prove that a relation is not a function, it is only necessary to find two distinct ordered pairs with the same first coordinate, and the pair $(0, 2)$ and $(0, -2)$ works to show that both of these relations fail to be functions. To again illustrate the difference between these relations, note that $(0, 1)$, $(0, 0.005)$, and $(0, -1.9)$ are all elements of the relation in Example 1d, but not of the relation in Example 1c.

Finally, the relation in Example 1e also fails to be a function. It is only necessary to think of two people who have the same mother to convince yourself of this fact.

In determining that the relation in Example 1b is a function, we noted that two distinct ordered pairs in the plane have the same first coordinate only if they are aligned vertically, and that no two points on the graph of the equation $-3x+7y=13$ have this property. We could also have used this criterion to determine that the relation in Example 1a is not a function, since in the graph of the relation the two ordered pairs $(-4,2)$ and $(-4,0)$ clearly lie on the same vertical line. This visual method of determining whether a relation is a function is quick and easily applied whenever an accurate graph of the relation is available to study. It is restated and given a name below.

The Vertical Line Test

If a relation can be graphed in the Cartesian plane, the relation is a function if and only if no vertical line passes through the graph more than once. If even *one* vertical line intersects the graph of the relation two or more times, the relation fails to be a function.

Note that vertical lines that miss the graph of a relation entirely don't prevent the relation from being a function; it is only the presence of a vertical line that hits the graph two or more times that indicates the relation isn't a function. The next example illustrates some more applications of the vertical line test.

example 2

a. The relation $R=\{(-3,2),(-1,0),(0,2),(2,-4),(4,0)\}$, graphed below, is a function. Any given vertical line in the plane either intersects the graph once or not at all.

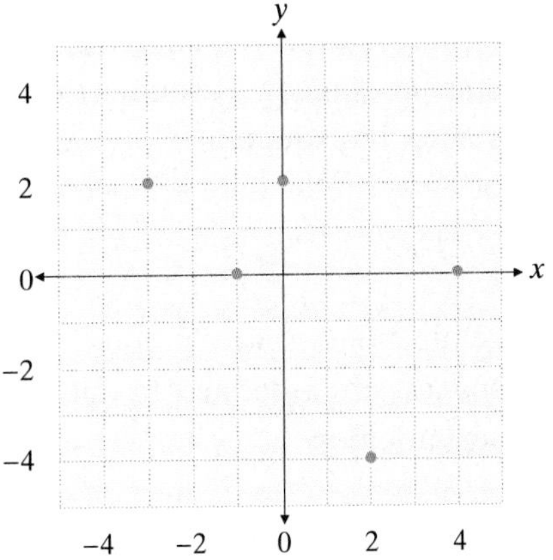

cont'd. on next page ...

b. The relation graphed below is not a function, as there are many vertical lines that intersect the graph more than once. The dashed line is one such vertical line.

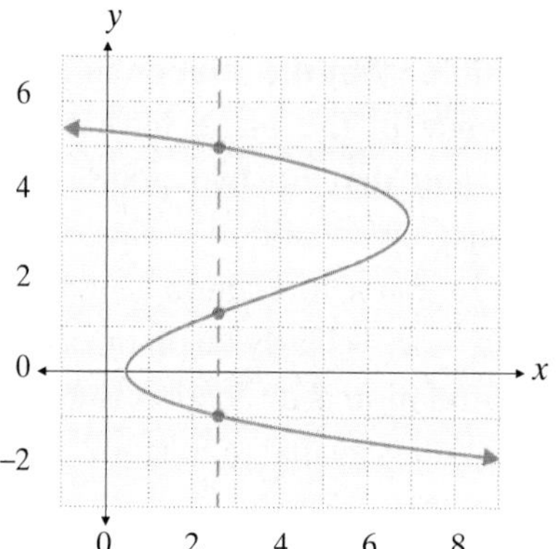

c. The relation graphed below is a function. In this case, every vertical line in the plane intersects the graph exactly once.

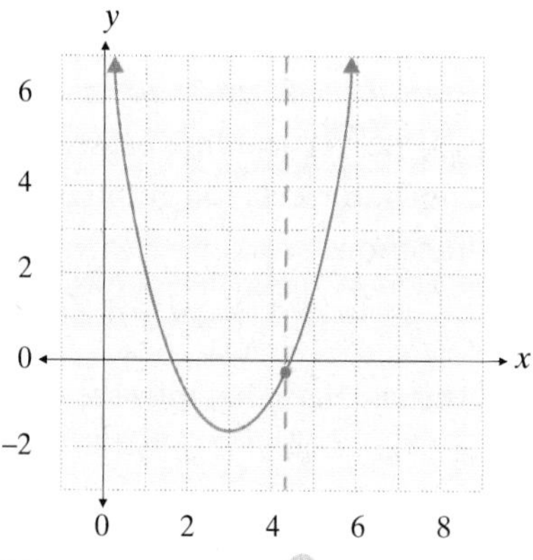

Topic 3: Functional Notation and Function Evaluation

Functions are useful in mathematics for a variety of reasons, all of which ultimately depend on the property that differentiates functions from other relations. Specifically, given an element of the domain of a function, the function relates it to a *unique* element of the range; there is no ambiguity in which element of the range to associate with a given domain element. Relations that are not functions fail this criterion of uniqueness. For instance, the equation $x = y^2$ describes a perfectly good relation, but for a given value of x, say $x = 9$, there are two elements of the range associated with it: both $(9, 3)$ and $(9, -3)$ are on the graph of $x = y^2$.

Because functions assign a unique element of the range to each element of the domain, functions are often defined by means of a notation not yet seen in this text, called functional notation.

Functional notation accomplishes a number of things at once: it gives a name to the function being defined, it provides a formula to work with when using the function, and it makes the solving of many, many mathematical problems far easier.

As an example of functional notation, consider the function defined by the graph of $10x - 5y = 15$. We know this represents a function because if we graph $10x - 5y = 15$ we obtain a non-vertical line, and such a line always passes the vertical line test.

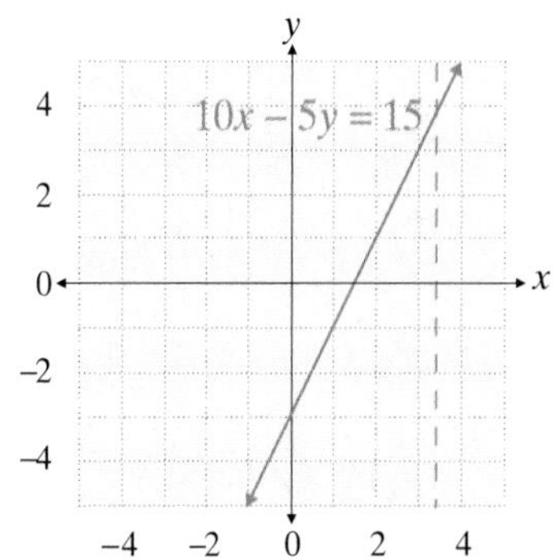

If we need to know which specific element of the range corresponds to a given element of the domain, say $x = 4$, we can substitute 4 for x in the equation and solve for y. But if we have to do this for more than one domain element, it is foolish to repeat the process over and over. Why not just solve for y once and for all? If we do this, we obtain the equivalent equation $y = 2x - 3$. Now, given any element of the domain (that is, given any value for x), we can determine the corresponding range element (the value for y). In this context, x is called the **independent variable** and y is called the **dependent variable**, as the value for y depends on the value for x.

In functional notation, the function defined by $10x - 5y = 15$ is written as $f(x) = 2x - 3$, where we have arbitrarily chosen to name the function f. The key point here is that we have used a mathematical formula to describe what the function named f does to whatever element of the domain it is given. The notation $f(x) = 2x - 3$ is read "f of x equals two times x, minus three," and indicates that when given a specific value for x, the function f returns two times that value, minus three. For instance, $f(4) = 2(4) - 3$, or $f(4) = 5$ (this last line is read "f of 4 equals 5"). Notice that the only difference between the equation $y = 2x - 3$ and the function $f(x) = 2x - 3$ is one of notation: we have replaced y with $f(x)$.

caution!

By far the most common error made when encountering functions for the first time is to think that $f(x)$ stands for the product of f and x. This is entirely wrong! While it is true that parentheses are often used to indicate multiplication, they are also used in defining functions. This is another example of the unfortunate reuse of symbols.

A word on the use of the variable x is in order here. In defining the function f as $f(x) = 2x - 3$, the critical idea being conveyed is the formula. We can use absolutely any symbol at all as a placeholder in defining the formula that we have named f. For instance, $f(n) = 2n - 3$, $f(z) = 2z - 3$, $f(\square) = 2\square - 3$ and $f(\ ) = 2(\ ) - 3$ all define exactly the same function: the function that multiplies what it is given by two and then subtracts three. The variable (or symbol) that is used in defining a given function is called its **argument**, and serves as nothing more than a placeholder.

example 3

Each of the following equations in x and y represents a function. Rewrite each one using functional notation, and then evaluate each function at $x = -3$.

a. $7x+3=2y-1$ **b.** $y-5=x^2$ **c.** $\sqrt{1-x}-2y=6$

Solutions:

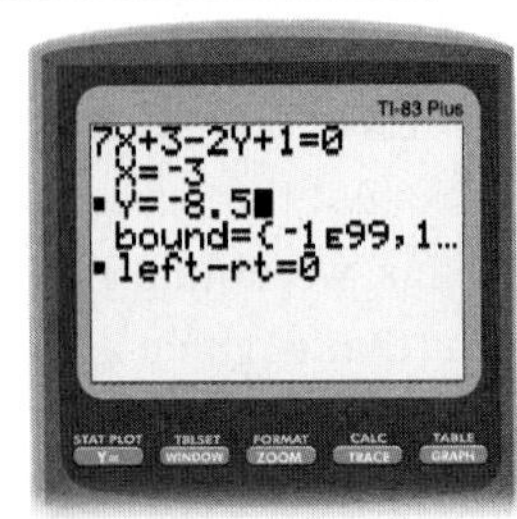

This screen was performed using equation solver.

a. $7x+3=2y-1$

$$7x-2y=-4$$
$$-2y=-7x-4$$
$$y=\frac{7}{2}x+2$$

The first step is to solve the equation for the dependent variable y.

$$f(x)=\frac{7}{2}x+2$$

We can name the function anything at all. Typical names of functions are f, g, h, etc.

$$f(-3)=\frac{7}{2}(-3)+2$$
$$=-\frac{17}{2}$$

We now evaluate f at -3 to obtain the desired answer.

b. $y-5=x^2$

Again, we begin by solving for y.

$$y=x^2+5$$
$$g(x)=x^2+5$$

To distinguish this function from the last, we have used a different name.

$$g(-3)=(-3)^2+5$$
$$=14$$

Replace the argument with -3 to finish.

c. $\sqrt{1-x}-2y=6$

$$-2y=6-\sqrt{1-x}$$
$$y=-3+\frac{\sqrt{1-x}}{2}$$
$$h(x)=-3+\frac{\sqrt{1-x}}{2}$$

We have solved for y and named the function h.

$$h(-3)=-3+\frac{\sqrt{1-(-3)}}{2}$$
$$=-3+\frac{2}{2}$$
$$=-2$$

The resulting evaluation tells us that h of -3 is equal to -2.

technology note

Functions and functional notation are immensely important not just as mathematical concepts but also in the way we use technology as mathematical aids. Graphing calculators and computer algebra systems all have methods for defining functions and then making use of them. The illustration below shows how the function $f(x)=x^3+7x^{3/2}-3x$ is defined in *Mathematica*, and how the function is then evaluated at 4 and 10. Graphing calculators and other computer algebra systems have their own particular syntax for defining and using functions. (*Consult your user's manual for the specific steps for your calculator or software, or see Appendix A for guidance on using Mathematica.*)

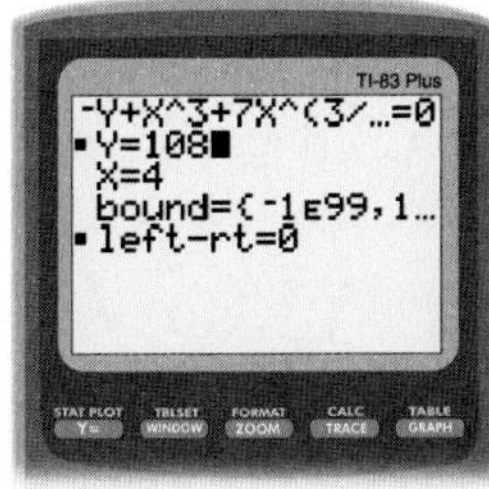

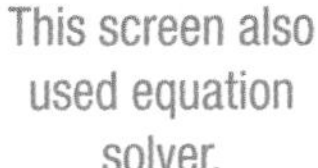
This screen also used equation solver.

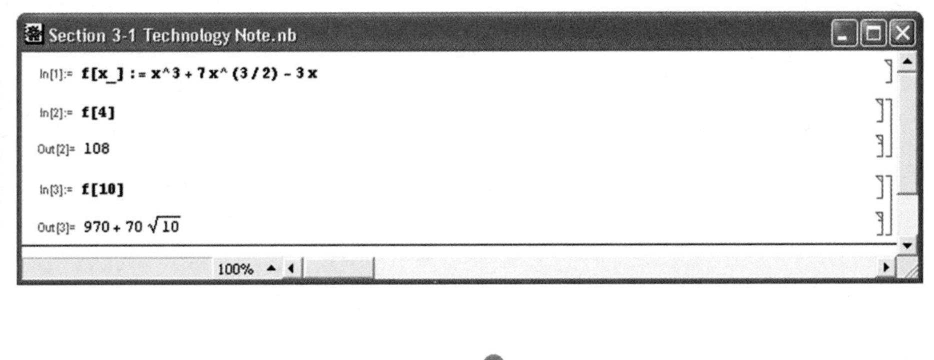

We will not always be replacing the arguments of functions with numbers. In many instances, we will have reason to replace the argument of a function with another variable or possibly a more complicated algebraic expression. Keep in mind that this just involves substituting something for the placeholder used in defining the function.

example 4

Given the function $f(x)=3x^2-2$, evaluate:

a. $f(a)$ **b.** $f(x+h)$ **c.** $\dfrac{f(x+h)-f(x)}{h}$

cont'd. on next page ...

Solutions:

a. $f(a) = 3a^2 - 2$

This is just a matter of replacing x by a.

b. $f(x+h) = 3(x+h)^2 - 2$

$= 3(x^2 + 2xh + h^2) - 2$

$= 3x^2 + 6xh + 3h^2 - 2$

In this problem, we replace x by $x + h$ and proceed to simplify the result.

c. $\dfrac{f(x+h) - f(x)}{h} = \dfrac{[3x^2 + 6xh + 3h^2 - 2] - [3x^2 - 2]}{h}$

$= \dfrac{6xh + 3h^2}{h}$

$= \dfrac{h(6x + 3h)}{h}$

$= 6x + 3h$

We can use the result from above in simplifying this expression. In the original expression, the h in the denominator is rewritten until, eventually, it cancels a factor of h in the numerator.

technology note

The illustration below shows how *Mathematica* can be used to carry out tasks like those seen in Example 4. The first command defines the function $f(x) = 2x^4 - 5x^2$, and the second command then simplifies the expression $\dfrac{f(x+h) - f(x)}{h}$. Other computer algebra systems and some graphing calculators can be used to perform the same task. (*Consult your user's manual for the specific steps for your calculator or software, or see Appendix A for guidance on using Mathematica.)*

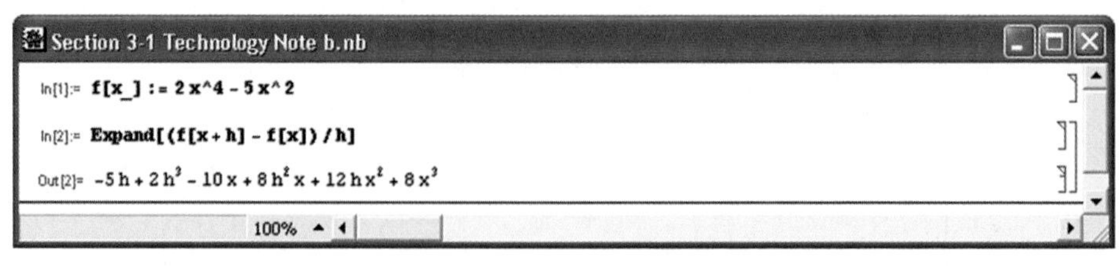

Section 3-1 Technology Note b.nb
In[1]:= f[x_] := 2 x^4 - 5 x^2
In[2]:= Expand[(f[x + h] - f[x]) / h]
Out[2]= -5 h + 2 h³ - 10 x + 8 h² x + 12 h x² + 8 x³
100%

There is one final piece of functional notation that is often encountered, especially in later math classes such as Calculus.

Domain and Codomain Notation

The notation $f: A \to B$ (read "f defined from A to B" or "f maps A to B") implies that f is a function from the set A to the set B. The symbols indicate that the domain of f is the set A, and that the range of f is a subset of the set B. In this context, the set B is often called the **codomain** of f.

Note that while the notation $f: A \to B$ implies the domain of f is the entire set A, there is no requirement for the range of f to be all of B. If it so happens that the range of f actually *is* the entire set B, f is said to be *onto* B (or, more formally, to be a *surjective* function). The next example illustrates how this notation is typically encountered, and also points out some of the subtleties inherent in these notions.

example 5

Identify the domain, the codomain, and the range of each of the following functions.

a. $f: \mathbb{R} \to \mathbb{R}$ by $f(x) = x^2$

b. $g: \mathbb{R} \to [0, \infty)$ by $g(x) = x^2$

c. $h: \mathbb{Z} \to \mathbb{Z}$ by $h(x) = x^2$

d. $j: \mathbb{N} \to \mathbb{R}$ by $j(x) = x^2$

Solutions:

a. The "$f: \mathbb{R} \to \mathbb{R}$" portion of the statement tells us that a function f on the real numbers is about to be defined, and that each value of the function will also be a real number. That is, the domain and codomain of f are both $\mathbb{R}$.

The "$f(x) = x^2$" portion tells us the details of how the function acts. Namely, it returns the square of each real number it is given. Since the square of any real number is nonnegative, and since every nonnegative real number is the square of some real number, the range of f is the interval $[0, \infty)$.

b. The function g is very similar to the function f in part a. The only difference is that the notation "$g: \mathbb{R} \to [0, \infty)$" tells us in advance that the codomain of g is the nonnegative real numbers. Note that the domain of g is $\mathbb{R}$ and the range of g is the same as the range of f. But since the range of g is the same as the codomain of g, the function g is said to be onto, or surjective.

cont'd. on next page ...

This points out that the quality of being "onto" depends entirely on how the codomain of the function is specified. If it is no larger than the range of the function, then the function is onto.

c. The function "$h: \mathbb{Z} \to \mathbb{Z}$ by $h(x) = x^2$" has a domain and codomain of $\mathbb{Z}$. But if we think about the result of squaring any given integer (positive or negative), we quickly see that the range of *h* is the set $\{0, 1, 4, 9, 16, 25, ...\}$. That is, the range of *h* consists of those integers which are squares of other integers. Since the range of *h* is not the same as the codomain, *h* is not onto.

d. The function "$j: \mathbb{N} \to \mathbb{R}$ by $j(x) = x^2$" is one final variation on the squaring function. The action of *j* is the same as that of the previous three functions, but this time the domain is specified to be the natural numbers (the positive integers) and the codomain is the entire set of real numbers. Since the range of *j* is the set $\{1, 4, 9, 16, 25, ...\}$, which is not the same as the codomain, *j* is not onto.

Topic 4: Implied Domain of a Function

Sometimes, especially in applications, the domain of a function under consideration is defined at the same time as the function itself. Often, however, no mention is made explicitly of the values that may be "plugged into" the function, and it is up to us to determine what the domain is. In these cases, the domain of the function is implied by the formula used in defining the function. It is assumed that the domain of the function consists of all those real numbers at which the function can be evaluated to obtain a real number. This means, for instance, that any values for the argument that result in division by zero or an even root of a negative number must be excluded from the domain.

example 6

Determine the domain of the following functions.

a. $f(x) = 5x - \sqrt{3-x}$

b. $g(x) = \dfrac{x-3}{x^2 - 1}$

Solutions:

a. The $5x$ in the formula $5x - \sqrt{3-x}$ can't cause any problems, as any real number can be safely multiplied by 5. However, for $\sqrt{3-x}$ to be a real number, we need $3 - x \geq 0$. Solving this inequality for *x* results in the statement $x \leq 3$. Using interval notation, the domain of the function *f* is the interval $(-\infty, 3]$.

b. The formula in this function is a rational expression, or a ratio of polynomials. A polynomial in x can be evaluated for any value of x, so the only potential problem we must guard against is division by zero. The denominator of this formula is zero if $x^2 - 1 = 0$, so we need to solve this quadratic equation and exclude the solutions from the domain of g. This equation can be solved by factoring: $(x+1)(x-1)=0$ tells us that we must exclude $x=-1$ and $x=1$ from the domain. In interval notation, the domain of g is $(-\infty, -1)\cup(-1, 1)\cup(1, \infty)$.

exercises

For each relation below, describe the domain and range. See Example 1.

1. $R=\{(-2, 5), (-2, 3), (-2, 0), (-2, -9)\}$

2. $S=\{(0, 0), (-5, 2), (3, 3), (5, 3)\}$

3. $A=\{(\pi, 2), (-2\pi, 4), (3, 0), (1, 7)\}$

4. $B=\{(3, 3), (-4, 3), (3, 8), (3, -2)\}$

5. $T=\{(x, y) \mid x\in\mathbb{Z} \text{ and } y=2x\}$

6. $U=\{(\pi, y) \mid y\in\mathbb{Q}\}$

7. $C=\{(x, 3x+4) \mid x\in\mathbb{Z}\}$

8. $D=\{(5x, 3y) \mid x\in\mathbb{Z} \text{ and } y\in\mathbb{Z}\}$

9. $3x-4y=17$

10. $x+y=0$

11. $x=|y|$

12. $y=x^2$

13. $y=-1$

14. $x=3$

15. $x=4x$

16. $y=7\pi^2$

17.

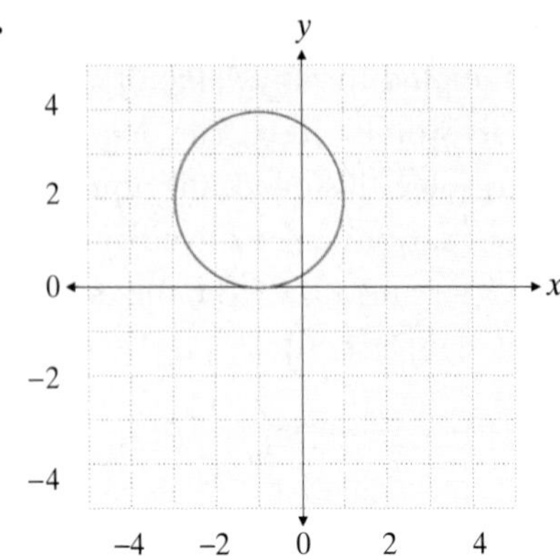

18.

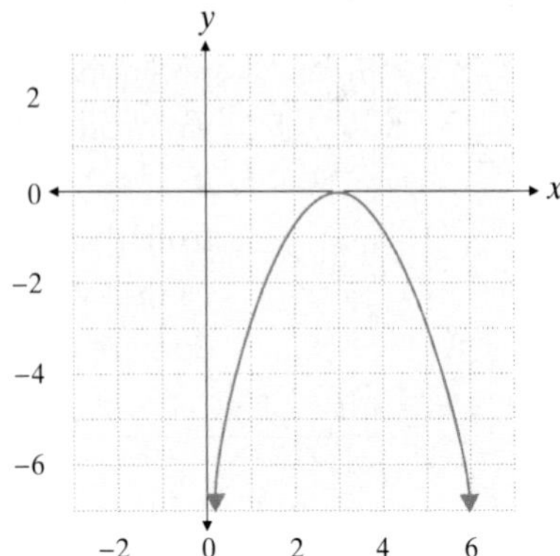

19.

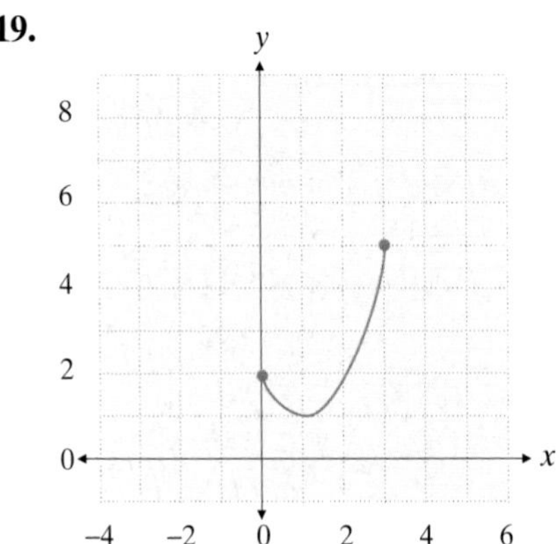

20.

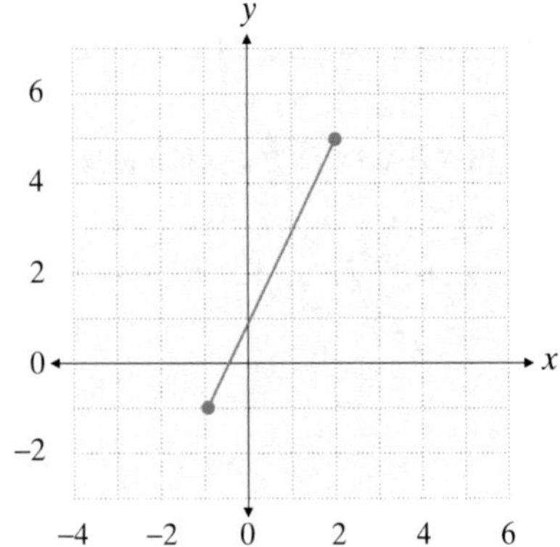

21.

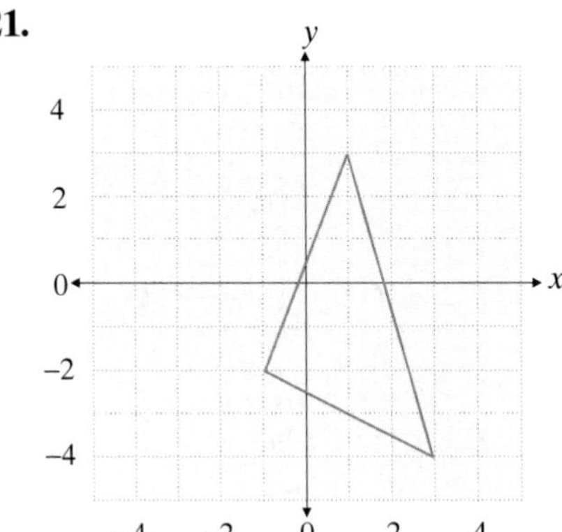

22.

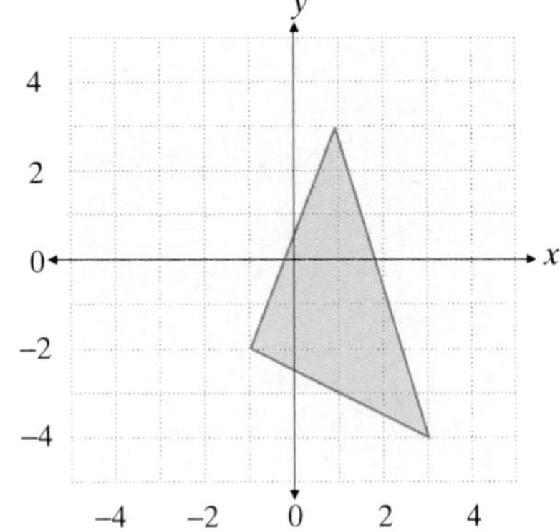

23. $V = \{(x, y) \mid x \text{ is the brother of } y\}$

24. $W = \{(x, y) \mid y \text{ is the daughter of } x\}$

Determine which of the relations below are functions. For those that are not, identify two ordered pairs with the same first coordinate. See Example 2.

25. $R = \{(-2, 5), (2, 4), (-2, 3), (3, -9)\}$

26. $S = \{(3, -2), (4, -2)\}$

27. $T = \{(-1, 2), (1, 1), (2, -1), (-3, 1)\}$

28. $U = \{(4, 5), (2, -3), (-2, 1), (4, -1)\}$

29.

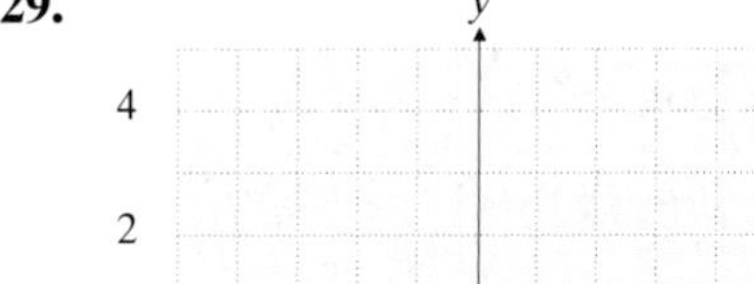

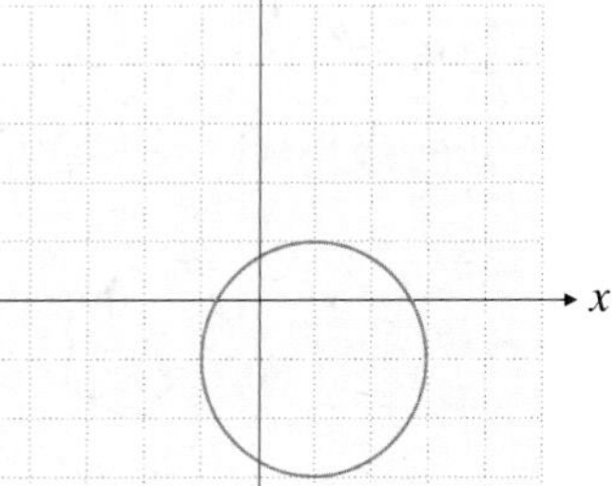

30.

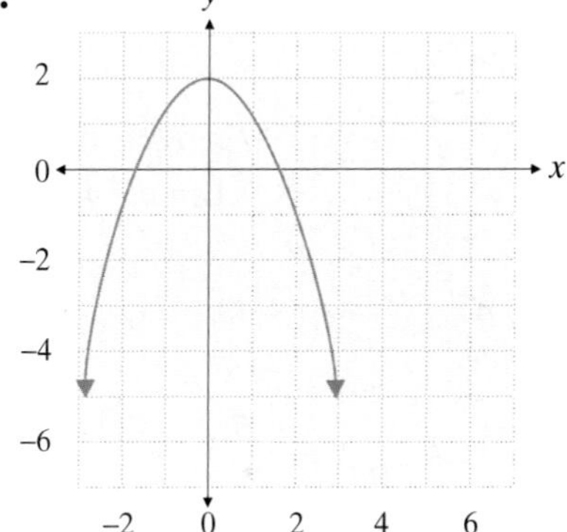

31.

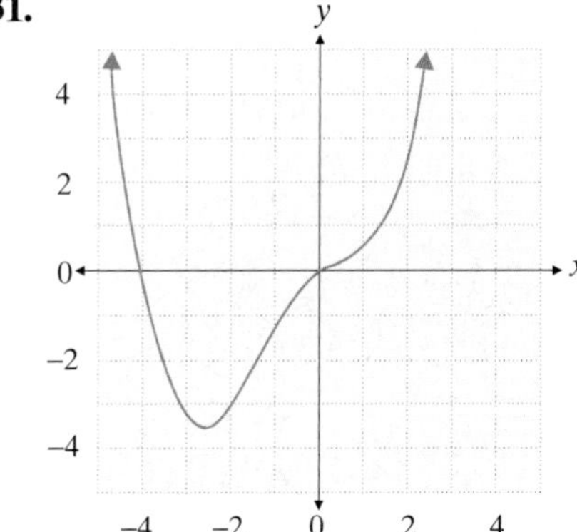

32.

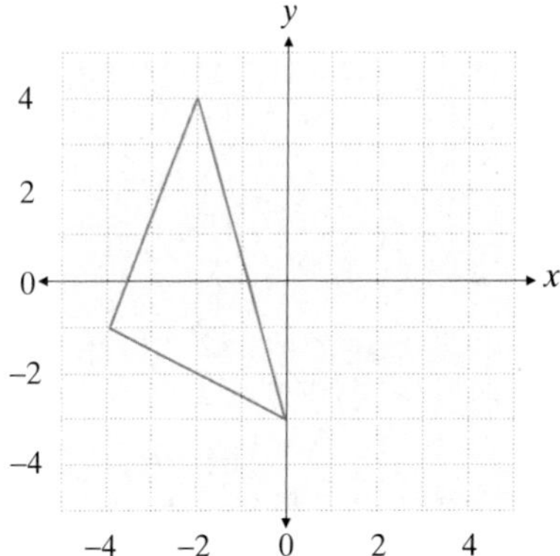

Determine whether each of the following relations is a function. If it is a function, give the relation's domain.

33. $y = \frac{1}{x}$ **34.** $x = y^2 - 1$ **35.** $x + y^2 = 0$ **36.** $y = 2x^2 - 4$ **37.** $y = \frac{x-1}{x+2}$

38. $x^2 + y^2 = 1$ **39.** $y = |x - 2|$ **40.** $y = x^3$ **41.** $y^2 - x^2 = 3$ **42.** $y = \sqrt{x-4}$

Rewrite each of the relations below as a function of x. Then evaluate the function at x = –1. See Example 3.

43. $6x^2 - x + 3y = x + 2y$

44. $2y - \sqrt[3]{x} = x - (x-1)^2$

45. $\dfrac{x+3y}{5} = 2$

46. $x^2 + y = 3 - 4x^2 + 2y$

47. $y - 2x^2 = -2(x + x^2 + 5)$

48. $\dfrac{9y+2}{6} = \dfrac{3x-1}{2}$

*For each function below, determine **(a)** f(x – 1), **(b)** f(x + a) – f(x), and **(c)** f(x²). See Example 4.*

49. $f(x) = x^2 + 3x$

50. $f(x) = \sqrt{x}$

51. $f(x) = 3x + 2$

52. $f(x) = -x^2 - 7$

53. $f(x) = 2(5 - 3x)$

54. $f(x) = 2x^2 + \sqrt[4]{x}$

*Find the value for each of the functions below for **(a)** f(0), **(b)** f(–2), **(c)** f(3), **(d)** f(x + h), and **(e)*** $\dfrac{f(x+h)-f(x)}{h}$.

55. $f(x) = x^2 - 4x + 5$

56. $f(x) = 5x^2 - 2$

57. $f(x) = \dfrac{3x-2}{2}$

58. $f(x) = 5x + 6$

59. $f(x) = \sqrt{x-8}$

60. $f(x) = x^2 + 3$

61. $f(x) = x^3$

62. $f(x) = \dfrac{x-3}{x+2}$

63. $f(x) = 6x - x^2$

Identify the domain, the codomain, and the range of each of the following functions. See Example 5.

64. $f : \mathbb{R} \to \mathbb{R}$ by $f(x) = 3x$

65. $g : \mathbb{Z} \to \mathbb{Z}$ by $g(x) = 3x$

66. $f : \mathbb{Z} \to \mathbb{Z}$ by $f(x) = x + 5$

67. $g : [0, \infty) \to \mathbb{R}$ by $g(x) = \sqrt{x}$

68. $h : \mathbb{N} \to \mathbb{N}$ by $h(x) = x + 5$

69. $h : \mathbb{N} \to \mathbb{R}$ by $h(x) = \dfrac{x}{2}$

Determine the implied domain of each of the following functions. See Example 6.

70. $f(x)=\sqrt{x-1}$

71. $g(x)=\sqrt[5]{x+3}-2$

72. $h(x)=\dfrac{3x}{x^2-x-6}$

73. $f(x)=(2x+6)^{1/2}$

74. $g(x)=\sqrt[4]{2x^2+3}$

75. $h(x)=\dfrac{3x^2-6x}{x^2-6x+9}$

technology exercises

Use a graphing calculator or a computer algebra system such as Mathematica to first define each of the following functions and then perform the indicated function evaluations. See the Technology Note after Example 3 for guidance.

76. $f(x)=\dfrac{7x^{5/3}-2x^{1/3}}{x^{1/2}}$; find $f(8)$ and $f(12)$

77. $g(x)=\sqrt{x^3-4x^2+2x+31}$; find $g(2)$ and $g(3)$

78. $f(x)=\dfrac{2x^5-9x^3+12}{4x^3-7x+6}$; find $f(-3)$ and $f(2)$

79. $g(x)=(5x^2-7x+1)^3$; find $g(-19)$ and $g(12)$

Use a graphing calculator, if possible, or a computer algebra system such as Mathematica to first define each of the following functions and then simplify the expression $\dfrac{f(x+h)-f(x)}{h}$. *See the Technology Note after Example 4 for guidance.*

80. $f(x)=x^3-1$

81. $f(x)=(x^2+1)^3$

82. $f(x)=3x^2-x$

83. $f(x)=\dfrac{1}{x-1}$

3.2 Linear and Quadratic Functions

TOPICS

1. Linear functions and their graphs
2. Quadratic functions and their graphs
3. Interlude: maximization/minimization problems

Topic 1: Linear Functions and Their Graphs

Much of the next several sections of this chapter will be devoted to gaining familiarity with some of the types of functions that commonly arise in mathematics. We will discuss two classes of functions in this section, beginning with the class of linear functions.

Recall that a linear equation in two variables is an equation whose graph consists of a straight line in the Cartesian plane. Similarly, a linear function (of one variable) is a function whose graph is a straight line. We can define such functions algebraically as follows.

Linear Functions

A **linear function** f of one variable, say the variable x, is any function that can be written in the form $f(x) = ax + b$, where a and b are real numbers. If $a \neq 0$, a function $f(x) = ax + b$ is also called a **first-degree function**.

In the last section, we learned that a function defined by an equation in x and y can be written in functional form by solving the equation for y and then replacing y with $f(x)$. Of course the process can be reversed, so the function $f(x) = ax + b$ appears in equation form as $y = ax + b$, and we recognize this as a linear equation written in slope-intercept form. This means the graph of the function $f(x) = ax + b$ is a straight line with a y-intercept of b (the constant term) and a slope of a (the coefficient of the variable x). For this reason, some texts define a linear function as any function that can be written in the form $f(x) = mx + b$, since m is the letter traditionally used to denote slope and b is commonly used to refer to the y-intercept.

The observations in the last paragraph give us a quickly applied method of graphing linear functions, but before we look at some examples a few words on graphing functions in general are in order. As we noted in Section 3.1, the graph of a function is a plot of all the ordered pairs that make up the function; that is, the graph of a function *f* is the plot of all the ordered pairs in the set $\{(x, y) | (x, y) \text{ is an element of } f\}$. We have a great deal of experience in plotting such sets if the ordered pairs are defined by an equation in *x* and *y*, but we have only plotted a few functions that have been defined with functional notation. But any function of *x* defined with functional notation can be written as an equation in *x* and *y* by replacing $f(x)$ with *y*, so the graph of a function *f* consists of a plot of the ordered pairs in the set $\{(x, f(x)) | x \in \text{domain of } f\}$.

Consider, for example, the function $f(x) = -3x + 5$. Figure 1 contains a table of four ordered pairs defined by the function, followed by a graph of the entire function with the four specific ordered pairs noted.

x	$f(x)$
-1	8
0	5
1	2
2	-1

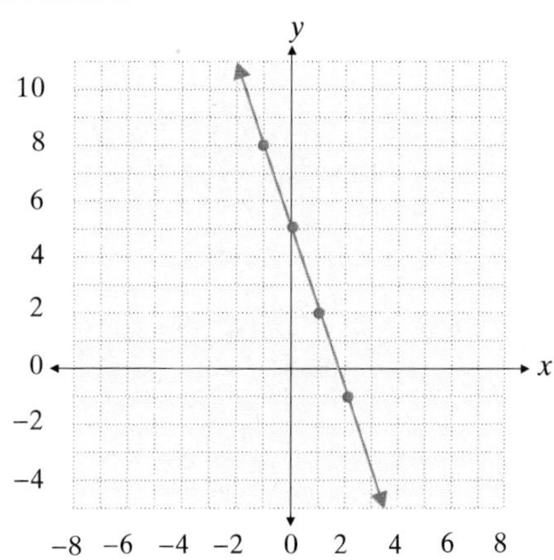

Figure 1: Graph of $f(x) = -3x + 5$

Again, note that every point on the graph of the function in Figure 1 is an ordered pair of the form $(x, f(x))$; we have simply highlighted four of them with green dots.

We could have graphed the function $f(x) = -3x + 5$ by noting that it is a straight line with a slope of -3 and a *y*-intercept of 5. We will use this approach in the following examples.

example 1

Graph the following linear functions.

a. $f(x) = \dfrac{4 + 6x}{2}$ **b.** $g(x) = 3$

cont'd. on next page ...

Solutions:

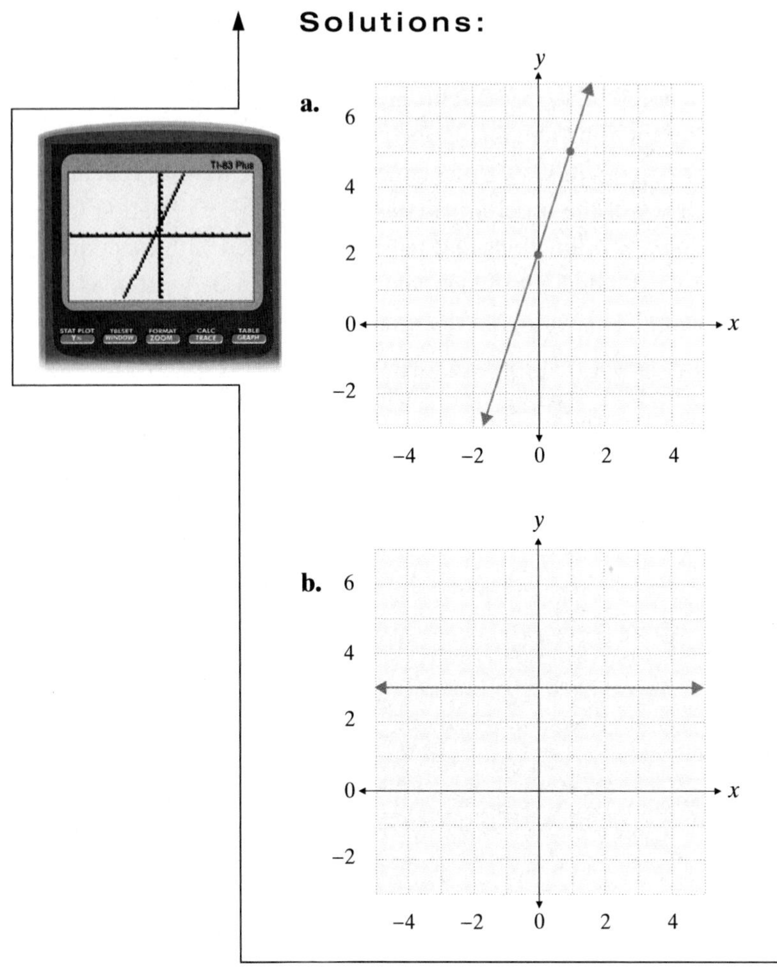

a. The function f can be rewritten as $f(x) = 3x + 2$, and in this form we recognize it as a line with a slope of 3 and a y-intercept of 2.

To graph the function, then, we can plot the ordered pair (0, 2) and locate another point on the line by moving up 3 units and over to the right 1 unit, giving us the ordered pair (1, 5). Once these two points have been plotted, connecting them with a straight line completes the process.

b. The graph of the function g is a straight line with a slope of 0 and a y-intercept of 3. A linear function with a slope of 0 is also called a **constant** function, as it turns any input into one fixed constant, in this case the number 3. The graph of a constant function is always a horizontal line.

Example 1b is an example of a function whose graph is a horizontal line. Of course, no function can have a vertical line as its graph, as a vertical line clearly can't pass the vertical line test. Thus, vertical lines can represent the graphs of equations, but not functions.

Topic 2: Quadratic Functions and Their Graphs

In Section 1.7, we learned how to solve quadratic equations in one variable. We will now study quadratic *functions* of one variable, and learn how to relate this new material to what we already know.

Quadratic Functions

A **quadratic**, or **second-degree**, **function** f of one variable, say the variable x, is any function that can be written in the form $f(x) = ax^2 + bx + c$, where a, b, and c are real numbers and $a \neq 0$.

The graph of any quadratic function is a roughly U-shaped curve known as a **parabola**. We will study parabolas in more generality in Chapter 9, but in this section we will learn how to graph parabolas as they arise in the context of quadratic functions.

The graph in Figure 2 is an example of a parabola. Specifically, it is the graph of the quadratic function $f(x) = x^2$, and the table that precedes the graph contains a few of the ordered pairs on the graph.

x	$f(x)$
−3	9
−1	1
0	0
2	4

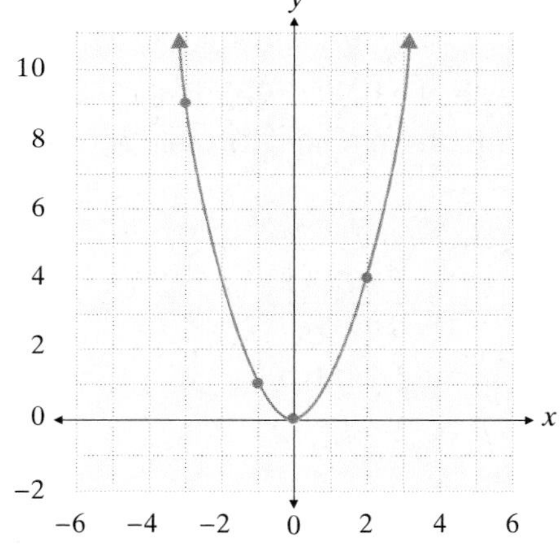

Figure 2: Graph of $f(x) = x^2$

As Figure 2 demonstrates, one key characteristic of a parabola is that it has one (and only one) point at which the graph "changes direction". This point is known as the **vertex**; scanning the plane from left to right, it is the point on a parabola where the graph stops going down and begins to go up (if the parabola opens upward), or else stops going up and begins to go down (if the parabola opens downward). Another characteristic of a parabola is that it is symmetric with respect to its **axis**, a straight line passing through the vertex and dividing the parabola into two halves that are mirror images of each other. This line is called the **axis of symmetry**. Every parabola that represents the graph of a quadratic function has a vertical axis, but we will see parabolas later in the text that have non-vertical axes. Finally, parabolas can be relatively skinny or relatively broad, meaning that the curve of the parabola at the vertex can range from very sharp to very flat.

Our knowledge of linear equations gives us a very satisfactory method for sketching the graphs of linear functions. Our goal now is to develop a method that allows us to sketch the graphs of quadratic functions. We *could*, given a quadratic function, simply plot a number of points and then connect them with a U-shaped curve, but this method has several drawbacks, chief among them being that it is slow and not very accurate. Ideally, we will develop a method that is fast and that pinpoints some of the important characteristics of a parabola (such as the location of its vertex) exactly.

We will develop our graphing method by working from the answer backward. We will first see what effects various mathematical operations have on the graphs of parabolas, and then see how this knowledge lets us graph a generic quadratic function.

To begin, let us say that the graph of the function $f(x)=x^2$, graphed in Figure 2, is the prototypical parabola. We already know its characteristics: its vertex is at the origin, its axis is the y-axis, it opens upward, and the sharpness of the curve at its vertex will serve as a convenient reference when discussing other parabolas. Now consider the function $g(x)=(x-3)^2$, obtained by replacing x in the formula for f with $x-3$. Both of the formulas x^2 and $(x-3)^2$ are non-negative for any value of x, and x^2 is equal to 0 when $x=0$. What value of x results in $(x-3)^2$ equaling 0? The answer is $x=3$. In other words, the point $(0,0)$ on the graph of f corresponds to the point $(3,0)$ on the graph of g. With this in mind, the table and graph in Figure 3 should not be too surprising.

x	$g(x)$
0	9
2	1
3	0
5	4

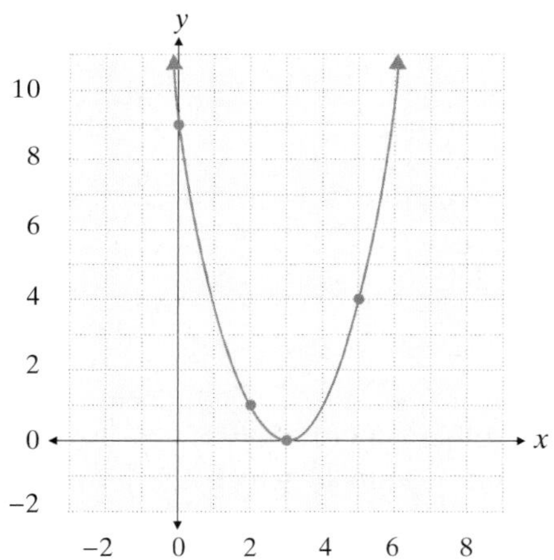

Figure 3: Graph of $g(x)=(x-3)^2$

Notice that the shape of the graph of g is identical to that of f, but it has been shifted over to the right by 3 units. This is our first example of how we can manipulate graphs of functions.

Consider now the function h obtained by replacing the x in x^2 by $x + 7$. As with the functions f and g, $h(x) = (x+7)^2$ is non-negative for all values of x, and only one value for x will return a value of 0: $h(-7) = 0$. Compare the table and graph in Figure 4 with those in Figures 2 and 3.

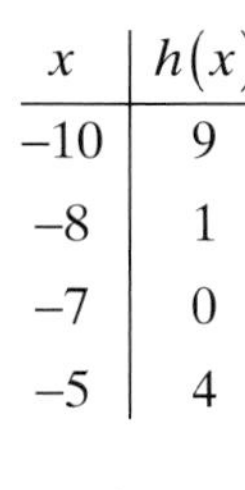

x	$h(x)$
–10	9
–8	1
–7	0
–5	4

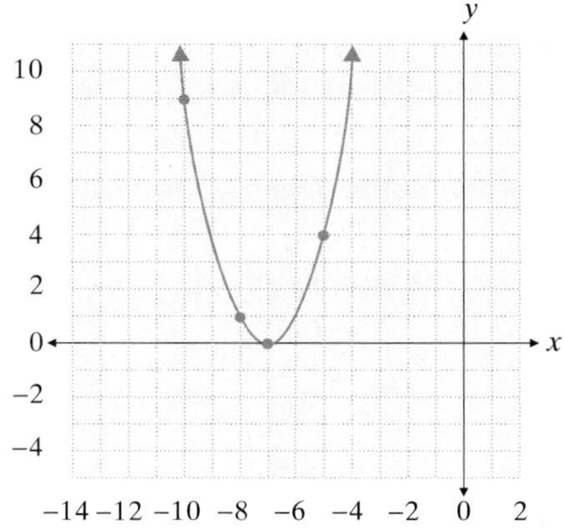

Figure 4: Graph of $h(x) = (x+7)^2$

We have thus demonstrated the ability to shift the prototypical parabola to the left and right. To sum up, the graph of $g(x) = (x-h)^2$ has the same shape as the graph of $f(x) = x^2$, but shifted h units to the right if h is positive and h units to the left if h is negative.

How do we shift a parabola up and down? To move the graph of $f(x) = x^2$ up by a fixed number of units, we need to add that number of units to the second coordinate of each ordered pair. Similarly, to move the graph down we need to subtract the desired number of units from each second coordinate. To demonstrate this, consider the table and graphs for the two functions $i(x) = x^2 + 5$ and $j(x) = x^2 - 2$ in Figure 5.

x	$i(x)$	$j(x)$
–3	14	7
–1	6	–1
0	5	–2
2	9	2

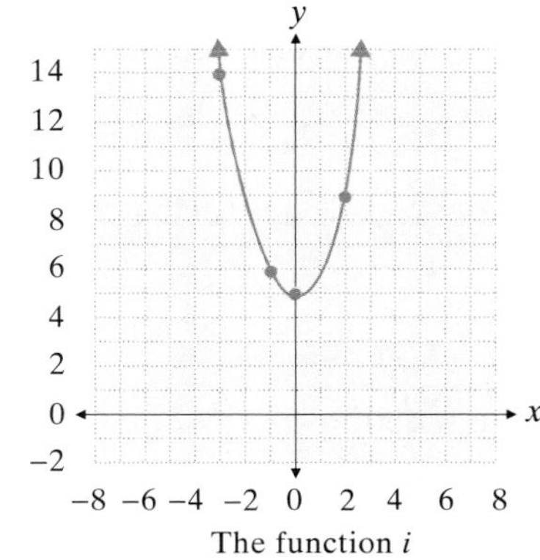

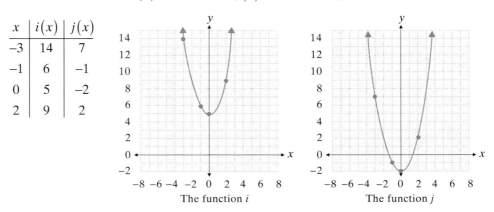

The function i

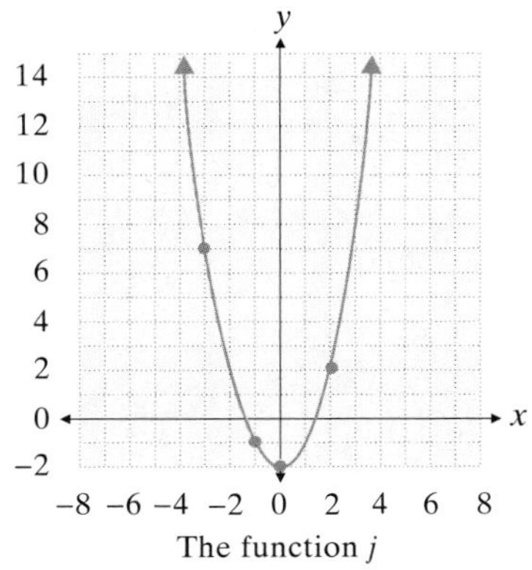

The function j

Figure 5: Graph of $i(x) = x^2 + 5$ and $j(x) = x^2 - 2$

Now, how do we make a parabola skinnier or broader? To make the prototypical parabola skinnier (or to make the curve at the vertex sharper), we need to stretch the graph vertically. That is, we want to leave the vertex untouched, but increase the second coordinates of all other points on the graph. We can do this by multiplying the formula x^2 by a constant a greater than 1 to obtain the formula ax^2. Multiplying the formula x^2 by a constant a that lies between 0 and 1 makes the parabola broader (that is, it makes the curve at the vertex flatter). Finally, multiplying x^2 by a negative constant a turns all of the non-negative outputs of f into non-positive outputs, resulting in a parabola that opens downward instead of upward.

Compare the graphs of $k(x)=6x^2$ and $l(x)=-\frac{1}{4}x^2$ in Figure 6 to the prototypical quadratic $f(x)=x^2$, which is shown as a green dotted line.

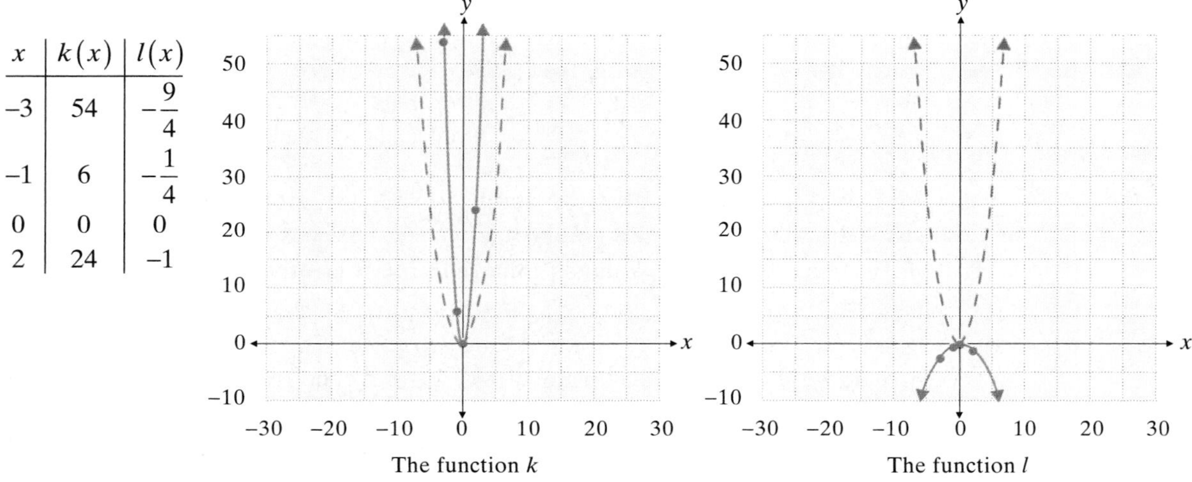

x	$k(x)$	$l(x)$
-3	54	$-\frac{9}{4}$
-1	6	$-\frac{1}{4}$
0	0	0
2	24	-1

Figure 6: Graph of $k(x)=6x^2$ and $l(x)=-\frac{1}{4}x^2$

The following box summarizes all of the above ways of altering the prototypical parabola $f(x)=x^2$.

Vertex Form of a Quadratic Function

The graph of the function $g(x)=a(x-h)^2+k$, where a, h, and k are real numbers and $a\neq 0$, is a parabola whose vertex is at (h, k). The parabola is narrower than $f(x)=x^2$ if $|a|>1$ and is broader than $f(x)=x^2$ if $0<|a|<1$. The parabola opens upward if a is positive, and downward if a is negative.

The question now is: given a quadratic function $f(x)=ax^2+bx+c$, how do we determine the location of its vertex, whether it opens upward or downward, and whether it is skinnier or broader than the prototypical parabola? The answer is to use the method of completing the square.

example 2

Sketch the graph of the function $f(x)=-x^2-2x+3$. Locate the vertex and the x-intercepts exactly.

Solution:

We identify the vertex of the function by completing the square on the expression $-x^2-2x+3$, as shown below.

$$\begin{aligned} f(x) &= -x^2-2x+3 \\ &= -\left(x^2+2x\right)+3 \\ &= -\left(x^2+2x+1\right)+1+3 \\ &= -(x+1)^2+4 \end{aligned}$$

The first step is to factor out the leading coefficient from the first two terms. In this problem, the leading coefficient is -1. We then complete the square on the x^2 and $2x$ terms, as shown. Note, though, that this amounts to adding -1 to the function, so we compensate by adding 1 as well.

Completing the square thus tells us that the vertex of the parabola lies at the point $(-1, 4)$. Note that we have to interpret the expression $-(x+1)^2+4$ as $-(x-(-1))^2+4$, so h is -1 and k is 4.

The instructions also ask us to identify the x-intercepts. The graph of any quadratic function will intersect the x-axis in 0, 1, or 2 places, depending on the orientation of the corresponding parabola. An x-intercept of the function f will be a point on the x-axis where $f(x)=0$, so we need to solve the equation $-x^2-2x+3=0$. This can be done by factoring:

$$\begin{aligned} -x^2-2x+3&=0 \\ x^2+2x-3&=0 \\ (x+3)(x-1)&=0 \\ x&=-3,\ 1. \end{aligned}$$

cont'd. on next page ...

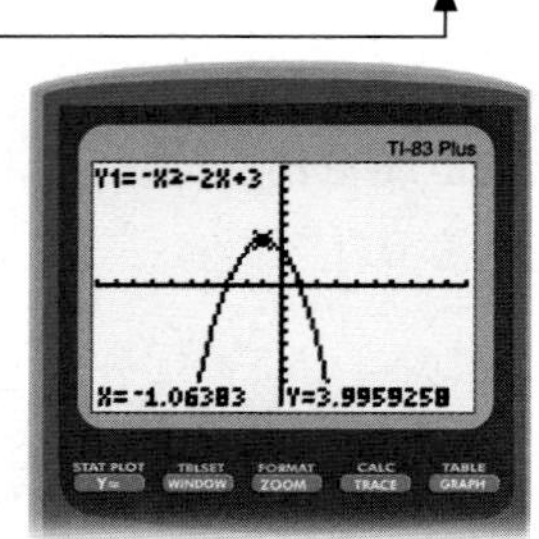

After graphing this equation in the usual manner by selecting Y= and then Graph, individual coordinates can be seen by selecting Trace and using the left and right arrows to move along the graph.

The vertex form of the function, $f(x) = -(x+1)^2 + 4$, tells us that this quadratic opens downward, has its vertex at (–1, 4), and is neither skinnier nor broader than the prototypical parabola. We now also know that it crosses the x-axis at –3 and 1. Putting this all together, we obtain the following graph.

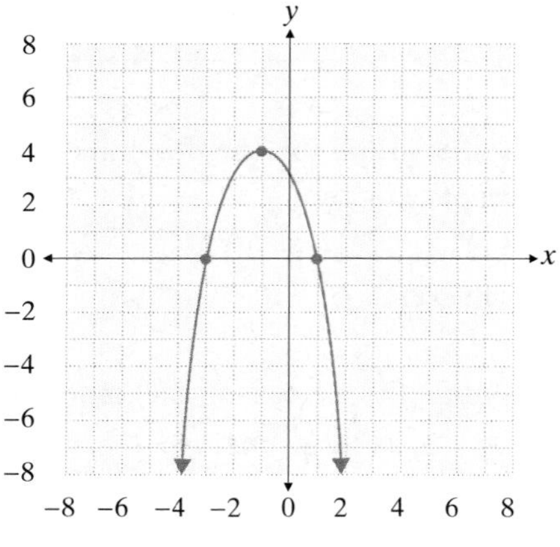

Topic 3: Interlude: Maximization/Minimization Problems

Many applications of mathematics involve determining the value (or values) of the variable x that return either the maximum possible value or the minimum possible value of some function $f(x)$. Such problems are called Max/Min problems for short. Examples from business include minimizing a cost function and maximizing a profit function. Examples from physics include maximizing a function that measures the height of a rocket as a function of time, and minimizing a function that measures the energy required by a particle accelerator.

We are now in a position to solve Max/Min problems if the function involved is a quadratic. Specifically, if the parabola corresponding to a given quadratic function opens upward, we know that the vertex is the lowest (minimum) point on the graph. Similarly, if the parabola opens downward, the vertex is the highest (maximum) point on the graph. In either case, we locate the vertex by the process of completing the square.

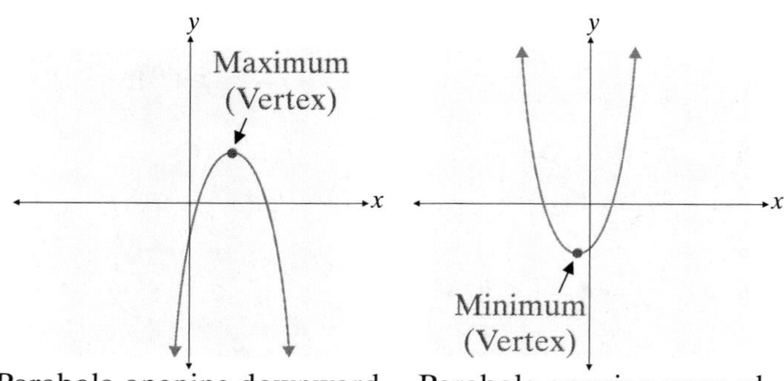

We can even shorten the process of locating the vertex of a parabola by completing the square on the generic quadratic $f(x) = ax^2 + bx + c$ once and for all. The work below should remind you of the derivation of the quadratic formula.

$$
\begin{aligned}
f(x) &= ax^2 + bx + c \\
&= a\left(x^2 + \frac{b}{a}x\right) + c \\
&= a\left(x^2 + \frac{b}{a}x + \frac{b^2}{4a^2}\right) - \frac{b^2}{4a} + c \\
&= a\left(x + \frac{b}{2a}\right)^2 + \frac{4ac - b^2}{4a}
\end{aligned}
$$

As always, we begin by factoring the leading coefficient a from the first two terms.

To complete the square, we add the square of half of $\frac{b}{a}$ inside the parentheses. This means we also have to subtract a times this quantity outside the parentheses.

The vertex is at $\left(-\frac{b}{2a}, \frac{4ac - b^2}{4a}\right)$.

An easier way to remember this formula is to realize that *every* point on the graph of a function f has the form $(x, f(x))$, so the vertex must lie at $\left(-\frac{b}{2a}, f\left(-\frac{b}{2a}\right)\right)$. That is, if $f(x) = ax^2 + bx + c$, then $f\left(-\frac{b}{2a}\right) = \frac{4ac - b^2}{4a}$. You should verify this claim.

Vertex of a Quadratic Function

The vertex of a quadratic function $f(x) = ax^2 + bx + c$ is

$$\left(-\frac{b}{2a}, f\left(-\frac{b}{2a}\right)\right).$$

Note, this point is the maximum of the graph if $a < 0$ and the minimum if $a > 0$.

example 3

A farmer plans to use 100 feet of spare fencing material to form a rectangular garden plot against the side of a long barn, using the barn as one side of the plot. How should he split up the fencing among the other three sides in order to maximize the area of the garden plot?

Solution:

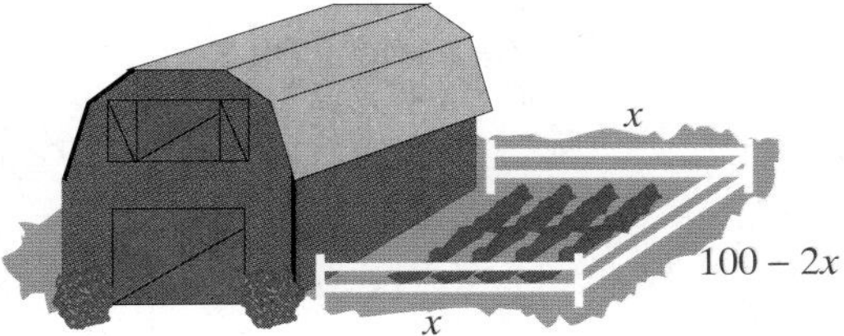

If we let x represent the length of one side of the plot, as shown in the diagram to the right, then the dimensions of the plot are x feet by $100 - 2x$ feet. The letter A, for area, is a reasonable name for the function that we wish to maximize in this problem, and so we want to find the maximum possible value of $A(x) = x(100 - 2x)$.

cont'd. on next page ...

If we multiply out the formula for A, we recognize it as the quadratic function $A(x) = -2x^2 + 100x$. This is a parabola opening downward, so the vertex will be the maximum point on the graph of A.

Using our work above, we know that the vertex of A is the ordered pair $\left(-\frac{100}{2(-2)}, A\left(-\frac{100}{2(-2)}\right)\right)$, or $(25, A(25))$. That is, to maximize area, we should let $x = 25$, and so $100 - 2x = 50$. The resulting maximum possible area, 25 feet × 50 feet, or 1250 square feet, is also the value $A(25)$.

exercises

Graph the following linear functions. See Example 1.

1. $f(x) = -5x + 2$
2. $g(x) = \frac{3x-2}{4}$
3. $h(x) = -x + 2$
4. $p(x) = -2$
5. $g(x) = 3 - 2x$
6. $r(x) = 2 - \frac{x}{5}$
7. $f(x) = -2(1-x)$
8. $a(x) = 3\left(1 - \frac{1}{3}x\right) + x$
9. $f(x) = 2 - 4x$
10. $g(x) = \frac{2x-8}{4}$
11. $h(x) = 5x - 10$
12. $k(x) = 3x - \frac{2+6x}{2}$
13. $m(x) = \frac{-x+25}{10}$
14. $q(x) = 1.5x - 1$
15. $w(x) = (x-2) - (2+x)$

Graph the following quadratic functions, locating the vertices and x-intercepts (if any) accurately. See Example 2.

16. $f(x) = (x-2)^2 + 3$
17. $g(x) = -(x+2)^2 - 1$
18. $h(x) = x^2 + 6x + 7$
19. $F(x) = 3x^2 + 2$
20. $G(x) = x^2 - x - 6$
21. $p(x) = -2x^2 + 2x + 12$
22. $q(x) = 2x^2 + 4x + 3$
23. $r(x) = -3x^2 - 1$
24. $s(x) = \frac{(x-1)^2}{4}$
25. $m(x) = x^2 + 2x + 4$
26. $n(x) = (x+2)(2-x)$
27. $p(x) = -x^2 + 2x - 5$
28. $f(x) = 4x^2 - 6$
29. $k(x) = 2x^2 - 4x$
30. $j(x) = (x+10)(x-2) + 36$

Solve the following maximization/minimization problems by analyzing the appropriate quadratic function. See Example 3.

31. Cindy wants to construct three rectangular dog-training arenas side by side, as shown, using a total of 400 feet of fencing. What should the overall length and width be in order to maximize the area of the three combined arenas? (Suggestion: let x represent the width, as shown, and find an expression for the overall length in terms of x.)

32. Among all the pairs of numbers with a sum of 10, find the pair whose product is maximum.

33. Among all rectangles that have a perimeter of 20, find the dimensions of the one whose area is largest.

34. Find the point on the line $2x + y = 5$ that is closest to the origin. (Hint: instead of trying to minimize the distance between the origin and points on the line, minimize the square of the distance.)

35. Among all the pairs of numbers (x, y) such that $2x + y = 20$, find the pair for which the sum of the squares is minimum.

36. A rancher has a rectangular piece of sheet metal that is 20 inches wide by 10 feet long. He plans to fold the metal into a three-sided channel and weld two other sheets of metal to the ends to form a watering trough 10 feet long, as shown. How should he fold the metal in order to maximize the volume of the resulting trough?

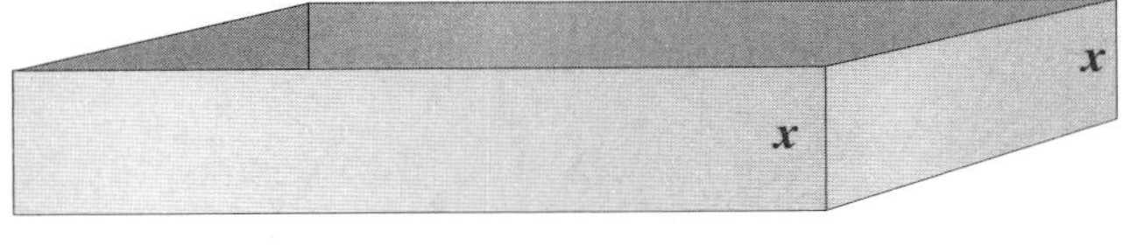

10 feet

37. Sitting in a tree, 48 feet above ground level, Sue shoots a pebble straight up with a velocity of 64 feet per second. As a function of time, t, the height of the pebble in feet is $h(t) = -16t^2 + 64t + 48$. What is the maximum height attained by the pebble?

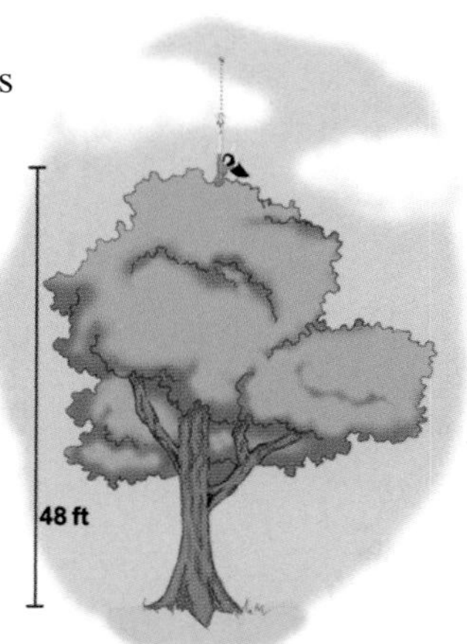

38. A ball is thrown upward with a velocity of 48 feet per second from the top of a 144-foot building. What is the maximum height of the ball? (Hint: Use $h(t) = -16t^2 + 48t + 144$ where the height h of the ball is a function of time t.)

39. A rock is thrown upward with a velocity of 80 feet per second from the top of a 64-foot high cliff. What is the maximum height of the rock? (Hint: Use $h(t) = -16t^2 + 80t + 64$ where the height h of the rock is a function of time t.)

40. Find a pair of numbers whose product is maximum when the pair of numbers has a sum of 16.

41. The back of George's property is a creek. George would like to enclose a rectangular area, using the creek as one side and fencing for the other three sides, to create a pasture for his two horses. If he has 300 feet of material, what is the maximum area the pasture can be?

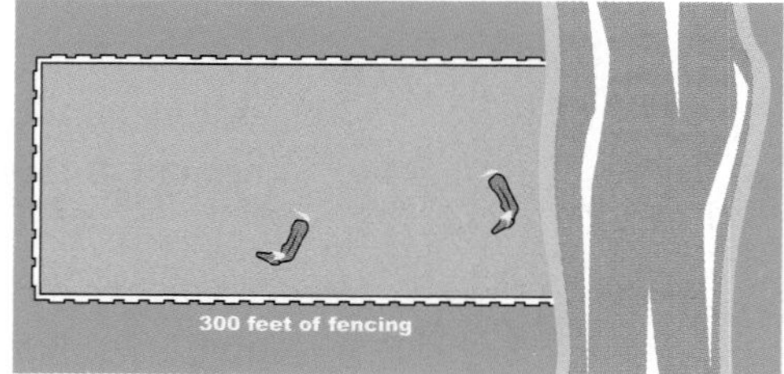

42. Find a pair of numbers whose product is maximum when two times the first number plus the second number is 48.

43. The total revenue for Thompsons' Studio Apartments is given as the function $R(x) = 100x - 0.1x^2$, where x is the number of rooms rented. What is the number of rooms rented that produces the maximum revenue?

44. The total revenue of Tran's Machinery Rental is given as the function $R(x) = 300x - 0.4x^2$, where x is the number of units rented. What is the number of units rented that produces the maximum revenue?

45. The total cost of producing a type of small car is given as the function $C(x) = 9000 - 135x + 0.045x^2$, where x is the number of cars produced. How many cars should be produced to incur a minimum cost?

46. The total cost of manufacturing a set of golf clubs is given as the function $C(x) = 800 - 10x + 0.20x^2$, where x is the number of sets of golf clubs produced. How many sets of golf clubs should be manufactured to incur minimum cost?

47. The owner of a parking lot is going to enclose a rectangular area with fencing using an existing fence as one of the sides. The owner has 220 feet of new fencing material (which is much less than the length of the existing fence). What is the maximum square feet that the area can be?

Match the following quadratic functions with their graphs.

48. $f(x) = (x - 4)^2$

49. $f(x) = x^2 + 1$

50. $f(x) = -(x + 2)^2 + 3$

51. $f(x) = x^2 + 2x$

52. $f(x) = -x^2 + 5$

53. $f(x) = x^2 - 2x - 3$

54. $f(x) = x^2 + 6x + 9$

55. $f(x) = x^2 + 2x + 2$

a.

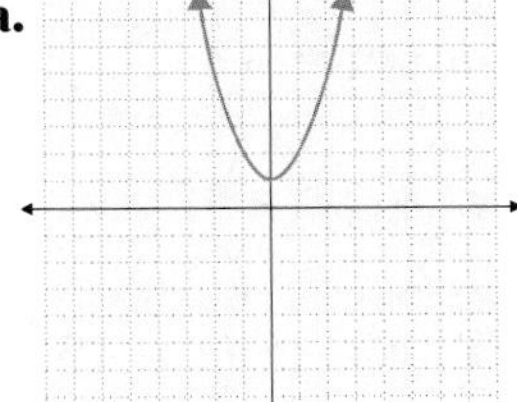

b.

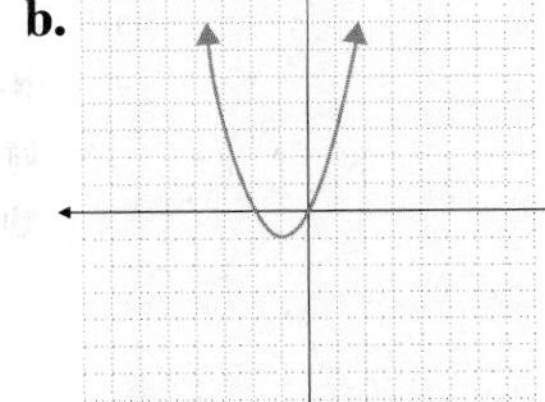

c.

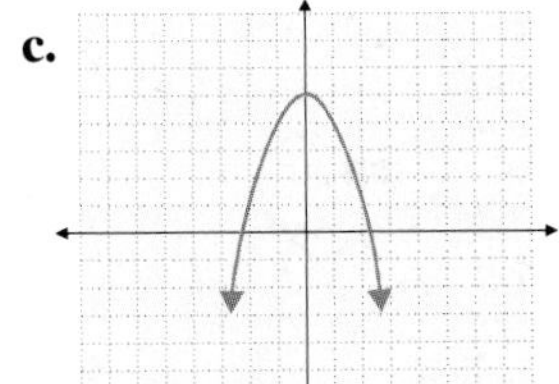

d.

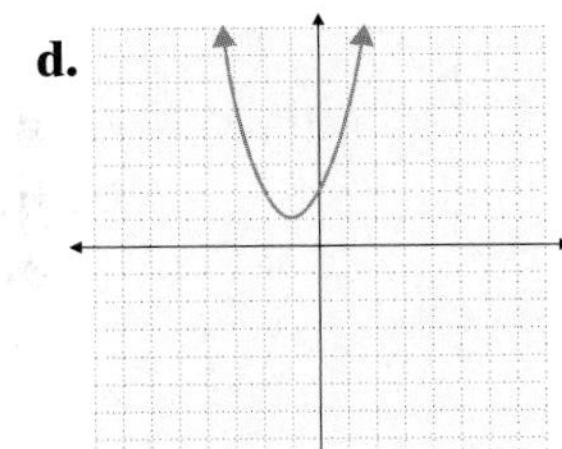

e.

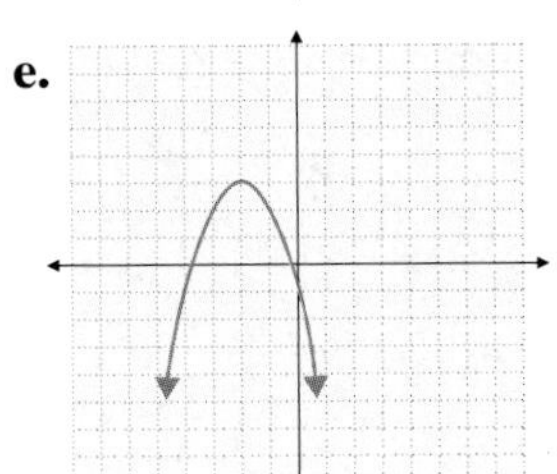

f.

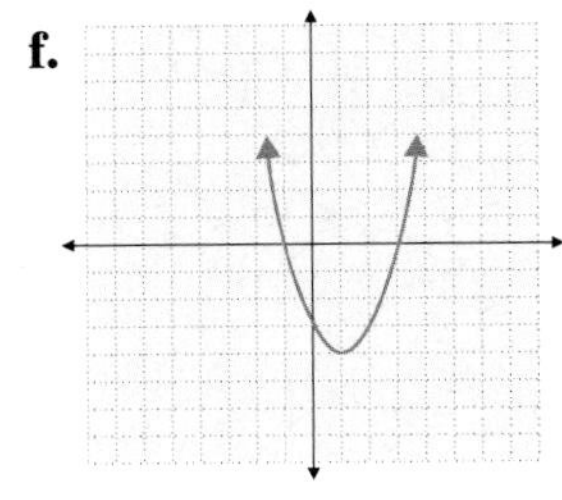

g.

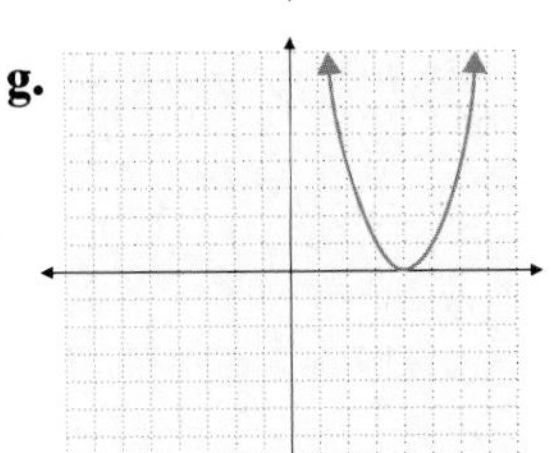

h.

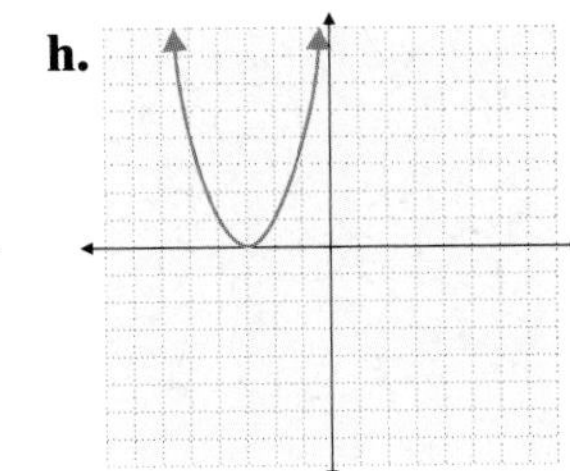

Use a graphing calculator to graph the following quadratic functions. Then determine the vertex and x-intercepts.

56. $f(x) = 2x^2 - 16x + 31$

57. $f(x) = -x^2 - 2x + 3$

58. $f(x) = x^2 - 8x - 20$

59. $f(x) = x^2 - 4x$

60. $f(x) = 25 - x^2$

61. $f(x) = 3x^2 + 18x$

62. $f(x) = x^2 + 2x + 1$

63. $f(x) = 3x^2 - 8x + 2$

64. $f(x) = -x^2 + 10x - 4$

65. $f(x) = \frac{1}{2}x^2 + x - 1$

Graph the equations a through c on the same set of x- and y-axes for each of the following problems. Compare and describe the differences of each graph as they compare to $y = x^2$.

66. **a.** $y = -x^2$ **b.** $y = -\frac{1}{2}x^2$ **c.** $y = -2x^2$

67. **a.** $y = x^2 + 2$ **b.** $y = x^2 - 2$ **c.** $y = x^2 + \frac{9}{2}$

68. **a.** $y = (x - 4)^2$ **b.** $y = (x + 4)^2$ **c.** $y = \left(x - \frac{3}{2}\right)^2$

69. **a.** $y = -(x + 3)^2 - 3$ **b.** $y = (x - 2)^2 + 4$ **c.** $y = \frac{1}{2}(x + 5)^2 + 5$

3.3 Other Common Functions

TOPICS

1. Functions of the form ax^n
2. Functions of the form $\frac{a}{x^n}$
3. Functions of the form $ax^{1/n}$
4. The absolute value function
5. The greatest integer function
6. Piecewise-defined functions

In Section 3.2, we investigated the behavior of linear and quadratic functions at great length. These are just two types of commonly occurring functions; there are many other functions that arise naturally in the course of solving various problems. In this section, we will briefly explore a variety of other classes of functions, with the aim of building up a useful portfolio of functions with which you are familiar.

Topic 1: Functions of the Form ax^n

We already know what the graph of any function of the form $f(x) = ax$ or $f(x) = ax^2$ looks like, as these are, respectively, simple linear and quadratic functions. What happens to the graphs as we increase the exponent, and consider functions of the form $f(x) = ax^3$, $f(x) = ax^4$, etc?

Roughly speaking, the behavior of a function of the form $f(x) = ax^n$, where a is a real number and n is a natural number, falls into one of two categories. Consider the graphs in Figure 1 on the next page.

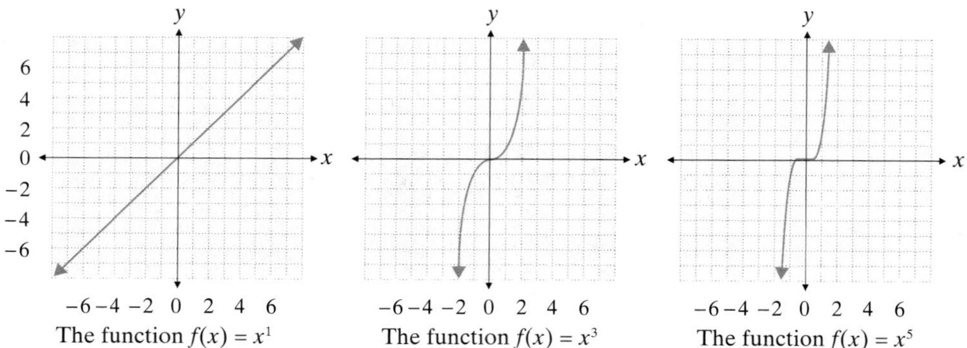

Figure 1: Odd Exponents

The three graphs in Figure 1 show the behavior of $f(x)=x^n$ for the first three odd exponents. Note that in each case, the domain and the range of the function are both the entire set of real numbers; the same is true for higher odd exponents as well. Now, consider the graphs in Figure 2:

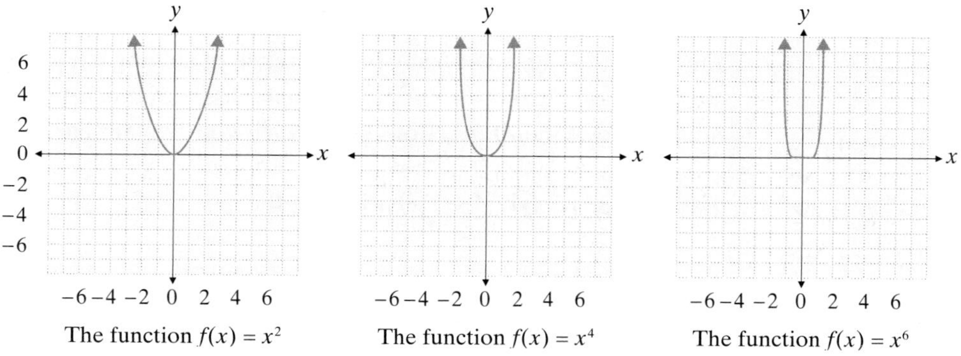

Figure 2: Even Exponents

These three functions are also similar to one another. The first one is the prototypical parabola that we studied in Section 3.2. The other two bear some similarity to parabolas, but are flatter near the origin and rise more steeply for $|x|>1$. For any function of the form $f(x)=x^n$ where n is an even natural number, the domain is the entire set of real numbers and the range is the interval $[0, \infty)$.

Multiplying a function of the form x^n by a constant a has the effect that we noticed in Section 3.2. If $|a|>1$, the graph of the function is stretched vertically; if $0<|a|<1$, the graph is compressed vertically; and if $a < 0$, the graph is reflected with respect to the x-axis. We can use this knowledge, along with a few specific points, to quickly sketch reasonably accurate graphs of any function of the form $f(x)=ax^n$.

example 1

Sketch the graphs of the following functions.

a. $f(x) = \dfrac{x^4}{5}$ **b.** $g(x) = -x^3$

Solutions:

a.

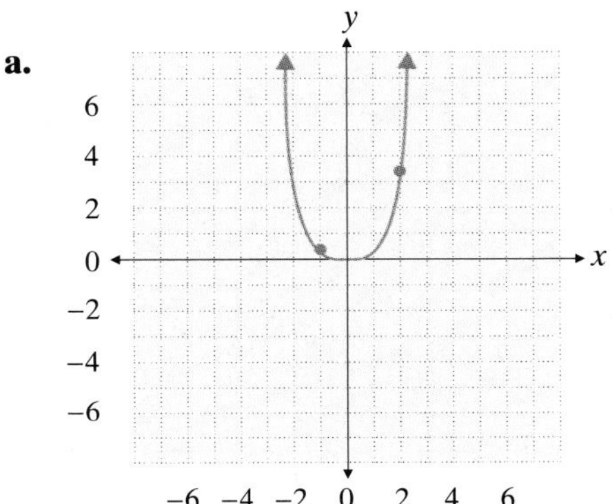

The graph of the function f will have the same basic shape as the function x^4, but compressed a bit vertically because of the factor of $\frac{1}{5}$. To make our sketch reasonably accurate, we can calculate the coordinates of a few points on the graph. The graph to the left illustrates that $f(-1) = \frac{1}{5}$ and that $f(2) = \frac{16}{5}$.

b.

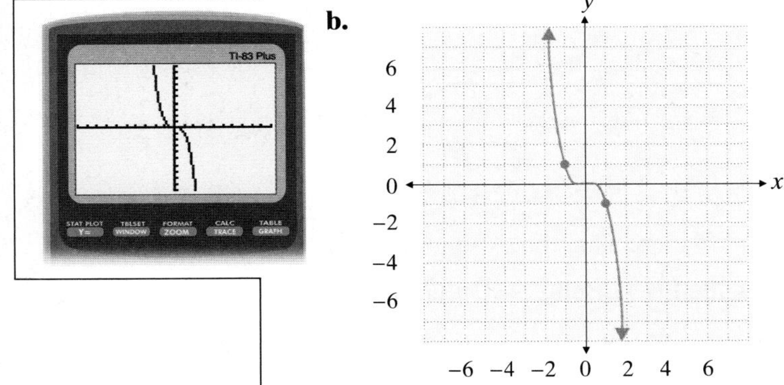

We know that the function g will have the same shape as the function x^3, but reflected with respect to the x-axis because of the factor of -1. The graph at left illustrates this point.

We have also plotted a few points on the graph of g, namely $(-1, 1)$ and $(1, -1)$, just to verify our analysis.

Topic 2: Functions of the Form $\dfrac{a}{x^n}$

We could also describe the following functions as having the form ax^{-n}, where a is a real number and n is a natural number. Again, the graphs of these functions fall roughly into two categories, as illustrated in Figures 3 and 4.

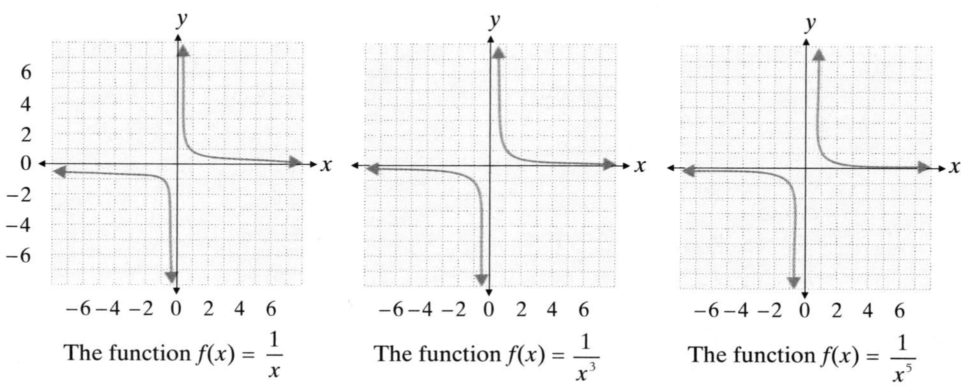

Figure 3: Odd Exponents

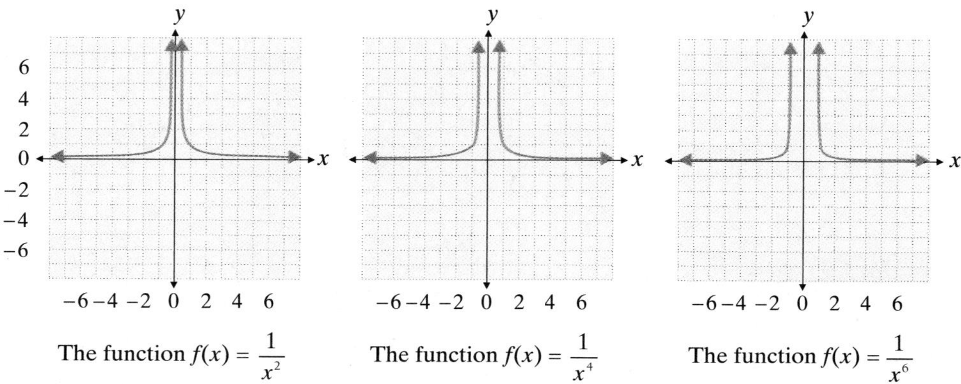

Figure 4: Even Exponents

As with functions of the form ax^n, increasing the exponent on functions of the form $\frac{a}{x^n}$ sharpens the curve of the graph near the origin. Note that the domain of any function of the form $f(x) = \frac{a}{x^n}$ is $(-\infty, 0) \cup (0, \infty)$, but that the range depends on whether n is even or odd. When n is odd, the range is also $(-\infty, 0) \cup (0, \infty)$, and when n is even the range is $(0, \infty)$.

example 2

Sketch the graph of the function $f(x) = -\frac{1}{4x}$.

Solution:

The graph of the function f is similar to that of the function $\frac{1}{x}$, with two differences. We obtain the formula $-\frac{1}{4x}$ by multiplying $\frac{1}{x}$ by $-\frac{1}{4}$, a negative number between −1 and 1. So one difference is that the graph of f is the reflection of $\frac{1}{x}$ with respect to the x-axis. The other difference is that the graph of f is compressed vertically, compared to $\frac{1}{x}$.

With the previous in mind, we can calculate the coordinates of a few points (such as $\left(-\frac{1}{4}, 1\right)$ and $\left(1, -\frac{1}{4}\right)$) and sketch the graph of f as shown below.

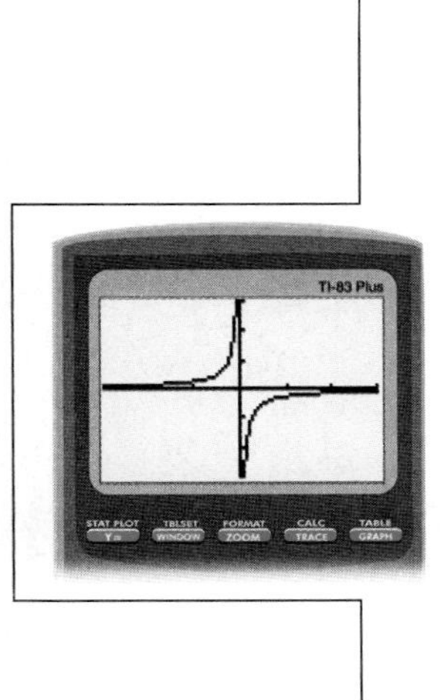

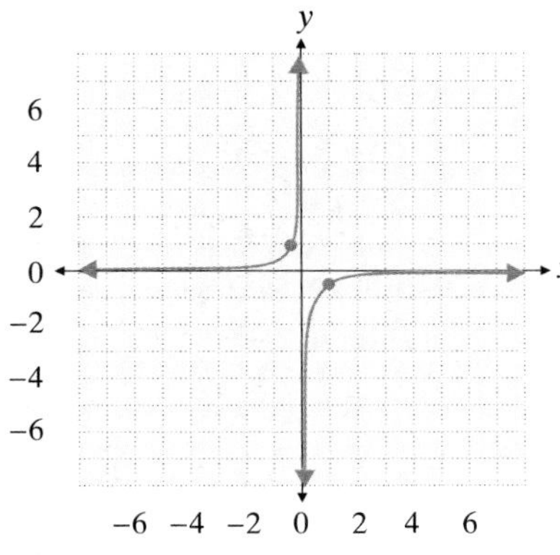

Topic 3: Functions of the Form $ax^{1/n}$

Using radical notation, these are functions of the form $a\sqrt[n]{x}$, where a is again a real number and n is a natural number. Square root and cube root functions, in particular, are commonly seen in mathematics.

Functions of this form again fall into one of two categories, depending on whether n is odd or even. To begin with, note that the domain and range are both the entire set of real numbers when n is odd, and that both are the interval $[0, \infty)$ when n is even. Figures 5 and 6 illustrate the two basic shapes of functions of this form.

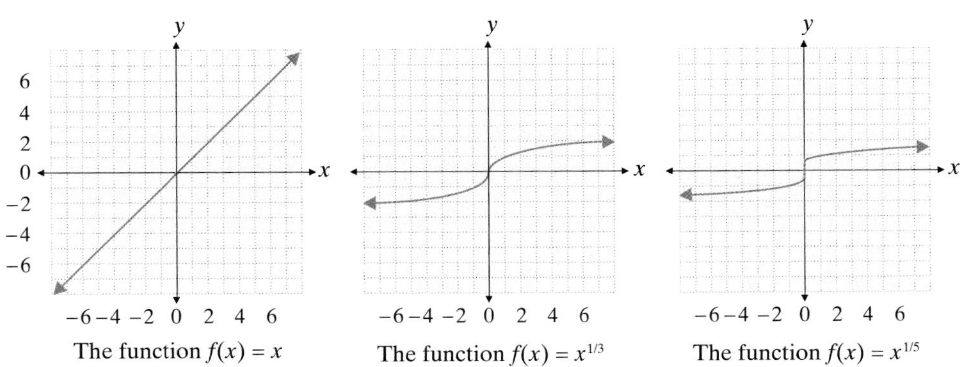

The function $f(x) = x$
The function $f(x) = x^{1/3}$
The function $f(x) = x^{1/5}$

Figure 5: Odd Roots

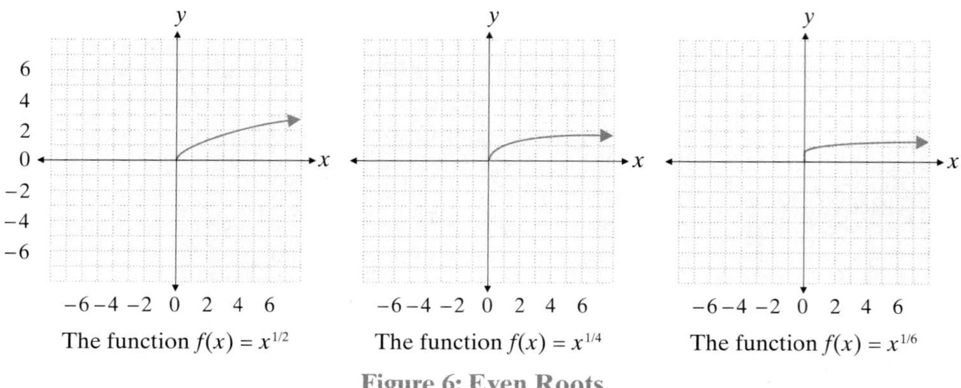

The function $f(x) = x^{1/2}$ The function $f(x) = x^{1/4}$ The function $f(x) = x^{1/6}$

Figure 6: Even Roots

At this point, you may be thinking that the graphs in Figures 5 and 6 appear familiar. The shapes in Figure 5 are the same as those seen in Figure 1, but rotated by 90 degrees and reflected with respect to the x-axis. Similarly, the shapes in Figure 6 bear some resemblance to those in Figure 2, except the graphs have been rotated and half of the graphs appear to have been erased. (These functions have a domain of $[0, \infty)$ since taking the even root of a negative number results in a nonreal number.) This resemblance is no accident, and is perhaps not surprising given that n^{th} roots undo n^{th} powers. We will explore this observation in much more detail in Section 3.6.

Topic 4: The Absolute Value Function

The Absolute Value Function

The basic absolute value function is $f(x)=|x|$. Note that for any value of x, $f(x)$ is non-negative, so the graph of f should lie on or above the x-axis. One way to determine its exact shape is to review the definition of absolute value:

$$|x|=\begin{cases} x & \text{if } x \geq 0 \\ -x & \text{if } x < 0 \end{cases}.$$

This means that for non-negative values of x, $f(x)$ is a linear function with a slope of 1, and for negative values of x, $f(x)$ is a linear function with a slope of −1. Both linear functions have a y-intercept of 0, so the complete graph of f is as shown in Figure 7.

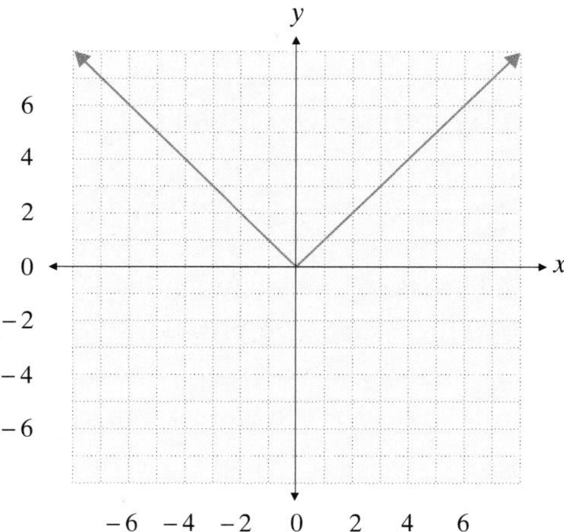

Figure 7: The Absolute Value Function

The effect of multiplying $|x|$ by a real number a is what we have come to expect: if $|a| > 1$ the graph is stretched vertically; if $0 < |a| < 1$, the graph is compressed vertically; and if a is negative the graph is reflected with respect to the x-axis.

example 3

Sketch the graph of the function $f(x) = -2|x|$.

To graph an absolute value, select **Y=** and enter the equation. To enter the absolute value, select **Math** and then use the right arrow to select `Num`. Press **1**, and you will see `abs` in your equation.

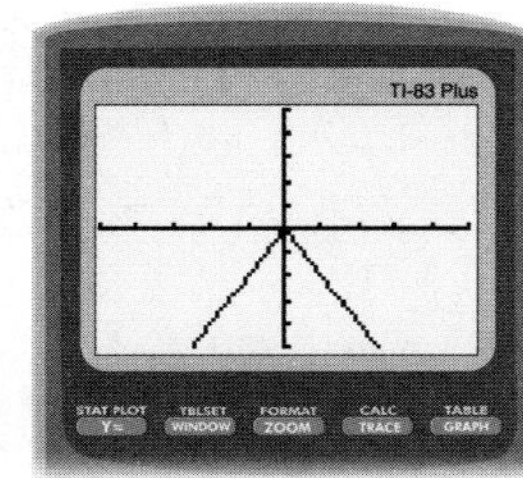

Solution:

The graph of f will be a stretched version of $|x|$, reflected with respect to the x-axis. As always, we can plot a few points to verify that our reasoning is correct. In the graph below, we have plotted the values of $f(-4)$ and $f(2)$.

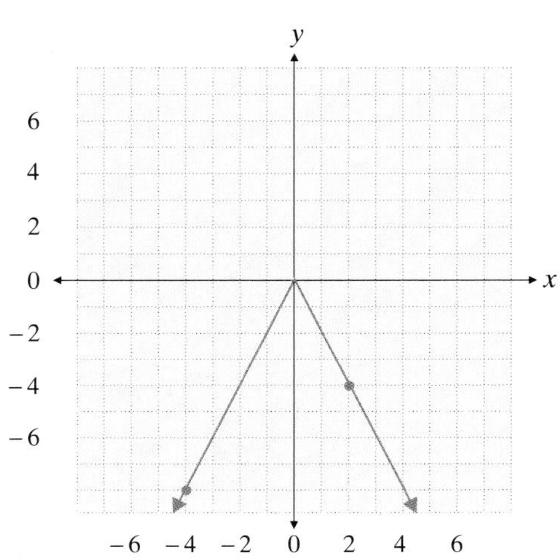

Topic 5: The Greatest Integer Function

The Greatest Integer Function

The greatest integer function, $f(x)=[\![x]\!]$, is a function commonly encountered in computer science applications. It is defined as follows: the **greatest integer of *x*** is the largest integer less than or equal to x. For instance, $[\![4.3]\!]=4$ and $[\![-2.9]\!]=-3$ (note that -3 is the largest integer to the left of -2.9 on the real number line).

Careful study of the greatest integer function reveals that its graph must consist of intervals where the function is constant, and that these portions of the graph must be separated by discrete "jumps," or breaks, in the graph. For instance, any value for x chosen from the interval $[1,2)$ results in $f(x)=1$, but $f(2)=2$. Similarly, any value for x chosen from the interval $[-3,-2)$ results in $f(x)=-3$, but $f(-2)=-2$.

Our graph of the greatest integer function must somehow indicate this repeated pattern of jumps. In cases like this, it is conventional to use an open circle on the graph to indicate that the function is either undefined at that point or is defined to be another value. Closed circles are used to emphasize that a certain point really does lie on the graph of the function. With these conventions in mind, the graph of the greatest integer function appears in Figure 8.

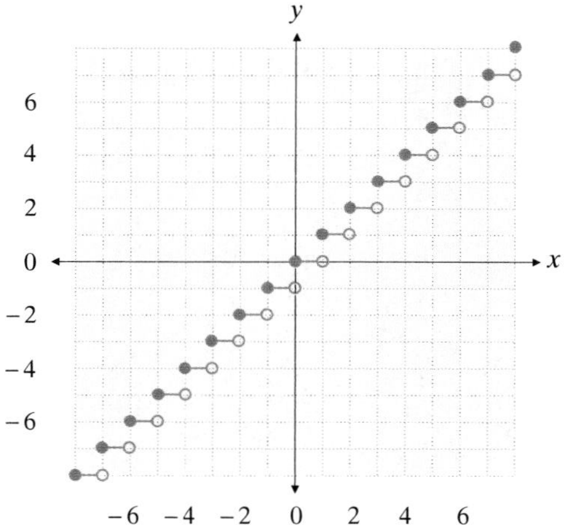

Figure 8: The Greatest Integer Function

Topic 6: Piecewise-Defined Functions

The last category of functions that we will discuss in this section is that of piecewise-defined functions.

Piecewise-Defined Functions

A **piecewise-defined** function is a function defined in terms of two or more formulas, each valid for its own unique portion of the real number line. In evaluating a piecewise-defined function f at a certain value for x, it is important to correctly identify which formula is valid for that particular value.

We have worked with one piecewise-defined function already; in evaluating the absolute value of x, we use one formula if x is greater than or equal to 0 and a different formula if x is less than 0. Keep the absolute value function in mind as you study the piecewise-defined function in Example 4.

example 4

Sketch the graph of the function $f(x)=\begin{cases} -2x-2 & \text{if } x \leq -1 \\ x^2 & \text{if } x > -1 \end{cases}$.

Solution:

The function f is a linear function on the interval $(-\infty,-1]$ and a quadratic function on the interval $(-1,\infty)$. To graph f, we graph each portion separately, making sure that each formula is applied only on the appropriate interval. The complete graph appears below, with the points $f(-4)=6$ and $f(2)=4$ noted in particular. Also note the use of a closed circle at $(-1,0)$ to emphasize that this point is part of the graph, and the use of an open circle at $(-1,1)$ to indicate that this point is not part of the graph. That is, the value of $f(-1)$ is 0, not 1.

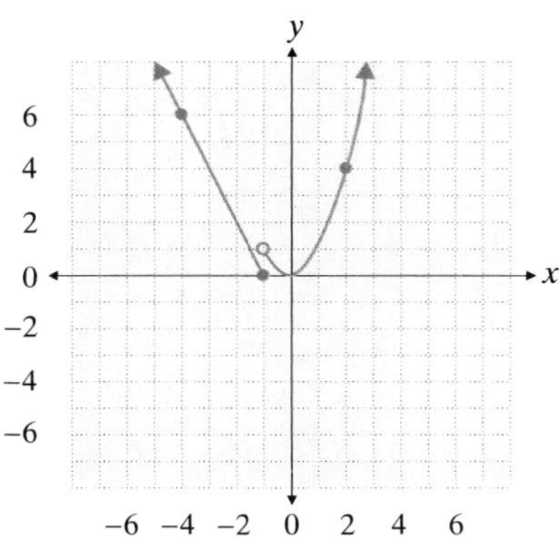

technology note

As their name implies, graphing calculators are very useful for sketching graphs of functions, though care must be used in interpreting the results. Computer algebra systems like *Mathematica* are also very good at graphing, and generally provide better quality sketches. The command in *Mathematica* for graphing a function is `Plot`, and its use in sketching the graph of $f(x)=x^3-2x^2+1$ is illustrated below.
(Consult your user's manual for instruction on how to use your graphing calculator or software package to graph functions, or see Appendix A for more guidance on using Mathematica.)

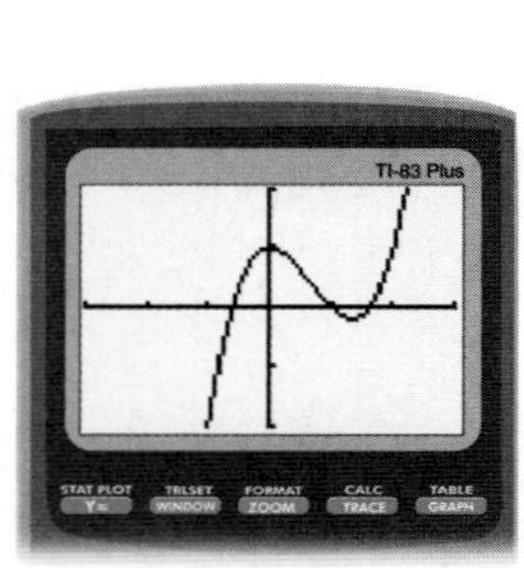

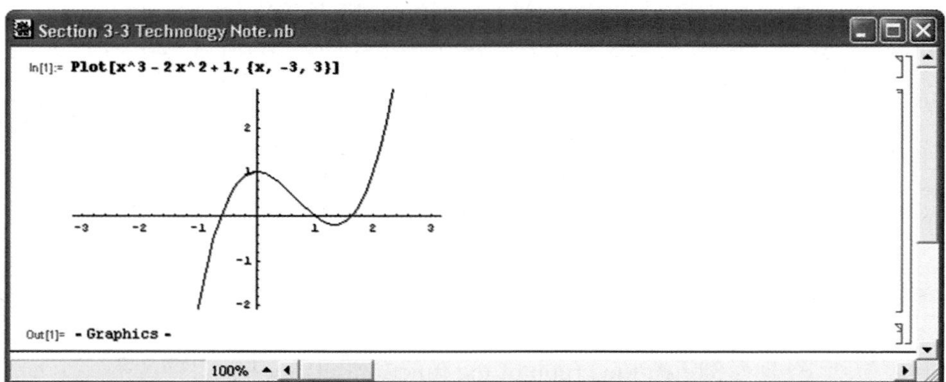

exercises

Sketch the graphs of the following functions. Pay particular attention to intercepts, if any, and locate these accurately. See Examples 1 through 4.

1. $f(x) = -x^3$

2. $g(x) = 2x^2$

3. $F(x) = \sqrt{x}$

4. $h(x) = \frac{1}{x}$

5. $p(x) = -\frac{2}{x}$

6. $q(x) = -\sqrt[3]{x}$

7. $G(x) = -|x|$

8. $k(x) = \frac{1}{x^3}$

9. $f(x) = 2[\![x]\!]$

10. $P(x) = -[\![x]\!]$

11. $G(x) = \frac{\sqrt{x}}{2}$

12. $H(x) = 0.5\sqrt[3]{x}$

13. $r(x) = 3|x|$

14. $p(x) = \frac{-1}{x^2}$

15. $W(x) = \frac{x^4}{16}$

16. $k(x) = \frac{x^3}{9}$

17. $h(x) = 2\sqrt[3]{x}$

18. $S(x) = \frac{4}{x^2}$

19. $d(x) = 2x^5$

20. $f(x) = -x^2$

21. $m(x) = \left[\!\!\left[\frac{x}{2}\right]\!\!\right]$

22. $r(x) = \frac{\sqrt[3]{x}}{3}$

23. $s(x) = -2|x|$

24. $t(x) = \frac{x^6}{4}$

25. $f(x) = \begin{cases} 3-x & \text{if } x < -2 \\ \sqrt[3]{x} & \text{if } x \geq -2 \end{cases}$

26. $g(x) = \begin{cases} -x^2 & \text{if } x \leq 1 \\ x^2 & \text{if } x > 1 \end{cases}$

27. $h(x) = \begin{cases} -|x| & \text{if } x < 2 \\ [\![x]\!] & \text{if } x \geq 2 \end{cases}$

28. $r(x) = \begin{cases} \frac{1}{x} & \text{if } x < 1 \\ -x & \text{if } x \geq 1 \end{cases}$

29. $p(x) = \begin{cases} x+1 & \text{if } x < -2 \\ x^3 & \text{if } -2 \leq x < 3 \\ -1-x & \text{if } x \geq 3 \end{cases}$

30. $q(x) = \begin{cases} -1 & \text{if } x \in \mathbb{Z} \\ 1 & \text{if } x \notin \mathbb{Z} \end{cases}$

31. $s(x) = \begin{cases} \frac{x^2}{3} & \text{if } x < 0 \\ -\frac{x^2}{3} & \text{if } x \geq 0 \end{cases}$

32. $t(x) = \begin{cases} x^4 & \text{if } x \leq 1 \\ [\![x]\!] & \text{if } x > 1 \end{cases}$

33. $u(x) = \begin{cases} [\![x]\!] & \text{if } x \leq 1 \\ 2x-2 & \text{if } x > 1 \end{cases}$

34. $v(x) = \begin{cases} x^2 & \text{if } -1 \leq x \leq 1 \\ |x| & \text{if } x < -1 \text{ or } x > 1 \end{cases}$

35. $M(x) = \begin{cases} x & \text{if } x \in \mathbb{Z} \\ -x & \text{if } x \notin \mathbb{Z} \end{cases}$

36. $N(x) = \begin{cases} x^2 & \text{if } x \in \mathbb{Z} \\ [\![x]\!] & \text{if } x \notin \mathbb{Z} \end{cases}$

technology exercises

Use a graphing calculator or a computer algebra system such as Mathematica to graph the following functions. Experiment by instructing your calculator or software to draw different portions of the plane until you obtain a sketch that seems to capture the meaningful parts of the graph. See the Technology Note after Example 4 for guidance.

37. $f(x) = 10x^5 - x^3$

38. $g(x) = x^5 + x^2$

39. $f(x) = x^3 - 5x^2 + x$

40. $g(x) = \sqrt{x} - x^2$

41. $f(x) = \sqrt{x} + 3x - 1$

42. $g(x) = x^4 - 3x^3 + 2$

3.4 Variation and Multi-Variable Functions

TOPICS

1. Direct variation
2. Inverse variation
3. Joint variation
4. Interlude: multi-variable functions

Topic 1: Direct Variation

A number of natural phenomena exhibit the mathematical property of variation: one quantity varies (or changes) as a result of a change in another quantity. One example is the electrostatic force of attraction between two oppositely charged particles, which varies in response to the distance between the particles. Another example is the distance traveled by a falling object, which varies as time increases. Of course, the principle underlying variation is that of functional dependence; in the first example, the force of attraction is a function of distance, and in the second example the distance traveled is a function of time.

We have now gained enough familiarity with functions that we can define the most common forms of variation.

Direct Variation

We say that y **varies directly as the nth power of** x (or that y is **proportional to the nth power of** x) if there is a non-zero constant k (called the **constant of proportionality**) such that

$$y = kx^n.$$

Many variation problems involve determining what, exactly, the constant of proportionality is in a given situation. This can be easily done if enough information is given about how the various quantities in the problem vary with respect to one another, and once k is determined many other questions can be answered. The following example illustrates the solution of a typical direct variation problem.

example 1

Hooke's Law says that the force exerted by the spring in a spring scale varies directly with the distance that the spring is stretched. If a 5 pound mass suspended on a spring scale stretches the spring 2 inches, how far will a 13 pound mass stretch it?

Solution:

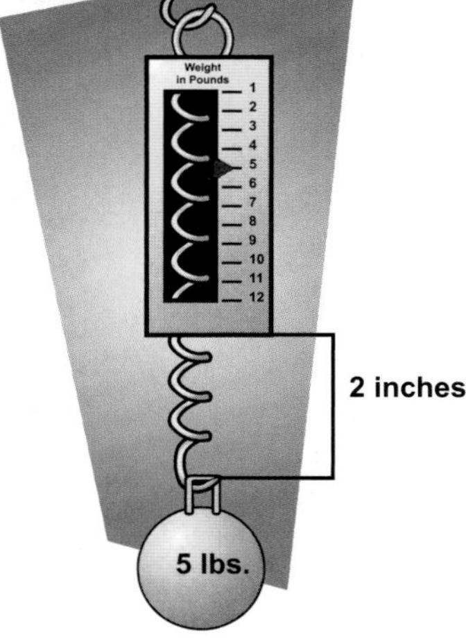

The first equation tells us that $F = kx$, where F represents the force exerted by the spring and x represents the distance that the spring is stretched. When a mass is suspended on a spring scale (and is stationary), the force exerted upward by the spring must equal the force downward due to gravity, so the spring exerts a force of 5 pounds when a 5 pound mass is suspended from it. So the second sentence tells us that

$$5 = 2k,$$

or $k = \frac{5}{2}$. We can now answer the question:

$$13 = \frac{5}{2}x$$
$$\frac{26}{5} = x.$$

So the spring stretches 5.2 inches when a 13 pound mass is suspended from it.

Topic 2: Inverse Variation

In many situations, an increase in one quantity results in a corresponding decrease in another quantity, and vice-versa. Again, this is a natural illustration of a functional relationship between quantities, and an appropriate name for this type of relationship is *inverse variation.*

Inverse Variation

We say that y **varies inversely as the nth power of** x (or that y is **inversely proportional to the nth power of** x) if there is a non-zero constant k such that

$$y = \frac{k}{x^n}.$$

The method of solving an inverse variation problem is identical to that seen in the first example. First, write an equation that expresses the nature of the relationship (including the as-yet-unknown constant of proportionality). Second, use the given information to determine the constant of proportionality. Third, use the knowledge gained to answer the question.

example 2

The weight of a person, relative to the Earth, is inversely proportional to the square of the person's distance from the center of the Earth. Using a radius for the Earth of 6370 kilometers, how much does a 180 pound man weigh when flying in a jet 9 kilometers above the Earth's surface?

Solution:

If we let W stand for the weight of a person and d the distance between the person and the Earth's center, the first sentence tells us that

$$W = \frac{k}{d^2}.$$

The second sentence gives us enough information to determine k. Namely, we know that $W = 180$ (pounds) when $d = 6370$ (kilometers). Solving the equation for k and substituting in the values that we know, we obtain

$$k = Wd^2 = (180)(6370)^2 \approx 7.3\times 10^9.$$

When the man is 9 kilometers above the Earth's surface, we know $d = 6379$, so the man's weight while flying is

$$\begin{aligned} W &= \frac{(180)(6370)^2}{(6379)^2} \\ &= 179.49 \text{ pounds.} \end{aligned}$$

Flying is not, therefore, a terribly effective way to lose weight.

Topic 3: Joint Variation

In more complicated situations, it may be necessary to identify more than two variables and to express how the variables relate to one another. And it may very well be the case that one quantity varies directly with respect to some variables and inversely with respect to others. For instance, the force of gravitational attraction F between two bodies of mass m_1 and mass m_2 varies directly as the product of the masses and inversely as the square of the distance between the masses: $F = \frac{km_1m_2}{d^2}$.

When one quantity varies directly as two or more other quantities, the word *jointly* is often used.

Joint Variation

We say that z **varies jointly as** x and y (or that z is **jointly proportional to** x **and** y) if there is a non-zero constant k such that

$$z = kxy.$$

If z varies **jointly as the** n^{th} **power of** x **and the** m^{th} **power of** y, we write

$$z = kx^n y^m.$$

example 3

The volume of a right circular cylinder varies jointly as the height and the square of the radius. Express this relationship in equation form.

Solution:

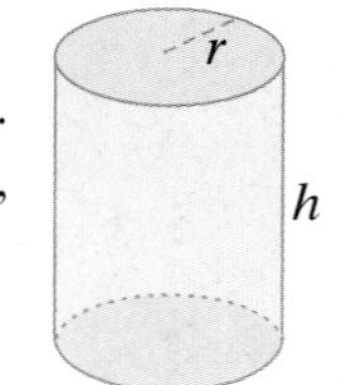

This simple problem merely asks for the form of the variation equation. If we let V stand for the volume of a right circular cylinder, r for its radius, and h for its height, we would write

$$V = kr^2h.$$

Of course, we are already familiar with this volume formula and know that the constant of proportionality is actually π, so we could provide more information and write

$$V = \pi r^2h.$$

Topic 4: Interlude: Multi-Variable Functions

The topic of variation provides an excellent opportunity to introduce functions that depend on two or more arguments. Examples abound in both pure and applied mathematics, and as you progress through Calculus and later math classes you will encounter such functions frequently.

Consider again how the force of gravity F between two objects depends on the masses m_1 and m_2 of the objects and the distance d between them:

$$F = \frac{km_1m_2}{d^2}.$$

If we change any of the three quantities m_1, m_2, and d, the force F changes in response. A slight extension of our familiar functional notation leads us then to express F as a function of m_1, m_2, and d and to write

$$F(m_1, m_2, d) = \frac{km_1m_2}{d^2}.$$

In fact, we can be a bit more precise and replace the constant of proportionality k with G, the Universal Gravitational Constant. Through many measurements in many different experiments, G has been determined to be approximately $6.67 \times 10^{-11}\ \mathrm{N} \cdot \mathrm{m}^2/\mathrm{kg}^2$ (N stands for the unit of force called the *Newton*; 1 Newton of force gives a mass of 1 kg an acceleration of 1 $\mathrm{m/s^2}$). If we use this value for G, we must be sure to measure the masses of the objects in kilograms and the distance between them in meters.

The next example illustrates an application of this function to which all of us on Earth can relate.

example 4

Determine the approximate force of gravitational attraction between the Earth and the Moon.

Solution:

The mass of the Earth is approximately 6.0×10^{24} kg and the mass of the Moon is approximately 7.4×10^{22} kg. The distance between these two bodies varies, but on average it is 3.8×10^{8} m. Using functional notation, we would write

cont'd on next page ...

$$F\left(6.0\times10^{24},\ 7.4\times10^{22},\ 3.8\times10^{8}\right)=\frac{\left(6.67\times10^{-11}\right)\left(6.0\times10^{24}\right)\left(7.4\times10^{22}\right)}{\left(3.8\times10^{8}\right)^{2}}$$
$$=2.1\times10^{20}\text{ N.}$$

It is this force of mutual attraction that keeps the Moon in orbit about the Earth.

exercises

Mathematical modeling is the process of finding a function that describes how quantities or variables relate to one another. The function is called the mathematical model. Find the mathematical model for each of the following verbal statements.

1. A varies directly as the product of b and h.
2. V varies directly as the product of four-thirds and r cubed.
3. W varies inversely as d squared.
4. P varies inversely as V.
5. r varies inversely as t.
6. S varies directly as the product of four and r squared.
7. x varies jointly as the cube of y and the square of z.
8. a varies jointly as the square of b and inversely to c.

Solve the following variation problems. See Examples 1, 2, and 3.

9. Suppose that y varies directly as the square root of x, and that $y=36$ when $x=16$. What is y when $x=20$?

10. Suppose that y varies inversely as the cube of x, and that $y=0.005$ when $x=10$. What is y when $x=5$?

11. Suppose that y varies directly as the cube root of x, and that $y=75$ when $x=125$. What is y when $x=128$?

12. Suppose that y is proportional to the 5th power of x, and that $y=96$ when $x=2$. What is y when $x=5$?

13. Suppose that y varies inversely as the square of x, and that $y=3$ when $x=4$. What is y when $x=8$?

14. Suppose that y is inversely proportional to the 4th power of x, and that $y = 1.5$ when $x = 4$. What is y when $x = 20$?

15. z varies directly as the square of x and inversely as y. If $z = 36$ when $x = 6$ and $y = 7$, what value does z have when $x = 12$ and $y = 21$?

16. Suppose that z varies jointly as the square of x and the cube of y, and that $z = 768$ when $x = 4$ and $y = 2$. What is z when $x = 3$ and $y = 2$?

17. Suppose that z is jointly proportional to x and y, and that $z = 90$ when $x = 1.5$ and $y = 3$. What is z when $x = .8$ and $y = 7$?

18. Suppose that z is jointly proportional to x and the cube of y, and that $z = 9828$ when $x = 13$ and $y = 6$. What is z when $x = 7$ and $y = 8$?

19. The distance that an object falls from rest, when air resistance is negligible, varies directly as the square of the time. A stone dropped from rest travels 144 feet in the first 3 seconds. How far does it travel in the first 4 seconds?

20. A record store manager observes that the number of CDs sold seems to vary inversely as the price per CD. If the store sells 840 CDs per week when the price per CD is \$15.99, how many does he expect to sell if he lowers the price to \$14.99?

21. A person's Body Mass Index (BMI) is used by physicians to determine if a patient's weight falls within reasonable guidelines relative to the patient's height. The BMI varies directly as a person's weight in pounds and inversely as the square of a person's height in inches. Given that a 6-foot tall man weighing 180 pounds has a BMI of 24.41, what is the BMI of a woman weighing 120 pounds with a height of 5 feet 4 inches?

22. The force necessary to keep a car from skidding as it travels along a circular arc varies directly as the product of the weight of the car and the square of the car's speed, and inversely as the radius of the arc. If it takes 241 pounds of force to keep a 2200 pound car moving 35 miles per hour on an arc whose radius is 750 feet, how many pounds of force would be required if the car were to travel 40 miles per hour?

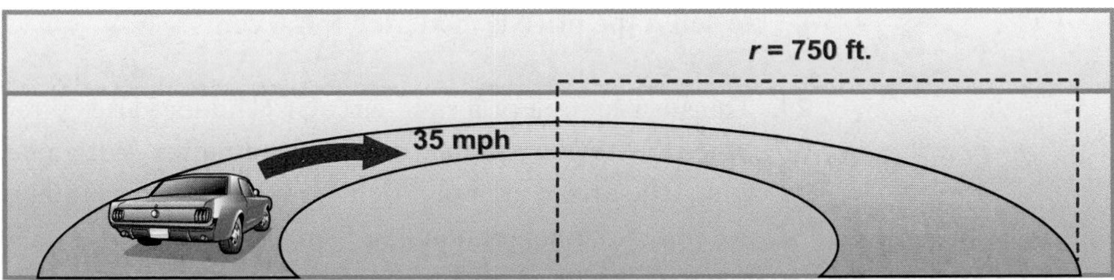

23. If a beam of width w, height h, and length l is supported at both ends, the maximum load that the beam can hold varies directly as the product of the width and the square of the height, and inversely as the length. A given beam 10 meters long with a width of 10 centimeters and a height of 5 centimeters can hold a load of 200 kilograms when the beam is supported at both ends. If the supports are moved inward so that the effective length of the beam is shorter, the beam can support more load. What should the distance between the supports be if the beam has to hold a load of 300 kilograms?

24. In a simple electric circuit connecting a battery and a light bulb, the current I varies directly with the voltage V but inversely with the resistance R. When a 1.5 volt battery is connected to a light bulb with resistance .3 ohms (Ω), the current that travels through the circuit is 5 amps. Find the current if the same light bulb is connected to a 6 volt battery.

25. The amount of time it takes for water to flow down a drainage pipe is inversely proportional to the square of the radius of the pipe. If a pipe of radius 1 cm can empty a sink in 25 seconds, find the radius of a pipe that would allow the sink to drain completely in 16 seconds.

26. The perimeter of a square varies directly as the length of the side of a square. If the perimeter of a square is 308 inches when one side is 77 inches, what is the perimeter of a square when the side is 133 inches?

27. The circumference of a circle varies directly as the diameter. A circular pizza slice has a length of 6.5 inches when the circumference of the pizza is 40.82 inches. What would the circumference of a pizza be if the pizza slice has a length of 5.5 inches?

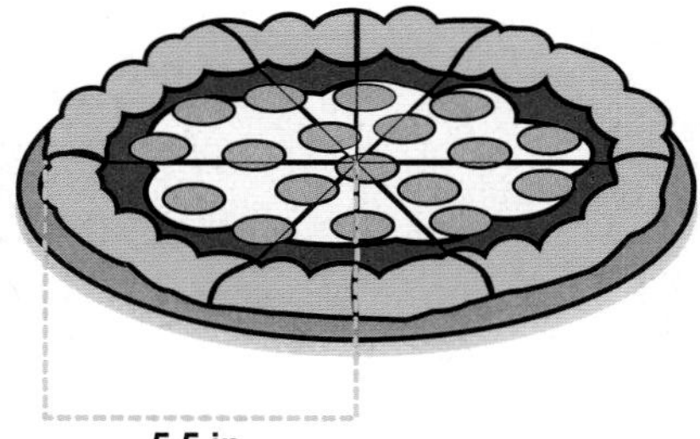

28. A hot dog vendor has determined that the number of hot dogs she sells a day is inversely proportional to the price she charges. The vendor wants to decide if increasing her price by 50 cents will drive away too many customers. On average, she sells 80 hot dogs a day at a price of \$3.50. How many hot dogs can she expect to sell if the price is increased by 50 cents?

29. The surface area of a right circular cylinder varies directly as the sum of the radius times the height and the square of the radius. With a height of 18 in. and a radius of 7 in., the surface area of a right circular cylinder is 1099 in^2. What would the surface area be if the height equaled 5 in. and the radius equaled 3.2 in.?

30. The velocity of an object undergoing no acceleration is inversely proportional to its time of travel. If an object traveling at the speed of light covers a distance of 8.9937×10^8 meters in 3 seconds, what is the value for the speed of light in meters per second?

31. In an electrical schematic, the voltage across a load is directly proportional to the power used by the load but inversely proportional to the current through the load. If a computer is connected to a wall outlet and the computer needs 18 volts to run and absorbs 54 watts of power, the current through the computer is 3 amps. Find the power absorbed by the computer if the same 18 volt computer is attached to a circuit with a loop current of .5 amps.

Express the indicated quantities as functions of the other variables. See Example 3.

32. A person's Body Mass Index (BMI) varies directly as a person's weight in pounds and inversely as the square of a person's height in inches. Given that a 6-foot tall man weighing 180 pounds has a BMI of 24.41, express BMI as a function of weight (w) and height (h).

33. The electric pressure varies directly with the square of the surface charge density (σ) and inversely with the permittivity (ε). If the surface charge density is 6 coulombs per unit area and the free space permittivity equals 3, the pressure is equal to 6 N/m^2. Express the electric pressure as a function of surface charge density and permittivity.

34. The volume of a right circular cylinder varies directly as the radius squared times the height of the cylinder. If the radius is 7 and the height is 4, the volume is equal to 615.44. Determine the expression of the volume of a right circular cylinder.

Find an equation for the relationship given and then use the equation to find the unknown value.

35. The variable a is proportional to $\sqrt{b}$. If $a = 15$ when $b = 9$, what is a when $b = 12$?

36. The variable a varies directly as b. If $a = 3$ when $b = 9$, what is a when $b = 7$?

37. The variable a varies directly as the square of b. If $a = 9$ when $b = 2$, what is a when $b = 4$?

38. The variable a is proportional to the square of b and inversely as the square root of c. If $a = 108$ when $b = 6$ and $c = 4$, what is a when $b = 4$ and $c = 9$?

39. The variable a varies jointly as b and c. If $a = 210$ when $b = 14$ and $c = 5$, what is the value of a when $b = 6$ and $c = 6$?

40. The variable a varies directly as the cube of b and inversely as c. If $a = 9$ when $b = 6$ and $c = 7$, what is the value of a when $b = 3$ and $c = 21$?

41. The price of gasoline purchased varies directly with the number of gallons of gas purchased. If 16 gallons of gas are purchased for \$34.40, what is the price of purchasing 20 gallons?

42. The illumination, I, of a light source varies directly as the intensity, i, and inversely as the square of the distance, d. If a light source with an intensity of 500 cp (candle-power) has an illumination of 20 fc (foot-candles) at a distance of 15 feet, what is the illumination at a distance of 20 feet?

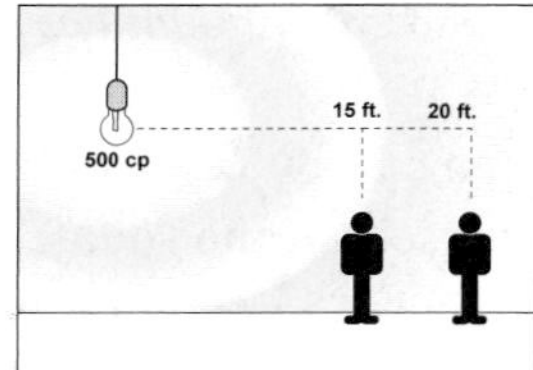

43. The force exerted by a spring varies directly with the distance that the spring is stretched. A hanging spring will stretch 9 cm if a weight of 15 grams is placed on the end of the spring. How far will the spring stretch if the weight is increased to 20 grams?

44. The volume of a cylinder varies jointly as its height and the square of its radius. If a cylinder has the measurements $V = 301.44$ cubic inches, $r = 4$ inches, and $h = 6$ inches, what is the volume of a cylinder that has a radius of 6 inches and a height of 8 inches?

45. The volume of a gas in a storage container varies inversely as the pressure on the gas. If the volume is 100 cubic centimeters under a pressure of 8 grams, what would the volume of the gas be if the pressure was decreased to 4 grams?

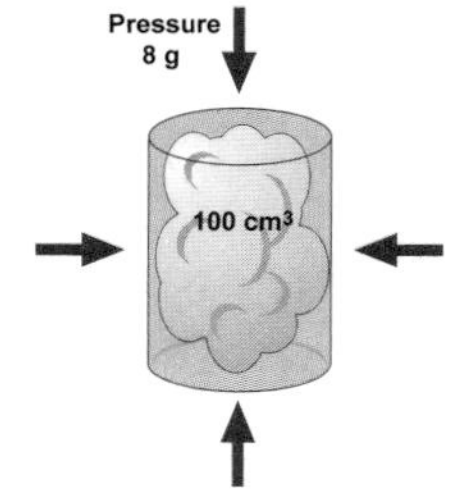

46. F is jointly proportional to a and b and varies inversely to c. If $F = 10$ when $a = 6$, $b = 5$, and $c = 2$, what is the value of F when $a = 12$, $b = 6$, and $c = 3$?

47. The resistance of a wire varies directly as its length and inversely as the square of the diameter. When a wire is 500 feet long and has a diameter of .015 in., it has a resistance of 20 ohms. What is the resistance of a wire that is 1200 feet long and has a diameter of .025 in.?

3.5 Transformations of Functions

TOPICS

1. Shifting, stretching, and reflecting graphs
2. Symmetry of functions and equations
3. Intervals of monotonicity

Topic 1: Shifting, Stretching, and Reflecting Graphs

Much of the material in this section was introduced in Section 3.2, in our discussion of quadratic functions. You may want to review the ways in which the prototypical quadratic function $f(x)=x^2$ can be shifted, stretched, and reflected as you work through the more general ideas here.

Horizontal Shifting

Let $f(x)$ be a function whose graph is known, and let h be a fixed real number. If we replace x in the definition of f by $x - h$, we obtain a new function $g(x)=f(x-h)$. The graph of g is the same shape as the graph of f, but shifted to the right by h units if $h > 0$ and shifted to the left by h units if $h < 0$.

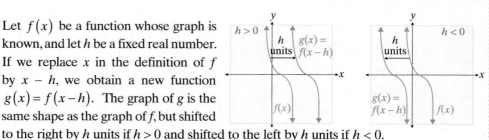

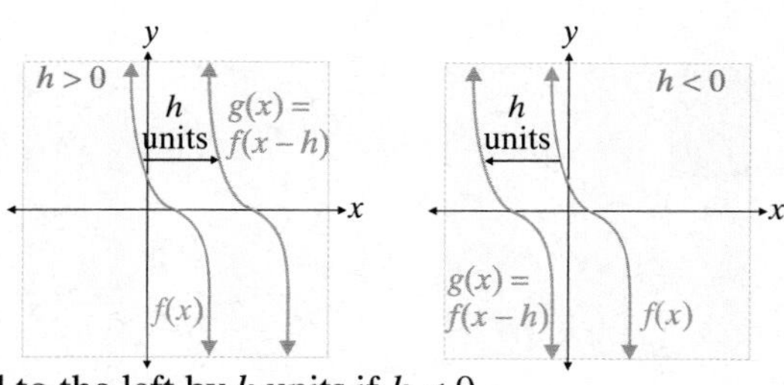

caution!

It is easy to forget that the minus sign in the expression $x - h$ is critical. It may help to remember a few specific examples: replacing x with $x - 5$ shifts the graph 5 units to the *right*, since 5 is positive. Replacing x with $x + 5$ shifts the graph 5 units to the *left*, since we have actually replaced x with $x - (-5)$. With practice, knowing the effect that replacing x with $x - h$ has on the graph of a function will become natural.

example 1

Sketch the graphs of the following functions.

a. $f(x) = (x+2)^3$ **b.** $g(x) = |x-4|$

Solutions:

a.

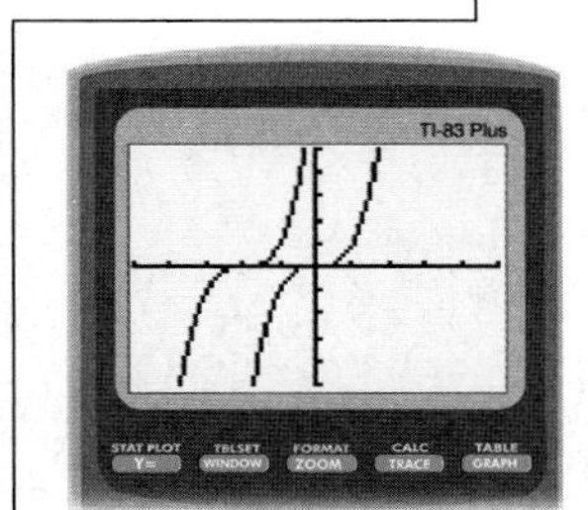

The screen above is the graph of both $y = x^3$ and $y = (x+2)^3$.

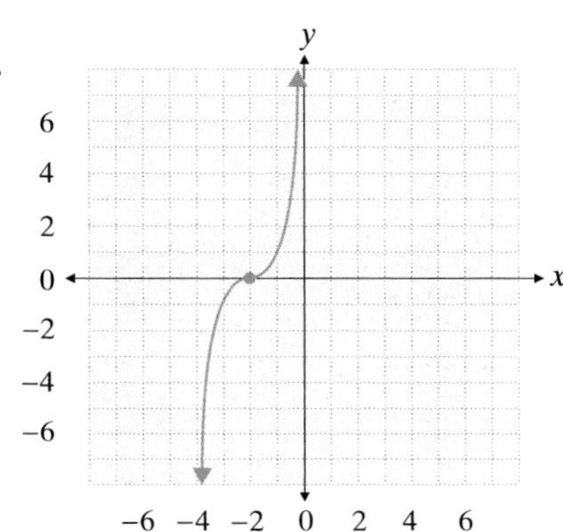

The shape of $(x+2)^3$ is the same as the shape of x^3, since one expression is obtained from the other by replacing x by $x + 2$. We simply draw the basic cubic shape (the shape of $y = x^3$) shifted to the left by 2 units. Note, for example, that $(-2, 0)$ is one point on the graph.

b.

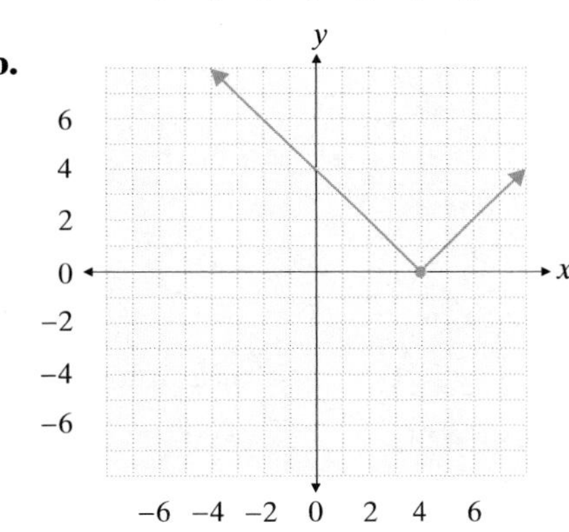

The graph of $g(x) = |x-4|$ has the same shape as the graph of the standard absolute value function, but shifted to the right by 4 units. Note, for example, that $(4, 0)$ lies on the graph of g.

Vertical Shifting

Let $f(x)$ be a function whose graph is known, and let k be a fixed real number. The graph of the function $g(x) = f(x) + k$ is the same shape as the graph of f, but shifted upward if $k > 0$ and downward if $k < 0$.

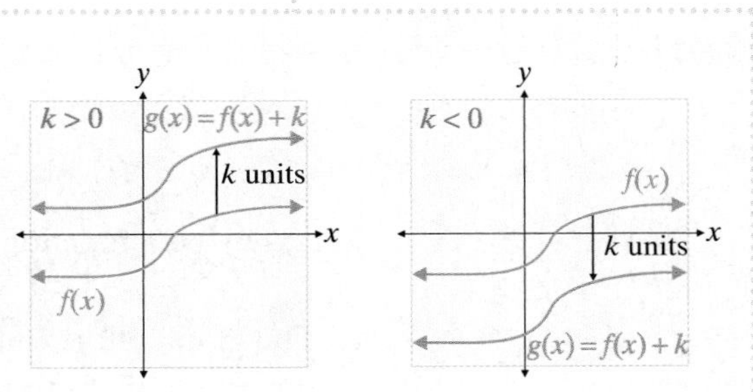

The effect of adding a constant k to a function can be most easily remembered by recalling that every point on the graph of a function f has the form $(x, f(x))$, so adding k shifts every point $(x, f(x))$ to $(x, f(x)+k)$. These new points are above the originals if k is positive and below the originals if k is negative.

example 2

Sketch the graphs of the following functions.

a. $f(x)=\frac{1}{x}+3$ **b.** $g(x)=\sqrt[3]{x}-2$

Solutions:

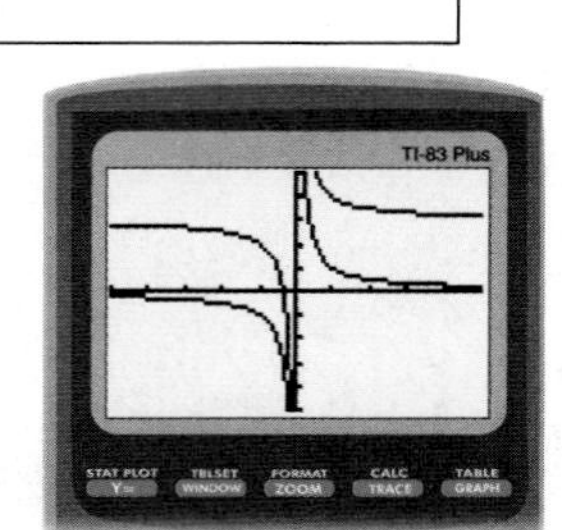

The screen above is the graph of both $y=\frac{1}{x}$ and $y=\frac{1}{x}+3.$

a.

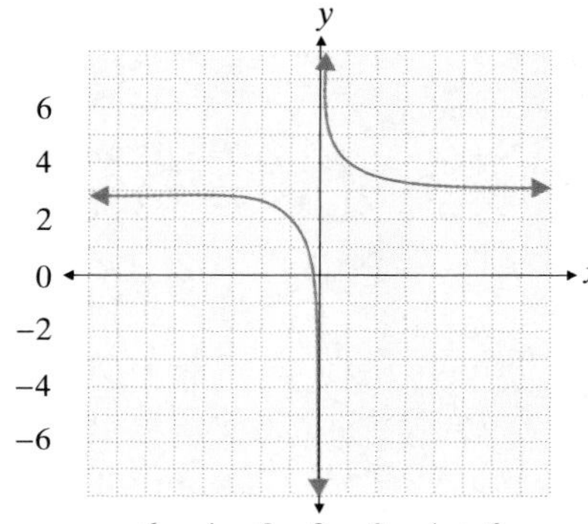

The graph of $f(x)=\frac{1}{x}+3$ is the graph of $y=\frac{1}{x}$ shifted up 3 units. Note that this doesn't affect the domain: the domain of f is $(-\infty, 0)\cup(0, \infty)$, the same as the domain of $y=\frac{1}{x}$. However, the range is affected. The range of f is $(-\infty, 3)\cup(3, \infty)$.

b.

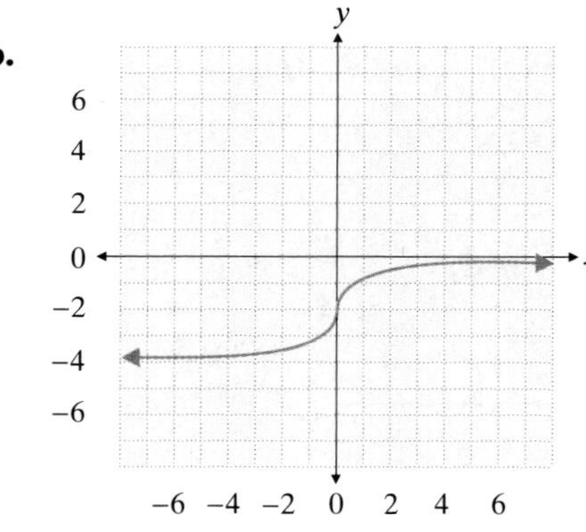

To graph $g(x)=\sqrt[3]{x}-2$, we shift the graph of $y=\sqrt[3]{x}$ down by 2 units.

Reflecting With Respect to the Axes

Let $f(x)$ be a function whose graph is known.

1. The graph of the function $g(x)=-f(x)$ is the reflection of the graph of f with respect to the x-axis.

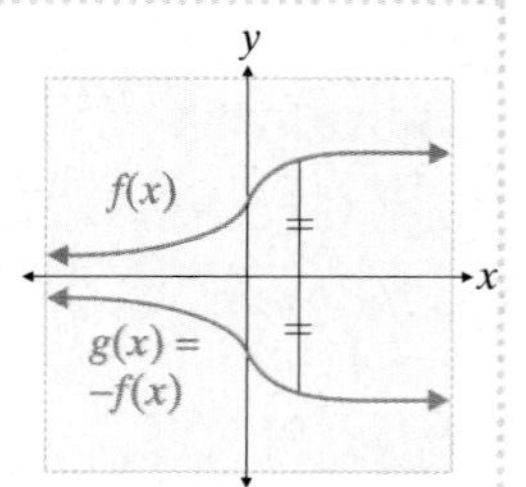

Reflecting With Respect to the Axes (cont'd.)

2. The graph of the function $g(x)=f(-x)$ is the reflection of the graph of f with respect to the y-axis.

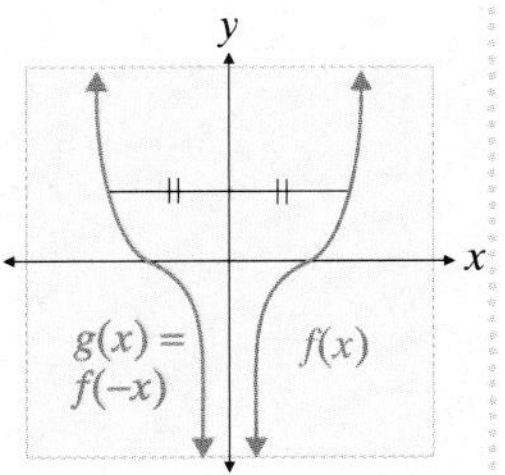

In other words, a function is reflected with respect to the x-axis by multiplying the entire function by -1, and reflected with respect to the y-axis by replacing x with $-x$.

example 3

Sketch the graphs of the following functions.

a. $f(x)=-x^2$ **b.** $g(x)=\sqrt{-x}$

Solutions:

a.

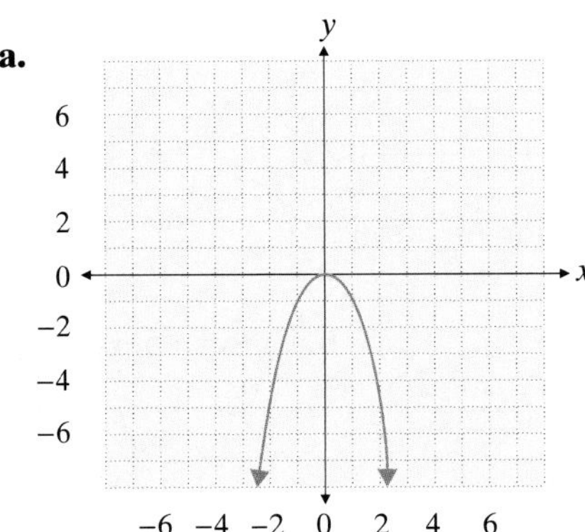

To graph $f(x)=-x^2$, we turn the graph of the prototypical parabola $y=x^2$ upside down. Note that the domain is still the entire real line, but the range of f is the interval $(-\infty, 0]$.

b.

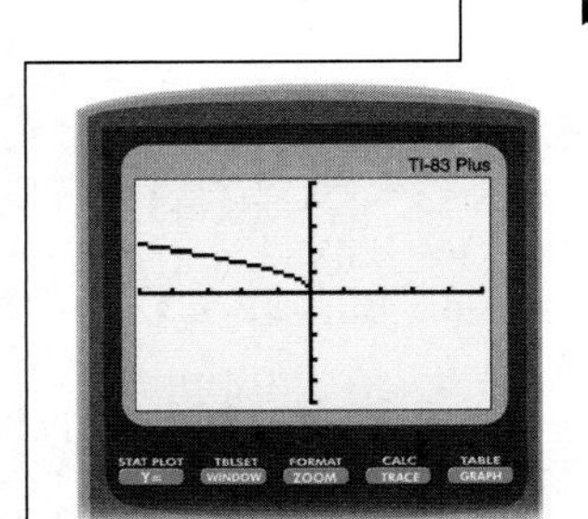

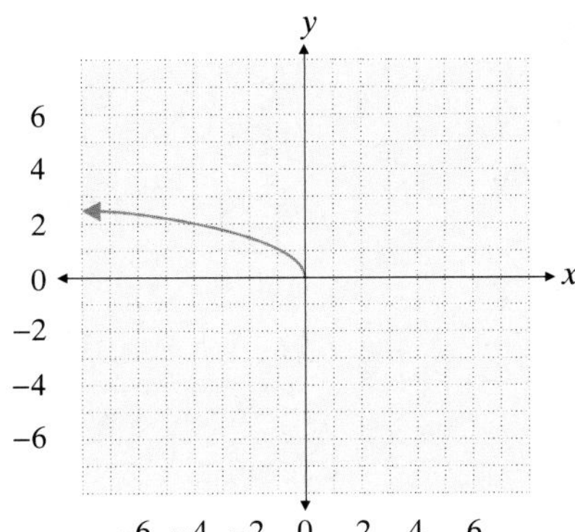

To graph $g(x)=\sqrt{-x}$, we reflect the graph of $y=\sqrt{x}$ with respect to the y-axis. Note that this changes the domain, but not the range. The domain of g is the interval $(-\infty, 0]$ and the range is $[0, \infty)$.

Stretching and Compressing

Let $f(x)$ be a function whose graph is known, and let a be a positive real number.

1. The graph of the function $g(x) = af(x)$ is *stretched* vertically compared to the graph of f if $a > 1$.

2. The graph of the function $g(x) = af(x)$ is *compressed* vertically compared to the graph of f if $0 < a < 1$.

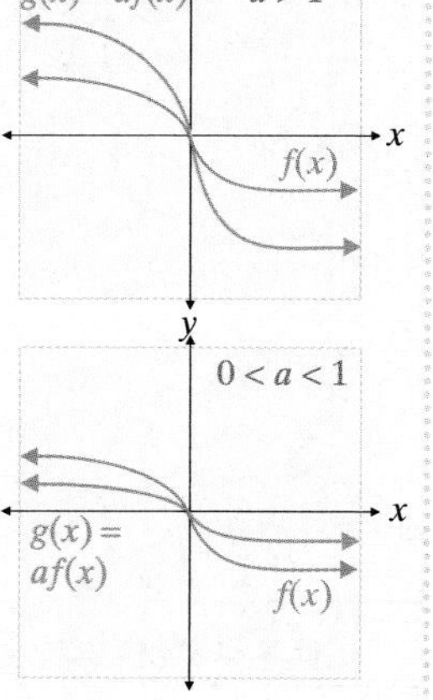

If the function g is obtained from the function f by multiplying f by a negative real number, think of the number as the product of –1 and a positive real number (namely, its absolute value). The box above tells us what multiplication by a positive constant does to a graph, and we already know that multiplying a function by –1 reflects the graph with respect to the x-axis.

example 4

Sketch the graphs of the following functions.

a. $f(x) = \dfrac{\sqrt{x}}{10}$ **b.** $g(x) = 5|x|$

Solutions:

a.

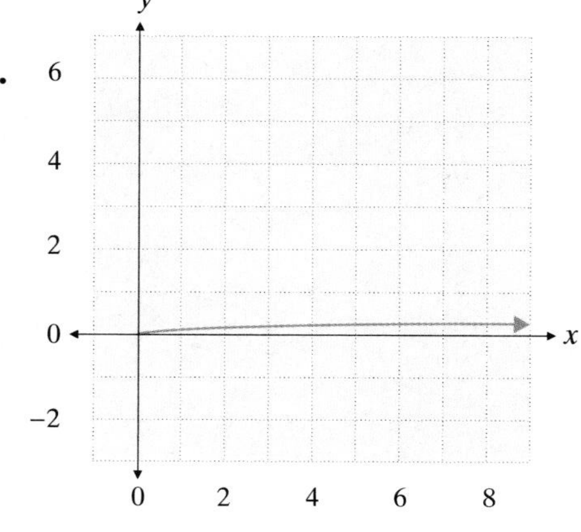

The graph of the function f is compressed considerably compared to the graph of $\sqrt{x}$, because of the factor of $\frac{1}{10}$. The shape is similar to the shape of $\sqrt{x}$, but all of the second coordinates have been multiplied by the factor of $\frac{1}{10}$, and are consequently smaller.

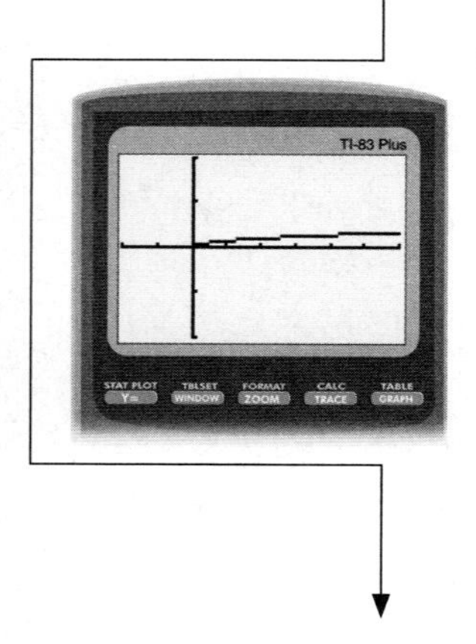

b.

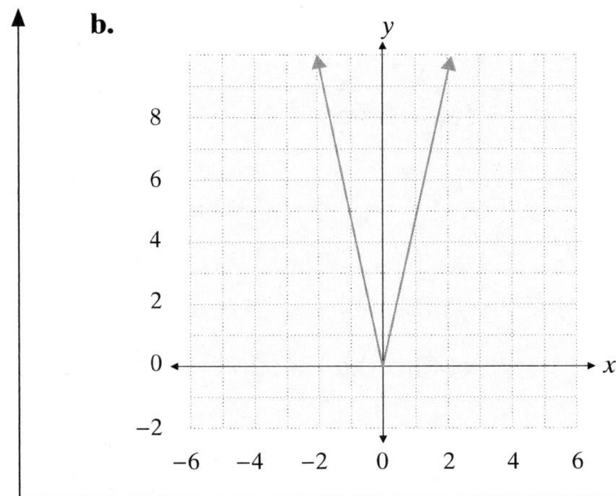

In contrast to the last example, the graph of $g(x)=5|x|$ is stretched compared to the standard absolute value function. Every second coordinate has been multiplied by a factor of 5, and is consequently larger.

We can now put all of the above together, and consider functions that have been derived through a sequence of transformations from simpler functions.

If a function g has been obtained from a simpler function f through a number of transformations, g can usually be understood by looking for the transformations in this order:

1. Horizontal shifts.
2. Stretching and compressing.
3. Reflections.
4. Vertical shifts.

Consider, for example, the function $g(x)=-2\sqrt{x+1}+3$. The function g has been "built up" from the basic square root function through a variety of transformations. First, $\sqrt{x}$ has been transformed into $\sqrt{x+1}$ by replacing x with $x+1$, and we know that this corresponds graphically to a shift leftward by 1 unit. Next, the function $\sqrt{x+1}$ has been multiplied by 2 to get the function $2\sqrt{x+1}$, and we know that this has the effect of stretching the graph of $\sqrt{x+1}$ vertically. The function $2\sqrt{x+1}$ has then been multiplied by –1, giving us $-2\sqrt{x+1}$, and the graph of this is the reflection of $2\sqrt{x+1}$ with respect to the x-axis. Finally, the constant 3 has been added to $-2\sqrt{x+1}$, shifting the entire graph upward by 3 units. These transformations are illustrated in order in Figure 1, culminating in the graph of $g(x)=-2\sqrt{x+1}+3$.

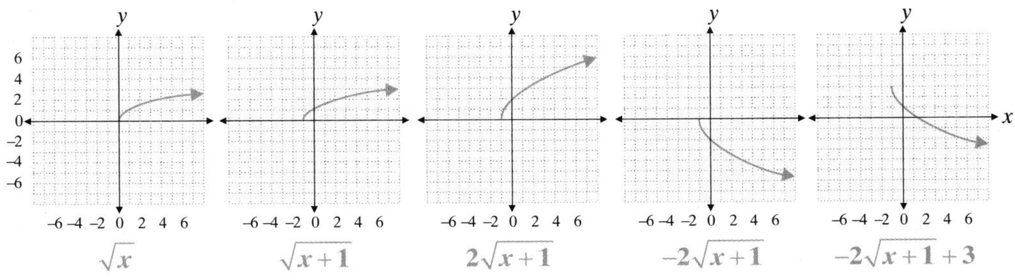

Figure 1: Building the Graph of $g(x)=-2\sqrt{x+1}+3$

example 5

Sketch the graph of the function $f(x)=\frac{1}{2-x}$.

Solution:

The less complex function that f is similar to is $\frac{1}{x}$. If we replace x by $x+2$ (shifting the graph 2 units to the left), we obtain the function $\frac{1}{x+2}$, which is closer to what we want. If we now replace x by $-x$, we have $\frac{1}{-x+2}$, which is the same as f. This last transformation reflects the graph of $\frac{1}{x+2}$ with respect to the y-axis. The entire sequence of transformations is shown below, ending with the graph of f.

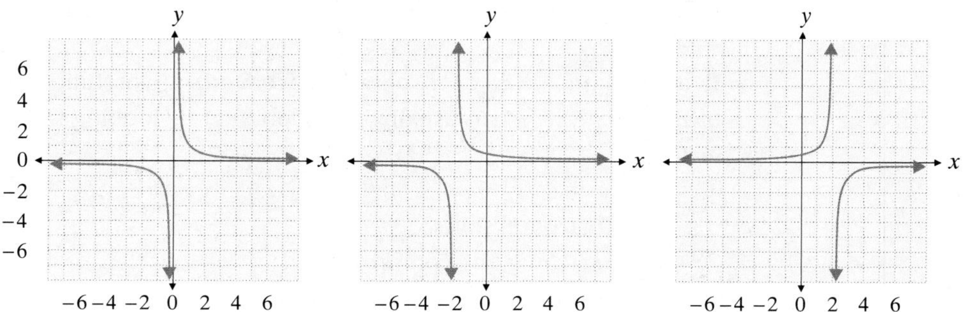

Note: An alternative approach to graphing $f(x)=\frac{1}{2-x}$ is to rewrite the function in the form $f(x)=-\frac{1}{x-2}$. In this form, the graph of f is the graph of $\frac{1}{x}$ shifted two units to the right, and then reflected with respect to the x-axis. The result is the same, as you should verify.

Topic 2: Symmetry of Functions and Equations

For some functions, replacing x with $-x$ has no effect, in the sense that the resulting function is identical to the original. We know that replacing x with $-x$ reflects the graph of a function with respect to the y-axis, so if f is one of these functions, the graph of f must be symmetric with respect to the y-axis. Algebraically, this means that $f(-x) = f(x)$ for all x in the domain of f.

y-axis Symmetry

The graph of a function f has **y-axis symmetry**, or is **symmetric with respect to the y-axis**, if $f(-x) = f(x)$ for all x in the domain of f. Such functions are called **even** functions.

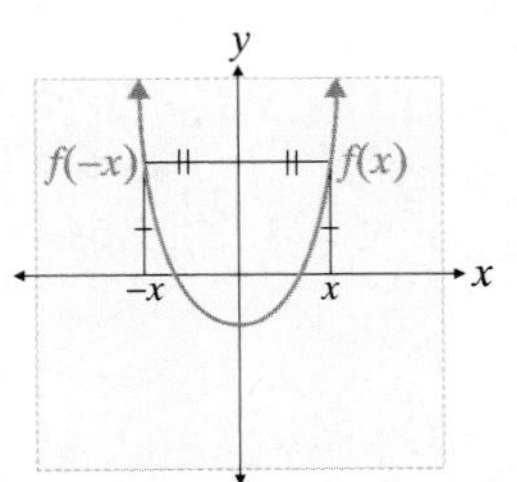

Functions whose graphs have y-axis symmetry are called even functions because polynomial functions with only even exponents form one large class of functions with this property. For instance, consider the function $f(x) = 7x^8 - 5x^4 + 2x^2 - 3$. This function is a polynomial of four terms, all of which have even degree (8, 4, 2, and 0, respectively). If we replace x with $-x$ and simplify the result, we obtain the function f again:

$$\begin{aligned} f(-x) &= 7(-x)^8 - 5(-x)^4 + 2(-x)^2 - 3 \\ &= 7x^8 - 5x^4 + 2x^2 - 3 \\ &= f(x). \end{aligned}$$

Be aware, however, that such polynomial functions are not the only even functions. We will see many more examples as we proceed.

There is another class of functions for which replacing x with $-x$ results in the exact opposite of the original function. That is, $f(-x) = -f(x)$ for all x in the domain. What does this mean geometrically? Suppose f is such a function, and that $(x, f(x))$ is a point on the graph of f. If we change the sign of both coordinates, we obtain a new point that is the reflection through the origin of the original.

For instance, if $(x, f(x))$ lies in the first quadrant, $(-x, -f(x))$ lies in the third, and if $(x, f(x))$ lies in the second quadrant, $(-x, -f(x))$ lies in the fourth. But since $f(-x) = -f(x)$, the point $(-x, -f(x))$ can be rewritten as $(-x, f(-x))$. Written in this form, we know that $(-x, f(-x))$ is a point on the graph of *f*, since *any* point of the form $(?, f(?))$ lies on the graph of *f*. So a function with the property $f(-x) = -f(x)$ has a graph that is symmetric with respect to the origin.

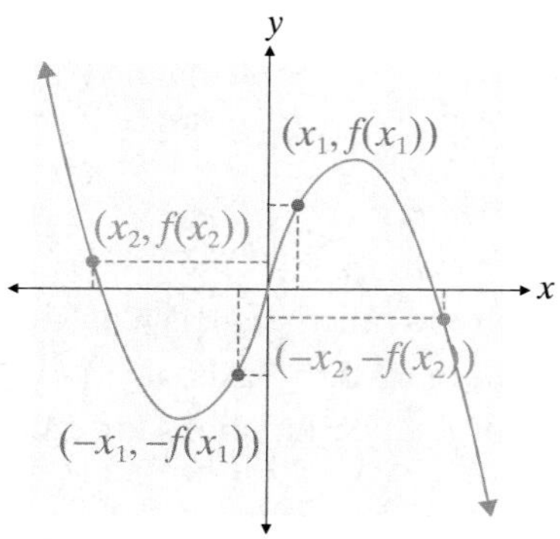

Origin Symmetry

The graph of a function *f* has **origin symmetry**, or is **symmetric with respect to the origin**, if $f(-x) = -f(x)$ for all x in the domain of *f*. Such functions are called **odd** functions.

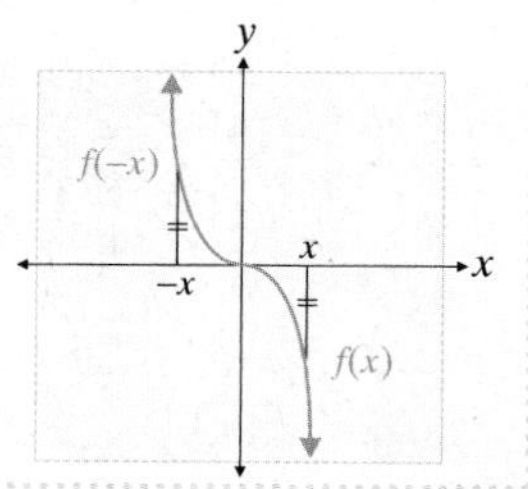

As you might guess, such functions are called odd because polynomial functions with only odd exponents serve as simple examples. For instance, the function $f(x) = -2x^3 + 8x$ is odd:

$$\begin{aligned} f(-x) &= -2(-x)^3 + 8(-x) \\ &= -2(-x^3) + 8(-x) \\ &= 2x^3 - 8x \\ &= -f(x). \end{aligned}$$

As far as functions are concerned, *y*-axis and origin symmetry are the two principal types of symmetry. You may have a sense, though, that something is missing. What about *x*-axis symmetry? It is certainly possible to draw a graph that displays *x*-axis symmetry; but unless the graph lies entirely on the *x*-axis, such a graph cannot represent a function. (Why not? Draw a variety of graphs that are symmetric with respect to the *x*-axis and convince yourself that they cannot be functions.)

This brings us back to relations. Recall that any equation in x and y defines a relation between the two variables. There are three principal types of symmetry that equations can possess.

Symmetry of Equations

We say that an equation in x and y is symmetric with respect to:

1. **The y-axis** if replacing x with $-x$ results in an equivalent equation.
2. **The x-axis** if replacing y with $-y$ results in an equivalent equation.
3. **The origin** if replacing x with $-x$ and y with $-y$ results in an equivalent equation.

Noting the symmetry of a function or an equation can serve as a useful aid in graphing. For instance, in graphing an even function it is only necessary to graph the part to the right of the y-axis, as the left half of the graph is the reflection of the right half with respect to the y-axis. Similarly, if a function is odd, the left half of its graph is the reflection of the right half through the origin.

example 6

Sketch the graphs of the following relations.

a. $f(x)=\frac{1}{x^2}$ **b.** $g(x)=x^3-x$ **c.** $x=y^2$

Solutions:

a.

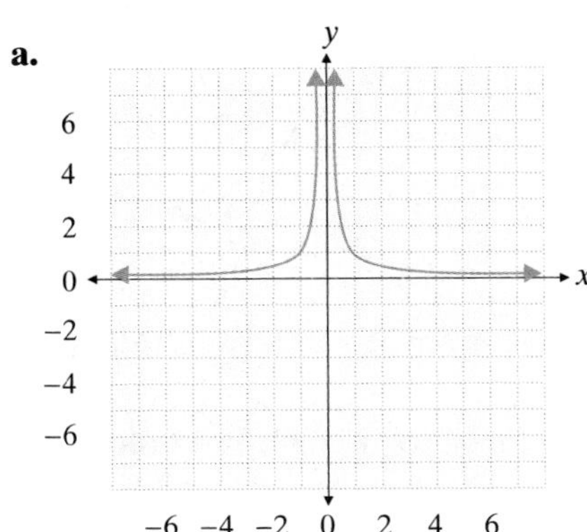

This relation is actually a function, one that we have already graphed in Section 3.3. Note that it is indeed an even function:

$$\begin{aligned} f(-x) &= \frac{1}{(-x)^2} \\ &= \frac{1}{x^2} \\ &= f(x). \end{aligned}$$

cont'd. on next page ...

b.

We do not quite have the tools yet to graph general polynomial functions, but $g(x)=x^3-x$ can be done. For one thing, g is odd: $g(-x)=-g(x)$ (as you should verify). If we now calculate a few values, such as $g(0)=0$, $g\left(\frac{1}{2}\right)=-\frac{3}{8}$, $g(1)=0$, and $g(2)=6$, and reflect these through the origin, we begin to get a good idea of the shape of g.

c.

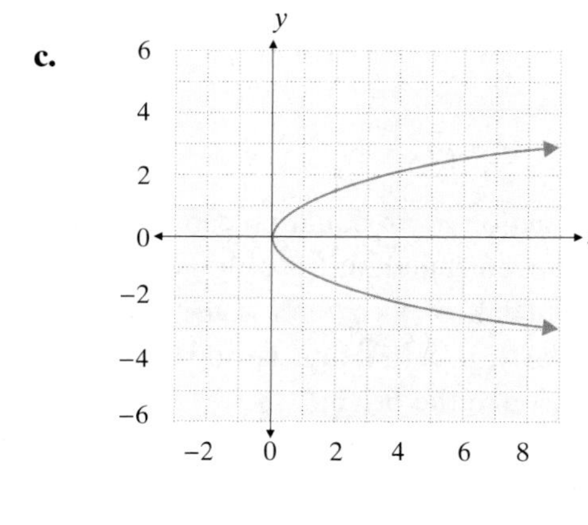

The equation $x=y^2$ does not represent a function, but it is a relation in x and y that has x-axis symmetry. If we replace y with $-y$ and simplify the result, we obtain the original equation:

$$x=(-y)^2$$
$$x=y^2.$$

The upper half of the graph is the function $y=\sqrt{x}$, so drawing this and its reflection gives us the complete graph of $x=y^2$.

Topic 3: Intervals of Monotonicity

In many applications, it is useful to identify the intervals on the x-axis for which a function f is increasing in value, decreasing in value, or remaining constant. In this context, we say that these are intervals on which f is **monotone**.

Increasing, Decreasing, and Constant

We say that a function f is:

1. **Increasing on an interval** if for any x_1 and x_2 in the interval with $x_1 < x_2$, it is the case that $f(x_1) < f(x_2)$.
2. **Decreasing on an interval** if for any x_1 and x_2 in the interval with $x_1 < x_2$, it is the case that $f(x_1) > f(x_2)$.
3. **Constant on an interval** if for any x_1 and x_2 in the interval, it is the case that $f(x_1) = f(x_2)$.

Determining the intervals of monotonicity of a function is a task that can be quite demanding, and in many cases is best tackled with the tools of calculus. We will look at some problems now in which algebra and our intuition will be sufficient.

example 7

Determine the intervals of monotonicity of the function $f(x)=(x-2)^2-1$.

Solution:

We know that the graph of f is the prototypical parabola shifted 2 units to the right and down 1 unit, as shown below.

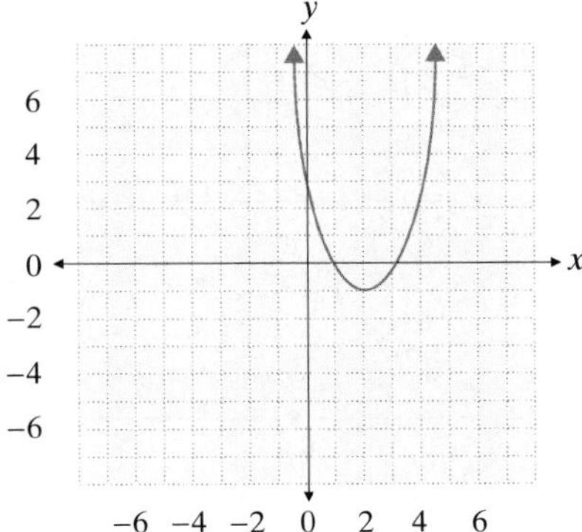

From the graph, we can see that f is decreasing on the interval $(-\infty, 2)$ and increasing on the interval $(2, \infty)$. Remember that these are intervals of the x-axis: if x_1 and x_2 are any two points in the interval $(-\infty, 2)$, with $x_1 < x_2$, then $f(x_1) > f(x_2)$. In other words, f is falling on this interval as we scan the graph from left to right. On the other hand, f is rising on the interval $(2, \infty)$ as we scan the graph from left to right.

example 8

The water level of a certain river varied over the course of a year as follows. In January, the level was 13 feet. From that level, the water increased linearly to a level of 18 feet in May. The water remained constant at that level until July, at which point it began to decrease linearly to a final level of 11 feet in December. Graph the water level as a function of time, and determine the intervals of monotonicity.

cont'd. on next page ...

Solution:

If we let 1 represent January through to 12 for December, the levels of monotonicity are as follows: increasing on $(1,5)$, constant on $(5,7)$, and decreasing on $(7,12)$. The graph of the water level as a function of the month is shown below.

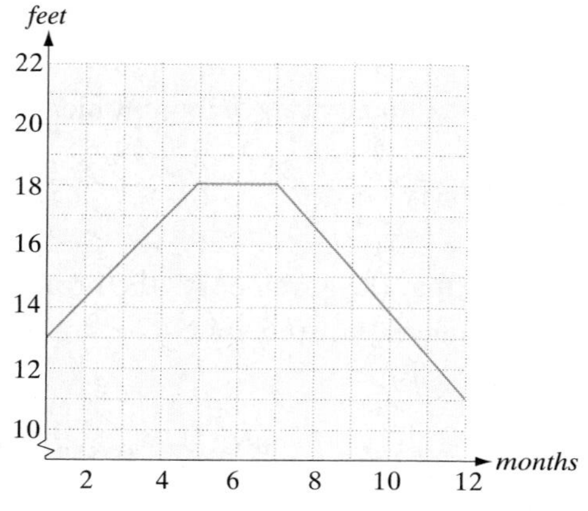

exercises

Sketch the graphs of the following functions by first identifying the more basic functions that have been shifted, reflected, stretched or compressed. Then determine the domain and range of each function. See Examples 1 through 5.

1. $f(x)=(x+2)^3$

2. $G(x)=|x-4|$

3. $p(x)=-(x+1)^2+2$

4. $g(x)=\sqrt{x+3}-1$

5. $q(x)=(1-x)^2$

6. $r(x)=-\sqrt[3]{x}$

7. $s(x)=\sqrt{2-x}$

8. $F(x)=\frac{|x+2|}{3}+3$

9. $w(x)=\frac{1}{(x-3)^2}$

10. $v(x)=\frac{1}{3x}-2$

11. $f(x)=\frac{1}{2-x}$

12. $k(x)=\sqrt{-x}+2$

13. $b(x)=[\![x-4]\!]+4$

14. $R(x)=4-|2x|$

15. $S(x)=(3-x)^3$

16. $g(x)=-\frac{1}{x+1}$

17. $h(x)=\frac{x^2}{2}-3$

18. $W(x)=1-|4-x|$

19. $g(x)=x^2-6x+9$ (Hint: find a better way to write the function.)

20. $h(x)=\frac{|x|}{x}$ (Hint: evaluate h at a few points to understand its behavior.)

21. $W(x)=\dfrac{x-1}{|x-1|}$ **22.** $S(x)=[\![x-2]\!]$ **23.** $V(x)=-3\sqrt{x-1}+2$

Write an equation for each of the functions described below.

24. Use the function $f(x) = x^2$. Move the function 3 units to the left and 4 units down.

25. Use the function $f(x) = x^2$. Move the function 4 units to the right and 2 units up.

26. Use the function $f(x) = x^2$. Move the function 6 units up and reflect across the x-axis.

27. Use the function $f(x) = x^2$. Move the function 2 units to the right and reflect across the y-axis.

28. Use the function $f(x) = x^3$. Move the function 1 unit to the left and reflect across the y-axis.

29. Use the function $f(x) = x^3$. Move the function 10 units to the right and 4 units up.

30. Use the function $f(x)=\sqrt{x}$. Move the function 5 units to the left and reflect across the x-axis.

31. Use the function $f(x)=\sqrt{x}$. Move the function 3 units down and reflect across the y-axis.

32. Use the function $f(x)=|x|$. Move the function 7 units to the left, reflect across the x-axis and reflect across the y-axis.

33. Use the function $f(x)=|x|$. Move the function 8 units to the right, 2 units up and reflect across the x-axis.

For each function below, determine if it is even, odd, or neither, and then graph the function. For each equation below, determine if it has y-axis symmetry, x-axis symmetry, origin symmetry, or none of the above, and then sketch the graph of the equation. See Example 6.

34. $f(x)=|x|+3$ **35.** $g(x)=x^3$ **36.** $h(x)=x^3-1$

37. $w(x) = \sqrt[3]{x}$

38. $x = -y^2$

39. $3y - 2x = 1$

40. $x + y = 1$

41. $F(x) = (x-1)^2$

42. $x = y^2 + 1$

43. $x = |2y|$

44. $g(x) = \dfrac{x^2}{5} - 5$

45. $s(x) = [\![x + \frac{1}{2}]\!]$

46. $m(x) = \sqrt[3]{x} - 1$

47. $xy = 2$

48. $x + y^2 = 3$

For each function below, find the intervals where the function is increasing, decreasing, or constant. See Examples 7 and 8.

49. $f(x) = (x+3)^2$

50. $g(x) = -|x-2|$

51. $h(x) = \dfrac{1}{x-1}$

52. $H(x) = \dfrac{1}{(x+3)^2}$

53. $G(x) = \sqrt{x+1}$

54. $F(x) = -2$

55. $p(x) = -30|x-1|$

56. $q(x) = (4-x)^2 + 1$

57. $r(x) = \dfrac{(x-7)^4}{-2} + 4$

58. $P(x) = \begin{cases} (x+3)^2 & \text{if } x < -1 \\ 1 & \text{if } x \geq -1 \end{cases}$

59. $Q(x) = \begin{cases} |x-1| & \text{if } x \leq 3 \\ 5-x & \text{if } x > 3 \end{cases}$

60. During the summer months, a swimming pool of one depth has varying water levels. In May the pool was 3.4 feet deep. After a steady increase due to the rain, the water level reached 4.9 feet in June. By July the water level decreased linearly to 4.2 feet. Knowing that the pool would be covered for the winter, the manager filled the pool in a linear fashion until it reached 5 feet in August. Graph the function as a function of time and determine the intervals of monotonicity.

61. The cost of a calling card is \$2 for the first minute and \$0.23 for each additional minute (even if the full minute is not used). A model of the function is given by $C(t) = 2 - 0.23[\![-t+1]\!]$ where t is the number of minutes the call lasted. Graph the function for an 8 minute, 35 second call and determine the intervals of monotonicity. How much did the call cost?

62. The rate of profit made by a hot dog vendor is given by the function

$$P(x) = \begin{cases} 2x - 3, & x \geq 0 \text{ and } x < 7 \\ \frac{1}{4}x^2, & x \geq 7 \end{cases}$$

where x is the number of hot dogs sold. Graph the function as a function of time and determine the intervals of monotonicity.

63. The rate of cost incurred by a newspaper stand is given by the function

$$P(x)=\begin{cases}-2\sqrt{x}+8, & x\geq 0 \text{ and } x<3\\ -x+8, & x\geq 3\end{cases}$$

where x is the number of newspapers sold. Graph the function as a function of time and determine the intervals of monotonicity.

technology exercises

Mentally sketch the graph of each of the following functions by identifying the basic shape that has been shifted, reflected, stretched, or compressed. Then use a graphing calculator or a computer algebra system (if available) to graph each function and check your reasoning. (See Appendix A for guidance on using Mathematica.)

64. $f(x)=-2(3-x)^3+5$ **65.** $f(x)=\dfrac{3}{x+5}-1$ **66.** $f(x)=\dfrac{-1}{(x-2)^2}-3$

67. $f(x)=-3|x+2|-4$ **68.** $f(x)=-\sqrt{1-x}+2$ **69.** $f(x)=\sqrt[3]{2+x}-1$

Write a possible equation for each of the functions depicted on the graphing calculator screens.

70.

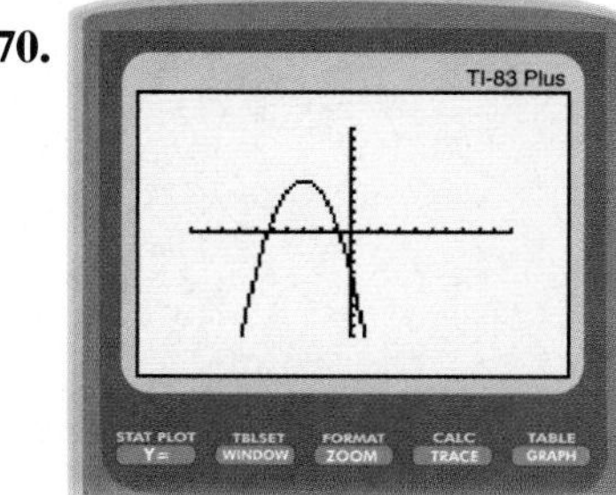

71.

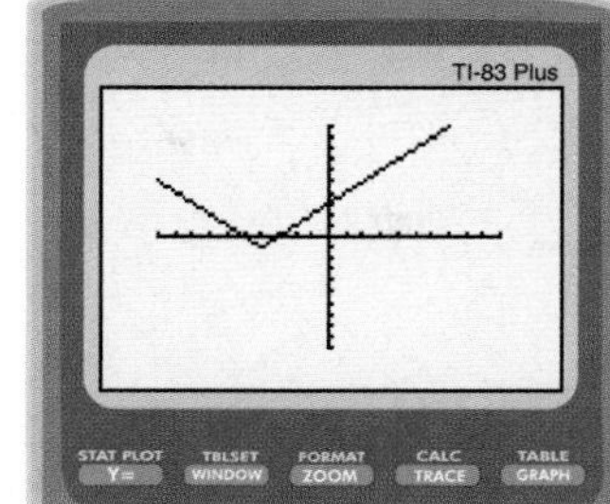

72.

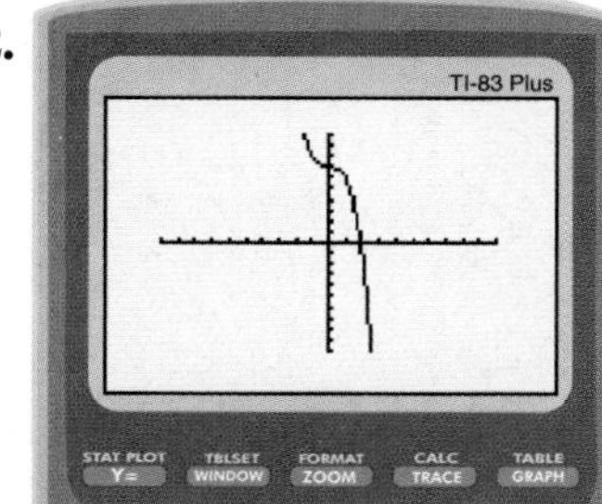

73.

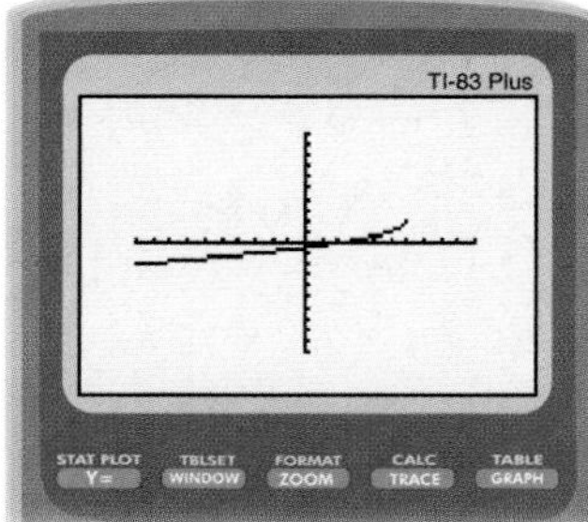

74. 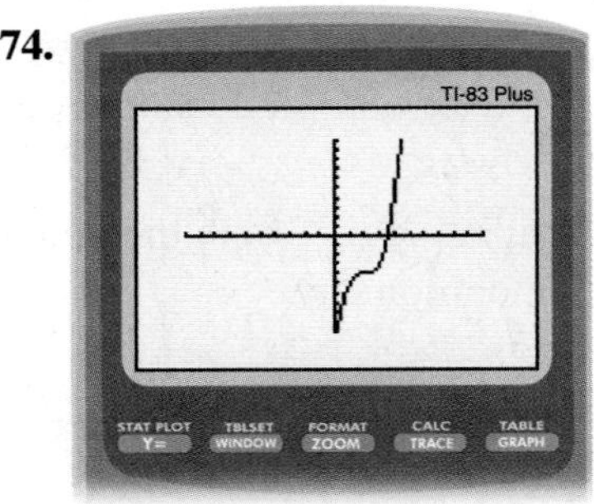
TI-83 Plus
STAT PLOT
Y=
TBLSET
WINDOW
FORMAT
ZOOM
CALC
TRACE
TABLE
GRAPH

75. 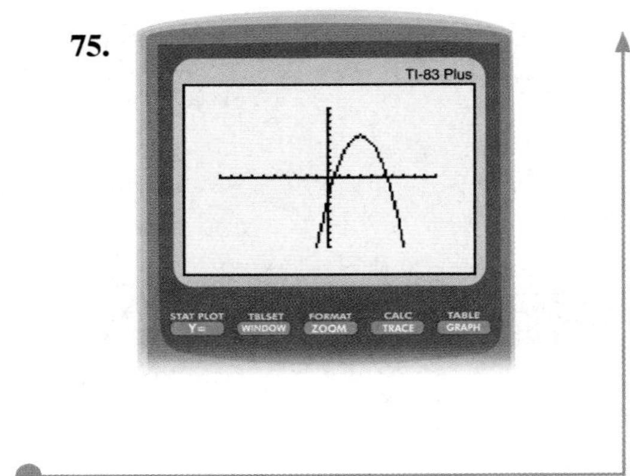
TI-83 Plus
STAT PLOT
Y=
TBLSET
WINDOW
FORMAT
ZOOM
CALC
TRACE
TABLE
GRAPH

3.6 Combining Functions

TOPICS

1. Combining functions arithmetically
2. Composing functions
3. Decomposing functions
4. Interlude: recursive graphics

Topic 1: Combining Functions Arithmetically

In Section 3.5, we gained considerable experience in building new functions from old by shifting, reflecting and stretching the old functions. In this section, we will explore other ways of building functions.

We begin with four arithmetic ways of combining two or more functions to obtain new functions. The four arithmetic operations are very familiar to you: addition, subtraction, multiplication, and division. The only possibly new idea is that we are applying these operations to functions, not numbers. But as we will see, the arithmetic combination of functions is based entirely on the arithmetic combination of numbers.

Addition, Subtraction, Multiplication, and Division of Functions

Let f and g be two functions. The **sum** $f + g$, **difference** $f - g$, **product** fg, and **quotient** $\frac{f}{g}$ are four new functions defined as follows:

1. $(f+g)(x)=f(x)+g(x)$.

2. $(f-g)(x)=f(x)-g(x)$.

3. $(fg)(x)=f(x)g(x)$.

4. $\left(\frac{f}{g}\right)(x)=\frac{f(x)}{g(x)}$, provided that $g(x)\neq 0$.

The domain of each of these new functions consists of the common elements of the domains of f and g individually, with the added condition that in the quotient function we have to omit those elements for which $g(x)=0$.

We have actually combined functions arithmetically already, though we didn't use the terminology of functions at the time, when we studied addition, subtraction, multiplication, and division of polynomials in Section 1.3.

With the above definition, we can determine the sum, difference, product, or quotient of two functions at one particular value for x, or find a formula for these new functions based on the formulas for f and g, if they are available.

example 1

Given that $f(-2)=5$ and $g(-2)=-3$, find $(f-g)(-2)$ and $\left(\frac{f}{g}\right)(-2)$.

Solution:

By the definition of the difference and quotient of functions,

$$\begin{aligned}(f-g)(-2) &= f(-2)-g(-2)\\ &= 5-(-3)\\ &= 8,\end{aligned}$$

and

$$\begin{aligned}\left(\frac{f}{g}\right)(-2) &= \frac{f(-2)}{g(-2)}\\ &= \frac{5}{-3}\\ &= -\frac{5}{3}.\end{aligned}$$

example 2

Given the two functions $f(x)=4x^2-1$ and $g(x)=\sqrt{x}$, find $(f+g)(x)$ and $(fg)(x)$.

Solution:

By the definition of the sum and product of functions,

$$\begin{aligned}(f+g)(x) &= f(x)+g(x)\\ &= 4x^2-1+\sqrt{x},\end{aligned}$$

and

$$\begin{aligned}(fg)(x) &= \left(4x^2-1\right)\left(\sqrt{x}\right)\\ &= 4x^{5/2}-x^{1/2}.\end{aligned}$$

The domain of both $f+g$ and fg is the interval $[0,\infty)$, even though the domain of f alone is the entire real line.

example 3

Based on the graphs of f and g below, determine the domain of $\frac{f}{g}$ and evaluate $\left(\frac{f}{g}\right)(1)$.

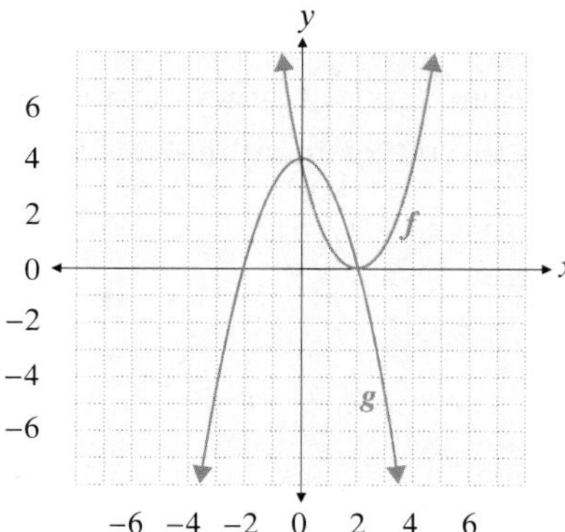

Solution:

Based on the graphs, it appears that g is 0 when $x = -2$ and $x = 2$. The domain of both f and g individually is the entire set of real numbers, so the domain of $\frac{f}{g}$ is $(-\infty,-2)\cup(-2,2)\cup(2,\infty)$. Also based on the graphs, it appears that $f(1)=1$ and $g(1)=3$, so $\left(\frac{f}{g}\right)(1)=\frac{1}{3}$.

Topic 2: Composing Functions

A fifth way of combining functions is to form the *composition* of one function with another. Informally speaking, this means to apply one function, say *f*, to the output of another function, say *g*. The symbol for composition is an open circle.

Composing Functions

Let f and g be two functions. The **composition** of f and g, denoted $f \circ g$, is the function defined by $(f \circ g)(x) = f(g(x))$. The domain of $f \circ g$ consists of all x in the domain of g for which $g(x)$ is in turn in the domain of f. The function $f \circ g$ is read "f composed with g", or "f of g."

Note that the order of f and g is important. In general, we can expect the function $f \circ g$ to be different from the function $g \circ f$. In formal terms, the composition of two functions, unlike the sum and product of two functions, is not commutative.

The diagram in Figure 1 is a sort of schematic of the composition of two functions. The ovals represent sets, with the leftmost oval being the domain of the function g. The arrows indicate the element that x is associated with by the various functions.

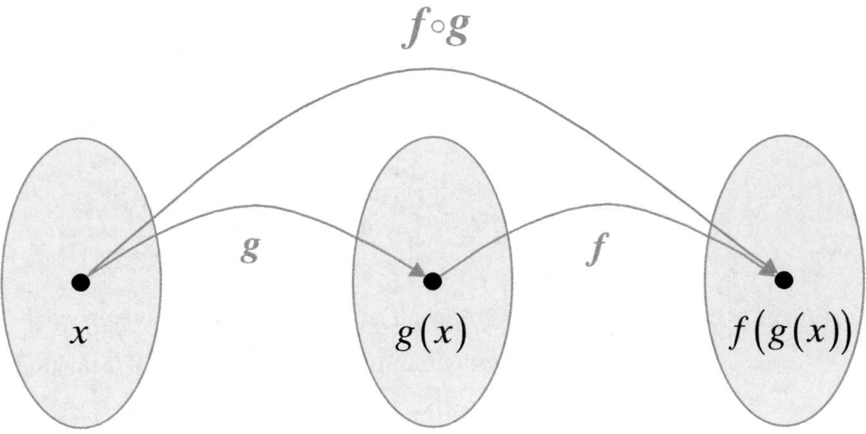

Figure 1: Composition of f and g

As with the four arithmetic ways of combining functions, we can evaluate the composition of two functions at a single point, or find a formula for the composition if we have been given formulas for the individual functions.

example 4

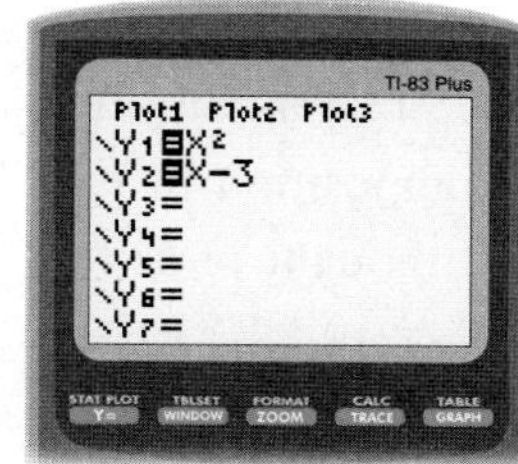

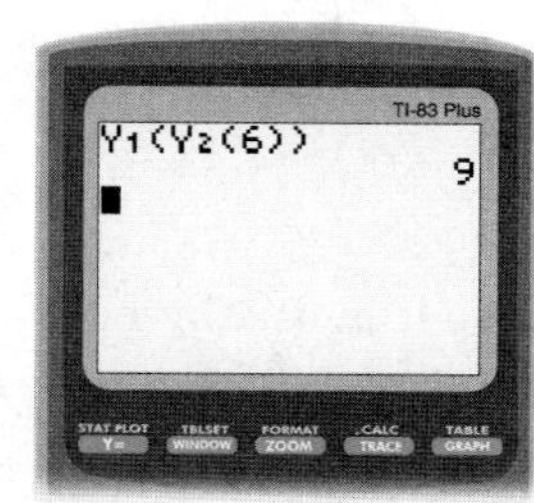

Enter the functions using **Y=**. To return to the main screen, select **2nd** and **Mode** **(Quit)**. To enter `Y1` and `Y2`, select **Vars**, use the right arrow to select `Y-Vars` and press **Enter**. Then **Enter** on the desired variable.

Given $f(x)=x^2$ and $g(x)=x-3$, find:

a. $(f\circ g)(6)$

b. $(f\circ g)(x)$

Solutions:

a. $g(6)=6-3=3$ and so $(f\circ g)(6)=f(g(6))=f(3)=3^2=9.$

b. To find the formula for $f\circ g,$ we simply apply the definition of composition and then simplify

$$\begin{aligned}(f\circ g)(x)&=f(g(x))\\&=f(x-3)\\&=(x-3)^2\\&=x^2-6x+9.\end{aligned}$$

Note that once we have found a formula for $f\circ g$, we have an alternative way of answering the first question: $(f\circ g)(6)=(6)^2-(6)(6)+9=9$.

example 5

Let $f(x)=x^2-4$ and $g(x)=\sqrt{x}$. Find formulas and state the domains for:

a. $f\circ g$

b. $g\circ f$

Solutions:

a. $$\begin{aligned}(f\circ g)(x)&=f\left(\sqrt{x}\right)\\&=\left(\sqrt{x}\right)^2-4\\&=x-4\end{aligned}$$

The answer at left may seem to indicate that the domain of $f\circ g$ is all real numbers, but this is incorrect. The domain of $f\circ g$ is actually the interval $[0,\infty)$, because only non-negative numbers can be plugged into g.

b. $$\begin{aligned}(g\circ f)(x)&=g\left(x^2-4\right)\\&=\sqrt{x^2-4}\end{aligned}$$

The domain of $g\circ f$ consists of all x for which $x^2-4\geq 0$, or $x^2\geq 4$. In interval form, the domain is the set $(-\infty,-2]\cup[2,\infty)$.

Topic 3: Decomposing Functions

Often, functions can be best understood by recognizing them as a composition of two or more simpler functions. We have already seen one instance of this: shifting, reflecting, stretching, and compressing can all be thought of as a composition of two or more functions. For example, the function $h(x)=(x-2)^3$ can be thought of as the composition of the functions $f(x)=x^3$ and $g(x)=x-2$. Note that

$$\begin{aligned} f(g(x)) &= f(x-2) \\ &= (x-2)^3 \\ &= h(x). \end{aligned}$$

To "decompose" a function into a composition of simpler functions, it is usually best to identify what the function does to its argument from the inside out. That is, identify the first thing that is done to the argument, then the second, and so on. Each action describes a less complex function, and can be identified as such. The composition of these functions, with the innermost function corresponding to the first action, the next innermost corresponding to the second action, and so on, is then equivalent to the original function.

Decomposition can often be done in several different ways. Consider, for example, the function $f(x)=\sqrt[3]{5x^2-1}$. The following table illustrates just some of the ways f can be written as a composition of functions. Be sure you understand how each of the different compositions is equivalent to f.

a. $g(x)=\sqrt[3]{x}$
$h(x)=5x^2-1$

$$\begin{aligned} g(h(x)) &= g(5x^2-1) \\ &= \sqrt[3]{5x^2-1} \\ &= f(x) \end{aligned}$$

b. $g(x)=\sqrt[3]{x-1}$
$h(x)=5x^2$

$$\begin{aligned} g(h(x)) &= g(5x^2) \\ &= \sqrt[3]{5x^2-1} \\ &= f(x) \end{aligned}$$

c. $g(x)=\sqrt[3]{x}$
$h(x)=5x-1$
$i(x)=x^2$

$$\begin{aligned} g(h(i(x))) &= g(h(x^2)) \\ &= g(5x^2-1) \\ &= \sqrt[3]{5x^2-1} \\ &= f(x) \end{aligned}$$

example 6

Decompose the function $f(x)=|x^2-3|+2$ into:

a. a composition of two functions.

b. a composition of three functions.

Solutions:

a. $g(x)=|x|+2$
$h(x)=x^2-3$

$$g(h(x))=g(x^2-3)$$
$$=|x^2-3|+2$$
$$=f(x)$$

b. $g(x)=x+2$
$h(x)=|x-3|$
$i(x)=x^2$

$$g(h(i(x)))=g(h(x^2))$$
$$=g(|x^2-3|)$$
$$=|x^2-3|+2$$
$$=f(x)$$

Topic 4: Interlude: Recursive Graphics

Recursion, in general, refers to using the output of a function as its input, and repeating the process a certain number of times. In other words, recursion refers to the composition of a function with itself, possibly many times. Recursion has many varied uses, one of which is a branch of mathematical art.

Some special nomenclature and notation have evolved to describe recursion. If f is a function, $f^2(x)$ is used in this context to stand for $f(f(x))$, or $(f \circ f)(x)$ (not $(f(x))^2$!) Similarly, $f^3(x)$ stands for $f(f(f(x)))$, or $(f \circ f \circ f)(x)$, and so on. The functions f^2, f^3, ... are called **iterates** of *f*, with f^n being the $\boldsymbol{n}^{\text{th}}$ **iterate** of *f*.

Some of the most famous recursively generated mathematical art is based on functions whose inputs and outputs are complex numbers. Recall from Section 1.4 that every complex number can be expressed in the form, $a + bi$ where a and b are real numbers and i is the imaginary unit. A one-dimensional coordinate system, such as the real number line, is insufficient to graph complex numbers, but complex numbers are easily graphed in a two-dimensional coordinate system.

> So Nat'ralists observe, A Flea
> Hath Smaller Fleas that on him prey
> And these have smaller Fleas to bite 'em
> And so proceed, ad infinitum.
>
> - Jonathan Swift

To graph the number $a + bi$, we treat it as the ordered pair (a, b) and plot the point (a,b) in the Cartesian plane, where the horizontal axis represents pure real numbers and the vertical axis represents pure imaginary numbers.

Benoit Mandelbrot used the function $f(z)=z^2+c$, where both z and c are variables representing complex numbers, to generate the image known as the Mandelbrot set in the 1970s. The basic idea is to evaluate the sequence of iterates $f(0)=0^2+c=c$, $f^2(0)=f(c)=c^2+c$, $f^3(0)=f(c^2+c)=(c^2+c)^2+c, \ldots$ for various complex numbers c and determine if the sequence of complex numbers stays close to the origin or not. Those complex numbers c that result in so-called "bounded" sequences are colored black, while those that lead to unbounded sequences are colored white. The author has used similar ideas to generate his own recursive art, as described below.

The image "i of the storm" reproduced here is based on the function $f(z)=\frac{(1-i)z^4+(7+i)z}{2z^5+6}$, where again z is a variable that will be replaced with complex numbers. The image is actually a picture of the complex plane, with the origin in the very center of the golden ring. The golden ring consists of those complex numbers that lie between 0.9 and 1.1 in distance from the origin. The rules for coloring other complex numbers in the plane are as follows: given an initial complex number z not on the gold ring, $f(z)$ is calculated. If the complex number $f(z)$ lies somewhere on the gold ring, the original number z is colored the deepest shade of green. If not, the iterate $f^2(z)$ is calculated. If this result lies in the gold ring, the original z is colored a bluish shade of green. If not, the process continues up to the 12th iterate $f^{12}(z)$, using a different color each time. If $f^{12}(z)$ lies in the gold ring, z is colored red, and if not the process halts and z is colored black.

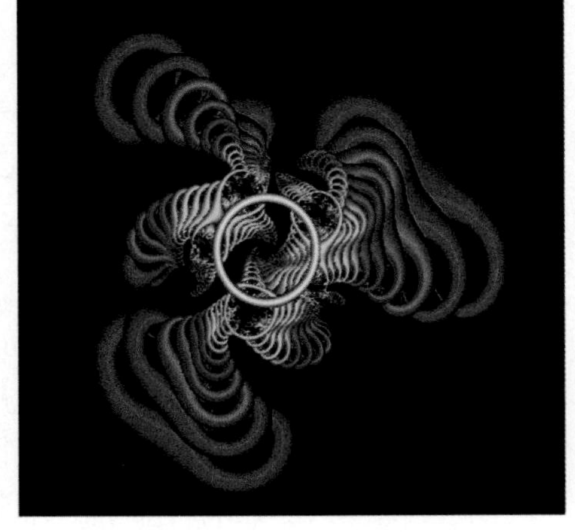

The idea of recursion can be used to generate any number of similar images, with the end result usually striking and often surprising even to the creator.

exercises

In each of the following problems, use the information given to determine ***(a)*** $(f+g)(-1)$ ***(b)*** $(f-g)(-1)$, ***(c)*** $(fg)(-1)$, *and* ***(d)*** $\left(\frac{f}{g}\right)(-1)$. *See Examples 1, 2, and 3.*

1. $f(-1)=-3$ and $g(-1)=5$

2. $f(-1)=0$ and $g(-1)=-1$

3. $f(x)=x^2-3$ and $g(x)=x$

4. $f(x)=\sqrt[3]{x}$ and $g(x)=x-1$

5. $f(-1)=15$ and $g(-1)=-3$

6. $f(x)=\dfrac{x+5}{2}$ and $g(x)=6x$

7. $f(x)=x^4+1$ and $g(x)=x^{11}+2$

8. $f(x)=\dfrac{6-x}{2}$ and $g(x)=\sqrt{\dfrac{x}{-4}}$

9. $f=\{(5,2),(0,-1),(-1,3),(-2,4)\}$ and $g=\{(-1,3),(0,5)\}$

10. $f=\{(3,15),(2,-1),(-1,1)\}$ and $g(x)=-2$

11.

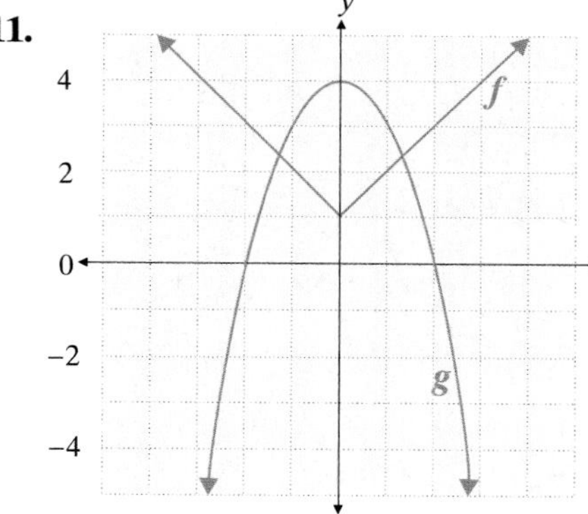

12.

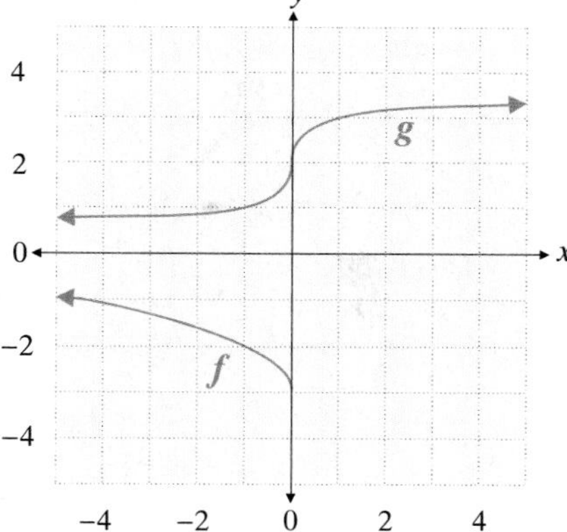

In each of the following problems, find ***(a)*** *the formula and domain for* $f+g$ *and* ***(b)*** *the formula and domain for* $\frac{f}{g}$. *See Examples 2 and 3.*

13. $f(x)=|x|$ and $g(x)=\sqrt{x}$

14. $f(x)=x^2-1$ and $g(x)=\sqrt[3]{x}$

15. $f(x)=x-1$ and $g(x)=x^2-1$

16. $f(x)=x^{3/2}$ and $g(x)=x-3$

17. $f(x)=3x$ and $g(x)=x^3-8$

18. $f(x)=x^3+4$ and $g(x)=\sqrt{x-2}$

19. $f(x)=-2x^2$ and $g(x)=[\![x+4]\!]$

20. $f(x)=6x-1$ and $g(x)=x^{2/3}$

Evaluate the following given $f(x)=\frac{1}{x^2}$ and $g(x)=2x+3$. See Examples 1, 2, and 3.

21. $(f+g)(-7)$ **22.** $(f+g)(10)$ **23.** $(f-g)(-5)$ **24.** $(f-g)(0)$

25. $(fg)(4)$ **26.** $(fg)(-3)$ **27.** $\left(\frac{f}{g}\right)(-2)$ **28.** $\left(\frac{f}{g}\right)(0)$

29. $\left(\frac{g}{f}\right)(1)$ **30.** $\left(\frac{g}{f}\right)(-6)$

In each of the following problems, use the information given to determine $(f \circ g)(3)$. See Example 4.

31. $f(-5)=2$ and $g(3)=-5$

32. $f(\pi)=\pi^2$ and $g(3)=\pi$

33. $f(x)=x^2-3$ and $g(x)=\sqrt{x}$

34. $f(x)=\sqrt{x^2-9}$ and $g(x)=1-2x$

35. $f(x)=2+\sqrt{x}$ and $g(x)=x^3+x^2$

36. $f(x)=x^{3/2}-3$ and $g(x)=\left[\!\left[\frac{3x}{2}\right]\!\right]$

37. $f(x)=\sqrt{x+6}$ and $g(x)=\sqrt{4x-3}$

38. $f(x)=\sqrt{\frac{3x}{14}}$ and $g(x)=x^4-x^3-x^2-x$

39.

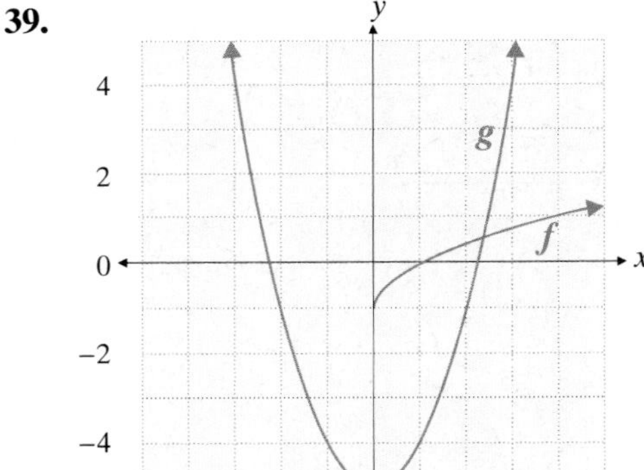

40.

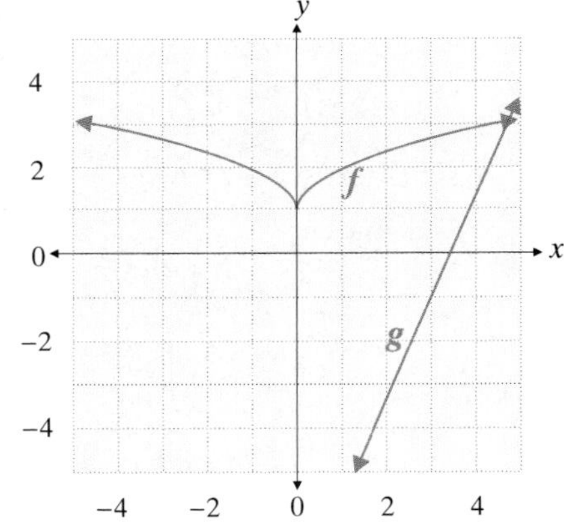

*In each of the following problems, find **(a)** the formula and domain for $f \circ g$ and **(b)** the formula and domain for $g \circ f$. See Example 5.*

41. $f(x)=\sqrt{x-1}$ and $g(x)=x^2$

42. $f(x)=\frac{1}{x}$ and $g(x)=x-1$

43. $f(x)=\frac{4x-2}{3}$ and $g(x)=\frac{1}{x}$

44. $f(x)=1-x$ and $g(x)=\sqrt{x}$

45. $f(x)=[\![x-3]\!]$ and $g(x)=x^3+1$

46. $f(x)=x^2+2x$ and $g(x)=x-3$

47. $f(x)=x^2+1$ and $g(x)=3x^2+5$

48. $f(x)=\sqrt{x}$ and $g(x)=2x$

49. $f(x)=\dfrac{1}{x+7}$ and $g(x)=\dfrac{2}{x}$

50. $f(x)=\dfrac{1}{x}$ and $g(x)=\dfrac{1}{x}$

51. $f(x)=x^2$ and $g(x)=3x+1$

52. $f(x)=\sqrt[3]{x}$ and $g(x)=x^3$

53. $f(x)=\sqrt{x-4}$ and $g(x)=x^2+2$

54. $f(x)=\dfrac{3}{1-x}$ and $g(x)=3x^2$

Write the following functions as a composition of two functions. Answers will vary. See Example 6.

55. $f(x)=\sqrt[3]{3x^2-1}$

56. $f(x)=\dfrac{2}{5x-1}$

57. $f(x)=|x-2|+3$

58. $f(x)=x+\sqrt{x+2}-5$

59. $f(x)=|x^3-5x|+7$

60. $f(x)=\dfrac{\sqrt{x-3}}{x^2-6x+9}$

Solve the following application problems.

61. The volume of a right circular cylinder is given by the formula $V=\pi r^2 h$. If the height h is three times the radius r, show the volume V as a function of r.

62. The surface area S of a wind sock is defined by the formula $S=\pi r^2+\pi rs$ where r is the radius of the wind sock and s is the length of the side of the wind sock. As the windsock is being knitted by an automated knitter, the length of the side s is increasing with time t defined by the formula $s(t)=\dfrac{1}{4}t^2,\ t\geq 0,$ find the surface area S of the wind sock as a function of time t.

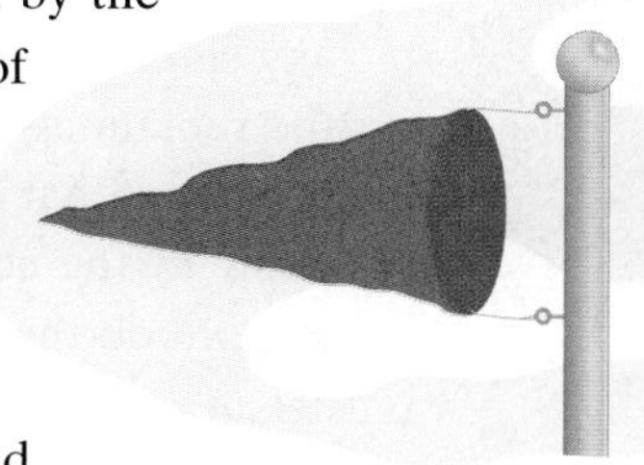

63. The volume V of the wind sock described in the previous question is given by the formula $V=\dfrac{1}{3}\pi r^2 h$ where r is the radius of the wind sock and h is the height of the wind sock. If the height h is increasing with time t defined by the formula $h(t)=\dfrac{1}{4}t^2,\ t\geq 0,$ find the volume V of the wind sock as a function of time t.

64. A widget factory produces n widgets in t hours of a single day. The number of widgets the factory produces is given by the formula $n(t)=10{,}000t-25t^2,\ 0\leq t\leq 9$. The cost c in dollars of producing n widgets is given by the formula $c(n)=2040+1.74n$. Find the cost c as a function of time t that the factory is producing the widgets.

65. Suppose that $M(x)$ represents the percent of income spent on a house mortgage in the year x and $C(x)$ represents the percent of income spent on car mortgage in the year x. If $I(x)$ represents the income in year x, determine the function L that represents the total mortgage expenses in year x.

66. Given two odd functions f and g, show that $f \circ g$ is also odd. Verify this fact with the particular functions $f(x) = \sqrt[3]{x}$ and $g(x) = \frac{-x^3}{3x^2 - 9}$. Recall that a function is odd if $f(-x) = -f(x)$ for all x in the domain of f.

67. Given two even functions f and g, show that the product is also even. Verify this fact with the particular functions $f(x) = 2x^4 - x^2$ and $g(x) = \frac{1}{x^2}$. Recall that a function is even if $f(-x) = f(x)$ for all x in the domain of f.

As mentioned in the Interlude, a given complex number c is said to be in the Mandelbrot set if, for the function $f(z) = z^2 + c$, the sequence of iterates $f(0), f^2(0), f^3(0), \ldots$ stays close to the origin (which is the complex number $0 + 0i$). It can be shown that if any single iterate falls more than 2 units in distance (magnitude) from the origin, then the remaining iterates will grow larger and larger in magnitude. In practice, computer programs that generate the Mandelbrot set calculate the iterates up to a pre-decided point in the sequence, such as $f^{50}(0)$, and if no iterate to this point exceeds 2 in magnitude the number c is admitted to the set. The magnitude of a complex number $a + bi$ is the distance between the point (a, b) and the origin, so the formula for the magnitude of $a + bi$ is $\sqrt{a^2 + b^2}$.

Use the above criterion to determine, without a calculator or computer, if the following complex numbers are in the Mandelbrot set or not.

68. $c = 0$ **69.** $c = 1$ **70.** $c = i$ **71.** $c = -1$ **72.** $c = 1 + i$

73. $c = -2$ **74.** $c = 1 - i$ **75.** $c = -1 - i$

technology exercises

A computer algebra system like *Mathematica* can be very useful in combining functions. The illustration below shows how the two functions $f(x) = x^2 - 3x - 5$ and $g(x) = (x-2)^3$ are defined in *Mathematica*, and then shows how to ask *Mathematica* for $(f+g)(7)$, $(fg)(x)$, $(f \circ g)(x)$, $(g \circ f)(x)$, and $(f \circ g)(-3)$. (Note how the composition of two functions is expressed in the *Mathematica* command).

Use a computer algebra system, if available, to evaluate $(f+g)(x)$, $(fg)(x)$, $(f \circ g)(x)$ *and* $(g \circ f)(x)$ *for each of the following pairs of functions. (See Appendix A for guidance on using Mathematica).*

Section 3-6 Technology Exercises.nb

```
In[1]:= f[x_] := x^2 - 3 x - 5
In[2]:= g[x_] := (x - 2) ^3
In[3]:= f[7] + g[7]
Out[3]= 148
In[4]:= Expand[f[x] * g[x]]
Out[4]= 40 - 36 x - 14 x^2 + 25 x^3 - 9 x^4 + x^5
In[5]:= Expand[f[g[x]]]
Out[5]= 83 - 228 x + 258 x^2 - 163 x^3 + 60 x^4 - 12 x^5 + x^6
In[6]:= Expand[g[f[x]]]
Out[6]= -343 - 441 x - 42 x^2 + 99 x^3 + 6 x^4 - 9 x^5 + x^6
In[7]:= f[g[-3]]
Out[7]= 15995
```

100%

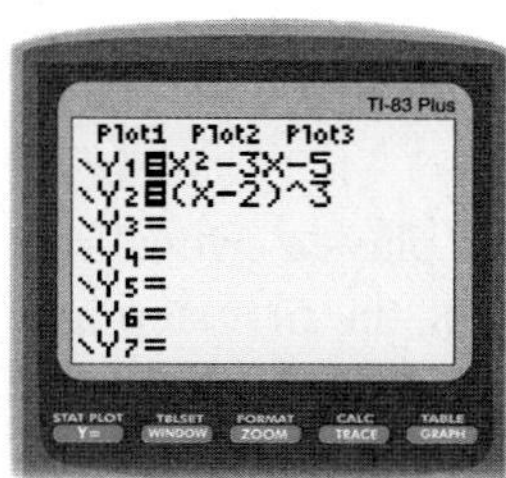

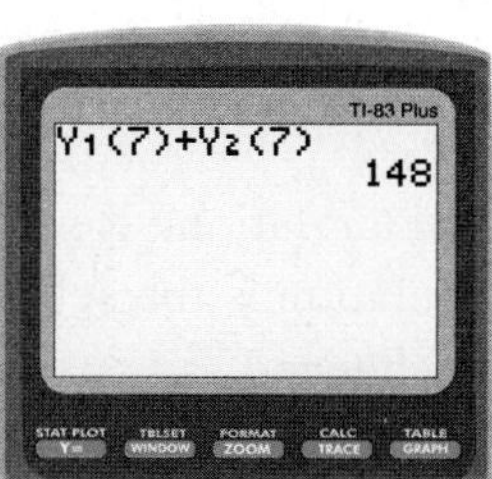

To enter Y_1 and Y_2, select **Vars** as described on page 265.

76. $f(x) = (3x+2)^2$ and $g(x) = \sqrt{x^2+5}$

77. $f(x) = \dfrac{1}{3x-5}$ and $g(x) = (x+2)^3$

78. $f(x) = \dfrac{x+1}{x-1}$ and $g(x) = \dfrac{x-1}{x}$

3.7 Inverses of Functions

TOPICS

1. Inverses of relations
2. Inverse functions and the horizontal line test
3. Finding inverse function formulas

Topic 1: Inverses of Relations

In many problems, "undoing" one or more mathematical operations plays a critical role in the solution process. For instance, to solve the equation $3x + 2 = 8$, the first step is to "undo" the addition of 2 on the left-hand side (by subtracting 2 from both sides) and the second step is to "undo" the multiplication by 3 (by dividing both sides by 3). In the context of more complex problems, the "undoing" process is often a matter of finding and applying the inverse of a function.

We begin our discussion with the more general idea of the inverse of a relation. Recall that a relation is just a set of ordered pairs; the inverse of a given relation is the set of these ordered pairs with the first and second coordinates of each exchanged.

Inverse of a Relation

Let R be a relation. The **inverse of *R***, denoted R^{-1}, is the set

$$R^{-1} = \{(b, a) | (a, b) \in R\}.$$

example 1

Determine the inverse of each of the following relations. Then graph each relation and its inverse, and determine the domain and range of both.

a. $R = \{(4, -1), (-3, 2), (0, 5)\}$ **b.** $y = x^2$

Solutions:

a. $R = \{(4, -1), (-3, 2), (0, 5)\}$

$R^{-1} = \{(-1, 4), (2, -3), (5, 0)\}$

$\text{Dom}(R) = \{4, -3, 0\}$ and $\text{Ran}(R) = \{-1, 2, 5\}$

$\text{Dom}(R^{-1}) = \{-1, 2, 5\}$ and $\text{Ran}(R^{-1}) = \{4, -3, 0\}$

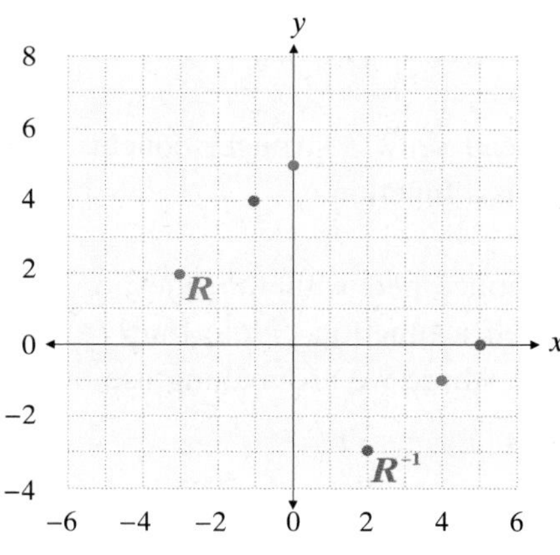

In the graph to the left, R is in blue and its inverse is in red. The relation R consists of three ordered pairs, and its inverse is simply these three ordered pairs with the coordinates exchanged. Note that the domain of R is the range of R^{-1}, and vice versa.

b. $R = \{(x, y) \mid y = x^2\}$

$R^{-1} = \{(x, y) \mid x = y^2\}$

$\text{Dom}(R) = \mathbb{R}$ and $\text{Ran}(R) = [0, \infty)$

$\text{Dom}(R^{-1}) = [0, \infty)$ and $\text{Ran}(R^{-1}) = \mathbb{R}$

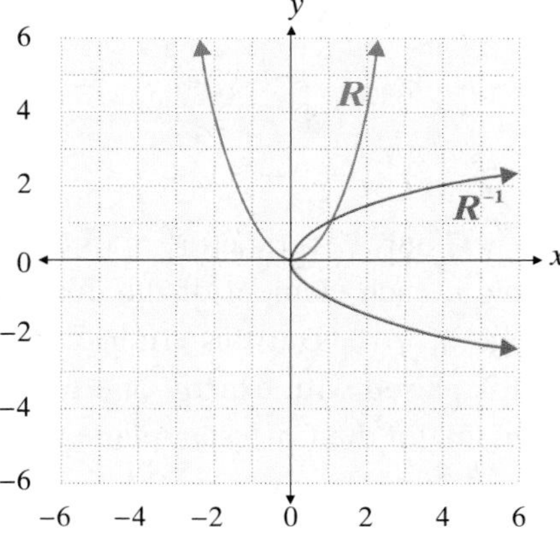

In this problem, R is described by the given equation in x and y. The inverse relation is the set of ordered pairs in R with the coordinates exchanged, so we can describe the inverse relation by just exchanging x and y in the equation, as shown at left.

Consider the graphs of the two relations and their respective inverses in Example 1. By definition, an ordered pair (b, a) lies on the graph of a relation R^{-1} if and only if (a, b) lies on the graph of R, so it shouldn't be surprising that the graphs of a relation and its inverse bear some resemblance to one another. Specifically, they are mirror images of one another with respect to the line $y = x$. If you were to fold the Cartesian plane in half along the line $y = x$ in the two examples above, you would see that the points in R and R^{-1} coincide with one another.

The two relations in Example 1 illustrate another important point. Note that in both cases, R is a function, as its graph passes the vertical line test. By the same criterion, R^{-1} in Example 1a is also a function, but R^{-1} in Example 1b is not. The conclusion to be drawn is that even if a relation is a function, its inverse may or may not be a function.

Topic 2: Inverse Functions and the Horizontal Line Test

With a bit more thought, we can draw a stronger conclusion from Example 1 about when the inverse of a relation is a function.

In practice, we will only be concerned with the question of when the inverse of a function f, denoted f^{-1}, is itself a function. Note that f^{-1} has already been defined: f^{-1} stands for the inverse of f, where we are making use of the fact that a function is also a relation.

caution!

We are faced with another example of reuse of notation. f^{-1} does *not* stand for $\frac{1}{f}$!

We use an exponent of –1 to indicate the reciprocal of a number or an algebraic expression, but when applied to a function or a relation it stands for the inverse relation.

Assume that f is a function. f^{-1} will only be a function in its own right if its graph passes the vertical line test; that is, only if each element of the domain of f^{-1} is paired with exactly one element of the range of f^{-1}. But this criterion is identical to saying that each element of the range of f is paired with exactly one element of the domain of f. In other words, every *horizontal* line in the Cartesian plane must intersect the graph of f no more than once. We say that functions meeting this condition pass the horizontal line test. See Figure 1.

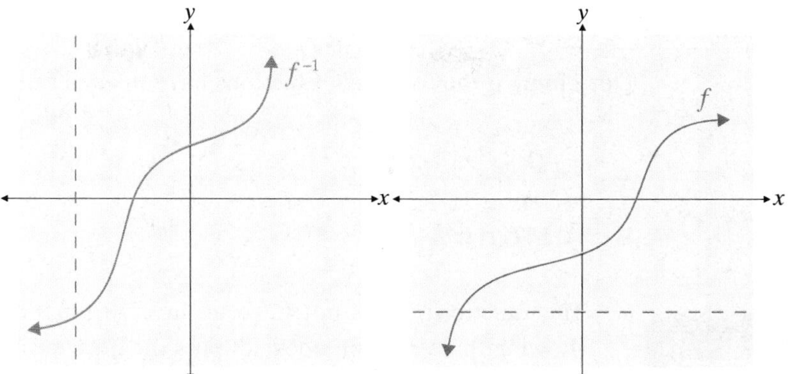

Figure 1: Vertical Line Test and Horizontal Line Test

The Horizontal Line Test

Let f be a function. We say that the graph of f passes the **horizontal line test** if every horizontal line in the plane intersects the graph no more than once.

Of course, the horizontal line test is only useful if the graph of f is available to study. We can also phrase the above condition in a non-graphical manner as follows. The inverse of f will only be a function if for every pair of distinct elements x_1 and x_2 in the domain of f, we have $f(x_1) \neq f(x_2)$. This criterion is important enough to merit a name.

One-to-One Functions

A function f is **one-to-one** if for every pair of distinct elements x_1 and x_2 in the domain of f, we have $f(x_1) \neq f(x_2)$. This means that every element of the range of f is paired with exactly one element of the domain of f.

To sum up: the inverse f^{-1} of a function f is also a function if and only if f is one-to-one. If the graph of f is available for examination, f is one-to-one if and only if its graph passes the horizontal line test.

If we now examine Example 1 again, we see that the function R in Example 1a is one-to-one, and so we know that its inverse is also a function. On the other hand, the function R in Example 1b is not one-to-one (plenty of horizontal lines pass through the graph twice), so its inverse is not a function.

example 2

Determine if the following functions have inverse functions.

a. $f(x)=|x|$

b. $g(x)=(x+2)^3$

Solutions:

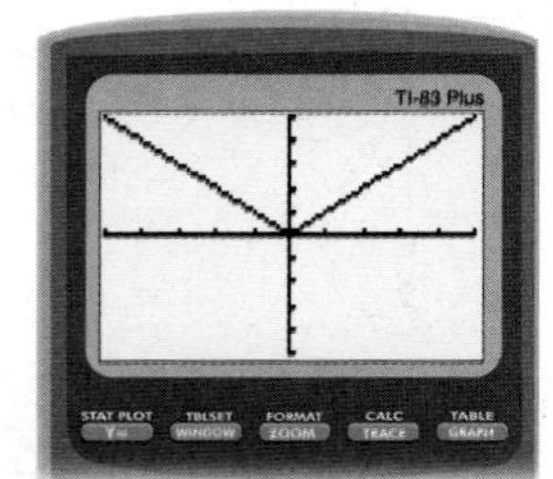

a. The function f does not have an inverse function, a fact easily demonstrated by showing that its graph does not pass the horizontal line test (recall that the graph of f is an upward-opening V shape). An algebraic proof that f does not have an inverse function is the following: even though $-3 \neq 3$, we have $f(-3)=f(3)$. Of course, there are an infinite number of pairs of numbers that show f is not one-to-one, but one such pair is all it takes to show that f does not have an inverse function.

b. The graph of g is the standard cubic shape shifted horizontally two units to the left, and the standard cubic shape passes the horizontal line test, so g has an inverse function. But again, it is good practice to prove this algebraically. Note how each line in the following argument implies the next line:

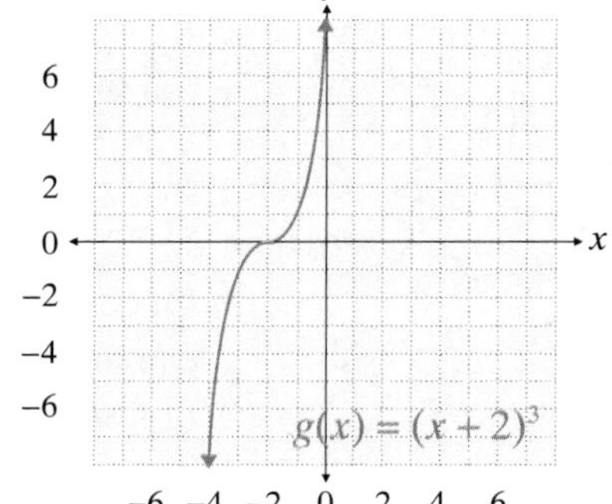

$$\begin{aligned} x_1 \neq x_2 &\Rightarrow x_1+2 \neq x_2+2 \\ &\Rightarrow (x_1+2)^3 \neq (x_2+2)^3 \\ &\Rightarrow g(x_1) \neq g(x_2). \end{aligned}$$

This argument shows that any two distinct elements of the domain of g lead to different values when plugged into g, so g is one-to-one and hence has an inverse function.

Consider the function in Example 2a again. As we noted, the function $f(x)=|x|$ is not one-to-one on its implied domain, and so cannot have an inverse function. However, if we *restrict* the domain of f explicitly by specifying that the domain is the interval $[0, \infty)$, the new function, with its restricted domain, is one-to-one and has an inverse function. Of course, this **restriction of domain** changes the function; in this case the graph of the new function is the right-hand half of the graph of the absolute value function.

Topic 3: Finding Inverse Function Formulas

In applying the notion of the inverse of a function, we will most often begin with a formula for f and want to find a formula for f^{-1}. This will allow us, for instance, to transform equations of the form

$$f(x) = y$$

into the form

$$x = f^{-1}(y).$$

Before we discuss the general algorithm for finding a formula for f^{-1}, consider the problem with which we began this section. If we define $f(x) = 3x + 2$, the equation $3x + 2 = 8$ can be written as $f(x) = 8$. Note that f is one-to-one, so f^{-1} does exist (as a function). If we can find a formula for f^{-1}, we can transform the equation into $x = f^{-1}(8)$. This is an overly complicated way to solve this equation, but it illustrates the point well.

What should the formula for f^{-1} be? Consider what f does to its argument. The first action is to multiply x by 3, and the second is to add 2. To "undo" f, we need to negate these two actions in reverse order: subtract 2 and then divide the result by 3. So,

$$f^{-1}(x) = \frac{x-2}{3}.$$

Applying this to the problem at hand, we obtain

$$x = f^{-1}(8) = \frac{8-2}{3} = 2.$$

This method of analyzing a function f and then finding a formula for f^{-1} by undoing the actions of f in reverse order is conceptually important, and works for simple functions. For other functions, however, the following algorithm may be necessary.

To Find a Formula for f^{-1}

Let f be a one-to-one function, and assume that f is defined by a formula. To find a formula for f^{-1}, perform the following steps:

1. Replace $f(x)$ in the definition of f with the variable y. The result is an equation in x and y that is solved for y at this point.
2. Interchange x and y in the equation.
3. Solve the new equation for y.
4. Replace the y in the resulting equation with $f^{-1}(x)$.

example 3

Find the inverse of each of the following functions.

a. $f(x)=(x-1)^3+2$ **b.** $g(x)=\dfrac{x-3}{2x+1}$

Solutions:

a.

$$f(x)=(x-1)^3+2$$
$$f^{-1}(x)=(x-2)^{1/3}+1$$

We can find the inverse of this function either by the algorithm, or by undoing the actions of f in reverse order. The function f subtracts 1 from x, cubes the result, and adds 2; its inverse will first subtract 2, take the cube root of the result, and then add 1.

b.

$$g(x)=\frac{x-3}{2x+1}$$

The inverse of the function g is most easily found by the algorithm.

$$y=\frac{x-3}{2x+1}$$

The first step is to replace $g(x)$ with y.

$$x=\frac{y-3}{2y+1}$$

The second step is to interchange x and y in the equation.

$$x(2y+1)=y-3$$
$$2xy+x=y-3$$
$$2xy-y=-x-3$$

We now have to solve the equation for y. We begin by clearing the equation of fractions, and then proceed to collect all the terms that contain y on one side.

$$y(2x-1)=-x-3$$
$$y=\frac{-x-3}{2x-1}$$

Factoring out the y on the left-hand side and dividing by $2x-1$ completes the process.

$$g^{-1}(x)=\frac{-x-3}{2x-1}$$

The last step is to name the formula g^{-1}.

Remember that the graphs of a relation and its inverse are mirror images of one another with respect to the line $y = x$; this is still true if the relations are functions. We can demonstrate this fact by graphing the function and its inverse from Example 3a above, as shown in Figure 2.

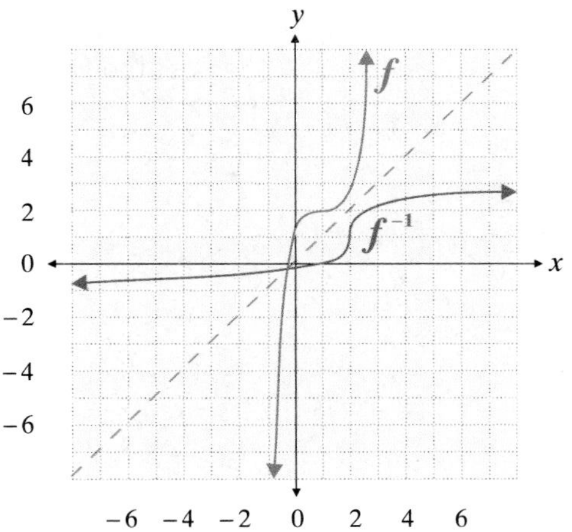

Figure 2: Graph of a Function and its Inverse

We can use the functions and their inverses from Example 3 to illustrate one last important point. The key characteristic of the inverse of a function is that it undoes the function. This means that if a function and its inverse are composed together, in either order, the resulting function has no effect on any allowable input; specifically,

Properties of Inverses

$$f\left(f^{-1}(x)\right)=x \text{ for all } x\in \operatorname{Dom}\left(f^{-1}\right), \text{ and}$$
$$f^{-1}\left(f(x)\right)=x \text{ for all } x\in \operatorname{Dom}(f).$$

For example, given $f(x)=(x-1)^3+2$ and $f^{-1}(x)=(x-2)^{1/3}+1$:

$$\begin{aligned}
f\left(f^{-1}(x)\right)&=f\left((x-2)^{1/3}+1\right)\\
&=\left((x-2)^{1/3}+1-1\right)^3+2\\
&=\left((x-2)^{1/3}\right)^3+2\\
&=x-2+2\\
&=x.
\end{aligned}$$

A similar calculation shows that $f^{-1}\left(f(x)\right)=x$, as you should verify.

As another example, consider $g(x)=\frac{x-3}{2x+1}$ and $g^{-1}(x)=\frac{-x-3}{2x-1}$:

$$\begin{aligned}
g^{-1}(g(x)) &= g^{-1}\left(\frac{x-3}{2x+1}\right)\\
&= \frac{-\frac{x-3}{2x+1}-3}{2\left(\frac{x-3}{2x+1}\right)-1}\\
&= \left(\frac{-\frac{x-3}{2x+1}-3}{2\left(\frac{x-3}{2x+1}\right)-1}\right)\left(\frac{2x+1}{2x+1}\right)\\
&= \frac{-x+3-6x-3}{2x-6-2x-1}\\
&= \frac{-7x}{-7}\\
&= x.
\end{aligned}$$

Similarly, $g(g^{-1}(x))=x$, as you should verify.

exercises

Graph the inverse of each of the following relations, and state its domain and range. See Example 1.

1. $R=\{(-4,2),(3,2),(0,-1),(3,-2)\}$

2. $S=\{(-3,-3),(-1,-1),(0,1),(4,4)\}$

3. $y=x^3$

4. $y=|x|+2$

5. $x=|y|$

6. $x=-\sqrt{y}$

7. $y=\frac{1}{2}x-3$

8. $y=-x+1$

9. $y=[\![x]\!]$

10. $T=\{(4,2),(3,-1),(-2,-1),(2,4)\}$

11. $x=y^2-2$

12. $y=2\sqrt{x}$

Determine if the following functions have inverse functions. If not, suggest a domain to restrict the function to so that it would have an inverse function (answers may vary). See Example 2.

13. $f(x)=x^2+1$ **14.** $g(x)=(x-2)^3-1$ **15.** $h(x)=\sqrt{x+3}$

16. $s(x)=\dfrac{1}{x^2}$ **17.** $G(x)=3x-5$ **18.** $F(x)=-x^2+5$

19. $r(x)=-\sqrt{x^3}$ **20.** $b(x)=[\![x]\!]$ **21.** $f(x)=x^2-4x$

22. $m(x)=\dfrac{13x-2}{4}$ **23.** $H(x)=|x-12|$ **24.** $p(x)=10-x^2$

Find the inverse of each of the following functions. See Example 3.

25. $f(x)=x^{1/3}-2$ **26.** $g(x)=4x-3$ **27.** $r(x)=\dfrac{x-1}{3x+2}$

28. $s(x)=\dfrac{1-x}{1+x}$ **29.** $F(x)=(x-5)^3+2$ **30.** $G(x)=\sqrt[3]{3x-1}$

31. $V(x)=\dfrac{x+5}{2}$ **32.** $W(x)=\dfrac{1}{x}$ **33.** $h(x)=x^{3/5}-2$

34. $A(x)=(x^3+1)^{1/5}$ **35.** $J(x)=\dfrac{2}{1-3x}$ **36.** $k(x)=\dfrac{x+4}{3-x}$

37. $h(x)=x^7+6$ **38.** $F(x)=\dfrac{3-x^5}{-9}$ **39.** $r(x)=\sqrt[5]{2x}$

40. $P(x)=(2+3x)^3$ **41.** $f(x)=3(2x)^{1/3}$ **42.** $q(x)=(x-2)^2+2$

In each of the following problems, show that $f(f^{-1}(x))=x$ *and that* $f^{-1}(f(x))=x$.

43. $f(x)=\dfrac{3x-1}{5}$ and $f^{-1}(x)=\dfrac{5x+1}{3}$ **44.** $f(x)=\sqrt[3]{x+2}-1$ and $f^{-1}(x)=(x+1)^3-2$

45. $f(x)=\dfrac{2x+7}{x-1}$ and $f^{-1}(x)=\dfrac{x+7}{x-2}$ **46.** $f(x)=x^2,\ x\geq 0$ and $f^{-1}(x)=\sqrt{x}$

47. $f(x)=2x-3$ and $f^{-1}(x)=\dfrac{x+3}{2}$ **48.** $f(x)=\sqrt[3]{x+1}$ and $f^{-1}(x)=x^3-1$

49. $f(x)=\dfrac{1}{x}$ and $f^{-1}(x)=\dfrac{1}{x}$ **50.** $f(x)=\dfrac{x-5}{2x+3}$ and $f^{-1}(x)=\dfrac{3x+5}{1-2x}$

51. $f(x)=(x-2)^2,\ x\geq 2$ and $f^{-1}(x)=\sqrt{x}+2,\ x\geq 0$

52. $f(x)=\dfrac{1}{1+x},\ x\geq 0$ and $f^{-1}(x)=\dfrac{1-x}{x},\ 0<x\leq 1$

*Match the following functions with the graphs of the inverses of the functions. The graphs are labeled **a** through **f**.*

53. $f(x) = x^3$

54. $f(x) = x - 5$

55. $f(x) = \sqrt{x-4}$

56. $f(x) = x^2$

57. $f(x) = \dfrac{x}{4}$

58. $f(x) = \sqrt[3]{x+1}$

a.

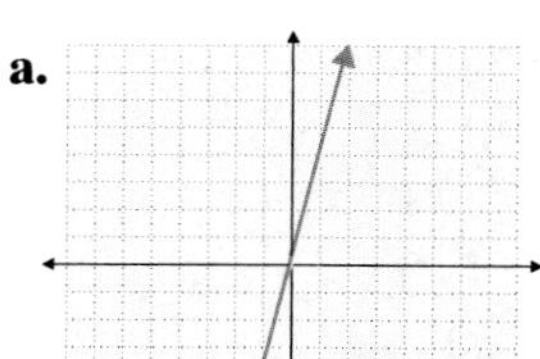

b.

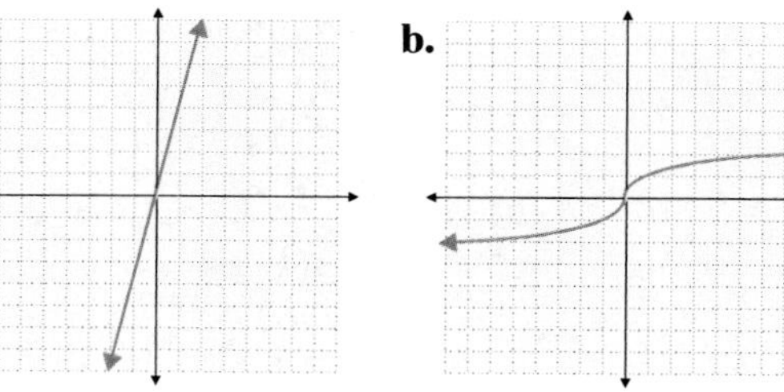

c.

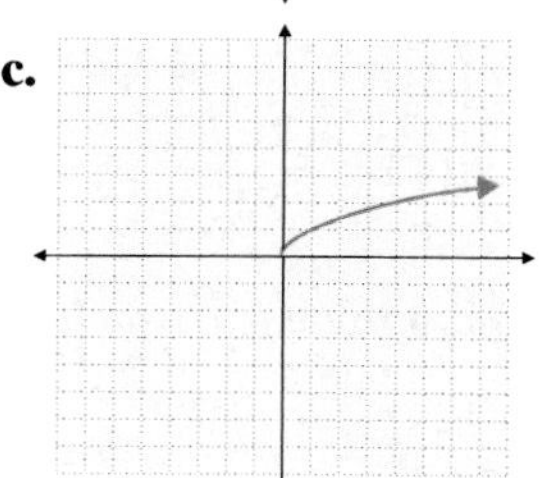

d.

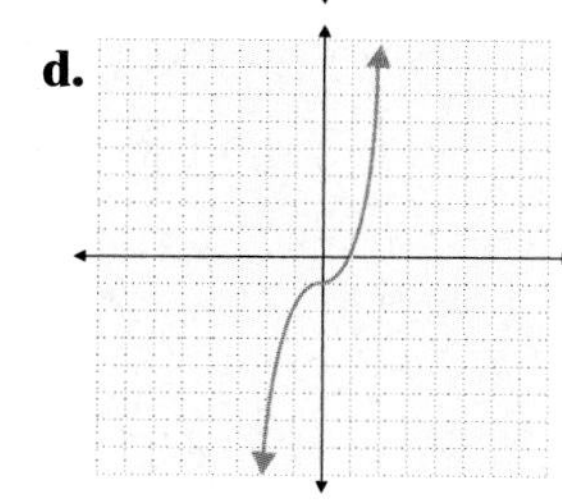

e.

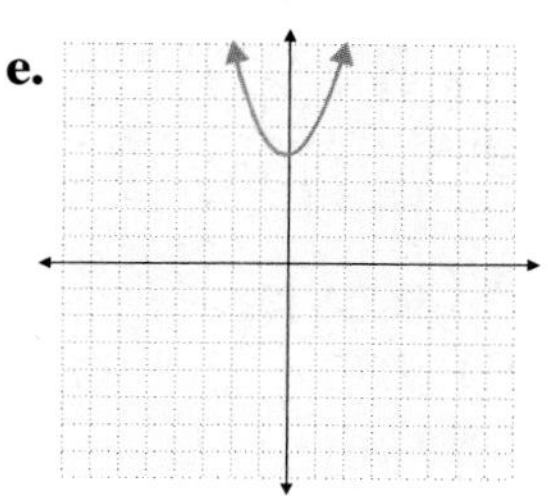

f.

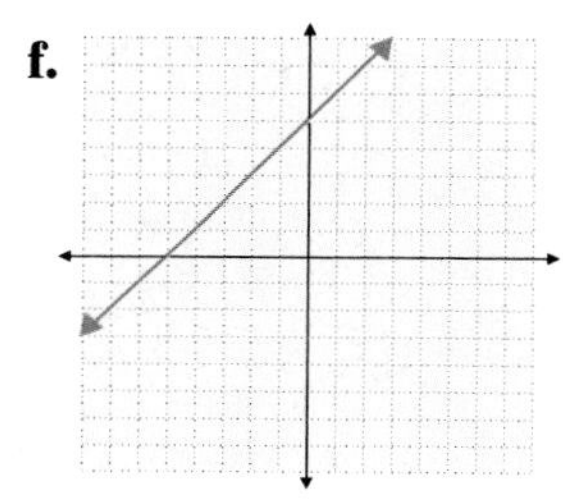

An inverse function can be used to encode and decode words and sentences by assigning each letter of the alphabet a numerical value ($A = 1, B = 2, C = 3, \ldots Z = 26$). Example: Use the function $f(x) = x^2$ to encode the word PRECALCULUS. The encoded message would be 256 324 25 9 1 144 9 441 144 441 361. The word can then be decoded by using the inverse function $f^{-1}(x) = \sqrt{x}$. The inverse values are 16 18 5 3 1 12 3 21 12 21 19 which translates back to the word PRECALCULUS. Encode or decode the following words using the numerical values ($A = 1, B = 2, C = 3, \ldots Z = 26$).

59. Encode the message SANDY SHOES using the function $f(x) = 4x - 3$.

60. Encode the message WILL IT RAIN TODAY using the function $f(x) = x^2 - 8$.

61. The following message was encoded using the function $f(x) = 8x - 7$. Decode the message.

41 137 65 145 9 33 33 169 113 89 89 33 193 9 1 89 89 1 105 25 57 113 137 145 33 145 57 113 33 145

62. The following message was encoded using the function $f(x) = 5x + 1$. Decode the message.

91 26 66 26 66 11 26 91 126 76 106 91 96 106 71 11 61 76 16 56

63. The following message was encoded using the function $f(x) = x^3$. Decode the message.

27 1 8000 27 512 1 12167 1 10648 125

64. The following message was encoded using the function $f(x) = -3 - 5x$. Decode the message.

−13 −28 −8 −18 −43 −33 −108 −73 −48 −73 −103 −43 −28 −98 −108 −73

A graphing calculator can be used to verify the inverse of a function. Use a graphing calculator to graph the following functions. Determine the domain and range of the inverse.

65. $f(x) = \sqrt{x+5}$ and $f^{-1}(x) = x^2 - 5$

66. $f(x) = -\sqrt{x^2 - 16}$ and $f^{-1}(x) = \sqrt{(-x)^2 + 16}$

67. $f(x) = x^2 + 3$ and $f^{-1}(x) = \sqrt{x-3}$

68. $f(x) = x^3 - 1$ and $f^{-1}(x) = \sqrt[3]{x+1}$

69. $f(x) = \dfrac{2x+1}{x-1}$ and $f^{-1}(x) = \dfrac{x+1}{x-2}$

70. $f(x) = \dfrac{4}{\sqrt{x}}$ and $f^{-1}(x) = \dfrac{16}{x^2}$

chapter three project

The Ozone Layer

As time goes on, there is continually increasing awareness, controversy, and legislation regarding the ozone layer and other environmental issues. The hole in the ozone layer over the South Pole disappears and reappears annually, and one model for its growth assumes the hole is circular and that its radius grows at a constant rate of 2.6 kilometers per hour.

1. Write the area of the circle as a function of the radius, r.
2. Assuming that t is measured in hours, that $t=0$ corresponds to the start of the annual growth of the hole, and that the radius of the hole is initially 0, write the radius as a function of time, t.
3. Write the area of the circle as a function of time, t.
4. What is the radius after 3 hours?
5. What is the radius after 5.5 hours?
6. What is the area of the circle after 3 hours?
7. What is the area of the circle after 5.5 hours?
8. What is the average rate of change of the area from 3 hours to 5.5 hours?
9. What is the average rate of change of the area from 5.5 hours to 8 hours?
10. Is the average rate of change of the area increasing or decreasing as time passes?

CHAPTER REVIEW

3.1 Relations and Functions

topics	pages	test exercises
Relations, domains, and ranges • The definition of *relation* as a set of ordered pairs • The definitions of *domain* and *range* as, respectively, the set of first coordinates and the set of second coordinates for a given relation • The correspondence between a relation and its graph in the Cartesian plane	p. 189 – 192	1 – 4
Functions and the vertical line test • The definition of a *function* as a relation in which every element of the domain is paired with *exactly* one element of the range • The meaning of the *vertical line test* as applied to the graph of a relation and in identifying functions	p. 192 – 194	1 – 4
Functional notation and function evaluation • The meaning of *functional notation* • Evaluation of a function for a given argument • The role of the argument as a placeholder in defining a function • The definition of *domain* and *codomain*	p. 194 – 200	5 – 8
Implied domain of a function • Determining the domain of a function when it is not stated explicitly	p. 200 – 201	1 – 4

3.2 Linear and Quadratic Functions

topics	pages	test exercises
Linear functions and their graphs • The definition of a *linear*, or *first-degree*, function: any function that can be written in the form $f(x) = ax + b$ • The graph of a linear function	p. 206 – 208	9 – 10

topics (continued)	pages	test exercises
Quadratic functions and their graphs • The definition of a *quadratic*, or *second-degree*, function: any function that can be written in the form $f(x) = ax^2 + bx + c$ • The graph of a quadratic function, including the location of the vertex and x- and y-intercepts • The *vertex form* of a quadratic function with center (h, k): $g(x) = a(x-h)^2 + k$	p. 208 – 214	11 – 12
Interlude: maximization/minimization problems • The role of completing the square in locating the maximum or minimum value of a quadratic function and the result: $\left(-\frac{b}{2a}, f\left(-\frac{b}{2a}\right)\right)$	p. 214 – 216	13

3.3 Other Common Functions

topics	pages	test exercises
Functions of the form ax^n • The basic form of the graph of ax^n when n is even • The basic form of the graph of ax^n when n is odd	p. 221 – 223	14
Functions of the form $\frac{a}{x^n}$ • The basic form of the graph of $\frac{a}{x^n}$ when n is even • The basic form of the graph of $\frac{a}{x^n}$ when n is odd	p. 223 – 225	15
Functions of the form $ax^{1/n}$ • The basic form of the graph of $ax^{1/n}$ when n is even • The basic form of the graph of $ax^{1/n}$ when n is odd	p. 225 – 226	16
The absolute value function • The basic form of the graph of the absolute value function	p. 226 – 227	17
The greatest integer function • The basic form of the graph of the greatest integer function	p. 228	18
Piecewise-defined functions • The definition of *piecewise-defined function*: a function defined in terms of two or more formulas, each valid for its own unique portion of the real number line	p. 229	19 – 20

3.4 Variation and Multi-Variable Functions

topics	pages	test exercises
Direct variation • Concept of *direct variation* ($y = kx^n$) • Determining the *constant of proportionality* (k)	p. 233 – 234	22
Inverse variation • Concept of *inverse variation* $\left(y = \frac{k}{x^n}\right)$	p. 234 – 235	23
Joint variation • Combination of direct and inverse variation in a single formula • Concept of *joint variation*	p. 236	21
Interlude: multi-variable functions • Expressing one quantity as a function of two or more variables • Evaluating multi-variable functions	p. 237 – 238	24

3.5 Transformations of Functions

topics	pages	test exercises
Shifting, stretching, and reflecting graphs • Replacing the argument x with $x - h$ in a function to shift its graph h units horizontally • Adding k to a function to shift its graph k units vertically • Multiplying a function by -1 to reflect its graph with respect to the x-axis • Replacing the argument x with $-x$ in a function to reflect its graph with respect to the y-axis • Multiplying a function by an appropriate constant to stretch or compress its graph	p. 244 – 250	25 – 26

topics (continued)	pages	test exercises
Symmetry of functions and equations • The meaning of *y-axis symmetry* $(f(-x) = f(x))$ • The meaning of *x-axis symmetry* $(f(-y) = f(y))$ • The meaning of *origin symmetry* $(f(-x) = -f(x))$ • The meaning of *even* (functions with y-axis symmetry) and *odd* (functions with origin symmetry) functions	p. 251 – 254	27 – 30
Intervals of monotonicity • The meaning of *increasing* $(f(x_1) < f(x_2))$, *decreasing* $(f(x_1) > f(x_2))$, and *constant* $(f(x_1) = f(x_2))$ as applied to the graph of a function	p. 254 – 256	31 – 32

3.6 Combining Functions

topics	pages	test exercises
Combining functions arithmetically • *Sums* $((f+g)(x)=f(x)+g(x))$, *differences* $((f-g)(x)=f(x)-g(x))$, *products* $((fg)(x)=f(x)g(x))$, and *quotients* $\left(\left(\frac{f}{g}\right)(x)=\frac{f(x)}{g(x)}\right)$ of functions, and how to evaluate such combinations • Identifying the domain of an arithmetic combination of functions	p. 261 – 263	33 – 34
Composing functions • The meaning of *composition* of functions $(f \circ g(x) = f(g(x)))$ • The domain of the composition of a function • Determining a formula for the composition of two functions, and evaluating a composition of functions for a given argument	p. 263 – 265	35 – 36
Decomposing functions • The decomposition of complicated functions into simpler functions	p. 266 – 267	37 – 38

topics (continued)	pages	test exercises
Interlude: recursive graphics • Recursion as an application of compositions, and the meaning of the iterates of a function	p. 267 – 268	39 – 40

3.7

Inverses of Functions

topics	pages	test exercises
Inverses of relations • The definition of the *inverse* $\left(R^{-1} = \{(b, a) \mid (a, b) \in R\}\right)$ of a relation, and the correspondence between the graph of a relation and the graph of its inverse	p. 274 – 276	41 – 42
Inverse functions and the horizontal line test • The concept of the *inverse* of a function, and the notation used (f^{-1}) • Identifying when a function has an inverse function • The correspondence between the *horizontal line test* and whether a function is *one-to-one* • Restriction of domain	p. 276 – 278	46
Finding inverse function formulas • The algorithmic approach to finding the inverse of a function • Constructing the formula for the inverse of a function by "undoing" the function step by step	p. 279 – 282	43 – 45

chapter test

Section 3.1

For each relation below, describe the domain and range and determine whether or not the relation is a function.

1. $R=\{(-2,9),(-3,-3),(-2,2),(-2,-9)\}$

2. $3x-4y=17$

3. $x=y^2-6$

4. $x=\sqrt{y-4}$

For each function below, determine **(a)** $f(x-1)$, **(b)** $f(x+a)-f(x)$ *and* **(c)** $f(x^2)$.

5. $f(x)=(x+5)(2x)$

6. $f(x)=\sqrt[3]{x}+6(x+4)$

Identify the domain, codomain, and range of each of the following functions.

7. $g:\mathbb{N}\to\mathbb{R}$ by $g(x)=\dfrac{3x}{4}$

8. $h:\mathbb{R}\to\mathbb{R}$ by $h(x)=5x+1$

Section 3.2

Graph the following linear functions.

9. $f(x)=7x-2$

10. $g(x)=\dfrac{2x-6}{3}$

Graph the following quadratic functions, locating the vertices and x-intercepts (if any) accurately.

11. $f(x)=(x-1)^2-1$

12. $g(x)=-x^2-6x-11$

13. Among all the pairs of numbers with a sum of 15, find the pair whose product is maximum.

Section 3.3

Sketch the graphs of the following functions. Pay particular attention to intercepts, if any, and locate these accurately.

14. $f(x)=4x^3$

15. $f(x)=\dfrac{-2}{x^2}$

16. $f(x)=\dfrac{\sqrt[3]{x}}{2}$

17. $f(x)=5|-x|$

18. $f(x)=\left[\!\left[\dfrac{2x}{3}\right]\!\right]$

19. $f(x)=\begin{cases} x^2 & \text{if } x<1 \\ \dfrac{1}{x} & \text{if } x\geq 1 \end{cases}$

20. $g(x)=\begin{cases} (x+1)^2-1 & \text{if } x\leq 0 \\ \sqrt[3]{x} & \text{if } x>0 \end{cases}$

Section 3.4

Solve the following variation problems.

21. Suppose that y varies jointly with the cube of x and the square root of z. If $y=270$ when $x=3$ and $z=25$, what is y when $x=2$ and $z=9$?

22. The distance that an object falls from rest, when air resistance is negligible, varies directly as the square of the time. A stone dropped from rest travels 400 feet in the first 5 seconds. How far did it travel in the first 2 seconds?

23. A video store manager observes that the number of videos rented seems to vary inversely as the price of a rental. If the store's customers rent 1050 videos per month when the price per rental is \$3.49, how many videos per month does he expect to rent if he lowers the price to \$2.99?

24. Determine the approximate distance between the Earth, with a mass of approximately 6.0×10^{24} kg, and an object which has a mass of 6.42×10^{22} kg if the gravitational force equals approximately 4.95×10^{21} N. Remember $F=\dfrac{km_1m_2}{d^2}$ and the Universal Gravitational Constant equals $6.67\times10^{-11}\,\text{Nm}^2/\text{kg}^2$.

Section 3.5

Sketch the graphs of the following functions by first identifying the more basic functions that have been shifted, reflected, stretched, or compressed. Then determine the domain and range of each function.

25. $f(x)=(x-1)^3+2$

26. $G(x)=4|x+3|$

Determine if the following functions are even, odd, or neither and then graph.

27. $f(x)=\dfrac{1}{3}x^3$

28. $g(x)=\sqrt{x}$

Determine if the following equations have y-axis symmetry, x-axis symmetry, origin symmetry, or no symmetry and then graph.

29. $y = |5x|$

30. $x^2 + y^2 = 25$

For each function below, find the intervals where the function is increasing, decreasing, or constant.

31. $f(x) = (x-2)^4 - 6$

32. $R(x) = \begin{cases} (x+2)^2 & \text{if } x < -1 \\ -x & \text{if } x \geq -1 \end{cases}$

Section 3.6

In each of the following problems, find ***(a)*** *the formula and domain for* $f+g$ *and* ***(b)*** *the formula and domain for* $\dfrac{f}{g}$.

33. $f(x) = x^2$ and $g(x) = \sqrt{x}$

34. $f(x) = \dfrac{1}{x-2}$ and $g(x) = \sqrt[3]{x}$

In each of the following problems, use the information given to determine ***(a)*** $(f \circ g)(x)$, ***(b)*** $(g \circ f)(x)$, *and* ***(c)*** $(f \circ g)(3)$.

35. $f(x) = -x + 1$ and $g(x) = -x - 1$

36. $f(x) = \dfrac{x^{-1}}{18} - 3$ and $g(x) = \dfrac{x-4}{x^3}$

Decompose the following functions. Answers will vary.

37. $f(x) = \dfrac{\sqrt{x+3}+2}{x^2+6x+9}$

38. $f(x) = |x+2| + x^2 + 4x + 4$

Determine if the following complex numbers are in the Mandelbrot set.

39. $c = -i$

40. $c = 2$

Section 3.7

Graph the inverse of each of the following relations and state the domain and range.

41. $R = \{(-3,5), (2,1), (0,-5), (-1,-2)\}$

42. $y = \dfrac{1}{3}x^2$

Find a formula for the inverse of each of the following functions.

43. $r(x) = \dfrac{2}{7x-1}$

44. $g(x) = \dfrac{4x-3}{x}$

45. $f(x) = x^{1/5} - 6$

In the following problem, show that $f\left(f^{-1}(x)\right)=x$ *and that* $f^{-1}\left(f(x)\right)=x$.

46. $f(x)=\dfrac{6x-7}{2-x}$ and $f^{-1}(x)=\dfrac{2x+7}{6+x}$

chapter FOUR

POLYNOMIAL FUNCTIONS

OVERVIEW

The information in this chapter will guide you in your work with polynomials – graphing, dividing, solving equations and inequalities, etc. Business applications may often make use of polynomial inequalities to find values for which various situations are profitable. On page 310 you will find a problem describing a skateboard company; given information about their revenue and expenses, you will be asked to find the number of skateboards they should manufacture in order to make a profit. You will master this type of problem using techniques for *Solving Polynomial Inequalities* as shown in Example 3 on page 306.

Introduction

In Chapter 3, we studied properties of functions in general, and learned the nomenclature and notation commonly used when working with functions and their graphs. In this chapter, we narrow our focus and concentrate on polynomial functions.

We have, of course, already seen many examples of polynomial functions, but have barely scratched the surface as far as obtaining a deep understanding of polynomials is concerned. We will soon see many mathematical methods that are peculiar to polynomials, and make many observations that are relevant only to polynomials. Some of these methods include polynomial division and the Rational Zero Theorem, and many of the observations point out the strong connection between factors of polynomials, solutions of polynomial equations, and (when appropriate) graphs of polynomial functions. Our deeper understanding of polynomial functions will then make solving polynomial inequalities a far easier task than it would be otherwise.

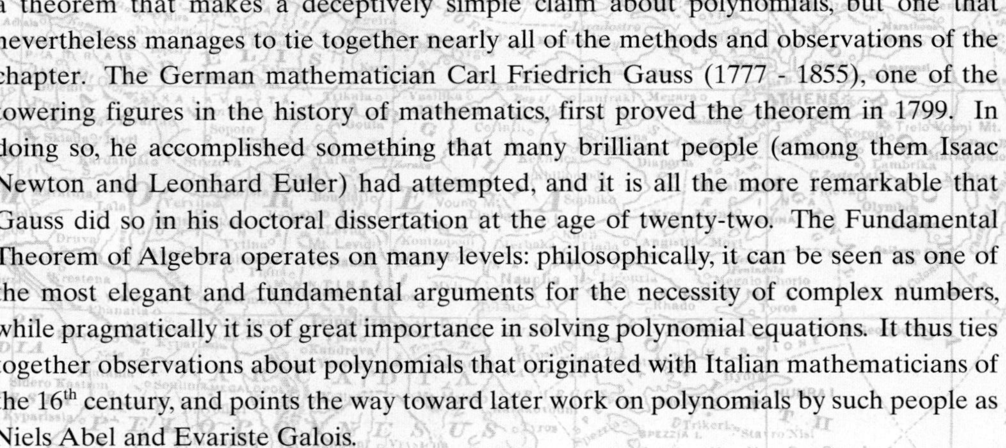

The discussion of polynomials concludes with the Fundamental Theorem of Algebra, a theorem that makes a deceptively simple claim about polynomials, but one that nevertheless manages to tie together nearly all of the methods and observations of the chapter. The German mathematician Carl Friedrich Gauss (1777 - 1855), one of the towering figures in the history of mathematics, first proved the theorem in 1799. In doing so, he accomplished something that many brilliant people (among them Isaac Newton and Leonhard Euler) had attempted, and it is all the more remarkable that Gauss did so in his doctoral dissertation at the age of twenty-two. The Fundamental Theorem of Algebra operates on many levels: philosophically, it can be seen as one of the most elegant and fundamental arguments for the necessity of complex numbers, while pragmatically it is of great importance in solving polynomial equations. It thus ties together observations about polynomials that originated with Italian mathematicians of the 16th century, and points the way toward later work on polynomials by such people as Niels Abel and Evariste Galois.

Gauss

The chapter ends with an opportunity to put to use many of the skills acquired thus far. An understanding of the subject of the last section, rational functions, depends not only on a knowledge of polynomials (of which rational functions are ratios), but also of *x*- and *y*-intercepts, factoring, and transformations of functions.

4.1 Introduction to Polynomial Equations and Graphs

TOPICS

1. Zeros of polynomials and solutions of polynomial equations
2. Graphing factored polynomials
3. Solving polynomial inequalities

Topic 1: Zeros of Polynomials and Solutions of Polynomial Equations

At this point, we have a firm grasp of how linear and quadratic polynomial functions behave, and we have tools guaranteed to solve all linear and quadratic equations. We have also studied some elementary higher-degree polynomial functions (those of the basic form ax^n for some natural number n). In this section, we begin a more complete exploration of higher-degree polynomials.

Ideally, by the end of this chapter our mastery of linear and quadratic polynomials would extend to all polynomials. Unfortunately, this is too much to expect: not surprisingly, the complexity of polynomial functions increases with the degree. This means, for instance, that higher-degree polynomials are generally more difficult to graph accurately, and that we cannot necessarily expect to find exact solutions to higher-degree polynomial equations. Nevertheless, there is much that we can say about a given polynomial function.

In order to make our discussion as general as possible, we will often refer to a generic n^{th} degree polynomial function $f(x) = a_n x^n + a_{n-1}x^{n-1} + \ldots + a_1 x + a_0$ where $a_n, a_{n-1}, \ldots, a_1, a_0$ all represent constants (possibly non-real complex number constants) and $a_n \neq 0$. We will begin by identifying the values of the variable that make a polynomial function equal to zero.

Zeros of a Polynomial

The number k (k may be a complex number) is said to be a **zero** of the polynomial function $f(x)=a_nx^n+a_{n-1}x^{n-1}+\ldots+a_1x+a_0$ if $f(k)=0$. This is also expressed by saying that k is a **root** or a **solution** of the equation $f(x)=0$.

If f is a polynomial with real coefficients, and if k is a real number zero of f, then the statement $f(k)=0$ means the graph of f crosses the x-axis at $(k,0)$. In this case, k may be referred to as an x-intercept of f.

The task of determining the zeros of a polynomial arises in many contexts, two of which are solving polynomial equations and graphing polynomials. We will discuss polynomial equations first.

Polynomial Equations

A **polynomial equation in one variable**, say the variable x, is an equation that can be written in the form

$$a_nx^n + a_{n-1}x^{n-1} + \ldots + a_1x + a_0 = 0,$$

where a_n, a_{n-1}, ..., a_1, a_0 are constants and n is a positive integer. Assuming $a_n \neq 0$, we say such an equation is of degree n. a_n is known as the **leading coefficient** and a_0 is called the **constant coefficient**.

Just as with linear and quadratic equations (which are polynomial equations of degree 1 and 2, respectively), a given polynomial equation may not initially appear as in the above definition. The first task, then, may be to add or subtract terms as necessary so that one side of the equation is zero. Once this is done, the zeros of the polynomial constitute the solutions of the equation.

example 1

Verify that the given values of x solve the corresponding polynomial equations.

a. $6x^2 - x^3 = 12 + 5x;\ x = 4$

b. $x^2 = 2x - 5;\ x = 1 + 2i$

c. $\dfrac{x}{1-i} = 3x^2;\ x = 0$

Solutions:

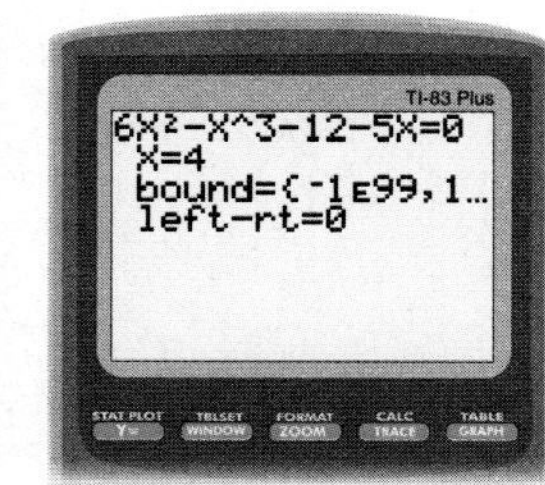

This screen was performed using equation solver.

a.

$$6x^2 - x^3 = 12 + 5x$$
$$6(4)^2 - (4)^3 \overset{?}{=} 12 + 5(4)$$
$$96 - 64 \overset{?}{=} 12 + 20$$
$$32 = 32$$

Verifying that 4 is a solution of the equation is just a matter of substituting 4 for x and simplifying both sides. This has also verified that 4 is a zero of the polynomial function $f(x) = -x^3 + 6x^2 - 5x - 12$. That is, $f(4) = 0$.

b.

$$x^2 = 2x - 5$$
$$x^2 - 2x + 5 = 0$$
$$x = \frac{2 \pm \sqrt{4-20}}{2}$$
$$x = \frac{2 \pm 4i}{2}$$
$$x = 1 \pm 2i$$

We could verify that $1+2i$ is a solution of the equation $x^2 = 2x - 5$ by substituting $1+2i$ for x and simplifying both sides, but since this is a quadratic equation we have the tools necessary to solve the equation for ourselves. The quadratic formula (see Section 1.7) tells us that $1+2i$ and $1-2i$ are the two solutions of the equation.

c.

$$\frac{x}{1-i} = 3x^2$$
$$\frac{0}{1-i} \overset{?}{=} 3(0)^2$$
$$0 = 0$$

Don't be misled by the complex number in the denominator; this is another polynomial equation, and verification that 0 is a solution is shown at left. For practice, verify that the polynomial $f(x) = 3x^2 - \dfrac{x}{1-i}$ could be rewritten as $f(x) = 3x^2 - \left(\dfrac{1}{2} + \dfrac{i}{2}\right)x$ (see Section 1.4 for review). Of course, 0 is a zero of f.

Topic 2: Graphing Factored Polynomials

Consider again the generic polynomial function $f(x) = a_n x^n + a_{n-1}x^{n-1} + \ldots + a_1 x + a_0$, where $a_n \neq 0$ and, for the moment, all of the coefficients $a_n,\ a_{n-1},\ \ldots,\ a_1,\ a_0$ are assumed to be real. Our goal is to be able to sketch the graph of such a function, paying particular attention to the behavior of f as $x \to -\infty$ and as $x \to \infty$, the y-intercept of f, and the x-intercepts of f (if there are any).

The behavior of a polynomial function as $x \to \pm\infty$ can be determined as follows. In brief, the graph of $f(x) = a_n x^n + a_{n-1}x^{n-1} + \ldots + a_1 x + a_0$ is similar to the graph of $a_n x^n$ for values of x that are very large in magnitude; that is, as $x \to -\infty$ and as $x \to \infty$, the leading term of $f(x) = a_n x^n + a_{n-1}x^{n-1} + \ldots + a_1 x + a_0$ dominates the behavior. Since we have already studied functions of the form $a_n x^n$, we know roughly how the sketch of an nth degree polynomial function should appear far away from the origin. Specifically, we know from Section 3.3 that

if n is even, $x^n \to \infty$ as $x \to -\infty$ and as $x \to \infty$,

and

if n is odd then $x^n \to -\infty$ as $x \to -\infty$ and $x^n \to \infty$ as $x \to \infty$.

We also know that

if a_n is positive, multiplying x^n by a_n merely compresses (if $0 < a_n < 1$) or stretches (if $a_n > 1$) the graph of x^n,

while

if a_n is negative, the graph of $a_n x^n$ is the reflection with respect to the x-axis of the graph of $|a_n| x^n$.

These observations are summarized below for a polynomial $f(x)$ of degree n.

End Behavior of a Polynomial

Given a polynomial $f(x) = a_n x^n + a_{n-1}x^{n-1} + \ldots + a_1 x + a_0$, we have

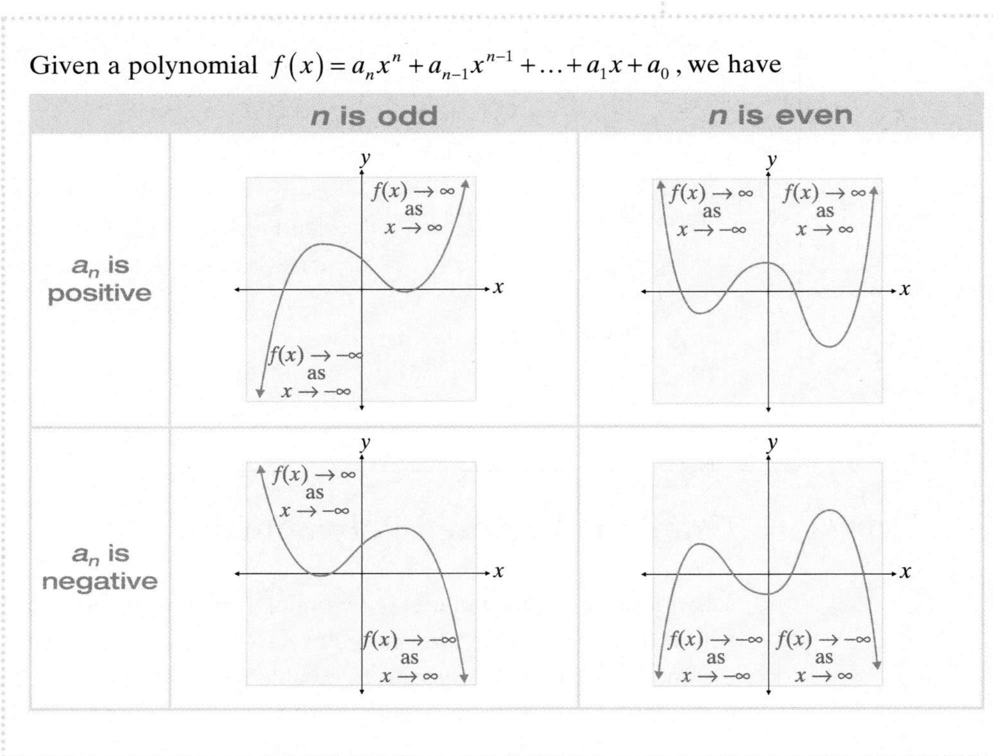

Figure 1 below contains a few examples of these observations.

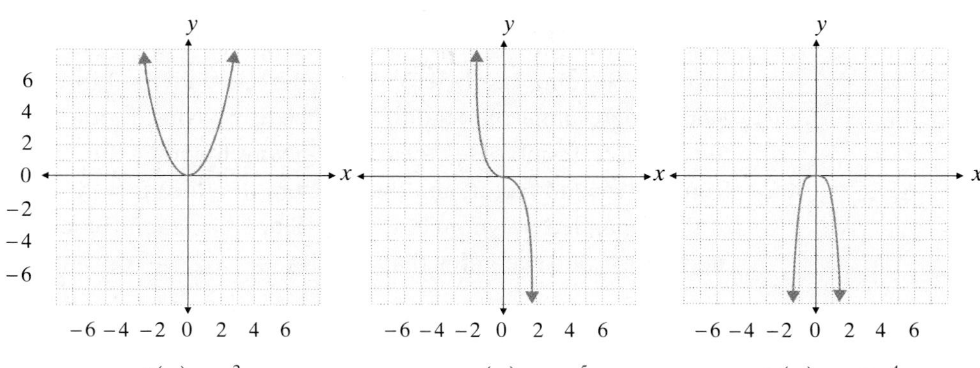

Figure 1: Examples of Behavior at $\pm\infty$

Near the origin, however, the graph of $f(x)=a_nx^n+a_{n-1}x^{n-1}+\ldots+a_1x+a_0$ is likely to be quite different from the graph of a_nx^n. For one thing, the y-intercept of f is $(0, a_0)$ and a_0 is possibly non-zero, whereas a_nx^n always crosses the y-axis at $(0,0)$. (Note that to determine the y-intercept of f, we simply evaluate f at 0; for our generic polynomial, $f(0)=a_0$.) Also, an nth degree polynomial may have, in general, a very different number of x-intercepts than the function a_nx^n (which has exactly one x-intercept, namely the origin). The y-intercept of a polynomial function is usually easy to determine, but finding the x-intercepts (if any exist) may require more effort.

At this point, the value of writing a polynomial function in factored form should become clear. If we are able to factor a given polynomial f into a product of linear factors, every linear factor with real coefficients will correspond to an x-intercept of the graph of f. To see why this is true, consider the polynomial function

$$f(x)=(3x-5)(x+2)(2x-6).$$

To determine the x-intercepts of this polynomial, we need to find all real numbers k for which $f(k)=0$; that is, we need to solve the equation

$$(3x-5)(x+2)(2x-6)=0.$$

Since the polynomial in this polynomial equation appears in factored form, the Zero-Factor Property (see Section 1.7) tells us immediately that the only solutions are those values of x for which $3x-5=0$, $x+2=0$, or $2x-6=0$. Solving these three linear equations gives us the x-coordinates of the three x-intercepts of f: $\left\{\frac{5}{3}, -2, 3\right\}$.

What else can we say about the graph of $f(x)=(3x-5)(x+2)(2x-6)$? If we were to multiply out the three linear factors, we would obtain a third degree polynomial. In order to know how the graph of *f* behaves as $x \to -\infty$ and as $x \to \infty$, we also need to know the leading coefficient. Instead of actually multiplying out *f* completely, we can quickly determine the leading coefficient by considering how the x^3 term arises. The third degree term comes from multiplying together the $3x$ from the first factor, the x from the second factor, and the $2x$ from the third factor. That is, the third degree term of the expanded polynomial is $6x^3$. Since the leading coefficient, 6, is positive, and the degree, 3, is odd, we know that $f(x) \to -\infty$ as $x \to -\infty$ and $f(x) \to \infty$ as $x \to \infty$.

The last observation we will make about the graph of *f* is that $f(0)=(0-5)(0+2)(0-6)$, or $f(0)=60$, so the *y*-intercept is $(0, 60)$. Putting all of the above together, along with a few more computed values of *f*, we obtain the sketch in Figure 2 below (note the difference in the horizontal and vertical scales).

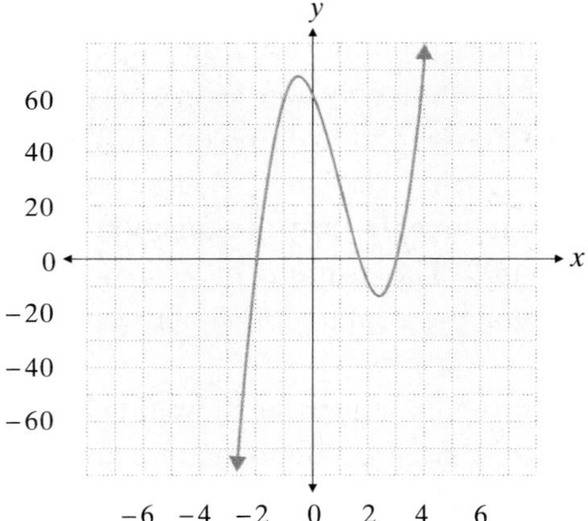

Figure 2: Graph of $f(x)=(3x-5)(x+2)(2x-6)$

The above method of graphing factored polynomials leaves unanswered one very important question: What do we do if we are given a polynomial in non-factored form? Since the only potentially difficult step in the method is determining the *x*-intercepts, this question is equivalent to: How do we solve the generic n^{th} degree polynomial equation $a_n x^n + a_{n-1}x^{n-1} + \ldots + a_1 x + a_0 = 0$? The answers to these questions will constitute much of the rest of this chapter, so for the remainder of this section all polynomials will either be given in factored form or else will be readily factored with the tools we already possess.

example 2

Sketch the graphs of the following polynomial functions, paying particular attention to the x-intercepts, the y-intercept, and the behavior as $x \to \pm\infty$.

Solutions:

a. $f(x) = -x(2x+1)(x-2)$ **b.** $g(x) = x^4 - 1$ **c.** $h(x) = x^2 + 2x - 3$

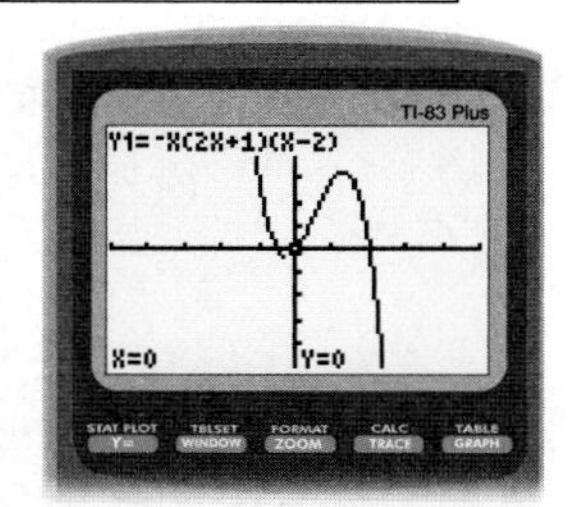

a.

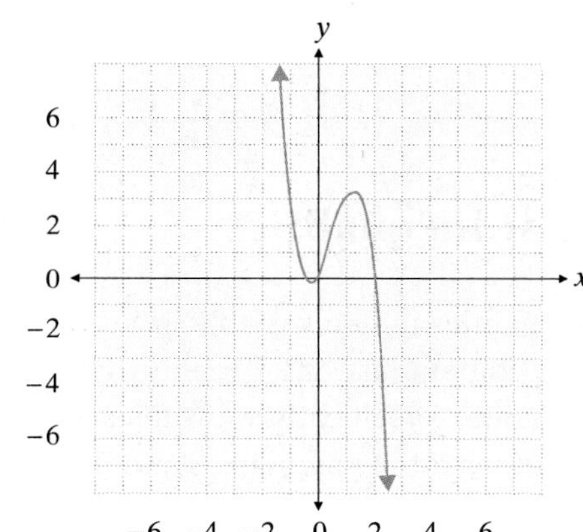

The solutions of $-x(2x+1)(x-2) = 0$ are 0, $-\frac{1}{2}$, and 2, and so these are the x-coordinates of the three x-intercepts.

The graph of f crosses the y-axis at $f(0) = 0$.

Finally, if we were to multiply out the three linear factors of f, the highest degree term would be $(-x)(2x)(x) = -2x^3$. The degree of f and the fact that the leading coefficient is negative indicates how f behaves far away from the origin.

b.

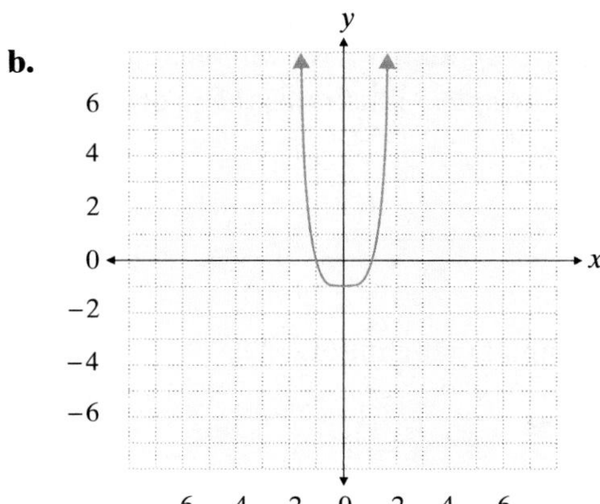

We know, from the discussion in Chapter 3, that the graph of $g(x) = x^4 - 1$ is the graph of x^4 shifted down one unit. But we can also determine the x-intercepts by solving the equation $x^4 - 1 = 0$. Since the polynomial is a difference of two squares, we can factor it to obtain the equation $(x^2 - 1)(x^2 + 1) = 0$, and then factor again to obtain $(x-1)(x+1)(x^2+1) = 0$. Only the two linear factors correspond to x-intercepts (the factor $x^2 + 1$ leads to zeros of i and $-i$, which are irrelevant for graphing purposes).

cont'd. on next page ...

c.

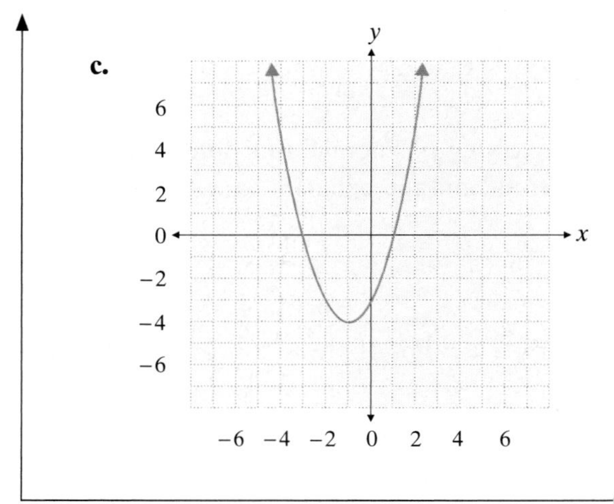

We can solve the quadratic equation $x^2+2x-3=0$ by factoring, which gives us $(x+3)(x-1)=0$, and hence x-intercepts of -3 and 1 for the function h. We also know that the shape of h is an upward opening parabola, and that the graph crosses the y-axis at $h(0)=-3$. We also know that the axis of symmetry must be a vertical line halfway between the x-intercepts, so the x-coordinate of the vertex must be -1. Note that $h(-1)=-4$.

Topic 3: Solving Polynomial Inequalities

We first encountered inequalities in one variable in Section 1.6. As in that section, the inequalities we study here contain just one variable, but the variable may now appear with exponents larger than one. Specifically, we will learn how to solve inequalities that would be polynomial equations if the inequality symbol were replaced by equality. The techniques learned here will be useful when we study rational functions later in this chapter, and will also aid your understanding of just what, exactly, the graph of a function tells us.

Every polynomial inequality can be rewritten in the form $f(x)<0$, $f(x)\le 0$, $f(x)>0$, or $f(x)\ge 0$, where f is a polynomial function, and this will be the key to solving the inequality. By graphing the polynomial f, we will be able to easily pick out the intervals that solve the inequality. Example 3 illustrates the process.

example 3

Solve the following polynomial inequalities.

a. $2x^2-3x>9$ **b.** $x^4-2x^2<-x^3$ **c.** $(x+3)(x+1)(x-2)\ge 0$

Solutions:

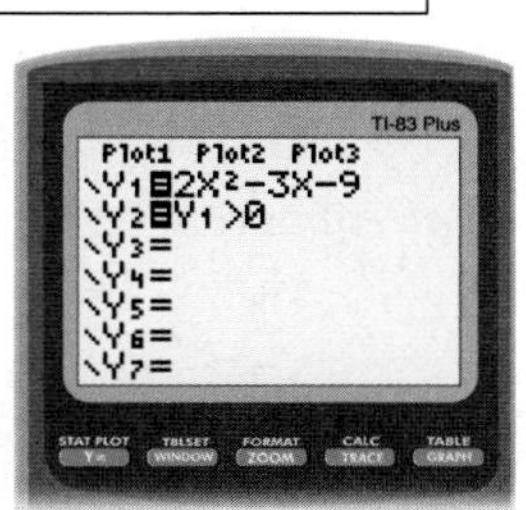

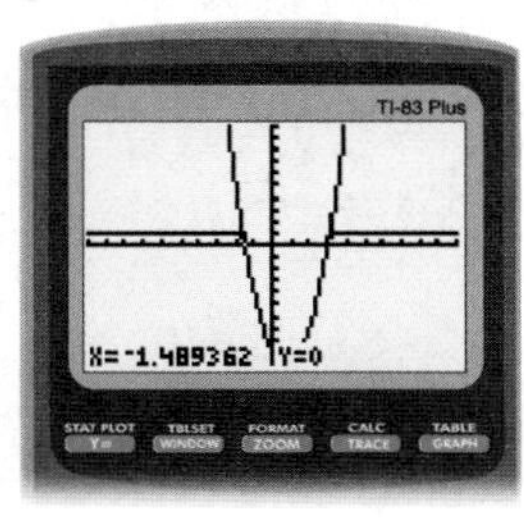

Before selecting the Y= screen, change the **Mode** from `Sequential` to `Simul`. To enter Y_1, select **Vars** (as described on pg. 265), and to enter >, select **2nd Math (Test)** and enter the desired symbol.

a.

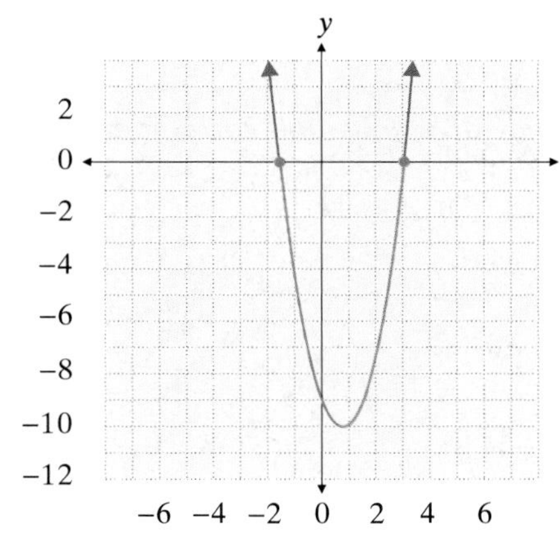

We need to identify the intervals for which $2x^2-3x-9>0$; that is, we need to locate the intervals on which the graph of the polynomial $f(x)=2x^2-3x-9$ is strictly above the x-axis. Factoring is the first step: $f(x)=(2x+3)(x-3)$. This tells us that $-\frac{3}{2}$ and 3 are the x-intercepts of this upward-opening parabola. After graphing, we see the solution (graphed in red) of the inequality is $\left(-\infty,-\frac{3}{2}\right)\cup(3,\infty)$.

b.

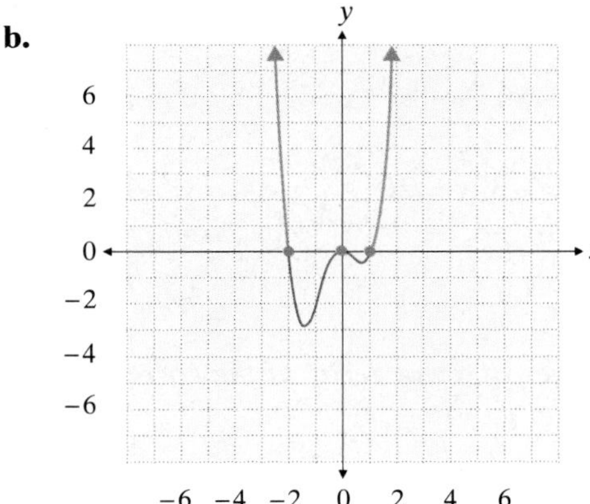

Again, we begin by writing the inequality in the form $x^4+x^3-2x^2<0$. The function $f(x)=x^4+x^3-2x^2$ can be factored more easily after first factoring out x^2, and we obtain $f(x)=x^2(x+2)(x-1)$. This means that the x-intercepts are 0, –2, and 1, but this function behaves a bit differently near 0 than others we have graphed so far. By plotting a few points near 0, we see that f touches the x-axis at this x-intercept, but does not cross. The solution of the inequality consists of those intervals where f is strictly below the x-axis: $(-2,0)\cup(0,1)$.

c.

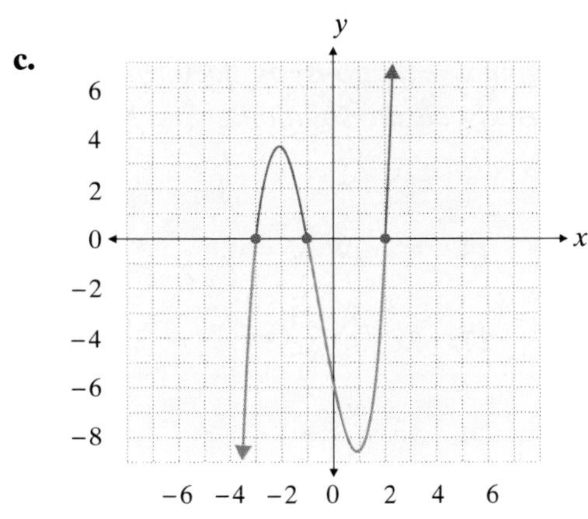

The polynomial in $(x+3)(x+1)(x-2)\geq 0$ is already factored, and the graph of the function $f(x)=(x+3)(x+1)(x-2)$ is as shown. The solution of the inequality consists of those intervals where the graph of f is above or on the x-axis. Hence, the solution is $[-3,-1]\cup[2,\infty)$. Note that –3, –1, and 2 are included in the solution set.

exercises

Verify that the given values of x solve the corresponding polynomial equations. See Example 1.

1. $9x^2 - 4x = 2x^3 + 15;\ x = -1$
2. $x^2 - 4x = -13;\ x = 2 - 3i$
3. $x^2 + 13 = 4x;\ x = 2 + 3i$
4. $3x^3 + (5 - 3i)x^2 = (2 + 5i)x - 2i;\ x = i$
5. $9x^2 - 4x = 2x^3 + 15;\ x = 3$
6. $9x^2 - 4x = 2x^3 + 15;\ x = \frac{5}{2}$
7. $3x^3 + (5 - 3i)x^2 = (2 + 5i)x - 2i;\ x = -2$
8. $23x^7 - 12x^5 = 63x^4 - 3x^2;\ x = 0$
9. $4x^5 - 8x^4 - 12x^3 = 16x^2 - 25x - 69;\ x = 3$
10. $x^2 - 4x - 12 = 0;\ x = 6$
11. $x^5 - 10x^4 - 80x^2 = 32 - 80x - 40x^3;\ x = 2$
12. $x^2 + 74 = 10x;\ x = 5 + 7i$
13. $4x^2 + 32x + (8 + i)x^3 = -8;\ x = 2i$
14. $8x - 17 = x^2;\ x = 4 - i$
15. $(5 - 3i)x - 3x = 4 - 6i;\ x = 2$
16. $x^6 - x^5 + 7x^4 + x^3 - 9x = -1;\ x = 1$

Determine if the given values of x are zeros of the corresponding polynomial equation. See Example 1.

17. $16x = x^3 + x^2 + 20;\ x = -5$
18. $x^4 - 13x^2 + 12 = -x^3 + x;\ x = -1$
19. $x^4 - 3x^3 - 10x^2 = 0;\ x = 2$
20. $4x^5 - 216x^2 = 36x^3 - 24x^4;\ x = -6$
21. $x^3 - 8ix + 30 = 15x + 2x^2 + 16i;\ x = -i$
22. $x^3 - 7x^2 + 4x - 28 = 0;\ x = 2i$

Solve the following polynomial equations by factoring and/or using the quadratic formula, making sure to identify all the solutions. See Example 2. See Section 1.7 for review, if necessary.

23. $x^3 - x^2 - 6x = 0$
24. $x^2 - 2x + 5 = 0$
25. $x^4 + x^2 - 2 = 0$
26. $2x^2 + 5x = 3$
27. $9x^2 = 6x - 1$
28. $x^4 - 8x^2 + 15 = 0$
29. $x^3 - x^2 = 72x$
30. $x^2 + 5x = -\frac{25}{4}$
31. $2x^2 + 5 = 11x$
32. $x^4 - 8x^3 + 25x^2 = 0$
33. $x^4 - 13x^2 + 36 = 0$
34. $x^4 + 7x^2 = 8$

For each of the following polynomials, determine the degree and the leading coefficient; then describe the behavior of the graph as $x \to \pm\infty$.

35. $p(x) = 2x^4 - 3x^3 - 6x^2 - x - 23$

36. $i(x) = 4x^7 + 5x^5 + 12$

37. $r(x) = (3x + 5)(x - 2)(2x - 1)(4x - 7)$

38. $h(x) = -6x^5 + 2x^3 - 7x$

39. $g(x) = (x - 5)^3(2x + 1)(-x - 1)$

40. $f(x) = -2(x + 4)(x - 4)(x^2)$

For each of the following polynomial functions, describe the behavior of its graph as $x \to \pm\infty$ and identify the x- and y-intercepts. Use this information to then sketch the graph of each polynomial. See Example 2.

41. $f(x) = (x-3)(x+2)(x+4)$

42. $g(x) = (3-x)(x+2)(x+4)$

43. $f(x) = (x-2)^2(x+5)$

44. $h(x) = -(x+2)^3$

45. $r(x) = x^2 - 2x - 3$

46. $s(x) = x^3 + 3x^2 + 2x$

47. $f(x) = -(x-2)(x+1)^2(x+3)$

48. $g(x) = (x-3)^5$

49. $f(x) = 5x^4 + 6x^3 - 25x^2 - 30x$

50. $p(x) = -x^5 + 3x^4 + 4x^3 - 12x^2$

51. $g(x) = (2x - 3)(x - 5)(1 - x)^2$

52. $h(x) = (5x + 6)^2(x + 2)^3$

*In Exercises 53 through 58, use the behavior as $x \to \pm\infty$ and the intercepts to match each polynomial with one of the graphs labeled **(a)** through **(f)**.*

53. $g(x) = (x+1)^2(x-3)^2$

54. $h(x) = 1 - (x + 2)^2$

55. $f(x) = (x-1)(x+2)(3-x)$

56. 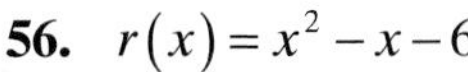 $r(x) = x^2 - x - 6$

57. 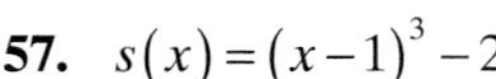 $s(x) = (x-1)^3 - 2$

58. $j(x) = (x - 2)^2(4x + 1)(x + 2)(x - 2)$

a. **b.**

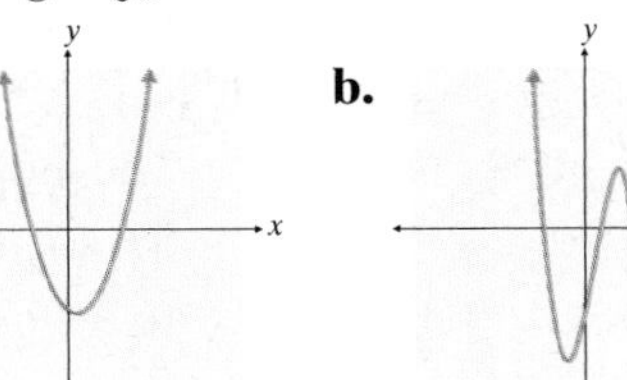

c. **d.**

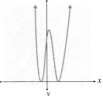

e.

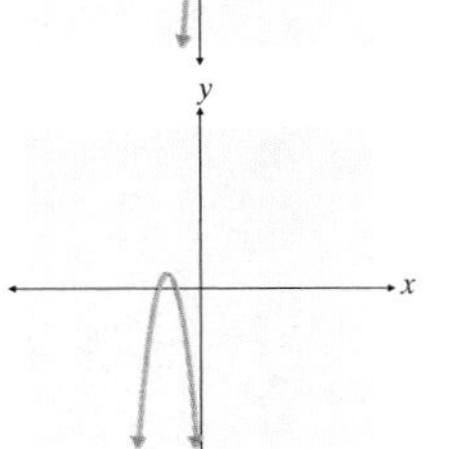

f.

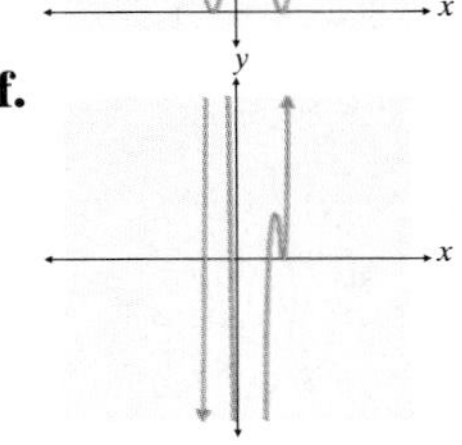

Solve the following polynomial inequalities. See Example 3.

59. $x^2 - x - 6 \le 0$

60. $x^2 > x + 6$

61. $(x+2)^2(x-1)^2 > 0$

62. $x^3 + 3x^2 + 2x < 0$

63. $(x-2)(x+1)(x+3) \ge 0$

64. $(x-1)(x+2)(3-x) \le 0$

65. $-x^3 - x^2 + 30x > 0$

66. $(x^2-1)(x-4)(x+5) \le 0$

67. $x^4 + x^2 > 0$

Solve the following application problems.

68. A small start-up skateboard company projects that the cost per month of manufacturing x skateboards will be $C(x) = 10x + 300$, and the revenue per month from selling x skateboards will be $r(x) = -x^2 + 50x$. Given that profit is revenue – cost, what value for x will give the company a non-negative profit?

69. The population of sea lions on an island is represented by the function $L(m) = 110m^2 - 0.35m^4 + 750$, where m is the number of months the sea lions have been observed on the island. Given this information, how many months until no more sea lions remain on the island?

70. An electronics company is deciding whether or not to begin producing CD-players. The company must determine if doing so is profitable enough to start producing CD-players. The profit function is modeled by the equation $P(x) = x + 0.27x^2 - 0.0015x^3 - 300$, where x is the number of CD-players produced. Given this equation, how many CD-players must the company produce to make a profit?

71. The population of mosquitoes in a city in Florida is modeled by the function $M(w) = 200w^2 - 0.01w^4 + 1200$, where w is the number of weeks since the town began spraying for mosquitoes. How many weeks will it take for all the mosquitoes to have died off?

4.2 Polynomial Division and the Division Algorithm

TOPICS

1. The Division Algorithm and the Remainder Theorem
2. Polynomial long division and synthetic division
3. Constructing polynomials with given zeros

Topic 1: The Division Algorithm and the Remainder Theorem

In Section 4.1, we made use of the factored form of a polynomial in graphing polynomials and in solving polynomial equations and inequalities. In this section, we will formalize the observations we made and introduce some useful techniques for working with polynomials.

We begin with a statement about polynomials that is the analog of an elementary number fact: If a natural number is divided by a smaller one, the result can be expressed as a whole number (called the quotient) plus a fraction less than 1 (the fraction may in fact be 0). The polynomial version of this is called the *Division Algorithm.*

The Division Algorithm

Let $p(x)$ and $d(x)$ be polynomials such that $d(x) \neq 0$ and with the degree of d less than or equal to the degree of p. Then there are unique polynomials $q(x)$ and $r(x)$, called the quotient and the remainder, respectively, such that

$$\underbrace{p(x)}_{\text{dividend}} = \underbrace{q(x)}_{\text{quotient}} \cdot \underbrace{d(x)}_{\text{divisor}} + \underbrace{r(x)}_{\text{remainder}}.$$

The degree of the remainder, r, is less than the degree of the divisor, d, or else the remainder is 0, in which case we say d divides evenly into the polynomial, p. If the remainder is 0, the two polynomials q and d are factors of p.

If we divide every term in the equation $p(x)=q(x)\cdot d(x)+r(x)$ by the polynomial d, we obtain a form that is very similar to the number fact mentioned above:

$$\frac{p(x)}{d(x)}=q(x)+\frac{r(x)}{d(x)}.$$

This fact may be stated in words as something like "If one polynomial is divided by another of smaller degree, the result is a polynomial plus, possibly, a ratio of two polynomials, the numerator of which has a smaller degree than the denominator." This version of the division algorithm will be very useful when we study rational functions later in Section 4.5.

In many cases, we will be interested in dividing a given polynomial p by a divisor of the form $d(x)=x-k$. The division algorithm tells us that the remainder is guaranteed to be either 0 (which is a polynomial of undefined degree), or else a polynomial of degree 0 (since the degree of d is 1). In either case, the remainder polynomial is guaranteed to be simply a number, so $p(x)=q(x)\cdot(x-k)+r$ where r is a constant.

What does this mean if k happens to be a zero of the polynomial p? If this is the case, $p(k)=0$, so

$$\begin{aligned}0 &= p(k)\\ &= q(k)\cdot(k-k)+r\\ &= r.\end{aligned}$$

In other words, if k is a zero of the polynomial p, $p(x)=q(x)\cdot(x-k)$. On the other hand, if $x-k$ divides a given polynomial p evenly, then we know $p(x)=q(x)\cdot(x-k)$ and hence $p(k)=q(k)\cdot(k-k)=0$, so k is a zero of p. Together, these two observations constitute a major tool that we use in graphing polynomials and in solving polynomial equations, summarized below for reference.

Zeros and Linear Factors

The number k is a zero of a polynomial $p(x)$ if and only if the linear polynomial $x-k$ is a factor of p. In this case, $p(x)=q(x)\cdot(x-k)$ for some quotient polynomial q. This also means that k is a solution of the polynomial equation $p(x)=0$, and if p is a polynomial with real coefficients and if k is a real number, then k is an x-intercept of p.

The above reasoning also leads to the more general conclusion called the *remainder theorem.*

The Remainder Theorem

If the polynomial $p(x)$ is divided by $x - k$, the remainder is $p(k)$. That is,

$$p(x) = q(x) \cdot (x - k) + p(k).$$

Topic 2: Polynomial Long Division and Synthetic Division

To make use of the division algorithm and the remainder theorem, we need to be able to actually divide one polynomial by another. Polynomial long division is the analog of numerical long division, and provides the means for dividing any polynomial by another of equal or smaller degree. Synthetic division is a shortcut that allows us to quickly divide a polynomial by a polynomial of the form $x - k$.

Polynomial long division is best explained with an example. As you study Example 1, notice the similarities between polynomial division and numerical division.

example 1

Divide the polynomial $6x^5 - 5x^4 + 10x^3 - 15x^2 + 9x - 19$ by the polynomial $2x^2 - x + 3$.

Solution:

$$\begin{array}{r} 3x^3 \\ 2x^2 - x + 3\overline{\big)6x^5 - 5x^4 + 10x^3 - 15x^2 + 9x - 19} \\ \underline{-\left(6x^5 - 3x^4 + 9x^3\right)} \\ -2x^4 + x^3 - 15x^2 + 9x - 19 \end{array}$$

The first step is to arrange the dividend and the divisor in descending order. The first term of the quotient is then the first term of the dividend divided by the first term of the divisor, giving us $3x^3$ in this case. Multiply $3x^3$ by $2x^2 - x + 3$ to obtain the polynomial $6x^5 - 3x^4 + 9x^3$, which we subtract from the dividend.

cont'd. on next page ...

$$\begin{array}{r}
3x^3 - x^2 \\
2x^2 - x + 3\overline{)6x^5 - 5x^4 + 10x^3 - 15x^2 + 9x - 19} \\
-\left(6x^5 - 3x^4 + 9x^3\right) \\
\hline
-2x^4 + x^3 - 15x^2 + 9x - 19 \\
-\left(-2x^4 + x^3 - 3x^2\right) \\
\hline
-12x^2 + 9x - 19
\end{array}$$

To determine the second term of the quotient, we repeat the process from the previous page, this time dividing $-2x^4$ by $2x^2$ to get $-x^2$ in the quotient. Multiply this by $2x^2 - x + 3$ to obtain the next polynomial we subtract, $-2x^4 + x^3 - 3x^2$.

$$\begin{array}{r}
3x^3 - x^2 - 6 \\
2x^2 - x + 3\overline{)6x^5 - 5x^4 + 10x^3 - 15x^2 + 9x - 19} \\
-\left(6x^5 - 3x^4 + 9x^3\right) \\
\hline
-2x^4 + x^3 - 15x^2 + 9x - 19 \\
-\left(-2x^4 + x^3 - 3x^2\right) \\
\hline
-12x^2 + 9x - 19 \\
-\left(-12x^2 + 6x - 18\right) \\
\hline
3x - 1
\end{array}$$

To complete the division, we repeat the procedure one more time. At this point, the process halts, as the degree of $3x - 1$ is smaller than the degree of the divisor.

We can summarize the result as:

$$\frac{6x^5 - 5x^4 + 10x^3 - 15x^2 + 9x - 19}{2x^2 - x + 3} = 3x^3 - x^2 - 6 + \frac{3x - 1}{2x^2 - x + 3}.$$

caution!

Although polynomial long division is a straightforward process, one common error is to forget to distribute the minus sign in each step as one polynomial is subtracted from the one above it. A good way to avoid this error is to put parentheses around the polynomial being subtracted, as in Example 1 above.

Synthetic division is a shortened version of polynomial long division, and can be used when the divisor is of the form $x - k$ for some constant k. Hence, we cannot use synthetic division when dividing by, for example, $x^2 - 1$, since x is raised to a power other than 1 or $3x + 5$, since the coefficient of x is not 1. Synthetic division does not do anything that long division can't do (and in fact is only applicable in certain circumstances), but the speed of synthetic division is often convenient.

Compare the division of $-2x^3+8x^2-9x+7$ by $x-2$ below, using long division on the left and synthetic division on the right.

$$
\begin{array}{r}
-2x^2+4x-1 \\
x-2\overline{\big)-2x^3+8x^2-9x+7} \\
\underline{-\left(-2x^3+4x^2\right)} \\
4x^2-9x+7 \\
\underline{-\left(4x^2-8x\right)} \\
-x+7 \\
\underline{-\left(-x+2\right)} \\
5
\end{array}
\qquad
\begin{array}{r|rrrr}
2 & -2 & 8 & -9 & 7 \\
 & & -4 & 8 & -2 \\
\hline
 & -2 & 4 & -1 & 5
\end{array}
$$

The increase in efficiency in synthetic division comes from omitting the variables in the division process. Instead of various powers of the variable, synthetic division uses a tabular arrangement to keep track of the coefficients of the dividend and, ultimately, the coefficients of the quotient and the remainder. In the example above, note that the numbers in red are the coefficients of the dividend and the numbers in blue are the coefficients of the quotient and remainder. More importantly, note that the number 2, which corresponds to k in the form $x - k$, appears without the minus sign. This is a very important and easily overlooked fact: When dividing by $x-k$ using synthetic division, the number that appears in the upper left spot is k.

Synthetic division is an iterative process based on simple numerical multiplication and addition. Once we have written k and the coefficients of the dividend as above, the synthetic division process begins by copying the leading coefficient of the dividend (-2 in the above example) in the first slot below the horizontal line. This number is then multiplied by the number k (which is 2 above) and the result is written directly below the second coefficient of the dividend (8 in the example). The two numbers in that column are then added and the result written in the second slot below the horizontal line. The process is repeated until the last column is completed, and the last number written down is the remainder. The other numbers in the bottom row constitute the coefficients of the quotient, which will be a polynomial of one degree less than the dividend. For instance, in our example above and to the right, the remainder is 5 and the quotient is $-2x^2+4x-1$, read from the bottom row.

Figure 1 is an illustration of the calculations described on the previous page, for the same division problem.

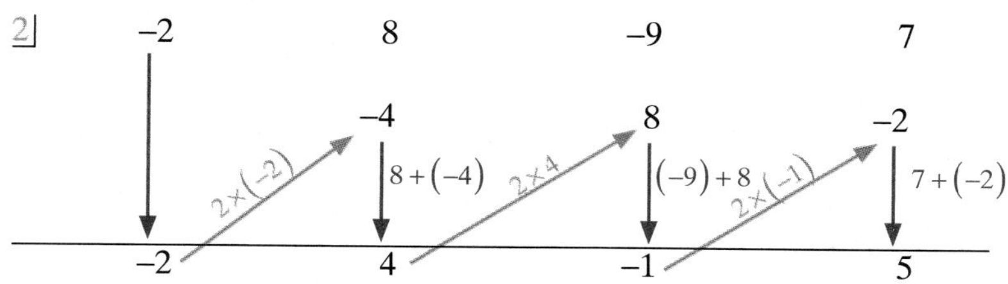

Figure 1: Synthetic Division

This process of synthetic division is summarized below.

Synthetic Division

Synthetic division can be used to divide $f(x)=a_nx^n+a_{n-1}x^{n-1}+\ldots+a_1x+a_0$ by $x-k$ as follows:

$$
\begin{array}{c|cccccccc}
k & a_n & a_{n-1} & a_{n-2} & \cdots & a_2 & a_1 & a_0 \\
 & & k\cdot b_n & k\cdot b_{n-1} & \cdots & k\cdot b_3 & k\cdot b_2 & k\cdot b_1 \\
\hline
 & b_n & b_{n-1} & b_{n-2} & \cdots & b_2 & b_1 & b_0
\end{array}
$$

where $b_n = a_n$

$b_i = a_i + k\cdot b_{i+1}$ for $i = 0, \ldots, n-1$

$b_0 =$ remainder (if $b_0 = 0$; then k is a zero of $f(x)$)

Because synthetic division is much faster than long division, it is very useful in determining if $x-k$ is a factor of a given polynomial. By the remainder theorem, synthetic division also provides a quick means of determining $p(k)$ for a given polynomial p, since $p(k)$ is the remainder when $p(x)$ is divided by $x-k$. This fact is used in Example 2.

example 2

For each polynomial p below, determine if the given k is a zero. If not, determine $p(k)$.

a. $p(x)=3x^8+9x^7-x^3-3x^2+x-1;\ k=-3$

b. $p(x)=-2x^4+11x^3-5x^2-3x+15;\ k=5$

Solutions:

a.

$\underline{-3\rvert}$	3	9	0	0	0	−1	−3	1	−1
		−9	0	0	0	0	3	0	−3
	3	0	0	0	0	−1	0	1	−4

First of all, note that it is essential that we place a number of 0's in certain slots in the first row, as these serve as placeholders for the missing terms of the dividend (namely, x^6, x^5, and x^4). The fact that the last number is non-zero means −3 is not a zero of p. What we *can* conclude is that $p(-3)=-4$.

$$\frac{3x^8+9x^7-x^3-3x^2+x-1}{x+3} = 3x^7-x^2+1-\frac{4}{x+3}$$

Note also that we can write the result of the division as a polynomial plus a rational function. And since $k=-3$, the divisor is $x-(-3)$, or $x+3$. Here we used the form $p(x)=q(x)+\frac{r(x)}{d(x)}$.

b.

$\underline{5\rvert}$	−2	11	−5	−3	15
		−10	5	0	−15
	−2	1	0	−3	0

In this case, the remainder is 0, and hence 5 is a zero of the polynomial p.

$$-2x^4+11x^3-5x^2-3x+15 = \left(-2x^3+x^2-3\right)(x-5)$$

Since the remainder is 0, we now know of two factors of p, as illustrated at left. Here we have $p(x)=q(x)\cdot d(x)$.

To convince yourself of the utility of synthetic division in evaluating polynomials, examine Example 2a again. The direct way to determine $p(-3)$ is, of course, to simplify the following: $p(-3)=3(-3)^8+9(-3)^7-(-3)^3-3(-3)^2+(-3)-1$. This is a tedious process, but the result, $3(6561)+9(-2187)-(-27)-3(9)+(-3)-1$, does indeed equal -4. Compare these calculations with the far simpler synthetic division used in the example.

Before leaving the subject of polynomial division, let us consider polynomials with complex coefficients and/or complex zeros. When graphing polynomials, we will be concerned with those that have only real coefficients, but complex zeros and coefficients may still arise in intermediate stages of the graphing process. And in solving polynomial equations, we have already seen (in the case of quadratic equations) that complex numbers may be the *only* solutions. For these reasons, it is important for us to be able to handle complex numbers as they arise.

example 3

a. Divide $p(x)=x^4+1$ by $d(x)=x^2+i$.

b. Perform the indicated division: $\dfrac{-3x^3+(5-2i)x^2+(-4+i)x+(1-i)}{x-1+i}$.

Solutions:

a.

$$\begin{array}{r} x^2-i \\ x^2+i\,\big)\overline{x^4+1} \\ \underline{-(x^4+ix^2)} \\ -ix^2+1 \\ \underline{-(-ix^2+1)} \\ 0 \end{array}$$

Note that the divisor is not of the form $x-k$, so long division is called for.

The procedure is the same as always, though complex number arithmetic is called for (for instance, we need to know $-i^2=1$). The remainder of 0 tells us that $x^4+1=(x^2+i)(x^2-i)$.

b.

$$\begin{array}{r|rrrr} 1-i & -3 & 5-2i & -4+i & 1-i \\ & & -3+3i & 3-i & -1+i \\ \hline & -3 & 2+i & -1 & 0 \end{array}$$

In this problem, the divisor can be written as $x-(1-i)$, so we can use synthetic division. Note that we have proved that $1-i$ is a zero of the numerator of the original fraction.

Topic 3: Constructing Polynomials with Given Zeros

The last topic in this section concerns reversing the division process. We now know well the connection between zeros and factors: k is a zero of the polynomial $p(x)$ if and only if $x-k$ is a factor of $p(x)$. We can make use of this fact to construct polynomials that have certain desired properties, as illustrated in Example 4.

example 4

Construct a polynomial that has the given properties.

a. Third degree, zeros of –3, 2, and 5, and goes to $-\infty$ as $x \to \infty$.

b. Fourth degree, zeros of –5, –2, 1, and 3, and a y-intercept of 15.

Solutions:

a.
$$p(x) = -(x+3)(x-2)(x-5)$$
$$= -x^3 + 4x^2 + 11x - 30$$

We begin by noting that $x+3$, $x-2$, and $x-5$ must be factors of the polynomial p we are to construct, since these factors give rise to the desired zeros. Further, there can be no other linear (or higher degree) factors, since the polynomial is to be third degree. But a cubic with a positive leading coefficient goes to ∞ as $x \to \infty$, so we must multiply the three linear factors by a negative constant to achieve the desired behavior. If we multiply by –1, the result is as shown at left.

b.
$$p(x) = a(x+5)(x+2)(x-1)(x-3)$$
$$p(0) = a(5)(2)(-1)(-3) = 30a$$
$$30a = 15 \Rightarrow a = \frac{1}{2}$$

$$p(x) = \frac{1}{2}(x+5)(x+2)(x-1)(x-3)$$
$$= \frac{1}{2}x^4 + \frac{3}{2}x^3 - \frac{15}{2}x^2 - \frac{19}{2}x + 15$$

The given zeros and the desired degree of the polynomial lead to the four linear factors shown at left. And as in the first problem, the linear factors can be multiplied by any non-zero constant a without affecting the zeros. Since we want the y-intercept to be 15, it must be the case that $30a = 15$, allowing us to solve for a.

If we multiply out the four linear factors and the constant, the result is as shown.

exercises

Use polynomial long division to rewrite each of the following fractions in the form $q(x)+\frac{r(x)}{d(x)}$, where $d(x)$ is the denominator of the original fraction, $q(x)$ is the quotient, and $r(x)$ is the remainder. See Example 1.

1. $\frac{6x^4-2x^3+8x^2+3x+1}{2x^2+2}$ **2.** $\frac{5x^2+9x-6}{x+2}$ **3.** $\frac{x^3-6x^2+12x-10}{x^2-4x+4}$

4. $\frac{7x^5-x^4+2x^3-x^2}{x^2+1}$ **5.** $\frac{4x^3-6x^2+x-7}{x+2}$ **6.** $\frac{x^3+2x^2-4x-8}{x-3}$

7. $\frac{3x^5+18x^4-7x^3+9x^2+4x}{3x^2-1}$ **8.** $\frac{9x^5-10x^4+18x^3-28x^2+x+3}{9x^2-x-1}$

9. $\frac{2x^3-3ix^2+11x+(1-5i)}{2x-i}$ **10.** $\frac{9x^3-(18+9i)x^2+x+(-2-i)}{x-2-i}$

11. $\frac{3x^3+ix^2+9x+3i}{3x+i}$ **12.** $\frac{35x^4+(14-10i)x^3-(7+4i)x^2+2ix}{7x-2i}$

13. $\frac{2x^5-5x^4+7x^3-10x^2+7x-5}{x^2-x+1}$ **14.** $\frac{14x^5-2x^4+27x^3-3x^2+9x}{2x^3+3x}$

15. $\frac{x^4+x^2-20x-8}{x-3}$ **16.** $\frac{2x^5-3x^2+1}{x^2+1}$

17. $\frac{9x^3+2x}{3x-5}$ **18.** $\frac{-4x^5+8x^3-2}{2x^3+x}$

Use synthetic division to determine if the given value for k is a zero of the corresponding polynomial. If not, determine $p(k)$. See Example 2.

19. $p(x)=32x^5-80x^4+80x^3-40x^2+10x+2;\ k=1$

20. $p(x)=32x^5-80x^4+80x^3-40x^2+10x+2;\ k=\frac{1}{2}$

21. $p(x)=12x^4-7x^3-32x^2-7x+6;\ k=2$

22. $p(x)=12x^4-7x^3-32x^2-7x+6;\ k=1$

23. $p(x)=12x^4-7x^3-32x^2-7x+6;\ k=\frac{1}{3}$

24. $p(x) = 2x^2 - (3-5i)x + (3-9i);\ k = -2$

25. $p(x) = 8x^4 - 2x + 6;\ k = 1$

26. $p(x) = x^4 - 1;\ k = 1$

27. $p(x) = x^5 + 32;\ k = -2$

28. $p(x) = 3x^5 + 9x^4 + 2x^2 + 5x - 3;\ k = -3$

29. $p(x) = 2x^2 - (3-5i)x + (3-9i);\ k = -3i$

30. $p(x) = x^2 - 6x + 13;\ k = 2$

31. $p(x) = x^2 - 6x + 13;\ k = 3 - 2i$

32. $p(x) = 3x^3 - 13x^2 - 28x - 12;\ k = -2$

33. $p(x) = 3x^3 - 13x^2 - 28x - 12;\ k = 6$

34. $p(x) = 2x^3 - 8x^2 - 23x + 63;\ k = 2$

35. $p(x) = 2x^3 - 8x^2 - 23x + 63;\ k = 5$

36. $p(x) = x^4 - 3x^3 - 3x^2 + 11x - 6;\ k = 1$

37. $p(x) = x^4 - 3x^3 - 3x^2 + 11x - 6;\ k = -2$

38. $p(x) = x^4 - 3x^3 - 3x^2 + 11x - 6;\ k = 3$

Use synthetic division to rewrite each of the following fractions in the form $q(x) + \frac{r(x)}{d(x)}$, *where* $d(x)$ *is the denominator of the original fraction,* $q(x)$ *is the quotient, and* $r(x)$ *is the remainder. See Examples 2 and 3.*

39. $\dfrac{x^3 + x^2 - 18x + 9}{x + 5}$

40. $\dfrac{-2x^5 + 4x^4 + 3x^3 - 7x^2 + 3x - 2}{x - 2}$

41. $\dfrac{x^8 + x^7 - 3x^3 - 3x^2 + 3}{x + 1}$

42. $\dfrac{x^8 - 5x^7 - 3x^3 + 15x^2 - 2}{x - 5}$

43. $\dfrac{4x^3-(16+4i)x^2+(14+4i)x+(-6-2i)}{x-3-i}$

44. $\dfrac{x^6-2x^5+2x^4+4x^2-8x+8}{x-1+i}$

45. $\dfrac{x^5-3x^4+x^3-5x^2+18}{x-2}$

46. $\dfrac{x^5-3x^4+x^3-5x^2+18}{x-3}$

47. $\dfrac{x^4+(i-1)x^3+(1-i)x^2+ix}{x+i}$

48. $\dfrac{x^6+8x^5+x^3+8x^2-14x-112}{x+8}$

Construct polynomial functions with the stated properties (answers may not be unique). See Example 4.

49. Second degree, zeros of –4 and 3, and goes to $-\infty$ as $x \to -\infty$.

50. Third degree, zeros of –2, 1, and 3, and a y-intercept of –12.

51. Second degree, zeros of $2-3i$ and $2+3i$, and a y-intercept of –13.

52. Third degree, zeros of $1-i$, $2+i$, and –1, and a leading coefficient of –2.

53. Fourth degree and a single x-intercept of 3.

54. Second degree, zeros of $-\dfrac{3}{4}$ and 2, and a y-intercept of 6.

55. Fourth degree, zeros of –3, –2, and 1, and a y-intercept of 18.

56. Third degree, zeros of 1, 2, and 3, and passes through the point $(4, 12)$.

technology exercises

Wise use of a computer algebra system can allow us to easily solve problems in cases where hand computation would be depressingly tedious. Suppose, for instance, we are asked to construct a fifth degree polynomial function that passes through the point $(2, 10)$ and has zeros of -4, -1, 3, $-3+i$, and $-3-i$. The illustration below shows how *Mathematica* can be used to define a function with the appropriate zeros (note that the zeros are unaffected by any non-zero choice for the constant a) and then further refined so as to pass through the required point.

If possible, make similar use of technology to construct polynomial functions with the stated properties.

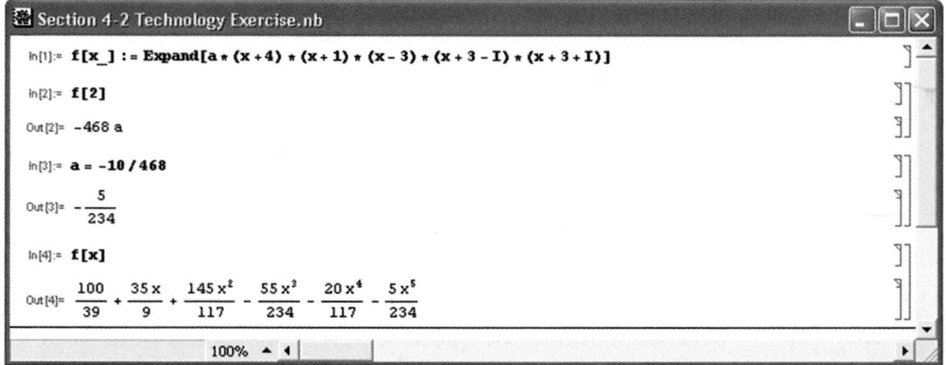

57. Fifth degree, zeros of $1+i$, $1-i$, $-3+2i$, $-3-2i$, and 7, and a y-intercept of 2.

58. Third degree, zeros of $2-5i$, $-1+3i$, and $3-i$, and with the property that $f(1) = -78-26i$.

59. Third degree, zeros of $2-5i$, $-1+3i$, and $3-i$, and with the property that $f(1+i) = 24+4i$.

4.3 Locating Real Zeros of Polynomials

TOPICS

1. The Rational Zero Theorem
2. Descartes' Rule of Signs
3. Bounds of real zeros
4. The Intermediate Value Theorem

Topic 1: The Rational Zero Theorem

For a given polynomial function $f(x)=a_nx^n+a_{n-1}x^{n-1}+\ldots+a_1x+a_0$, we know that the number k is a zero if and only if $x-k$ is a factor of f. Furthermore, if $x-k$ is a factor of f, we can use either polynomial long division or synthetic division to actually divide f by $x-k$ and find the quotient polynomial q, allowing us to write $f(x)=(x-k)\cdot q(x)$. What we are lacking, at this point, is a method for determining the zeros of f. This means, in effect, that the skills we have acquired in the last two sections cannot be put to use until we have some way of locating a zero k as a starting point.

If there were a method guaranteed to find all the zeros of a given polynomial, we would be able to solve any polynomial equation and (if the coefficients of the polynomial were all real) we would be able to sketch with fair accuracy the graph of any polynomial. As hinted at in Section 4.1, however, no such method exists. In fact, it can be proved (using ideas from a branch of mathematics called *abstract algebra*) that there is no formula based on elementary mathematical operations that identifies all the zeros of an arbitrary polynomial of degree five or higher. What we do have, though, are some tools that give us hints about where to look for zeros of a given polynomial. One such tool is the Rational Zero Theorem.

The Rational Zero Theorem

If $f(x) = a_n x^n + a_{n-1}x^{n-1} + \ldots + a_1 x + a_0$ is a polynomial with integer coefficients, then any rational zero of f must be of the form $\frac{p}{q}$, where p is a factor of the constant term a_0 and q is a factor of the leading coefficient a_n.

The proof of this statement is outlined in Exercise 57 following this section.

Before demonstrating how the Rational Zero Theorem is used, some cautions on what it doesn't do are in order. First, it doesn't necessarily find even a single zero of a polynomial; instead, it identifies a list of rational numbers that could *potentially* be zeros. In order to determine if any of the potential zeros actually *are* zeros, we have to test them. Second, the theorem says nothing about irrational zeros or complex zeros. If a given polynomial has some zeros that are either irrational or else have a non-zero imaginary part, we must resort to other means to find them.

example 1

For each of the polynomials that follow, list all of the potential rational zeros. Then write the polynomial in factored form and identify the actual zeros.

a. $f(x) = 2x^3 + 5x^2 - 4x - 3$ **b.** $g(x) = 27x^4 - 9x^3 - 33x^2 - x - 4$

Note in the table below that the y value is zero for both $x = -3$ and $x = 1$.

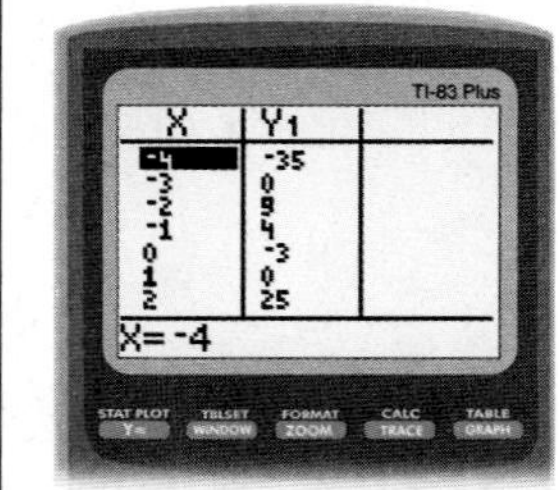

Solutions:

a. Factors of $a_0 = 3$: $\pm\{1, 3\}$

Factors of $a_3 = 2$: $\pm\{1, 2\}$

Possible rational zeros: $\pm\left\{1, 3, \frac{1}{2}, \frac{3}{2}\right\}$

We begin by listing the factors of the constant term, −3, and the leading coefficient, 2. Note that each number has two positive factors and two negative factors; this fact is denoted with the ± symbol in front of each set. By the Rational Zero Theorem, any rational zero must be one of the eight numbers generated by dividing factors of a_0 by factors of a_3.

cont'd. on next page ...

$$\begin{array}{r|rrrr} 1 & 2 & 5 & -4 & -3 \\ & & 2 & 7 & 3 \\ \hline & 2 & 7 & 3 & 0 \end{array}$$

$$\begin{aligned} f(x) &= (x-1)(2x^2+7x+3) \\ &= (x-1)(2x+1)(x+3) \end{aligned}$$

Actual zeros: $\left\{1, -\frac{1}{2}, -3\right\}$

We can then use synthetic division to identify which of the potential zeros actually *are* zeros. In the work at left, we first discover that 1 is a zero and then finish by factoring the trinomial $2x^2+7x+3$. You may have to try several potential zeros before finding an actual zero.

b. Factors of $a_0 = 4$: $\pm\{1, 2, 4\}$

Factors of $a_4 = 27$: $\pm\{1, 3, 9, 27\}$

Possible rational zeros:

$$\pm\left\{1, 2, 4, \frac{1}{3}, \frac{2}{3}, \frac{4}{3}, \frac{1}{9}, \frac{2}{9}, \frac{4}{9}, \frac{1}{27}, \frac{2}{27}, \frac{4}{27}\right\}$$

$$\begin{array}{r|rrrrr} -1 & 27 & -9 & -33 & -1 & -4 \\ & & -27 & 36 & -3 & 4 \\ \hline \frac{4}{3} & 27 & -36 & 3 & -4 & 0 \\ & & 36 & 0 & 4 & \\ \hline & 27 & 0 & 3 & 0 & \end{array}$$

$$\begin{aligned} g(x) &= (x+1)\left(x-\frac{4}{3}\right)(27x^2+3) \\ &= 27(x+1)\left(x-\frac{4}{3}\right)\left(x-\frac{i}{3}\right)\left(x+\frac{i}{3}\right) \\ &= (x+1)(3x-4)(3x-i)(3x+i) \end{aligned}$$

Actual zeros: $\left\{-1, \frac{4}{3}, \frac{i}{3}, -\frac{i}{3}\right\}$

In this problem, our list of potential rational zeros contains 24 candidates (12 positive and 12 negative). Although the task of trying so many may seem daunting, the Rational Zero Theorem has at least eliminated all rational numbers except these 24. The work at left illustrates that two of the 24 numbers, -1 and $\frac{4}{3}$, are actual zeros of g. (Note how the result of the first synthetic division is used in the second synthetic division.)

Once we have found two zeros and factored out the corresponding linear factors, by synthetic division, we are left with the task of factoring $27x^2+3$. By solving the equation $27x^2+3=0$, we find $27x^2+3 = 27\left(x-\frac{i}{3}\right)\left(x+\frac{i}{3}\right)$.

If we wish, we can distribute the three factors of 3 in 27 among the last three linear factors to obtain factors without fractions, as shown. Of course, this doesn't affect the four zeros of g.

Topic 2: Descartes' Rule of Signs

Descartes' Rule of Signs is another tool to aid us in our search for zeros. Unlike the Rational Zeros Theorem, Descartes' Rule doesn't identify candidates; instead, it gives us guidelines on how many positive real zeros and how many negative real zeros we can expect a given polynomial to have. Again, this tool tells us nothing about zeros with non-zero imaginary parts.

Descartes' Rule of Signs

Let $f(x) = a_n x^n + a_{n-1}x^{n-1} + \ldots + a_1 x + a_0$ be a polynomial with real coefficients, and assume $a_0 \neq 0$. A *variation in sign* of *f* is a change in the sign of one coefficient of *f* to the next, either from positive to negative or vice versa.

1. The number of *positive real zeros* of *f* is either the number of variations in sign of $f(x)$ or is less than this number by a positive even integer.
2. The number of *negative real zeros* of *f* is either the number of variations in sign of $f(-x)$ or is less than this number by a positive even integer.

Note that in order to apply Descartes' Rule of Signs, it is critical to first write the terms of the polynomial in descending order. Note also that unless the number of variations in sign is 0 or 1, the rule does not give us a definitive answer for the number of zeros to expect. For instance, if the number of variations in sign of $f(x)$ is 4, we know only that there will be 4, 2, or 0 positive real zeros.

example 2

Use Descartes' Rule of Signs to determine the possible number of positive and negative real zeros of each of the following polynomials. Then use the Rational Zeros Theorem and other means to find the zeros, if possible.

a. $f(x) = 2x^3 + 3x^2 - 14x - 21$

b. $g(x) = 3x^3 - 10x^2 + \frac{51}{4}x - \frac{13}{4}$

cont'd. on next page ...

Solutions:

a. $f(x) = 2x^3 + 3x^2 \underset{\text{sign change}}{} -14x \; -21$

$f(-x) = 2(-x)^3 + 3(-x)^2 - 14(-x) - 21$

$= -2x^3 \underset{\text{sign change}}{} +3x^2 \; +14x \underset{\text{sign change}}{} -21$

There is only one variation in sign in the function $f(x)$, and so we know there is exactly one positive real zero. We next simplify the function $f(-x)$ and determine that it has two variations in sign. Thus, f has either 2 or 0 negative real zeros.

Factors of -21: $\pm\{1, 3, 7, 21\}$

Factors of 2: $\pm\{1, 2\}$

Possible rational zeros: $\pm\left\{1, 3, 7, 21, \frac{1}{2}, \frac{3}{2}, \frac{7}{2}, \frac{21}{2}\right\}$

$$\begin{array}{r|rrrr} -\frac{3}{2} & 2 & 3 & -14 & -21 \\ & & -3 & 0 & 21 \\ \hline & 2 & 0 & -14 & 0 \end{array}$$

$f(x) = \left(x + \frac{3}{2}\right)\left(2x^2 - 14\right)$

$= 2\left(x + \frac{3}{2}\right)\left(x - \sqrt{7}\right)\left(x + \sqrt{7}\right)$

Zeros: $\left\{-\frac{3}{2}, \sqrt{7}, -\sqrt{7}\right\}$

We can now use the Rational Zero Theorem to try to actually find the zeros. The calculations at left show there are 16 rational numbers that are possible zeros of f, and trial and error eventually leads to the discovery that $-\frac{3}{2}$ is actually a zero. Now, since we have found 1 negative zero, we know there must be a second.

Synthetic division allows us to factor f as shown, and as always we know we can factor the quadratic quotient. In this case, we can solve the equation $2x^2 - 14 = 0$ fairly easily (though the quadratic formula could also be used). Note that the end result is indeed 1 positive zero and 2 negative zeros.

b. $g(x) = 3x^3 - 10x^2 + \frac{51}{4}x - \frac{13}{4}$

$g(-x) = 3(-x)^3 - 10(-x)^2 + \frac{51}{4}(-x) - \frac{13}{4}$

$= -3x^3 - 10x^2 - \frac{51}{4}x - \frac{13}{4}$

There are 3 variations in sign in the function $g(x)$, so there are either 3 or 1 positive real zeros. And since $g(-x)$ has no variations in sign, $g(x)$ has no negative real zeros.

$$g(x)=\frac{1}{4}\left(12x^3-40x^2+51x-13\right)$$

Factors of -13: $\pm\{1, 13\}$

Factors of 12: $\pm\{1, 2, 3, 4, 6, 12\}$

Possible rational zeros:

$$\pm\left\{1, 13, \frac{1}{2}, \frac{13}{2}, \frac{1}{3}, \frac{13}{3}, \frac{1}{4}, \frac{13}{4}, \frac{1}{6}, \frac{13}{6}, \frac{1}{12}, \frac{13}{12}\right\}$$

Before applying the Rational Zero Theorem, we have to make one small adjustment. The theorem requires that all the coefficients of the polynomial be integers, and g does not meet this condition. However, we can factor out $\frac{1}{4}$ as shown and apply the theorem to $12x^3-40x^2+51x-13$, since the zeros of this polynomial are the same as the zeros of g (multiplying a function by a non-zero constant has no effect on the zeros.)

$$\begin{array}{r|rrrr} \frac{1}{3} & 12 & -40 & 51 & -13 \\ & & 4 & -12 & 13 \\ \hline & 12 & -36 & 39 & 0 \end{array}$$

$$g(x)=\frac{1}{4}\left(x-\frac{1}{3}\right)\left(12x^2-36x+39\right)$$

$$=\frac{12}{4}\left(x-\frac{1}{3}\right)\left(x-\left(\frac{3-2i}{2}\right)\right)\left(x-\left(\frac{3+2i}{2}\right)\right)$$

Zeros: $\left\{\frac{1}{3}, \frac{3-2i}{2}, \frac{3+2i}{2}\right\}$

After a trial and error search, synthetic division shows that $\frac{1}{3}$ is an actual zero of g. At left, g is shown in completely factored form (don't forget to include the factor of $\frac{1}{4}$). We can factor 12 from $12x^2-36x+39$, and apply the quadratic formula to the result keeping in mind that we must multiply the product of the two linear factors that result by 12 to obtain the correct leading coefficient.

Topic 3: Bounds of Real Zeros

Although the Rational Zero Theorem and Descartes' Rule of Signs are useful for determining the zeros of a polynomial, Examples 1 and 2 illustrate that more guidance would certainly be welcome, especially guidance that reduces the number of potential zeros that must be tested by trial and error. The theorem on the following page does just that.

Upper and Lower Bounds of Zeros

Assume $f(x)$ is a polynomial with real coefficients, a positive leading coefficient, and degree ≥ 1. Let a and b be fixed numbers, with $a < 0 < b$. Then:

1. No real zero of f is larger than b (we say b is an *upper bound* of the zeros of f) if the last row in the synthetic division of $f(x)$ by $x-b$ contains no negative numbers. That is, b is an upper bound of the zeros if the quotient and remainder have no negative coefficients when $f(x)$ is divided by $x-b$.
2. No real zero of f is smaller than a (we say a is a *lower bound* of the zeros of f) if the last row in the synthetic division of $f(x)$ by $x-a$ has entries that alternate in sign (0 can be counted as either positive or negative, as necessary).

Example 3 revisits the polynomial $f(x)=2x^3+3x^2-14x-21$ that we studied in Example 2a, and illustrates the use of the above theorem.

example 3

Use synthetic division to identify upper and lower bounds of the real zeros of the polynomial $f(x)=2x^3+3x^2-14x-21$.

Solution:

$$\begin{array}{r|rrrr} 2 & 2 & 3 & -14 & -21 \\ & & 4 & 14 & 0 \\ \hline & 2 & 7 & 0 & -21 \end{array}$$

We can begin by trying out any positive number whatsoever as an upper bound. The work at left shows that 2 is not necessarily an upper bound, as the last row contains a negative number.

$$\begin{array}{r|rrrr} 3 & 2 & 3 & -14 & -21 \\ & & 6 & 27 & 39 \\ \hline & 2 & 9 & 13 & 18 \end{array}$$

The number 3, however, is an upper bound according to the theorem, as all of the coefficients in the last row are non-negative. In other words, all real zeros of f (including any irrational zeros) are less than or equal to 3.

$$\begin{array}{r|rrrr} -3 & 2 & 3 & -14 & -21 \\ & & -6 & 9 & - \\ \hline & 2 & -3 & -5 & - \end{array}$$

The work at left tests whether -3 is a lower bound, according to the theorem. The synthetic division has not been completed, because as soon as the signs in the last row cease to alternate, the theorem doesn't apply.

$$\begin{array}{r|rrrr} -4 & 2 & 3 & -14 & -21 \\ & & -8 & 20 & -24 \\ \hline & 2 & -5 & 6 & -45 \end{array}$$

When we try a lower number, we find that -4 is a lower bound, as the signs in the last row alternate. Remember that if a 0 appears, it can be counted as either positive or negative, whichever leads to a sequence of alternating signs.

To summarize, we now know that all real zeros of f lie in the interval $[-4, 3]$, meaning there is no need to test the numbers $\left\{7, 21, \frac{7}{2}, \frac{21}{2}, -7, -21, -\frac{21}{2}\right\}$ out of the list of potential rational zeros $\pm\left\{1, 3, 7, 21, \frac{1}{2}, \frac{3}{2}, \frac{7}{2}, \frac{21}{2}\right\}$. We have already determined in Example 2 that the zeros of f are $\left\{-\frac{3}{2}, \sqrt{7}, -\sqrt{7}\right\}$, all of which do indeed lie in the interval $[-4, 3]$.

caution!

Don't read more into the Upper and Lower Bounds Theorem than is actually there. For instance, –3 actually *is* a lower bound of the zeros of $f(x) = 2x^3 + 3x^2 - 14x - 21$, but the theorem is not powerful enough to indicate this. The work in Example 3 shows that –4 is a lower bound, but the theorem fails to spot the fact that –3 is a better lower bound. The tradeoff for this weakness in the theorem is that it is quickly and easily applied.

Topic 4: The Intermediate Value Theorem

The last technique for locating zeros that we study makes use of a property of polynomials called *continuity*. Although continuity of functions will not be discussed in this text, one consequence of continuity is that the graph of a continuous function has no "breaks" in it. That is, assuming that the function can be graphed at all, it can be drawn without lifting the drawing tool.

Intermediate Value Theorem

Assume that $f(x)$ is a polynomial with real coefficients, and that a and b are real numbers with $a < b$. If $f(a)$ and $f(b)$ differ in sign, then there is at least one point c such that $a < c < b$ and $f(c) = 0$. That is, at least one zero of f lies between a and b.

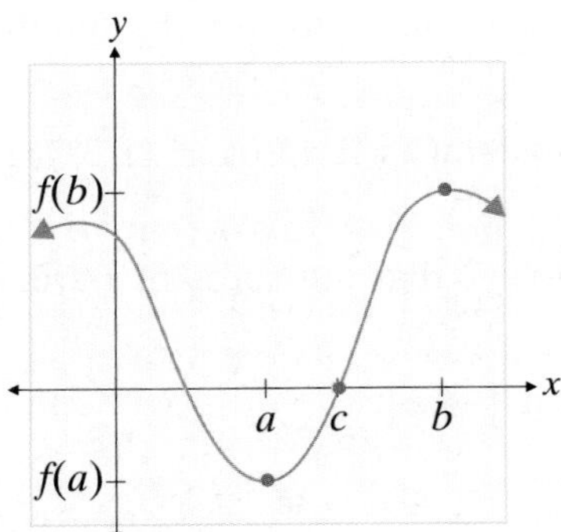

We can use the Intermediate Value Theorem to prove that a zero of a given polynomial must lie in a particular interval, and then through repeated application to "home in" on the zero. Example 4 illustrates this process.

example 4

a. Show that $f(x) = x^3 + 3x - 7$ has a zero between 1 and 2.

b. Find an approximation of the zero to the nearest tenth.

Solutions:

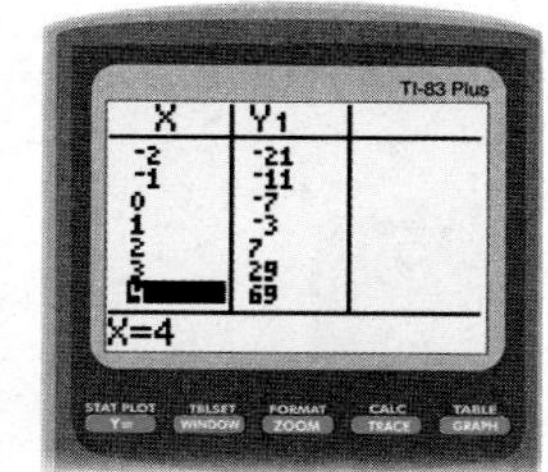

This table was reached by entering the equation in Y= and selecting 2nd Graph (Table). Notice that the signs of the y values change between $x = 1$ and $x = 2$, indicating that there is a zero there.

a.

$$\begin{array}{r|rrrr} 1 & 1 & 0 & 3 & -7 \\ & & 1 & 1 & 4 \\ \hline & 1 & 1 & 4 & -3 \end{array}$$

$$\begin{array}{r|rrrr} 2 & 1 & 0 & 3 & -7 \\ & & 2 & 4 & 14 \\ \hline & 1 & 2 & 7 & 7 \end{array}$$

For the function $f(x) = x^3 + 3x - 7$, we could determine $f(1)$ and $f(2)$ by direct computation, but for many polynomials it is quicker to use synthetic division as illustrated at left. Remember that the remainder upon dividing $f(x)$ by $x - k$ is $f(k)$; thus, $f(1) = -3$ and $f(2) = 7$. The critical point is that these two values are opposite in sign, so a zero of f must lie between 1 and 2.

b.

$$\begin{array}{r|rrrr} 1.5 & 1 & 0 & 3 & -7 \\ & & 1.5 & 2.25 & 7.875 \\ \hline & 1 & 1.5 & 5.25 & 0.875 \end{array}$$

Since we know f has a zero between 1 and 2, we might guess 1.5 as a first approximation for the zero. If we continue to use synthetic division as shown, we find $f(1.5) = 0.875$, which tells us that the actual zero of f must lie between 1 and 1.5. We might further guess that the zero is closer to 1.5 than it is to 1, since $f(1.5)$ is closer to 0 than is $f(1)$.

$$\begin{aligned} f(1.4) &= (1.4)^3 + 3(1.4) - 7 \\ &= -0.056 \end{aligned}$$

With this in mind, we calculate $f(1.4)$. Again, we can use synthetic division or direct computation as shown at left. The result is negative, so we now know the actual zero must lie between 1.4 and 1.5.

$$\begin{aligned} f(1.45) &= (1.45)^3 + 3(1.45) - 7 \\ &= 0.398625 \end{aligned}$$

The value of f midway between 1.4 and 1.5 is positive, so the zero must lie between 1.4 and 1.45 (and in fact is probably closer to 1.4). In any event, we have shown that the zero, to the nearest tenth, is 1.4. (Continuing the process would lead to the better approximation 1.41).

technology note

Graphing calculators, while not as powerful as computer algebra systems such as *Mathematica*, are nonetheless useful aids in finding real zeros of polynomials. Their graphing capability can be used, for example, to quickly eliminate some candidates for zeros given by the Rational Zero Theorem. And the "Zoom" feature available on most graphing calculators is a convenient tool when applying the Intermediate Value Theorem (see your owner's manual for help with your specific calculator). The illustration below shows successive screen shots of the graph of $f(x)=x^3+3x-7$ as we "zoom in" on the approximate zero identified in Example 4.

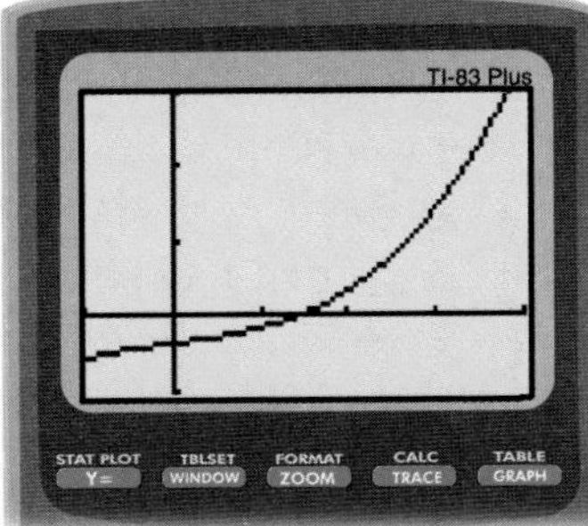

y-axis: $[-20, 60]$; x-axis: $[-1, 4]$
y scale = 20; x scale = 1

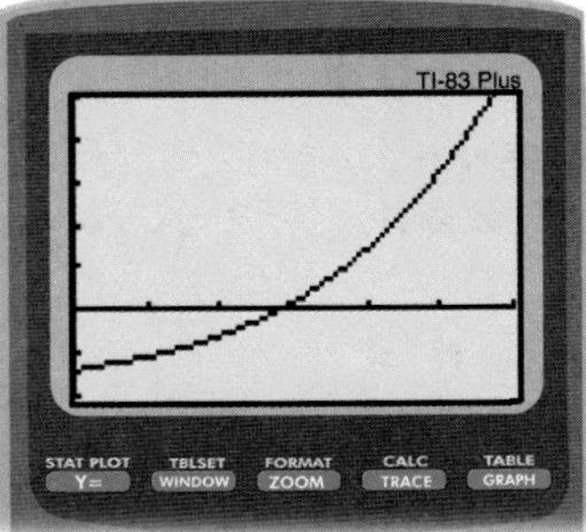

y-axis: $[-10, 25]$; x-axis: $[0, 3]$
y scale = 5; x scale = 0.5

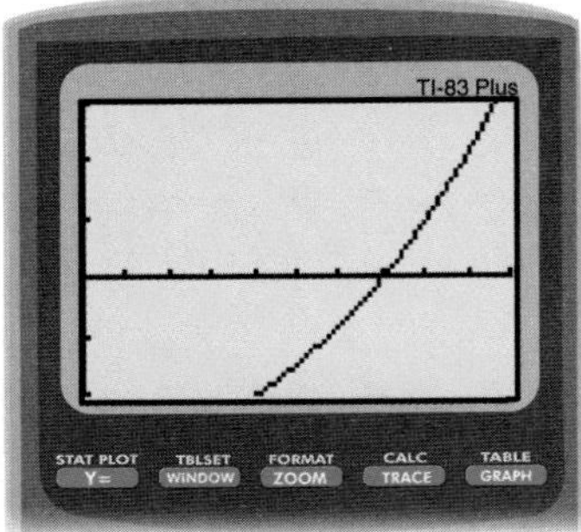

y-axis: $[-4, 6]$; x-axis: $[0, 2]$
y scale = 2; x scale = 0.2

exercises

Throughout these exercises, a graphing calculator or a computer algebra system may be used as an additional tool in identifying zeros, if permitted by your instructor.

List all of the potential rational zeros of the following polynomials. Then use polynomial division and the quadratic formula, if necessary, to identify the actual zeros. See Example 1.

1. $f(x)=3x^3+5x^2-26x+8$

2. $g(x)=-2x^3+11x^2+x-30$

3. $p(x)=x^4-5x^3+10x^2-20x+24$

4. $h(x)=x^3-3x^2+9x+13$

5. $q(x)=x^3-10x^2+23x-14$

6. $r(x)=x^4+x^3+23x^2+25x-50$

7. $s(x)=2x^3-9x^2+4x+15$

8. $t(x)=x^3-6x^2+13x-20$

9. $j(x)=3x^4-3$

10. $k(x)=x^4-10x^2+24$

Using the Rational Zero Theorem or your answers to the preceding problems, solve the following polynomial equations.

11. $x^4+x-2=-2x^4+x+1$

12. $x^4+10=10x^2-14$

13. $x^3-3x^2+9x+13=0$

14. $3x^3+5x^2=26x-8$

15. $x^4+10x^2-20x=5x^3-24$

16. $-2x^3+11x^2+x=30$

17. $2x^3-12x^2+26x=40$

18. $2x^3+9x^2+4x=15$

19. $x^4+x^3+23x^2=50-25x$

20. $x^3+23x=10x^2+14$

Use Descartes' Rule of Signs to determine the possible number of positive and negative real zeros of each of the following polynomials. See Example 2.

21. $f(x)=x^3+8x^2+17x+10$

22. $f(x)=x^3+2x^2-5x-6$

23. $f(x)=x^3-6x^2+3x+10$

24. $g(x)=x^3+6x^2+11x+6$

25. $f(x) = x^4 - 5x^3 - 2x^2 + 40x - 48$

26. $g(x) = x^3 + 3x^2 + 3x + 9$

27. $f(x) = x^4 - 25$

28. $g(x) = x^4 - 7x^3 + 5x^2 + 31x - 30$

29. $f(x) = 5x^5 - x^4 + 2x^3 + x - 9$

30. $g(x) = -6x^7 - x^5 - 7x^3 - 2x$

Use synthetic division to identify upper and lower bounds of the real zeros of the following polynomials (answers may vary). See Example 3.

31. $f(x) = x^3 + 4x^2 - x - 4$

32. $f(x) = 2x^3 - 3x^2 - 8x - 3$

33. $f(x) = x^3 - 6x^2 + 3x + 10$

34. $g(x) = x^3 + 6x^2 + 11x + 6$

35. $f(x) = x^4 - 5x^3 - 2x^2 + 40x - 48$

36. $g(x) = x^3 + 3x^2 + 3x + 9$

37. $f(x) = x^4 - 25$

38. $g(x) = x^4 - 7x^3 + 5x^2 + 31x - 30$

39. $f(x) = 2x^3 - 7x^2 - 28x - 12$

40. $g(x) = x^5 + x^4 - 9x^3 - x^2 + 20x - 12$

Using your answers to the preceding problems, polynomial division, and the quadratic formula, if necessary, find all of the zeros of the following polynomials.

41. $f(x) = x^3 + 4x^2 - x - 4$

42. $f(x) = 2x^3 - 3x^2 - 8x - 3$

43. $f(x) = x^3 - 6x^2 + 3x + 10$

44. $g(x) = x^3 + 6x^2 + 11x + 6$

45. $f(x) = x^4 - 5x^3 - 2x^2 + 40x - 48$

46. $g(x) = x^3 + 3x^2 + 3x + 9$

47. $f(x) = x^4 - 25$

48. $g(x) = x^4 - 7x^3 + 5x^2 + 31x - 30$

49. $f(x) = 2x^3 - 7x^2 - 28x - 12$

50. $g(x) = x^5 + x^4 - 9x^3 - x^2 + 20x - 12$

Use the Intermediate Value Theorem to show that each of the following polynomials has a real zero between the indicated values. See Example 4.

51. $f(x) = 5x^3 - 4x^2 - 31x - 6$; -2 and -1

52. $f(x) = x^4 - 9x^2 - 14$; 1 and 4

53. $f(x) = x^4 + 2x^3 - 10x^2 - 14x + 21$; 2 and 3

54. $f(x) = -x^3 + 2x^2 + 13x - 26$; -4 and -3

55. $14x + 10x^2 = x^4 + 2x^3 + 21$; 2 and 3 **56.** $x^3 - 2x^2 = 13(x-2)$; -4 and -3

57. Construct a proof of the Rational Zero Theorem by following the suggested steps.

a. Assuming that $\frac{p}{q}$ is a zero of the polynomial $f(x) = a_n x^n + a_{n-1}x^{n-1} + \ldots + a_1 x + a_0$, show that the equation $a_n\left(\frac{p}{q}\right)^n + a_{n-1}\left(\frac{p}{q}\right)^{n-1} + \ldots + a_1\left(\frac{p}{q}\right) + a_0 = 0$ can be written in the form $a_n p^n + a_{n-1}p^{n-1}q + \ldots + a_1 pq^{n-1} = -a_0 q^n$.

b. It can be assumed that $\frac{p}{q}$ is written in lowest terms (that is, the greatest common divisor of p and q is 1). By examining the left-hand side of the last equation above, show that p must be a divisor of the right-hand side, and hence a factor of a_0.

c. By rearranging the equation so that all terms with a factor of q are on one side, use a similar argument to show that q must be a factor of a_n.

Using any of the methods discussed in this section as guides, find all of the real zeros of the following functions.

58. $f(x) = 3x^3 - 18x^2 + 9x + 30$

59. $f(x) = -4x^3 - 19x^2 + 29x - 6$

60. $f(x) = 3x^5 + 7x^4 + 12x^3 + 28x^2 - 15x - 35$

61. $f(x) = 2x^4 + 5x^3 - 9x^2 - 15x + 9$

62. $f(x) = -15x^4 + 44x^3 + 15x^2 - 72x - 28$

63. $f(x) = 2x^4 + 13x^3 - 23x^2 - 32x + 20$

64. $f(x) = 3x^4 + 7x^3 - 25x^2 - 63x - 18$

65. $f(x) = x^5 + 7x^4 + 5x^3 - 43x^2 - 42x + 72$

66. $f(x) = 2x^5 - 3x^4 - 47x^3 + 103x^2 + 45x - 100$

67. $f(x) = x^6 - 125x^4 + 4804x^2 - 57{,}600$

Using any of the methods discussed in this section as guides, solve the following equations.

68. $x^3 + 6x^2 + 11x = -6$

69. $x^3 - 7x = 6(x^2 - 10)$

70. $x^3 + 9x^2 = 2x + 18$

71. $6x^3 + 14 = 41x^2 + 9x$

72. $4x^3 = 18x^2 + 106x + 48$

73. $3x^3 + 15x^2 - 6x = 72$

74. $8x^4 + 24 + 8x = 2x^3 + 38x^2$

75. $x^4 + 7x^2 = 3x^3 + 21x$

76. $6x^6 - 10x^5 - 9x^4 + 27x^3 = 20x^2 + 18x - 30$

77. $4x^5 - 5x^4 + 20x^2 = 6x^3 + 25x + 30$

4.4 The Fundamental Theorem of Algebra

TOPICS

1. The Fundamental Theorem of Algebra
2. Multiple zeros and their geometric meaning
3. Conjugate pairs of zeros
4. Summary of polynomial methods

Topic 1: The Fundamental Theorem of Algebra

We are now ready to tie together all that we have learned about polynomials, and we begin with a powerful but deceptively simple-looking statement called the Fundamental Theorem of Algebra.

The Fundamental Theorem of Algebra

If f is a polynomial of degree n, with $n \geq 1$, then f has at least one zero. That is, the equation $f(x) = 0$ has at least one root, or solution. It is important to note that the zero of f, and consequently the solution of $f(x) = 0$, may be a non-real complex number.

Mathematicians began to suspect the truth of this statement in the first half of the 17th century, but a convincing proof did not appear until the 22-year old German mathematician Carl Friedrich Gauss (1777 - 1855) provided one in his doctoral dissertation of 1799.

Although the proof of the Fundamental Theorem of Algebra is beyond the scope of this book, we can use it to prove a consequence that summarizes much of the previous three sections. The following theorem has great implications in solving polynomial equations and in graphing real-coefficient polynomial functions.

The Linear Factors Theorem

Given the polynomial $f(x) = a_n x^n + a_{n-1}x^{n-1} + \ldots + a_1 x + a_0$, where $n \geq 1$ and $a_n \neq 0$, f can be factored as $f(x) = a_n(x - c_1)(x - c_2)\cdots(x - c_n)$, where $c_1, c_2, \ldots, c_n$ are constants (possibly non-real complex constants and not necessarily distinct). In other words, an nth degree polynomial can be factored as a product of n linear factors.

Proof: The Fundamental Theorem of Algebra tells us that $f(x)$ has at least one zero; call it c_1. From our work in Section 4.2 then, we know that $x - c_1$ is a factor of *f*, and we can write

$$f(x) = (x - c_1)q_1(x),$$

where $q_1(x)$ is a polynomial of degree $n-1$. Note that the leading coefficient of q_1 must be a_n, since we divided *f* by $x - c_1$, a polynomial with a leading coefficient of 1.

If the degree of q_1 is 0 (that is, if $n = 1$), then $q_1(x) = a_n$ and $f(x) = a_n(x - c_1)$. If this is not the case, then q_1 is of degree 1 or larger, and by the Fundamental Theorem of Algebra q_1 itself has at least one zero; call it c_2. By the same reasoning then, we can write

$$f(x) = (x - c_1)(x - c_2)q_2(x),$$

where q_2 is a polynomial of degree $n-2$, also with leading coefficient a_n. We can perform this process a total of n times (and no more), at which point we have the desired result:

$$f(x) = a_n(x - c_1)(x - c_2)\cdots(x - c_n).$$

caution!

Keep in mind that some of the linear factors thus obtained may be identical, and that some of the constants $c_1, c_2, \ldots, c_n$ may be non-real complex numbers. The meaning of these possible outcomes will be explored shortly.

The Linear Factors Theorem tells us that, in theory, any polynomial can be written as a product of linear factors; what it does *not* do is tell us how to determine the linear factors. To accomplish that task, we must still rely on the techniques developed in Section 4.3.

In the case where all of the coefficients of $f(x) = a_n x^n + a_{n-1}x^{n-1} + \ldots + a_1 x + a_0$ are real, the Linear Factors Theorem tells us that the graph of f has *at most n x*-intercepts (and can only have *exactly n x*-intercepts if all n zeros are real and distinct). Indirectly, the theorem tells us something more: the graph of f can have at most $n - 1$ *turning points*. A **turning point** of a graph is a point where the graph changes behavior from decreasing to increasing or vice versa. These facts are summarized in the box below.

Interpreting the Linear Factors Theorem

The graph of an nth degree polynomial function has *at most n x*-intercepts and *at most* $n-1$ turning points.

Figure 1 illustrates that the degree of a polynomial merely gives us an upper bound on the number of x-intercepts and turning points. Note that the graph of the 4th degree polynomial f has just two x-intercepts and only one turning point, while the graph of the 3rd degree polynomial g has three x-intercepts and two turning points.

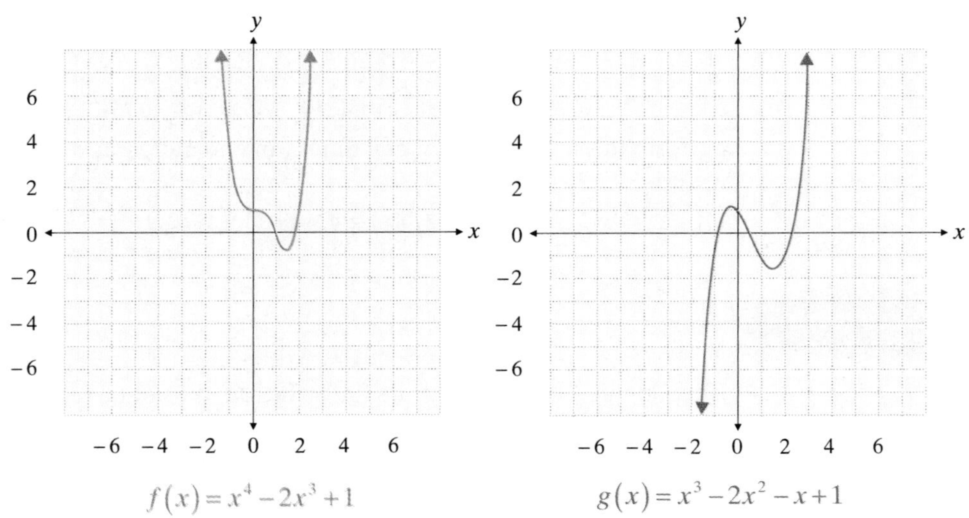

$f(x) = x^4 - 2x^3 + 1$ $g(x) = x^3 - 2x^2 - x + 1$

Figure 1: x-Intercepts and Turning Points

Topic 2: Multiple Zeros and Their Geometric Meaning

If the linear factor $x-c$ appears k times in the factorization of a polynomial, we say the number c is a *zero of multiplicity k*. If we are graphing a polynomial f for which c is a real zero of multiplicity k, then c is certainly an x-intercept of the graph of f, but the behavior of the graph near c depends on whether k is even or odd. We have already seen numerous examples of this. Consider, for instance, the function $f(x)=(x-1)^3$: clearly, 1 is the only zero of f, and it is evidently a zero of multiplicity 3. (Note that f is a 3rd degree polynomial, so if it had been given to us in non-factored form, we would still know it could be written as a product of three linear factors.) From our work in Chapter 3, we know that the graph of f is the prototypical cubic shape shifted to the right by 1 unit, as shown in Figure 2.

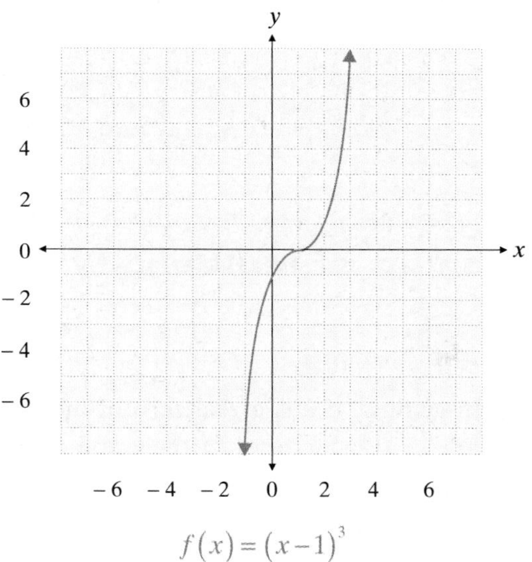

Figure 2: Zero of Multiplicity 3

Compare the behavior of $f(x)=(x-1)^3$ near its zero to the behavior of the function $g(x)=(x+2)^4$ near its own zero of –2, a zero of multiplicity 4. Figure 3 contains the graph of g.

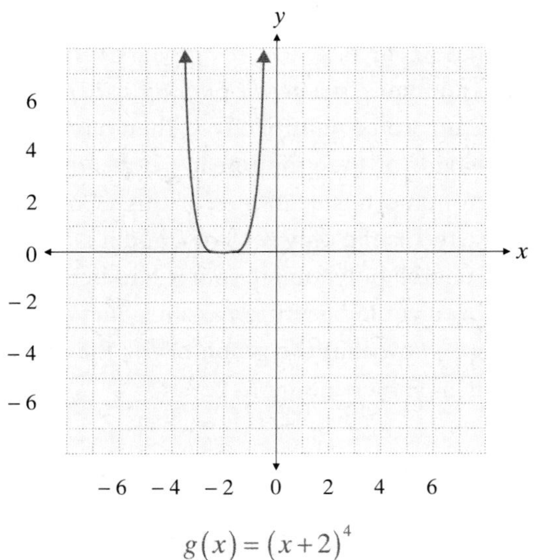

$g(x) = (x+2)^4$

Figure 3: Zero of Multiplicity 4

Geometric Meaning of Multiplicity

In general, if c is a real zero of multiplicity k of a polynomial f (alternatively, if $(x-c)^k$ is a factor of f), the graph of f will touch the x-axis at $(c, 0)$ and:

- cross through the x-axis if k is odd;
- stay on the same side of the x-axis if k is even.

Further, if $k \geq 2$, the graph of f will "flatten out" near $(c, 0)$.

With an understanding of how a zero's multiplicity affects a polynomial, constructing a reasonably accurate sketch of the graph becomes easier. Example 1 illustrates this point.

example 1

Sketch the graph of the polynomial $f(x)=(x+2)(x+1)^2(x-3)^3$.

Solution:

$f(x)\to\infty$ as $x\to\pm\infty$

The polynomial f has degree 6, and so we know the behavior of the graph as $x\to-\infty$ and $x\to\infty$.

$$\begin{aligned} f(0) &= (0+2)(0+1)^2(0-3)^3 \\ &= (2)(1)(-27) \\ &= -54 \end{aligned}$$

The y-intercept is easy to determine, as always.

f crosses at –2 and 3, but not at –1, and flattens out at –1 and 3

We can also determine at which of the x-intercepts the graph actually crosses the axis.

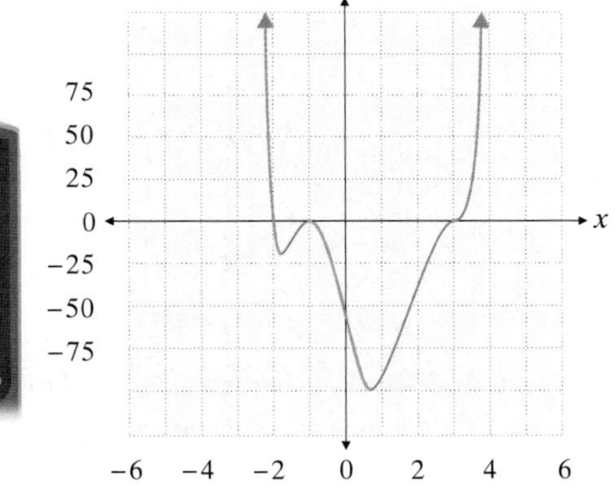

The sketch at left takes into account all of the above observations. Note that the graph flattens out at –1 and at 3, but not at –2. Note also the extreme difference in the scales of the two axes.

To fill in portions of the graph between zeros accurately, we must still compute a few values of the function. For instance, $f(1)=-96$ and $f(2)=-36$. This also gives us a way to double-check our analysis of the behavior on either side of a zero.

Topic 3: Conjugate Pairs of Zeros

If all the coefficients of a given polynomial $f(x)=a_nx^n+a_{n-1}x^{n-1}+\ldots+a_1x+a_0$ are real, then f is a function that transforms real numbers into other real numbers and consequently f can be graphed in the Cartesian plane. Nonetheless, it is very possible that some of the constants $c_1, c_2, \ldots, c_n$ in the factored form of f, $f(x)=a_n(x-c_1)(x-c_2)\cdots(x-c_n)$, might be non-real complex numbers. If this happens, such complex numbers must occur in pairs, as described on the next page.

The Conjugate Roots Theorem

Let $f(x) = a_n x^n + a_{n-1}x^{n-1} + \ldots + a_1 x + a_0$ be a polynomial with only real coefficients. If the complex number $a + bi$ is a zero of f, then so is its complex conjugate $a - bi$. In terms of the linear factors of f, this means that if $x - (a + bi)$ is a factor of f, then so is $x - (a - bi)$.

We can make use of this fact in several ways. For instance, if we are given one non-real zero of a real-coefficient polynomial, we automatically know a second zero. We may also find the theorem useful when constructing polynomials with certain specified properties.

Another consequence of the Conjugate Roots Theorem is that every polynomial with real coefficients can be factored into a product of linear factors and irreducible quadratic factors with real coefficients. Each irreducible quadratic factor with real coefficients corresponds to the product of two factors of the form $x - (a + bi)$ and $x - (a - bi)$.

example 2

Given that $4 - 3i$ is a zero of the polynomial $f(x) = x^4 - 8x^3 + 200x - 625$, factor f completely.

Solution:

Since $4 - 3i$ is a zero of f, we know that $4 + 3i$ is a zero as well. This gives us two ways to proceed: we could divide f by $x - (4 - 3i)$ and then divide the result by $x - (4 + 3i)$ (this would be most efficiently done with synthetic division), or we could multiply $x - (4 - 3i)$ and $x - (4 + 3i)$ and divide f by their product (using polynomial long division). In either case, we will be left with a quadratic polynomial that we know we can factor.

If we take the second approach, the first step is as follows:

$$\begin{aligned}(x - (4 - 3i))(x - (4 + 3i)) &= (x - 4 + 3i)(x - 4 - 3i)\\ &= x^2 - 4x \cancel{- 3ix} - 4x + 16 \cancel{+ 12i} \cancel{+ 3ix} \cancel{- 12i} - 9i^2\\ &= x^2 - 8x + 25\end{aligned}$$

Now, we divide f by this product:

$$\begin{array}{r}
x^2 - 25 \\
x^2 - 8x + 25 \overline{\big) x^4 - 8x^3 + 200x - 625} \\
\underline{-\left(x^4 - 8x^3 + 25x^2\right)} \\
-25x^2 + 200x - 625 \\
\underline{-\left(-25x^2 + 200x - 625\right)} \\
0
\end{array}$$

The quotient, $x^2 - 25$, is a difference of two squares and is easily factored, giving us our final result:

$$f(x) = (x - 4 + 3i)(x - 4 - 3i)(x - 5)(x + 5).$$

example 3

Construct a 4th degree real-coefficient polynomial function f with zeros of 2, –5, and $1+i$ such that $f(1) = 12$.

Solution:

As $1+i$ is to be one of the zeros and f is to have only real coefficients, $1-i$ must be a zero as well. Based on this, f must be of the form

$$f(x) = a\big(x - (1+i)\big)\big(x - (1-i)\big)(x-2)(x+5)$$

for some real constant a. Of course, a must be chosen so that $f(1) = 12$. In order to do this, we might begin by multiplying out $\big(x-(1+i)\big)\big(x-(1-i)\big)$:

$$\begin{aligned}
\big(x-(1+i)\big)\big(x-(1-i)\big) &= (x - 1 - i)(x - 1 + i) \\
&= x^2 - x \cancel{+ix} - x + 1 \cancel{-i} \cancel{-ix} \cancel{+i} - i^2 \\
&= x^2 - 2x + 2.
\end{aligned}$$

Thus,

$$\begin{aligned}
f(1) &= a\left(1^2 - 2(1) + 2\right)(1-2)(1+5) \\
&= a(1)(-1)(6) \\
&= -6a.
\end{aligned}$$

cont'd. on next page ...

Since we want $f(1) = 12$, we need $12 = -6a$ or $a = -2$. In factored form, then, the polynomial is

$$f(x) = -2(x-1-i)(x-1+i)(x-2)(x+5),$$

which, multiplied out, is $f(x) = -2x^4 - 2x^3 + 28x^2 - 52x + 40$.

Topic 4: Summary of Polynomial Methods

All of the methods that you have learned in this chapter may be useful in solving a given polynomial problem, whether the problem focuses on graphing a polynomial function, solving a polynomial equation, or solving a polynomial inequality. Now that all of the methods have been introduced, it makes sense to summarize them and see how they contribute to the big picture.

Recall that, in general, an nth degree polynomial function has the form

$$f(x) = a_n x^n + a_{n-1} x^{n-1} + \ldots + a_1 x + a_0,$$

where $a_n \neq 0$ and any (or all) of the coefficients may be non-real complex numbers. Keep in mind that it only makes sense to talk about graphing f in the Cartesian plane if all of the coefficients are real. Similarly, a polynomial inequality in which f appears on one side only makes sense if all the coefficients are real. For this reason, most of the polynomials in this text have only real coefficients.

Nevertheless, complex numbers arise in a very natural way when working with polynomials, as some of the numbers $c_1, c_2, \ldots, c_n$ in the factored form of f,

$$f(x) = a_n(x-c_1)(x-c_2)\cdots(x-c_n),$$

may be non-real even if all of $a_1, a_2, \ldots, a_n$ are real. The fact that f can, in principle, be factored is a direct consequence of the Fundamental Theorem of Algebra.

Factoring f into a product of linear factors as shown is the central point in solving a polynomial equation and (when the coefficients of f are real) in graphing a polynomial and solving a polynomial inequality. In particular, the observations summarized on the following page are of value:

- The solutions of the polynomial equation $f(x)=0$ are the numbers $c_1, c_2, \ldots, c_n$.
- When $a_1, a_2, \ldots, a_n$ are all real, the x-intercepts of the graph of f are the real numbers in the list $c_1, c_2, \ldots, c_n$. Keep in mind that if a given c_i appears in the list k times, it is a *zero of multiplicity k*. If an x-intercept of f is of multiplicity k, the behavior of f near that x-intercept depends on whether k is even or odd. Any non-real zeros in the list $c_1, c_2, \ldots, c_n$ will appear in conjugate pairs.
- When $a_1, a_2, \ldots, a_n$ are all real, the solution of the polynomial inequality $f(x) > 0$ consists of all the open intervals on the x-axis where the graph of f lies strictly above the x-axis. The solution of $f(x) < 0$ consists of all the open intervals where the graph of f lies strictly below the x-axis. The solutions of $f(x) \geq 0$ and $f(x) \leq 0$ consist of closed intervals.

The remaining topics discussed in this chapter are observations and techniques that aid us in filling in the details of the big picture. They are:

- The observation that the degree of a polynomial and the sign of its leading coefficient tell us how the graph of the polynomial behaves as $x \to -\infty$ and as $x \to \infty$.
- The observation that the graph of f crosses the y-axis at the easily computed point $(0, f(0))$.
- The technique of polynomial long division, useful in dividing one polynomial by another of the same or smaller degree.
- The technique of synthetic division, a shortcut that applies when dividing a polynomial by a polynomial of the form $x - k$. Recall that the remainder of this division is the value $f(k)$.
- The Rational Zero Theorem, which provides a list of potential rational zeros for polynomials with integer coefficients.
- Descartes' Rule of Signs, which provides guidance on the number of positive and negative real zeros that a real-coefficient polynomial might have.
- The Upper and Lower Bounds rule, which indicates an interval in which to search for all the zeros of a real-coefficient polynomial.
- The Intermediate Value Theorem, which can be used to "home in" on a real zero of a given polynomial.

As you solve various polynomial problems, try to keep the big picture in mind. Often, it is useful to literally keep a picture, namely the graph of the polynomial, in mind even if the problem does not specifically involve graphing.

exercises

Throughout these exercises, a graphing calculator or a computer algebra system may be used as an additional tool in identifying zeros and in checking your graphing, if permitted by your instructor.

Sketch the graph of each factored polynomial. See Example 1.

1. $f(x)=(x+1)^4(x-2)^3(x-1)$

2. $g(x)=-x^3(x-1)(x+2)^2$

3. $f(x)=-x(x+2)(x-1)^2$

4. $g(x)=(x+2)(x-1)^3$

5. $f(x)=(x-1)^4(x-2)(x-3)$

6. $g(x)=(x+1)^2(x-2)^3$

Use all available methods (e.g. the Rational Zero Theorem, Descartes' Rule of Signs, polynomial division, etc.) to factor each of the following polynomials completely, and then sketch the graph of each one. See Example 1.

7. $f(x)=x^5+4x^4+x^3-10x^2-4x+8$

8. $p(x)=2x^3-x^2-8x-5$

9. $s(x)=-x^4+2x^3+8x^2-10x-15$

10. $f(x)=-x^3+6x^2-12x+8$

11. $H(x)=x^4-x^3-5x^2+3x+6$

12. $h(x)=x^5-11x^4+46x^3-90x^2+81x-27$

Use all available methods (e.g. the Rational Zero Theorem, Descartes' Rule of Signs, polynomial division, etc.) to solve each polynomial equation. Use the Linear Factors Theorem to make sure you find the appropriate number of solutions, counting multiplicity.

13. $x^5+4x^4+x^3=10x^2+4x-8$

14. $x^4+15=2x^3+8x^2-10x$

15. $x^4+x^3+3x^2+5x-10=0$

16. $x^3-9x^2=30-28x$

17. $x^5+x^4-x^3+7x^2-20x+12=0$

18. $2x^4-5x^3-2x^2+15x=0$

Use all available methods (in particular, the Conjugate Roots Theorem, if applicable) to factor each of the following polynomials completely, making use of the given zero if one is given. See Example 2.

19. $f(x) = x^4 - 9x^3 + 27x^2 - 15x - 52$; $3 - 2i$ is a zero.

20. $g(x) = x^3 - (1-i)x^2 - (8-i)x + (12-6i)$; $2-i$ is a zero.

21. $f(x) = x^3 - (2+3i)x^2 - (1-3i)x + (2+6i)$; 2 is a zero.

22. $p(x) = x^4 - 2x^3 + 14x^2 - 8x + 40$; $2i$ is a zero.

23. $f(x) = x^4 - 3x^3 + 5x^2 - x - 10$

24. $g(x) = x^6 - 8x^5 + 25x^4 - 40x^3 + 40x^2 - 32x + 16$

25. $n(x) = x^4 - 4x^3 + 6x^2 + 28x - 91$; $2+3i$ is a zero.

26. $G(x) = x^4 - 14x^3 + 98x^2 - 686x + 2401$; $7i$ is a zero.

27. $r(x) = x^4 + 7x^3 - 41x^2 + 33x$

28. $d(x) = x^5 - x^4 - 18x^3 + 18x^2 + 81x - 81$

29. $P(x) = x^3 - 6x^2 + 28x - 40$

30. $g(x) = x^6 - x^4 - 16x^2 + 16$

Construct polynomial functions with the stated properties. See Example 3.

31. Third degree, only real coefficients, –1 and $5+i$ are two of the zeros, y-intercept is –52.

32. Fourth degree, only real coefficients, $\sqrt{7}$ and $i\sqrt{5}$ are two of the zeros, y-intercept is –35.

33. Fifth degree, 1 is a zero of multiplicity 3, –2 is the only other zero, leading coefficient is 2.

34. Fifth degree, only real coefficients, 0 is the only real zero, $1+i$ is a zero of multiplicity 1, leading coefficient is 1.

35. Fourth degree, only real coefficients, x-intercepts are 0 and 6, $-2i$ is a zero, leading coefficient is 3.

36. Fifth degree, -2 is a zero of multiplicity 2, another integer is a zero of multiplicity 3, y-intercept is -108, leading coefficient is 1.

Solve the following application problem.

37. An open-top box is to be constructed from a 10" by 18" sheet of tin by cutting out squares from each corner as shown and then folding up the sides. Let $V(x)$ denote the volume of the resulting box.

a. Write $V(x)$ as a product of linear factors.

b. For which values of x is $V(x)=0$?

c. Which of your answers from part **b** are physically possible?

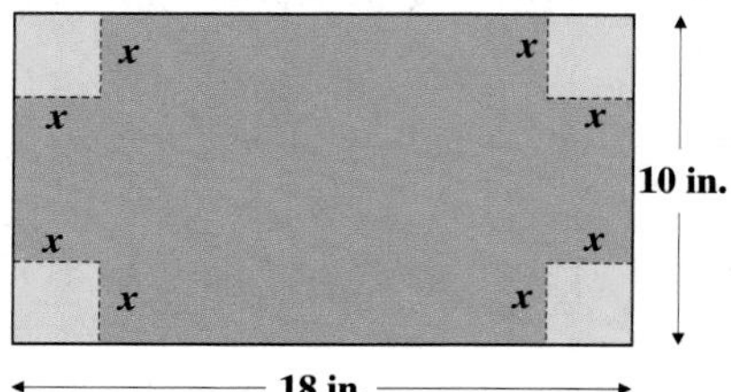

38. Assume $f(x)$ is an nth degree polynomial with real coefficients. Explain why the following statement is true: If n is even, the number of turning points is odd and if n is odd, the number if turning points is even.

4.5 Rational Functions and Rational Inequalities

TOPICS

1. Definitions and useful notation
2. Vertical asymptotes
3. Horizontal and oblique asymptotes
4. Graphing rational functions
5. Solving rational inequalities

Topic 1: Definitions and Useful Notation

Our study of polynomials leads naturally to a study of rational functions, as rational functions are ratios of polynomials. But the behavior of rational functions can be significantly more complex than that of polynomials, and it makes sense to begin by introducing some ideas and notation that will aid us.

Rational Functions

A **rational function** is a function that can be written in the form

$$f(x)=\frac{p(x)}{q(x)},$$

where $p(x)$ and $q(x)$ are both polynomial functions and $q(x)\neq 0$. Of course, even though q is not allowed to be identically zero, there will often be values of x for which $q(x)$ is zero, and at these values the fraction is undefined. Consequently, the **domain of f** is the set $\{x\in\mathbb{R}\,|\,q(x)\neq 0\}$.

We have already seen a few rational functions and are familiar with their graphs. For instance, in Chapter 3 we studied the basic reciprocal function – one of the simplest rational functions – and modifications of it. You already know how to construct the graphs, shown in Figure 1, of the following three rational functions:

$$f(x)=\frac{1}{x},$$

$$g(x)=\frac{1}{x-2},$$

$$h(x)=\frac{1}{x+1}+2=\frac{1}{x+1}+\frac{2x+2}{x+1}=\frac{2x+3}{x+1}$$

(Note that h can be written as a ratio of polynomials, so h is a rational function.)

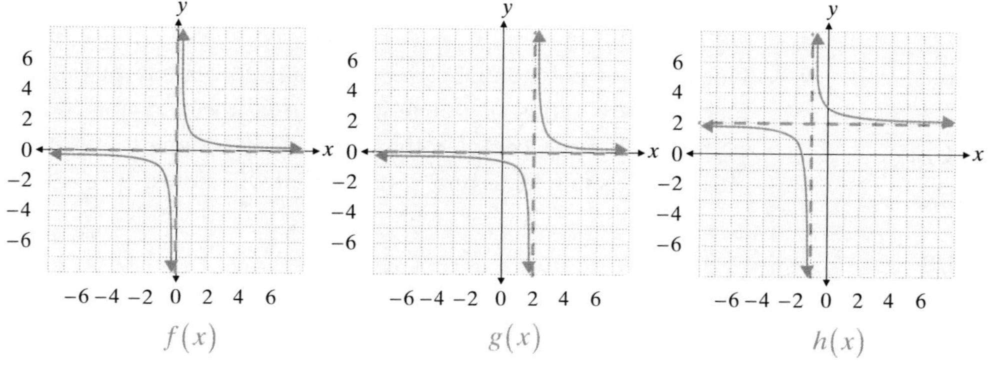

Figure 1: Graphs of Three Rational Functions

Each of the three graphs in Figure 1 is augmented by two dashed lines, one vertical and one horizontal. It is important to realize that in each graph the two dashed lines are not part of the function: they are examples of asymptotes, and they serve as guides to understanding the function. Roughly speaking, an asymptote is an auxiliary line that the graph of a function approaches. Three kinds of asymptotes will appear in our study of rational functions: vertical, horizontal, and oblique.

Vertical Asymptotes

The vertical line $x = c$ is a **vertical asymptote** of a function f if $f(x)$ increases in magnitude without bound as x approaches c. Examples of vertical asymptotes appear in Figure 2. The graph of a rational function cannot intersect a vertical asymptote.

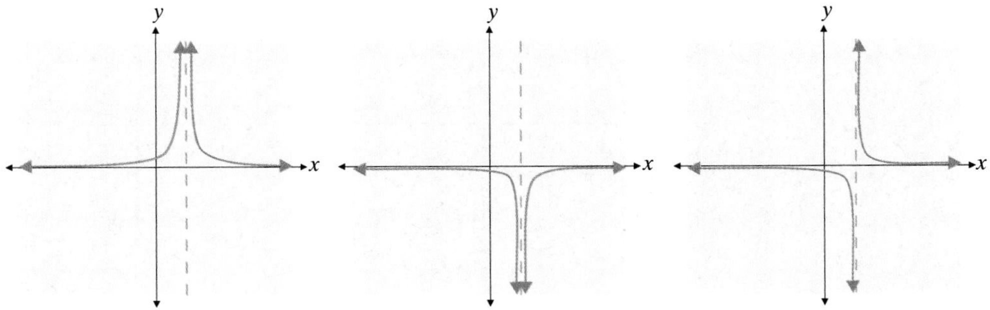

Figure 2: Vertical Asymptotes

Horizontal Asymptotes

The horizontal line $y = c$ is a **horizontal asymptote** of a function f if $f(x)$ approaches the value c as $x \to -\infty$ or as $x \to \infty$. Examples of horizontal asymptotes appear in Figure 3. The graph of a rational function may intersect a horizontal asymptote near the origin, but will eventually approach the asymptote from one side only as x increases in magnitude.

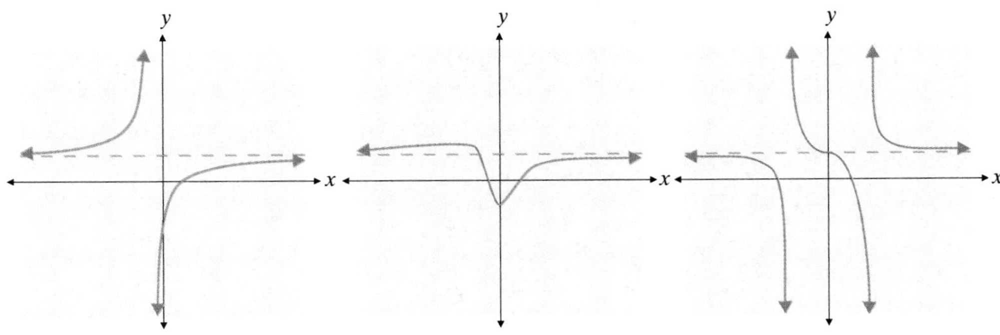

Figure 3: Horizontal Asymptotes

Oblique Asymptotes

A non-vertical, non-horizontal line may also be an asymptote of a function f. Examples of **oblique** (or **slant**) **asymptotes** appear in Figure 4. Again, the graph of a rational function may intersect an oblique asymptote.

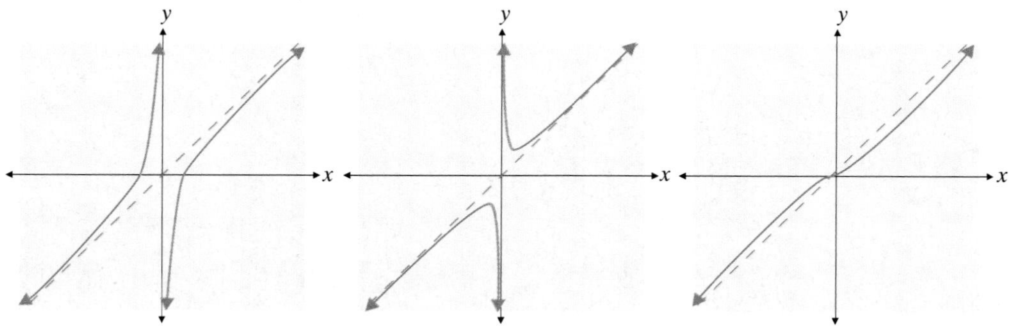

Figure 4: Oblique Asymptotes

As Figures 2, 3, and 4 illustrate, the behavior of rational functions with respect to asymptotes can vary considerably. In order to describe the behavior of a given rational function more easily, we introduce the following notation.

Asymptote Notation

The notation $x \to c^-$ is used in describing the behavior of a graph as x approaches the value c from the left (the negative side).

The notation $x \to c^+$ is used in describing behavior as x approaches c from the right (the positive side).

The notation $x \to c$ is used in describing behavior that is the same on both sides of c.

Figure 5 illustrates how the above notation can be used to describe the behavior of functions.

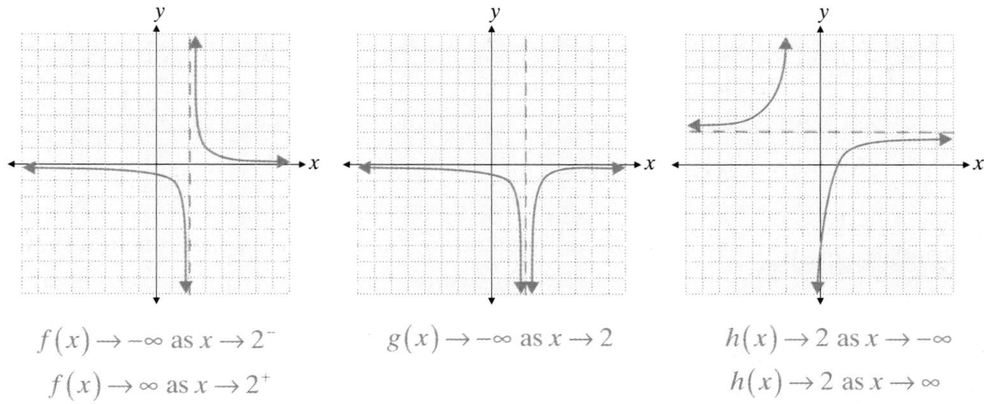

$f(x) \to -\infty$ as $x \to 2^-$
$f(x) \to \infty$ as $x \to 2^+$

$g(x) \to -\infty$ as $x \to 2$

$h(x) \to 2$ as $x \to -\infty$
$h(x) \to 2$ as $x \to \infty$

Figure 5: Asymptote Notation

Topic 2: Vertical Asymptotes

With the above notation and figures as background, we are ready to delve into the details of identifying asymptotes for rational functions.

Equations for Vertical Asymptotes

If the rational function $f(x) = \frac{p(x)}{q(x)}$ has been written in reduced form (so that p and q have no common factors), the vertical line $x = c$ is a vertical asymptote for f if and only if c is a real zero of the polynomial q. In other words, f has vertical asymptotes at the x-intercepts of q.

Note that the numerator of a rational function is irrelevant in locating the vertical asymptotes, assuming that all common factors in the fraction have been canceled.

example 1

Find the equations for the vertical asymptotes of the following functions.

a. $f(x)=\dfrac{x^2-x}{x-1}$ **b.** $g(x)=\dfrac{x^2+1}{x^2+2x-15}$

Solutions:

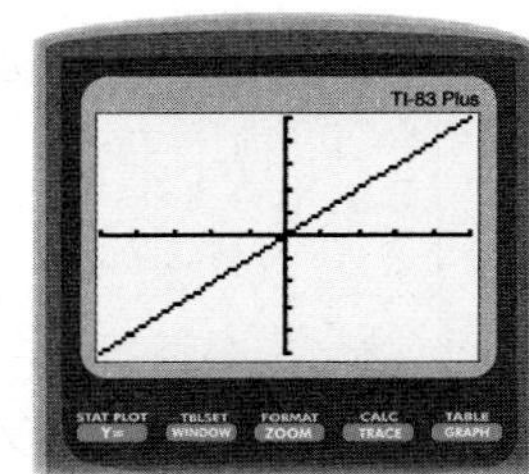

a. One of the first steps in analyzing a rational function should always be determining the domain. Since the denominator of f is zero when $x=1$, the domain of f is the set $\{x\in\mathbb{R}\,|\,x\neq 1\}$; that is, f is defined for all real numbers except 1. Once the domain has been noted, we can proceed to look for common factors to cancel. In this case, a factor of $x-1$ can be canceled from the numerator and denominator, leaving $f(x)=x$ (but only for $x\neq 1$). Since the reduced form of f has no denominator, f has no vertical asymptotes.

b. As usual, in order to determine the domain of the rational function we have to factor the denominator. And after making note of the domain, we will look for common factors to cancel, so we may as well factor the numerator, if possible:

$$g(x)=\frac{x^2+1}{x^2+2x-15}=\frac{x^2+1}{(x+5)(x-3)}.$$

What we see in this example is that the domain is $\{x\in\mathbb{R}\,|\,x\neq -5 \text{ and } x\neq 3\}$ and, since the numerator is a sum of two squares and cannot be factored over the real numbers, the fraction is automatically in reduced form. Thus, the equations of the two vertical asymptotes are $x=-5$ and $x=3$.

Topic 3: Horizontal and Oblique Asymptotes

To determine horizontal and oblique asymptotes, we are interested in the behavior of a function $f(x)$ as $x\to-\infty$ and as $x\to\infty$. If f is a rational function, f is a ratio of two polynomials p and q, so we can begin by considering the effect p and q have on one another. There are many possibilities, but in order to develop some intuition about rational functions, consider the examples on the following page:

1) $f(x)=\dfrac{6x+1}{5x^2-2x+3}$. Any polynomial of degree greater than or equal to 1 gets larger and larger in magnitude (without bound) as $x \to \pm\infty$, and that is true of the numerator and denominator of this function. The critical point, though, is that the degree of the denominator is larger than the degree of the numerator, and hence the behavior of the denominator dominates. As $x \to -\infty$ and as $x \to \infty$, the increasingly large values of the denominator overpower the large values of the numerator, and the fraction as a whole approaches 0. (Think about the result of dividing a fixed number, no matter how large, by a much larger number.) In symbols, $f(x) \to 0$ as $x \to \pm\infty$.

2) $f(x)=\dfrac{6x^2-3x+2}{3x^2+5x-17}$. In this case, the degrees of the numerator and denominator are the same, and neither dominates the other as $x \to \pm\infty$. We know, however, that in any given polynomial the highest power term dominates the rest far away from the origin. In this case, the numerator "looks" more and more like $6x^2$ and the denominator "looks" more and more like $3x^2$ as $x \to -\infty$ and as $x \to \infty$. As a whole, then, the fraction approaches $\dfrac{6x^2}{3x^2}=\dfrac{6}{3}=2$ far away from the origin. In symbols, $f(x) \to 2$ as $x \to \pm\infty$.

3) $f(x)=\dfrac{7x^3+2x-1}{x^2+4x}$. A comparison of the degrees indicates that the numerator of this fraction dominates the denominator far away from the origin. From our work with polynomials, we know that dividing a 3rd degree polynomial by a 2nd degree polynomial results in a 1st degree quotient with a possible remainder. As we will see, the quotient (whose graph is a straight line since its degree is 1) is the oblique asymptote of *f*.

4) $f(x)=\dfrac{3x^5-2x^3+7x^2-1}{4x^3+19x^2-3x+5}$. If the degree of the numerator is larger than the degree of the denominator by more than 1 (as in this case), the rational function gets larger without bound as $x \to \pm\infty$, but does not approach any straight line asymptote.

The following box summarizes the situation as far as horizontal and oblique asymptotes are concerned.

Equations for Horizontal and Oblique Asymptotes

Let $f(x)=\dfrac{p(x)}{q(x)}$ be a rational function, where p is an nth degree polynomial with leading coefficient a_n and q is an mth degree polynomial with leading coefficient b_m. Then:

1. If $n < m$, the horizontal line $y = 0$ (the x-axis) is the horizontal asymptote for f.
2. If $n = m$, the horizontal line $y=\dfrac{a_n}{b_m}$ is the horizontal asymptote for f.
3. If $n = m + 1$, the line $y = g(x)$ is an oblique asymptote for f, where g is the quotient polynomial obtained by dividing p by q. (The remainder polynomial is irrelevant.)
4. If $n > m + 1$, there is no straight line horizontal or oblique asymptote for f.

example 2

Find the equation for the horizontal or oblique asymptote of the following functions.

a. $g(x)=\dfrac{x^2+1}{x^2+2x-15}$ **b.** $h(x)=\dfrac{x^3+x^2+2x+2}{x^2+9}$

Solutions:

a. Because the numerator and denominator of g have the same degree, the asymptote is the horizontal line $y = 1$ (since 1 is the ratio of the leading coefficients).

b. Because the degree of the numerator is one more than the degree of the denominator, we know h has an oblique asymptote. To get the equation for it, we need to do polynomial division:

$$\begin{array}{r} x+1 \\ x^2+9\overline{)x^3+x^2+2x+2} \\ -\left(x^3+0x^2+9x\right) \\ \hline x^2-7x+2 \\ -\left(x^2+0x+9\right) \\ \hline -7x-7 \end{array}$$

This tells us that

$$h(x)=x+1+\frac{-7x-7}{x^2+9},$$

but we only need the quotient, $x + 1$, for the task at hand. The equation for the oblique asymptote is $y = x + 1$.

Topic 4: Graphing Rational Functions

Much of our experience in graph-sketching will be useful as we graph rational functions. But in addition to the standard steps of identifying the x-intercepts (if any), the y-intercept (if there is one), useful symmetry, and so on, we will make use of asymptotes when graphing rational functions. The following is a list of suggested steps in such work.

Graphing Steps

Given a rational function f,

1. Factor the denominator in order to determine the domain of f.
2. Factor the numerator as well and cancel any common factors.
3. Examine the remaining factors in the denominator to determine the equations for any vertical asymptotes.
4. Compare the degrees of the numerator and denominator to determine if there is a horizontal or oblique asymptote, and proceed to find the equation for such.
5. Determine the y-intercept, assuming 0 is in the domain of f.
6. Determine the x-intercepts, if there are any, by setting the numerator of the reduced fraction equal to 0.
7. Determine if there is any symmetry that can be used as an aid.
8. Plot enough points to determine the behavior of f between x-intercepts and between vertical asymptotes.

The work in Example 3 will illustrate these steps, using three functions that we have already studied in part.

example 3

Sketch the graphs of the following rational functions.

a. $f(x)=\dfrac{x^2-x}{x-1}$ **b.** $g(x)=\dfrac{x^2+1}{x^2+2x-15}$ **c.** $h(x)=\dfrac{x^3+x^2+2x+2}{x^2+9}$

Solutions:

a.

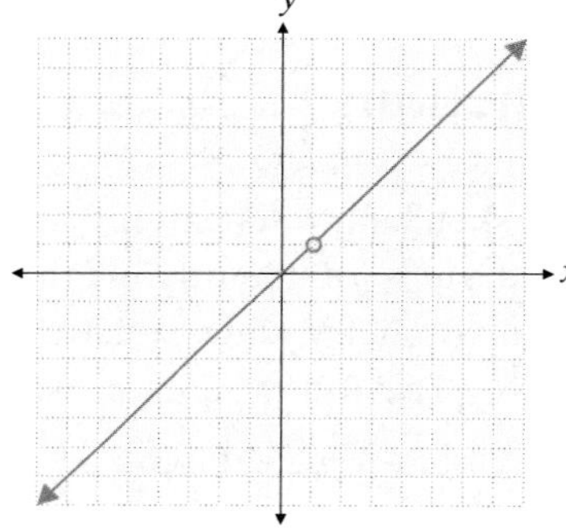

We have already noted, in Example 1a, that the domain of f is all real numbers except for 1, and that after canceling common factors the function reduces to $f(x)=x$. Graphing f then is simply a matter of graphing the straight line with a slope of 1 and a y-intercept of 0, and then noting that f is not defined when $x=1$. As usual, we denote this with an open circle at the point of the graph where $x=1$.

b.

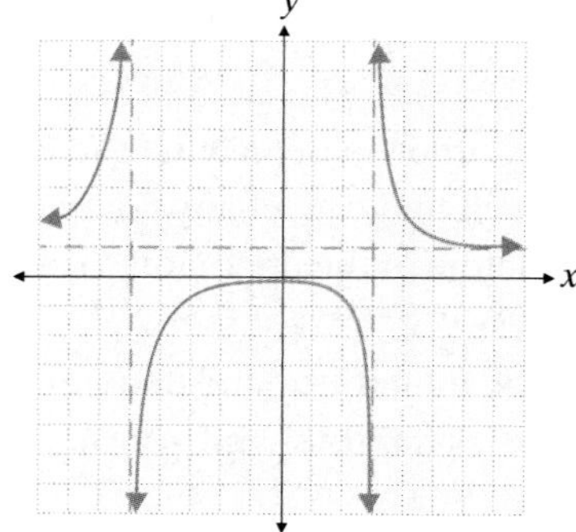

In Example 1b, we determined that the function g has the vertical asymptotes $x=-5$ and $x=3$, and in Example 2a we determined it has the horizontal asymptote $y=1$. These have been drawn at left.

The y-intercept can be calculated by finding the value of the function at 0: $g(0)=-\frac{1}{15}$. Note that g has no x-intercepts, as there is no real number solution to $x^2+1=0$. The remainder of the work consists in plotting enough points to allow us to sketch the function as shown.

c.

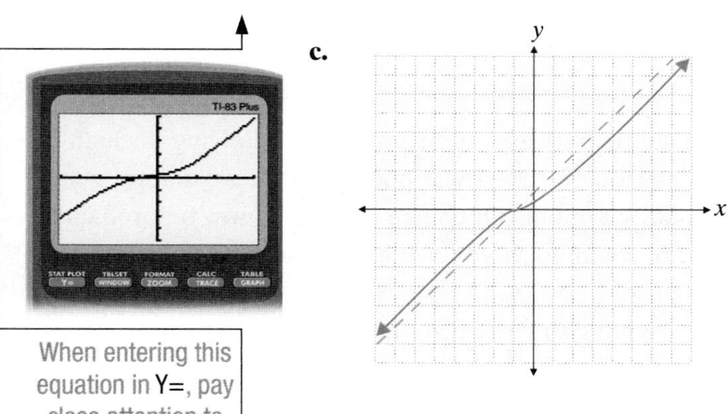

When entering this equation in Y=, pay close attention to the placement of parentheses. Then select Graph.

In Example 2b, we determined that h has an oblique asymptote of $y = x + 1$. Note that h has no vertical asymptotes, as the denominator of h does not factor over the real numbers.

The function crosses the y-axis at $\frac{2}{9}$, but to determine the x-intercept(s) we have to work a bit harder. One of the potential rational zeros of the numerator, -1, actually is a zero, and the numerator factors as $(x+1)(x^2+2)$, so -1 is the only x-intercept.

Topic 5: Solving Rational Inequalities

Rational inequalities are inequalities that can be written with a rational function on one side and 0 on the other, and this definition is the key to their solution. For instance, if we were asked to solve the inequality

$$\frac{x^2+1}{x^2+2x-15} > 0,$$

we would only have to examine our graph in Example 3b and note that the only intervals on which the function $g(x) = \frac{x^2+1}{x^2+2x-15}$ is positive are $(-\infty, -5)$ and $(3, \infty)$, and so the solution of the inequality is the union of these two intervals.

Forgetting this observation can lead to disasters. Suppose, for example, we tried to solve the inequality

$$\frac{x}{x+2} < 3$$

by multiplying through by $x + 2$ and thus clearing the inequality of fractions. The result would be

$$x < 3x + 6,$$

a simple linear inequality. We can solve this to obtain $x > -3$, or the interval $(-3, \infty)$. But does this really work? The number $-\frac{5}{2}$ is in this interval, but

$$\frac{-\frac{5}{2}}{-\frac{5}{2}+2} = 5,$$

which is not less than 3. Also, note that -4 solves the inequality, but -4 is not in the interval $(-3, \infty)$.

What went wrong? The answer is subtle but very important. By multiplying through by $x + 2$, we implicitly made the assumption that $x + 2$ was positive, since we did not worry about possibly reversing the inequality symbol. Of course we don't know beforehand if $x + 2$ is positive or not, since we are trying to solve for the variable x.

To solve this inequality correctly, we have to write it in the standard form of a rational inequality, as shown in Example 4.

example 4

Solve the rational inequality $\frac{x}{x+2} < 3$.

Solution:

We begin by subtracting 3 from both sides and writing the left-hand side as a single rational function, giving us:

$$\frac{-2x-6}{x+2} < 0.$$

We can factor -2 from the numerator of the fraction and then divide both sides by -2 (remembering to reverse the inequality symbol) to obtain the simpler inequality

$$\frac{x+3}{x+2} > 0.$$

Thus, the solution of the original inequality will consist of those intervals for which the rational function $f(x) = \frac{x+3}{x+2}$ is positive. (Alternatively, we can recognize that the only way for the fraction to be positive is if the factor $x + 3$ in the numerator and the factor $x + 2$ in the denominator have the same sign. We could check the intervals $(-\infty, -3), (-3, -2)$, and $(-2, \infty)$ one by one to see when the factors agree in sign).

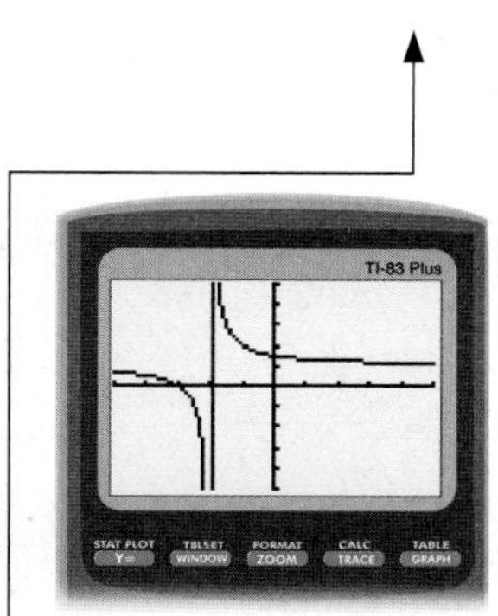

The graph of $f(x)=\dfrac{x+3}{x+2}$ is as shown to the right, and from the graph we can see that the solution (shown in red) to the inequality is $(-\infty,-3)\cup(-2,\infty)$.

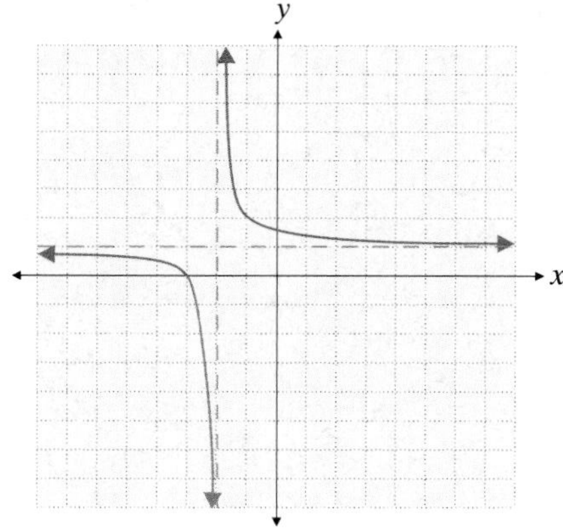

exercises

Find equations for the vertical asymptotes, if any, for each of the following rational functions. See Example 1.

1. $f(x)=\dfrac{5}{x-1}$
2. $f(x)=\dfrac{x^2+3}{x+3}$
3. $f(x)=\dfrac{x^2-4}{x+2}$
4. $f(x)=\dfrac{x^2-4}{2x-x^2}$
5. $f(x)=\dfrac{x+2}{x^2-9}$
6. $f(x)=\dfrac{x^2-2x-3}{2x^2-5x-3}$
7. $f(x)=\dfrac{2x^2+2x-4}{x^2+2x+1}$
8. $f(x)=\dfrac{-3x+5}{x-2}$
9. $f(x)=\dfrac{3x^2+1}{x-2}$
10. $f(x)=\dfrac{x^3-27}{x^2+5}$
11. $f(x)=\dfrac{x^2+5}{x^3-27}$
12. $f(x)=\dfrac{x^2+2x}{x+1}$
13. $f(x)=\dfrac{x^2-1}{x^2-8x+7}$
14. $f(x)=\dfrac{2x^2+7x-14}{2x^2+7x-15}$
15. $f(x)=\dfrac{x^3-6x^2+11x-6}{x^3+8}$
16. $f(x)=\dfrac{x^2-2x-15}{x-5}$
17. $f(x)=\dfrac{x^2-16}{x^2-4}$
18. $f(x)=\dfrac{x^2+4x+4}{x^2+x-2}$

Find equations for the horizontal or oblique asymptotes, if any, for each of the following rational functions. See Example 2.

19. $f(x)=\dfrac{5}{x-1}$

20. $f(x)=\dfrac{x^2+3}{x+3}$

21. $f(x)=\dfrac{x^2-4}{x+2}$

22. $f(x)=\dfrac{x^2-4}{2x-x^2}$

23. $f(x)=\dfrac{x+2}{x^2-9}$

24. $f(x)=\dfrac{x^2-2x-3}{2x^2-5x-3}$

25. $f(x)=\dfrac{2x^2+2x-4}{x^2+2x+1}$

26. $f(x)=\dfrac{-3x+5}{x-2}$

27. $f(x)=\dfrac{3x^2+1}{x-2}$

28. $f(x)=\dfrac{x^3-27}{x^2+5}$

29. $f(x)=\dfrac{x^2+5}{x^3-27}$

30. $f(x)=\dfrac{x^2+2x}{x+1}$

31. $f(x)=\dfrac{x^2-81}{x^3+7x-12}$

32. $f(x)=\dfrac{x^3-3x^2+2x}{x-7}$

33. $f(x)=\dfrac{x^2-9x+4}{x+2}$

34. $f(x)=\dfrac{-x^5+2x^2}{5x^5+3x^3-7}$

35. $f(x)=\dfrac{5x^2-x+12}{x-1}$

36. $f(x)=\dfrac{2x^2-5x+6}{x-3}$

Sketch the graphs of the following rational functions, making use of your work in the problems above and additional information about intercepts and any other points that may be useful. See Example 3.

37. $f(x)=\dfrac{5}{x-1}$

38. $f(x)=\dfrac{x^2+3}{x+3}$

39. $f(x)=\dfrac{x^2-4}{x+2}$

40. $f(x)=\dfrac{x^2-4}{2x-x^2}$

41. $f(x)=\dfrac{x+2}{x^2-9}$

42. $f(x)=\dfrac{x^2-2x-3}{2x^2-5x-3}$

43. $f(x)=\dfrac{2x^2+2x-4}{x^2+2x+1}$

44. $f(x)=\dfrac{-3x+5}{x-2}$

45. $f(x)=\dfrac{3x^2+1}{x-2}$

46. $f(x)=\dfrac{x^3-27}{x^2+5}$

47. $f(x)=\dfrac{x^2+5}{x^3-27}$

48. $f(x)=\dfrac{x^2+2x}{x+1}$

Solve the following rational inequalities. See Example 4.

49. $2x<\dfrac{4}{x+1}$

50. $\dfrac{5}{x-2}\geq\dfrac{3x}{x-2}$

51. $\dfrac{5}{x-2}>\dfrac{3}{x+2}$

52. $\dfrac{x}{x^2-x-6}\leq\dfrac{-1}{x^2-x-6}$

53. $\dfrac{x}{x^2-x-6}\leq\dfrac{-2}{x^2-x-6}$

54. $x>\dfrac{1}{x}$

55. $\dfrac{4}{x-3} \le \dfrac{4}{x}$

56. $\dfrac{x-7}{x-3} \ge \dfrac{x}{x-1}$

57. $\dfrac{x}{x^2+3x+2} > \dfrac{1}{x^2+3x+2}$

58. $\dfrac{1}{x-4} \ge \dfrac{1}{x+1}$

59. $\dfrac{x}{x+1} \ge \dfrac{x+1}{x}$

60. $\dfrac{x}{x^2-2x-3} > \dfrac{3}{x^2-2x-3}$

technology exercises

As you know by now, graphing calculators and computer algebra systems can be very useful tools for analyzing functions. In particular, several of the *Mathematica* commands you have seen previously are useful in understanding rational functions. The simple `Plot` command can be used to obtain a graph of a rational function, though always with the caution that such mechanically derived plots can be misleading. The `Factor` command can be used to factor the numerator and denominator of a rational function, and is thus helpful in identifying the domain, the vertical asymptotes, and the x-intercepts. The illustration below shows the use of these two commands in understanding one particular rational function (comments in red are the conclusions that can be drawn from *Mathematica*'s output).

If available, make similar use of technology to plot, identify the domain, and identify the vertical asymptotes of the rational functions that follow.

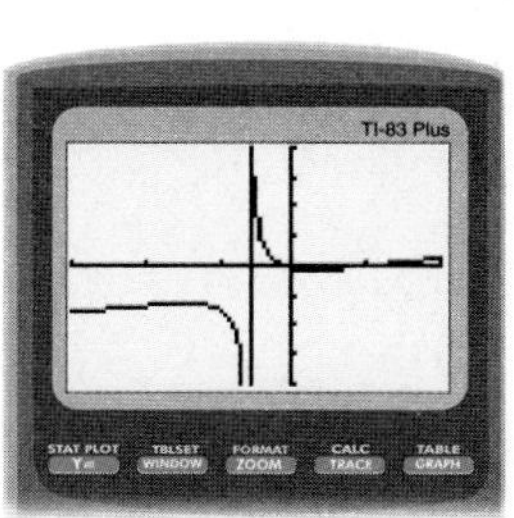
TI-83 Plus

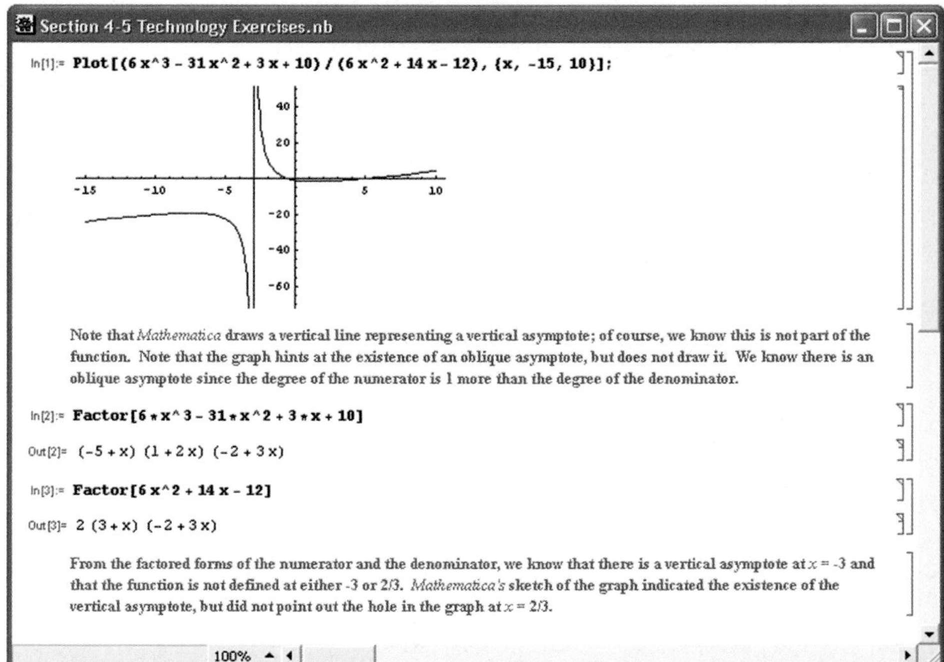
Section 4-5 Technology Exercises.nb
In[1]:= Plot[(6x^3 - 31x^2 + 3x + 10) / (6x^2 + 14x - 12), {x, -15, 10}];
40
20
-15
-10
-5
5
10
-20
-40
-60
Note that Mathematica draws a vertical line representing a vertical asymptote; of course, we know this is not part of the function. Note that the graph hints at the existence of an oblique asymptote, but does not draw it. We know there is an oblique asymptote since the degree of the numerator is 1 more than the degree of the denominator.
In[2]:= Factor[6*x^3 - 31*x^2 + 3*x + 10]
Out[2]= (-5 + x) (1 + 2 x) (-2 + 3 x)
In[3]:= Factor[6 x^2 + 14 x - 12]
Out[3]= 2 (3 + x) (-2 + 3 x)
From the factored forms of the numerator and the denominator, we know that there is a vertical asymptote at x = -3 and that the function is not defined at either -3 or 2/3. Mathematica's sketch of the graph indicated the existence of the vertical asymptote, but did not point out the hole in the graph at x = 2/3.
100%

cont'd. on next page ...

61. $f(x)=\dfrac{8x^3+30x^2-47x+15}{4x^3-8x^2-15x+9}$

62. $f(x)=\dfrac{8x^3+30x^2-47x+15}{4x^2+5x-6}$

63. $f(x)=\dfrac{15x^3-61x^2+2x+8}{5x^2+8x-4}$

64. $f(x)=\dfrac{5x^2+8x-4}{15x^3-61x^2+2x+8}$

65. $f(x)=\dfrac{6x^3-11x^2-26x+40}{6x^2+5}$

66. $f(x)=\dfrac{6x^3-11x^2-26x+40}{6x^3-5x^2-49x+60}$

chapter four project

Polynomial Functions

Ace Automobiles is an international auto manufacturer that has just finished the design for a new sports car. In order to produce this car, they must lease a new assembly plant, purchase new robotic equipment, and hire new staff.

Ace has projected the monthly costs of these expenditures for budgeting purposes. According to their estimates, the plant lease and all utilities will cost \$72,000; the depreciation on the new equipment will be \$130,000; salaries will cost \$480,000; and all other combined overhead will be \$47,000. Thus the monthly total for all expenses will be \$729,000.

The cost of raw materials to produce each car will be \$4500.

The cost function per month for manufacturing x new cars will be

$$C(x) = 4500x + 729{,}000.$$

The revenue per month from selling x cars will be

$$r(x) = 13{,}500x.$$

1. How many cars per month must the new plant make in order to show a profit, if profit is revenue – cost?
2. If the plant manager decides to increase staff and create two shifts, resulting in an additional \$300,000 in salaries with all other expenses remaining unchanged, how many cars would the plant then have to produce in a month to be profitable?
3. Assume now that the plant at full capacity (including the second shift) can produce a maximum of 75 cars per month. What minimum price per car must Ace Automobiles set in order to maintain a profit?

CHAPTER REVIEW

4.1 Introduction to Polynomial Equations and Graphs

topics	pages	test exercises
Zeros of polynomials and solutions of polynomial equations • The connection between *zeros* of polynomials and *roots* or *solutions* of polynomial equations • The connection between x-intercepts and real zeros of polynomials with real coefficients	p. 299 – 301	1 – 4
Graphing factored polynomials • The geometric meaning of linear factors of a polynomial • Behavior of a polynomial as $x \to \pm\infty$ based on its degree and the sign of its leading coefficient	p. 301 – 306	5 – 7
Solving polynomial inequalities • The use of the graph of a polynomial in solving a polynomial inequality	p. 306 – 307	8 – 10

4.2 Polynomial Division and the Division Algorithm

topics	pages	test exercises
The Division Algorithm and the Remainder Theorem • The meaning of the terms *quotient*, *divisor*, *dividend*, and *remainder* as applied to polynomial division • The *Division Algorithm* and what it implies about the degree of the remainder ($p(x) = q(x)d(x) + r(x)$; the degree of the remainder, r, is less than the degree of the divisor, d) • The connection between zeros of a polynomial and its linear factors (if k is a zero of a polynomial p, then $x + k$ is a factor of p) • The *Remainder Theorem* ($p(x) = q(x)(x - k) + p(k)$)	p. 311 – 313	11 – 18
Polynomial long division and synthetic division • The method of *polynomial long division* • The method of *synthetic division* • Using synthetic division to evaluate polynomials for given values	p. 313 – 318	11 – 16

topics (continued)	pages	test exercises
Constructing polynomials with given zeros • Using the theorems to construct polynomials with desired properties	p. 318 – 319	17 – 18

4.3 Locating Real Zeros of Polynomials

topics	pages	test exercises
The Rational Zero Theorem • Using the *Rational Zero Theorem* to identify possible zeros	p. 324 – 326	19 – 20
Descartes' Rule of Signs • Using *Descartes' Rule of Signs* to identify the possible number of positive and negative real zeros	p. 327 – 329	21 – 22
Bounds of real zeros • Determining *upper* and *lower bounds* for the real zeros of a polynomial	p. 329 – 332	23 – 24
The Intermediate Value Theorem • Using the *Intermediate Value Theorem* to approximate real zeros of a polynomial	p. 332 – 334	25 – 26

4.4 The Fundamental Theorem of Algebra

topics	pages	test exercises
The Fundamental Theorem of Algebra • The *Fundamental Theorem of Algebra* and what it says about solutions of polynomial equations • The *Linear Factors Theorem* and what it says about the number of zeros of a polynomial based on its degree • How to determine the maximum number of x-intercepts and *turning points* of the graph of a polynomial	p. 338 – 340	27 – 28
Multiple zeros and their geometric meaning • The meaning of *multiplicity* as applied to zeros of a polynomial • The geometric implications of zeros of a given multiplicity	p. 341 – 343	27 – 28, 31

topics (continued)	pages	test exercises
Conjugate pairs of zeros • The *Conjugate Roots Theorem* and what it says about polynomials with real coefficients • Using the Conjugate Roots Theorem to construct polynomials with desired properties	p. 343 – 346	29 – 31
Summary of polynomial methods • Using all the tools of the chapter to graph polynomials, factor polynomials, and solve polynomial equations and inequalities	p. 346 – 348	27 – 32

4.5 Rational Functions and Rational Inequalities

topics	pages	test exercises
Definitions and useful notation • The definition of a *rational function*, and how to determine domains of rational functions • The meaning of *vertical*, *horizontal*, and *oblique asymptotes*	p. 351 – 355	33 – 40
Vertical asymptotes • How to find the vertical asymptotes for a given rational function	p. 355 – 356	33 – 34
Horizontal and oblique asymptotes • How to determine if a rational function has a horizontal or oblique asymptote based on the degrees of the numerator and denominator • How to find the horizontal asymptote of a given rational function, if there is one • How to find the oblique asymptote of a given rational function, if there is one	p. 356 – 359	35 – 36
Graphing rational functions • How to use knowledge of a rational function's domain, intercepts, asymptotes, and symmetry to sketch its graph	p. 359 – 361	37 – 38
Solving rational inequalities • How to use graphing knowledge to solve *rational inequalities*	p. 361 – 363	39 – 40

chapter test

Section 4.1

Verify that the given values of x solve the corresponding polynomial equations.

1. $4x^3 - 5x^2 = -3x + 18;\ x = 2$

2. $x^2 - 6x = -13;\ x = 3 + 2i$

3. $x^3 + x = 6x^2 - 164;\ x = 5 - 4i$

4. $x^3 + (1 + 4i)x = (7 - 2i)x^2 - 2i + 36;\ x = -2i$

For each of the following polynomial functions, describe the behavior of its graph as $x \to \pm\infty$ and identify the x- and y-intercepts. Use this information to then sketch the graph of each polynomial.

5. $f(x) = (x + 1)^2(x - 2)$

6. $g(x) = (x + 4)(x - 3)(x + 6)$

7. $s(x) = x^3 + 5x^2 + 6x$

Solve the following polynomial inequalities.

8. $2x^2 + 15 \le 11x$

9. $(x - 3)^2(x + 1)^2 > 0$

10. $(x - 4)(x + 2)(x^2 - 1) \le 0$

Section 4.2

Use polynomial long division to rewrite each of the following fractions in the form $q(x) + \frac{r(x)}{d(x)}$, where $d(x)$ is the denominator of the original fraction, $q(x)$ is the quotient, and $r(x)$ is the remainder.

11. $\frac{8x^4 - 6x^3 + 2x^2 + 3x + 4}{2x^2 - 1}$

12. $\frac{11x^2 + 2x - 5}{x - 3}$

13. $\frac{x^4 - 3x^2 + x - 8}{x^2 + 3x + 2}$

Use synthetic division to determine if the given value for k is a zero of the corresponding polynomial. If not, determine $p(k)$.

14. $p(x) = 6x^5 - 23x^4 - 95x^3 + 70x^2 + 204x - 72;\ k = 1$

15. $p(x) = 48x^4 + 10x^3 - 51x^2 - 10x + 3;\ k = \frac{1}{6}$

16. $p(x) = 18x^5 - 87x^4 + 110x^3 - 28x^2 - 16x + 3;\ k = \frac{2}{3}$

Construct a polynomial that has the given properties.

17. Third-degree; zeros of $-1, 4$ and -5; and goes to $-\infty$ as $x \to \infty$

18. Second-degree; zeros of 3 and -2; and y-intercept of -6

Section 4.3

List all of the potential rational zeros of the following polynomials. Then use polynomial division and the quadratic formula, if necessary, to identify the actual zeros.

19. $f(x) = x^4 + 3x^3 - 3x^2 - 11x - 6$

20. $g(x) = 2x^3 - 11x^2 + 18x - 9$

Use Descartes' Rule of Signs to determine the possible number of positive and negative real zeros of each of the following polynomials.

21. $f(x) = 2x^4 - 3x^3 - x^2 + 3x + 10$

22. $g(x) = x^6 - 4x^5 - 2x^4 + x^3 - 6x^2 - 11x + 6$

Use synthetic division to identify integer upper and lower bounds of the real zeros of the following polynomials.

23. $f(x) = 2x^3 - 11x^2 + 3x + 36$

24. $g(x) = 4x^3 - 16x^2 - 79x - 35$

Use the Intermediate Value Theorem to show that each of the following polynomials has a real zero between the indicated values.

25. $f(x) = x^3 + 3x^2 - 4x - 10$; 1 and 2

26. $f(x) = -x^4 + 5x^3 - x^2 - 10x - 2$; -2 and -1

Section 4.4

Use all available methods (e.g. the Rational Zero Theorem, Descartes' Rule of Signs, polynomial division, etc.) to solve each polynomial equation. Use the Linear Factors Theorem to make sure you find the appropriate number of solutions, counting multiplicity.

27. $3x^5 + x^4 + 5x^3 = x^2 + 28x + 20$

28. $8x^5 + 12x^4 - 18x^3 - 35x^2 = 18x + 3$

Use all available methods (in particular, the Conjugate Roots Theorem, if applicable) to factor each of the following polynomials completely, making use of the given zero.

29. $f(x) = 14x^4 - 109x^3 + 296x^2 - 321x + 70;\ 2 + i$ is a zero.

30. $f(x) = x^4 - 5x^3 + 19x^2 - 125x - 150;\ -5i$ is a zero.

Construct polynomial functions with the stated properties.

31. Fourth degree, only real coefficients, $\frac{1}{2}$ and $1 + 2i$ are two of the zeros, y-intercept is -30, leading coefficient is 2.

32. Fifth degree, only real coefficients, -1 is a zero of multiplicity 3, $\sqrt{6}$ is a zero, y-intercept is -6, leading coefficient is 1.

Section 4.5

Find equations for the vertical asymptotes, if any, for each of the following rational functions.

33. $f(x) = \dfrac{4}{2x - 5}$

34. $f(x) = \dfrac{x^2 - 3x + 2}{x - 1}$

Find equations for the horizontal or oblique asymptotes, if any, for each of the following rational functions.

35. $f(x) = \dfrac{2x^3 + 5x^2 - 1}{x^2 - 2x}$

36. $f(x) = \dfrac{x^2 - x + 8}{3x^2 - 7}$

Sketch the graphs of the following rational functions.

37. $f(x) = \dfrac{x^2 + 5}{x + 5}$

38. $f(x) = \dfrac{x + 2}{x^2 + 7x}$

Solve the following rational inequalities.

39. $\dfrac{7}{x + 3} \geq \dfrac{2x}{x + 3}$

40. $\dfrac{x}{x^2 - 5x + 6} < \dfrac{3}{x^2 - 5x + 6}$

chapter

FIVE

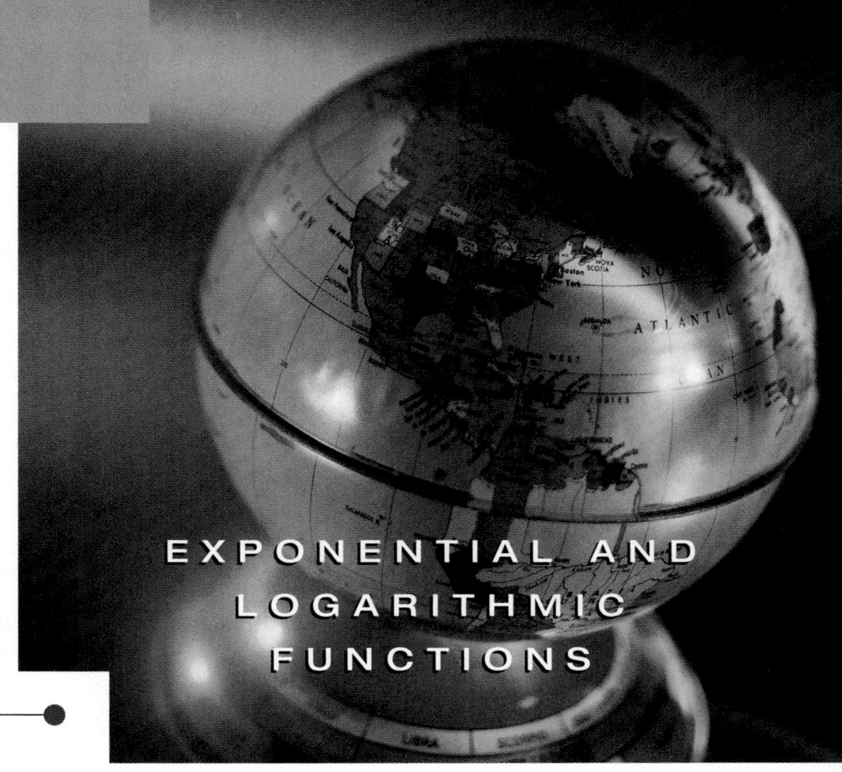

EXPONENTIAL AND LOGARITHMIC FUNCTIONS

OVERVIEW

By the end of this chapter you should be able to estimate the behavior of growing populations, interest-earning accounts, and decaying elements. On page 441 you will find a problem regarding world population. Given information about current population and growth rate, you will be asked to calculate the time it will take to reach a population of 20 billion (disregarding complicated issues of limited resources to support such a population). You will master this type of problem using tools such as the *Summary of Logarithmic Properties* on page 430.

Introduction

This chapter introduces two entirely new classes of functions, both of which are of enormous importance in many natural and man-made contexts. As we will see, the two classes of functions are inverses of one another, though historically exponential and logarithmic functions were developed independently and for unrelated reasons.

We will begin with a study of exponential functions. These are functions in which the variable appears in the exponent while the base is a constant, just the opposite of what we have seen so often in the individual terms of polynomials. As with many mathematical concepts, the argument can be made that exponential functions exist in the natural world independently of mankind, and that consequently mathematicians have done nothing more than observe (and formalize) what there is to be seen. Exponential behavior is exhibited, for example, in the rate at which radioactive substances undergo decay, in how the temperature of an object changes when placed in an environment held at a constant temperature, and in the fundamental principles of population growth. But exponential functions also arise in discussing such man-made phenomena as the growth of investment funds.

In fact, we will use the formula for compound interest to motivate the introduction of the most famous and useful base for exponential functions, the irrational constant e (the first few digits of which are 2.718281828459…). The Swiss mathematician Leonhard Euler (1707 - 1783), who identified many of this number's unique properties (such as the fact that $e = 1 + \frac{1}{1} + \frac{1}{1 \cdot 2} + \frac{1}{1 \cdot 2 \cdot 3} + \ldots$), was one of the first to recognize the fundamental importance of e, and in fact is responsible for the choice of the letter e as its symbol. The constant e also arises very naturally in the context of calculus, but that discussion must wait for a later course.

Napier

Logarithms are inverses, in the function sense, of exponentials, but historically their development was for very different reasons. Much of the development of logarithms is due to John Napier (1550 - 1617), a Scottish writer of political and religious tracts and an amateur mathematician and scientist. It was the goal of simplifying computations of large numbers that led him to devise early versions of logarithmic functions, to construct what today would be called tables of logarithms, and to design a prototype of what would eventually become a slide rule. Of course, today it is not necessary to resort to logarithms in order to carry out difficult computations, but the properties of logarithms make them invaluable in solving certain equations. Further, the fact that they are inverses of exponential functions means they have just as much a place in the natural world.

5.1 Exponential Functions and Their Graphs

TOPICS

1. Definition and classification of exponential functions
2. Graphing exponential functions
3. Solving elementary exponential equations

Topic 1: Definition and Classification of Exponential Functions

We have studied extensively simple functions such as $f(x)=x^2$ and $g(x)=x^3$, and are familiar with radical functions such as $h(x)=x^{1/3}$. We have also explored more exotic species of functions, such as rational functions, step functions, and the like. To this point, though, we have not encountered a function in which the variable appears in the exponent. That will change in this chapter.

An *exponential function* is, informally, a function in which a constant is raised to a variable power. As a class, exponential functions are extremely important because of the large number of diverse situations in which they arise. Examples include radioactive decay, population growth, compound interest, spread of epidemics, rates of temperature change, and many others. The formal definition follows.

Exponential Functions

Let a be a fixed positive real number not equal to 1. The exponential function with base a is the function

$$f(x)=a^x.$$

Note that for any such constant a, a^x is defined for all real numbers x; consequently, the domain of $f(x) = a^x$ is the set of all real numbers. How about the range of f? This question requires a bit more thought to answer.

First, recall that if a is any non-zero number, a^0 is defined to be 1. This means the y-intercept of any exponential function, regardless of the base, is the point $(0,1)$. Aside from this commonality, exponential functions fall into two classes, depending on whether a lies between 0 and 1 or if a is larger than 1. (Incidentally, the only reason we don't allow a to equal 1 is because the result is too boring: $1^x = 1$ for all x. And the reason we don't allow a to be negative is because a^x would not be real for some values of x; consider, for example, what would happen if $a = -1$ and $x = \frac{1}{2}$.)

Consider the following calculations for two sample exponential functions:

x	$f(x) = \left(\frac{1}{3}\right)^x$	$g(x) = 2^x$
−2	$f(-2) = \left(\frac{1}{3}\right)^{-2} = 9$	$g(-2) = \frac{1}{4}$
−1	$f(-1) = 3$	$g(-1) = \frac{1}{2}$
1	$f(1) = \frac{1}{3}$	$g(1) = 2$
2	$f(2) = \frac{1}{9}$	$g(2) = 4$

Note that the values of f decrease as x increases while the values of g do just the opposite. Note also that for any x, both $f(x)$ and $g(x)$ are positive. A bit of experimentation will convince you that any exponential function with a base between 0 and 1 will behave like the function f above, while any exponential function with a base larger than 1 will behave like g. We have terminology to describe this: f is an example of a *decreasing* function and g is an example of an *increasing* function. If we plot a few more points to be sure of the behavior and then fill in the gaps, we get the graphs of f and g that appear in Figure 1.

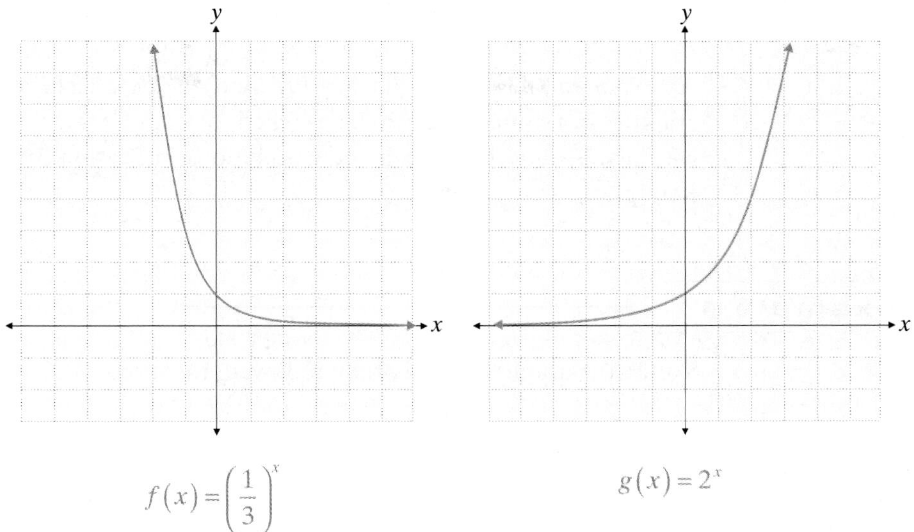

Figure 1: Two Exponential Functions

The above graphs suggest that the range of an exponential function is $(0, \infty)$, and that is indeed the case. Note that the base is immaterial: the range of a^x is the positive real numbers for any allowable base (that is, for any positive a not equal to 1).

The following table summarizes what we have learned.

Behavior of Exponential Functions

Given a positive real number a not equal to 1, the function $f(x) = a^x$ is:

- a decreasing function if $0 < a < 1$, with $f(x) \to \infty$ as $x \to -\infty$ and $f(x) \to 0$ as $x \to \infty$;
- an increasing function if $a > 1$, with $f(x) \to 0$ as $x \to -\infty$ and $f(x) \to \infty$ as $x \to \infty$.

In either case, the point $(0, 1)$ lies on the graph of f, the domain of f is the set of real numbers, and the range of f is the set of positive real numbers.

Topic 2: Graphing Exponential Functions

An exponential function, like any function, can be transformed in ways that result in the graph being shifted, reflected, stretched, or compressed. You may want to review the basic ideas underlying such transformations in Section 3.5 as you study the following examples.

example 1

Sketch the graphs of each of the following functions.

a. $f(x) = 2^{-x}$

b. $g(x) = -3^{x} + 1$

c. $r(x) = 3^{x-2} - 1$

d. $s(x) = \left(\frac{1}{2}\right)^{1-x}$

Solutions:

a.

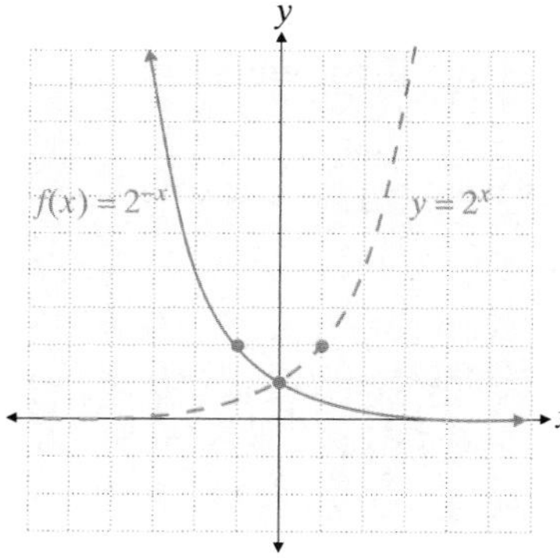

The graph of $f(x) = 2^{-x}$ is the reflection of the graph of $y = 2^{x}$ with respect to the y-axis, as x has been replaced by $-x$. In the picture at left, the dashed green line is the graph of 2^{x} while the blue line, its reflection, is the graph of $f(x) = 2^{-x}$.

There is another way to think about the function $f(x) = 2^{-x}$. Note that $2^{-x} = \left(2^{-1}\right)^{x} = \left(\frac{1}{2}\right)^{x}$, so f is actually a decreasing exponential function.

b.

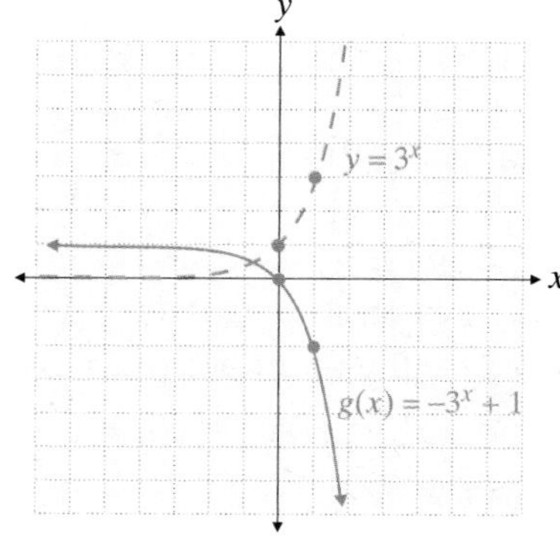

The graph of $y = 3^{x}$ is the dashed green line at left (note that $(0, 1)$ and $(1, 3)$, two easily determined points, lie on the graph).

The effect of multiplying a function by -1 is, as always, the reflection of the graph with respect to the x-axis, and the effect of adding 1 to a function is a vertical upward shift of the graph. The blue line at left is thus the graph of $g(x) = -3^{x} + 1$.

c.

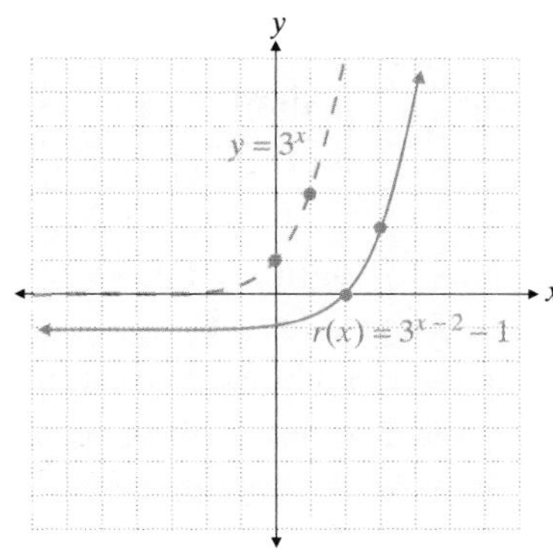

The graph of the function $r(x)=3^{x-2}-1$ is also closely related to the graph of $y=3^x$, again shown as a dashed green line at left. This time, though, x has been replaced by $x-2$, resulting in a shift to the right of 2 units. Subtracting 1 from 3^{x-2} then moves the graph down by 1 unit.

Note that $3^{x-2}=3^x\cdot 3^{-2}=\left(\frac{1}{9}\right)\left(3^x\right)$, so this function could have been presented as $r(x)=\frac{3^x}{9}-1$.

d.

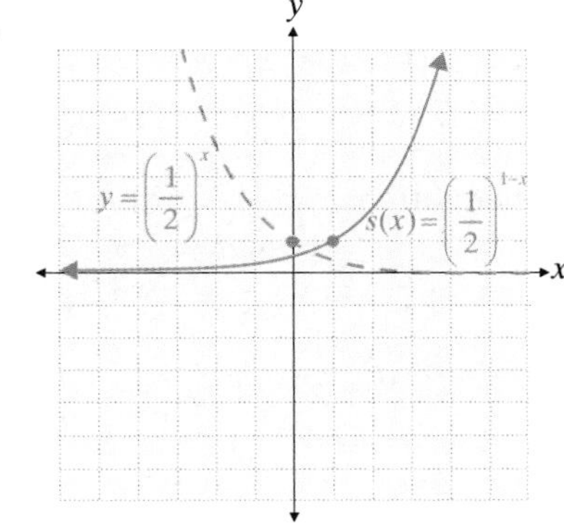

One way to analyze the function s is to relate it to the function $y=\left(\frac{1}{2}\right)^x$, the graph of which is the dashed green line at left. Replacing x with $x+1$ shifts the graph 1 unit to the left, and replacing the x with $-x$ then reflects the result with respect to the y-axis.

Another way to analyze the function is to recognize that $s(x)=\left(\frac{1}{2}\right)\left(2^x\right)$, the graph of which is a compression of the graph of 2^x.

Topic 3: Solving Elementary Exponential Equations

An exponential equation in one variable is, as you might expect, an equation in which the variable appears as an exponent. In other words, an exponential function (or some variant of one) appears as a term in the equation, so it is natural to consider such equations now.

We are not yet ready to tackle exponential equations in full generality; this must wait until we have discussed a class of functions called *logarithms* (which will happen in Section 5.3). But we *are* ready to solve exponential equations that, through algebraic manipulation, can be put into the form

$$a^x=a^b,$$

where a is an allowable base for an exponential function (positive and not equal to 1) and b is a constant.

You might guess, just from the form of the equation $a^x = a^b$, that the solution is $x = b$. This guess is correct, but its justification is somewhat subtle. Note, for instance, that the solution of the (non-exponential) equation $1^x = 1^3$ is not simply $x = 3$: any real number x solves the equation because both sides reduce to 1.

The reason that the single value b is the solution of an exponential equation of the form $a^x = a^b$ is that the exponential function $f(x) = a^x$ is one-to-one (its graph passes the horizontal line test). Recall from Section 3.7 that if g is a one-to-one function, then the only way for $g(x_1)$ to equal $g(x_2)$ is if $x_1 = x_2$. In the case of the function $f(x) = a^x$, the equation $a^x = a^b$ is equivalent to the statement $f(x) = f(b)$, and this implies $x = b$, since f is one-to-one.

An exponential equation may not appear in the simple form $a^x = a^b$ initially. Example 2 illustrates the sorts of steps often necessary in solving exponential equations.

example 2

Solve the following exponential equations.

a. $25^x - 125 = 0$ **b.** $8^{y-1} = \frac{1}{2}$ **c.** $\left(\frac{2}{3}\right)^x = \frac{9}{4}$

Solutions:

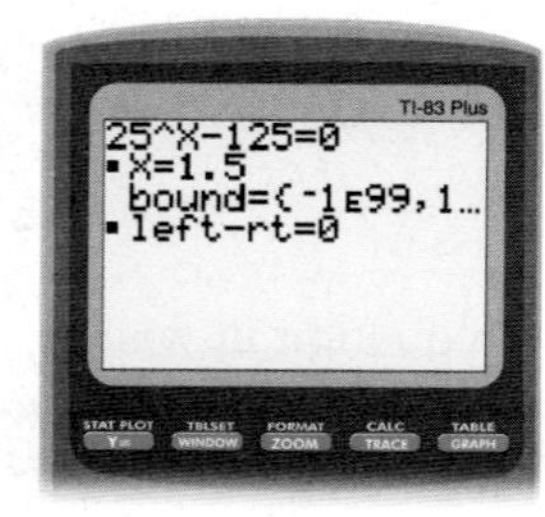

a.
$$25^x - 125 = 0$$
$$25^x = 125$$
$$\left(5^2\right)^x = 5^3$$
$$5^{2x} = 5^3$$
$$2x = 3$$
$$x = \frac{3}{2}$$

We may as well start by isolating the term with the variable on one side, with the hope that both sides can then be rewritten with the same base. Doing so requires finding a number for which both 25 and 125 are powers; the number 5 meets this criterion. The remainder of the work involves using the properties of exponents to write both sides with a base of 5, and then equating the powers. The result is a simple linear equation. Verify that $\frac{3}{2}$ does indeed solve the original equation.

b. $8^{y-1} = \frac{1}{2}$

$\left(2^3\right)^{y-1} = 2^{-1}$

$2^{3y-3} = 2^{-1}$

$3y - 3 = -1$

$3y = 2$

$y = \frac{2}{3}$

Again, we need to rewrite both sides so that they have the same base. Since 8 and $\frac{1}{2}$ can both be written as 2 raised to an appropriate power, we do so, writing 8 as 2^3 and $\frac{1}{2}$ as 2^{-1}.

After applying the properties of exponents to simplify both sides, we equate the exponents and solve for y. You should again verify that this solution works in the original equation.

c. $\left(\frac{2}{3}\right)^x = \frac{9}{4}$

$\left(\frac{2}{3}\right)^x = \left(\frac{3}{2}\right)^2$

$\left(\frac{2}{3}\right)^x = \left(\frac{2}{3}\right)^{-2}$

$x = -2$

The choice of base may not be as obvious in this problem, but 3 and 2 seem likely to appear in some way. If we initially write the right-hand side as shown in the second step to the left, we can then make the two bases equal by making the change shown in the third step.

After making both bases the same, the solution is clear.

exercises

Sketch the graphs of the following functions. See Example 1.

1. $f(x) = 4^x$

2. $g(x) = (.5)^x$

3. $s(x) = 3^{x-2}$

4. $f(x) = \left(\frac{1}{3}\right)^{x+1}$

5. $r(x) = 5^{x-2} + 3$

6. $h(x) = 1 - 2^{x+1}$

7. $f(x) = 2^{-x}$

8. $r(x) = 3^{2-x}$

9. $g(x) = 3\left(2^{-x}\right)$

10. $h(x) = 2^{2x}$

11. $s(x) = (.2)^{-x}$

12. $f(x) = \frac{1}{2^x} + 1$

13. $g(x) = 3 - 2^{-x}$

14. $r(x) = \frac{1}{2^{3-x}}$

15. $h(x) = \left(\frac{1}{2}\right)^{5-x}$

16. $m(x) = 3^{2x+1}$

17. $p(x) = 2 - 4^{2-x}$

18. $q(x) = 5^{3-2x}$

Solve the following exponential equations. See Example 2.

19. $5^x = 125$

20. $3^{2x-1} = 27$

21. $9^{2x-5} = 27^{x-2}$

22. $10^x = .01$

23. $4^{-x} = 16$

24. $\left(\frac{2}{3}\right)^{x+3} = \left(\frac{9}{4}\right)^{-x}$

25. $5^x = .2$

26. $7^{x^2+3x} = \frac{1}{49}$

27. $3^{x^2+4x} = 81^{-1}$

28. $\left(\frac{1}{2}\right)^{x-3} = \left(\frac{1}{4}\right)^{x-5}$

29. $64^{x+(7/6)} = 2$

30. $6^{2x} = 36^{2x-3}$

31. $4^{2x-5} = 8^{x/2}$

32. $\left(\frac{2}{5}\right)^{2x+4} = \left(\frac{4}{25}\right)^{11}$

33. $4^{4x-7} = \frac{1}{64}$

34. $-10^x = -.001$

35. $3^x = 27^{x+4}$

36. $1000^{-x} = 10^{x-8}$

37. $2^x = \left(\frac{1}{2}\right)^{13}$

38. $1^{3x-7} = 4^{2-x}$

39. $5^{3x-1} = 625^x$

40. $2^{x+1} = 64^3$

41. $\frac{1}{5}^{x-4} = 625^{1/2}$

42. $4^{3x+2} = \frac{1}{4}^{-2x}$

43. $\left(e^{x+2}\right)^3 = \left(e^x\right)\frac{1}{\left(e^{3x}\right)}$

44. $3^{2x-7} = 81^{x/2}$

Match the graphs of the following functions to the appropriate equation.

45. $f(x) = 2^{3x}$ **46.** $h(x) = 5^x - 1$ **47.** $g(x) = 2\left(4^{x-1}\right)$

48. $p(x) = 1 - 2^{-x}$ **49.** $f(x) = 6^{4-x}$ **50.** $r(x) = \dfrac{1}{3^x}$

51. $m(x) = -2 + 2^{-3x}$ **52.** $g(x) = \left(\dfrac{1}{4}\right)^{1+x}$ **53.** $h(x) = 3^{\frac{1}{2}x}$

54. $s(x) = 1^x - 4$

a.

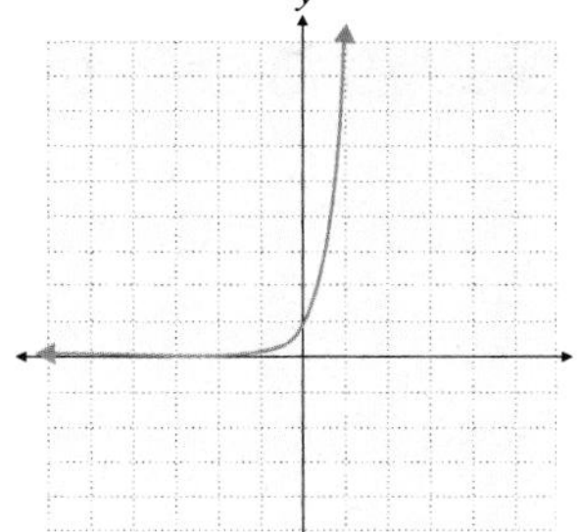

b.

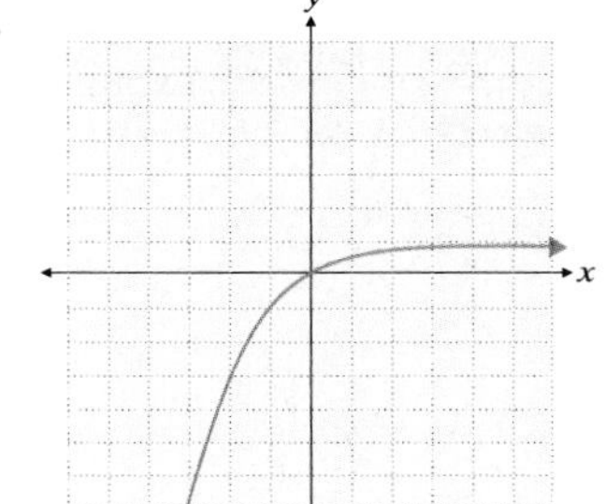

c.

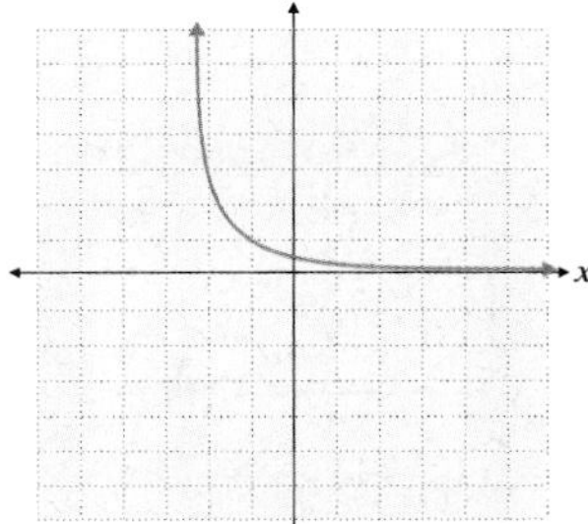

d.

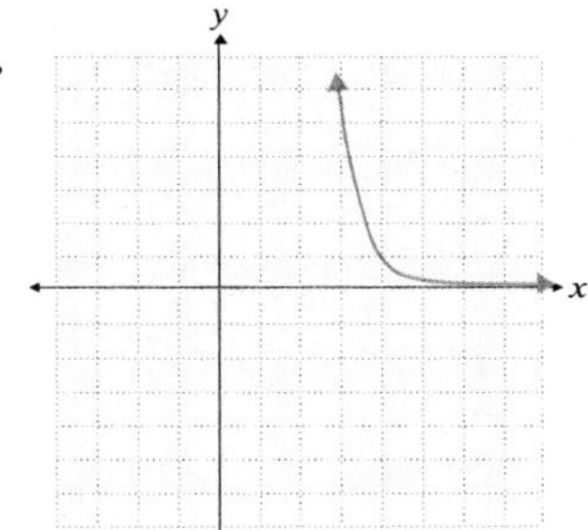

e.

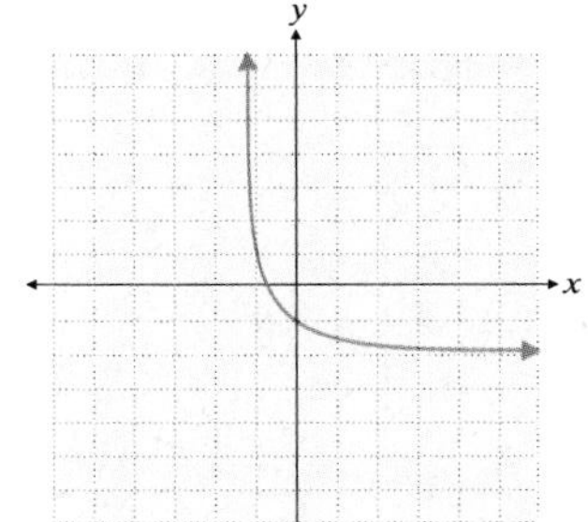

f.

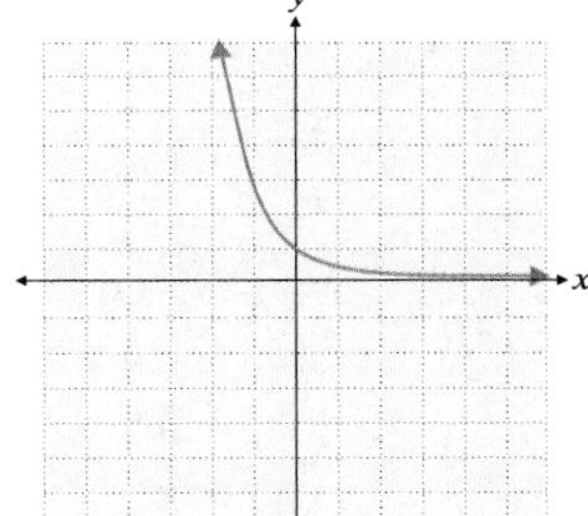

g.

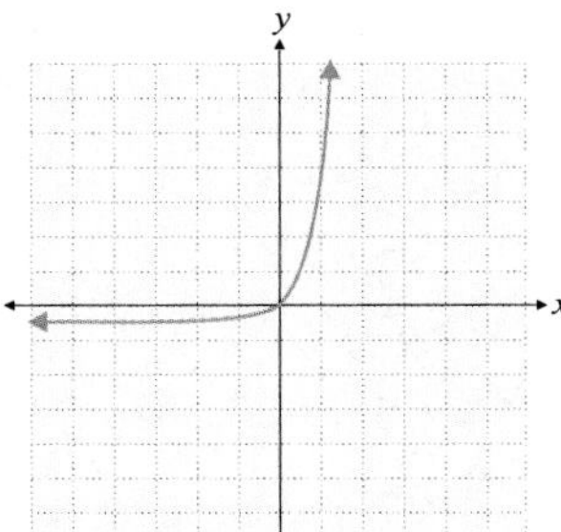

h.

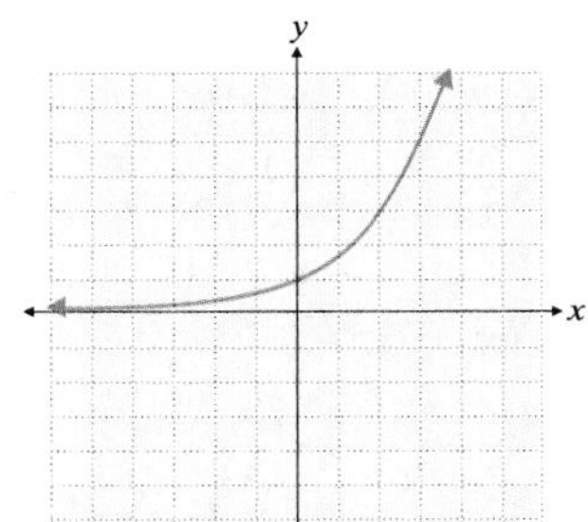

i.

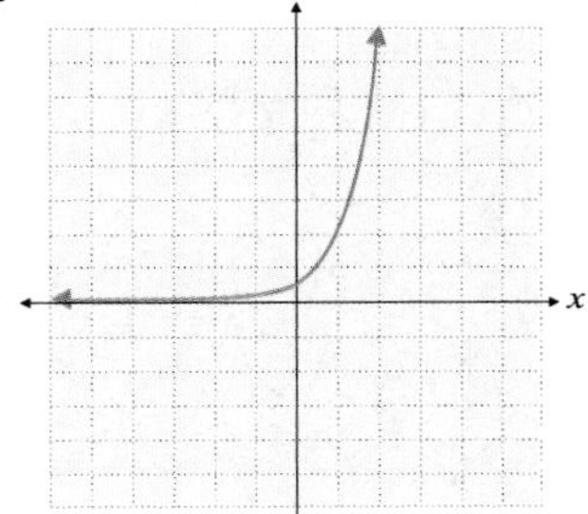

j.

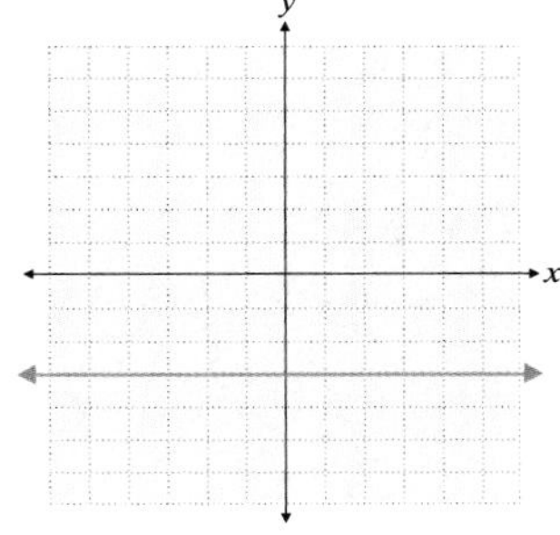

technology exercises

Graphing calculators and computer algebra systems generally produce very nice graphs of exponential functions, provided you specify the plotting "window" appropriately. The graphs below were generated using the simple `Plot` command in *Mathematica*.

Use a graphing calculator or a computer algebra system, if available, to check your graphs of the functions that follow.

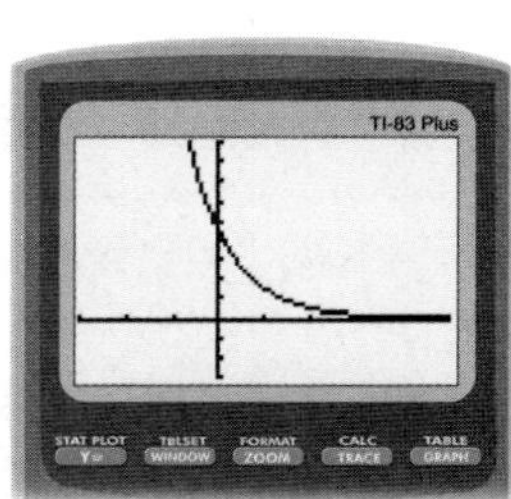

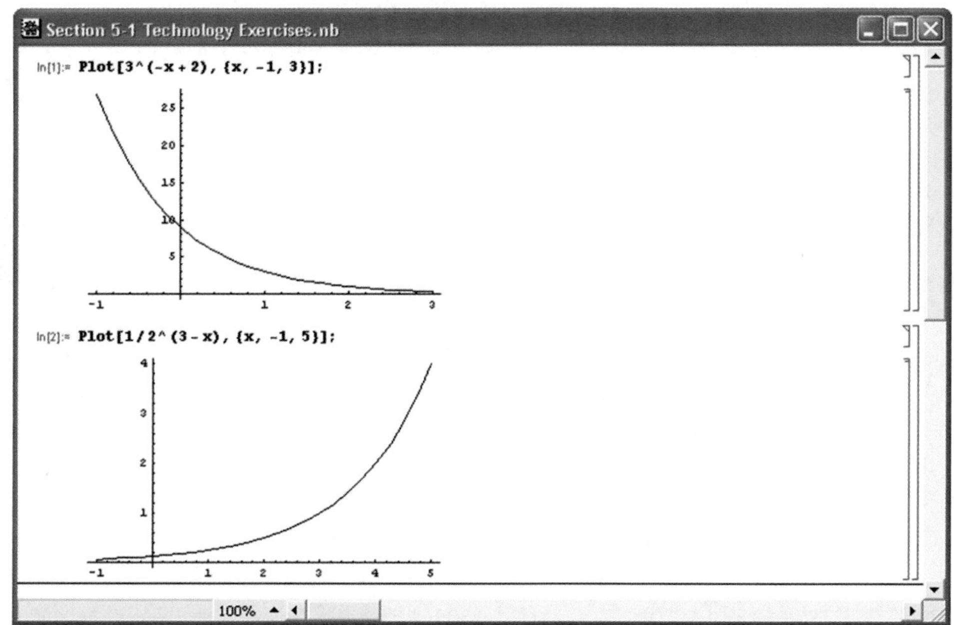

55. $f(x) = 2^{5-x}$

56. $g(x) = \dfrac{3}{2^{x-1}}$

57. $h(x) = 0.4^{x+5}$

58. $f(x) = 2.1^{3-x}$

59. $g(x) = \dfrac{1}{3^{x+2}}$

60. $h(x) = \left(\dfrac{2}{3}\right)^{x+2}$

5.2 Applications of Exponential Functions

TOPICS

1. Models of population growth
2. Radioactive decay
3. Compound interest and the number e

Topic 1: Models of Population Growth

Exponential functions arise naturally in a wide array of situations. In this section, we will study a few such situations in some detail, beginning with population models.

Over the centuries many people, working in such areas as mathematics, biology, and sociology, have created mathematical models of population. The models have represented many types of populations, such as people in a given city or country, wolves in a wildlife habitat, or number of bacteria in a Petri dish. In general, such models can be quite complex, depending on factors like availability of food, space constraints, and effects of disease and predation. The length of time the model is intended to correspond to in reality also plays a large role in the mathematical details of the model. But at their most basic, many population models assume population growth displays exponential behavior.

The reason for this is that the growth of a population usually depends to a large extent on the number of members capable of producing more members. For instance, in the simple case of bacteria in a Petri dish, the more bacteria there are the faster the population will grow, as there are more bacteria cloning themselves. (This assumes an abundant food supply and no constraints on population from lack of space, but at least initially this is often the case.) In any situation where the rate of growth of a population is proportional to the size of the population, the population will grow exponentially.

Example 1 illustrates this with a specific model of bacterial population growth.

example 1

A biologist is culturing bacteria in a large Petri dish. She begins with 1000 bacteria, and supplies sufficient food so that for the first five hours the bacteria population grows exponentially. If the population doubles every hour, how many bacteria are there two and a half hours after she begins?

Solution:

We are seeking a function that describes the population size. The function will depend on the variable t (for time), and the situation suggests we should measure t in hours. If we name the function P (for Population), and if we assume that $t = 0$ corresponds to the moment when the biologist begins the culturing process, we can determine a few values of P:

t	$P(t)$
0	$P(0) = 1000$
1	$P(1) = 2000$
2	$P(2) = 4000$
$\vdots$	$\vdots$
5	$P(5) = 32{,}000$

That is, the population at time 0 is 1000 bacteria, the population at time 1 (hour) is 2000 bacteria, and the population doubles every hour thereafter up to at least $t = 5$ (hours).

But we have been asked to calculate $P(2.5)$, and this is not so easily done. A good starting point, though, is the fact that since the population growth is exponential, we know

$$P(t) = P_0 a^t$$

for some constant P_0 and some base a. In other words, the population function is an exponential function (with some as-yet-unknown base), possibly multiplied by a constant (labeled P_0 here). We just have to determine P_0 and a.

Note that $P(0) = P_0 a^0 = P_0$, and we already know that $P(0) = 1000$, so it must be the case that $P_0 = 1000$. In fact, this is the rationale for the name of this constant: P_0 is the population at time 0.

Now we know that $P(t) = 1000a^t$, and we have only the task of determining a. But since we know $P(1) = 2000$, and from the above $P(1) = 1000a^1 = 1000a$, we have the equation $2000 = 1000a$, or $a = 2$. Thus, $P(t) = 1000(2^t)$.

We can now use a calculator to determine that $P(2.5) = 1000(2^{2.5}) \approx 5657$.

Topic 2: Radioactive Decay

In contrast to populations (at least healthy populations), radioactive substances diminish with time. To be exact, the mass of a radioactive substance decreases over time as the substance decays into other elements. Since exponential functions with a base between 0 and 1 are decreasing functions, this suggests that radioactive decay is modeled by

$$A(t) = A_0 a^t,$$

where $A(t)$ represents the amount of a given substance at time t, A_0 is the amount at time $t = 0$, and a is a number between 0 and 1.

The fact of radioactive decay is important (indirectly) in using radioactivity for power generation, and important again in working out the details of storing spent radioactive fuel rods. Radioactive decay also has other uses, one of which goes by the name of *radio-carbon dating*. This is a technique used by archaeologists, anthropologists, and others to estimate how long ago an organism died. The method depends on the fact that living organisms constantly absorb molecules of the radioactive substance carbon-14 while alive, but the intake of carbon-14 ceases once the organism dies. It is believed that the percentage of carbon-14 on Earth (relative to other isotopes of carbon) has been relatively constant over time, so the first step in the method is to determine the percentage of carbon-14 in the remains of a given organism. By comparing this (smaller) percentage to the percentage found in living tissue, an estimate of when the organism died can then be made.

The mathematics of the age estimation depends on the fact that half of a given mass of carbon-14 decays over a period of 5728 years. This is known as the *half-life* of carbon-14; every radioactive substance has a half-life, and the half-life is usually an important issue when working with such substances. In the case of carbon-14, the half-life of 5728 years means that if, for instance, an organism contained 12 grams of carbon-14 at death, it would contain 6 grams after 5728 years, 3 grams after another 5728 years, and so on. (In exponential growth, there is the related concept of *doubling-time*, which is the length of time needed for the function to double in value. The doubling-time for the bacteria population in Example 1 is 1 hour.)

example 2

Determine the base a so that the function $A(t) = A_0 a^t$ accurately describes the decay of carbon-14 as a function of t years.

Solution:

Note that $A(0) = A_0 a^0 = A_0$, so A_0 represents the amount of carbon-14 at time $t = 0$. Since we are seeking a general formula, we don't know what A_0 is specifically; that is, the value of A_0 will vary depending on the details of the situation. But we can still determine the base constant a.

What we know is that half of the original amount of carbon-14 decays over a period of 5728 years, so $A(5728)$ will be half of A_0. This gives us the equation

$$A_0\left(a^{5728}\right) = \frac{A_0}{2}.$$

To solve this for a, we can first divide both sides by A_0; the fact that A_0 then disappears from the equation just emphasizes that its exact value is irrelevant for the task at hand. So we have the equation

$$a^{5728} = \frac{1}{2}.$$

At this point, a calculator is called for, as we need to take the 5728th root of both sides. This gives us the value for a that we seek:

$$a = \left(\frac{1}{2}\right)^{1/5728} \approx 0.999879.$$

The function is thus $A(t) = A_0(0.999879)^t$ (using our approximate value for a). We can now verify that the function behaves as expected by evaluating A at various multiples of the half-life for carbon-14, as shown.

$$A(1 \cdot 5728) = A_0(0.999879)^{5728} \approx 0.5A_0$$

$$A(2 \cdot 5728) = A_0(0.999879)^{11456} \approx 0.25A_0$$

$$A(3 \cdot 5728) = A_0(0.999879)^{17184} \approx 0.125A_0$$

This is our first encounter with radioactive decay and radio-carbon dating. We will come back to the topic in a later section and refine our understanding.

Topic 3: Compound Interest and the Number e

It is difficult to think of an application of exponential functions that is of more immediate importance to more people than compound interest. Virtually everyone living today eventually has some experience, positive or negative, with the notion of compound interest. The positive experiences derive from investing money in various ways, while the negative (but usually unavoidable) experiences derive from car loans, mortgages, credit card charges, and the like.

The basic compound interest formula can be easily understood by considering what happens when money is invested in a simple savings account. Typically, a savings account is set up to pay interest at an annual rate of r (which we will represent in decimal form) compounded n times a year. For instance, a bank may offer an annual interest rate of 5% (meaning $r = 0.05$) compounded monthly (so $n = 12$). Compounding, in this context, is the act of calculating the interest earned on an investment and adding that amount to the investment. An investment in a monthly-compounded account will have interest added to it twelve times over the course of a year.

Suppose an amount of P (for Principal) dollars is invested in a savings account at an annual rate of r compounded n times per year. What we are seeking is a formula for the amount of money, which we will call A, in the account after t years. If we say that a period is the length of time between compoundings, interest is calculated at the rate of $\frac{r}{n}$ per period (for instance, if $r = 0.05$ and $n = 12$, interest is earned at a rate of $\frac{0.05}{12} \approx 0.00417$ per month). Table 1 illustrates how compounding increases the amount in the account over the course of several periods.

Period	Amount Accumulated
0	$A = P$
1	$A = P\left(1+\frac{r}{n}\right)$
2	$A = P\left(1+\frac{r}{n}\right)\left(1+\frac{r}{n}\right) = P\left(1+\frac{r}{n}\right)^2$
3	$A = P\left(1+\frac{r}{n}\right)^2\left(1+\frac{r}{n}\right) = P\left(1+\frac{r}{n}\right)^3$
$\vdots$	$\vdots$
k	$A = P\left(1+\frac{r}{n}\right)^{k-1}\left(1+\frac{r}{n}\right) = P\left(1+\frac{r}{n}\right)^k$

Table 1: Effect of Compounding on an Investment of *P* Dollars

Notice that the accumulation at the end of each period is the accumulation at the end of the preceding period multiplied by $\left(1+\frac{r}{n}\right)$; that is, the account has the amount from the end of the preceding period plus that amount multiplied by $\frac{r}{n}$. The amount A is then a function of t, the number of years the account stays active. Since the number of investment periods in t years is nt, we obtain the following formula.

Compound Interest Formula

An investment of P dollars at an annual interest rate of r, compounded n times per year, has an accumulated value after t years of

$$A(t) = P\left(1+\frac{r}{n}\right)^{nt}.$$

example 3

Sandy invests $10,000 dollars in a savings account earning 4.5% annual interest, compounded quarterly. What is the value of her investment after three and a half years?

Solution:

The first step is to determine the values of the various pieces of the compound interest formula. We know that $P = 10{,}000$, $r = 0.045$ (remember to express the interest rate in decimal form), $n = 4$ (the account is compounded four times a year), and $t = 3.5$ (the length of time concerned, in years). Thus, the accumulation after three and a half years is

$$\begin{aligned} A(3.5) &= 10{,}000\left(1+\frac{0.045}{4}\right)^{(4)(3.5)} \\ &= 10{,}000(1.01125)^{14} \\ &= \$11{,}695.52. \end{aligned}$$

The compound interest formula can also be used to determine the interest rate of an existing savings account, as demonstrated in Example 4.

example 4

Nine months after depositing \$520.00 in a monthly-compounded savings account, Frank checks his balance and finds the account has \$528.84. Being the forgetful type, he can't quite remember what the annual interest rate for his account is supposed to be, but he sees that the bank is advertising a rate of 2.5% for new accounts. Should he close out his existing account and open a new one?

Solution:

≈Your Bank≈

Introducing 2.5% Annual Interest Rate on all new accounts!

As in Example 3, we will begin by identifying the known quantities in the compound interest formula: note that $P = 520$, $n = 12$ (12 compoundings per year), and $t = 0.75$ (nine months is three-quarters of a year). Further, the accumulation A at the time in question is \$528.84. This gives us the equation

$$528.84 = 520\left(1 + \frac{r}{12}\right)^{(12)(0.75)}$$

to solve for r, the annual interest rate. After dividing both sides by 520, we can use a calculator to take the 9th root of both sides and solve for r, as shown:

$$528.84 = 520\left(1 + \frac{r}{12}\right)^{9}$$

$$1.017 = \left(1 + \frac{r}{12}\right)^{9}$$

$$1.001875 \approx 1 + \frac{r}{12}$$

$$0.001875 \approx \frac{r}{12}$$

$$0.0225 \approx r.$$

Thus, his current savings account is paying an annual interest rate of 2.25%, and he would gain a slight advantage by switching to a new account.

The compound interest formula serves as a convenient means of introducing the extremely important irrational number e. Just as π is the symbol conventionally used for the irrational number obtained by dividing the circumference of a circle by its diameter, e is the symbol traditionally used to denote what is undoubtedly the most useful base for an exponential function, and it is consequently called the **natural base**.

A complete understanding of the preceding statement requires a discussion of some concepts encountered in calculus, but we'll see many purely algebraic uses of e in this text. The first arises from considering what happens to the compound interest formula when the number of compoundings per year increases without bound. That is, we will examine the formula as $n \to \infty$.

In order to do this easily, we need to perform some algebraic manipulation of the formula, as follows:

$$\begin{aligned} A(t) &= P\left(1+\frac{r}{n}\right)^{nt} \\ &= P\left(1+\frac{1}{m}\right)^{rmt} \qquad \left(\text{we have replaced } \frac{r}{n} \text{ with } \frac{1}{m}; \text{ therefore } n = rm\right) \\ &= P\left(\left(1+\frac{1}{m}\right)^{m}\right)^{rt}. \end{aligned}$$

Although the manipulation may appear somewhat strange, it has accomplished the important task of isolating the part of the formula that changes as $n \to \infty$. First, note that the change of variables made in the second step implies $m = \frac{n}{r}$, and so letting $n \to \infty$ means that $m \to \infty$ as well. Since every other quantity in the formula remains fixed, we only have to understand what happens to

$$\left(1+\frac{1}{m}\right)^{m}$$

as m grows without bound.

Unfortunately, this is no trivial undertaking. You might think that letting m get larger and larger would make the expression grow larger and larger, as m is the exponent in the expression and the base, for any positive value of m, is larger than 1. But at the same time, the base approaches 1 as m increases without bound, and 1 raised to any power is simply 1. It turns out that these two effects balance one another out, with the result that

$$\left(1+\frac{1}{m}\right)^{m} \to e \approx 2.7183 \text{ as } m \to \infty.$$

As far as compound interest is concerned, this means that if P dollars is invested in an account that is compounded continuously (the interest is calculated at every instant and added to the account), the accumulation in the account is determined by the following formula.

Continuous Compounding Formula

If P dollars is invested for t years in a continuously compounded account with an annual interest rate of r, the accumulation is:

$$A(t) = Pe^{rt}.$$

example 5

If Sandy (last seen in Example 3) has the option of investing her $10,000 in a continuously compounded account earning 4.5% annual interest for three and a half years, what will her accumulation be?

Solution:

The relevant formula is applied:

$$\begin{aligned} A(3.5) &= 10{,}000e^{(0.045)(3.5)} \\ &= 10{,}000e^{0.1575} \\ &= \$11705.81. \end{aligned}$$

Such an account thus earns $10.29 more than the quarterly compounded account in Example 3.

As we will see in Section 4 of this chapter, all exponential functions can be expressed with the natural base e (or any other legitimate base, for that matter). For instance, the formula for the radioactive decay of carbon-14, using the base e, is

$$A(t) = A_0e^{-0.000121t}.$$

You should verify that this version of the decay formula does indeed give the same values for $A(t)$ as the version derived in Example 2.

exercises

The following problems all involve working with exponential functions similar to those encountered in this section. See Examples 1 through 5 for guidance as you work through the problems.

1. The half-life of radium is approximately 1600 years.

a. Determine a so that $A(t) = A_0 a^t$ describes the amount of radium left after t years, where A_0 is the amount at time $t = 0$.

b. How much of a 1 gram sample of radium would remain after 100 years?

c. How much of a 1 gram sample of radium would remain after 1000 years?

2. The radioactive element Polonium-210 has a relatively short half-life of 140 days, and one way to model the amount of Polonium-210 remaining after t days is with the function $A(t) = A_0 e^{-0.004951t}$, where A_0 is the mass at time $t = 0$ (note that $A(140) = \frac{A_0}{2}$). What percentage of the original mass of a sample of Polonium-210 remains after one year?

3. A certain species of fish is to be introduced into a new man-made lake, and wildlife experts estimate the population will grow according to $P(t) = (1000)2^{t/3}$, where t represents the number of years from the time of introduction.

a. What is the doubling-time for this population of fish?

b. How long will it take for the population to reach 8000 fish, according to this model?

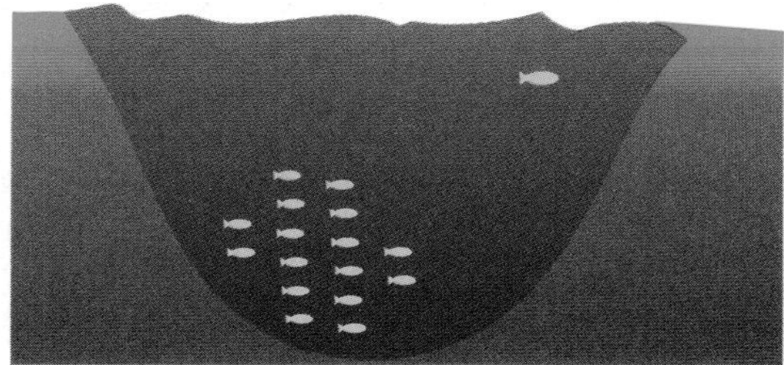

4. The population of a certain inner-city area is estimated to be declining according to the model $P(t) = 237{,}000e^{-0.018t}$, where t is the number of years from the present. What does this model predict the population will be in ten years?

5. In an effort to control vegetation overgrowth, 100 rabbits are released in an isolated area that is free of predators. After one year, it is estimated that the rabbit population has increased to 500. Assuming exponential population growth, what will the population be after another 6 months?

6. Assuming a current world population of 6 billion people, an annual growth rate of 1.9% per year, and a worst-case scenario of exponential growth, what will the world population be in: **a.** 10 years? **b.** 50 years?

7. Madiha has \$3500 that she wants to invest in a simple savings account for two and a half years, at which time she plans to close out the account and use the money as a down payment on a car. She finds one local bank offering an annual interest rate of 2.75% compounded monthly, and another bank offering an annual interest rate of 2.7% compounded daily (365 times per year). Which bank should she choose?

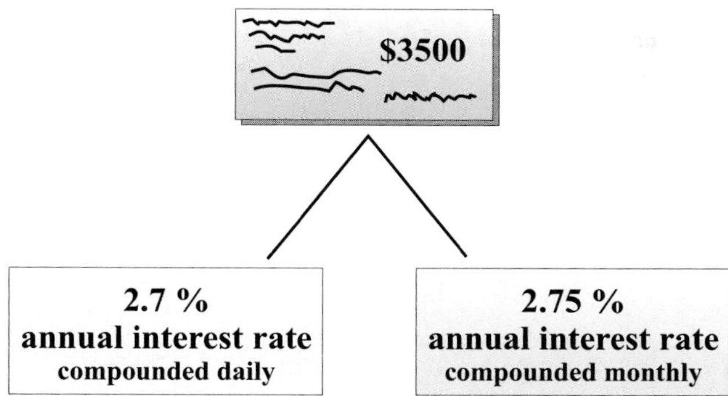

8. Madiha from the last problem does some more searching and finds an on-line bank offering an annual rate of 2.75% compounded continuously. How much more money will she earn over two and a half years if she chooses this bank rather than the local bank offering the same rate compounded monthly?

9. Tom hopes to earn \$1000 in interest in three years time from \$10,000 that he has available to invest. To decide if it's feasible to do this by investing in a simple monthly-compounded savings account, he needs to determine the annual interest rate such an account would have to offer for him to meet his goal. What would the annual rate of interest have to be?

10. An investment firm claims that its clients usually double their principal in five years time. What annual rate of interest would a savings account, compounded monthly, have to offer in order to match this claim?

11. The function $C(t) = C(1 + r)^t$ models the rise in the cost of a product that has a cost of C today, subject to an average yearly inflation rate of r for t years. If the average annual rate of inflation over the next decade is assumed to be 3%, what will the inflation-adjusted cost of a \$100,000 house be in 10 years?

12. Given the inflation model $C(t) = C(1 + r)^t$ (see the previous problem), and given that a loaf of bread that currently sells for \$3.60 sold for \$3.10 six years ago, what has the average annual rate of inflation been for the past six years?

13. The function $N(t) = \dfrac{10{,}000}{1 + 999e^{-t}}$ models the number of people in a small town who have caught the flu t weeks after the initial outbreak.

a. How many people were ill initially?

b. How many people have caught the flu after 8 weeks?

c. Describe what happens to the function $N(t)$ as $t \to \infty$.

14. The concentration $C(t)$ of a certain drug in the bloodstream after t minutes is given by the formula $C(t) = 0.05\left(1 - e^{-0.2t}\right)$. What is the concentration after 10 minutes?

15. Carbon-11 has a radioactive half-life of approximately 20 minutes, and is useful as a diagnostic tool in certain medical applications. Because of the relatively short half-life, time is a crucial factor when conducting experiments with this element.

a. Determine a so that $A(t) = A_0 a^t$ describes the amount of carbon-11 left after t minutes, where A_0 is the amount at time $t = 0$.

b. How much of a 2 kg sample of carbon-11 would be left after 30 minutes?

c. How much of a 2 kg sample of carbon-11 would be left after 6 hours?

16. Charles has recently inherited \$8000 which he wants to deposit in a savings account. He has determined that his two best bets are an account that compounds annually at a rate of 3.20% and an account that compounds continuously at an annual rate of 3.15%. Which account would pay Charles more interest?

17. Marshall invests \$1250 in a mutual fund which boasts a 5.7% annual return compounded semiannually (twice a year). After 3 and a half years, Marshall decides to withdraw his money.

a. How much is in his account?

b. How much has he made in interest from his investment?

18. Adam is working in a lab testing bacteria populations. After starting out with a population of 375 bacteria, he observes the change in population and notices that the population doubles every 27 minutes.

a. Find the equation for the population P in terms of time t in minutes, rounding a to the nearest thousandth.

b. Find the population after 2 hours.

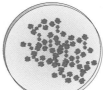

At time 0 After 27 minutes

19. A new virus has broken out in isolated parts of Africa, and is spreading exponentially through tribal villages. The growth of this new virus can be mapped using the following formula where P stands for the number of people in a village and d stands for the number of days since the virus first appeared. According to this equation, how many people in a tribe of 300 will be infected after 5 days?

$$V = P\left(1 - e^{-0.18d}\right)$$

20. A new hybrid car is equipped with a battery that is meant to improve gas mileage by making the car run on electric power as well as gasoline. The batteries in these new cars must be changed out every so often to ensure proper operation. The power in the battery decreases according to the exponential equation below. After 30 days (d), how much power in watts (w) is left in the battery?

$$w(d) = 40e^{-0.06d}$$

21. A young economics student has come across a very profitable investment scheme in which his money will accrue interest according to the equation listed below. If this student invests $1250 into this lucrative endeavor, how much money will he have after 24 months? I represents the investment and m represents the number of months the money has been invested for.

$$C = Ie^{0.08m}$$

22. A family releases a couple of pet rabbits into the wild. Upon being released the rabbits begin to reproduce at an exponential rate. The population increase follows the following formula. After 2 years how large is the rabbit population where n stands for the initial rabbit population (2) and m stands for the number of months?

$$P = ne^{0.5m}$$

23. Inside a business network, an email worm was downloaded by an employee. This worm goes through the infected computer's address book and sends itself to all the listed email addresses. This worm very rapidly works its way through the network following the equation below where C is the number of computers in the network and h is the number of hours after its discovery. After only 8 hours, how many computers has the worm infected if there are 150 computers in the network?

$$W = C\left(1 - e^{-0.12h}\right)$$

24. A construction crew has been assigned to build an apartment complex. The work of the crew can be modeled using the exponential formula below where A is the total number of apartments to be built, w is the number of weeks, and F is the number of finished apartments. Out of a total of 100 apartments, how many apartments have been finished after 4 weeks of work?

$$F = A\left(1 - e^{-0.1w}\right)$$

technology note

Scientific and graphing calculators usually have a button labeled e^x, to be used in evaluating or graphing exponential functions with the base e. Computer algebra systems such as *Mathematica* are similarly equipped. In *Mathematica*, the expression e^x can be entered as either **E^x** or **Exp[x]**. The example below shows how the function $f(x) = e^{3x}$ is defined, graphed, and then evaluated for two different values in *Mathematica*.

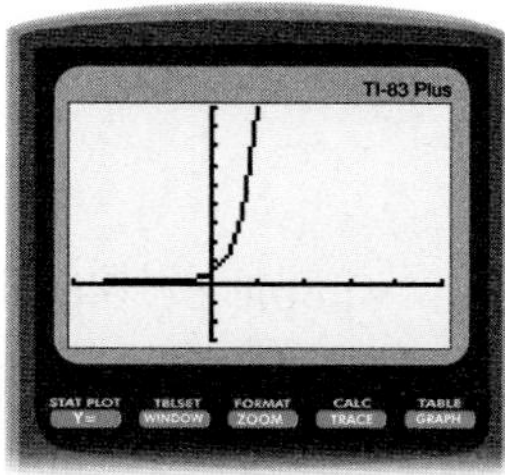

Window:
$x = [-3, 5]$
$y = [-6, 18]$

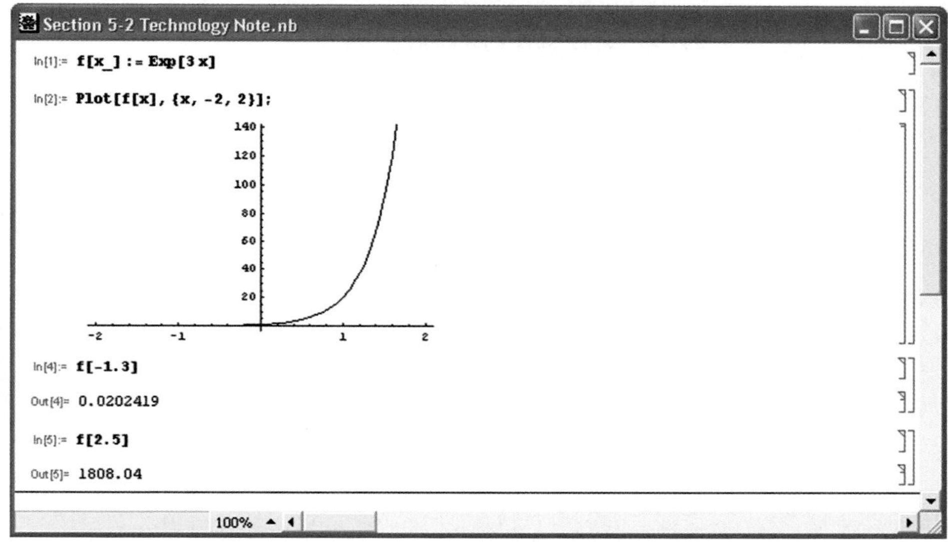

5.3 Logarithmic Functions and Their Graphs

TOPICS

1. Definition of logarithmic functions
2. Graphing logarithmic functions
3. Evaluating elementary logarithmic expressions
4. Common and natural logarithms

Topic 1: Definition of Logarithmic Functions

The need for logarithmic functions can be demonstrated with a few sample problems, some easily solved and some, at present, not.

Solvable	Not easily solvable yet
We can solve the equation $$2^x = 8$$ by writing 8 as 2^3 and equating exponents. This is an example of an elementary exponential equation such as we learned to solve in Section 1 of this chapter.	We cannot solve the equation $$2^x = 9$$ in the same way, even though this equation is only slightly different. All we can say at the moment is that x must be a bit larger than 3.
If we know A, P, n, and t, we can solve the compound interest equation $$A = P\left(1+\frac{r}{n}\right)^{nt}$$ for the annual interest rate r, as we did in Example 4 of the last section.	If we know A, P, and t, it is not so easy to determine the annual interest rate r from $$A = Pe^{rt},$$ the continuous compounding equation.

In both of the first two equations, the variable x appears in the exponent (making them exponential equations), but only in the first equation can x be extricated from the exponent with the tools we already have. In the second pair of equations, we are able to solve for the interest rate r in the first case but it is difficult in the continuous-compounding case, and for the same reason: the variable is inconveniently "stuck" in the exponent. What is needed in such situations is a way of undoing the exponentiation.

This might seem familiar. We have "undone" functions before, and we have done so (whether we realized it at the time or not) by finding inverses of functions. In this case, we need to find the inverse of the generic exponential function $f(x) = a^x$. Happily, we know at the outset that $f(x) = a^x$ has an inverse, because the graph of f passes the horizontal line test (regardless of whether $0 < a < 1$ or $a > 1$).

If we apply the algorithm for finding inverses of functions, our work looks something like the following:

$f(x) = a^x$	The first step is to rewrite the function as an equation, simply by replacing $f(x)$ with y. We then interchange x and y and proceed to solve for y.
$y = a^x$	
$x = a^y$	
$y = ?$	At this point, we are stuck again. What is y?

No concept or notation that we have encountered up to this point allows us to solve the equation $x = a^y$ for the variable y, and this is ultimately the reason for introducing the new class of functions called *logarithms*.

Logarithmic Functions

Let a be a fixed positive real number not equal to 1. The **logarithmic function with base *a*** is defined to be the inverse of the exponential function with base a, and is denoted $\log_a x$. In symbols, if

$$f(x) = a^x, \text{ then } f^{-1}(x) = \log_a x.$$

In equation form, the definition of logarithm means that the equations

$$x = a^y \text{ and } y = \log_a x$$

are equivalent. Note that a is the base in both equations: either the base of the exponential function or the base of the logarithmic function.

Since logarithmic functions are an entirely new class of functions, the first task is to study them and build an intuitive understanding of their behavior. For instance, we can determine that the domain of $\log_a x$ (for any allowable a) is the positive real numbers, because the positive real numbers constitute the range of a^x (remember that the domain of f^{-1} is the range of f). Similarly, the range of $\log_a x$ is the entire set of real numbers, because the real numbers make up the domain of a^x. But probably the best way to become familiar with logarithmic functions is to gain experience with their graphs.

Topic 2: Graphing Logarithmic Functions

Recall that the graphs of a function and its inverse are reflections of one another with respect to the line $y = x$. Since exponential functions come in two forms ($0 < a < 1$ and $a > 1$), logarithmic functions fall into two similar categories. In both of the graphs in Figure 1, the dashed green line is the graph of an exponential function representative of its class, and the solid blue line is the corresponding logarithmic function.

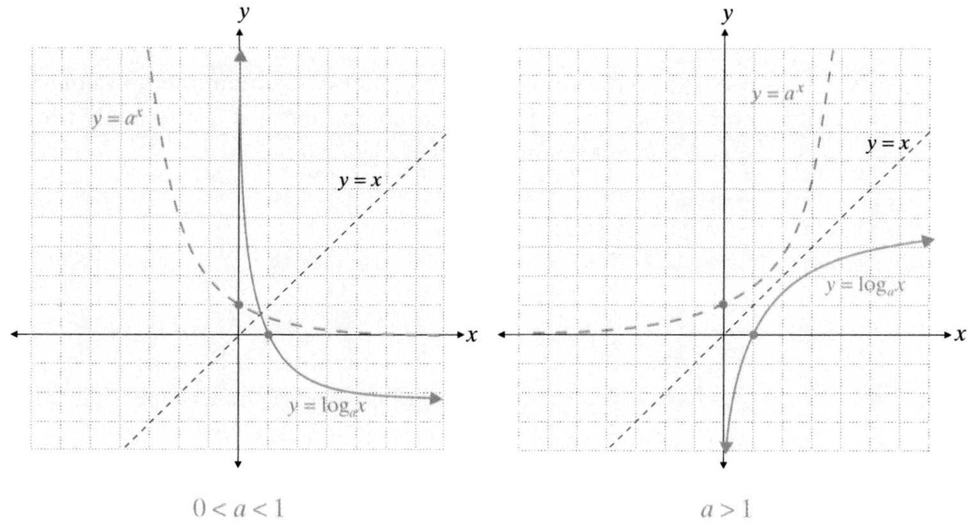

Figure 1: The Two Classes of Logarithmic Functions

Note that the domain of each of the logarithmic functions in Figure 1 is indeed $(0, \infty)$, and that the range in each case is $(-\infty, \infty)$. Also note that the y-axis is a vertical asymptote for both, and that neither has a horizontal asymptote.

Of course, we can shift, reflect, stretch and compress the graphs of logarithms as we can any other function. The techniques we learned in Section 3.5 are all we need to do the work in Example 1.

example 1

Sketch the graphs of the following functions.

a. $f(x) = \log_3(x + 2) + 1$ **b.** $g(x) = \log_2(-x - 1)$ **c.** $h(x) = \log_{1/2}(x) - 2$

Solutions:

a.

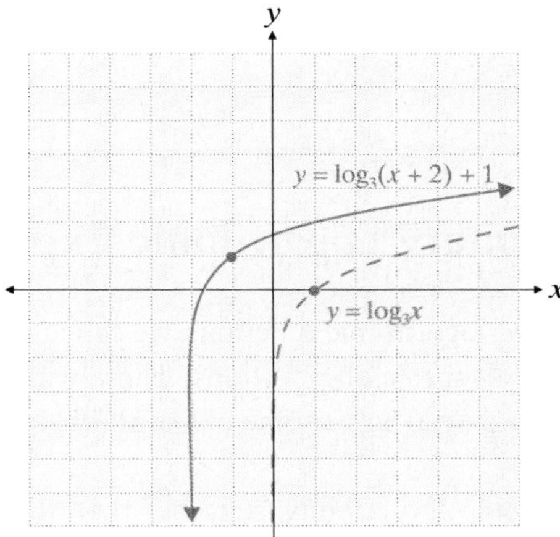

The dashed green line at left is the graph of the function $y = \log_3 x$, the shape of which is the basis for the graph of f. We know several points on the graph of $\log_3 x$ exactly; for instance, $(1, 0)$ and $(3, 1)$ are on the graph of $\log_3 x$ because $(0, 1)$ and $(1, 3)$ are on the graph of 3^x.

To graph f, we then shift everything 2 units to the left and 1 unit up.

b.

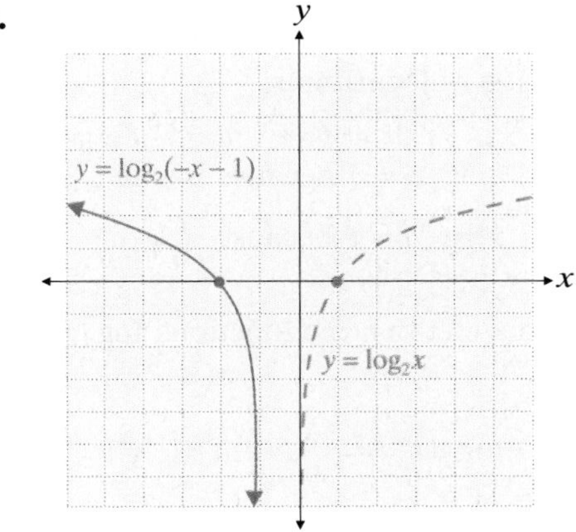

Again, we start with the basic shape of the graph, and then worry about the transformations. The basic shape of the graph of g is the same as the shape of $y = \log_2 x$, drawn as a dashed green line at left. As in part a., some points on $\log_2 x$ are easily determined, such as $(1, 0)$ and $(2, 1)$.

To obtain g from $\log_2 x$, the variable x is replaced with $x - 1$, which shifts the graph 1 unit to the right, and then x is replaced by $-x$, which reflects the graph with respect to the y-axis.

cont'd. on next page ...

c.

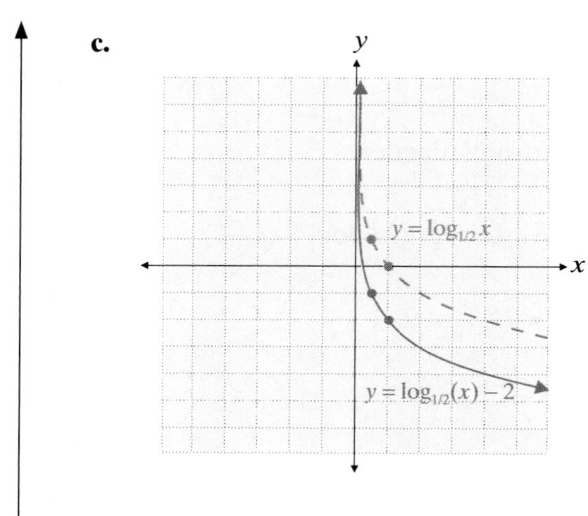

The basic logarithmic function $y = \log_{1/2} x$ is a decreasing function, as the base is between 0 and 1. Note that $(1, 0)$ and $\left(\frac{1}{2}, 1\right)$ are two points on the dashed green line, the graph of $\log_{1/2} x$. If we then shift the graph 2 units down, we obtain the graph of the function h.

Topic 3: Evaluating Elementary Logarithmic Expressions

Now that we have graphed logarithmic functions, we can augment our understanding of their behavior with a few algebraic observations. These will enable us to evaluate some logarithmic expressions and even solve some elementary logarithmic equations.

First, our work in Example 1 certainly suggests that the point $(1, 0)$ is always on the graph of $\log_a x$ for any allowable base a, and this is indeed the case. A similar observation is that $(a, 1)$ is always on the graph of $\log_a x$. These two facts are nothing more than restatements of two corresponding facts about exponential functions, and are a consequence of the definition of logarithms:

$$\log_a 1 = 0 \text{ because } a^0 = 1.$$
$$\log_a a = 1 \text{ because } a^1 = a.$$

More generally, we can use the fact that the functions $\log_a x$ and a^x are inverses of one another to write:

$$\log_a \left(a^x\right) = x \text{ and } a^{\log_a x} = x.$$

Example 2 makes use of these facts in evaluating some specific logarithms.

example 2

Evaluate the following logarithmic expressions.

a. $\log_5 25$ **b.** $\log_{1/2} 2$ **c.** $\log_\pi \sqrt{\pi}$

d. $\log_{17} 1$ **e.** $\log_{16} 4$ **f.** $\log_{10}\left(\frac{1}{100}\right)$

Solutions:

a. $\log_5 25 = \log_5 5^2 = 2$ Equivalent exponential equation: $25 = 5^2$

b. $\log_{1/2} 2 = \log_{1/2}\left(\frac{1}{2}\right)^{-1} = -1$ Equivalent exponential equation: $2 = \left(\frac{1}{2}\right)^{-1}$

c. $\log_\pi \sqrt{\pi} = \log_\pi \pi^{1/2} = \frac{1}{2}$ Equivalent exponential equation: $\sqrt{\pi} = \pi^{1/2}$

d. $\log_{17} 1 = 0$ Equivalent exponential equation: $1 = 17^0$

e. $\log_{16} 4 = \log_{16} 16^{1/2} = \frac{1}{2}$ Equivalent exponential equation: $4 = 16^{1/2}$

f. $\log_{10}\left(\frac{1}{100}\right) = \log_{10} 10^{-2} = -2$ Equivalent exponential equation: $\frac{1}{100} = 10^{-2}$

example 3

Use the elementary properties of logarithms to solve the following equations.

a. $\log_6(2x) = -1$ **b.** $3^{\log_{3x} 2} = 2$ **c.** $\log_2 8^x = 5$

Solutions:

a. $\log_6(2x) = -1$

$$2x = 6^{-1}$$

$$2x = \frac{1}{6}$$

$$x = \frac{1}{12}$$

One technique that we will find very fruitful for solving logarithmic equations is to convert the equation to the equivalent exponential form. The second equation at left means exactly the same thing as the first.

Note that $\log_6\left(\frac{2}{12}\right) = \log_6\left(6^{-1}\right) = -1$.

b. $3^{\log_{3x} 2} = 2$

$$\log_{3x} 2 = \log_3 2$$

$$3x = 3$$

$$x = 1$$

This time, we have converted from the exponential form to the logarithmic form and then equated the bases. We could also have noted from the start that the base of the logarithm must be the same as the base of the exponential function for the statement to be true.

c. $\log_2 8^x = 5$

$$8^x = 2^5$$

$$\left(2^3\right)^x = 2^5$$

$$2^{3x} = 2^5$$

$$3x = 5$$

$$x = \frac{5}{3}$$

The exponential form of the equation, shown as the second line, is of a form that we have solved before. We know that if we can write both sides with the same base, the exponents must be equal.

Note that $\log_2\left(8^{5/3}\right) = \log_2\left(8^{1/3}\right)^5 = \log_2\left(2^5\right) = 5.$

Topic 4: Common and Natural Logarithms

We have already mentioned the fact that the number e, called the natural base, plays a fundamental role in many important real-world situations, so it is perhaps not surprising that the logarithmic function with base e is worthy of special attention. For historical reasons, the logarithmic function with base 10 is also singled out for further mention.

Common and Natural Logarithms

- The function $\log_{10} x$ is called the **common logarithm**, and is usually written as $\log x$.
- The function $\log_e x$ is called the **natural logarithm**, and is usually written as $\ln x$.

Another way in which these particular logarithms are special is that most calculators, if they are capable of calculating logarithms at all, are only equipped to evaluate common and natural logarithms. Such calculators normally have a button labeled LOG for the common logarithm and a button labeled LN for the natural logarithm.

Properties of Natural Logarithms

$\ln x = y \Leftrightarrow e^y = x$

Properties	Reasons
1. $\ln 1 = 0$	Raise e to the power 0 to get 1.
2. $\ln e = 1$	Raise e to the power 1 to get e.
3. $\ln e^x = x$	Raise e to the power x to get e^x.
4. $e^{\ln x} = x$	$\ln x$ is the power to which e must be raised to get x.

example 4

Evaluate the following logarithmic expressions.

a. $\ln\sqrt[3]{e}$ **b.** $\log 1000$ **c.** $\ln(4.78)$ **d.** $\log(10.5)$

Solutions:

a. $\ln\sqrt[3]{e} = \ln e^{1/3} = \dfrac{1}{3}$

No calculator is necessary for this problem, just an application of an elementary property of logarithms.

b. $\log 1000 = \log 10^3 = 3$

Again, no calculator is required.

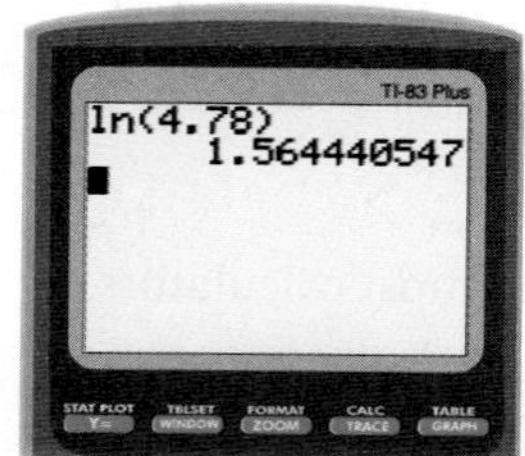

c. $\ln(4.78) \approx 1.564$

This time a calculator is needed and only an approximate answer can be given. Be sure to use the correct logarithm.

d. $\log(10.5) \approx 1.021$

Again, we must use a calculator, though we can say beforehand that the answer should be only slightly larger than 1, as $\log 10 = 1$ and 10.5 is only slightly larger than 10.

exercises

Sketch the graphs of the following functions. See Example 1.

1. $f(x) = \log_3(x-1)$ **2.** $g(x) = \log_5(x+2) - 1$ **3.** $r(x) = \log_{1/2}(x-3)$

4. $p(x) = 3 - \log_2(x+1)$ **5.** $q(x) = \log_3(2-x)$ **6.** $s(x) = \log_{1/3}(5-x)$

7. $h(x) = \log_7(x-3) + 3$ **8.** $m(x) = \log_{1/2}(1-x)$ **9.** $f(x) = \log_3(6-x)$

10. $p(x) = 4 - \log_{10}(x+3)$ **11.** $s(x) = -\log_{1/3}(-x)$ **12.** $g(x) = \log_5(2x) - 1$

Evaluate the following logarithmic expressions without the use of a calculator. See Example 2.

13. $\log_7 \sqrt{7}$

14. $\log_{1/2} 4$

15. $\log_9 \left(\frac{1}{81}\right)$

16. $\log_3 27$

17. $\log_{27} 3$

18. $\log_3 (\log_{27} 3)$

19. $\ln e^{2.89}$

20. $\log 0.0001$

21. $\log_a a^{5/3}$

22. $\ln\left(\frac{1}{e}\right)$

23. $\log\left(\log\left(10^{10}\right)\right)$

24. $\log_3 1$

25. $\ln \sqrt[5]{e}$

26. $\log_{1/16} 4$

27. $\log_8 4^{\log 1000}$

Use the elementary properties of logarithms to solve the following equations. See Example 3.

28. $\log_{16} x = \frac{3}{4}$

29. $\log_{16} x^{1/2} = \frac{3}{4}$

30. $\log_{16} x = -\frac{3}{4}$

31. $\log_5 5^{\log_3 x} = 2$

32. $\log_a a^{\log_b x} = 0$

33. $\log_3 9^{2x} = -2$

34. $\log_{1/3} 3^x = 2$

35. $\log_7 (3x) = -1$

36. $4^{\log_3 x} = 0$

37. $\log_9 (2x - 1) = 2$

38. $\log_x \left(\log_{1/2} \frac{1}{4}\right) = 1$

39. $6^{\log_x e^2} = e$

Solve the following logarithmic equations, using a calculator if necessary to evaluate the logarithms. See Example 4.

40. $\log(3x) = 2.1$

41. $\log x^2 = -2$

42. $\ln(x + 1) = 3$

43. $\ln 2x = -1$

44. $\ln e^x = 5.6$

45. $\ln\left(\ln x^2\right) = 0$

46. $\log 19 = 3x$

47. $\log e^x = 5.6$

48. $\log 300^{\log x} = 9$

49. $\log x^{10} = 10$

50. $\log\left(\log\left(x - 2\right)\right) = 1$

Match the graph of the appropriate equation to the logarithmic function.

51. $f(x) = \log_2 x - 1$ **52.** $f(x) = \log_2 (2 - x)$ **53.** $f(x) = \log_2 (-x)$

54. $f(x) = \log_2 (x - 3)$ **55.** $f(x) = 1 - \log_2 x$ **56.** $f(x) = -\log_2 x$

57. $f(x) = -\log_2 (-x)$ **58.** $f(x) = \log_2 x$ **59.** $f(x) = \log_2 x + 3$

a.

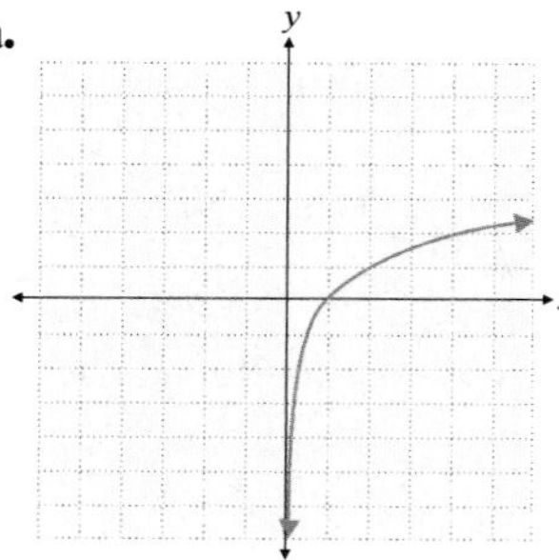

b.

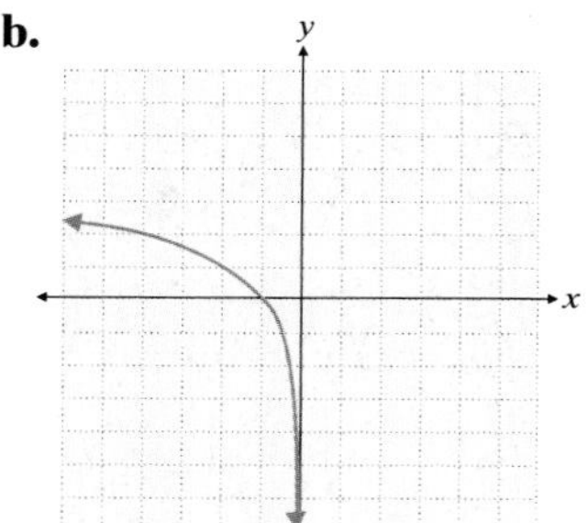

c.

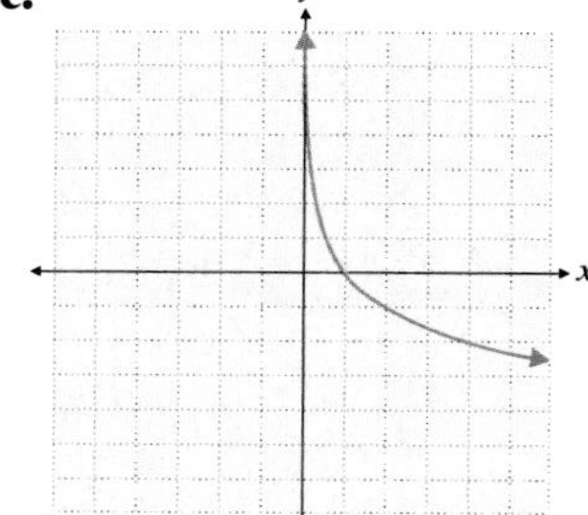

d.

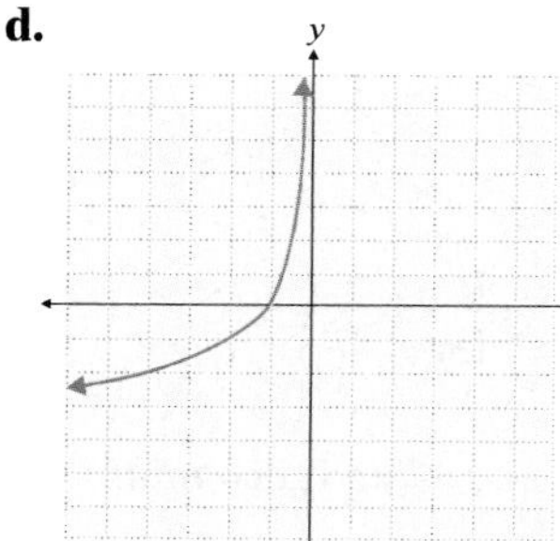

e.

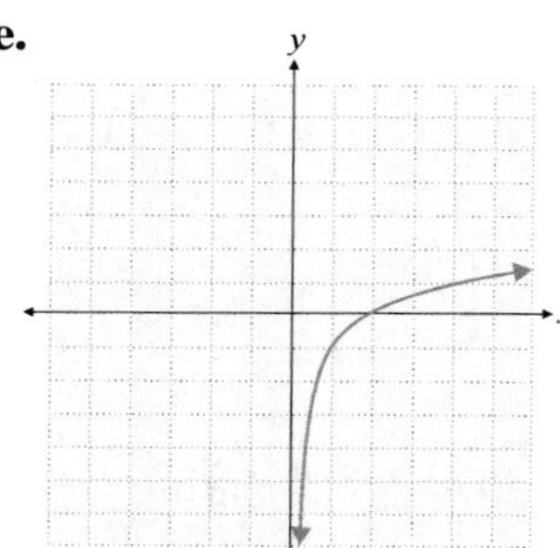

f.

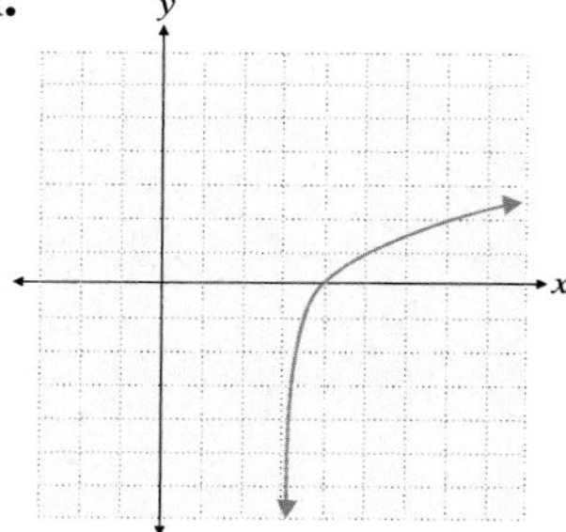

g.

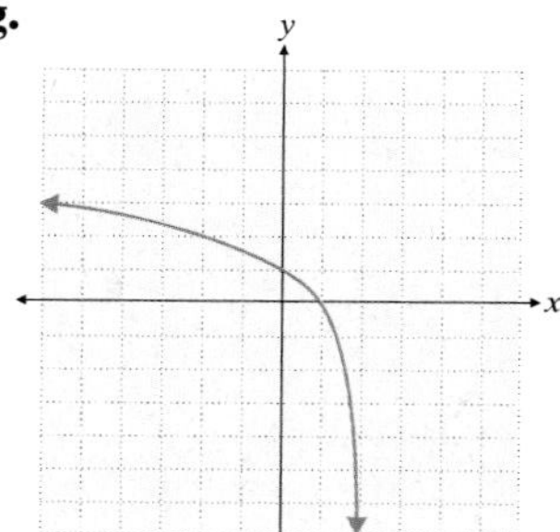

h.

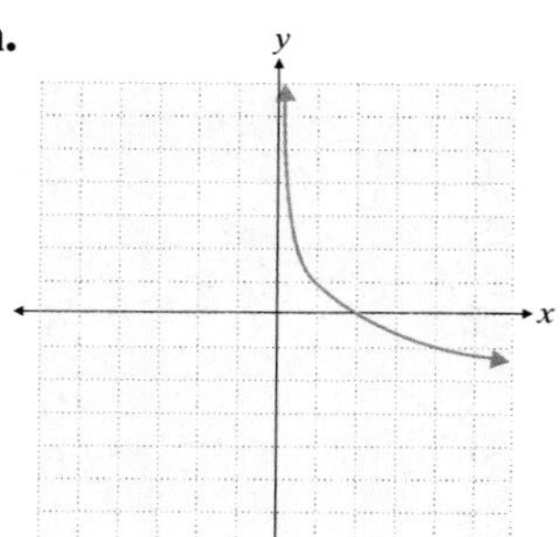

i.

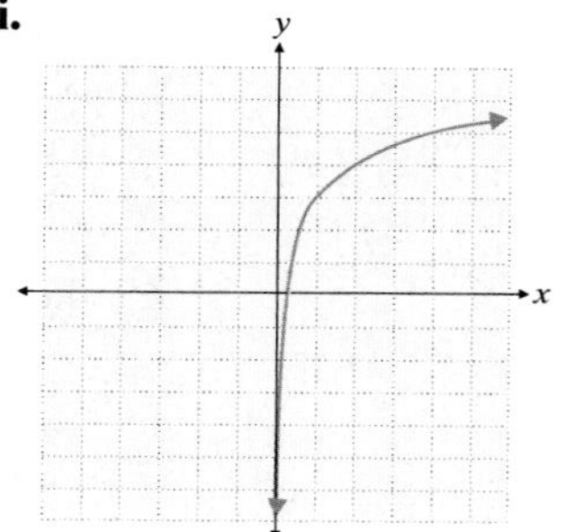

Write the following equations in logarithmic terms.

60. $625 = 5^4$

61. $216 = 6^3$

62. $x^3 = 27$

63. $b^2 = 3.2$

64. $4.2^3 = C$

65. $1.3^2 = V$

66. $4^x = 31$

67. $16^{2x} = 215$

68. $4x^{\sqrt{3}} = 13$

69. $e^x = \pi$

70. $2^{e^x} = 11$

71. $4^e = N$

Write the following logarithmic equations as exponential equations.

72. $\log_3 81 = 4$

73. $\log_2 \frac{1}{8} = -3$

74. $\log_b 4 = \frac{1}{2}$

75. $\log_y 9 = 2$

76. $\log_2 15 = b$

77. $\log_5 8 = d$

78. $\log_5 W = 12$

79. $\log_7 T = 6$

80. $\log_\pi 2x = 4$

81. $\log_{\sqrt{3}} 2\pi = x$

82. $\ln 2 = x$

83. $\ln 5x = 3$

technology exercises

As already mentioned, most calculators can only calculate natural logarithms and common logarithms directly. The computer algebra system *Mathematica*, in contrast, can evaluate logarithms of any allowable base. The command to evaluate $\log_2 15$, for example, is `Log[2., 15]` (the decimal point tells *Mathematica* to give us a decimal approximation of the answer). If only one number is supplied to the `Log` function, *Mathematica* assumes the base is e and calculates the natural logarithm of the given number. The illustration below shows the result of asking *Mathematica* to determine $\log_2 15$ and $\ln 9$.

Use a computer algebra system, if available, to evaluate the expressions that follow. (We will see in the next section how a calculator can be used to evaluate such expressions.)

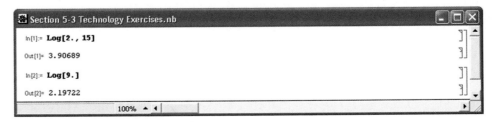

85. $\log_7 45$

86. $\log_{1/2} 9$

87. $\log_5 30$

88. $\log_{0.9} 2$

89. $\log_{1/10}\left(\frac{1}{9}\right)$

90. $\log_{4.5} 12$

91. $\log_{2.3} 10$

92. $\log_{19} 5$

5.4 Properties and Applications of Logarithms

TOPICS

1. Properties of logarithms
2. The change of base formula
3. Applications of logarithmic functions

Topic 1: Properties of Logarithms

In the last section, we introduced logarithmic functions and made note of some of their elementary properties. Our motivation was our inability, at that time, to solve certain equations. Let us reconsider the two sample problems that initiated our discussion of logarithms and see if we have made progress. We'll begin with the continuous-compounding interest problem.

example 1

Anne reads an ad in the paper for a new bank in town. The bank is advertising "Continuously-Compounded Savings Accounts!" in an attempt to attract customers, but fails to mention the annual interest rate. Curious, she goes to the bank and is told by an account agent that if she were to invest, say, \$10,000 in an account, her money would grow to \$10,202.01 in a year's time. But, strangely, the agent also refuses to divulge the yearly interest rate. What rate is the bank offering?

Solution:

We need to solve the equation $A = Pe^{rt}$ for r, given that $A = 10,202.01$, $P = 10,000$, and $t = 1$. We are now able to do so:

$$10,202.01 = 10,000e^{r(1)}$$

$$1.020201 = e^{r} \quad \text{Divide by 10,000.}$$

$$\ln(1.020201) = r \quad \text{Convert to logarithmic form. Recall } \ln(e^{r}) = r.$$

$$r = 0.0199 \approx 0.02 \quad \text{Use a calculator to determine } r.$$

cont'd. on next page ...

To solve the equation, we convert the exponential equation to its logarithmic form. Note that we use the natural logarithm since the base of the exponential function is e. We can then use a calculator to evaluate r, and discover that the annual interest rate is 2%.

example 2

Solve the equation $2^x = 9$.

Solution:

To solve this problem, we convert the equation to logarithmic form to obtain the solution $x = \log_2 9$. The only problem with this answer is that we still don't know anything about x in decimal form, other than that it is bound to be slightly more than 3.

As Example 2 shows, our ability to work with logarithms leaves something to be desired. In this section we will derive some more important properties of logarithms that will allow us to solve more complicated equations, as well as provide a decimal approximation to the solution of $2^x = 9$.

The following three properties of logarithmic functions are a consequence of corresponding properties of exponential functions, a not so surprising fact considering how logarithms are defined.

Properties of Logarithms

In the following, assume that a (the logarithmic base) is a fixed positive real number not equal to 1, that x and y represent positive real numbers, and that r represents any real number.

1. $\log_a(xy) = \log_a x + \log_a y$ ("the log of a product is the sum of the logs")
2. $\log_a\left(\frac{x}{y}\right) = \log_a x - \log_a y$ ("the log of a quotient is the difference of the logs")
3. $\log_a(x^r) = r\log_a x$ ("the log of something raised to a power is the power times the log")

We will illustrate the correspondence between these properties and the more familiar properties of exponents by proving the first one. The proofs of the second and third will be left as exercises.

Proof: We will start with the right-hand side of the statement, and convert the expressions to exponential form. To do this, let $m = \log_a x$ and $n = \log_a y$. The equivalent exponential form of these two equations is:

$$x = a^m \text{ and } y = a^n.$$

Since we are interested in the product xy, note that

$$xy = a^m a^n = a^{m+n}.$$

The statement $xy = a^{m+n}$ can be converted to logarithmic form, giving us

$$\log_a(xy) = m + n.$$

If we now refer back to the original definition of m and n, we have achieved our goal:

$$\log_a(xy) = \log_a x + \log_a y.$$

The properties of logarithms may appear strange at first, but with time they will come to seem as natural as the properties of exponents. In fact, they *are* the properties of exponents, simply restated in logarithmic form.

caution!

Errors in working with logarithms often arise from imperfect recall of the properties on the previous page. It is common to be misled, for example, into making mistakes of the sort shown on the next page. Note carefully the differences (and tempting similarities) between the incorrect and correct statements.

Incorrect Statements	Correct Statements
$\log_a(x+y) = \log_a x + \log_a y$	$\log_a(xy) = \log_a x + \log_a y$
$\log_a(xy) = (\log_a x)(\log_a y)$	$\log_a(xy) = \log_a x + \log_a y$
$\dfrac{\log_a x}{\log_a y} = \log_a x - \log_a y$	$\log_a\left(\dfrac{x}{y}\right) = \log_a x - \log_a y$
$\dfrac{\log_a x}{\log_a y} = \log_a\left(\dfrac{x}{y}\right)$	$\log_a\left(\dfrac{x}{y}\right) = \log_a x - \log_a y$

In some situations, we will find it useful to use properties of logarithms to decompose a complicated expression into a sum or difference of simpler expressions, while in other situations we will do just the reverse, combining a sum or a difference of logarithms into one logarithm. Examples 3 and 4 illustrate the process.

example 3

Use the properties of logarithms to expand the following expressions as much as possible (that is, decompose the expressions into sums or differences of the simplest possible terms).

a. $\log_4\left(64x^3\sqrt{y}\right)$ **b.** $\log_a \sqrt[3]{\dfrac{xy^2}{z^4}}$ **c.** $\log\left(\dfrac{2.7\times10^4}{x^{-2}}\right)$

Solutions:

a.
$$\begin{aligned}\log_4\left(64x^3\sqrt{y}\right) &= \log_4 64 + \log_4 x^3 + \log_4 \sqrt{y}\\ &= \log_4 4^3 + \log_4 x^3 + \log_4 y^{1/2}\\ &= 3 + 3\log_4 x + \frac{1}{2}\log_4 y\end{aligned}$$

We can initially write the expression as a sum of three log terms. We evaluate the first, and the second and third can be rewritten by bringing the exponent down in front.

b.
$$\begin{aligned}\log_a \sqrt[3]{\frac{xy^2}{z^4}} &= \log_a\left(\frac{xy^2}{z^4}\right)^{1/3}\\ &= \frac{1}{3}\log_a\left(\frac{xy^2}{z^4}\right)\\ &= \frac{1}{3}\left(\log_a x + \log_a y^2 - \log_a z^4\right)\\ &= \frac{1}{3}\left(\log_a x + 2\log_a y - 4\log_a z\right)\end{aligned}$$

Before we can decompose the fraction, we must apply the third property of logarithms and eliminate the radical. Once the exponent of $\frac{1}{3}$ has been brought to the front, the fraction can be written in simpler form as shown. Note that the base is immaterial: a can represent any legitimate logarithmic base.

c. $\log\left(\frac{2.7\times10^4}{x^{-2}}\right) = \log(2.7) + \log(10^4) - \log x^{-2}$

$= \log(2.7) + 4 + 2\log x$

$\approx 4.43 + 2\log x$

Recall that if the base is not explicitly written, it is assumed to be 10. This is a convenient base when dealing with scientific notation. Note that we can use a calculator to approximate $\log(2.7)$.

example 4

Use the properties of logarithms to condense the following expressions as much as possible (that is, rewrite the expressions as a sum or difference of as few logarithms as possible).

a. $2\log_3\left(\frac{x}{3}\right) - \log_3\left(\frac{1}{y}\right)$ **b.** $\ln x^2 - \frac{1}{2}\ln y + \ln 2$ **c.** $\log_b 5 + 2\log_b x^{-1}$

Solutions:

a. $2\log_3\left(\frac{x}{3}\right) - \log_3\left(\frac{1}{y}\right) = \log_3\left(\frac{x}{3}\right)^2 + \log_3\left(\frac{1}{y}\right)^{-1}$

$= \log_3\left(\frac{x^2}{9}\right) + \log_3 y$

$= \log_3\left(\frac{x^2 y}{9}\right)$

Note that before a sum or difference of log terms can be combined, they have to have the same coefficient. This is usually straightforward to arrange, as the coefficient can be moved up into the exponent, as shown. Note also that the order in which the properties are applied can vary, though all lead to the final answer.

b. $\ln x^2 - \frac{1}{2}\ln y + \ln 2 = \ln x^2 - \ln y^{1/2} + \ln 2$

$= \ln\left(\frac{x^2}{y^{1/2}}\right) + \ln 2$

$= \ln\left(\frac{2x^2}{y^{1/2}}\right)$ or $\ln\left(\frac{2x^2}{\sqrt{y}}\right)$

Again, we begin by rewriting each term so that its coefficient is 1 or –1; the sums or differences that result can then be combined.

The final answer can be written in several different ways, two of which are shown at left.

c. $\log_b 5 + 2\log_b x^{-1} = \log_b 5 + \log_b x^{-2}$

$= \log_b 5x^{-2}$ or $\log_b\left(\frac{5}{x^2}\right)$

The base is immaterial; the properties of logarithms hold for any legitimate base.

Topic 2: The Change of Base Formula

The properties we just derived can be used to provide an answer to a question about logarithms that has been left hanging to this point. A specific illustration of the question arose in Example 2: How do we determine the decimal form of a number like $\log_2 9$?

Paradoxically enough, to answer this question we will undo our work in Example 2. In the following, we assign a variable to the quantity $\log_2 9$, convert the resulting logarithmic equation into exponential form, take the natural logarithm of both sides, and then solve for the variable.

$$
\begin{aligned}
x &= \log_2 9 && \text{Let } x \text{ stand for the number } \log_2 9. \\
2^x &= 9 && \text{Convert to exponential form.} \\
\ln\left(2^x\right) &= \ln 9 && \text{Take the natural log of both sides.} \\
x \ln 2 &= \ln 9 && \text{Apply a property of logarithms.} \\
x &= \frac{\ln 9}{\ln 2} && \text{Solve for } x. \\
&\approx 3.17
\end{aligned}
$$

Of course, there is nothing special about the natural logarithm, at least as far as this problem is concerned. However, if a calculator is to be used to approximate the number $\log_2 9$, there are (for most calculators) only two good choices: the natural log and the common log. If we had done the work above with the common logarithm, the final answer would have been the same. That is,

$$\frac{\log 9}{\log 2} \approx 3.17.$$

More generally, a logarithm with base b can be converted to a logarithm with base a through the same reasoning, as summarized below.

Change of Base Formula

Let a and b both be positive real numbers, neither of them equal to 1, and let x be a positive real number. Then

$$\log_b x = \frac{\log_a x}{\log_a b}.$$

example 5

Evaluate the following logarithmic expressions, using the base of your choice.

a. $\log_7 15$ **b.** $\log_{1/2} 3$ **c.** $\log_\pi 5$

Solutions:

a. $$\log_7 15 = \frac{\ln 15}{\ln 7} \approx 1.392$$

The calculation at left uses the natural logarithm, but the common logarithm achieves the same result. Use a calculator to verify that $7^{1.392} \approx 15$.

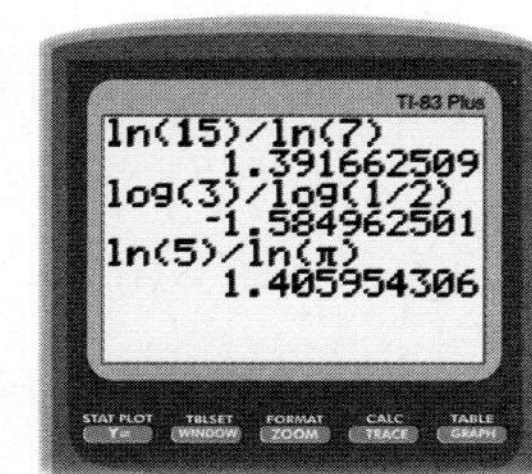

b. $$\log_{1/2} 3 = \frac{\log 3}{\log\left(\frac{1}{2}\right)} \approx -1.585$$

Since the base of the original logarithm is between 0 and 1, we expect the answer to be negative, as it is.

c. $$\log_\pi 5 = \frac{\ln 5}{\ln \pi} \approx 1.406$$

Although a base of π is not often encountered, the original expression is legitimate, and the change of base formula allows us to evaluate it.

Topic 3: Applications of Logarithmic Functions

Logarithms appear in many different contexts and have a wide variety of uses. This is due partly to the fact that logarithmic functions are the inverses of exponential functions, and partly to the logarithmic properties we have discussed. In fact the one individual who can be most credited for "inventing" logarithms, John Napier of Scotland (1550 - 1617), was inspired in his work by the very utility of what we now call logarithmic properties.

Computationally, logarithms are useful because they relocate exponents as coefficients, thus making them more easily worked with. Consider, for example, a very large number such as 3×10^{17} or a very small number such as 6×10^{-9}. The common logarithm (used because it has a base of 10) expresses these numbers on a more comfortable scale:

$$\log\left(3\times10^{17}\right) = \log 3 + \log\left(10^{17}\right) = 17 + \log 3 \approx 17.477$$
$$\log\left(6\times10^{-9}\right) = \log 6 + \log\left(10^{-9}\right) = -9 + \log 6 \approx -8.222.$$

Napier, working long before the advent of electronic calculating devices, devised logarithms in order to take advantage of this property. Although his particular motivation is not as meaningful today, it is still convenient in many instances to express quantities in terms of their logarithms. We will illustrate this with the next three examples.

example 6

In chemistry, the concentration of hydronium ions in a solution determines its acidity. Since the concentrations are small numbers, but can vary over many orders of magnitude, it is convenient to express acidity in terms of the pH scale, as follows.

The **pH** of a solution is defined to be $-\log\left[H_3O^+\right]$, where $\left[H_3O^+\right]$ is the concentration of hydronium ions in moles/liter. Solutions with a pH less than 7 are said to be *acidic*, while those with a pH greater than 7 are *basic*.

If a sample of orange juice is determined to have a $\left[H_3O^+\right]$ concentration of 1.58×10^{-4} moles/liter, what is its pH?

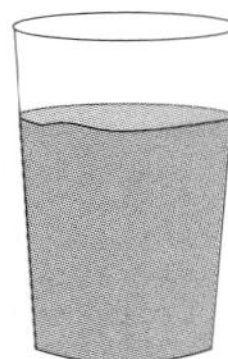

1.58×10^{-4} moles / liter

Solution:

Applying the above formula (and using a calculator), the pH is equal to

$$\text{pH} = -\log\left(1.58 \times 10^{-4}\right) = -(-3.80) = 3.8.$$

After doing one calculation, the reason for the minus sign in the formula is more apparent. By multiplying the log of the concentration by –1, the pH of a solution is necessarily positive, and convenient for comparative purposes.

example 7

The energy released during earthquakes can vary greatly, but logarithms provide a convenient way to analyze and compare the intensity of earthquakes.

Earthquake intensity is measured on the *Richter scale* (named for the American seismologist Charles Richter, 1900 - 1985). In the formula that follows, I_0 is the intensity of a just-discernable earthquake, I is the intensity of an earthquake being analyzed, and R is its ranking on the Richter scale:

$$R = \log\left(\frac{I}{I_0}\right).$$

By this measure, earthquakes range from a classification of Small $(R < 4.5)$, to Moderate $(4.5 \leq R < 5.5)$, to Large $(5.5 \leq R < 6.5)$, to Major $(6.5 \leq R < 7.5)$, and finally to Greatest $(7.5 \leq R)$.

Note that the base 10 logarithm means that every increase of 1 unit on the Richter scale corresponds to an increase by a factor of 10 in the intensity, and a barely discernable earthquake has a rank of 0 $(\log(1) = 0)$.

The January 2001 earthquake in the state of Gujarat in India was $80{,}000{,}000$ times as intense as a 0-level earthquake. What was the Richter ranking of this devastating event?

INDIA

Solution:

If we let I denote the intensity of the Gujarat earthquake, then $I = 80{,}000{,}000 I_0$, so

$$\begin{aligned} R &= \log\left(\frac{80{,}000{,}000 I_0}{I_0}\right) \\ &= \log\left(8 \times 10^7\right) \\ &= \log(8) + 7 \\ &\approx 7.9. \end{aligned}$$

The Gujarat earthquake thus fell in the category of Greatest on the Richter scale.

example 8

Sound intensity is another quantity that varies greatly, and the measure of how the human ear perceives intensity, in units called *decibels*, is very similar to the measure of earthquake intensity.

In the **decibel** formula that follows, I_0 is the intensity of a just-discernable sound, I is the intensity of the sound being analyzed, and D is its decibel level:

$$D = 10\log\left(\frac{I}{I_0}\right).$$

Decibel levels range from 0 for a barely discernable sound, to 40 for the level of normal conversation, to 80 for heavy traffic, to 120 for a loud rock concert, and finally (as far as humans are concerned) to around 160, at which point the eardrum is likely to rupture.

Given that $I_0 = 10^{-12}$ watts/meter^2, what is the decibel level of jet airliner's engines at a distance of 45 meters, for which the sound intensity is 50 watts/meter^2?

Solution:

$$\begin{aligned} D &= 10\log\left(\frac{50}{10^{-12}}\right) \\ &= 10\log\left(5\times 10^{13}\right) \\ &= 10\left(13+\log 5\right) \\ &\approx 137\text{ dB} \end{aligned}$$

In other words, the sound level would probably not be literally ear-splitting, but it would be very painful.

exercises

Use properties of logarithms to expand the following expressions as much as possible. Simplify any numerical expressions that can be evaluated without a calculator. See Example 3.

1. $\log_5\left(125x^3\right)$

2. $\ln\left(\frac{x^2y}{3}\right)$

3. $\ln\left(\frac{e^2p}{q^3}\right)$

4. $\log(100x)$

5. $\log_9 9xy^{-3}$

6. $\log_6 \sqrt[3]{\frac{p^2}{q}}$

7. $\ln\left(\frac{\sqrt{x^3}pq^5}{e^7}\right)$

8. $\log_a \sqrt[5]{\frac{a^4b}{c^2}}$

9. $\log\left(\log\left(100x^3\right)\right)$

10. $\log_3(9x+27y)$

11. $\log\left(\frac{10}{\sqrt{x+y}}\right)$

12. $\ln\left(\ln\left(e^{ex}\right)\right)$

13. $\log_2\left(\frac{y^2+z}{16x^4}\right)$

14. $\log\left(\log\left(100{,}000^{2x}\right)\right)$

15. $\log_b \sqrt{\frac{x^4y}{z^2}}$

16. $\ln\left(7x^2-42x+63\right)$

17. $\log_b ab^2c^b$

18. $\ln\left(\ln\left(e^{e^x}\right)\right)$

Use properties of logarithms to condense the following expressions as much as possible, writing each answer as a single term with a coefficient of 1. See Example 4.

19. $\log_5 x - 2\log_5 y$

20. $\log_5\left(x^2-25\right)-\log_5(x-5)$

21. $\ln\left(x^2y\right)-\ln y-\ln x$

22. $\frac{1}{3}\log_2 x+\log_2(x+3)$

23. $\frac{1}{5}\left(\log_7\left(x^2\right)-\log_7(pq)\right)$

24. $\ln 3+\ln p-2\ln q$

25. $2\left(\log_5\sqrt{x}-\log_5 y\right)$

26. $\log(x-10)-\log x$

27. $2\log a^2b-\log\frac{1}{b}+\log\frac{1}{a}$

28. $3\left(\ln\sqrt[3]{e^2}-\ln xy\right)$

29. $\log x-\log y$

30. $\log_2(4x)-\log_2 x$

31. $\log_5 20-\log_5 5$

32. $\log 30-\log 2-\log 5$

33. $\ln 15+\ln 3$

34. $\ln 8-\ln 4+\ln 3$

35. $0.5\log_3 16-\log_3 4$

36. $3\log_7 2-2\log_7 4$

37. $0.25\ln 81+\ln 4$

38. $2(\log 4-\log 1+\log 2)$

39. $\log 11+0.5\log 9-\log 3$

40. $3\log_4(x^2)+\log_4(x^6)$

41. $\log_8(2x^2-2y)-0.25\log_8 16$

42. $\log_{3x} x^2+\log_{3x} 18-\log_{3x} 6$

*Use the properties of logarithms to write each of the following as a single term that does **not** contain a logarithm.*

43. $5^{2\log_5 x}$

44. $10^{\log y^2-3\log x}$

45. $e^{2-\ln x+\ln p}$

46. $e^{5(\ln\sqrt[5]{3}+\ln x)}$

47. $10^{\log x^3-4\log y}$

48. $a^{\log_a b+4\log_a\sqrt{a}}$

49. $10^{2\log x}$

50. $10^{4\log x-2\log x}$

51. $\log_4 16\cdot\log_x x^2$

52. $e^{\ln x+2+\ln x^2}$

53. $4^{\log_4(3x)+0.5\log_4(16x^2)}$

54. $4^{2\log_2 6-\log_2 9}$

Evaluate the following logarithmic expressions to two decimals. See Example 5.

55. $\log_4 17$

56. $2\log_{1/3} 5$

57. $\log_9 8$

58. $\log_2 0.01$

59. $\log_{12} 10.5$

60. $\log(\ln 2)$

61. $\log_6 3^4$

62. $\log_7 14.3$

63. $\log_{1/2}\pi^{-2}$

64. $\log_{1/5} 626$

65. $\ln(\log 123)$

66. $\log_{17} 0.041$

67. $\log 16$

68. $\log_3 9$

69. $\log_5 20$

70. $\log_8 26$

71. $\log_4 0.25$

72. $\log_{1.8} 9$

73. $\log_{2.5} 34$

74. $\log_{0.5} 10$

75. $\log_4 2.9$

76. $\log_{0.4} 14$

77. $\log_{0.2} 17$

78. $\log_{0.16} 2.8$

Without using a calculator, evaluate the following expressions.

79. $\log_4 16$

80. $\log_5 25^3$

81. $\ln e^4+\ln e^3$

82. $\log_4\frac{1}{64}$

83. $\ln e^{1.5}-\log_4 2$

84. $\log_2 8^{(2\log_2 4-\log_2 4)}$

Use your knowledge of logarithms to answer the following questions. See Examples 6, 7, and 8.

85. A certain brand of tomato juice has a $\left[H_3O^+\right]$ concentration of 3.16×10^{-6} moles/liter. What is the pH of this brand?

86. One type of detergent, when added to neutral water with a pH of 7, results in a solution with a $\left[H_3O^+\right]$ concentration that is 5.62×10^{-4} times weaker than that of the water. What is the pH of the solution?

87. What is the concentration of $\left[H_3O^+\right]$ in lemon juice with a pH of 3.2?

88. The 1994 Northridge, California earthquake measured 6.7 on the Richter scale. What was the intensity, relative to a 0-level earthquake, of this event?

89. How much stronger was the 2001 Gujarat earthquake (7.9 on the Richter scale) than the 1994 Northridge earthquake described in Exercise 88?

90. A construction worker operating a jackhammer would experience noise with an intensity of 20 watts/meter^2 if it weren't for ear-protection. Given that $I_0 = 10^{-12}$ watts/meter^2, what is the decibel level for such noise?

91. A microphone picks up the sound of a thunderclap, and measures its decibel level as 105. Given that $I_0 = 10^{-12}$ watts/meter^2, with what sound intensity did the thunderclap reach the microphone?

92. Matt, a lifeguard, has to make sure that the pH of the swimming pool stays between 7.2 and 7.6. If the pH is out of this range, he has to add chemicals which alter the pH level of the pool. If Matt measures the $\left[H_3O^+\right]$ concentration in the swimming pool to be 2.40×10^{-8} moles/liter, what is the pH? Does he need to change the pH by adding chemicals to the water?

93. The intensity of a cat's soft purring is measured to be 2.19×10^{-11}. Given that $I_0 = 10^{-12}$ watts/meter^2, what is the decibel level of this noise?

5.5 Exponential and Logarithmic Equations

TOPICS

1. Converting between exponential and logarithmic forms
2. Further applications of exponential and logarithmic equations
3. Interlude: analysis of a stock market investment

Topic 1: Converting Between Exponential and Logarithmic Forms

At this point, we have all the tools we need to solve the most common sorts of exponential and logarithmic equations. All that is left is to develop our skill in using the tools.

We have already solved many exponential and logarithmic equations, using the more elementary facts about exponential and logarithmic functions to quickly obtain solutions. But many equations require a bit more work to solve. While there is no algorithm to follow in dealing with more complicated equations, you may find it useful to keep the following general approach in mind: If a given equation doesn't yield a solution easily, try converting it from exponential form to logarithmic form or vice versa, as the occasion warrants.

Of course, all of the well-known properties of exponents and their logarithmic counterparts are of great use too. For reference, the logarithmic properties that we have noted over the last two sections are reproduced here, in brief form.

Summary of Logarithmic Properties

1. The equations $x = a^y$ and $y = \log_a x$ are equivalent, and are, respectively, the exponential form and the logarithmic form of the same statement.
2. The inverse of the function $f(x) = a^x$ is $f^{-1}(x) = \log_a x$, and vice versa.
3. A consequence of the last point is that $\log_a(a^x) = x$ and $a^{\log_a x} = x$. In particular, $\log_a 1 = 0$ and $\log_a a = 1$.
4. $\log_a(xy) = \log_a x + \log_a y$ ("the log of a product is the sum of the logs")
5. $\log_a\left(\frac{x}{y}\right) = \log_a x - \log_a y$ ("the log of a quotient is the difference of the logs")
6. $\log_a(x^r) = r\log_a x$ ("the log of something raised to a power is the power times the log")

The next several examples illustrate typical uses of the properties, and the desirability of converting between the exponential and the logarithmic forms of an equation.

Steps for Solving Exponential Equations

1. Isolate the exponential expression on one side of the equation.
2. Take the logarithm of each side.
3. Bring the exponent down in front.
4. Solve the resulting equation.

example 1

Solve the equation $3^{2-5x} = 11$. Express the answer exactly and as a decimal approximation.

Solution:

This problem is a good illustration of an exponential equation that is not easily solved in its original form (because the two sides of the equation do not have the same base), but is less complex as a logarithmic equation. The work on the next page begins with the conversion of the equation into logarithmic form and then proceeds with the solution of the resulting linear equation.

$$
\begin{aligned}
3^{2-5x} &= 11 \\
\ln\left(3^{2-5x}\right) &= \ln 11 && \text{Take the log of both sides (any base).} \\
(2-5x)\ln 3 &= \ln 11 && \text{Bring the exponent down in front.} \\
2-5x &= \frac{\ln 11}{\ln 3} && \text{Solve the linear equation.} \\
-5x &= \frac{\ln 11}{\ln 3} - 2 \\
x &= -\frac{\ln 11}{5\ln 3} + \frac{2}{5} && \text{An exact form of the answer.}
\end{aligned}
$$

The exact form of the answer is not unique. For one thing, a logarithm of any base may be used in the second line above, as any logarithm allows the variable to be brought down from the exponent. And if elegance is the only goal, using a logarithm with base 3 is undeniably best, as shown:

$$
\begin{aligned}
3^{2-5x} &= 11 \\
2-5x &= \log_3 11 \\
-5x &= \log_3 11 - 2 \\
x &= \frac{2-\log_3 11}{5}
\end{aligned}
$$

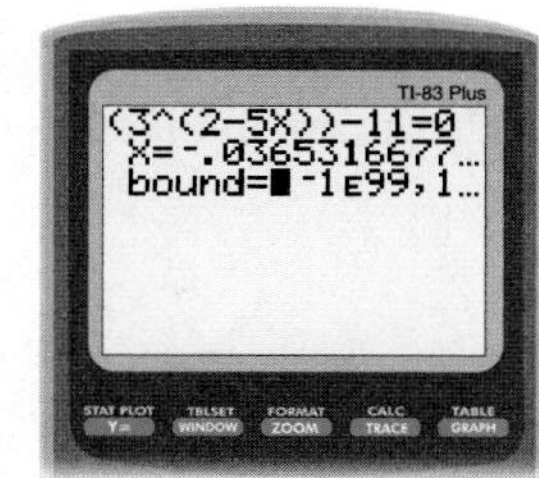

But for most purposes, either the natural or common logarithm is the most efficient choice, as the answer can then be approximated with a calculator:

$$x = -\frac{\ln 11}{5\ln 3} + \frac{2}{5} \approx -0.037.$$

It can be verified with a calculator that this is indeed the solution.

example 2

Solve the equation $5^{3x-1} = 2^{x+3}$. Express the answer exactly and as a decimal approximation.

Solution:

As in the first example, taking a logarithm of both sides is the key. For variety, we will use the common logarithm this time.

cont'd. on next page ...

$$\log\left(5^{3x-1}\right) = \log\left(2^{x+3}\right)$$ Take the log of both sides.

$$(3x-1)\log 5 = (x+3)\log 2$$ Use a property of logarithms.

$$3x\log 5 - \log 5 = x\log 2 + 3\log 2$$ Multiply out.

$$3x\log 5 - x\log 2 = 3\log 2 + \log 5$$ Collect terms with x on one side.

$$x(3\log 5 - \log 2) = \log 8 + \log 5$$ Factor out x.

$$x = \frac{\log 40}{\log 125 - \log 2}$$ Proceed to simplify the answer.

$$x = \frac{\log 40}{\log\left(\frac{125}{2}\right)} \approx 0.892$$

Again, the exact form of the answer could appear in many different ways, depending on the base of the logarithm chosen and the logarithmic properties used in simplifying the answer.

Steps for Solving Logarithmic Equations

1. Isolate the logarithmic terms on one side of the equation, this may involve combining terms first.
2. Write both sides in exponential form.
3. Solve the resulting equation.

example 3

Solve the equation $\log_5 x = \log_5(2x+3) - \log_5(2x-3)$.

Solution:

This is an example of a logarithmic equation that is not easily solved in its logarithmic form. But once a few properties of logarithms have been exploited, the equation can be rewritten in a very familiar form.

$$\log_5 x - \log_5(2x+3) + \log_5(2x-3) = 0$$
$$\log_5\left(\frac{x(2x-3)}{2x+3}\right) = 0$$
$$\frac{x(2x-3)}{2x+3} = 5^0$$
$$\frac{x(2x-3)}{2x+3} = 1$$
$$x(2x-3) = 2x+3$$
$$2x^2 - 3x - 2x - 3 = 0$$
$$2x^2 - 5x - 3 = 0$$

The original logarithmic equation has been turned into a quadratic equation, which we now proceed to solve.

$$2x^2 - 5x - 3 = 0$$
$$(2x+1)(x-3) = 0$$
$$x = -\frac{1}{2} \text{ or } x = 3$$

But one last and crucial step remains. While the two solutions above unquestionably solve the quadratic equation, remember that the goal is to solve the initial logarithmic equation. As we have seen with some other types of equations, the process of solving logarithmic equations can introduce extraneous solutions. If we check our two potential solutions in the original equation, we quickly discover that only one of them is valid.

$$\log_5\left(-\frac{1}{2}\right) \stackrel{?}{=} \log_5\left(2\left(-\frac{1}{2}\right)+3\right) - \log_5\left(2\left(-\frac{1}{2}\right)-3\right)$$

$-\frac{1}{2}$ is not a solution, as $\log_5\left(-\frac{1}{2}\right)$ is not defined.

$$\log_5(3) \stackrel{?}{=} \log_5(2(3)+3) - \log_5(2(3)-3)$$
$$\log_5 3 \stackrel{?}{=} \log_5 9 - \log_5 3$$
$$\log_5 3 \stackrel{?}{=} \log_5\left(\frac{9}{3}\right)$$

$\log_5 3 = \log_5 3$ The solution set of the equation is thus $\{3\}$.

The last example illustrates again the importance of verifying that potential solutions actually solve the original problem. In the case of logarithmic equations, always keep in mind that logarithms of negative numbers are not defined.

example 4

Solve the equation $\log_7(3x-2)=2$.

Solution:

This equation is more quickly solved than the previous one. Once the equation is rewritten in its exponential form, the solution is apparent. We can also verify that the resulting solution actually solves the original equation.

$$\begin{aligned}\log_7(3x-2)&=2\\3x-2&=7^2\\3x&=51\\x&=17\end{aligned}$$

Topic 2: Further Applications of Exponential and Logarithmic Equations

We will conclude our discussion of exponential and logarithmic equations by revisiting some important applications.

example 5

Rita is saving up money for a down payment on a new car. She currently has \$5500, but knows she can get a loan at a lower interest rate if she can put down \$6000. If she invests her \$5500 in a money market account that earns 4.8% annually, compounded monthly, how long will it take her to accumulate the \$6000?

Solution:

We need to solve the compound interest formula, which in this case takes the form

$$6000=5500\left(1+\frac{0.048}{12}\right)^{12t}.$$

Since we are looking for a numerical value for t, the number of years Rita needs to invest her money, we should apply either the natural log or the common log to both sides of the equation, as follows.

$$\frac{6000}{5500}=\left(1+\frac{0.048}{12}\right)^{12t}$$

$$\ln\left(\frac{6000}{5500}\right)=\ln\left(1+\frac{0.048}{12}\right)^{12t}$$

$$\ln\left(\frac{6000}{5500}\right)=(12t)\ln(1.004)$$

$$\frac{\ln(6000/5500)}{12\ln(1.004)}=t$$

$$t\approx 1.82 \text{ years}$$

In other words, it will take a bit less than a year and 10 months for the \$5500 to grow to \$6000 (meaning, probably, that Rita will want to speed up the process by investing additional principal as she is able to).

example 6

We have already discussed how the radioactive decay of carbon-14 can be used to arrive at age estimates of carbon-based fossils, and we constructed an exponential function describing the rate of decay. A much more common form of the function is

$$A(t)=A_0e^{-0.000121t},$$

where $A(t)$ is the mass of carbon-14 remaining after t years, and A_0 is the mass of carbon-14 initially. Use this formula to determine:

a. The half-life of carbon-14 (that is, the number of years required for half of a given mass of carbon-14 to decay); and

b. The number of years necessary for 2.3 grams of carbon-14 to decay to 1.5 grams.

cont'd. on next page ...

79. Wayne has \$12,500 in a high interest savings account at 3.66% annual interest compounded monthly. Assuming he makes no deposits or withdrawals, how long will it take for his investment to grow to \$15,000?

80. Ben and Casey both open money market accounts with 4.9% annual interest compounded continuously. Ben opens his account with \$8700 while Casey opens her account with \$3100.
a. How long will it take Ben's account to reach \$10,000?
b. How long will it take Casey's account to reach \$10,000?
c. How much money will be in Ben's account after the time found in part b?

81. Cesium-137 has a half-life of approximately 30 years. How long would it take for 160 grams of cesium-137 to decay to 159 grams?

82. A chemist, running tests on an unknown sample from an illegal waste dump, isolates 50 grams of what she suspects is a radioactive element. In order to help identify the element, she would like to know its half-life. She determines that after 40 days only 44 grams of the original element remains. What is the half-life of this mystery element?

technology exercises

The Interlude of this section concludes with a polynomial equation of degree $n+1$, where n is the number of months during which an investor makes a fixed periodic investment. In practice, n is typically large enough to make solving the polynomial equation a nontrivial task, and a computer algebra system such as *Mathematica* is very useful.

Suppose, for instance, that an investor makes a periodic investment of \$100.00 per month and has an accumulation of \$1300.00 after one year. Since only \$1200.00 was invested over the 12-month period, the investor clearly made money. But at what rate? As described in the Interlude, it would be nice to determine the equivalent annual rate of interest, compounded monthly, that would have generated the same return. The *Mathematica* commands below illustrate how this can be done. First the constants A and P, representing the total accumulation and the monthly investment respectively, are defined. Next, the command **NSolve** is used to solve the 13th degree polynomial equation. **NSolve** is used in the same way that **Solve** is used, but returns only numerical approximations to the solutions of the given equation and is a better choice in this case because of the high degree of the equation.

We know, from our study of polynomial functions in Chapter 4, that a 13th degree polynomial equation can have, potentially, 13 distinct solutions. The particular equation in this example does indeed have 13 different solutions, only three of which are real numbers. *Mathematica* returns a list of all 13 solutions, with the single negative solution listed first, the ten complex solutions next, and the two positive solutions listed last. As expected, $x=1$ is one solution, and the second positive real solution is slightly larger than 1. When we use this second positive solution to solve for r, we obtain an annual monthly-compounded interest rate of 14.7%.

```
Section 5-5 Technology Exercises.nb

In[1]:= A = 1300
Out[1]= 1300

In[2]:= P = 100
Out[2]= 100

In[3]:= NSolve[x^13 - (A / P + 1) * x + A / P == 0, x]
Out[3]= {{x → -1.30311}, {x → -1.13687 - 0.630111 ⅈ}, {x → -1.13687 + 0.630111 ⅈ},
        {x → -0.683116 - 1.09374 ⅈ}, {x → -0.683116 + 1.09374 ⅈ}, {x → -0.064897 - 1.26924 ⅈ},
        {x → -0.064897 + 1.26924 ⅈ}, {x → 0.54846 - 1.11259 ⅈ}, {x → 0.54846 + 1.11259 ⅈ},
        {x → 0.981856 - 0.668899 ⅈ}, {x → 0.981856 + 0.668899 ⅈ}, {x → 1.}, {x → 1.01225}}

In[4]:= Solve[1 + r / 12 == 1.01225, r]
Out[4]= {{r → 0.147}}

100%
```

What if the accumulation at the end of one year had been \$1100.00 instead of \$1300.00? This means the investment lost money, so the equivalent annual interest rate should be negative. The *Mathematica* output below confirms this; note that again there are only two positive real solutions to the equation, but this time we use the one slightly less than 1 to determine that $r = -16.2\%$.

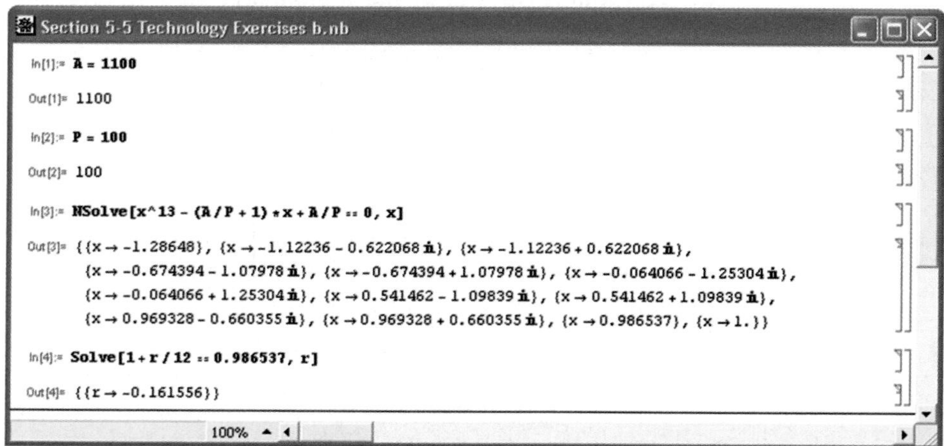

Use a computer algebra system, if available, to determine the equivalent monthly-compounded interest rate in the following scenarios. In each problem, P represents the principal invested each month and A is the accumulated value at the end of n months.

83. $n = 24$, $P = \$50.00$, and $A = \$1275.00$

84. $n = 120$, $P = \$100.00$, and $A = \$15,000.00$

85. $n = 12$, $P = \$50.00$, and $A = \$590.00$

86. $n = 24$, $P = \$75.00$, and $A = \$2000.00$

chapter five project

Exponential Functions

Computer viruses have cost US companies billions of dollars in damages and lost revenues over the last few years. One factor that makes computer viruses so devastating is the rate at which they spread. A virus can potentially spread across the world in a matter of hours depending on its characteristics, and whom it attacks.

Consider the growth of the following virus. A new virus has been created and is distributed to 100 computers in a company via a corporate E-mail. From these workstations the virus continues to spread. Let $t = 0$ be the time of the first 100 infections, and at $t = 17$ minutes the population of infected computers grows to 200. Assume the anti-virus companies are not able to identify the virus or slow its progress for 24 hours, allowing the virus to grow exponentially.

1. What will the population of the infected computers be after 1 hour?
2. What will the population be after 1 hour 30 minutes?
3. What will the population be after a full 24 hours?

Suppose another virus is developed and released on the same 100 computers. This virus grows according to $P(t) = (100)2^{\frac{t}{2}}$, where t represents the number of hours from the time of introduction.

4. What is the doubling-time for this virus?
5. How long will it take for the virus to infect 2000 computers, according to this model?

CHAPTER REVIEW

5.1 Exponential Functions and Their Graphs

topics	pages	test exercises
Definition and classification of exponential functions • The definition of *exponential functions* • The two classes of exponential functions, and their behavior	p. 377 – 379	1 – 5
Graphing exponential functions • Graphing transformations of exponential functions	p. 380 – 381	1 – 2
Solving elementary exponential equations • The definition of an exponential equation • Using the one-to-one nature of exponential functions to solve simple exponential equations	p. 381 – 383	3 – 5

5.2 Applications of Exponential Functions

topics	pages	test exercises
Models of population growth • Modeling population growth with exponential functions • Using population information to construct a corresponding exponential function	p. 387 – 388	8
Radioactive decay • Modeling radioactive decay with exponential functions • The use of exponential functions in radio-carbon dating • Using half-life measurements and other data to construct exponential functions that model radioactive decay	p. 389 – 390	7

topics (continued)	pages	test exercises
Compound interest and the number e • The meaning of *compounding* in investment contexts • The use of the basic *compound interest formula* • Extending the compound interest formula to the *continuous compounding* case • The meaning of the *natural base* e, and its approximate value	p. 391 – 395	6

5.3 Logarithmic Functions and Their Graphs

topics	pages	test exercises
Definition of logarithmic functions • Understanding the need for *logarithmic functions* • The connection between a^x and $\log_a x$	p. 402 – 404	9 – 18
Graphing logarithmic functions • The two classes of logarithmic functions, and their behavior	p. 404 – 406	9 – 10
Evaluating elementary logarithmic expressions • Evaluating simple logarithmic expressions manually	p. 406 – 408	11 – 16
Common and natural logarithms • The definition of the *common* and *natural* logarithms • Using a calculator to evaluate common and natural logarithms	p. 409 – 410	17 – 18

5.4 Properties and Applications of Logarithms

topics	pages	test exercises
Properties of logarithms • Basic properties of logarithmic functions, and how they derive from properties of exponents • Using the properties of logarithms to expand a single complicated logarithmic expression into a sum of simpler logarithmic expressions • Using the properties of logarithms to combine sums of logarithmic expressions into a single logarithmic expression	p. 415 – 419	19 – 22
The change of base formula • The derivation of the change of base formula and its use	p. 420 – 421	23 – 24
Applications of logarithmic functions • The historical use of logarithms in simplifying computation • The use of logarithms in defining pH, earthquake intensity, and noise levels	p. 421 – 424	25 – 26

5.5 Exponential and Logarithmic Equations

topics	pages	test exercises
Converting between exponential and logarithmic forms • Using the definition of logarithms and their properties to convert between the exponential and logarithmic forms of a given equation	p. 429 – 434	27 – 30
Further applications of exponential and logarithmic equations • Using logarithms to solve exponential equations	p. 434 – 436	31 – 32
Interlude: analysis of a stock market investment • The meaning of *dollar-cost averaging*, and how to determine the equivalent annual rate of interest for an investment	p. 436 – 438	

chapter test

Section 5.1

Sketch the graphs of the following functions.

1. $f(x) = \left(\frac{1}{2}\right)^{x-1} + 3$

2. $r(x) = 2^{-x+4} - 2$

Solve the following elementary exponential equations.

3. $3^{3x-5} = 81$

4. $\left(\frac{2}{5}\right)^{-4x} = \left(\frac{25}{4}\right)^{x-1}$

5. $10{,}000^{x} = 10^{-2x-12}$

Section 5.2

Solve the following problems involving exponential functions.

6. Melissa has recently inherited \$15,000 which she wants to deposit into a savings account for 10 years. She has determined that her two best bets are an account that compounds annually at a rate of 3.95% and an account that compounds continuously at an annual rate of 3.85%. Which account would pay Melissa more interest?

7. Bill has come upon a 37 gram sample of iodine-131. He isolates the sample and waits for 2 weeks. After this time period, only 11 grams of iodine-131 remains. Determine the equation modeling the radioactive decay of this substance.

8. Katherine is working in a lab testing bacteria populations. Starting out with a population of 870 bacteria, she notices that the population doubles every 22 minutes. Find **(a)** the equation for the population P in terms of time t in minutes, and **(b)** the population after 68 minutes.

Section 5.3

Sketch the graphs of the following functions.

9. $f(x) = \log_4(3 - x) + 2$

10. $g(x) = 1 - \log_5(2x)$

Evaluate the following logarithmic expressions without the use of a calculator.

11. $\log_{27} 9^{\log 1000}$

12. $\log_{1/3} 9$

13. $\log_4\left(\frac{1}{64}\right)$

Use the elementary properties of logarithms to solve the following equations.

14. $\log_6 6^{\log_5 x} = 3$

15. $\log_9 x^{1/2} = \frac{3}{4}$

16. $\log_x \left(\log_{1/2} \frac{1}{16} \right) = 2$

Solve the following equations, using a calculator if necessary to evaluate the logarithms.

17. $\log 27 = 5x$

18. $\ln(2x - 1) = 3$

Section 5.4

Use properties of logarithms to expand the following expressions as much as possible. Simplify any numerical expressions that can be evaluated without a calculator.

19. $\log \sqrt{\frac{x^3}{4\pi^5}}$

20. $\ln\left(\frac{\sqrt{a^5}mn^2}{e^5} \right)$

Use properties of logarithms to condense the following expressions as much as possible, writing each answer as a single term with a coefficient of 1.

21. $\frac{1}{3}\left(\log_2\left(a^5\right) - \log_2\left(bc^3\right)\right)$

22. $\ln 4 - \ln x^2 - 7\ln y$

Evaluate the following logarithmic expressions.

23. $\log_7 18$

24. $3 \log_{1/4} 6$

Use your knowledge of logarithmic functions to answer the following application questions.

25. Cameron had front row tickets to a recent rock concert. The noise intensity in the front row was 10^{-1} W/m^2. Given that $I_0 = 10^{-12}$ W/m^2, what was the decibel level Cameron experienced the night of the concert?

26. The largest earthquake since 1900 occurred May 22, 1960 in Chile, measuring 9.5 on the Richter Scale. What was the intensity, relative to a 0-level earthquake?

Section 5.5

Solve the following exponential and logarithmic equations. When appropriate, write the answer as both an exact expression and as a decimal approximation.

27. $e^{8-5x} = 16$

28. $10^{6/x} = 321$

29. $\ln(x+1) + \ln(x-1) = \ln(x+5)$

30. $\log_2(x+3) + \log_2(x+4) = \log_2(3x+8)$

Use your knowledge of exponential and logarithmic functions to answer the following questions.

31. Rick puts \$6500 in a high interest money market account at 4.36% annual interest compounded monthly. Assuming he makes no deposits or withdrawals, how long will it take for his investment to grow to \$7000?

32. Sodium-24 has a half-life of approximately 15 hours. How long would it take for 350 grams of sodium-24 to decay to 12 grams?

chapter SIX

OVERVIEW

By the end of this chapter you should be able to evaluate and graph trigonometric functions. Trigonometric functions can be used to describe the behavior of many naturally occurring phenomena, such as waves and harmonic motion. On page 515, you will be asked to find the period of oscillation of a cord swinging from the light bulb of a ceiling fan. You will master this type of problem using the definition of simple harmonic motion and frequency on page 508.

Introduction

This chapter is the first of three dealing almost exclusively with an area of mathematics called *trigonometry*, an area which is big enough to study on its own (as it often is) but which is also intrinsically related to the algebraic and geometric concepts that make up the rest of this text. This close association will become even more apparent when you study Calculus, so its inclusion in a book preparing you for Calculus is very much appropriate.

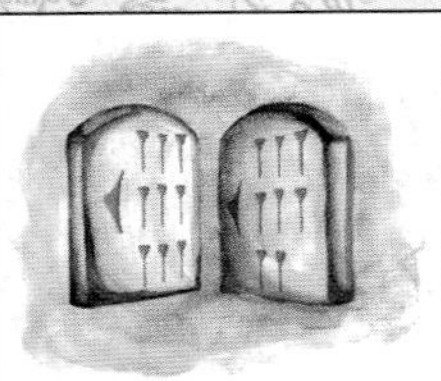

Babylonian numbers

The early history of trigonometry is not quite so well-documented as that of geometry, but archeologists have unearthed clay tablets indicating that Babylonian mathematicians around 2000 BC were already developing ideas that we would classify today as trigonometric. And that Babylonian heritage is of more than academic interest. We owe to the Babylonians of the 1st millennium BC our *degree* unit of angle measure. Although many competing arguments have been proposed as to why the Babylonians fixed on 360° as being the measure of one full rotation, there is no doubt that the convention began with them. (Some of the competing arguments are built around such things as connections between a full circle and the calendar, the fact that 360 can be factored many different ways, and the fact that Babylonians apparently divided the day into 12 "hours" of 30 parts each.)

The word "trigonometry" itself is Greek, and translates roughly as "measurement of triangles." From the start, trigonometry found important applications in astronomy, navigation, and surveying, and those applications have only grown in importance over time. Initially, trigonometry focused on ratios of side-lengths of triangles, and that perspective is alive and strong still. But with the development of Calculus in the 17th century, mathematicians also began to view the trigonometric relations as functions of real numbers. This secondary perspective lends itself to applications involving rotations or oscillations, and trigonometry quickly became an indispensable tool in engineering, explanations of wave propagation, and modern signal processing.

As with every topic in this text, try to keep the historical background in mind as you learn the material. Mathematics is not immune to societal and other pressures, and many aspects of trigonometry's history demonstrate this. The presentation in this text draws upon more than 2000 years of development, and would be very unfamiliar to early users of trigonometry. Relatively recent developments in technology have also had a profound effect on the way trigonometry is taught and learned – if you find yourself tiring at some point while studying this chapter, comfort yourself with the thought that dreaded "trig tables" are a thing of the past! Learning how to use them once constituted a large part of trigonometry, but calculators and computer software make such tedium unnecessary now. (If you have no idea what a "trig table" is, and want to subject yourself to a lecture on how kids today have it too easy, ask someone who learned trigonometry prior to the mid-1970's for an explanation.)

6.1 Radian and Degree Measure of Angles

TOPICS

1. The unit circle and angle measure
2. Conversion between degrees and radians
3. Commonly encountered angles
4. Arc length and angular speed
5. Area of a circular sector

Topic 1: The Unit Circle and Angle Measure

Trigonometry is, at heart, the study of angles. Although much of trigonometry can be discussed without reference to angle measure (and in fact early Greek mathematicians did just that), we will find it very useful to have a method of describing the size of angles. As it turns out, there are two common ways to measure angles (as well as a number of less common ways). The method of measuring angles in terms of *degrees* is one that you are probably very familiar with – references to degree measure occur in all sorts of non-mathematical contexts. But, as mentioned in the introduction to this chapter, the definition of degree has more of a cultural basis than a mathematical basis, so it shouldn't be too surprising that there is a way of measuring angles that makes more mathematical sense. That more mathematically useful way is called *radian measure.*

Radian Measure

Let θ (the Greek letter *theta*) be an angle at the center of a circle of radius 1 (the unit circle), as shown in the diagram. The measure of θ in **radians** (abbreviated as **rad**) is the length of that portion of the circle *subtended* by θ (that is, the portion of the circumference shown in red). Note that the unit of length measurement is immaterial. As long as the circle has a radius of 1 (unit), the length of the subtended portion of the circle (in the same units) is defined to be the radian measure of the angle.

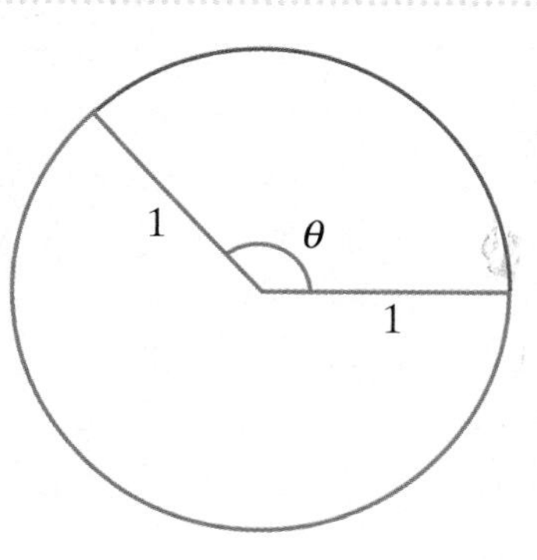

Of course, in general an angle θ is defined by any two rays R_1 and R_2 sharing a common origin, as shown in Figure 1. We can associate a sign with the measure of θ by designating one ray, say R_1, as the **initial side** and the other ray, R_2, as the **terminal side**. If θ is defined by a counterclockwise rotation from the initial side to the terminal side, we say θ has **positive measure**, and if θ is defined by a clockwise rotation we say it has **negative measure**. In Figure 1, the red angle has positive measure while the blue angle is negative.

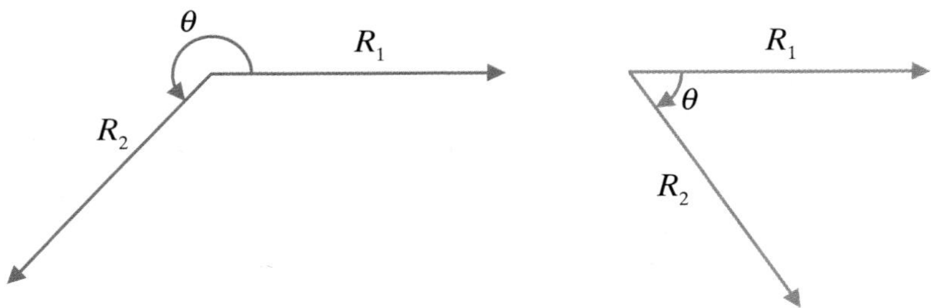

Figure 1: Positive and Negative Angle Measure

In order to use our definition of radian to measure an angle defined by two rays, as in Figure 1, we place the vertex of the angle at the center of a **unit circle** (a circle of radius 1) and measure the length of the arc between the initial and terminal sides of the angle. Further, we say the angle is in **standard position** if its vertex is located at the origin of the Cartesian plane and its initial side lies along the positive x-axis. In this case, the unit circle is then the graph of the equation $x^2 + y^2 = 1$. Figure 2 illustrates the second angle from Figure 1 placed in standard position.

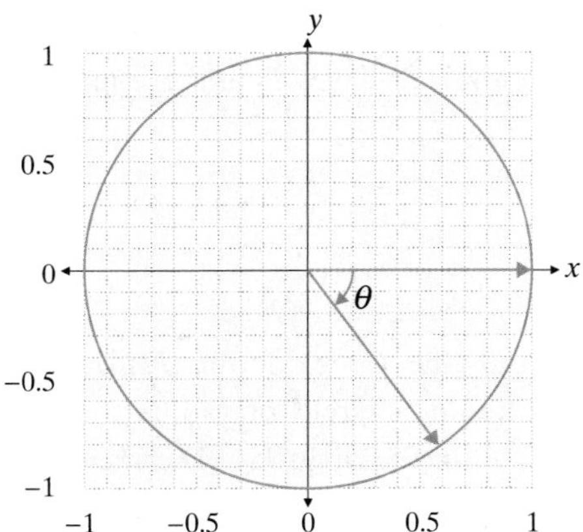

Figure 2: Standard Position of an Angle

Topic 2: Conversion Between Degrees and Radians

Once radian measure has been defined, the next order of business is to acquire a degree of familiarity with its use. To begin with, we want to be able to translate between degree measure and radian measure of an angle.

This is easily done if we recall the formula for the circumference C of a circle of radius r: $C = 2\pi r$. For the unit circle under discussion, $r = 1$ so $C = 2\pi$. If we think of the entire circumference as being the portion of the unit circle subtended by an angle of 360° (that is, an angle whose terminal side and initial side coincide), we have just determined that an angle of 360° corresponds to 2π radians. Using this as a starting point, we see that an angle of 180° corresponds to π radians (such an angle subtends half the circumference of the unit circle) and an angle of 90° corresponds to $\frac{\pi}{2}$ radians (that is, a right angle has a measure of $\frac{\pi}{2}$ radians). From the equation $180° = \pi$ rad, we can derive the conversion formulas that follow.

Degrees	Radians
360°	2π
180°	π
90°	$\frac{\pi}{2}$

Conversion Formulas

Since $180° = \pi$ rad, we know that $1° = \frac{\pi}{180}$ rad and $\left(\frac{180}{\pi}\right)^{\circ} = 1$ rad. Multiplying both sides of these equations by an arbitrary quantity x, we have:

1. $x^{\circ} = (x)\left(\frac{\pi}{180}\right)$ rad, and

2. $x \text{ rad} = x\left(\frac{180}{\pi}\right)^{\circ}$.

In particular, note that 1 rad $\approx 57.296°$, so an angle of 1 rad cuts off a bit less than one-sixth of a circle (an angle of 60° cuts off exactly one-sixth of a circle).

example 1

Convert the following angle measures as directed.

a. Express $\frac{\pi}{3}$ rad in degrees. **b.** Express 270° in radians.

c. Express –2 rad in degrees.

Solutions:

a. $\frac{\pi}{3} \text{ rad} = \left(\frac{\pi}{3}\right)\left(\frac{180}{\pi}\right)^{\circ} = 60^{\circ}.$

b. $270^{\circ} = (270)\left(\frac{\pi}{180}\right) \text{rad} = \frac{3\pi}{2} \text{ rad}.$

c. $-2 \text{ rad} = (-2)\left(\frac{180}{\pi}\right)^{\circ} \approx -114.592^{\circ}.$

Before continuing, a note on terminology: Whenever an angle is measured in degrees, its measure will appear followed by the degree symbol (°); angles measured in radians will either appear with the abbreviation "rad" afterward or, more commonly, with no notation at all. It is a reflection of the importance of radian measure in mathematics that if no indication of the method of measurement appears, we are to assume the angle is measured in radians.

Topic 3: Commonly Encountered Angles

It is tempting, when teaching or learning a new area of mathematics, to restrict attention to examples that are artificially "nice." That is, examples in which complicated terms in the accompanying equations either never appear or else conveniently cancel, and examples in which the final answer is suspiciously devoid of ugly fractions and approximations. With this in mind, you might dismiss the angles in the following discussion as unrealistically pleasant to work with. But the justification for studying the angles of 30°, 45°, and 60° is that, first, they actually do appear fairly frequently in real life and, second, they are undeniably useful in building an understanding of trigonometric functions. We will encounter them repeatedly in the sections that follow.

At the moment, we are primarily interested in determining the radian measures that correspond to these common angles, but that is a simple matter of applying the appropriate conversion formula. While we have them before us, therefore, we will also note how these angles relate to one another in the context of triangles. This knowledge will prove to be useful very soon.

First, the radian equivalents:

$$30^\circ = (30)\left(\frac{\pi}{180}\right) = \frac{\pi}{6}$$

$$45^\circ = (45)\left(\frac{\pi}{180}\right) = \frac{\pi}{4}$$

$$60^\circ = (60)\left(\frac{\pi}{180}\right) = \frac{\pi}{3}$$

and, while we're at it:

$$90^\circ = (90)\left(\frac{\pi}{180}\right) = \frac{\pi}{2}.$$

(Remember that the absence of notation following an angle means the angle is measured in radians.)

The triangle connection comes from an application of the Pythagorean Theorem. Recall that if a and b are the lengths of the two legs of a right triangle and if c is the length of the hypotenuse, then $a^2 + b^2 = c^2$. Recall also that the sum of the angles of a triangle is always 180°, or π radians. So a triangle with one vertex of measure $\frac{\pi}{6}$ and a second vertex of measure $\frac{\pi}{3}$ must have a right angle for the third vertex, and similarly for a triangle with two vertices of measure $\frac{\pi}{4}$. These observations are illustrated in Figure 3.

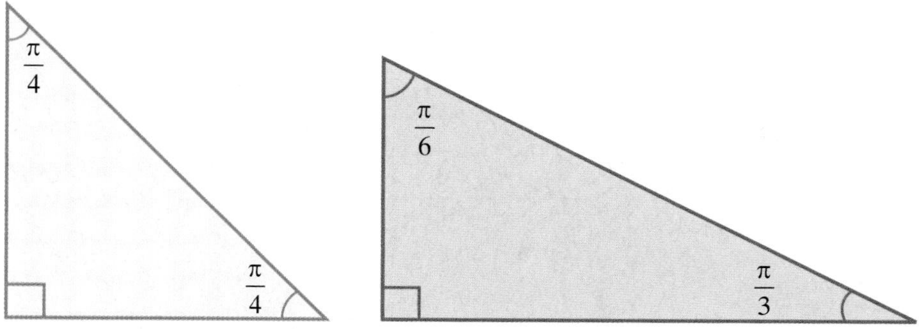

Figure 3: Common Triangles

Now, suppose the triangle on the left in Figure 3 has legs of length 1. Pythagoras' Theorem tells us then that the length of the hypotenuse is $\sqrt{1^2 + 1^2} = \sqrt{2}$. In general, any triangle with two angles of measure $\frac{\pi}{4}$ will be a right triangle with two legs of equal length, and the ratios of the lengths of the sides will be $1:1:\sqrt{2}$.

We have to work slightly harder to figure out the ratios of the lengths of the sides of the triangle on the right (the $30°-60°-90°$ triangle, in degree terms). Note that if we join the triangle with its mirror image, we obtain an equilateral triangle (since all the angles will measure 60°) as shown in Figure 4. This means that the length of the shorter leg of the original triangle must be half of the length of the hypotenuse. So if we assume the shorter leg has a length of 1, the hypotenuse has a length of two and the Pythagorean Theorem tells us that the longer leg has a length of $\sqrt{2^2-1^2}=\sqrt{3}$. In general, the ratio of the short leg to the long leg to the hypotenuse of such a triangle is $1:\sqrt{3}:2$.

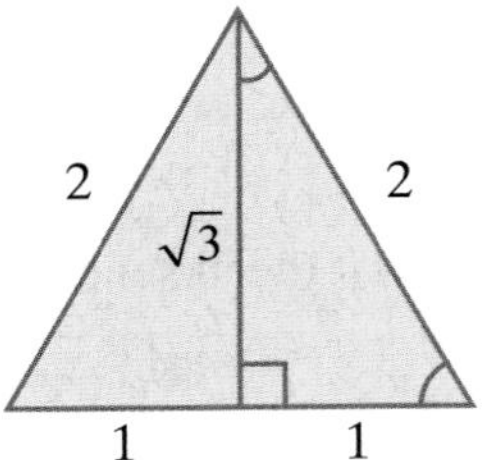

Figure 4: Doubling the $\frac{\pi}{6}-\frac{\pi}{3}-\frac{\pi}{2}$ Triangle

example 2

Use the information in each diagram to determine the radian measure of the indicated angle.

a.

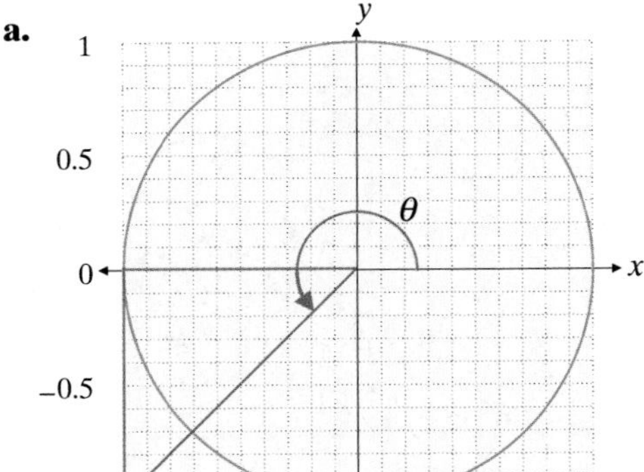

b.

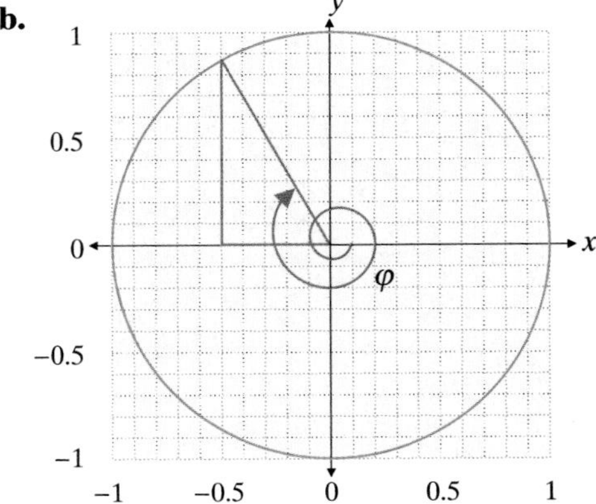

Solutions:

a. The angle θ is in standard position, and is positive. It can also be seen that the measure of the angle is π radians plus a bit more, where the "bit more" comes from the angle whose initial side is the negative x-axis and whose terminal side contains the hypotenuse of the red $\frac{\pi}{4}-\frac{\pi}{4}-\frac{\pi}{2}$ triangle. So the "bit more" must be $\frac{\pi}{4}$ radians, and the angle θ has measure $\pi+\frac{\pi}{4}=\frac{5\pi}{4}$.

b. The angle φ (the Greek letter *phi*) is also in standard position, but its measure is negative. It is defined by beginning at the positive x-axis and rotating -2π radians (that is, going full-circle in the clockwise direction), continuing for another $-\pi$ radians (another half-circle), and then continuing on for a bit more. This time, the angle corresponding to the "bit more" has its initial side on the negative x-axis and terminal side on the hypotenuse of a $\frac{\pi}{6}-\frac{\pi}{3}-\frac{\pi}{2}$ triangle. We know the triangle must be of this sort because its hypotenuse has length 1 (do you see why?) and its shorter leg has length $\frac{1}{2}$, so the ratio of the shorter leg to the hypotenuse is $1:2$. Hence the "bit more" must have measure $-\frac{\pi}{3}$ and altogether the measure of φ is $-2\pi-\pi-\frac{\pi}{3}=-\frac{10\pi}{3}$.

Topic 4: Arc Length and Angular Speed

The advantages of radian measure over degree measure will appear repeatedly over the next several chapters (and later in Calculus). The first advantage is actually just a restatement of the definition of radian. Recall that the radian measure of an angle is related to that portion of the unit circle cut off (or subtended) by the angle when the angle is placed at the center of the circle; the length of the subtended arc is an example of *arc length*. In other words, if θ is a central angle of a unit circle, then $\frac{\theta}{2\pi}$ is the fraction of the circle's circumference subtended by θ. More generally, if θ is a central angle of a circle of radius r, the length s of the portion of the circle subtended by θ is the same fraction multiplied by the circumference.

Arc Length Formula

Given a circle of radius r, the length s of the arc subtended by a central angle θ is given by:

$$s=\left(\frac{\theta}{2\pi}\right)(2\pi r)$$
$$=r\theta$$

(In the figure on the right, the unit circle is drawn in black, while the larger circle has radius r.)

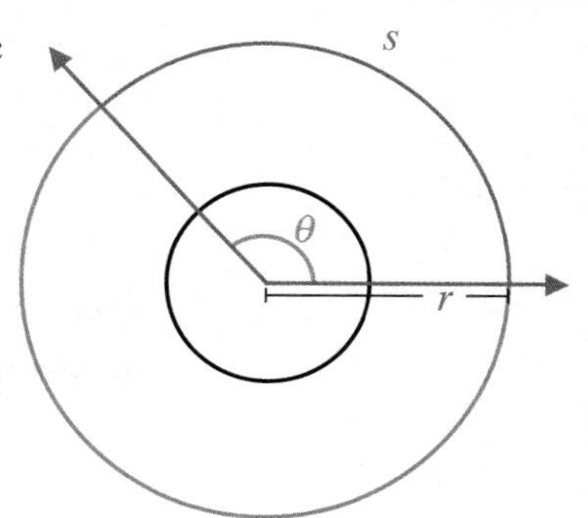

example 3

The Galapagos Islands lie almost exactly on the equator, and are located at 90° West Longitude. Suppose a ship sails along the equator from the Galapagos to the International Date Line (at 180° Longitude). How far does the ship travel? (Assume the Earth's radius is 6370 kilometers.)

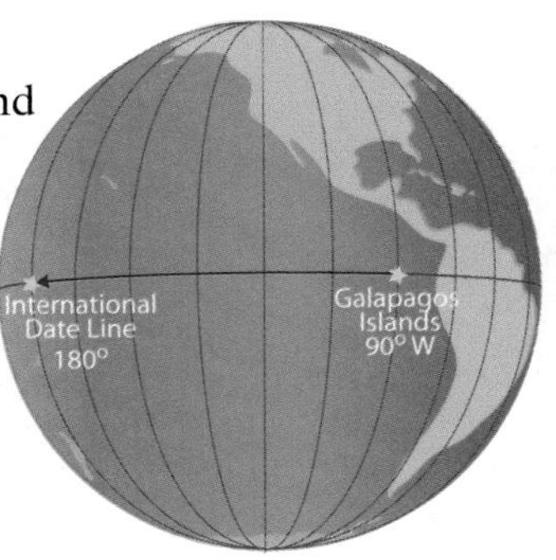

Solution:

The critical observation is that the ship travels one-quarter of the Earth's circumference (from 90° West of 0° Longitude to 180° from 0°), so the angle (at the center of the Earth) described by the ship's path is $\frac{\pi}{2}$. Using the arc-length formula, the distance traveled is

$$\begin{aligned} s &= (6370)\left(\frac{\pi}{2}\right) \\ &\approx 10{,}006 \text{ km.} \end{aligned}$$

Now that we have a convenient formula for arc-length, we can easily determine the speed with which an object traverses a given arc. For instance, if we are told that the ship in Example 3 takes 15 days to make its journey, we can calculate that its average speed is

$$\frac{10{,}006}{(15)(24)} \approx 27.8 \frac{\text{km}}{\text{hr}}.$$

Often, information about rate of travel along a circle's circumference is given in terms of *angular speed*, which is a measure of the angle traversed over time.

Angular Speed and Linear Speed

If an object moves along an arc of a circle defined by a central angle θ in time t, the object is said to have an **angular speed** ω (the Greek letter "omega") given by:

$$\omega = \frac{\theta}{t}.$$

If the circle has a radius of r, the distance traveled in time t is the arc-length s, and the **linear speed** v is given by:

$$v = \frac{s}{t} = \frac{r\theta}{t} = r\omega.$$

example 4

Suppose an ant crawls along the rim of a circular glass with radius 2 inches, and traverses the arc indicated in red below in 20 seconds. What are the angular and linear speeds of the ant, and how far does it travel?

Solution:

As in Example 2b, we can determine that the triangle shown in blue is a $30° - 60° - 90°$ right triangle, and its vertex at the origin must have measure $\frac{\pi}{3}$. This means the ant describes an angle of $\theta = \frac{\pi}{2} + \frac{\pi}{6} = \frac{2\pi}{3}$ as it walks, so its angular speed is

$$\omega = \frac{\theta}{t} = \frac{\frac{2\pi}{3}}{20} = \frac{\pi}{30}\frac{\text{rad}}{\text{s}}.$$

Given that the radius of the glass is 2 inches, the linear speed of the ant is

$$v = r\omega = 2\left(\frac{\pi}{30}\right) = \frac{\pi}{15} \approx 0.21\frac{\text{in}}{\text{s}}.$$

Finally, the distance the ant travels is $s = r\theta = (2)\left(\frac{2\pi}{3}\right) = \frac{4\pi}{3}$ in. or approximately 4.19 in.

caution!

The arc-length and angular speed formulas, as well as the area formula that follows, are only true for angles measured in radians. Equivalent but less convenient formulas can be derived for angles measured in terms of degrees.

Topic 5: Area of a Circular Sector

We will close this section with one last example of the value of measuring angles in radians.

A **sector** of a circle is the portion of a circle between two radii. The area of a sector, then, can range from 0 to πr^2 square units, where the radius of the circle is assumed to be r units. Of course, the two radii defining a given sector can also be taken to be the initial and terminal sides of a central angle θ. Just as $\frac{\theta}{2\pi}$ represents the fraction of a circle's circumference subtended by θ, this same ratio represents the portion of a circle's area contained in the sector of angular size θ. This gives us the following formula.

Sector Area Formula

The area A of a sector with a central angle of θ in a circle of radius r is

$$A = \left(\frac{\theta}{2\pi}\right)\left(\pi r^2\right) = \frac{r^2\theta}{2}.$$

example 5

Determine the areas of the sectors defined by the given radii and angles.

a. Circle of radius 3 cm, central angle of 52°

b. Circle of radius $\frac{1}{2}$ ft., central angle of $\frac{4\pi}{3}$

Solutions:

a. In order to use the above sector area formula, the first step is to convert the angle measure to radians:

$$52^\circ = (52)\left(\frac{\pi}{180}\right) = \frac{13\pi}{45}.$$

Now, the formula is easily applied:

$$A = \frac{\left(3^2\right)\left(\frac{13\pi}{45}\right)}{2} = \frac{13\pi}{10} \approx 4.08 \text{ cm}^2.$$

b. Since the angle is given in radians, we have immediately:

$$A = \frac{\left(\frac{1}{2}\right)^2\left(\frac{4\pi}{3}\right)}{2} = \frac{\pi}{6} \approx 0.52 \text{ ft}^2.$$

exercises

In questions 1 – 10, convert the radian measure to degrees. See Example 1.

1. $\frac{5\pi}{4}$ **2.** $\frac{\pi}{180}$ **3.** $\frac{-3\pi}{8}$ **4.** $\frac{-7\pi}{6}$ **5.** $\frac{2\pi}{3}$

6. $\frac{7\pi}{20}$ **7.** $\frac{5\pi}{6}$ **8.** $\frac{11\pi}{10}$ **9.** $\frac{-9\pi}{4}$ **10.** $\frac{-5\pi}{3}$

In questions 11 – 20, convert the degree measure to radians. See Example 1.

11. 47° **12.** 93° **13.** 132° **14.** 154° **15.** 148°

16. 120° **17.** 480° **18.** 520° **19.** 125° **20.** 90°

Convert the following angle measures as directed. See Example 1.

21. Express $\frac{3\pi}{2}$ in degrees.

22. Express $-\frac{9\pi}{4}$ in degrees.

23. Express 3π in degrees.

24. Express $\frac{\pi}{12}$ in degrees.

25. Express $-\frac{2\pi}{5}$ in degrees.

26. Express $\frac{2\pi}{3}$ in degrees.

27. Express 20° in radians.

28. Express 340° in radians.

29. Express –144° in radians.

30. Express 66° in radians.

31. Express 30° in radians.

32. Express 180° in radians.

The unit circle shown below shows several angles in radians or degrees. Fill in the corresponding radian or degree for questions 33 – 44.

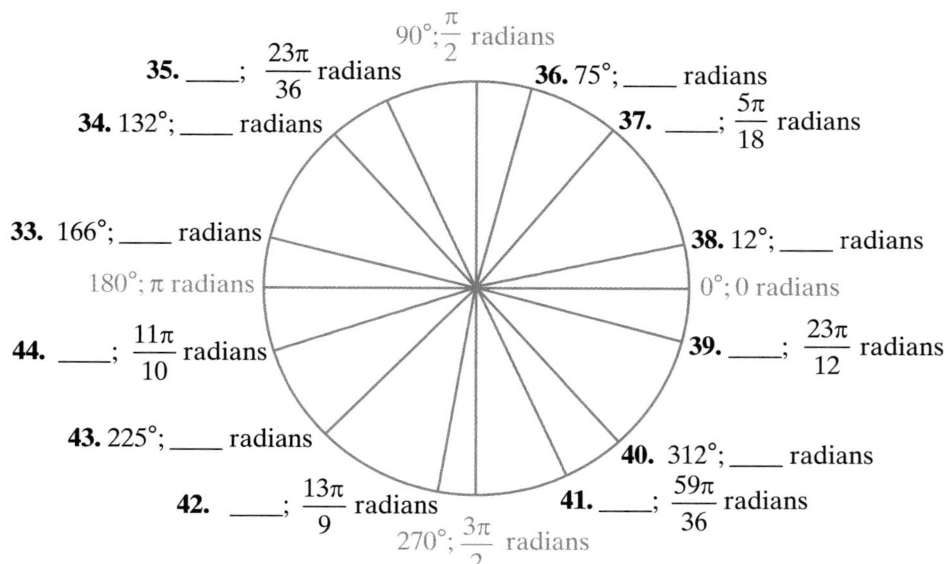

Use the information in each diagram to determine the radian measure of the indicated angle. See Example 2.

45.

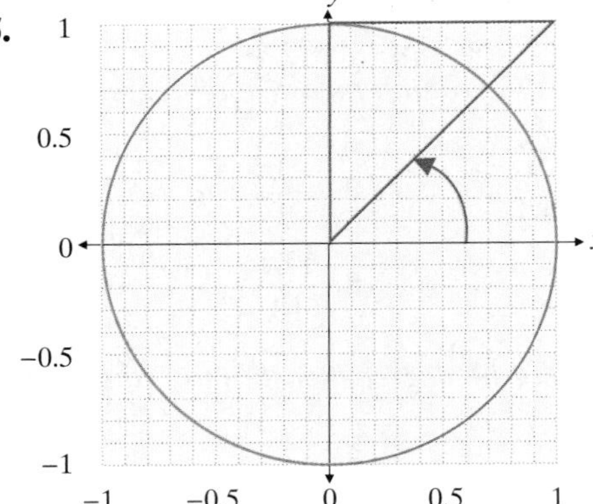

46.

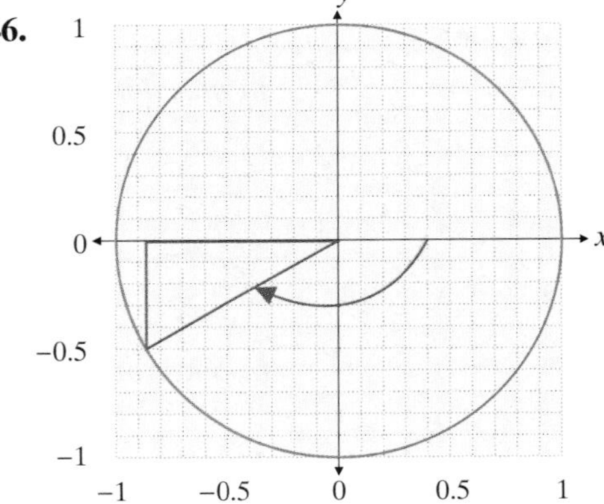

47.

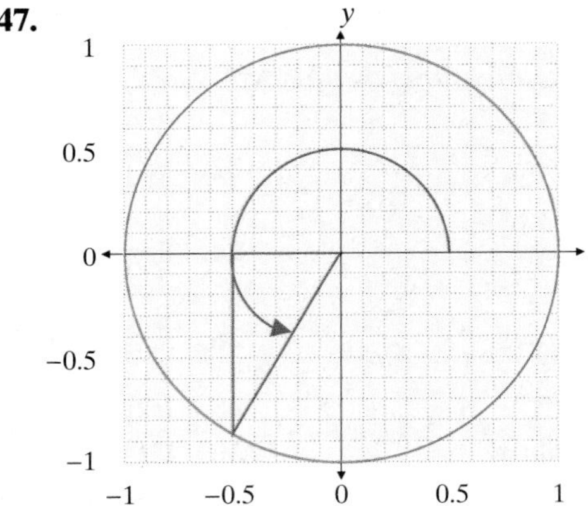

48.

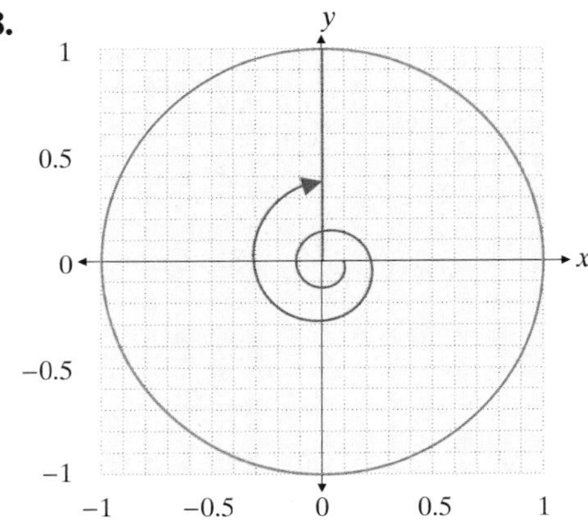

49.

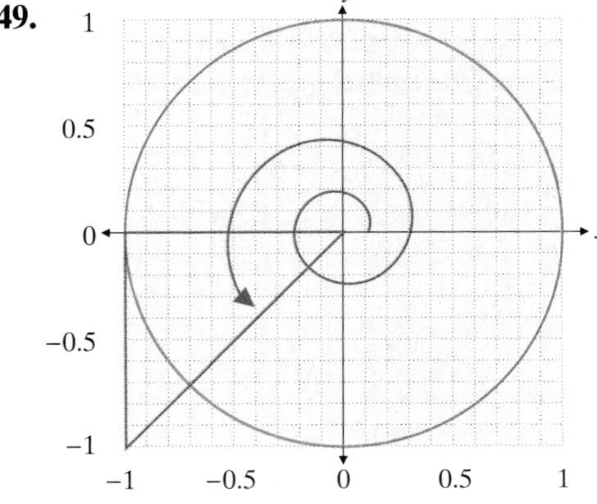

50.

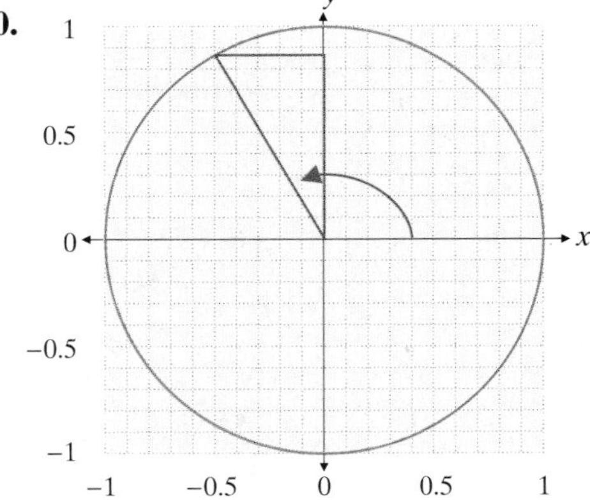

Sketch the indicated angles. See Example 2.

51. $\frac{5\pi}{2}$ **52.** $-60°$ **53.** $210°$

54. $-\frac{\pi}{3}$ **55.** $\frac{7\pi}{4}$ **56.** $120°$

Find the arc length of a circle with the given radius r and central angle θ. Give the answer in the given unit of measure and round decimals off to the nearest hundredth.

57. $r = 4$ in.; $\theta = 1$

58. $r = 9$ cm; $\theta = \dfrac{\pi}{2}$

59. $r = 15$ feet; $\theta = \dfrac{\pi}{4}$

60. $r = 80$ km; $\theta = 180°$

61. $r = 16.5$ m; $\theta = 30°$

62. $r = 7$ feet; $\theta = 90°$

Find the indicated arc length in each of the following problems. See Example 3. (Round your answers to the nearest hundredth.)

63. Given a circle of radius 5 inches, find the length of the arc subtended by a central angle of 17° (hint: convert to radians first).

64. Given a circle of radius 22.5 cm, find the length of the arc subtended by a central angle of 3π.

65. Given a circle with a diameter of 6 feet, find the length of the arc subtended by a central angle of 68° (hint: convert to radians first).

66. Given a circle of radius 7 meters, find the length of the arc subtended by a central angle of $\dfrac{7\pi}{8}$.

67. Assuming that Columbia, SC and Daytona Beach, FL have the same longitude (81° W), use a radius of 6370 kilometers for the Earth and the following to find the distance between the two cities.

City	Latitude
Columbia, SC	34° N
Daytona Beach, FL	29.25° N

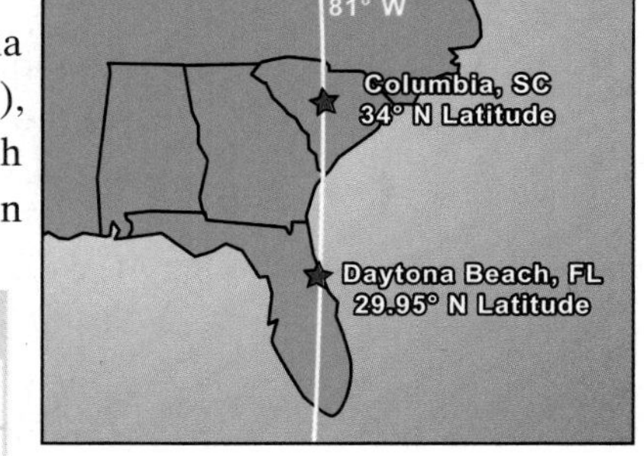

68. Given that two cities on the equator are 100 miles apart and have the same latitude (that is, one is due west of the other), what is the difference in their longitudes? Use a value of 3960 miles for the radius of the earth.

69. Using a radius of 1.2 cm for the average eyeball, find the central angle formed to meet the edges of an iris (the colored portion of the eye) with an arc length of 9 mm.

70. Find the distance between Denver, CO and Roswell, NM which lie on the same longitude. The latitude of Denver is 39.75° N and the latitude of Roswell is 33.3° N. Use a radius of 3960 miles for the Earth.

71. Find the distance between Atlanta, Georgia and Cincinnati, Ohio which lie on the same longitude. The latitude of Atlanta is 33.67° N and the latitude of Cincinnati is 39.17° N. Assume the Earth's radius is 6370 kilometers.

72. Find the distance between Greenwich, England and Valencia, Spain which lie on the same longitude. The latitude of Greenwich is 51.48° N and the latitude of Valencia is 39.47° N. Assume the Earth's radius is 6370 kilometers.

73. Find the distance between La Paz, Bolivia and Caracas, Venezuela which lie on the same longitude. The latitude of La Paz is 16.50° S and the latitude of Caracas is 10.52° N. Assume the Earth's radius is 6370 kilometers.

74. Find the distance between Bucharest, Romania and Johannesburg, South Africa which lie on the same longitude. The latitude of Bucharest is 44.43° N and the latitude of Johannesburg is 26.21° S. Assume the Earth's radius is 6370 kilometers.

Find the radian measure of the central angle θ given the radius r and the arc length s transcribed by θ. Leave all answers in fraction form.

75. $r = 14$ ft.; $s = 63$ ft. **76.** $r = 16$ in.; $s = 6$ in. **77.** $r = 23.5$ dm; $s = 10.5$ dm

78. $r = 13$ cm; $s = 130$ cm **79.** $r = 2$ km; $s = 22.5$ km **80.** $r = 33$ ft.; $s = 11$ ft.

The following problems ask you to determine the angular and/or linear speeds of various objects. See Example 4. (Round your answers to the nearest hundredth.)

81. An industrial circular saw blade has a 10-inch radius and spins at 1000 rpm. Find **(a)** the angular speed of a tooth of the blade in rad/min., and **(b)** the linear speed of the tooth in feet per second.

82. The Earth takes roughly 23 hours and 56 minutes to rotate once about its axis. Using a radius for the earth of 3960 miles, what is the linear speed in miles per hour (relative to the center of the Earth) of a person standing on the equator? (Ignore, for the purposes of this problem, such motion as the rotation of the Earth about the Sun.)

83. A stationary exercise bike is ridden at a constant speed, causing the wheel to spin at a rate of 50 revolutions per minute. If a tack becomes lodged in the tire of radius 14 inches, find **(a)** the angular speed of the tack in rad/min., and **(b)** the linear speed of the tack in feet per minute.

84. A horse is tethered and urged to trot such that it completes a circular path every 5 seconds. If the rope which tethers it is 20 feet long, what is the linear speed of the horse in miles per hour?

The following problems ask you to calculate the area of a sector of a circle. See Example 5.

85. Find the area of the green shaded portion of the circle below:

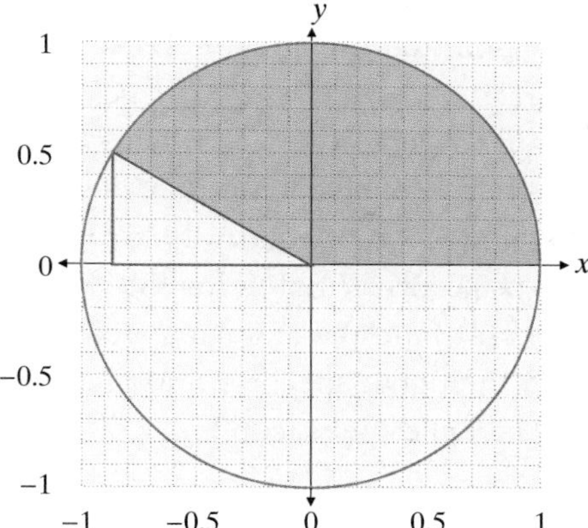

86. Find the area of the green shaded portion of the circle below:

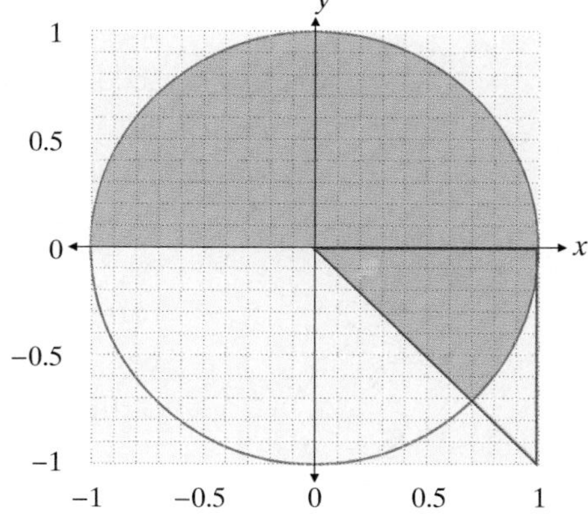

87. Find the area of the sector of a circle of radius 7 cm with a central angle of 70°.

88. Find the area of the sector of a circle of radius 3.5 ft. with a central angle of 27°.

89. Find the area of the sector of a circle of radius 4 m with a central angle of $\frac{3\pi}{5}$.

90. Find the area of the sector of a circle of radius 16 inches with a central angle of 138°.

91. Find the area of the sector of a circle of radius 20 ft. with a central angle of $\frac{\pi}{2}$.

92. Find the area of the sector of a circle of radius 19 km with a central angle of 5.31°.

93. A pie of radius 5 inches is cut into 8 equal pieces. What is the area of each piece?

94. The minute hand of a clock extends out to the edge of the clock's face, which is a circle of radius 2 inches. What area does the minute hand sweep out between 9:05 and 9:25?

95. The circular spinner for a board game is divided into 6 equal wedges, each of a different color. If the radius is 5 cm, what area is encompassed by 2 wedges?

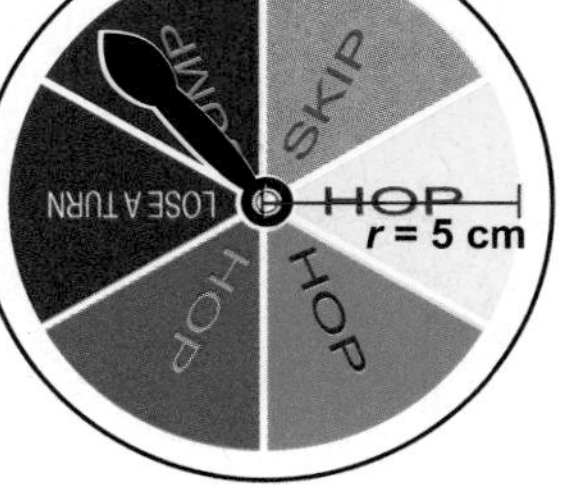

96. The floppy disk drive (FDD) was invented in 1967 to store information for computer users. The first floppy drive used an 8-inch disk and had a radius of 3.91 inches. The drive motor would spin at 300 rotations per minute (RPM). **(a)** Find the angular speed of the 8-inch disk in radians per second. **(b)** Find the linear speed of a particular point on the circumference of the 8-inch disk in inches per second.

97. The 8-inch floppy disk drive evolved into a smaller 5.25-inch disk that was used in the personal computers (PC) in the early 1980's. The 5.25-inch disk had a radius of 2.53 inches. The usual drive motor for the 5.25-inch disk would spin at 360 rotations per minute. **(a)** Find the angular speed of the 5.25-inch disk in radians per second. **(b)** Find the linear speed of a particular point on the circumference of the 5.25-inch disk in inches per second.

98. Two gears are rotating to turn a conveyor belt. The smaller gear rotates 80° as the larger gear rotates 50°. If the larger gear has a radius of 18.7 in., what is the radius of the smaller gear?

99. Two water mills are on display at a local museum. The smaller water mill rotates counterclockwise and turns the larger water mill in a clockwise direction. If the smaller water mill has a radius of 5.23 ft. and the larger water mill has radius of 8.16 ft., what is the degree of rotation of the larger wheel when the smaller rotates 60°?

6.2 Trigonometric Functions of Acute Angles

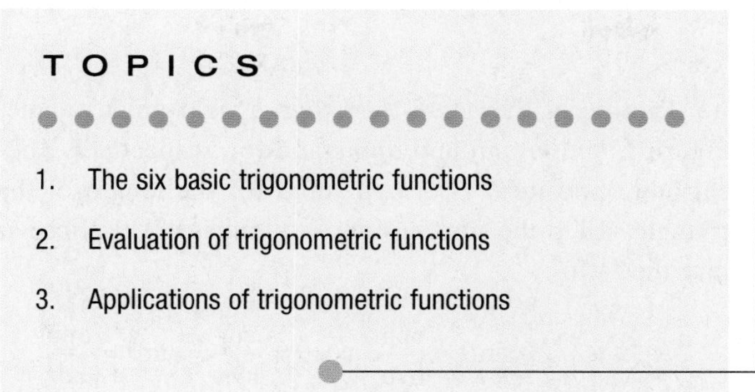

TOPICS
1. The six basic trigonometric functions
2. Evaluation of trigonometric functions
3. Applications of trigonometric functions

Topic 1: The Six Basic Trigonometric Functions

Now that we are equipped with a useful way of measuring angles, we can proceed to define the basic trigonometric functions that will dominate the discussion in this and the next two chapters.

It is worthwhile to reflect again on the fact that the material we are studying was developed by countless individuals over the span of several thousand years, and that the cultures they lived in and the problems they hoped to solve varied greatly. Much of the early impetus in developing trigonometry came from astronomy, but by the time of the Renaissance the utility of trigonometry in navigation, surveying, and engineering was thoroughly well-recognized. That utility has only increased with time, along with the fields in which trigonometric skill is necessary. We will find trigonometry particularly useful, for example, in solving plane geometry problems.

With the benefit of thousands of years of development behind us, we will take an approach to defining the six basic trigonometric functions that calls upon the work of many different eras. That is, our treatment of trigonometry would not be immediately recognizable to, say, early Greek mathematicians or to Italian mathematicians of the 15th and 16th centuries. Instead, our definitions will be motivated by the desire to make the trigonometric functions most readily useful in a wide variety of applications.

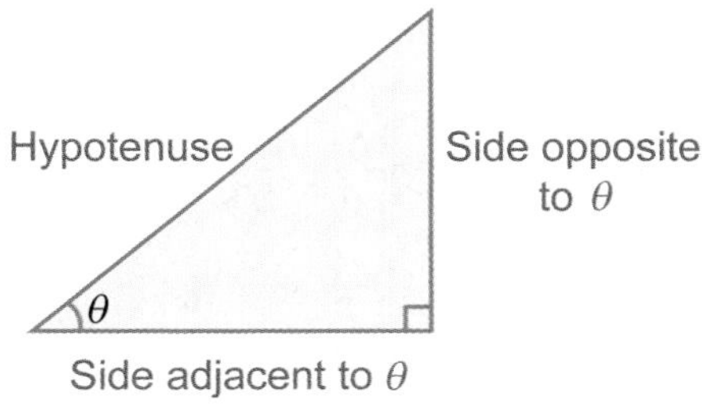

Figure 1: Legs Labeled Relative to θ

To begin, consider a right triangle such as the one shown in the figure to the left. We will define six functions which are functions of the angle θ. In order to do so, we label the two legs of the triangle as **adjacent** to and **opposite** of θ, as shown.

Sine, Cosine, and Tangent

Assume θ is one of the acute (less than a right angle) angles in a right triangle, as in Figure 1, and let *adj* and *opp* stand for, respectively, the lengths of the sides adjacent and opposite θ. Let *hyp* stand for the length of the hypotenuse of the right triangle. Then the **sine**, **cosine**, and **tangent** of θ, abbreviated $\sin\theta$, $\cos\theta$ and $\tan\theta$, are the ratios:

$$\sin\theta = \frac{\text{opp}}{\text{hyp}}, \quad \cos\theta = \frac{\text{adj}}{\text{hyp}}, \quad \tan\theta = \frac{\text{opp}}{\text{adj}}.$$

(Note that sine, cosine, and tangent are indeed functions of the angle θ, and to be consistent with our functional notation we would expect to see, for example, $\sin(\theta) = \frac{\text{opp}}{\text{hyp}}$. By convention, though, the parentheses around the argument are omitted unless called for to make the meaning clear.)

Incidentally, the name sine appears to have evolved through a complicated history of abbreviations and mistranslations, beginning with an Arabic word for "half-chord." Our name for the ratio comes from the Latin word *sinus*, which means "bay," but the reference to water is entirely accidental. On the other hand, *cosine* and *tangent*, along with the three functions still to be defined, have meaningful names. More on the subject of names will appear soon.

The remaining three basic trigonometric functions are reciprocals of the first three, as follows:

Cosecant, Secant, and Cotangent

Again, assume θ is one of the acute angles in a right triangle, as in Figure 1. Then the **cosecant**, **secant**, and **cotangent** of θ, abbreviated $\csc\theta$, $\sec\theta$, and $\cot\theta$, are the reciprocals, respectively, of $\sin\theta$, $\cos\theta$, and $\tan\theta$. That is,

$$\csc\theta = \frac{1}{\sin\theta} = \frac{\text{hyp}}{\text{opp}}, \quad \sec\theta = \frac{1}{\cos\theta} = \frac{\text{hyp}}{\text{adj}}, \quad \cot\theta = \frac{1}{\tan\theta} = \frac{\text{adj}}{\text{opp}}.$$

example 1

Use the information contained in the two figures below to determine the values of the six trigonometric functions of θ.

a.

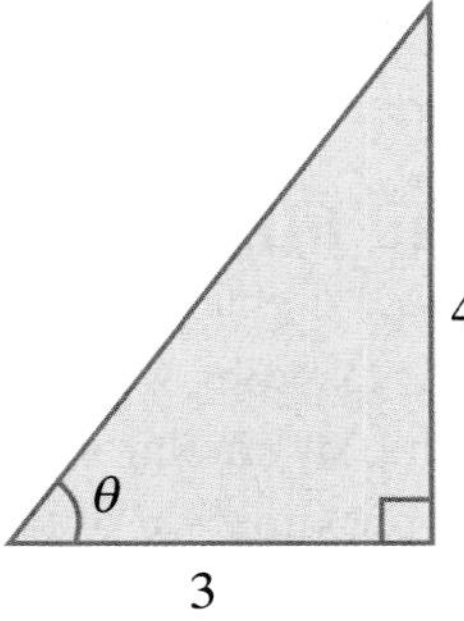

b.

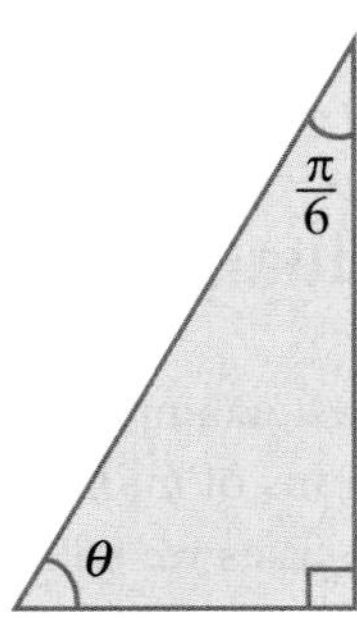

Solutions:

a. With the information given, we can determine tan θ and cot θ without effort (make sure you see why this is so). In order to evaluate the remaining four trigonometric functions at θ, all we need to do is determine the length of the hypotenuse. By the Pythagorean Theorem,

$$\begin{aligned}(\text{hyp})^2 &= (\text{adj})^2 + (\text{opp})^2 \\ &= 3^2 + 4^2 \\ &= 25\end{aligned}$$

so hyp = 5. Thus:

$$\sin\theta = \frac{\text{opp}}{\text{hyp}} = \frac{4}{5},\ \cos\theta = \frac{\text{adj}}{\text{hyp}} = \frac{3}{5},\ \tan\theta = \frac{\text{opp}}{\text{adj}} = \frac{4}{3}$$

and

$$\csc\theta = \frac{1}{\sin\theta} = \frac{5}{4},\ \sec\theta = \frac{1}{\cos\theta} = \frac{5}{3},\ \cot\theta = \frac{1}{\tan\theta} = \frac{3}{4}.$$

b. Since the triangle pictured contains a right angle and an angle of $\frac{\pi}{6}$ radians, the angle θ must have measure $\frac{\pi}{3}$; in degree terms, this is a $30° - 60° - 90°$ triangle. Such triangles always have sides in the ratio $1 : \sqrt{3} : 2$, so even though we are not given the length of any side, we can still evaluate all six trigonometric functions at θ. For example, if the shorter leg (the leg adjacent to θ) is assumed to have a length of a, then the other leg must have a length of $a\sqrt{3}$ and the hypotenuse must have a length of $2a$. So,

$$\sin\theta = \frac{\text{opp}}{\text{hyp}} = \frac{a\sqrt{3}}{2a} = \frac{\sqrt{3}}{2}.$$

cont'd. on next page ...

Similarly,

$$\cos\theta = \frac{\text{adj}}{\text{hyp}} = \frac{1}{2}, \ \tan\theta = \frac{\text{opp}}{\text{adj}} = \sqrt{3},$$

$$\csc\theta = \frac{1}{\sin\theta} = \frac{2}{\sqrt{3}}, \ \sec\theta = \frac{1}{\cos\theta} = 2, \ \cot\theta = \frac{1}{\tan\theta} = \frac{1}{\sqrt{3}}.$$

Topic 2: Evaluation of Trigonometric Functions

The last example introduces the sort of reasoning we can use to evaluate trigonometric functions of many angles, and we will employ similar methods often in what follows. In other cases we will want a numerical approximation of the value of some trigonometric function, and a calculator will prove to be very useful.

example 2

Evaluate the tangent and secant of $\theta = \frac{\pi}{4}$.

Solution:

With practice, you'll be able to determine $\tan\frac{\pi}{4}$ and $\sec\frac{\pi}{4}$ mentally, but initially it's very useful to draw a picture in order to visualize what is being asked. Since we are working with an angle of $\frac{\pi}{4}$ radians (45°), the remaining angle of our right triangle must be the same size. We've already noted that the sides of such a triangle have lengths in the ratio $1:1:\sqrt{2}$, so we can draw a triangle such as:

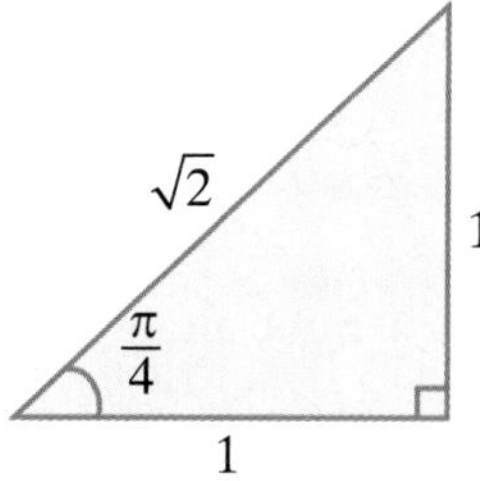

Note that the lengths of the sides have been arbitrarily chosen to be 1, 1, and $\sqrt{2}$. Any three numbers in the ratio of $1:1:\sqrt{2}$ could be used (the common factor will cancel out in the evaluation of any trigonometric function), so we may as well use these relatively simple numbers.

Now it is straightforward to note that

$$\tan\frac{\pi}{4} = \frac{1}{1} = 1$$

and

$$\sec\frac{\pi}{4} = \frac{\sqrt{2}}{1} = \sqrt{2}.$$

Up to this point in this chapter, a calculator has not been needed to perform any of the evaluations. But if we are asked to evaluate an expression such as sin 56.4° (we may encounter such an expression while solving a real-world application), some technological assistance is called for.

Fortunately, calculators (and computer software) that are equipped to handle trigonometric functions are readily available and easily used. However, it is important to remember that angles can be measured in terms of either degrees or radians, and *you* are responsible for putting the calculator in the correct mode (degree or radian) before performing the evaluation. This warning deserves to be repeated:

caution!

Before using a calculator to evaluate a given trigonometric expression, determine whether the angle in the expression is measured in degrees or radians. Then put the calculator in the appropriate mode prior to the evaluation.

example 3

Use a calculator to evaluate the following expressions.

a. $\sin 56.4°$ **b.** $\cot \dfrac{5\pi}{11}$

Solutions:

a. Refer to the user's manual to determine how to put your calculator in degree mode. Typically, there is a button labeled "mode" and pressing it leads to the option of choosing either "degree" or "radian." Once the calculator is in the correct mode, press the "sin" button, enter 56.4 on the number pad, and press the "=" or "Enter" button. The answer, rounded off to 4 decimal places, is .8329. The exact number of digits on your display will depend on the calculator and its current settings. If your display reads –.1481, your calculator is in the incorrect (radian) mode for this problem.

Make sure your MODE is set to Radian, then enter 1 / TAN (5π / 11). Your screen should appear as follows.

b. The absence of the degree symbol in the expression $\dfrac{5\pi}{11}$ tells us that the angle is measured in radians, so the first step is to place your calculator in radian mode. Next, recall that cotangent is the reciprocal of tangent. Most calculators don't have buttons specifically for the cotangent, secant, and cosecant functions; if you need to evaluate an expression containing one of these functions, take the reciprocal of, respectively, the tangent, cosine, or sine of the given angle.

cont'd. on next page ...

In this case, use your calculator to confirm that

$$\tan\frac{5\pi}{11} \approx 6.9552,$$

and therefore

$$\cot\frac{5\pi}{11} = \frac{1}{\tan\frac{5\pi}{11}} \approx .1438.$$

As mentioned at the start of this section, trigonometry has been around for several thousand years. For all but the last few decades of that very long history, users of trigonometry did not have the option of being able to punch a few buttons on a calculator in order to perform their calculations. In the not too distant past, a large part of trigonometry consisted of teaching students how to use tables of pre-determined evaluations (so-called "trig tables"). Thankfully, we are past the need for such instruction; however, one legacy of the pre-calculator days of trigonometry lives on and needs to be discussed. That legacy concerns notation.

Today, decimal notation most naturally suits the use of calculators. When calculations were done by hand, however, angles were more commonly expressed in the "degrees, minutes, seconds" (DMS) notation, and we still encounter this notation frequently in some contexts (surveying and astronomy, to name two). We need, therefore, to be able to convert from the DMS notation to decimal notation.

Degree, Minute, Second Notation

In the context of angle measure,

$$1' = \text{one minute} = \left(\frac{1}{60}\right)\left(1^\circ\right)$$

and

$$1'' = \text{one second} = \left(\frac{1}{60}\right)\left(1'\right) = \left(\frac{1}{3600}\right)\left(1^\circ\right).$$

For instance, an angle given as $14^\circ 37' 23''$ ("14 degrees, 37 minutes, 23 seconds") can also be written in decimal form (rounded to four decimal places) as 14.6231°, since

$$14^\circ 37' 23'' = 14 + \frac{37}{60} + \frac{23}{3600} \approx 14.6231^\circ.$$

Topic 3: Applications of Trigonometric Functions

One way or another, most applications of trigonometry involve using given information about a triangle to determine something else about the triangle. At this point, the triangles we work with are all right triangles, and a basic knowledge of the trigonometric functions and the Pythagorean Theorem suffice as tools. The process of determining unknown angles and/or dimensions from known data in such cases is often termed **solving right triangles**. In later sections, we will consider arbitrary triangles and will enlarge our collection of tools with a variety of trigonometric identities and theorems.

example 4

Before cutting down a dead tree in your yard, you very sensibly decide to determine its height. Backing up 40 feet from the tree (which rises straight up from level ground), you use a *theodolite* (a surveyor's instrument that accurately measures angles) and note that the angle between the ground and the top of the tree is $61°55'39''$. How tall is the tree?

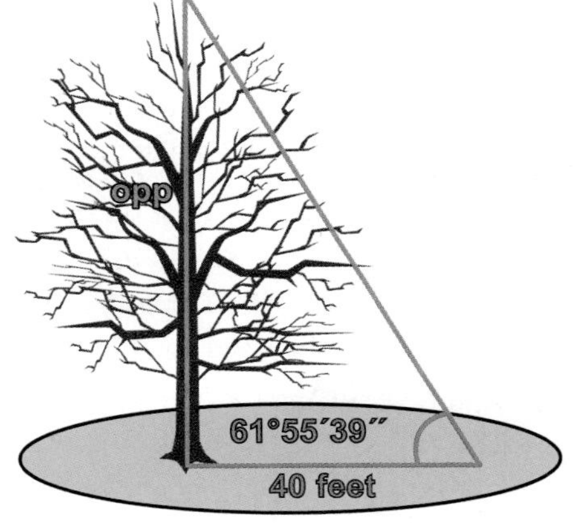

Solution:

As in so many problems, a picture is of great help. Note that this problem indeed involves solving a right triangle: we know the measure of an angle and the length of the angle's adjacent leg, and we want the length of the opposite leg. This observation gives us the best clue as to which trigonometric function to use; since tangent and cotangent are the two which don't depend on the length of the hypotenuse, chances are one of these is a good choice.

Note also that we have to convert the theodolite's reading into decimal form before using a calculator:

$$61°55'39'' = 61 + \frac{55}{60} + \frac{39}{3600} \approx 61.9275°.$$

Now we can use the figure to see that

$$\tan 61.9275° = \frac{\text{opp}}{40},$$

so

$$\text{opp} = 40 \tan 61.9275° \approx 75.$$

That is, the tree is approximately 75 feet tall.

example 5

The manufacturer of a certain brand of 16-foot ladder recommends that, when in use, the angle between the ground and the ladder should equal 75°. What distance should the foot of the ladder be from the base of the wall it is leaning against?

Solution:

Since we are given information about an angle, its adjacent side, and the hypotenuse of a right triangle, cosine is the logical trigonometric function to use in solving this problem (equivalently, secant could be used, but calculators are equipped with a "cos" and not a "sec" button so our current technology tends to lead to the use of cosine).

We want to determine the length of the adjacent side when the ladder is resting against the wall with its recommended angle of 75°, and we note that

$$\cos 75^\circ = \frac{\text{adj}}{16}.$$

This gives us

$$\text{adj} = 16\cos 75^\circ \approx 4.14 \text{ feet},$$

or a bit less than 4 feet, 2 inches.

In many surveying problems, it is frequently necessary to determine the height of some distant object when it is impossible or impractical to measure how far away the object is. One way to determine the height anyway begins with the diagram in Figure 2. In the diagram, assume that distance d and angles α and β can be measured, but that distance x is unknown. How can we determine height h?

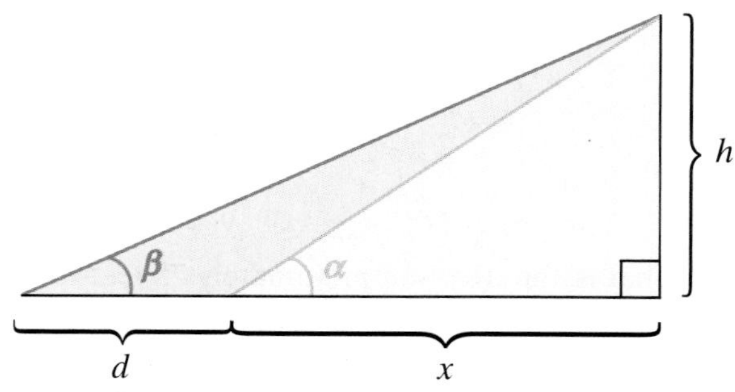

Figure 2: Determining h from Two Angles

There are two right triangles apparent in the diagram, and since we know or desire to know something about an angle and its opposite and adjacent sides in each triangle, the tangent function looks promising. We'll begin with the two trigonometric relations

$$\tan\alpha = \frac{h}{x}$$

and

$$\tan\beta = \frac{h}{x+d}.$$

The premise is that we want to find a formula for h in terms of α, β and d alone, since we don't know the length x. We can start by solving the two equations above for h to get $h = x\tan\alpha$ and $h = (x + d)\tan\beta$. These two equations actually are an example of a *system of two equations* in the two unknowns x and h, and we will study such systems in detail in Chapter 10. For our present purposes, we'll note that if we can solve one equation for x and use the result in the other equation, we'll obtain a single equation in which h is the only unknown (this is called the *method of substitution*). For instance, if we solve the first equation for x,

$$x = \frac{h}{\tan\alpha},$$

and make that substitution for x in the second equation, we obtain

$$h = \left(\frac{h}{\tan\alpha} + d\right)\tan\beta.$$

Now we can solve this equation for h:

$$h = \left(\frac{h}{\tan\alpha} + d\right)\tan\beta$$

$$h = \frac{h\tan\beta}{\tan\alpha} + d\tan\beta \qquad \text{Multiply out.}$$

$$h\tan\alpha = h\tan\beta + d\tan\alpha\tan\beta \qquad \text{Clear fractions by multiplying by } \tan\alpha.$$

$$h\tan\alpha - h\tan\beta = d\tan\alpha\tan\beta \qquad \text{Isolate terms with } h \text{ on one side.}$$

$$h(\tan\alpha - \tan\beta) = d\tan\alpha\tan\beta \qquad \text{Factor out } h.$$

$$h = \frac{d\tan\alpha\tan\beta}{\tan\alpha - \tan\beta}. \qquad \text{Divide to solve for } h.$$

This formula is well-known to surveyors, though it often appears in the slightly more appealing form obtained by dividing the numerator and denominator of the fraction on the right by tan α tan β:

$$\begin{aligned} h &= \frac{d \tan\alpha \tan\beta}{\tan\alpha - \tan\beta} \\ &= \frac{d}{\dfrac{\tan\alpha}{\tan\alpha\tan\beta} - \dfrac{\tan\beta}{\tan\alpha\tan\beta}} \\ &= \frac{d}{\dfrac{1}{\tan\beta} - \dfrac{1}{\tan\alpha}} \\ &= \frac{d}{\cot\beta - \cot\alpha}. \end{aligned}$$

example 6

Approached from one direction, Mt. Baldy rises out of a perfectly level desert plain. A surveyor standing in the desert some distance from the mountain measures the angle of elevation between the desert floor and the top of the mountain to be $60^\circ 1' 16''$. She then backs up 1000 feet and determines the new angle of elevation to be $56^\circ 3' 23''$. How high above the desert plain does Mt. Baldy rise?

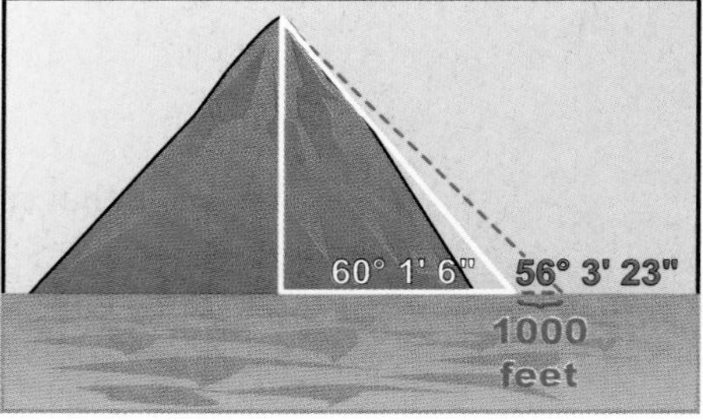

Solution:

Using the notation of the derivation above, we are given $\alpha = 60^\circ 1' 6''$, $\beta = 56^\circ 3' 23''$, and $d = 1000$. Converting to decimal notation, $\alpha \approx 60.0183^\circ$ and $\beta \approx 56.0564^\circ$, so

$$\begin{aligned} h &= \frac{1000}{\cot 56.0564^\circ - \cot 60.0183^\circ} \\ &\approx \frac{1000}{.6731 - .5769} \\ &\approx 10{,}395 \text{ feet.} \end{aligned}$$

exercises

Use the information contained in the figures to determine the values of the six trigonometric functions of θ. Rationalize all denominators. See Example 1.

1.

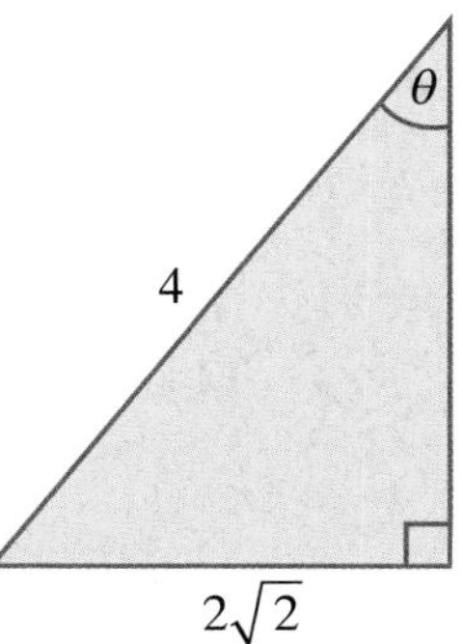

2.

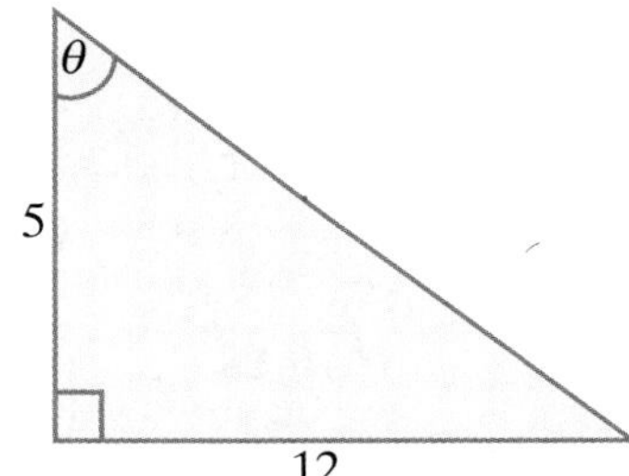

3.

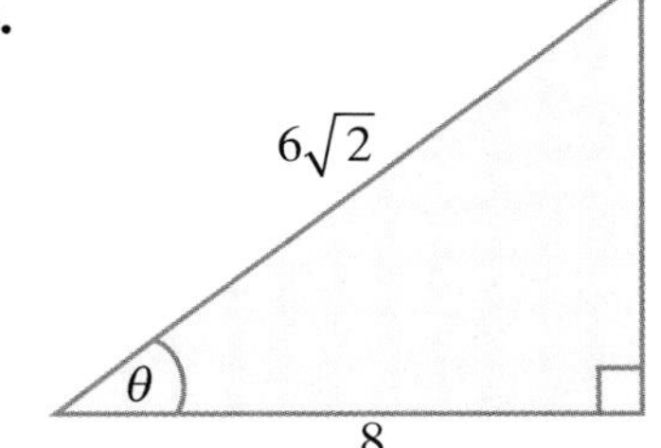

4.

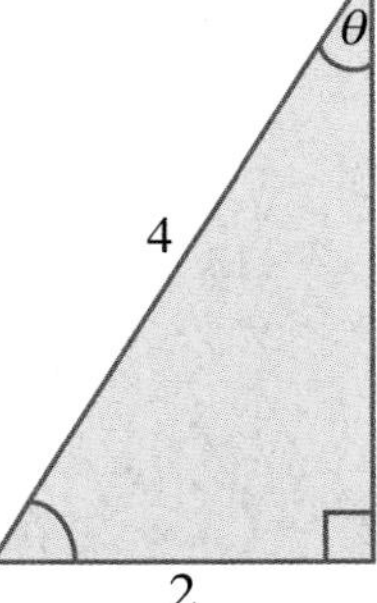

5.

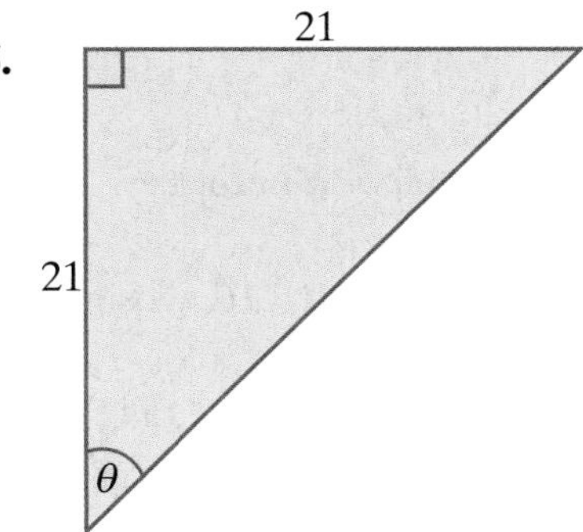

6.

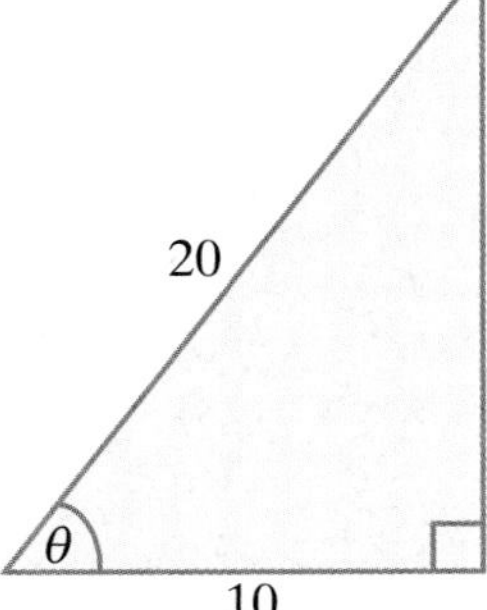

7.

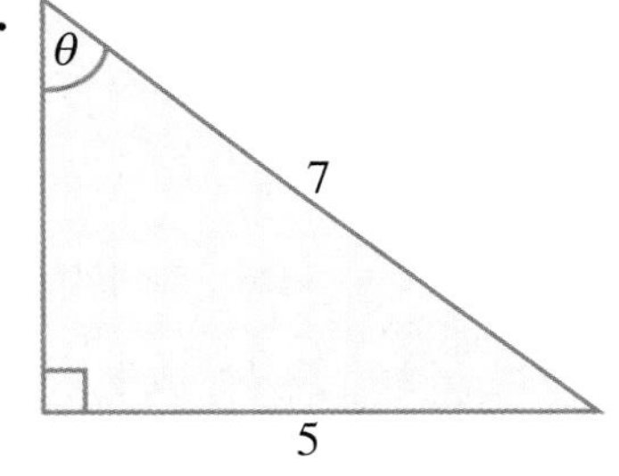

8.

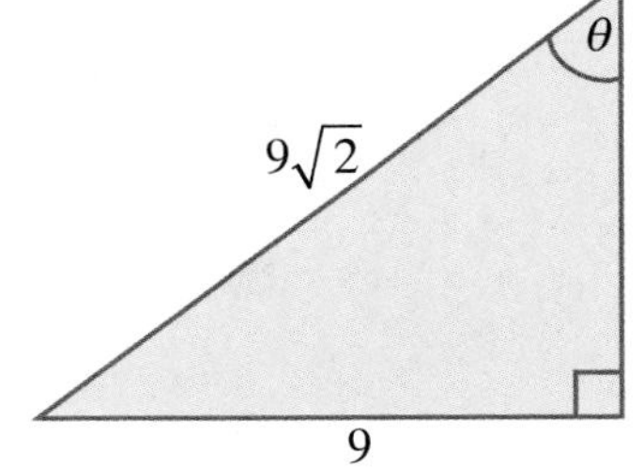

9.

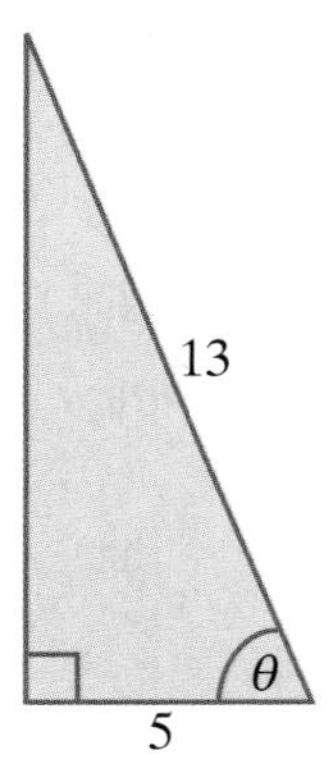

10.

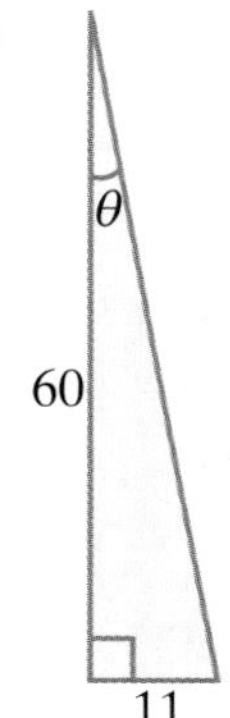

11.

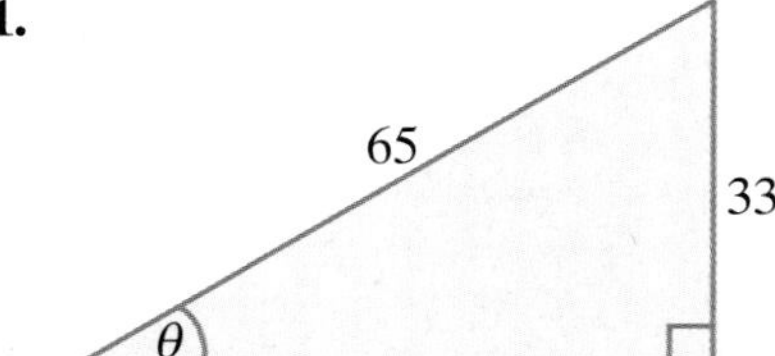

12.

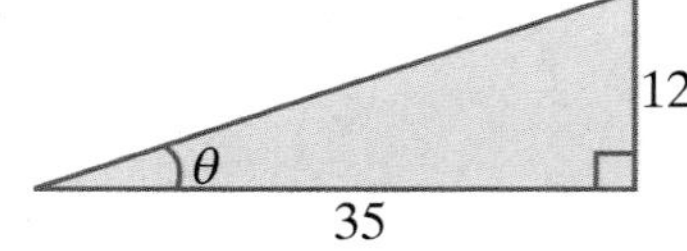

13.

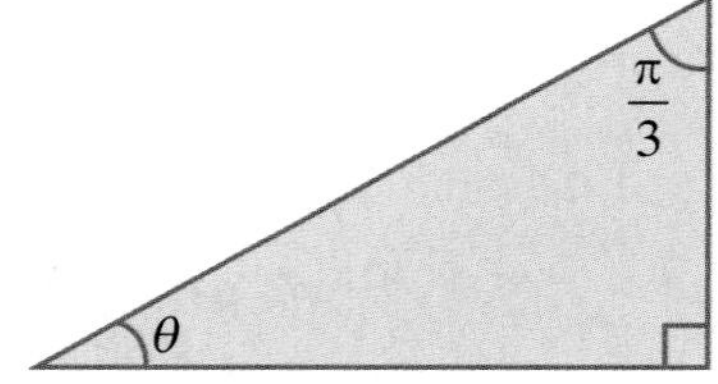

14.

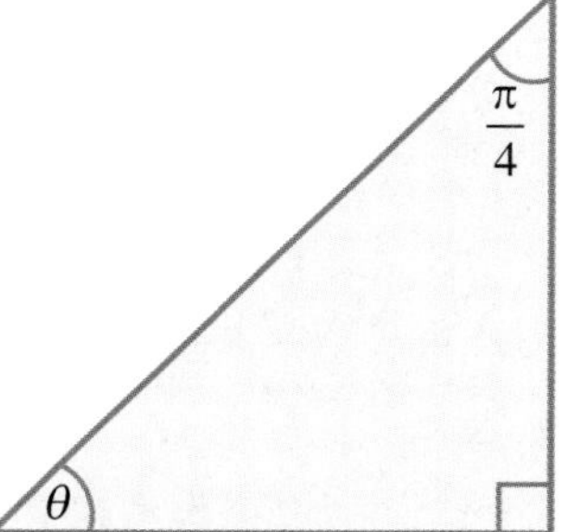

Evaluate the expressions, using a calculator if necessary. See Examples 2 and 3.

15. sine and cosecant of $\frac{\pi}{4}$

16. cosine and tangent of $\frac{\pi}{7}$

17. $\sec 60^\circ$

18. $\tan 71^\circ$ and $\cot 71^\circ$

19. $\csc \frac{\pi}{6}$

20. sine of $\frac{3\pi}{5}$

21. secant and tangent of 5°

22. cosine of 28.37°

23. cotangent of $\frac{\pi}{3}$

24. $\sin \frac{2\pi}{5}$ and $\cos \frac{2\pi}{5}$

25. $\tan 87.2^\circ$

26. $\csc 54^\circ$

Use a graphing calculator to evaluate the following expressions. Round off answers to four decimal places. See Example 3.

27. $\sin 94°$ **28.** $\cos 72°$ **29.** $\tan 146°$ **30.** $\csc 17°$

31. $\sec 88°$ **32.** $\cot 159°$ **33.** $\tan \dfrac{4\pi}{5}$ **34.** $\cos \dfrac{11\pi}{4}$

35. $\sin \dfrac{\pi}{8}$ **36.** $\cot \dfrac{7\pi}{3}$ **37.** $\sec \dfrac{8\pi}{3}$ **38.** $\csc \dfrac{7\pi}{11}$

Convert each expression from degrees, minutes, seconds (DMS) notation to decimal notation. Round off answers to four decimal places.

39. $38°54'19''$ **40.** $256°12'1''$ **41.** $124°78'90''$ **42.** $5°8'3650''$

43. $920°99'56''$ **44.** $90°30'600''$

Determine the value of the given trigonometric expression given the value of another trigonometric expression. Round off answers to four decimal places.

45. Find the $\sin \theta$, if the $\csc \theta = 8.7$.

46. Find the $\cos \theta$, if the $\sec \theta = \dfrac{-7}{4}$.

47. Find the $\tan \theta$, if the $\cot \theta = \dfrac{\sqrt{15}}{3}$.

48. Find the $\cot \theta$, if the $\tan \theta = 2.5$.

49. Find the $\sec \theta$, if the $\cos \theta = 0.2$.

50. Find the $\csc \theta$, if the $\sin \theta = -\dfrac{1}{5}$.

Determine whether the following statements are true or false. Use a graphing calculator when necessary.

51. If $\sin \theta = 0.8$, then $\csc \theta = 1.25$.

52. If $\cos \theta = 0.96$, then $\sec \theta = 1\dfrac{1}{24}$.

53. If $\tan \theta = 4\dfrac{4}{9}$, then $\cot \theta = 0.225$.

54. If $\sin \theta = 0.5625$, then $\csc \theta = 2.48$.

55. If $\cos \theta = .75$, then $\sec \theta = \dfrac{8}{3}$.

56. If $\tan \theta = 0.2540$, then $\cot \theta = 3.937$.

Use an appropriate trigonometric function and a calculator if necessary to solve the following problems. See Examples 4 and 5.

57. A hang glider wants to determine if a certain vertical cliff is a suitable height for her lift-off. From a distance of 40 yards, she measures the angle from the ground to its tip as $80°55'24''$. How high is the cliff in feet?

58. A mahi-mahi is hooked on 70 feet of fishing line, 10 feet of which is above the surface of the water. The angle of depression from the water's surface to the line is $40°$. How deep is the fish?

59. A filing cabinet is 3 feet and 4 inches tall from the floor. If a piece of string is stretched from the top of the cabinet to a point on the floor and the angle between the string and the floor is 11°, what is the length of the string?

60. A tree being cut down makes a 70° angle with the ground when the tip of the tree is directly above a spot that is 40 feet from the base of the tree. Find the height of the tree.

61. Stephen is standing 15 yards from a stream, but instead of walking directly towards the stream, he decides to take a more scenic (though straight-line) path to the stream. If the angle between the scenic route and the stream is 18°, how far did Stephen walk?

62. The builder of a parking garage wants to build a ramp at an angle of 16° that covers a horizontal span of 40 feet. What is the vertical rise of the ramp?

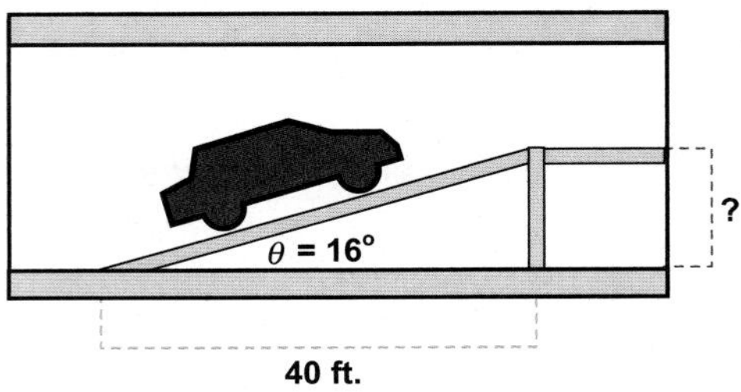

63. A kitesurfer's lines are 20 m long and make an angle of 37° with the ocean while heading away from the beach under current wind conditions. How high above the water is the kite flying?

64. An anthropologist studying a tribe of indigenous people wants to know the dimensions of their stone-hewn temple. After walking 15 meters from the structure, she measures the angle to its top to be 53°. What is the height of the temple?

65. A radio tower has a 64-foot shadow cast by the sun. If the angle from the tip of the shadow to the top of the tower is 78.5°, what is the height of the radio tower?

66. A ladder is propped up to a barn at an 80° angle. If the ladder is 22 feet long, what is the approximate height where the top of the ladder touches the barn?

67. The ramp of a moving truck touches the ground 12 feet away from the end of the truck. If the ramp makes an angle of 30° relative to the ground, what is the length of the ramp?

68. The angle of elevation of a flying kite is 61°7′21″. If the other end of the 40-foot long string attached to the kite is tied to the ground, what is the approximate height of the kite?

69. A length of rope is attached from the top of a dock to the rope tie device located on the underneath of the boat at the water's surface. The rope is 33 feet in length and has an angle of elevation relative to the surface of the water of 12°. How high above the water does the dock sit?

Use the formula from Example 6 to solve the following problems.

70. A surveyor wants to find the width of a river without crossing it. He sights an abandoned tire on the opposite bank (the banks are straight and parallel) and measures the angle from where he stands relative to the shore to be 31°. After walking precisely 15 feet away from the tire, he measures the same angle to be 13.5°. How wide is the river?

71. A drawbridge operator in a control room observes a sailboat approaching and finds the angle of depression to the boat to be 9°. Twenty minutes later, the angle to the same boat is 19°. If the sailboat has traveled 68.2 m, how high above water is the control room?

72. A birdwatcher discovers a hawk's nest in a tree some distance away. She wants to determine its height, so she measures the angle from the level ground to the nest at 40°. After approaching 25 feet closer to the tree, she finds the same angle to be 52.5°. How high does the nest sit, in feet?

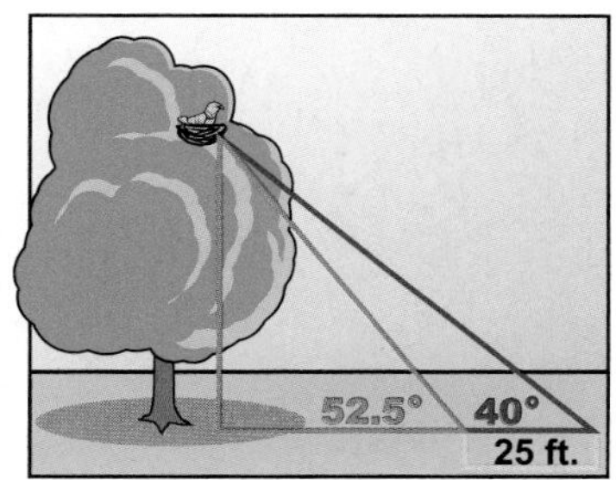

73. A surveyor standing some distance from a plateau measures the angle of elevation from the ground to the top of the plateau to be 46°57′12″. The surveyor then walks forward 800 feet and measures the angle of elevation to be 55°37′70″. What is the height of the plateau?

74. A surveyor standing some distance from a hill measures the angle of elevation from the ground to the top of the hill to be 83°45′97″. The surveyor then steps back 300 feet and measures the angle of elevation to be 75°44′16″. What is the height of the hill?

6.3 Trigonometric Functions of Any Angle

TOPICS

1. Extending the domains of the trigonometric functions
2. Evaluation using reference angles
3. Relationships between trigonometric functions

Topic 1: Extending the Domains of the Trigonometric Functions

The definitions given in the last section of the six trigonometric functions implicitly assumed that the angle under discussion was greater than 0 and less than $\frac{\pi}{2}$ radians. With just a little bit of extrapolation, the domain of definition of each function can be extended slightly as indicated in the table below:

Function	Initial Extended Domain (interval from which θ can be chosen)
$\sin\theta$	$\left[0, \frac{\pi}{2}\right]$ or $0 \le \theta \le \frac{\pi}{2}$
$\cos\theta$	$\left[0, \frac{\pi}{2}\right]$ or $0 \le \theta \le \frac{\pi}{2}$
$\tan\theta$	$\left[0, \frac{\pi}{2}\right)$ or $0 \le \theta < \frac{\pi}{2}$
$\csc\theta$	$\left(0, \frac{\pi}{2}\right]$ or $0 < \theta \le \frac{\pi}{2}$
$\sec\theta$	$\left[0, \frac{\pi}{2}\right)$ or $0 \le \theta < \frac{\pi}{2}$
$\cot\theta$	$\left(0, \frac{\pi}{2}\right]$ or $0 < \theta \le \frac{\pi}{2}$

The reasoning behind the extrapolation is as follows. Suppose that (x, y) is a point anywhere in the first quadrant of the Cartesian plane (including possibly on the x-axis or the y-axis). By drawing a line segment from the origin to (x, y) and another line segment from (x, y) vertically down to the x-axis (that is, the point $(x, 0)$), we obtain a right triangle with θ defined as pictured in Figure 1, where $r = \sqrt{x^2 + y^2}$.

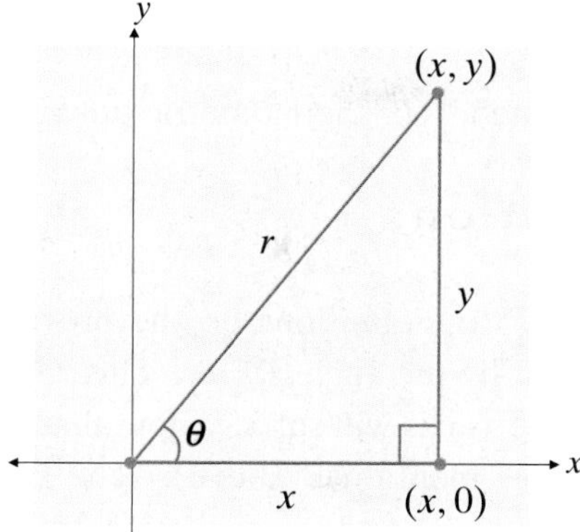

Figure 1: Trigonometric Functions in the First Quadrant

If either x or y is 0 (that is, if the point (x, y) lies on one of the coordinate axes), the triangle formed is *degenerate*. This is not a moral condemnation – it just means that two of the edges coincide so the figure doesn't appear to be a triangle. But as we will see, degenerate triangles won't affect our extended definitions of the trigonometric functions.

Recall that the original definitions of the functions were in terms of the hypotenuse and opposite and adjacent sides of a right triangle. Referring to Figure 1, we can rephrase and extend the definitions as follows:

$\sin\theta = \dfrac{y}{r}$	This now defines $\sin\theta$ for $0 \le \theta \le \dfrac{\pi}{2}$.
$\cos\theta = \dfrac{x}{r}$	This now defines $\cos\theta$ for $0 \le \theta \le \dfrac{\pi}{2}$.
$\tan\theta = \dfrac{y}{x}$ (for $x \ne 0$)	Note that the restriction $x \ne 0$ means $\theta \ne \dfrac{\pi}{2}$.
$\csc\theta = \dfrac{r}{y}$ (for $y \ne 0$)	Note that the restriction $y \ne 0$ means $\theta \ne 0$.
$\sec\theta = \dfrac{r}{x}$ (for $x \ne 0$)	Note that the restriction $x \ne 0$ means $\theta \ne \dfrac{\pi}{2}$.
$\cot\theta = \dfrac{x}{y}$ (for $y \ne 0$)	Note that the restriction $y \ne 0$ means $\theta \ne 0$.

Note that for a given angle θ in one of the intervals on the previous page, there are an infinite number of points (x, y) in the first quadrant that could be used in order to complete the picture in Figure 1 (any point lying along the ray rotated θ from the positive x-axis will work). But because all of the possible triangles thus formed are similar, any one of them suffices to define a given trigonometric function.

example 1

Evaluate all six trigonometric functions at $\theta = 0$, if possible.

Solution:

To use the definitions on the previous page, we need to pick a point along the ray defined by the angle $\theta = 0$. This ray lies along the positive x-axis, so any point on the positive x-axis will suffice (note that the angle $\theta = 0$ leads to one of the two degenerate triangles mentioned above). The first point on the positive x-axis that may come to mind is (1, 0), and this will certainly work. For this point, $x = 1$, $y = 0$, and $r = 1$. Now:

$$\sin 0 = \frac{0}{1} = 0, \ \cos 0 = \frac{1}{1} = 1, \ \tan 0 = \frac{0}{1} = 0$$

and

$$\csc 0 \text{ is undefined, } \sec 0 = \frac{1}{1} = 1, \ \cot 0 \text{ is undefined.}$$

The association between an angle θ and a point (x, y) on the ray defined by θ actually allows us to extend the domains of the trigonometric functions to a far greater extent. The table below simply repeats, for reference, the definitions on the previous page.

Extending the Domains of the Trig Functions

Given an angle θ in standard position, let (x, y) be any point (other than the origin) on the terminal side of the angle. Letting $r = \sqrt{x^2 + y^2}$ we define:

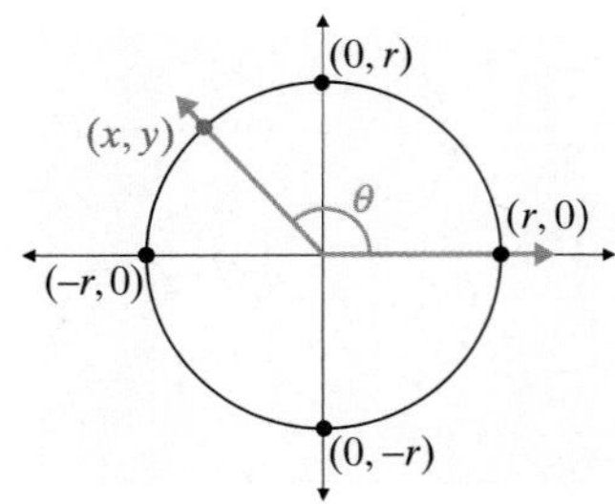

$\sin\theta = \dfrac{y}{r}$	$\cos\theta = \dfrac{x}{r}$	$\tan\theta = \dfrac{y}{x}$ (for $x \neq 0$)
$\csc\theta = \dfrac{r}{y}$ (for $y \neq 0$)	$\sec\theta = \dfrac{r}{x}$ (for $x \neq 0$)	$\cot\theta = \dfrac{x}{y}$ (for $y \neq 0$)

In other words,

- sin and cos are defined for all real numbers;
- tan and sec are defined for all real numbers except $\dfrac{\pi}{2} + n\pi$; and
- cot and csc are defined for all real numbers except $n\pi$

for some integer n.

We have now defined $\sin\theta$ and $\cos\theta$ for any real number θ, and the other four trigonometric functions have been defined for *nearly* any real number θ. The exact meaning of this last sentence will become clear as you study the remaining examples in this chapter.

example 2

Determine the values of the six trigonometric functions of the angle θ.

a. $\theta = -\frac{5\pi}{2}$ **b.** $\theta = 120^\circ$

Solution:

a. Recall that a negative angle corresponds to a clockwise rotation from the positive x-axis. The angle $\theta = -\frac{5\pi}{2}$ indicates one full revolution clockwise (-2π radians) plus another quarter revolution clockwise ($-\frac{\pi}{2}$ radians), resulting in the terminal side of the angle pointing straight down along the negative y-axis as shown. To evaluate the six trigonometric functions, we just need to choose a point on the terminal side of the angle; although it's a somewhat boring choice, the point $(0, -1)$ is probably easiest to work with (note that this gives us $r = 1$). Now:

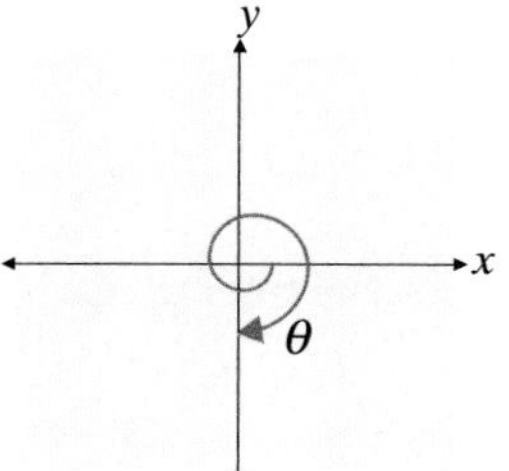

$$\sin\left(-\frac{5\pi}{2}\right) = \frac{-1}{1} = -1,\ \cos\left(-\frac{5\pi}{2}\right) = \frac{0}{1} = 0,\ \tan\left(-\frac{5\pi}{2}\right) = \frac{-1}{0} \text{ is undefined}$$

and

$$\csc\left(-\frac{5\pi}{2}\right) = \frac{1}{-1} = -1,\ \sec\left(-\frac{5\pi}{2}\right) = \frac{1}{0} \text{ is undefined},\ \cot\left(-\frac{5\pi}{2}\right) = \frac{0}{-1} = 0.$$

b. The angle $\theta = 120^\circ$ is 30° more than a (counter-clockwise oriented) right angle, leading to the triangle shown at right. Once we realize we are dealing with a $30^\circ - 60^\circ - 90^\circ$ triangle, we know that the ratios of the side-lengths must be $1:\sqrt{3}:2$ and we can easily locate the point $\left(-1, \sqrt{3}\right)$ on the terminal side of the angle.

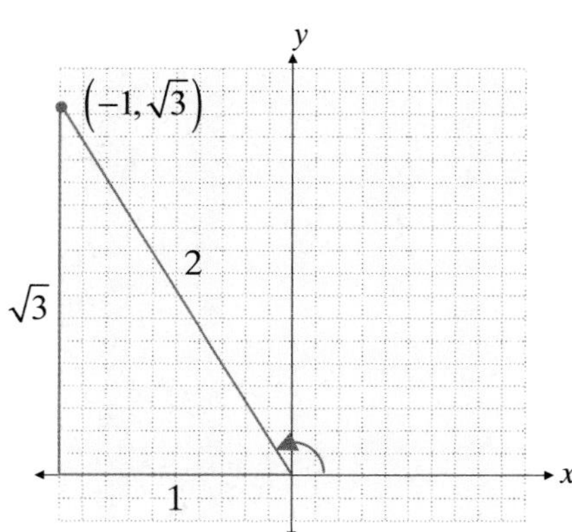

cont'd. on next page ...

The rest is straightforward. Using the point $\left(-1, \sqrt{3}\right)$ and the value $r = 2$ we determine that

$$\sin 120^\circ = \frac{\sqrt{3}}{2}, \ \cos 120^\circ = \frac{-1}{2}, \ \tan 120^\circ = \frac{\sqrt{3}}{-1} = -\sqrt{3}$$

and

$$\csc 120^\circ = \frac{2}{\sqrt{3}}, \ \sec 120^\circ = \frac{2}{-1} = -2, \ \cot 120^\circ = \frac{-1}{\sqrt{3}}.$$

Topic 2: Evaluation Using Reference Angles

The last example hinted at the fact that the evaluation of a trigonometric function at a non-acute angle can be related to its evaluation at an angle in the interval $\left[0, \frac{\pi}{2}\right]$. Such angles are called *reference angles*, and the precise definition is as follows.

Reference Angle

Given an angle θ in standard position, the **reference angle** θ' associated with it is the angle formed by the x-axis and the terminal side of θ. Reference angles are always greater than or equal to 0 and less than or equal to $\frac{\pi}{2}$ radians $\left(0 \le \theta' \le \frac{\pi}{2}\right)$.

Figure 2 illustrates four ways in which a given angle θ can relate to its reference angle θ'.

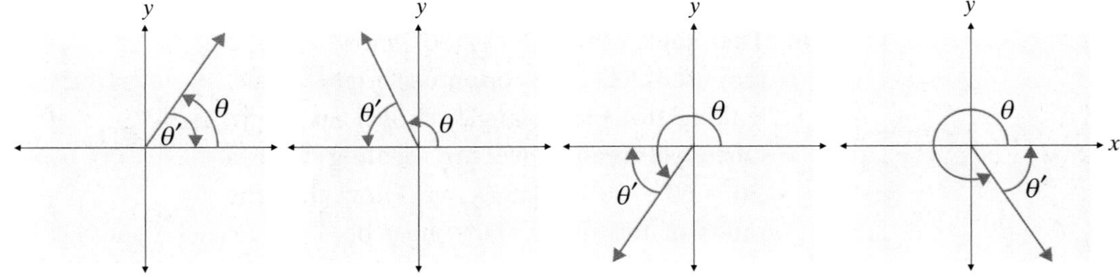

Figure 2: Angles and Associated Reference Angles

example 3

Determine the reference angle associated with each of the following angles.

a. $\theta = \dfrac{9\pi}{8}$ **b.** $\varphi = -655^\circ$

Solutions:

a. The terminal side of $\theta = \dfrac{9\pi}{8}$ lies in the third quadrant, so the reference angle is determined by it and the negative x-axis. Specifically,

$$\theta' = \frac{9\pi}{8} - \pi = \frac{\pi}{8}.$$

b. The terminal side of $\varphi = -655^\circ$ lies in the first quadrant (note that the angle describes one complete revolution and more than three-quarters of a second clockwise revolution around the origin). An additional clockwise rotation of 65° would result in two full revolutions, so that is the reference angle. That is,

$$2(360^\circ) - 655^\circ = 65^\circ.$$

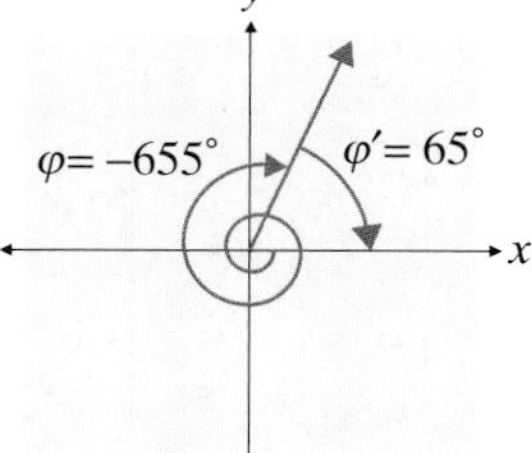

The value of reference angles lies in the fact that a trigonometric function evaluated at a given angle will be the same as the function evaluated at the reference angle, except possibly for sign. We implicitly used this fact in Example 2, and the reason why it's true is not hard to see. Since we have defined the trigonometric functions in terms of the coordinates of a point chosen on the terminal side of an angle, the value doesn't depend at all on how many revolutions around the origin (or in which direction) the angle describes. In Example 2b, for instance, the key step lay in determining that the reference angle for 120° is 60°. This led to the construction of a $30^\circ - 60^\circ - 90^\circ$ triangle and the easy evaluation of all six trigonometric functions.

Sign is the one thing that may differ between the value of a trigonometric function at θ and the value of the same function at the reference angle θ'. By thinking about the signs of the x and y coordinates of points in the four quadrants, it's easy to see that all trig functions are positive in the first quadrant, that sine (and its reciprocal cosecant) are positive in the second quadrant, that tangent (and its reciprocal cotangent) are positive in the third quadrant, and that cosine (and its reciprocal secant) are positive in the fourth quadrant. A bit of propaganda has evolved as a mnemonic to help students remember this: "All Students Take Calculus" may remind you that, beginning in the first quadrant, the functions that are positive are:

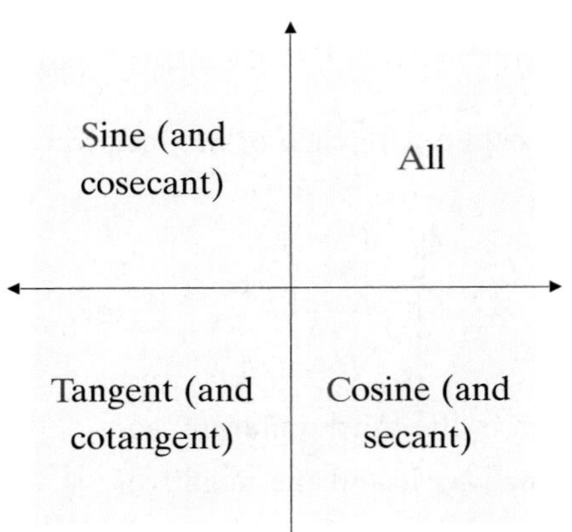

Signs of the Trigonometric Functions

Quadrant	Positive	Negative
I	all	none
II	sin, csc	cos, sec, tan, cot
III	tan, cot	sin, csc, cos, sec
IV	cos, sec	sin, csc, tan, cot

example 4

Evaluate the following:

a. $\cos \frac{7\pi}{6}$ **b.** $\tan(-225°)$

Solutions:

a. The terminal side of $\frac{7\pi}{6}$ lies in the third quadrant and its reference angle is $\frac{\pi}{6}$. $\cos \frac{\pi}{6} = \frac{\sqrt{3}}{2}$, but cosine is negative in the third quadrant, so $\cos \frac{7\pi}{6} = -\frac{\sqrt{3}}{2}$.

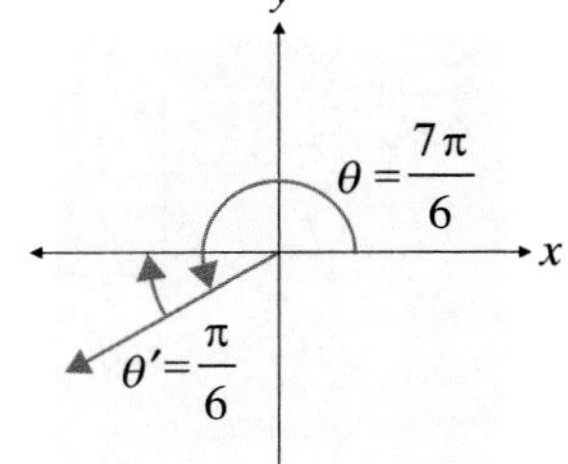

b. The terminal side of $-225°$ lies in the second quadrant and its reference angle is $45°$. Since tangent is negative in the second quadrant, $\tan(-225°) = -\tan(45°) = -1$.

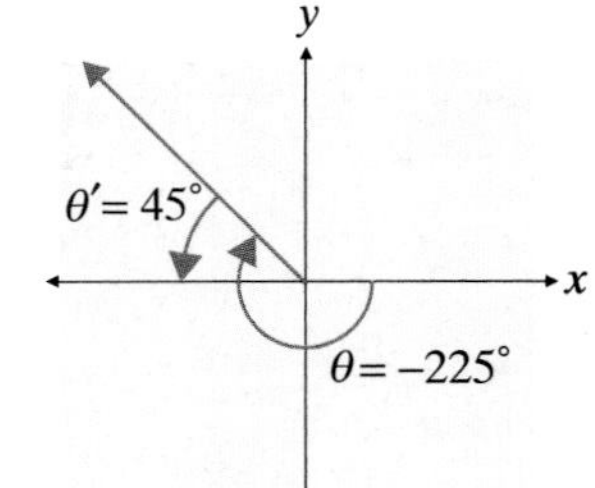

Topic 3: Relationships Between Trigonometric Functions

If you have been studying the definitions and examples in this chapter carefully, you may be starting to develop the feeling that there is a great deal of redundancy in trigonometry. This is no illusion; as specific examples, we know now that there are two ways to measure angles and that three of the trigonometric functions are just reciprocals of the other three. There are other redundancies that are a bit more subtle, but you may have developed a sense of them already. We will end this section with a discussion of the qualities of trigonometric functions that lead to this feeling.

The preceding paragraph should not be taken to mean, however, that some of what you are learning in this chapter is pointless. Take, for example, the fact that cosecant, secant, and cotangent are simply reciprocals of sine, cosine, and tangent, and therefore seem unnecessary. In Calculus, you'll encounter problems that are more easily stated and solved in terms of, for example, secant rather than cosine. This argument (that a problem is easier to solve with the choice of one function over another) is potent, and can't be disregarded. Why discard something if its existence makes life easier?

Another reason, if not a justification, for the existence of all six trigonometric functions is historical. This history is worth spending a paragraph or two on just for the light it sheds on the nomenclature of trigonometry, but as it turns out the nomenclature in turn leads to some useful facts.

Starting with a circle of radius 1, construct the lines as shown in Figure 3. The words *secant* and *tangent* are derived from Latin names for, respectively, the lines $\overline{OB}$ and $\overline{AB}$ (we use "tangent" in everyday language to describe a situation where one object is just touching another). If the line passing through C and D were continued down until it intersected the circle again, it would form a *chord*; as it is, $\overline{CD}$ forms a half-chord and *sine* is the word that evolved to denote its length.

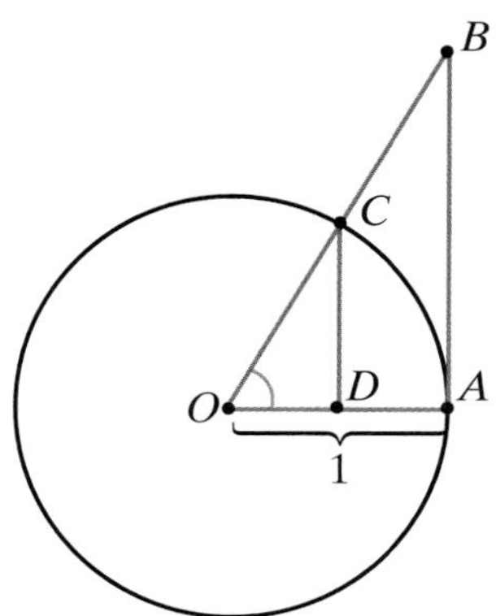

Figure 3: Etymology of the Trigonometric Functions

Make sure you see how the lengths of the line segments $\overline{OB}$, $\overline{AB}$, and $\overline{CD}$ represent, respectively, the secant, tangent, and sine of the central angle shown in blue (hint: it's important to remember that $\overline{OA}$ and $\overline{OC}$ have length 1). The next step is to see how these three functions lead to the remaining three trigonometric functions (cosine, cosecant, and cotangent).

Complementary angles are two angles which, when put together, form a right angle. That is, the sum of their measures is $\frac{\pi}{2}$ (or 90° in terms of degree measure). The cosine, cosecant, and cotangent of an angle θ are, respectively, the sine, secant, and tangent of the *complement* of θ. In terms of formulas, this gives us our first set of identities.

Cofunction Identities

Given an angle θ (measured in radians), $\frac{\pi}{2}-\theta$ is the measure of its complement. Thus:

$$\cos\theta = \sin\left(\frac{\pi}{2}-\theta\right), \ \csc\theta = \sec\left(\frac{\pi}{2}-\theta\right), \text{ and } \cot\theta = \tan\left(\frac{\pi}{2}-\theta\right).$$

We will encounter many more identities in the sections to come. In general, trigonometric identities are equations that are useful in simplifying or evaluating expressions. For reference purposes, we list here another set of three identities that you know well by now:

Reciprocal Identities

For a given angle θ for which both sides of the equation make sense,

$$\csc\theta = \frac{1}{\sin\theta}, \ \sec\theta = \frac{1}{\cos\theta}, \text{ and } \cot\theta = \frac{1}{\tan\theta}.$$

We'll finish this initial list of identities with two that you may have already noted. Recall our definitions of tangent and cotangent in terms of a point (x, y) on the terminal side of an angle θ:

$$\tan\theta = \frac{y}{x} \text{ and } \cot\theta = \frac{x}{y}.$$

Since $\sin\theta = \frac{y}{r}$ and $\cos\theta = \frac{x}{r}$, the following identities are apparent.

Quotient Identities

For a given angle θ for which both sides of the equation make sense,

$$\tan\theta = \frac{\sin\theta}{\cos\theta} \text{ and } \cot\theta = \frac{\cos\theta}{\sin\theta}.$$

example 5

Express each of the following in terms of the appropriate cofunction, and verify the equivalence of the two expressions.

a. $\cos\left(-\frac{5\pi}{11}\right)$

b. $\cot 195°$

Solutions:

a. $\cos\left(-\frac{5\pi}{11}\right) = \sin\left(\frac{\pi}{2} - \left(-\frac{5\pi}{11}\right)\right)$ The cosine of an angle is equal to the sine of the complement of the angle.

$= \sin\left(\frac{11\pi}{22} + \frac{10\pi}{22}\right)$ Simplify the argument.

$= \sin\left(\frac{21\pi}{22}\right).$ Now use a calculator to verify that $\cos\left(-\frac{5\pi}{11}\right)$ and $\sin\left(\frac{21\pi}{22}\right)$ are both approximately 0.1423.

b. $\cot 195° = \tan\left(90° - 195°\right)$

$= \tan\left(-105°\right).$

The same relation between cotangent and tangent applies, though the angles are measured in degrees in this example. Verify that both sides are approximately 3.7321.

We'll conclude this section with examples that illustrate another way to use the relationships between trigonometric functions.

example 6

Given that $\cos\theta = -\frac{\sqrt{3}}{2}$ and $\tan\theta$ is negative, determine θ and $\tan\theta$.

Solution:

Since $\cos\theta$ is negative, we know the terminal side of θ must lie in either the second or third quadrant (remember: "All Students Take Calculus" reminds us the cosine is only positive in the first and fourth quadrants). Given that $\tan\theta$ is also negative, the terminal side of θ must lie in the second quadrant (tangent is positive in the third). A diagram is helpful at this point:

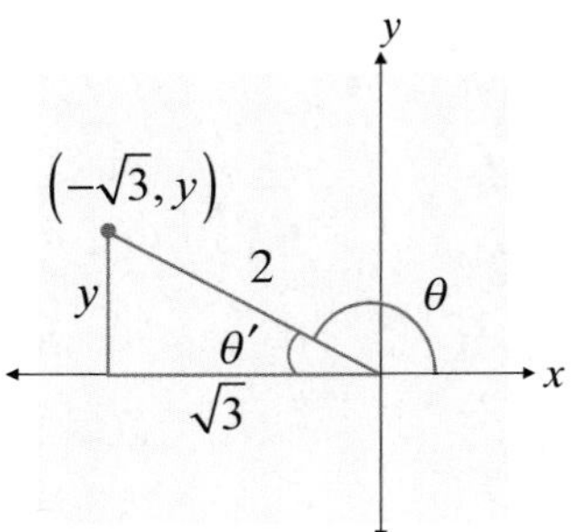

In the diagram, we've drawn θ with its terminal side in Quadrant II and we've drawn the reference angle θ'. We've also noted the relative magnitudes of the adjacent leg and hypotenuse of the right triangle, as indicated by the fact that $\cos\theta = -\frac{\sqrt{3}}{2}$. Of course, the actual lengths don't have to be $\sqrt{3}$ and 2, respectively, but they do have to be some multiple of $\sqrt{3}$ and 2, so we may as well use the simplest choice.

What we were not given initially is the length labeled "y" and the actual angle θ. But from the diagram, we can now recognize that the triangle is a familiar one and that $\theta' = \frac{\pi}{6}$, so $\theta = \frac{5\pi}{6}$. Finally, it must be the case that $y = 1$ and therefore $\tan\theta = -\frac{1}{\sqrt{3}}$.

example 7

Given that $\cot\theta = 0.4$ and that θ lies in the first quadrant, determine $\sin\theta$.

Solution:

All trigonometric functions are ratios, so it will probably be useful to express cotangent as a fraction. The result will help us construct a right triangle that relates to the given information. To that end, note that $\cot\theta = 0.4 = \frac{4}{10} = \frac{2}{5}$. If we take the numerator and denominator as the lengths of the adjacent and opposite sides of a right triangle, we are led to the diagram on the next page.

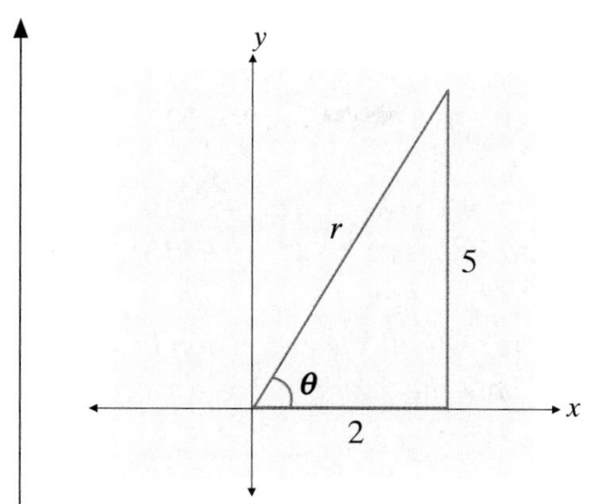

Now we can use the Pythagorean Theorem to determine that $r = \sqrt{4+25} = \sqrt{29}$, and so $\sin\theta = \dfrac{5}{\sqrt{29}}$.

exercises

Evaluate all six trigonometric functions at the given θ, using a calculator if necessary. See Examples 1 and 2.

1. $\theta = 45°$ **2.** $\theta = \dfrac{\pi}{2}$ **3.** $\theta = 60°$ **4.** $\theta = \dfrac{3\pi}{4}$ **5.** $\theta = \dfrac{5\pi}{2}$

6. $\theta = -520°$ **7.** $\theta = 305°$ **8.** $\theta = -1105°$ **9.** $\theta = 6\pi$ **10.** $\theta = 670°$

11. $\theta = \dfrac{3\pi}{2}$ **12.** $\theta = -215°$ **13.** $\theta = \dfrac{5\pi}{4}$ **14.** $\theta = 780°$ **15.** $\theta = -445°$

Determine the reference angle associated with the given angle. See Example 3.

16. $\varphi = 98°$ **17.** $\theta = \dfrac{9\pi}{2}$ **18.** $\varphi = -60°$ **19.** $\theta = \dfrac{5\pi}{4}$ **20.** $\theta = \dfrac{5\pi}{2}$

21. $\varphi = 313°$ **22.** $\theta = \dfrac{7\pi}{6}$ **23.** $\varphi = -168°$ **24.** $\theta = \dfrac{6\pi}{5}$ **25.** $\varphi = 216°$

26. $\theta = \dfrac{3\pi}{2}$ **27.** $\varphi = -330°$ **28.** $\theta = \dfrac{7\pi}{4}$ **29.** $\varphi = 718°$ **30.** $\varphi = 105°$

Determine which of the four quadrants angle θ is located in.

31. $\sin\theta > 0$ and $\tan\theta < 0$ **32.** $\sin\theta < 0$ and $\cos\theta > 0$ **33.** $\tan\theta > 0$ and $\sec\theta > 0$

34. $\cos\theta > 0$ and $\cot\theta < 0$ **35.** $\sec\theta < 0$ and $\csc\theta < 0$ **36.** $\cot\theta > 0$ and $\csc\theta > 0$

37. $\cot\theta > 0$ and $\cos\theta < 0$ **38.** $\sin\theta > 0$ and $\sec\theta < 0$

Match the angle θ in questions 39 – 48 with the correct reference angle θ' in choices a – c. Answers will be used more than once.

a. $\theta' = 30°$ **b.** $\theta' = 45°$ **c.** $\theta' = 60°$

39. $\theta = 300°$ **40.** $\theta = 150°$ **41.** $\theta = -135°$ **42.** $\theta = 210°$ **43.** $\theta = -120°$

44. $\theta = 315°$ **45.** $\theta = 510°$ **46.** $\theta = 600°$ **47.** $\theta = 855°$ **48.** $\theta = 480°$

*In each of the following problems, **(a)** rewrite the expression in terms of the given angle's reference angle, and then **(b)** evaluate the result, using a calculator if necessary. See Example 4.*

49. $\tan 98°$ **50.** $\sin\frac{9\pi}{2}$ **51.** $\cos -60°$ **52.** $\tan\frac{5\pi}{4}$ **53.** $\cos\frac{5\pi}{2}$

54. $\sin 313°$ **55.** $\cos\frac{7\pi}{6}$ **56.** $\tan -168°$ **57.** $\cos\frac{6\pi}{5}$ **58.** $\sin 216°$

59. $\tan\frac{3\pi}{2}$ **60.** $\cos -330°$ **61.** $\sin\frac{7\pi}{4}$ **62.** $\tan 718°$ **63.** $\sin 105°$

Use the appropriate identity to answer the following questions. Choose only one answer per question.

64. Which choice is equivalent to $\sin 18°$?

a. $\tan 72°$ **b.** $\cos 72°$ **c.** $\csc 72°$ **d.** $\sec 162°$ **e.** $\cos 162°$

65. Which choice is equivalent to $\sec \frac{\pi}{6}$?

a. $\csc \frac{\pi}{3}$ **b.** $\cos \frac{\pi}{2}$ **c.** $\sin \frac{\pi}{6}$ **d.** $\cos \frac{\pi}{3}$ **e.** $\tan \frac{\pi}{6}$

66. Which choice is equivalent to $\tan \frac{\pi}{12}$?

a. $\sin \frac{\pi}{2}$ **b.** $\cos \frac{\pi}{12}$ **c.** $\cot \frac{\pi}{2}$ **d.** $\cot \frac{\pi}{12}$ **e.** $\cot \frac{5\pi}{12}$

67. Which choice is equivalent to $\cos 87°$?

a. $\sin 93°$ **b.** $\cos 93°$ **c.** $\sin 273°$ **d.** $\sec 3°$ **e.** $\sin 3°$

Express each of the following in terms of the appropriate cofunction, and verify the equivalence of the two expressions. See Example 5.

68. $\cot 135°$ **69.** $\sec\frac{\pi}{2}$ **70.** $\sin -60°$ **71.** $\cos\left(-\frac{3\pi}{4}\right)$ **72.** $\csc\frac{5\pi}{6}$

73. $\cot 313°$ **74.** $\cos\left(\frac{-3\pi}{6}\right)$ **75.** $\csc -168°$ **76.** $\sin\left(\frac{-4\pi}{5}\right)$ **77.** $\sec 216°$

78. $\csc\frac{3\pi}{2}$ **79.** $\cos -15°$ **80.** $\cot\frac{\pi}{4}$ **81.** $\tan -105°$ **82.** $\sec 105°$

Using a graphing calculator, determine the tangent and cotangent for each question. Round each answer to three decimal places.

83. $\sin\theta = 0.978$ and $\cos\theta = 0.208$

84. $\sin\theta = 0.588$ and $\cos\theta = -0.809$

85. $\sin\theta = -0.966$ and $\cos\theta = -0.259$

86. $\sin\theta = -0.866$ and $\cos\theta = -0.5$

87. $\sin\theta = -0.699$ and $\cos\theta = 0.743$

88. $\sin\theta = -0.995$ and $\cos\theta = -0.105$

The three cofunction identities presented in this section have three companion identities, as follows:

$$\sin\theta = \cos\left(\frac{\pi}{2}-\theta\right),\ \sec\theta = \csc\left(\frac{\pi}{2}-\theta\right),\ \text{and}\ \tan\theta = \cot\left(\frac{\pi}{2}-\theta\right).$$

89. Prove these three identities. (Hint: For the first identity, begin with the observation that $\sin\theta = \sin\left(\frac{\pi}{2}-\left(\frac{\pi}{2}-\theta\right)\right)$ and then apply one of the three original cofunction identities.)

Use the given information about each angle to evaluate the expressions. See Examples 6 and 7.

90. Given that $\cos\theta = \frac{\sqrt{12}}{4}$ and $\tan\theta$ is negative, determine θ and $\tan\theta$.

91. Given that $\csc\theta = 0.6$ and θ lies in the second quadrant, determine $\cot\theta$.

92. Given that $\tan\theta = \frac{\sqrt{3}}{3}$ and $\sin\theta$ is positive, determine θ and $\sin\theta$.

93. Given that $\sec\theta = 0.3$ and θ lies in the fourth quadrant, determine $\csc\theta$.

Express each of the following in terms of the appropriate cofunction. Check the term given and the cofunction found with your calculator to verify the expressions are equivalent.

94. $\sin\frac{7\pi}{4}$

95. $\csc\frac{8\pi}{3}$

96. $\cot\frac{3\pi}{4}$

97. $\cos -\frac{5\pi}{3}$

98. $\tan 15°$

99. $\sec -315°$

6.4 Graphs of Trigonometric Functions

TOPICS

1. Graphing the basic trigonometric functions
2. Periodicity and other observations
3. Graphing transformed trigonometric functions
4. Interlude: damped harmonic motion

Topic 1: Graphing the Basic Trigonometric Functions

In the preceding sections, you have been exposed to some of the highlights of more than 2000 years of thought regarding trigonometry and its uses. With the fundamentals out of the way, it's time to reflect on a subtle but important point.

Most of the applications of the trigonometric functions implicitly view them as either functions of (acute) angles or as functions of real numbers. (Incidentally, the trigonometric functions can be extended even further to be functions of complex numbers, but that discussion will have to wait for a later course.) We've had quite a bit of experience with applications of the first sort; chances are, if you sketch a triangle in the course of solving a problem, you're using the "angle" point of view. In this section, we will concentrate on the second point of view.

To emphasize the theme of this section, we will now frequently use x to represent the argument of a given trigonometric function, and x will be a variable standing for (nearly) any real number. To start with, we know from Section 3 that sine and cosine are functions of any real number x, so the first order of business will be to graph them as functions defined on the real line. The process will be familiar to you – we have taken similar steps to graph the functions we encountered in Chapters 3, 4, and 5.

Recall that the most elementary approach to graphing a function for the first time is to plot points. That is, evaluate a given function $f(x)$ for a sufficient number of values x so that, when all the calculated points of the form $(x, f(x))$ are plotted, a reasonable guess for the graph of f can be constructed. We'll do this for the two functions sine and cosine at the same time.

For instance, for $x = 0$ we know that $\sin 0 = 0$ and $\cos 0 = 1$. We could use a calculator to confirm these facts, but it's better to *understand* why the values are what they are. In this context, letting $x = 0$ tells us to consider a degenerate triangle whose initial and terminal sides both lie along the positive x-axis, and the definitions of sine and cosine in Section 3 then lead to

$$\sin 0 = \frac{0}{1} = 0 \text{ and } \cos 0 = \frac{1}{1} = 1.$$

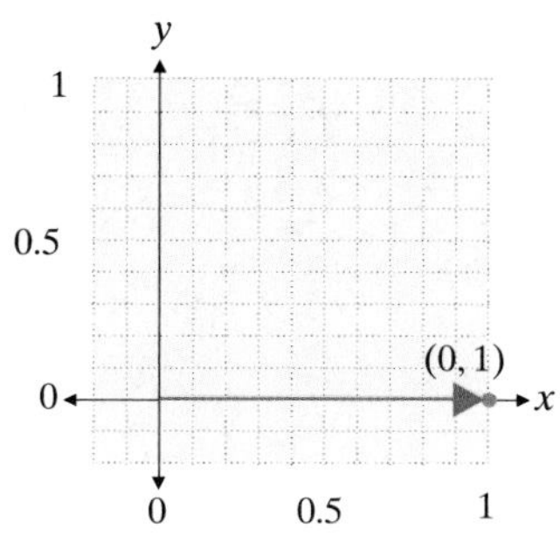

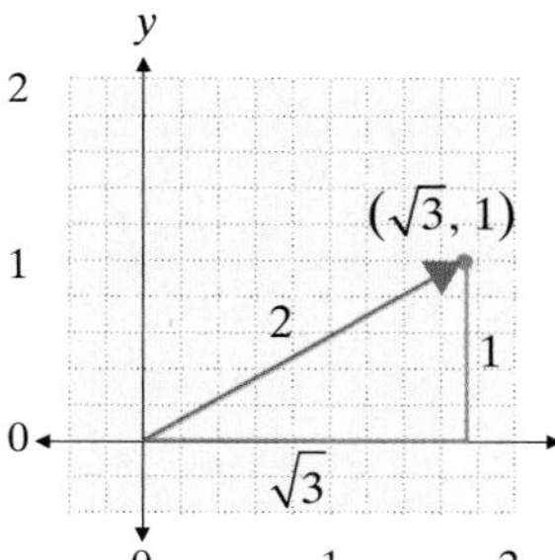

Similarly, for $x = \frac{\pi}{6}$ we know

$$\sin \frac{\pi}{6} = \frac{1}{2} \text{ and } \cos \frac{\pi}{6} = \frac{\sqrt{3}}{2}.$$

If we let x take on other easily-worked-with values between 0 and 2π, we obtain the following table:

x	$\sin x$	$\cos x$
0	0	1
$\frac{\pi}{6}$	$\frac{1}{2}$	$\frac{\sqrt{3}}{2}$
$\frac{\pi}{4}$	$\frac{1}{\sqrt{2}}$	$\frac{1}{\sqrt{2}}$
$\frac{\pi}{3}$	$\frac{\sqrt{3}}{2}$	$\frac{1}{2}$
$\frac{\pi}{2}$	1	0
$\frac{2\pi}{3}$	$\frac{\sqrt{3}}{2}$	$-\frac{1}{2}$
$\frac{3\pi}{4}$	$\frac{1}{\sqrt{2}}$	$-\frac{1}{\sqrt{2}}$
$\frac{5\pi}{6}$	$\frac{1}{2}$	$-\frac{\sqrt{3}}{2}$

x	$\sin x$	$\cos x$
π	0	-1
$\frac{7\pi}{6}$	$-\frac{1}{2}$	$-\frac{\sqrt{3}}{2}$
$\frac{5\pi}{4}$	$-\frac{1}{\sqrt{2}}$	$-\frac{1}{\sqrt{2}}$
$\frac{4\pi}{3}$	$-\frac{\sqrt{3}}{2}$	$-\frac{1}{2}$
$\frac{3\pi}{2}$	-1	0
$\frac{5\pi}{3}$	$-\frac{\sqrt{3}}{2}$	$\frac{1}{2}$
$\frac{7\pi}{4}$	$-\frac{1}{\sqrt{2}}$	$\frac{1}{\sqrt{2}}$
$\frac{11\pi}{6}$	$-\frac{1}{2}$	$\frac{\sqrt{3}}{2}$

Figure 1: Selected Values of Sine and Cosine

This table represents the values of sine and cosine as x assumes the value of convenient angles, beginning with $x = 0$ and rotating around counter-clockwise for one full circle. Of course, there's no need to let x take on higher values, or negative numbers for

that matter, since the above table will simply repeat. For instance, $\sin 2\pi = \sin 0$ and $\sin\left(-\frac{\pi}{6}\right) = \sin\left(\frac{11\pi}{6}\right)$. This observation is a recognition of the *periodicity* of sine and cosine; we say that these two functions both have a period of 2π. We will discuss the periodicity of all trigonometric functions more thoroughly soon.

Figure 2 shows the result of plotting the calculated points for sine, with the red curve drawn to smoothly pass through the points.

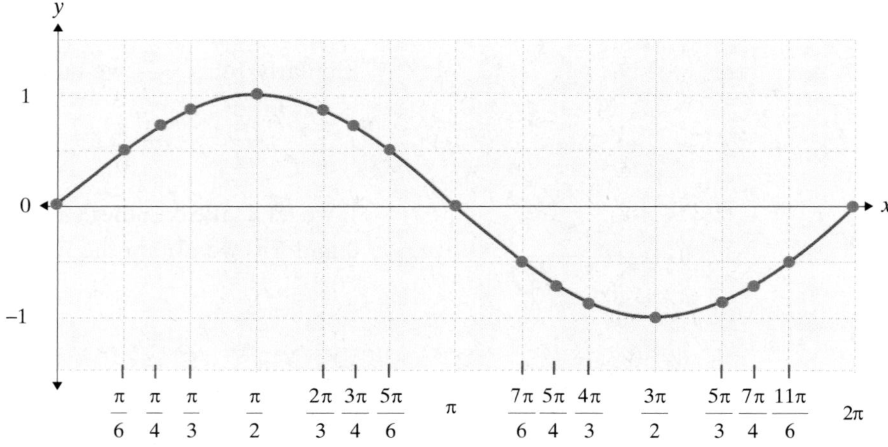

Figure 2: Graph of Sine between 0 and 2π

Similarly, Figure 3 contains the result of plotting the calculated points for cosine.

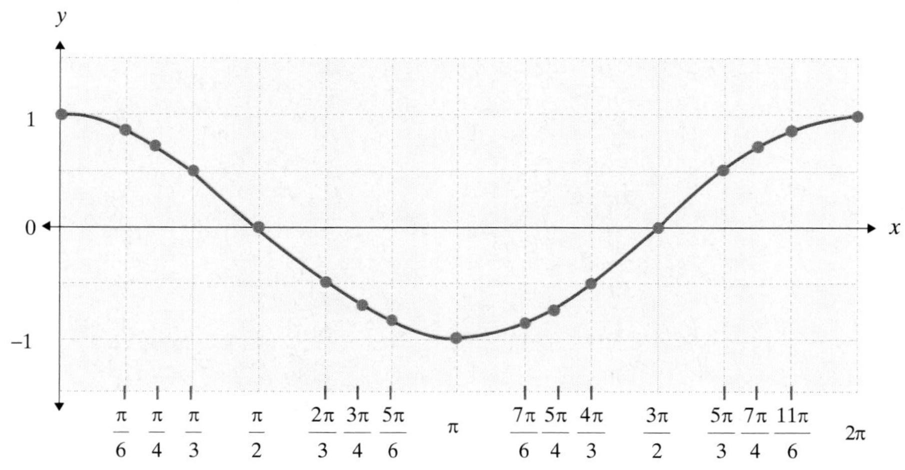

Figure 3: Graph of Cosine between 0 and 2π

One fact that leaps out from a glance at the graphs is that both sine and cosine take on values only between –1 and 1. This makes perfect sense, of course, since the length of the hypotenuse of a right triangle is always greater than or equal to the length of either leg, but the graphs drive this point home in a way that tables of figures don't. Another fact that is starting to appear is that sine and cosine seem to have similar shapes, one shifted horizontally with respect to the other. This is clearer if we extend the graphs to more of the real line, making use of each function's periodicity.

In Figure 4, both functions are graphed over a longer interval of the x-axis, with sine in blue and cosine in red.

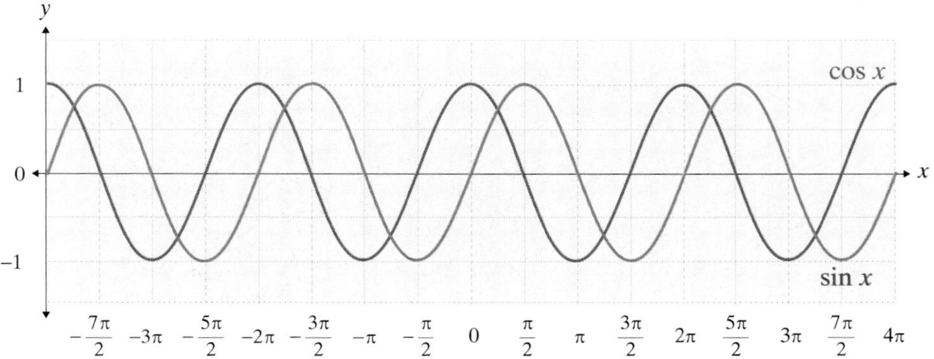

Figure 4: Graphs of Sine and Cosine

In the examples that follow, we will take similar steps to construct sketches of graphs of various functions. As with nearly all of our graphing, however, our intent is not to compete with graphing calculators and computer software for the most accurate pictures. Instead, our goal is to understand the behavior of functions and be able to pick out important qualities like periodicity, symmetry, the existence of asymptotes, and so on.

example 1

Sketch the graph of the cosecant function.

Solution:

Since cosecant is the reciprocal of sine, the data in Figure 1 and the graph in Figure 4 are sufficient to give us the understanding we seek. First, we note that $\sin x = 0$ when x is any multiple of π, so csc x is undefined for the same values of x. That is, the domain of the cosecant function is all real numbers *except* $\pm\{0, \pi, 2\pi, 3\pi, \ldots\}$. We knew this already – this corresponds to the restriction in the general definition of cosecant given in Section 6.3. Second, in the interval $[0, \pi]$, the graph of sine is symmetric with respect to

cont'd. on next page ...

the line $x = \dfrac{\pi}{2}$, so the graph of cosecant will have the same property. Similar statements hold for other intervals. And since sine is non-negative (and less than or equal to 1) for all x in $[0, \pi]$, the reciprocal of sine will also be non-negative (but greater than or equal to 1) for the same x.

Since $\sin x \to 0$ as x approaches any multiple of π, we know that cosecant will have vertical asymptotes at these points. Extending the reasoning in the last paragraph, we also know that $\csc x \geq 1$ on the intervals $\ldots, (-4\pi, -3\pi), (-2\pi, -\pi), (0, \pi), (2\pi, 3\pi), \ldots$ and that $\csc x \leq -1$ on the intervals $\ldots, (-3\pi, -2\pi), (-\pi, 0), (\pi, 2\pi), (3\pi, 4\pi), \ldots$. Putting all this together, we obtain the following sketch:

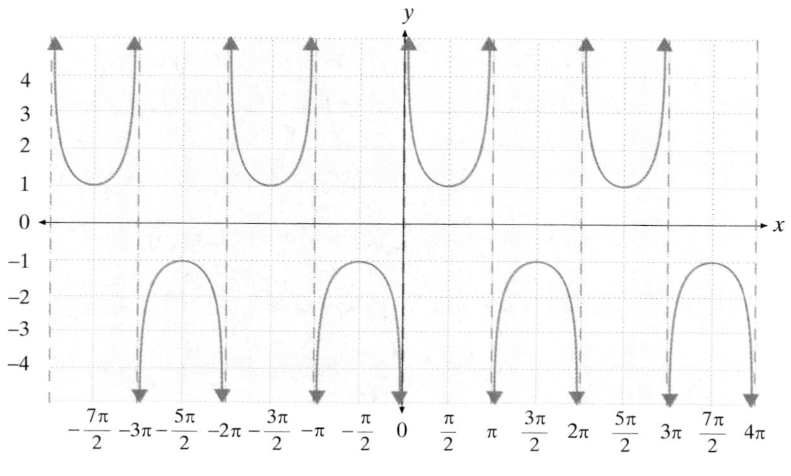

Topic 2: Periodicity and Other Observations

With a few graphs to look back on, we'll now formally define what we mean by periodicity.

Period of a Function

A function f is said to be **periodic** if there is a positive number p such that

$$f(x+p) = f(x)$$

for all x in the domain of f. The smallest such number p is called the **period** of f.

For instance, we know that $\sin(x+2\pi)=\sin(x)$, $\cos(x+2\pi)=\cos(x)$, and $\csc(x+2\pi)=\csc(x)$. It's also true that $\sin(x+2n\pi)=\sin(x)$ for any integer n, but 2π is the smallest positive constant p for which $\sin(x+p)=\sin(x)$, so the period of sine (and cosine and cosecant) is 2π.

example 2

Determine the period of the secant, tangent, and cotangent functions.

Solution:

As you no doubt expect after studying cosecant, the period of secant is the same as the period of cosine, since secant is the reciprocal of cosine. We can prove this algebraically as follows:

$$\sec(x+2\pi)=\frac{1}{\cos(x+2\pi)}=\frac{1}{\cos(x)}=\sec(x),$$

so the period of secant is no larger than 2π. And if there were a smaller positive number p for which $\sec(x+p)=\sec(x)$, then cosine would necessarily have the same period p, contradicting what we know about the period of cosine. Note, however, that secant is not defined for all real numbers; the domain of secant is all real numbers except where cosine $= 0$, namely $\pm\left\{\frac{\pi}{2}, \frac{3\pi}{2}, \frac{5\pi}{2}, \ldots\right\}$.

We'll have to work a bit harder to determine the period of tangent. A reasonable guess might be 2π, since four of the trigonometric functions have period 2π, but we'll see that this is wrong. Recall the general definition of the tangent function: given an angle θ and any point (x, y) on the terminal side of θ (other than the origin, of course),

$$\tan\theta=\frac{y}{x}.$$

If $\theta\in\left[0, \frac{\pi}{2}\right)$, $\tan\theta$ is positive because both y and x are positive. If $\theta\in\left[\pi, \frac{3\pi}{2}\right)$, $\tan\theta$ is again positive because y and x are both negative. If the terminal side of θ lies in the second or fourth quadrant, $\tan\theta$ is negative. This alone is enough to tell us that the period of tangent can't be less than π (do you see why?). But more precisely, consider what the tangent function does to a given angle θ and to $\theta+\pi$, as illustrated in Figure 5:

cont'd. on next page ...

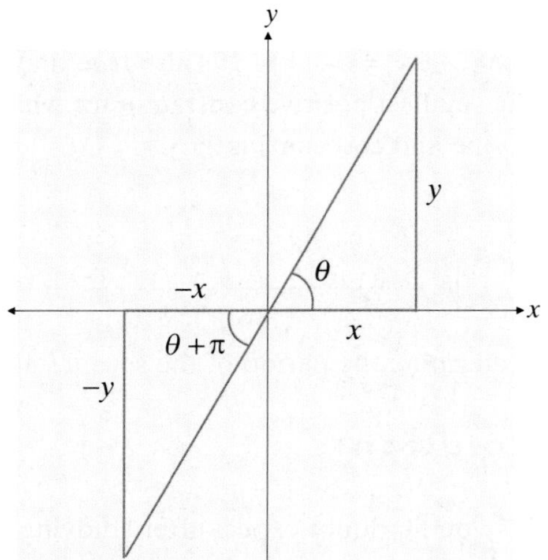

Figure 5: Tangent of an Angle and the Angle plus π

Since $\frac{-y}{-x} = \frac{y}{x}$ (with the restriction $x \neq 0$), we see that $\tan(\theta + \pi) = \tan(\theta)$. This, along with the observation in the preceding paragraph, tells us that the period of tangent is π, and consequently the period of cotangent is also π.

With the information in the last example for inspiration, we'll proceed to get a better understanding of tangent.

example 3

Sketch the graph of the tangent function.

Solution:

First, we'll recap what we know: tangent is not defined for multiples of $\frac{\pi}{2}$, its period is π, and it is non-negative on the interval $\left[0, \frac{\pi}{2}\right)$ and non-positive on the interval $\left(-\frac{\pi}{2}, 0\right]$. And we can either draw some triangles or use the identity $\tan x = \frac{\sin x}{\cos x}$ to easily calculate some values:

x	$-\frac{\pi}{3}$	$-\frac{\pi}{4}$	$-\frac{\pi}{6}$	0	$\frac{\pi}{6}$	$\frac{\pi}{4}$	$\frac{\pi}{3}$
tan x	$-\sqrt{3}$	-1	$-\frac{1}{\sqrt{3}}$	0	$\frac{1}{\sqrt{3}}$	1	$\sqrt{3}$

Now we can proceed to plot the above points to sketch the graph of tangent in the interval $\left(-\frac{\pi}{2}, \frac{\pi}{2}\right)$, and we can use its periodicity to extend the graph over a larger interval:

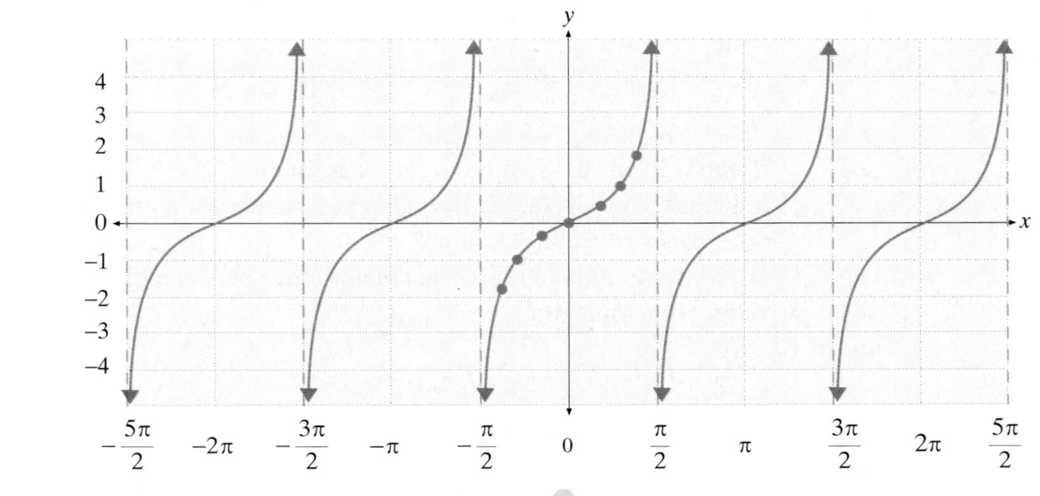

In Chapter 3 we defined the terms *even* and *odd* as applied to functions. Recall that a function *f* is *odd* if $f(-x) = -f(x)$ for all *x* in the domain, and *even* if $f(-x) = f(x)$ for all *x* in the domain. Geometrically, this means that the graph of *f* is either symmetric with respect to the origin or the *y*-axis, respectively. We've constructed enough graphs now to note that cosine and its reciprocal, secant, are even functions, while sine, cosecant, tangent, and cotangent are all odd. These facts are summarized below.

Even/Odd Identities

$\sin(-x) = -\sin x$	$\cos(-x) = \cos x$	$\tan(-x) = -\tan x$
$\csc(-x) = -\csc x$	$\sec(-x) = \sec x$	$\cot(-x) = -\cot x$

Topic 3: Graphing Transformed Trigonometric Functions

In actual use, a trigonometric function is unlikely to appear in its pristine form. That is, in order to be useful in solving a problem, a given function will probably have to be modified somewhat. Geometrically, this means that the graph of a function will appear stretched, compressed, reflected, or shifted relative to the graph of its fundamental form. We will finish out this section with a discussion of the more common transformations.

We have had quite a bit of experience with transformations of functions in general, but in the context of trigonometry some additional nomenclature is useful. We'll begin with a definition of one particularly useful term.

Amplitude of Sine and Cosine Curves

Given a fixed real number a, the **amplitude** of the function $f(x) = a\sin x$ or the function $g(x) = a\cos x$ is the value $|a|$. As we know, the multiplication of $\sin x$ or $\cos x$ by a stretches (or compresses, if $-1 < a < 1$) the graph vertically by a factor of $|a|$, so the amplitude represents the distance between the x-axis and the maximum value of the function.

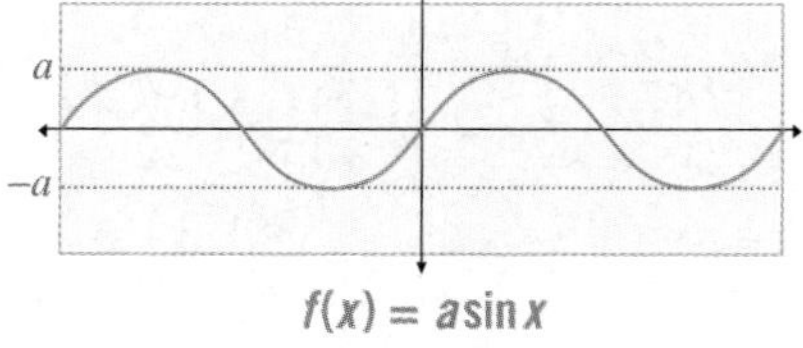

$f(x) = a\sin x$

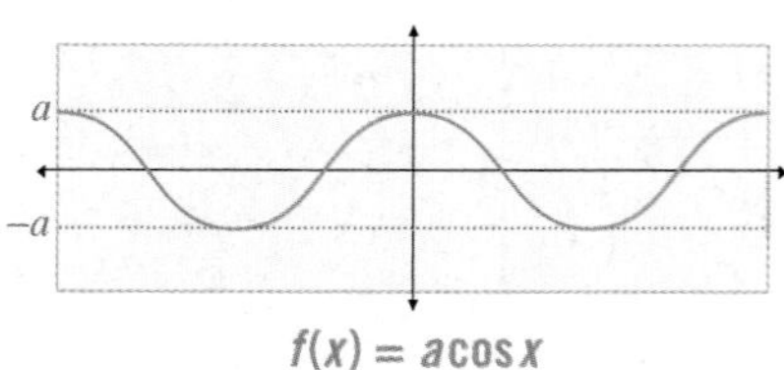

$f(x) = a\cos x$

As we have seen, the shapes of the sine and cosine curves are identical; one is merely shifted horizontally with respect to the other. For this reason, both graphs are said to be *sinusoidal*, so we have now defined the amplitude of a sinusoidal curve. Such curves arise in numerous physical situations, often when an object is displaying *simple harmonic motion (SHM)*.

Simple Harmonic Motion and Frequency

If an object is oscillating and its displacement from some mid-point at time t can be described by either $f(t) = a\sin\omega t$ or $g(t) = a\cos\omega t$, the object is said to be in **simple harmonic motion**. In either case, a is a real number and ω ("omega") is a positive real number. The amplitude of the object (the maximum displacement from rest) is $|a|$ and the **frequency** of the object's motion is $\frac{\omega}{2\pi}$.

Frequency is a measure of the number of times an object goes through one complete cycle of motion in one unit of time. If time t is being measured in seconds, the frequency is measured in terms of *cycles per second*, or **Hertz (Hz)**.

example 4

A first approximation to the motion of an object suspended at the end of a spring and set into vertical oscillation is given by $y = a\cos\omega t$, where y is the displacement of the object above or below its rest position and a and ω are as defined on the previous page. Suppose several potatoes are dumped into the basket of a grocer's scale, which then proceeds to bounce up and down with a frequency of 3 Hz. Given that the distance traveled between peaks of the basket's oscillation is 8 centimeters, find a mathematical model for the basket's motion.

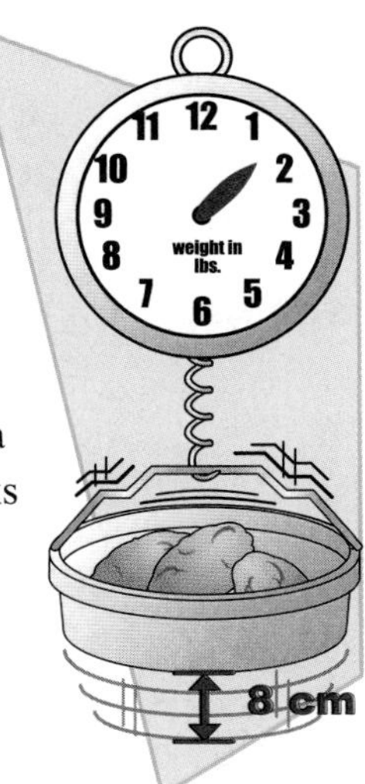

Solution:

From the given information, we know that our model must describe an object traveling 4 centimeters above and 4 centimeters below its rest position; in other words, the amplitude must be 4. We have a choice to make, however: we can define our coordinate system so that a positive displacement is either up or down. The most natural choice in this problem is probably to choose positive displacement as being in the upward direction, so we would like to arrange it so that $y = -4$ when $t = 0$. That is, a good model of the basket's oscillation will have the basket starting out 4 centimeters below its midpoint position at the moment ($t = 0$) when the potatoes are dumped in. This means we want to set $a = -4$.

A frequency of 3 Hz means that the basket makes 3 complete up-and-down cycles every second. This means that $\omega = 6\pi$, and so our model is $y = -4\cos 6\pi t$. To see how our model relates to the physical situation, consider the following graph:

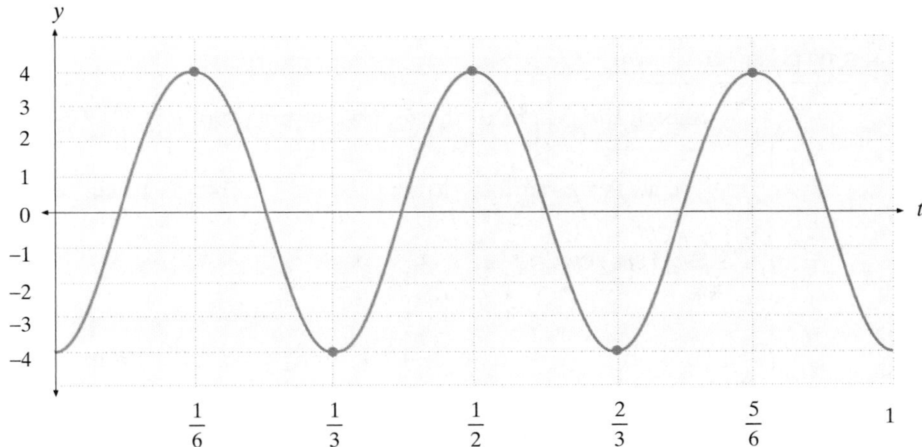

From the graph, it's clear that the basket really does make 3 complete cycles during the course of the first second, that it starts out 4 centimeters below the mid-point position, and that its maximum displacement above mid-point is also 4 centimeters.

Of course, our experience tells us that the simple model seen in Example 4 isn't very accurate over a long period of time – eventually the basket settles down and ceases its oscillatory motion. We'll return to this problem at the end of this section, but first we'll define a few more useful terms.

Period Revisited

We know that sine and cosine (and their reciprocals) go through one complete cycle over an interval of length 2π. The modified functions

$$f(x) = \sin bx \text{ and } g(x) = \cos bx$$

both have a period of $\frac{2\pi}{b}$. Note then that the period is the reciprocal of the frequency. This makes sense, since the frequency measures how many cycles occur over an interval of length 1, while the period is the length of the interval required for one complete cycle.

Notice in Example 4 that $b = 6\pi$, so the period is $\frac{2\pi}{6\pi} = \frac{1}{3}$.

Replacing the variable x with bx thus has the effect of stretching or compressing the graph of $\sin x$ or $\cos x$ horizontally, just as multiplying either function by a constant a has the same effect vertically. One further effect we need to be able to achieve is a horizontal shifting to the left or right. We've actually already learned how to do this in general: replacing the variable x with $x - c$ shifts a graph to the right if c is positive and to the left if c is negative.

example 5

Sketch the graphs of the following functions.

a. $f(x) = \sin 2\pi x$ **b.** $g(x) = \sin(x - \pi)$

Solutions:

a.

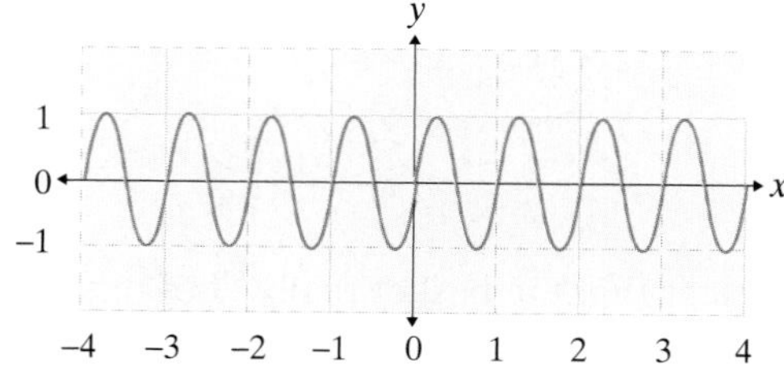

The period of $f(x) = \sin 2\pi x$ is $\frac{2\pi}{2\pi} = 1$ but this is the only difference between the function f and $\sin x$.

b.

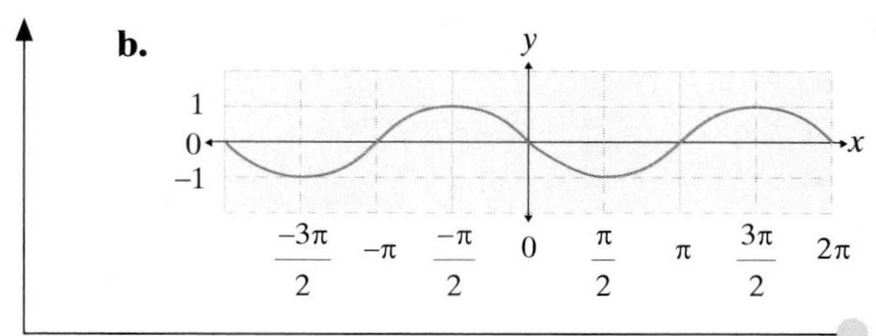

The graph of $g(x) = \sin(x - \pi)$ is simply the usual sine shape shifted π units to the right.

In the context of trigonometry, shifting a function to the left or right is called a **phase shift**, and in general it may occur in combination with a change in period and amplitude. For instance, the function $f(x) = 3\sin(5\pi x - 4\pi)$ has an amplitude of 3, a phase shift, and an altered period. One way to determine the phase shift and period precisely is to rewrite the function as follows:

$$\begin{aligned} f(x) &= 3\sin(5\pi x - 4\pi) \\ &= 3\sin\left[5\pi\left(x - \frac{4}{5}\right)\right]. \end{aligned}$$

The function $3\sin 5\pi x$ has a period of $\frac{2}{5}$, but when x is replaced with $x - \frac{4}{5}$, we know the graph of $3\sin 5\pi x$ gets shifted to the right by $\frac{4}{5}$ units.

Another way to determine details about the period and phase shift of $f(x) = 3\sin(5\pi x - 4\pi)$ is to relate the beginning and end of one cycle of the function to the beginning and end of one cycle of sin x. That is, we know that a cycle of f will begin when $x = 0$ and will end when $x = 2\pi$.

Solving these two equations for x gives us:

$$\begin{aligned} 5\pi x - 4\pi &= 0 & \qquad 5\pi x - 4\pi &= 2\pi \\ 5\pi x &= 4\pi & 5\pi x &= 6\pi \\ x &= \frac{4}{5} & x &= \frac{6}{5} \end{aligned}$$

So one complete cycle of f occurs over the interval between $x = \frac{4}{5}$ and $x = \frac{6}{5}$, telling us again that the period is $\frac{2}{5}$.

These observations are summarized on the next page for the functions sine and cosine. The same sort of analysis will prove to be just as useful for the remaining trigonometric functions, however.

Amplitude, Period, and Phase Shift Combined

Given constants a, $b > 0$, and c, the functions

$$f(x) = a\sin(bx - c) \text{ and } g(x) = a\cos(bx - c)$$

have **amplitude** $|a|$, **period** $\frac{2\pi}{b}$, and a **phase shift** of $\frac{c}{b}$. The left end-point of one cycle of either function is $\frac{c}{b}$ and the right end-point is $\frac{c}{b} + \frac{2\pi}{b}$.

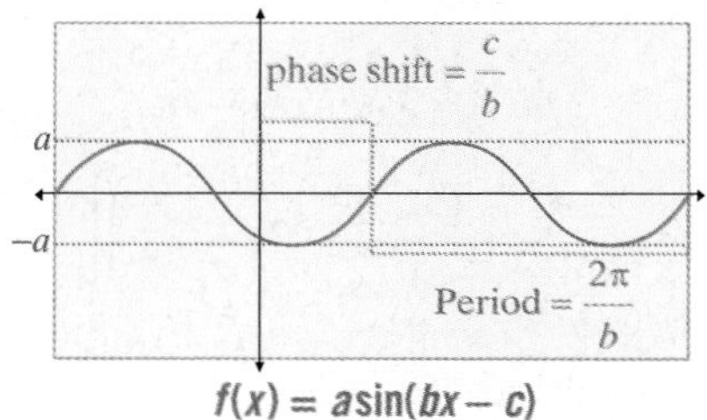

$f(x) = a\sin(bx - c)$

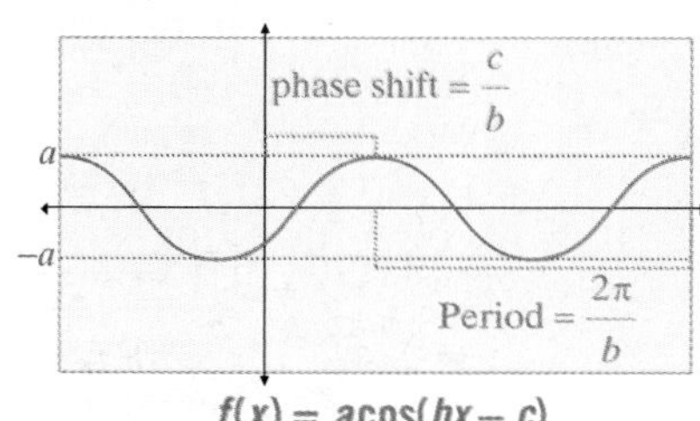

$f(x) = a\cos(bx - c)$

example 6

Sketch the graph of $f(x) = -2\sec\left(\pi x + \frac{\pi}{2}\right)$.

Solution:

The starting point is the secant function. We know its period is 2π and that it has vertical asymptotes at $\pm\left\{\frac{\pi}{2}, \frac{3\pi}{2}, \frac{5\pi}{2}, \ldots\right\}$. It completes one cycle between $\frac{\pi}{2}$ and $\frac{5\pi}{2}$, so these will be convenient values to use in determining the left and right end-points of one cycle of f:

$$\pi x + \frac{\pi}{2} = \frac{\pi}{2} \qquad \pi x + \frac{\pi}{2} = \frac{5\pi}{2}$$
$$\pi x = 0 \qquad \pi x = 2\pi$$
$$x = 0 \qquad x = 2$$

Note that this implicitly tells us that the function f has a period of 2.

The factor of –2 in front of secant stretches the graph vertically by a factor of 2 and reflects it with respect to the x-axis. For instance, where $\sec\left(\pi x + \frac{\pi}{2}\right)$ has values of 1 and –1, the function f will have values of, respectively, –2 and 2. Putting this all together, our sketch is as shown on the next page:

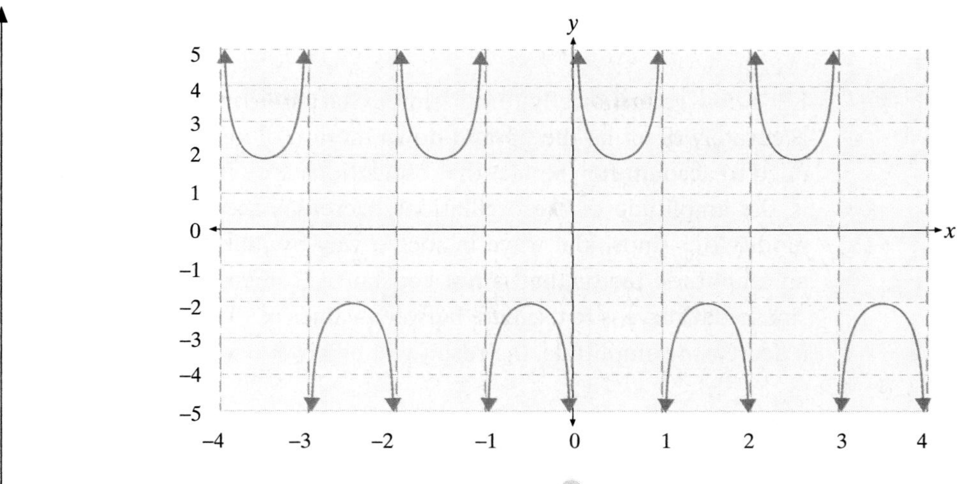

example 7

Sketch the graph of $g(x) = 1 + \sin\left(x - \frac{\pi}{4}\right)$.

Solution:

The graph of g is the graph of $\sin x$ shifted to the right by $\frac{\pi}{4}$ units and up by 1 (recall that adding a constant to a function merely shifts the graph up or down, according to whether the constant is positive or negative). Neither the amplitude nor period have changed, however. Our sketch is thus:

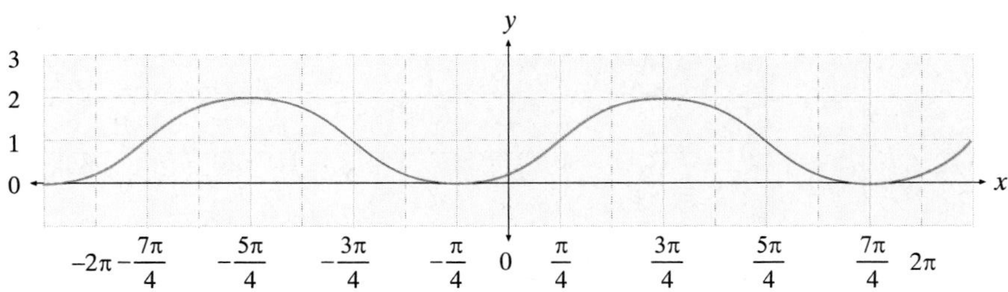

Topic 4: Interlude: Damped Harmonic Motion

Let's now return briefly to our simple harmonic motion problem. If we want to more accurately describe the up-and-down motion of an object suspended from a spring, we need to account for the fact that oscillations are often *damped* as time progresses. That is, the amplitude of the oscillations decreases according to some rule. We can easily modify our sinusoidal wave in such a way by multiplying a sine or cosine function by an amplitude factor that is not constant. Remember: the fundamental behavior of a sinusoidal curve is to oscillate between values of –1 and 1. If we multiply such a wave by a decreasing amplitude, the result will be a wave whose oscillations diminish over time.

example 8

Sketch the graph of $f(t) = -4e^{-t}\cos 6\pi t$.

Solution:

We've already graphed the function $f(t) = -4\cos 6\pi t$ (this function was our simple model for the motion of the grocer's basket in Example 4). The factor of e^{-t} provides the desired damping effect. In the figure below, the blue curves are the graphs of $4e^{-t}$ and $-4e^{-t}$, and are included to show how they describe the "envelope" of amplitude modulation. The result is that the magnitude of the displacement of the grocer's basket decreases over time. Notice, however, the period is unaffected.

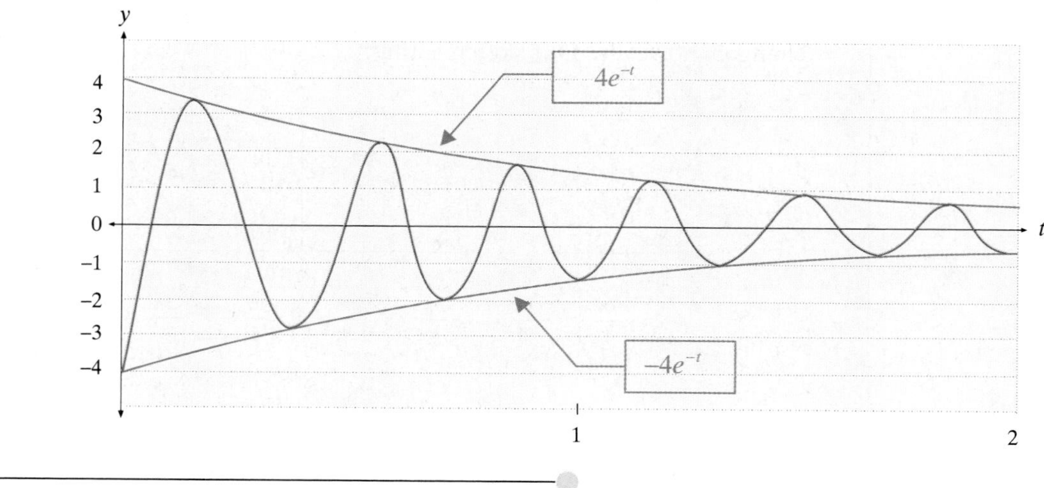

In the Exercises to follow, you'll be asked to sketch graphs of similar products of damping factors and trigonometric functions.

exercises

Given the information contained in Examples 1, 2, and 3, construct sketches of the following functions.

1. cotangent

2. secant

Use the given information to construct functions modeling the described behavior. See Example 4.

3. A baby is playing with a toy attached above his head on a coiled spring. The baby pulls the toy down a distance of 3 inches from its equilibrium position, and then releases it. The time for one oscillation is 2 seconds. Find the amplitude and period, then give the function for its displacement.

4. A pull cord for a lighted ceiling fan is swinging back and forth. The end of the cord swings a total distance of 4 inches from end to end and reaches a speed of 9 inches/s at the midpoint. Find the period of oscillation.

5. Marcel is bouncing a basketball at 10 ft./s. The distance from the ground to his waist is approximately 3 feet. Find the amplitude and period, then give the function for its displacement.

Determine the amplitude, period, and phase shift of each of the following trigonometric equations.

6. $y = 2\cos x$

7. $y = \frac{3}{2}\sin x$

8. $y = 5 + 4\cos x$

9. $y = \sin(x - 5)$

10. $y = -\sin x$

11. $2y = \cos x$

12. $y = -3\cos(x + 7)$

13. $y = \frac{2}{3}\sin x$

14. $y = 2\sin(2x)$

15. $y = -3\cos\left(\frac{1}{2}x\right)$

16. $2y = 3\sin(\pi\theta)$

17. $y = \cos(3\pi\theta - 2)$

18. $y = 0.5\sin(8x + 1)$

19. $y = 7\cos\left(x \cdot \frac{\pi}{2} + \frac{3}{2}\right)$

20. $5y = 8\cos(2\pi x + 4)$

21. $y = 2 - \frac{3}{4}\sin(-3 + x)$

Sketch the graphs of the following functions. See Examples 5 and 6.

22. $f(x)=\cos \pi x$

23. $g(x)=-2\sin 5x$

24. $f(x)=\csc \frac{3\pi}{4}x$

25. $g(x)=3\sin(x-2\pi)$

26. $g(x)=\tan\left(3\pi x-\frac{\pi}{2}\right)$

27. $f(x)=-5\cot \pi x$

28. $g(x)=\sin\left(x-\frac{\pi}{4}\right)$

29. $f(x)=4\cos\left(\frac{3x}{2}+\frac{\pi}{2}\right)$

30. $f(x)=\frac{1}{2}\tan\left(\frac{1}{2}x-\frac{\pi}{3}\right)$

31. $g(x)=2\cos(4x-2)$

32. $f(x)=\cos(x-\pi)$

33. $g(x)=3\sin 4x$

34. $g(x)=\csc\left(\frac{3\pi}{2}x-\frac{1}{2}\right)$

35. $f(x)=-\sin 2\pi x$

36. $f(x)=5\tan\left(3\pi-\frac{\pi}{2}x\right)$

Sketch the graphs of the following functions. See Example 7.

37. $g(x)=1+\sin(x-2\pi)$

38. $f(x)=2-\cos 2\pi x$

39. $f(x)=4+\csc\left(1-\frac{5\pi}{4}x\right)$

40. $g(x)=5-2\sin\left(x-\frac{\pi}{2}\right)$

41. $g(x)=1+\tan\left(\pi x-\frac{\pi}{4}\right)$

42. $f(x)=-3+5\cos x$

43. $g(x)=2-\sin\left(2x-\frac{\pi}{4}\right)$

44. $f(x)=4+\tan\left(x+\frac{3\pi}{2}\right)$

45. $f(x)=\frac{1}{2}-5\sin\left(\frac{1}{2}x-\frac{\pi}{2}\right)$

46. $g(x)=1-\frac{1}{4}\cos\left(\frac{1}{4}x-\frac{\pi}{2}\right)$

47. $g(x)=2+\frac{5}{6}\sec\left(\frac{1}{2}x-\pi x\right)$

48. $f(x)=\frac{1}{2}\tan\left(\frac{3}{4}x-2\pi\right)+3$

Sketch the following functions modeling damped harmonic motion. See Example 8.

49. $g(t)=-2e^{-t}\cos 5\pi t$

50. $f(t)=e^{-t}\sin\frac{3\pi}{4}t$

51. $g(t)=e^{t}\sin\left(3t-\frac{\pi}{2}\right)$

52. $g(t)=3e^{-t}\cos\left(5t-\frac{\pi}{2}\right)$

53. $f(t) = -3 + 5e^{-t}\cos t$

54. $f(t) = -5e^{t}\cos\dfrac{3\pi}{2}t$

55. $f(t) = \dfrac{1}{2}e^{-t}\sin\left(\dfrac{5}{6}t - 4\pi\right) + 2$

56. $g(t) = 2 + e^{-t}\sin\left(t - \dfrac{\pi}{4}\right)$

6.5 Inverse Trigonometric Functions

TOPICS

1. The definitions of inverse trigonometric functions
2. Evaluation of inverse trigonometric functions
3. Applications of inverse trigonometric functions

Topic 1: The Definitions of Inverse Trigonometric Functions

The rationale for the inverse trigonometric functions is the rationale for inverses of functions in general. In many situations, we will want to find an angle having a certain specified property, and our method will be to "undo" the action of a given trigonometric function. As a simple example, suppose we need to find an acute angle θ for which $\sin\theta = \frac{1}{2}$. Our experience is sufficient for this task; we've worked with nice angles enough to recognize that $\sin\frac{\pi}{6} = \frac{1}{2}$, so it must be the case that $\theta = \frac{\pi}{6}$. But what if we seek an angle φ for which $\sin\varphi = 0.7$? The problem is similar, but we don't yet have a way to determine φ.

Recall from Chapter 3, however, that a function will have an inverse only if it is one-to-one. Recall also, if the graph of the function is available, this means the graph must pass the horizontal line test; this is something the trigonometric functions markedly fail to do. Fortunately, there is a way out. By restricting the domain of a trigonometric function wisely, we can make it one-to-one and thus invertible. We will go through the process step by step for the sine function and then briefly show how the other trigonometric functions are dealt with similarly.

There are many ways in which we could restrict the domain of sine in order to make it one-to-one, but we are guided also by the desire to not lose more than we have to in the restriction. For instance, we could specify that we will only define sine over the interval $\left[0, \frac{\pi}{2}\right]$, but by doing so we prevent the newly-defined function from ever taking on a negative value (note that $0 \le \sin x \le 1$ for $x \in \left[0, \frac{\pi}{2}\right]$). Figure 1 indicates that $\left[-\frac{\pi}{2}, \frac{\pi}{2}\right]$ is the largest interval containing $\left[0, \frac{\pi}{2}\right]$ that we could choose for the restricted domain; the bold red portion of the graph is one-to-one and takes on all values between –1 and 1.

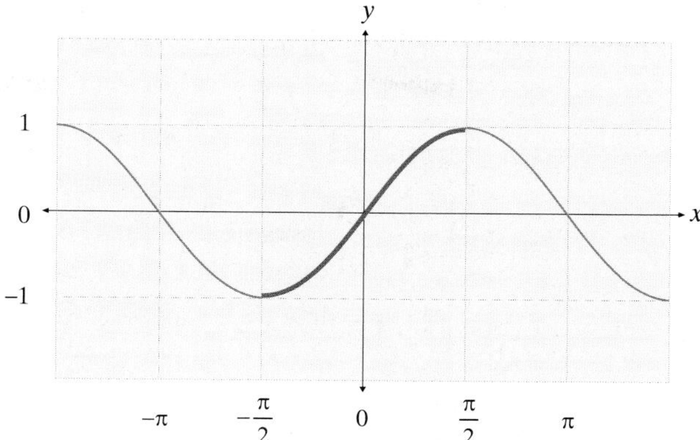

Figure 1: Restricting the Domain of Sine

In practice, context will tell us whether we want to think of sine as being defined on the entire real line or only over the interval $\left[-\frac{\pi}{2}, \frac{\pi}{2}\right]$, but the biggest hint will be whether we need to apply the inverse of the sine function. If so, the restricted domain for sine is called for.

Two notations are commonly used for the inverse trigonometric functions. In the case of sine, $y = \sin x$ is equivalent to the equations

$$x = \arcsin y \text{ and } x = \sin^{-1} y.$$

The arcsine notation derives from the fact that $\arcsin y$ is the length of the arc (on the unit circle) corresponding to the angle x. The $\sin^{-1} y$ notation is in keeping with our use of f^{-1} to stand for the inverse of the function f.

caution!

But in using this notation, remember:

$$\sin^{-1} y \neq \frac{1}{\sin y}$$

In order to avoid this possible source of confusion, some texts use only arcsine for the inverse sine function.

In summary,

Arcsine

Given $x \in \left[-\frac{\pi}{2}, \frac{\pi}{2}\right]$, **arcsine** is defined by either of the following:

$$y = \sin x \Leftrightarrow x = \arcsin y \text{ and } y = \sin x \Leftrightarrow x = \sin^{-1} y.$$

In words, arcsin y is the angle whose sine is y. Since the (restricted) domain of sine is $\left[-\frac{\pi}{2}, \frac{\pi}{2}\right]$ and its range is $[-1, 1]$, the domain of arcsine is $[-1, 1]$ and its range is $\left[-\frac{\pi}{2}, \frac{\pi}{2}\right]$.

The best way to finish up this introduction to arcsine is with a graph of the function. In Chapter 3 we saw that the graphs of a function and its inverse are reflections of one another with respect to the line $y = x$, and this is all we need in order to generate Figure 2.

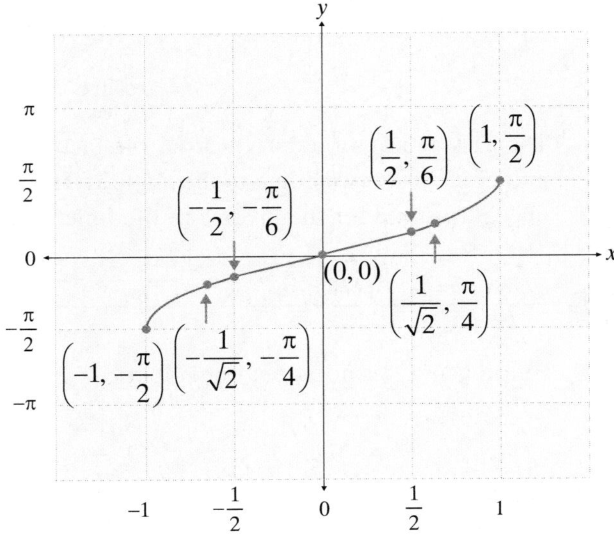

Figure 2: Graph of Arcsine

example 1

With the derivation of arcsine as a guide, construct a definition of arccosine and plot the resulting function.

Solution:

As with sine, we first need to restrict the domain of cosine to an interval over which cosine is one-to-one. Picture the graph of cosine in your mind (or refer to Figure 4 in Section 6.4 for a refresher). Most people would probably say that the natural choice for the restricted domain is the interval $[0, \pi]$, and this is indeed the convention. This is all we need in order to make our definition: Given $x \in [0, \pi]$, **arccosine** (with its two notations) is defined by

$$y = \cos x \Leftrightarrow x = \arccos y \text{ and } y = \cos x \Leftrightarrow x = \cos^{-1} y.$$

To graph the arccosine, we simply reflect the restricted graph of cosine with respect to the line $y = x$. Since the (restricted) domain of cosine is $[0, \pi]$ and the range is $[-1, 1]$, we know that the domain of arccosine will be $[-1, 1]$ and its range will be $[0, \pi]$. This knowledge serves as a good way to double-check our graph of arccosine.

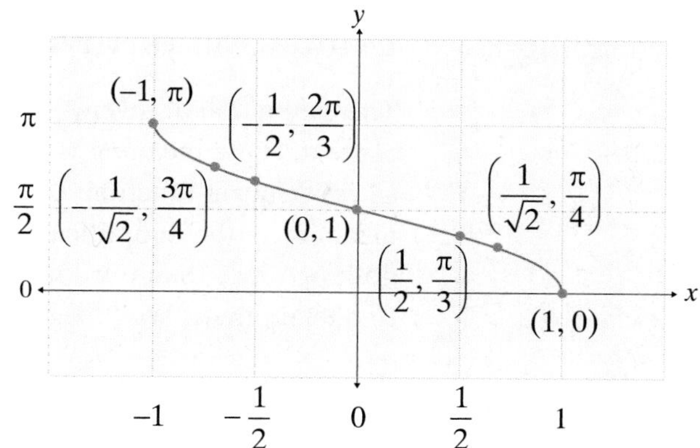

Arctangent is the third commonly encountered inverse function, and its definition and graph are arrived at in a similar manner. The box below summarizes facts about the definitions, domains, and ranges of arcsine, arccosine, and arctangent, and Figure 3 illustrates the graph of arctangent. Note the horizontal asymptotes in the graph of arctangent, corresponding to the vertical asymptotes in the graph of tangent.

Arcsine, Arccosine, and Arctangent

Function	Notation 1	Notation 2	Domain (y)	Range (x)
Inverse Sine	$\arcsin y = x \Leftrightarrow y = \sin x$	$\sin^{-1} y = x \Leftrightarrow y = \sin x$	$[-1, 1]$	$\left[-\frac{\pi}{2}, \frac{\pi}{2}\right]$
Inverse Cosine	$\arccos y = x \Leftrightarrow y = \cos x$	$\cos^{-1} y = x \Leftrightarrow y = \cos x$	$[-1, 1]$	$[0, \pi]$
Inverse Tangent	$\arctan y = x \Leftrightarrow y = \tan x$	$\tan^{-1} y = x \Leftrightarrow y = \tan x$	$(-\infty, \infty)$	$\left(-\frac{\pi}{2}, \frac{\pi}{2}\right)$

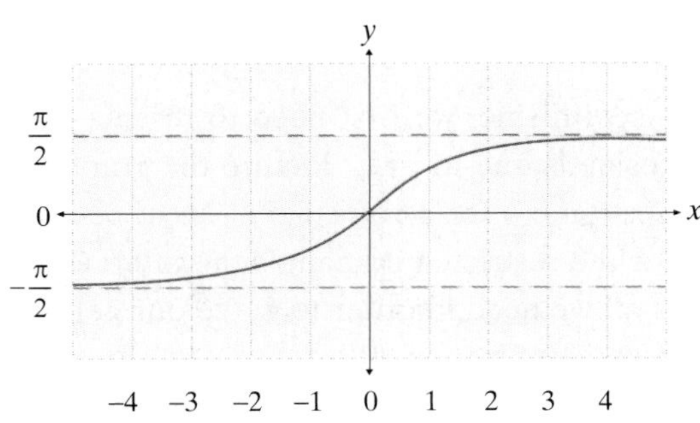

Figure 3: Graph of Arctangent

Topic 2: Evaluation of Inverse Trigonometric Functions

The evaluation of inverse trigonometric functions can take several forms, depending on context. One meaning is the actual numerical evaluation of an expression containing an inverse trig function; this may or may not require the use of a calculator. Another meaning is the simplification of expressions containing inverse trig functions, using nothing more than our knowledge of how functions and their inverses behave relative to one another. We'll begin with some numerical examples.

example 2

Evaluate the following expressions, using a calculator if necessary.

a. $\cos^{-1}(-0.3)$ **b.** $\tan^{-1}(-1)$ **c.** $\sin^{-1} 2.3$

Solutions:

a. Evaluation of $\cos^{-1}(-0.3)$ calls for technology of some sort. The labels on calculators vary, but many calculators have a $\cos^{-1}$ function accessed by first pressing a button labeled "2nd" or "Inv" and then the COS button. Remember, however, that the unit of angle measure is assumed to be radians unless directed otherwise, so before proceeding make sure your calculator is in radian mode. One possible sequence of keystrokes to evaluate $\cos^{-1}(-0.3)$ is

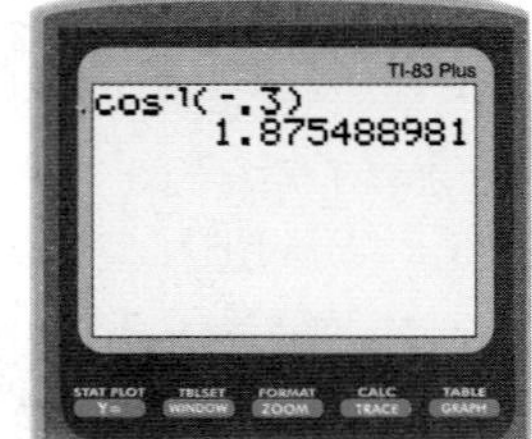

The result you see should be approximately 1.875 (and remember, this is 1.875 radians).

Using the software package *Mathematica*, the command to use is `Arccos`.

b. There is no need to use a calculator to evaluate $\tan^{-1}(-1)$. A glance at the graph of arctangent tells us that $\tan^{-1}(-1)$ is a negative number apparently halfway between 0 and $-\frac{\pi}{2}$ (that is, a negative angle whose terminal side is in the fourth quadrant). By drawing the appropriate right triangle, we can verify that it is indeed the case that $\tan\left(-\frac{\pi}{4}\right)=-1$, so $\tan^{-1}(-1)=-\frac{\pi}{4}$. However, if we do use a calculator the answer we get is $\approx -.7854$; it is then up to us to recognize this as approximately $-\frac{\pi}{4}$.

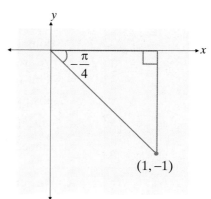

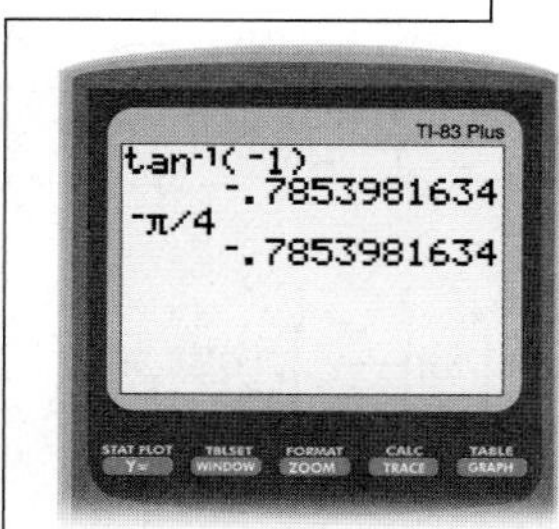

c. The domain of $\sin^{-1}$ is $[-1, 1]$, so the short answer is that $\sin^{-1} 2.3$ cannot be evaluated. After all, how could there be an angle whose sine is more than 1? For the purposes of this text, the answer is indeed simply that $\sin^{-1} 2.3$ is not defined. As a teaser toward more advanced mathematics, however, you may recall an aside at the start of Section 6.4 that the domains of the trigonometric functions can be extended to include complex numbers. Under those conditions, $\sin^{-1} 2.3$ actually has a complex number value (with non-zero imaginary part).

The single most important attribute of the inverse trigonometric functions is that they reverse the action of the functions they are associated with; remember that, in general, $f^{-1}(f(x))=x$ and $f(f^{-1}(x))=x$. But these statements are only true if all of the expressions contained in them make sense. A solid understanding of domains and ranges prevents potential errors, as shown in the next example.

example 3

Evaluate the following expressions, if possible.

a. $\arcsin\left(\sin\frac{3\pi}{4}\right)$ **b.** $\cos\left(\cos^{-1}(-0.2)\right)$ **c.** $\tan^{-1}\left(\tan\frac{7\pi}{6}\right)$

Solutions:

a. The potential error in this problem is to assume that $\arcsin\left(\sin\frac{3\pi}{4}\right)=\frac{3\pi}{4}$, since arcsin and sin are inverse functions of one another. But $\frac{3\pi}{4}$ lies outside the range of arcsin, which is $\left[-\frac{\pi}{2},\frac{\pi}{2}\right]$ so we know this can't be the answer. The key is to evaluate the expressions individually:

$$\sin\left(\frac{3\pi}{4}\right)=\frac{1}{\sqrt{2}}$$

and then

$$\arcsin\left(\frac{1}{\sqrt{2}}\right)=\frac{\pi}{4}.$$

cont'd. on next page ...

b. The number –0.2 lies in the domain of arccosine, and all real numbers lie in the domain of cosine, so all the parts of the expression $\cos\left(\cos^{-1}(-0.2)\right)$ make sense and we are safe in stating $\cos\left(\cos^{-1}(-0.2)\right) = -0.2$. If we wanted to explore the expression a bit further, we could note that, from the graph of arccosine, $\cos^{-1}(-0.2)$, is some positive number (and a calculator tells us it is approximately 1.8), and further that $\cos^{-1}(-0.2)$ must be greater than $\frac{\pi}{2}$ since $\cos\left(\cos^{-1}(-0.2)\right)$ is negative. This is indeed the case: $\cos^{-1}(-0.2)$ is approximately 101.5°.

c. We run into the same problem with $\tan^{-1}\left(\tan\frac{7\pi}{6}\right)$ as with $\arcsin\left(\sin\frac{3\pi}{4}\right)$, but we will present a slightly different way of thinking about the resolution here. Instead of evaluating $\tan\frac{7\pi}{6}$ literally, consider only the steps involved in doing so. The first step is to determine that the reference angle for $\frac{7\pi}{6}$ is $\frac{\pi}{6}$, and the second step is to note that $\tan\frac{7\pi}{6}$ and $\tan\frac{\pi}{6}$ have the same sign (the terminal side of $\frac{7\pi}{6}$ is in the third quadrant, and tangent is positive there). So $\tan^{-1}\left(\tan\frac{7\pi}{6}\right) = \tan^{-1}\left(\tan\frac{\pi}{6}\right) = \frac{\pi}{6}$, as $\frac{\pi}{6}$ lies in the range of $\tan^{-1}$.

The last example demonstrated the evaluation of compositions of trig functions with their inverses, but of course other compositions are possible. In many cases, a picture aids greatly in the computation.

example 4

Evaluate the following expressions.

a. $\tan\left(\sin^{-1}\left(-\frac{4}{5}\right)\right)$ **b.** $\cos(\arctan 0.4)$

Solutions:

a. Remember that the range of arcsin is $\left[-\frac{\pi}{2}, \frac{\pi}{2}\right]$, and in particular that $\sin^{-1}\left(-\frac{4}{5}\right)$ will lie between $-\frac{\pi}{2}$ and 0 (the graph tells us that arcsin of a negative number is negative). If we let $\theta = \sin^{-1}\left(-\frac{4}{5}\right)$ then $\sin\theta = -\frac{4}{5}$ and we can sketch the triangle shown on the next page to illustrate the relationship between θ and the given numbers:

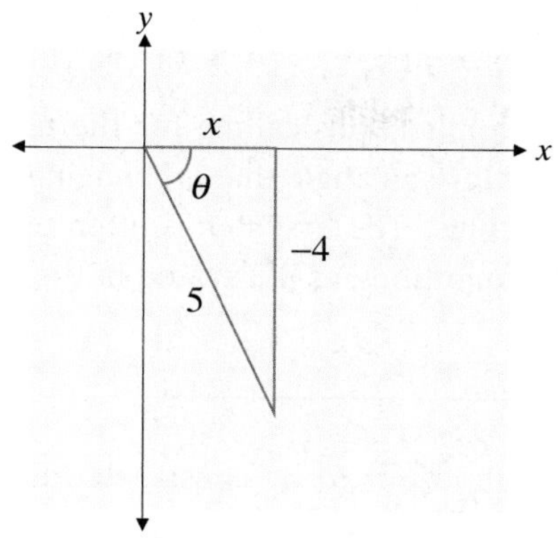

Of course, Pythagoras' Theorem allows us to calculate x:

$$x=\sqrt{5^2-(-4)^2}=\sqrt{9}=3.$$

Now we can see that $\tan\theta=-\frac{4}{3}$, so

$$\tan\left(\sin^{-1}\left(-\frac{4}{5}\right)\right)=-\frac{4}{3}.$$

b. We can employ the same method and let $\theta=\arctan 0.4$. This leads to

$$\tan\theta=0.4=\frac{4}{10}=\frac{2}{5}$$

and then to the sketch

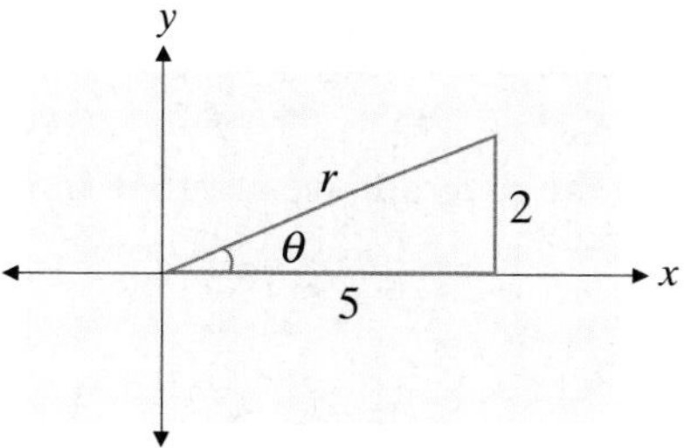

Pythagoras' Theorem gives us $r=\sqrt{5^2+2^2}=\sqrt{29}$, and so $\cos(\arctan 0.4)=\frac{5}{\sqrt{29}}$.

Topic 3: Applications of Inverse Trigonometric Functions

Many applications calling for the use of inverse trigonometric functions are dynamic, and feature an angle that is changing over time; you will encounter many such problems in Calculus. The first step is often to determine a formula for a given angle in terms of other quantities, as illustrated in the next examples.

example 5

A lighthouse is to be constructed half a mile from a long, straight reef, as shown. In order to ensure the light illuminates certain portions of the reef within specified lengths of time, the engineer needs a formula for θ in terms of x. Find such a formula.

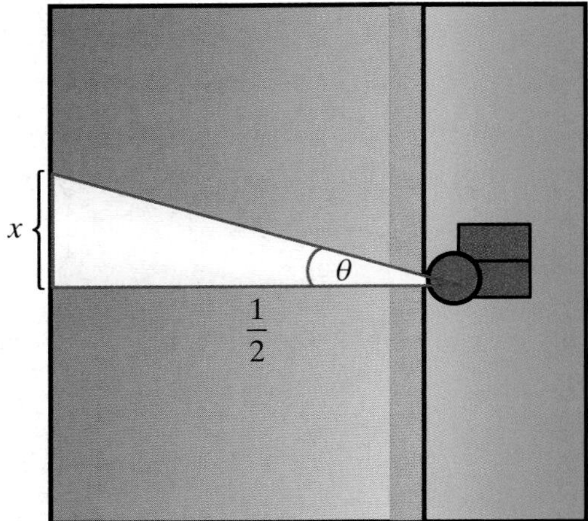

Solution:

From the diagram, we see that $\tan\theta = \frac{x}{\frac{1}{2}} = 2x$, so the formula for θ is simply

$$\theta = \tan^{-1} 2x.$$

example 6

Express $\sin\left(\cos^{-1} 2x\right)$ as an algebraic function of x, assuming $-\frac{1}{2} \le x \le \frac{1}{2}$.

Solution:

Let $\theta = \cos^{-1} 2x$. Then $\cos\theta = 2x$ and we are led to consider a sketch like the one on the next page:

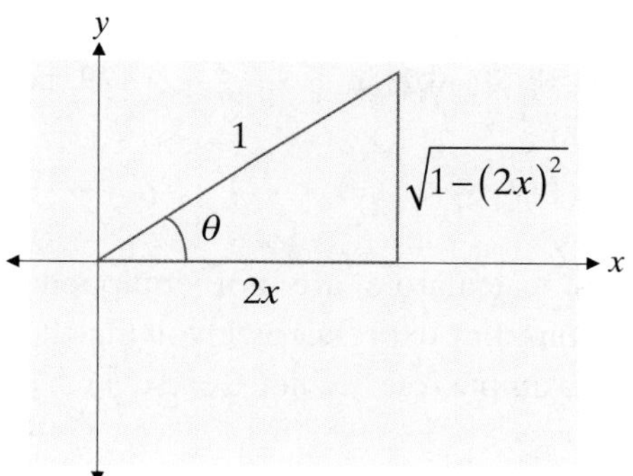

In the sketch, we have chosen the simplest lengths for the adjacent side and the hypotenuse that make $\cos\theta = 2x$, though of course any positive multiple of these lengths would also work. And as always, once the lengths of two sides of the right triangle have been determined, Pythagoras' Theorem provides the length of the third side. Now we can refer to the sketch to see that

$$\sin\left(\cos^{-1} 2x\right) = \sin\theta = \frac{\sqrt{1-4x^2}}{1} = \sqrt{1-4x^2}.$$

exercises

Construct sketches of the following functions. See Example 1.

1. arccosecant **2.** arcsecant **3.** arccotangent

Solve each equation for y without the use of a calculator.

4. $y = \sin^{-1}(-1)$ **5.** $y = \cos^{-1}\left(\frac{\sqrt{2}}{2}\right)$ **6.** $y = \tan^{-1} 1$ **7.** $y = \cot^{-1}\left(-\frac{\sqrt{3}}{3}\right)$

8. $y = \sec^{-1}\left(\frac{2\sqrt{3}}{3}\right)$ **9.** $y = \csc^{-1}(-2)$ **10.** $y = \arcsin 0$ **11.** $y = \arccos(-1)$

12. $y = \arctan\left(-\sqrt{3}\right)$ **13.** $y = \operatorname{arccot}\left(-\sqrt{3}\right)$ **14.** $y = \operatorname{arcsec} 2$ **15.** $y = \operatorname{arccsc}\left(\sqrt{2}\right)$

16. $y = \operatorname{arccot}(-1)$ **17.** $y = \tan^{-1}\left(\frac{\sqrt{3}}{3}\right)$ **18.** $y = \cos^{-1}\left(-\frac{1}{2}\right)$ **19.** $y = \csc^{-1} 2$

20. $y = \arcsin\left(-\frac{1}{2}\right)$ **21.** $y = \sec^{-1}(-1)$ **22.** $y = \operatorname{arccsc} 1$ **23.** $y = \arctan 0$

24. $y = \sin^{-1}\left(\frac{\sqrt{2}}{2}\right)$ **25.** $y = \arccos\left(-\frac{\sqrt{2}}{2}\right)$ **26.** $y = \operatorname{arcsec}(-2)$ **27.** $y = \cot^{-1}\left(-\sqrt{3}\right)$

Evaluate the following expressions, using a calculator if necessary. See Example 2.

28. $\sin^{-1}(-0.2)$ **29.** $\cos^{-1} 4$ **30.** $\sin^{-1}(-0.9)$

31. $\tan^{-1} 5$ **32.** $\cos^{-1}(-0.4)$ **33.** $\tan^{-1} 0.8$

Most calculators are not equipped with arccosecant, arcsecant, and arccotangent buttons, but expressions involving these functions can still be evaluated. To evaluate, for example, $\csc^{-1} x$, let $\theta = \csc^{-1} x$. Then

$$\csc\theta = x$$
$$\frac{1}{\sin\theta} = x$$
$$\sin\theta = \frac{1}{x}$$
$$\theta = \sin^{-1}\left(\frac{1}{x}\right).$$

Use the above method to evaluate the following expressions.

34. $\csc^{-1} 5$ **35.** $\sec^{-1}(-0.5)$ **36.** $\cot^{-1} 150$

37. $\cot^{-1}(-0.2)$ **38.** $\csc^{-1}(-8.9)$ **39.** $\sec^{-1} 2$

Evaluate the following expressions, if possible. See Example 3.

40. $\cos^{-1}\left(\cos\frac{2\pi}{4}\right)$ **41.** $\sin^{-1}\left(\sin\frac{3\pi}{2}\right)$ **42.** $\tan(\tan^{-1} 0.5)$

43. $\sin^{-1}\left(\sin\frac{7\pi}{6}\right)$ **44.** $\cos(\cos^{-1} -0.8)$ **45.** $\tan^{-1}\left(\tan\frac{5\pi}{4}\right)$

Evaluate the following expressions. See Example 4.

46. $\sin(\arctan 0.4)$ **47.** $\sin^{-1}\left(\cos\frac{3\pi}{2}\right)$ **48.** $\cos(\tan^{-1} 0.5)$

49. $\arcsin(\tan 1)$ **50.** $\tan(\cos^{-1} -0.8)$ **51.** $\tan^{-1}(\cos 5)$

Find the value of each expression without using a calculator.

52. $\sin\left(\arctan\sqrt{3}\right)$ **53.** $\cos\left(\sec^{-1} -2\right)$ **54.** $\tan(\text{arccot } 1)$

55. $\csc\left(\arccos -\frac{\sqrt{3}}{2}\right)$ **56.** $\tan\left(\sin^{-1} -\frac{\sqrt{2}}{2}\right)$ **57.** $\sec\left(\csc^{-1}\frac{2\sqrt{3}}{3}\right)$

58. $\cos\left(\cot^{-1} -1\right)$

59. $\sec\left(\arcsin -\frac{1}{2}\right)$

60. $\cot\left(\text{arcsec}\ \sqrt{2}\right)$

61. $\cot\left(\text{arccsc}\ -2\right)$

62. $\sin\left(\cos^{-1}\frac{\sqrt{2}}{2}\right)$

63. $\sec\left(\tan^{-1} -\frac{\sqrt{3}}{3}\right)$

64. $\sec\left(\arccos -\frac{\sqrt{2}}{2}\right)$

65. $\tan\left(\csc^{-1} -2\right)$

66. $\sin\left(\text{arcsec}\ \sqrt{2}\right)$

67. $\csc\left(\cot^{-1}\sqrt{3}\right)$

68. $\cot\left(\sin^{-1} -\frac{\sqrt{2}}{2}\right)$

69. $\cos\left(\arctan -\frac{\sqrt{3}}{3}\right)$

Use inverse trigonometric functions to solve the following problems. See Example 5.

70. Kim is watching a space shuttle launch from an observation spot two miles away from the launch pad. Find the angle of elevation to the shuttle for each of the following heights.

a. 0.5 miles **b.** 2 miles **c.** 2.8 miles

71. Jesse is rowing in the men's singles race. The length of the oar from the side of the shell to the water is 7 feet. At what angle is the oar from the side of the boat when the blade is at the following distance from the boat?

a. 2 feet **b.** 3 feet **c.** 5 feet

Express the following functions as purely algebraic functions. See Example 6.

72. $\tan\left(\cos^{-1} x\right)$

73. $\cot\left(\sin^{-1}\frac{2}{x}\right)$

74. $\sec\left(\tan^{-1} 3x\right)$

75. $\tan\left(\sin^{-1}\frac{x}{\sqrt{x^2+3}}\right)$

76. $\sin\left(\sec^{-1} x\right)$

77. $\cos\left(\tan^{-1}\frac{x}{4}\right)$

Using a graphing calculator, find the measure of θ in degrees. Remember to make sure your calculator is in the correct mode.

78. $\theta = \sin^{-1} 0.74184113$ **79.** $\theta = \arctan(-0.258416)$ **80.** $\theta = \text{arccsc}\ 1.847526$
81. $\theta = \sec^{-1}(-1.1224539)$ **82.** $\theta = \cot^{-1} 0.57496998$

Using a graphing calculator, find the value of θ in radians. Remember to make sure your calculator is in the correct mode.

83. $\theta = \arccos(-0.1115598)$ **84.** $\theta = \text{arccot}\ 1.547773$ **85.** $\theta = \tan^{-1} 5.999999$
86. $\theta = \csc^{-1}(-1.333333)$ **87.** $\theta = \arcsin 0.65937229$

Sketch a graph of the following functions. Then graph the functions using a graphing calculator to check your answer.

88. $f(x) = \sin^{-1}(x-3)$ **89.** $f(x) = \sec^{-1}(2x)$ **90.** $f(x) = \arctan\left(\frac{x}{2}\right)$
91. $f(x) = 2\arccos x$

chapter six project

Trigonometric Applications

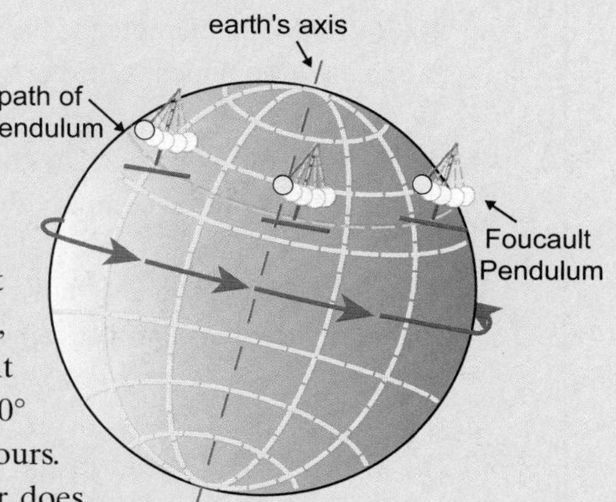

At the Paris Observatory in 1851, Jean Foucault used a long pendulum to prove that the earth is rotating. As it swings, the pendulum appears to change its path. However, it is not the pendulum that changes path, but the room rotating underneath it. At the North Pole the earth revolves 360° underneath the pendulum over 24 hours. The path of a pendulum at the equator does not revolve at all; instead, the pendulum travels in a huge circle while the earth spins. At points between the two, the pendulum cannot show how far it travels, but it can show how much the earth is revolving underneath it. To calculate how much the earth revolves in a particular location, use the following equation:

$$\text{degrees of revolution} = 360° \sin(\text{latitude of location})$$

Location	Latitude	Location	Latitude
United Nations, NY	40°44′58″	St. Isaac's Cathedral, Russia	59°53′02″
California Academy of Sciences	37°46′12″	Paris Observatory, France	48°48′58″
Smithsonian, Washington, DC	38°53′19″		

1. In a 24 hour period, how many degrees does the earth revolve at the above locations?

Your university has decided to install a 50 ft. Foucault Pendulum in your science building and asked you to make sure that there is enough room.

2. If the pendulum swings a total of 16°, how long is the arc traced in the air by the tip of the pendulum during one swing?

3. The school plans to build a small wall encircling the swinging pendulum. What should the diameter of the circle be if they want the tip of the pendulum to come within 6 inches of the wall?

4. When the pendulum reaches the farthest point from the center, how much higher will the tip be compared to when it is at the center?

5. If the science center only has room for a circular wall of diameter 12 feet, how many degrees can the pendulum swing and still stay 6 inches from the wall?

6. The Foucault Pendulum in the United Nations building has a length of 75 feet and a period of 10 seconds. Assuming simple harmonic motion and that at $t = 0$ the pendulum is at its farthest distance away (6 feet from the center of the circle), what function models the motion of the pendulum? Graph this function.

CHAPTER REVIEW

6.1 Radian and Degree Measure of Angles

topics	pages	test exercises
The unit circle and angle measure • The definition of *radian* (length of the portion of the circle subtended by θ) • Definition of initial side and terminal side of an angle	p. 455 – 456	3 – 6
Conversion between degrees and radians • Converting from degrees to radians: $x^\circ = (x)\left(\frac{\pi}{180}\right)\text{rad}$ • Converting from radians to degrees: $x \text{ rad} = x\left(\frac{180}{\pi}\right)^\circ$	p. 457 – 458	1 – 2
Commonly encountered angles • 30, 45, 60, and 90 and the ratios of sides of triangles with common angles	p. 458 – 461	3 – 6
Arc length and angular speed • The definition of (fraction of a circle's circumference subtended by θ) and formula for *arc length*, s ($s = r\theta$) • The definition of (speed of an object moving along the arc of a circle defined by a central angle θ and time t) and formula for *angular speed*, ω $\left(\omega = \frac{\theta}{t}\right)$ • The definition of (the distance traveled in time t around an arc length s of a circle with radius r) and formula for *linear speed*, v $\left(v = \frac{s}{t} = \frac{r\theta}{t} = r\omega\right)$	p. 461 – 463	7 – 10
Area of a circular sector • The definition of *sector* (the portion of the circle between two radii) • Finding the area of a sector: $A = \left(\frac{\theta}{2\pi}\right)\left(\pi r^2\right) = \frac{r^2\theta}{2}$	p. 464	11 – 12

6.2 Trigonometric Functions of Acute Angles

topics	pages	test exercises
The six basic trigonometric functions • Using the sides of a triangle to define the six basic trigonometric functions: $\sin\theta = \frac{\text{opp}}{\text{hyp}},\ \cos\theta = \frac{\text{adj}}{\text{hyp}},\ \tan\theta = \frac{\text{opp}}{\text{adj}}$ $\csc\theta = \frac{1}{\sin\theta} = \frac{\text{hyp}}{\text{opp}},\ \sec\theta = \frac{1}{\cos\theta} = \frac{\text{hyp}}{\text{adj}},\ \cot\theta = \frac{1}{\tan\theta} = \frac{\text{adj}}{\text{opp}}$	p. 471 – 474	13 – 14
Evaluation of trigonometric functions • Find the value of each of the trigonometric functions given the value of the angle • Degree, minute, second notation: $1' = \text{one minute} = \left(\frac{1}{60}\right)(1^\circ)$ $1'' = \text{one second} = \left(\frac{1}{60}\right)(1') = \left(\frac{1}{3600}\right)(1^\circ)$	p. 474 – 476	15 – 16
Applications of trigonometric functions • Using trigonometric functions and the Pythagorean Theorem to solve application problems • Solving surveying problems and the formula: $h = \frac{d}{\cot\beta - \cot\alpha}$	p. 477 – 480	17 – 20

6.3 Trigonometric Functions of Any Angle

topics	pages	test exercises
Extending the domains of trigonometric functions • Using $r = \sqrt{x^2 + y^2}$ where (x, y) is a point on the terminal side of an angle to extend the domains of the trig functions: $\sin\theta = \frac{y}{r},\ \cos\theta = \frac{x}{r},\ \tan\theta = \frac{y}{x}\ (\text{for } x \neq 0),$ $\csc\theta = \frac{r}{y}\ (\text{for } y \neq 0),\ \sec\theta = \frac{r}{x}\ (\text{for } x \neq 0),\ \cot\theta = \frac{x}{y}\ (\text{for } y \neq 0)$	p. 486 – 490	21 – 22

topics (continued)	pages	test exercises
Evaluation using reference angles • The definition of *reference angle* (angle formed by the x-axis and the terminal side of the given angle) • Using reference angles to evaluate a given angle	p. 490 – 492	23 – 26
Relationships between trigonometric functions • Cofunction identities: $\cos\theta = \sin\left(\frac{\pi}{2} - \theta\right)$, $\csc\theta = \sec\left(\frac{\pi}{2} - \theta\right)$, and $\cot\theta = \tan\left(\frac{\pi}{2} - \theta\right)$ • Reciprocal identities: $\csc\theta = \frac{1}{\sin\theta}$, $\sec\theta = \frac{1}{\cos\theta}$, and $\cot\theta = \frac{1}{\tan\theta}$ • Quotient identies: $\tan\theta = \frac{\sin\theta}{\cos\theta}$ and $\cot\theta = \frac{\cos\theta}{\sin\theta}$ • Using trig identities to find the value of an angle or another trig function	p. 493 – 497	27 – 30

6.4

Graphs of Trigonometric Functions

topics	pages	test exercises
Graphing the basic trigonometric functions • Plotting known points to determine the general shape of the trig graphs	p. 500 – 504	31
Periodicity and other observations • The definition of the *period* of a function (the smallest number p such that $f(x + p) = f(x)$) • Even trig functions: $\cos x$ and $\sec x$ • Odd trig functions: $\sin x$, $\csc x$, $\tan x$, and $\cot x$	p. 504 – 507	32
Graphing transformed trigonometric functions • The definition of the *amplitude* of sine and cosine • The definition of *simple harmonic motion* (oscillating motion that can be described by $f(t) = a \sin \omega t$ or $g(t) = a \cos \omega t$) • The definition of *frequency* (the number of times an object goes through one complete cycle of motion in one unit of time, $\frac{\omega}{2\pi}$) • The definition of *period* for a function of the form $f(x) = \sin bx$ or $g(x) = \cos bx$: $\frac{2\pi}{b}$ • Sketching transformed graphs	p. 507 – 513	33 – 36

topics (continued)	pages	test exercises
Interlude: damped harmonic motion • Graphing functions representing damped harmonic motion	p. 514	37 – 38

6.5 Inverse Trigonometric Functions

topics	pages	test exercises
The definitions of inverse trigonometric functions • Restricting the domain of trig functions so the inverse is defined • Inverse notation for trig functions • Graphing inverse trig functions	p. 518 – 522	39
Evaluation of inverse trigonometric functions • Numerical evaluations of inverse trig functions • Simplifying expressions involving inverse trig functions • Evaluating compositions of trig functions and inverse trig functions	p. 522 – 525	40 – 47
Applications of inverse trigonometric functions • Determining a formula for a given angle in terms of other quantities	p. 526– 527	48 – 49

chapter test

Section 6.1

Convert the following angle measures as directed.

1. Express $\frac{4\pi}{3}$ in degrees.

2. Express $-219°$ in radians.

Use the information in each diagram to determine the radian measure of the indicated angle.

3.

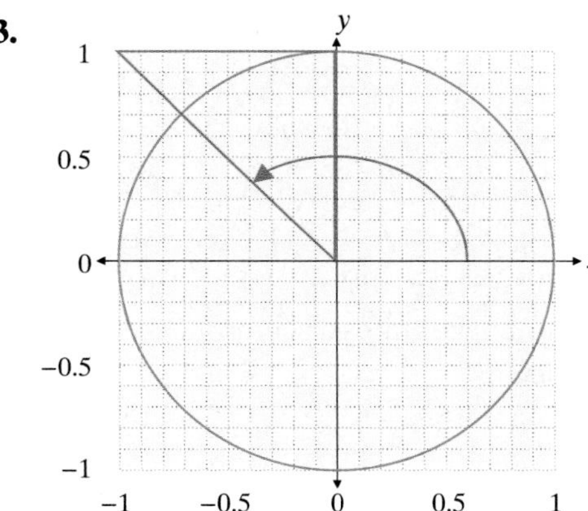

4.

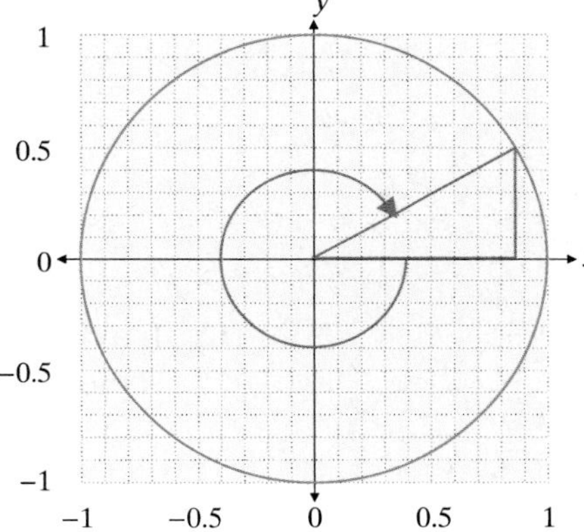

Sketch the indicated angles.

5. $-\frac{5\pi}{3}$

6. $175°$

Find the indicated arc length in each of the following problems.

7. Given a circle with a diameter of 1.5 m, find the length of the arc subtended by a central angle of 92° (hint: convert to radians first).

8. Given that two cities are 1200 miles apart and have the same longitude (that is, one is due south of the other), what is the difference in their latitudes? Use a value of 3960 miles for the radius of the earth.

The following problem asks you to determine the angular and/or linear speeds of various objects.

9. A DVD has a 2-inch radius and spins at 600 rpm. Find **(a)** the angular speed of a mark on its edge in rad/min, and **(b)** the linear speed of the mark in inches per second.

10. A round metal gear has a radius of 3 inches and spins at 90 rpm. What is the angular speed of a notch on the edge of the gear?

The following problems ask you to calculate the area of a sector of a circle.

11. Find the area of the green shaded portion of the circle below:

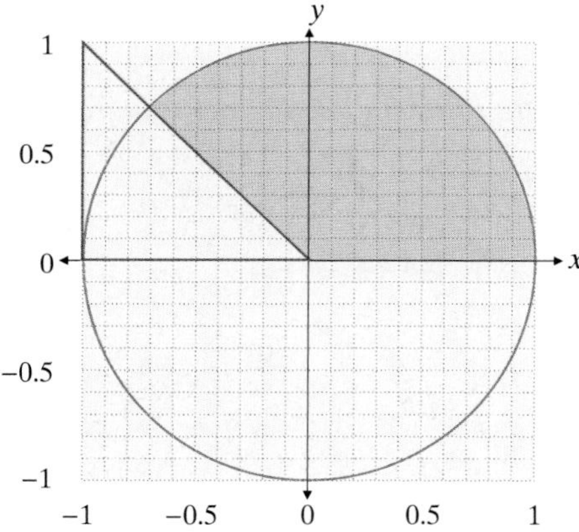

12. If a circle of radius 7 cm contains a sector marked by a central angle of −172°, what is the area of the sector?

Section 6.2

Use the information contained in the figures to determine the values of the six trigonometric functions of θ.

13.

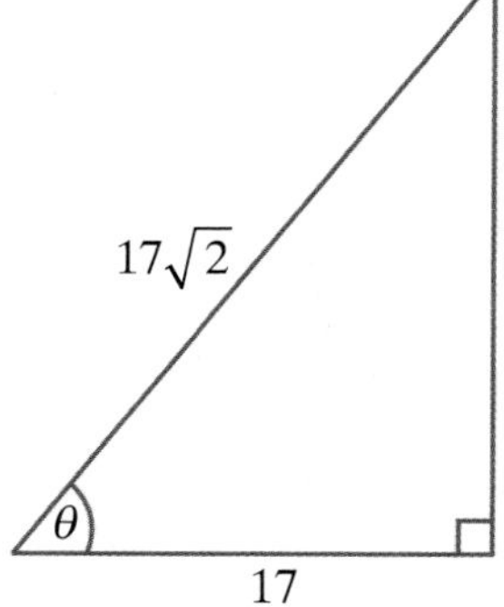

14.

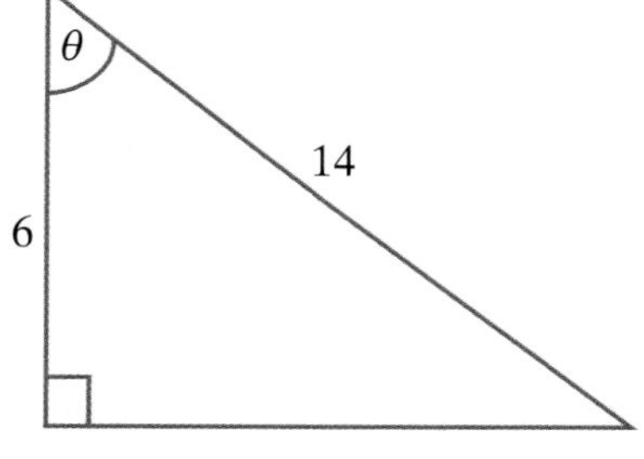

Evaluate the expressions, using a calculator if necessary.

15. tan 63° and cot 63°

16. sine and cosecant of $\frac{5\pi}{6}$

Use an appropriate trigonometric function and a calculator if necessary to solve the following problems.

17. A tourist is going to ride a cable car up a mountain near Lake Tahoe. He knows that the elevation at the base of the mountain is 4530 feet, and that the top of the mountain reaches 6976 feet. If the angle of ascent for the cable car line is 55° 33′ 20″, how far does each car travel?

18. A professor uses a laser pointer while giving a Power Point® lecture. If the laser makes an angle of 13° with the horizontal when she is pointing at a spot on screen that is 3.5 feet above the horizontal, how far is she standing from the screen?

Use the formula $h = \dfrac{d}{\cot\beta - \cot\alpha}$ *to solve the following problems.*

19. To determine the height of a skyscraper, a surveyor stands down the street and measures the angle to its highest point to be 52°. He then travels 45 feet closer to the building and finds the same angle to be 44°. How tall is the building?

20. A camera operator sits in a stationary crane, 40 feet above ground. He films a scene below of an automobile approaching. At the beginning of the scene, the angle of depression to the driver is 23°. After a direct approach towards the crane, the scene ends with the camera at an angle of 47°. How far did the vehicle travel?

Section 6.3

Evaluate the six trigonometric functions for the angles given below, using a calculator if necessary.

21. $\theta = -125°$

22. $\theta = 5\pi$

Determine the reference angle associated with the given angle.

23. $\varphi = 86°$

24. $\theta = \dfrac{3\pi}{2}$

*In each of the following problems, **(a)** rewrite the expression in terms of the given angle's reference angle, and then **(b)** evaluate the result, using a calculator if necessary.*

25. $\sin 179°$

26. $\cos -\dfrac{9\pi}{4}$

Express each of the following in terms of the appropriate cofunction, and verify the equivalence of the two expressions.

27. $\tan -\frac{\pi}{5}$

28. $\csc 52°$

Use the given information about each angle to evaluate the expressions.

29. Given that $\tan\theta = \frac{1}{2}$ and $\sin\theta$ is negative, determine $\sin\theta$.

30. Given that $\csc\theta = 0.8$ and θ lies in the fourth quadrant, determine $\cot\theta$.

Section 6.4

31. Construct a sketch of the sine function.

32. Construct a function modeling the following behavior: A clock's pendulum spans a distance of 3 inches from one end of its swing to the other. The time for one oscillation is one second.

Sketch the graphs of the following functions.

33. $f(x) = \csc 2\pi x$

34. $g(x) = -2\tan 3x$

35. $g(x) = 2 + \sin(x - \pi)$

36. $f(x) = 1 - \cos 3\pi x$

Sketch the following functions modeling damped harmonic motion.

37. $g(t) = -e^{-t+2}\sin 3\pi t$

38. $f(t) = \frac{1}{2}e^{-t}\cos(t + 2\pi) - 1$

Section 6.5

39. Sketch the arcsecant function.

Evaluate the following expressions, using a calculator if necessary.

40. $\tan^{-1}(0.6)$

41. $\cos^{-1} -2$

Convert the following expressions to ones that can be entered in a calculator, then evaluate.

42. $\sec^{-1}(-5.3)$

43. $\cot^{-1} 127$

Evaluate the following expressions, if possible.

44. $\cos^{-1}\left(\cos\frac{\pi}{2}\right)$

45. $\tan(\tan^{-1} 0.75)$

46. $\cos(\tan^{-1} -0.4)$

47. $\sin^{-1}\left(\cos\frac{3\pi}{4}\right)$

Express the following functions as purely algebraic functions.

48. $\sin(\sec^{-1} 2x)$

49. $\tan\left(\sin^{-1}\frac{x}{\sqrt{x^2+4}}\right)$

Introduction

The last chapter introduced the trigonometric functions and their basic properties and uses. This chapter now delves deeper into how the functions relate to one another; this deeper understanding will allow us to simplify unwieldy expressions, solve trigonometric equations, and extend our grasp of trigonometry yet further.

The relationships between the trigonometric functions fall into a category of equations called *identities*, the formal definition of which was given in Chapter 1. Briefly, a trigonometric identity (also sometimes referred to as a trigonometric formula) is an equation that is always true. As such, there is no need to solve an identity – instead, identities are used as tools in accomplishing other tasks. Identities began to be used soon after the appearance of trigonometric functions. By the second century AD, Ptolomy, the Greek astronomer and geographer, was using identities in his work with the chords of a circle, the accepted form of trigonometry at the time. For his early work entitled *Almagest*, he used the sum and difference formulas and something that resembled the modern day half angle formula in order to update the current chord tables to an accuracy of three decimal places.

Identities must either be discovered or verified, and both processes call for the application of algebra. Further, the use of an identity usually calls for algebraic skill; some of the uses we will explore include simplifying expressions, evaluating trigonometric functions exactly, and solving *conditional* equations (which are equations that are *not* always true). In fact, the underlying theme of this chapter is the marriage of trigonometry and algebra. Whereas the last chapter focused almost exclusively on the basics of trigonometry, this chapter and the next bring algebra back into the discussion. The union of trigonometry and algebra culminates in the last section of the chapter, where you will see that such algebraic methods as factoring and solving for specific terms are critical in being able to solve trigonometric equations.

Viéte

The concept of mixing algebra and trigonometry did not happen overnight. For centuries mathematicians separated the realms of real numbers and geometric ideas. It was not until the sixteenth century that mathematicians finally began to use algebra for abstract quantities rather than simply for concrete values. This transition was thanks in part to François Viéte (1540 - 1603), who was the first to consistently apply algebraic methods to his work in trigonometry. He introduced the sum-to-product formulas, the law of tangents, and a recurrence formula that allows $\cos(nx)$ to be presented in terms of the cosine of lower multiples of x. Although Viéte's breakthrough application of algebra to trigonometry may seem remarkably basic, it is good to remember that many important mathematical discoveries come from examining a problem in a new, unexpected, and sometimes simple way.

7.1 Fundamental Identities and Their Uses

TOPICS

1. Previously encountered identities
2. Simplifying trigonometric expressions
3. Verifying trigonometric identities
4. Interlude: trigonometric substitutions

Topic 1: Previously Encountered Identities

Chapter 6 presented the foundations of trigonometry, and concentrated largely on the geometric and functional aspects of the material. This chapter focuses more on the relationships between trigonometric functions and on the marriage of algebra and trigonometry. It does so by introducing and then using statements of equality known as *trigonometric identities*. We begin with a review of the identities already seen, though three of the equations below have only been alluded to in passing.

Identities Already Seen

Reciprocal Identities

$$\csc x = \frac{1}{\sin x} \qquad \sec x = \frac{1}{\cos x} \qquad \cot x = \frac{1}{\tan x}$$

$$\sin x = \frac{1}{\csc x} \qquad \cos x = \frac{1}{\sec x} \qquad \tan x = \frac{1}{\cot x}$$

Quotient Identities

$$\tan x = \frac{\sin x}{\cos x} \qquad \cot x = \frac{\cos x}{\sin x}$$

cont'd. on next page ...

Identities Already Seen (cont'd.)

Cofunction Identities

$$\cos x = \sin\left(\frac{\pi}{2} - x\right) \qquad \csc x = \sec\left(\frac{\pi}{2} - x\right) \qquad \cot x = \tan\left(\frac{\pi}{2} - x\right)$$

$$\sin x = \cos\left(\frac{\pi}{2} - x\right) \qquad \sec x = \csc\left(\frac{\pi}{2} - x\right) \qquad \tan x = \cot\left(\frac{\pi}{2} - x\right)$$

Period Identities

$$\sin(x + 2\pi) = \sin x \qquad \cos(x + 2\pi) = \cos x$$

$$\csc(x + 2\pi) = \csc x \qquad \sec(x + 2\pi) = \sec x$$

$$\tan(x + \pi) = \tan x \qquad \cot(x + \pi) = \cot x$$

Even/Odd Identities

$$\sin(-x) = -\sin x \qquad \cos(-x) = \cos x \qquad \tan(-x) = -\tan x$$

$$\csc(-x) = -\csc x \qquad \sec(-x) = \sec x \qquad \cot(-x) = -\cot x$$

Pythagorean Identities

$$\sin^2 x + \cos^2 x = 1 \qquad \tan^2 x + 1 = \sec^2 x \qquad 1 + \cot^2 x = \csc^2 x$$

The Pythagorean Identities have not explicitly appeared yet, but they are based on an idea we have used frequently. Their names allude to the familiar Pythagorean Theorem, and the first identity follows from consideration of a diagram such as that in Figure 1.

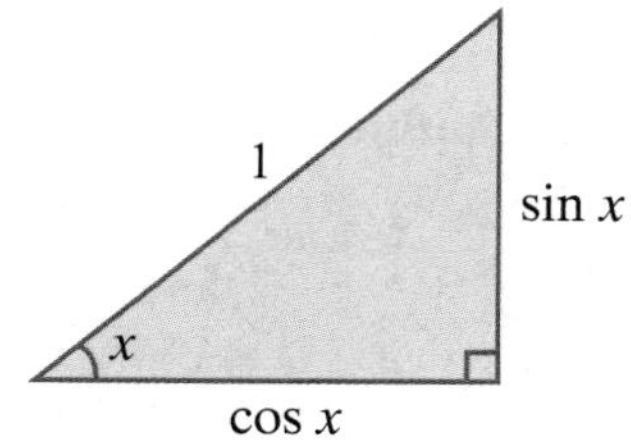

Figure 1: Derivation of $\sin^2 x + \cos^2 x = 1$

The diagram assumes that x is a real number between 0 and $\frac{\pi}{2}$, but a similar diagram can be drawn given any real number (recall that the use of reference angles makes this possible). If a right triangle with a hypotenuse of length 1 is drawn, then the legs of the right triangle must be of length $\sin x$ and $\cos x$ (be sure you see how this follows from the definitions of sine and cosine). The Pythagorean Theorem then leads to the first Pythagorean Identity; dividing through by $\cos^2 x$ and $\sin^2 x$ leads to, respectively, the second and third Pythagorean Identities.

The ideas behind the identities can lead to statements that appear, superficially, to be different. For instance, we know that the 2π-periodicity of sine makes all of the following statements true:

$$\sin(x-6\pi)=\sin x, \quad \sin(x+4\pi)=\sin(x+2\pi), \quad \text{and} \quad \sin(x+2\pi)=\sin(x-2\pi).$$

Deeper contemplation of the graph of sine leads to the similar identity

$$\sin x=-\sin(x+\pi).$$

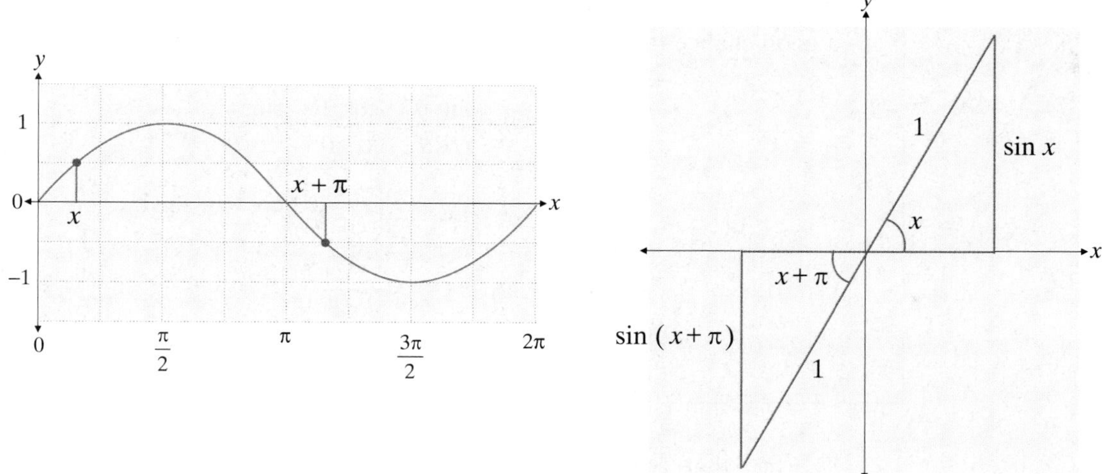

Similar statements based on periodicity can be deduced for the other trigonometric functions.

Topic 2: Simplifying Trigonometric Expressions

One common use of trigonometric identities is in simplifying expressions. Frequently, the first answer obtained in solving a trigonometry-related problem is unnecessarily complicated, and the judicious use of identities leads to a simpler form of the answer. This happens often, for instance, in Calculus problems.

example 1

Simplify the expression $\cos\theta + \sin\theta\tan\theta$.

Solution:

Skill in using identities to simplify trigonometric expressions only comes with practice, but the guiding principle is to rewrite the expression in such a way that the use of one or more identities becomes apparent. In the case of the expression $\cos\theta + \sin\theta\tan\theta$, a good way to begin is to rewrite it in terms of only sine and cosine, since three distinct functions is unnecessary.

$$\begin{aligned}\cos\theta + \sin\theta\tan\theta &= \cos\theta + \sin\theta\left(\frac{\sin\theta}{\cos\theta}\right)\\ &= \cos\theta + \frac{\sin^2\theta}{\cos\theta}\end{aligned}$$

Now, the presence of $\sin^2\theta$ should remind you of the first Pythagorean Identity (one that you'll find yourself using very frequently). Remember that $\sin^2\theta + \cos^2\theta = 1$, so if some other term can be rewritten in such a way that $\sin^2\theta$ is added to $\cos^2\theta$, there is a good chance that the identity applies. In the problem at hand,

$$\begin{aligned}\cos\theta + \frac{\sin^2\theta}{\cos\theta} &= \frac{\cos^2\theta}{\cos\theta} + \frac{\sin^2\theta}{\cos\theta}\\ &= \left(\frac{1}{\cos\theta}\right)\left(\cos^2\theta + \sin^2\theta\right)\\ &= \frac{1}{\cos\theta}\\ &= \sec\theta.\end{aligned}$$

The fact that $\cos\theta + \sin\theta\tan\theta = \sec\theta$ is not at all obvious initially, but is also not at all surprising. Such equivalences between relatively complicated and relatively simple expressions occur frequently in trigonometry, and it's usually worth spending some amount of time to see if a complicated expression can be simplified. After all, if the expression $\cos\theta + \sin\theta\tan\theta$ were the answer to a real-world problem, and if the next step was to evaluate the expression for numerous values of θ, it would certainly be easier to simply evaluate $\sec\theta$ instead.

example 2

Simplify the expression $\cot\alpha + \dfrac{\sin\alpha}{1+\cos\alpha}$.

Solution:

As in Example 1, rewriting the expression in terms of only sine and cosine is a good way to start. And as in so many problems, finding a common denominator and combining the resulting fractions is a promising way to proceed. After that, the remaining steps suggest themselves clearly:

$$\cot\alpha + \frac{\sin\alpha}{1+\cos\alpha} = \frac{\cos\alpha}{\sin\alpha} + \frac{\sin\alpha}{1+\cos\alpha}$$

$$= \left(\frac{\cos\alpha}{\sin\alpha}\right)\left(\frac{1+\cos\alpha}{1+\cos\alpha}\right) + \left(\frac{\sin\alpha}{1+\cos\alpha}\right)\left(\frac{\sin\alpha}{\sin\alpha}\right)$$

Multiply by appropriate factors to achieve a common denominator.

$$= \frac{\cos\alpha + \cos^2\alpha + \sin^2\alpha}{\sin\alpha(1+\cos\alpha)}$$

Apply the first Pythagorean Identity.

$$= \frac{\cos\alpha + 1}{\sin\alpha(1+\cos\alpha)}$$

Cancel the common factors.

$$= \frac{1}{\sin\alpha}$$

$$= \csc\alpha.$$

Topic 3: Verifying Trigonometric Identities

The identities we have seen to this point are all very fundamental, and in fact some are merely restatements of definitions. Nevertheless, they are also very useful and serve as good examples of the class of equations known as identities. Recall from Section 1.5 that an identity is an equation that is true for any (allowable) value of the variable; this would be a good time to review the list at the start of this section and verify that those equations indeed fit this description.

Of course, most equations that we encounter in algebra are not identities, and in fact the goal of much of our work is to determine exactly which values of the variable(s) make the equation true. In Section 1.5 non-identity equations were labeled *conditional*, but in practice the adjective is usually dropped since most equations we encounter are of this sort. We have already solved many simple conditional trigonometric equations, but we will study such equations in much greater depth in Section 7.4. At that time, we will see that trigonometric identities are very useful in determining the solutions of trigonometric equations.

Before we get to that point, though, we need to build up our repertoire of identities, and one step in doing so is to verify that a proposed identity really is true for all values of the variable. This is called *verifying an identity*, and the process is often very similar to using identities to simplify expressions. While there is no guaranteed method to use in such verification, there are some general guidelines to follow.

Guidelines for Verifying Trigonometric Identities

1. **Work with one side at a time.** Choose one side of the equation to work with and simplify it. The more complicated side is usually the best choice. The goal is to transform it into the other side.

2. **Apply trigonometric identities as appropriate.** To do so, it will probably be necessary to combine fractions, add or subtract terms, factor expressions, or use other algebraic manipulations.

3. **Rewrite in terms of sine and cosine if necessary.** If you are stuck, expressing everything in terms of sine and cosine often leads to inspiration.

example 3

Verify the identity $2\csc^2 x = \dfrac{1}{1-\cos x} + \dfrac{1}{1+\cos x}$.

Solution:

The right-hand side is more complicated, so we'll begin with it. Clearly, combining the two fractions is a good way to begin, especially as we can see that a denominator of $1-\cos^2 x$ will eventually appear.

$$\frac{1}{1-\cos x} + \frac{1}{1+\cos x} = \frac{1+\cos x + 1 - \cos x}{(1-\cos x)(1+\cos x)}$$

Modify each fraction in order to obtain a common denominator, and combine.

$$= \frac{2}{1-\cos^2 x}$$

Multiply out the denominator.

$$= \frac{2}{\sin^2 x}$$

Apply the first Pythagorean Identity.

$$= 2\csc^2 x$$

In verifying some identities, it may be easiest to simplify both sides of the equation individually. The goal in this case is to achieve a single simpler expression which is equivalent to both sides of the original equation.

example 4

Verify the identity $\frac{\tan^2 x}{1+\sec x} = \frac{1-\cos x}{\cos x}$.

Solution:

It's not immediately clear that either side is more complicated than the other, so we can try simplifying both. Beginning with the left-hand side, we can use another of the Pythagorean Identities as follows:

$$\frac{\tan^2 x}{1+\sec x} = \frac{\sec^2 x - 1}{1+\sec x}$$

Use the second Pythagorean Identity to rewrite the numerator.

$$= \frac{(\sec x - 1)(\sec x + 1)}{1+\sec x}$$

Factor the difference of two squares.

$$= \sec x - 1.$$

Cancel the common factors.

The right-hand side is more easily dealt with:

$$\frac{1-\cos x}{\cos x} = \frac{1}{\cos x} - \frac{\cos x}{\cos x}$$

$$= \sec x - 1.$$

Break the single fraction into two and simplify.

Since both sides of the original equation are equivalent to $\sec x - 1$, the identity is true.

example 5

Verify the identity $\frac{\cos\varphi \cot\varphi}{1-\sin\varphi} - 1 = \csc\varphi$.

Solution:

The left-hand side is clearly more complicated, and there are several ways of beginning the process of simplifying it. First, rewriting the numerator of the fraction in terms of sine and cosine looks promising, as a factor of $\cos^2\varphi$ would then appear. Second, obtaining a denominator of $1-\sin^2\varphi$ in the fraction would be easily done and would allow one of the Pythagorean Identities to apply. We'll try both ideas:

cont'd. on next page ...

$$\frac{\cos\varphi\cot\varphi}{1-\sin\varphi}-1=\frac{\cos\varphi\left(\dfrac{\cos\varphi}{\sin\varphi}\right)}{1-\sin\varphi}-1$$

Rewrite cotangent in terms of sine and cosine.

$$=\left(\frac{\cos^2\varphi}{\sin\varphi(1-\sin\varphi)}\right)\left(\frac{1+\sin\varphi}{1+\sin\varphi}\right)-1$$

Multiply appropriately in order to obtain $1-\sin^2\varphi$ in the denominator.

$$=\left(\frac{\cos^2\varphi}{\sin\varphi\left(1-\sin^2\varphi\right)}\right)(1+\sin\varphi)-1$$

$$=\left(\frac{\cos^2\varphi}{\sin\varphi\cos^2\varphi}\right)(1+\sin\varphi)-1$$

Apply the first Pythagorean Identity.

$$=\frac{1+\sin\varphi}{\sin\varphi}-1$$

Cancel common factors.

$$=\frac{1}{\sin\varphi}+1-1$$

Break apart fraction and simplify.

$$=\csc\varphi$$

Topic 4: Interlude: Trigonometric Substitutions

In several classes of Calculus problems, it is very convenient to be able to replace certain algebraic expressions with trigonometric expressions. Most often, these replacements depend on one of the Pythagorean Identities, and the work involved is reminiscent of that in other problems in this section. The example below illustrates a typical trigonometric substitution.

example 6

Use the substitution $\sin\theta=\frac{x}{2}$ to write $\sqrt{4-x^2}$ as a trigonometric expression. Assume $0\le\theta\le\frac{\pi}{2}$.

Solution:

Although it is not necessary for the task at hand, a diagram motivating the substitution may be helpful. The triangle below illustrates the geometric relation between θ and the various algebraic expressions:

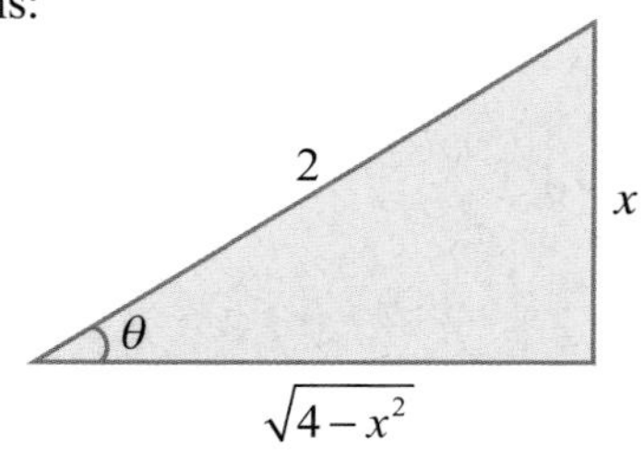

The suggested substitution can be rewritten as $x = 2\sin\theta$, and so we obtain

$$\begin{aligned}\sqrt{4-x^2} &= \sqrt{4-(2\sin\theta)^2}\\ &= \sqrt{4-4\sin^2\theta}\\ &= 2\sqrt{1-\sin^2\theta}\\ &= 2\cos\theta.\end{aligned}$$

exercises

Use trigonometric identities to simplify the expressions. There may be more than one correct answer. See Examples 1 and 2.

1. $\tan x \csc x$

2. $\dfrac{1}{\tan^2\theta+1}$

3. $\dfrac{\tan t}{\sec t}$

4. $\cot^2 x - \cot^2 x\cos^2 x$

5. $\sin(-x)\tan x$

6. $\dfrac{1}{\sec^2 x} + \sin x\cos\left(\dfrac{\pi}{2}-x\right)$

7. $\sin(\alpha+2\pi)\sec\alpha$

8. $\sin t\,(\csc t - \sin t)$

9. $\cos y\,(1+\tan^2 y)$

10. $\dfrac{1}{\cos x\csc(-x)}$

11. $\dfrac{1-\tan^2 x}{\cot^2 x - 1}$

12. $\dfrac{\sin\beta\tan\left(\dfrac{\pi}{2}-\beta\right)}{\cos\beta}$

Verify the identities. See Examples 3, 4, and 5.

13. $(1-\cos\theta)(1+\cos\theta) = \sin^2\theta$

14. $\csc x - \sin x = \cos x\cot x$

15. $\sec^2 y - \tan^2 y = \sec y\cos(-y)$

16. $\dfrac{\cos\beta\cot\beta}{1-\sin\beta} - 1 = \csc\beta$

17. $\dfrac{\sin\left(\dfrac{\pi}{2}-x\right)}{\cos\left(\dfrac{\pi}{2}-x\right)} = \cot x$

18. $\dfrac{\sec^2\theta}{\tan\theta} = \sec\theta\csc\theta$

19. $\dfrac{1}{\tan x} + \tan x = \dfrac{\sec^2 x}{\tan x}$

20. $\sin^2 t + \sin^2\left(\dfrac{\pi}{2}-t\right) = 1$

21. $\dfrac{1}{\sin(\theta+2\pi)+1} + \dfrac{1}{\csc(\theta+2\pi)+1} = 1$

22. $3 + \cot^2\alpha = 2 + \csc^2\alpha$

23. $\sin^2 x - \sin^4 x = \cos^2(-x) - \cos^4(-x)$

24. $\cot\left(\frac{\pi}{2} - \beta\right)\cot\beta = 1$

25. $\dfrac{\cos\left(\frac{\pi}{2} - \alpha\right)}{\csc\alpha} - 1 = \sin\alpha\cot(-\alpha)\cos(-\alpha)$

Use the suggested substitution to rewrite the given expression as a trigonometric expression. See Example 6.

26. $\sqrt{x^2 + 1},\ x = \tan\theta$

27. $\sqrt{x^2 - 16},\ x = 4\sec\theta$

28. $\sqrt{9 - x^2},\ \cos\theta = \frac{x}{3}$

29. $\sqrt{4x^2 + 100},\ \cot\theta = \frac{x}{5}$

30. $\sqrt{64 - x^2},\ x = 8\sin\theta$

31. $\sqrt{x^2 - 4},\ x = 2\csc\theta$

32. $\sqrt{x^2 + 25},\ \tan\theta = \frac{x}{5}$

33. $\sqrt{144 - 9x^2},\ x = 4\cos\theta$

Show how the identities below follow from the first Pythagorean Identity.

34. $\tan^2 x + 1 = \sec^2 x$

35. $1 + \cot^2 x = \csc^2 x$

7.2 Sum and Difference Identities

TOPICS

1. The identities
2. Using sum and difference identities for exact evaluation
3. Applications of sum and difference identities

Topic 1: The Identities

This section and the next will introduce new identities, selected proofs of the identities, and many examples of how the identities can be used. The first group of identities concerns functions acting on sums or differences of angles.

Sum and Difference Identities

Sine Identities

$$\sin(u+v) = \sin u \cos v + \cos u \sin v \qquad \sin(u-v) = \sin u \cos v - \cos u \sin v$$

Cosine Identities

$$\cos(u+v) = \cos u \cos v - \sin u \sin v \qquad \cos(u-v) = \cos u \cos v + \sin u \sin v$$

Tangent Identities

$$\tan(u+v) = \frac{\tan u + \tan v}{1 - \tan u \tan v} \qquad \tan(u-v) = \frac{\tan u - \tan v}{1 + \tan u \tan v}$$

We will prove two of these identities here, and indicate in the Exercises how the other four identities can then be derived.

Consider the unit circle in Figure 1 on the following page. In the figure, angle v has been placed in standard position and angle u has been positioned in such a way that angle $u + v$ is in standard position. In addition, the negative of angle u has also been drawn in standard position.

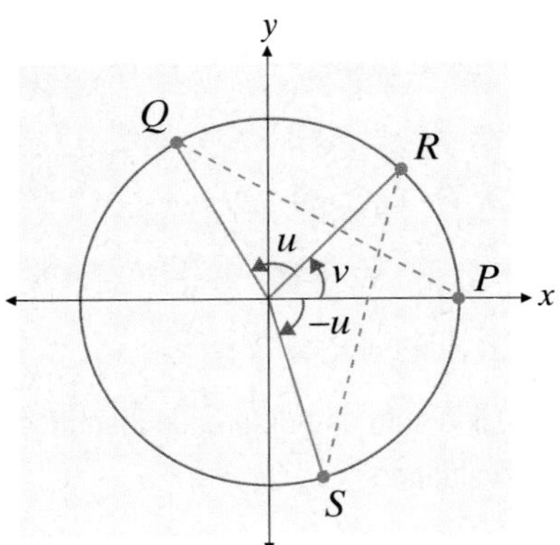

Figure 1: Derivation of Cosine Sum Identity

Using the fact that the radius of the circle is 1, we can easily identify the coordinates of the points P, Q, R, and S:

$$P = (1, 0),\ Q = \left(\cos(u+v),\ \sin(u+v)\right),\ R = \left(\cos v,\ \sin v\right),\ \text{and}\ S = \left(\cos u,\ -\sin u\right).$$

Note that for point S, we have used the fact that $\sin(-u) = -\sin u$. Note also that the chord $\overline{PQ}$ has the same length as the chord $\overline{RS}$, since the subtended angles both have magnitude $u + v$. Since we know the coordinates of the endpoints of the chords, we can use the Distance Formula to obtain the equation

$$\sqrt{\left(\cos(u+v)-1\right)^2 + \left(\sin(u+v)-0\right)^2} = \sqrt{\left(\cos v - \cos u\right)^2 + \left(\sin v + \sin u\right)^2}\ .$$

The square roots are easily eliminated by squaring both sides, and we can then proceed to expand the squared terms:

$$\cos^2(u+v) - 2\cos(u+v) + 1 + \sin^2(u+v) =$$
$$\cos^2 v - 2\cos u\cos v + \cos^2 u + \sin^2 v + 2\sin u\sin v + \sin^2 u.$$

By now, you should be attuned to the presence of $\sin^2(\ )$ and $\cos^2(\ )$ terms; every pair of these having the same argument can be replaced with 1. Making that replacement three times in the preceding equation gives us

$$2 - 2\cos(u+v) = 2 - 2\cos u\cos v + 2\sin u\sin v,$$

and subtracting 2 from both sides and then dividing by −2 yields

$$\cos(u+v) = \cos u\cos v - \sin u\sin v.$$

Now that we have proved one of the identities, the rest follow relatively quickly. In particular, the difference identity for cosine is very easily proved by replacing v with $-v$ in the sum identity and making use of the fact that cosine is an even function and sine is odd:

$$\begin{aligned}\cos(u-v) &= \cos u\cos(-v) - \sin u\sin(-v)\\ &= \cos u\cos v + \sin u\sin v.\end{aligned}$$

Topic 2: Using Sum and Difference Identities for Exact Evaluation

It was fairly easy, back in Chapter 6, to determine the exact values of the trigonometric functions acting on the angles $\frac{\pi}{6}, \frac{\pi}{4},$ and $\frac{\pi}{3}$. It may seem odd, therefore, that these are still the only (acute) angles for which we can perform exact evaluation. To this point, for instance, our only option for evaluating $\sin 75^\circ$ has been to use a calculator and note that $\sin 75^\circ \approx 0.9659$. The sum and difference identities extend our ability to obtain exact values greatly, as seen in the next example.

example 1

Determine the exact value of $\sin 75^\circ$.

Solution:

All problems of this sort will call for us to express the given angle in terms of angles about which we know more. In this case, we'll use the fact that $75^\circ = 45^\circ + 30^\circ$:

$$\begin{aligned}\sin 75^\circ &= \sin\left(45^\circ + 30^\circ\right)\\ &= \sin 45^\circ \cos 30^\circ + \cos 45^\circ \sin 30^\circ\\ &= \left(\frac{1}{\sqrt{2}}\right)\left(\frac{\sqrt{3}}{2}\right) + \left(\frac{1}{\sqrt{2}}\right)\left(\frac{1}{2}\right)\\ &= \frac{\sqrt{3}+1}{2\sqrt{2}}.\end{aligned}$$

You can now easily verify that this exact value is approximately 0.9659.

example 2

Determine the exact value of $\cos 75^\circ$.

Solution:

The purpose of this example is twofold. The first is to point out that we are starting to build up a significant collection of tools and knowledge. The second is to emphasize that two identical answers may appear, superficially, to be different. We could certainly use the sum identity for cosine to evaluate $\cos 75^\circ$, taking steps very similar to those in Example 1. Using this approach, we would obtain

cont'd. on next page ...

$$\cos 75^\circ = \frac{\sqrt{6}-\sqrt{2}}{4}.$$

But, coming immediately after the evaluation of $\sin 75^\circ$, it would also make sense to use the identity $\cos^2 x + \sin^2 x = 1$ to obtain

$$\begin{aligned}\cos 75^\circ &= \sqrt{1-\sin^2 75^\circ}\\ &= \sqrt{1-\left(\frac{\sqrt{3}+1}{2\sqrt{2}}\right)^2}\\ &= \sqrt{1-\frac{4+2\sqrt{3}}{8}}\\ &= \sqrt{\frac{2-\sqrt{3}}{4}}\\ &= \frac{\sqrt{2-\sqrt{3}}}{2}.\end{aligned}$$

Are these two answers actually the same?

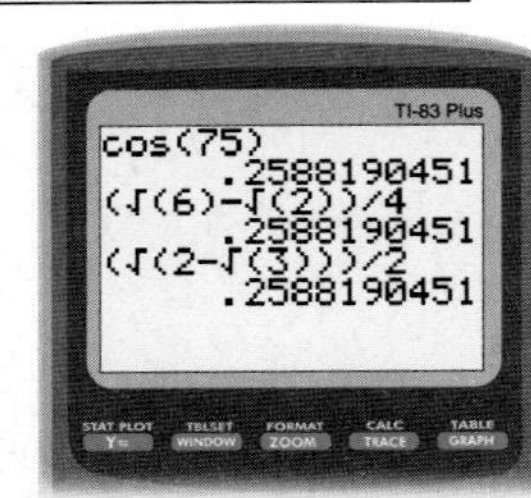

You can use a calculator to determine that both are approximately 0.2588, but it is more convincing to prove the equivalence mathematically. A common way of doing this is to set the two answers equal to one another and to simplify the expressions in the resulting equation until either an obviously true or obviously false statement is achieved. Keep in mind, however, that the "equality" below is actually a question initially; we won't know that the two expressions actually are equal until we achieve an obviously true equation.

$\frac{\sqrt{6}-\sqrt{2}}{4} \stackrel{?}{=} \frac{\sqrt{2-\sqrt{3}}}{2}$	Use the two expressions to form an "equation."
$\sqrt{6}-\sqrt{2} \stackrel{?}{=} 2\sqrt{2-\sqrt{3}}$	Multiply both sides by 4.
$6-2\sqrt{12}+2 \stackrel{?}{=} 4\left(2-\sqrt{3}\right)$	Square both sides.
$8-4\sqrt{3} = 8-4\sqrt{3}$	Simplify to obtain an obviously true statement.

example 3

Determine the exact value of $\tan \frac{\pi}{12}$.

Solution:

To make use of previous results, we need to note that $\frac{\pi}{12} = \frac{\pi}{3} - \frac{\pi}{4}$. So

$$\begin{aligned}\tan \frac{\pi}{12} &= \tan\left(\frac{\pi}{3} - \frac{\pi}{4}\right) \\ &= \frac{\tan \frac{\pi}{3} - \tan \frac{\pi}{4}}{1 + \tan \frac{\pi}{3} \tan \frac{\pi}{4}} \\ &= \frac{\sqrt{3} - 1}{1 + \left(\sqrt{3}\right)(1)} \\ &= \frac{\sqrt{3} - 1}{\sqrt{3} + 1}.\end{aligned}$$

Although the need arises less often, the sum and difference identities can also be used in reverse.

example 4

Determine the exact value of $\sin 80^\circ \cos 20^\circ - \cos 80^\circ \sin 20^\circ$.

Solution:

The key step is to recognize the expression as the difference identity for sine. We can then proceed as follows:

$$\begin{aligned}\sin 80^\circ \cos 20^\circ - \cos 80^\circ \sin 20^\circ &= \sin\left(80^\circ - 20^\circ\right) \\ &= \sin 60^\circ \\ &= \frac{\sqrt{3}}{2}.\end{aligned}$$

Topic 3: Applications of Sum and Difference Identities

The truth of the cofunction identities can be easily seen through the use of a unit circle diagram, as in the argument at the start of this section. But the difference identities can also be used to furnish a very quick algebraic verification, as in the next example.

example 5

Use a difference identity to verify that $\sin\left(\frac{\pi}{2}-x\right)=\cos x$.

Solution:

$$\begin{aligned}\sin\left(\frac{\pi}{2}-x\right)&=\sin\frac{\pi}{2}\cos x-\cos\frac{\pi}{2}\sin x\\&=(1)(\cos x)-(0)(\sin x)\\&=\cos x\end{aligned}$$

The identities in this section are useful in simplifying trigonometric functions of sums or differences no matter how unwieldy they may initially appear to be.

example 6

Express $\sin\left(\tan^{-1}x+\cos^{-1}x\right)$ as an algebraic function of x.

Solution:

From our previous experience with compositions of trig functions and inverse trig functions, we know the general approach to take; the only additional complication here is that the argument of sine is a sum. If we let $u=\tan^{-1}x$ and $v=\cos^{-1}x$, then $\tan u=x$ and $\cos v=x$ and we are led to consider two right triangles such as those below.

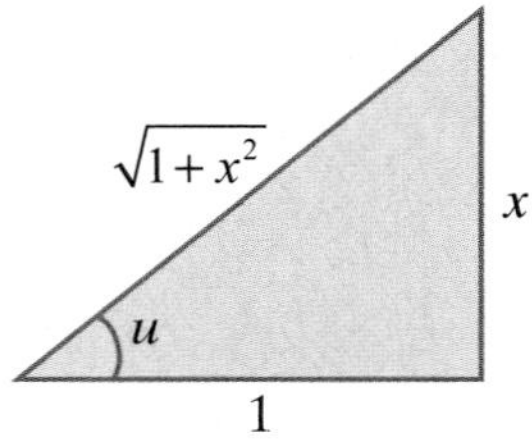

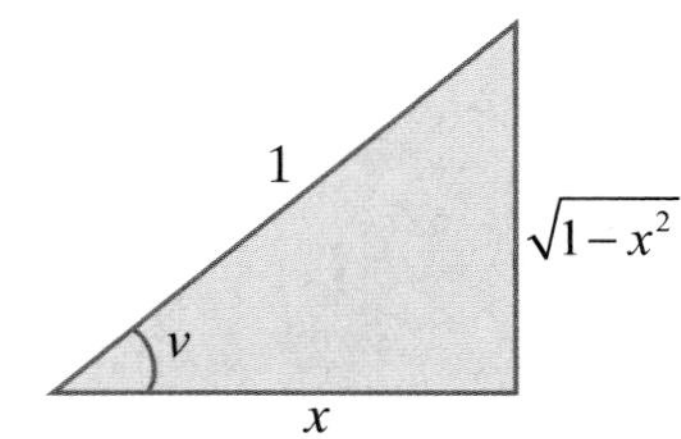

We can now apply the appropriate sum identity:

$$\begin{aligned}\sin\left(\tan^{-1}x+\cos^{-1}x\right)&=\sin(u+v)\\&=\sin u\cos v+\cos u\sin v\\&=\left(\frac{x}{\sqrt{1+x^2}}\right)(x)+\left(\frac{1}{\sqrt{1+x^2}}\right)\left(\sqrt{1-x^2}\right)\\&=\frac{x^2}{\sqrt{1+x^2}}+\frac{\sqrt{1-x^2}}{\sqrt{1+x^2}}.\end{aligned}$$

technology note

Computer algebra systems, such as *Mathematica*, and some calculators are capable of simplifying expressions such as $\sin\left(\tan^{-1}x+\cos^{-1}x\right)$. The command to do this in *Mathematica* is `TrigExpand`; its use is demonstrated in the figure below.

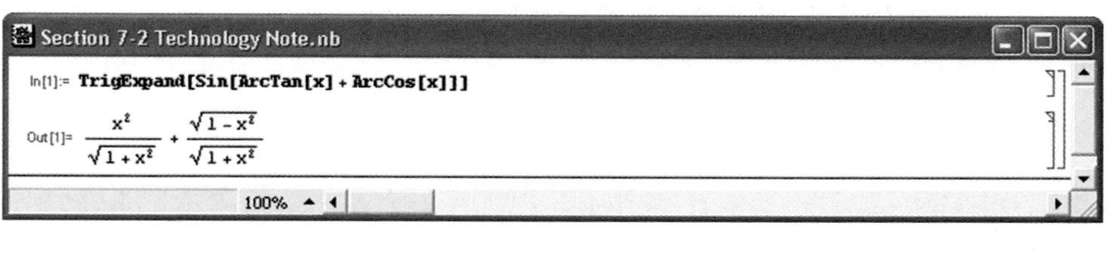

The next example is one that you will encounter again in Calculus.

example 7

Show that $\dfrac{\sin(x+h)-\sin x}{h}=\sin x\left(\dfrac{\cos h-1}{h}\right)+\cos x\left(\dfrac{\sin h}{h}\right)$.

Solution:

$$\begin{aligned}\frac{\sin(x+h)-\sin x}{h}&=\frac{\sin x\cos h+\cos x\sin h-\sin x}{h}\\&=\frac{\sin x\cos h-\sin x}{h}+\frac{\cos x\sin h}{h}\\&=\sin x\left(\frac{\cos h-1}{h}\right)+\cos x\left(\frac{\sin h}{h}\right)\end{aligned}$$

The last example of an application of the sum/difference identities leads to a surprising conclusion. As you know, the graph of cosine has the same shape as the graph of sine shifted to the left by $\frac{\pi}{2}$ (one of the cofunction identities is the algebraic form of this statement). But the connection between cosine and sine goes deeper; as it turns out, a *sum* of a sine term and a cosine term can be rephrased in terms of sine or cosine alone. This is true even if the sine and cosine functions have different amplitudes.

To see this, consider the expression $A\sin x + B\cos x$, a sum of a sine curve with amplitude A and a cosine curve with amplitude B. This has some resemblance to the right-hand side of the sum identity for sine, but only if A can be replaced with a cosine term and B with a sine term. That is the inspiration for the following:

$$A\sin x + B\cos x = \sqrt{A^2 + B^2}\left(\frac{A}{\sqrt{A^2 + B^2}}\sin x + \frac{B}{\sqrt{A^2 + B^2}}\cos x\right).$$

If an angle φ exists for which

$$\cos\varphi = \frac{A}{\sqrt{A^2 + B^2}} \text{ and } \sin\varphi = \frac{B}{\sqrt{A^2 + B^2}},$$

then the sum identity gives us

$$\begin{aligned} A\sin x + B\cos x &= \sqrt{A^2 + B^2}\left(\cos\varphi\sin x + \sin\varphi\cos x\right) \\ &= \sqrt{A^2 + B^2}\sin\left(x + \varphi\right). \end{aligned}$$

Does such a φ exist? Figure 2 to the right makes it clear that φ is easily found.

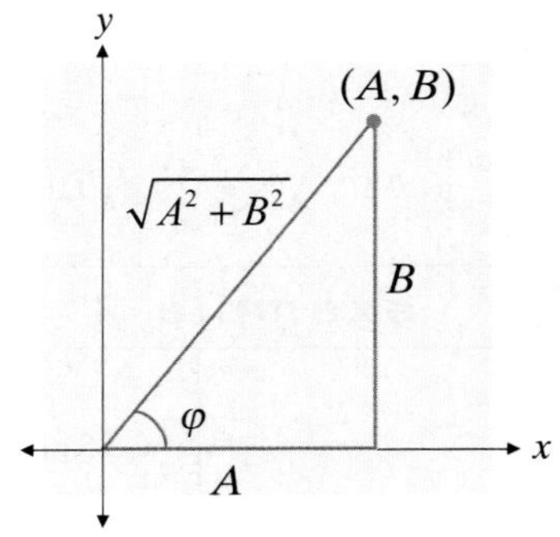

Figure 2: Determining the Phase Shift φ

Sum of Sines and Cosines

$$\begin{aligned} A\sin x + B\cos x &= \sqrt{A^2 + B^2}\left(\frac{A}{\sqrt{A^2 + B^2}}\sin x + \frac{B}{\sqrt{A^2 + B^2}}\cos x\right) \\ &= \sqrt{A^2 + B^2}\sin(x + \varphi) \end{aligned}$$

where $\cos\varphi = \frac{A}{\sqrt{A^2 + B^2}}$ and $\sin\varphi = \frac{B}{\sqrt{A^2 + B^2}}$.

example 8

Express the function $f(x) = \sin x - \sqrt{3}\cos x$ in terms of a single sine function, and graph the result.

Solution:

Since $A = 1$ and $B = -\sqrt{3}$, $\sqrt{A^2 + B^2} = 2$ and φ must satisfy

$$\cos\varphi = \frac{1}{2} \text{ and } \sin\varphi = \frac{-\sqrt{3}}{2}.$$

This means φ lies in the fourth quadrant, and as a negative angle it can be expressed as $\varphi = -\frac{\pi}{3}$ (we could also write $\varphi = \frac{5\pi}{3}$). Using the above derivation,

$$f(x) = 2\sin\left(x - \frac{\pi}{3}\right),$$

and the graph of this function is shown below.

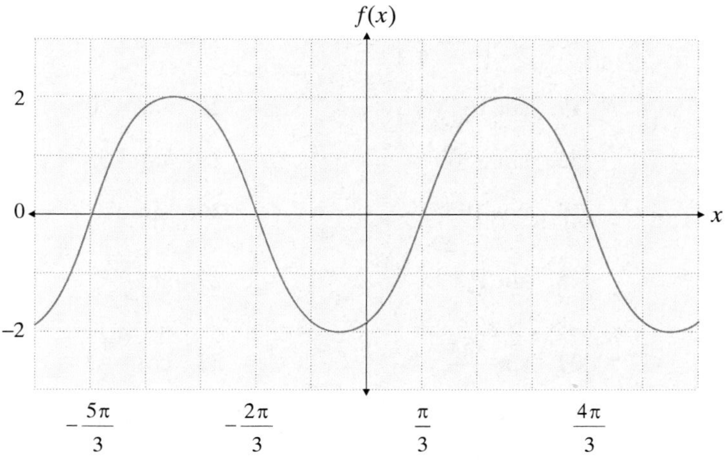

exercises

Use the sum and difference identities to determine the exact value of each of the following expressions. See Examples 1, 2, and 3.

1. $\cos\left(\frac{\pi}{4}+\frac{\pi}{3}\right)$

2. $\sin\left(\frac{\pi}{6}+\frac{3\pi}{4}\right)$

3. $\tan\left(\frac{4\pi}{3}+\frac{5\pi}{4}\right)$

4. $\sin\left(\frac{2\pi}{3}+\frac{\pi}{4}\right)$

5. $\cos\left(\frac{7\pi}{6}-\frac{\pi}{6}\right)$

6. $\tan\left(\frac{\pi}{3}-\frac{3\pi}{4}\right)$

7. $\cos\left(\frac{4\pi}{3}+\frac{5\pi}{3}\right)$

8. $\tan\left(\frac{4\pi}{3}-\frac{5\pi}{4}\right)$

9. $\cos\left(\frac{7\pi}{6}-\frac{5\pi}{3}\right)$

10. $\cos\left(\frac{7\pi}{6}+\frac{5\pi}{3}\right)$

11. $\sin\left(\frac{7\pi}{4}+\frac{5\pi}{4}\right)$

12. $\cos\left(\frac{\pi}{3}+\frac{11\pi}{6}\right)$

13. $\cos\left(\frac{\pi}{4}-\frac{\pi}{6}\right)$

14. $\sin\left(\frac{5\pi}{4}-\frac{\pi}{3}\right)$

15. $\sin\left(\frac{7\pi}{4}+\frac{2\pi}{3}\right)$

16. $\cos\left(\frac{\pi}{3}-\frac{\pi}{4}\right)$

17. $\tan 75°$

18. $\tan 15°$

19. $\sin 165°$

20. $\tan 150°$

21. $\cos -15°$

22. $\sin -30°$

23. $\tan 255°$

24. $\cos 135°$

25. $\cos 195°$

26. $\sin 270°$

27. $\cos 165°$

28. $\sin 315°$

29. $\sin\frac{\pi}{12}$

30. $\tan\frac{5\pi}{12}$

31. $\sin\frac{-5\pi}{6}$

32. $\cos\frac{7\pi}{12}$

33. $\cos\frac{25\pi}{12}$

34. $\sin\frac{13\pi}{12}$

35. $\sin\frac{11\pi}{12}$

36. $\tan\frac{7\pi}{12}$

37. $\cos\frac{-7\pi}{6}$

38. $\cos\frac{-\pi}{3}$

39. $\sin\frac{5\pi}{12}$

40. $\tan\frac{\pi}{12}$

Find the sum or difference for each given question.

41. $\sin\alpha = \frac{4}{5}$ and $\sin\beta = \frac{5}{13}$. Both α and β are in quadrant I. Find $\cos(\alpha-\beta)$.

42. $\sin\alpha = -\frac{15}{17}$ and $\cos\beta = -\frac{3}{5}$. Both α and β are in quadrant III. Find $\sin(\alpha-\beta)$.

43. $\cos\alpha = -\frac{15}{17}$ and $\cos\beta = -\frac{3}{5}$. α is in quadrant II and β is in quadrant III. Find $\sin(\alpha+\beta)$.

44. $\cos\alpha = -\frac{24}{25}$ and $\sin\beta = \frac{5}{13}$. α is in quadrant III and β is in quadrant I. Find $\cos(\alpha+\beta)$.

45. $\cos\alpha = \frac{2}{5}$ and $\cos\beta = \frac{1}{5}$. Both α and β are in quadrant I. Find $\sin(\beta-\alpha)$.

46. $\cos\alpha = -\frac{2}{3}$ and $\sin\beta = -\frac{2\sqrt{2}}{3}$. α is in quadrant III and β is in quadrant IV. Find $\tan(\alpha+\beta)$.

Use the sum and difference identities to rewrite each of the following expressions as a trigonometric function of a single number, and then evaluate the result. See Example 4.

47. $\sin 15^\circ \cos 30^\circ + \cos 15^\circ \sin 30^\circ$

48. $\cos\frac{5\pi}{12}\cos\frac{2\pi}{3} + \sin\frac{5\pi}{12}\sin\frac{2\pi}{3}$

49. $\dfrac{\tan 100^\circ + \tan 35^\circ}{1-\tan 100^\circ \tan 35^\circ}$

50. $\sin 125^\circ \cos 35^\circ - \cos 125^\circ \sin 35^\circ$

51. $\dfrac{\tan\frac{5\pi}{16} - \tan\frac{\pi}{16}}{1+\tan\frac{5\pi}{16}\tan\frac{\pi}{16}}$

52. $\cos 15^\circ \cos 15^\circ - \sin 15^\circ \sin 15^\circ$

53. $\sin 70^\circ \cos 80^\circ + \cos 70^\circ \sin 80^\circ$

54. $\cos\frac{\pi}{5}\cos\frac{3\pi}{10} - \sin\frac{\pi}{5}\sin\frac{3\pi}{10}$

55. $\cos 182^\circ \cos 47^\circ + \sin 182^\circ \sin 47^\circ$

56. $\dfrac{\tan\frac{5\pi}{12} + \tan\frac{3\pi}{4}}{1-\tan\frac{5\pi}{12}\tan\frac{3\pi}{4}}$

57. $\dfrac{\tan 70^\circ - \tan 10^\circ}{1+\tan 70^\circ \tan 10^\circ}$

58. $\sin\frac{5\pi}{12}\cos\frac{\pi}{12} - \cos\frac{5\pi}{12}\sin\frac{\pi}{12}$

Use the sum and difference identities to verify the following identities. See Example 5.

59. $\tan\left(\frac{\pi}{2}-\theta\right) = \cot\theta$ (Hint: use sin and cos)

60. $\cos^2 u - \sin^2 v = \cos(u+v)\cos(u-v)$

61. $\cos\left(\frac{3\pi}{2}-\alpha\right) = -\sin\alpha$

62. $\sin(\beta-\theta) + \sin(\beta+\theta) = 2\sin\beta\cos\theta$

63. $\tan\left(\alpha-\frac{5\pi}{4}\right) = \dfrac{\tan\alpha - 1}{1+\tan\alpha}$

64. $\sec\left(\frac{\pi}{2}-u\right) = \csc u$

65. $\tan(\pi + 2\pi) = 0$

66. $\sin\left(\frac{5\pi}{6}+\theta\right)=\frac{1}{2}\left(\cos\theta-\sqrt{3}\sin\theta\right)$

67. $\sin(u + v)\sin(u - v) = \sin^2 u - \sin^2 v$

68. $\cos\left(\frac{7\pi}{4}-\beta\right)=\frac{\sqrt{2}}{2}(\cos\beta-\sin\beta)$

Using a graphing calculator, determine whether the following identities are true or false. (Hint: Graph both expressions on each side of the equality separately and determine if the graphs coincide.)

69. $\sin\left(\frac{\pi}{2}-\theta\right)=\cos\theta$

70. $\cos\left(\theta-\frac{3\pi}{2}\right)=-\sin\theta$

71. $\cot(\pi+\theta)=-\tan\theta$

72. $\sec\left(\frac{\pi}{2}-\theta\right)=\csc\theta$

73. $\sin\left(\frac{\pi}{6}+\theta\right)=\frac{1}{2}\left(\cos\theta+\sqrt{3}\sin\theta\right)$

74. $\frac{1+\tan\theta}{1-\tan\theta}=-\tan\theta$

Express each of the following as an algebraic function of x. See Example 6.

75. $\sin(\sin^{-1} 2x + \cos^{-1} 2x)$

76. $\sin(\arctan 2x - \arccos 2x)$

77. $\cos(\arctan 2x - \arcsin x)$

78. $\cos(\cos^{-1} x - \sin^{-1} x)$

79. $\cos(\arccos x + \arcsin 2x)$

80. $\sin(\sin^{-1} x - \cos^{-1} x)$

Express each of the following functions in terms of a single sine function, and graph the result. See Example 8.

81. $f(x)=\sin x+\cos x$

82. $g(x)=\sin x+\sqrt{3}\cos x$

83. $h(\beta) = \sin 2\beta - \cos 2\beta$

84. $f(\theta)=-\sqrt{3}\sin\theta+\cos\theta$

85. $g(u) = 5\sin 5u + 12\cos 5u$

86. $h(v)=8\cos\frac{v}{2}+6\sin\frac{v}{2}$

87. Use a cofunction identity to prove the sum and difference identities for sine. (Hint: note that $\sin(u+v)=\cos\left(\frac{\pi}{2}-(u+v)\right)=\cos\left(\left(\frac{\pi}{2}-u\right)-v\right)$ and apply the difference identity for cosine.)

88. Given sum identities for sine and cosine, prove the sum identity for tangent.

89. Prove or disprove that $\sin(u + v) + \sin(u - v) = 2 \sin u \cos v$.

90. Prove or disprove that $\dfrac{\cos(u+v)}{\cos u \cos v} = \tan u + \tan v$.

91. Prove or disprove that $\dfrac{\cos(u-v)}{\cos(u+v)} = 2\tan u \tan v$.

92. Prove or disprove that $\dfrac{\sin(u+v)}{\sin(u-v)} = \dfrac{\tan u + \tan v}{\tan u - \tan v}$.

93. Use the sine and cosine difference formulas to prove $\tan(u-v) = \dfrac{\tan u - \tan v}{1 + \tan u \tan v}$.

7.3 Product-Sum Identities

TOPICS

1. Double-angle identities
2. Power-reducing identities
3. Half-angle identities
4. Product-to-sum and sum-to-product identities

Topic 1: Double-Angle Identities

We will enlarge our repertoire greatly in this section, introducing and using four useful classes of identities. We will also prove a few selected identities in order to demonstrate typical methods of proof.

Our first class of identities contains those in which the argument of a given trigonometric function is twice an angle, and the identities are thus commonly called *double-angle identities*.

Double-Angle Identities

Sine Identity: $\sin 2u = 2\sin u \cos u$

Cosine Identities:

$$\begin{aligned}\cos 2u &= \cos^2 u - \sin^2 u \\ &= 2\cos^2 u - 1 \\ &= 1 - 2\sin^2 u\end{aligned}$$

Tangent Identity: $\tan 2u = \dfrac{2\tan u}{1-\tan^2 u}$

These identities are easily verified using the sum identities of the previous section. We will prove two of the cosine identities here, and leave the remainder of the proofs as exercises. For the first, note that

$$\begin{aligned}\cos 2u &= \cos(u+u) \\ &= \cos u \cos u - \sin u \sin u \\ &= \cos^2 u - \sin^2 u.\end{aligned}$$

And for the second cosine identity, we can apply a Pythagorean Identity:

$$\begin{aligned}\cos 2u &= \cos^2 u - \sin^2 u \\ &= \cos^2 u - \left(1 - \cos^2 u\right) \\ &= 2\cos^2 u - 1.\end{aligned}$$

example 1

Given that $\cos x = -\frac{2}{\sqrt{5}}$ and that sin x is positive, determine cos $2x$, sin $2x$, and tan $2x$.

Solution:

First, note that since cos x is negative and sin x is positive, we know that x lies in quadrant II. The Pythagorean Theorem (or the first Pythagorean Identity) then tells us that $\sin x = \frac{1}{\sqrt{5}}$, as follows:

$$\sin x = \sqrt{1 - \left(-\frac{2}{\sqrt{5}}\right)^2} = \sqrt{1 - \frac{4}{5}} = \frac{1}{\sqrt{5}}.$$

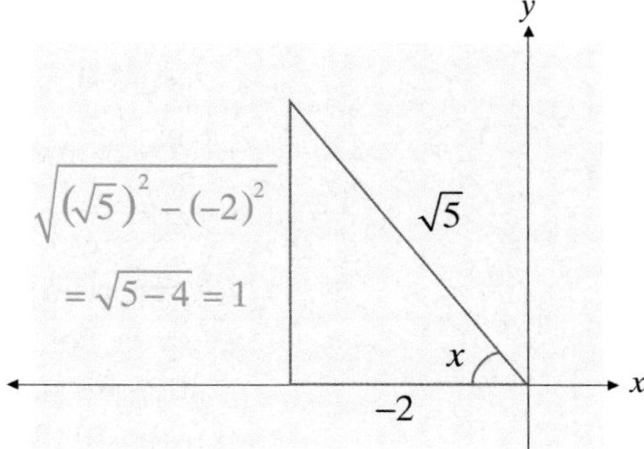

We can now easily apply two of the double-angle identities:

$$\cos 2x = \cos^2 x - \sin^2 x = \left(-\frac{2}{\sqrt{5}}\right)^2 - \left(\frac{1}{\sqrt{5}}\right)^2 = \frac{4}{5} - \frac{1}{5} = \frac{3}{5}$$

and

$$\sin 2x = 2\sin x \cos x = 2\left(\frac{1}{\sqrt{5}}\right)\left(-\frac{2}{\sqrt{5}}\right) = -\frac{4}{5}.$$

We can now most easily determine tan $2x$ by noting that $\tan 2x = \frac{\sin 2x}{\cos 2x} = -\frac{4}{3}$, but we could also have determined that $\tan x = -\frac{1}{2}$ and used the tangent double-angle identity as follows:

$$\tan 2x = \frac{2\tan x}{1 - \tan^2 x} = \frac{2\left(-\frac{1}{2}\right)}{1 - \left(-\frac{1}{2}\right)^2} = \frac{-1}{\frac{3}{4}} = -\frac{4}{3}.$$

Of course, the double-angle identities (in conjunction with others) can be used to prove new identities, as shown in the next example.

example 2

Prove that $\sin 3x = 3\sin x - 4\sin^3 x$.

Solution:

$\sin 3x = \sin(2x + x)$	Rewrite the argument as a sum.
$= \sin 2x \cos x + \cos 2x \sin x$	Apply the sum identity for sine.
$= 2\sin x \cos x \cos x + (1 - 2\sin^2 x)\sin x$	Apply two of the double-angle identities.
$= 2\sin x \cos^2 x + \sin x - 2\sin^3 x$	
$= 2\sin x(1 - \sin^2 x) + \sin x - 2\sin^3 x$	Apply the first Pythagorean Identity.
$= 2\sin x - 2\sin^3 x + \sin x - 2\sin^3 x$	
$= 3\sin x - 4\sin^3 x$	

Not surprisingly, the same sort of argument can be used on any expression of the form sin nx or cos nx, where n is a positive integer. In fact, cos nx can be rewritten as an nth-degree polynomial in cos x (the polynomials are called *Chebyshev polynomials*, after the 19th-century Russian Pafnuty Lvovich Chebyshev, a mathematician famous for, among many other things, the variety of spellings of his last name). A similar statement is almost true for sin nx; see the Exercises for more on this topic.

Topic 2: Power-Reducing Identities

The double-angle identities just introduced are useful when sin x, cos x, or tan x is known and an expression containing sin nx, cos nx, or tan nx (for some positive integer n) must be dealt with. But they are also useful when faced with an expression containing $\sin^2 x$, $\cos^2 x$, or $\tan^2 x$. Quite often in calculus, for instance, it's very helpful to be able to lower the power of a trigonometric expression – that is, to be able to replace, say, $\cos^2 x$, with an expression containing only cosine to the first power. The following three identities follow quickly from the double-angle identities and accomplish the power-reducing task.

Power-Reducing Identities

Sine Identity: $\sin^2 x = \dfrac{1 - \cos 2x}{2}$

Cosine Identity: $\cos^2 x = \dfrac{1 + \cos 2x}{2}$

Tangent Identity: $\tan^2 x = \dfrac{1 - \cos 2x}{1 + \cos 2x}$

The first identity is easily proved by beginning with one of the double-angle identities for cosine:

$$\cos 2x = 1 - 2\sin^2 x$$
$$2\sin^2 x = 1 - \cos 2x$$
$$\sin^2 x = \frac{1-\cos 2x}{2}.$$

We begin with the one double-angle identity that contains $\sin^2 x$ and no other squared terms, and then proceed to solve for $\sin^2 x$.

Of course, the power-reducing identities can be used to reduce powers other than 2, as shown in the next example.

example 3

Express $\sin^5 x$ in terms containing only first powers of sine and cosine.

Solution:

The procedure is to break the expression down into pieces to which the power-reducing identities apply, repeating the process if necessary:

$$\begin{aligned}\sin^5 x &= \sin^2 x \sin^2 x \sin x\\ &= \left(\frac{1-\cos 2x}{2}\right)\left(\frac{1-\cos 2x}{2}\right)\sin x\\ &= \left(\frac{1-2\cos 2x+\cos^2 2x}{4}\right)\sin x\\ &= \left(\frac{1}{4}\right)\left(\sin x - 2\cos 2x \sin x + \cos^2 2x \sin x\right)\\ &= \left(\frac{1}{4}\right)\left(\sin x - 2\cos 2x \sin x + \left(\frac{1+\cos 4x}{2}\right)\sin x\right)\\ &= \left(\frac{1}{4}\right)\left(\sin x - 2\cos 2x \sin x + \frac{\sin x}{2} + \frac{\cos 4x \sin x}{2}\right)\\ &= \left(\frac{1}{4}\right)\left(\frac{3\sin x}{2} - 2\cos 2x \sin x + \frac{\cos 4x \sin x}{2}\right).\end{aligned}$$

Topic 3: Half-Angle Identities

We have reached the point where we have enough identities to accomplish most of the common trigonometric tasks that may present themselves. For instance, if we needed the exact evaluation of $\sin\frac{\pi}{8}$, $\cos\frac{\pi}{8}$, or $\tan\frac{\pi}{8}$, we could start with an appropriate power-reducing identity and make use of the fact that twice $\frac{\pi}{8}$ is $\frac{\pi}{4}$, one of the "nice" angles. The box below contains general identities relating sine, cosine, and tangent of half an angle to expressions involving the whole angle. After proving one of the identities, we will proceed to determine the exact values of $\sin\frac{\pi}{8}$, $\cos\frac{\pi}{8}$, and $\tan\frac{\pi}{8}$.

Half-Angle Identities

Sine Identity: $\sin\frac{x}{2} = \pm\sqrt{\frac{1-\cos x}{2}}$

Cosine Identity: $\cos\frac{x}{2} = \pm\sqrt{\frac{1+\cos x}{2}}$

Tangent Identities: $\tan\frac{x}{2} = \frac{1-\cos x}{\sin x} = \frac{\sin x}{1+\cos x}$

The proofs of the first two half-angle identities follow from the corresponding power-reducing identities, with the sign of the result being determined by the quadrant in which $\frac{x}{2}$ lies (see the Exercises for guidance in the proofs). For the tangent half-angle identity, we begin by replacing x with $\frac{x}{2}$ in the power-reducing identity and proceed as follows:

$$\tan^2\frac{x}{2} = \frac{1-\cos 2\left(\frac{x}{2}\right)}{1+\cos 2\left(\frac{x}{2}\right)}$$

Apply the tangent power-reducing identity with an argument of $\frac{x}{2}$.

$$\tan\frac{x}{2} = \pm\sqrt{\frac{1-\cos x}{1+\cos x}}$$

Simplify the right-hand side and take the square root of both sides.

$$\tan\frac{x}{2} = \pm\sqrt{\left(\frac{1-\cos x}{1+\cos x}\right)\left(\frac{1-\cos x}{1-\cos x}\right)}$$

Multiply the numerator and denominator by $1-\cos x$.

$$\tan\frac{x}{2} = \pm\sqrt{\frac{(1-\cos x)^2}{1-\cos^2 x}}$$

Expand the denominator of the right-hand side and apply the Pythagorean Identity.

$$\tan\frac{x}{2} = \pm\frac{|1-\cos x|}{|\sin x|}$$

Simplify the radical.

All that remains is to simplify the formula a bit. First, note that $1 - \cos x$ is always non-negative, so the absolute value in the numerator is unnecessary. The next step is to note that $\tan\frac{x}{2}$ and $\sin x$ have the same sign for any value of x; the verification of this is left to the reader (consider individually the four cases of $0 \le \frac{x}{2} < \frac{\pi}{2}$, $\frac{\pi}{2} < \frac{x}{2} \le \pi$, $-\frac{\pi}{2} < \frac{x}{2} \le 0$ and $-\pi \le \frac{x}{2} < -\frac{\pi}{2}$). This means that if the absolute value in the denominator is removed, then the $\pm$ in the formula can also be discarded, since its only purpose is to remind us to choose the correct sign depending on the quadrant in which $\frac{x}{2}$ lies. We are thus led to the desired result; the second formula in the tangent half-angle identity is left as an exercise.

example 4

Determine the exact values of $\sin\frac{\pi}{8}$, $\cos\frac{\pi}{8}$, and $\tan\frac{\pi}{8}$.

Solution:

Since $\cos\frac{\pi}{4} = \sin\frac{\pi}{4} = \frac{\sqrt{2}}{2}$, we can easily determine that

$$\sin\frac{\pi}{8} = \sqrt{\frac{1-\cos\frac{\pi}{4}}{2}} = \sqrt{\frac{1-\frac{\sqrt{2}}{2}}{2}} = \sqrt{\frac{2-\sqrt{2}}{4}} = \frac{1}{2}\sqrt{2-\sqrt{2}},$$

$$\cos\frac{\pi}{8} = \sqrt{\frac{1+\cos\frac{\pi}{4}}{2}} = \sqrt{\frac{1+\frac{\sqrt{2}}{2}}{2}} = \sqrt{\frac{2+\sqrt{2}}{4}} = \frac{1}{2}\sqrt{2+\sqrt{2}},$$

and

$$\tan\frac{\pi}{8} = \frac{1-\cos\frac{\pi}{4}}{\sin\frac{\pi}{4}} = \frac{1-\frac{\sqrt{2}}{2}}{\frac{\sqrt{2}}{2}} = \frac{2\sqrt{2}-2}{2} = \sqrt{2}-1.$$

Of course, we could also have used the fact that $\tan x = \frac{\sin x}{\cos x}$ for the last calculation. Verifying that the two seemingly different answers obtained are actually equivalent is left to the reader.

example 5

Determine the exact value of $\cos 105^\circ$.

Solution:

Once we note that 105° is half of 210°, and recall that the reference angle of 210° is 30° the calculation is straightforward. We must also be careful, however, to insert a minus sign before the radical, as 105° lies in the second quadrant and $\cos 105^\circ$ is therefore negative.

$$\cos 105^\circ = -\sqrt{\frac{1+\cos 210^\circ}{2}} = -\sqrt{\frac{1-\frac{\sqrt{3}}{2}}{2}} = -\frac{1}{2}\sqrt{2-\sqrt{3}}$$

Topic 4: Product-to-Sum and Sum-to-Product Identities

The final set of identities in this section is easily verified using the sum and difference identities, and the proofs are left to the reader. The first box below contains identities that allow us to rewrite certain products of trigonometric functions as sums or differences of functions.

Product-to-Sum Identities

$$\sin x\cos y = \frac{1}{2}\left[\sin(x+y)+\sin(x-y)\right] \qquad \cos x\sin y = \frac{1}{2}\left[\sin(x+y)-\sin(x-y)\right]$$

$$\sin x\sin y = \frac{1}{2}\left[\cos(x-y)-\cos(x+y)\right] \qquad \cos x\cos y = \frac{1}{2}\left[\cos(x+y)+\cos(x-y)\right]$$

example 6

Express $\sin^5 x$ in terms containing only first powers of sine.

Solution:

In Example 3, we used a power-reducing identity to determine that

$$\sin^5 x = \left(\frac{1}{4}\right)\left(\frac{3\sin x}{2} - 2\cos 2x \sin x + \frac{\cos 4x \sin x}{2}\right).$$

We can now use one of the product-to-sum identities to rewrite two of the terms in this expression, as follows:

$$\begin{aligned}
\sin^5 x &= \left(\frac{1}{4}\right)\left[\frac{3\sin x}{2} - 2\cos 2x \sin x + \frac{\cos 4x \sin x}{2}\right] \\
&= \left(\frac{1}{4}\right)\left[\frac{3\sin x}{2} - 2\left(\frac{1}{2}(\sin 3x - \sin x)\right) + \frac{1}{2}\left(\frac{1}{2}(\sin 5x - \sin 3x)\right)\right] \\
&= \frac{3}{8}\sin x - \frac{1}{4}\sin 3x + \frac{1}{4}\sin x + \frac{1}{16}\sin 5x - \frac{1}{16}\sin 3x \\
&= \frac{5}{8}\sin x - \frac{5}{16}\sin 3x + \frac{1}{16}\sin 5x.
\end{aligned}$$

Consider, now, the first product-to-sum identity, with u and v in place of x and y:

$$\sin u \cos v = \frac{1}{2}\left[\sin(u+v) + \sin(u-v)\right].$$

If we replace u with $\frac{x+y}{2}$ and v with $\frac{x-y}{2}$, we obtain:

$$\sin\left(\frac{x+y}{2}\right)\cos\left(\frac{x-y}{2}\right) = \frac{1}{2}(\sin x + \sin y).$$

Multiplying this last equation through by 2 gives us an identity that allows us to rewrite a sum of sine functions as a product of a sine and cosine function, and it is the first of the four sum-to-product identities in the box on the following page.

Sum-to-Product Identities

$$\sin x + \sin y = 2\sin\left(\frac{x+y}{2}\right)\cos\left(\frac{x-y}{2}\right) \qquad \sin x - \sin y = 2\cos\left(\frac{x+y}{2}\right)\sin\left(\frac{x-y}{2}\right)$$

$$\cos x + \cos y = 2\cos\left(\frac{x+y}{2}\right)\cos\left(\frac{x-y}{2}\right) \qquad \cos x - \cos y = -2\sin\left(\frac{x+y}{2}\right)\sin\left(\frac{x-y}{2}\right)$$

example 7

Verify the identity $\dfrac{\sin 2x + \sin 4x}{\cos 2x + \cos 4x} = \tan 3x$.

Solution:

We can apply two of the sum-to-product identities to easily verify the statement:

$$\begin{aligned}\frac{\sin 2x + \sin 4x}{\cos 2x + \cos 4x} &= \frac{2\sin\left(\frac{6x}{2}\right)\cos\left(\frac{-2x}{2}\right)}{2\cos\left(\frac{6x}{2}\right)\cos\left(\frac{-2x}{2}\right)} \\ &= \frac{\sin 3x}{\cos 3x} \\ &= \tan 3x.\end{aligned}$$

exercises

Use the information given in each problem to determine cos 2x, sin 2x, and tan 2x. See Example 1.

1. $\sin x = \frac{3}{5}$ and $\cos x$ is positive

2. $\tan x = -4$ and $\sin x$ is negative

3. $\cos x = -\frac{2}{\sqrt{6}}$ and $\sin x$ is positive

4. $\sin x = \frac{1}{\sqrt{5}}$ and $\tan x$ is positive

5. $\tan x = \frac{1}{\sqrt{3}}$ and $\cos x$ is negative

6. $\cos x = -3$ and $\tan x$ is negative

Verify the following trigonometric identities. See Example 2.

7. $\tan 3x = \dfrac{3\tan x - \tan^3 x}{1 - 3\tan^2 x}$

8. $\sin 2x = \dfrac{2\tan x}{1 + \tan^2 x}$

9. $\dfrac{\sin 4x - \sin 2x}{\cos 4x + \cos 2x} = \tan x$

10. $\dfrac{\sin 3x}{\sin x} = 3 - 4\sin^2 x$

11. $2\sin^2 3x = 1 - \cos 6x$

12. $\sin 3x = 3\sin x \cos^2 x - \sin^3 x$

Use a power-reducing identity to rewrite the expressions as directed. See Example 3.

13. Rewrite $\sin^3 x$ in terms containing only first powers of sine and cosine.

14. Rewrite $\sin^4 x$ in terms containing only first powers of cosine.

15. Rewrite $\sin^4 x \cos^2 x$ in terms containing only first powers of cosine.

16. Rewrite $\cos^3 x \sin^2 x$ in terms containing only first powers of cosine.

17. Rewrite $\tan^4 x \sin x$ in terms containing only first powers of sine and cosine.

18. Rewrite $\sin^8 x$ in terms containing only first powers of cosine.

Determine the exact values of each of the following expressions. See Examples 4 and 5.

19. $\sin\left(\dfrac{3\pi}{8}\right)$

20. $\tan 112°30'$

21. $\cos\left(-\dfrac{\pi}{12}\right)$

22. $\tan\dfrac{7\pi}{12}$

23. $\sin 75°$

24. $\cos 165°$

Use the product-to-sum identities to rewrite the expressions as a sum or difference. See Example 6.

25. $\sin 3x \cos 3x$

26. $\cos\dfrac{\pi}{4}\sin\dfrac{\pi}{4}$

27. $5\cos 105° \sin 15°$

28. $2\cos 75° \cos 45°$

29. $\sin(x + y)\sin(x - y)$

30. $\sin\dfrac{5\pi}{6}\cos\dfrac{\pi}{6}$

31. $\sin\dfrac{5\pi}{4}\sin\dfrac{2\pi}{3}$

32. $\cos\beta\cos 3\beta$

33. $2\cos\dfrac{\pi}{3}\sin\dfrac{\pi}{6}$

Use the sum-to-product identities to rewrite the expressions as a product. See Example 7.

34. $\sin\frac{\pi}{4}+\sin\frac{3\pi}{4}$

35. $\sin 6x + \sin 2x$

36. $\cos 60° + \cos 30°$

37. $\cos 3\beta - \cos \beta$

38. $\sin\pi - \sin\frac{\pi}{2}$

39. $\sin 135° - \sin 15°$

40. $\cos 6x - \cos 2x$

41. $\cos\frac{7\pi}{6} - \cos\frac{\pi}{4}$

42. $\sin(\pi + \theta) + \sin(\pi - \theta)$

43. Two of the double-angle identities were proved in this chapter. Prove the remaining three double-angle identities.

44. The power-reducing identity for sine was proved in this chapter. Prove the remaining two power-reducing identities.

45. Prove the half-angle identities for sine and cosine by replacing x with $\frac{x}{2}$ in an appropriately chosen identity.

46. As mentioned in this chapter, $\cos(nx)$ can be expressed as a polynomial of degree n in $\cos x$; such polynomials are called Chebyshev polynomials. For $\sin(nx)$, the equivalent rewriting is a product of $\sin x$ and a polynomial of degree $n - 1$ in $\cos x$. Expand $\sin(nx)$ and $\cos(nx)$ for $n = 2$, 3, and 4 and compare the results.

7.4 Trigonometric Equations

TOPICS

1. Applying algebraic techniques to trigonometric equations
2. Using inverse trigonometric functions

Topic 1: Applying Algebraic Techniques to Trigonometric Equations

The majority of the trigonometric equations we have seen thus far in this chapter have been identities, and the goal has been to either verify the identity or use it in simplifying a trigonometric expression. In this section, we will study equations that are conditional – that is, equations which are not true for all allowable values of the variable. The goal, of course, is to identify exactly those values of the variable that *do* make a given equation true.

The identities we have learned in the first three sections of this chapter will prove to be very useful again, especially when combined with standard algebraic techniques covered in previous chapters. It is important to remember, however, that the solution to an equation consists of *all* values that make the equation true. Since trigonometric functions are periodic, it is not at all unusual for a trigonometric equation to have an infinite number of solutions, and the goal is to find and describe all of them.

One common approach in solving trigonometric equations, especially if only one trigonometric function is involved, is to isolate the trigonometric expression on one side of the equation. Our knowledge of the behavior of trigonometric functions will then often allow us to describe the complete solution set.

example 1

Solve the equation $3-6\cos x=0$.

Solution:

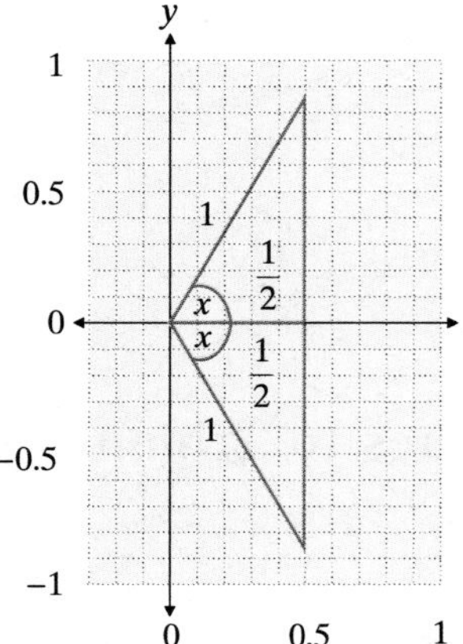

Isolating cos x is easily accomplished:

$$3-6\cos x=0$$
$$3=6\cos x$$
$$\cos x=\frac{1}{2}.$$

This tells us several things. First, the fact that cos x is positive means that x lies in the first or fourth quadrant. This means that $0\le x\le\frac{\pi}{2}$ or $-\frac{\pi}{2}\le x\le 0$ (we could also use the range $\frac{3\pi}{2}\le x\le 2\pi$ to describe fourth-quadrant angles). We can then sketch a triangle or two, if necessary, to remind ourselves that $\cos\frac{\pi}{3}=\frac{1}{2}$ and $\cos\left(-\frac{\pi}{3}\right)=\frac{1}{2}$.

Then, since cosine is 2π-periodic, we can describe the complete solution set with the two equations

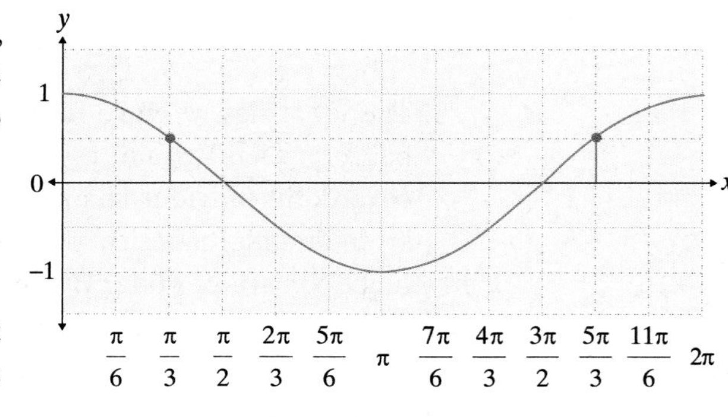

$$x=\frac{\pi}{3}+2n\pi \text{ and } x=-\frac{\pi}{3}+2n\pi,$$

where n is any integer. Note that this set could be described, less precisely, as

$$\left\{\ldots-\frac{7\pi}{3},-\frac{5\pi}{3},-\frac{\pi}{3},\frac{\pi}{3},\frac{5\pi}{3},\frac{7\pi}{3},\ldots\right\}.$$

example 2

Solve the equation $\tan^2 x - 1 = 2$.

Solution:

The trigonometric function can again be isolated fairly easily:

$$\tan^2 x - 1 = 2$$
$$\tan^2 x = 3$$
$$\tan x = \pm\sqrt{3}.$$

As in Example 1, the task now is to identify *all* the angles which solve the last equation (which, because of the plus-or-minus symbol, is actually two equations). Since tangent has a period of π, we'll be done if we find all the solutions in an interval of length π and add to them all integer multiples of π. For instance, in the interval $\left[-\frac{\pi}{2}, \frac{\pi}{2}\right]$, $\tan x = \sqrt{3}$ when $x = \frac{\pi}{3}$ and $\tan x = -\sqrt{3}$ when $x = -\frac{\pi}{3}$. So the complete solution set is described by

$$x = \frac{\pi}{3} + n\pi \text{ and } x = -\frac{\pi}{3} + n\pi.$$

The last example included a squared trigonometric term, but the equation was easily solved by taking the square root of both sides. More complicated quadratic-type trigonometric equations can be solved by factoring, by completing the square, or by the quadratic formula.

example 3

Solve the equation $\sin^2 x - \sin x = \sin x + 3$.

Solution:

The equation $\sin^2 x - \sin x = \sin x + 3$ is quadratic in sin x; that is, if sin x were replaced by x, it would simply be a quadratic equation. This observation provides the key to its solution: we will use our previous experience with quadratic equations to solve for sin x and then use our knowledge of trigonometry to solve for x.

$$\sin^2 x - \sin x = \sin x + 3$$
$$\sin^2 x - 2\sin x - 3 = 0$$
$$(\sin x - 3)(\sin x + 1) = 0$$

Remember that the first step in solving quadratic equations by factoring or by the quadratic formula is to collect all the terms on one side. In this problem, we can easily factor the resulting left-hand side.

cont'd. on next page ...

As in any quadratic equation solved by factoring, the two factors now lead to two equations:

$\sin x = 3$ or $\sin x = -1$

no real solution $\qquad x = \frac{3\pi}{2} + 2n\pi$

The first equation has no real number solution, as 3 lies outside the range of the sine function. The second equation has an infinite number of solutions.

The examples in this section so far have involved only one trigonometric function. If an equation initially appears with two or more functions, one approach to solving it is to use an identity to obtain a new equation in only one function.

example 4

Solve the equation $-2\cos^2 x + 1 = \sin x$.

Solution:

The presence of a $\cos^2 x$ or $\sin^2 x$ term is almost always an indication that the first Pythagorean Identity may be useful:

$$-2\cos^2 x + 1 = \sin x$$
$$-2\left(1 - \sin^2 x\right) + 1 = \sin x$$

Use the first Pythagorean Identity to rewrite the equation in terms of sine alone.

$$2\sin^2 x - \sin x - 1 = 0$$
$$(2\sin x + 1)(\sin x - 1) = 0.$$

Solve the resulting quadratic in sine.

As in the last example, we have two equations to solve:

$2\sin x + 1 = 0$ or $\sin x - 1 = 0$

$\sin x = -\frac{1}{2}$ $\qquad \sin x = 1$

$x = -\frac{5\pi}{6} + 2n\pi, -\frac{\pi}{6} + 2n\pi$ $\qquad x = \frac{\pi}{2} + 2n\pi$

The solutions to the first equation are angles lying in the third and fourth quadrants, while the solution to the second equation is a right angle. Three equations are thus necessary to describe the solution set.

There is another way to consider the equation $-2\cos^2 x + 1 = \sin x$ that is very instructive, though usually not sufficient to clearly identify the solutions. The equation is satisfied at those values of x for which the graphs of the functions $f(x) = -2\cos^2 x + 1$ and $g(x) = \sin x$ coincide. Figure 1 on the next page contains the graphs of these two functions over the interval $[-\pi, \pi]$ (f is shown in red and g is shown in blue); note that the two graphs intersect at exactly the points found above.

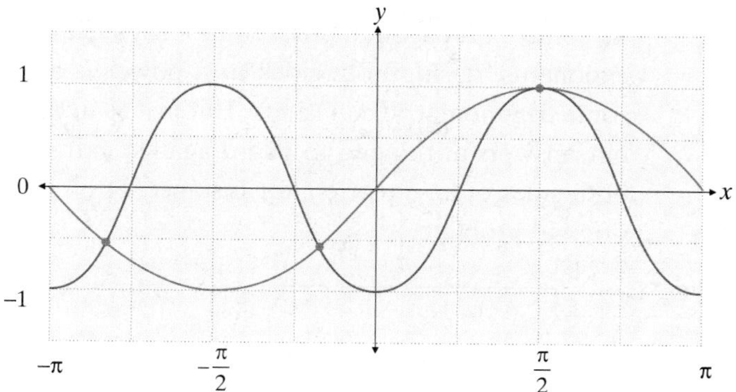

Figure 1: Graphical Representation of Solution to $-2\cos^2 x + 1 = \sin x$

example 5

Solve the equation $\sin x = \cos x$ in the interval $[0, 2\pi)$.

Solution:

We have used the graphs of sine and cosine to such an extent that the solutions to this equation should come as no surprise. Graphically, we are looking for those x's at which sine and cosine coincide. The graph below illustrates the two points in the interval $[0, 2\pi)$ where this happens:

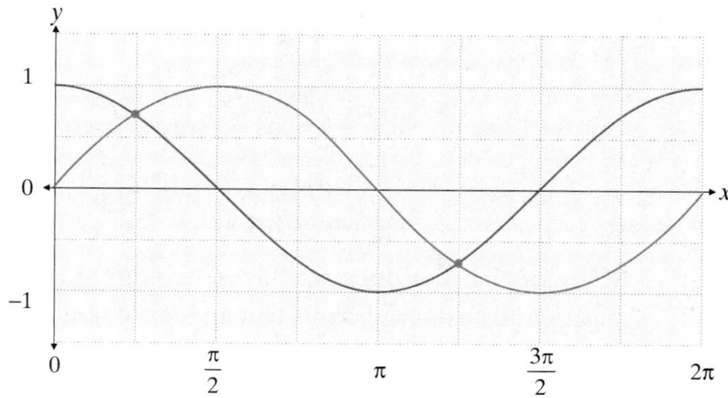

From the graph, it appears that the solution of the equation is the set $\left\{\frac{\pi}{4}, \frac{5\pi}{4}\right\}$, and this is indeed the case. Algebraically, we could also solve the equation $\sin x = \cos x$ by dividing both sides by cos x and making use of our knowledge of the tangent function. Note, though, that we can only do this after we determine that we won't lose any solutions in the process – dividing by cos x is only a legitimate step if cos x is non-zero. But when cosine *is* zero, sine is not, so we can safely discard those values of x as they can't possibly be part of the solution set. After the division, we have the equation $\tan x = 1$, and the solution set of this equation in the interval $[0, 2\pi)$ is $\left\{\frac{\pi}{4}, \frac{5\pi}{4}\right\}$.

In some non-trigonometric equations, such as those involving radicals or fractional exponents, raising both sides to a power is a useful technique. The same is true for some trigonometric equations. But just as in the non-trigonometric case, we must check our answers at the end to guard against extraneous solutions; remember that squaring both sides of an equation, for instance, often introduces solutions that do not solve the original problem.

example 6

Solve the equation $\sin x - 1 = \cos x$ in the interval $[0, 2\pi)$.

Solution:

$$\sin x - 1 = \cos x$$

$$(\sin x - 1)^2 = \cos^2 x$$

$$\sin^2 x - 2\sin x + 1 - \cos^2 x = 0$$

$$\sin^2 x - 2\sin x + 1 - (1 - \sin^2 x) = 0$$

$$2\sin^2 x - 2\sin x = 0$$

$$2(\sin x)(\sin x - 1) = 0$$

$$\sin x = 0 \text{ and } \sin x = 1$$

$$x = 0, \pi \text{ or } x = \frac{\pi}{2}$$

$$\sin 0 - 1 \overset{?}{=} \cos 0 \quad \sin\frac{\pi}{2} - 1 \overset{?}{=} \cos\frac{\pi}{2} \quad \sin\pi - 1 \overset{?}{=} \cos\pi$$

$$-1 \neq 1 \qquad 0 = 0 \qquad -1 = -1$$

By squaring both sides of the equation, we can apply an identity and rewrite the equation in terms of a single trigonometric function.

The new equation is quadratic in sine, and can be solved by factoring.

At this point, there are three potential solutions. However, inserting the potential solutions into the original equation reveals that the solution set is actually $\left\{\frac{\pi}{2}, \pi\right\}$.

Many of the identities we have encountered contain arguments that are multiples of angles, and such arguments can appear in conditional equations as well.

example 7

Solve the equation $\frac{\sec 3t}{2} - 1 = 0$.

Solution:

$$\frac{\sec 3t}{2} - 1 = 0$$

$$\sec 3t = 2$$

$$3t = \frac{\pi}{3} + 2n\pi \quad \text{and} \quad 3t = -\frac{\pi}{3} + 2n\pi$$

$$t = \frac{\pi}{9} + \frac{2n\pi}{3} \quad \text{and} \quad t = -\frac{\pi}{9} + \frac{2n\pi}{3}$$

Begin by isolating the single trigonometric term in the equation.

There are two solutions of the equation in any interval of length 2π.

Divide both sides (including the $2n\pi$ term) by 3.

Topic 2: Using Inverse Trigonometric Functions

It is, of course, no accident that the examples so far in this section have had "nice" solutions based on the commonly-encountered angles of $\frac{\pi}{6}$, $\frac{\pi}{4}$, $\frac{\pi}{3}$, and $\frac{\pi}{2}$. The preceding examples have been undeniably convenient for demonstrating the sorts of techniques used in solving trigonometric equations. But real-world problems do not, necessarily, give rise to equations with such easily described solutions. It is in solving these more unwieldy problems that inverse trigonometric functions are most useful.

example 8

Solve the equation $\tan^2 x + 2\tan x = 3$ on the interval $\left(-\frac{\pi}{2}, \frac{\pi}{2}\right)$.

Solution:

$$\tan^2 x + 2\tan x = 3$$

The equation is quadratic in tangent, and can be solved by factoring.

$$\tan^2 x + 2\tan x - 3 = 0$$

$$(\tan x + 3)(\tan x - 1) = 0$$

$$\tan x = -3 \text{ or } \tan x = 1$$

$$x = \tan^{-1}(-3) \text{ and } x = \frac{\pi}{4}$$

At this point, one of the two resulting equations can be solved easily. The other solution is $\tan^{-1}(-3)$, which is approximately -1.24905.

example 9

Solve the equation $6\sin x - 2 = \sin x$ on the interval $[0, 2\pi)$.

Solution:

$$6\sin x - 2 = \sin x$$

The trigonometric function can be easily isolated.

$$5\sin x = 2$$

$$\sin x = 0.4$$

$$x = \sin^{-1} 0.4 \text{ and } x = \pi - \sin^{-1} 0.4$$

Note that since $\sin x$ is positive, x lies in the first or second quadrant. The solution in the first quadrant is $\sin^{-1} 0.4$, and the solution in the second quadrant has $\sin^{-1} 0.4$ as its reference angle.

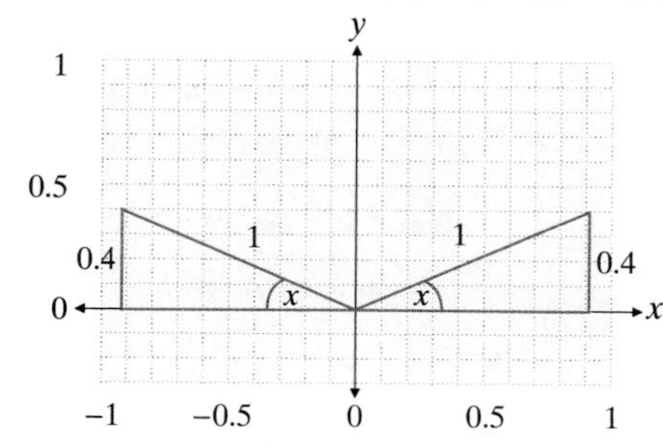

exercises

Use trigonometric identities and algebraic methods, as necessary, to solve the following trigonometric equations. See Examples 1 through 7.

1. $2\sin x + 1 = 0$

2. $4\sin^2 x + 2 = 3$

3. $\sqrt{2} - 2\cos x = 0$

4. $4 \cos x = 2$

5. $2\cos x - \sqrt{3} = 0$

6. $\sin 2x = \sqrt{3}\cos 2x$

7. $-\dfrac{1}{\sqrt{48}\sin x} = \dfrac{1}{8}$

8. $\sin^2 x - \sin x = 2 \sin x - 2$

9. $\sqrt{3}\tan x + 1 = -2$

10. $(3 \tan^2 x - 1)(\tan^2 x - 3) = 0$

11. $\sec^2 x - 1 = 0$

12. $\sin^2 x = \sin^2 x + \cos^2 x$

13. $\sec x + \tan x = 1$

14. $\cos x + \sin x \tan x = 2$

15. $\sin^2 x + \cos^2 x + \tan^2 x = 0$

16. $\cos^2 x = 3 \sin^2 x$

17. $2 \cos^2 x - 3 = 5 \cos x$

18. $\sin^3 x = \sin x$

19. $\dfrac{\cot 2x}{\sqrt{3}} = -1$

20. $\cos x - 1 = \sin x$

Use trigonometric identities, algebraic methods, and inverse trigonometric functions, as necessary, to solve the following trigonometric equations on the interval $[0, 2\pi]$. See Examples 8 and 9.

21. $2\sin^2 x + 7\sin x = 4$

22. $\tan^2 x = \tan x + 6$

23. $2 \cos^2 x - 1 = 0$

24. $\sec^2 x - 3 = -\tan x$

25. $0.05 \sin^3 x = 0.1 \sin x$

26. $12 \sin x - 1 = 8 \sin x$

27. $2 \cos^2 x + 11 \cos x = -5$

28. $10 \sin x - 16 = -11 + 18 \sin x$

29. $\sec^2 x - 2 + 8 \tan x = 42$

30. $\sin(-x) = 5 \sin x - 3$

Verify that the x-values given are solutions to the given equation.

31. $2 \cos x + 1 = 0$;
a. $x = \dfrac{2\pi}{3}$ **b.** $x = \dfrac{4\pi}{3}$

32. $3 \sec^2 x - 4 = 0$
a. $x = \dfrac{\pi}{6}$ **b.** $x = \dfrac{5\pi}{6}$

33. $2 \sin^2 x - \sin x - 1 = 0$
a. $x = \dfrac{\pi}{2}$ **b.** $x = \dfrac{7\pi}{6}$

34. $\tan^2 3x = 3$
a. $x = \dfrac{\pi}{9}$ **b.** $x = \dfrac{2\pi}{9}$

35. $\csc^4 x - 4\csc^2 x = 0$

a. $x = \frac{\pi}{6}$ **b.** $x = \frac{5\pi}{6}$

36. $3\cot^2 x - 1 = 0$

a. $x = \frac{\pi}{3}$ **b.** $x = \frac{2\pi}{3}$

37. $2\cot x + 1 = -1$

a. $x = \frac{3\pi}{4}$ **b.** $x = \frac{7\pi}{4}$

38. $\csc^2 x = 2\cot x$

a. $x = \frac{\pi}{4}$ **b.** $x = \frac{5\pi}{4}$

39. $2\sec x + 1 = \sec x + 3$

a. $x = \frac{\pi}{3}$ **b.** $x = \frac{5\pi}{3}$

40. $2\sin^2 2x = 1$

a. $x = \frac{\pi}{8}$ **b.** $x = \frac{3\pi}{8}$

Use a graphing calculator to approximate the solutions of the given equation for the interval $[0, 2\pi)$. Round answers to four decimal places.

41. $x\tan x - 3 = 0$

42. $2\sin x + \cos x = 0$

43. $2\cos^2 x - \sin x = 0$

44. $\cot^2 x - \sec^2 x = 0$

45. $2\sin x - \csc^2 x = 0$

46. $2\sin x = 1 - 2\cos x$

47. $\log x = -\sin x$

48. $\sin\left(\frac{x}{2}\right) = 2\cos(2x)$

Determine if the value given is a solution to the trigonometric function. If the value of x is not a solution, give all solutions to the equation.

49. $2\cos x = -1;\ x = \frac{4\pi}{3} + 2n\pi$

50. $\tan 3x(\tan x - 1) = 0;\ x = \frac{\pi}{4} + n\pi$

51. $3\sec^2 x = 4;\ x = \frac{\pi}{6} + n\pi$

52. $\sin^2 x - 3\cos^2 x = 0;\ x = \frac{\pi}{3} + 2n\pi$

53. $\sqrt{3}\csc x = 2;\ x = \frac{2\pi}{3} + 2n\pi$

54. $2\sin^2 x - 1 = 0;\ x = \frac{\pi}{4} + n\pi$

55. $\tan x = -\sqrt{3};\ x = \frac{\pi}{6} + n\pi$

56. $\tan^2 3x - 3 = 0;\ x = \frac{2\pi}{9} + \frac{n\pi}{3}$

57. $3\cot^2 x = 1;\ x = \frac{\pi}{3} + n\pi$

58. $\cos 2x(2\cos x + 1) = 0;\ x = \frac{5\pi}{6} + n\pi$

Solve the following quadratic-like equations in the interval $[0, 2\pi)$.

59. $2\sin^2 x - \sin x - 1 = 0$

60. $2\sin^2 x + 3\cos x - 3 = 0$

61. $\sin x - \cos x - 1 = 0$

62. $\tan x + \sqrt{3} = \sec x$

63. $\cos 2x - \cos x = 0$

64. $2\cos^2 x - \sqrt{3}\cos x = 0$

65. $\csc^2 x - 2\cot x = 0$

Solve the following equations in the interval [0°, 360°). Give the exact solutions when appropriate, otherwise, round answers to the nearest tenth.

66. $\sin^2 x \cos x - \cos x = 0$

67. $\cos^2 x = \sin^2 x$

68. $\tan x = \cot x$

69. $2 \sin x = \csc x + 1$

70. $\sec^2 x - 2 \tan x = 4$

71. $\sin^2 x = 2 \sin x - 3$

72. $2 \cos^2 x - 1 = -2 \cos x$

73. $2\sin x \cot x + \sqrt{3}\cot x - 2\sqrt{3}\sin x - 3 = 0$

Solve the algebraic and trigonometric equations given. Restrict the solutions to the trigonometric equations to the interval [0, 2π).

74. $6s^2 - 13s + 6 = 0;\ 6\cos^2 t - 13\cos t + 6 = 0$

75. $s^2 + s - 12 = 0;\ \sin^2 t + \sin t - 12 = 0$

76. $2s^2 + 7s - 15 = 0;\ 2\tan^2 t + 7\tan t - 15 = 0$

77. $4s^2 - 4s - 1 = 0;\ 4\cos^2 t - 4\cos t - 1 = 0$

78. While working Exercises 74 – 77, what did you observe as the maximum number of real solutions the algebraic equations can have?

79. While working Exercises 74 – 77, what did you observe as the maximum number of real solutions the trigonometric equations can have in the interval $[0, 2\pi)$?

80. Use your graphing calculator to solve the equation $2\sin(2x) = \sqrt{3}$ on the interval $[0°, 360°)$. (Remember to set your MODE to `Degree`.)

81. Use your graphing calculator to solve the equation $\sin(3x) - \frac{1}{2} = 0$ on the interval $[0°, 360°)$. (Remember to set your MODE to `Degree`.)

82. Use your graphing calculator to solve the equation $2\sin(4x) - 1 = 0$ on the interval $[0°, 360°)$. (Remember to set your MODE to `Degree`.)

83. While working with the trigonometric equations in Exercises 80 – 82, what did you observe about the solutions to the equation of the form $y = \sin(ax)$ in the interval $[0°, 360°)$?

84. An arrow is shot by an archer at an angle of θ in reference to the horizontal. The initial velocity is $v_0 = 100$ feet per second. If the arrow hits a target 300 feet from where the archer is standing, determine θ for the range given by $r = \frac{1}{32}v_0^2 \sin 2\theta$.

85. A baseball leaves a bat at an angle of θ in reference to the horizontal. The initial velocity is $v_0 = 95$ feet per second. The ball is caught 160 feet from where it is hit. What is the value of θ if the range is given by the projectile $r = \frac{1}{32}v_0^2 \sin 2\theta$?

chapter seven project

Trigonometric Identities

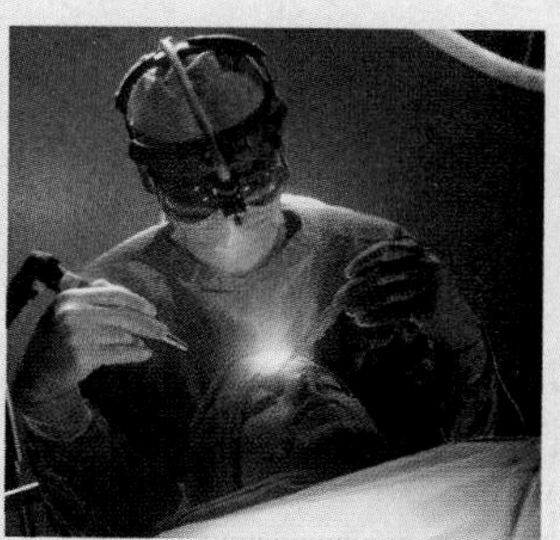

Lasers are used in everything from CD players to scanners at the check-out counter, to intricate medical surgeries, to weaponry. A beam of light from a flashlight shines for a couple hundred yards, but a laser's narrow band of light can be reflected off the moon and detected on Earth. A laser has this ability because its light is *coherent*. Coherent light means that each light wave has exactly the same amplitude, direction, and phase. Coherence reflects the superposition principle which states that when combining two waves, the resulting wave is the sum of the two individual waves.

Let's examine how the superposition principle works:

1. Consider two waves with a difference in displacement of $\frac{\pi}{2}$:

$$y_1 = 2\sin(kx - \omega t)$$
$$y_2 = 2\sin\left(kx - \omega t + \frac{\pi}{2}\right)$$

Using a trigonometric identity, add these two waves to find the equation of their superposition. What is the amplitude of the resulting wave?

2. For the set of equations:

$$y_1 = A\sin(kx - \omega t)$$
$$y_2 = A\sin(kx - \omega t + \delta)$$

a. For what values of δ would the amplitude be the largest? (This happens when two waves are coherent and it is called constructive interference.)

b. What is the smallest amplitude possible for $y_1 + y_2$?

c. For what values of δ would the smallest amplitude occur? (This is called destructive interference.)

3. Graph the following equations:

$$y_1 = 3\sin t$$
$$y_2 = 3\sin\left(t + \frac{\pi}{3}\right)$$

Now graph $y_1 + y_2$. Discuss the relationship between the three graphs.

CHAPTER REVIEW

7.1 Fundamental Identities and Their Uses

topics	pages	test exercises
Previously encountered identities • Reciprocal identities, Quotient identities, Cofunction identities, Period identities, Even identities, Odd identities, and Pythagorean identities	p. 545 – 547	7
Simplifying trigonometric expressions • Using trigonometric identities to simplify more complicated expressions	p. 547 – 549	1 – 2
Verifying trigonometric identities • Guidelines for verifying trigonometric identities: 1) Work with one side at a time; 2) Apply trigonometric identities as appropriate; 3) Rewrite in terms of sine and cosine if necessary	p. 549 – 552	3 – 4
Interlude: trigonometric substitutions • Replacing algebraic expressions with trigonometric expressions	p. 552 – 553	5 – 6

7.2 Sum and Difference Identities

topics	pages	test exercises
The identities • Sine Identities: $\sin(u+v)=\sin u\cos v+\cos u\sin v$ and $\sin(u-v)=\sin u\cos v-\cos u\sin v$ • Cosine Identities: $\cos(u+v)=\cos u\cos v-\sin u\sin v$ and $\cos(u-v)=\cos u\cos v+\sin u\sin v$ • Tangent Identities: $\tan(u+v)=\frac{\tan u+\tan v}{1-\tan u\tan v}$ and $\tan(u-v)=\frac{\tan u-\tan v}{1+\tan u\tan v}$	p. 555 – 556	8 – 15
Using sum and difference identities for exact evaluation	p. 557 – 559	8 – 9
Applications of sum and difference identities	p. 560 – 563	10 – 15

7.3 Product-Sum Identities

topics	pages	test exercises
Double-angle identities • Sine Identity: $\sin 2u = 2\sin u\cos u$ • Cosine Identities: $\cos 2u = \cos^2 u - \sin^2 u$ $= 2\cos^2 u - 1$ $= 1 - 2\sin^2 u$ • Tangent Identity: $\tan 2u = \frac{2\tan u}{1-\tan^2 u}$	p. 568 – 570	16 – 17
Power-reducing identities • Sine Identity: $\sin^2 x = \frac{1-\cos 2x}{2}$ • Cosine Identity: $\cos^2 x = \frac{1+\cos 2x}{2}$ • Tangent Identity: $\tan^2 x = \frac{1-\cos 2x}{1+\cos 2x}$	p. 570 – 571	18
Half-angle identities • Sine Identity: $\sin\frac{x}{2} = \pm\sqrt{\frac{1-\cos x}{2}}$ • Cosine Identity: $\cos\frac{x}{2} = \pm\sqrt{\frac{1+\cos x}{2}}$ • Tangent Identity: $\tan\frac{x}{2} = \frac{1-\cos x}{\sin x} = \frac{\sin x}{1+\cos x}$	p. 572 – 574	19 – 20
Product-to-sum and sum-to-product identities • Product-to-sum identities: $\sin x\cos y = \frac{1}{2}\left[\sin(x+y)+\sin(x-y)\right]$ $\cos x\sin y = \frac{1}{2}\left[\sin(x+y)-\sin(x-y)\right]$ $\sin x\sin y = \frac{1}{2}\left[\cos(x-y)-\cos(x+y)\right]$ $\cos x\cos y = \frac{1}{2}\left[\cos(x+y)+\cos(x-y)\right]$	p. 574 – 576	21 – 24

topics (continued)	pages	test exercises
• Sum-to-product identities: $\sin x+\sin y=2\sin\left(\frac{x+y}{2}\right)\cos\left(\frac{x-y}{2}\right)$ $\sin x-\sin y=2\cos\left(\frac{x+y}{2}\right)\sin\left(\frac{x-y}{2}\right)$ $\cos x+\cos y=2\cos\left(\frac{x+y}{2}\right)\cos\left(\frac{x-y}{2}\right)$ $\cos x-\cos y=-2\sin\left(\frac{x+y}{2}\right)\sin\left(\frac{x-y}{2}\right)$		

7.4 Trigonometric Equations

topics	pages	test exercises
Applying algebraic techniques to trigonometric equations • Using trigonometric identities and algebraic techniques to solve trigonometric equations	p. 579 – 584	25 – 32
Using inverse trigonometric functions • Using inverse trigonometric functions to solve trigonometric equations	p. 585	29 –32

chapter test

Section 7.1

Use trigonometric identities to simplify the expressions. There may be more than one correct answer.

1. $\sin^2(-x) - 5\sec^2 x \cot^2 x$

2. $\dfrac{\tan\beta\cos(\beta)}{\sin(-\beta)}$

Verify the identities.

3. $1 + \sin^2\theta = \csc^2\theta - \cot^2\theta + \sin^2\theta$

4. $(\sin^2 x)(\csc^2 x - 1) = \cos^2 x$

Use the suggested substitution to rewrite the given expression as a trigonometric expression. For all problems, assume $0 \le \theta \le \dfrac{\pi}{2}$.

5. $\sqrt{x^2 + 169}$, $\dfrac{x}{13} = \tan\theta$

6. $\sqrt{128 - 2x^2}$, $x = 8\sin\theta$

Show how the identity below follows from the first Pythagorean Identity.

7. $\tan^2 x + 1 = \sec^2 x$

Section 7.2

Use the sum and difference identities to determine the exact value of each of the following expressions.

8. $\sin\dfrac{25\pi}{12}$

9. $\tan 165°$

Use the sum and difference identities to rewrite each of the following expressions as a trigonometric function of a single number, and then evaluate the result.

10. $\cos 131° \cos 28° + \sin 131° \sin 28°$

11. $\dfrac{\tan\dfrac{5\pi}{8} + \tan\dfrac{\pi}{4}}{1 - \tan\dfrac{5\pi}{8}\tan\dfrac{\pi}{4}}$

Use the sum and difference identities to verify the following identities.

12. $\csc\left(\frac{3\pi}{4}+\theta\right)=\sec\left(\frac{\pi}{4}+\theta\right)$

13. $\cos^2\frac{\pi}{2}-\sin^2\frac{3\pi}{4}=\cos\frac{5\pi}{4}\cos\left(-\frac{\pi}{4}\right)$

Express the following as an algebraic function of x.

14. $\sin(\tan^{-1}x+\sin^{-1}2x)$

Express the following function in terms of sine functions only.

15. $h(\beta)=2\sqrt{3}\sin(4\beta)+2\cos(4\beta)$

Section 7.3

Use the information given to determine cos 2x, sin 2x, and tan 2x.

16. $\sin x=\frac{1}{3}$ and $\cos x$ is negative

Verify the following trigonometric identity.

17. $\tan 4x=\dfrac{4\tan x-4\tan^3 x}{1-6\tan^2 x+\tan^4 x}$

Use a power-reducing identity to rewrite the expression as directed.

18. Rewrite $\cos^5 x$ in terms containing only first powers of sine and cosine.

Determine the exact values of each of the following expressions.

19. $\sin\frac{11\pi}{12}$

20. $\cos 15°$

Use the Product-to-Sum Identities to rewrite the expressions as a sum or difference.

21. $\sin 5x\cos 5x$

22. $\cos\frac{3\pi}{5}\sin\frac{3\pi}{5}$

Use the Sum-to-Product Identities to rewrite the expressions as a product.

23. $\sin\frac{5x}{6}+\sin\frac{x}{6}$

24. $\cos 5x+\cos 3x$

Section 7.4

Use trigonometric identities and algebraic methods, as necessary, to solve the following trigonometric equations.

25. $4\sin x\cos x = \sqrt{3}$

26. $-\dfrac{2\cos^2\theta}{3} = \sin\theta - 1$

27. $2\cos^8 x = 4\cos^6 x$

28. $\tan\theta + 1 = \sqrt{3} + \sqrt{3}\cot\theta$

Use trigonometric identities, algebraic methods, and inverse trigonometric functions, as necessary, to solve the following trigonometric equations.

29. $5\sin^2 x + 4\sin x = 6$

30. $3\cos^2 x + 8\cos x = -4$

31. $4\sec^2 x - 4 + 16\tan x = 24$

32. $8\sin x - 4 = 16\sin x$

chapter EIGHT

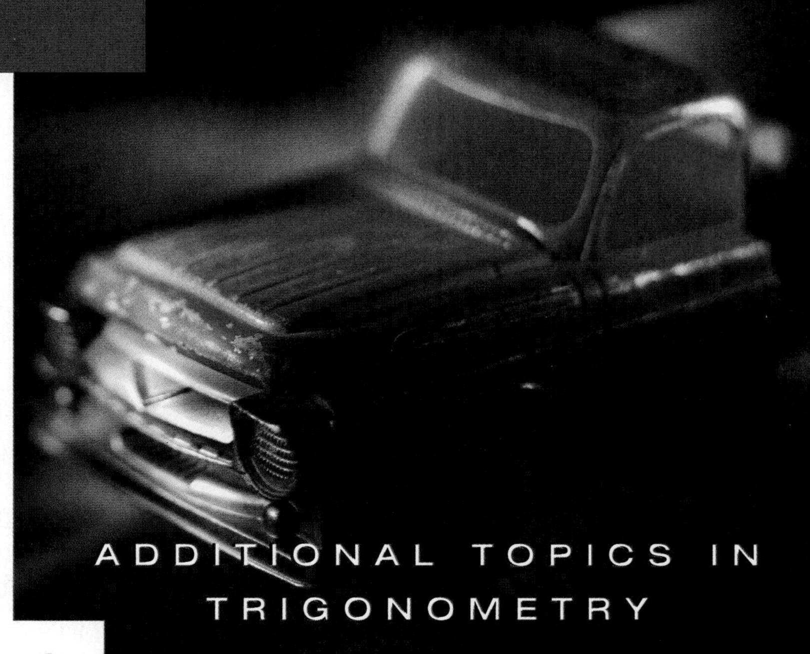

ADDITIONAL TOPICS IN TRIGONOMETRY

OVERVIEW

By the end of this chapter, you should be able to perform many operations with vectors, including addition, scalar multiplication, and a special kind of vector multiplication called dot product. Vectors have many uses in the physical sciences due to their ability to concisely describe and perform calculations about the forces acting around us. On page 669, you will find a problem about a boy playing with his toy truck in which you will be asked to find the total force acting upon the truck. You will master this type of problem using tools such as *Vector Operations Using Components* on page 661.

Introduction

This is the last of three chapters devoted to the introduction and use of trigonometric concepts. Chapter 6 introduced the nomenclature of trigonometry and the six fundamental trigonometric functions, and Chapter 7 focused on a deeper understanding of how those functions behave. The goal of this chapter is to show how trigonometry relates to a surprisingly large number of topics seen elsewhere in this text, even in situations which would at first seem to have nothing in common with trigonometry.

The applications of trigonometry seen in this chapter vary widely. As a preview, here is a cursory list of the uses that will be developed:

- New theorems and relations that are exceedingly useful in surveying, navigation, and other practical concerns
- New ways to measure areas of triangles
- A second planar coordinate system that simplifies the graphing and solving of some kinds of equations
- A second way of describing curves in the plane that allows us to answer questions that have been awkward up to this point
- A second way of representing complex numbers that simplifies multiplication and division, and offers an easy method for the calculation and visualization of roots
- A very powerful method of representing *directed magnitudes*, allowing us to mathematically describe quantities in which the direction that a magnitude is applied is of critical importance
- A new mathematical operation that greatly simplifies the solving of problems involving, for example, force and work

Hamilton

As usual, the concepts presented come from many different eras and cultures. One of the formulas for triangular area is named for a first century AD mathematician from Alexandria, but was probably known by mathematicians from several different areas a century or two before. The method of representing directed magnitudes owes much to the work of the 19th-century Irish mathematician William Rowan Hamilton. And the use of trigonometry in handling complex numbers developed over the course of several centuries and through the work of many mathematicians working toward many widely differing goals.

8.1 The Law of Sines and the Law of Cosines

TOPICS

1. The Law of Sines and its use
2. The Law of Cosines and its use
3. Areas of triangles

Topic 1: The Law of Sines and its Use

Chapter 6 introduced the trigonometric functions and demonstrated how they are used to solve a variety of problems involving triangles. Chapter 7 then explored the properties of the trigonometric functions and developed relationships and identities that greatly extended their usefulness. But the triangles analyzed to this point have all been right triangles, and even our more general treatment of the trigonometric functions has drawn, explicitly or implicitly, upon properties of right triangles. In this section we will expand the class of triangles we can analyze to include *oblique* triangles – those that have no right angle.

Most of the problems in this section contain a step in which a triangle must be *solved*; that is, given some information about a triangle's sides and/or angles, we will need to determine something else about the triangle. Note that there are six fundamental quantities that can be measured for any triangle: the lengths of the three sides, and the sizes of the three angles. We will see that if we know the length of any one side and any two of the other five quantities, we can determine the remaining three quantities.

In keeping with tradition, we will let A, B, and C denote the measures of the three angles of a given triangle and a, b, and c the lengths of the sides, with the side of length a opposite the angle of measure A and similarly for b and c. We will also follow the convention of letting a letter stand for both a quantity and the leg or angle it measures. Figure 1 illustrates a typical oblique triangle and labeling of its sides and angles.

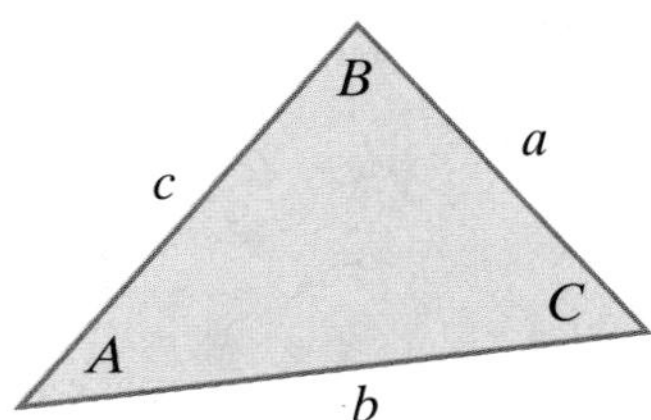

Figure 1: Oblique Triangle with Sides and Angles Labeled

In describing what we know about a triangle, we will let S represent knowledge about a leg and A knowledge about an angle, so for instance the notation SAS is shorthand for a situation in which we know the lengths of two legs and the measure of the angle between them. The table below categorizes the cases in which the two laws, known as the Law of Sines and the Law of Cosines, allow us to solve a given triangle.

Use of the Law of Sines and the Law of Cosines

Law of Sines	Law of Cosines
Two angles and a side (AAS or ASA)	Two sides and the included angle (SAS)
Two sides and a non-included angle (SSA)	Three sides (SSS)

We are now ready to state and use the Law of Sines.

The Law of Sines

Given a triangle with sides and angles labeled according to the convention on the previous page,

$$\frac{\sin A}{a} = \frac{\sin B}{b} = \frac{\sin C}{c}.$$

Note that since A, B, and C represent measures of angles in a triangle, all three must lie between 0 and π radians; hence, $\sin A$, $\sin B$, and $\sin C$ are all non-zero and the Law of Sines can also be written in the form

$$\frac{a}{\sin A} = \frac{b}{\sin B} = \frac{c}{\sin C}.$$

While the Law of Sines is typically applied to oblique triangles, its truth relies on decomposing a triangle into two right triangles. For instance, if angles A and C are acute, we can construct the altitude h and determine its value two different ways, as shown in Figure 2.

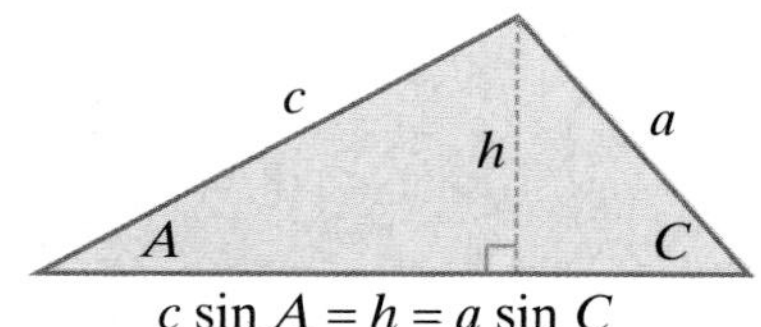

Figure 2: The Law of Sines (Acute Case)

The exact same relation is true if one of the angles, say A, is obtuse, but this realization depends on recalling that $\sin A = \sin(\pi - A)$ (that is, the sine of an angle in the second quadrant is equal to the sine of the reference angle). Figure 3 illustrates this case.

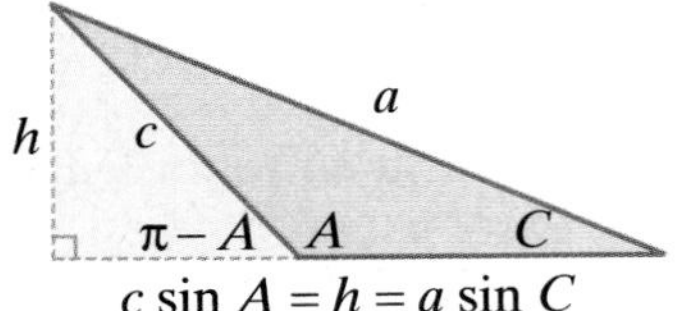

Figure 3: The Law of Sines (Obtuse Case)

In either case, dividing both sides of the equation $c\sin A = a\sin C$ by ac leads to the Law of Sines as it relates to A, C, a, and c. Similar diagrams provide the rest of the law.

The next few Examples demonstrate the use of the Law of Sines in the situations where it is applicable.

example 1

(***AAS* situation**) Sarah is piloting a hot-air balloon, and finds herself becalmed directly above a long straight road. She notices mile-markers on the road, and determines the angle of depression to the two markers as shown below. How far is she from marker A? What is her altitude?

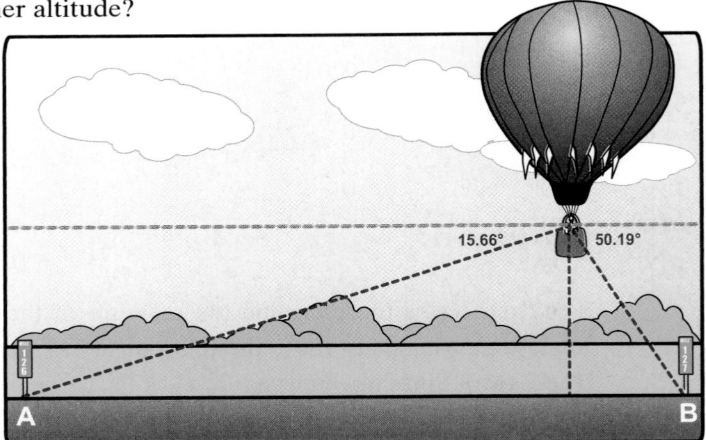

Solution:

To begin, note that the vertex at mile marker B also has measure 50.19° (this follows from the equality of opposite angles formed by a line cutting two parallel lines). Further, the vertex at Sarah's position has measure $180^\circ - 15.66^\circ - 50.19^\circ = 114.15^\circ$. Expressing the one length that we know in feet, we have the following information:

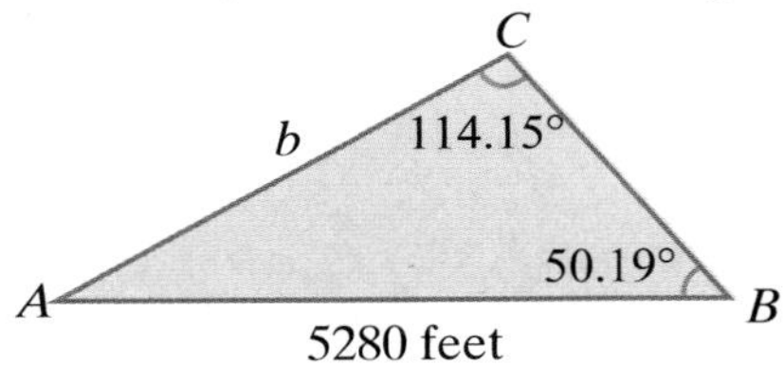

cont'd. on next page ...

We can now use the Law of Sines to obtain the equation

$$\frac{b}{\sin 50.19^\circ} = \frac{5280}{\sin 114.15^\circ},$$

which we can easily solve for length b to get $b = \left(\sin 50.19^\circ\right)\left(\frac{5280}{\sin 114.15^\circ}\right) \approx 4445$ feet.

To determine Sarah's altitude h, we can use the fact that the measure of the vertex at marker A is 15.66° and observe that

$$\sin 15.66^\circ = \frac{h}{4445}.$$

Solving this for h, we obtain $h \approx 1200$ feet.

example 2

(*ASA* situation) *A* surveyor has the task of determining the dimensions of the triangular plot of land shown below. He has already measured the length of the short edge of the plot, and has also determined the measure of two of the vertices. What are the lengths of the other two edges?

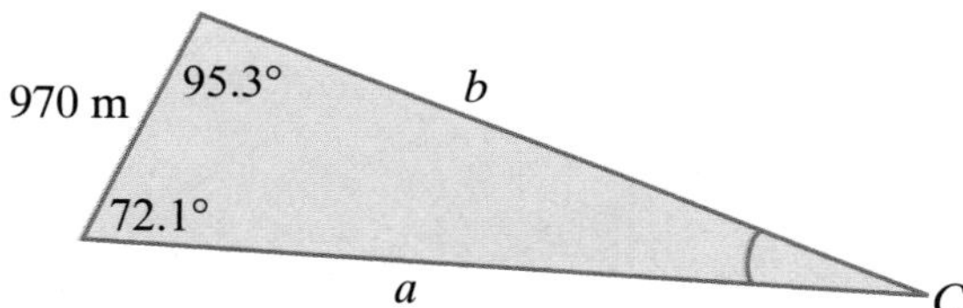

Solution:

The first step is to determine the measure of the third vertex. Since the sum of the angles in a triangle is 180°, the third angle has measure $180^\circ - 95.3^\circ - 72.1^\circ$, or 12.6°. The Law of Sines now tells us

$$\frac{970}{\sin 12.6^\circ} = \frac{a}{\sin 95.3^\circ} \quad \text{and} \quad \frac{970}{\sin 12.6^\circ} = \frac{b}{\sin 72.1^\circ}.$$

Solving for a and b, we obtain $a \approx 4428$ meters and $b \approx 4231$ meters.

As Examples 1 and 2 demonstrated, problems in which two angles and a side are known are readily dealt with. Usually, the application of some simple facts about triangles and angles, followed by the Law of Sines, allows us to completely determine the dimensions of a triangle in the AAS and ASA cases. Unfortunately, the same is not true for the SSA case. Given two sides a and b and the angle A opposite a, there may be no triangle, exactly one triangle, or two triangles that satisfy the conditions. To see why this is so, consider the four possibilities that may occur when A is acute:

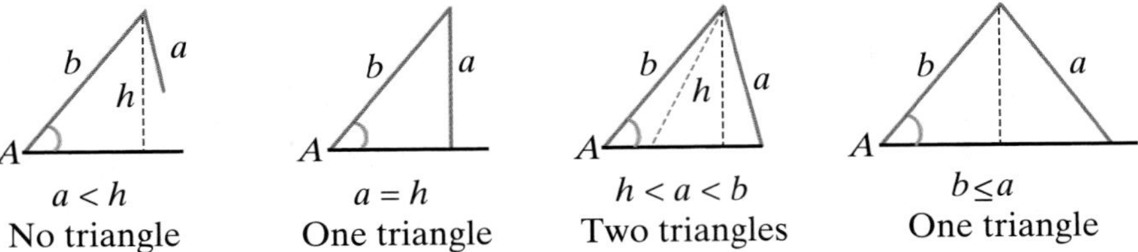

Figure 4: *SSA* Case with Acute Angle *A*

As Figure 4 shows, the size of a in relation to h and b, where $h = b\sin A$, determines whether a triangle exists which fits the given information, and whether the triangle is unique if it does exist. There are fewer possibilities if A is obtuse, but the ambiguity is still present. Figure 5 illustrates the obtuse case.

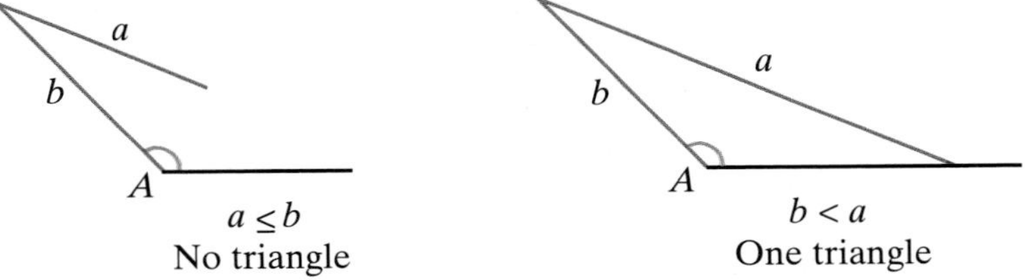

Figure 5: *SSA* Case with Obtuse Angle *A*

Because of the uncertainty over whether a unique triangle exists in the SSA case, it is sometimes called the **ambiguous case**. In practice, the ambiguity is resolved through consideration of additional information.

example 3

Construct a triangle, if possible, for which $A = \frac{\pi}{6}$, b = 12 cm, and a = 7 cm.

Solution:

A sketch is often useful in order to get an initial idea of whether such a triangle is possible. In this case, a rough sketch might look something like the following:

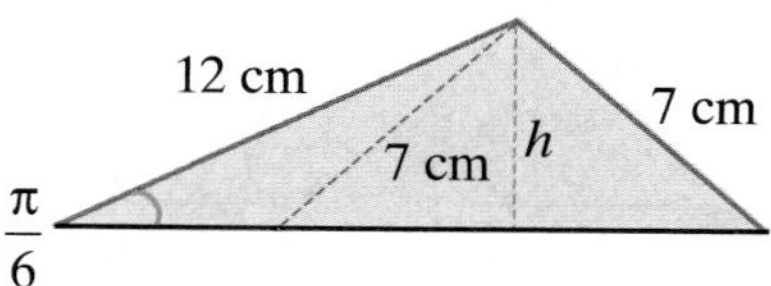

Such a sketch doesn't prove anything, but it certainly hints at the possibility of two triangles satisfying the given information, and the sketch also reminds us of the criterion to check: whether $h < a < b$. Since $h = 12\sin\left(\frac{\pi}{6}\right) = 6$, and since $6 < 7 < 12$, we know two such triangles really do exist. The remaining task is to completely determine the dimensions of both triangles.

Triangle 1: We will arbitrarily designate the triangle in which B is acute as Triangle 1. The Law of Sines tells us that

$$\frac{\sin\frac{\pi}{6}}{7} = \frac{\sin B}{12},$$

and so $\sin B = \frac{6}{7}$. Thus, $B = \sin^{-1}\left(\frac{6}{7}\right) \approx 1.03$ (remember that this is in radians). This means that angle C has measure $\pi - \frac{\pi}{6} - \sin^{-1}\left(\frac{6}{7}\right)$, or approximately 1.59, and a second application of the Law of Sines gives us

$$\frac{c}{\sin 1.59} = \frac{7}{\sin\frac{\pi}{6}},$$

so $c \approx 14.0$ cm.

Triangle 2: In the second triangle, B is the obtuse angle for which $\sin B = \frac{6}{7}$; that is, the reference angle of B is 1.03, and hence $B \approx \pi - 1.03 = 2.11$ (a quick check with a calculator will verify that $\sin 2.11 \approx \frac{6}{7}$). In this case, $C \approx \pi - \frac{\pi}{6} - 2.11 = 0.51$, and hence

$$\frac{c}{\sin 0.51} = \frac{7}{\sin\frac{\pi}{6}}.$$

Solving this for c gives us $c \approx 6.8$ cm.

example 4

Construct a triangle, if possible, for which $A = 75^\circ$, $b = 15$ units, and $a = 10$ units.

Solution:

A sufficiently accurate sketch may lead you to conclude that no such triangle is possible. The proof of this fact follows from noting that $h = 15\sin 75^\circ \approx 14.5$, and that $a < h$. In other words, a is too short to reach side c and form a triangle, so no triangle satisfies the given conditions.

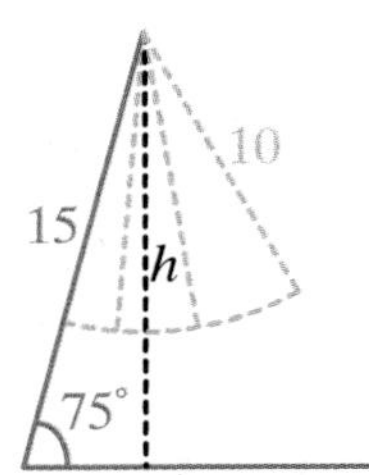

Topic 2: The Law of Cosines and its Use

The Law of Sines is of no use if we are given information about three sides of a triangle or two sides and the included angle of a triangle. Fortunately, the Law of Cosines handles the *SSS* and *SAS* cases easily.

The Law of Cosines

Given a triangle ABC, with sides labeled conventionally, the following are all true:

$$a^2 = b^2 + c^2 - 2bc\cos A$$

$$b^2 = a^2 + c^2 - 2ac\cos B$$

$$c^2 = a^2 + b^2 - 2ab\cos C$$

The similarity of these statements to the Pythagorean Theorem is no coincidence, and you should verify that the Law of Cosines, when applied to a right triangle, reduces to the Pythagorean Theorem. The Law of Cosines is thus an extension of the simpler theorem, but its proof depends on constructing an auxiliary right triangle to which we can apply the Pythagorean Theorem. To prove the first statement, for example, position the triangle under consideration so that angle A is in standard position in the plane, as shown in Figure 6.

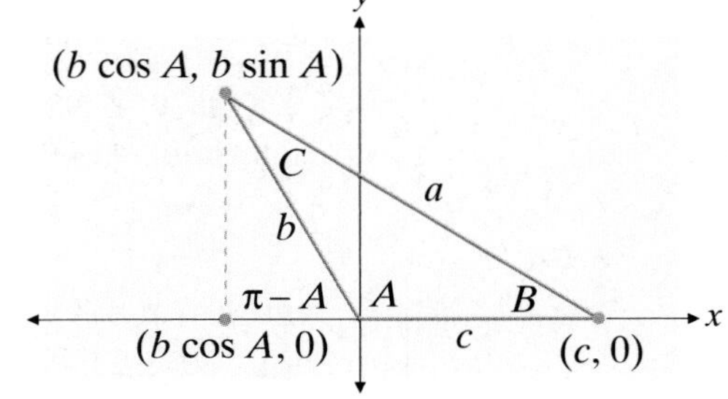

Figure 6: Proving the Law of Cosines

Applying the Pythagorean Theorem to the right triangle formed by the points $(c, 0)$, $(b\cos A, b\sin A)$ and $(b\cos A, 0)$ we obtain

$$\begin{aligned}a^2 &= (c - b\cos A)^2 + (b\sin A)^2 \\ &= c^2 - 2bc\cos A + b^2\cos^2 A + b^2\sin^2 A \\ &= b^2(\cos^2 A + \sin^2 A) + c^2 - 2bc\cos A \\ &= b^2 + c^2 - 2bc\cos A.\end{aligned}$$

The other two parts of the Law of Cosines are proved similarly.

In solving triangles, it's useful to remember a simple fact: the longest side of a triangle is opposite the largest angle. This is important especially when working backward to determine an angle from the sine of the angle, as seen in Example 5 below. And to avoid possible confusion when applying the Law of Cosines to solve for an angle, it's a good idea to solve for the largest angle first. If the largest angle is obtuse, the other two angles must be acute (a triangle can have at most one obtuse angle). Of course, if the largest angle is acute, this also means the other two angles (which are by definition smaller) must be acute.

example 5

(***SSS* situation**) Determine the three angles for a triangle in which $a = 3$ inches, $b = 5$ inches, and $c = 7$ inches.

Solution:

Guided by the observations above, we'll solve for C, the largest angle, first. By the Law of Cosines,

$$\begin{aligned}7^2 &= 3^2 + 5^2 - 2(3)(5)\cos C \\ 49 &= 34 - 30\cos C\end{aligned}$$

and so $\cos C = \frac{(49-34)}{(-30)} = -\frac{1}{2}$. Since $\cos C$ is negative, we know C is obtuse, and a calculator tells us that $C = \cos^{-1}(-0.5) = 120°$. (We can also determine C without a calculator by first determining C's reference angle; since $\cos 60° = 0.5$, we know that $C = 180° - 60° = 120°$.)

We can now use either Law to determine another angle. Using the Law of Sines to determine A, we have

$$\frac{\sin A}{3} = \frac{\sin 120°}{7}$$

and so $\sin A \approx 0.37$. Using a calculator, this means $A \approx 21.79°$ (and we know that A is not $180° - 21.79°$ because we have already determined that A is acute). We can now easily determine that $B \approx 180° - 120° - 21.79° = 38.21°$.

example 6

(***SAS* situation**) The course of a sailboat race instructs the sailors to head due east 11 kilometers to the first buoy. They are then to veer off to port by $20°$ and proceed for another 15 kilometers, at which point they should find the second buoy. What bearing would a sailor take to go from the starting point directly to the second buoy?

Solution:

A picture is definitely in order:

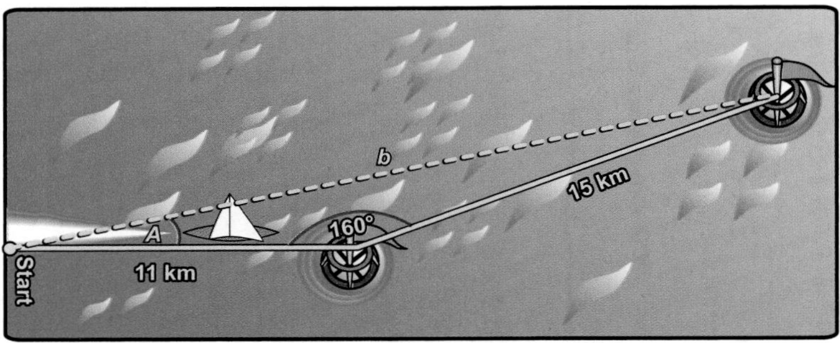

We wish to determine A, but neither Law allows us to do so directly. However, the Law of Cosines will allow us to determine the third side, which we have labeled b in the diagram:

$$\begin{aligned} b^2 &= 11^2 + 15^2 - 2(11)(15)\cos 160° \\ &= 121 + 225 - 330\cos 160° \\ &\approx 656.10. \end{aligned}$$

This gives us $b \approx 25.61$ kilometers, and we can now apply the Law of Sines to determine $\sin A$:

$$\frac{\sin A}{15} = \frac{\sin 160°}{25.61}.$$

This gives us $\sin A \approx 0.20$ and thus $A \approx 11.56°$ North of East.

Topic 3: Areas of Triangles

Look again at the proof of the Law of Sines. In that discussion, we determined the height of a given triangle two different ways, both of which involved the sine of an angle. The two formulas in one particular case are depicted again for reference in Figure 7.

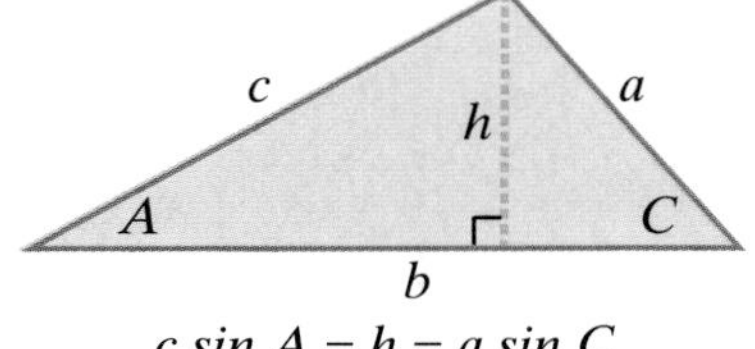

$c \sin A = h = a \sin C$

Figure 7: Two Formulas for Height

Recall that the area of a triangle is given by the formula $\left(\frac{1}{2}\right)(\text{base})(\text{height})$; in Figure 7, the length of the base is b, and we have two ways of determining the height. However, multiplying the base b by either expression for height results in a product of the lengths of two sides and the sine of the included angle. This same sort of product occurs if we instead consider one of the other sides of the triangle to be the base. In summary, we have derived the following.

Area of a Triangle (Sine Formula)

The area of a triangle is one-half the product of the lengths of any two sides and the sine of their included angle. That is,

$$\text{Area} = \frac{1}{2}ab\sin C = \frac{1}{2}bc\sin A = \frac{1}{2}ac\sin B.$$

Remarkably, there is also a formula for triangular area that does not depend on knowing *any* of the three angles. The formula is usually called Heron's Formula (Heron was an Alexandrian mathematician of the first century AD), but there is evidence that Archimedes (287 BC - 212 BC) and mathematicians of other cultures had also discovered it.

Area of a Triangle (Heron's Formula)

Given a triangle with sides a, b, and c, let $s = \dfrac{a+b+c}{2}$. Then

$$\text{Area} = \sqrt{s(s-a)(s-b)(s-c)}.$$

Our knowledge of trigonometry and the two Laws of this section allows us to derive Heron's Formula. Beginning with the Sine Formula for area, followed by one of the Pythagorean Identities and some factoring, we know

$$\begin{aligned}\text{Area}^2 &= \frac{1}{4}a^2b^2\sin^2 C \\ &= \frac{1}{4}a^2b^2\left(1-\cos^2 C\right) \\ &= \frac{1}{4}a^2b^2\left(1-\cos C\right)\left(1+\cos C\right).\end{aligned}$$

By the Law of Cosines, we can rewrite the last two factors as follows.

$$1-\cos C = 1-\frac{a^2+b^2-c^2}{2ab} \qquad 1+\cos C = 1+\frac{a^2+b^2-c^2}{2ab}$$
$$=\frac{2ab-a^2-b^2+c^2}{2ab} \qquad =\frac{2ab+a^2+b^2-c^2}{2ab}$$
$$=\frac{c^2-\left(a^2-2ab+b^2\right)}{2ab} \qquad =\frac{\left(a^2+2ab+b^2\right)-c^2}{2ab}$$
$$=\frac{c^2-(a-b)^2}{2ab} \qquad =\frac{(a+b)^2-c^2}{2ab}$$
$$=\frac{(c+a-b)(c-a+b)}{2ab} \qquad =\frac{(a+b+c)(a+b-c)}{2ab}$$

Replacing the above expressions for the original factors, we now have

$$\begin{aligned}\text{Area}^2 &= \frac{1}{4}a^2b^2\left(\frac{(c+a-b)(c-a+b)}{2ab}\right)\left(\frac{(a+b+c)(a+b-c)}{2ab}\right)\\ &= \frac{1}{16}(a+b+c)(c-a+b)(c+a-b)(a+b-c)\\ &= \left(\frac{a+b+c}{2}\right)\left(\frac{-a+b+c}{2}\right)\left(\frac{a-b+c}{2}\right)\left(\frac{a+b-c}{2}\right)\\ &= s(s-a)(s-b)(s-c).\end{aligned}$$

Taking the square root of both sides completes the process.

example 7

A businessman has an opportunity to buy a triangular plot of land in a part of town known colloquially as "Five Points," as five roads intersect there to form five equal angles. The property has 147 feet of frontage on one road, and 207 feet of frontage on the other. What is the square footage of the property?

Solution:

Since the roads intersect to form five equal angles, the included angle of the lot must be 72°. The Sine Formula for area then quickly gives us

$$\begin{aligned}\text{Area} &= \frac{1}{2}(147)(207)\left(\sin 72^\circ\right)\\ &\approx 14,470 \text{ ft}^2.\end{aligned}$$

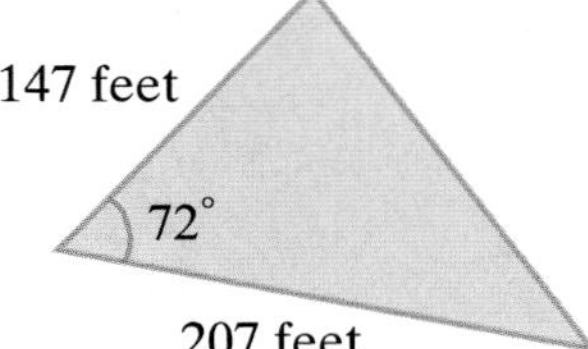

example 8

A set-designer is putting together a backdrop for a play, and one element of the scene is a large triangular piece of wood. The edges of the triangle are of lengths 4 meters, 7 meters, and 9 meters. She wants to know the square area of the triangle in order to estimate the amount of paint needed to cover it.

Solution:

Heron's Formula is easily applied:

$$s = \frac{4+7+9}{2} = 10$$

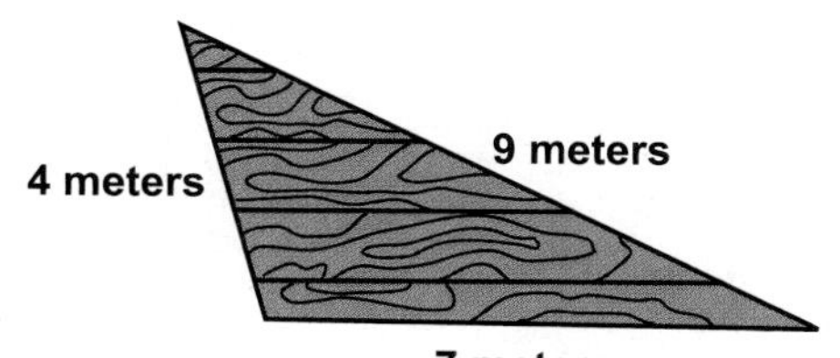

and so

$$\begin{aligned}\text{Area} &= \sqrt{(10)(10-4)(10-7)(10-9)}\\ &= \sqrt{(10)(6)(3)(1)}\\ &= \sqrt{180}\\ &= 6\sqrt{5}\\ &\approx 13.4 \text{ m}^2.\end{aligned}$$

exercises

Solve the following problems. See Examples 1 and 2.

1. A plane flies 730 miles from Charleston, SC to Cleveland, OH with a bearing of N 30° W (30° West of North). The plane then flies from Cleveland to Dallas, TX at a S 42° W bearing (42° West of South). How far is Dallas from Charleston (assume Dallas and Charleston are at the same latitude)?

2. A telephone pole was recently hit by a car and now leans 6° from the vertical. A point 40 feet away from the base of the pole has an angle of elevation of 36° to the top of the pole. How tall is the pole?

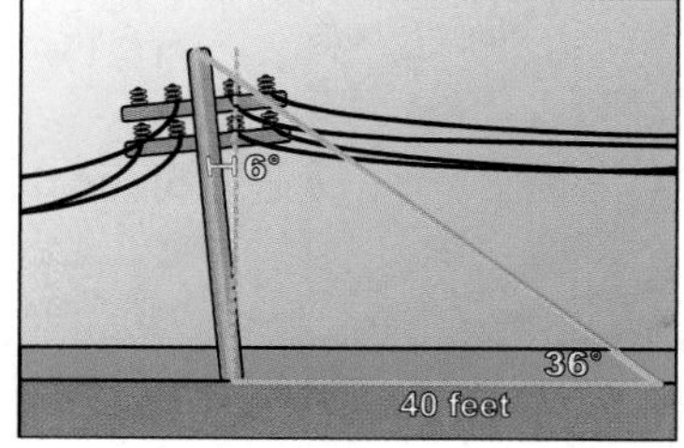

3. Jack wants to build a tree house. His parents worry that he is building it too high. If Jack's dad is looking at the tree house location from a 70° angle and then moves back 10 feet so he can see it at a 50° angle, how high is the tree house location?

4. Brandy is flying an airplane and is descending at a 10° angle towards a runway. If she can see a lake behind her that is 500 feet away from the runway at a 50° angle, how much further does she have to fly till she lands?

5. John's lizard ran out of the house. It ran 20 feet, turned and ran 30 feet, and then turned 140° to face the house. How far away from the house is John's lizard?

6. Kristin is playing miniature golf. She hits the ball and it bounces off a brick, making a 110° angle. Her ball comes to a stop 8.0 feet away at a 20° angle from where it started. How far did the ball travel?

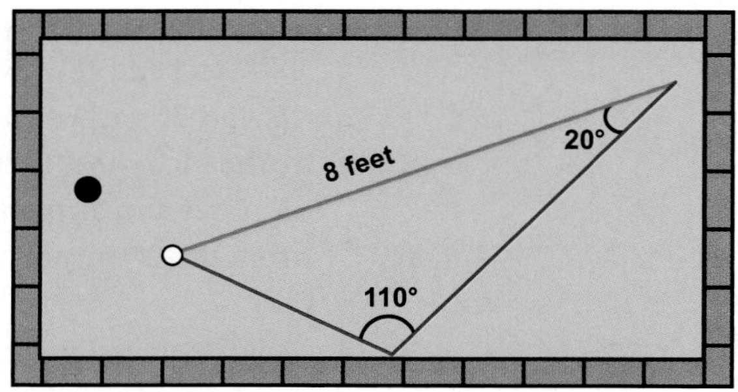

7. Janet is racing her friend Susan. Susan runs 10° away from Janet for 2000 feet. If she has to turn at a 50° angle to get back to Janet's path, how much shorter was Janet's run?

8. Two pieces of mail blew out into the yard. When Bob went to pick them up, he walked 10 feet to the first piece, turned 40° and walked to the next piece, and finally turned 150° to walk back to where he started. How far did Bob walk?

9. A bridge is suspended over a gorge shaped like an upside down isosceles triangle. If the bridge makes an 80° angle with the side of the gorge and the bridge is 1000 feet long, how deep is the gorge?

10. Brittany and Jim are playing catch. They are standing 30 feet away from each other. Ryan wants to join them and stands at a 50° angle away from Jim and at a 70° angle away from Brittany. How far away is Ryan from Jim and Ryan from Brittany?

11. Alan is golfing and sets up for a long drive. He slices it and hits a tree 80 feet away. The ball ricochets off of it at a 110° angle and comes to a stop 20° away from the direction he hit it. How far from Alan did the ball land?

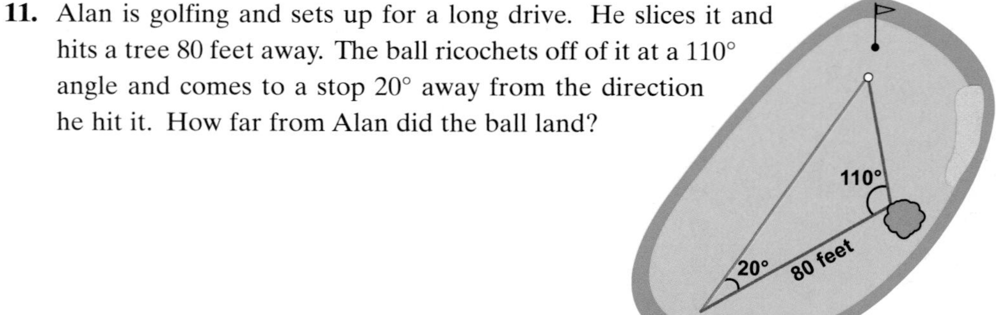

12. A ping pong net has become bent at a 70° angle instead of a 90° angle. The bottom of the net is 4.5 feet away from the end of the table. If the top of the net is 4.35 feet away from the end of the table, how high is the net?

13. An airplane has to fly between 3 airports. The trip from the 1st to the 2nd is 120 miles. After landing at the 2nd airport, the airplane must turn 140° to head toward the 3rd airport. At the 3rd airport it must turn 100° to head to the first airport. How far does the airplane have to travel from the 2nd airport to the 3rd?

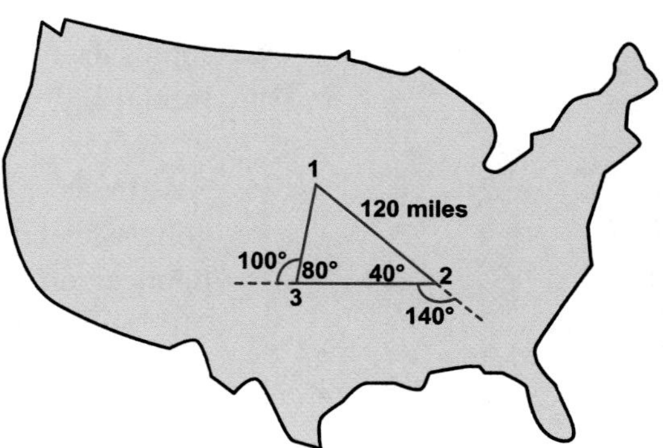

14. An extremely limber girl can bend over backwards at her waist making a 60° with her back. If her legs are 3.5 feet long and the distance between her hands and feet is 4 feet, what is the distance between her waist and hands?

Solve for the remaining angle and sides of the triangles. See Examples 1 and 2.

15. $A = 30°,\ B = 45°,\ a = 3$
16. $A = 60°,\ B = 40°,\ a = 2$
17. $A = 70°,\ B = 50°,\ b = 4$
18. $A = 100°,\ B = 20°,\ b = 1$
19. $B = 70°,\ C = 30°,\ c = 2$
20. $B = 120°,\ C = 40°,\ b = 6$
21. $A = 20°,\ B = 10°,\ a = 2$
22. $B = 100°,\ C = 30°,\ a = 3$

Solve for the remaining angles and side of any triangle that can be created. See Examples 3 and 4.

23. $A = 40°,\ a = 2,\ b = 4$
24. $A = 40°,\ a = 4,\ b = 4$
25. $C = 45°,\ a = 2,\ c = 4$
26. $C = 140°,\ b = 1,\ c = 9$
27. $A = 60°,\ a = 5,\ c = 6$
28. $B = 80°,\ a = 2,\ b = 6$
29. $B = 50°,\ b = 2,\ c = 5$
30. $B = 110°,\ a = 1,\ b = 8$

Construct a triangle, if possible, using the following information. Round the missing lengths and angles to two decimal places. See Examples 3 and 4.

31. $A = 60°,\ a = 10,\ b = 6$
32. $C = 42°,\ b = 9,\ c = 3$
33. $B = 13.2°,\ A = 63.7°,\ b = 21.2$
34. $A = 6°23',\ B = 64°15',\ c = 2.5$
35. $C = 100°,\ a = 18.1,\ c = 20.4$
36. $A = 108°,\ a = 9,\ b = 8.9$
37. $C = 24°,\ b = 2.4,\ c = 1.5$
38. $B = 16.9°,\ A = 29.7°,\ b = 17.8$
39. $A = 46°53',\ B = 74°13',\ c = 3.1$
40. $C = 116°,\ a = 24.1,\ c = 25$
41. $A = 30°,\ a = 15,\ b = 13$
42. $C = 74°,\ b = 4.5,\ c = 23$

Solve for the remaining angles and side of the triangles. See Examples 5 and 6.

43. $A = 60°,\ b = 3,\ c = 7$

44. $A = 40°,\ b = 2,\ c = 3$

45. $B = 50°,\ a = 4,\ c = 6$

46. $B = 45°,\ a = 5,\ c = 4$

47. $C = 30°,\ a = 8,\ b = 6$

48. $A = 110°,\ b = 2,\ c = 1$

49. $C = 70°,\ a = 5,\ b = 7$

50. $B = 100°,\ a = 1,\ c = 3$

Solve for the angles of the given triangles. See Examples 5 and 6.

51. $a = 3,\ b = 4,\ c = 2$

52. $a = 5,\ b = 2,\ c = 6$

53. $a = 8,\ b = 6,\ c = 3$

54. $a = 9,\ b = 4,\ c = 7$

55. $a = 5,\ b = 5,\ c = 5$

56. $a = 6,\ b = 4,\ c = 7$

57. $a = 5,\ b = 3,\ c = 4$

58. $a = 7,\ b = 2,\ c = 8$

Solve the following problems. See Examples 5 and 6.

59. A log is seen floating down a stream. The log is first spotted 10 feet away. Ten seconds later the log is 70 feet away making a 60° angle between the two sightings. How far did the log travel?

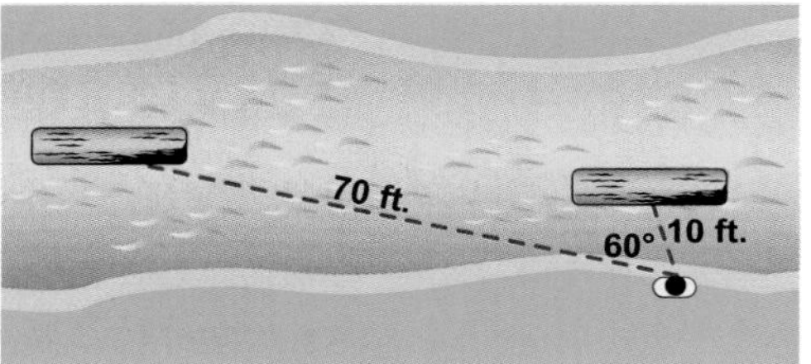

60. A bullet is fired and ricochets off a metal sign 100 feet away making an 80° angle as it speeds toward a tree where it embeds itself. If the sign and tree are 60 feet apart, how far did the bullet stop from where it was fired?

61. Astronomers once thought the sun revolved around the earth. The sun is 9.3×10^7 miles away and moves 15° across the sky in an hour. Assuming the sun travels in a straight line, how far would it have had to travel?

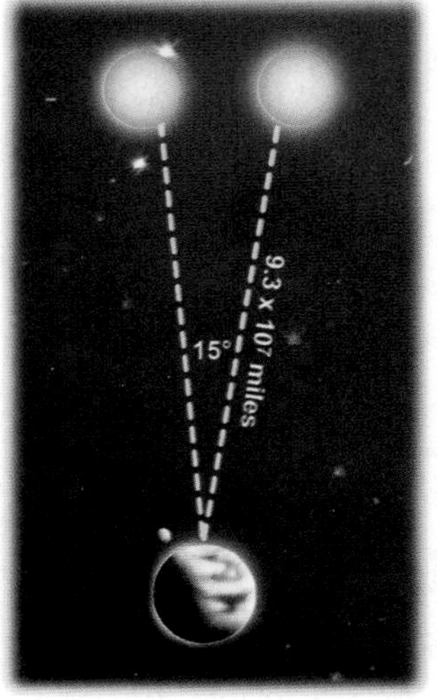

62. A pitcher 60 feet away throws a baseball to Joey. Joey bunts the ball at a 20° angle away from the pitcher. If the ball travels 50 feet, how far does the pitcher have to run to pick up the ball?

63. Nick is surfing a wave that carries him for 20 feet. He executes a sharp spray making a 100° angle. He rides the wave for 5 feet more before he topples into the water. How far is Nick from where he started?

64. A farmer puts a piece of fence across an inside corner of his barn to make a pen for his chickens. The lengths of the sides of the pen are 7 feet, 5 feet, and 8 feet. What are the respective angles?

65. Teresa wants to make a picture frame with two 5 inch and two 12 inch pieces of wood. If the diagonal length is 13 inches, what do the inside angles have to be for the two imaginary triangles?

66. Brian is up to bat. He hits the ball straight at the pitcher 60 feet away. The ball ricochets off the pitcher's head at a 100° angle and comes to rest 40 feet away from the pitcher. How far did the ball travel away from Brian?

67. A plane took off and ascended for 1000 feet before leveling off. The plane flew for 500 feet which put it 1480 feet directly away from where it started. What angle did the plane make when it leveled off?

Construct a triangle, if possible, using the following information and the Law of Cosines. See Examples 5 and 6.

68. $A = 65°,\ c = 13,\ b = 7$

69. $C = 35°,\ b = 12,\ a = 14$

70. $B = 24.2°,\ a = 13.3,\ c = 21.2$

71. $C = 46°7',\ a = 27.8,\ b = 19.4$

72. $A = 103°,\ c = 8, b = 6.3$

73. $C = 75°4',\ b = 15.4,\ a = 16.8$

74. $b = 12,\ c = 9,\ a = 15$

75. $c = 4.78,\ b = 16.46,\ a = 16.54$

76. $b = 4.2,\ a = 7.6,\ c = 9.2$

77. $b = 6.84,\ c = 10.87,\ a = 7.37$

78. $a = 76.45,\ b = 94.45,\ c = 84.42$

79. $a = 5,\ b = 10,\ c = 7$

Find the area of the triangle using the following information. See Examples 7 and 8.

80. $A = 131°,\ b = 10,\ c = 25$

81. $B = 60°7',\ c = 18,\ a = 6$

82. $C = 103°,\ a = 10,\ b = 2$

83. $B = 54°,\ a = 10,\ c = 7$

84. $A = 67°49',\ c = 4.2,\ b = 9.5$

85. $C = 46°,\ b = 20,\ a = 19$

86. $A = 86°,\ b = 24,\ c = 28$

87. $b = 12,\ c = 18,\ a = 15$

Solve the following problems. See Examples 7 and 8.

88. Bob wants to build an ice skating rink in his backyard, but his wife says he can only use the part beyond the wood-chipped path running through their yard. How large would his rink be if it is triangular-shaped with sides of length 20 feet, 23 feet, and 32 feet?

89. Nancy wants to plant wildflowers between the two intersecting paths in her garden. If the paths intersect at a 72° angle and she wants the flowers to extend 12 feet down one path and 15 down the other, how large is the area she wants to plant?

90. The U.S.S. Cyclops mysteriously disappeared somewhere in the Bermuda Triangle in 1910. Miami, Florida; San Juan, Puerto Rico; and the Bermudas are generally accepted as the three points of the triangle. The distance from Miami to San Juan and Miami to the Bermudas is 908.2 nautical miles and the distance from San Juan to the Bermudas is 839.1 nautical miles. How large an area must be searched to look for the remains of the missing ship?

91. An A-frame house overlooking the Atlantic Ocean has windows entirely covering one end. If the roof intersects at a 54° angle and the roof is 21 feet long from peak to ground, how much area do the windows cover?

92. Brian just bought a used sailboat, but it needs a new sail. The dimensions of the sail are 11 feet × 12 feet × 7 feet.

a. What are the three angles of the sail?

b. How much fabric would a sail of this size require?

c. Suppose he plans to make the sail three different colors by dividing the largest angle so the 12-foot side is split into three sections of 5 feet (blue), 4 feet (yellow), and 3 feet (red), respectively. How much fabric of each color would he need?

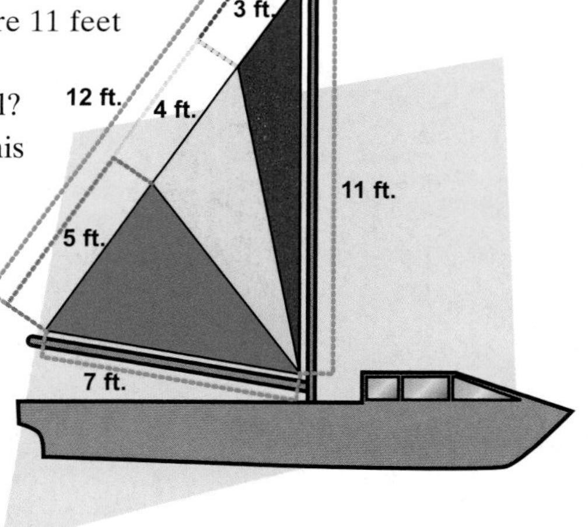

93. Any regular (all sides are equal) n-sided polygon can be divided into n equal triangles by drawing a line from each vertex to the center of the polygon. A pentagon would be divided as shown in the figure.

a. If each side has a length of 6 inches, what would the area of the given pentagon be?

b. Using a similar method, what would be the area of an octagon with sides of length 11 inches?

c. What would be the area of a five pointed star where each line segment has a length of 8 inches?

8.2 Polar Coordinates and Polar Equations

TOPICS

1. The polar coordinate system
2. Coordinate conversion
3. The form of polar equations
4. Graphing polar equations

Topic 1: The Polar Coordinate System

The Cartesian coordinate system was introduced in Chapter 2, and it has served us well to this point. But there are many situations for which a rectangular coordinate system is not the most natural choice. Some planar images and some equations in two variables have a symmetry which is awkward to express in terms of the familiar x and y coordinates. Polar coordinates provide an alternative framework for these cases.

The **polar coordinate system**, like the Cartesian coordinate system, serves as a means of locating points in the plane, and both systems are centered at a point O called the **origin**, sometimes referred to as the **pole** in the polar system. Starting from the origin O, a ray (or half-line) called the **polar axis** is drawn; in practice, the polar axis is usually drawn extending horizontally to the right, so that it corresponds with the positive x-axis in the Cartesian system. Now, given any point P in the plane other than the origin, the line segment $\overline{OP}$ has a unique positive length; we will label this length r (as in *radius*). Finally, we let θ denote the angle, measured counterclockwise, between the polar axis and the segment $\overline{OP}$, and we say (r, θ) are **polar coordinates** of the point P. The origin is the unique point for which $r = 0$ and for which the angle θ is irrelevant; the coordinates $(0, \theta)$ refer to O for any angle θ. Figure 1 illustrates the process of determining r and θ for several points.

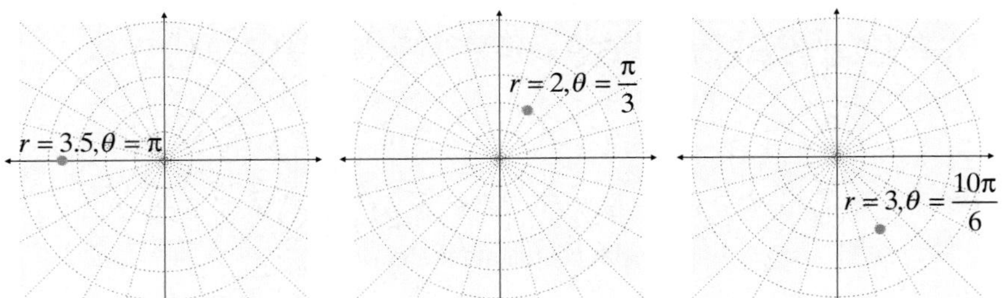

Figure 1: The Polar Coordinate System

Before proceeding further, it is important to recognize one very critical difference between Cartesian and polar coordinates. A given point P corresponds to unique Cartesian coordinates (x, y), but the polar coordinates (r, θ) of P as described above are only one of an infinite number of ways of specifying P in the polar system. Our familiarity with trigonometric functions indicates one reason for this: $(r, \theta + 2n\pi)$ also represents P, since θ and $\theta + 2n\pi$ have the same terminal sides for any integer n. Further, $(-r, \theta + (2n + 1)\pi)$ also represents P for any integer n, given the interpretation that $-r$ indicates travel in the opposite direction through the origin. These observations are illustrated in Figure 2 with alternate descriptions of the points from Figure 1.

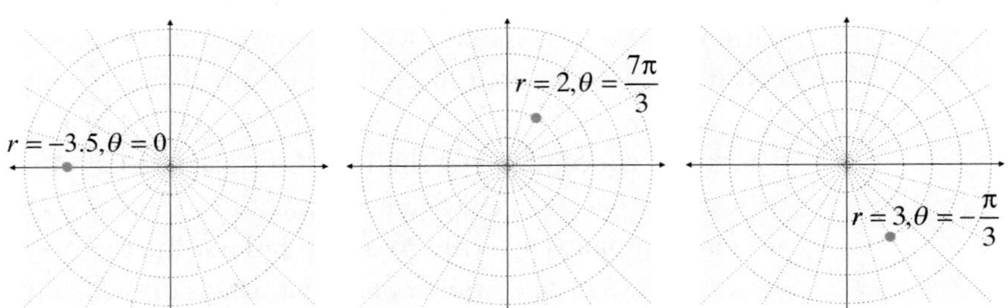

Figure 2: Alternate Polar Coordinates

example 1

Plot the points given by the following polar coordinates.

a. $\left(2, \frac{3\pi}{4}\right)$ **b.** $\left(3.5, -\frac{5\pi}{2}\right)$ **c.** $\left(-1, \frac{4\pi}{3}\right)$

Solutions:

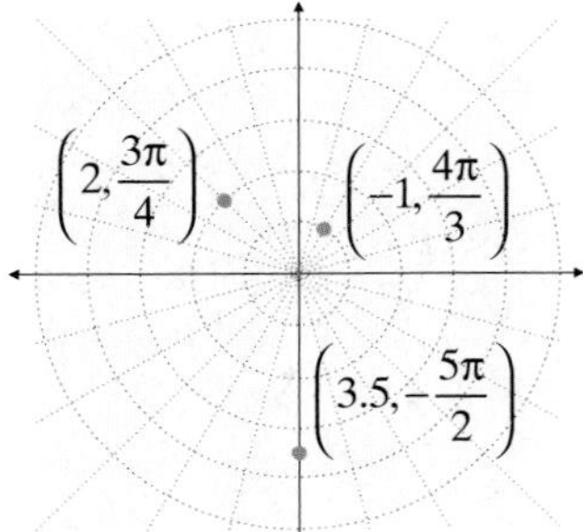

Topic 2: Coordinate Conversion

Since we have two systems by which to specify points in the plane, it should come as no surprise that we will occasionally need to be able to translate information from one system to the other. As we will soon see, this will be especially useful when faced with an equation which is awkward to graph in one coordinate system, but straightforward in the other. Fortunately, converting from Cartesian coordinates to polar coordinates, and vice-versa, is easily accomplished. To do so, we will assume the polar axis is aligned with the positive x-axis, and that a fixed point P has Cartesian coordinates (x, y) and polar coordinates (r, θ). Then, as seen in Figure 3, $r^2 = x^2 + y^2$ and

$$\cos\theta = \frac{x}{r}, \ \sin\theta = \frac{y}{r}, \text{ and } \tan\theta = \frac{y}{x}.$$

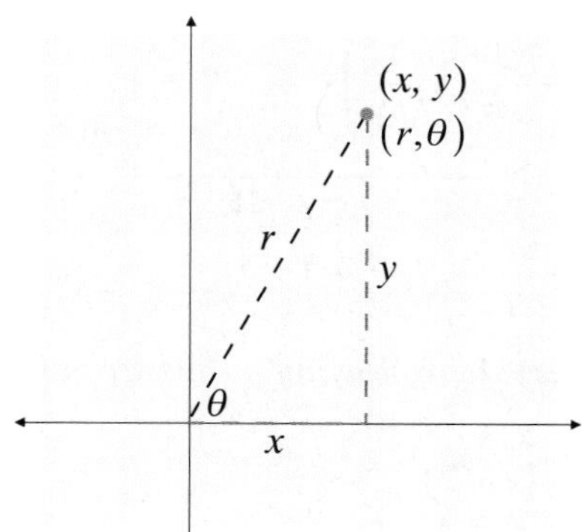

Figure 3: Coordinate Conversion

These relations, and the diagram in Figure 3, should seem very familiar – we encountered them first in defining the trigonometric functions. They are restated as conversion formulas in the box below.

Coordinate Conversion

Converting from polar to Cartesian coordinates: Given (r, θ), x and y are defined by

$$x = r\cos\theta \text{ and } y = r\sin\theta.$$

Converting from Cartesian to polar coordinates: Given (x, y), r and θ are defined by

$$r^2 = x^2 + y^2 \text{ and } \tan\theta = \frac{y}{x} \quad (x \neq 0).$$

Make note of the quadrant of the original Cartesian coordinate when converting, to be sure the polar coordinates fall in the same quadrant. (See Example 3.)

example 2

Convert the following points from polar to Cartesian coordinates.

a. $\left(2, -\frac{\pi}{3}\right)$ **b.** $\left(-3, \frac{\pi}{4}\right)$

Solutions:

a. $x = 2\cos\left(-\frac{\pi}{3}\right) = 1$ and $y = 2\sin\left(-\frac{\pi}{3}\right) = -\sqrt{3}$, so the Cartesian coordinates are $\left(1, -\sqrt{3}\right)$.

b. $x = -3\cos\left(\frac{\pi}{4}\right) = -\frac{3}{\sqrt{2}}$ and $y = -3\sin\left(\frac{\pi}{4}\right) = -\frac{3}{\sqrt{2}}$, so the Cartesian coordinates are $\left(-\frac{3}{\sqrt{2}}, -\frac{3}{\sqrt{2}}\right)$.

example 3

Convert the following points from Cartesian to polar coordinates.

a. $(-3, 2)$ **b.** $\left(\sqrt{3}, -1\right)$

Solutions:

a. To avoid careless error, make sure you have some rough idea of what the answer should be before doing any calculations. Since $(-3, 2)$ is in the second quadrant, one possible conversion will lead to $\frac{\pi}{2} < \theta < \pi$. To get the exact angle, we use the fact that $\tan\theta = -\frac{2}{3}$, so $\theta = \tan^{-1}\left(-\frac{2}{3}\right) \approx 2.55$ (remember, this is in radians). Depending on your calculator, though, you may have found $\tan^{-1}\left(-\frac{2}{3}\right) \approx -0.59$, an angle in the fourth quadrant; if so, it is up to you to remember to either add π in order to get an angle in the second quadrant, or to use a negative value for r. The radius is more easily determined: $r = \sqrt{(-3)^2 + (2)^2} = \sqrt{13}$. The two answers we have found are thus $\left(\sqrt{13}, 2.55\right)$ and $\left(-\sqrt{13}, -0.59\right)$.

b. Polar coordinates of $\left(\sqrt{3},-1\right)$ are more easily determined. The point lies in the fourth quadrant, so the two most obvious conversions will lead to either $\frac{3\pi}{2}<\theta<2\pi$ or $-\frac{\pi}{2}<\theta<0$. The coordinates $\left(\sqrt{3},-1\right)$ give rise to a familiar triangle with convenient angles, and it's probably easiest to use $\theta=-\frac{\pi}{6}$. The radius is $r=2$, so one set of polar coordinates for the point is $\left(2,-\frac{\pi}{6}\right)$.

Topic 3: The Form of Polar Equations

In the most general terms, a *polar equation* is an equation in the variables r and θ that defines a relationship between these polar coordinates (just as equations in x and y define a relationship between rectangular coordinates). A solution to a polar equation is thus an ordered pair (r, θ) that makes the equation true, and the graph of a polar equation consists of all such ordered pairs. Many polar equations can be written in the form $r=f(\theta)$; in such a form, the distance from the origin is expressed as a function of the angle θ and the graph of the equation is usually fairly easy to determine. We will soon study many examples of polar equations, but first we will examine the effect of translating an equation from one coordinate system to another.

The basic tools for such translation are the formulas for coordinate conversion used in Examples 2 and 3. Translating from rectangular coordinates to polar coordinates is particularly straightforward: simply replace every occurrence of x with $r\cos\theta$ and every occurrence of y with $r\sin\theta$. Translation in the opposite direction may require significantly more effort.

example 4

Rewrite the equation $x^2-2x+y^2=0$ in polar form.

Solution:

Making the appropriate substitutions in the equation $x^2-2x+y^2=0$ and simplifying, we obtain

$$\begin{aligned}(r\cos\theta)^2-2(r\cos\theta)+(r\sin\theta)^2&=0\\ r^2\cos^2\theta-2r\cos\theta+r^2\sin^2\theta&=0\\ r^2\left(\cos^2\theta+\sin^2\theta\right)-2r\cos\theta&=0\\ r^2-2r\cos\theta&=0.\end{aligned}$$

example 5

Rewrite the equation $2r = \sec\theta$ in rectangular coordinates.

Solution:

One good way to begin is to rewrite the equation as follows:

$$2r = \sec\theta$$

$$2r = \frac{1}{\cos\theta}$$

$$r\cos\theta = \frac{1}{2}.$$

Now we recognize the term on the left-hand side as x, and we see the equation is $x = \frac{1}{2}$.

Topic 4: Graphing Polar Equations

Gaining a sense of confidence in your ability to graph polar equations calls for nothing more than familiarity with a number of examples. As with equations in rectangular coordinates, we will begin with some very simple, but illustrative, examples.

example 6

Sketch the graphs of the following polar equations, and then convert the equations to rectangular coordinates.

a. $r = 3$ **b.** $\theta = \frac{2\pi}{3}$

Solutions:

a. Since θ doesn't appear in the equation $r = 3$, θ is allowed to take on any value. The equation thus describes all points (r, θ) for which $r = 3$; that is, a circle of radius 3. The equation in rectangular coordinates is $x^2 + y^2 = 9$.

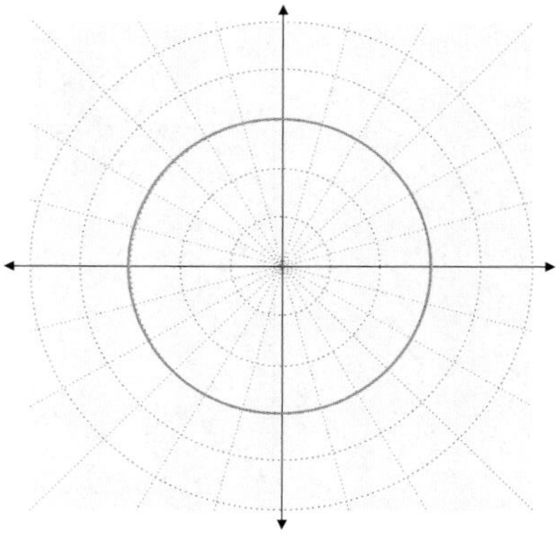

b. In this equation, r is allowed to take on any value. But every point (r, θ) that satisfies the equation must have $\theta = \frac{2\pi}{3}$. The graph is thus a straight line passing through the origin, and $\frac{y}{x} = \tan\theta = \tan\left(\frac{2\pi}{3}\right) = -\sqrt{3}$, so $y = -\sqrt{3}x$.

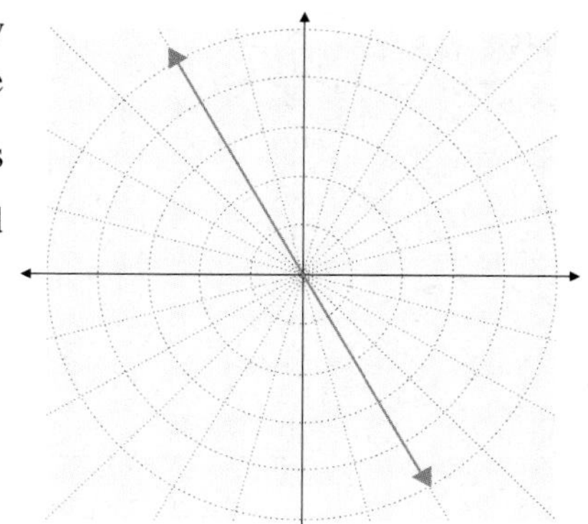

Of course, in general we expect an equation in polar coordinates to contain both r's and θ's. As with rectangular coordinates, the most basic approach to sketching the graph of a polar equation is to plot some representative points and to connect the points as seems appropriate. This method, applied judiciously and perhaps in combination with some algebra, will take us far.

example 7

Sketch the graph of the equation $r = 2\cos\theta$.

Solution:

We can begin by calculating some values of r for given θ's:

θ	0	$\frac{\pi}{6}$	$\frac{\pi}{4}$	$\frac{\pi}{3}$	$\frac{\pi}{2}$	$\frac{2\pi}{3}$	$\frac{3\pi}{4}$	$\frac{5\pi}{6}$
r	2	$\sqrt{3}$	$\sqrt{2}$	1	0	-1	$-\sqrt{2}$	$-\sqrt{3}$

Now if we plot the pairs (r, θ) from the table, we obtain the points in red shown at right. These certainly appear to lie along the circumference of a circle, so the points have been connected with the curve in blue. If we convert the equation into rectangular coordinates as shown on the next page, we see that the graph indeed is a circle of radius 1 centered at $x = 1, y = 0$.

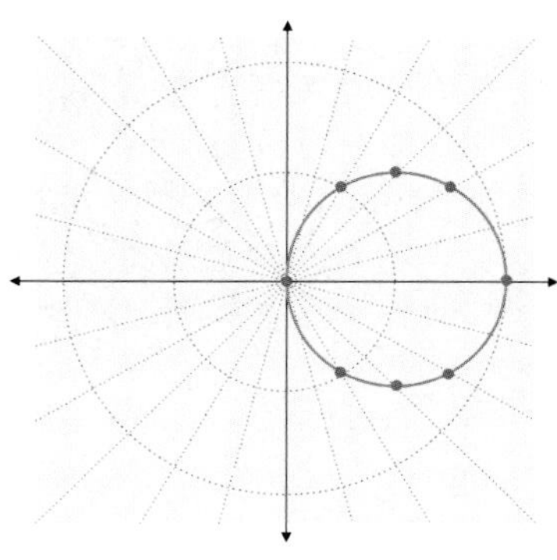

cont'd. on next page ...

$$
\begin{aligned}
r &= 2\cos\theta \\
r^2 &= 2r\cos\theta \\
x^2 + y^2 &= 2x \\
x^2 - 2x + y^2 &= 0 \\
x^2 - 2x + 1 + y^2 &= 1 \\
(x-1)^2 + y^2 &= 1
\end{aligned}
$$

We have used symmetry in the past as an aid in graphing functions and equations, and the concept is no less useful in polar coordinates. Consider the three types of symmetry illustrated in Figure 4.

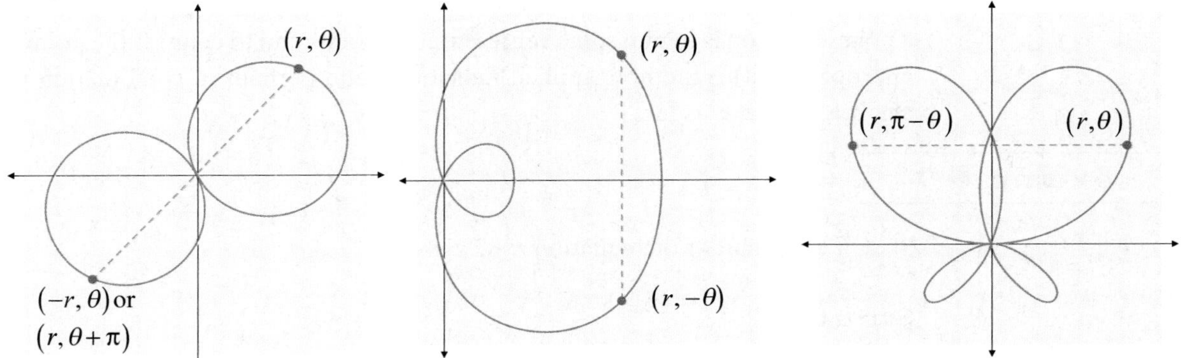

Figure 4: Symmetry in Polar Coordinates

Algebraically, symmetry can be recognized with the following tests.

Symmetry of Polar Equations

An equation in r and θ is symmetric with respect to:

1. **The pole** if replacing r with $-r$ (or replacing θ with $\theta+\pi$) results in an equivalent equation.
2. **The polar axis** if replacing θ with $-\theta$ results in an equivalent equation.
3. **The line** $\boldsymbol{\theta = \frac{\pi}{2}}$ if replacing θ with $\pi - \theta$ results in an equivalent equation.

We will conclude this section with a catalog of some polar equations that arise frequently enough to have been given names. Exploring a few of these further will give us the opportunity to apply the above symmetry tests and gain more familiarity with graphing in polar coordinates.

Common Polar Equations and Graphs

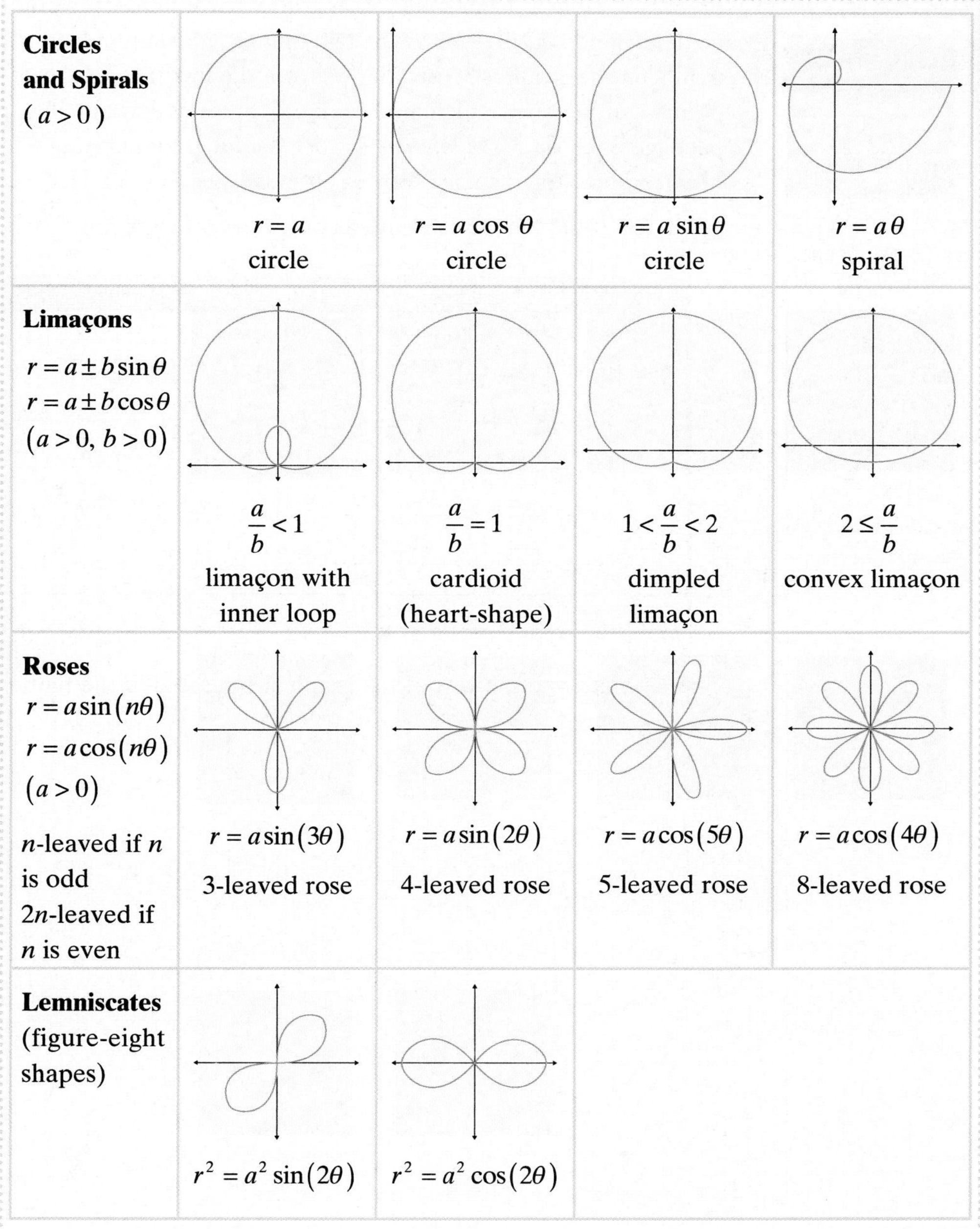

Circles and Spirals $(a>0)$	$r=a$ circle	$r=a\cos\theta$ circle	$r=a\sin\theta$ circle	$r=a\theta$ spiral
Limaçons $r=a\pm b\sin\theta$ $r=a\pm b\cos\theta$ $(a>0,\ b>0)$	$\frac{a}{b}<1$ limaçon with inner loop	$\frac{a}{b}=1$ cardioid (heart-shape)	$1<\frac{a}{b}<2$ dimpled limaçon	$2\leq\frac{a}{b}$ convex limaçon
Roses $r=a\sin(n\theta)$ $r=a\cos(n\theta)$ $(a>0)$ n-leaved if n is odd $2n$-leaved if n is even	$r=a\sin(3\theta)$ 3-leaved rose	$r=a\sin(2\theta)$ 4-leaved rose	$r=a\cos(5\theta)$ 5-leaved rose	$r=a\cos(4\theta)$ 8-leaved rose
Lemniscates (figure-eight shapes)	$r^2=a^2\sin(2\theta)$	$r^2=a^2\cos(2\theta)$		

example 8

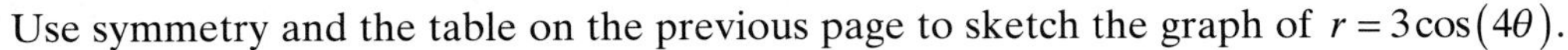

Use symmetry and the table on the previous page to sketch the graph of $r = 3\cos(4\theta)$.

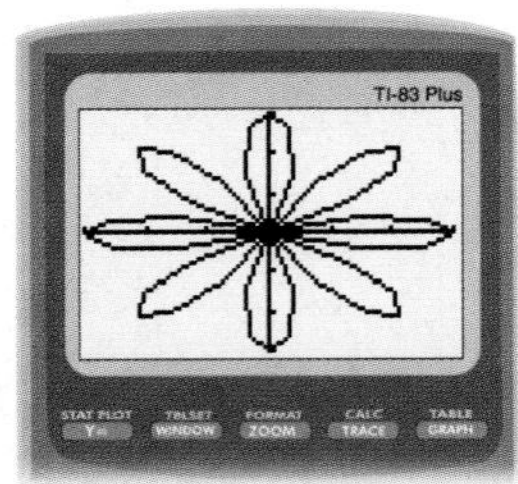

Make sure `Pol` is selected under the **Mode** menu. Press **Y=** and enter `3cos(4θ)`. Then press **Window** and set the parameters as: `θmax=2π`, `Xmin=-3`, `Xmax=3`, `Ymin=-3`, and `Ymax=3`. Finally, press **Graph** and the screen should appear as above.

Solution:

The equation $r = 3\cos(4\theta)$ possesses all three kinds of symmetry, and reference to the preceding table indicates the graph is an 8-leaved rose. Several aspects of the graph can be determined easily, such as the fact that the maximum distance between the origin and points on the graph is 3; this follows from the fact that $-1 \le \cos(4\theta) \le 1$ for all θ. Some points can be easily calculated by hand, but a programmable calculator or a software package such as *Mathematica* is very useful in generating a large number of points. *Mathematica*, for instance, can easily generate the polar coordinates of points on the graph for $0 \le \theta \le 2\pi$ in increments of $\frac{\pi}{16}$, as shown below:

Section 8-2 Example 8.nb

In[1]:= `Table[{3 Cos[4 θ], θ}, {θ, 0, 2 Pi, Pi / 16}]`

Out[1]= $\{\{3, 0\}, \{\frac{3}{\sqrt{2}}, \frac{\pi}{16}\}, \{0, \frac{\pi}{8}\}, \{-\frac{3}{\sqrt{2}}, \frac{3\pi}{16}\}, \{-3, \frac{\pi}{4}\}, \{-\frac{3}{\sqrt{2}}, \frac{5\pi}{16}\}, \{0, \frac{3\pi}{8}\},$
$\{\frac{3}{\sqrt{2}}, \frac{7\pi}{16}\}, \{3, \frac{\pi}{2}\}, \{\frac{3}{\sqrt{2}}, \frac{9\pi}{16}\}, \{0, \frac{5\pi}{8}\}, \{-\frac{3}{\sqrt{2}}, \frac{11\pi}{16}\}, \{-3, \frac{3\pi}{4}\}, \{-\frac{3}{\sqrt{2}}, \frac{13\pi}{16}\},$
$\{0, \frac{7\pi}{8}\}, \{\frac{3}{\sqrt{2}}, \frac{15\pi}{16}\}, \{3, \pi\}, \{\frac{3}{\sqrt{2}}, \frac{17\pi}{16}\}, \{0, \frac{9\pi}{8}\}, \{-\frac{3}{\sqrt{2}}, \frac{19\pi}{16}\}, \{-3, \frac{5\pi}{4}\},$
$\{-\frac{3}{\sqrt{2}}, \frac{21\pi}{16}\}, \{0, \frac{11\pi}{8}\}, \{\frac{3}{\sqrt{2}}, \frac{23\pi}{16}\}, \{3, \frac{3\pi}{2}\}, \{\frac{3}{\sqrt{2}}, \frac{25\pi}{16}\}, \{0, \frac{13\pi}{8}\},$
$\{-\frac{3}{\sqrt{2}}, \frac{27\pi}{16}\}, \{-3, \frac{7\pi}{4}\}, \{-\frac{3}{\sqrt{2}}, \frac{29\pi}{16}\}, \{0, \frac{15\pi}{8}\}, \{\frac{3}{\sqrt{2}}, \frac{31\pi}{16}\}, \{3, 2\pi\}\}$

100%

These points, and the rest of the graph, are plotted in the figure below.

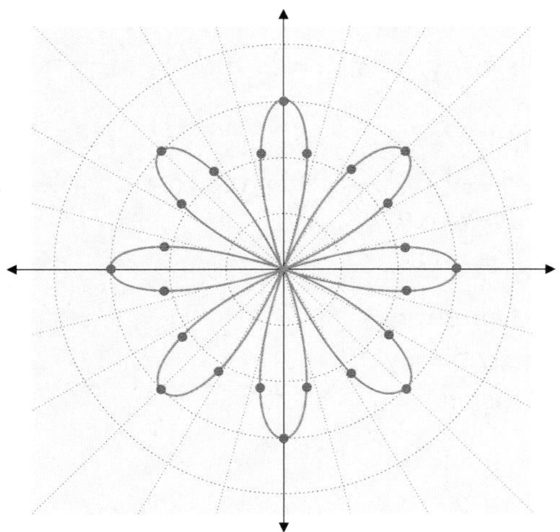

example 9

Use symmetry and the table on page 625 to sketch the graph of $r = 1 + 2\cos\theta$.

Make sure **Mode** is set to `Pol`. Press **Y=** and enter `1+2cos(θ)`. Then press **Window** and set the parameters as: `θmax=2π`, `Xmin=-3`, `Xmax=3`, `Ymin=-2`, and `Ymax=2`. Finally, press **Graph** and the screen should appear as above.

Solution:

The graph of $r = 1 + 2\cos\theta$ is a limaçon with an inner loop. Because

$$1 + 2\cos\theta = 1 + 2\cos(-\theta)$$

(since cosine is an even function), the graph is symmetric with respect to the polar axis. The graph and some points on the graph are as follows:

θ	r	θ	r
0	3	$\frac{4\pi}{3}$	0
$\frac{\pi}{3}$	2	$\frac{3\pi}{2}$	1
$\frac{\pi}{2}$	1	$\frac{5\pi}{3}$	2
$\frac{2\pi}{3}$	0	2π	3
π	-1		

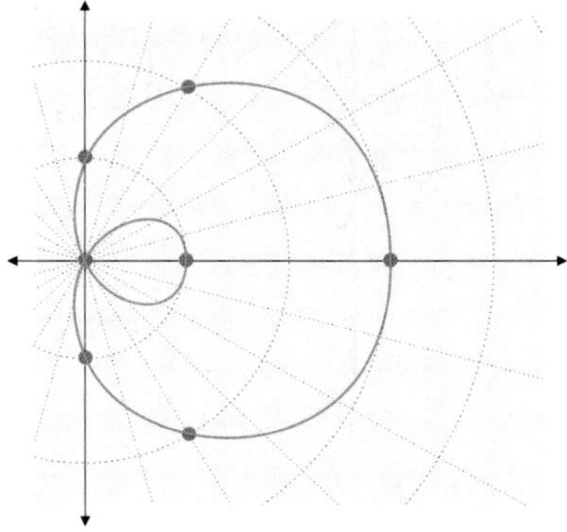

technology note

Graphing calculators and computer algebra systems, such as *Mathematica*, can be very useful in constructing accurate graphs of polar equations, though their capabilities are limited and vary considerably. Often, a polar equation must be solved for r in order to use the graphing feature of a particular technology. Figure 5 on the next page illustrates how *Mathematica* is used to graph a cardioid and a lemniscate. Note that in order to graph the lemniscate, we have had to solve the equation $r^2 = 4\sin(2\theta)$ for r and then instruct *Mathematica* to graph both the positive and negative square roots that result.

cont'd. on next page ...

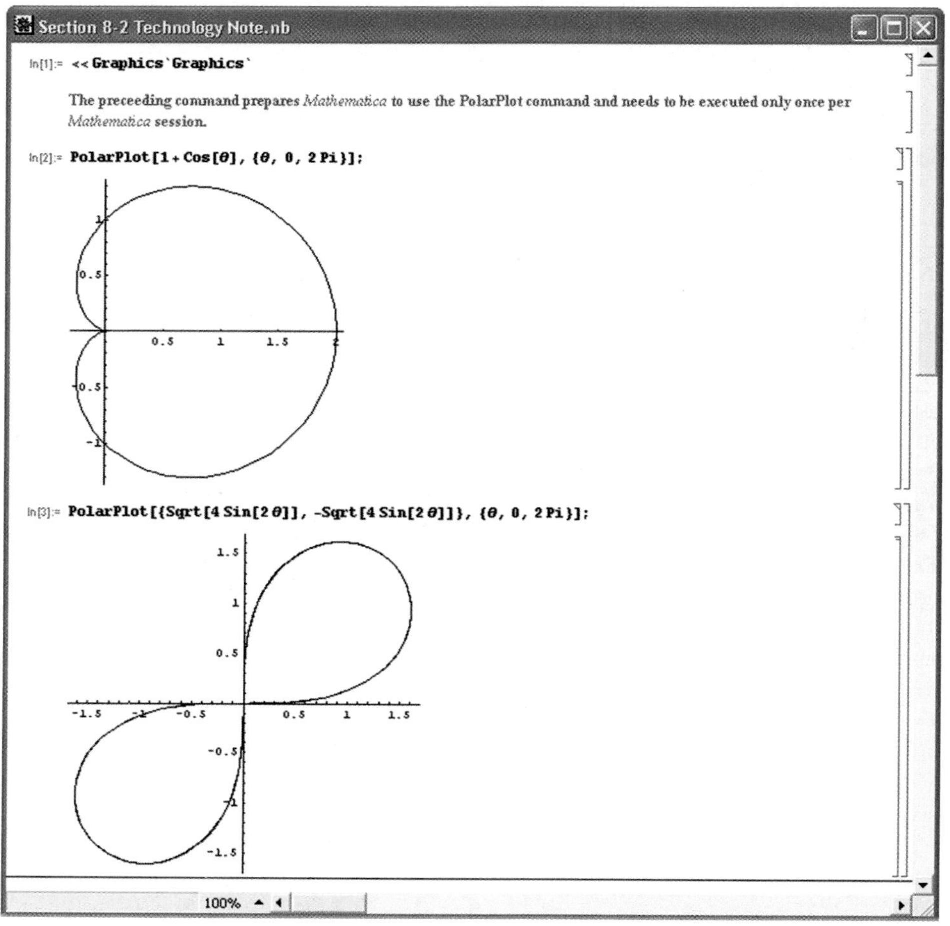

Figure 5: Graphing Polar Equations in *Mathematica*

exercises

Plot the point given by the polar coordinates. See Example 1.

1. $\left(-1, \frac{5\pi}{4}\right)$ **2.** $\left(-5, \frac{3\pi}{2}\right)$ **3.** $\left(\frac{1}{4}, \frac{-7\pi}{6}\right)$

4. $\left(\sqrt{3}, \frac{-\pi}{3}\right)$ **5.** $\left(4\frac{8}{9}, -\pi\right)$ **6.** $\left(\frac{7}{\sqrt{2}}, \frac{\pi}{2}\right)$

Convert the point from a polar to a Cartesian coordinate. See Example 2.

7. $\left(5, \frac{7\pi}{4}\right)$ **8.** $(0, 2\pi)$ **9.** $\left(6.25, \frac{-3\pi}{4}\right)$

10. $\left(-2.25, \frac{\pi}{4}\right)$ **11.** $\left(3, \frac{-5\pi}{6}\right)$ **12.** $\left(-11, \frac{10\pi}{12}\right)$

Convert the point from a Cartesian to a polar coordinate. See Example 3.

13. $(-3, 0)$ **14.** $\left(-6, \sqrt{3}\right)$ **15.** $(12, -1)$

16. $(8, 0)$ **17.** $\left(-\sqrt{3}, 9\right)$ **18.** $(-5, -5)$

Rewrite the rectangular equation in polar form. See Example 4.

19. $x^2 + y^2 = 25$ **20.** $x^2 + y^2 = 81$ **21.** $x = 12$

22. $y = 16$ **23.** $y = x$ **24.** $y = b$

25. $x = 16a$ **26.** $x^2 + y^2 = a$ **27.** $x^2 + y^2 = 4ax$

28. $x^2 + y^2 = 4ay$ **29.** $y^2 - 4 = 4x$ **30.** $x^2 + y^2 = 36a^2$

Rewrite the polar equation in rectangular form. See Example 5.

31. $r = 5\cos\theta$ **32.** $r = 8\sin\theta$ **33.** $r = 7$

34. $\theta = \frac{\pi}{6}$ **35.** $18r = 9\csc\theta$ **36.** $r = 2\sec\theta$

37. $r^2 = \sin 2\theta$ **38.** $r = \frac{2}{1-\cos\theta}$ **39.** $r = \frac{12}{4\sin\theta + 7\cos\theta}$

40. $r = \frac{16}{4+4\sin\theta}$

Rewrite the polar equation in rectangular form and sketch the graph. See Examples 6 and 7.

41. $r = 3$ **42.** $r = 6$ **43.** $\theta = \frac{5\pi}{6}$

44. $\theta = \frac{\pi}{4}$ **45.** $r = 7\sec\theta$ **46.** $r = 2\csc\theta$

Sketch a graph of the polar equation. See Examples 8 and 9.

47. $r = 4$

48. $r = 5$

49. $\theta = \frac{4\pi}{3}$

50. $\theta = \frac{-\pi}{3}$

51. $r = 6\cos\theta$

52. $r = 2\sin\theta$

53. $r = 3 - 3\sin\theta$

54. $r = 6 + 5\cos\theta$

55. $r = 7(1 + \cos\theta)$

56. $r = 2(1 - 2\sin\theta)$

57. $r = 4 - 3\sin\theta$

58. $r = 3 + 4\sin\theta$

59. $r = 3\sin 3\theta$

60. $r = 5\sin 3\theta$

61. $r = 2\sin 2\theta$

62. $r = 4\sin 2\theta$

63. $r = 5\cos 5\theta$

64. $r = 4\cos 5\theta$

65. $r = 4\cos 4\theta$

66. $r = 3\cos 4\theta$

67. $r^2 = 16\sin 2\theta$

68. $r^2 = 9\cos 2\theta$

8.3 Parametric Equations

TOPICS

1. Sketching parametric curves
2. Eliminating the parameter
3. Constructing parametric equations

Topic 1: Sketching Parametric Curves

Many curves in the plane are most naturally defined in terms of a variable called a **parameter**, a variable different from those representing the coordinates. For instance, if (x, y) denotes the position of a thrown object, both x and y may be thought of as functions of a third variable t, representing time. In fact, thinking of the position (x, y) as a function of time t provides more information than a simple equation relating x and y: by calculating (x, y) for various values of t, we can determine not just the shape of the object's flight, but *where* the object is at any given time. If we have determined exactly how x and y depend on t, so that

$$x = f(t) \text{ and } y = g(t)$$

for some functions f and g, we have determined **parametric equations** for the curve traced out by (x, y).

The example provided by a thrown object is important enough to warrant further study, and a thorough analysis will lead to our first illustration of how parametric curves are sketched. In Section 1.7, we saw that the height of an object, moving only under the influence of the force of gravity g, is given by the formula

$$h(t) = -\frac{1}{2}gt^2 + v_0 t + h_0,$$

where v_0 is the initial (vertical) velocity and h_0 is the initial height of the object. In this formula, t represents the variable time; in the derivation that follows, t will retain this meaning, but we will now also consider t as the parameter describing the object's travel.

Only strictly vertical travel was considered in Section 1.7, but we now have the tools to analyze more realistic scenarios. We will set up our coordinate system so that the object starts off at position $(0, h_0)$ at time $t = 0$, and we will assume that it has an initial velocity of v_0 at an angle θ, as shown in Figure 1.

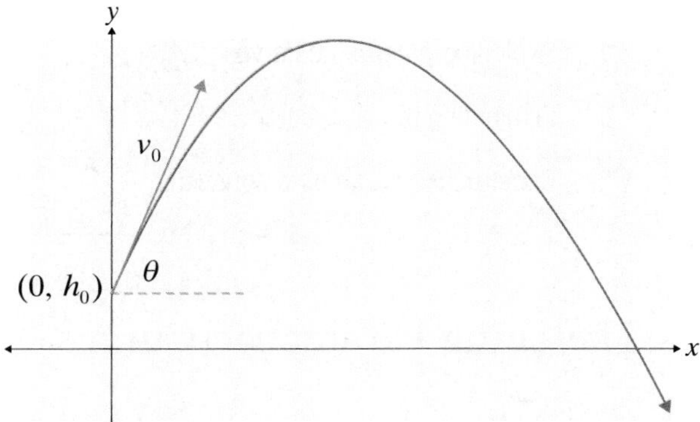

Figure 1: Describing an Object's Travel Parametrically

It is important to realize that v_0 is the size of the initial velocity in the direction θ, and that therefore the initial *vertical* velocity is $v_0 \sin\theta$ and the initial *horizontal* velocity is $v_0 \cos\theta$. If we let (x, y) stand for the position of the object at time t, the formula from Section 1.7 tells us

$$y = -\frac{1}{2}gt^2 + (v_0 \sin\theta)t + h_0,$$

so we have succeeded in writing y as a function of t. The situation in the horizontal direction is different. Because there is no force affecting the object's horizontal velocity over time (we are assuming there is no frictional force present), the horizontal displacement is simply the product of the horizontal velocity and time (remember, $\text{distance} = \text{rate} \times \text{time}$). That is,

$$x = (v_0 \cos\theta)t.$$

We now have parametric equations defining the path of the object's travel. A specific application of these general equations follows.

example 1

Suppose a baseball is hit 3 feet above the ground, and that it leaves the bat at a speed of 100 miles an hour at an angle of 20° from the horizontal. Construct a set of parametric equations describing the ball's flight, and sketch a graph of the ball's travel.

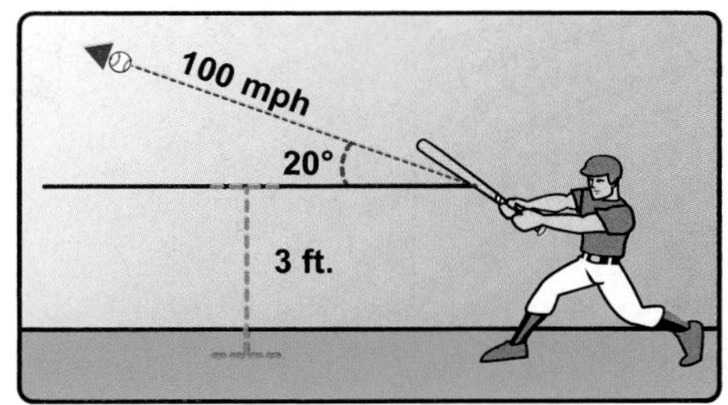

Solution:

With the information given, it's appropriate to work in units of feet and seconds. Recall that $g = 32 \text{ ft./s}^2$, and that all the quantities must be expressed in the same units. Hence:

$$h_0 = 3 \text{ feet}$$

and

$$v_0 = 100 \ \frac{\text{miles}}{\text{hour}} = \left(\frac{100 \text{ miles}}{\text{hour}}\right)\left(\frac{5280 \text{ feet}}{1 \text{ mile}}\right)\left(\frac{1 \text{ hour}}{3600 \text{ seconds}}\right) \approx 146.7 \ \frac{\text{feet}}{\text{second}}.$$

We can now determine the parametric equations:

$$\begin{aligned} y &\approx -\frac{1}{2}(32)t^2 + (146.7)\left(\sin 20^\circ\right)t + 3 \\ &= -16t^2 + 50.2t + 3 \end{aligned}$$

and

$$\begin{aligned} x &\approx (146.7)\left(\cos 20^\circ\right)t \\ &= 137.9t. \end{aligned}$$

For the purposes of this application, the part of the curve we are interested in begins with $t = 0$, the moment the bat hits the ball. Note that at this time, $x = 0$ and $y = 3$; that is, the ball starts off at the point $(0, 3)$, as we intended. As time moves on, the x-coordinate increases and the y-coordinate at first increases and then decreases. Considering the physical situation giving rise to the equations, we are probably only interested in the curve up to the point where $y = 0$ (when the ball hits the ground). If we calculate (x, y) for values of t from 0 to 3.5, in half-second increments, we obtain the following table:

t	0	0.5	1.0	1.5
(x, y)	$(0, 3)$	$(69.0, 24.1)$	$(137.9, 37.2)$	$(206.9, 42.3)$

t	2.0	2.5	3.0	3.5
(x, y)	$(275.8, 39.4)$	$(344.8, 28.5)$	$(413.7, 9.6)$	$(482.7, -17.3)$

cont'd. on next page ...

We can now plot these points and connect them, as usual, with a smooth curve. The resulting graph is:

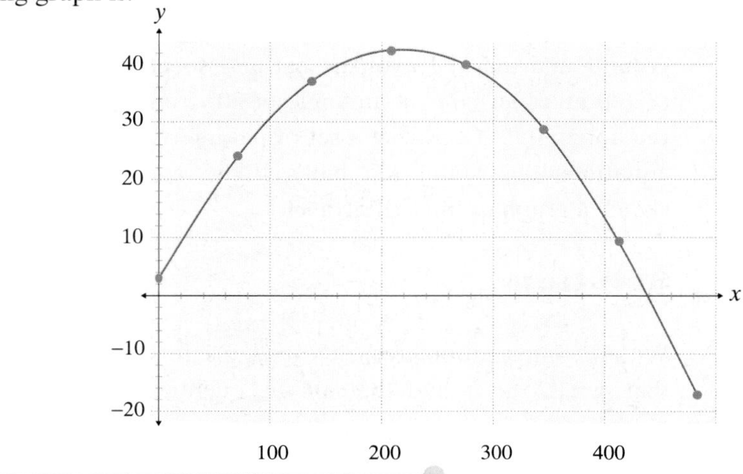

We have completed the tasks set in Example 1, but there are still some advantages to parametric curves to point out. First, curves described by parametric equations automatically possess an **orientation**, which is the direction traveled along the curve as the parameter increases. If the parameter represents time, as in Example 1, the orientation has an obvious physical meaning: it is the direction the ball travels as it traces out the curve. Second, the parametric description of the curve allows many questions to be easily answered. Consider, for example, the following question.

example 2

Suppose the ball in Example 1 has been hit toward a 10-foot high fence that is 400 feet from home plate. Will the ball clear the fence?

Solution:

The graph of the ball's flight makes it appear that the answer may be yes, but pictures can be deceptive and a graph is certainly not a proof. However, we can easily resolve the situation in two steps: first, determine the time when the ball is 400 feet from home plate, and then calculate the height of the ball at that time. We determine the time by setting the equation for x equal to 400 and solving for t:

$$\begin{aligned} 137.9t &= 400 \\ t &= 2.9 \text{ seconds.} \end{aligned}$$

Then we evaluate the height y at this time:

$$\begin{aligned} y &= -16(2.9)^2 + 50.2(2.9) + 3 \\ &= 14.0 \text{ feet.} \end{aligned}$$

So the ball does indeed clear the fence.

Topic 2: Eliminating the Parameter

While the parametric form of a curve is very useful in answering some questions and for some graphing purposes, it should not be assumed that one form or the other is better for *all* purposes. In some cases, it is convenient to convert parametric equations into a single equation relating x and y; we will call such equations rectangular equations, to make the distinction with parametric equations.

The process of converting from parametric form to rectangular form is called, appropriately enough, **eliminating the parameter**. The most common method is to solve one of the equations for the parameter and to substitute the result for the parameter in the other equation.

example 3

Sketch the graph described by the parametric equations $x=\frac{1}{\sqrt{t-1}}+1$ and $y=\frac{1}{t-1}$ by eliminating the parameter.

Solution:

If we solve the second equation for t, the result is as follows:

$$\begin{aligned} y &= \frac{1}{t-1} \\ t-1 &= \frac{1}{y} \\ t &= \frac{1}{y}+1. \end{aligned}$$

Substituting this into the first equation and simplifying, we obtain an equation in x and y.

$$\begin{aligned} x &= \frac{1}{\sqrt{\left(\frac{1}{y}+1\right)-1}}+1 \\ &= \frac{1}{\sqrt{\frac{1}{y}}}+1 \\ &= \sqrt{y}+1 \end{aligned}$$

We are more accustomed to graphing such equations in x and y if they are solved for y, and doing so in this case leads to a curve we know well:

$$\begin{aligned} x &= \sqrt{y}+1 \\ x-1 &= \sqrt{y} \\ (x-1)^2 &= y. \end{aligned}$$

cont'd. on next page ...

As we know, the graph of this last equation is an upward-opening parabola with vertex at (1, 0). But before sketching this curve and proclaiming it as the answer, look again at the original parametric equations. The formulas in t are only valid for $t > 1$ (because of the presence of $\sqrt{t-1}$), and this restriction on t leads to corresponding restrictions on x and y. If we consider all possible values of t bigger than 1, we see that the restriction on y is simply $y > 0$. However, as $t \to \infty$, $x \to 1$ and as $t \to 1$, $x \to \infty$. In particular, note that there is no value of t for which $x = 1$ (or anything smaller than 1). The graph defined by the original parametric equations is thus only half of the parabola, as shown above.

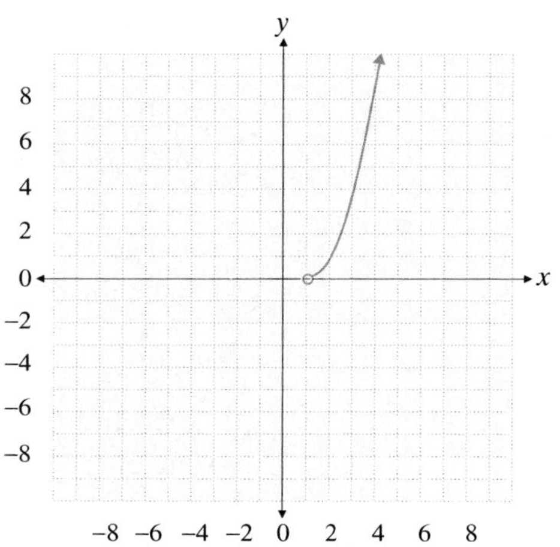

In addition to time t, angle measure θ is often used as a parameter. In fact, we have encountered parametric equations in θ already, though this was not clear at the time. Consider the following familiar example.

example 4

Sketch the graph described by the parametric equations $x = 5\cos\theta$ and $y = 5\sin\theta$ for $0 \le \theta \le 2\pi$.

Solution:

We have many ways of determining the shape of this curve. One way is to recognize the equations $x = 5\cos\theta$ and $y = 5\sin\theta$ as specific instances of the formulas $x = r\cos\theta$ and $y = r\sin\theta$ that relate rectangular and polar coordinates. If we make this connection, we see immediately that $r = 5$ and the curve is thus a circle of radius 5 centered at the origin.

But we don't have to rely on this insight in order to obtain the answer. Another approach is to eliminate the parameter as in the last example, and again we have several options. One is to solve for θ in one equation and substitute the result in the other equation:

$$y = 5\sin\theta$$

$$\sin\theta = \frac{y}{5}$$

$$\theta = \sin^{-1}\left(\frac{y}{5}\right)$$

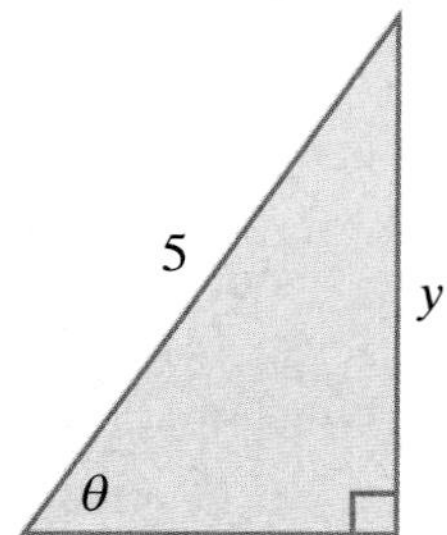

and so

$$x = 5\cos\left(\sin^{-1}\left(\frac{y}{5}\right)\right)$$
$$= 5\left(\frac{\sqrt{25-y^2}}{5}\right)$$
$$= \sqrt{25-y^2}.$$

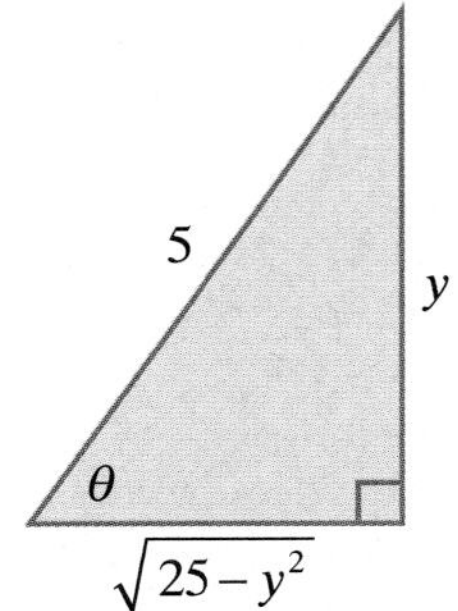

This last equation is more meaningful if we square both sides and rearrange terms slightly:

$$x = \sqrt{25-y^2}$$
$$x^2 = 25 - y^2$$
$$x^2 + y^2 = 25.$$

Using this method, we arrive at the rectangular form for the equation of a circle of radius 5 centered at the origin.

Another method of eliminating the parameter is indirect, but very useful. If we square both sides of each of the parametric equations, we can apply Pythagoras' Theorem as follows:

$$x^2 = 25\cos^2\theta \text{ and } y^2 = 25\sin^2\theta$$

so

$$x^2 + y^2 = 25\cos^2\theta + 25\sin^2\theta$$
$$= 25\left(\cos^2\theta + \sin^2\theta\right)$$
$$= 25.$$

Using any method, we arrive at the graph sketched at right.

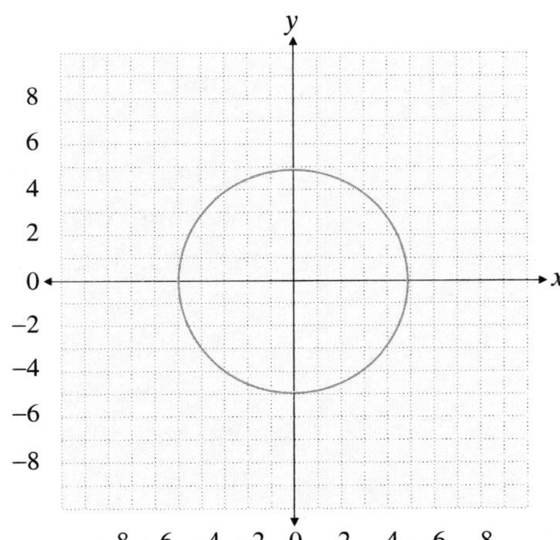

Topic 3: Constructing Parametric Equations

As we saw in Examples 1 and 2, the parametric form of an equation can prove to be very useful in answering some questions. This naturally leads us to the question of *how* parametric equations describing a given curve can be constructed. We will consider several methods.

example 5

Construct parametric equations describing the graph of $y = x^2 + 3$.

Solution:

One hallmark of parametric descriptions of graphs is that they are not unique. Geometrically, the reason for this is that a given curve can be traced out by two parameters at different "speeds" or even in opposite directions. We will illustrate this point with two solutions to the problem in this example.

Solution 1: One parameterization of the graph is achieved by letting $x = t$ and thereby defining $y = t^2 + 3$. As t increases, the parabola defined by the parametric equations is traced out from left to right. Plot a few points to verify this claim.

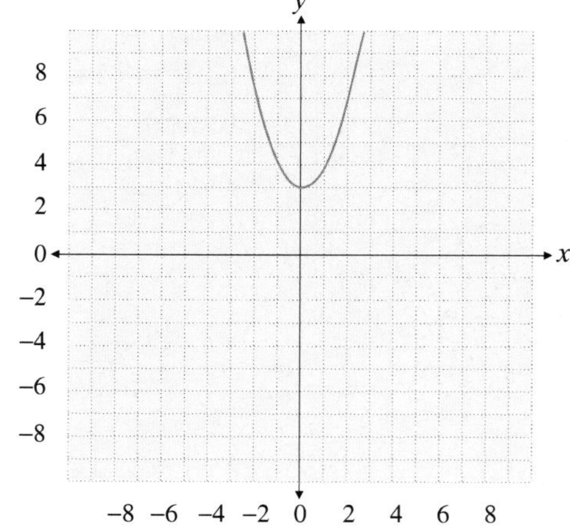

Solution 2: One of an infinite number of alternative solutions, and one that traces out the graph in the opposite direction, is achieved by letting $x = 5 - t$ (there is no special significance to the number 5 here). Now, as t increases, x decreases and so the points on the parabola will be traced out from right to left. We know this to be true even before determining that $y = x^2 + 3 = (5 - t)^2 + 3 = t^2 - 10t + 28$. Again, plot a few points to verify the claim.

example 6

A **cycloid** is the curve (in red below) traced out by a point on a circle as it rolls along a straight line. Fix a point P on a circle of radius a, and assume that P lies initially at the origin and that the circle rolls to the right as shown. Find parametric equations describing the cycloid traced out by P.

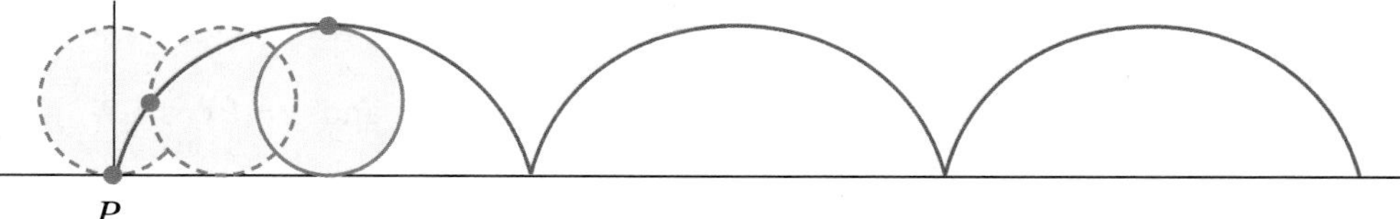

Solution:

The first step is to make a wise choice of parameter. To this end, let θ measure the angle between the ray extending straight down from the circle's center and the ray extending from the circle's center through the point P. Then when the circle is in the initial position, the two rays coincide and $\theta = 0$. As the circle rolls to the right, θ increases; when the circle is in the position of the second circle shown above, $\theta = \frac{\pi}{2}$, and $\theta = \pi$ when the point P reaches its topmost position the first time (the third circle shown above).

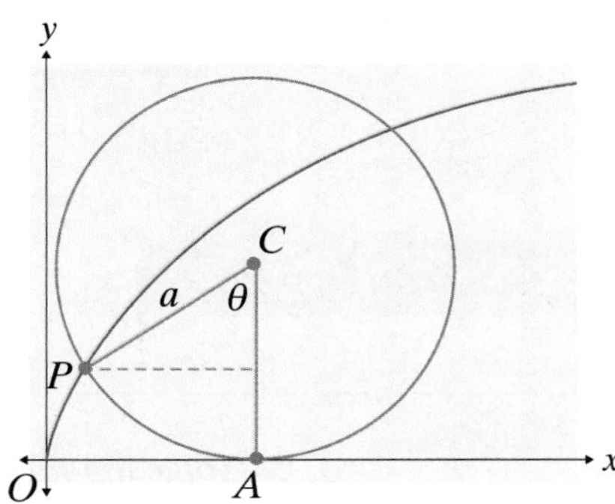

Consider the enlarged diagram at right. If we let (x, y) denote the coordinates of the point P, our goal is to write x and y as functions of θ. Our first observation is that the line segment $\overline{OA}$ must have the same length as the arc $\widehat{PA}$ (since the circle rolls without slipping); the arc-length formula tells us the length of $\widehat{PA}$ is $a\theta$, so this is the length of $\overline{OA}$ as well. This means the coordinates of point C are $(a\theta, a)$ and that the coordinates of point A are $(a\theta, 0)$. The dashed line in the diagram has length $a\sin\theta$, and the other leg of the right triangle has length $a\cos\theta$. Putting this all together, we have

$$x + a\sin\theta = a\theta \text{ and } y + a\cos\theta = a.$$

Solving for x and y, we obtain the desired parametric equations as shown below (verifying that the equations above remain true as θ takes on larger values is left to the reader):

$$x = a(\theta - \sin\theta) \text{ and } y = a(1 - \cos\theta).$$

technology note

Many graphing calculators and computer algebra systems can generate parametrically-defined curves. In *Mathematica* the command to do so is **ParametricPlot**. Figure 2 below illustrates the use of the command to graph a curve known as a **Lissajous figure**, after the 19th-century mathematician Jules-Antoine Lissajous. In general, a Lissajous figure is the shape of the curve defined by the parametric equations

$$x = A\sin\omega_1 t \text{ and } y = B\cos\omega_2 t,$$

where A, B, ω_1 and ω_2 are real constants. Since the values of sine and cosine always lie between –1 and 1, the Lissajous figure will lie inside the rectangle defined by $-A \le x \le A$ and $-B \le y \le B$.

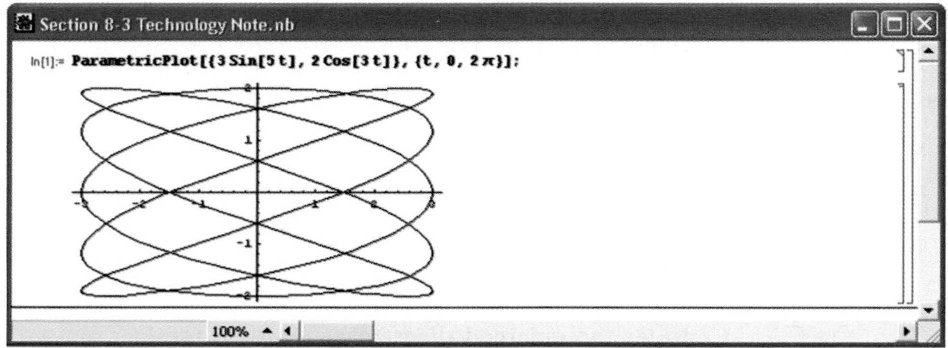

Figure 2: A Lissajous Figure

exercises

Solve the following problems concerning parametric equations. See Examples 1 and 2.

1. Given the parametric equations $x = 5 + t$ and $y = \frac{\sqrt{t}}{t-2}$, construct a table of the points (x, y) that result from t values from zero to six, and then sketch the curve.

2. Given the parametric equations $x = \frac{\tan\theta}{2}$ and $y = \cos^2\theta + 3$, construct a table of the points (x, y) that result from the values $\theta = 0, \frac{\pi}{6}, \frac{\pi}{3}, \frac{\pi}{2}, \frac{2\pi}{3}, \frac{5\pi}{6}$, and π. Using these values, sketch the graph of the equations.

3. Suppose that a circus performer is shot from a cannon at a rate of 80 mph, at an angle of 60° from the horizontal. The cannon sits on a platform 10 feet above ground.
   a. Construct a set of parametric equations describing the performer's stunt.
   b. Sketch a graph of his flight.
   c. How high is the acrobat 1.5 seconds after leaving the cannon?
   d. How far from the cannon should a landing net be placed, if it is placed at ground level?
   e. At what time t will the performer land in the net?
   f. If a 12 ft. high wall of flames is placed 70 feet from the cannon, will he clear it unharmed?

80 mph

60°

10 ft.

4. On his morning paper route, John throws a newspaper from his car window 3.5 feet from the ground. The paper has an initial velocity of 10 ft./s, and is tossed at an angle of 10° from the horizontal.
   a. Construct a set of parametric equations modeling the path of the newspaper.
   b. Sketch a graph of the paper's path.

5. François shoots a basketball at an angle of 48° from the horizontal. It leaves his hands 7 feet from the ground with a velocity of 21 ft./s.
   a. Construct a set of parametric equations describing the shot.
   b. Sketch a graph of the basketball's flight.
   c. If the goal is 15 ft. away and 11 ft. high, will he make the shot?

Sketch the graphs of the following parametric equations by eliminating the parameter. See Examples 3 and 4.

6. $x = 3(t+1)$ and $y = 2t$

7. $x = \sqrt{t-2}$ and $y = 3t - 2$

8. $x = 1 + t$ and $y = \frac{t-3}{2}$

9. $x = |t+3|$ and $y = t - 5$

10. $x = \frac{t}{4}$ and $y = t^2$

11. $x = \frac{t}{t+2}$ and $y = \sqrt{t}$

12. $x = \sqrt{t+3}$ and $y = t + 3$

13. $x = \frac{2}{|t-3|}$ and $y = 2t - 1$

14. $x = \cos\theta$ and $y = 2\sin\theta$

15. $x = 3\sin\theta - 1$ and $y = \frac{\cos\theta}{2}$

16. $x = 1 - \sin\theta$ and $y = \sin\theta - 1$

17. $x = 2\cos\theta$ and $y = 3\cos\theta$

18. $x = 2\sin\theta + 2$ and $y = 2\cos\theta + 2$

19. $x = \sin\theta$ and $y = 4 - 3\cos\theta$

Construct parametric equations describing the graphs of the following equations. See Example 5.

20. $y = (x+1)^2$

21. $y = 5x - 2$

22. $y = -x^2 - 5$

23. $x^2 + \dfrac{y^2}{4} = 1$

24. $x = y^2 + 4$

25. $y = x^2 + 1$

26. $y = \dfrac{1}{x}$

27. $x = 4y - 6$

28. $y = |x - 1|$

29. $x = 2(y - 3)$

30. $y^2 = 1 - x^2$

31. $x = \dfrac{1}{3y}$

32. $y = x^2 - x - 6$

Construct parametric equations for the line with the given attributes.

33. Slope –2, passing through (–5, –2)

34. Slope $\frac{1}{4}$, passing through (10, 12)

35. Slope 3, passing through (7, 2)

36. Passing through (0, 0) and (7, 4)

37. Passing through (6, –3) and (2, 3)

38. Passing through (12, 3) and (–4, –5)

Construct parametric equations for the circle with the given attributes.

39. Center (0, 0), radius 1

40. Center (–4, 2), radius 3

41. Center (7, –5), radius 4

42. Center (0, –2), radius 6

Construct parametric equations for the ellipse with the given attributes.

43. Vertices $(\pm 3, 0)$, Foci $(\pm 2, 0)$

44. Vertices $(\pm 6, 1)$, Foci $(\pm 5, 1)$

45. Vertices (5, 2) and (5, –4), Foci (5, 0) and (5, –2)

46. Vertices (7, 3) and (7, –3), Foci (7, 1) and (7, –1)

Construct parametric equations for the hyperbola with the given attributes.

47. Vertices $(\pm 3, 0)$, Foci $(\pm 4, 0)$

48. Vertices $(\pm 2, 0)$, Foci $(\pm 7, 0)$

49. Vertices $(0, \pm 3)$, Foci $(0, \pm 5)$

50. Vertices $(0, \pm 3)$, Foci $(0, \pm 6)$

51. A wheel of radius 12 inches rolls along a flat surface in a straight line. There is a fixed point P that initially lies at the point (0, 0). Find parametric equations describing the cycloid traced out by P.

52. A ball is rolled on the floor in a straight line from one person to another person. The ball has a radius of 3 cm and there is a fixed point P located on the ball. Let the person rolling the ball represent the origin. Find parametric equations describing the cycloid traced out by P.

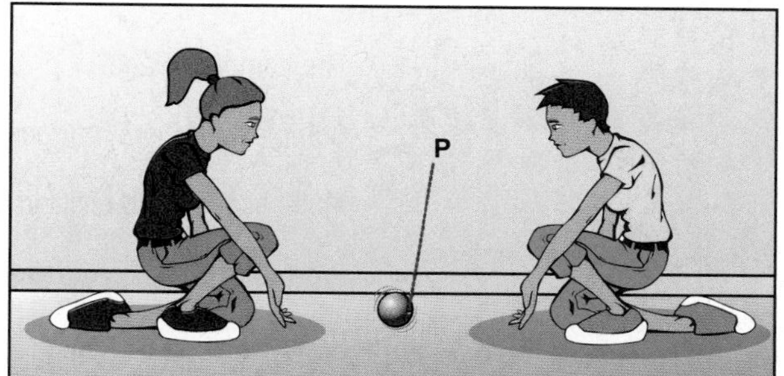

8.4 Trigonometric Form of Complex Numbers

TOPICS

1. The complex plane
2. Complex numbers in trigonometric form
3. Multiplication and division of complex numbers
4. Powers of complex numbers
5. Roots of complex numbers

Topic 1: The Complex Plane

Complex numbers were introduced in Section 1.4, and at that point the most basic arithmetic operations were extended from the field of real numbers to the field of complex numbers. Since then you have been exposed to many algebraic and trigonometric concepts, some of which can now be put to use to enlarge our understanding of complex numbers and expand our ability to work with them.

The first idea we can put to use is that of a two-dimensional plane. We will use the plane as a way to visualize complex numbers, exactly as we use the real line to visualize real numbers; this use was foreshadowed in the interlude on recursive graphics in Section 3.6. We need a two-dimensional framework in order to plot complex numbers since complex numbers consist of two independent components – the real part and the imaginary part. In fact, for the purposes of graphing, we identify a given complex number $z = a + bi$ with the ordered pair (a, b). In this context, we label the horizontal axis of a rectangular coordinate system the **real axis**, and we label the vertical axis the **imaginary axis**; the plane as a whole is called the **complex plane**. Figure 1 illustrates how several complex numbers are graphed.

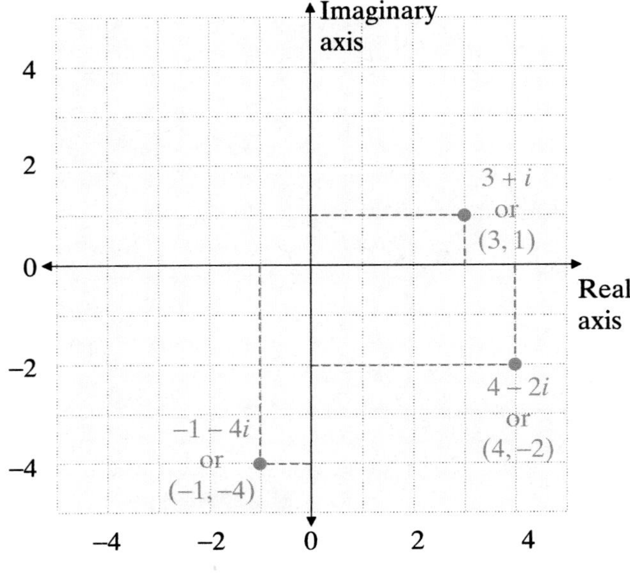

Figure 1: The Complex Plane

The association of a complex number $z = a + bi$ with the ordered pair (a, b) in the plane gives rise immediately to a way to measure the size of z.

Magnitude of a Complex Number

We say the **magnitude** of a complex number $z = a + bi$, also known as the **modulus** or **absolute value** of z, is the real number

$$|z| = \sqrt{a^2 + b^2},$$

and we use $|z|$ to denote this non-negative quantity.

The formula for $|z|$ is, of course, nothing more than the formula for the distance between (a, b) and the origin of the plane, so the magnitude of a complex number is a measure of how distant it is from the complex number $0 + 0i$ (which we usually just write as 0). Note also that if z is real, $|z|$ has the same meaning as the absolute value of a real number.

example 1

Determine the magnitudes of the following complex numbers.

a. $-2 + 5i$ **b.** $1 - 3i$

Solutions:

a. $|-2 + 5i| = \sqrt{(-2)^2 + (5)^2} = \sqrt{29}$

b. $|1 - 3i| = \sqrt{(1)^2 + (-3)^2} = \sqrt{10}$

example 2

Graph the regions of the complex plane defined by the following.

a. $\{z \mid |z| \le 1\}$

b. $\{z \mid |z| \le 1 \text{ and } \operatorname{Im}(z) \ge 0\}$

Solutions:

a. In words, $\{z \mid |z| \le 1\}$ is the set of all complex numbers with magnitude less than or equal to 1. The region appears at right.

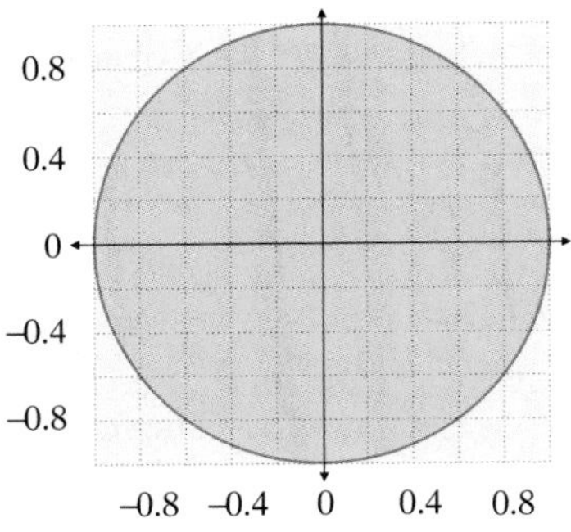

b. The region $\{z \mid |z| \le 1 \text{ and } \operatorname{Im}(z) \ge 0\}$ consists of only those complex numbers with magnitude less than or equal to 1 and non-negative imaginary parts. The graph of the region is a half-circle as shown.

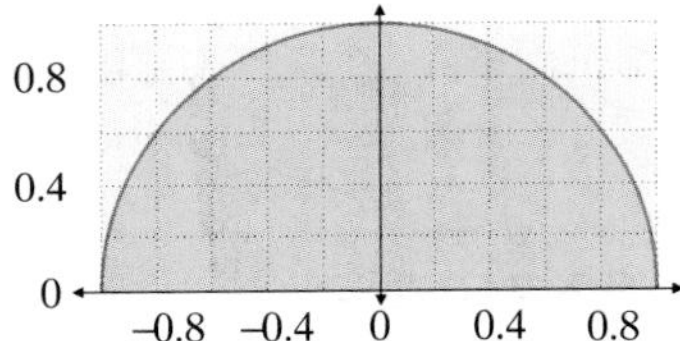

Topic 2: Complex Numbers in Trigonometric Form

We can also put our knowledge of trigonometry to use in describing complex numbers, with the result that many operations, such as multiplication, division, and the taking of roots, are made much easier.

To see how trigonometry applies, consider how the graph of a complex number gives rise to an angle θ in a very natural way, as shown below. Given $z = a + bi$, we know

$$\sin\theta = \frac{b}{|z|} \text{ and } \cos\theta = \frac{a}{|z|},$$

so $a = |z|\cos\theta$ and $b = |z|\sin\theta$. This leads to the following definition.

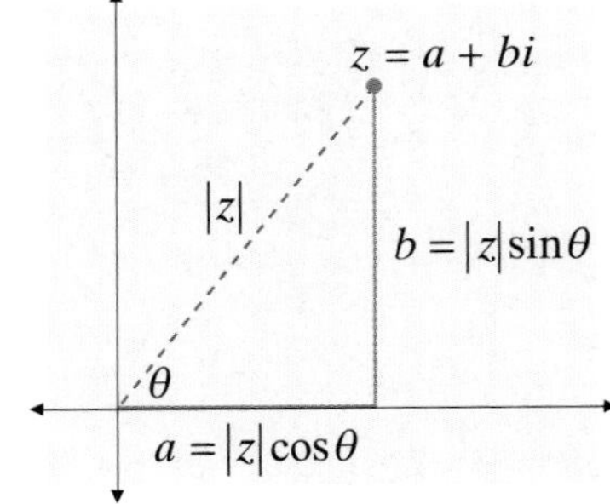

Trigonometric Form of $z = a + bi$

Given $z = a + bi$, the **trigonometric form** of z is given by

$$z = |z|\cos\theta + i|z|\sin\theta$$
$$= |z|(\cos\theta + i\sin\theta),$$

where $|z| = \sqrt{a^2 + b^2}$ and θ satisfies $\tan\theta = \frac{b}{a}$. The angle θ is called the **argument** of z.

In this way, we also relate the graphing of complex numbers to the polar coordinates discussed in Section 8.2. And as mentioned in that section, the angle θ corresponding to a specific point in the plane is not unique; however, any two arguments for a given complex number will differ by a multiple of 2π.

example 3

Write each of the following complex numbers in trigonometric form.

a. $-1+i$ **b.** $3-4i$

Solutions:

a. $|-1+i| = \sqrt{1+1} = \sqrt{2}$ and the argument of $-1+i$ is $\frac{3\pi}{4}$, so in trigonometric form $-1+i$ is written $\sqrt{2}\left(\cos\frac{3\pi}{4} + i\sin\frac{3\pi}{4}\right)$.

b. The magnitude of $3-4i$ is 5, but the argument is not so neatly expressed. In degrees, $\theta = \tan^{-1}\left(-\frac{4}{3}\right) \approx -53.1^\circ$. So $3-4i \approx 5\left(\cos\left(-53.1^\circ\right) + i\sin\left(-53.1^\circ\right)\right)$.

example 4

Write the complex number $2\left(\cos\frac{\pi}{6} + i\sin\frac{\pi}{6}\right)$ in standard form.

Solution:

$$2\left(\cos\frac{\pi}{6} + i\sin\frac{\pi}{6}\right) = 2\left(\frac{\sqrt{3}}{2} + i\frac{1}{2}\right) = \sqrt{3} + i.$$

Although it is beyond the scope of this text to prove, there is a very famous and useful identity that relates exponential functions and arguments of complex numbers. The identity is called **Euler's Formula**, and states: $e^{i\theta} = \cos\theta + i\sin\theta$. In the context of the trigonometric form of complex numbers, we can use this to write

$$z = |z|(\cos\theta + i\sin\theta) = |z|e^{i\theta}.$$

We will return to this elegant formula soon.

Topic 3: Multiplication and Division of Complex Numbers

The remainder of this section will show how, as promised, the trigonometric form of complex numbers simplifies certain computations. We start with multiplication and division of complex numbers.

Given two complex numbers $z_1 = |z_1|(\cos\theta_1 + i\sin\theta_1)$ and $z_2 = |z_2|(\cos\theta_2 + i\sin\theta_2)$, we can use our basic knowledge of complex multiplication to write the product as

$$\begin{aligned} z_1 z_2 &= |z_1|(\cos\theta_1 + i\sin\theta_1)|z_2|(\cos\theta_2 + i\sin\theta_2) \\ &= |z_1||z_2|\left[(\cos\theta_1\cos\theta_2 - \sin\theta_1\sin\theta_2) + i(\sin\theta_1\cos\theta_2 + \cos\theta_1\sin\theta_2)\right]. \end{aligned}$$

The combinations of sines and cosines that result should look familiar: they appear in the sum and difference trigonometric identities of Section 7.2. Applying these, we can simplify the above product to obtain

$$z_1 z_2 = |z_1||z_2|\left[\cos(\theta_1 + \theta_2) + i\sin(\theta_1 + \theta_2)\right].$$

A similar result can be obtained for the division of complex numbers, but the verification will be left as an exercise. The box below summarizes the observations.

Product and Quotient of Complex Numbers

Given $z_1 = |z_1|(\cos\theta_1 + i\sin\theta_1)$ and $z_2 = |z_2|(\cos\theta_2 + i\sin\theta_2)$,

$$z_1 z_2 = |z_1||z_2|\left[\cos(\theta_1 + \theta_2) + i\sin(\theta_1 + \theta_2)\right]$$

and

$$\frac{z_1}{z_2} = \frac{|z_1|}{|z_2|}\left[\cos(\theta_1 - \theta_2) + i\sin(\theta_1 - \theta_2)\right] \quad (z_2 \neq 0).$$

example 5

Use the product formula on the previous page to find the product of $z_1 = 3\left(\cos\frac{5\pi}{6} + i\sin\frac{5\pi}{6}\right)$ and $z_2 = 2\left(\cos\frac{2\pi}{3} + i\sin\frac{2\pi}{3}\right)$.

Solution:

$$\begin{aligned} z_1 z_2 &= (3)(2)\left(\cos\left(\frac{5\pi}{6} + \frac{2\pi}{3}\right) + i\sin\left(\frac{5\pi}{6} + \frac{2\pi}{3}\right)\right) \\ &= 6\left(\cos\left(\frac{9\pi}{6}\right) + i\sin\left(\frac{9\pi}{6}\right)\right) \\ &= 6\left(\cos\left(\frac{3\pi}{2}\right) + i\sin\left(\frac{3\pi}{2}\right)\right) \\ &= -6i. \end{aligned}$$

example 6

Use the quotient formula on the previous page to divide $z_1 = 6\left(\cos 117^\circ + i\sin 117^\circ\right)$ by $z_2 = 2\left(\cos 72^\circ + i\sin 72^\circ\right)$.

Solution:

$$\begin{aligned} \frac{z_1}{z_2} &= \frac{6\left(\cos 117^\circ + i\sin 117^\circ\right)}{2\left(\cos 72^\circ + i\sin 72^\circ\right)} \\ &= 3\left(\cos\left(117^\circ - 72^\circ\right) + i\sin\left(117^\circ - 72^\circ\right)\right) \\ &= 3\left(\cos 45^\circ + i\sin 45^\circ\right) \\ &= \frac{3}{\sqrt{2}} + i\frac{3}{\sqrt{2}}. \end{aligned}$$

To remember the product and quotient formulas just derived, it's useful to note that multiplication of complex numbers corresponds to addition of the arguments, while division of complex numbers corresponds to subtraction of the arguments (along with a respective product or quotient of the magnitudes). This correspondence should seem familiar: it's exactly the behavior of multiplication and division of exponential functions. This similarity is no accident, as we see if we rephrase multiplication and division using Euler's Formula as shown on the next page.

$$z_1 z_2 = \left(|z_1| e^{i\theta_1}\right)\left(|z_2| e^{i\theta_2}\right) = |z_1||z_2| e^{i\theta_1} e^{i\theta_2} = |z_1||z_2| e^{i(\theta_1+\theta_2)}$$

and

$$\frac{z_1}{z_2} = \frac{|z_1| e^{i\theta_1}}{|z_2| e^{i\theta_2}} = \left(\frac{|z_1|}{|z_2|}\right)\left(\frac{e^{i\theta_1}}{e^{i\theta_2}}\right) = \frac{|z_1|}{|z_2|} e^{i(\theta_1-\theta_2)}.$$

This behavior of multiplication and division of complex numbers also manifests itself graphically, as seen in the next example.

example 7

Plot the two complex numbers $2e^{i\left(\frac{\pi}{3}\right)}$ and $3e^{i\left(\frac{5\pi}{6}\right)}$ and then, on the same graph, plot their product.

Solution:

First note that, by the above formula, $\left(2e^{i\left(\frac{\pi}{3}\right)}\right)\left(3e^{i\left(\frac{5\pi}{6}\right)}\right) = 6e^{i\left(\frac{7\pi}{6}\right)}$. We can now use the magnitudes and arguments of the three complex numbers as their polar coordinates and construct the graph below. Note in particular how the argument of the product is the sum of the arguments of the two given complex numbers.

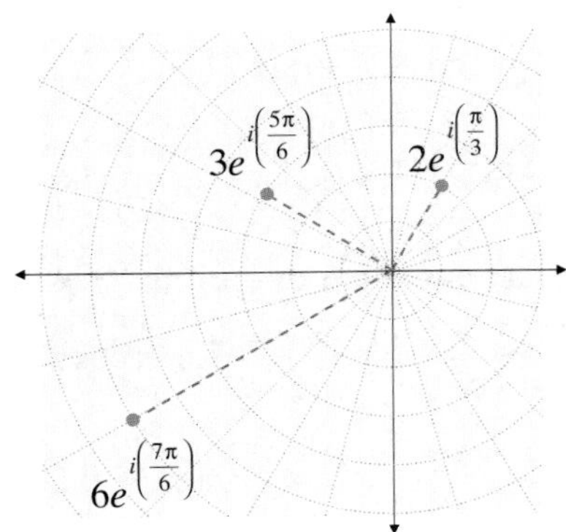

Topic 4: Powers of Complex Numbers

The product formula can be used to greatly speed up the process of calculating powers of complex numbers. For instance, if $z = |z|(\cos\theta + i\sin\theta)$, then

$$\begin{aligned} z^2 &= \left[|z|(\cos\theta + i\sin\theta)\right]^2 \\ &= |z|^2(\cos 2\theta + i\sin 2\theta). \end{aligned}$$

Similarly,

$$\begin{aligned} z^3 &= z^2 z \\ &= |z|^2 (\cos 2\theta + i\sin 2\theta)|z|(\cos\theta + i\sin\theta) \\ &= |z|^3 (\cos 3\theta + i\sin 3\theta). \end{aligned}$$

The pattern that is developing is even more apparent if we use Euler's Formula:

$$\begin{aligned} z^n &= \left(|z|e^{i\theta}\right)^n \\ &= |z|^n e^{in\theta}. \end{aligned}$$

These observations are named after the early 18th-century mathematician Abraham DeMoivre.

Powers of Complex Numbers (DeMoivre's Theorem)

Given $z = |z|(\cos\theta + i\sin\theta)$ and a positive integer n,

$$z^n = |z|^n (\cos n\theta + i\sin n\theta).$$

Using Euler's Formula, this appears in the form

$$z^n = |z|^n e^{in\theta}.$$

example 8

Use DeMoivre's Theorem to calculate $\left(\sqrt{3} - i\right)^7$.

Solution:

In one form,

$$\begin{aligned} \left(\sqrt{3} - i\right)^7 &= 2^7 e^{7i\left(-\frac{\pi}{6}\right)} \\ &= 128e^{i\left(-\frac{7\pi}{6}\right)} \\ &= 128e^{i\left(\frac{5\pi}{6}\right)}. \end{aligned}$$

In the alternative form, $\left(\sqrt{3} - i\right)^7 = 128\left(\cos\frac{5\pi}{6} + i\sin\frac{5\pi}{6}\right)$.

Topic 5: Roots of Complex Numbers

DeMoivre's Theorem can be used to determine roots of complex numbers. The first step is to realize that if $z = |z|(\cos\theta + i\sin\theta)$ is a non-zero complex number and if n is a positive integer, z has n distinct n^{th} roots. This follows from an application of the Fundamental Theorem of Algebra: the nth-degree polynomial equation $w^n = z$ has n solutions (here, w is a complex variable).

One n^{th} root of z is easily determined. If we let $w_0 = |z|^{1/n}\left(\cos\frac{\theta}{n} + i\sin\frac{\theta}{n}\right)$, then by DeMoivre's Theorem

$$\begin{aligned} w_0{}^n &= \left(|z|^{\frac{1}{n}}\right)^n \left(\cos\left(n\frac{\theta}{n}\right) + i\sin\left(n\frac{\theta}{n}\right)\right) \\ &= |z|(\cos\theta + i\sin\theta) \\ &= z, \end{aligned}$$

so w_0 is indeed an nth root. But as we know, replacing θ with $\theta + 2k\pi$ results in an equivalent way of writing z. This observation leads to the following:

Roots of Complex Numbers

Given the non-zero complex number $z = |z|(\cos\theta + i\sin\theta)$ and the positive integer n,

$$w_k = |z|^{\frac{1}{n}}\left[\cos\left(\frac{\theta + 2k\pi}{n}\right) + i\sin\left(\frac{\theta + 2k\pi}{n}\right)\right]$$

defines n distinct n^{th} roots of z for $k = 0, 1, \ldots, n-1$. Alternatively,

$$w_k = |z|^{\frac{1}{n}} e^{i\left(\frac{\theta + 2k\pi}{n}\right)}.$$

example 9

Find all the fifth roots of 1, and graph their locations in the complex plane.

Solution:

The easiest way to describe the five fifth roots uses the Euler formulation. Since 1 can be written as e^{0i} we know $\theta = 0$. Thus we have:

$$\left\{(1)^{\frac{1}{5}}e^{i\left(\frac{0+2(0)\pi}{5}\right)}, (1)^{\frac{1}{5}}e^{i\left(\frac{0+2(1)\pi}{5}\right)}, (1)^{\frac{1}{5}}e^{i\left(\frac{0+2(2)\pi}{5}\right)}, (1)^{\frac{1}{5}}e^{i\left(\frac{0+2(3)\pi}{5}\right)}, (1)^{\frac{1}{5}}e^{i\left(\frac{0+2(4)\pi}{5}\right)}\right\}$$

which simplifies to

$$\left\{1, e^{i\left(\frac{2\pi}{5}\right)}, e^{i\left(\frac{4\pi}{5}\right)}, e^{i\left(\frac{6\pi}{5}\right)}, e^{i\left(\frac{8\pi}{5}\right)}\right\}.$$

The graph of the five roots appears below. Note that they are evenly spaced around a circle of radius 1.

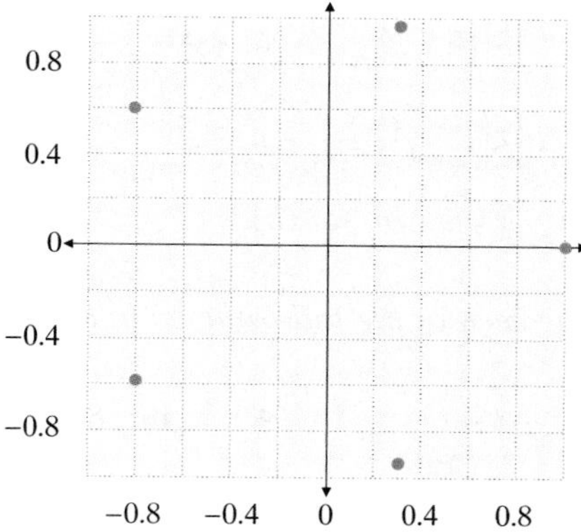

example 10

Find all the fourth roots of $-1-i\sqrt{3}$.

Solution:

Again, the roots are most easily written using Euler's formula, and we begin by writing $-1-i\sqrt{3}$ as $2e^{i\left(\frac{4\pi}{3}\right)}$. Then the four fourth roots are:

$$\left\{2^{\frac{1}{4}}e^{i\left(\frac{\pi}{3}\right)}, 2^{\frac{1}{4}}e^{i\left(\frac{5\pi}{6}\right)}, 2^{\frac{1}{4}}e^{i\left(\frac{4\pi}{3}\right)}, 2^{\frac{1}{4}}e^{i\left(\frac{11\pi}{6}\right)}\right\}.$$

exercises

Graph and determine the magnitudes of the following complex numbers. See Example 1.

1. $3+5i$ **2.** $-1+3i$ **3.** $2-4i$

4. $-6-i$ **5.** $4+4i$ **6.** $5+2i$

Sketch z_1, z_2, z_1+z_2, and z_1z_2 on the same complex plane.

7. $z_1=7+2i, z_2=-2+3i$ **8.** $z_1=-1-2i,\ z_2=4+4i$

9. $z_1=3+i, z_2=5-i$ **10.** $z_1=5-2i,\ z_2=-1+5i$

Graph the regions of the complex plane defined by the following. See Example 2.

11. $\{z \mid |z|<3\}$ **12.** $\{z \mid 1\le|z|\le4\}$

13. $\{z \mid |z|\ge1\}$ **14.** $\{z=a+bi \mid a>1,\ b>2\}$

15. $\{z=a+bi \mid a\ge b\}$ **16.** $\{z \mid |z|=4\}$

Write each of the following complex numbers in trigonometric form. See Example 3.

17. $-3-i$ **18.** $5-3i$ **19.** $1+2i$

20. $3+\sqrt{3}i$ **21.** $4+2i$ **22.** $5-\sqrt{2}i$

23. $\sqrt{2}-\sqrt{2}i$ **24.** $2\sqrt{3}-2i$ **25.** $3+4\mathrm{i}$

26. $1+i$ **27.** $4-4\sqrt{3}i$ **28.** $2\sqrt{2}-i$

Write each of the complex numbers in standard form. See Example 4.

29. $3\left(\cos\frac{5\pi}{6}+i\sin\frac{5\pi}{6}\right)$ **30.** $\cos\frac{\pi}{3}+i\sin\frac{\pi}{3}$

31. $2\left(\cos\frac{4\pi}{3}+i\sin\frac{4\pi}{3}\right)$ **32.** $6(\cos\pi+i\sin\pi)$

33. $5\left(\cos\frac{3\pi}{4}+i\sin\frac{3\pi}{4}\right)$ **34.** $\cos\frac{5\pi}{4}+i\sin\frac{5\pi}{4}$

35. $\frac{3}{2}\left(\cos150^\circ+i\sin150^\circ\right)$ **36.** $4\left(\cos210^\circ+i\sin210^\circ\right)$

37. $5\left[\cos\left(78^\circ\,20'\right)+i\sin\left(78^\circ\,20'\right)\right]$ **38.** $3\left[\cos\left(121^\circ\,40'\right)+i\sin\left(121^\circ\,40'\right)\right]$

Perform the following operations and show the answer in both trigonometric form and standard form. See Examples 5 and 6.

39. $\left[4\left(\cos 60° + i\sin 60°\right)\right]\left[4\left(\cos 330° + i\sin 330°\right)\right]$

40. $\left[3\left(\cos 180° + i\sin 180°\right)\right]\left[4\left(\cos 30° + i\sin 30°\right)\right]$

41. $\left[\sqrt{2}\left(\cos\frac{5\pi}{4} + i\sin\frac{5\pi}{4}\right)\right]\left[3\sqrt{3}\left(\cos\frac{\pi}{6} + i\sin\frac{\pi}{6}\right)\right]$

42. $\left[10\left(\cos\frac{5\pi}{6} + i\sin\frac{5\pi}{6}\right)\right]\left[6\left(\cos\frac{2\pi}{3} + i\sin\frac{2\pi}{3}\right)\right]$

43. $(-1+3i)\left(\sqrt{3}+i\right)$

44. $2i(4+5i)$

45. $\dfrac{6\left(\cos 225° + i\sin 225°\right)}{3\left(\cos 45° + i\sin 45°\right)}$

46. $\dfrac{8\left(\cos 135° + i\sin 135°\right)}{4\left(\cos 30° + i\sin 30°\right)}$

47. $\dfrac{10\left(\cos\frac{5\pi}{3} + i\sin\frac{5\pi}{3}\right)}{3\left(\cos\frac{7\pi}{6} + i\sin\frac{7\pi}{6}\right)}$

48. $\dfrac{12\left(\cos\frac{10\pi}{3} + i\sin\frac{10\pi}{3}\right)}{6(\cos 2\pi + i\sin 2\pi)}$

49. $\dfrac{-i}{1+i}$

50. $\dfrac{-2-2i}{4+3i}$

51. $\dfrac{2e^{\frac{2\pi}{3}i}}{e^{\frac{\pi}{4}i}}$

52. $\left(e^{210° i}\right)\left(e^{90° i}\right)$

Plot both complex numbers and, on the same graph, plot their product. See Example 7.

53. $\left[4\left(\cos\frac{5\pi}{6} + i\sin\frac{5\pi}{6}\right)\right]\left[2(\cos\pi + i\sin\pi)\right]$

54. $\left[5\left(\cos\frac{3\pi}{2} + i\sin\frac{3\pi}{2}\right)\right]\left[3\left(\cos\frac{5\pi}{6} + i\sin\frac{5\pi}{6}\right)\right]$

55. $(2-5i)\left(\sqrt{2}+2i\right)$

56. $(-4-2i)\left(-\sqrt{3}+4i\right)$

57. $\left(2e^{\frac{\pi}{3}i}\right)\left(3e^{\frac{5\pi}{4}i}\right)$

58. $\left(5e^{\frac{5\pi}{3}i}\right)\left(e^{\pi i}\right)$

Use DeMoivre's Theorem to calculate the following. See Example 8.

59. $\left(1-\sqrt{3}i\right)^5$ **60.** $\left(\frac{\sqrt{2}}{2}+\frac{\sqrt{2}}{2}i\right)^{22}$ **61.** $(5+3i)^{17}$

62. $\left(-\sqrt{3}+i\right)^{13}$ **63.** $\left(\cos\frac{\pi}{4}+i\sin\frac{\pi}{4}\right)^8$ **64.** $\left[2\left(\cos 135^\circ+i\sin 135^\circ\right)\right]^4$

Find the indicated roots of the following and graphically represent each set in the complex plane. See Examples 9 and 10.

65. The fourth roots of –1.

66. The cube roots of $64i$.

67. The square roots of $2\sqrt{3}+2i$.

68. The fourth roots of $-1-i$.

69. The fourth roots of 256.

70. The fourth roots of $16\left(\cos\frac{4\pi}{3}+i\sin\frac{4\pi}{3}\right)$.

71. The square roots of $4\left(\cos 120^\circ+i\sin 120^\circ\right)$.

Solve the following equations. See Examples 9 and 10.

72. $z^3-i=0$ **73.** $z^2-4\sqrt{3}-4i=0$ **74.** $z^4+81i=0$

75. $z^5+32=0$ **76.** $z^3+4\sqrt{2}-i=0$ **77.** $z^2+25i=0$

8.5 Vectors in the Cartesian Plane

TOPICS

1. Vector terminology
2. Basic vector operations
3. Component form of a vector
4. Vector applications

Topic 1: Vector Terminology

Many quantities are defined by their size. For example, length, area, mass, price, and temperature are fully determined by a single number; such numbers, representing only magnitude, are called **scalars**. Other quantities, however, cannot be adequately described by a single number. Force and velocity, for instance, possess both a *magnitude* and a *direction*, and a complete description of these quantities must somehow include both. Such *directed magnitudes* are called **vectors**.

We will introduce vectors in the setting of the two-dimensional plane, but the study of vectors in general constitutes an enormous area of mathematics, and vectors are easily extended into spaces of any dimension. In the plane, vectors are often represented as directed line segments (informally known as "arrows"). Such a directed line segment begins at a point P called the **initial point** and ends at a point Q called the **terminal point**, and the notation $\overrightarrow{PQ}$ is used to refer to the directed line segment. A subtle but very important point, though, is that a vector is characterized *entirely* by its direction and its magnitude, not by its initial and terminal points. That is, for a specific pair of points P and Q, $\overrightarrow{PQ}$ is only one way of depicting the vector it represents. We will use bold lowercase letters to denote vectors in general, and an expression such as $\mathbf{u} = \overrightarrow{PQ}$ means that $\mathbf{u}$ is a vector whose length is the same as the length of the line segment $\overrightarrow{PQ}$ and whose direction is defined by the initial point P and the terminal point Q. To make this important point clear, Figure 1 illustrates five different ways of depicting the one vector $\mathbf{u}$.

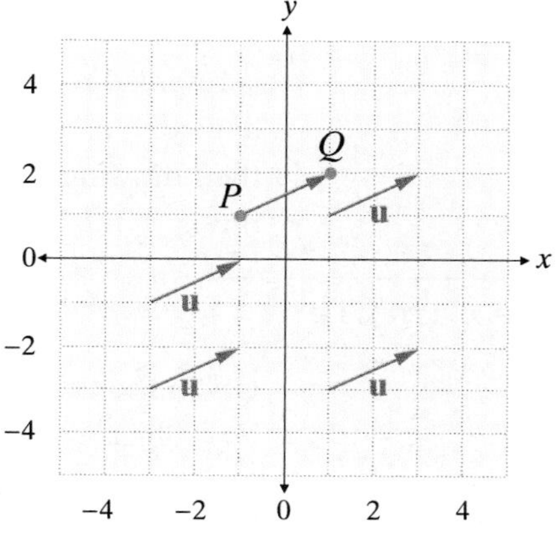

Figure 1: Five Depictions of u

Since vectors are characterized entirely by length and direction, we say that two vectors **u** and **v** with the same length and the same direction are **equal** (no matter where they might actually be depicted, if they are graphed) and we write **u** = **v**. The **length** of a vector may also be referred to as its **magnitude** or **norm**, and is denoted $\|\mathbf{u}\|$. In Figure 1, the length of **u** is the same as the length of the line segment from $(-1, 1)$ to $(1, 2)$, and the familiar distance formula tells us

$$\|\mathbf{u}\| = \|\overline{PQ}\| = \sqrt{(1-(-1))^2 + (2-1)^2} = \sqrt{5}.$$

Magnitude of a Vector

The length of a vector is called its **magnitude** and is denoted $\|\mathbf{u}\|$.

$$\|\mathbf{u}\| = \sqrt{a^2 + b^2}$$

where a is the horizontal displacement and b is the vertical displacement of the vector.

Topic 2: Basic Vector Operations

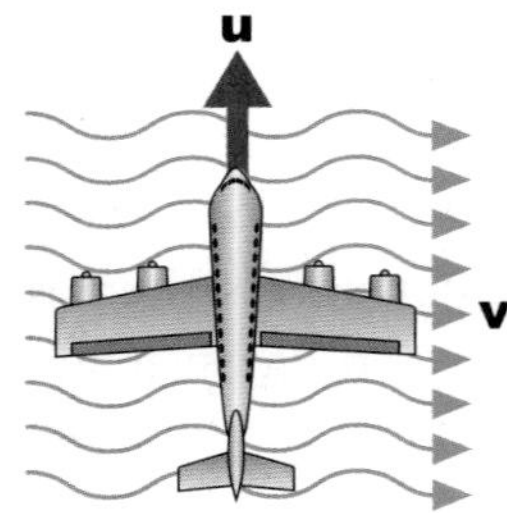

The applications of vectors point the way toward the basic ways that vectors can be combined. Consider the following scenario: wind with a velocity of **v** is affecting the flight of a plane that would be flying, if the wind were absent, with a velocity of **u**. What is the actual velocity of the plane given the presence of the wind? Both the plane and the wind possess, individually, a speed and a direction; the combination of the plane's velocity and the wind's velocity results in a third velocity that represents the actual progress of the plane.

The use of vectors facilitates such calculations. **Vector addition** is the operation called for in the above scenario; the plane's actual velocity is given by **u** + **v**, and this expression is referred to as the **sum** of the two vectors. Graphically, **u** + **v** can be represented by depicting **u**, placing **v** so that its initial point coincides with the terminal point of **u**, and then drawing a directed line segment from the initial point of **u** to the terminal point of **v**.

example 1

Given the two vectors **u** and **v** depicted at right, construct a graphical representation of **u** + **v**.

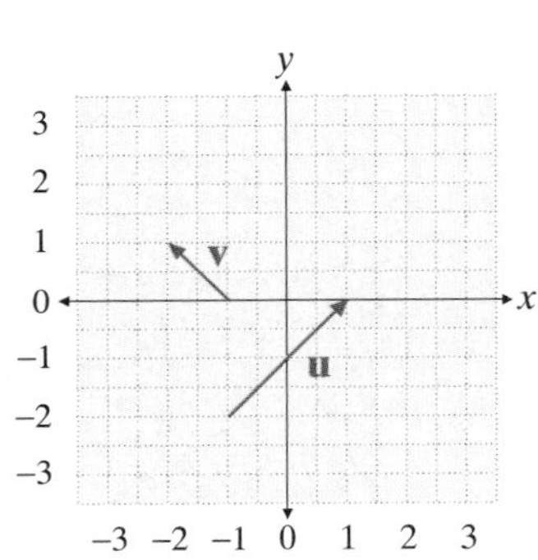

Solution:

The simplest way to proceed is to leave **u** alone and to redraw **v** so that its initial point is the terminal point of **u**. If we then construct a directed line segment from **u**'s initial point to **v**'s terminal point, the result will depict **u** + **v**. This has been done in the graph to the right, with **u** + **v** shown in blue.

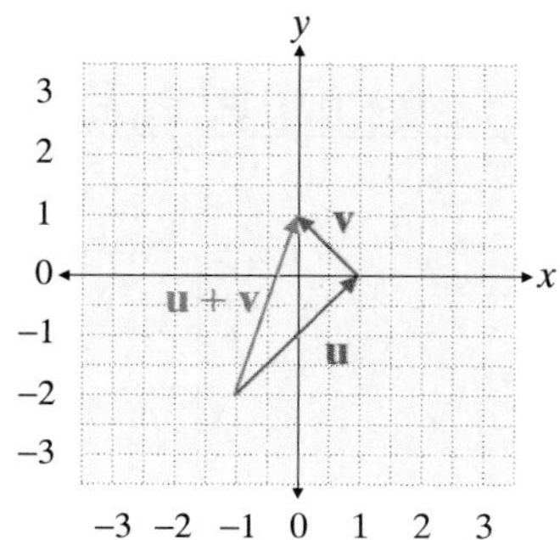

Now consider a second scenario. A person is attempting to paddle a kayak upstream, and her velocity would be **u** if the water wasn't flowing. But suppose the water *is* flowing, with a magnitude equal to the kayaker's but the exact opposite direction. If we let **v** denote the water's velocity, then **u** + **v** = **0**, where **0** is the **zero vector**. The zero vector is the unique vector which has a magnitude of 0, and in the context of vectors its function is equivalent to the number 0 on the real line (or the number 0 on the complex plane). Further, it makes sense to write **v** = –**u** and to say that –**u** is the *additive inverse* of the vector **u**.

Consider one last scenario. Suppose you are walking along a hallway with a velocity **u**, and realize you are late for a meeting. You break into a slow run, moving three times faster than you were before. Your direction is unchanged, but your magnitude is three times as large; a convenient way to denote your new velocity is 3**u**, i.e. the scalar 3 times the vector **u**. Negative numbers can multiply vectors too: (–1)**u** means the same as –**u** and represents a vector with the same magnitude as **u** but the opposite direction.

In all, the set of vectors we are describing possess properties that are very familiar to us from our study of real and complex numbers. The complete set of such properties appears below.

Properties of Vector Addition and Scalar Multiplication

Assume **u**, **v**, and **w** represent vectors, while a and b represent scalars. Then the following hold:

Vector Addition Properties	Scalar Multiplication Properties
$\mathbf{u}+\mathbf{v}=\mathbf{v}+\mathbf{u}$	$a(\mathbf{u}+\mathbf{v})=a\mathbf{u}+a\mathbf{v}$
$\mathbf{u}+(\mathbf{v}+\mathbf{w})=(\mathbf{u}+\mathbf{v})+\mathbf{w}$	$(a+b)\mathbf{u}=a\mathbf{u}+b\mathbf{u}$
$\mathbf{u}+\mathbf{0}=\mathbf{u}$	$(ab)\mathbf{u}=a(b\mathbf{u})=b(a\mathbf{u})$
$\mathbf{u}+(-\mathbf{u})=\mathbf{0}$	$1\mathbf{u}=\mathbf{u}$, $0\mathbf{u}=\mathbf{0}$, and $a\mathbf{0}=\mathbf{0}$
	$\lVert a\mathbf{u}\rVert=\lvert a\rvert\lVert\mathbf{u}\rVert$

example 2

Given the two vectors **u** and **v** depicted at right, construct graphical representations of $-2\mathbf{u}$, $3\mathbf{u} + \mathbf{v}$, and $\mathbf{u} - \mathbf{v}$.

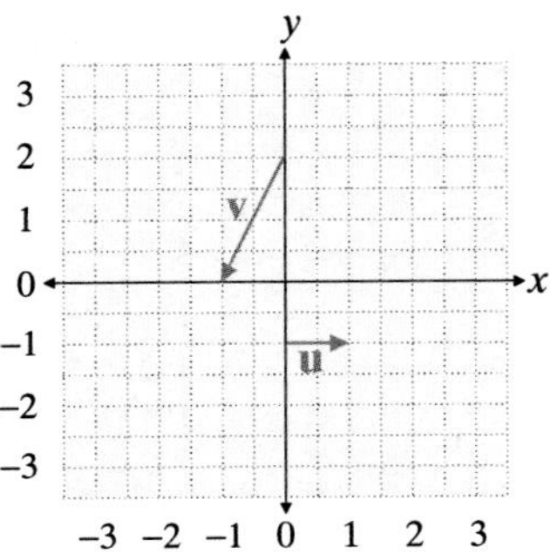

Solution:

Remember that the vectors can be placed anywhere on the plane, as long as they have the correct magnitude and direction. With a bit of planning, all three answers can be graphed in the same image, as shown to the right. Note that $\mathbf{u} - \mathbf{v}$ means the same thing as $\mathbf{u} + (-\mathbf{v})$.

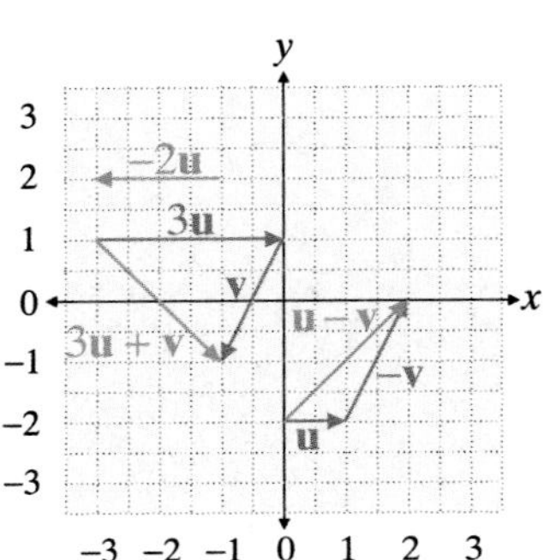

Topic 3: Component Form of a Vector

Our graphical work with vectors to this point has illustrated the important differences between scalars and vectors, and it has also begun to show how vectors can be used to solve certain problems. But in order to be really useful, we need a way to treat vectors in a more analytical fashion; that is, we need a form that can be used in equations.

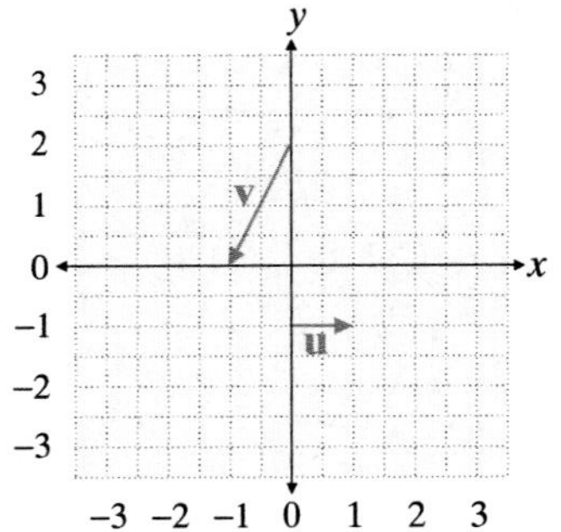

The **component form** of a vector is one such form. Consider again the two vectors **u** and **v** from the last example. One way of completely characterizing **u** is to note that the horizontal displacement between the initial and terminal points is 1, while the vertical displacement is 0. The notation for this observation is $\mathbf{u} = \langle 1, 0 \rangle$; in a similar fashion, we would write $\mathbf{v} = \langle -1, -2 \rangle$. The similarity between this notation and ordered pair notation is no accident, as the order in which the components appear is critical in both cases. Note, however, the very important distinction between vectors and points in the plane: $(-1, -2)$ refers to the one single point that is 1 unit to the left and 2 units down from the origin, while $\langle -1, -2 \rangle$ refers to the vector whose horizontal displacement is -1 and whose vertical displacement is -2. A depiction of $\mathbf{v} = \langle -1, -2 \rangle$ can be placed anywhere in the plane, as long as the displacement between the initial point and the terminal point is correct.

Determining the components of a vector from its initial point P and its terminal point Q is simply a matter of calculating the displacements. If $P=(x_1,y_1)$ and $Q=(x_2,y_2)$, then

$$\overrightarrow{PQ}=\langle x_2-x_1,y_2-y_1\rangle.$$

The component form of a vector provides a simple and precise way of performing vector operations, as shown below.

Vector Operations Using Components

Given two vectors $\mathbf{u}=\langle u_1,u_2\rangle$ and $\mathbf{v}=\langle v_1,v_2\rangle$ and a scalar a,

$$\mathbf{u}+\mathbf{v}=\langle u_1+v_1,u_2+v_2\rangle$$

$$a\mathbf{u}=\langle au_1,au_2\rangle$$

and

$$\|\mathbf{u}\|=\sqrt{u_1^{\,2}+u_2^{\,2}}$$

example 3

Given the two vectors **u** and **v** depicted at right, determine the component forms of –2**u**, 3**u** + **v**, and **u** – **v**, and then find the magnitudes of **u** and **v**.

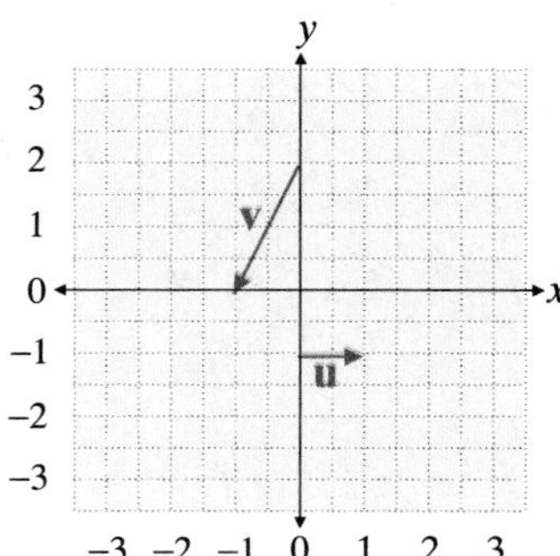

Solution:

In the discussion above, we determined that **u** = ⟨1, 0⟩ and **v** = ⟨–1, –2⟩; note how these components can be determined by identifying (0, –1) as the initial point of **u**, (1, –1) as the terminal point of **u**, and (0, 2) and (–1, 0) as the initial and terminal points of **v**.

The remainder of the work is straightforward:

$$-2\mathbf{u}=\langle(-2)(1),(-2)(0)\rangle=\langle -2,0\rangle,$$

$$3\mathbf{u}+\mathbf{v}=\langle 3,0\rangle+\langle -1,-2\rangle=\langle 2,-2\rangle,$$

$$\mathbf{u}-\mathbf{v}=\langle 1,0\rangle-\langle -1,-2\rangle=\langle 2,2\rangle,$$

$$\|\mathbf{u}\|=\sqrt{1^2+0^2}=1,$$

$$\|\mathbf{v}\|=\sqrt{(-1)^2+(-2)^2}=\sqrt{5}.$$

The vector **u** in the preceding two examples is just one example of a **unit vector**, which is any vector with a length of 1. There are, of course, an infinite number of unit vectors (think of any arrow of length 1 pointing in an arbitrary direction), but two of them are given special names:

$$\mathbf{i} = \langle 1, 0\rangle \text{ and } \mathbf{j} = \langle 0, 1\rangle.$$

In other words, **i** and **j** are the two unit vectors parallel to the coordinate axes. These two unit vectors are useful in some contexts because *any* vector in the plane can be written in terms of **i** and **j**, as follows:

$$\langle a,b\rangle = a\langle 1,0\rangle + b\langle 0,1\rangle = a\mathbf{i} + b\mathbf{j}$$

The last expression in the equation above is an example of a **linear combination**; any sum of scalar multiples of vectors constitutes a linear combination of the vectors, and the properties of linear combinations turn out to be very important in the study of vectors.

example 4

Let $\mathbf{u} = \langle -5, 3\rangle$. Find (**a**) a unit vector pointing in the same direction as **u**, and (**b**) the linear combination of **i** and **j** that is equivalent to **u**.

Solutions:

a. The key in constructing a unit vector with the same direction as **u** is to multiply **u** by an appropriate scalar a such that $\|a\mathbf{u}\| = 1$. We can solve this for a as follows:

$$\|a\mathbf{u}\| = 1$$

$$|a|\|\mathbf{u}\| = 1$$

$$a = \frac{1}{\|\mathbf{u}\|}.$$

Note that the absolute value sign around a has been dropped in the third equation, since we know in advance that we want to multiply **u** by a positive number. In our case, we have $\|\mathbf{u}\| = \sqrt{25+9} = \sqrt{34}$, so the desired unit vector is

$$\left(\frac{1}{\sqrt{34}}\right)\langle -5,3\rangle = \left\langle -\frac{5}{\sqrt{34}}, \frac{3}{\sqrt{34}}\right\rangle.$$

b. The second task is fairly straightforward: $\mathbf{u} = \langle -5,3\rangle = -5\mathbf{i} + 3\mathbf{j}$.

In some cases, information about a vector's magnitude and its angle with respect to the positive x-axis may be given to us. If we need to work instead with the components of the vector, our knowledge of trigonometry comes to the rescue. The right triangle in Figure 2 is familiar from our study of sine and cosine.

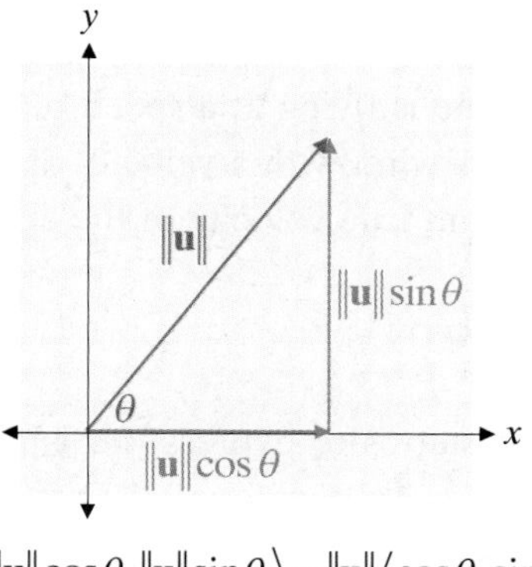

$$\mathbf{u} = \left\langle \|\mathbf{u}\|\cos\theta, \|\mathbf{u}\|\sin\theta \right\rangle = \|\mathbf{u}\|\left\langle \cos\theta, \sin\theta \right\rangle$$

Figure 2: Components Derived from Magnitude and Angle

example 5

A baseball is hit, and leaves the bat at a speed of 100 mph and at an angle of 20° from the horizontal. Express this velocity in vector form.

Solution:

The speed is the vector's magnitude, and $\theta = 20^\circ$. So the ball's velocity as it leaves the bat is the vector

$$100\left\langle \cos 20^\circ, \sin 20^\circ \right\rangle \approx \langle 93.97, 34.20 \rangle.$$

Topic 4: Vector Applications

Let us return to an application that served as the inspiration for vector addition.

example 6

An airplane is flying at a speed of 200 miles per hour at a bearing of N 33° E when it encounters wind with a velocity of 35 miles per hour at a bearing of N 47° W. What is the resultant true velocity of the airplane?

Solution:

The first step is to express the plane's velocity and the wind's velocity as vectors. A bearing of N 33° E means 33° East of North, but it is more convenient to think of this as 57° North of East (in keeping with our convention of measuring angles relative to the positive x-axis). Similarly, the wind's bearing of N 47° W equates to a bearing of 137° as measured from due East. If we let **p** denote the plane's velocity and **w** the wind's velocity, we have:

$$\mathbf{p} = 200\left\langle \cos 57^\circ, \sin 57^\circ \right\rangle \approx \left\langle 108.9, 167.7 \right\rangle$$

and

$$\mathbf{w} = 35\left\langle \cos 137^\circ, \sin 137^\circ \right\rangle \approx \left\langle -25.6, 23.9 \right\rangle.$$

The plane's true velocity is now simply $\mathbf{p} + \mathbf{w} = \langle 83.3, 191.6 \rangle$. It may also be useful to determine that the speed of the plane is now $\|\mathbf{p} + \mathbf{w}\| = \sqrt{(83.3)^2 + (191.6)^2} \approx 208.9$ miles per hour, and that its bearing is 66.5° North of East. This last angle is derived from the fact that $\tan\theta = \dfrac{191.6}{83.3}$, so $\theta = \tan^{-1}\dfrac{191.6}{83.3} \approx 66.5^\circ$. The diagram below illustrates the three vectors in this problem.

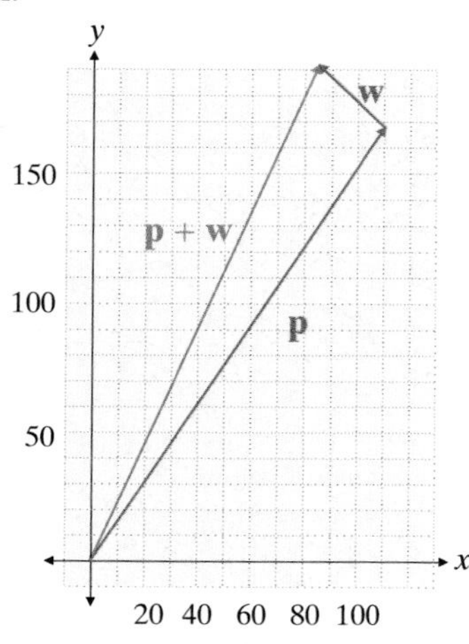

example 7

A cat is slowly pushing a 5 pound plant across a table, with the intention of knocking it off the edge (determining why cats feel the need to do so is beyond the scope of this text). The cat is pushing with a force of 1 pound. What is the total force being applied to the plant?

Solution:

Weight is itself a force – it is the force due to gravity that the Earth exerts on an object. Forces exerted on an object are added as vectors, and the result is the total applied force. If we let $\mathbf{F}_1$ denote the weight of the plant and $\mathbf{F}_2$ the force exerted by the cat, the force on the plant is

$$\mathbf{F}_1 + \mathbf{F}_2 = \langle 0,-5 \rangle + \langle 1,0 \rangle = \langle 1,-5 \rangle.$$

The magnitude of this total force is $\sqrt{1+25} \approx 5.1$ pounds, and the diagram below illustrates the situation (note the difference in scale between the axes).

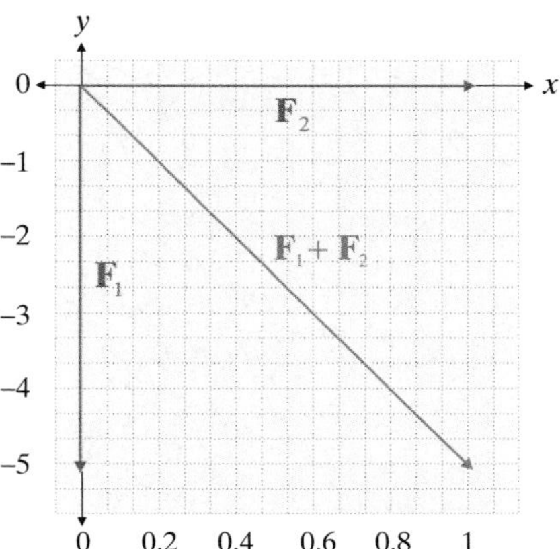

exercises

Use the figure to sketch a graph for the specified vector. See Examples 1 and 2.

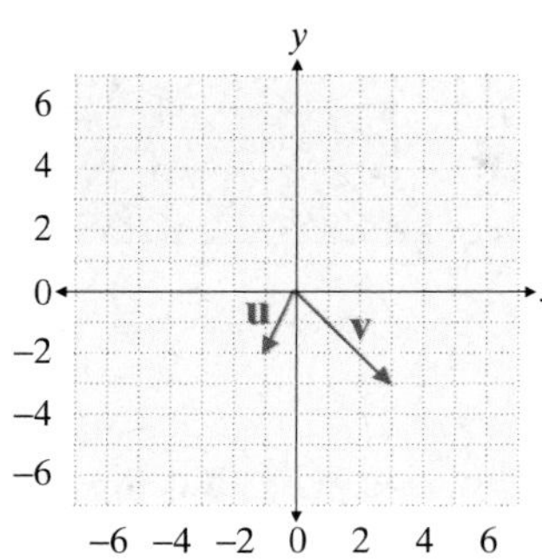

1. $-\mathbf{u}$

2. $2\mathbf{u} + \mathbf{v}$

3. $3\mathbf{v}$

4. $-\frac{1}{2}\mathbf{u} - \mathbf{v}$

5. $2\mathbf{u} - 2\mathbf{v}$

6. $\mathbf{u} + 3\mathbf{v}$

Find the component form and the magnitude of vector **v** *for each of the following. See Example 3.*

7.

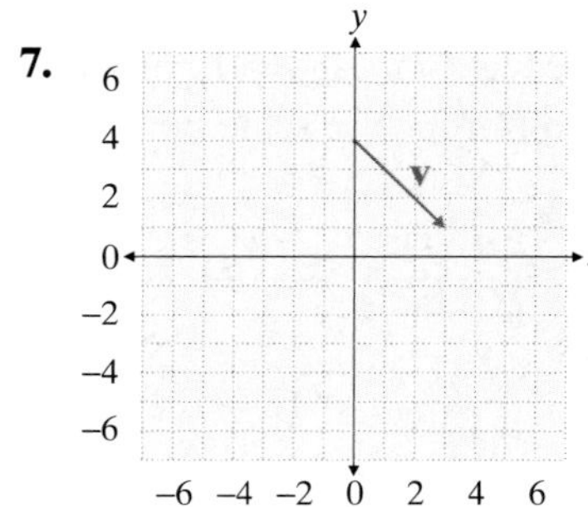

8.

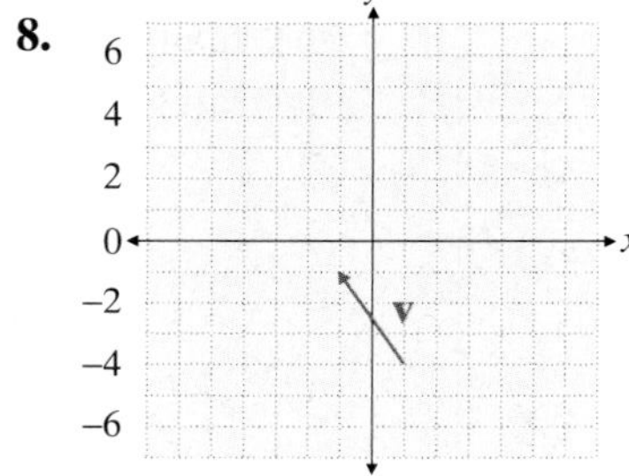

9.

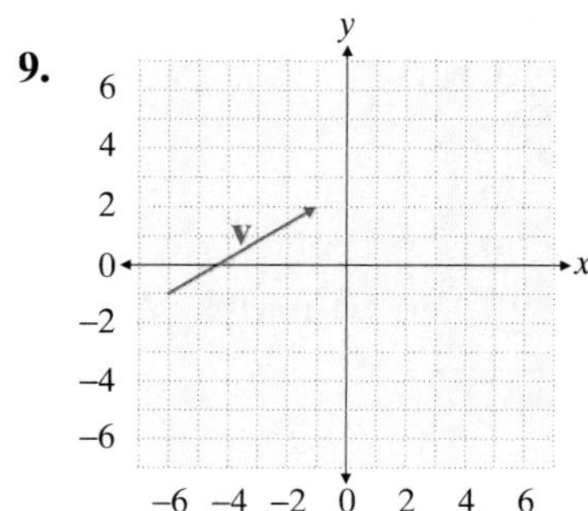

10.

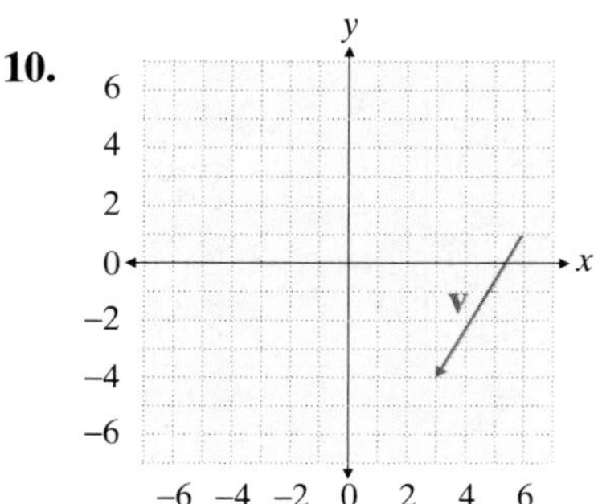

Find the component form and the magnitude of the vectors defined by the points given below. Assume the first point given is the initial point and the second point given is the terminal point. See Example 3.

11. $(-2, 4), (3, 3)$

12. $(1, 6), (2, 3)$

13. $(5, -2), (-2, 5)$

14. $(4, 0), (-1, 7)$

15. $(3, 4), (-1, -2)$

16. $(1, -6), (0, 0)$

For each of the following, calculate (a) $2\mathbf{u} + \mathbf{v}$*, (b)* $-\mathbf{u} + 3\mathbf{v}$*, and (c)* $-2\mathbf{v}$*. See Example 3.*

17. $\mathbf{u} = \langle -2, 4 \rangle, \mathbf{v} = \langle 2, 0 \rangle$

18. $\mathbf{u} = \langle 4, 1 \rangle, \mathbf{v} = \langle 2, 5 \rangle$

19. $\mathbf{u} = \langle 2, 0 \rangle, \mathbf{v} = \langle -3, 4 \rangle$

20. $\mathbf{u} = \langle 1, 3 \rangle, \mathbf{v} = \langle 4, 4 \rangle$

21. $\mathbf{u} = \langle -1, -4 \rangle, \mathbf{v} = \langle -3, -2 \rangle$

22. $\mathbf{u} = \langle 0, -5 \rangle, \mathbf{v} = \langle -1, 2 \rangle$

For each of the following graphs, determine the component forms of $-\mathbf{u}$*,* $2\mathbf{u} - \mathbf{v}$*, and* $\mathbf{u} + \mathbf{v}$ *and find the magnitudes of* $\mathbf{u}$ *and* $\mathbf{v}$*. See Example 3.*

23.

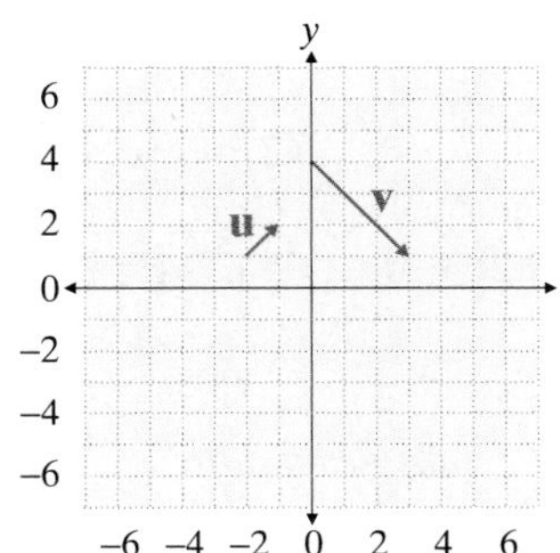

24.

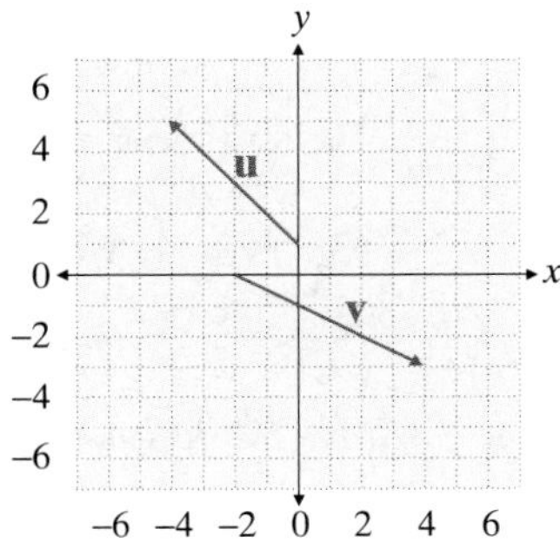

25.

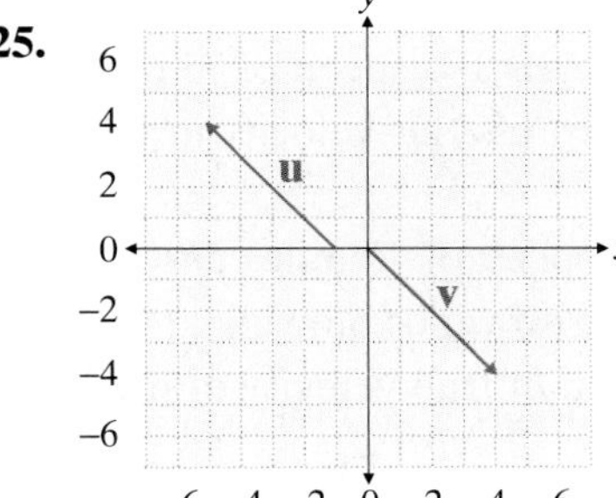

26.

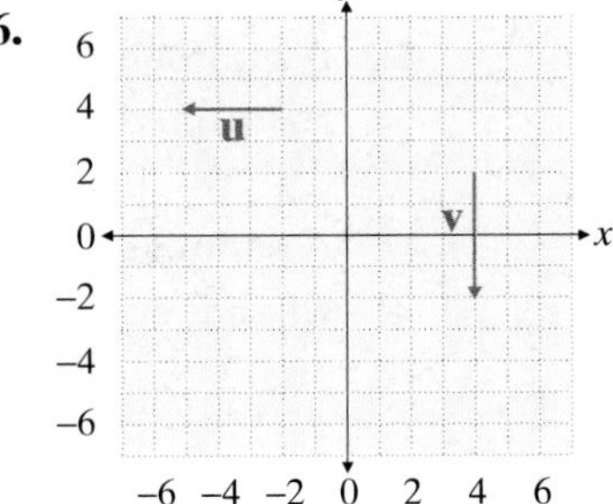

Given the vector $\mathbf{u}$*, find (a) a unit vector pointing in the same direction as* $\mathbf{u}$*, and (b) the linear combination of* $\mathbf{i}$ *and* $\mathbf{j}$ *that is equivalent to* $\mathbf{u}$*. See Example 4.*

27. $\mathbf{u} = \langle 6, -3 \rangle$

28. $\mathbf{u} = \langle 1, 4 \rangle$

29. $\mathbf{u} = \langle -5, -1 \rangle$

30. $\mathbf{u} = \langle -4, 3 \rangle$

31. $\mathbf{u} = \langle 2, 3 \rangle$

32. $\mathbf{u} = \langle 5, 2 \rangle$

Find the magnitude and direction angle of the vector **v**.

33. $\mathbf{v} = 5\left(\cos 30^\circ \mathbf{i} + \sin 30^\circ \mathbf{j}\right)$

34. $\mathbf{v} = 7\left(\cos 45^\circ \mathbf{i} + \sin 45^\circ \mathbf{j}\right)$

35. $\mathbf{v} = 4\mathbf{i} + 3\mathbf{j}$

36. $\mathbf{v} = -2\mathbf{i} - 2\mathbf{j}$

Find the component form of **v** *given its magnitude and the angle it makes with the positive x-axis. See Example 5.*

37. $\|\mathbf{v}\| = 6,\ \theta = 30^\circ$

38. $\|\mathbf{v}\| = \frac{5}{2},\ \theta = 0^\circ$

39. $\|\mathbf{v}\| = 18,\ \theta = 135^\circ$

40. $\|\mathbf{v}\| = 3\sqrt{3},\ \theta = 90^\circ$

41. $\|\mathbf{v}\| = 1,\ \theta = 120^\circ$

42. $\|\mathbf{v}\| = 4\sqrt{2},\ \theta = 45^\circ$

43. $\|\mathbf{v}\| = 4$, **v** in the direction of $2\mathbf{i} + 3\mathbf{j}$

44. $\|\mathbf{v}\| = 7$, **v** in the direction of $\mathbf{i} + 4\mathbf{j}$

Solve the following problems. See Examples 6 and 7.

45. A spitball is fired into the air at a speed of 4 ft./sec. and at an angle of 30° from the horizontal. Express this velocity in vector form.

46. A golf ball is driven into the air at a speed of 75 miles per hour and at an angle of 50° from the horizontal. Express this velocity in vector form.

47. A sailboat is traveling at a speed of 45 miles per hour with a bearing of N 59° W, when it encounters a front with winds blowing at 15 miles per hour with a bearing of S 3° E. What is the resultant true velocity of the sailboat?

48. An underwater missile is traveling at a speed of 350 miles per hour and bearing of S 17° W, when it meets a current traveling at 44 miles per hour in the direction of N 61° W. What is the resultant true velocity of the underwater missile?

49. Prometheus is slowly pushing a 1235 pound boulder across a flat plain with a force of 150 pounds. What is the total force being applied to the boulder?

50. A boy is pushing a toy truck across the floor. If the toy weighs 3 pounds and the boy is exerting half a pound of pressure on the toy, what is the total force being applied to the toy truck?

8.6 The Dot Product and its Uses

TOPICS

1. The dot product
2. Projections of vectors
3. Applications of the dot product

Topic 1: The Dot Product

This last section introduces a vector operation that is distinct from the vector addition and scalar multiplication that you learned in Section 8.5. The reason for introducing this third operation is simply its usefulness in performing a variety of tasks.

The operation goes by a variety of names; it is often referred to as the **dot product** of two vectors, due to the notation used, but it is also often called the **scalar product** because it is an operation that turns two vectors into a scalar. More advanced texts often refer to the operation as the **inner product**, to differentiate it from a number of other products that can be defined on vectors (including one called, as you might have guessed, the *outer product*).

The Dot Product in the Plane

Given two vectors $\mathbf{u} = \langle u_1, u_2 \rangle$ and $\mathbf{v} = \langle v_1, v_2 \rangle$, the **dot product** $\mathbf{u} \cdot \mathbf{v}$ of the two vectors is the scalar defined by

$$\mathbf{u} \cdot \mathbf{v} = u_1 v_1 + u_2 v_2.$$

Note that the dot product of two vectors is, indeed, a scalar, and that it can be positive, 0, or negative. Its calculation is simple, assuming the components of the two vectors are known.

example 1

Calculate each of the following dot products.

a. $\langle -5, 2\rangle \cdot \langle 3, -1\rangle$ **b.** $\langle -5, 2\rangle \cdot \langle 2, 5\rangle$ **c.** $\langle -5, 2\rangle \cdot \langle -5, 2\rangle$

Solutions:

a. $\langle -5, 2\rangle \cdot \langle 3, -1\rangle = (-5)(3)+(2)(-1) = -15-2 = -17$

b. $\langle -5, 2\rangle \cdot \langle 2, 5\rangle = (-5)(2)+(2)(5) = -10+10 = 0$

c. $\langle -5, 2\rangle \cdot \langle -5, 2\rangle = (-5)(-5)+(2)(2) = 25+4 = 29$

A number of properties of the dot product can be proven. The verifications of the following statements are left as exercises.

Elementary Properties of the Dot Product

Given two vectors $\mathbf{u} = \langle u_1, u_2\rangle$ and $\mathbf{v} = \langle v_1, v_2\rangle$ and a scalar a, the following hold:

1. $\mathbf{u} \cdot \mathbf{v} = \mathbf{v} \cdot \mathbf{u}$

2. $\mathbf{0} \cdot \mathbf{u} = 0$

3. $\mathbf{u} \cdot (\mathbf{v} + \mathbf{w}) = \mathbf{u} \cdot \mathbf{v} + \mathbf{u} \cdot \mathbf{w}$

4. $a(\mathbf{u} \cdot \mathbf{v}) = (a\mathbf{u}) \cdot \mathbf{v} = \mathbf{u} \cdot (a\mathbf{v})$

5. $\mathbf{u} \cdot \mathbf{u} = \|\mathbf{u}\|^2$

example 2

Given that the dot product of a vector **u** with itself is 8, determine the magnitude of **u**.

Solution:

Since $\|\mathbf{u}\|^2 = \mathbf{u} \cdot \mathbf{u} = 8,$ it follows that $\|\mathbf{u}\| = 2\sqrt{2}.$

There is another property, however, whose truth is not so immediately clear. It often goes by the name of the Dot Product Theorem.

The Dot Product Theorem

Let two nonzero vectors **u** and **v** be depicted so that their initial points coincide, and let θ be the smaller of the two angles formed by **u** and **v** (so $0 \le \theta \le \pi$). Then

$$\mathbf{u} \cdot \mathbf{v} = \|\mathbf{u}\|\|\mathbf{v}\|\cos\theta.$$

The proof of this statement makes good use of the Law of Cosines and several of the elementary properties listed above. Starting with the two vectors **u** and **v** drawn as directed, the third side of a triangle is formed by **u** – **v** as shown in Figure 1.

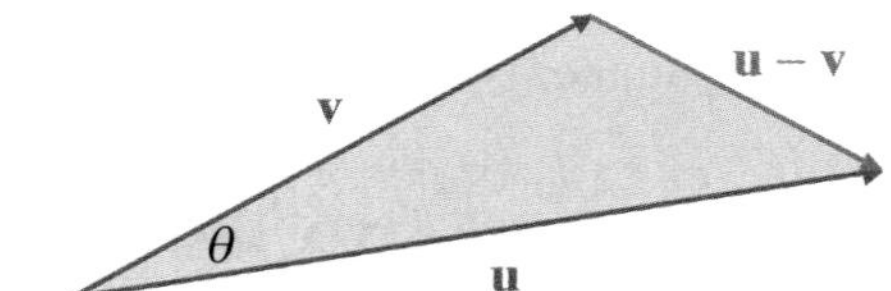

Figure 1: Proving the Dot Product Theorem

By the Law of Cosines, then, $\|\mathbf{u}-\mathbf{v}\|^2 = \|\mathbf{u}\|^2 + \|\mathbf{v}\|^2 - 2\|\mathbf{u}\|\|\mathbf{v}\|\cos\theta$. But we have a second way of expressing $\|\mathbf{u}-\mathbf{v}\|^2$:

$$\begin{aligned}\|\mathbf{u}-\mathbf{v}\|^2 &= (\mathbf{u}-\mathbf{v})\cdot(\mathbf{u}-\mathbf{v}) \\ &= \mathbf{u}\cdot\mathbf{u} - \mathbf{u}\cdot\mathbf{v} - \mathbf{v}\cdot\mathbf{u} + \mathbf{v}\cdot\mathbf{v} \\ &= \|\mathbf{u}\|^2 - 2(\mathbf{u}\cdot\mathbf{v}) + \|\mathbf{v}\|^2.\end{aligned}$$

Equating these two different ways of expressing $\|\mathbf{u}-\mathbf{v}\|^2$ and canceling like terms, we obtain

$$-2\|\mathbf{u}\|\|\mathbf{v}\|\cos\theta = -2(\mathbf{u}\cdot\mathbf{v}),$$

and dividing both sides by –2 then gives us the desired result.

While the Dot Product Theorem is occasionally useful in calculating the dot product of two given vectors, it is more often used to determine the angle between two vectors.

example 3

Find the angle between the vector $\mathbf{u} = \langle -3, 1\rangle$ and the vector $\mathbf{v} = \langle 2, 5\rangle$.

Solution:

By the Dot Product Theorem,

$$\cos\theta = \frac{\mathbf{u}\cdot\mathbf{v}}{\|\mathbf{u}\|\|\mathbf{v}\|} = \frac{\langle -3, 1\rangle\cdot\langle 2, 5\rangle}{\left(\sqrt{10}\right)\left(\sqrt{29}\right)} = \frac{-1}{\sqrt{290}}$$

and so

$$\theta = \cos^{-1}\left(-\frac{1}{\sqrt{290}}\right) \approx 93.4^\circ.$$

The Dot Product Theorem also makes it clear that pairs of vectors for which θ is a right angle are easily identified by the dot product. If $\theta = \frac{\pi}{2}$ for two given vectors $\mathbf{u}$ and $\mathbf{v}$, then $\cos\theta = 0$ and hence the dot product is 0. We turn this into a definition as follows:

Orthogonal Vectors

Two nonzero vectors $\mathbf{u}$ and $\mathbf{v}$ are said to be **orthogonal** (or **perpendicular**) if

$$\mathbf{u}\cdot\mathbf{v} = 0.$$

This is equivalent to saying that the angle θ defined by the two vectors is a right angle.

example 4

Find a vector orthogonal to the vector $\mathbf{u} = \langle 7, -2\rangle$.

Solution:

There are an infinite number of vectors $\mathbf{v}$ for which $\mathbf{u}\cdot\mathbf{v} = 0$, and hence an infinite number of correct answers. One such is $\mathbf{v} = \langle 2, 7\rangle$, and another is $\mathbf{v} = \langle -4, -14\rangle$. Any nonzero multiple of $\langle 2, 7\rangle$ is orthogonal to $\mathbf{u}$.

Topic 2: Projections of Vectors

The solution of many problems depends on being able to decompose a vector into a sum of two vectors in a specific way. Usually, the task amounts to writing a vector $\mathbf{u}$ as a sum $\mathbf{w}_1 + \mathbf{w}_2$ where $\mathbf{w}_1$ is parallel to a second vector $\mathbf{v}$ and $\mathbf{w}_2$ is orthogonal to $\mathbf{v}$.

In the language of vectors, we say $\mathbf{w}_1$ is the **projection of u onto v**, and write $\mathbf{w}_1 = \text{proj}_{\mathbf{v}}\mathbf{u}$. If we can determine $\text{proj}_{\mathbf{v}}\mathbf{u}$, then we can easily find $\mathbf{w}_2$ by noting that

$$\mathbf{w}_2 = \mathbf{u} - \text{proj}_{\mathbf{v}}\mathbf{u},$$

and since $\mathbf{w}_2$ is perpendicular to $\mathbf{v}$, some books refer to $\mathbf{w}_2$ as $\text{perp}_{\mathbf{v}}\mathbf{u}$.

The first step is to find a formula for $\text{proj}_{\mathbf{v}}\mathbf{u}$ in terms of $\mathbf{u}$ and $\mathbf{v}$. From our study of trigonometry and the right triangle in the diagram, we know that

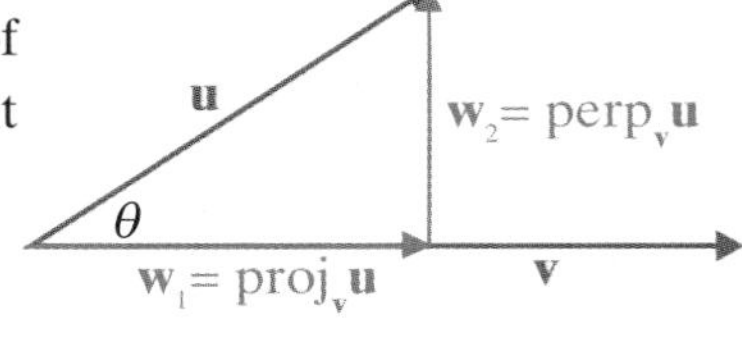

$$\cos\theta = \frac{\|\mathbf{w}_1\|}{\|\mathbf{u}\|},$$

so

$$\|\mathbf{w}_1\| = \|\mathbf{u}\|\cos\theta.$$

We can now apply the Dot Product Theorem and write

$$\|\mathbf{w}_1\| = \|\mathbf{u}\|\cos\theta = \|\mathbf{u}\|\left(\frac{\mathbf{u}\cdot\mathbf{v}}{\|\mathbf{u}\|\|\mathbf{v}\|}\right) = \frac{\mathbf{u}\cdot\mathbf{v}}{\|\mathbf{v}\|}.$$

This, however, is only the *magnitude* of the projection of $\mathbf{u}$ onto $\mathbf{v}$. In order to construct the actual vector $\mathbf{w}_1$, we need to multiply this magnitude by a unit vector pointing in the same direction as $\mathbf{v}$:

$$\mathbf{w}_1 = \left(\frac{\mathbf{u}\cdot\mathbf{v}}{\|\mathbf{v}\|}\right)\left(\frac{\mathbf{v}}{\|\mathbf{v}\|}\right) = \left(\frac{\mathbf{u}\cdot\mathbf{v}}{\|\mathbf{v}\|^2}\right)\mathbf{v}.$$

This is summarized in the following box.

Projection of u onto v

Let $\mathbf{u}$ and $\mathbf{v}$ be nonzero vectors. The **projection of u onto v** is the vector

$$\text{proj}_{\mathbf{v}}\mathbf{u} = \left(\frac{\mathbf{u}\cdot\mathbf{v}}{\|\mathbf{v}\|^2}\right)\mathbf{v}.$$

example 5

Find the projection of $\mathbf{u} = \langle 2, 4 \rangle$ onto $\mathbf{v} = \langle 7, -1 \rangle$, and then write $\mathbf{u}$ as a sum of two orthogonal vectors, one of which is $\text{proj}_{\mathbf{v}}\mathbf{u}$.

Solution:

First,

$$\begin{aligned} \text{proj}_{\mathbf{v}}\mathbf{u} &= \left(\frac{\langle 2, 4 \rangle \cdot \langle 7, -1 \rangle}{\left\| \langle 7, -1 \rangle \right\|^2} \right) \langle 7, -1 \rangle \\ &= \left(\frac{14 - 4}{\left(\sqrt{49+1}\right)^2} \right) \langle 7, -1 \rangle \\ &= \left(\frac{1}{5} \right) \langle 7, -1 \rangle \\ &= \left\langle \frac{7}{5}, -\frac{1}{5} \right\rangle. \end{aligned}$$

Now, we can determine that $\text{perp}_{\mathbf{v}}\mathbf{u} = \langle 2, 4 \rangle - \left\langle \frac{7}{5}, -\frac{1}{5} \right\rangle = \left\langle \frac{3}{5}, \frac{21}{5} \right\rangle$.

So, we can write $\mathbf{u}$ as the sum of the two orthogonal vectors $\text{proj}_{\mathbf{v}}\mathbf{u}$ and $\text{perp}_{\mathbf{v}}\mathbf{u}$.

$$\mathbf{u} = \text{proj}_{\mathbf{v}}\mathbf{u} + \text{perp}_{\mathbf{v}}\mathbf{u} = \left\langle \frac{7}{5}, -\frac{1}{5} \right\rangle + \left\langle \frac{3}{5}, \frac{21}{5} \right\rangle = \langle 2, 4 \rangle = \mathbf{u}$$

The graphs of $\mathbf{u}$ and $\mathbf{v}$ are shown here in red, while the orthogonal decomposition of $\mathbf{u}$ appears in blue.

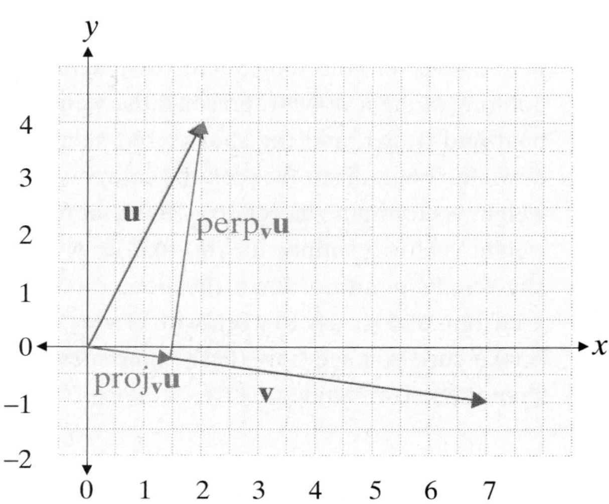

Topic 3: Applications of the Dot Product

The dot product and associated concepts are very useful in answering some common questions. We will close this section with a few examples.

example 6

A boat and trailer, which together weigh 650 pounds, are to be pulled up a boat ramp that has an incline of 30°. What force is required to merely prevent the boat and trailer from rolling down the ramp?

Solution:

As seen in Example 7 of Section 8.5, weight is simply an example of force: it is the force that the Earth exerts on an object through gravity. To gain some insight into this problem, suppose the question were "What is the force needed to lift the boat and trailer straight up?" The answer would be that 650 pounds of force is needed to counter the pull of gravity, so anything greater than 650 pounds of force would result in the boat and trailer being lifted.

But the actual situation isn't so extreme. Only the portion of the combined weight that is parallel to the boat ramp must be countered. That is, we are looking for the component of the 650 pounds that is directed at an angle of 30° from the horizontal. The picture at right illustrates the vectors under discussion.

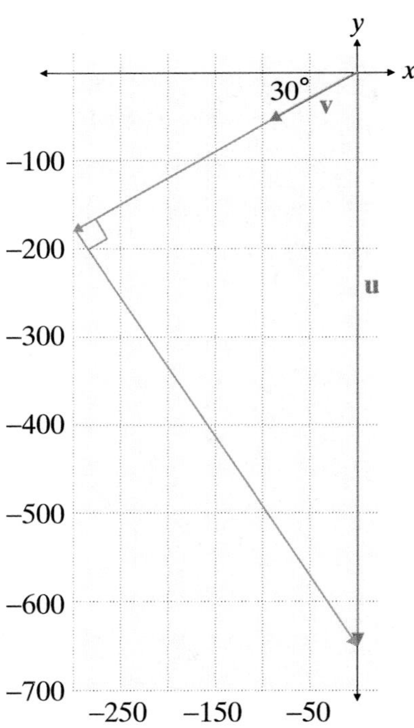

In the diagram, the long vector of magnitude 650 pointing straight down represents the weight of the boat and trailer, and the shorter red vector points down the boat ramp. We seek the component of the weight vector in the direction of the boat ramp. The weight vector is simply $\mathbf{u} = \langle 0, -650 \rangle$. A vector of any length pointing down the boat ramp can be used; one that is easy to work with is $\mathbf{v} = \langle -\sqrt{3}, -1 \rangle$ (make sure you see how these components follow from the known angle of 30°).

Now,

$$\begin{aligned}
\text{proj}_{\mathbf{v}}\mathbf{u} &= \left(\frac{\langle 0, -650\rangle \cdot \langle -\sqrt{3}, -1\rangle}{\left\|\langle -\sqrt{3}, -1\rangle\right\|^2}\right)\langle -\sqrt{3}, -1\rangle \\
&= \left(\frac{(-650)(-1)}{\left(\sqrt{3+1}\right)^2}\right)\langle -\sqrt{3}, -1\rangle \\
&= \left(\frac{650}{4}\right)\langle -\sqrt{3}, -1\rangle \\
&= \left\langle -\frac{325\sqrt{3}}{2}, \frac{-325}{2}\right\rangle.
\end{aligned}$$

The magnitude of this projection, which is 325 pounds, is thus the force that must be exerted to keep the boat and trailer stationary. Anything more than this will allow the boat and trailer to be pulled up the ramp.

Work has a technical meaning in physics and engineering, and the dot product is very useful in calculating how much work is required to accomplish a task. Technically, **work** is the application of a force through a certain distance. For instance, applying a force of 375 pounds to pull the boat and trailer of Example 6 a distance of 20 feet up the ramp results in a certain amount of work being done.

It is important to realize, though, that only the component of the force applied in the direction of motion contributes to the work done. Fortunately, we can use the dot product to make this calculation. If we let $\mathbf{D}$ represent the vector of the motion (so that $\|\mathbf{D}\|$ is the distance traveled) and $\mathbf{F}$ the force applied (not necessarily in the same direction as $\mathbf{D}$), then the component of the force in the direction of motion is $\|\mathbf{F}\|\cos\theta$, where θ is, as usual, the angle between the two vectors. The work done is then the product of $\|\mathbf{D}\|$ and $\|\mathbf{F}\|\cos\theta$, that is, $W = \|\mathbf{D}\|\|\mathbf{F}\|\cos\theta$. But this last expression should look familiar – it appears in the Dot Product Theorem, allowing us to write the simpler formula below:

$$W = \mathbf{F}\cdot\mathbf{D}.$$

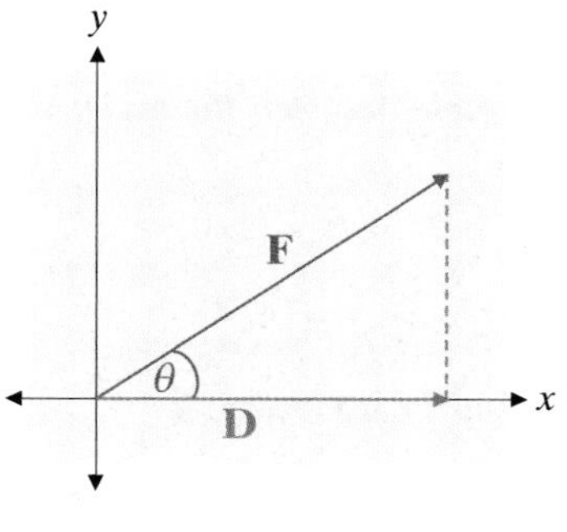

example 7

A child pulls a wagon along a sidewalk, exerting a force of 15 pounds on the handle of the wagon. The handle is at an angle of 40° to the horizontal. If the child pulls the wagon a distance of 50 feet, what work has been done?

Solution:

We start by defining the force and distance vectors:

$$\mathbf{F} = 15\langle \cos 40°, \sin 40° \rangle \text{ and } \mathbf{D} = \langle 50, 0 \rangle.$$

The calculation is now straightforward:

$$W = 15\langle \cos 40°, \sin 40° \rangle \cdot \langle 50, 0 \rangle \approx (15)(0.766)(50) + (15)(0.643)(0) = 574.5 \text{ foot-pounds.}$$

exercises

Calculate each of the following dot products. See Example 1.

1. $\langle 4, 3 \rangle \cdot \langle 5, -1 \rangle$

2. $\langle 2, 4 \rangle \cdot \langle -1, -1 \rangle$

3. $\langle 3, 5 \rangle \cdot \langle 2, 0 \rangle$

4. $\langle -1, 6 \rangle \cdot \langle 6, 1 \rangle$

5. $\langle 2, 2 \rangle \cdot \langle 2, 2 \rangle$

6. $\langle 1, 2 \rangle \cdot \langle 3, 4 \rangle$

7. $\langle -4, 3 \rangle \cdot \langle 2, 3 \rangle$

8. $\langle -2, -4 \rangle \cdot \langle 6, 2 \rangle$

9. $\mathbf{u} = 5\mathbf{i} + \mathbf{j}, \mathbf{v} = -2\mathbf{i} + 3\mathbf{j}$

10. $\mathbf{u} = \mathbf{i} - 5\mathbf{j}, \mathbf{v} = -2\mathbf{i} - 4\mathbf{j}$

Find the indicated quantity given $\mathbf{u} = \langle -2, 3 \rangle$ and $\mathbf{v} = \langle 4, 4 \rangle$. See Example 1.

11. $\mathbf{v} \cdot \mathbf{v}$

12. $4\mathbf{u} \cdot \mathbf{v}$

13. $\|\mathbf{u}\| + 2$

14. $(\mathbf{u} \cdot \mathbf{v})2\mathbf{v}$

Find the magnitude of $\mathbf{u}$ using the dot product. See Example 2.

15. $\mathbf{u} = \langle 6, -1 \rangle$

16. $\mathbf{u} = \langle 10, 3 \rangle$

17. $\mathbf{u} = 2\mathbf{i} + 7\mathbf{j}$

18. $\mathbf{u} = -3\mathbf{i} + 4\mathbf{j}$

Find the angle between the given vectors. See Example 3.

19. $\mathbf{u} = \langle -2, 3 \rangle, \mathbf{v} = \langle 1, 0 \rangle$

20. $\mathbf{u} = \langle 5, 4 \rangle, \mathbf{v} = \langle 3, 2 \rangle$

21. $\mathbf{u} = \langle 3, 5 \rangle, \mathbf{v} = \langle 4, 4 \rangle$

22. $\mathbf{u} = \langle -4, 2 \rangle, \mathbf{v} = \langle 1, 5 \rangle$

23. $\mathbf{u} = -\mathbf{i} + 2\mathbf{j}, \mathbf{v} = 3\mathbf{i} - 3\mathbf{j}$

24. $\mathbf{u} = 5\mathbf{i} + 2\mathbf{j}, \mathbf{v} = 4\mathbf{i} + \mathbf{j}$

25. $\mathbf{u} = \cos\left(\frac{3\pi}{4}\right)\mathbf{i} + \sin\left(\frac{3\pi}{4}\right)\mathbf{j},\ \mathbf{v} = \cos\left(\frac{\pi}{2}\right)\mathbf{i} + \sin\left(\frac{\pi}{2}\right)\mathbf{j}$

26. $\mathbf{u} = \cos\left(\frac{\pi}{4}\right)\mathbf{i} + \sin\left(\frac{\pi}{4}\right)\mathbf{j},\ \mathbf{v} = \cos\left(\frac{5\pi}{6}\right)\mathbf{i} + \sin\left(\frac{5\pi}{6}\right)\mathbf{j}$

Use vectors to find the interior angles of the triangles given the following sets of vertices.

27. $(3, 3), (4, 2), (-1, -6)$

28. $(0, 0), (0, 5), (3, 6)$

29. $(-2, -1), (2, 4), (-4, 5)$

30. $(6, 3), (-5, 2), (-6, 1)$

Find $\mathbf{u} \cdot \mathbf{v}$ *where* θ *is the angle between* $\mathbf{u}$ *and* $\mathbf{v}$*. See Example 3.*

31. $\|\mathbf{u}\| = 25, \|\mathbf{v}\| = 5, \theta = 120^\circ$

32. $\|\mathbf{u}\| = 4, \|\mathbf{v}\| = 64, \theta = \frac{\pi}{6}$

33. $\|\mathbf{u}\| = 16, \|\mathbf{v}\| = 4, \theta = \frac{3\pi}{4}$

34. $\|\mathbf{u}\| = 9, \|\mathbf{v}\| = 10, \theta = \frac{2\pi}{3}$

Find <u>two</u> vectors orthogonal to the given vector. See Example 4. Answers may vary.

35. $\mathbf{u} = \langle 3, -3 \rangle$

36. $\mathbf{u} = \langle 4, 1 \rangle$

37. $\mathbf{u} = \langle 2, -6 \rangle$

38. $\mathbf{u} = \langle 5, 4 \rangle$

Determine whether $\mathbf{u}$ *and* $\mathbf{v}$ *are orthogonal, parallel, or neither. See Example 4.*

39. $\mathbf{u} = \langle 2, -3 \rangle, \mathbf{v} = \langle 1, 6 \rangle$

40. $\mathbf{u} = \langle -12, 30 \rangle, \mathbf{v} = \left\langle \frac{1}{2}, -\frac{5}{4} \right\rangle$

41. $\mathbf{u} = 2\mathbf{i} - 2\mathbf{j}, \mathbf{v} = -\mathbf{i} - \mathbf{j}$

42. $\mathbf{u} = \mathbf{i}, \mathbf{v} = -2\mathbf{i} + 2\mathbf{j}$

Find the projection of $\mathbf{u}$ *onto* $\mathbf{v}$*, and then write* $\mathbf{u}$ *as a sum of two orthogonal vectors, one of which is* $\text{proj}_{\mathbf{v}}\mathbf{u}$*. See Example 5.*

43. $\mathbf{u} = \langle 1, 3 \rangle, \mathbf{v} = \langle 4, 2 \rangle$

44. $\mathbf{u} = \langle 2, 2 \rangle, \mathbf{v} = \langle 1, -7 \rangle$

45. $\mathbf{u} = \langle 3, -5 \rangle, \mathbf{v} = \langle 6, 2 \rangle$

46. $\mathbf{u} = \langle 0, 3 \rangle, \mathbf{v} = \langle 2, 6 \rangle$

47. $\mathbf{u} = \langle -3, -3 \rangle, \mathbf{v} = \langle -4, -1 \rangle$

48. $\mathbf{u} = \langle 4, 2 \rangle, \mathbf{v} = \langle 1, 5 \rangle$

Find the work done on a particle moving from J to K if the magnitude and direction of the force are given by $\mathbf{v}$*. See Example 7.*

49. $J = (1, 4), K = (5, 6), \mathbf{v} = \langle 2, 3 \rangle$

50. $J = (-3, 2), K = (0, 5), \mathbf{v} = \langle 4, 2 \rangle$

51. $J = (3, 0), K = (-4, -2), \mathbf{v} = -\mathbf{i} + 2\mathbf{j}$

52. $J = (3, -3), K = (5, 1), \mathbf{v} = 6\mathbf{i} - 3\mathbf{j}$

Solve the following problems. See Examples 6 and 7.

53. A truck with a gross weight of 25,000 pounds is parked on an 8° slope. What force is required to prevent the truck from rolling down the hill?

54. A child sits in his go-cart at the start position of a race atop a hill. If the hill has a slope of 3°, and the child and go-cart have a total weight of 250 pounds, what force is required to keep them stationary at the start position?

55. A woman on skis holds herself stationary, with the use of her ski poles, on a ski slope that is 45° from the horizontal. If the woman and her skis have a total weight of 155 pounds, what is the force required to prevent her from sliding down the slope?

56. A child pulls a sled over the snow, exerting a force of 25 pounds on the attached rope. The rope is 35° to the horizontal. If the child pulls the sled a distance of 80 feet, what work has been done?

57. The world's strongest man pulls a log 200 feet, and the tension in the cable connecting the man and log is 3000 pounds. What is the work being done if the cable is being held 15° to the horizontal?

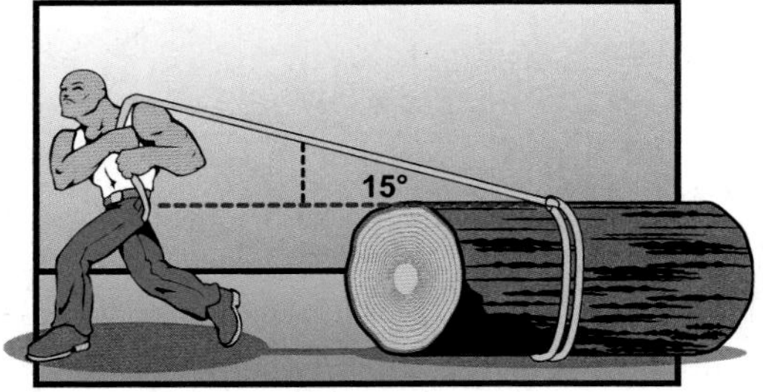

58. A recreational vehicle pulls a passenger car behind it, exerting 1250 pounds on the attach point. The point of attachment is 30° to the horizontal. If the RV pulls the car a distance of 2 miles, what work has been done?

chapter eight project

Trigonometric Applications

Built by King Khufu from 2589 - 2566 BC to serve as his tomb, the Great Pyramid of Giza covers 13 acres and weighs more than 6.5 million tons. The Great Pyramid is the oldest and the only remaining wonder of the 7 Wonders of the Ancient World. Even using modern technology, engineers in the 21st century would have difficulties recreating this impressive structure. Today, we can only put forth theories as to how this amazing structure was created.

When first built, the Egyptians called the Great Pyramid Ikhet, which means Glorious Light, for the sides of the pyramid were covered in highly polished limestone which would have shone brightly under the hot Egyptian Sun.

1. When built, the length of each side of the base was 754 ft. and the distance from each corner of the base to the peak was 718 ft. What was the surface area of the four sides?

The Egyptians quarried most of the stone locally, but they also floated huge granite blocks down the Nile River from Aswan.

2. Your barge with the latest shipment of granite for King Khufu is quickly approaching Giza. The river is flowing at 195 yards per minute. You command your oarsmen to start rowing towards shore at a 65° angle from the direction of the current. They can row 260 yards per minute.
   - **a.** What is the resultant true velocity of the barge?
   - **b.** The Nile River is 840 yards wide near Giza. If the boat is in the center of the river and the dock is 750 yards ahead, will they hit the bank before or after it? By how much will they miss it? [Hint: (velocity)(time) = distance]

3. Once the boat reaches the shore, the granite stones need to be moved into place.
   - **a.** If a granite block weighs 8300 pounds and 12 of your men are each pulling on it (in the same direction) with a force of 115 pounds, what is the total force being applied?

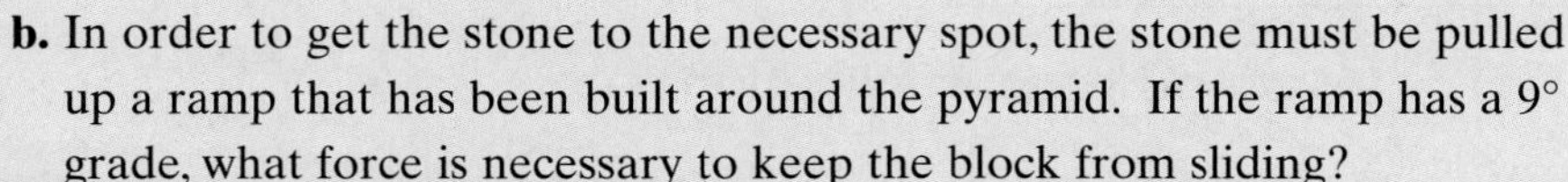

b. In order to get the stone to the necessary spot, the stone must be pulled up a ramp that has been built around the pyramid. If the ramp has a 9° grade, what force is necessary to keep the block from sliding?

c. If the top of the pyramid is currently 320 feet above the desert, how long does the ramp have to be?

d. How much work is it for the 12 men to drag the stone to the top of the pyramid? (Remember, in this case work is done in both the horizontal and vertical directions.)

CHAPTER REVIEW

8.1 The Law of Sines and the Law of Cosines

topics	pages	test exercises
The Law of Sines and its Use • Use of the Law of Sines: AAS, ASA, or SSA • Use of the Law of Cosines: SAS or SSS • The Law of Sines: $\frac{\sin A}{a} = \frac{\sin B}{b} = \frac{\sin C}{c}$ • Solving applications using the Law of Sines • SSA Case: If A is acute and $a < h$, then no triangle exists, if A is acute and $a = h$, then one triangle exists, if A is acute and $h < a < b$, then two triangles exist, if A is acute and $b \le a$, then one triangle exists, if A is obtuse and $a \le b$, then no triangle exists, if A is obtuse and $b < a$, then one triangle exists.	p. 599 – 605	2 – 5
The Law of Cosines and its Use • The Law of Cosines: $a^2 = b^2 + c^2 - 2bc\cos A$ $b^2 = a^2 + c^2 - 2ac\cos B$ $c^2 = a^2 + b^2 - 2ab\cos C$ • Solving applications using the Law of Cosines	p. 605 – 607	1, 6 – 9
Areas of triangles • Area of a triangle (Sine Formula): Area $= \frac{1}{2}ab\sin C = \frac{1}{2}bc\sin A = \frac{1}{2}ac\sin B$. • Area of a triangle (Heron's Formula): Area $= \sqrt{s(s-a)(s-b)(s-c)}$ where $s = \frac{a+b+c}{2}$.	p. 607 – 610	

8.2 Polar Coordinates and Polar Equations

topics	pages	test exercises
The polar coordinate system • Polar coordinates: (r, θ) where r is the distance from the origin to the given point and θ is the angle, measured counterclockwise from the polar axis to a line segment connecting the origin to the given point.	p. 617 – 618	

topics (continued)	pages	test exercises
Coordinate conversion • Converting from polar to Cartesian coordinates: $x = r\cos\theta$ and $y = r\sin\theta$. • Converting from Cartesian to polar coordinates: $r^2 = x^2 + y^2$ and $\tan\theta = \frac{y}{x}$ $(x \neq 0)$.	p. 619 – 621	10 – 13
The form of polar equations • Polar equation: an equation in the variables r and θ that defines a relationship between these coordinates • Solution to a polar equation: an ordered pair (r, θ) that makes the equation true	p. 621 – 622	14 – 17
Graphing polar equations • Symmetry of polar equations: pole symmetry, polar axis symmetry, and symmetry with respect to the line $\theta = \frac{\pi}{2}$ • Common polar equations and their graphs	p. 622 – 628	18 – 19

8.3 Parametric Equations

topics	pages	test exercises
Sketching parametric curves • Describing an object's travel parametrically	p. 631 – 634	20 – 23
Eliminating the parameter • Eliminating the parameter: the process of converting from parametric form to rectangular form	p. 635 – 637	20 – 23
Constructing parametric equations • How parametric equations describing a given curve are constructed	p. 638– 640	24 – 27

8.4 Trigonometric Form of Complex Numbers

topics	pages	test exercises
The complex plane • Coordinates of a complex number $z = a + bi$: (a, b) • Horizontal axis = real axis; vertical axis = imaginary axis • Complex plane: plane created by the real and imaginary axes. • Magnitude of a complex number: $\lvert z \rvert = \sqrt{a^2 + b^2}$	p. 644 – 646	28 – 33
Complex numbers in trigonometric form • Trigonometric form of $z = a + bi$: $\lvert z \rvert(\cos\theta + i\sin\theta)$ where $\lvert z \rvert = \sqrt{a^2 + b^2}$ and $\tan\theta = \frac{b}{a}$. • Complex number written using Euler's formula: $z = \lvert z \rvert e^{i\theta}$	p. 646 – 648	34 – 37
Multiplication and division of complex numbers • Multiplication: $z_1 z_2 = \lvert z_1 \rvert \lvert z_2 \rvert [\cos(\theta_1 + \theta_2) + i\sin(\theta_1 + \theta_2)]$ • Division: $\frac{z_1}{z_2} = \frac{\lvert z_1 \rvert}{\lvert z_2 \rvert}[\cos(\theta_1 - \theta_2) + i\sin(\theta_1 - \theta_2)]$	p. 648 – 650	38 – 41
Powers of complex numbers • DeMoivre's Theorem: $z^n = \lvert z \rvert^n(\cos n\theta + i\sin n\theta) = \lvert z \rvert^n e^{in\theta}$	p. 650 – 651	42 – 43
Roots of complex numbers • Roots of complex numbers: $w_k = \lvert z \rvert^{\frac{1}{n}}\left[\cos\left(\frac{\theta + 2k\pi}{n}\right) + i\sin\left(\frac{\theta + 2k\pi}{n}\right)\right] = \lvert z \rvert^{\frac{1}{n}} e^{i\left(\frac{\theta + 2k\pi}{n}\right)}$	p. 652 – 653	44 – 47

8.5 Vectors in the Cartesian Plane

topics	pages	test exercises
Vector terminology • Vector: quantity possessing magnitude and direction, often represented by a directed line segment. The point where the directed line segment begins is called the initial point and the point where it ends is called the terminal point . • Magnitude: length of the vector, denoted $\lVert \mathbf{u} \rVert$.	p. 657 – 658	48 – 49

topics (continued)	pages	test exercises
Basic vector operations • Properties of vector addition • Properties of scalar multiplication	p. 658 – 660	50 – 53
Component form of a vector • Component form: $\mathbf{v} = \langle x_2 - x_1, y_2 - y_1 \rangle$ • Vector operations using components • Unit vector: any vector with length 1 such as $\mathbf{i} = \langle 1, 0 \rangle$ and $\mathbf{j} = \langle 0, 1 \rangle$ • $\langle a, b \rangle = a\mathbf{i} + b\mathbf{j}$ • Linear combinations: any sum of scalar multiples of vectors • $\mathbf{u} = \lVert \mathbf{u} \rVert \langle \cos\theta, \sin\theta \rangle$	p. 660 – 663	52 – 59
Vector applications	p. 664 – 665	60 – 61

8.6 The Dot Product and its Uses

topics	pages	test exercises
The dot product • Dot product: $\mathbf{u} \cdot \mathbf{v} = u_1 v_1 + u_2 v_2$. • Properties of the dot product • The dot product theorem: $\mathbf{u} \cdot \mathbf{v} = \lVert \mathbf{u} \rVert \lVert \mathbf{v} \rVert \cos\theta$. • Orthogonal vectors: two vectors are orthogonal if $\mathbf{u} \cdot \mathbf{v} = 0$.	p. 670 – 673	62 – 69
Projections of vectors • Projection of **u** onto **v**: $\text{proj}_{\mathbf{v}}\mathbf{u} = \left(\dfrac{\mathbf{u} \cdot \mathbf{v}}{\lVert \mathbf{v} \rVert^2} \right) \mathbf{v}$.	p. 674 – 675	70 – 71
Applications of the dot product • Work $= \mathbf{F} \cdot \mathbf{D}$	p. 676 – 678	72 – 75

chapter test

Section 8.1

Solve the following problem.

1. The base of a 25 ft. ladder is positioned 7 ft. away from an office building situated on a slight hill, and the ladder and ground form a 62° angle. At what angle and at what height does the ladder touch the building?

Construct a triangle, if possible, using the following information.

2. $A = 30^\circ, B = 45^\circ, b = 4$
3. $a = 15, c = 13, C = 57^\circ$
4. $A = 74^\circ 20', C = 37^\circ, c = 23$
5. $b = 8, c = 13, C = 78^\circ$

Construct a triangle, if possible, using the following information and the Law of Cosines.

6. $A = 62^\circ, b = 8, c = 10$
7. $B = 94^\circ 7', a = 6, c = 14$
8. $a = 9, b = 2.5, c = 7.3$
9. $a = 10.8, b = 13.4, c = 6$

Section 8.2

Convert the points from polar to Cartesian coordinates.

10. $\left(-3.45, \frac{\pi}{3}\right)$
11. $\left(7, \frac{7\pi}{6}\right)$

Convert the points from Cartesian to polar coordinates.

12. $\left(-\sqrt{3}, -1\right)$
13. $(10, 12)$

Rewrite the rectangular equation in polar form.

14. $x^2 + y^2 = 16a^2$
15. $x^2 + y^2 = 9ax$

Rewrite the polar equation in rectangular form.

16. $r = 2\cos\theta$
17. $r = \dfrac{16}{4\cos\theta + 4\sin\theta}$

Sketch a graph of the polar equation.

18. $r = 4\sin 3\theta$
19. $r^2 = 25\cos 2\theta$

Section 8.3

Sketch the graphs of the following parametric equations by eliminating the parameter.

20. $x = \frac{1}{36t}$ and $y = t^2$

21. $x = t + 5$ and $y = |t - 2|$

22. $x = \frac{3}{4t - 2}$ and $y = 2t - 2$

23. $x = 4\sin\theta$ and $y = \cos\theta + 1$

Construct parametric equations describing the graphs of the following equations.

24. $y^2 = x^2 + 4$

25. $6x = 2 - y$

Construct parametric equations for the line or conic with the given attributes.

26. Line: Passing through (14, 4) and (–3, –8)

27. Circle: Center (1, 1), radius 1

Section 8.4

Graph and determine the magnitudes of the following complex numbers.

28. $5 + 2i$

29. $-3 + 3i$

Sketch z_1, z_2, $z_1 + z_2$, and z_1z_2 on the same complex plane.

30. $z_1 = -2 - 3i$, $z_2 = 6 + 3i$

31. $z_1 = 4 + 2i$, $z_2 = -5 + i$

Graph the regions of the complex plane defined by the following.

32. $\{z \mid 2 \le |z| \le 3\}$

33. $\{z = a + bi \mid a > 2,\ b > 3\}$

Write each of the following complex numbers in trigonometric form.

34. $2\sqrt{3} - 3i$

35. $1 + 4i$

Write each of the following complex numbers in standard form.

36. $4\left(\cos\frac{7\pi}{4} + i\sin\frac{7\pi}{4}\right)$

37. $3\left(\cos 60^\circ + i\sin 60^\circ\right)$

Perform the following operations and show the answer in both trigonometric form and standard form.

38. $\left[\sqrt{3}\left(\cos\frac{2\pi}{3}+i\sin\frac{2\pi}{3}\right)\right]\left[4\sqrt{3}\left(\cos\frac{7\pi}{6}+i\sin\frac{7\pi}{6}\right)\right]$

39. $\frac{5\left(\cos 240^\circ+i\sin 240^\circ\right)}{\left(\cos 120^\circ+i\sin 120^\circ\right)}$

40. $\frac{-\sqrt{3}+i}{1-i\sqrt{3}}$

41. $\left(12e^{35^\circ i}\right)\left(2e^{280^\circ i}\right)$

Use DeMoivre's Theorem to calculate the following.

42. $\left(1+\sqrt{3}i\right)^6$

43. $\left[3\left(\cos 240^\circ+i\sin 240^\circ\right)\right]^{11}$

Find the indicated roots of the following and graphically represent each set in the complex plane.

44. The square roots of $-144i$.

45. The cube roots of $125\left(\cos\frac{7\pi}{4}+i\sin\frac{7\pi}{4}\right)$.

Solve the following equations.

46. $z^4-1+i=0$

47. $z^3+4\sqrt{2}-4i\sqrt{2}=0$

Section 8.5

Find the component form and the magnitude of the vector **v** *for each of the following, assuming the first point given is the initial point and the second point given is the terminal point.*

48. $(-1, 0), (4, -5)$

49. $(6, 5), (-4, -1)$

*For each of the following, calculate and graph (a) 2***u** *+* **v***, (b) –***u** *+ 3***v***, and (c) –2***v***.*

50. $\mathbf{u}=\langle 1, 3\rangle, \mathbf{v}=\langle -5, 2\rangle$

51. $\mathbf{u}=\langle 1, -1\rangle, \mathbf{v}=\langle 4, -3\rangle$

For each of the following graphs, determine the component forms of $-\mathbf{u}$, $2\mathbf{u}-\mathbf{v}$, *and* $\mathbf{u}+\mathbf{v}$ *and find the magnitudes of* **u** *and* **v**.

52. 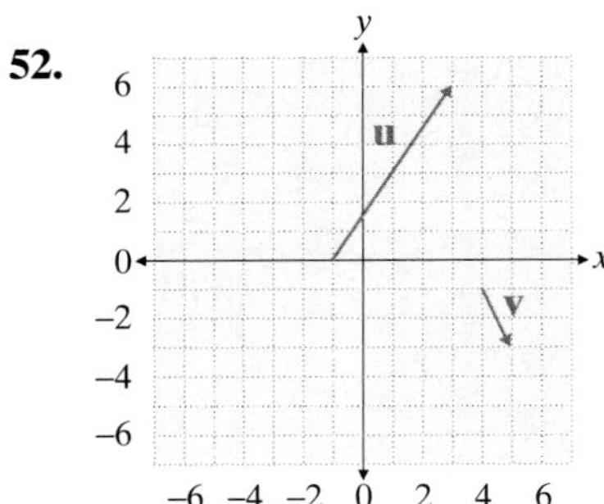

53. 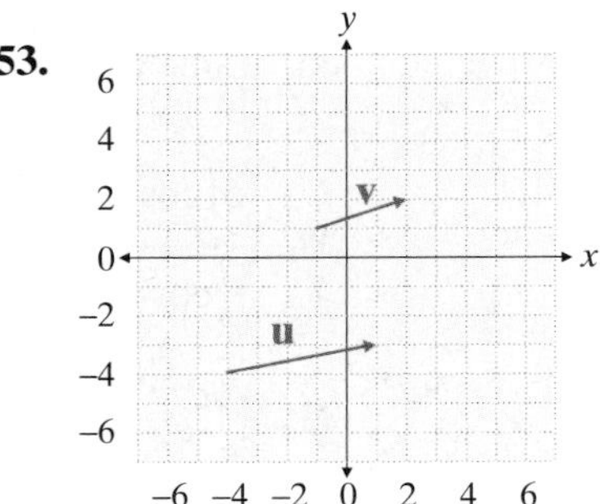

Given the vector **u**, *find (a) a unit vector pointing in the same direction as* **u**, *and (b) the linear combination of* **i** *and* **j** *that is equivalent to* **u**.

54. $\mathbf{u}=\langle -4, 5\rangle$

55. $\mathbf{u}=\langle 6, 3\rangle$

Find the magnitude and direction angle of the vector **v**.

56. $\mathbf{v}=4\left(\cos 135^\circ\,\mathbf{i}+\sin 135^\circ\,\mathbf{j}\right)$

57. $\mathbf{v}=5\mathbf{i}-\mathbf{j}$

Find the component form of **v** *given its magnitude and the angle it makes with the positive x-axis.*

58. $\|\mathbf{v}\|=2\sqrt{2},\ \theta=60^\circ$

59. $\|\mathbf{v}\|=6$, **v** in the direction of $3\mathbf{i}-4\mathbf{j}$

Solve the following problems.

60. A golf ball is driven into the air at a speed of 90 miles per hour and an angle of 45° from the horizontal. Express this velocity in vector form.

61. A sailboat is traveling at a speed of 55 miles per hour with a bearing of W 66° N, when it encounters a front with winds blowing at 20 miles per hour with a bearing of S 10° W. What is the resultant true velocity of the sailboat?

Section 8.6

Find the indicated quantity given $\mathbf{u} = \langle 1, -4 \rangle$ *and* $\mathbf{v} = \langle 2, 5 \rangle$.

62. $3\mathbf{u} \cdot \mathbf{v}$

63. $(\mathbf{u} \cdot \mathbf{v})3\mathbf{v}$

Find the magnitude of **u** *using the dot product.*

64. $\mathbf{u} = \langle -2, -3 \rangle$

65. $\mathbf{u} = -\mathbf{i} - 3\mathbf{j}$

Find the angle between the given vectors.

66. $\mathbf{u} = \langle 5, -5 \rangle, \mathbf{v} = \langle 1, 4 \rangle$

67. $\mathbf{u} = \cos\left(\frac{\pi}{4}\right)\mathbf{i} + \sin\left(\frac{\pi}{4}\right)\mathbf{j}, \ \mathbf{v} = \cos\left(\frac{2\pi}{3}\right)\mathbf{i} + \sin\left(\frac{2\pi}{3}\right)\mathbf{j}$

Find $\mathbf{u} \cdot \mathbf{v}$ *where* θ *is the angle between* **u** *and* **v**.

68. $\|\mathbf{u}\| = 16, \|\mathbf{v}\| = 2, \theta = 60^\circ$

69. $\|\mathbf{u}\| = 8, \|\mathbf{v}\| = 9, \theta = \frac{2\pi}{3}$

Find the projection of **u** *onto* **v**, *and then write* **u** *as a sum of two orthogonal vectors, one of which is* $proj_{\mathbf{v}}\mathbf{u}$.

70. $\mathbf{u} = \langle 2, 3 \rangle, \mathbf{v} = \langle -1, 5 \rangle$

71. $\mathbf{u} = \langle 4, -1 \rangle, \mathbf{v} = \langle 2, 2 \rangle$

Find the work done in a particle moving from J to K if the magnitude and direction of the force are given by **v**.

72. $J = (2, 4), K = (3, 6), \mathbf{v} = \langle 1, 3 \rangle$

73. $J = (-5, 3), K = (0, 4), \mathbf{v} = \langle 5, 6 \rangle$

74. A truck with a gross weight of 33,000 pounds is parked on a 6° slope. What force is required to prevent the truck from rolling down the hill?

75. The world's strongest man pulls a log 160 feet, and the tension in the cable connecting the man and log is 2650 pounds. What is the work being done if the cable is being held 10° to the horizontal?

chapter NINE

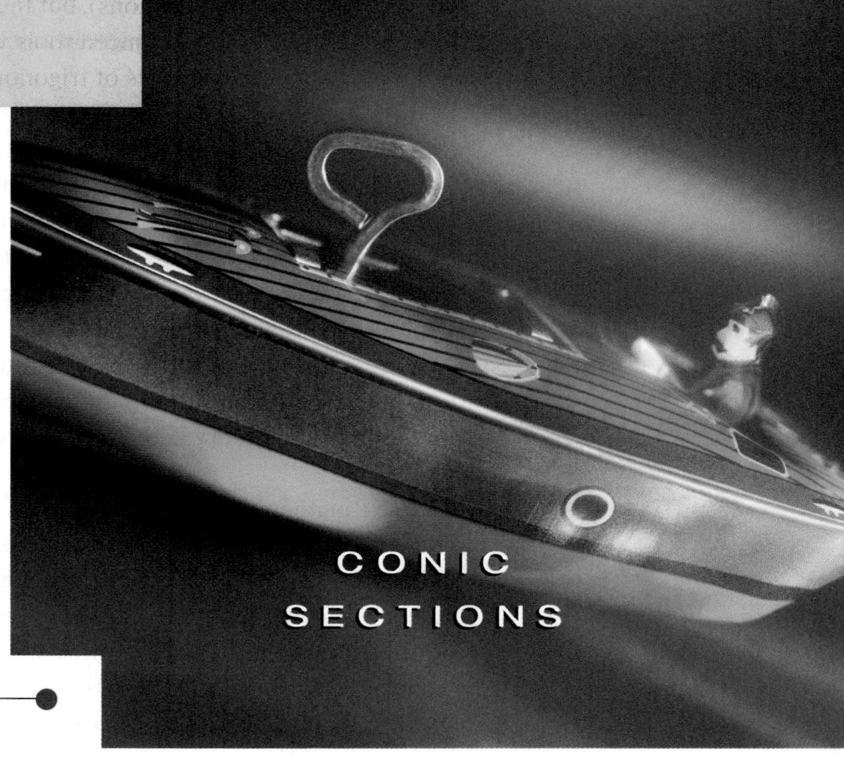

OVERVIEW

By the end of this chapter you should be able to graph hyperbolas, find their vertices, and even describe them with equations. One interesting application which makes use of hyperbolas and their characteristics is the "LORAN" radio-communications system. On page 728 you will find a problem regarding the position of a ship at sea. Given information about the distances between transmitters and the ship, you will use hyperbolas to discover an equation describing its possible locations. You will master this type of problem using tools such as the *Standard Form of a Hyperbola* on page 724.

those conic sections for which $B = 0$; we will then see the effect of allowing B to be nonzero in Section 9.4.

Algebraic Definition of Conics

A conic section described by an equation of the form $Ax^2 + Cy^2 + Dx + Ey + F = 0$, where at least one of the two coefficients A and C is not equal to 0, is:

1. an **ellipse** if the product AC is positive;
2. a **parabola** if the product AC is 0;
3. a **hyperbola** if the product AC is negative.

Finally, the three conic sections are characterized by certain geometric properties that are similar in nature. We will make use of these properties in the work to follow.

Geometric Properties of Conics

1. An ellipse consists of the set of points in the plane for which the sum of the distances d_1 and d_2 to two *foci* (plural of *focus*) is a fixed constant. See the first diagram in Figure 2.
2. A parabola consists of the set of points that are the same distance d from a line (called the *directrix*), and a point not on the line (called the *focus*). See the second diagram in Figure 2.
3. A hyperbola consists of the set of points for which the magnitude of the difference of the distances d_1 and d_2 to two *foci* is fixed. See the third diagram in Figure 2.

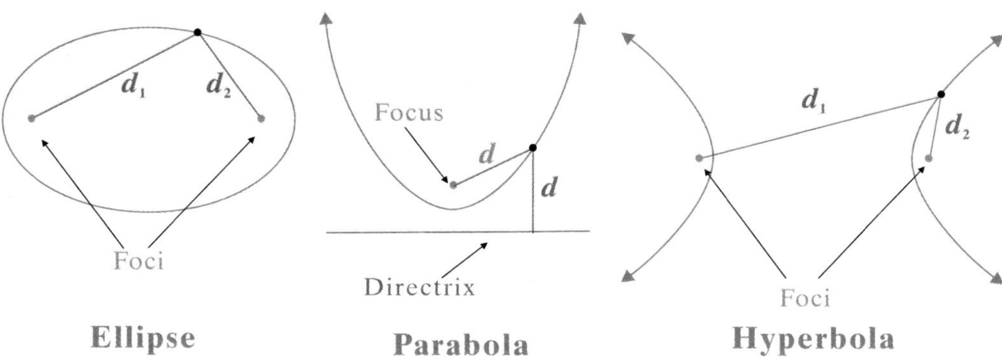

Figure 2: Properties of Conics

Topic 2: Derivation of the Standard Form of an Ellipse

Our goal is to use the properties of the conic sections to derive useful forms of their equations. Specifically, we want to be able to easily construct the graph of a conic section from its equation and, reversing the process, be able to find the equation for a conic section with known properties. We begin with the ellipse.

In order to simplify our work, we will initially assume we are working with an ellipse centered at the origin with foci at $(-c, 0)$ and $(c, 0)$; after deriving a useful form of the equation for this ellipse, it will be an easy matter to generalize it for an arbitrary ellipse. What we seek is an equation in x and y that identifies those points (x, y) on the ellipse, and the only property we need is that the sum $d_1 + d_2$ from Figure 3 is fixed. To make our algebra easier, we will denote this sum as $2a$; that is, $d_1 + d_2 = 2a$.

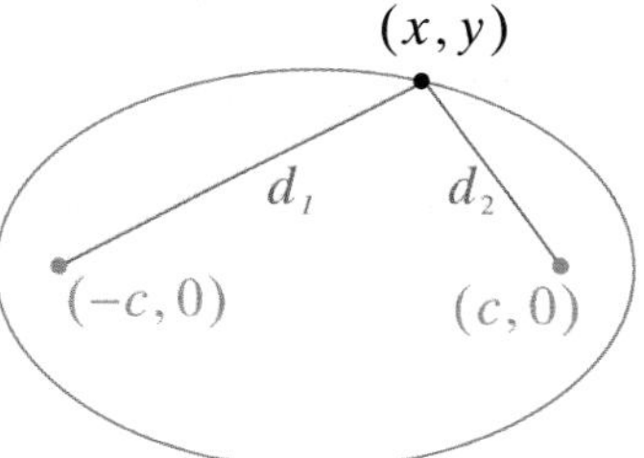

Figure 3: Parts of an Ellipse

The rest of the derivation consists of applying the distance formula from Section 2.1 followed by algebraic simplification. Using the distance formula to find d_1 and d_2, we know that

$$\sqrt{(x+c)^2 + y^2} + \sqrt{(x-c)^2 + y^2} = 2a.$$

In one sense, we have achieved our goal, as this is indeed an equation in x and y that describes all the points on the ellipse. What we desire, though, is to make the equation easier to use and interpret. We can do this by eliminating the radicals in the equation (as you learned to do in Section 1.7) and by ultimately renaming some of the constants that appear in the equation. We begin by isolating one radical, squaring both sides, and eliminating some terms:

$$\left(\sqrt{(x+c)^2 + y^2}\right)^2 = \left(2a - \sqrt{(x-c)^2 + y^2}\right)^2$$

$$(x+c)^2 + y^2 = 4a^2 - 4a\sqrt{(x-c)^2 + y^2} + (x-c)^2 + y^2$$

$$\cancel{x^2} + 2cx + \cancel{c^2} + \cancel{y^2} = 4a^2 - 4a\sqrt{(x-c)^2 + y^2} + \cancel{x^2} - 2cx + \cancel{c^2} + \cancel{y^2}.$$

There is still one radical left, so we isolate it and repeat the process:

$$4cx - 4a^2 = -4a\sqrt{(x-c)^2 + y^2}$$

$$a^2 - cx = a\sqrt{(x-c)^2 + y^2}$$

$$\left(a^2 - cx\right)^2 = a^2\left((x-c)^2 + y^2\right)$$

$$a^4 - \cancel{2a^2cx} + c^2x^2 = a^2x^2 - \cancel{2a^2cx} + a^2c^2 + a^2y^2$$

$$a^4 - a^2c^2 = a^2x^2 - c^2x^2 + a^2y^2.$$

This form of the equation has the virtue of no square root terms, but we can make a superficial change to improve the appearance even more. From the diagram in Figure 3, it's clear that $2a$ is larger than $2c$; that is, even in the extreme case of very skinny ellipses, the sum $d_1 + d_2$ (which we have called $2a$) will approach $2c$ but will always be larger than $2c$. We can conclude that $a > c$, so $a^2 - c^2$ is a positive number. All of this is simply to justify renaming $a^2 - c^2$ as b^2. In other words, we replace the constant $a^2 - c^2$ with the simpler symbol b^2, and we are safe in replacing $a^2 - c^2$ with what is clearly a non-negative number because we have shown that $a^2 - c^2$ is positive.

Thus:

$$a^4 - a^2c^2 = a^2x^2 - c^2x^2 + a^2y^2$$

$$a^2\left(a^2 - c^2\right) = \left(a^2 - c^2\right)x^2 + a^2y^2$$

$$a^2b^2 = b^2x^2 + a^2y^2.$$

If we write this as $b^2x^2 + a^2y^2 - a^2b^2 = 0$, we have a simpler equation that does indeed describe an ellipse by the algebraic definition (since the product of the coefficients of x^2 and y^2 is positive). For our purposes, we will find it useful to make one more cosmetic adjustment and divide the equation by a^2b^2:

$$b^2x^2 + a^2y^2 = a^2b^2$$

$$\frac{x^2}{a^2} + \frac{y^2}{b^2} = 1.$$

Before generalizing this equation to arbitrary ellipses, let's examine its utility with an example.

example 1

Graph the equation $\frac{x^2}{16}+\frac{y^2}{9}=1$, and determine the foci of the resulting ellipse.

Solution:

This equation is in the form derived above, so we can see by inspection that $a = 4$ and $b = 3$. We can calculate the x and y intercepts of such an equation: when $y = 0$ it must be the case that $x^2 = 16$, so $x = \pm 4$, and when $x = 0$ the equation reduces to $y^2 = 9$, so $y = \pm 3$. So $(-4, 0)$ and $(4, 0)$ are the two x-intercepts and $(0, -3)$ and $(0, 3)$ are the two y-intercepts. Further, these four points are extremes of the graph, as any x-value larger than 4 in magnitude leads to imaginary values for y, and any y-value larger than 3 in magnitude leads to imaginary values for x.

Before plotting these intercepts and filling in the rest of the ellipse, we will use the relation $a^2 - c^2 = b^2$ from the derivation above to determine c, and hence the location of the two foci. Rearranging the terms of the equation, we obtain $c^2 = a^2 - b^2$, and so

$$c^2 = 16 - 9 = 7$$
$$c = \pm\sqrt{7}.$$

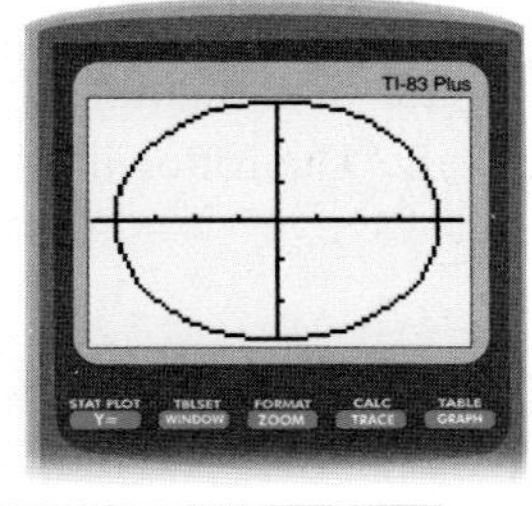

To display the graph of an ellipse, select the blue **Apps** button. Select `3:Conics`, then `2:Ellipse`. Now select **1** to enter a horizontal ellipse, and plug in the values from the equation. To view the graph, select **Graph**. (You may need to visit the Texas Instruments website to download the "Conic Graphing" application.)

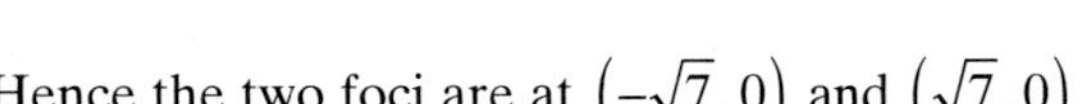

Hence the two foci are at $\left(-\sqrt{7}, 0\right)$ and $\left(\sqrt{7}, 0\right)$.

Putting these pieces together, and connecting the four extreme points with an elliptically shaped curve, we obtain the picture below.

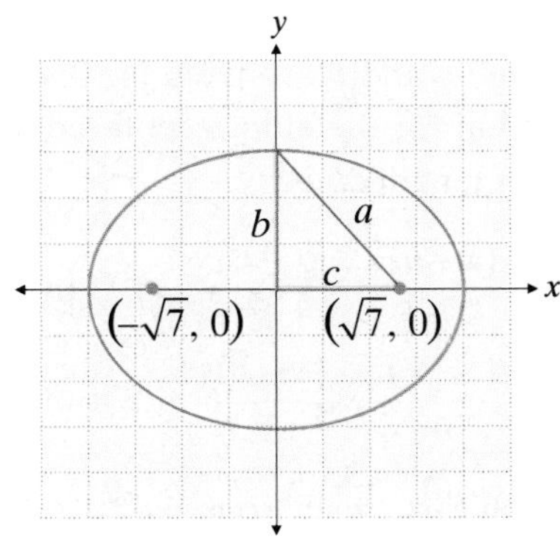

Topic 3: Using the Standard Form of an Ellipse

We have nearly, but not quite, derived the general form of an ellipse. At this point, we know that

$$\frac{x^2}{a^2}+\frac{y^2}{b^2}=1$$

describes an ellipse centered at the origin that is elongated horizontally. If the foci of an origin-centered ellipse are instead at $(0, c)$ and $(0, -c)$, so that the ellipse is elongated vertically, then the equation is

$$\frac{x^2}{b^2}+\frac{y^2}{a^2}=1,$$

where we are again assuming $a > b$ and $c^2 = a^2 - b^2$. The last task is to come up with the equation for an ellipse that is centered at the point (h, k). But we know how to do this: replacing x with $x - h$ in an equation shifts the graph of the equation h units horizontally, and replacing y with $y - k$ shifts the graph k units vertically.

To simplify our discussion, we will say the **major axis** of an ellipse is the line segment extending from one extreme point of the ellipse to the other and passing through the two foci and the center. The two ends of the major axis are the **vertices** of the ellipse. The **minor axis** is the line segment that also passes through the center and is perpendicular to the major axis; it spans the distance between the other two extremes of the ellipse. Thus, $2a$ and $2b$ are the lengths of the major and minor axes, respectively. The following box uses these terms to summarize our work.

Standard Form of an Ellipse

Assume a and b are fixed positive numbers with $a > b$. The standard form of the equation for the ellipse centered at (h, k) with major axis of length $2a$ and minor axis of length $2b$ is:

- $\frac{(x-h)^2}{a^2}+\frac{(y-k)^2}{b^2}=1$ if the major axis is horizontal;
- $\frac{(x-h)^2}{b^2}+\frac{(y-k)^2}{a^2}=1$ if the major axis is vertical.

In both cases, the two foci of the ellipse are located on the major axis c units away from the center of the ellipse, where $c^2 = a^2 - b^2$.

The next two examples will demonstrate how we can use the standard form to perform two different tasks.

example 2

Graph the equation $25x^2 + 4y^2 + 100x - 24y + 36 = 0$, and determine the foci of the resulting ellipse.

Solution:

The equation $25x^2 + 4y^2 + 100x - 24y + 36 = 0$ definitely represents an ellipse, as the product of the coefficients of x^2 and y^2 is positive, but the equation in this form tells us almost nothing about the specifics of the ellipse. We have encountered similar situations previously, and a familiar technique will assist us here just as it has in the past. That technique is the method of completing the square, as shown below.

$$25x^2 + 4y^2 + 100x - 24y + 36 = 0$$
$$25\left(x^2 + 4x\right) + 4\left(y^2 - 6y\right) = -36$$
$$25\left(x^2 + 4x + 4\right) + 4\left(y^2 - 6y + 9\right) = -36 + 100 + 36$$
$$25(x+2)^2 + 4(y-3)^2 = 100$$
$$\frac{(x+2)^2}{4} + \frac{(y-3)^2}{25} = 1$$

We begin by rearranging the terms as shown. Recall that the method involves completing the square (for both x and y in this case) in order to obtain perfect square trinomials. This requires adding 4 to the expression $x^2 + 4x$ and 9 to $y^2 - 6y$, and compensating by adding numbers to the right-hand side as well.

Note that the last step in the work above is to divide both sides by 100 in order to obtain the number 1 on the right-hand side. At this point, we have an equivalent equation that is in the standard form for an ellipse.

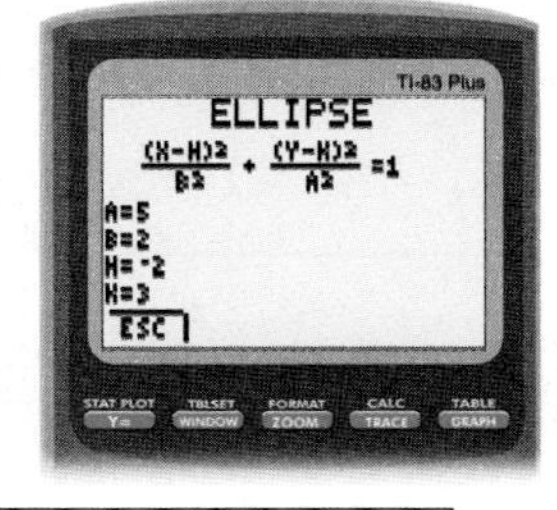

This ellipse is graphed in the same manner as the previous one, except the second equation form was selected for a vertical ellipse. Note that A and B are now switched. Also, remember that the value for H in this case is negative.

From the standard form, we can tell that the center of the ellipse is (–2, 3), that the ellipse is elongated vertically (since the denominator under the y term is larger than that under the x term), and that $a = 5$ and $b = 2$. So the length of the major (vertical) axis is 10 and the length of the minor axis is 4. This gives us enough information to sketch the graph at right.

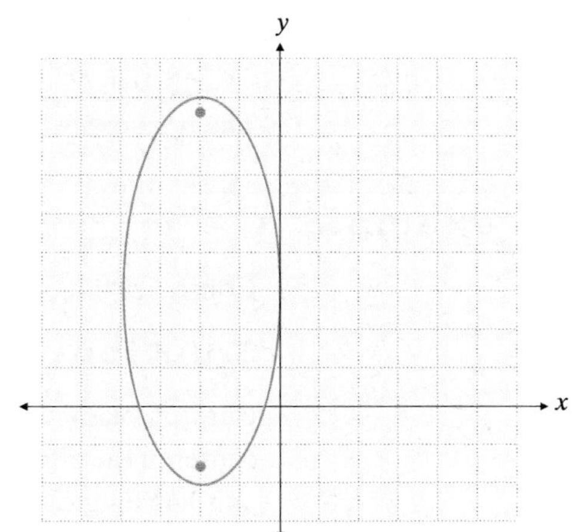

To determine the locations of the two foci, we solve the equation $c^2 = a^2 - b^2 = 21$ for c, obtaining $c = \pm\sqrt{21} \approx \pm 4.6$. So the two foci are approximately 4.6 units above and below the center of the ellipse, at (–2, 7.6) and (–2, –1.6).

technology note

A computer algebra system, such as *Mathematica*, can sketch the graphs of equations as well as functions, and therefore can be used to construct graphs of ellipses. Such graphs can't be relied on to identify features of an ellipse such as the lengths of the major and minor axes or the locations of the foci, but they can be useful in checking your work. In *Mathematica*, the command used to graph an equation is **`ImplicitPlot`**. To use the command, you must first execute the command **`<<Graphics`ImplicitPlot``** (this extends the basic capabilities of *Mathematica* and must be executed only one time in a *Mathematica* session and only if you intend to use **`ImplicitPlot`**). The use of these commands is illustrated below in sketching the graph of $x^2-6x+2y^2+8y=3$.

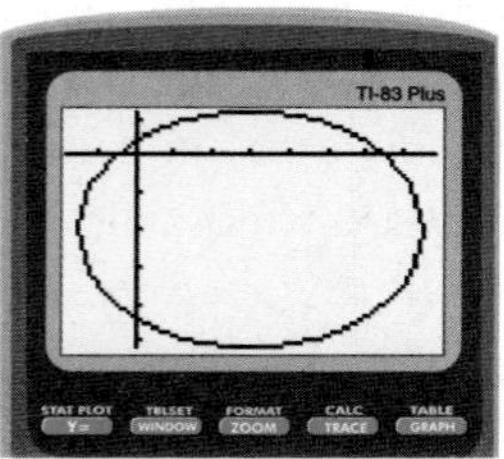

Again, select **Apps**, `3:Conics`, and `2:Ellipse`. This graph uses equation 1 for a horizontal ellipse.

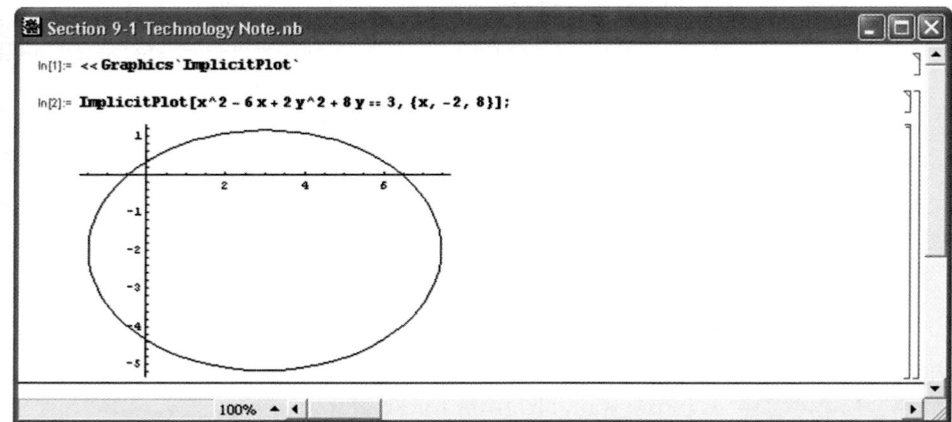

The next example reverses the process, and is faster.

example 3

Construct the equation, in standard form, for the ellipse whose graph is given below.

Solution:

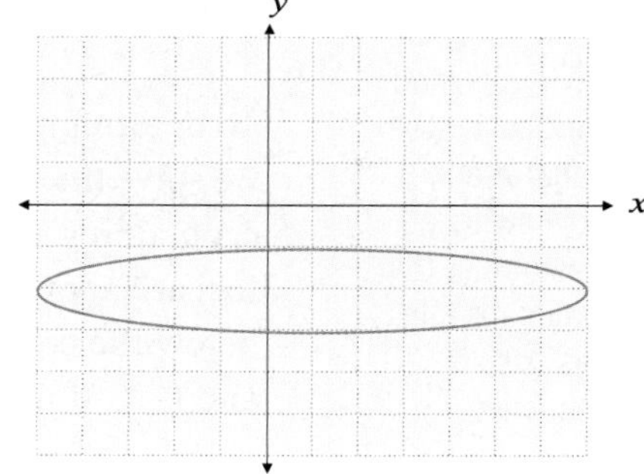

Careful examination of the graph shows that the center of the ellipse is at $(1,-2)$, and that the major axis has length 12 and the minor axis has length 2. This means that $a=6$ and $b=1$, and since the ellipse is elongated horizontally, the equation must be $\frac{(x-1)^2}{36}+(y+2)^2=1$.

Example 2 demonstrated that the technique of graphing ellipses bears much similarity to that of graphing circles, and this similarity warrants further investigation. To begin, we might wonder about the result if the numbers a and b in the standard form of an ellipse happen to be the same. If $a = b = r$, then the standard form is

$$\frac{(x-h)^2}{r^2} + \frac{(y-k)^2}{r^2} = 1,$$

which, if we multiply through by r^2, gives us $(x-h)^2 + (y-k)^2 = r^2$, the standard form of a circle. So a circle, not surprisingly, is an ellipse whose major and minor axes are the same length.

A circle, viewed as a type of ellipse, has only one focus, and that focus coincides with the center of the circle. Ellipses that are nearly circular have two foci, but they are relatively close to the center, while narrow ellipses have foci far away from the center. This gives us a convenient way to characterize the relative "skinniness" of an ellipse.

Eccentricity of an Ellipse

Given an ellipse with major and minor axes of lengths $2a$ and $2b$, respectively, the **eccentricity** of the ellipse, often given the symbol e, is defined by

$$e = \frac{c}{a} = \frac{\sqrt{a^2 - b^2}}{a}.$$

If $e = 0$, then $c = 0$ and the ellipse is actually a circle. At the other extreme, c may be close to (but cannot equal) a, in which case e is close to 1. An eccentricity close to 1 indicates a relatively narrow ellipse.

Topic 4: Interlude: Planetary Orbits

Johannes Kepler (1571-1630) was the German-born astronomer/mathematician who first demonstrated that the planets in our solar system follow elliptical orbits. He did this by laboring, over a period of twenty-one years, to mathematically model the astronomical observations of his predecessor Tycho Brahe. The ultimate result was Kepler's Three Laws of Planetary Motion.

I. The planets orbit the sun in elliptical paths, with the sun at one focus of each orbit.

II. A line segment between the sun and a given planet sweeps over equal areas of space in equal intervals of time.

III. The square of the time needed for a planet's complete revolution about the sun is proportional to the cube of the orbit's semimajor (half of the major) axis.

You are probably already familiar with the fact that the shapes of the various planetary orbits in our solar system vary widely, and that these differences greatly influence seasons on the planets. For instance, Pluto has a much more eccentric orbit than Earth, in both the informal and formal meaning of the term. The eccentricity of Pluto's orbit is approximately 0.248, while the eccentricity of Earth's orbit is approximately 0.017. The eccentricities can be easily calculated from astronomical data, but we can also use the eccentricity of a planet's orbit to answer particular questions, as follows.

example 4

Given that the furthest Earth gets from the sun is approximately 94.56 million miles, and that the eccentricity of Earth's orbit is approximately 0.017, estimate the closest approach of the Earth to the sun.

Solution:

Kepler's first law states that each planet follows an elliptical orbit, and that the sun is positioned at one focus of the ellipse. Thus the Earth is furthest from the sun when Earth is at the end of the ellipse's major axis on the other side of the center from the sun.

Using our standard terminology, this means that

$$a + c = 94.56 \text{ (million miles).}$$

By the definition of eccentricity, we also know that

$$0.017 = \frac{c}{a},$$

or $c = 0.017a$. Combining these pieces of information, we get $1.017a = 94.56$, leading to $a = 92.98$ million miles. From this, we know $c = (0.017)(92.98) = 1.58$ million miles. The closest approach to the sun must be $a - c$, which is thus approximately 91.4 million miles.

exercises

Find the center, foci, and vertices of each ellipse that the equation describes. See Examples 1 and 2.

1. $\frac{(x-5)^2}{4} + \frac{(y-2)^2}{25} = 1$

2. $\frac{(x+3)}{9} + \frac{(y+1)}{16} = 1$

3. $(x+2)^2 + 3(y+5)^2 = 9$

4. $4(x-4)^2 + (y-2)^2 = 8$

5. $x^2 + 6x + 2y^2 - 8y + 13 = 0$

6. $2x^2 + y^2 - 4x + 4y - 10 = 0$

7. $4x^2 + y^2 + 40x - 2y + 85 = 0$

8. $x^2 + 2y^2 - 6x + 16y + 37 = 0$

9. $x^2 + 3y^2 + 8x - 12y + 1 = 0$

10. $4x^2 + 3y^2 - 8x + 18y + 19 = 0$

11. $x^2 - 4x + 5y^2 - 1 = 0$

12. $x^2 + 4y^2 + 24y + 28 = 0$

Match the corresponding equation to the appropriate graph. See Examples 1 and 2.

13. $\dfrac{(x-3)^2}{9}+\dfrac{(y-2)^2}{25}=1$

14. $\dfrac{(x-1)^2}{4}+\dfrac{y^2}{16}=1$

15. $x^2+\dfrac{(y-3)^2}{4}=1$

16. $\dfrac{x^2}{16}+\dfrac{y^2}{9}=1$

17. $\dfrac{(x-2)^2}{9}+\dfrac{(y+2)^2}{9}=1$

18. $\dfrac{x^2}{4}+\dfrac{y^2}{16}=1$

19. $\dfrac{(x-1)^2}{4}+y^2=1$

20. $\dfrac{(x+2)^2}{9}+\dfrac{(y-1)^2}{4}=1$

a.

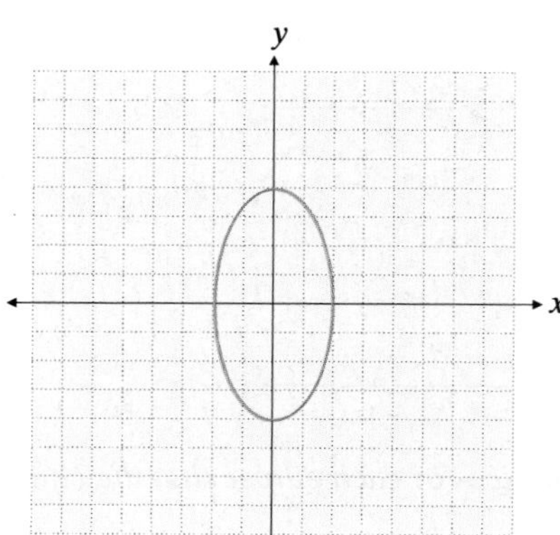

b.

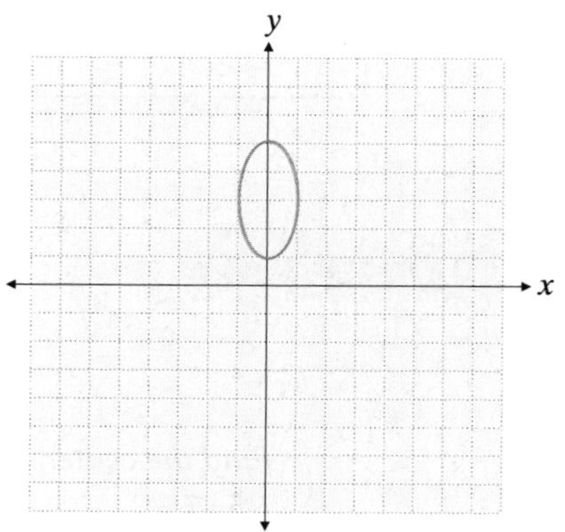

c.

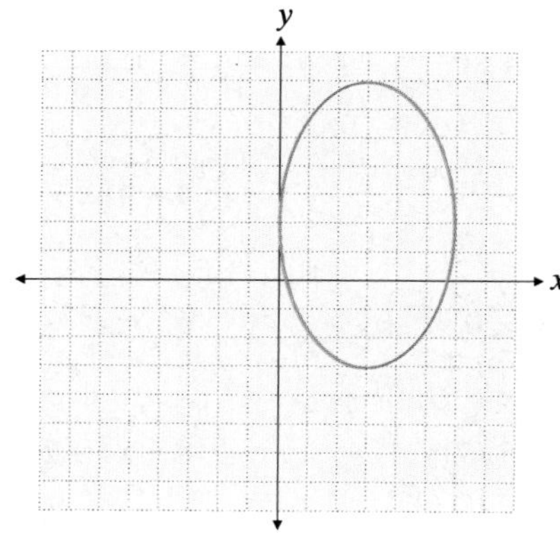

d.

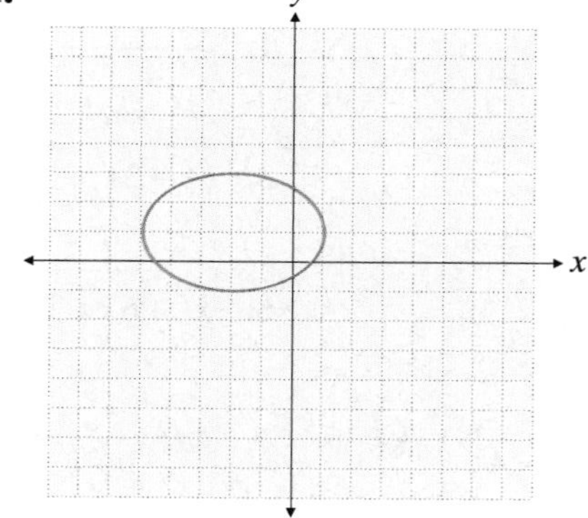

e.

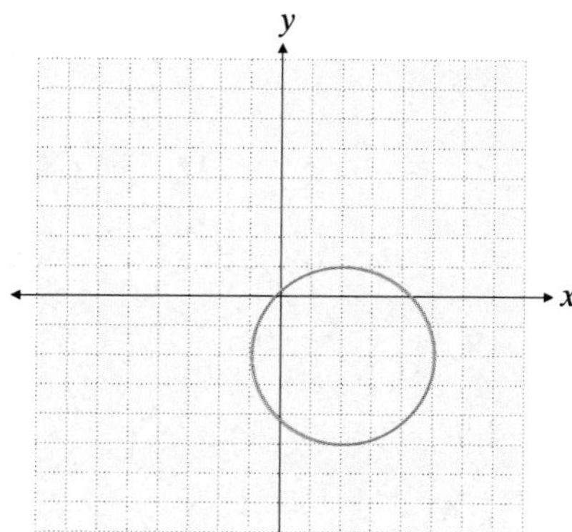

f.

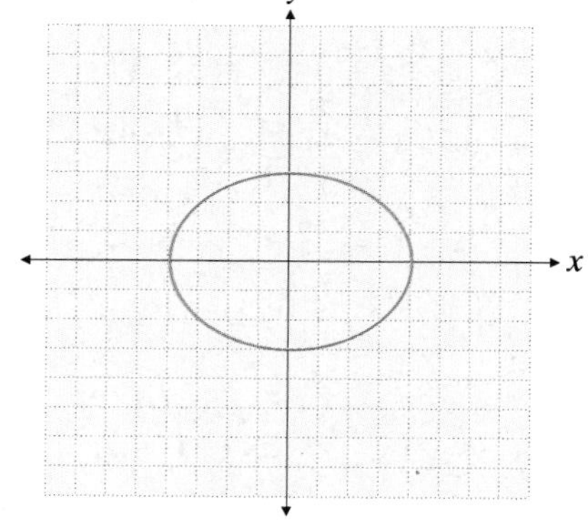

g.

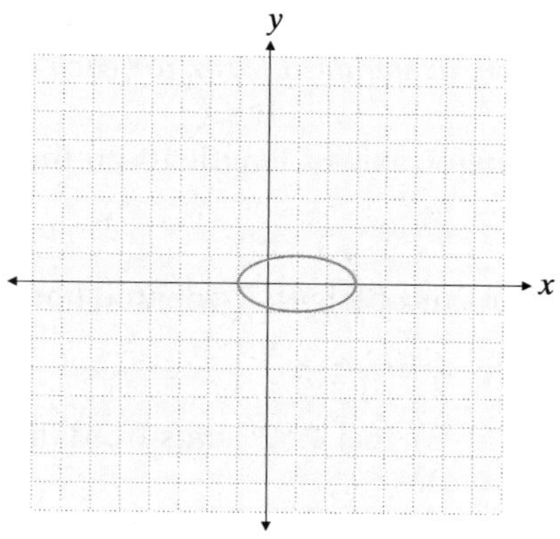

h.

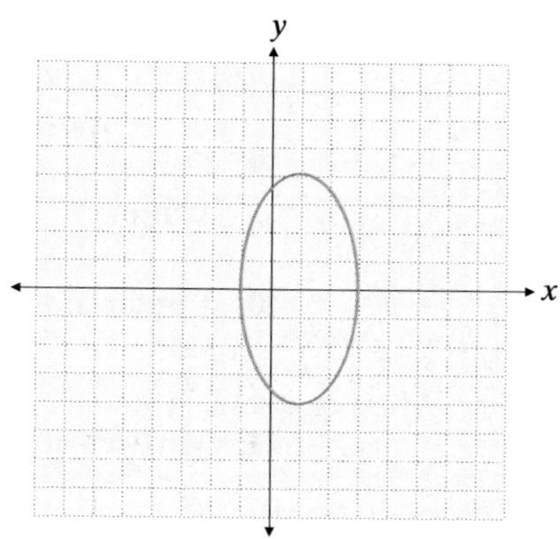

Sketch the graphs of the following ellipses, and determine the coordinates of the foci. See Examples 1 and 2.

21. $\frac{(x-3)^2}{9}+\frac{(y+1)^2}{1}=1$

22. $\frac{(x+5)^2}{4}+\frac{(y+2)^2}{16}=1$

23. $\frac{(x-3)^2}{9}+\frac{(y-4)^2}{4}=1$

24. $\frac{x^2}{25}+\frac{(y-3)^2}{16}=1$

25. $(x-1)^2+\frac{(y-4)^2}{4}=1$

26. $\frac{(x-4)^2}{16}+\frac{(y-4)^2}{4}=1$

27. $\frac{(x+1)^2}{25}+\frac{(y+5)^2}{4}=1$

28. $\frac{(x-2)^2}{9}+\frac{(y+1)^2}{9}=1$

29. $\frac{(x+2)^2}{16}+\frac{(y+1)^2}{9}=1$

30. $\frac{x^2}{25}+(y+2)^2=1$

31. $9x^2+16y^2+18x-64y=71$

32. $9x^2+4y^2-36x-24y+36=0$

33. $16x^2+y^2+160x-6y=-393$

34. $25x^2+4y^2-100x+8y+4=0$

35. $4x^2+9y^2+40x+90y+289=0$

36. $16x^2+y^2-64x+6y+57=0$

37. $4x^2+y^2+4y=0$

38. $9x^2+4y^2+108x-32y=-352$

In each of the following problems, an ellipse is described either by picture or by properties it possesses. Find the equation, in standard form, for each ellipse. See Example 3.

39. Center at the origin, major axis of length 10 on the y-axis, foci 3 units from the center.

40. Center at $(-2, 3)$, major axis of length 8 oriented horizontally, minor axis of length 4.

41. Vertices at $(1, 4)$ and $(1, -2)$, foci $2\sqrt{2}$ units from the center.

42. Vertices at $(5, -1)$ and $(1, -1)$, minor axis of length 2.

43. Foci at $(0, 0)$ and $(6, 0)$, $e = \frac{1}{2}$.

44. Vertices at $(-1, 4)$ and $(-1, 0)$, $e = 0$.

45. Vertices at $(-2, -1)$ and $(-2, -5)$, minor axis of length 2.

46. Vertices at $(-4, 6)$ and $(-14, 6)$, $e = \frac{2}{5}$.

47. Vertices at $(1, 3)$ and $(9, 3)$, one of the foci at $(6, 3)$.

48. Foci at $(2, -4)$ and $(2, -8)$, minor axis of length 6.

49.

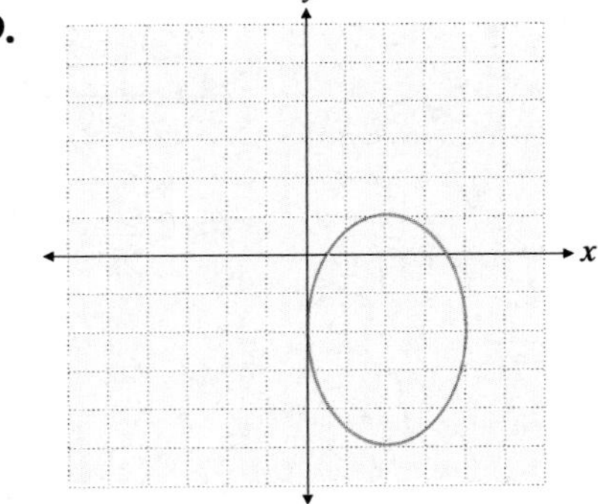

50.

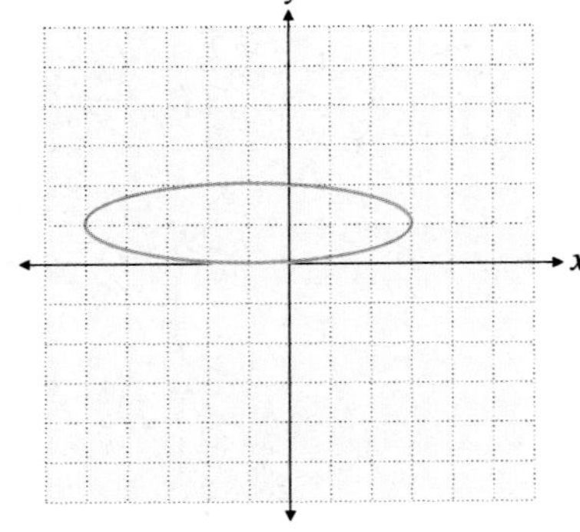

51.

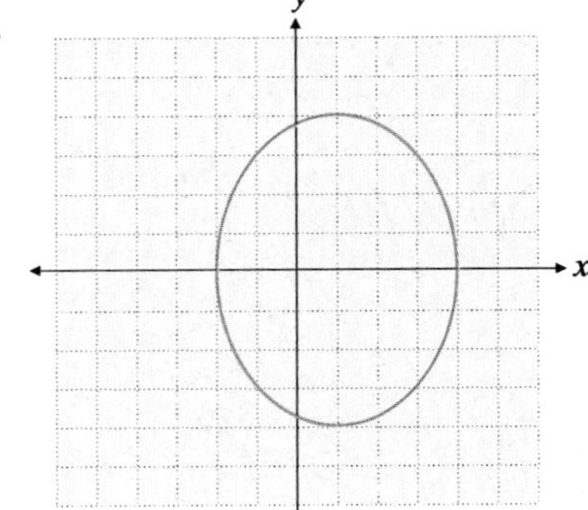

52.

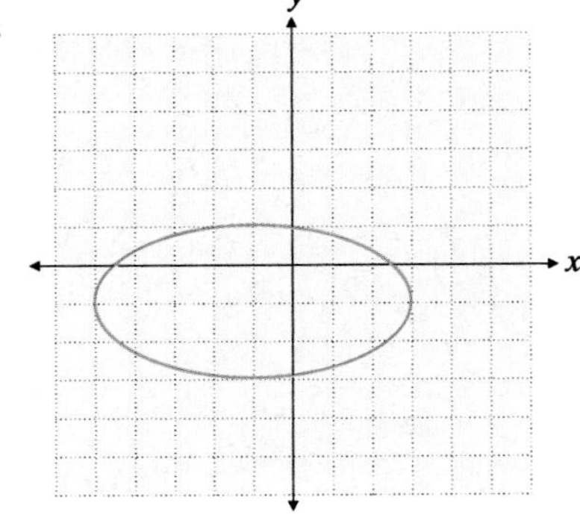

Find the eccentricity and the lengths of the minor and major axes of the following ellipses.

53. $\frac{x^2}{100}+\frac{y^2}{144}=1$ **54.** $\frac{x^2}{64}+\frac{y^2}{9}=1$ **55.** $x^2+9y^2=36$

56. $25x^2+4y^2=100$ **57.** $4x^2+16y^2=16$ **58.** $5x^2+8y^2=40$

59. $20x^2+10y^2=40$ **60.** $\frac{1}{4}x^2+\frac{1}{12}y^2=\frac{1}{2}$ **61.** $x^2=49-7y^2$

Use your knowledge of ellipses to answer the following questions. See Example 4.

62. The orbit of Halley's Comet is an ellipse with an eccentricity of 0.967. Its closest approach to the sun is approximately 54,591,000 miles. What is the furthest Halley's Comet ever gets from the sun?

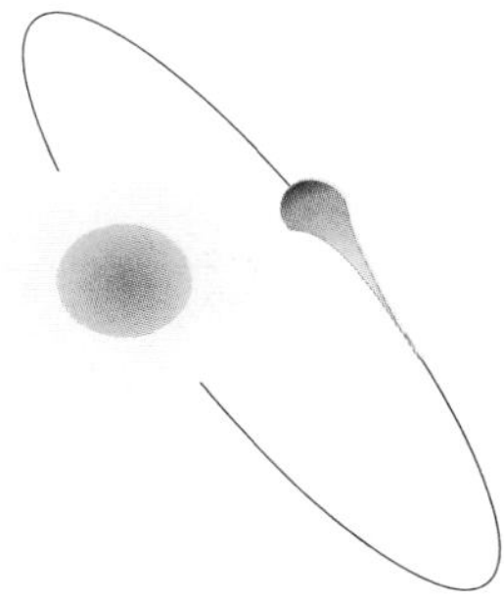

63. Pluto's closest approach to the sun is approximately 4.43×10^9 kilometers, and its maximum distance from the sun is approximately 7.37×10^9 kilometers. What is the eccentricity of Pluto's orbit?

64. The archway supporting a bridge over a road is in the shape of half an ellipse. The archway is 60 feet wide and is 15 feet tall at the middle. A large truck is 10 feet wide and 14 feet, 9 inches tall. Is the truck capable of passing under the archway?

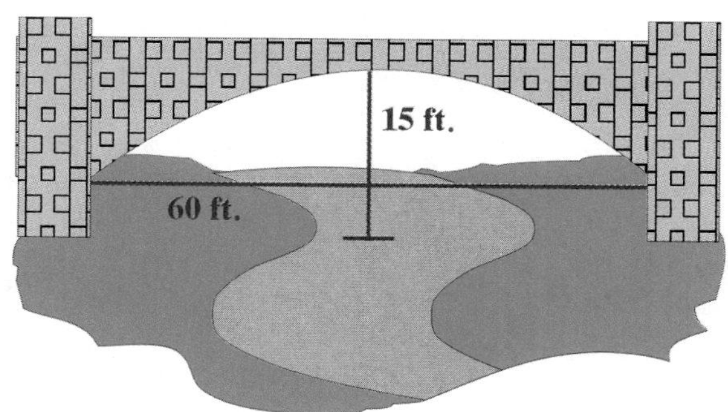

65. Use the information given in Example 4 to determine the length of the minor axis of the ellipse formed by Earth's orbit around the sun.

66. Since the sum of the distances from each of the two foci to any point on an ellipse is constant, we can draw an ellipse using the following method. Tack the ends of a length of string at two points (the foci) and, keeping the string taut by pulling outward with the tip of a pencil, trace around the foci to form an ellipse (the total length of the string remains constant). If you want to create an ellipse with a major axis of length 5 cm and a minor axis of length 3 cm, how long should your string be and how far apart should you place the tacks? Use the relationships of distances and formulas that you have learned in this section.

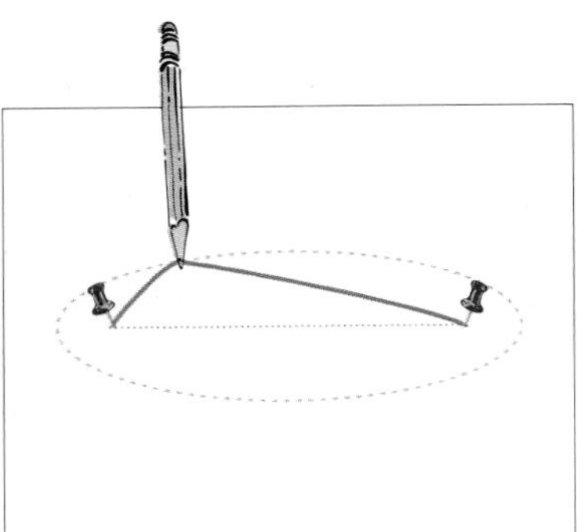

technology exercises

Use a computer algebra system such as Mathematica (or a suitably equipped graphing calculator) to graph the following equations. See the Technology Note after Example 2 for guidance.

67. $15x^2 + 9y^2 + 150x - 36y = -276$

68. $5x^2 + 12y^2 - 20x + 144y + 392 = 0$

69. $3x^2 + 2y^2 = 3 - 18x$

70. $2x^2 + 5y^2 = 70y - 205$

9.2 The Parabola

TOPICS

1. Derivation of the standard form of a parabola (as a conic section)
2. Using the standard form of a parabola
3. Interlude: parabolic mirrors

Topic 1: Derivation of the Standard Form of a Parabola (as a Conic Section)

We have already studied, in Section 3.2, parabolas as they arise in the context of quadratic functions. Specifically, we know that any function of the form

$$f(x) = a(x-h)^2 + k$$

describes a vertically-oriented parabola in the plane (opening upward or downward, depending on the sign of a) whose vertex is at (h, k). (We also know how to use the method of completing the square to transform a function $f(x) = ax^2 + bx + c$ into the more useful form above; we will have occasion to use this skill again in this section.) The material in this section does not replace what we have learned. Instead, viewing parabolas as one of the three varieties of conic sections broadens our understanding of them, and allows us to work with parabolic curves that are not defined by functions.

Recall from Section 9.1 that a principal geometric property of parabolas is that each point on a given parabola is equidistant from a fixed point called the *focus* and a fixed line called the *directrix*. Just as with ellipses, we will derive a useful form of the equation for a parabola from one geometric property. The form that we seek will allow us to quickly sketch fairly accurate graphs of parabolas if we are given the equation, and in turn construct equations for a parabola if we are given some geometric information about it.

Following convention, we will let p denote the distance between the vertex of the parabola and the focus, and hence also the distance between the vertex and the directrix, as shown in Figure 1. And as in Section 9.1, we will begin our derivation by assuming the parabola is oriented vertically and that the vertex is at the origin. After our initial work, it will be a relatively simple matter to generalize to horizontally-oriented parabolas and parabolas whose vertices are not at the origin.

Given the information above, we know the focus of our parabola is at $(0, p)$ and the equation for the directrix is $y = -p$. The equation below then characterizes all those points that are of equal distance d from the focus and the directrix:

$$\sqrt{x^2+(y-p)^2}=\sqrt{(x-x)^2+(y+p)^2}.$$

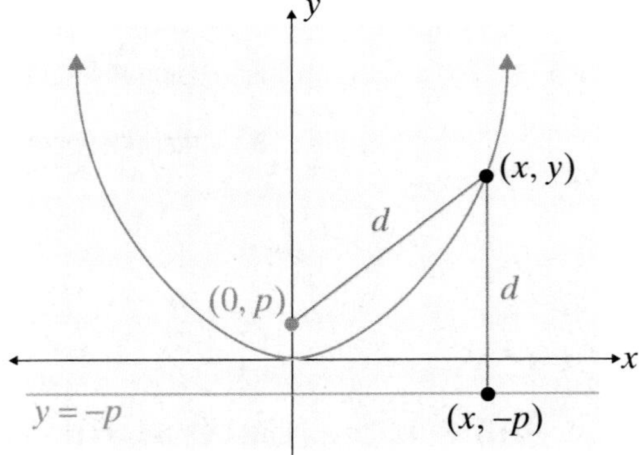

Figure 1: Parts of a Parabola

One way to make the equation look better is to square both sides. We can then proceed to eliminate some terms and combine others to arrive at a useful form:

$$\begin{aligned} x^2+(y-p)^2 &= (y+p)^2 \\ x^2+y^2-2py+p^2 &= y^2+2py+p^2 \\ x^2 &= 4py. \end{aligned}$$

The conclusion is that any equation that can be written in the above form describes a parabola whose vertex is at the origin and whose focus is at $(0, p)$. This is true even if p is negative; if this is the case then the parabola opens downward.

As we have seen in other contexts, replacing x with y and vice-versa reflects a graph about the line $y = x$, so the equation $y^2 = 4px$ describes a horizontally-oriented parabola with vertex at $(0, 0)$. Swapping x and y reflects the focus and directrix of a parabola as well, so that the focus of $y^2 = 4px$ is at $(p, 0)$ and the directrix is the vertical line $x = -p$. And, as always, replacing x with $x - h$ and y with $y - k$ shifts a graph h units horizontally and k units vertically.

In summary:

Standard Form of a Parabola

Assume p is a fixed non-zero real number. The **standard form** of the equation for the parabola with vertex at (h,k) is:

- $(x-h)^2 = 4p(y-k)$ if the parabola is vertically-oriented.
  If $p > 0$, the parabola opens up.
  If $p < 0$, the parabola opens down.
  The focus is at $(h, k+p)$.
  The equation of the directrix is $y = k-p$.
- $(y-k)^2 = 4p(x-h)$ if the parabola is horizontally-oriented.
  If $p > 0$, the parabola opens right.
  If $p < 0$, the parabola opens left.
  The focus is at $(h+p, k)$.
  The equation of the directrix is $x = h-p$.

As you know, parabolas can be relatively wide or relatively narrow. The constant p in the standard form is the parameter that determines the degree of flatness of the parabola, a fact that will be visually clear to you if you draw several parabolas whose foci are of varying distances from their vertices.

Topic 2: Using the Standard Form of a Parabola

Putting the standard form of a parabola to use usually requires the use of some familiar algebraic techniques, as the next several examples demonstrate.

example 1

Graph the equation $-y^2+2x+2y+5=0$, and determine the focus and directrix of the resulting parabola.

Solution:

We know the equation $-y^2+2x+2y+5=0$, represents a parabola, as the product of the coefficients of x^2 and y^2 is 0 (that is, only one of the variables appears to the second power), but the equation in this form is difficult to graph. We can put the equation into the standard form (as a conic) by completing the square with respect to y (the squared variable) and then identifying p.

cont'd. on next page ...

$$-y^2+2x+2y+5=0$$
$$y^2-2y=2x+5$$
$$y^2-2y+1=2x+5+1$$
$$(y-1)^2=2(x+3)$$
$$(y-1)^2=4\left(\frac{1}{2}\right)(x+3)$$

We begin by rearranging the terms as shown. In order to complete the square, we have to add 1 to both sides.

The last step is to figure out what value the constant p has in the equation. To do this, we need to put the right-hand side into the form $4p(x-h)$, as shown.

Now that the equation is in standard form, by inspection we can tell that the vertex is at (–3, 1) and that $p=\frac{1}{2}$. Since the focus is p units to the right of the vertex and the directrix is a vertical line p units to the left of the vertex, we obtain

$$\text{focus: } \left(-\frac{5}{2},1\right)$$
$$\text{directrix: } x=-\frac{7}{2}.$$

The only remaining task is to sketch the graph. We can begin by plotting the vertex at (–3, 1). We can plot two other points by noting that the two points whose x-coordinates are $-\frac{5}{2}$ will be 1 unit away from the directrix; therefore they must lie 1 unit away from the focus as well. That is, $\left(-\frac{5}{2},0\right)$ and $\left(-\frac{5}{2},2\right)$ must also lie on the parabola. This gives us some idea of the "flatness" of the parabola.

To duplicate the graph below, select **Apps**, then `3:Conics` and `4:Parabola`. Select equation **1** for a horizontal parabola, and plug in the values from the equation. To view the graph, select **Graph**.

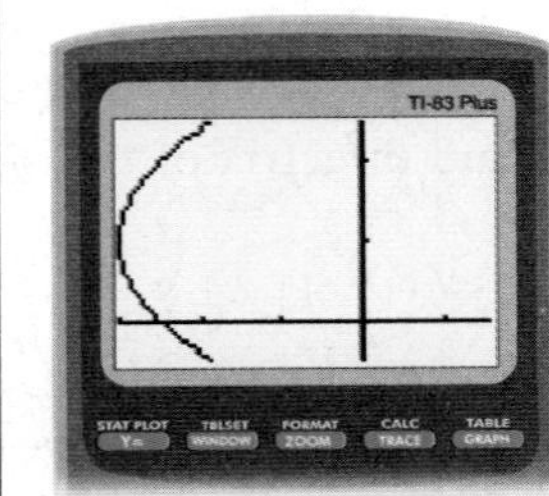

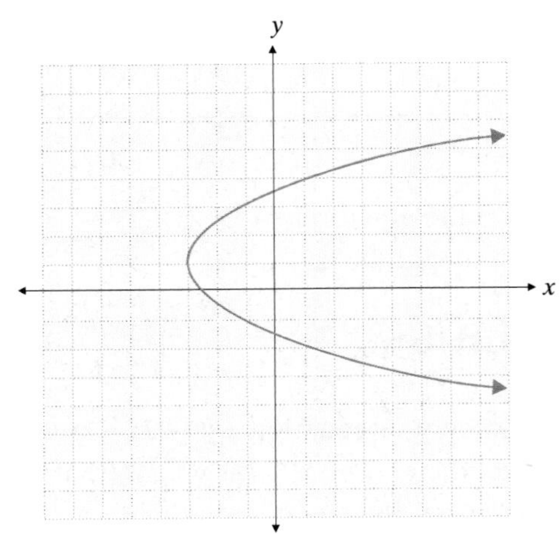

The last step in Example 1 warrants further discussion. Because every point on a given parabola is equidistant from the parabola's focus and its directrix, the two points that are each $2p$ units away from the directrix and from the focus can be plotted as shown in Figure 2.

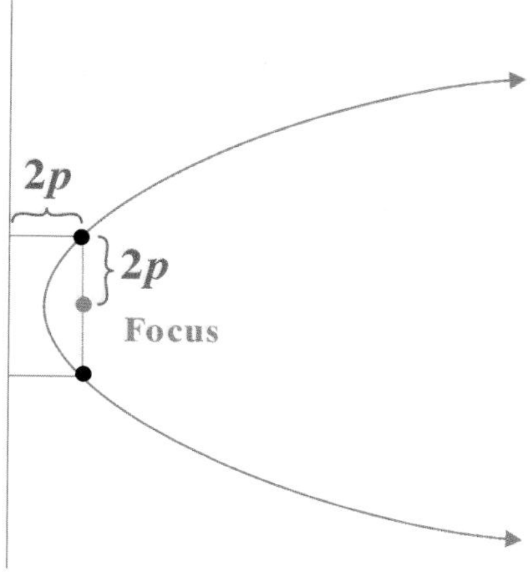

Figure 2: Two Easily Plotted Points

Example 2 asks for an equation based on some geometric information.

example 2

Given that the directrix of a parabola is the line $y = 2$, that $p = -1$, and that the parabola is symmetric with respect to the line $x = 2$, find the equation that defines the parabola.

Solution:

From the given information, we can immediately draw the picture at right to check our work. We know the parabola is oriented vertically because the directrix is horizontal, and from the equation for the line of symmetry we know the x-coordinate of the vertex (and the focus) must be 2. Moving down 1 unit from the directrix (as p is negative) puts the vertex at (2, 1). So the equation must be:

$$(x-2)^2 = -4(y-1).$$

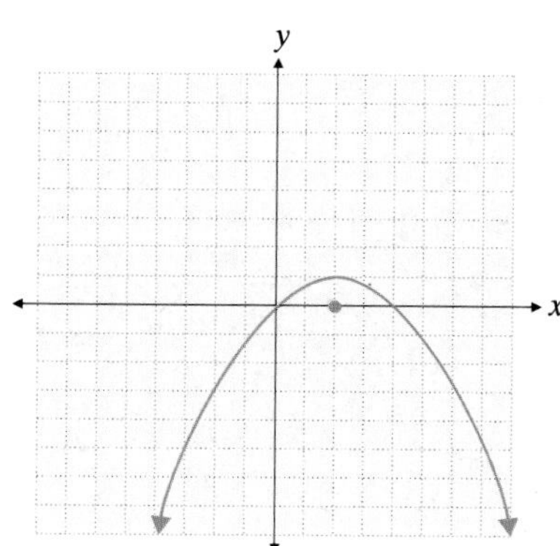

Topic 3: Interlude: Parabolic Mirrors

Parabolas arise naturally in many diverse contexts, and knowledge of parabolas is a key to solving many practical applications. We have already seen one very important application: the path described by a thrown object (in the absence of friction and other forces) is a parabola. More exactly, we know that the height $h(t)$ of an object with initial velocity v_0 and initial height h_0 is

$$h(t) = -\frac{1}{2}gt^2 + v_0t + h_0,$$

where g, a constant, is the acceleration due to gravity (see Section 1.7 for details). This is a classic example of a quadratic function, and we experience the fact that a parabola is the shape of its graph whenever we see, for example, a thrown baseball.

But parabolas also arise in contexts where a quadratic function may not be the most natural way to describe them or may simply be irrelevant. In some applications, the geometric properties of parabolas are the key issue. For instance, the fact that every point on a parabola is equidistant from its focus and its directrix has the consequence that the opposite angles shown in Figure 3 are equal to one another at each point. That is, if we imagine rays of light emanating from the focus of the parabola, each ray reflected from the inner surface of the parabola is parallel to the axis of symmetry.

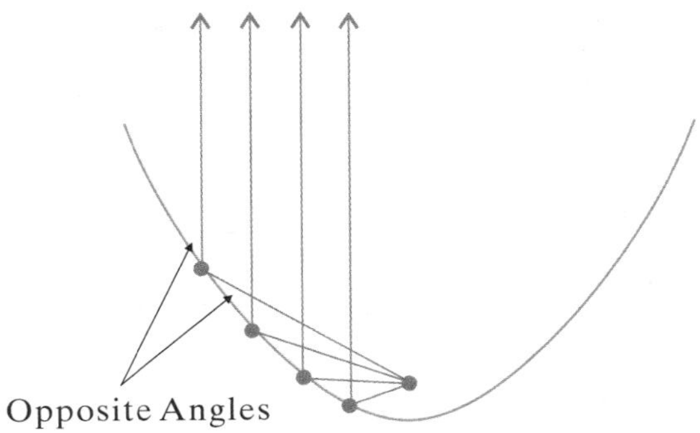

Figure 3: Equal Angles of Reflection in a Parabola

This property can be used in a variety of concrete applications. If a parabola is rotated about its axis of symmetry, a three-dimensional shape called a *paraboloid* is the result. A *parabolic mirror* is then made by coating the inner surface of a paraboloid with a reflecting material. Parabolic mirrors are the basis of searchlights and vehicle headlights, with a light source placed at the focus of the paraboloid, and are also the basis of one design of telescope, in which incoming (parallel) starlight is reflected to an eyepiece at the focus.

example 3

The Hale Telescope at the Mount Palomar observatory in California is a very large reflecting telescope. The paraboloid is the top surface of a large cylinder of Pyrex glass 200 inches in diameter. Along the outer rim, the cylinder of Pyrex is approximately 26.8 inches thick, while at the center the Pyrex is 23 inches thick. Given this, where is the focus of the parabolic mirror located?

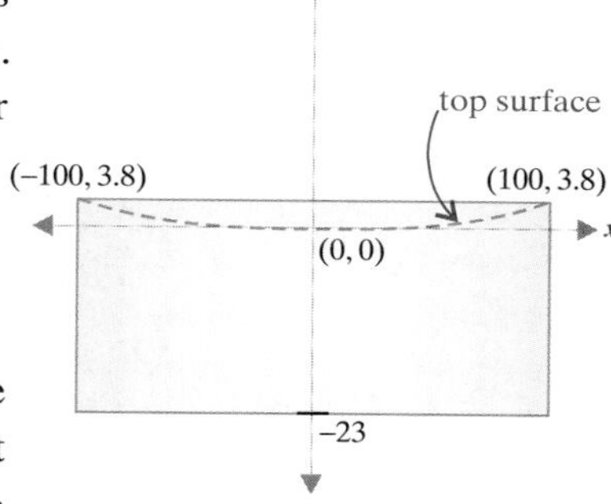

Solution:

In order to make the math as easy as possible, we can locate the origin of our coordinate system at the vertex of a parabolic cross-section of the Hale mirror, and we can assume the parabola opens upward. What we know, then, is that the equation $x^2 = 4py$ describes the shape of the cross-section for some p. We will be able to locate the focus of the parabola if we can determine p.

$(0, 0)$ is one point (namely the vertex) on the graph of $x^2 = 4py$. What we have been told will allow us to find two other points on the graph. Since the difference in thickness of the mirror between the center and the outer rim is 3.8 inches, and since the mirror has a diameter of 200 inches, the two points $(-100, 3.8)$ and $(100, 3.8)$ must also lie on the graph. Using either one of these points in the equation $x^2 = 4py$, we can solve for p:

$$\begin{aligned}(100)^2 &= 4p(3.8)\\ 10000 &= 15.2p\\ p &\approx 657.9 \text{ inches}\\ p &\approx 54.8 \text{ feet.}\end{aligned}$$

We know that the focus of a parabola is p units from the vertex, so the focus of the Hale Telescope is nearly 55 feet from the mirror.

exercises

Graph the following parabolas and determine the focus and directrix of each. See Example 1.

1. $(x+1)^2 = 4(y-3)$ **2.** $(y-4)^2 = -2(x-1)$ **3.** $y^2 - 4y = 8x + 4$

4. $y^2 + 2y + 12x + 37 = 0$ **5.** $x^2 - 8y = 6x - 1$ **6.** $x^2 + 6x + 8y = -17$

7. $(y-1)^2 = 8(x+3)$ **8.** $(x-2)^2 = 4(y+1)$ **9.** $(y+1)^2 = -12(x+1)$

10. $x^2 + 2x + 8y = 31$ **11.** $y^2 + 6y - 2x + 13 = 0$ **12.** $x^2 - 2x - 4y + 13 = 0$

13. $y^2 = 6x$ **14.** $x^2 = 2y$ **15.** $x^2 = 7y$

16. $x^2 = -5y$ **17.** $y = -12x^2$ **18.** $x = -4y^2$

19. $x = \frac{1}{6}y^2$ **20.** $\frac{1}{5}x = -y^2$ **21.** $y^2 + 16x = 0$

22. $-6x - 2y^2 = 0$ **23.** $4y + 2x^2 = 4$ **24.** $2y^2 - 10x = 10$

Find the equation, in standard form, for the parabola with the given properties or with the given graph. See Example 2.

25. Focus at $(-2, 1)$, directrix is the y-axis.

26. Focus at $(-2, 1)$, directrix is the x-axis.

27. Vertex at $(3, -1)$, focus at $(3, 1)$.

28. Symmetric with respect to the line $y = 1$, directrix is the line $x = 2$, and $p = -3$.

29. Vertex at $(3, -2)$, directrix is the line $x = -3$.

30. Vertex at $(7, 8)$, directrix is the line $x = \frac{27}{4}$.

31. Focus at $\left(-3, -\frac{3}{2}\right)$, directrix is the line $y = -\frac{1}{2}$.

32. Vertex at (3, 16), focus at (3, 11).

33. Vertex at (–4, 3), focus at $\left(-\frac{3}{2}, 3\right)$.

34. Symmetric with respect to the x-axis, focus at (–3, 0), and $p = 2$.

35.

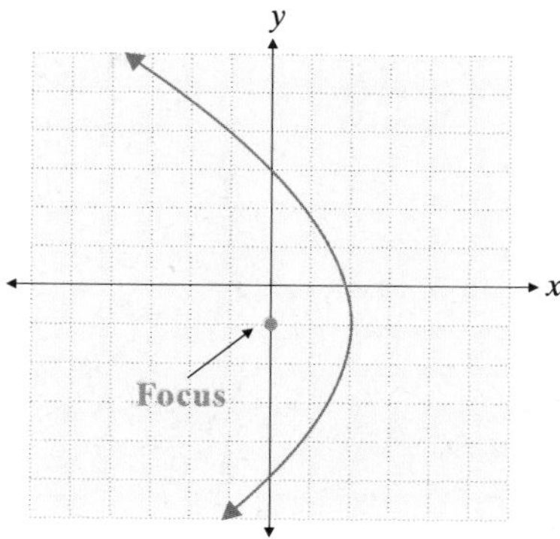

36.

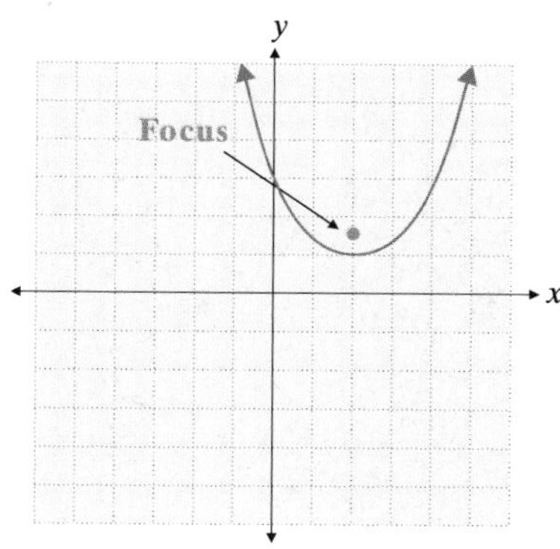

Match the following equations to the appropriate graph.

37. $(x+2)^2 = 3(y-1)$ **38.** $(y-1)^2 = 2(x+2)$ **39.** $y^2 = 4(x+1)$

40. $x^2 = 2(y+1)$ **41.** $(x-1)^2 = -(y-2)$ **42.** $(y+2)^2 = 3x$

43. $(x-2)^2 = 4y$ **44.** $y^2 = -2(x+1)$

a.

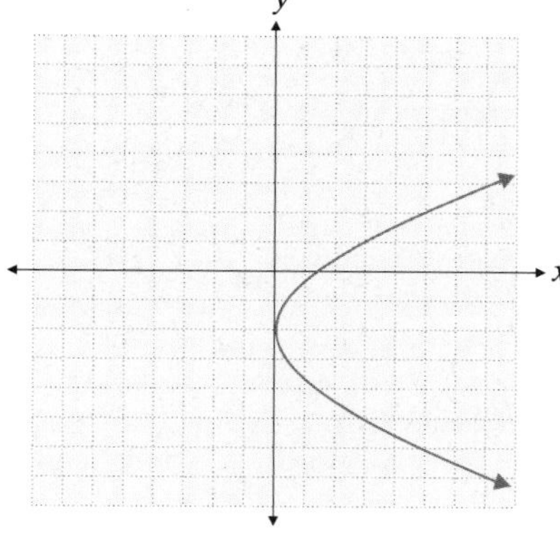

b.

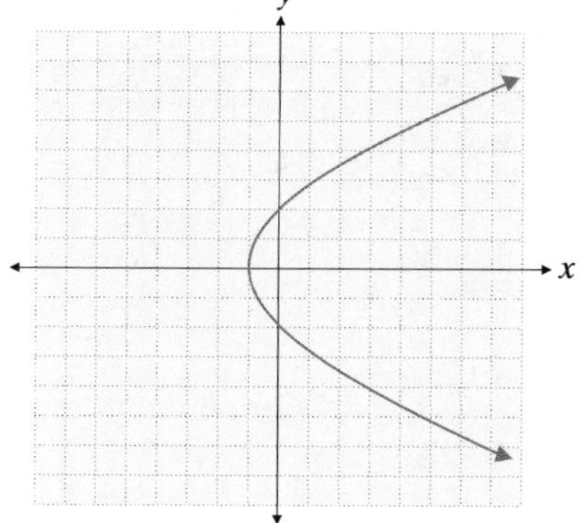

c.

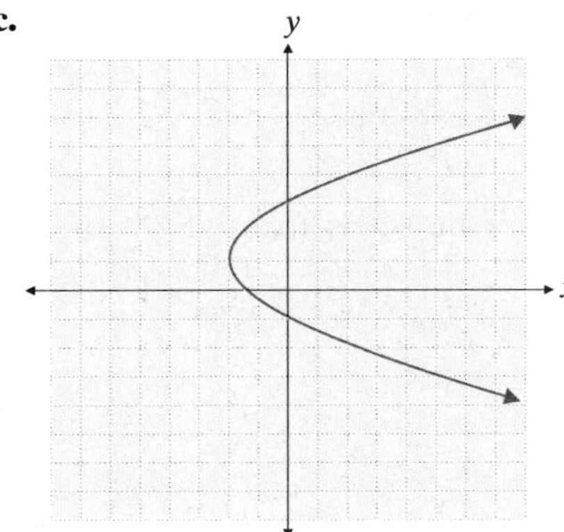

d.

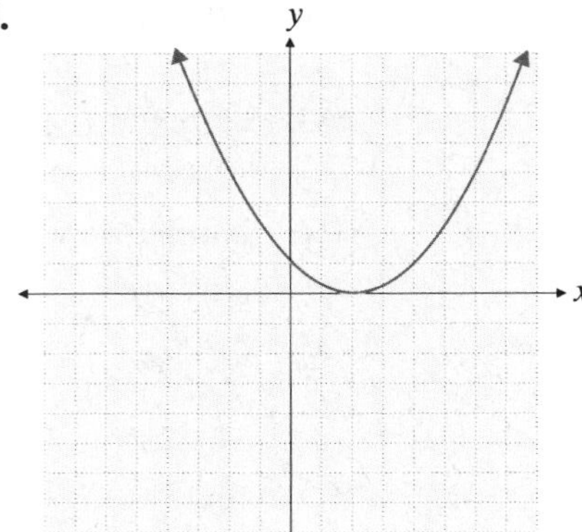

e.

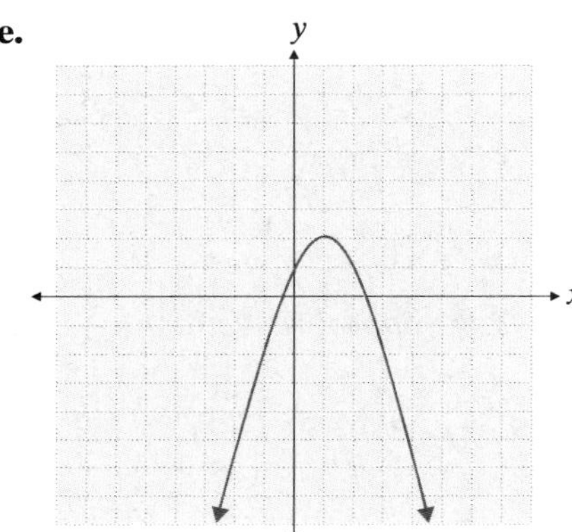

f.

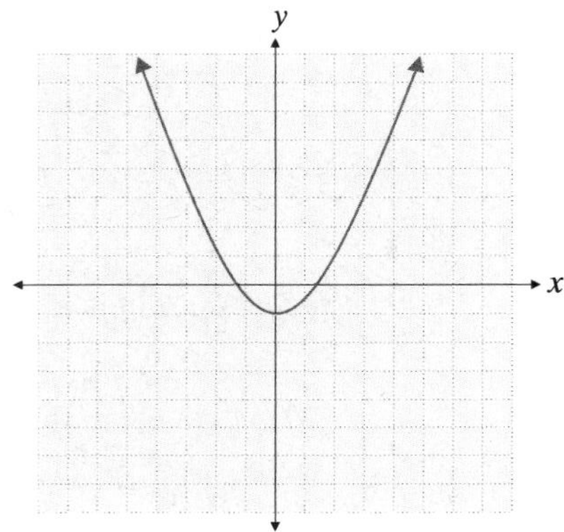

g.

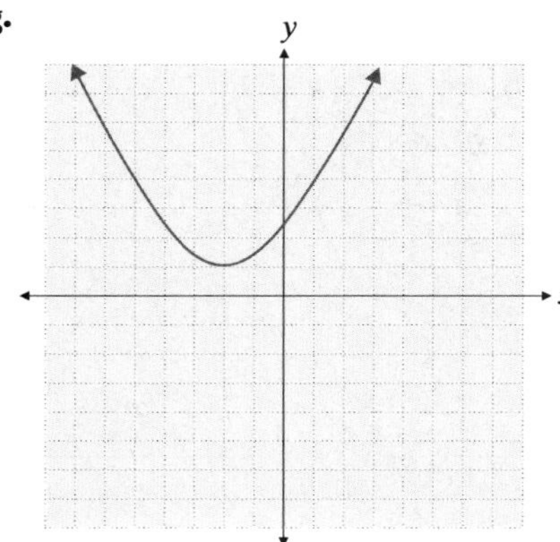

h.

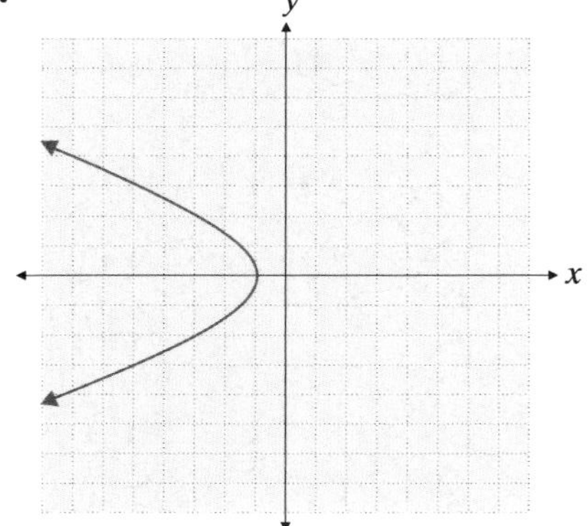

Graph the following parabolas.

45. $x^2 = -4(y+1)$ **46.** $y^2 = -2(x+3)$ **47.** $(x-5)^2 = 2y$

48. $x^2 - 4x - 8y - 12 = 0$ **49.** $y^2 + 2y - 2x - 5 = 0$ **50.** $-x^2 + 6x - 4y + 1 = 0$

51. $(y-1)^2 = 6(x+2)$ **52.** $(x-3)^2 = -10(y+2)$ **53.** $(y+3)^2 = 5(x-1)$

Use your knowledge of parabolas to answer the following questions. See Example 3.

54. One design for a solar furnace is based on the paraboloid formed by rotating the parabola $x^2 = 8y$ around its axis of symmetry. The object to be heated in the furnace is then placed at the focus of the paraboloid (assume that x and y are in units of feet). How far from the vertex of the paraboloid is the hottest part of the furnace?

55. A certain brand of satellite dish antenna has a diameter of 6 feet and a depth of 1 foot, as shown. How far from the vertex of the dish should the receiver of the antenna be placed?

$d = 6$ ft.
depth = 1 ft.

56. A spotlight is made by placing a strong light bulb inside a reflective paraboloid formed by rotating the parabola $x^2 = 6y$ around its axis of symmetry (assume that x and y are in units of inches). In order to have the brightest, most concentrated light beam, how far from the vertex should the bulb be placed?

technology exercises

Use a computer algebra system such as Mathematica (or a suitably equipped graphing calculator) to graph the following equations. See the Technology Note after Example 2 in Section 9.1 for guidance.

57. $3x^2 - 4y + 24x = -56$

58. $y^2 + 2y = 8x - 41$

59. $y^2 - 6y + 4x = -17$

60. $x^2 - 6x + 12y + 21 = 0$

9.3 The Hyperbola

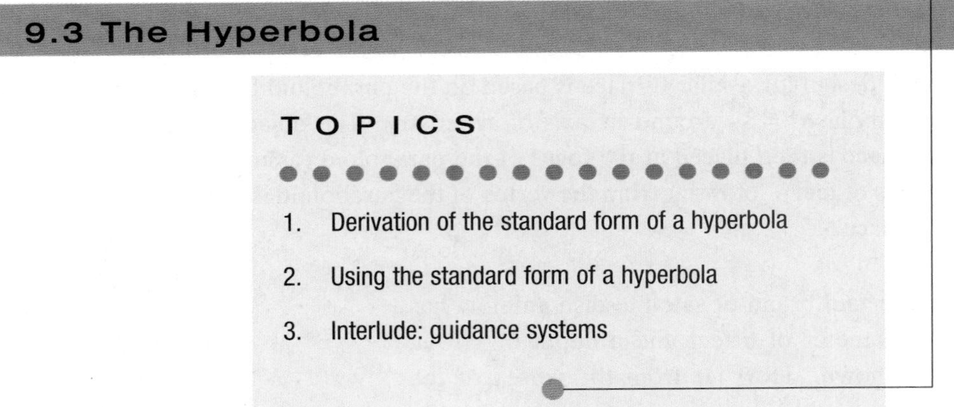

TOPICS
1. Derivation of the standard form of a hyperbola
2. Using the standard form of a hyperbola
3. Interlude: guidance systems

Topic 1: Derivation of the Standard Form of a Hyperbola

The pattern for exploring hyperbolas, the last of the three varieties of conic sections, will be familiar to you by now. We begin with a characteristic geometric property of hyperbolas, and use this to derive a useful form of the equation for a hyperbola. This we will call the standard form.

The characteristic geometric property of hyperbolas is similar to the one for ellipses. Recall that the points of an ellipse are those for which the sum of the distances to two foci is a fixed constant. For the points on a hyperbola, the magnitude of the *difference* of the distances to two foci is a fixed constant. This results in two disjoint pieces, called *branches*, of a hyperbola. The point halfway between the two branches is the center of the hyperbola, and the two points on the hyperbola closest to the center are the two *vertices*. Figure 1 illustrates how these parts of a hyperbola relate to one another.

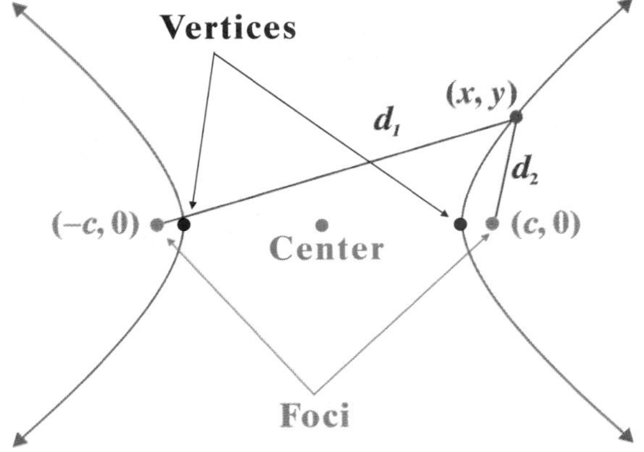

Figure 1: Parts of a Hyperbola

To begin, we will work with a hyperbola oriented as in Figure 1, with the center at the origin and the two foci at $(-c, 0)$ and $(c, 0)$. We know that for every point (x, y) on the hyperbola, the quantity $|d_1 - d_2|$ is fixed; in order to simplify the algebra, we will let $2a$ denote this fixed quantity. Using the distance formula for d_1 and d_2, then, we have:

$$\left|\sqrt{(x+c)^2 + (y-0)^2} - \sqrt{(x-c)^2 + (y-0)^2}\right| = 2a.$$

As in our work with ellipses, we wish to manipulate this equation into a more useful form. Because of the absolute value symbols, we know that the one equation above actually represents two equations, as the left-hand side represents either $d_1 - d_2$ or $d_2 - d_1$ depending on which quantity is non-negative. Fortunately, the two cases quickly converge after we isolate either one of the radicals and square both sides. In what follows, d_1 has been isolated and both sides squared:

$$\left(\sqrt{(x+c)^2 + (y-0)^2}\right)^2 = \left(2a + \sqrt{(x-c)^2 + (y-0)^2}\right)^2$$

$$(x+c)^2 + y^2 = 4a^2 + 4a\sqrt{(x-c)^2 + y^2} + (x-c)^2 + y^2.$$

This should look very familiar; our work so far closely parallels the derivation of the standard form for an ellipse. In fact, if we follow the same steps we will arrive at the form

$$\frac{x^2}{a^2} - \frac{y^2}{c^2 - a^2} = 1$$

and as with ellipses we will make a change of variables to improve the appearance. Note that a (the distance of either vertex from the center) is less than c (the distance of either focus from the center), so $c^2 - a^2$ is positive and we are safe in renaming $c^2 - a^2$ as b^2. This changes the appearance of the above equation to

$$\frac{x^2}{a^2} - \frac{y^2}{b^2} = 1,$$

with the understanding that $b^2 = c^2 - a^2$ (we will find it more useful to write this as $c^2 = a^2 + b^2$ later, so we can solve for c).

To generalize this equation, we make the usual observations that swapping x and y reflects the graph in Figure 1 with respect to the line $y = x$ (giving us a hyperbola with foci on the y-axis) and replacing x with $x - h$ and y with $y - k$ moves the center to (h, k).

The following box summarizes our work.

Standard Form of a Hyperbola

The **standard form** of the equation for the hyperbola with center at (h,k) is:

- $\dfrac{(x-h)^2}{a^2} - \dfrac{(y-k)^2}{b^2} = 1$ if the foci are aligned horizontally;
- $\dfrac{(y-k)^2}{a^2} - \dfrac{(x-h)^2}{b^2} = 1$ if the foci are aligned vertically.

In either case, the foci are located c units away from the center, where $c^2 = a^2 + b^2$, and the vertices are located a units away from the center.

The standard form for a hyperbola is undeniably useful, as it tells us by inspection where the center is and hence where the two vertices are (as well as the foci if we wish). But this knowledge alone leaves a lot of uncertainty about the shape of the hyperbola. We always have the option of calculating and then plotting additional points to help us fill in the graph, but fortunately there is a more efficient aid. The analysis that follows shows that the branches of a hyperbola approach two *asymptotes* far away from the center.

Consider again a hyperbola centered at the origin with foci aligned horizontally, as in Figure 1. Then for some pair of constants a and b, the hyperbola is described by

$$\frac{x^2}{a^2} - \frac{y^2}{b^2} = 1.$$

To understand how this hyperbola behaves far away from the center, we can solve the equation for y:

$$y^2 = b^2\left(\frac{x^2}{a^2} - 1\right).$$

We are wondering how y behaves when x is large in magnitude, but the answer is not clear from the equation in this form. As is often the case, a little algebraic manipulation sheds light on the issue. Factoring out the fraction from the parentheses gives us the equation

$$y^2 = \frac{b^2x^2}{a^2}\left(1 - \frac{a^2}{x^2}\right),$$

and taking the square root of both sides leads to

$$y = \pm \frac{b}{a}x\sqrt{1-\frac{a^2}{x^2}}.$$

The advantage of this last form is that as x goes to ∞ or $-\infty$, it is clear that the radicand approaches 1. Thus, y gets closer and closer to the value

$$\frac{b}{a}x \text{ or } -\frac{b}{a}x$$

for values of x that are large in magnitude. In other words, the two straight lines

$$y = \frac{b}{a}x \text{ and } y = -\frac{b}{a}x$$

(which intersect at the center) are the asymptotes of the hyperbola.

For hyperbolas whose foci are aligned vertically, the equations for the asymptotes are

$$x = \frac{b}{a}y \text{ and } x = -\frac{b}{a}y;$$

that is, x and y exchange places. If we solve these last two equations for y, and also consider hyperbolas not centered at the origin, we obtain the general situation described below.

Asymptotes of Hyperbolas

- The asymptotes of the hyperbola $\frac{(x-h)^2}{a^2} - \frac{(y-k)^2}{b^2} = 1$ are the two lines $y-k = \frac{b}{a}(x-h)$ and $y-k = -\frac{b}{a}(x-h)$.

- The asymptotes of the hyperbola $\frac{(y-k)^2}{a^2} - \frac{(x-h)^2}{b^2} = 1$ are the two lines $y-k = \frac{a}{b}(x-h)$ and $y-k = -\frac{a}{b}(x-h)$.

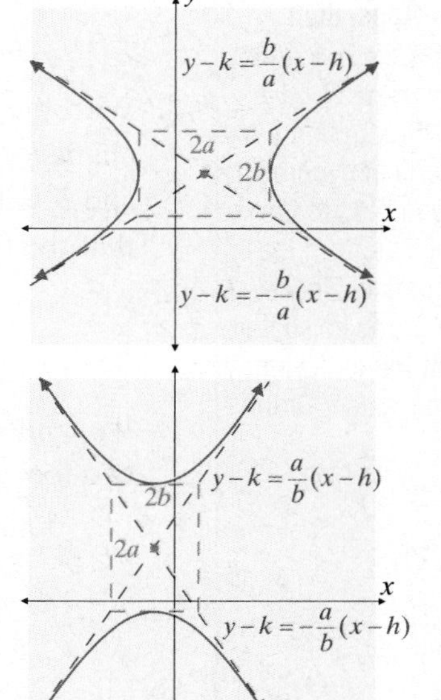

Topic 2: Using the Standard Form of a Hyperbola

It's time to see how the above analysis can be put to use.

example 1

Graph the equation $-16x^2+9y^2+96x+18y=279$, indicating the asymptotes of the resulting hyperbola.

Solution:

The fact that $-16x^2+9y^2+96x+18y=279$ represents a hyperbola follows from the fact that the product of the coefficients of x^2 and y^2 is negative. As with many ellipse and parabola problems, though, we will need to use the method of completing the square to write the equation in standard form and thus identify details of the hyperbola.

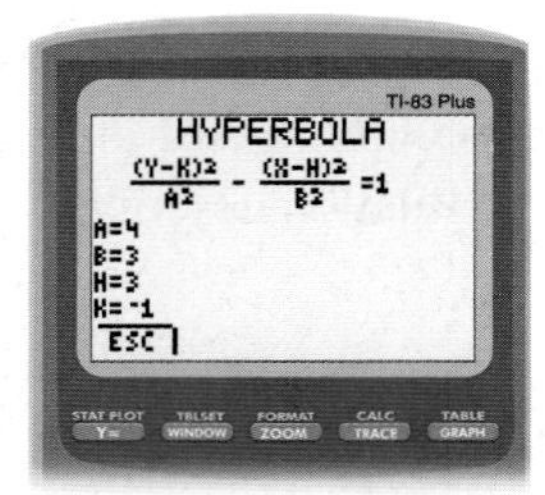

To graph a hyperbola, select **Apps**, then `3:Conics` and `3:Hyperbola`. Select equation **2** for a vertical hyperbola, and plug in the values from the equation. To view the graph as shown on the following page, select **Graph**.

$$-16x^2+9y^2+96x+18y=279$$
$$-16\left(x^2-6x\right)+9\left(y^2+2y\right)=279$$
$$-16\left(x^2-6x+9\right)+9\left(y^2+2y+1\right)=279-144+9$$
$$-16(x-3)^2+9(y+1)^2=144$$
$$\frac{(y+1)^2}{16}-\frac{(x-3)^2}{9}=1$$

We begin by factoring out the coefficients of x^2 and y^2 from the terms containing x and y, respectively. We then add the appropriate constant to complete each perfect square trinomial. Remember to compensate by adding the appropriate numbers to the right-hand side as well. The last step is to divide by 144.

With the equation in standard form, we now know that the foci of the hyperbola are aligned vertically (since the positive fraction on the left is in the variable y), that $a = 4$ and $b = 3$, and that the center of the hyperbola is at $(3, -1)$. This tells us immediately that the two vertices of the hyperbola must lie at

$$(3, -1-4) = (3, -5) \text{ and } (3, -1+4) = (3, 3).$$

We could also, if we wish, determine the foci of the hyperbola: $c^2 = a^2+b^2 = 25$ so the two foci are at

$$(3, -1-5) = (3, -6) \text{ and } (3, -1+5) = (3, 4).$$

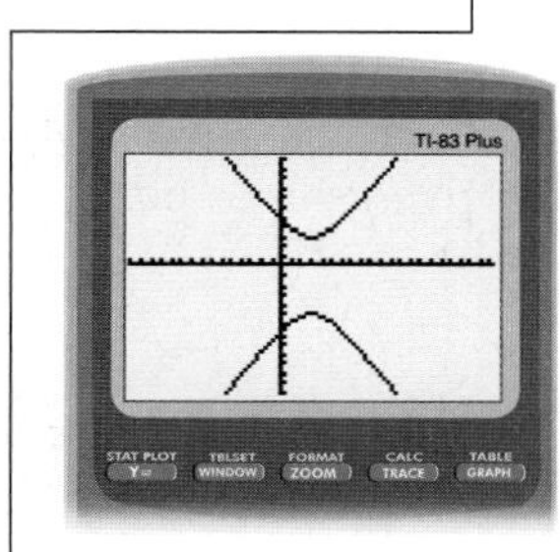

But the best aid in graphing the hyperbola will be the two asymptotes. The diagram at right illustrates a way to sketch the asymptotes of a hyperbola. We know that both asymptotes pass through the center of the hyperbola, and we know that one has a slope of $\frac{4}{3}$ and the other a slope of $-\frac{4}{3}$. The dashed-line rectangle drawn in the figure is centered at $(3, -1)$, with the bottom and top edges 4 units away from the center and the left and right edges 3 units away.

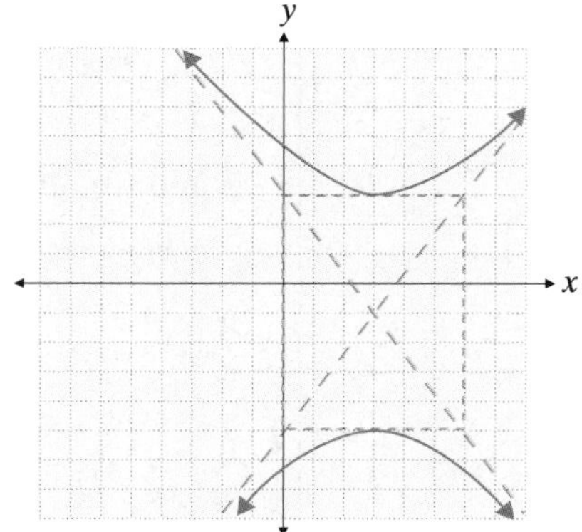

In other words, the rectangle has a height of $2a$ and a width of $2b$ (if the hyperbola was oriented horizontally, the width of the corresponding rectangle would be $2a$ and the height would be $2b$). Two dashed lines drawn through diagonally opposite corners of this rectangle then correspond to the two asymptotes, and the hyperbola can be drawn accordingly.

example 2

Given that the asymptotes of a certain hyperbola have slopes of $\frac{1}{2}$ and $-\frac{1}{2}$ and that the vertices of the hyperbola are at $(-1, 0)$ and $(7, 0)$, find the equation (in standard form) for the hyperbola.

Solution:

The fact that the vertices (and hence the foci) are aligned horizontally tells us immediately that we are going to be constructing an equation of the form

$$\frac{(x-h)^2}{a^2} - \frac{(y-k)^2}{b^2} = 1.$$

Our only remaining task is to determine h, k, a, and b. Since the vertices are 8 units apart, we know that $2a = 8$, or $a = 4$. We also know that the center lies halfway between the vertices, and that halfway point is $(3, 0)$. All that remains is to determine b.

cont'd. on next page ...

As the foci are aligned horizontally, we know the asymptotes are of the form

$$y-k=\frac{b}{a}(x-h) \text{ and } y-k=-\frac{b}{a}(x-h),$$

and we already know a, h, and k. Specifically, we know that the two given slopes must correspond to the fractions $\frac{b}{a}$ and $-\frac{b}{a}$. That is,

$$\frac{1}{2}=\frac{b}{a}, \text{ so } \frac{1}{2}=\frac{b}{4}, \text{ and so } b=2.$$

The equation we seek is thus

$$\frac{(x-3)^2}{16}-\frac{y^2}{4}=1.$$

A sketch of the hyperbola, along with its asymptotes, appears at right.

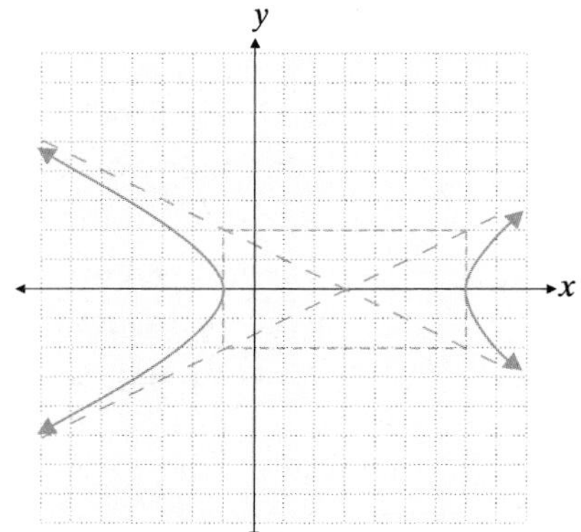

Topic 3: Interlude: Guidance Systems

Hyperbolas, like ellipses and parabolas, arise in a wide variety of contexts. For instance, hyperbolas are seen in architecture and structural engineering (think of the shape of a nuclear power plant's cooling towers) and in astronomy (comets that make a single pass through our solar system don't have elliptical orbits, but instead trace one branch of a hyperbola).

One of the most natural applications of hyperbolas concerns guidance systems, such as LORAN (LOng RAnge Navigation). LORAN is a radio-communication system that can be used to determine the location of a ship at sea, and the basis of LORAN is an understanding of hyperbolic curves.

Consider a situation in which two land-based radio transmitters, located at sites A and B in Figure 2, send out a pulse signal simultaneously. A receiver on a ship, located at C, would receive the two signals at slightly different times due to the difference in the distances the signal must travel. Since the times for signal travel are proportional to the respective distances d_1 and d_2, the difference in time between receipt of the two signals is proportional to $|d_1-d_2|$. In other words, a person on the ship could determine $|d_1-d_2|$ by noting the time-delay in receiving the two signals.

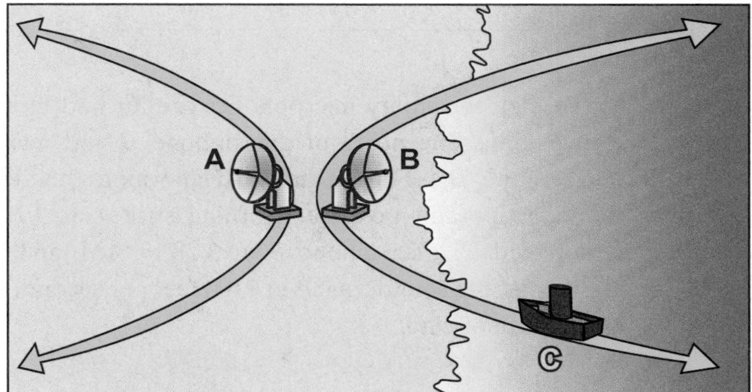

Figure 2: One Set of Transmitters

But as Figure 2 indicates, any point along the hyperbola (only one branch of which is relevant as far as the ship is concerned) could correspond to the position of the ship, as $|d_1 - d_2|$ is the same for all the points on the hyperbola. One way to actually determine the location of the ship is to perform the same computations for another pair of simultaneous signals sent out from two additional transmitters, located at A' and B' in Figure 3, and note the point of intersection of the two hyperbolas.

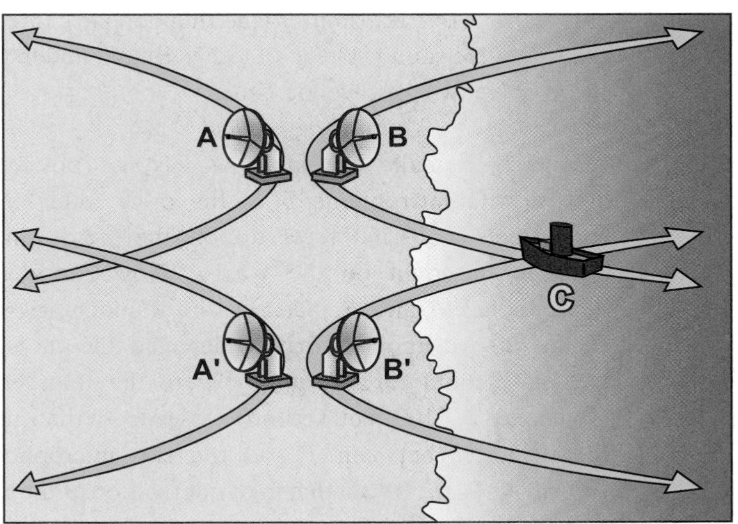

Figure 3: Two Sets of Transmitters

example 3

Three high-sensitivity microphones are located in a forest preserve, with microphone A two miles due north of microphone B and microphone C two miles due east of microphone B. During an early morning thunderstorm, microphone A detects a thunderclap (and possible lightning strike) at 3:28:15 AM. The same thunderclap is detected by microphone B at 3:28:19 AM, and by microphone C at 3:28:25 AM. Assuming that sound travels at 1100 feet per second, graphically approximate the source of the thunderclap.

Solution:

This situation is similar to the LORAN navigational system just described, but in this case there is one transmitter (the thunderclap), three receivers (the microphones), and the waves are sound instead of radio. But an understanding of how hyperbolas relate to the problem will resolve the issue.

In order to make the math as easy as possible, we will set up a coordinate system with microphone B at the origin and microphone A located 10,560 units (two miles = 10,560 feet) up on the y-axis and microphone C located 10,560 units to the right on the x-axis. Now consider the thunderclap in relation to microphones A and B. Clearly, the thunder originated at a point closer to A than B, but all we know is that A detected the thunder 4 seconds earlier than B. If we let T stand for the point where the thunder originated, then T is 4400 feet (4 seconds × 1100 feet/second) closer to A than to B; if we let d_1 and d_2 represent the distances between T and the two microphones A and B, respectively, then $|d_1 - d_2| = 4400$. Recall that in our derivation of the standard form, we let $|d_1 - d_2| = 2a$, and a is the distance from the center of a hyperbola to either one of its vertices. Hence, $2a = 4400$, or $a = 2200$. We now know that T lies somewhere along a hyperbola with foci at the two microphones, center at $(0, 5280)$ (the halfway point between A and B), and distance from the center to either vertex equal to 2200 feet. The equation for such a hyperbola is

$$\frac{(y-5280)^2}{(2200)^2} - \frac{x^2}{b^2} = 1,$$

with the constant b still undetermined. But since c is the distance from the center to either focus, and since $b^2 = c^2 - a^2$, we know $b^2 = (5280)^2 - (2200)^2 = 23{,}038{,}400$ and so the equation is

$$\frac{(y-5280)^2}{(2200)^2} - \frac{x^2}{23{,}038{,}400} = 1.$$

We can now do a similar analysis for microphones B and C, knowing that the difference in time is 6 seconds and hence $|d_1 - d_2| = 6600$. This means that for the hyperbola with foci at B and C, $a = 3300$ and $b^2 = c^2 - a^2 = 16{,}988{,}400$. Note that the center of this hyperbola is $(5280, 0)$, so the equation for the possible location of T relative to these two microphones is

$$\frac{(x-5280)^2}{(3300)^2} - \frac{y^2}{16{,}988{,}400} = 1.$$

We are now ready to graph these two hyperbolas and approximate the source of the thunder. Figure 4 contains the graph, marked off in square-mile grids, with the three microphones marked in green.

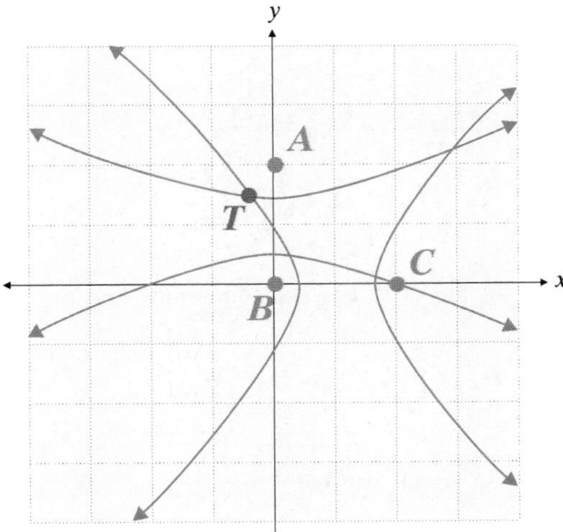

Figure 4: Three Microphones and a Possible Lightning Strike

Figure 4 indicates that there are four points of intersection of the two hyperbolas. How do we know which one corresponds to T? The point we are looking for should be closest to A and next closest to B, and this is the point marked in red. The coordinates of this point are approximately $(-1646, 7606)$, and this is the place to look for lightning damage.

exercises

Sketch the graphs of the following hyperbolas, using asymptotes as guides. Determine the coordinates of the foci in each case. See Example 1.

1. $\frac{(x+3)^2}{4}-\frac{(y+1)^2}{9}=1$

2. $\frac{(y-2)^2}{25}-\frac{(x+2)^2}{9}=1$

3. $4y^2-x^2-24y+2x=-19$

4. $x^2-9y^2+4x+18y-14=0$

5. $9x^2-25y^2=18x-50y+241$

6. $9x^2-16y^2+116=36x+64y$

7. $\frac{x^2}{16}-\frac{(y-2)^2}{4}=1$

8. $\frac{(y-1)^2}{9}-(x+3)^2=1$

9. $9y^2-25x^2-36y-100x=289$

10. $9x^2+18x=4y^2+27$

11. $9x^2-16y^2-36x+32y-124=0$

12. $x^2-y^2+6x-6y=4$

Find the center, foci, and vertices of each hyperbola that the equation describes.

13. $\frac{(x+3)^2}{4}-\frac{(y-2)^2}{9}=1$

14. $\frac{(y-2)^2}{16}-\frac{(x+1)^2}{9}=1$

15. $3(x-1)^2-(y+4)^2=9$

16. $(y-2)^2-2(x-4)^2=4$

17. $(x+2)^2-5(y-1)^2=25$

18. $6(y+2)^2-(x+1)^2=12$

19. $2x^2+12x-y^2-2y+9=0$

20. $y^2-9x^2+6y+72x-144=0$

21. $x^2-4y^2-2x=0$

22. $4y^2-x^2+32y+2x+47=0$

23. $4x^2-y^2-64x+10y+167=0$

24. $4x^2-9y^2-36y-72=0$

Find the equation, in standard form, for the hyperbola with the given properties or with the given graph. See Example 2.

25. Foci at $(-3, 0)$ and $(3, 0)$ and vertices at $(-2, 0)$ and $(2, 0)$.

26. Foci at $(1, 5)$ and $(1, -1)$ and vertices at $(1, 3)$ and $(1, 1)$.

27. Asymptotes of $y = \pm 2x$ and vertices at $(0, -1)$ and $(0, 1)$.

28. Asymptotes of $y = \pm(x - 2) + 1$ and vertices at $(-1, 1)$ and $(5, 1)$.

29. Foci at $(2, 4)$ and $(-2, 4)$ and asymptotes of $y = \pm 3x + 4$.

30. Foci at $(-1, 3)$ and $(-1, -1)$ and asymptotes of $y = \pm(x + 1) + 1$.

31. Foci at $(2, 5)$ and $(10, 5)$ and vertices at $(3, 5)$ and $(9, 5)$.

32. Foci at $(7, 4)$ and $(7, -4)$ and vertices at $(7, 1)$ and $(7, -1)$.

33. Asymptotes of $y = \pm(2x + 8) + 3$ and vertices at $(-6, 3)$ and $(-2, 3)$.

34. Asymptotes of $y = \pm \frac{4}{3}x - 3$ and vertices at $(0, -7)$ and $(0, 1)$.

35.

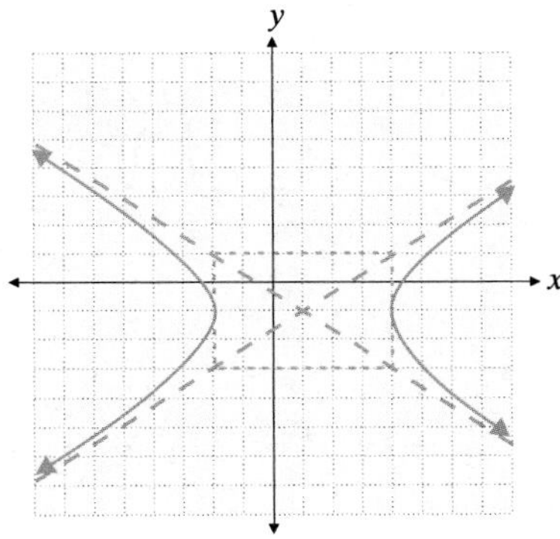

36.

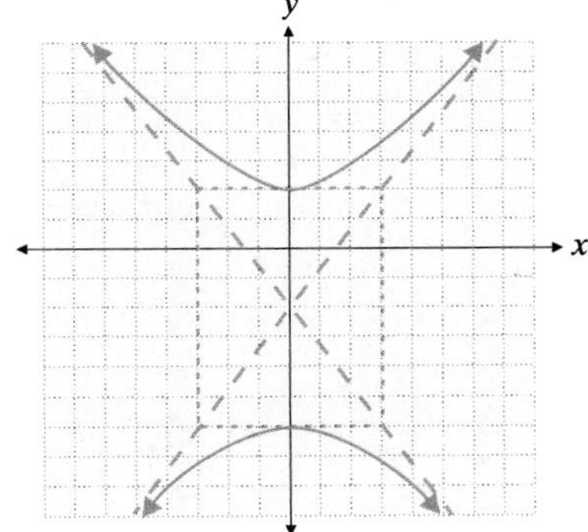

37.

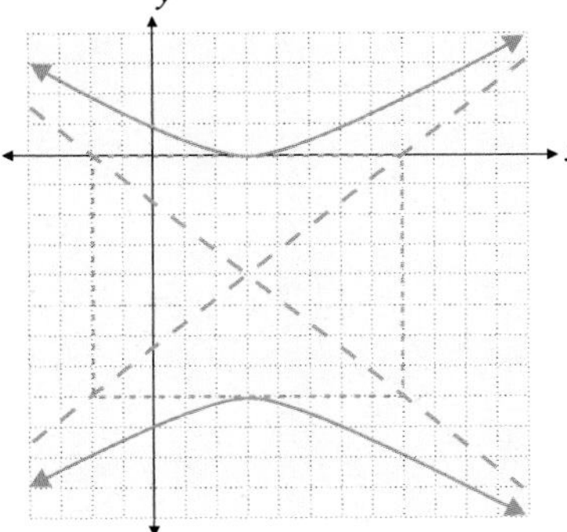

38.

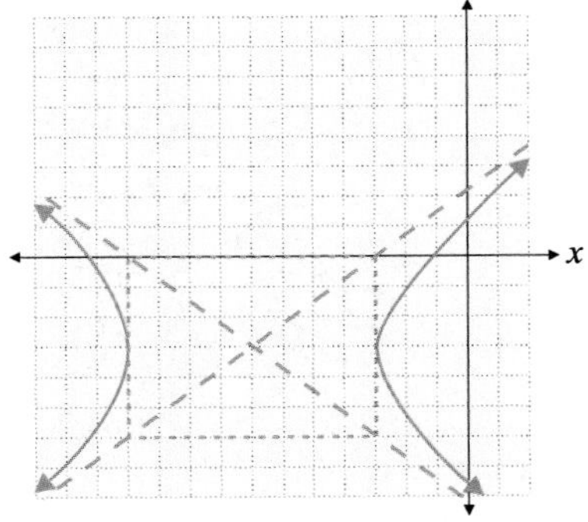

Match the corresponding equation to the appropriate graph.

39. $\frac{x^2}{9} - y^2 = 1$

40. $y^2 - \frac{x^2}{4} = 1$

41. $x^2 - \frac{(y-3)^2}{4} = 1$

42. $\frac{(x-3)^2}{4} - \frac{(y+1)^2}{9} = 1$

43. $(y+2)^2 - \frac{(x-2)^2}{4} = 1$

44. $\frac{x^2}{9} - \frac{(y+2)^2}{4} = 1$

45. $\frac{y^2}{4} - (x-1)^2 = 1$

46. $x^2 - y^2 = 1$

a.

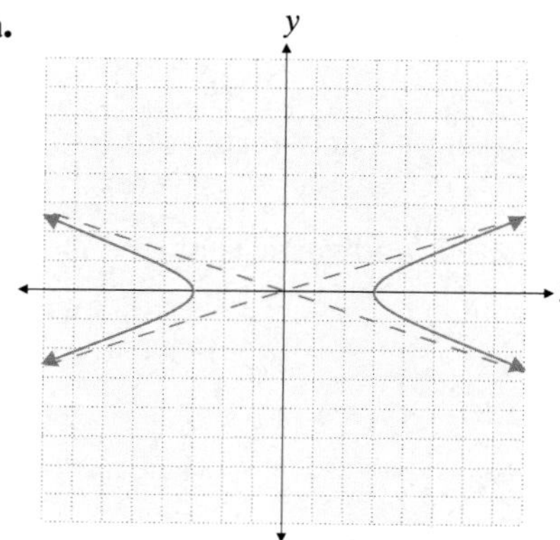

b.

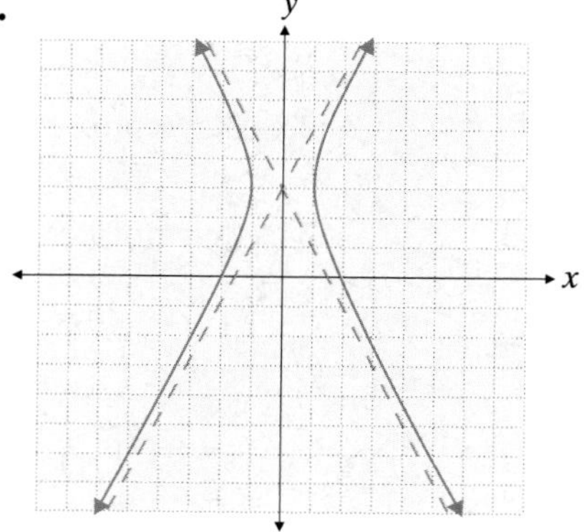

c.

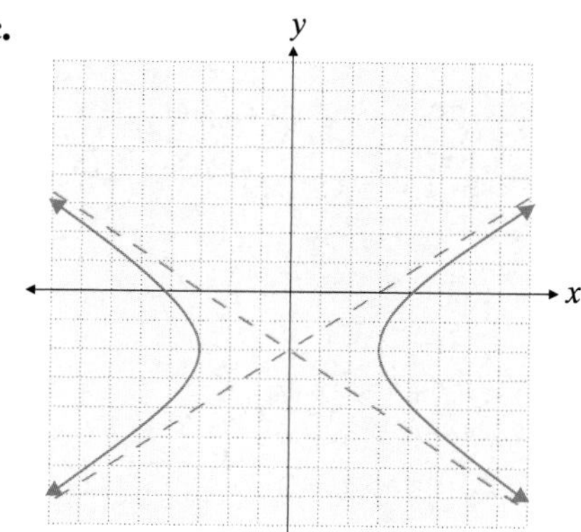

d.

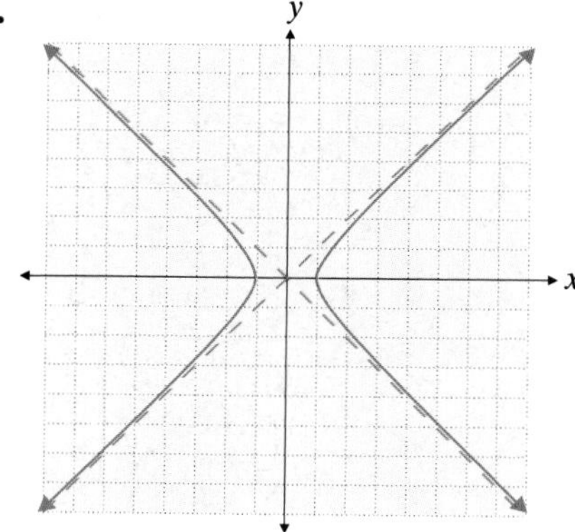

e.

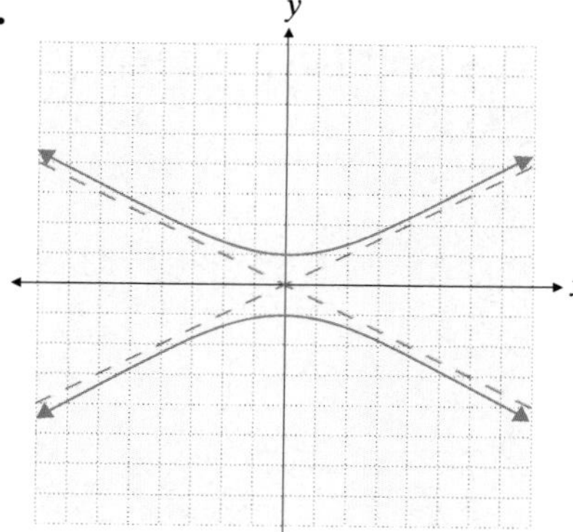

f.

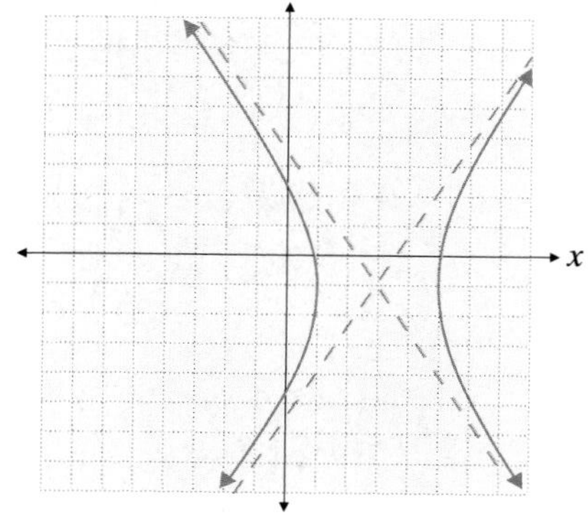

g.

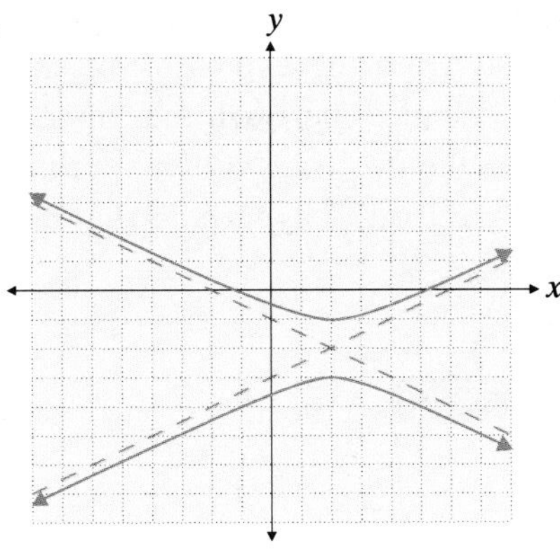

h.

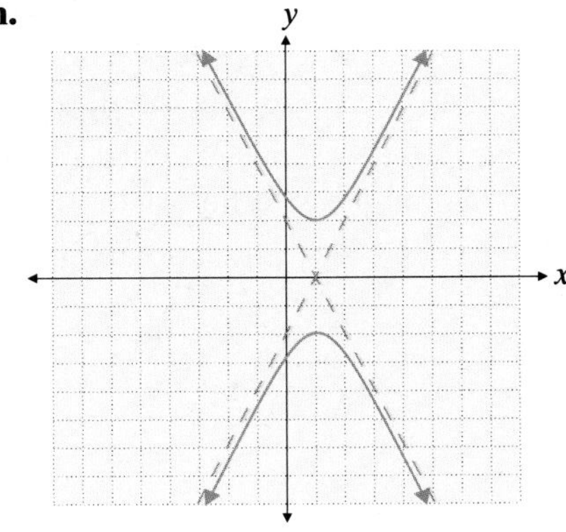

Using a graphing calculator, graph the following.

47. $x^2 - 6y^2 = 15$

48. $4y^2 - 9x^2 = 18$

49. $3y^2 - 18x^2 = 36$

50. $x^2 - 6 = 3y^2$

51. $(y+2)^2 - 20 = 5x^2$

52. $(x+5)^2 = 3(y-2)^2 + 15$

Use your knowledge of hyperbolas to answer the following questions. See Example 3.

53. As mentioned in this section, some comets trace one branch of a hyperbola through the solar system, with the sun at one focus. Suppose a comet is spotted that appears to be headed straight for Earth, as shown in the figure below. As the comet gets closer, however, it becomes apparent that it will pass between the Earth, which lies at the center of the hyperbolic path of the comet, and the sun. In the end, the closest the comet comes to Earth is 60,000,000 miles. Using an estimate of 94,000,000 miles for the distance from the Earth to the sun, and positioning the Earth at the origin of a coordinate system, find the equation for the path of the comet.

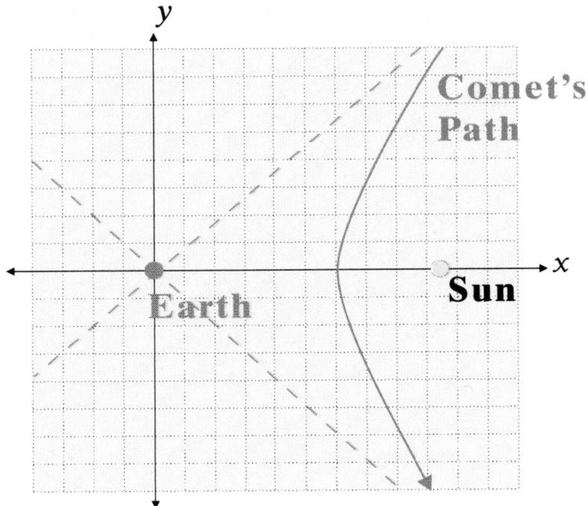

54. Suppose two LORAN radio transmitters are 26 miles apart. A ship at sea receives signals sent simultaneously from the two transmitters and is able to determine that the difference in the distances between the ship and each of the transmitters is 24 miles. By positioning the two transmitters on the y-axis, each 13 miles from the origin, find the equation for the hyperbola that describes the set of possible locations for the ship.

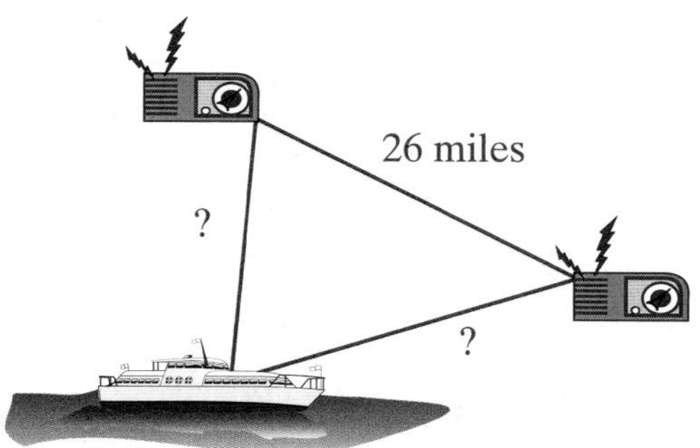

technology exercises

Use a computer algebra system such as Mathematica (or a suitably equipped graphing calculator) to graph the following equations. See the Technology Note after Example 2 in Section 9.1 for guidance.

55. $x^2 - 2y^2 = 4x + 12y + 26$

56. $2x^2 - y^2 + 12x + 2y = 3$

57. $5x^2 - y^2 + 20x = 10y + 25$

58. $x^2 - 5y^2 = 14x + 20y - 4$

9.4 Rotation of Conics

TOPICS

1. Rotation formulas
2. Rotation invariants

Topic 1: Rotation Formulas

In the first three sections of this chapter, we studied equations of the form

$$Ax^2 + Cy^2 + Dx + Ey + F = 0.$$

The choice of letters for the coefficients is traditional, and certainly hints at a missing term. As mentioned briefly in Section 9.1, the most general form of a conic in Cartesian coordinates is

$$Ax^2 + Bxy + Cy^2 + Dx + Ey + F = 0$$

and we are now ready to explore the consequences of adding a nonzero xy term to the equation.

Geometrically, the condition $B \neq 0$ leads to a conic that is not oriented vertically or horizontally. The ellipses, parabolas, and hyperbolas graphed so far have all been aligned so that the axis (or axes in the case of ellipses and hyperbolas) of symmetry are parallel to the x-axis or y-axis. If $B \neq 0$, the graph is a conic section that has been rotated through some angle θ. We can exploit this fact to make the graphing of a rotated conic section relatively easy. Our method will be to introduce a new set of coordinate axes (i.e. a second Cartesian plane) rotated by an acute angle θ with respect to the original coordinate axes, and then to graph the conic in the new coordinate system. Algebraically, the goal is to begin with an equation of the form

$$Ax^2 + Bxy + Cy^2 + Dx + Ey + F = 0$$

and define a new set of coordinate axes x' and y' in which the equation has the form

$$A'x'^2 + C'y'^2 + D'x' + E'y' + F' = 0.$$

That is, the coefficient B' in the new coordinate system is 0, and hence the graphing techniques of the first three sections of this chapter apply.

We begin with a picture. Figure 1 is an illustration of two rectangular coordinate systems, with the new system rotated by an acute angle θ with respect to the original.

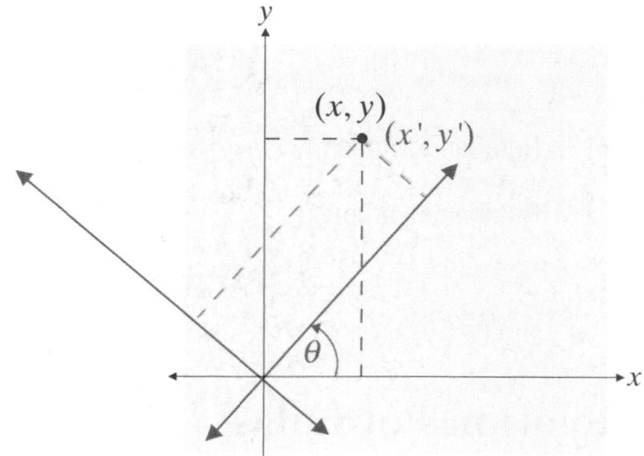

Figure 1: Two Coordinate Systems

The point in the figure has two sets of coordinates, corresponding to the two coordinate planes. However, the distance r between the origin and the point is the same in both, and this fact and the introduction of the angle θ' shown in Figure 2 allow us to begin to relate the two sets of coordinates.

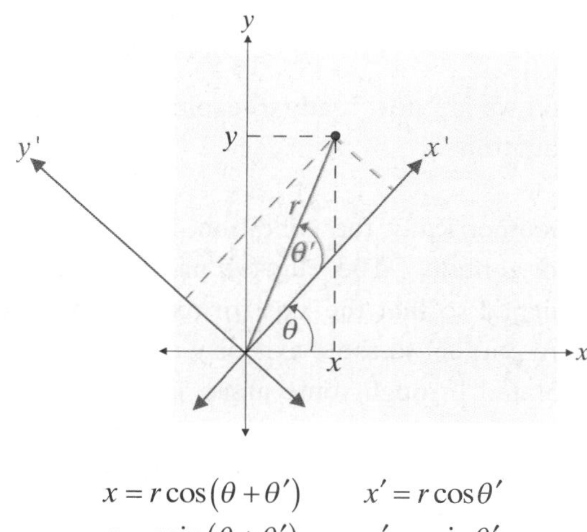

$$x = r\cos(\theta + \theta') \qquad x' = r\cos\theta'$$
$$y = r\sin(\theta + \theta') \qquad y' = r\sin\theta'$$

Figure 2: Relation Between x', y', x, and y

We can now apply one of the trigonometric identities from Chapter 7:

$$\begin{aligned} x &= r\cos(\theta + \theta') \\ &= r(\cos\theta\cos\theta' - \sin\theta\sin\theta') \\ &= r\cos\theta'\cos\theta - r\sin\theta'\sin\theta \\ &= x'\cos\theta - y'\sin\theta. \end{aligned}$$

Similarly, $y = x'\sin\theta + y'\cos\theta$. We will also need to be able to express x' and y' in terms of x and y, and we can do so through clever application of another trigonometric identity. By multiplying the two equations we have just derived by $\cos\theta$ and $\sin\theta$, respectively, and then adding the results, we obtain:

$$\begin{aligned}
&\left(x\cos\theta = x'\cos^2\theta - y'\sin\theta\cos\theta\right)\\
+&\left(y\sin\theta = x'\sin^2\theta + y'\sin\theta\cos\theta\right)\\
\hline
x\cos\theta + y\sin\theta &= x'\left(\cos^2\theta + \sin^2\theta\right)\\
&= x'.
\end{aligned}$$

A similar manipulation allows us to express y' in terms of x and y. All four relations are summarized below.

Rotation Relations

Given a rectangular coordinate system with axes x' and y' rotated by angle θ with respect to axes x and y, as in Figure 1, the two sets of coordinates (x', y') and (x, y) for the same point are related by:

$$\begin{aligned}
x &= x'\cos\theta - y'\sin\theta \qquad & x' &= x\cos\theta + y\sin\theta\\
y &= x'\sin\theta + y'\cos\theta \qquad & y' &= -x\sin\theta + y\cos\theta
\end{aligned}$$

example 1

Given that $\theta = \frac{\pi}{6}$, find the $x'y'$-coordinates of the point with xy-coordinates $(-1, 5)$.

Solution:

Using the above relations,

$$\begin{aligned}
x' &= (-1)\cos\frac{\pi}{6} + 5\sin\frac{\pi}{6}\\
&= (-1)\left(\frac{\sqrt{3}}{2}\right) + (5)\left(\frac{1}{2}\right)\\
&= \frac{5-\sqrt{3}}{2}
\end{aligned}$$

cont'd. on next page ...

and
$$y' = -(-1)\sin\frac{\pi}{6} + 5\cos\frac{\pi}{6}$$
$$= \left(\frac{1}{2}\right) + (5)\left(\frac{\sqrt{3}}{2}\right)$$
$$= \frac{1+5\sqrt{3}}{2}.$$

The $x'y'$-coordinates are thus $\left(\frac{5-\sqrt{3}}{2}, \frac{1+5\sqrt{3}}{2}\right)$.

We are familiar with the graph of $y = \frac{1}{x}$ from our work in Chapters 3 and 4. The next example gives us another perspective on this equation, written in the form $xy = 1$.

example 2

Use the rotation $\theta = 45^\circ$ to show that the graph of $xy = 1$ is a hyperbola.

Solution:

Using the angle $\theta = 45^\circ$ in the rotation relations, we convert the equation as follows:

$$xy = 1$$
$$\left(x'\cos 45^\circ - y'\sin 45^\circ\right)\left(x'\sin 45^\circ + y'\cos 45^\circ\right) = 1$$
$$\left(\frac{x'}{\sqrt{2}} - \frac{y'}{\sqrt{2}}\right)\left(\frac{x'}{\sqrt{2}} + \frac{y'}{\sqrt{2}}\right) = 1$$
$$\frac{x'^2}{2} - \frac{x'y'}{2} + \frac{x'y'}{2} - \frac{y'^2}{2} = 1$$
$$\frac{x'^2}{2} - \frac{y'^2}{2} = 1.$$

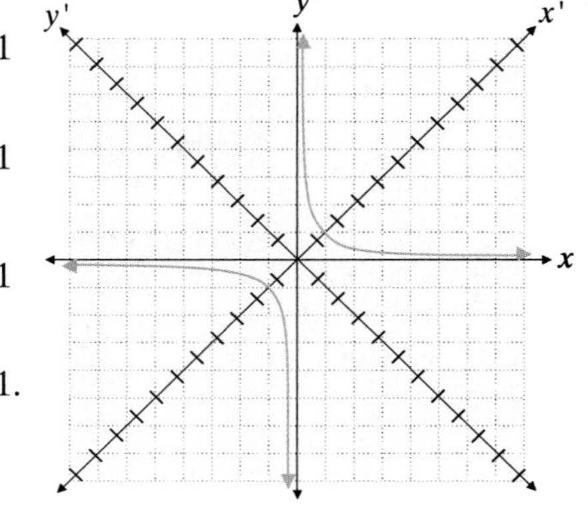

We recognize this last equation as a hyperbola in the $x'y'$-plane, with center at the origin and vertices $\sqrt{2}$ away. The asymptotes are $y' = \pm x'$, which correspond to the x- and y-axes.

Remember that the goal in general is to determine θ so that the conversion of the equation

$$Ax^2 + Bxy + Cy^2 + Dx + Ey + F = 0,$$

has no $x'y'$ term. Example 2 will serve as the inspiration; by replacing x and y with the corresponding x' and y' expressions and simplifying the result, we can derive a formula for the appropriate angle θ. We begin with the replacements:

$$\begin{aligned}&A(x'\cos\theta - y'\sin\theta)^2 + B(x'\cos\theta - y'\sin\theta)(x'\sin\theta + y'\cos\theta)\\&+C(x'\sin\theta + y'\cos\theta)^2 + D(x'\cos\theta - y'\sin\theta)\\&+E(x'\sin\theta + y'\cos\theta) + F = 0.\end{aligned}$$

When the left-hand side of this equation is expanded and like terms collected, the result is an equation of the form

$$A'x'^2 + B'x'y' + C'y'^2 + D'x' + E'y' + F' = 0$$

where

$$\begin{aligned}A' &= A\cos^2\theta + B\cos\theta\sin\theta + C\sin^2\theta\\B' &= 2(C-A)\cos\theta\sin\theta + B\left(\cos^2\theta - \sin^2\theta\right)\\C' &= A\sin^2\theta - B\cos\theta\sin\theta + C\cos^2\theta\\D' &= D\cos\theta + E\sin\theta\\E' &= -D\sin\theta + E\cos\theta\\F' &= F.\end{aligned}$$

Since we want $B' = 0,$ this gives us an equation in θ to solve. To do so, we will use the double-angle formulas for both sine and cosine in reverse:

$$\begin{aligned}2(C-A)\cos\theta\sin\theta + B\left(\cos^2\theta - \sin^2\theta\right) &= 0\\(C-A)\sin 2\theta + B\cos 2\theta &= 0\\B\cos 2\theta &= (A-C)\sin 2\theta\\\frac{\cos 2\theta}{\sin 2\theta} &= \frac{A-C}{B}\\\cot 2\theta &= \frac{A-C}{B}.\end{aligned}$$

This result is summarized on the following page.

Elimination of the xy Term

The graph of the equation $Ax^2+Bxy+Cy^2+Dx+Ey+F=0$ in the xy-plane is the same as the graph of the equation $A'x'^2+C'y'^2+D'x'+E'y'+F'=0$ in the $x'y'$-plane, where the angle of rotation θ between the two coordinate systems satisfies

$$\cot 2\theta = \frac{A-C}{B}.$$

Although formulas relating the primed coefficients to the unprimed coefficients were derived in the preceding discussion, in practice it is easier to simply determine θ and use the rotation relations to convert equations, as shown in this next example.

example 3

Graph the conic section $x^2+2\sqrt{3}xy+3y^2+\sqrt{3}x-y=0$ by first determining the appropriate angle θ by which to rotate the axes.

Solution:

By the formula above,

$$\cot 2\theta = \frac{1-3}{2\sqrt{3}} = -\frac{1}{\sqrt{3}}.$$

Since the angle θ is to be acute, it must be the case that $2\theta = \frac{2\pi}{3}$ and hence $\theta = \frac{\pi}{3}$. By the rotation relations, then,

$$\begin{aligned} x &= x'\cos\frac{\pi}{3} - y'\sin\frac{\pi}{3} \\ &= \frac{1}{2}x' - \frac{\sqrt{3}}{2}y' \end{aligned}$$

and

$$\begin{aligned} y &= x'\sin\frac{\pi}{3} + y'\cos\frac{\pi}{3} \\ &= \frac{\sqrt{3}}{2}x' + \frac{1}{2}y'. \end{aligned}$$

Making these substitutions into the equation, we obtain:

$$\left(\frac{1}{2}x' - \frac{\sqrt{3}}{2}y'\right)^2 + 2\sqrt{3}\left(\frac{1}{2}x' - \frac{\sqrt{3}}{2}y'\right)\left(\frac{\sqrt{3}}{2}x' + \frac{1}{2}y'\right) + 3\left(\frac{\sqrt{3}}{2}x' + \frac{1}{2}y'\right)^2$$
$$+\sqrt{3}\left(\frac{1}{2}x' - \frac{\sqrt{3}}{2}y'\right) - \left(\frac{\sqrt{3}}{2}x' + \frac{1}{2}y'\right) = 0.$$

Multiplying out and collecting like terms in this equation results in

$$y' = 2x'^2,$$

a much simpler equation. We recognize this as a parabola with vertex at the origin. The graph of this equation in the $x'y'$-plane is easily sketched:

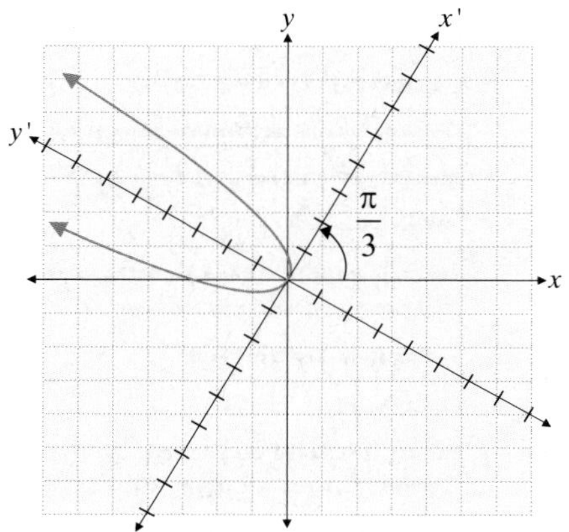

technology note

Graphing calculators and computer algebra systems, such as *Mathematica*, can be very useful in sketching accurate graphs of conic sections, though their capabilities vary widely. Software such as *Mathematica* can also greatly speed up the process of converting an equation from one rectangular coordinate system to another. Figure 3 illustrates the use of *Mathematica* in first converting the equation from Example 3 and then graphing the original equation.

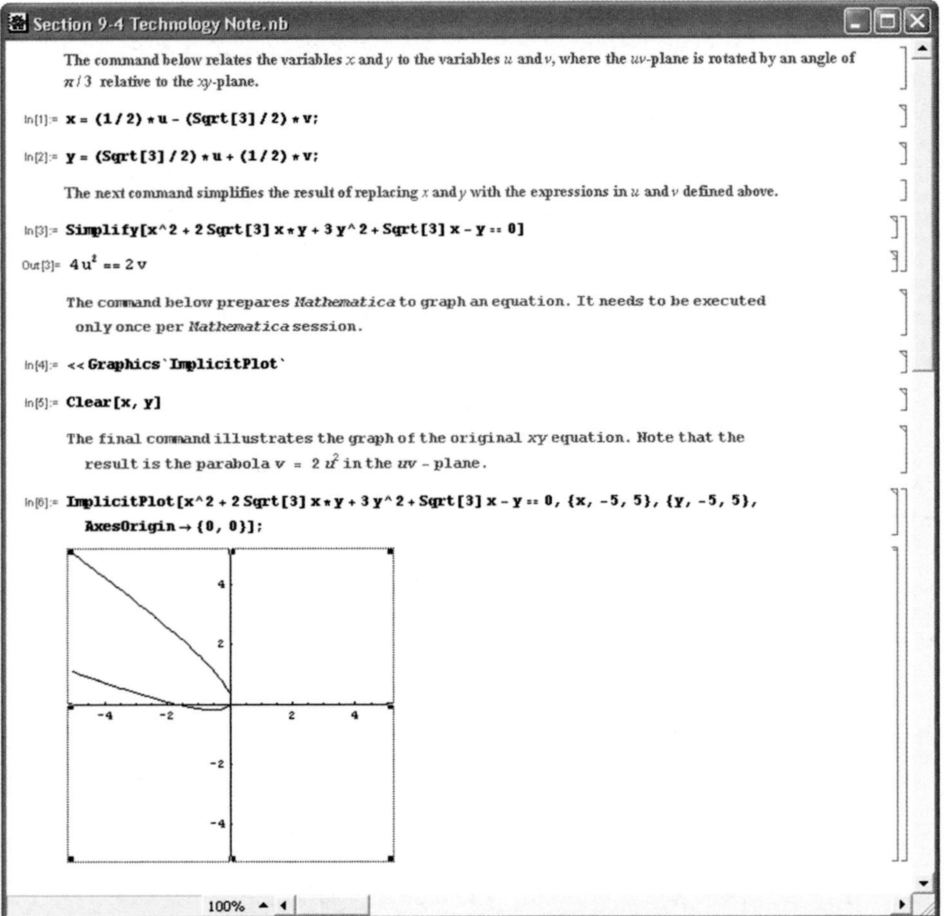

Figure 3: Using *Mathematica* to Graph a Rotated Conic

Topic 2: Rotation Invariants

The relationships we derived between A, B, C, D, E, F and $A', B', C', D', E',$ and F' lead to some interesting observations, one of which we have an immediate use for. Since $F' = F,$ we say the constant term is *invariant* under rotation. Slightly less obviously, $A' + C' = A + C,$ so $A + C$ is also said to be an invariant.

Less obvious still is the very important fact that $B'^2 - 4A'C' = B^2 - 4AC.$ This invariant is called the **discriminant** of the conic section, and its sign (except in the case of degenerate conics) identifies the graph of the equation as an ellipse, a parabola, or a hyperbola. The discriminant thus generalizes the role that the product AC played in Sections 9.1, 9.2, and 9.3.

Classifying Conics

Assuming the graph of the equation $Ax^2 + Bxy + Cy^2 + Dx + Ey + F = 0$ is a non-degenerate conic section, it is classified by its discriminant as follows:

1. **Ellipse** if $B^2 - 4AC < 0$
2. **Parabola** if $B^2 - 4AC = 0$
3. **Hyperbola** if $B^2 - 4AC > 0$

example 4

Classify the conic section $13x^2 + 10xy + 13y^2 + 42\sqrt{2}x - 6\sqrt{2}y + 18 = 0$ and then sketch its graph.

Solution:

The discriminant is $10^2 - 4(13)(13) = -576.$ Since this result is negative, the conic section is an ellipse. The next step is to determine the rotation angle θ:

$$\cot 2\theta = \frac{13-13}{10} = 0$$

$$\Rightarrow 2\theta = \frac{\pi}{2}$$

$$\Rightarrow \theta = \frac{\pi}{4}.$$

cont'd. on next page ...

The rotation relations are thus

$$x = x'\cos\frac{\pi}{4} - y'\sin\frac{\pi}{4}$$
$$= \frac{x'}{\sqrt{2}} - \frac{y'}{\sqrt{2}}$$

and

$$y = x'\sin\frac{\pi}{4} + y'\cos\frac{\pi}{4}$$
$$= \frac{x'}{\sqrt{2}} + \frac{y'}{\sqrt{2}}.$$

Making these substitutions in the original equation gives us

$$13\left(\frac{x'}{\sqrt{2}} - \frac{y'}{\sqrt{2}}\right)^2 + 10\left(\frac{x'}{\sqrt{2}} - \frac{y'}{\sqrt{2}}\right)\left(\frac{x'}{\sqrt{2}} + \frac{y'}{\sqrt{2}}\right) + 13\left(\frac{x'}{\sqrt{2}} + \frac{y'}{\sqrt{2}}\right)^2$$
$$+42\sqrt{2}\left(\frac{x'}{\sqrt{2}} - \frac{y'}{\sqrt{2}}\right) - 6\sqrt{2}\left(\frac{x'}{\sqrt{2}} + \frac{y'}{\sqrt{2}}\right) + 18 = 0$$

which, when multiplied out and simplified, reduces to

$$9x'^2 + 4y'^2 + 18x' - 24y' + 9 = 0.$$

In order to easily graph this ellipse in the $x'y'$-plane, we follow the usual completing-the-square process:

$$9\left(x'^2 + 2x'\right) + 4\left(y'^2 - 6y'\right) = -9$$
$$9\left(x'^2 + 2x' + 1\right) + 4\left(y'^2 - 6y' + 9\right) = -9 + 9 + 36$$
$$9(x'+1)^2 + 4(y'-3)^2 = 36$$
$$\frac{(x'+1)^2}{4} + \frac{(y'-3)^2}{9} = 1.$$

We can now easily graph this ellipse whose center, in the $x'y'$-plane, is $(-1, 3)$ and whose minor and major axes have lengths 4 and 6, respectively.

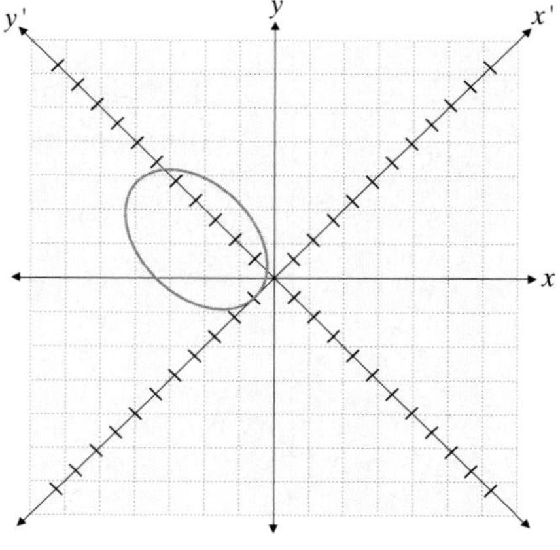

exercises

In the exercises below, the coordinates of a point are given. Find the $x'y'$-coordinates of each point for the given rotation angle θ. See Example 1.

1. $(8, 6)$, $\theta = 30°$

2. $(-5, 1)$, $\theta = \frac{\pi}{3}$

3. $\left(\frac{-1}{2}, \frac{-1}{8}\right)$, $\theta = \frac{\pi}{4}$

4. $(2.7, 5)$, $\theta = 15°$

5. $(13, -4)$, $\theta = 78°$

6. $\left(\frac{12}{\sqrt{18}}, \frac{240}{\sqrt{1152}}\right)$, $\theta = 45°$

7. $\left(3.65, \frac{3}{8}\right)$, $\theta = \frac{\pi}{6}$

8. $\left(\frac{3+\sqrt{48}}{-\left(2\sqrt{12}+3\right)}, \frac{\sqrt{4096}}{8\sqrt{25}\cdot\frac{1}{5}\sqrt{64}}\right)$, $\theta = \frac{\pi}{2}$

Use the discriminant to determine whether the equation of the given conic represents an ellipse, a parabola, or a hyperbola. See Example 4.

9. $2x^2 - 3xy + 2y^2 - 2x = 0$

10. $3x^2 + 7xy + 5y^2 - 6x + 7y + 15 = 0$

11. $3x^2 + 8xy + 4y^2 - 7 = 0$

12. $5x^2 + 6xy - 3y^2 - 9 = 0$

13. $-2x^2 - 8xy + 2y^2 + 2y + 5 = 0$

14. $3x^2 - 6xy + 3y^2 + 3x - 9 = 0$

15. $x^2 - xy + 4y^2 + 2x - 3y + 1 = 0$

16. $x^2 - 4xy + 4y^2 + 2x + 3y - 1 = 0$

Use the discriminant to classify each of the following conic sections. Then determine the angle θ that will allow you to convert the equation and eliminate the xy-term. Finally, sketch the conic section. See Examples 2, 3, and 4.

17. $xy = 2$

18. $xy - 4 = 0$

19. $x^2 + 2xy + y^2 - x + y = 0$

20. $7x^2 + 5\sqrt{3}xy + 2y^2 = 14$

21. $22x^2 + 6\sqrt{3}xy + 16y^2 - 49 = 276$

22. $2\sqrt{3}x^2 - 6xy + \sqrt{3}x - 9y = 0$

23. $34x^2 + 8\sqrt{3}xy + 42y^2 = 1380$

24. $xy + x - 4y = 6$

Use a graphing calculator or computer software to construct the graphs of the following conic sections.

25. $x^2+6xy+y^2=18$ **26.** $x^2-4xy+3y^2=12$ **27.** $9x^2+14xy-9y^2=15$

28. $36x^2-19xy+8y^2=72$ **29.** $40x^2+20xy+10y^2+(2\sqrt{2}-6)x-(4\sqrt{2}+8)y=90$

30. $72x^2+19xy+4y^2=20$ **31.** $48x^2+15xy+7y^2=28$ **32.** $72x^2+18xy-9y^2=14$

Match the equation with its corresponding graph.

33. $3x^2+2xy+y^2-10=0$ **34.** $x^2-4xy+4y^2+5\sqrt{5}y+1=0$

35. $xy-1=0$ **36.** $x^2+y^2-16x+39=0$

37. $x^2-y^2-16=0$ **38.** $3x^2+8xy+4y^2-7=0$

39. $4xy-9=0$ **40.** $x^2-6xy+9y^2-2y+1=0$

a.

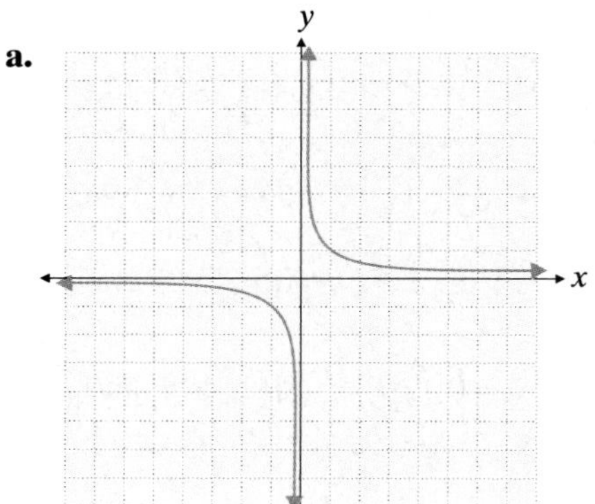

b.

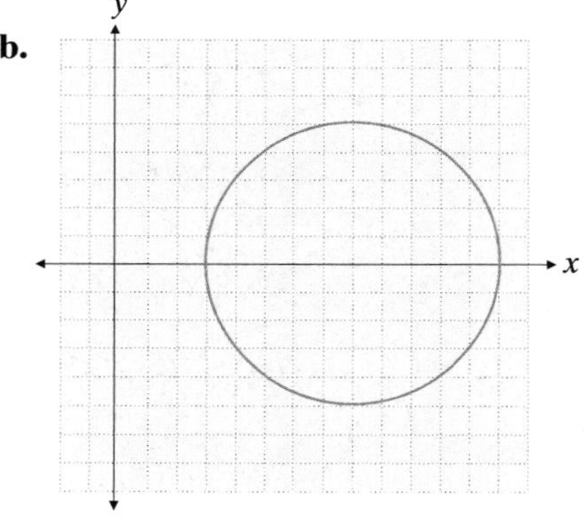

c.

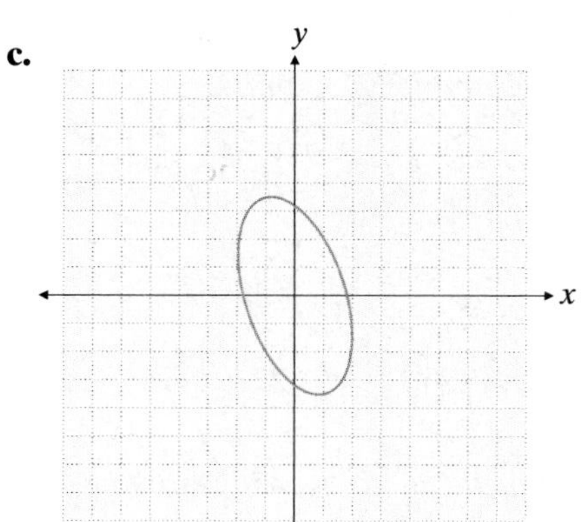

d.

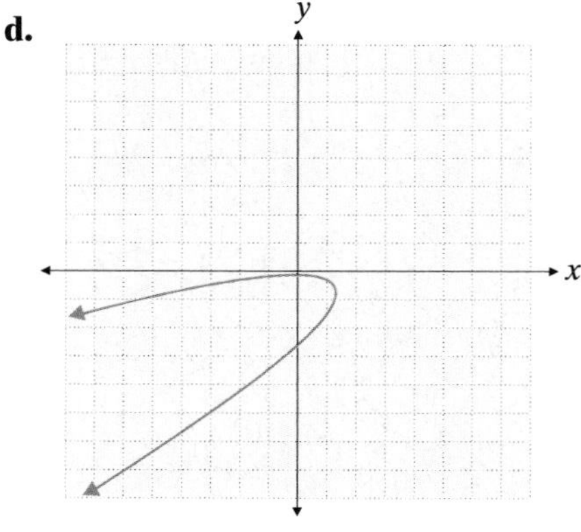

e.

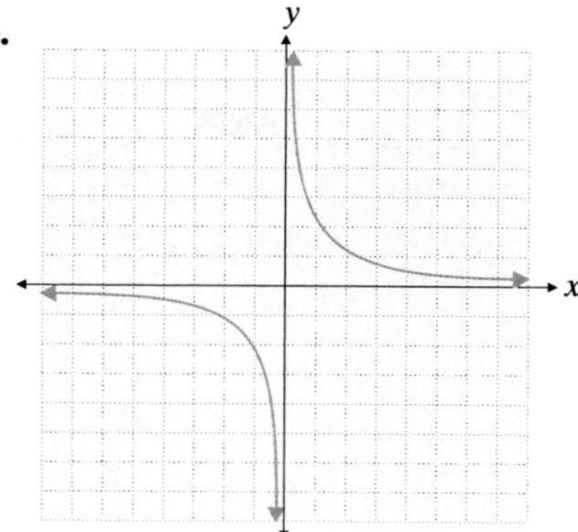

f.

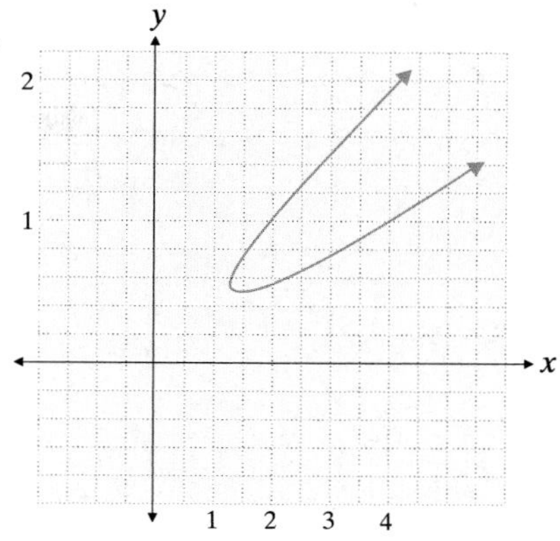

g.

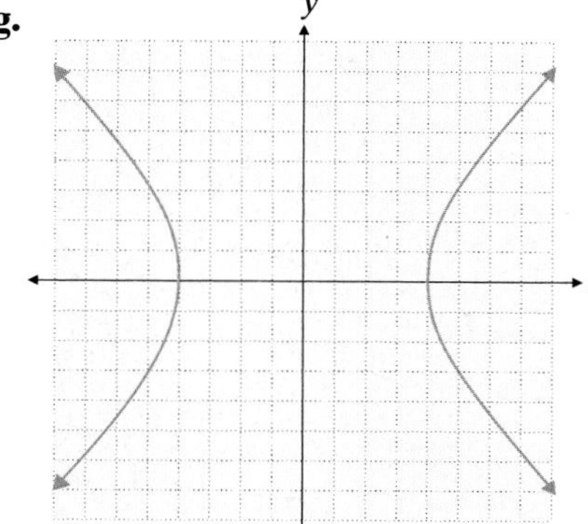

h.

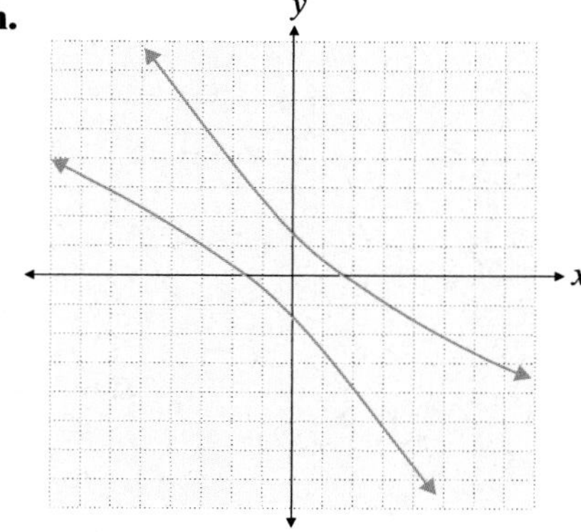

Use your knowledge of rotation of conics to answer the following questions.

41. You have just used the rotation of axes to rotate the x- and y-axes until they were parallel to the axes of the conic. The resulting equation in the $x'y'$-plane is of the form

$$A'(x')^2 + Bx'y' + E'y' + F' = 0.$$

What is wrong with the resulting equation?

42. What must the angle of rotation (θ) be if the coefficients of x^2 and y^2 are equal and $B \neq 0$? Support you answer.

43. Using the equation $7x^2 - 6\sqrt{3}xy + 13y^2 - 16 = 0$:

a. Show that the rotation invariant $F = F'$ is true.

b. Show that the rotation invariant $A + C = A' + C'$ is true.

c. Show that the rotation invariant $B^2 - 4AC = (B')^2 - 4A'C'$ is true.

9.5 Polar Form of Conic Sections

TOPICS

1. The focus/directrix definition of conic sections
2. Using the polar form of conic sections

The first three sections of this chapter dealt with the three varieties of conic sections individually, describing their geometric properties and the formulation of their associated equations in rectangular coordinates. This individual attention is useful when introducing conic sections, and rectangular coordinates are undeniably useful in working with conics to solve certain applications. But polar coordinates provide us with an alternative approach to the study of conic sections, one which possesses important advantages of its own.

Topic 1: The Focus/Directrix Definition of Conic Sections

Probably the most striking virtue of the polar coordinate formulation of conics is that the equations for the three types of conics all have the same form. The equations are easily identified, and the magnitude of a parameter e, called the *eccentricity,* determines whether the conic is an ellipse, a parabola, or a hyperbola. You first encountered eccentricity in the discussion of ellipses – the use of e in this section is an extension of the original use. An additional characteristic of the polar form of conics is that all three varieties are defined in terms of a *focus* and a *directrix*; previously, the directrix only made an appearance in the discussion of parabolas.

Throughout this section, we will assume a point F, called the **focus**, lies at the origin of the plane and that a line L, called the **directrix**, lies d units away from the focus. Until we discuss rotated conics in polar form, the directrix will be oriented either vertically or horizontally, so the equation for the directrix will be $x = -d$, $x = d$, $y = -d$, or $y = d$. We will let $D(P, F)$ denote the distance between a variable point P and the fixed focus F, and we will let $D(P, L)$ denote the shortest distance between a variable point P and the fixed directrix L. The **eccentricity** e will be a fixed positive number for any given conic, and we will make frequent use of the polar coordinates (r,θ) of points in the plane.

Focus/Directrix Description of Conics

Let $D(P, F)$ denote the distance between a variable point P and the fixed focus F; and let $D(P, L)$ denote the shortest distance between P and the fixed directrix L. A conic section consists of all points P in the plane which satisfy the equation

$$\frac{D(P,F)}{D(P,L)} = e,$$

where e is a fixed positive constant.

The conic is:

an ellipse if $0 < e < 1$,
a parabola if $e = 1$, and
a hyperbola if $e > 1$.

In words, a conic section consists of all those points for which the ratio in the distance to the focus and the distance to the directrix is a fixed constant e. This common definition for all three types of conic sections is pleasantly unified. Figure 1 illustrates three conic sections that share the directrix $x = -1$ but with three different eccentricities. In all three graphs, the directrix and the focus appear in green.

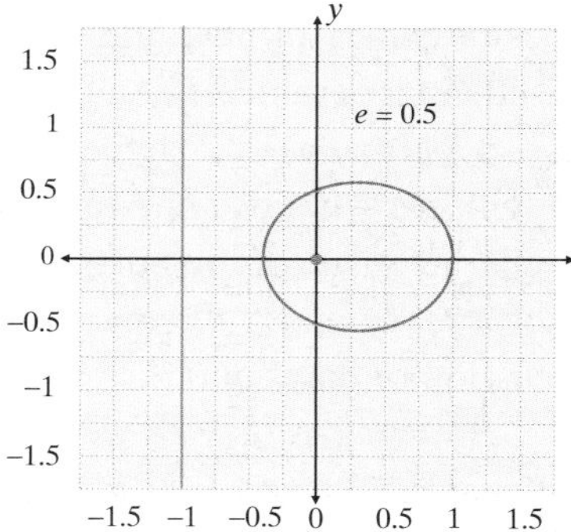

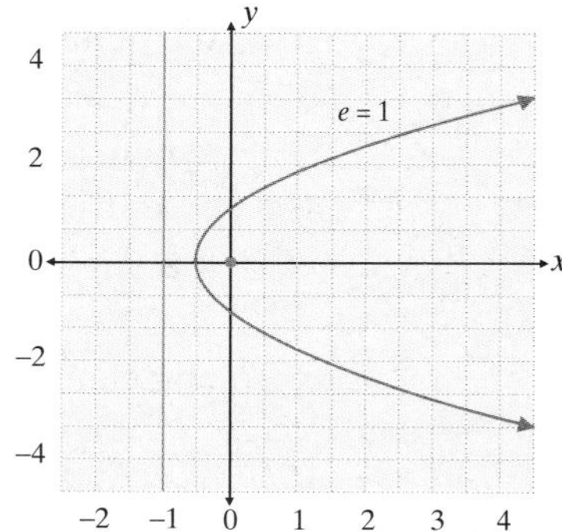

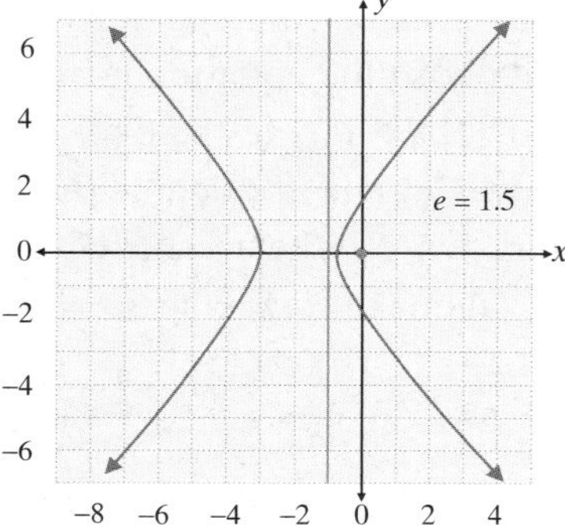

Figure 1: Common Directrix and Focus, Three Eccentricities

The next step is to use the focus/directrix description to develop the form of the polar equations for conics. If we let (r,θ) denote a point P on a given conic section, then $D(P,F)=r$. To determine $D(P, L)$, consider the diagram in Figure 2, which depicts an ellipse and a directrix of the form $x = d$.

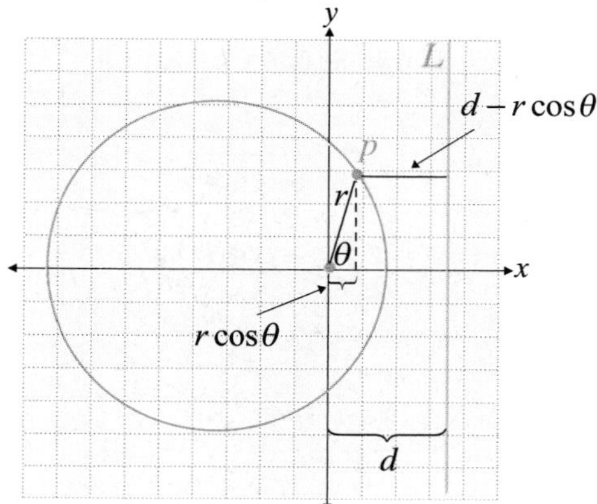

Figure 2: Determining $D(P, L)$

From Figure 2, we see that $D(P,L)=d-r\cos\theta$, so the equation in polar coordinates is

$$\frac{r}{d-r\cos\theta}=e.$$

This equation will be more useful if we solve for r:

$$\begin{aligned} r &= e(d-r\cos\theta) \\ r &= ed-er\cos\theta \\ r+er\cos\theta &= ed \\ r(1+e\cos\theta) &= ed \\ r &= \frac{ed}{1+e\cos\theta}. \end{aligned}$$

The derivations of the equations for a directrix of the form $x = -d$, $y = -d$, or $y = d$ are very similar, and will be left to the reader. The four possible forms and their geometric meanings are described in the following box.

Polar Form of Conic Sections

In polar coordinates, a conic section with focus at the origin and either a horizontal or vertical directrix has one of the following forms:

Conic Section Equation	Directrix
$r = \dfrac{ed}{1+e\cos\theta}$	Vertical Directrix $x = d$
$r = \dfrac{ed}{1-e\cos\theta}$	Vertical Directrix $x = -d$
$r = \dfrac{ed}{1+e\sin\theta}$	Horizontal Directrix $y = d$
$r = \dfrac{ed}{1-e\sin\theta}$	Horizontal Directrix $y = -d$

example 1

Identify the variety of each of the following conics, and determine the equation for the directrix in each case.

a. $r = \dfrac{15}{5-3\cos\theta}$ **b.** $r = \dfrac{2}{3+5\sin\theta}$

Solutions:

a. All we need to do is determine the two constants e and d, and we can accomplish this with simple algebraic manipulation. In order to have the correct form, the constant term in the denominator must be a 1. Thus:

$$r = \frac{15}{5-3\cos\theta}$$
$$= \frac{15\left(\frac{1}{5}\right)}{(5-3\cos\theta)\left(\frac{1}{5}\right)}$$
$$= \frac{3}{1-\left(\frac{3}{5}\right)\cos\theta}$$

This tells us that $e = \frac{3}{5}$, so the conic is an ellipse. To determine the constant d, we need to write the numerator as a product of e and d. Since the numerator is 3 and since we now know $e = \frac{3}{5}$, we can do this as follows:

$$r = \frac{3}{1-\left(\frac{3}{5}\right)\cos\theta} = \frac{\left(\frac{3}{5}\right)(5)}{1-\left(\frac{3}{5}\right)\cos\theta}.$$

Hence, $d = 5$. The last observation is that the trigonometric function in the denominator is cosine and that the sign between the two terms in the denominator is negative. By the guidelines for conics in polar form, this tells us that $x = -5$ is the equation for the directrix.

b. We use the same methods as above to determine first e and then d:

$$r = \frac{2}{3+5\sin\theta} = \frac{\frac{2}{3}}{1+\left(\frac{5}{3}\right)\sin\theta} = \frac{\left(\frac{5}{3}\right)\left(\frac{2}{5}\right)}{1+\left(\frac{5}{3}\right)\sin\theta}.$$

From this form, we can easily see that $e = \frac{5}{3}$ and that the directrix is $y = \frac{2}{5}$. Thus, the conic section is a hyperbola and the directrix is horizontal and above the x-axis.

example 2

Construct a polar equation for a leftward-opening parabola with focus at the origin and directrix 2 units from the focus.

Solution:

Since we are discussing a parabola, $e = 1$. If the parabola is to open to the left, the directrix must be oriented vertically and must lie to the right of the focus, so $d = 2$ and the trigonometric function in the denominator must be cosine. Thus, the equation is

$$r = \frac{2}{1+\cos\theta}.$$

The graph of the parabola appears below, with the focus and directrix shown in green:

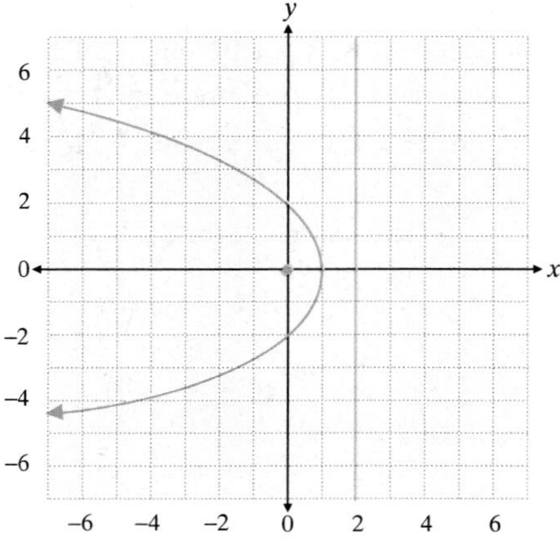

Topic 2: Using the Polar Form of Conic Sections

Example 1 illustrated how the two constants e and d can be easily determined from the equation for a conic in polar form. In order to get an even better understanding of a given conic section, it may be useful to determine the constants a, b, and c as well. These constants have the same meaning in this context as they did in Sections 9.1 and 9.3; that is, $2a$ is the distance between vertices of an ellipse or a hyperbola, $2b$ is the length of the minor axis of an ellipse or the width of the auxiliary rectangle that aids in sketching the asymptotes of a hyperbola, and $2c$ is the distance between foci of an ellipse or a hyperbola.

We begin with a. In Section 9.3, eccentricity was defined as $e = \frac{c}{a}$, and this relation still holds for the expanded definition of eccentricity that we now have. In originally discussing ellipses, we noted that $b^2 = a^2 - c^2$, and in originally discussing hyperbolas we

noted that $b^2 = c^2 - a^2$; we can combine these two statements by noting that b is non-negative and that $b^2 = \left|a^2 - c^2\right|$. Note that if $e = 1$, $a = c$ and, hence, $b = 0$; this should not be surprising, as b played no role in the original discussion of parabolas.

When graphing an ellipse or a hyperbola described in polar form, it is computationally easy to determine a. Using the above observations, it is then easy to determine c and b. This process is illustrated in the next several examples.

example 3

Sketch the graph of the conic section $r = \dfrac{15}{5 - 3\cos\theta}$.

Solution:

This is the first equation from Example 1, and we have already seen that the equation can be written in the form

$$r = \frac{\left(\frac{3}{5}\right)(5)}{1 - \left(\frac{3}{5}\right)\cos\theta}.$$

So $e = \frac{3}{5}$, $d = 5$, and the directrix is the line $x = -5$. This tells us that the graph is an ellipse and that the major axis is oriented horizontally (perpendicular to the directrix). The entire graph is traced out as θ increases from 0 to 2π, with the right vertex corresponding to $\theta = 0$ and the left vertex corresponding to $\theta = \pi$ (halfway around the ellipse). When $\theta = 0$, $r = \frac{15}{2}$ and when $\theta = \pi$, $r = \frac{15}{8}$. In rectangular coordinates, the coordinates of the right vertex are $\left(\frac{15}{2}, 0\right)$ and the coordinates of the left vertex are $\left(-\frac{15}{8}, 0\right)$ (remember that $\theta = \pi$ and a positive r corresponds to a point left of the origin).

Now that we know the coordinates of the two vertices, we can determine that

$$2a = \frac{15}{2} + \frac{15}{8} = \frac{75}{8}$$

and so $a = \frac{75}{16}$. Since $e = \frac{3}{5}$ and $c = ea$, this means

$$c = \left(\frac{3}{5}\right)\left(\frac{75}{16}\right) = \frac{45}{16}.$$

Finally, from the relation $b^2 = \left|a^2 - c^2\right|$, we can determine that $b = \frac{15}{4}$.

cont'd. on next page ...

With this knowledge of a and b, we can sketch the graph below:

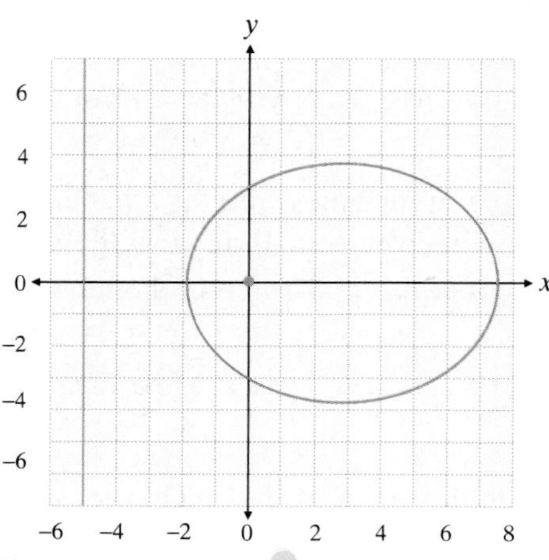

technology note

Computer algebra systems and graphing calculators with the capability to plot equations in polar mode are very convenient for constructing sketches of conic sections. In Mathematica, the command for such a graph is `PolarPlot`. Its use is illustrated in the next example.

example 4

Sketch the graph of the conic section $r = \dfrac{2}{3+5\sin\theta}$.

Solution:

This equation is the second from Example 1, and we already know $e = \frac{5}{3}$ and that $y = \frac{2}{5}$ is the equation for the directrix. This tells us that the graph is a hyperbola and that the vertices are aligned vertically (perpendicular again to the directrix). One vertex corresponds to $\theta = \frac{\pi}{2}$ while the other vertex corresponds to $\theta = \frac{3\pi}{2}$ (it might also be useful to note that the lower branch of the hyperbola crosses the x-axis at $\theta = 0$ and $\theta = \pi$). When $\theta = \frac{\pi}{2}$, $r = \frac{1}{4}$ and when $\theta = \frac{3\pi}{2}$, $r = -1$ (make note of the negative sign on r). This means that the rectangular coordinates of the two vertices are $\left(0, \frac{1}{4}\right)$ and $(0, 1)$, so $2a = \frac{3}{4}$ and $a = \frac{3}{8}$. From this, $c = ea = \frac{5}{8}$ and the relation $b^2 = \left|a^2 - c^2\right|$ leads to $b = \frac{1}{2}$.

Our work in Section 9.3 tells us the equations for the asymptotes of the hyperbola are $y-\frac{5}{8}=\pm\frac{3}{4}x$. Mathematica has been used to plot the hyperbola; the crossed straight lines running through $\left(0,\frac{5}{8}\right)$ are actually artifacts of the graphing algorithm, but they approximate the asymptotes.

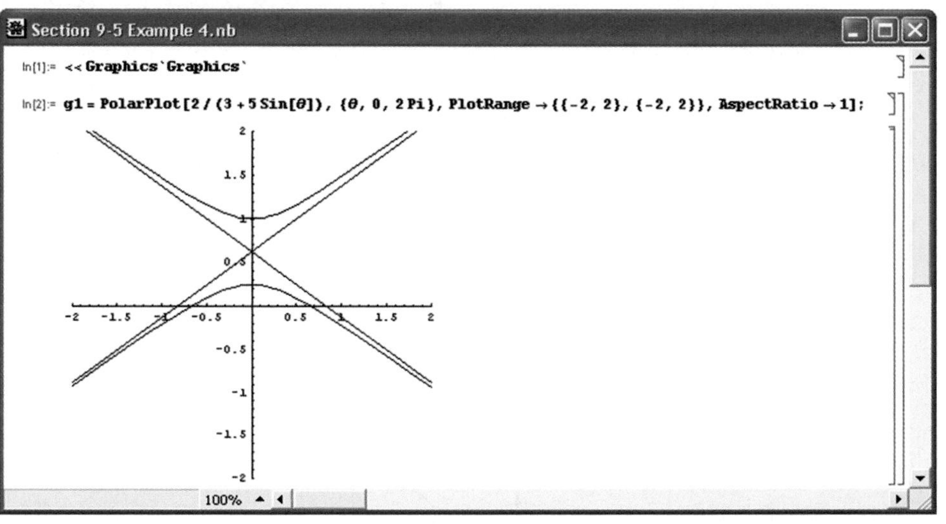

We will conclude this section with an example of rotation of conics in polar form. In general, the graph of an equation $r=f(\theta-\varphi)$ is the rotation of the graph of $r=f(\theta)$ by the angle φ counterclockwise. This makes rotation in polar coordinates particularly easy to handle.

example 5

Sketch the graph of the conic section $r=\dfrac{2}{1+\cos\left(\theta-\frac{\pi}{6}\right)}$.

Solution:

We constructed the equation $r=\dfrac{2}{1+\cos\theta}$ in Example 2, so we know its graph is a parabola opening to the left with directrix $x=2$. The graph of $r=\dfrac{2}{1+\cos\theta}$ is repeated to the right:

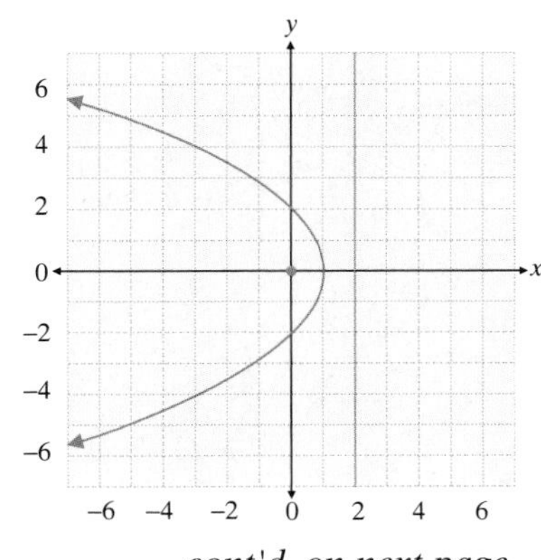

cont'd. on next page ...

The graph of $r = \dfrac{2}{1+\cos\left(\theta - \dfrac{\pi}{6}\right)}$ is simply the same shape rotated $\dfrac{\pi}{6}$ radians counterclockwise:

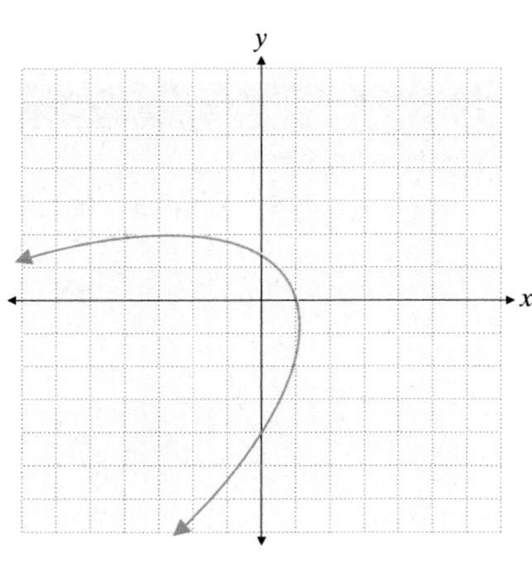

exercises

Match the graph with its corresponding polar equation. See Examples 2 and 3.

1. $r = \dfrac{3}{4-\cos(\theta)}$

2. $r = \dfrac{9}{6-2\sin(\theta)}$

3. $r = \dfrac{3}{3+4\sin(\theta)}$

4. $r = \dfrac{1}{2+2\cos(\theta)}$

5. $r = \dfrac{6}{1+3\sin(\theta)}$

6. $r = \dfrac{6}{1+3\cos(\theta)}$

a.

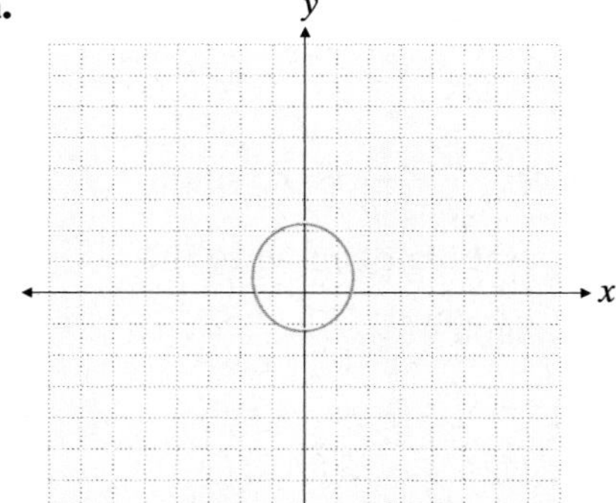

b.

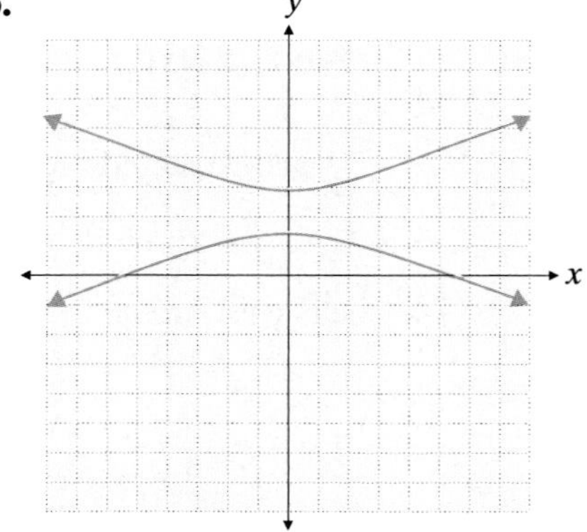

c.

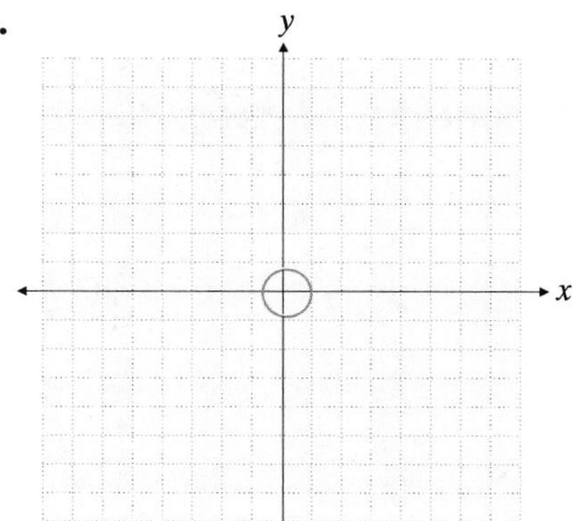

d.

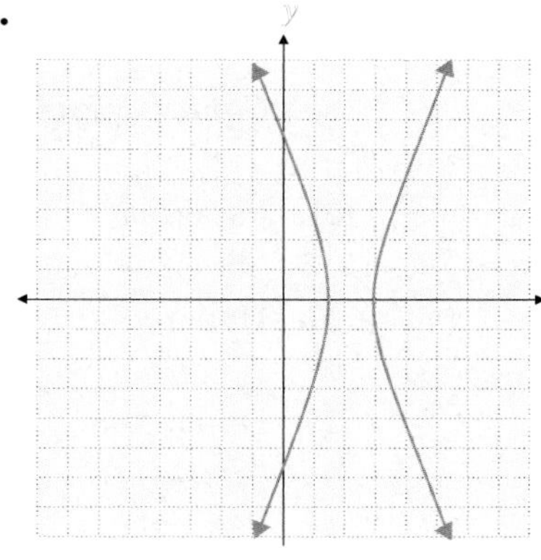

e.

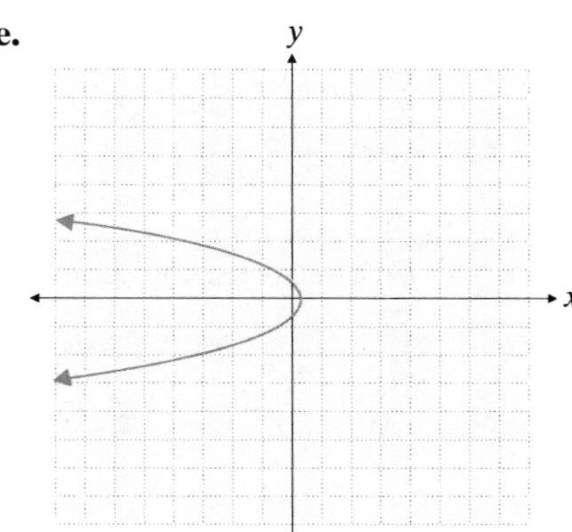

f.

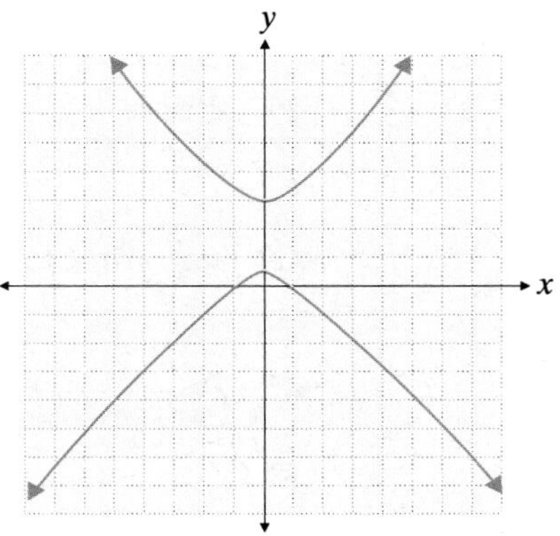

Identify each conic section and find the equation for its directrix. See Example 1.

7. $r = \dfrac{7}{1+6\sin(\theta)}$

8. $r = \dfrac{2}{1-\sin(\theta)}$

9. $r = \dfrac{3}{4-\cos(\theta)}$

10. $r = \dfrac{4}{2-2\cos(\theta)}$

11. $r = \dfrac{1}{1+3\cos(\theta)}$

12. $r = \dfrac{7}{3+2\sin(\theta)}$

13. $r = \dfrac{5}{2+\cos(\theta)}$

14. $r = \dfrac{3}{4-3\sin(\theta)}$

15. $r = \dfrac{6}{3-5\cos(\theta)}$

16. $r = \dfrac{8}{5-6\sin(\theta)}$

17. $r = \dfrac{3}{2+2\sin(\theta)}$

18. $r = \dfrac{-1}{3+4\cos(\theta)}$

19. $r = \dfrac{4}{6-7\cos(\theta)}$

20. $r = \dfrac{9}{5-4\sin(\theta)}$

Construct polar equations for the conic sections described below. See Example 2.

	Conic	*Eccentricity*	*Directrix*
21.	Parabola	$e=1$	$x=-2$
22.	Hyperbola	$e=2$	$x=-3$
23.	Hyperbola	$e=4$	$y=-\frac{3}{4}$
24.	Parabola	$e=1$	$x=2$
25.	Ellipse	$e=\frac{1}{4}$	$x=12$
26.	Ellipse	$e=\frac{1}{2}$	$y=8$

Sketch the graphs of the following conic sections. See Examples 3, 4, and 5.

27. $r=\frac{5}{1+3\cos(\theta)}$

28. $r=\frac{3}{2+\sin(\theta)}$

29. $r=\frac{4}{1-2\sin(\theta)}$

30. $r=\frac{6}{2-4\cos(\theta)}$

31. $r=\frac{9}{3-2\cos(\theta)}$

32. $r=\frac{5}{3+\sin(\theta)}$

33. $r=\frac{4}{1+2\cos(\theta)}$

34. $r=\frac{4}{2+2\sin(\theta)}$

35. $r=\frac{2}{1+\cos\left(\theta-\frac{\pi}{4}\right)}$

36. $r=\frac{4}{2+2\sin\left(\theta-\frac{\pi}{3}\right)}$

Use a graphing calculator or computer software program to construct the graphs of the following conic sections. See Examples 4 and 5.

37. $r = \dfrac{-3}{4 - 9\cos(\theta)}$

38. $r = \dfrac{9}{-4 + \frac{3}{2}\sin(\theta)}$

39. $r = \dfrac{-11}{3 - \cos(\theta)}$

40. $r = \dfrac{2}{10 + 4\sin(\theta)}$

41. $r = \dfrac{3}{7 + 3\cos(\theta)}$

42. $r = \dfrac{2}{2 + 3\cos\left(\theta - \frac{\pi}{4}\right)}$

43. $r = \dfrac{-7}{5 + 3\sin\left(\theta - \frac{\pi}{6}\right)}$

44. $r = \dfrac{5}{-2 - 4\sin\left(\theta + \frac{2\pi}{3}\right)}$

45. $r = \dfrac{4}{-3 - 2\cos\left(\theta + \frac{\pi}{3}\right)}$

46. $r = \dfrac{1}{1 + 4\sin\left(\theta + \frac{\pi}{6}\right)}$

Use your knowledge of conics to answer the following questions.

47. The planets of our solar system follow elliptical orbits with the sun located at one of the foci. If we assume the sun is located at the pole and the major axes of these elliptical orbits lie along the polar axis, the orbits of the planets can be expressed by the polar equation

$$r = \frac{\left(1 - e^2\right)a}{1 - e\cos\theta}$$

where e is the eccentricity. Verify the above equation.

48. Using the equation from Exercise 47, answer the following:

a. Show that the shortest distance from the sun to a planet, called the *perihelion*, is $r = a(1 - e)$.

b. Show that the longest distance from the sun to a planet, called the *aphelion*, is $r = a(1 + e)$.

c. Uranus is approximately 2.74×10^9 km away from the sun at perihelion and 3.00×10^9 km at aphelion. Find the eccentricity of Uranus' orbit.

d. The eccentricity of Neptune's path is 0.0113 and $a = 4.495 \times 10^9$. Determine the perihelion and aphelion distances for Neptune.

chapter nine project

Constructing a Bridge

Plans are in process to develop an uninhabited coastal island into a new resort. Before development can begin, a bridge must be constructed joining the island to the mainland.

Two possibilities are being considered for the support structure of the bridge. The archway could be built as a parabola, or in the shape of a semi-ellipse.

Assume all measurements that follow refer to dimensions at high tide. The county building inspector has deemed that in order to establish a solid foundation, the space between supports must be at least 300 ft. and the height at the center of the arch should be 80 ft. There is a commercial fishing dock located on the mainland whose fishing vessels travel constantly along this intercoastal waterway. The tallest of these ships requires a 60 ft. clearance to pass comfortably beneath the bridge. With these restrictions, the width of a channel with a minimum height of 60 ft. has to be determined for both possible shapes of the bridge to confirm that it will be suitable for the water traffic beneath it.

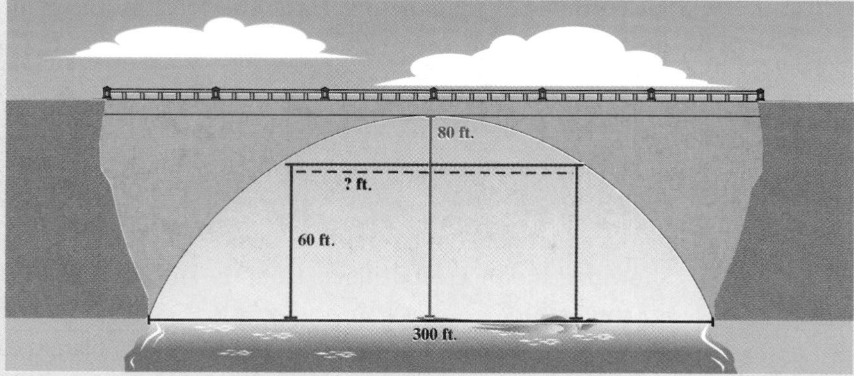

1. Find the equation of a parabola that will fit these constraints.
2. How wide is the channel with a minimum of 60 ft. vertical clearance for the parabola in question 1?
3. Find the equation of a semi-ellipse that will fit these constraints.

4. How wide is the channel with a minimum of 60 ft. vertical clearance for the semi-ellipse in question 3?
5. Which of these bridge designs would you choose, and why?
6. Suppose the tallest fishing ship installs a new antenna which raises the center height by 12 ft. How far off of center (to the left or right) can the ship now travel and still pass under the bridge without damage to the antenna:
   a. For the parabola?
   b. For the semi-ellipse?

CHAPTER REVIEW

9.1 The Ellipse

topics	pages	test exercises
Geometric and algebraic meaning of the conic sections • The geometric definitions of the three varieties of conic sections: *ellipses*, *parabolas*, and *hyperbolas*. • How to identify, algebraically, equations whose graphs are ellipses, parabolas, or hyperbolas • The meaning of the *foci* (plural of *focus*) and *vertices* of an ellipse	p. 695 – 696	1 – 2
Derivation of the standard form of an ellipse • How the standard form of the equation for an ellipse relates to its geometric properties	p. 697 – 699	1 – 2
Using the standard form of an ellipse • The meaning of an ellipse's *major* and *minor axes* • Using the standard form of an ellipse to sketch its graph • How to construct the equation for an ellipse with prescribed properties • The geometric and algebraic meaning of elliptical *eccentricity*	p. 700 – 703	3 – 6
Interlude: planetary orbits • Using knowledge of ellipses to answer questions about planetary orbits	p. 703 – 705	

9.2 The Parabola

topics	pages	test exercises
Derivation of the standard form of a parabola • How the standard form of the equation for a parabola relates to its geometric properties • The meaning of the *focus*, *vertex*, and *directrix* of a parabola	p. 711 – 713	7 – 8
Using the standard form of a parabola • Using the standard form of a parabola to sketch its graph • How to construct the equation for a parabola with prescribed properties	p. 713 – 715	9 – 11
Interlude: parabolic mirrors • The use of a parabola's focusing property, and how to answer questions related to this use	p. 716 – 717	12

9.3 The Hyperbola

topics	pages	test exercises
Derivation of the standard form of a hyperbola • How the standard form of the equation for a hyperbola relates to its geometric properties • The meaning of the *foci, vertices,* and *asymptotes* of a hyperbola	p. 722 – 725	13 – 14
Using the standard form of a hyperbola • Using the standard form of a hyperbola to sketch its graph • How to construct the equation for a hyperbola with prescribed properties	p. 726 – 728	15 – 18
Interlude: guidance systems • How properties of hyperbolas are used in designing guidance systems	p. 728 – 731	

9.4 Rotation of Conics

topics	pages	test exercises
Rotation formulas • How to determine a new set of coordinates to allow a rotation of the conic sections • How to relate the two sets of coordinates • How to eliminate the *xy* term	p. 737 – 744	19 – 24
Rotation invariants • How to classify conics by their invariants • The meaning of the *discriminant*	p. 745 – 746	25 – 30

9.5 Polar Form of Conic Sections

topics	pages	test exercises
The focus/directrix definition of conic sections • How to determine the polar form of conic sections • The meaning of the *focus, directrix,* and *eccentricity* of the polar form of conic sections	p. 751 – 756	31 – 37
Using the polar form of conic sections • Using the polar form to find the distance between vertices, the length of the minor axis, or the width of the auxiliary rectangle to find the asymptotes	p. 756 – 760	38 – 43

chapter test

Section 9.1

Sketch the graphs of the following ellipses, and determine the coordinates of the foci.

1. $\frac{(x+1)^2}{9}+\frac{(y-2)^2}{16}=1$

2. $x^2+9y^2-6x+18y=-9$

In each of the following problems, an ellipse is described by properties it possesses. Find the equation, in standard form, for each ellipse.

3. Center at $(-1, 4)$, major axis is vertical and of length 8, foci $\sqrt{7}$ units from the center.

4. Foci at $(1,2)$ and $(7,2)$, $e=\frac{1}{2}$.

5. Vertices at $\left(\frac{7}{2},-1\right)$ and $\left(\frac{1}{2},-1\right)$, $e=0$.

6. Vertices at $(1,-8)$ and $(1,2)$, minor axis of length 6.

Section 9.2

Graph the following parabolas and determine the focus and directrix of each.

7. $(y+1)^2=-12(x+3)$

8. $x^2-8x+2y+14=0$

Find the equation, in standard form, for the parabola with the given properties.

9. Vertex at $(-2,3)$, directrix is the line $y=2$.

10. Vertex at $(5,-3)$, focus at $(5,1)$.

11. Focus at $(3,-1)$, directrix is the line $x=2$.

Use your knowledge of parabolas to answer the following question.

12. A motorcycle headlight is made by placing a strong light bulb inside a reflective paraboloid formed by rotating the parabola $x^2 = 5y$ around its axis of symmetry (assume that x and y are in units of inches). In order to have the brightest, most concentrated light beam, how far from the vertex should the bulb be placed?

Section 9.3

Sketch the graphs of the following hyperbolas, using asymptotes as guides. Determine the coordinates of the foci in each case.

13. $\frac{(y+2)^2}{9} - \frac{(x-2)^2}{16} = 1$

14. $9x^2 - 4y^2 + 54x - 8y + 41 = 0$

Find the equation, in standard form, for the hyperbola with the given properties.

15. Vertices at $(4, -1)$ and $(-2, -1)$ and foci at $(5, -1)$ and $(-3, -1)$.

16. Asymptotes of $y = \pm\frac{5}{2}(x+1) - 2$ and vertices at $(-3, -2)$ and $(1, -2)$.

17. Foci at $(-1, -2)$ and $(-1, 8)$ and asymptotes of $y = \pm\left(\frac{3}{4}x + \frac{3}{4}\right) + 3$.

18. Asymptotes of $y = \pm(3x - 6) + 2$ and vertices at $(2, -1)$ and $(2, 5)$.

Section 9.4

In the exercises below, the xy-coordinates of a point are given. Find the $x'y'$-coordinates of each point for the given rotation angle θ.

19. $(-8, 7), \theta = \frac{\pi}{4}$

20. $(22, 86), \theta = \frac{\pi}{3}$

21. $(4.6, -8.9), \theta = 53°$

22. $(2\sqrt{3}, 6\sqrt{3}), \theta = 30°$

23. $(5,-32.1), \theta = 2.7°$

24. $(3,\sqrt{3}), \theta = \frac{\pi}{6}$

Use the discriminant to classify each of the following conic sections. Then determine the angle θ that will allow you to convert the equation and eliminate the xy-term. Finally, sketch the conic section.

25. $xy - 6 = 0$

26. $44x^2 + 12\sqrt{3}xy + 32y^2 - 582 = 718$

27. $10x^2 + 2\sqrt{3}xy + 12y^2 - y = 0$

28. $10\sqrt{3}x^2 + 42xy - 4\sqrt{3}y^2 = 187\sqrt{3}$

29. $x^2 + 2xy + y^2 + x - y = 0$

30. $3x^2 + 6xy + 3y^2 + 9x - 9y = \frac{36}{\sqrt{2}}$

Section 9.5

Identify each conic section and find the equation for its directrix.

31. $r = \frac{8}{1+2\sin\theta}$

32. $r = \frac{5}{4-8\sin\theta}$

33. $r = \frac{3}{7+6\sin\theta}$

34. $r = \frac{6}{9-9\cos\theta}$

35. $r = \frac{7}{4+4\sin\theta}$

36. $r = \frac{4}{6-3\cos\theta}$

37. $r = \frac{7}{5+2\cos\theta}$

Construct polar equations for the conic sections described below.

	Eccentricity	Directrix
38.	$e = 1$	$x = -3$
39.	$e = 4$	$y = 3$
40.	$e = \frac{1}{5}$	$y = -15$
41.	$e = \frac{1}{4}$	$x = 16$
42.	$e = 1$	$y = -7$
43.	$e = 9$	$x = \frac{1}{3}$

chapter TEN

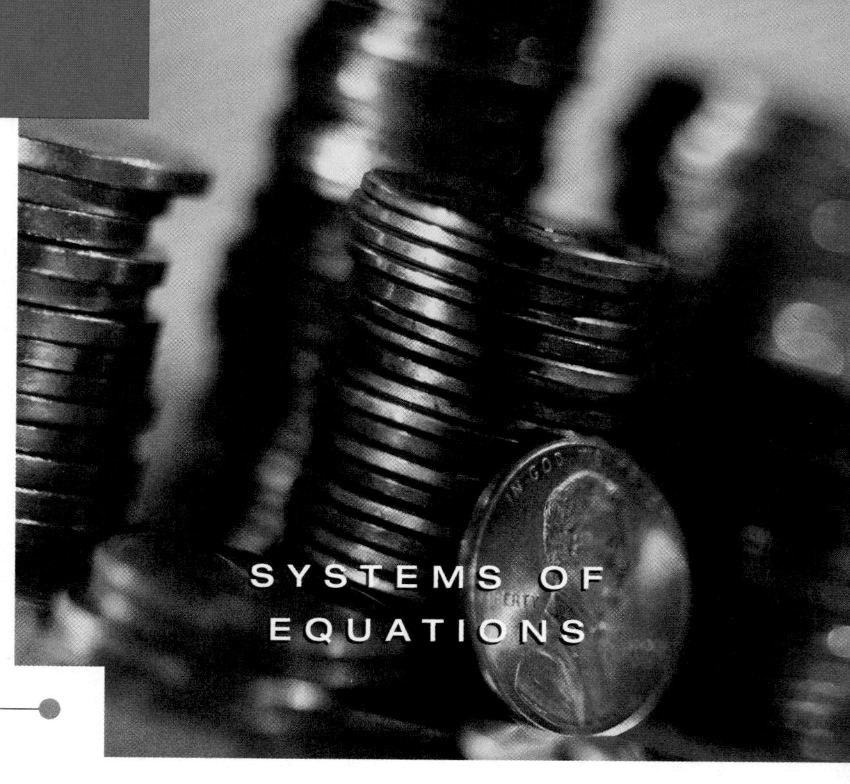

OVERVIEW

By the end of this chapter you should be able to solve both linear and nonlinear systems of equations. Given two or more equations in the same variables, you will find solutions for the variables involved that solve all the equations simultaneously. On page 789 you will find a problem involving an application of such a system in which a variety of loose coins is found. You will be given the total value of the coins and asked to calculate how many of the total number are nickels and how many are pennies. You will master this type of problem using tools such as the method for *Solving Systems by Elimination* as shown in Section 10.1, Example 3, on page 780.

Introduction

In this chapter, we return to the study of linear equations, arguably the simplest class of equations. But as we will soon see, our understanding of linear equations has room for growth in many directions. In fact, the material in this chapter serves as a good illustration of how, given incentive and opportunity, mathematicians extend ideas and techniques beyond the familiar.

In the particular case of linear equations, this extension has taken the form of:

1. Considering equations containing more variables (we have already seen the first step in this process when we moved from linear equations in one variable to linear equations in two).

2. Trying to find solutions that satisfy more than one linear equation at a time; such sets of equations are called *systems*.

3. Making use of elementary methods to solve systems, if possible, and developing entirely new methods, if necessary or desirable.

The third point above constitutes the bulk of this chapter. We start off, in our quest to solve systems of linear equations, by using some intuitive methods that work admirably if the equations contain only a few variables and if the system has only a few equations. Several centuries ago, however, mathematicians began to realize the limitations of such simple methods. As the problems that people wanted to solve led to systems of many variables and equations, refinements were made to the elementary methods of solution and, eventually, entirely new techniques were developed. The two refinements we will study are called *Gaussian elimination* and *Gauss-Jordan elimination*. Some of the new techniques that were developed for solving large systems of equations make use of *determinants of matrices* and *matrix inverses*, two concepts that can serve as an introduction to higher mathematics.

Gauss

The notion of the inverse of a matrix (a matrix is a rectangular array of numbers) closes out our discussion of solution methods, but in fact it brings us full circle and allows us to write systems of linear equations as single matrix equations. This illustrates how mathematicians, in extending our reach in terms of problem solving and in constructing new mathematics, are guided by the achievements of the past. Sections 10.6 and 10.7 then demonstrate two important uses of linear systems of equations, and the chapter ends with an introduction to systems of nonlinear equations.

10.1 Solving Systems by Substitution and Elimination

TOPICS

1. Definition and classification of linear systems of equations
2. Solving systems by substitution
3. Solving systems by elimination
4. Larger systems of equations
5. Applications of systems of equations

Topic 1: Definition and Classification of Linear Systems of Equations

Many problems of a mathematical nature are most naturally described by two or more equations in two or more variables. When the equations are all linear, such a collection of equations is called a *linear system of equations*, or sometimes *simultaneous linear equations*. The word *simultaneous* refers to the goal of identifying all the associated values – if any – of all the variables that solve all of the equations simultaneously.

We have some incidental familiarity with linear systems of equations from our work in graphing solutions of multiple linear inequalities in Section 2.5. In that section, we were concerned with identifying the regions of the plane that simultaneously satisfy collections of inequalities joined by the word "and" or the word "or." But the boundary line for the solution set of a given linear inequality in two variables is the graph of a linear equation, so we were effectively graphing solutions to systems of equations at the same time. This point is illustrated by the graphs in Figure 1, which pictorially identify the three possible varieties of systems of two linear equations in two variables.

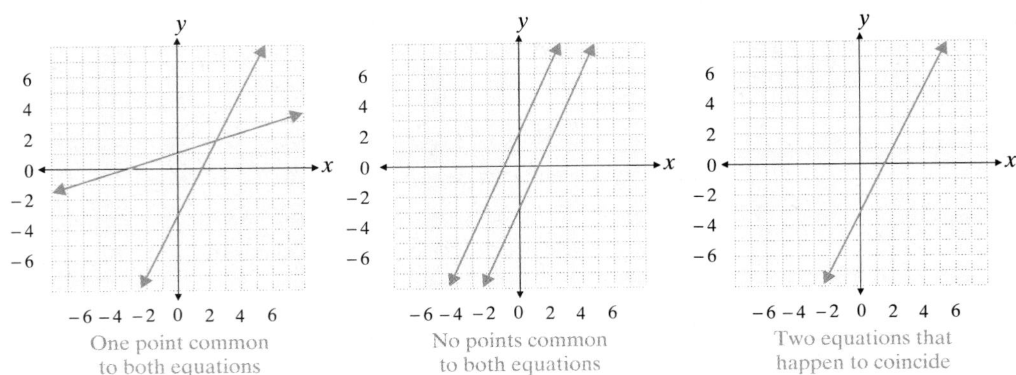

Figure 1: The Varieties of a Pair of Linear Equations in the Plane

In the first graph of Figure 1, the two lines intersect in exactly one point, and the ordered pair associated with that point is consequently a solution to both linear equations at the same time (of course, each equation individually has an infinite number of solutions). In the second graph, the two lines are parallel, and the corresponding system of equations has no solution as there is no point lying on both lines. In the final graph, the two lines actually coincide and appear as one, meaning that any ordered pair that solves one of the equations in the system solves the other as well, so the system has an infinite number of solutions.

We will soon encounter systems consisting of more than two equations and/or more than two variables, but larger systems will have one important similarity to the two-variable, two-equation systems in Figure 1: every linear system will have exactly one solution, no solution, or an infinite number of solutions. This fact is worthy of a definition.

Varieties of Linear Systems

- A linear system of equations with no solution is said to be **inconsistent**.
- A linear system of equations that has at least one solution is, naturally enough, called **consistent**. Any linear system with more than one solution must have an infinite number, and is said to be **dependent**.

Of course, the goal is to develop a systematic and effective method of solving linear systems. Because linear systems of equations arise in so many different contexts, and are of great importance both theoretically and practically, many solution methods have been devised. In fact, the foundation of the branch of mathematics called *linear algebra* is the study of linear systems and their solutions. In this text, we will use solution methods that fall into four broad classes, the first two of which are fairly easy to apply when the number of equations and variables is small.

Topic 2: Solving Systems by Substitution

The solution method of substitution hinges on solving one equation in a system for one of the variables, and substituting the result for that variable in the remaining equations. This can be a time-consuming process if the system is large (meaning more than a few equations and more than a few variables), and in fact the task may have to be repeated many times. But it is a very natural method to use for some small systems, as illustrated in Example 1.

example 1

Use the method of substitution to solve the system $\begin{cases} 2x - y = 1 \\ x + y = 5 \end{cases}$.

Solution:

Either equation can be solved for either variable, and the choice doesn't affect the final answer. One option is to solve the second equation for x. In the work that follows, we do exactly that, and then substitute the result in the first equation, giving us one equation in the variable y. Once we have solved for y, we can then go back and determine x.

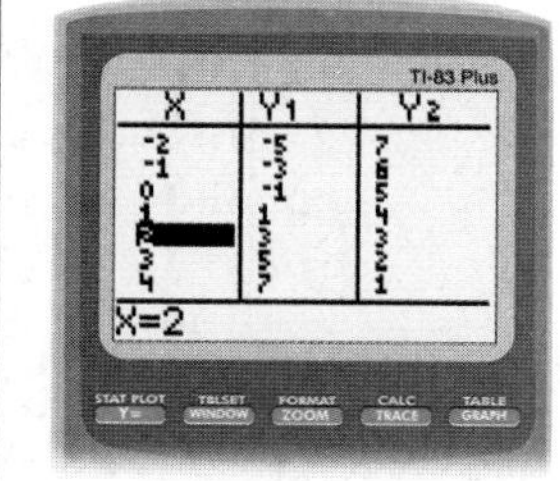

Notice that the y values are both "3" at $x = 2$, after entering the two equations in Y= and selecting 2nd Graph (Table).

$x = -y + 5$	Solve equation 2 for x.
$2(-y+5) - y = 1$	Substitute the result in equation 1.
$-2y + 10 - y = 1$	Multiply out.
$-3y = -9$	Proceed to solve for y.
$y = 3$, so $x = -3 + 5$	Use y to then solve for x.
Solution is $(2, 3)$.	The ordered pair $(2, 3)$ solves both equations.

Although it is not a necessary step in the solution process, it is worthwhile to graph a few systems and their solutions. Note how the graph below corresponds to the system and the ordered pair that we have found.

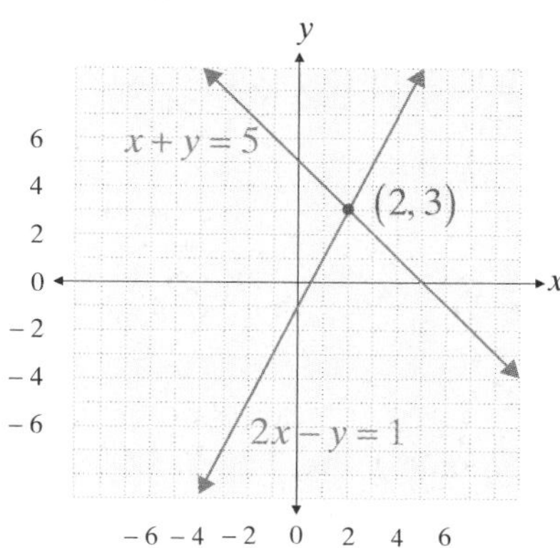

technology note

Computer software and graphing calculators are convenient for graphing systems of equations and can be used to quickly get a sense of where (or whether) to expect a solution. In *Mathematica*, any number of equations (linear or otherwise) may be plotted on the same graph. The first step is to execute the command `<<Graphics`ImplicitPlot`` as shown in the figure below (execute this command just once per *Mathematica* session). The syntax for plotting the two equations $4x - y = 3$ and $x + 3y = 2$ is then shown in the second command. The part reading `{x,-5,5}` tells *Mathematica* to display the portion of the plane from –5 to 5 on the x-axis, and the part reading `{y,-5,5}` does the same thing for the y-axis.

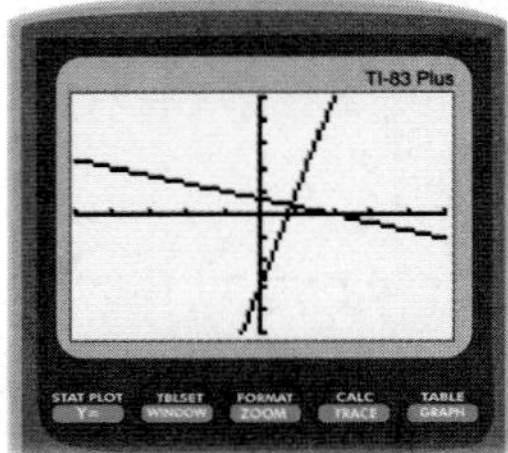

Remember to first solve for y before entering the equations in Y= and selecting Graph.

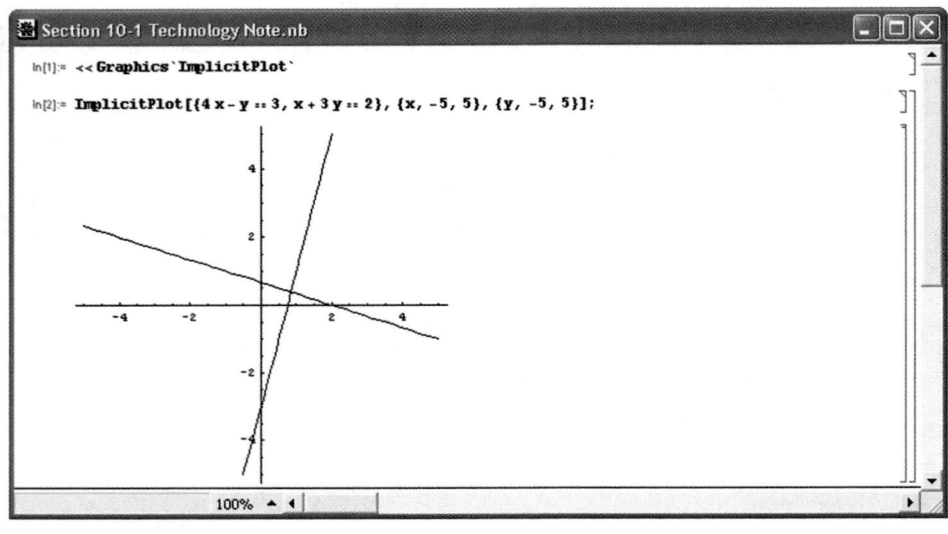

As Figure 1 illustrates, though, a single solution to a system is just one of three possible outcomes. A natural question, therefore, is: How does the solution process differ when the system is inconsistent or dependent?

example 2

Use the method of substitution to solve the system $\begin{cases} -2x+6y=6 \\ x+3=3y \end{cases}$.

Solution:

$$x = 3y-3 \quad \text{Solve equation 2 for } x.$$
$$-2(3y-3)+6y=6 \quad \text{Substitute the result in equation 1.}$$
$$-6y+6+6y=6 \quad \text{Multiply out and proceed to simplify.}$$
$$0=0 \quad \text{True, and trivial, statement.}$$

The last equation, obtained from equation 1 of the system after replacing x with (according to equation 2) an equivalent expression, is always true. Another way of saying this is that for any value of y, the corresponding value $3y - 3$ for x results in an ordered pair (x, y) that solves both equations. Since there are an infinite number of solutions, then, the system is dependent; in the two-variable, two-equation case, this means the graphs of the two equations coincide exactly. In fact, the first equation is simply a rearranged multiple of the second equation.

The question of how to describe the infinite number of solutions remains. One way is with a picture: since the graphs of the two equations are the same line in the plane, the graph of either one represents all the ordered pairs that solve the system. A more accurate method is to describe, with a formula, all the ordered pairs that make up the solution set. From the discussion above, we know one such description is

$$\{(3y-3,\ y) \mid y \in \mathbb{R}\},$$

or simply $\{(3y-3,\ y)\}$. If we had solved either equation for y instead of x, we would have obtained the alternative but equivalent set

$$\left\{\left(x,\ \frac{x+3}{3}\right) \middle| x \in \mathbb{R}\right\}.$$

We will soon encounter an example of the third possibility, a system with no solution.

Topic 3: Solving Systems by Elimination

In some systems, the expressions obtained by solving one equation for one variable are all somewhat unwieldy. (In practice, this often means that awkward fractions appear, no matter which variable is solved for.) In such cases, the solution method of elimination may be a more efficient choice. The elimination method is often a better choice for larger systems, as well.

The **method of elimination** is based on the goal of eliminating one variable in one equation by adding two equations together, and in fact some texts call this the *addition method*. The phrase "adding two equations together" means making use of the fact that if $A = B$ and $C = D$, then $A + C = B + D$. If the equation $A + C = B + D$ has fewer variables than either of the original two, then progress has been made. In fact, if the system is of the two-variable, two-equation variety, the new equation is ready to be solved for its remaining variable, and the solution of the system is then straightforward to find. The following examples illustrate the use of this method.

example 3

Use the method of elimination to solve the system $\begin{cases} 5x+3y=-7 \\ 7x-6y=-20 \end{cases}$.

Solution:

Although the method of substitution could indeed be used, solving for either x or y in either equation introduces some fractions that would be nice to avoid, if possible. Note, though, that the coefficient of y in the second equation is an integer multiple of the coefficient of y in the first. In particular, if we multiply all the terms in the first equation by 2, the coefficients of y will be negatives of one another; adding the two equations will then produce a third equation in which y does not appear.

In order to keep track of the steps we use, it is helpful to annotate our work with reminders of how, exactly, we are modifying the appearance of the system. Each step must result, of course, in an equivalent system: we are not changing the problem, only the way it looks. (The ultimate goal is to change the appearance of the system so that we can "read off" the answer.) In the work that follows, the arrow and notation in green indicates that we have modified the system by multiplying equation 1 by the constant 2.

Notice that the y values are both "1" at $x = -2$, after entering the two equations in Y= and selecting 2nd Graph (Table).

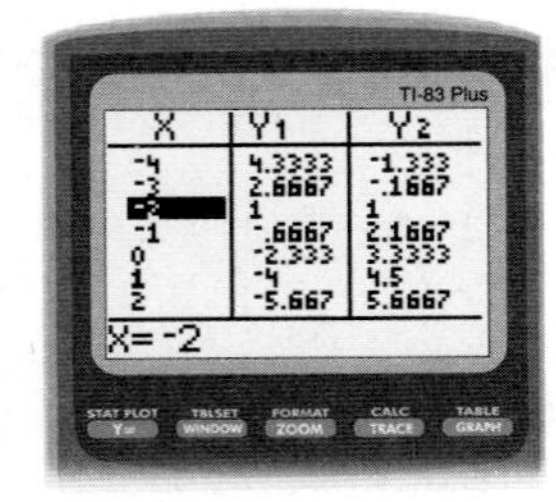

$$\begin{cases} 5x+3y=-7 \\ 7x-6y=-20 \end{cases} \xrightarrow{2E_1} \begin{cases} 10x+6y=-14 \\ 7x-6y=-20 \end{cases}$$

If we now add the left-hand sides of the two equations, and equate this with the sum of the right-hand sides, we get:

$$17x = -34,$$

or $x = -2$. We can now return to any of the other equations to determine y. For instance, equation 1 of the original system tells us that $5(-2)+3y=-7$, or $y = 1$. The ordered pair $(-2, 1)$ is thus the solution of the system. (Note that using the second equation gives the same y-value, and is a good way to check our work.)

example 4

Use the method of elimination to solve the system $\begin{cases} 2x-3y=3 \\ 3x-\frac{9}{2}y=11 \end{cases}$.

Solution:

Instead of multiplying just one equation by a constant, it is probably easier to eliminate x from the system by modifying both equations, as shown:

$$\begin{cases} 2x-3y=3 \\ 3x-\frac{9}{2}y=11 \end{cases} \xrightarrow[-2E_2]{3E_1} \begin{cases} 6x-9y=9 \\ -6x+9y=-22 \end{cases}$$

Although the intent of the above operation was to obtain coefficients of x that were negatives of one another, we have achieved the same thing for y. If we now add the equations, the result is $0=-13$, a false statement. This means that no ordered pair solves both equations, and the system is inconsistent. Graphically, the two lines defined by the equations are parallel, as shown to the right.

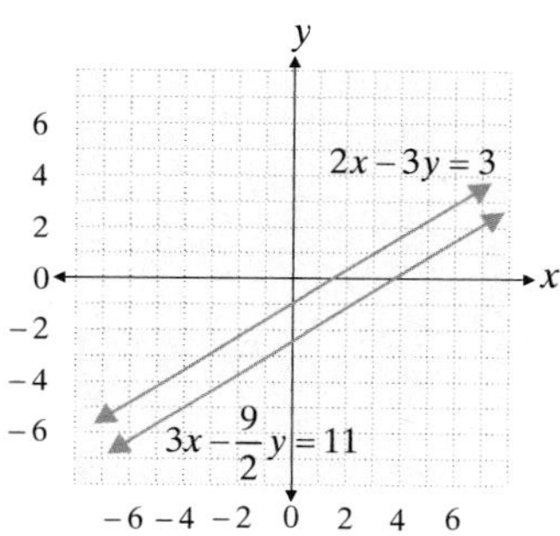

Topic 4: Larger Systems of Equations

Algebraically, larger systems of equations can be dealt with much the same way as the two-variable, two-equation systems that we have studied (though the number of steps needed to obtain a solution might increase). Geometrically, however, larger systems mean something quite different.

For example, if an equation contains three variables, say x, y, and z, a given solution of the equation must consist of an *ordered triple* of numbers, not an ordered pair. And a graphical representation of the ordered triple requires three coordinate axes, as opposed to two. This leads to the concept of three-dimensional space and to the association of points in space with ordered triples. Figure 2 is an illustration of the way in which the positive x, y, and z axes are typically represented on a two-dimensional surface, such as a piece of paper, a computer monitor, or a blackboard. The negative portion of each of the axes is not drawn, and the three axes meet at the origin (the point with (0, 0, 0) as its coordinates) at mutual right angles. As an illustration of how ordered triples appear in $\mathbb{R}^3$, the point $(1,2,4)$ is plotted (the thin colored lines are drawn merely for reference and are not part of the plot).

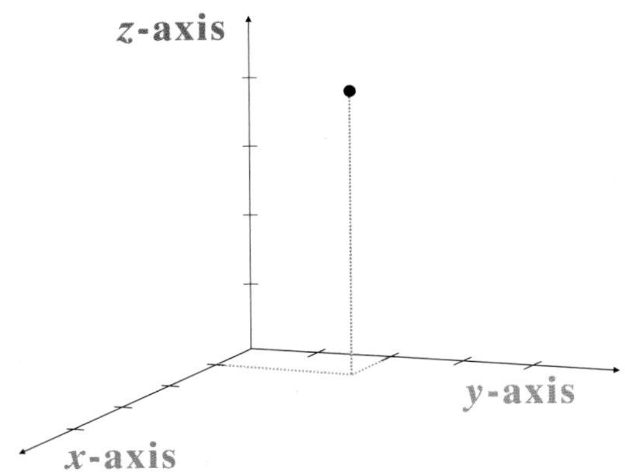

Figure 2: A Depiction of $\mathbb{R}^3$ and the Point (1, 2, 4)

For our purposes, we are interested in how the graph of a linear equation in three variables appears in $\mathbb{R}^3$. The answer is that any equation of the form

$$Ax + By + Cz = D$$

where not all of A, B, and C are 0, depicts a plane in three-dimensional space. A linear system of equations in three variables will thus describe a collection of planes in $\mathbb{R}^3$, one plane per equation. If a linear system of three variables contains three equations, then, it is possible for the three planes to intersect in exactly one point. It is also possible, however, for two of the planes to intersect in a line while the third plane contains no point of that line, for the three planes to intersect in a common line, for two or three of the planes to be parallel, or for two or three of the planes to coincide. In other words, the addition of another variable and another equation leads to many more possibilities than we have seen thus far. But it is still the case that a linear system has no solution, exactly one solution, or an infinite number of solutions. Figure 3 illustrates some of the possible configurations of a three-variable, three-equation linear system.

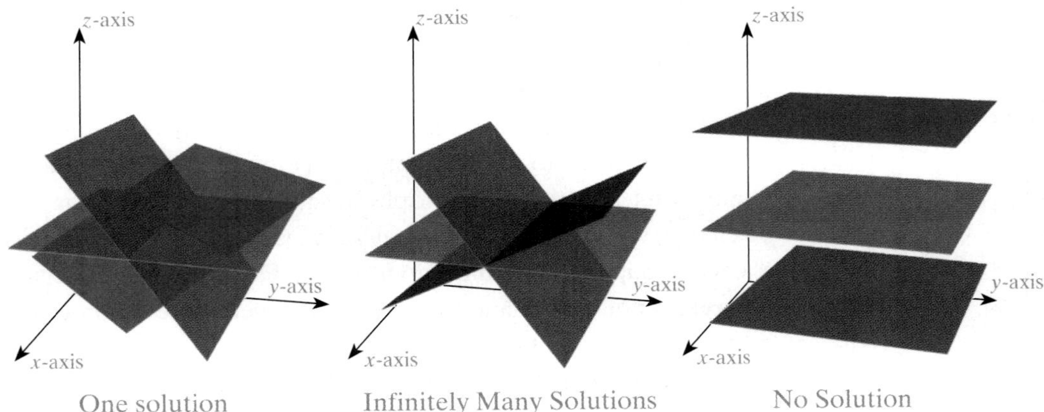

Figure 3: Some Possible 3-Variable, 3-Equation Systems

In Example 5, we will use the method of elimination to find the solution of a linear system in three variables.

example 5

Solve the system $\begin{cases} 3x-5y+z=-10 \\ -x+2y-3z=-7 \\ x-y-5z=-24 \end{cases}$.

Solution:

In general, a good approach to solving a large system of equations is to try to eliminate a variable and obtain a smaller system. In this problem, we know how to handle any 2-variable system that results from eliminating one of the original three variables.

There are many possible ways to proceed. One option is to use the second equation (or a multiple of it) to eliminate x when we add it to the first and third equations. This is what we have done below in order to obtain the two equations in green.

Using E_1+E_2: $\begin{cases} 3x-5y+z=-10 \\ -x+2y-3z=-7 \end{cases} \xrightarrow{3E_2} \begin{cases} 3x-5y+z=-10 \\ -3x+6y-9z=-21 \end{cases}$

$$y-8z=-31$$

Using E_2+E_3: $\begin{cases} -x+2y-3z=-7 \\ x-y-5z=-24 \end{cases}$

$$y-8z=-31$$

The two new equations are identical, and tell us that if an ordered triple (x, y, z) is to solve the system, it must be the case that $y = 8z - 31$. We can now use any equation that contains x to determine the relation between x and z. For instance, the third equation in the system tells us that $x = y + 5z - 24$, or

$$x=(8z-31)+5z-24=13z-55.$$

One description of the solution set is thus $\{(13z-55, 8z-31, z) \mid z \in \mathbb{R}\}$. This is a one-dimensional figure in $\mathbb{R}^3$, indicating that the three planes in the system intersect in a line.

We will soon see an example of a 3-variable system whose solution is a single point.

technology note

Computer software and many calculators are able to solve systems of equations, and can be used to check your understanding of the material in this chapter. The command in *Mathematica* for solving a system of equations is the `Solve` command, as illustrated below. In the first instance, *Mathematica* has been used to verify our solution from Example 5; the warning in blue indicates that one of the variables (z in this case) is free to take on any value. In the second instance, a very similar system of equations has a solution consisting of the single point (–2, 3, 1).

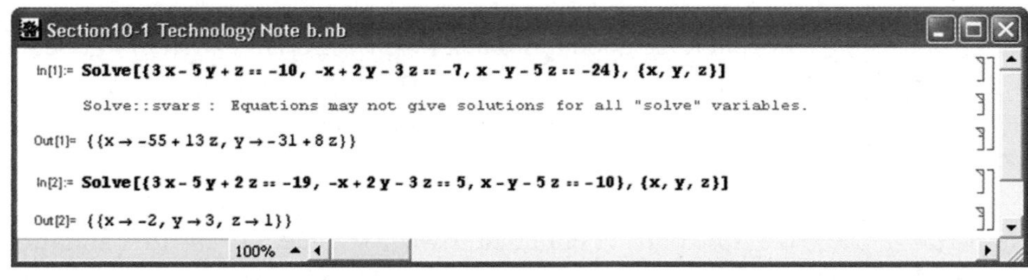

Section10-1 Technology Note b.nb
In[1]:= Solve[{3 x - 5 y + z == -10, -x + 2 y - 3 z == -7, x - y - 5 z == -24}, {x, y, z}]
Solve::svars : Equations may not give solutions for all "solve" variables.
Out[1]= {{x → -55 + 13 z, y → -31 + 8 z}}
In[2]:= Solve[{3 x - 5 y + 2 z == -19, -x + 2 y - 3 z == 5, x - y - 5 z == -10}, {x, y, z}]
Out[2]= {{x → -2, y → 3, z → 1}}
100%

Topic 5: Applications of Systems of Equations

Many applications that we have previously analyzed mathematically are more naturally stated in terms of two or more equations. Consider, for example, the following mixture problem.

example 6

A foundry needs to produce 75 tons of an alloy that is 34% copper. It has supplies of 9% copper alloy and 84% copper alloy. How many tons of each must be mixed to obtain the desired result?

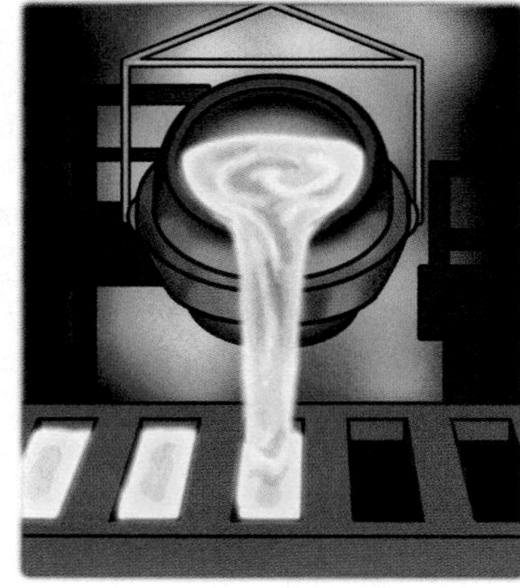

Solution:

To solve this problem without using a system of equations, it is necessary to arrive at one equation in one variable. But the work in doing this involves, essentially, stating the problem in system form.

If we let x represent the number of tons of 9% copper alloy needed, and y the number of tons of 84% copper alloy needed, one immediate fact is that $x + y = 75$; that is, we require 75 tons of alloy in the end. Another fact is that 9% of the x tons and 84% of the y tons represent the total amount of copper, and this amount must equal 34% of 75 tons; the second equation is thus $0.09x + 0.84y = 0.34(75)$. This gives us a system that can be solved by elimination:

$$\begin{cases} x + y = 75 \\ 0.09x + 0.84y = 0.34(75) \end{cases} \xrightarrow[100E_2]{-9E_1} \begin{cases} -9x - 9y = -675 \\ 9x + 84y = 2550 \end{cases}$$

Adding these two equations gives us the equation $75y = 1875$, or $y = 25$. From this, we deduce that $x = 50$, so 50 tons of 9% alloy and 25 tons of 84% alloy are needed.

The last example involves three variables, and is definitely most easily stated and solved in system form.

example 7

If the ages of three girls, Xenia, Yolanda, and Zsa Zsa, are added, the result is 30. The sum of Xenia's and Yolanda's ages is Zsa Zsa's age, while Xenia's age subtracted from Yolanda's is half of Zsa Zsa's age a year ago. How old are the three?

Solution:

Conveniently enough, x, y, and z are natural names for Xenia's, Yolanda's, and Zsa Zsa's ages. The first sentence tells us that $x + y + z = 30$, the second that $x + y = z$, and the third that $y - x = \frac{(z-1)}{2}$. To make the work easier, these equations can be modified as shown:

cont'd. on next page ...

$$\begin{cases} x+y+z=30 \\ x+y=z \\ y-x=\dfrac{z-1}{2} \end{cases} \xrightarrow{2E_3} \begin{cases} x+y+z=30 \\ x+y-z=0 \\ -2x+2y-z=-1 \end{cases}$$

From this, we see that the sum of the first and second equations results in $2x + 2y = 30$, or $x + y = 15$, and the sum of the first and third equations is $-x + 3y = 29$. The method of elimination is the best choice for solving the new system

$$\begin{cases} x+y=15 \\ -x+3y=29, \end{cases}$$

as the sum of the two equations gives us $4y = 44$, or $y = 11$. With this knowledge, the first of the last two equations tells us $x = 4$, and then any equation that includes z can be used to determine that $z = 15$. If this solution is interpreted geometrically, it means that the ordered triple (4, 11, 15) lies on the three planes described by the three equations. In context, it means that Xenia is 4, Yolanda is 11, and Zsa Zsa is 15.

exercises

Use the method of substitution to solve the following systems of equations. If a system is dependent, express the solution set in terms of one of the variables. See Examples 1 and 2.

1. $\begin{cases} 2x-y=-12 \\ 3x+y=-13 \end{cases}$

2. $\begin{cases} 2x-4y=-6 \\ 3x-y=-4 \end{cases}$

3. $\begin{cases} 3y=9 \\ x+2y=11 \end{cases}$

4. $\begin{cases} -3x-y=2 \\ 9x+3y=-6 \end{cases}$

5. $\begin{cases} 2x+y=-2 \\ -4x-2y=5 \end{cases}$

6. $\begin{cases} 5x-y=-21 \\ 9x+2y=-34 \end{cases}$

7. $\begin{cases} 2x-y=-3 \\ -4x+2y=6 \end{cases}$

8. $\begin{cases} 3x+6y=-12 \\ 2x+4y=-8 \end{cases}$

9. $\begin{cases} 2x+5y=33 \\ 3x=-3 \end{cases}$

10. $\begin{cases} 5x+2y=8 \\ 2x+y=6 \end{cases}$

11. $\begin{cases} -2x+y=5 \\ 9x-2y=5 \end{cases}$

12. $\begin{cases} 3x+y=4 \\ -2x+3y=1 \end{cases}$

13. $\begin{cases} 4x-y=-1 \\ -8x+2y=2 \end{cases}$

14. $\begin{cases} 4x-2y=3 \\ -2x+y=-7 \end{cases}$

15. $\begin{cases} 9x-y=-1 \\ 3x+2y=44 \end{cases}$

16. $\begin{cases} 3x+y=9 \\ 2x-y=6 \end{cases}$ **17.** $\begin{cases} x-2y=3 \\ 3x+y=30 \end{cases}$ **18.** $\begin{cases} 6x=18 \\ 2x-3y=-12 \end{cases}$

19. $\begin{cases} 4x-4y=12 \\ x+y=7 \end{cases}$ **20.** $\begin{cases} x-y=0 \\ -4x+5y=1 \end{cases}$ **21.** $\begin{cases} 2x+y=4 \\ 3x-y=6 \end{cases}$

22. $\begin{cases} 3x+y=4 \\ -x-y=4 \end{cases}$ **23.** $\begin{cases} 3x+4y=5 \\ 2x+2y=3 \end{cases}$ **24.** $\begin{cases} 5x+7y=-25 \\ x+14y=-68 \end{cases}$

25. $\begin{cases} 7x-5y=11 \\ x+5y=13 \end{cases}$ **26.** $\begin{cases} 3x-2y=12 \\ 4x+y=5 \end{cases}$ **27.** $\begin{cases} 4x-2y=6 \\ -3x+2y=2 \end{cases}$

28. $\begin{cases} 0.5x+3y=4 \\ 2x-6y=-14 \end{cases}$ **29.** $\begin{cases} 7x+2y=5 \\ 14x+6y=8 \end{cases}$ **30.** $\begin{cases} 3x-2y=1 \\ 4x-2y=4 \end{cases}$

31. $\begin{cases} 5x+13y=38 \\ -20x+36y=-64 \end{cases}$ **32.** $\begin{cases} 38x-14y=64 \\ 64x-27y=77 \end{cases}$ **33.** $\begin{cases} 28x+29y=-72 \\ 13x+25y=-218 \end{cases}$

34. $\begin{cases} 2x-y+3z=12 \\ 3x-2y+z=11 \\ 5x-7y-2z=11 \end{cases}$ **35.** $\begin{cases} 6x+4y+3z=-11 \\ 5x+y+2z=-13 \\ x-4y+5z=7 \end{cases}$ **36.** $\begin{cases} x+3y-z=7 \\ 5x-2y+8z=6 \\ 4x+y+9z=23 \end{cases}$

37. $\begin{cases} 2x-5y+7z=-16 \\ 8x-4y+5z=0 \\ x+2y-10z=10 \end{cases}$ **38.** $\begin{cases} 5x+9y+3z=-1 \\ x-17y-9z=17 \\ 11x+4y+z=32 \end{cases}$ **39.** $\begin{cases} 8x-y+13z=-13 \\ 6x+5y+4z=19 \\ 13x+19y-11z=36 \end{cases}$

Use the method of elimination to solve the following systems of equations. If a system is dependent, express the solution set in terms of one of the variables. See Examples 3 and 4.

40. $\begin{cases} 2x-3y=8 \\ 8x+5y=-2 \end{cases}$ **41.** $\begin{cases} -2x+3y=13 \\ 4x+2y=-18 \end{cases}$ **42.** $\begin{cases} 5x+7y=1 \\ -2x+3y=-12 \end{cases}$

43. $\begin{cases} x+2y=17 \\ 3x+4y=39 \end{cases}$ **44.** $\begin{cases} 5x-10y=9 \\ -x+2y=-3 \end{cases}$ **45.** $\begin{cases} -2x-2y=4 \\ 3x+3y=-6 \end{cases}$

46. $\begin{cases} \frac{2}{3}x+y=-3 \\ 3x+\frac{5}{2}y=-\frac{7}{2} \end{cases}$ **47.** $\begin{cases} \frac{x}{5}-y=-\frac{11}{5} \\ \frac{x}{4}+y=4 \end{cases}$ **48.** $\begin{cases} \frac{2}{3}x+2y=1 \\ x+3y=0 \end{cases}$

49. $\begin{cases} 4x+y=11 \\ 3x-2y=0 \end{cases}$ **50.** $\begin{cases} 7x+8y=-3 \\ -5x-4y=9 \end{cases}$ **51.** $\begin{cases} -x-5y=-6 \\ \frac{3}{5}x+3y=1 \end{cases}$

52. $\begin{cases} -2x-y=9 \\ 4x+2y=1 \end{cases}$ **53.** $\begin{cases} -2x+4y=6 \\ 3x-y=-4 \end{cases}$ **54.** $\begin{cases} 5x-6y=-1 \\ -4x+3y=-10 \end{cases}$

55. $\begin{cases} 5x-3y=-4 \\ -5x+6y=13 \end{cases}$ **56.** $\begin{cases} 3x+4y=36 \\ 3x-2y=18 \end{cases}$ **57.** $\begin{cases} 7x-3y=11 \\ 9x+6y=24 \end{cases}$

58. $\begin{cases} 3x-2y=1 \\ 9x+4y=43 \end{cases}$ **59.** $\begin{cases} -7x+4y=-1 \\ 14x-7y=7 \end{cases}$ **60.** $\begin{cases} 3x-3y=-21 \\ 6x+6y=6 \end{cases}$

61. $\begin{cases} 2x+5y=-11 \\ 3x+4y=-13 \end{cases}$ **62.** $\begin{cases} 5x-7y=-49 \\ 3x+9y=-3 \end{cases}$ **63.** $\begin{cases} 3x+17y=9 \\ 7x-23y=21 \end{cases}$

64. $\begin{cases} 4x+9y=0 \\ 5x+3y=33 \end{cases}$ **65.** $\begin{cases} 13x+15y=-14 \\ 19x+11y=56 \end{cases}$ **66.** $\begin{cases} 23x-5y=244 \\ -31x+29y=-84 \end{cases}$

67. $\begin{cases} 2x-4y+3z=11 \\ 3x-8y+6z=17 \\ x-4y=-3 \end{cases}$ **68.** $\begin{cases} 4x-3y+6z=5 \\ 2x+3y=7 \\ -6x+8y-7z=-4 \end{cases}$ **69.** $\begin{cases} 4x-3y+5z=13 \\ 7x+9y-2z=19 \\ 8x-5y=-2 \end{cases}$

70. $\begin{cases} x-3y+5z=6 \\ x-2y+z=5 \\ -4x+5y-6z=-17 \end{cases}$ **71.** $\begin{cases} 7x-5y=19 \\ 3x-10y+7z=37 \\ 3x+4y+8z=26 \end{cases}$ **72.** $\begin{cases} 12x+4y+5z=44 \\ -9x-5y+2z=23 \\ 7x+9y+7z=25 \end{cases}$

Use any convenient method to solve the following systems of equations. If a system is dependent, express the solution set in terms of one or more of the variables, as appropriate. See Example 5.

73. $\begin{cases} x-y+4z=-4 \\ 4x+y-2z=-1 \\ -y+2z=-3 \end{cases}$ **74.** $\begin{cases} x+2y=-1 \\ y+3z=7 \\ 2x+5z=21 \end{cases}$ **75.** $\begin{cases} x+y=4 \\ y+3z=-1 \\ 2x-2y+5z=-5 \end{cases}$

76. $\begin{cases} 2x-y=0 \\ 5x-3y-3z=5 \\ 2x+6z=-10 \end{cases}$ **77.** $\begin{cases} 3x-y+z=2 \\ -6x+2y-2z=-4 \\ -3x+y-z=-2 \end{cases}$ **78.** $\begin{cases} 2x-3y=-2 \\ x-4y+3z=0 \\ -2x+7y-5z=0 \end{cases}$

79. $\begin{cases} 3x - y + z = 2 \\ -6x + 2y - 2z = 1 \\ 5x + 2y - 3z = 2 \end{cases}$ **80.** $\begin{cases} 4x - y + 5z = 6 \\ 4x - 3y - 5z = -14 \\ -2x - 5z = -8 \end{cases}$ **81.** $\begin{cases} 3x + 8z = 3 \\ x + 3z = 1 \end{cases}$

82. $\begin{cases} x + 2y + z = 8 \\ 2x - 3y - 4z = -16 \\ x - 5y + 5z = 6 \end{cases}$ **83.** $\begin{cases} 2x - 7y - 4z = 7 \\ -x + 4y + 2z = -3 \\ 3y - 4z = -1 \end{cases}$ **84.** $\begin{cases} 4x + 4y - 2z = 6 \\ x - 5y + 3z = -2 \\ -2x - 2y + z = 3 \end{cases}$

85. $\begin{cases} 2x + 3y + 4z = 1 \\ 3x - 4y + 5z = -5 \\ 4x + 5y + 6z = 5 \end{cases}$ **86.** $\begin{cases} x - 4y + 2z = -1 \\ 2x + y - 3z = 10 \\ -3x + 12y - 6z = 3 \end{cases}$ **87.** $\begin{cases} x + 2y + 3z = 29 \\ 2x - y - z = -2 \\ 3x + 2y - 6z = -8 \end{cases}$

88. $\begin{cases} 5x - 2y + z = 14 \\ 8x + 4y = 12 \\ 9x = 18 \end{cases}$ **89.** $\begin{cases} 2x + 5y = 6 \\ 3y + 8z = -6 \\ x + 4y = -5 \end{cases}$ **90.** $\begin{cases} 4x + 3y + 4z = 5 \\ 5x - 6y - 2z = -12 \\ 5z = 20 \end{cases}$

91. $\begin{cases} 9x + 4y - 8z = -4 \\ -6x + 3y - 9z = -9 \\ 8y - 3z = 18 \end{cases}$ **92.** $\begin{cases} 7x + 5y - 9z = -2 \\ 6x + 4y - 7z = 0 \\ x - 18y - 7z = 7 \end{cases}$ **93.** $\begin{cases} 3x - 7y + z = 4 \\ 5x + 2y - 9z = 2 \\ -2x + 7y - z = 1 \end{cases}$

94. $\begin{cases} -5x + 2y - 20z = 14 \\ 2x - 3y + 10z = -19 \\ 7x + 4y - 7z = -7 \end{cases}$ **95.** $\begin{cases} 3y + 8z = -11 \\ 5x - 7y - 6z = -10 \\ 15x - 7y - 2z = 4 \end{cases}$ **96.** $\begin{cases} 9x + 5y - z = -42 \\ 6x - 2y + 3z = -21 \\ -2x + 6y - 5z = -5 \end{cases}$

97. $\begin{cases} 21x - 7y + 51z = 141 \\ 13x + 9y - 5z = -19 \\ 19x - 8y + 23z = 30 \end{cases}$ **98.** $\begin{cases} 5x - 10y + 11z = 19 \\ 27x + 9y + 7z = -44 \\ 2x + 19y - 4z = -3 \end{cases}$ **99.** $\begin{cases} -23x + 17y - 7z = -51 \\ -13x + 25y - 11z = 45 \\ 51x - 21y - 28z = -58 \end{cases}$

Use a system of equations to solve the following problems. See Examples 6 and 7.

100. Karen empties out her purse and finds 45 loose coins, consisting entirely of nickels and pennies. If the total value of the coins is \$1.37, how many nickels and how many pennies does she have?

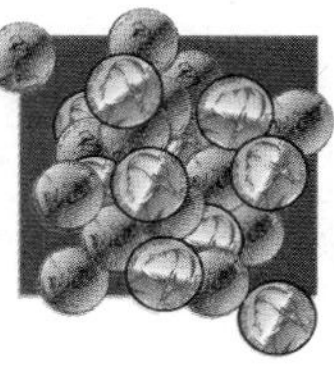

101. What choice of $a, b,$ and c will force the graph of the polynomial $f(x) = ax^2 + bx + c$ to have a y-intercept of 5 and to pass through the points $(1, 3)$ and $(2, 0)$?

102. Eliza's mother is 20 years older than Eliza, but 3 years younger than Eliza's father. Eliza's father is 7 years younger than three times Eliza's age. How old is Eliza?

103. An investor decides at the beginning of the year to invest some of his cash in an account paying 8% annual interest, and to put the rest in a stock fund that ends up earning 15% over the course of the year. He puts \$2000 more in the first account than in the stock fund, and at the end of the year he finds he has earned \$1310 in interest. How much money was invested at each of the two rates?

104. A tour organizer is planning on taking a group of 40 people to a musical. Balcony tickets cost $29.95 and regular tickets cost $19.95, and she collects a total of $1048.00 from her group to buy the tickets. How many people chose to sit in the balcony?

105. How many ounces each of a 12% alcohol solution and a 30% alcohol solution must be combined to obtain 60 ounces of an 18% solution?

106. Jack and Tyler went shopping for summer clothes. Shirts were $12.47 each, including tax, and shorts were $17.23 per pair, including tax. Jack and Tyler spent a total of $156.21 on 11 items. How many shirts and pairs of shorts did they buy?

107. Three years ago, Bob was twice as old as Marla. Fifteen years ago, Bob was three times as old as Marla. How old is Bob?

108. Deyanira empties out her pockets and finds 35 coins consisting of nickels and pennies. If the total value of the coins is $1.27, how many nickels and how many pennies does she have?

109. If an investor has invested $1000 in stocks and bonds, how much has he invested in stocks if he invested four times more in stocks than in bonds?

110. Twelve years ago, Jim was twice as old as Kristin. Sixteen years ago, Jim was three times older than Kristin. How old is Jim?

111. A movie brought in $740 in ticket sales one day. Tickets during the day cost $5 and tickets at night cost $7. If 120 tickets were sold, how many were sold during the day?

112. A computer has 24 screws in its case. If there are 7 times more slotted screws than thumb screws, how many thumb screws are in the computer?

113. Jael has $10,000 she would like to invest. She has narrowed her options down to a certificate of deposit paying 5% annually, bonds paying 4% annually, and stocks with an expected annual rate of return of 13.5%. If she wants to invest twice as much in the stocks as in the certificate of deposit and she wants to earn $1000 in interest by the end of the year, how much should she invest in each type of investment?

114. Lea ordered fruit baskets for three of her coworkers. One containing 5 apples, 2 oranges, and 1 mango cost $6.81. Another containing 2 mangos, 8 oranges, and 3 apples cost $11.88. The third contained 4 apples, 4 oranges, and 4 mangos and cost $11.04. How much did each type of fruit cost?

10.2 Matrix Notation and Gaussian Elimination

TOPICS

Topic 1: Linear Systems, Matrices, and Augmented Matrices

The scheduling of planes and passengers at an airport and the modeling of materials at the atomic level, to name just two, are applications of linear systems with literally thousands of variables. The solution methods of substitution and elimination, learned in the last section, are inadequate for dealing with such large systems. More sophisticated and powerful methods, several of which we will learn in this text, are all based on matrices of numbers associated with a given system. We will begin the discussion with a definition of the relevant terms.

Matrices and Matrix Notation

A **matrix** is a rectangular array of numbers, called **elements** or **entries** of the matrix. As the numbers are in a rectangular array, they naturally form **rows** and **columns**. It is often important to determine the size of a given matrix; we say that a matrix with m rows and n columns is an $m \times n$ matrix (read "m by n"), or of *order* $m \times n$. By convention, the number of rows is always stated first.

Matrices are often labeled with capital letters. The same letter in lower-case, with a pair of subscripts attached, is usually used to refer to its individual elements. For instance, if A is a matrix, a_{ij} refers to the element in the ith row and the jth column (again, rows are stated before columns).

example 1

Given the matrix $A = \begin{bmatrix} -27 & 0 & 1 \\ 5 & -\pi & 13 \end{bmatrix}$, determine:

a. The order of A. **b.** The value of a_{13}. **c.** The value of a_{21}.

Solutions:

a. A has 2 rows and 3 columns, and is thus a 2×3 matrix.
b. The value of the entry in the first row and third column is 1.
c. The value of the entry in the second row and first column is 5.

technology note

To enter matrices, select 2nd x^{-1} (Matrx). Use the right arrow to highlight `Edit`, and press 1 for matrix A. Enter the desired dimensions and values.

Computer algebra systems and many calculators understand matrix notation and operations. In *Mathematica*, for instance, the matrix A from Example 1 is defined as shown below. Note that A is entered as a list consisting of two sub-lists; each sub-list corresponds to a row in the matrix, so in this example there are two sub-lists of three elements each. The command **`MatrixForm`** is used to view the matrix as a rectangular array of numbers. The last two commands below illustrate how the values a_{13} and a_{21} are referred to in *Mathematica*.

```
Section 10-2 Technology Note.nb
In[1]:= A = {{-27, 0, 1}, {5, -Pi, 13}}
Out[1]= {{-27, 0, 1}, {5, -π, 13}}
In[2]:= MatrixForm[A]
Out[2]//MatrixForm=
  (-27  0   1)
  ( 5  -π  13)
In[3]:= A[[1, 3]]
Out[3]= 1
In[4]:= A[[2, 1]]
Out[4]= 5
100%
```

The following application of matrices to solve linear systems is based on a slight refinement of the method of elimination. The first step is to construct the augmented matrix associated with a given system.

Augmented Matrices

The **augmented matrix** of a linear system of equations is a matrix consisting of the coefficients of the variables (retaining the appropriate order), with an adjoined column consisting of the constants from the right-hand side of the system, assuming the equations have been written in standard form. (Informally, an equation is in *standard form* when it has been simplified and all variables appear on the left of the equation and the constant appears on the right.) The matrix of coefficients and the column of constants are customarily separated by a vertical bar.

For example, the augmented matrix for the system $\begin{cases} 3x-7y=-4 \\ -x+4y=9 \end{cases}$ is $\left[\begin{array}{cc|c} 3 & -7 & -4 \\ -1 & 4 & 9 \end{array}\right]$.

Of course, it may be necessary to rearrange the terms of a system in order to construct its augmented matrix.

example 2

Construct the augmented matrix for the linear system $\begin{cases} \dfrac{2x-6y}{2}=3-z \\ z-x+5y=12 \\ x+3y-2=2z \end{cases}$.

Solution:

The first step is to write each equation in standard form:

$$\begin{cases} \dfrac{2x-6y}{2}=3-z \\ z-x+5y=12 \\ x+3y-2=2z \end{cases} \longrightarrow \begin{cases} x-3y+z=3 \\ -x+5y+z=12 \\ x+3y-2z=2 \end{cases}$$

To create this screen, use the Matrx operation as described on page 792.

We can now read off the coefficients and constants to construct the augmented matrix:

$$\left[\begin{array}{ccc|c} 1 & -3 & 1 & 3 \\ -1 & 5 & 1 & 12 \\ 1 & 3 & -2 & 2 \end{array}\right]$$

Topic 2: Gaussian Elimination and Row Echelon Form

Consider the following augmented matrix:

$$\left[\begin{array}{ccc|c} 1 & 2 & -2 & 11 \\ 0 & 1 & -1 & 3 \\ 0 & 0 & 1 & -1 \end{array}\right].$$

We can translate this back into system form to obtain

$$\begin{cases} x+2y-2z=11 \\ \quad\quad y-z=3 \\ \quad\quad\quad z=-1 \end{cases},$$

and we note that this system can be solved when we *back-substitute* the last equation $z=-1$ into the second equation to obtain $y-(-1)=3$, or $y=2$, and then back-substitute again in the first equation to obtain $x+2(2)-2(-1)=11$, or $x=5$.

Of course, we have no reason to expect systems to be so singularly nice. The point of *Gaussian elimination* (named after the ubiquitous Carl Friedrich Gauss) is that it transforms an arbitrary augmented matrix into a form like the one above, and the solution of the corresponding system then solves the original system as well. The technical name for an augmented matrix in the form shown above is *row echelon form.*

Row Echelon Form

A matrix is in **row echelon form** if:

1. The first non-zero entry in each row is 1.
2. Every entry in the column below each such 1 (called a **leading 1**) is 0, and each leading 1 appears further to the right than leading 1's in previous rows.
3. All rows consisting entirely of 0's (if there are any) appear at the bottom.

The operations that make up Gaussian elimination are called elementary row operations. There are three such matrix operations, all based on very familiar facts concerning equations. Namely, that

- interchanging the order of two equations in a system has no effect on the system,
- all the terms in an equation can be multiplied by a non-zero constant, and finally
- two equations in a system can be added together.

The matrix form of these facts, along with the symbols we will use to annotate our work, appear below.

Elementary Row Operations

Assume A is an augmented matrix corresponding to a given system of equations. Each of the following operations on A results in the augmented matrix of an equivalent system. In the notation, R_i refers to row i of the matrix A.

1. Rows i and j can be interchanged. (Denoted $R_i \leftrightarrow R_j$)
2. Each entry in row i can be multiplied by a non-zero constant c. (Denoted cR_i)
3. Row j can be replaced with the sum of itself and a constant multiple of row i. (Denoted $cR_i + R_j$)

In summary, then, Gaussian elimination refines the method of elimination by removing the need to write variables at each step and by providing a framework for systematic application of row operations. Examples 3 and 4 illustrate its use in practice.

example 3

Use Gaussian elimination to solve the system $\begin{cases} -2x+y-5z=-6 \\ x+2y-z=-8 \\ 3x-y+2z=2 \end{cases}$.

Solution:

We will first construct the augmented matrix for the system, and then transform it into row echelon form. A good rule of thumb is to proceed by columns: after getting a leading 1 in the first row, use row operations to obtain 0's below it, and then repeat the process with each successive column.

$$\left[\begin{array}{ccc|c} -2 & 1 & -5 & -6 \\ 1 & 2 & -1 & -8 \\ 3 & -1 & 2 & 2 \end{array}\right] \xrightarrow{R_1 \leftrightarrow R_2} \left[\begin{array}{ccc|c} 1 & 2 & -1 & -8 \\ -2 & 1 & -5 & -6 \\ 3 & -1 & 2 & 2 \end{array}\right] \xrightarrow[-3R_1+R_3]{2R_1+R_2} \left[\begin{array}{ccc|c} 1 & 2 & -1 & -8 \\ 0 & 5 & -7 & -22 \\ 0 & -7 & 5 & 26 \end{array}\right]$$

cont'd. on next page ...

$$\xrightarrow{\frac{1}{5}R_2}\left[\begin{array}{ccc|c} 1 & 2 & -1 & -8 \\ 0 & 1 & -\frac{7}{5} & -\frac{22}{5} \\ 0 & -7 & 5 & 26 \end{array}\right]\xrightarrow{7R_2+R_3}\left[\begin{array}{ccc|c} 1 & 2 & -1 & -8 \\ 0 & 1 & -\frac{7}{5} & -\frac{22}{5} \\ 0 & 0 & -\frac{24}{5} & -\frac{24}{5} \end{array}\right]\xrightarrow{-\frac{5}{24}R_3}\left[\begin{array}{ccc|c} 1 & 2 & -1 & -8 \\ 0 & 1 & -\frac{7}{5} & -\frac{22}{5} \\ 0 & 0 & 1 & 1 \end{array}\right]$$

The third row in the last matrix tells us that $z = 1$. When we back-substitute this in the second row, we obtain

$$y-\frac{7}{5}(1)=-\frac{22}{5},$$

so $y = -3$. From the first row, then, we obtain

$$x+2(-3)-1(1)=-8,$$

or $x = -1$. Thus, the ordered triple $(-1,-3,1)$ solves the system of equations.

example 4

Use Gaussian elimination to solve the system $\begin{cases} x-3y+2z=1 \\ 2x+y-z=0 \\ -3x-5y+4z=3 \end{cases}$.

Solution:

We construct the system's augmented matrix and proceed as before.

$$\left[\begin{array}{ccc|c} 1 & -3 & 2 & 1 \\ 2 & 1 & -1 & 0 \\ -3 & -5 & 4 & 3 \end{array}\right]\xrightarrow[3R_1+R_3]{-2R_1+R_2}\left[\begin{array}{ccc|c} 1 & -3 & 2 & 1 \\ 0 & 7 & -5 & -2 \\ 0 & -14 & 10 & 6 \end{array}\right]$$

We might, however, notice at this point that the -14 in the third row, second column, is most easily changed to 0 if we add 2 times the second row to the third row. Of course, we eventually want a 1 in the second row, second column, but if we wish to avoid fractions as long as possible we can delay, for one step, dividing each entry in the second row by 7.

$$\left[\begin{array}{ccc|c} 1 & -3 & 2 & 1 \\ 0 & 7 & -5 & -2 \\ 0 & -14 & 10 & 6 \end{array}\right]\xrightarrow{2R_2+R_3}\left[\begin{array}{ccc|c} 1 & -3 & 2 & 1 \\ 0 & 7 & -5 & -2 \\ 0 & 0 & 0 & 2 \end{array}\right]$$

In fact, we see now that we can stop at this point. Although the matrix is not yet in row echelon form, we already know there is no solution to the system. The third row of the matrix corresponds to the equation $0 = 2$, and we know that a false statement of this sort means the solution set of the system is the empty set.

Topic 3: Gauss-Jordan Elimination and Reduced Row Echelon Form

The process of back-substitution, once a matrix has been put in row echelon form, amounts to eliminating all but one variable per equation, beginning with the last equation and working backward. Once this is realized, it is natural to ask: Why not accomplish this while the system is still in matrix form?

Gauss-Jordan elimination is the answer to this question. Just as Gaussian elimination is a refinement of the method of elimination, Gauss-Jordan elimination, co-named for Wilhelm Jordan (1842 - 1899), is a refinement of Gaussian elimination. The goal in the method is to put a given matrix into *reduced row echelon form.*

Reduced Row Echelon Form

A matrix is said to be in **reduced row echelon form** if:

1. It is in row echelon form.
2. Each entry **above** a leading 1 is also 0.

Consider, for instance, the last matrix obtained in Example 3. If we begin with that matrix and eliminate entries (that is, convert to 0) above the leading 1 in the third column, and then do the same in the second column, we obtain a matrix in reduced row echelon form:

$$\left[\begin{array}{ccc|c} 1 & 2 & -1 & -8 \\ 0 & 1 & -\frac{7}{5} & -\frac{22}{5} \\ 0 & 0 & 1 & 1 \end{array}\right] \xrightarrow[R_3+R_1]{\frac{7}{5}R_3+R_2} \left[\begin{array}{ccc|c} 1 & 2 & 0 & -7 \\ 0 & 1 & 0 & -3 \\ 0 & 0 & 1 & 1 \end{array}\right] \xrightarrow{-2R_2+R_1} \left[\begin{array}{ccc|c} 1 & 0 & 0 & -1 \\ 0 & 1 & 0 & -3 \\ 0 & 0 & 1 & 1 \end{array}\right]$$

If we now write this in system form, we have

$$\begin{cases} x = -1 \\ y = -3, \\ z = 1 \end{cases}$$

which is equivalent to the original system, but in a form that tells us the solution of the system.

example 5

Use Gauss-Jordan elimination to solve the system $\begin{cases} x-2y+3z=-5 \\ 2x+3y-z=1 \\ -x-5y+4z=-6 \end{cases}$.

Solution:

$$\left[\begin{array}{ccc|c} 1 & -2 & 3 & -5 \\ 2 & 3 & -1 & 1 \\ -1 & -5 & 4 & -6 \end{array}\right] \xrightarrow[R_1+R_3]{-2R_1+R_2} \left[\begin{array}{ccc|c} 1 & -2 & 3 & -5 \\ 0 & 7 & -7 & 11 \\ 0 & -7 & 7 & -11 \end{array}\right] \xrightarrow{R_2+R_3} \left[\begin{array}{ccc|c} 1 & -2 & 3 & -5 \\ 0 & 7 & -7 & 11 \\ 0 & 0 & 0 & 0 \end{array}\right]$$

We can pause at this intermediate stage and note that the row of 0's at the bottom corresponds to the true statement $0 = 0$, indicating that this system has an infinite number of solutions. But we don't yet have a good way of describing the infinite number of ordered triples that make up the solution set, so we continue:

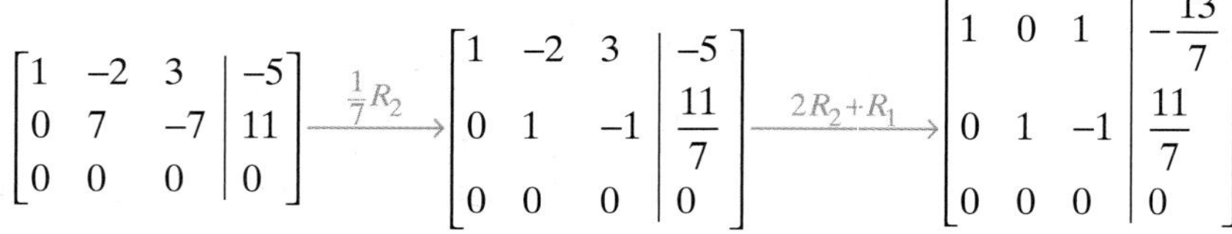

$$\left[\begin{array}{ccc|c} 1 & -2 & 3 & -5 \\ 0 & 7 & -7 & 11 \\ 0 & 0 & 0 & 0 \end{array}\right] \xrightarrow{\frac{1}{7}R_2} \left[\begin{array}{ccc|c} 1 & -2 & 3 & -5 \\ 0 & 1 & -1 & \frac{11}{7} \\ 0 & 0 & 0 & 0 \end{array}\right] \xrightarrow{2R_2+R_1} \left[\begin{array}{ccc|c} 1 & 0 & 1 & -\frac{13}{7} \\ 0 & 1 & -1 & \frac{11}{7} \\ 0 & 0 & 0 & 0 \end{array}\right]$$

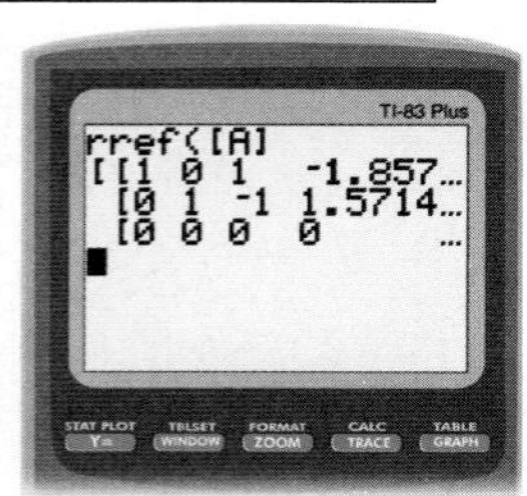

To find the reduced row echelon form of a matrix, enter it using the **Matrx** operation as described on pg. 792. Then **Quit (2nd Mode)**, select **Matrx**, and use the right arrow to highlight `Math`. Select `B↓rref`. This enters the command; to use it on the matrix you created, select **Matrx** then **1**.

The last matrix is in reduced row echelon form, as every leading 1 has 0's both below and above it. The corresponding system of equations is

$$\begin{cases} x+z=-\dfrac{13}{7} \\ y-z=\dfrac{11}{7} \end{cases},$$

and we can thus describe the solution set as $\left\{\left(-z-\frac{13}{7}, z+\frac{11}{7}, z\right) \middle| z \in \mathbb{R}\right\}$.

exercises

1. Let $A = \begin{bmatrix} 4 & -1 \\ 0 & 3 \\ 9 & -5 \end{bmatrix}$. Determine the following, if possible:

a. The order of A. **b.** The value of a_{12}. **c.** The value of a_{23}.

2. Let $B = \begin{bmatrix} -7 & 2 & 11 \end{bmatrix}$. Determine the following, if possible:

a. The order of B. **b.** The value of b_{12}. **c.** The value of b_{31}.

3. Let $C = \begin{bmatrix} 1 & 0 \\ 5 & -3 \\ 2 & 9 \\ \pi & e \\ 10 & -7 \end{bmatrix}$. Determine the following, if possible:

a. The order of C. **b.** The value of c_{23}. **c.** The value of c_{51}.

4. Let $D = \begin{bmatrix} -8 & 13 & -1 \\ 0 & 6 & 3 \\ 0 & -9 & 0 \end{bmatrix}$. Determine the following, if possible:

a. The order of D. **b.** The value of d_{23}. **c.** The value of d_{33}.

5. Let $E = \begin{bmatrix} -443 & 951 & 165 & 274 \\ 286 & -653 & 812 & -330 \\ 909 & 377 & 429 & -298 \end{bmatrix}$. Determine the following, if possible:

a. The order of E. **b.** The value of e_{42}. **c.** The value of e_{21}.

6. Let $A = \begin{bmatrix} 9 & 5 & 0 \\ 7 & 4 & 2 \end{bmatrix}$. Determine the following, if possible:

a. The order of A. **b.** The value of a_{22}. **c.** The value of a_{13}.

7. Let $B=\begin{bmatrix} 8 & 1 \\ 3 & 0 \\ 6 & 7 \end{bmatrix}$. Determine the following, if possible:

a. The order of B. **b.** The value of b_{12}. **c.** The value of b_{13}.

8. Let $C=\begin{bmatrix} 65 & 32 & 91 & 45 \\ 23 & 18 & 75 & 47 \\ 8 & 63 & 28 & 31 \end{bmatrix}$. Determine the following, if possible:

a. The order of C. **b.** The value of c_{43}. **c.** The value of c_{23}.

9. Let $D=\begin{bmatrix} 4 & 9 & 7 & 1 & 8 \\ 5 & 3 & 0 & 2 & 6 \end{bmatrix}$. Determine the following, if possible:

a. The order of D. **b.** The value of d_{21}. **c.** The value of d_{24}.

Construct the augmented matrix that corresponds to each of the following systems of equations. See Example 2. (Answers may appear in slightly different, but equivalent, forms.)

10. $\begin{cases} 5x+\dfrac{y-z}{2}=3 \\ 7(z-x)+y-2=0 \\ x-(4-z)=y \end{cases}$

11. $\begin{cases} \dfrac{2-3x}{2}=y \\ 3z+2(x+y)=0 \\ 2x-y=2(x-3z) \end{cases}$

12. $\begin{cases} 2(z+3)-x+y=z \\ -3(x-2y)-1=5z \\ \dfrac{x}{3}-(y-2z)=x \end{cases}$

13. $\begin{cases} \dfrac{12x-1}{5}+\dfrac{y}{2}=\dfrac{3z}{2} \\ y-(x+3z)=-(1-y) \\ 2x-2-z-2y=7x \end{cases}$

14. $\begin{cases} 4x+5y-3z=8 \\ 7x-2y+9=3 \\ 5x-6y+3z=0 \end{cases}$

15. $\begin{cases} \dfrac{3x+4y}{2}-3z=6 \\ 3(x-2y+9z)=0 \\ 2x+6y=3-z \end{cases}$

16. $\begin{cases} \dfrac{2x-4y}{3}=2z \\ 8x=2(y-3z)+7 \\ 3x=2y \end{cases}$

17. $\begin{cases} y-2z+4=3x \\ \dfrac{x}{2}-4y-1=z \\ 3(-y+z)-1=0 \end{cases}$

18. $$\begin{cases} \dfrac{2(2x-y)}{3}+z=7 \\ 4=\dfrac{3}{-x+y+3z} \\ 4x-8y+4=9x \end{cases}$$

19. $$\begin{cases} 0.5x-14y=\dfrac{z}{4}-8 \\ \dfrac{x}{5}-y+\dfrac{z}{4}=\dfrac{y}{6}-3 \\ \dfrac{2}{3}\left(\dfrac{4}{y-x-1}\right)=\dfrac{5}{z} \end{cases}$$

Fill in the blanks by performing the indicated row operations.

20. $\left[\begin{array}{cc|c} 3 & 2 & -7 \\ 1 & 3 & 5 \end{array}\right] \xrightarrow{-3R_2+R_1} \underline{?}$

21. $\left[\begin{array}{cc|c} 2 & -5 & 3 \\ -4 & 3 & -1 \end{array}\right] \xrightarrow[2R_1+R_2]{} \underline{?}$

22. $\left[\begin{array}{cc|c} 4 & 2 & -8 \\ 3 & -9 & 0 \end{array}\right] \xrightarrow[\frac{1}{3}R_2]{\frac{1}{2}R_1} \underline{?}$

23. $\left[\begin{array}{cc|c} 9 & -2 & 7 \\ 1 & 3 & -2 \end{array}\right] \xrightarrow{R_1 \leftrightarrow R_2} \underline{?}$

24. $\left[\begin{array}{cc|c} 4 & 1 & 5 \\ 3 & 6 & 0 \end{array}\right] \xrightarrow{2R_1} \underline{?}$

25. $\left[\begin{array}{cc|c} 8 & -2 & -4 \\ 3 & -1 & 7 \end{array}\right] \xrightarrow[-2R_2]{} \underline{?}$

26. $\left[\begin{array}{cc|c} 9 & 12 & -6 \\ 15 & -3 & 0 \end{array}\right] \xrightarrow{-\frac{1}{3}R_1} \underline{?}$

27. $\left[\begin{array}{cc|c} 4 & 12 & -6 \\ 7 & 3 & 9 \end{array}\right] \xrightarrow[\frac{1}{2}R_1+R_2]{} \underline{?}$

28. $\left[\begin{array}{cc|c} 3 & 0 & 1 \\ 5 & 7 & -2 \end{array}\right] \xrightarrow[3R_1+R_2]{} \underline{?}$

29. $\left[\begin{array}{cc|c} 8 & -2 & 10 \\ 9 & -3 & 0 \end{array}\right] \xrightarrow[-\frac{2}{3}R_2]{\frac{1}{2}R_1} \underline{?}$

30. $\left[\begin{array}{ccc|c} 5 & 2 & 9 & 7 \\ 1 & 3 & -5 & 0 \\ 2 & -4 & 1 & 8 \end{array}\right] \xrightarrow[-R_1+R_3]{2R_2} \underline{?}$

31. $\left[\begin{array}{ccc|c} 6 & -2 & 5 & 14 \\ -7 & 19 & 2 & 3 \\ -9 & 11 & -4 & 7 \end{array}\right] \xrightarrow[0.5R_3]{3R_1} \underline{?}$

32. $\left[\begin{array}{ccc|c} 5 & 3 & 13 & 15 \\ 17 & 9 & -8 & -14 \\ 4 & -11 & 19 & 8 \end{array}\right] \xrightarrow[-2R_2+R_3]{} \underline{?}$

33. $\left[\begin{array}{ccc|c} 8 & 11 & 18 & 2 \\ 14 & 33 & -3 & -5 \\ -9 & 21 & 12 & 9 \end{array}\right] \xrightarrow[-2R_3+R_2]{\frac{1}{3}R_3+R_1} \underline{?}$

34. $\left[\begin{array}{ccc|c} 1 & 3 & -2 & 4 \\ 3 & -1 & 8 & 2 \\ -5 & 0 & 2 & 7 \end{array}\right] \xrightarrow[5R_1+R_3]{-3R_1+R_2} \underline{?}$

35. $\left[\begin{array}{ccc|c} 2 & 3 & -3 & 5 \\ 1 & 1 & 3 & 4 \\ 3 & 3 & 9 & 12 \end{array}\right] \xrightarrow[-3R_2+R_3]{-2R_2+R_1} \underline{?}$

36. $\left[\begin{array}{cc|c} -3 & 2 & 2 \\ 5 & -4 & 1 \end{array}\right] \xrightarrow[2R_1+R_2]{} \underline{?}$

37. $\left[\begin{array}{cc|c} -5 & 20 & -15 \\ 2 & -12 & 5 \end{array}\right] \xrightarrow[\frac{1}{2}R_2]{\frac{1}{5}R_1} \underline{?}$

38. $\left[\begin{array}{ccc|c} 2 & 2 & 3 & 7 \\ -3 & 2 & 8 & -2 \\ 1 & 5 & 2 & 6 \end{array}\right] \xrightarrow[3R_3+R_2]{-2R_3+R_1} \underline{?}$

39. $\left[\begin{array}{ccc|c} 1 & 5 & -9 & 11 \\ 1 & 4 & -1 & 4 \\ 4 & 3 & 5 & 45 \end{array}\right] \xrightarrow[-4R_1+R_3]{-R_1+R_2} \underline{?}$

Use Gaussian elimination and back-substitution to solve the following systems of equations. See Examples 3 and 4.

40. $\begin{cases} 2x-4y=-6 \\ 3x-y=-4 \end{cases}$

41. $\begin{cases} 2x-5y=11 \\ 3x+2y=7 \end{cases}$

42. $\begin{cases} 5x-y=-21 \\ 9x+2y=-34 \end{cases}$

43. $\begin{cases} x-4y=-11 \\ 7x-y=4 \end{cases}$

44. $\begin{cases} x+2y=17 \\ 3x+4y=39 \end{cases}$

45. $\begin{cases} 2x+6y=4 \\ -4x-7y=7 \end{cases}$

46. $\begin{cases} 3x-2y=5 \\ -5x+4y=-3 \end{cases}$

47. $\begin{cases} 2x+y=-2 \\ -4x-2y=5 \end{cases}$

48. $\begin{cases} 6x-16y=10 \\ -3x+8y=4 \end{cases}$

49. $\begin{cases} 2x-3y=0 \\ 5x+y=17 \end{cases}$

50. $\begin{cases} 6x+3y=3 \\ x+y=3 \end{cases}$

51. $\begin{cases} 3x+6y=-12 \\ 2x+4y=-8 \end{cases}$

52. $\begin{cases} 4x+5y=9 \\ 8x+3y=-17 \end{cases}$

53. $\begin{cases} \frac{2}{3}x+2y=1 \\ x+3y=0 \end{cases}$

54. $\begin{cases} 13x-17y=-3 \\ -19x+15y=-35 \end{cases}$

55. $\begin{cases} 3x-9y-7z=-9 \\ 5x+11y-z=17 \\ -4x-8y+7z=5 \end{cases}$

56. $\begin{cases} 8x-y+5z=-8 \\ 11x-2y+9z=-9 \\ 7x-3y+13z=4 \end{cases}$

57. $\begin{cases} 17x+13y+8z=46 \\ -12x+3y+28z=-19 \\ 14x+5y-15z=-15 \end{cases}$

Use Gauss-Jordan elimination to solve the following systems of equations. See Example 5.

58. $\begin{cases} 2x-3y=8 \\ 8x+5y=-2 \end{cases}$

59. $\begin{cases} \frac{2}{3}x+y=-3 \\ 3x+\frac{5}{2}y=-\frac{7}{2} \end{cases}$

60. $\begin{cases} 3y=9 \\ x+2y=11 \end{cases}$

61. $\begin{cases} 6x+2y=-4 \\ -9x-3y=6 \end{cases}$

62. $\begin{cases} 3y=6 \\ 5x+2y=4 \end{cases}$

63. $\begin{cases} 3x+8y=-4 \\ x+2y=-2 \end{cases}$

64. $\begin{cases} -3x+2y=5 \\ 5x-2y=1 \end{cases}$

65. $\begin{cases} 9x-11y=10 \\ -4x+3y=-12 \end{cases}$

66. $\begin{cases} 9x-15y=-6 \\ -3x+11y=-10 \end{cases}$

67. $\begin{cases} 3x - 8y = 7 \\ 18x - 35y = -23 \end{cases}$

68. $\begin{cases} 4x + y - 3z = -9 \\ 2x - 3z = -19 \\ 7x - y - 4z = -29 \end{cases}$

69. $\begin{cases} -5x + 9y + 3z = 1 \\ 3x + 2y - 6z = 9 \\ x + 4y - z = 16 \end{cases}$

70. $\begin{cases} 2x - y = 0 \\ 5x - 3y - 3z = 5 \\ 2x + 6z = -10 \end{cases}$

71. $\begin{cases} x + y = 4 \\ y + 3z = -1 \\ 2x - 2y + 5z = -5 \end{cases}$

72. $\begin{cases} 2x - 3y = -2 \\ x - 4y + 3z = 0 \\ -2x + 7y - 5z = 0 \end{cases}$

73. $\begin{cases} 3x + 8z = 3 \\ -3x - 7z = -3 \\ x + 3z = 1 \end{cases}$

74. $\begin{cases} 3x - y + z = 2 \\ -6x + 2y - 2z = 1 \\ 5x + 2y - 3z = 2 \end{cases}$

75. $\begin{cases} x + 2y = -1 \\ y + 3z = 7 \\ 2x + 5z = 21 \end{cases}$

76. $\begin{cases} 2x + 8y - z = -5 \\ -5x + 3y + 4z = -6 \\ x - 4y - 5z = -8 \end{cases}$

77. $\begin{cases} 7x - 8y + 2z = -2 \\ 5x - 3y - z = -3 \\ 8x + y - 3z = 7 \end{cases}$

78. $\begin{cases} 8x + 14y - 3z = 3 \\ -6x + 2y + 7z = -13 \\ 8x + 19y + 3z = 11 \end{cases}$

79. $\begin{cases} 8x + 5y + 3z = -2 \\ 12x - y - 18z = 1 \\ 7x + 6y + 10z = 19 \end{cases}$

80. $\begin{cases} 4x + 8y + 7z = 27 \\ -2x + 9y - 8z = -15 \\ 9x + 13y + 7z = -33 \end{cases}$

81. $\begin{cases} w - x + 2z = 9 \\ 2w + 3y = -1 \\ -2w - 5y - z = 0 \\ x + 2y = -4 \end{cases}$

82. $\begin{cases} 3w - x + 5y + 3z = 2 \\ -4w - 10y - 2z = 10 \\ w - x + 2z = 7 \\ 4w - 2x + 5y + 5z = 9 \end{cases}$

10.3 Determinants and Cramer's Rule

TOPICS

1. Determinants and their evaluation
2. Using Cramer's rule to solve linear systems

Topic 1: Determinants and Their Evaluation

As far as solving linear systems is concerned, Gaussian elimination is certainly an improvement over the methods of substitution and elimination. But the improvement is due primarily to the systematic approach that Gaussian elimination, and in particular Gauss-Jordan elimination, encourages, and not to any significantly deeper understanding of systems. Cramer's rule, named after the Swiss mathematician Gabriel Cramer (1704 - 1752), is the first of two solution methods we will study that truly brings something new to the discussion.

Cramer's rule relies on the computation of a number, called the *determinant*, associated with every square (that is, $n \times n$) matrix, and we accordingly begin with a definition of determinant.

Determinant of a 2×2 Matrix

The **determinant** of the matrix $A = \begin{bmatrix} a_{11} & a_{12} \\ a_{21} & a_{22} \end{bmatrix}$, denoted $|A|$, is $a_{11}a_{22} - a_{21}a_{12}$.

Of course, this only defines determinants for matrices of one size, and unfortunately the simplicity of the definition does not carry over to larger sizes. The good news, however, is that determinants of 3×3, 4×4, and larger matrices can all be related, ultimately, to determinants of 2×2 matrices. We will do a few calculations before considering larger matrices.

example 1

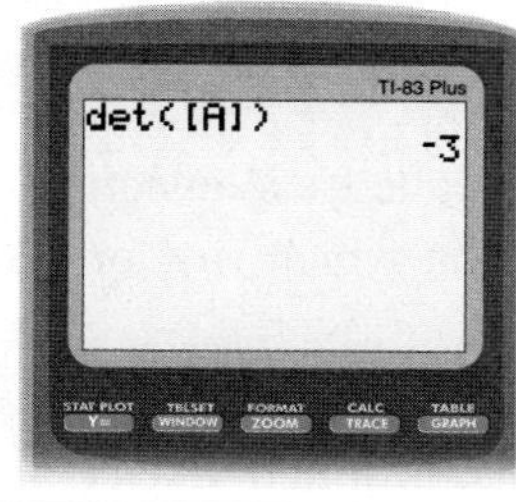

To find the determinant of a matrix, enter it using the **Matrx** operation (pg. 792). Then **Quit (2nd Mode)**, select **Matrx**, and use the right arrow to highlight `Math`. Select `1:det`. This enters the command; to use it on the matrix you created, select **Matrx** then **1**.

Evaluate the determinants of each of the following matrices:

a. $A=\begin{bmatrix}-2 & 3\\ -1 & 3\end{bmatrix}$ **b.** $B=\begin{bmatrix}2 & -3\\ -4 & 6\end{bmatrix}$

Solutions:

a. $|A|=(-2)(3)-(-1)(3)=-6+3=-3$

b. $|B|=(2)(6)-(-4)(-3)=12-12=0$

The most elementary evaluation of a 3×3 determinant depends on first calculating three 2×2 determinants, and then combining these three numbers in a certain way. Similarly, the most basic evaluation of a 4×4 determinant requires the evaluation of four 3×3 determinants, each of which in turn depends on three 2×2 determinants. Fortunately, as we will soon see, there are some useful properties of determinants that allow us to reduce the required number of computations, but we need a few more definitions first.

Minors and Cofactors

Let A be an $n\times n$ matrix, and let i and j be two fixed numbers each between 1 and n.

- The **minor** of the element a_{ij} is the determinant of the $(n-1)\times(n-1)$ matrix formed from A by deleting its ith row and j^{th} column.
- The **cofactor** of the element a_{ij} is $(-1)^{i+j}$ times the minor of a_{ij}. Thus, if $i+j$ is even, the cofactor of a_{ij} is the same as the minor of a_{ij}; if $i+j$ is odd, the cofactor is the negative of the minor.

The following matrix of signs may help you remember which minors to multiply by –1: if a_{ij} is in a position occupied by a – sign, change the sign of the minor of a_{ij} to obtain the cofactor.

$$\begin{bmatrix}+ & - & + & \cdots\\ - & + & - & \cdots\\ + & - & + & \cdots\\ \vdots & \vdots & \vdots & \ddots\end{bmatrix}$$

Determinant of an $n \times n$ Matrix

Evaluation of an $n \times n$ determinant is accomplished by what is called **expansion** along a fixed row or column. The end result does not depend on which row or column is chosen.

- To expand along the ith row, each element of that row is multiplied by its cofactor, and the n products are then added.
- To expand along the jth column, each element of that column is multiplied by its cofactor, and the n products are then added.

Note that, in either case, a total of n determinants, each $(n-1)\times(n-1)$, must potentially be computed. A wise choice of row or column will often reduce the number of determinants that must actually be evaluated, however.

It is time to put the definitions and techniques to use.

example 2

Evaluate the determinants of the following matrices. In each case, minimize the required number of computations by carefully choosing a row or column to expand along.

a. $A = \begin{bmatrix} -1 & 3 & 2 \\ -2 & 0 & 0 \\ 4 & 1 & 5 \end{bmatrix}$ **b.** $B = \begin{bmatrix} 4 & -2 & 3 & 0 \\ 2 & 1 & -1 & 3 \\ 3 & 0 & 1 & 1 \\ 2 & -2 & 0 & 0 \end{bmatrix}$

Solutions:

a. When calculating determinants, a row or column with many zeros is generally a good choice for the expansion, as we know beforehand the result of multiplying zero by any number. Thus, we will expand along the second row of A. In the computations on the next page, each entry of the second row appears in black, while its corresponding cofactor appears in green. Note the minus sign that appears in front of the first and third 2×2 determinant; the change in sign is part of the definition of the cofactor. (The cofactors corresponding to the zero entries are written for reference only; we will not bother to compute them.) Make sure you understand how each 2×2 matrix is generated by deleting the appropriate row and column of the original 3×3 matrix.

$$\begin{vmatrix} -1 & 3 & 2 \\ -2 & 0 & 0 \\ 4 & 1 & 5 \end{vmatrix} = (-2)\left(-\begin{vmatrix} 3 & 2 \\ 1 & 5 \end{vmatrix}\right) + (0)\left(\begin{vmatrix} -1 & 2 \\ 4 & 5 \end{vmatrix}\right) + (0)\left(-\begin{vmatrix} -1 & 3 \\ 4 & 1 \end{vmatrix}\right)$$

$$= (-2)(-13) + 0 + 0 = 26$$

We have the answer then: $|A| = 26$. But to illustrate the fact that the initial choice of row or column to expand along does not affect the answer, it might be good to evaluate the determinant a second way. If we expand along the first column, our work appears as follows:

$$\begin{vmatrix} -1 & 3 & 2 \\ -2 & 0 & 0 \\ 4 & 1 & 5 \end{vmatrix} = (-1)\left(\begin{vmatrix} 0 & 0 \\ 1 & 5 \end{vmatrix}\right) + (-2)\left(-\begin{vmatrix} 3 & 2 \\ 1 & 5 \end{vmatrix}\right) + (4)\left(\begin{vmatrix} 3 & 2 \\ 0 & 0 \end{vmatrix}\right)$$

$$= (-1)(0) + (-2)(-13) + (4)(0) = 26$$

b. The fourth row and the fourth column both contain two zeros, so either is a good choice as the basis for the expansion. If we use the fourth column, we obtain:

$$|B| = (0)\left(-\begin{vmatrix} 2 & 1 & -1 \\ 3 & 0 & 1 \\ 2 & -2 & 0 \end{vmatrix}\right) + (3)\left(\begin{vmatrix} 4 & -2 & 3 \\ 3 & 0 & 1 \\ 2 & -2 & 0 \end{vmatrix}\right) + (1)\left(-\begin{vmatrix} 4 & -2 & 3 \\ 2 & 1 & -1 \\ 2 & -2 & 0 \end{vmatrix}\right) + (0)\left(\begin{vmatrix} 4 & -2 & 3 \\ 2 & 1 & -1 \\ 3 & 0 & 1 \end{vmatrix}\right)$$

We now have two 3×3 determinants to evaluate, so our work continues:

$$|B| = (3)\left(\begin{vmatrix} 4 & -2 & 3 \\ 3 & 0 & 1 \\ 2 & -2 & 0 \end{vmatrix}\right) + (1)\left(-\begin{vmatrix} 4 & -2 & 3 \\ 2 & 1 & -1 \\ 2 & -2 & 0 \end{vmatrix}\right)$$

$$= (3)\left((2)\left(\begin{vmatrix} -2 & 3 \\ 0 & 1 \end{vmatrix}\right) + (-2)\left(-\begin{vmatrix} 4 & 3 \\ 3 & 1 \end{vmatrix}\right)\right) + (-1)\left((2)\left(\begin{vmatrix} -2 & 3 \\ 1 & -1 \end{vmatrix}\right) + (-2)\left(-\begin{vmatrix} 4 & 3 \\ 2 & -1 \end{vmatrix}\right)\right)$$

$$= (3)\big((2)(-2) + (-2)(5)\big) + (-1)\big((2)(-1) + (-2)(10)\big)$$

$$= (3)(-14) + (-1)(-22)$$

$$= -20$$

(Note that in both cases, the 3×3 determinants were evaluated by expanding around the third row, and those cofactors that were to be multiplied by zero were omitted.)

The screen below was produced using the **Matrx** operation as described on page 792 and the `det` feature as described on page 805.

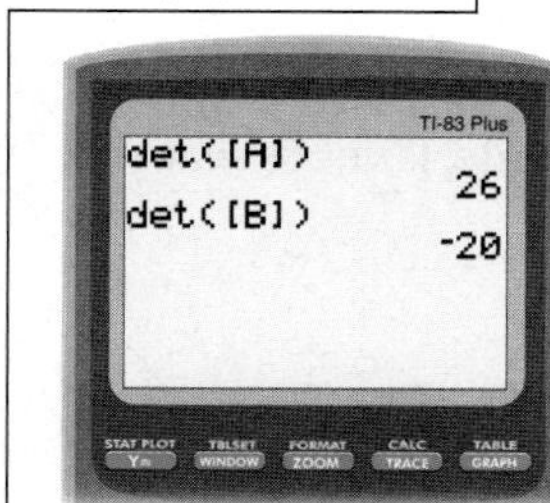

The last example shows that any tricks that might reduce the number of computations necessary would certainly be welcome. There are three such tricks, each of them due to a property of determinants.

Properties of Determinants

1. A constant can be factored out of each of the terms in a given row or column when computing determinants. For example,

$$\begin{vmatrix} 2 & -1 \\ 15 & 5 \end{vmatrix} = 5\begin{vmatrix} 2 & -1 \\ 3 & 1 \end{vmatrix} \text{ and } \begin{vmatrix} 4 & 7 \\ 12 & 9 \end{vmatrix} = 4\begin{vmatrix} 1 & 7 \\ 3 & 9 \end{vmatrix}$$

2. Interchanging two rows or two columns changes the determinant by a factor of –1. For example,

$$\begin{vmatrix} 2 & -1 \\ 15 & 5 \end{vmatrix} = -\begin{vmatrix} 15 & 5 \\ 2 & -1 \end{vmatrix} \text{ and } \begin{vmatrix} 3 & -2 \\ 7 & 1 \end{vmatrix} = -\begin{vmatrix} -2 & 3 \\ 1 & 7 \end{vmatrix}$$

3. The determinant is unchanged by adding a multiple of one row (or column) to another row (or column). For example,

$$\begin{vmatrix} 3 & -2 \\ 1 & -1 \end{vmatrix} \overset{-3R_2+R_1}{=} \begin{vmatrix} 0 & 1 \\ 1 & -1 \end{vmatrix}.$$

It is a good idea to annotate, as shown in green above, applications of this last property.

To demonstrate the utility of these properties, in particular the third one, let us revisit the 4×4 determinant from the last example.

example 3

Evaluate the determinant of the matrix $B = \begin{bmatrix} 4 & -2 & 3 & 0 \\ 2 & 1 & -1 & 3 \\ 3 & 0 & 1 & 1 \\ 2 & -2 & 0 & 0 \end{bmatrix}$.

Solution:

We will make use of the properties of determinants to try to obtain rows or columns with as many zeros as possible. In this problem, one application of the third property results in a column with only one non-zero entry, as shown:

$$\begin{vmatrix} 4 & -2 & 3 & 0 \\ 2 & 1 & -1 & 3 \\ 3 & 0 & 1 & 1 \\ 2 & -2 & 0 & 0 \end{vmatrix} \overset{-3R_3+R_2}{=} \begin{vmatrix} 4 & -2 & 3 & 0 \\ -7 & 1 & -4 & 0 \\ 3 & 0 & 1 & 1 \\ 2 & -2 & 0 & 0 \end{vmatrix}$$

We proceed to expand around the fourth column (do not forget the minus sign as part of the cofactor):

$$\begin{vmatrix} 4 & -2 & 3 & 0 \\ -7 & 1 & -4 & 0 \\ 3 & 0 & 1 & 1 \\ 2 & -2 & 0 & 0 \end{vmatrix} = (1)\left(-\begin{vmatrix} 4 & -2 & 3 \\ -7 & 1 & -4 \\ 2 & -2 & 0 \end{vmatrix}\right)$$

We can continue to apply the third property of determinants to simplify the evaluation of the 3×3 determinant. In what follows, the second column is replaced by the sum of the first and second columns, and then the determinant is evaluated by expanding around the third row.

$$\begin{aligned} |B| = -\begin{vmatrix} 4 & -2 & 3 \\ -7 & 1 & -4 \\ 2 & -2 & 0 \end{vmatrix} &\overset{C_1+C_2}{=} -\begin{vmatrix} 4 & 2 & 3 \\ -7 & -6 & -4 \\ 2 & 0 & 0 \end{vmatrix} \\ &= -(2)\left(\begin{vmatrix} 2 & 3 \\ -6 & -4 \end{vmatrix}\right) \\ &= (-2)(10) \\ &= -20 \end{aligned}$$

Notice that we have reduced the work to the evaluation of only one 3×3 determinant, which in turn involved evaluating only one 2×2 determinant.

technology note

Clearly, evaluating determinants of large matrices can be a time-consuming and error-prone process, and technology offers welcome assistance. Computer algebra systems and many calculators are able to calculate determinants of matrices up to a certain size (the exact size depends on the technology being used). In *Mathematica*, the command to calculate the determinant of a matrix is `Det`. In the illustration below, the matrix B from Example 3 is first defined (refer to the Technology Note in Section 10.2 for another example of defining matrices). The second command simply returns the matrix B in the familiar rectangular form, and allows us to verify that we defined B correctly. The third command then verifies the value for the determinant that we found manually.

The screens below were produced using the **Matrx** operation as described on page 792 and the `det` feature as described on page 805.

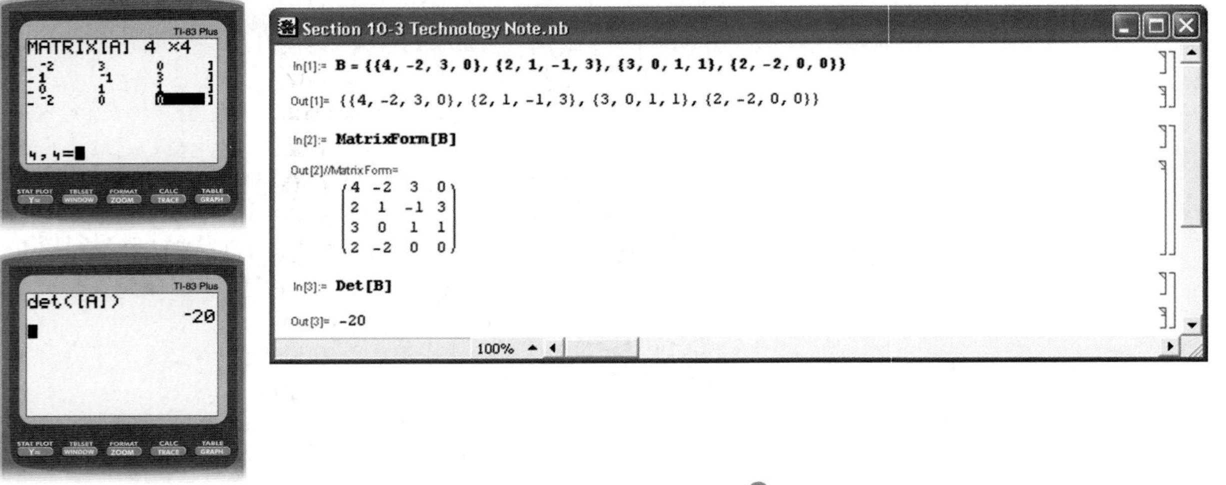

Topic 2: Using Cramer's Rule to Solve Linear Systems

To motivate the form of Cramer's rule, let us solve a 2-variable, 2-equation linear system once and for all. To do this, we note that any such system can be put into the form

$$\begin{cases} ax + by = e \\ cx + dy = f \end{cases}$$

where a, b, c, d, e, and f are all constants. If we can solve this system for x and y, then we will have a formula for the solution of any such system.

The methods we have at our disposal, at the moment, are substitution, elimination, Gaussian elimination, and Gauss-Jordan elimination. Any method can be used; if we choose the method of elimination, we can obtain an equation in x alone as follows:

$$\begin{cases} ax+by=e \\ cx+dy=f \end{cases} \xrightarrow[-bE_2]{dE_1} \begin{cases} adx+bdy=ed \\ -bcx-bdy=-bf \end{cases}$$
$$(ad-bc)x=ed-bf$$

This equation can then be solved for x to obtain

$$x=\frac{ed-bf}{ad-bc}.$$

Similarly, the system can be solved for y to obtain

$$y=\frac{af-ce}{ad-bc}.$$

These formulas are worthwhile on their own, but their true value becomes apparent when we express them using determinants. Note that the above formulas are equivalent to:

$$x=\frac{\begin{vmatrix} e & b \\ f & d \end{vmatrix}}{\begin{vmatrix} a & b \\ c & d \end{vmatrix}} \text{ and } y=\frac{\begin{vmatrix} a & e \\ c & f \end{vmatrix}}{\begin{vmatrix} a & b \\ c & d \end{vmatrix}}.$$

The denominator in both formulas is the determinant of the coefficient matrix, the square matrix consisting of the coefficients of the variables. The numerators differ, but there is an easily remembered pattern that describes them. If we let D_x and D_y represent the numerators of the formulas for x and y, respectively, then D_x is the determinant of the coefficient matrix with the first column (the x-column) replaced by the column of constants, and D_y is the determinant of the coefficient matrix with the second column replaced by the column of constants. If we then define D to be the determinant of the coefficient matrix, we obtain Cramer's rule for the 2-variable, 2-equation case:

$$x=\frac{D_x}{D} \text{ and } y=\frac{D_y}{D}.$$

Before looking at an application of the rule, and before generalizing the rule to larger systems, we should make one more observation. Whenever a fraction appears in our work, it is appropriate to pause and ask if the expression in the denominator might ever be zero, and what it means if in fact this happens. In this case, there are certainly numbers a, b, c, and d for which $D=0$; clearly, in such a system, we can not solve for x and y by Cramer's rule. The meaning of $D=0$ is that the system is either dependent or inconsistent. If both D_x and D_y are also zero, the system is dependent. If at least one of D_x and D_y is non-zero, the system has no solution.

example 4

Use Cramer's rule to solve the following systems.

a. $\begin{cases} 4x - 5y = 3 \\ -3x + 7y = 1 \end{cases}$ **b.** $\begin{cases} -x + 2y = -1 \\ 3x - 6y = 3 \end{cases}$

Solutions:

a. $D = \begin{vmatrix} 4 & -5 \\ -3 & 7 \end{vmatrix} = 28 - 15 = 13$

$D_x = \begin{vmatrix} 3 & -5 \\ 1 & 7 \end{vmatrix} = 21 - (-5) = 26$

$D_y = \begin{vmatrix} 4 & 3 \\ -3 & 1 \end{vmatrix} = 4 - (-9) = 13$

We will learn something about the system merely by calculating D, so that is what we do first. The fact that D is non-zero tells us there is a single solution to the system; we then find D_x and D_y in order to determine the solution.

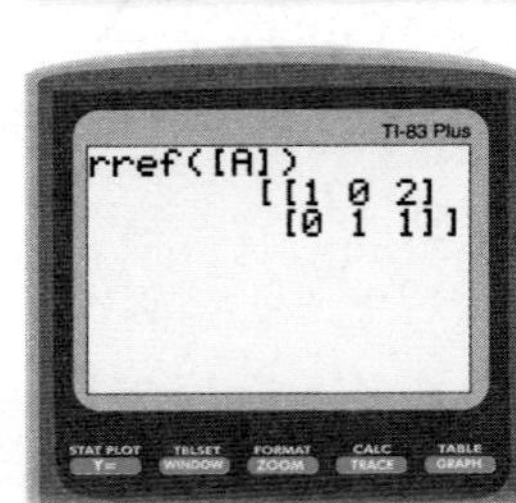

$x = \frac{26}{13} = 2$ and $y = \frac{13}{13} = 1$

So the single solution is $(2, 1)$.

Cramer's rule gives us the ordered pair (2, 1) as the single point that solves both equations.

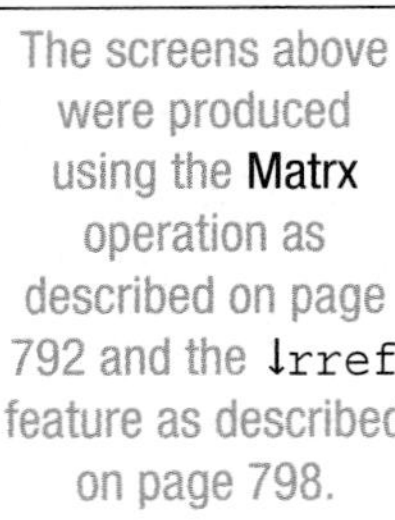

The screens above were produced using the Matrx operation as described on page 792 and the ↓rref feature as described on page 798.

b. $D = \begin{vmatrix} -1 & 2 \\ 3 & -6 \end{vmatrix} = 6 - 6 = 0$

$D_x = \begin{vmatrix} -1 & 2 \\ 3 & -6 \end{vmatrix} = 6 - 6 = 0$

$D_y = \begin{vmatrix} -1 & -1 \\ 3 & 3 \end{vmatrix} = -3 - (-3) = 0$

Solution set is $\{(2y+1, y) \mid y \in \mathbb{R}\}$.

Again, we calculate D first. The fact that $D = 0$ means either the system has no solution or it has an infinite number of solutions. Since both D_x and D_y also turn out to be zero, we know the system is dependent.

The infinite number of ordered pairs that solve the system can be described by solving either equation for either variable, as has been done at left.

Cramer's rule can be extended to solve any linear system of n equations in n variables, as shown in the box on the next page. This makes it appear to be an extremely powerful method, and indeed Cramer's rule is remarkable for its succinctness. Keep in mind, however, that using Cramer's rule to solve an n-equation, n-variable system entails calculating $n + 1$ determinants, of $n \times n$ size, so it is important to make liberal use of the labor-saving properties of determinants.

Cramer's Rule

A linear system of n equations in the n variables $x_1, x_2, \ldots, x_n$ can be written in the form

$$\begin{cases} a_{11}x_1 + a_{12}x_2 + \ldots + a_{1n}x_n = b_1 \\ a_{21}x_1 + a_{22}x_2 + \ldots + a_{2n}x_n = b_2 \\ \vdots \\ a_{n1}x_1 + a_{n2}x_2 + \ldots + a_{nn}x_n = b_n \end{cases}.$$

The solution of the system is given by the n formulas $x_1 = \frac{D_{x_1}}{D}$, $x_2 = \frac{D_{x_2}}{D}$, $\ldots$, $x_n = \frac{D_{x_n}}{D}$, where D is the determinant of the coefficient matrix and D_{x_i} is the determinant of the same matrix with the ith column replaced by the column of constants $b_1, b_2, \ldots, b_n$. If $D = 0$ and if each $D_{x_i} = 0$ as well, for $i = 1, \ldots, n$, the system is dependent and has an infinite number of solutions. If $D = 0$ and at least one of the D_{x_i}'s is non-zero, the system has no solution.

example 5

Use Cramer's rule to solve the system $\begin{cases} 3x - 2y - 2z = -1 \\ 3y + z = -7. \\ x + y + 2z = 0 \end{cases}$

Solution:

Note how the properties of determinants are used to simplify each calculation. In each case, the row or column used for expansion is written in green.

$$D = \begin{vmatrix} 3 & -2 & -2 \\ 0 & 3 & 1 \\ 1 & 1 & 2 \end{vmatrix} \overset{-3R_3+R_1}{=} \begin{vmatrix} 0 & -5 & -8 \\ 0 & 3 & 1 \\ 1 & 1 & 2 \end{vmatrix} = (1)\left(\begin{vmatrix} -5 & -8 \\ 3 & 1 \end{vmatrix}\right) = 19$$

$$D_x = \begin{vmatrix} -1 & -2 & -2 \\ -7 & 3 & 1 \\ 0 & 1 & 2 \end{vmatrix} \overset{-2C_2+C_3}{=} \begin{vmatrix} -1 & -2 & 2 \\ -7 & 3 & -5 \\ 0 & 1 & 0 \end{vmatrix} = (1)\left(-\begin{vmatrix} -1 & 2 \\ -7 & -5 \end{vmatrix}\right) = -19$$

cont'd. on next page ...

$$D_y = \begin{vmatrix} 3 & -1 & -2 \\ 0 & -7 & 1 \\ 1 & 0 & 2 \end{vmatrix} \overset{-2C_1+C_3}{=} \begin{vmatrix} 3 & -1 & -8 \\ 0 & -7 & 1 \\ 1 & 0 & 0 \end{vmatrix} = (1)\left(\begin{vmatrix} -1 & -8 \\ -7 & 1 \end{vmatrix}\right) = -57$$

$$D_z = \begin{vmatrix} 3 & -2 & -1 \\ 0 & 3 & -7 \\ 1 & 1 & 0 \end{vmatrix} \overset{-C_1+C_2}{=} \begin{vmatrix} 3 & -5 & -1 \\ 0 & 3 & -7 \\ 1 & 0 & 0 \end{vmatrix} = (1)\left(\begin{vmatrix} -5 & -1 \\ 3 & -7 \end{vmatrix}\right) = 38$$

We know immediately upon calculating D that the system has a unique solution, as $D \neq 0$. And after determining D_x, D_y, and D_z, we know the solution is the single ordered triple

$$(x, y, z) = \left(\frac{-19}{19}, \frac{-57}{19}, \frac{38}{19}\right) = (-1, -3, 2).$$

Why does Cramer's rule work? Once we have the formulas, we can use the properties of determinants (specifically, the first and third properties) to verify that they work. Consider again the generic n-equation, n-variable system

$$\begin{cases} a_{11}x_1 + a_{12}x_2 + \ldots + a_{1n}x_n = b_1 \\ a_{21}x_1 + a_{22}x_2 + \ldots + a_{2n}x_n = b_2 \\ \quad\vdots \\ a_{n1}x_1 + a_{n2}x_2 + \ldots + a_{nn}x_n = b_n \end{cases}.$$

If we assume that the system has a unique solution, which we will denote $(x_1, x_2, \ldots, x_n)$, then we can rewrite D_{x_1} as follows:

$$D_{x_1} = \begin{vmatrix} b_1 & a_{12} & \ldots & a_{1n} \\ b_2 & a_{22} & \ldots & a_{2n} \\ \vdots & \vdots & \ldots & \vdots \\ b_n & a_{n2} & \ldots & a_{nn} \end{vmatrix} = \begin{vmatrix} a_{11}x_1 + a_{12}x_2 + \ldots + a_{1n}x_n & a_{12} & \ldots & a_{1n} \\ a_{21}x_1 + a_{22}x_2 + \ldots + a_{2n}x_n & a_{22} & \ldots & a_{2n} \\ \vdots & \vdots & \ldots & \vdots \\ a_{n1}x_1 + a_{n2}x_2 + \ldots + a_{nn}x_n & a_{n2} & \ldots & a_{nn} \end{vmatrix}$$

We can now use the third property of determinants $n - 1$ times to simplify the first column. We do this by multiplying the second column by $-x_2$ and adding it to the first, multiplying the third column by $-x_3$ and adding it to the first, and so on. After doing this, we obtain

$$D_{x_1} = \begin{vmatrix} a_{11}x_1 & a_{12} & \ldots & a_{1n} \\ a_{21}x_1 & a_{22} & \ldots & a_{2n} \\ \vdots & \vdots & \ldots & \vdots \\ a_{n1}x_1 & a_{n2} & \ldots & a_{nn} \end{vmatrix} = x_1 \begin{vmatrix} a_{11} & a_{12} & \ldots & a_{1n} \\ a_{21} & a_{22} & \ldots & a_{2n} \\ \vdots & \vdots & \ldots & \vdots \\ a_{n1} & a_{n2} & \ldots & a_{nn} \end{vmatrix} = x_1 D.$$

Note that we have used the first property of determinants to factor x_1 out of the determinant. We can now solve this equation for x_1 (since $D \neq 0$) to obtain

$$x_1 = \frac{D_{x_1}}{D},$$

and we can then repeat the process for the remaining variables $x_2, x_3, \ldots, x_n$.

exercises

Evaluate the following determinants. See Example 1.

1. $\begin{vmatrix} 4 & -3 \\ 1 & 2 \end{vmatrix}$ **2.** $\begin{vmatrix} 5 & -2 \\ 5 & -2 \end{vmatrix}$ **3.** $\begin{vmatrix} 0 & 3 \\ -5 & 2 \end{vmatrix}$ **4.** $\begin{vmatrix} 34 & -2 \\ 17 & -1 \end{vmatrix}$ **5.** $\begin{vmatrix} a & x \\ x & b \end{vmatrix}$ **6.** $\begin{vmatrix} 5x & 2 \\ -x & 1 \end{vmatrix}$

7. $\begin{vmatrix} -2 & 2 \\ -2 & -2 \end{vmatrix}$ **8.** $\begin{vmatrix} ac & 2ad \\ bc & db \end{vmatrix}$ **9.** $\begin{vmatrix} -1 & 2 \\ 3 & 4 \end{vmatrix}$ **10.** $\begin{vmatrix} w & x \\ y & z \end{vmatrix}$ **11.** $\begin{vmatrix} -2 & 9 \\ 5 & -3 \end{vmatrix}$ **12.** $\begin{vmatrix} 2y & 3x \\ y-1 & x^2 \end{vmatrix}$

Solve for x by finding the determinant.

13. $\begin{vmatrix} x-2 & 2 \\ 2 & x+1 \end{vmatrix} = 0$ **14.** $\begin{vmatrix} x+7 & -2 \\ 9 & x-2 \end{vmatrix} = 0$ **15.** $\begin{vmatrix} x+1 & 8 \\ 1 & x+3 \end{vmatrix} = 0$

16. $\begin{vmatrix} x-8 & 11 \\ -2 & x+5 \end{vmatrix} = 0$ **17.** $\begin{vmatrix} x+6 & 2 \\ -1 & x+3 \end{vmatrix} = 0$ **18.** $\begin{vmatrix} x-4 & -4 \\ 3 & x+9 \end{vmatrix} = 0$

19. $\begin{vmatrix} x+5 & 3 \\ 3 & x-3 \end{vmatrix} = 0$ **20.** $\begin{vmatrix} x+3 & 6 \\ 5 & x+7 \end{vmatrix} = 0$ **21.** $\begin{vmatrix} x-3 & 2 \\ 1 & x-4 \end{vmatrix} = 0$

Use the matrix $A = \begin{bmatrix} 2 & -1 & 5 \\ 0 & 1 & 3 \\ 1 & 0 & -2 \end{bmatrix}$ *to evaluate the following:*

22. The minor of a_{12}. **23.** The cofactor of a_{12}. **24.** The minor of a_{22}.

25. The cofactor of a_{22}. **26.** The cofactor of a_{32}. **27.** The cofactor of a_{33}.

28. The minor of a_{13}. **29.** The cofactor of a_{21}. **30.** The cofactor of a_{31}.

Find the determinant of the matrix by the method of expansion by cofactors along the given row or column.

31. $\begin{vmatrix} 4 & 5 & 3 \\ -1 & 2 & 7 \\ 11 & 6 & 2 \end{vmatrix}$ Expand along Row 3

32. $\begin{vmatrix} 8 & 2 & 0 \\ 3 & 4 & 7 \\ 1 & 0 & 2 \end{vmatrix}$ Expand along Column 1

33. $\begin{vmatrix} 5 & 8 & 5 \\ 0 & -6 & 3 \\ 2 & 4 & -1 \end{vmatrix}$ Expand along Row 1

34. $\begin{vmatrix} -4 & 2 & 1 \\ 9 & 12 & 8 \\ 0 & 6 & -3 \end{vmatrix}$ Expand along Column 1

35. $\begin{vmatrix} 13 & 0 & -7 \\ 4 & 2 & 3 \\ 1 & 4 & 0 \end{vmatrix}$ Expand along Row 2

36. $\begin{vmatrix} 7 & 0 & 1 \\ 2 & 5 & 3 \\ 8 & 6 & 2 \end{vmatrix}$ Expand along Column 3

37. $\begin{vmatrix} 8 & 0 & -7 & 5 \\ 4 & -2 & 3 & 3 \\ -1 & 1 & 0 & 2 \\ 2 & 0 & 6 & 0 \end{vmatrix}$ Expand along Row 4

38. $\begin{vmatrix} 4 & -2 & 9 & 2 \\ 7 & 0 & 1 & 7 \\ -6 & 3 & 0 & 1 \\ 3 & 1 & 2 & 0 \end{vmatrix}$ Expand along Column 2

Evaluate the following determinants. In each case, minimize the required number of computations by carefully choosing a row or column to expand along, and use the properties of determinants to simplify the process. See Examples 2 and 3.

39. $\begin{vmatrix} 2 & 0 & 1 \\ -5 & 1 & 0 \\ 3 & -1 & 1 \end{vmatrix}$

40. $\begin{vmatrix} 12 & 3 & 1 \\ 1 & 1 & -1 \\ 0 & 2 & 0 \end{vmatrix}$

41. $\begin{vmatrix} 12 & 3 & 6 \\ 2 & 2 & -4 \\ 0 & 2 & 0 \end{vmatrix}$

42. $\begin{vmatrix} 2 & 1 & -3 & 0 \\ 1 & -2 & 1 & 0 \\ 0 & 1 & 0 & 1 \\ 2 & 0 & 1 & 1 \end{vmatrix}$

43. $\begin{vmatrix} x & x & x & x \\ 0 & x & x & x \\ 0 & 0 & x & x \\ 0 & 0 & 0 & x \end{vmatrix}$

44. $\begin{vmatrix} x & 0 & 0 & 0 \\ 0 & x & 0 & 0 \\ 0 & 0 & x & 0 \\ 0 & 0 & 0 & x \end{vmatrix}$

45. $\begin{vmatrix} 0 & 2 & 0 & 0 \\ -2 & -4 & 5 & 9 \\ 1 & 3 & -1 & 1 \\ 0 & 7 & 0 & 2 \end{vmatrix}$

46. $\begin{vmatrix} 1 & 2 & 3 \\ 4 & 5 & 6 \\ 7 & 8 & 9 \end{vmatrix}$

47. $\begin{vmatrix} x & x & 0 & 0 \\ yz & x^3 & z & x^4 \\ z & xy & x & 0 \\ x^2 & 0 & 0 & 0 \end{vmatrix}$

Using a graphing calculator or other computational aid, find the determinant of the matrix.

48. $\begin{vmatrix} 0.1 & 0.4 & -0.7 \\ 0.3 & -0.1 & 0.2 \\ 0.5 & -0.2 & 0.3 \end{vmatrix}$

49. $\begin{vmatrix} 0.1 & 0.3 & 0.1 \\ 0.2 & -0.2 & -0.1 \\ -0.1 & -0.4 & 0.5 \end{vmatrix}$

50. $\begin{vmatrix} 2.2 & 0.3 & -1.7 \\ 0.4 & -0.2 & 0.1 \\ 0.2 & 0.3 & -1.6 \end{vmatrix}$

51. $\begin{vmatrix} 3.1 & 0.6 & -1.1 \\ 1.2 & 5.2 & -7.3 \\ -0.1 & -4.1 & 6.5 \end{vmatrix}$

52. $\begin{vmatrix} 13 & 23 & -21 \\ 17 & -32 & 14 \\ 15 & 12 & -16 \end{vmatrix}$

53. $\begin{vmatrix} 25 & 32 & 17 \\ -13 & 14 & -24 \\ 16 & 26 & 36 \end{vmatrix}$

Use Cramer's rule to solve the following systems. See Examples 4 and 5.

54. $\begin{cases} 2x-3y=8 \\ 8x+5y=-2 \end{cases}$

55. $\begin{cases} 5x+7y=9 \\ 2x+3y=-7 \end{cases}$

56. $\begin{cases} 5x-10y=9 \\ -x+2y=-3 \end{cases}$

57. $\begin{cases} -2x-2y=4 \\ 3x+3y=-6 \end{cases}$

58. $\begin{cases} \frac{2}{3}x+y=-3 \\ 3x+\frac{5}{2}y=-\frac{7}{2} \end{cases}$

59. $\begin{cases} \frac{2}{3}x+2y=1 \\ x+3y=0 \end{cases}$

60. $\begin{cases} -4x+y=1 \\ 7x+2y=407 \end{cases}$

61. $\begin{cases} 5x-4y=-49 \\ 24x-19y=179 \end{cases}$

62. $\begin{cases} 23x+21y=-4 \\ x-3y=-8 \end{cases}$

63. $\begin{cases} -5x+10y=3 \\ \frac{7}{2}x-7y=20 \end{cases}$

64. $\begin{cases} x+2y=-1 \\ y+3z=7 \\ 2x+5z=21 \end{cases}$

65. $\begin{cases} 2x-y=0 \\ 5x-3y-3z=5 \\ 2x+6z=-10 \end{cases}$

66. $\begin{cases} 3x+8z=3 \\ -3x-7z=-3 \\ x+3z=1 \end{cases}$

67. $\begin{cases} 3w-x+5y+3z=2 \\ -4w-10y-2z=10 \\ w-x+2z=7 \\ 4w-2x+5y+5z=9 \end{cases}$

68. $\begin{cases} 2w+x-3y=3 \\ w-2x+y=1 \\ x+z=-2 \\ y+z=0 \end{cases}$

69. $\begin{cases} 3w-2x+y-5z=-1 \\ w+x-y+4z=2 \\ 4w-x-z=1 \\ 5w-x=9 \end{cases}$

70. $\begin{cases} 2w-3x+4y-z=21 \\ w+5x=2 \\ -2x+3y+z=12 \\ -3w+4z=-5 \end{cases}$

71. $\begin{cases} w-x+y-z=2 \\ 2w-x+3y=-5 \\ x-2z=7 \\ 3w+4x=-13 \end{cases}$

Use a graphing calculator or other computational aid and Cramer's rule to solve the systems of equations.

72. $\begin{cases} x-2y+3z=9 \\ -x+3y=-4 \\ 2x-5y+5z=17 \end{cases}$

73. $\begin{cases} 2x+4y+z=1 \\ x-2y-3z=2 \\ x+y-z=-1 \end{cases}$

74. $\begin{cases} x+y+z+w=6 \\ 2x+3y-w=0 \\ -3x+4y+z+2w=4 \\ x+2y-z+w=0 \end{cases}$

10.4 The Algebra of Matrices

TOPICS

1. Matrix addition
2. Scalar multiplication
3. Matrix multiplication
4. Interlude: transition matrices

Topic 1: Matrix Addition

As you continue on in mathematics, you will encounter many sets of objects that can be combined in ways similar to the ways real numbers can be combined. We have, in fact, seen several examples already in this text: addition, subtraction, multiplication, and division are all defined on the set of complex numbers and on real-valued functions. Loosely speaking, an *algebra* is a collection of objects together with one or more methods of combining the objects. In this section, we will see that a very natural algebra exists for sets of matrices. We start by defining what it means to add two matrices.

Matrix Addition

Two matrices A and B can be added to form the new matrix $A + B$ only if A and B are of the same order. The addition is performed by adding corresponding entries of the two matrices together; that is, the element in the ith row and the jth column of $A + B$ is given by $a_{ij} + b_{ij}$.

It is very important to note the restriction on the order of the two matrices in the above definition; a matrix with m rows and n columns can only be added to another matrix with m rows and n columns. It is, in fact, a hallmark of matrix algebra that the orders of the matrices under discussion must always be considered. This is true even for the concept of *matrix equality*, which we now define so we can use the equality symbol in our work with matrices.

Matrix Equality

Two matrices A and B are said to be **equal**, denoted $A = B$, if they are of the same order and corresponding entries of A and B are equal.

example 1

To add matrices, use the **Matrx** operation (pg. 792) to define matrices A and B. Then **Quit (2nd Mode)**; enter A by selecting **Matrx 1**, enter "+", and then select **Matrx 2** (B).

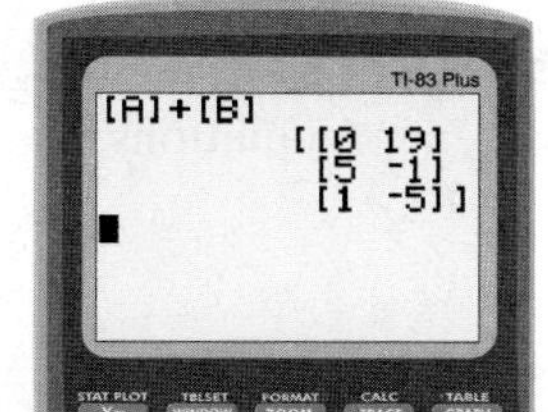

Perform the indicated addition, if possible.

a. $\begin{bmatrix} -3 & 2 \\ 0 & -5 \\ 11 & -9 \end{bmatrix} + \begin{bmatrix} 3 & 17 \\ 5 & 4 \\ -10 & 4 \end{bmatrix}$ **b.** $\begin{bmatrix} 2 & -5 \\ 1 & 0 \\ 0 & 3 \\ -7 & 10 \end{bmatrix} + \begin{bmatrix} 2 & 1 & 0 & -7 \\ -5 & 0 & 3 & 10 \end{bmatrix}$

Solutions:

a. Both matrices are 3×2, so the sum is defined and

$$\begin{bmatrix} -3 & 2 \\ 0 & -5 \\ 11 & -9 \end{bmatrix} + \begin{bmatrix} 3 & 17 \\ 5 & 4 \\ -10 & 4 \end{bmatrix} = \begin{bmatrix} 0 & 19 \\ 5 & -1 \\ 1 & -5 \end{bmatrix}.$$

b. The first matrix is 4×2 and the second is 2×4, so the sum cannot be performed.

Of course, the entries in matrices do not all have to be constants. In many applications of matrices, we will find it convenient to represent some of the entries initially as variables, with the intent of eventually solving for the variables. We can solve some examples of such *matrix equations* now, with nothing other than matrix addition and matrix equality defined.

example 2

Determine the values of the variables that will make each of the following statements true.

a. $\begin{bmatrix} -3 & a & b \\ -2 & a+b & 5 \end{bmatrix} = \begin{bmatrix} c & 3 & 7 \\ -2 & d & 5 \end{bmatrix}$ **b.** $\begin{bmatrix} 3x \\ 4 \end{bmatrix} + \begin{bmatrix} -y \\ 2x \end{bmatrix} = \begin{bmatrix} 13 \\ 7y \end{bmatrix}$

Solutions:

a. Solving for the four variables $a, b, c,$ and d in the equation

$$\begin{bmatrix} -3 & a & b \\ -2 & a+b & 5 \end{bmatrix} = \begin{bmatrix} c & 3 & 7 \\ -2 & d & 5 \end{bmatrix}$$

is just a matter of comparing the entries one-by-one. Beginning with the first rows, we note that $c = -3,\ a = 3,$ and $b = 7$. Of course, we have no control over the first and third entries in the second row of each matrix; if the constants in these two positions were not equal, there would be no way to make the matrix equality true. The only variable left to solve for is d, and since we have already determined a and b, it must be the case that $d = 10$.

b. Each matrix consists of only two entries, and after performing the matrix addition on the left and comparing corresponding entries, we arrive at the system of equations

$$\begin{cases} 3x - y = 13 \\ 4 + 2x = 7y \end{cases}.$$

We know a number of ways of solving such a system. If we choose to use Cramer's rule, we first rewrite the system as

$$\begin{cases} 3x - y = 13 \\ 2x - 7y = -4 \end{cases}$$

from which we can determine

$$D = \begin{vmatrix} 3 & -1 \\ 2 & -7 \end{vmatrix} = (3)(-7) - (2)(-1) = -19,$$

$$D_x = \begin{vmatrix} 13 & -1 \\ -4 & -7 \end{vmatrix} = (13)(-7) - (-4)(-1) = -95, \text{ and}$$

$$D_y = \begin{vmatrix} 3 & 13 \\ 2 & -4 \end{vmatrix} = (3)(-4) - (2)(13) = -38,$$

$$\text{so } x = \frac{D_x}{D} = \frac{-95}{-19} = 5 \text{ and } y = \frac{D_y}{D} = \frac{-38}{-19} = 2.$$

Topic 2: Scalar Multiplication

In the context of matrix algebra, a *scalar* is a real number (or, more generally, a complex number), and *scalar multiplication* refers to the product of a real number and a matrix. At first glance, it may seem strange to combine two very different sorts of objects (a scalar and a matrix), but the notion arises naturally from considering such matrix sums as

$$\begin{bmatrix} -5 & 2 \\ 1 & -3 \\ -2 & 7 \end{bmatrix} + \begin{bmatrix} -5 & 2 \\ 1 & -3 \\ -2 & 7 \end{bmatrix} = \begin{bmatrix} -10 & 4 \\ 2 & -6 \\ -4 & 14 \end{bmatrix}.$$

Since the result of adding a matrix to itself has the effect of doubling each entry, we are led to write

$$2\begin{bmatrix} -5 & 2 \\ 1 & -3 \\ -2 & 7 \end{bmatrix} = \begin{bmatrix} -10 & 4 \\ 2 & -6 \\ -4 & 14 \end{bmatrix}.$$

Extending this idea to arbitrary scalars leads to the following definition.

Scalar Multiplication

If A is an $m \times n$ matrix and c is a scalar, cA stands for the $m \times n$ matrix for which each entry is c times the corresponding entry of A. In other words, the entry in the ith row and j^{th} column of cA is ca_{ij}.

example 3

Given the matrices

$$A = \begin{bmatrix} -1 & 6 & 2 \\ -8 & 0 & 1 \end{bmatrix} \text{ and } B = \begin{bmatrix} 0 & -3 & 4 \\ 1 & -2 & 6 \end{bmatrix},$$

write $-3A + 2B$ as a single matrix.

Solution:

Our work consists of two tasks.

cont'd. on next page ...

The first is to multiply each entry of either A or B by its corresponding scalar, and the second is to then add the resulting matrices together:

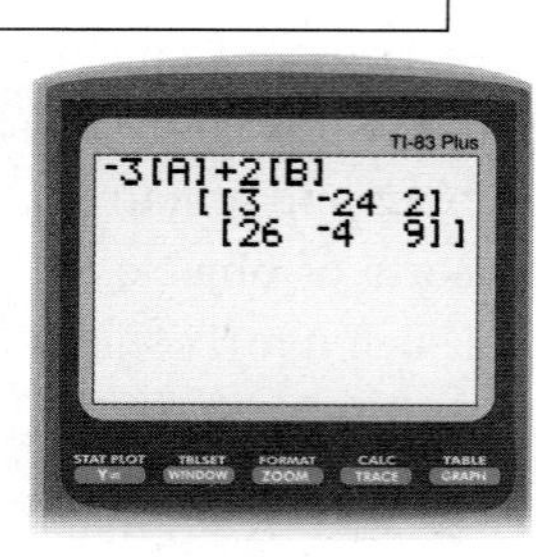

Enter the matrices in A and B using the **Matrx** command. Then **Quit** and enter the expression, selecting each matrix as desired through the **Matrx** menu.

$$\begin{aligned}-3A+2B&=-3\begin{bmatrix}-1 & 6 & 2\\ -8 & 0 & 1\end{bmatrix}+2\begin{bmatrix}0 & -3 & 4\\ 1 & -2 & 6\end{bmatrix}\\ &=\begin{bmatrix}3 & -18 & -6\\ 24 & 0 & -3\end{bmatrix}+\begin{bmatrix}0 & -6 & 8\\ 2 & -4 & 12\end{bmatrix}\\ &=\begin{bmatrix}3 & -24 & 2\\ 26 & -4 & 9\end{bmatrix}\end{aligned}$$

Remember that our overall goal is to develop an algebra of matrices, and we can use the algebra of real numbers as a model. Since subtraction of real numbers is defined in terms of addition $\left(a-b=a+(-b)\right)$, we can define matrix subtraction in the same way: if A and B are two matrices of the same order, then

$$A-B=A+(-B).$$

example 4

Perform the indicated subtraction: $\begin{bmatrix}3 & -5 & 2\end{bmatrix}-\begin{bmatrix}-2 & -5 & 3\end{bmatrix}$.

Solution:

As both matrices are of order 1×3, we know the subtraction is possible. The computation consists of subtracting entries in the second matrix from the corresponding entries in the first:

$$\begin{bmatrix}3 & -5 & 2\end{bmatrix}-\begin{bmatrix}-2 & -5 & 3\end{bmatrix}=\begin{bmatrix}3-(-2) & -5-(-5) & 2-3\end{bmatrix}=\begin{bmatrix}5 & 0 & -1\end{bmatrix}.$$

Topic 3: Matrix Multiplication

Unlike the definitions of matrix addition and scalar multiplication, both of which probably seem fairly natural and expected, the definition of matrix multiplication may at first be confusing and appear unnecessarily complicated. For one thing, matrix multiplication does *not* refer to multiplying corresponding entries of two matrices together. Such a definition would be undeniably simpler than the actual definition, but matrix multiplication is defined the way it is because of the demands of applications of matrices.

One of the best ways of motivating the definition of matrix multiplication requires a bit of preamble. To begin with, it is useful to consider matrices as functions. Note that the system of equations

$$\begin{cases} x' = ax + by \\ y' = cx + dy \end{cases}$$

can be viewed as a function that transforms the ordered pair (x, y) into the ordered pair (x', y'), and that the function is characterized entirely by the matrix

$$A = \begin{bmatrix} a & b \\ c & d \end{bmatrix}.$$

Similarly, the system

$$\begin{cases} x' = ex + fy \\ y' = gx + hy \end{cases}$$

is a function that transforms ordered pairs into ordered pairs, and it is associated with the matrix

$$B = \begin{bmatrix} e & f \\ g & h \end{bmatrix}.$$

Consider now the result of plugging the output of the first function into the second function (that is, of *composing* the two functions together so that B acts on the result of A). If we let (x', y') denote the output of the first function, we obtain the new ordered pair (x'', y'') given by

$$\begin{cases} x'' = ex' + fy' \\ y'' = gx' + hy' \end{cases} \text{ or } \begin{cases} x'' = e(ax + by) + f(cx + dy) \\ y'' = g(ax + by) + h(cx + dy) \end{cases}.$$

We can change the way the last system is written to obtain

$$\begin{cases} x'' = (ea + fc)x + (eb + fd)y \\ y'' = (ga + hc)x + (gb + hd)y \end{cases}$$

so the composition of the two functions, in the order BA, is characterized entirely by the matrix

$$\begin{bmatrix} ea + fc & eb + fd \\ ga + hc & gb + hd \end{bmatrix}.$$

Note that the entries of this last matrix are constructed by a combination of products and sums, where the products in each sum are between elements of the rows of B with elements of the columns of A. For example, the entry in the second row of the first column is the sum of the products of the second row of B with the first column of A. This is the basis of our formal definition of matrix multiplication.

Matrix Multiplication

Two matrices A and B can be multiplied together, resulting in a new matrix denoted AB, only if the length of each row of A (the matrix on the left) is the same as the length of each column of B (the matrix on the right). That is, if A is of order $m \times n$, the product AB is only defined if B is of order $n \times p$. The order of AB in this case will be $m \times p$.

If we let c_{ij} denote the entry in the ith row and j^{th} column of AB, c_{ij} is obtained from the ith row of A and the jth column of B by the formula

$$c_{ij} = a_{i1}b_{1j} + a_{i2}b_{2j} + \ldots + a_{in}b_{nj}.$$

caution!

Unlike numerical multiplication, matrix multiplication is not commutative. That is, given two matrices A and B, AB in general is not equal to BA. As an illustration of this fact, suppose A is a 3×4 matrix and B is a 4×2 matrix. Then AB is defined (and is of order 3×2), but BA does not even exist. We will see many more examples of the non-commutativity of matrix multiplication.

example 5

Given the two matrices $A = \begin{bmatrix} 2 & -3 \\ 4 & -1 \\ 1 & 0 \end{bmatrix}$ and $B = \begin{bmatrix} 5 & 0 & -2 \\ -4 & 1 & 3 \end{bmatrix}$, find the following products.

a. AB **b.** BA

Solutions:

a. A is of order 3×2 and B is of order 2×3, so AB is defined and will be of order 3×3. Remember that each entry of AB is formed from a row of A and a column of B. For instance, the first-row, first-column entry of AB is calculated as follows:

$$AB = \begin{bmatrix} 2 & -3 \\ 4 & -1 \\ 1 & 0 \end{bmatrix}\begin{bmatrix} 5 & 0 & -2 \\ -4 & 1 & 3 \end{bmatrix} = \begin{bmatrix} (2)(5)+(-3)(-4)=22 & ? & ? \\ ? & ? & ? \\ ? & ? & ? \end{bmatrix}$$

Similarly, the entry in the third row and second column of AB comes from the third row of A and the second column of B:

$$AB=\begin{bmatrix}2 & -3\\ 4 & -1\\ 1 & 0\end{bmatrix}\begin{bmatrix}5 & 0 & -2\\ -4 & 1 & 3\end{bmatrix}=\begin{bmatrix}? & ? & ?\\ ? & ? & ?\\ ? & (1)(0)+(0)(1)=0 & ?\end{bmatrix}.$$

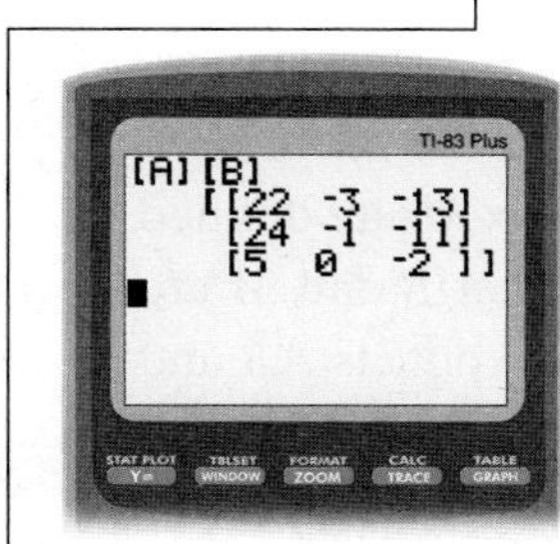

Continuing the process, we obtain

$$AB=\begin{bmatrix}2 & -3\\ 4 & -1\\ 1 & 0\end{bmatrix}\begin{bmatrix}5 & 0 & -2\\ -4 & 1 & 3\end{bmatrix}=\begin{bmatrix}22 & -3 & -13\\ 24 & -1 & -11\\ 5 & 0 & -2\end{bmatrix}.$$

b. From the orders of the two matrices, we know BA will be a 2×2 matrix. Note that each entry of BA is a sum of three products, the first of which is

$$BA=\begin{bmatrix}5 & 0 & -2\\ -4 & 1 & 3\end{bmatrix}\begin{bmatrix}2 & -3\\ 4 & -1\\ 1 & 0\end{bmatrix}=\begin{bmatrix}(5)(2)+(0)(4)+(-2)(1)=8 & ?\\ ? & ?\end{bmatrix}.$$

Continuing, the final answer is

$$BA=\begin{bmatrix}5 & 0 & -2\\ -4 & 1 & 3\end{bmatrix}\begin{bmatrix}2 & -3\\ 4 & -1\\ 1 & 0\end{bmatrix}=\begin{bmatrix}8 & -15\\ -1 & 11\end{bmatrix}.$$

technology note

Computer algebra systems and sufficiently advanced calculators are all capable of performing matrix addition, scalar multiplication, and matrix multiplication, though the exact notation used varies from one technology to another. In the illustration below, the matrices A and B from Example 5, along with a third matrix M, are defined in *Mathematica*. The expression $3B - 2M$ is then evaluated (note that B and M are the same size, so this expression makes sense), and then the matrix products AB and BA are evaluated. The notation for matrix multiplication in *Mathematica* is a period between the two matrices.

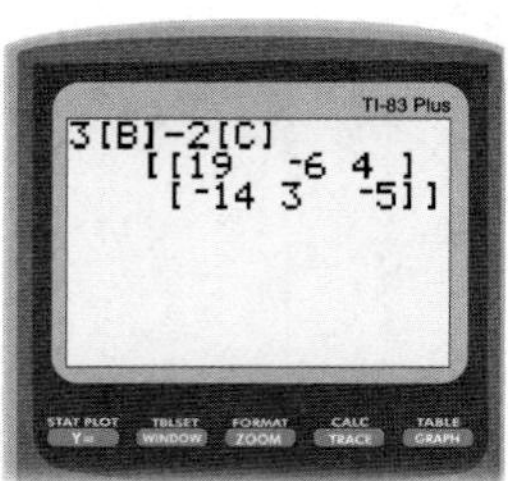

Note the matrix M has been renamed as C for compatibility on the TI-83 Plus.

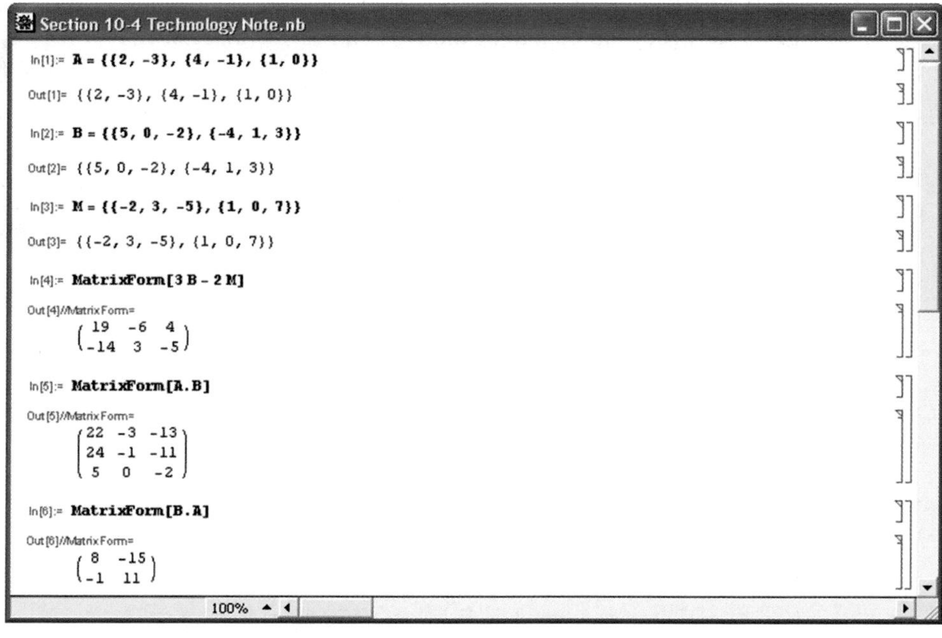

Now that matrix multiplication has been defined, it can be used in another way to illustrate the fact that matrices can be considered as functions. As already mentioned, the matrix

$$A = \begin{bmatrix} a & b \\ c & d \end{bmatrix}$$

completely characterizes the function that transforms (x, y) into (x', y') in the system

$$\begin{cases} x' = ax + by \\ y' = cx + dy \end{cases}.$$

What we mean by this is that the four numbers a, b, c, and d, along with knowledge of their placement in the matrix, suffice to describe the function. This is even more clear if we associate the ordered pair (x, y) with the 2×1 matrix

$$\begin{bmatrix} x \\ y \end{bmatrix}.$$

Then the matrix product

$$\begin{bmatrix} a & b \\ c & d \end{bmatrix}\begin{bmatrix} x \\ y \end{bmatrix},$$

which has the "look" of a function acting on its argument, gives us the expressions in the system. Verify for yourself that

$$\begin{bmatrix} a & b \\ c & d \end{bmatrix}\begin{bmatrix} x \\ y \end{bmatrix} = \begin{bmatrix} ax + by \\ cx + dy \end{bmatrix}.$$

Topic 4: Interlude: Transition Matrices

In a variety of important applications, matrices can be used to model how the "state" of a situation changes over time. To illustrate the idea, we will consider one such situation in some detail.

Suppose a new grocery store opens in a small town, with the intent of competing with the one existing store. The new store, which we will call store A, begins an aggressive advertising campaign, with the result that every month 45% of the customers of the existing store (store B) decide to start shopping at store A. Store B responds with a strong appeal to win back customers, however, and every month 30% of the customers of store A return to store B. In this highly dynamic situation, a number of questions may arise in the minds of the managers of the two stores, among them:

1. Given the known number of customers one month, how many customers can be expected the following month?

2. Can a certain percentage of the total number of the town's shoppers be expected in the long run? If so, what is that percentage?

To answer the first question, let a denote the number of customers of store A in a given month, and let b denote the number of customers of store B in the same month. In this simple example, we will assume that $a + b$, the total number of grocery store customers in the town, remains fixed over time.

In the following month, store A gains $0.45b$ customers (45% of its competitor's customers) and retains $0.7a$ of its existing customers. At the same time, store B gains $0.3a$ new customers (the 30% of A's clientele that switch back to store B), and retains $0.55b$ existing customers. In equation form, the number of customers of each store the following month are given by the two expressions:

$$\begin{cases} 0.7a + 0.45b \\ 0.3a + 0.55b \end{cases}.$$

Matrix multiplication allows us to write this information as

$$\begin{bmatrix} 0.7 & 0.45 \\ 0.3 & 0.55 \end{bmatrix}\begin{bmatrix} a \\ b \end{bmatrix},$$

and if we let P stand for the 2×2 matrix of percentages, we can think of P as a function that transforms one month's distribution of customers into the next month's. For instance, if store A has 360 customers and store B has 640 customers one month, then the following month their respective number of customers will be 540 and 460, as

$$\begin{bmatrix} 0.7 & 0.45 \\ 0.3 & 0.55 \end{bmatrix}\begin{bmatrix} 360 \\ 640 \end{bmatrix} = \begin{bmatrix} 540 \\ 460 \end{bmatrix}.$$

In such situations, a matrix like P is called a **transition matrix**, as it characterizes the transition of the system from one state to the next. Transition matrices are characterized by the fact that all of their entries are positive and the sum of the entries in each column is 1.

To answer the second question, we might continue the above specific example and ask how the 540 and 460 customers of the respective stores realign themselves one month later. Note that

$$\begin{bmatrix} 0.7 & 0.45 \\ 0.3 & 0.55 \end{bmatrix}\begin{bmatrix} 540 \\ 460 \end{bmatrix} = \begin{bmatrix} 585 \\ 415 \end{bmatrix},$$

meaning that at the end of the second month, store A has 585 customers and store B has 415. But we can obtain the same result by applying the matrix P^2 to the original distribution of 360 and 640 customers, as

$$P^2\begin{bmatrix} 360 \\ 640 \end{bmatrix} = P\left(P\begin{bmatrix} 360 \\ 640 \end{bmatrix}\right) = P\begin{bmatrix} 540 \\ 460 \end{bmatrix}.$$

In other words, we can form the composition of P with itself to obtain a function that transforms one month's distribution of customers into the distribution two months later. Note that

$$P^2 = \begin{bmatrix} 0.625 & 0.5625 \\ 0.375 & 0.4375 \end{bmatrix}$$

and that

$$\begin{bmatrix} 0.625 & 0.5625 \\ 0.375 & 0.4375 \end{bmatrix}\begin{bmatrix} 360 \\ 640 \end{bmatrix} = \begin{bmatrix} 585 \\ 415 \end{bmatrix}.$$

At this point, a calculator or computer software with matrix capability may be useful. We can use either aid to calculate high powers of P to see if there is a long-term trend in the distribution of the customers. For instance, to six decimal places,

$$P^5 = \begin{bmatrix} 0.600391 & 0.599414 \\ 0.399609 & 0.400586 \end{bmatrix},$$

and after a certain point calculators and software will round off the entries and give us a result of

$$P^n = \begin{bmatrix} 0.6 & 0.6 \\ 0.4 & 0.4 \end{bmatrix}$$

for large n (the value for n at which this happens will vary depending on the technology used). This means that, after a few months, store A can count on roughly 60% of the town's customers and store B can count on roughly 40% (the actual identities of the customers will, of course, keep changing from month to month, but the relative proportions will have stabilized). We can verify that the situation is stable by applying the transition matrix to an assumed 1000 customers split 60/40:

$$\begin{bmatrix} 0.7 & 0.45 \\ 0.3 & 0.55 \end{bmatrix}\begin{bmatrix} 600 \\ 400 \end{bmatrix} = \begin{bmatrix} 600 \\ 400 \end{bmatrix}.$$

You will see another approach to determining the stable long-term state in the exercises.

exercises

Given $A = \begin{bmatrix} 3 & -2 \\ 1 & 0 \\ 0 & 5 \end{bmatrix}$, $B = \begin{bmatrix} 4 & -5 \\ 3 & 0 \\ -2 & 2 \end{bmatrix}$, $C = \begin{bmatrix} 2 & -1 \\ 6 & 10 \\ -3 & 7 \end{bmatrix}$, *and* $D = \begin{bmatrix} 3 & 2 & 5 \\ -2 & -4 & 1 \end{bmatrix}$, *determine the following, if possible. See Examples 1 through 4.*

1. $3A - B$ **2.** $B - 2D$ **3.** $3C$ **4.** $\frac{1}{2}D$ **5.** $3D + C$ **6.** $A + B + C$

7. $2A + 2B$ **8.** $\frac{3}{2}B + \frac{1}{2}C$ **9.** $C - 3A$ **10.** $3C - A$ **11.** $4A - 3D$ **12.** $2(A - 3B)$

Determine values of the variables that will make the following equations true, if possible. See Examples 1 through 4.

13. $\begin{bmatrix} a & 2b & c \end{bmatrix} + 3\begin{bmatrix} a & 2 & -c \end{bmatrix} = \begin{bmatrix} 8 & 2 & 2 \end{bmatrix}$

14. $\begin{bmatrix} w & 5x \\ 2y & z \end{bmatrix} - 5\begin{bmatrix} w & x \\ y & -z \end{bmatrix} = \begin{bmatrix} w+5 & 0 \\ 6 & 1 \end{bmatrix}$

15. $\begin{bmatrix} 3x \\ 2y \end{bmatrix} + \begin{bmatrix} x \\ -y \\ z \end{bmatrix} = \begin{bmatrix} 4 \\ 0 \\ 2 \end{bmatrix}$

16. $\begin{bmatrix} 2a & 3b & c \end{bmatrix} = \begin{bmatrix} 4 \\ 3 \\ 0 \end{bmatrix}$

17. $\begin{bmatrix} x \\ 3x \end{bmatrix} - \begin{bmatrix} y \\ 2y \end{bmatrix} = \begin{bmatrix} 5 \\ 20 \end{bmatrix}$

18. $7\begin{bmatrix} -1 \\ y \end{bmatrix} = \begin{bmatrix} 2x \\ 5x \end{bmatrix} + 3\begin{bmatrix} y \\ 1 \end{bmatrix}$

19. $2\begin{bmatrix} x \\ 2y \end{bmatrix} - 3\begin{bmatrix} 5y \\ -3x \end{bmatrix} = \begin{bmatrix} -9 \\ 31 \end{bmatrix}$

20. $2\begin{bmatrix} 3r & s & 2t \end{bmatrix} - \begin{bmatrix} r & s & t \end{bmatrix} = \begin{bmatrix} 15 & 3 & 9 \end{bmatrix}$

21. $2\begin{bmatrix} 2x^2 & x \\ 7x & 4 \end{bmatrix} - \begin{bmatrix} 5x \\ x-2 \end{bmatrix} = \begin{bmatrix} 2x & 0 \\ 6 & x^2 \end{bmatrix}$

22. $\begin{bmatrix} -x \\ 3 \end{bmatrix} - 5\begin{bmatrix} 2 \\ y \end{bmatrix} = \begin{bmatrix} -2y \\ 3x \end{bmatrix}$

Evaluate the following matrix products, if possible. See Example 5.

23. $\begin{bmatrix} 3 & -2 & 1 \end{bmatrix}\begin{bmatrix} 5 & -1 \\ 0 & 3 \\ 9 & 4 \end{bmatrix}$

24. $\begin{bmatrix} 0 & -8 \\ 5 & 6 \end{bmatrix}\begin{bmatrix} 3 & 7 \end{bmatrix}$

25. $\begin{bmatrix} 3 & 7 \end{bmatrix}\begin{bmatrix} 0 & -8 \\ 5 & 6 \end{bmatrix}$

26. $\begin{bmatrix} 5 & 0 & -3 \end{bmatrix}\begin{bmatrix} 4 \\ 2 \\ -6 \end{bmatrix}$

27. $\begin{bmatrix} 3 & 9 & -4 \\ 0 & 0 & 2 \\ 5 & -2 & 7 \end{bmatrix}\begin{bmatrix} 3 & 2 \\ 2 & 1 \end{bmatrix}$

28. $\begin{bmatrix} 4 \\ 2 \\ -6 \end{bmatrix}\begin{bmatrix} 5 & 0 & -3 \end{bmatrix}$

Given $A = \begin{bmatrix} -3 & 1 \\ 2 & 3 \end{bmatrix}$, $B = \begin{bmatrix} 8 & -5 \end{bmatrix}$, $C = \begin{bmatrix} 4 \\ 7 \\ -2 \end{bmatrix}$, *and* $D = \begin{bmatrix} -5 & 4 \\ -1 & -1 \end{bmatrix}$, *determine the following, if possible. See Example 5.*

29. AB **30.** BA **31.** $BA + B$ **32.** A^2 **33.** C^2 **34.** CB

35. D^2 **36.** $CD + C$ **37.** DA **38.** AD **39.** DB **40.** $(BD)A$

The following questions refer to the discussion of transition matrices in this section.

41. Suppose that each month 20% of store B's customers switch to store A, and 10% of store A's customers switch back to store B. At the start of January, store A has 250 customers and store B has 750. How many customers can each store expect at the start of February? At the start of March?

42. Given the percentages stated in the last problem, what long-term proportion of the town's customers can each store expect? (A technological aid may be used to compute high powers of the transition matrix, or you can use the method described below.)

43. Suppose P is a 2×2 transition matrix, and we want to determine the effect of applying high powers of P to the matrix $\begin{bmatrix} x \\ y \end{bmatrix}$, where $x + y$ is a fixed constant, say c. (In our competing store situation, $x + y = 1000$.) If the long-term behavior approaches a steady-state, as in our two-store example, then there is some value for x and some value for y such that $x + y = c$ and $P\begin{bmatrix} x \\ y \end{bmatrix} = \begin{bmatrix} x \\ y \end{bmatrix}$.

In other words, once the steady-state has been reached, applying the matrix P to it has no effect on the state.

We can use this fact to actually solve for x and y as follows. Given the matrix

$$P = \begin{bmatrix} 0.7 & 0.45 \\ 0.3 & 0.55 \end{bmatrix},$$

write the equation $P\begin{bmatrix} x \\ y \end{bmatrix} = \begin{bmatrix} x \\ y \end{bmatrix}$ in system form. You should find that the two equations that result are actually identical. But if we now also use the fact that $x + y = 1000$, we can solve for the variables and find that $x = 600$ and $y = 400$. Verify that this is indeed the case.

10.5 Inverses of Matrices

TOPICS

1. The matrix form of a linear system
2. Finding the inverse of a matrix
3. Using matrix inverses to solve linear systems

Topic 1: The Matrix Form of a Linear System

In the last section, we noted that matrix multiplication allows us to think of matrices as functions. For example, if we express the ordered pair (x, y) as a 2×1 matrix, then the linear system

$$\begin{cases} ax+by=e \\ cx+dy=f \end{cases}$$

can be written as

$$\begin{bmatrix} a & b \\ c & d \end{bmatrix}\begin{bmatrix} x \\ y \end{bmatrix}=\begin{bmatrix} e \\ f \end{bmatrix}.$$

While it can be verified (after multiplying out the left-hand side) that the matrix equation is equivalent to the system of equations above it, this is not, by any means, a trivial observation. For one thing, the matrix form of a linear system is more efficient: it converts a system of any number of equations into a single matrix equation. More importantly, it uses the function interpretation of a matrix to express a system of equations in a form reminiscent of linear equations of a single variable. And since the generic linear equation $ax = b$ can be solved by dividing both sides by a,

$$ax=b \Leftrightarrow x=\frac{b}{a} \text{ (assuming } a \neq 0\text{)}$$

it is tempting to solve

$$\begin{bmatrix} a & b \\ c & d \end{bmatrix}\begin{bmatrix} x \\ y \end{bmatrix}=\begin{bmatrix} e \\ f \end{bmatrix}$$

by "dividing" both sides by the 2×2 matrix of coefficients. Unfortunately, while we can add, subtract, and multiply matrices (at least when the dimensions match up appropriately), we do not have any way yet of making sense of "matrix division."

We will return to this thought soon, but first we will see how some specific linear systems appear in matrix form.

example 1

Write each linear system as a matrix equation.

a. $\begin{cases} -3x+5y=2 \\ x-4y=-1 \end{cases}$ **b.** $\begin{cases} 3y-x=-2 \\ 4-z+y=5 \\ z-3x+3=y-x \end{cases}$

Solutions:

a. Since the system is in standard form, we can just read off the coefficients of x and y to form the equation

$$\begin{bmatrix} -3 & 5 \\ 1 & -4 \end{bmatrix}\begin{bmatrix} x \\ y \end{bmatrix}=\begin{bmatrix} 2 \\ -1 \end{bmatrix}.$$

Verify for yourself that $\begin{bmatrix} -3 & 5 \\ 1 & -4 \end{bmatrix}\begin{bmatrix} x \\ y \end{bmatrix}=\begin{bmatrix} -3x+5y \\ x-4y \end{bmatrix}$.

b. While we could, with care, determine by inspection what the coefficient matrix of the system is, writing each equation in standard form first is probably wise:

$$\begin{cases} 3y-x=-2 \\ 4-z+y=5 \\ z-3x+3=y-x \end{cases} = \begin{cases} -x+3y=-2 \\ y-z=1 \\ -2x-y+z=-3 \end{cases}$$

Now we can read off the coefficients to form the matrix equation

$$\begin{bmatrix} -1 & 3 & 0 \\ 0 & 1 & -1 \\ -2 & -1 & 1 \end{bmatrix}\begin{bmatrix} x \\ y \\ z \end{bmatrix}=\begin{bmatrix} -2 \\ 1 \\ -3 \end{bmatrix}.$$

Topic 2: Finding the Inverse of a Matrix

In order to solve matrix equations like the two we obtained in Example 1, we need a way to "undo" the matrix of coefficients that appears in front of the column of variables. To get a better understanding of what, exactly, we are seeking, consider the following matrix products:

$$\begin{bmatrix}1 & 0\\0 & 1\end{bmatrix}\begin{bmatrix}x\\y\end{bmatrix} \text{ and } \begin{bmatrix}1 & 0 & 0\\0 & 1 & 0\\0 & 0 & 1\end{bmatrix}\begin{bmatrix}x\\y\\z\end{bmatrix}.$$

If these matrix products appeared as the left-hand side of matrix equations, the equations would correspond to solutions of linear systems:

$$\begin{bmatrix}1 & 0\\0 & 1\end{bmatrix}\begin{bmatrix}x\\y\end{bmatrix}=\begin{bmatrix}a\\b\end{bmatrix} \text{ corresponds to } \begin{cases}x=a\\y=b\end{cases}$$

and

$$\begin{bmatrix}1 & 0 & 0\\0 & 1 & 0\\0 & 0 & 1\end{bmatrix}\begin{bmatrix}x\\y\\z\end{bmatrix}=\begin{bmatrix}a\\b\\c\end{bmatrix} \text{ corresponds to } \begin{cases}x=a\\y=b.\\z=c\end{cases}$$

Such matrices, composed of 0's and 1's, deserve a name.

Identity Matrices

The $n\times n$ **identity matrix**, denoted I_n (or by just I when there is no possibility of confusion), is the $n\times n$ matrix consisting of 1's on the *main diagonal* and 0's everywhere else. The **main diagonal** consists of those entries in the first row-first column, the second row-second column, and so on down to the nth row-n^{th} column. Every identity matrix, then, has the form

$$I=\begin{bmatrix}1 & 0 & 0 & \cdots & 0\\0 & 1 & 0 & \cdots & 0\\0 & 0 & 1 & \cdots & 0\\\vdots & \vdots & \vdots & \ddots & \vdots\\0 & 0 & 0 & \cdots & 1\end{bmatrix}.$$

The identity matrix (of a certain size) serves as the multiplicative identity on the set of appropriate matrices. This means that if matrices A and B are such that the matrix products are defined, then $AI=A$ and $IB=B$. In this sense, I serves the same purpose as the number 1 in the set of real numbers.

(Although we have no immediate use for it, every set of matrices of a certain fixed order contains an *additive identity*. It is the matrix of the same order consisting entirely of 0's, and it is easy to see that if it is added to any matrix A, the result is A. The 0 matrices, then, play the same role as the number 0 in the set of real numbers.)

We know that a linear system of n equations and n variables can be expressed as a matrix equation $AX = B$, where A is an $n \times n$ matrix composed of coefficients, X is an $n \times 1$ matrix composed of the n variables, and B is an $n \times 1$ matrix composed of the constants from the right-hand sides of the equations. If we could find a matrix, which we will label A^{-1}, with the property that $A^{-1}A = I$, then we could use A^{-1} to "undo" the matrix A as desired. We call the matrix A^{-1}, if it exists, the *inverse* of A; the situation is completely analogous to the fact that $\frac{1}{a}$ is the (multiplicative) inverse of the real number a. The formal definition follows.

The Inverse of a Matrix

Let A be an $n \times n$ matrix. If there exists an $n \times n$ matrix A^{-1} such that

$$A^{-1}A = I_n \text{ and } AA^{-1} = I_n,$$

we call A^{-1} the **inverse** of A.

As the following example illustrates, we already possess the knowledge of how to find the inverse of a matrix.

example 2

Find the inverse of the matrix $A = \begin{bmatrix} 2 & -3 \\ -1 & 2 \end{bmatrix}$.

Solution:

If we let $A^{-1} = \begin{bmatrix} w & x \\ y & z \end{bmatrix}$, then one equation that must be satisfied is $AA^{-1} = I$, or

$$\begin{bmatrix} 2 & -3 \\ -1 & 2 \end{bmatrix}\begin{bmatrix} w & x \\ y & z \end{bmatrix} = \begin{bmatrix} 1 & 0 \\ 0 & 1 \end{bmatrix}.$$

cont'd. on next page ...

If we carry out the matrix product, we see that we need to solve the equation

$$\begin{bmatrix} 2w-3y & 2x-3z \\ -w+2y & -x+2z \end{bmatrix} = \begin{bmatrix} 1 & 0 \\ 0 & 1 \end{bmatrix},$$

which, if we equate columns on each side, means we need to solve the two linear systems

$$\begin{cases} 2w-3y=1 \\ -w+2y=0 \end{cases} \text{ and } \begin{cases} 2x-3z=0 \\ -x+2z=1 \end{cases}.$$

Of course, we have many methods for solving such systems. One efficient approach is to use Gauss-Jordan elimination, and to solve both systems at the same time by forming a new kind of augmented matrix. Since the coefficients in front of the variables are the same, if we put the matrix

$$\left[\begin{array}{cc|cc} 2 & -3 & 1 & 0 \\ -1 & 2 & 0 & 1 \end{array}\right]$$

into reduced row-echelon form, the numbers that appear in the third column will solve the first system and the numbers that appear in the fourth column will solve the second system.

$$\left[\begin{array}{cc|cc} 2 & -3 & 1 & 0 \\ -1 & 2 & 0 & 1 \end{array}\right] \xrightarrow{R_1 \leftrightarrow R_2} \left[\begin{array}{cc|cc} -1 & 2 & 0 & 1 \\ 2 & -3 & 1 & 0 \end{array}\right] \xrightarrow{2R_1+R_2} \left[\begin{array}{cc|cc} -1 & 2 & 0 & 1 \\ 0 & 1 & 1 & 2 \end{array}\right]$$

$$\xrightarrow{-R_1} \left[\begin{array}{cc|cc} 1 & -2 & 0 & -1 \\ 0 & 1 & 1 & 2 \end{array}\right] \xrightarrow{2R_2+R_1} \left[\begin{array}{cc|cc} 1 & 0 & 2 & 3 \\ 0 & 1 & 1 & 2 \end{array}\right]$$

Enter the matrix using the **Matrx** operation (pg. 792). Then **Quit**, enter matrix *A* through the **Matrx** menu, and select x^{-1}. Be careful... calculators may find the "inverse" of non-invertible matrices!

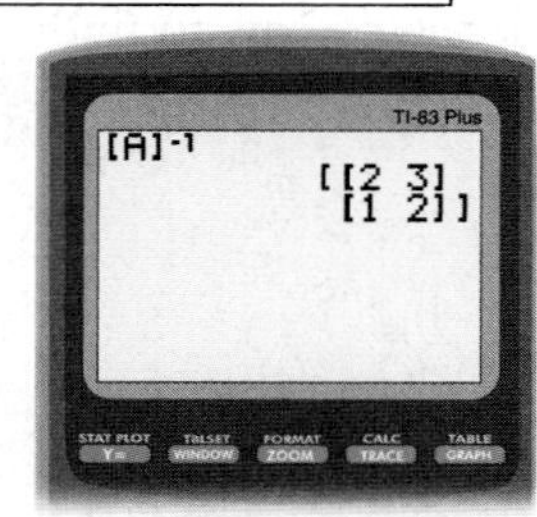

This tells us that $w = 2$ and $y = 1$ (from the third column) and $x = 3$ and $z = 2$ (from the fourth column). So

$$A^{-1} = \begin{bmatrix} 2 & 3 \\ 1 & 2 \end{bmatrix}.$$

We can now verify that

$$\begin{bmatrix} 2 & -3 \\ -1 & 2 \end{bmatrix}\begin{bmatrix} 2 & 3 \\ 1 & 2 \end{bmatrix} = \begin{bmatrix} 1 & 0 \\ 0 & 1 \end{bmatrix} \text{ and also } \begin{bmatrix} 2 & 3 \\ 1 & 2 \end{bmatrix}\begin{bmatrix} 2 & -3 \\ -1 & 2 \end{bmatrix} = \begin{bmatrix} 1 & 0 \\ 0 & 1 \end{bmatrix},$$

so we have indeed found A^{-1}.

Once we understand how the above method works, it becomes clear that we can omit the intermediate step of constructing the systems of equations and skip to the process of putting the appropriate augmented matrix into reduced row-echelon form.

Finding the Inverse of a Matrix

Let A be an $n\times n$ matrix. The inverse of A can be found by:

1. Forming the augmented matrix $[A \mid I]$, where I is the $n\times n$ identity matrix.
2. Using Gauss-Jordan elimination to put $[A \mid I]$ into the form $[I \mid B]$, if possible.
3. Defining A^{-1} to be B.

If it is not possible to put $[A \mid I]$ into reduced row-echelon form, then A does not have an inverse, and we say A is not *invertible*.

We will find the inverses of several more matrices soon, but the last sentence in the above box merits further discussion. Since it is certainly possible that a linear system has either no solution or an infinite number of solutions, we should not be surprised by the possibility that some matrices have no inverses. In such cases, fortunately, it becomes obvious during the process of trying to find the inverse that there is not one. We will examine such a situation in Example 3.

One last comment before we look at another example: there is a shortcut for finding inverses of 2×2 matrices that saves significant time. The shortcut also quickly identifies those 2×2 matrices that are invertible.

Inverse of a 2×2 Matrix

Let $A=\begin{bmatrix} a & b \\ c & d \end{bmatrix}$. Then $A^{-1}=\frac{1}{|A|}\begin{bmatrix} d & -b \\ -c & a \end{bmatrix}$, where $|A|=ad-bc$ is the determinant of A. Since $|A|$ appears in the denominator of a fraction, A^{-1} fails to exist if $|A|=0$.

Finding the inverse of a 3×3 or larger matrix requires more work than the above shortcut, but there is one similarity. Regardless of the order of a matrix, if its determinant is zero, the matrix has no inverse.

example 3

Find the inverses of the following matrices, if possible.

a. $A=\begin{bmatrix}3 & -5\\ 2 & 1\end{bmatrix}$ **b.** $B=\begin{bmatrix}2 & 4 & 2\\ -1 & 5 & -1\\ 3 & 1 & 3\end{bmatrix}$

Solutions:

a. Using the shortcut,

$$A^{-1}=\frac{1}{3-(-10)}\begin{bmatrix}1 & 5\\ -2 & 3\end{bmatrix}=\begin{bmatrix}\frac{1}{13} & \frac{5}{13}\\ \frac{-2}{13} & \frac{3}{13}\end{bmatrix}.$$

We can verify our work as follows:

$$\begin{bmatrix}3 & -5\\ 2 & 1\end{bmatrix}\begin{bmatrix}\frac{1}{13} & \frac{5}{13}\\ \frac{-2}{13} & \frac{3}{13}\end{bmatrix}=\begin{bmatrix}1 & 0\\ 0 & 1\end{bmatrix}=\begin{bmatrix}\frac{1}{13} & \frac{5}{13}\\ \frac{-2}{13} & \frac{3}{13}\end{bmatrix}\begin{bmatrix}3 & -5\\ 2 & 1\end{bmatrix}$$

b. A quick examination of B reveals two identical columns, so from the properties of determinants we know $|B|=0$. Based on the comment immediately preceding this example, then, we expect that B is not invertible, but we will begin the process of inverting B anyway to see what happens.

$$\left[\begin{array}{ccc|ccc}2 & 4 & 2 & 1 & 0 & 0\\ -1 & 5 & -1 & 0 & 1 & 0\\ 3 & 1 & 3 & 0 & 0 & 1\end{array}\right]\xrightarrow[3R_2+R_3]{2R_2+R_1}\left[\begin{array}{ccc|ccc}0 & 14 & 0 & 1 & 2 & 0\\ -1 & 5 & -1 & 0 & 1 & 0\\ 0 & 16 & 0 & 0 & 3 & 1\end{array}\right]$$

$$\xrightarrow{R_1\leftrightarrow R_2}\left[\begin{array}{ccc|ccc}-1 & 5 & -1 & 0 & 1 & 0\\ 0 & 14 & 0 & 1 & 2 & 0\\ 0 & 16 & 0 & 0 & 3 & 1\end{array}\right]\xrightarrow[\frac{1}{14}R_2]{-R_1}\left[\begin{array}{ccc|ccc}1 & -5 & 1 & 0 & -1 & 0\\ 0 & 1 & 0 & \frac{1}{14} & \frac{1}{7} & 0\\ 0 & 16 & 0 & 0 & 3 & 1\end{array}\right]$$

$$\xrightarrow{-16R_2+R_3}\left[\begin{array}{ccc|ccc}1 & -5 & 1 & 0 & -1 & 0\\ 0 & 1 & 0 & \frac{1}{14} & \frac{1}{7} & 0\\ 0 & 0 & 0 & -\frac{8}{7} & \frac{5}{7} & 1\end{array}\right]$$

At this point, we can stop. Once the first three entries of any row are 0 in the 3×3 matrix, there is no way to put the matrix into reduced row-echelon form. Thus, B has no inverse.

technology note

Computer algebra systems and sufficiently-advanced calculators are able to determine inverses of matrices. In *Mathematica*, the command to do so is **Inverse**. The illustration below shows first the definition of a 3×3 matrix A and then the evaluation of its inverse. Note that the output of the **Inverse** command is a list of sub-lists, which is the way that *Mathematica* works with matrices internally. To see A^{-1} in rectangular form, the command **MatrixForm** is used along with the **Inverse** command.

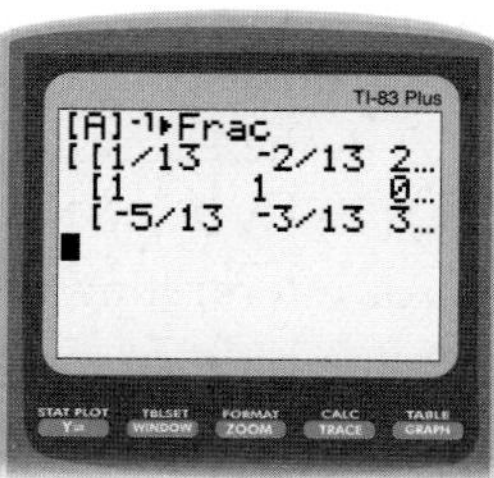

Enter the matrix and select x^{-1} as described on page 836. To show the answer in fraction form, select Math and 1 (Frac.)

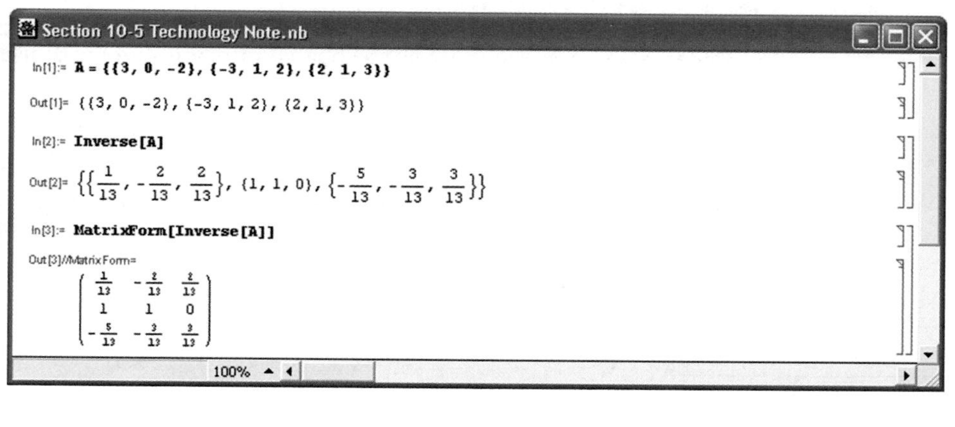

Topic 3: Using Matrix Inverses to Solve Linear Systems

We have now assembled all the tools we need to solve linear systems by the inverse matrix method.

Given a linear system of n equations in n variables, the first step is to write the system in matrix form as $AX = B$, where A is the $n \times n$ matrix of coefficients, X is the $n \times 1$ matrix of variables and B is the $n \times 1$ matrix of constants. The next step is to construct A^{-1}, if A^{-1} exists. There are now two ways of interpreting the meaning of the final step, though the two interpretations entail identical computations.

Interpretation 1: As we know, multiplying A by A^{-1} results in the $n \times n$ identity matrix, just as multiplying a non-zero number a by $\frac{1}{a}$ gives us the multiplicative identity 1. So if we multiply both sides of the equation $AX = B$ by A^{-1}, we obtain

$$\begin{aligned} A^{-1}AX &= A^{-1}B \\ IX &= A^{-1}B \\ X &= A^{-1}B. \end{aligned}$$

The entries in the $n \times 1$ matrix $A^{-1}B$ are thus the sought-after values for the variables listed in the $n \times 1$ matrix X.

Interpretation 2: The more sophisticated interpretation is to use the fact that A^{-1} is the inverse function of the function A to rewrite the equation $AX = B$ in the equivalent form $X = A^{-1}B$. This is an illustration of the generic statement that $f(x) = y$ if and only if $x = f^{-1}(y)$, if f indeed has an inverse function. The advantage of interpreting the solution process this way is that it emphasizes the point that a given matrix A may not have an inverse. The ramifications of this are explored in more detail in such classes as linear algebra.

example 4

Solve the following systems by the inverse matrix method.

a. $\begin{cases} 4x - 5y = 3 \\ -3x + 7y = 1 \end{cases}$ **b.** $\begin{cases} -x + 2y = 3 \\ 3x - 6y = -5 \end{cases}$

Solutions:

a. $\begin{cases} 4x - 5y = 3 \\ -3x + 7y = 1 \end{cases}$

$$\begin{bmatrix} 4 & -5 \\ -3 & 7 \end{bmatrix}\begin{bmatrix} x \\ y \end{bmatrix} = \begin{bmatrix} 3 \\ 1 \end{bmatrix}$$

The first step is to write the system in matrix form. Since both equations are in standard form, this is a straightforward task.

$$\begin{bmatrix} 4 & -5 \\ -3 & 7 \end{bmatrix}^{-1} = \frac{1}{28-15}\begin{bmatrix} 7 & 5 \\ 3 & 4 \end{bmatrix}$$

$$= \begin{bmatrix} \frac{7}{13} & \frac{5}{13} \\ \frac{3}{13} & \frac{4}{13} \end{bmatrix}$$

The main step in the inverse matrix method is to find the inverse of the coefficient matrix A. Since the matrix is of order 2×2 in this problem, we can use the shortcut to obtain the inverse quickly. Of course, we can check our work at this stage by making sure that the product of the original matrix and its inverse is the identity matrix.

$$\begin{bmatrix} x \\ y \end{bmatrix} = \begin{bmatrix} \frac{7}{13} & \frac{5}{13} \\ \frac{3}{13} & \frac{4}{13} \end{bmatrix} \begin{bmatrix} 3 \\ 1 \end{bmatrix}$$

$$= \begin{bmatrix} \frac{21}{13} + \frac{5}{13} \\ \frac{9}{13} + \frac{4}{13} \end{bmatrix}$$

$$= \begin{bmatrix} 2 \\ 1 \end{bmatrix}$$

Solution is (2, 1).

We can now rewrite the equation $AX = B$ in the form $X = A^{-1}B$, and proceed to carry out the matrix multiplication on the right to get the answer. Once we have $x = 2$ and $y = 1$, we can verify that this is indeed the solution of the system.

b. $\begin{cases} -x + 2y = 3 \\ 3x - 6y = -5 \end{cases}$

$$\begin{bmatrix} -1 & 2 \\ 3 & -6 \end{bmatrix} \begin{bmatrix} x \\ y \end{bmatrix} = \begin{bmatrix} 3 \\ -5 \end{bmatrix}$$

Again, we start by writing the system in matrix form.

$$\begin{bmatrix} -1 & 2 \\ 3 & -6 \end{bmatrix}^{-1} = \frac{1}{6-6} \begin{bmatrix} -6 & -2 \\ -3 & -1 \end{bmatrix}$$

In this problem, we note immediately that the coefficient matrix has no inverse, so we know that either the system has no solution or else an infinite number of solutions.

$$D_x = \begin{vmatrix} 3 & 2 \\ -5 & -6 \end{vmatrix} = -18 - (-10) \neq 0$$

Since $D_x \neq 0$, Cramer's rule tells us that this system has no solution.

No Solution.

We have now reached the end of our list of solution methods for linear systems; there is more to say on the topic, but further discussion is best left for a more advanced course. One question that might have occurred to you, though, has not yet been adequately addressed. That question is: How should the method of solution be chosen when faced with a specific system of equations?

Unfortunately, there is no simple answer. To begin with, if the system is small, or if it looks especially simple, then the method of substitution or the method of elimination may be the quickest route to a solution. Generally speaking, the larger and/or the more complicated the system, the likelier it is that one of the more advanced methods will be preferable. But if you find that your first choice of method leads to a computational headache, do not be reluctant to stop and try another method.

In practical applications, it is often the case that a number of closely related systems must be solved, and the only difference in the systems lies in the constants on the right-hand sides of the equations. That is, the coefficient matrix is the same for all the systems. In this case, the inverse matrix method is almost certainly the best choice, as illustrated below.

example 5

Solve the following three linear systems.

$$\begin{cases} x-4y+z=-6 \\ -2x+y=5 \\ 3y-z=3 \end{cases} \qquad \begin{cases} x-4y+z=-17 \\ -2x+y=2 \\ 3y-z=14 \end{cases} \qquad \begin{cases} x-4y+z=3 \\ -2x+y=5 \\ 3y-z=-5 \end{cases}$$

Solution:

The coefficient matrix is the same for the three systems, and this means we can avoid unnecessary computation if we use the inverse matrix method. We begin by finding the inverse of the coefficient matrix.

$$\left[\begin{array}{ccc|ccc} 1 & -4 & 1 & 1 & 0 & 0 \\ -2 & 1 & 0 & 0 & 1 & 0 \\ 0 & 3 & -1 & 0 & 0 & 1 \end{array}\right] \xrightarrow{2R_1+R_2} \left[\begin{array}{ccc|ccc} 1 & -4 & 1 & 1 & 0 & 0 \\ 0 & -7 & 2 & 2 & 1 & 0 \\ 0 & 3 & -1 & 0 & 0 & 1 \end{array}\right]$$

$$\xrightarrow{\frac{1}{3}R_3} \left[\begin{array}{ccc|ccc} 1 & -4 & 1 & 1 & 0 & 0 \\ 0 & -7 & 2 & 2 & 1 & 0 \\ 0 & 1 & -\frac{1}{3} & 0 & 0 & \frac{1}{3} \end{array}\right] \xrightarrow[7R_3+R_2]{4R_3+R_1} \left[\begin{array}{ccc|ccc} 1 & 0 & -\frac{1}{3} & 1 & 0 & \frac{4}{3} \\ 0 & 0 & -\frac{1}{3} & 2 & 1 & \frac{7}{3} \\ 0 & 1 & -\frac{1}{3} & 0 & 0 & \frac{1}{3} \end{array}\right]$$

$$\xrightarrow{-3R_2} \left[\begin{array}{ccc|ccc} 1 & 0 & -\frac{1}{3} & 1 & 0 & \frac{4}{3} \\ 0 & 0 & 1 & -6 & -3 & -7 \\ 0 & 1 & -\frac{1}{3} & 0 & 0 & \frac{1}{3} \end{array}\right] \xrightarrow{R_2 \leftrightarrow R_3} \left[\begin{array}{ccc|ccc} 1 & 0 & -\frac{1}{3} & 1 & 0 & \frac{4}{3} \\ 0 & 1 & -\frac{1}{3} & 0 & 0 & \frac{1}{3} \\ 0 & 0 & 1 & -6 & -3 & -7 \end{array}\right]$$

$$\xrightarrow[\frac{1}{3}R_3+R_2]{\frac{1}{3}R_3+R_1} \left[\begin{array}{ccc|ccc} 1 & 0 & 0 & -1 & -1 & -1 \\ 0 & 1 & 0 & -2 & -1 & -2 \\ 0 & 0 & 1 & -6 & -3 & -7 \end{array}\right]$$

$$A^{-1} = \begin{bmatrix} -1 & -1 & -1 \\ -2 & -1 & -2 \\ -6 & -3 & -7 \end{bmatrix}$$

The majority of the work has now been done, and we can solve the three systems quickly by multiplying by the three matrices consisting of the right-hand-side constants:

$$\begin{bmatrix} x \\ y \\ z \end{bmatrix} = \begin{bmatrix} -1 & -1 & -1 \\ -2 & -1 & -2 \\ -6 & -3 & -7 \end{bmatrix}\begin{bmatrix} -6 \\ 5 \\ 3 \end{bmatrix} = \begin{bmatrix} -2 \\ 1 \\ 0 \end{bmatrix}$$

$$\begin{bmatrix} x \\ y \\ z \end{bmatrix} = \begin{bmatrix} -1 & -1 & -1 \\ -2 & -1 & -2 \\ -6 & -3 & -7 \end{bmatrix}\begin{bmatrix} -17 \\ 2 \\ 14 \end{bmatrix} = \begin{bmatrix} 1 \\ 4 \\ -2 \end{bmatrix}$$

$$\begin{bmatrix} x \\ y \\ z \end{bmatrix} = \begin{bmatrix} -1 & -1 & -1 \\ -2 & -1 & -2 \\ -6 & -3 & -7 \end{bmatrix}\begin{bmatrix} 3 \\ 5 \\ -5 \end{bmatrix} = \begin{bmatrix} -3 \\ -1 \\ 2 \end{bmatrix}$$

exercises

Write each of the following systems of equations as a single matrix equation. See Example 1.

1. $\begin{cases} 14x - 5y = 7 \\ x + 9y = 2 \end{cases}$

2. $\begin{cases} x - 5 = 9y \\ 3y - 2x = 8 \end{cases}$

3. $\begin{cases} \dfrac{3x - 8y}{5} = 2 \\ y - 2 = 0 \end{cases}$

4. $\begin{cases} x - 7y = 5 \\ \dfrac{6 + x}{2} = 3y - 2 \end{cases}$

5. $\begin{cases} 4x = 3y - 9 \\ 13 - 2x = -4y \end{cases}$

6. $\begin{cases} -\dfrac{7}{3}y = \dfrac{5 - x}{6} \\ x - 5(y - 3) = -2 \end{cases}$

7. $\begin{cases} -6 - 2y = x \\ 9x + 14 = 3y \end{cases}$

8. $\begin{cases} x - y = 5 \\ 2 - z = x \\ z - 3y = 4 \end{cases}$

9. $\begin{cases} 3x_1 - 7x_2 + x_3 = -4 \\ x_1 - x_2 = 2 \\ 8x_2 + 5x_3 = -3 \end{cases}$

10. $\begin{cases} x_3 = x_2 \\ x_2 = x_1 \\ x_1 = x_3 \end{cases}$

11. $\begin{cases} 2x - y = -3z \\ y - x = 17 \\ 2 + z + 4x = 5y \end{cases}$

12. $\begin{cases} 2x_1 - 3x_3 = 7 \\ x_2 - 10x_3 = 0 \\ 2x_1 - x_2 + x_3 = 1 \end{cases}$

Find the inverse of each of the following matrices, if possible. See Examples 2 and 3.

13. $\begin{bmatrix} 0 & 4 \\ -5 & -1 \end{bmatrix}$

14. $\begin{bmatrix} -2 & -2 \\ -1 & 2 \end{bmatrix}$

15. $\begin{bmatrix} 3 & 4 \\ -4 & -5 \end{bmatrix}$

16. $\begin{bmatrix} -1 & -1 \\ -\frac{1}{4} & -\frac{1}{2} \end{bmatrix}$

17. $\begin{bmatrix} -\frac{1}{5} & 0 \\ \frac{1}{5} & \frac{1}{2} \end{bmatrix}$

18. $\begin{bmatrix} -7 & 2 \\ 7 & -2 \end{bmatrix}$

19. $\begin{bmatrix} -2 & -4 & -2 \\ 1 & -4 & 1 \\ 4 & -3 & 4 \end{bmatrix}$

20. $\begin{bmatrix} -3 & 0 & -4 \\ 2 & 5 & 4 \\ 1 & -5 & -2 \end{bmatrix}$

21. $\begin{bmatrix} -\frac{5}{11} & -\frac{8}{11} & 1 \\ \frac{13}{11} & \frac{12}{11} & -2 \\ -\frac{2}{11} & -\frac{1}{11} & 0 \end{bmatrix}$

22. $-\frac{1}{31}\begin{bmatrix} 17 & -8 & -2 \\ 1 & 5 & 9 \\ -6 & 1 & 8 \end{bmatrix}$

23. $\begin{bmatrix} -1 & 2 & -1 \\ 0 & 3 & -1 \\ 0 & 4 & -1 \end{bmatrix}$

24. $\begin{bmatrix} -1 & 0 & -1 \\ -\frac{3}{2} & \frac{1}{2} & -\frac{3}{2} \\ -\frac{1}{2} & 0 & -\frac{1}{4} \end{bmatrix}$

25. $\begin{bmatrix} -\frac{6}{5} & -\frac{2}{5} & -1 \\ \frac{3}{5} & \frac{1}{5} & 1 \\ 1 & 0 & 1 \end{bmatrix}$

26. $\begin{bmatrix} 2 & -2 & 1 \\ -2 & 2 & -3 \\ 1 & 0 & 2 \end{bmatrix}$

27. $\begin{bmatrix} 0 & 1 & 1 \\ 1 & 1 & 0 \\ 0 & 1 & 2 \end{bmatrix}$

28. $\begin{bmatrix} 9 & 8 & 7 \\ 6 & 5 & 4 \\ 3 & 2 & 1 \end{bmatrix}$

29. $\begin{bmatrix} \frac{2}{3} & \frac{8}{9} & \frac{1}{9} \\ -\frac{1}{3} & \frac{2}{9} & -\frac{2}{9} \\ -\frac{1}{3} & -\frac{7}{9} & -\frac{2}{9} \end{bmatrix}$

30. $\begin{bmatrix} -3 & -3 & -4 \\ 0 & \frac{1}{4} & \frac{1}{2} \\ 2 & 2 & 3 \end{bmatrix}$

Solve the following systems by the inverse matrix method, if possible. If the inverse matrix method does not apply, use any other method to determine if the system is inconsistent or dependent. See Example 4.

31. $\begin{cases} -2x-2y=9 \\ -x+2y=-3 \end{cases}$

32. $\begin{cases} 3x+4y=-2 \\ -4x-5y=9 \end{cases}$

33. $\begin{cases} -2x+3y=1 \\ 4x-6y=-2 \end{cases}$

34. $\begin{cases} -2x+4y=5 \\ x-4y=-3 \end{cases}$

35. $\begin{cases} -5x=10 \\ 2x+2y=-4 \end{cases}$

36. $\begin{cases} -3x+y=2 \\ 9x-3y=5 \end{cases}$

37. $\begin{cases} -x+2y-z=6 \\ 3y-z=-1 \\ 4y-z=5 \end{cases}$ **38.** $\begin{cases} x-4z=8 \\ -3x+2y=-2 \\ -2x+4z=4 \end{cases}$ **39.** $\begin{cases} -2x-4y-2z=1 \\ x-4y+z=0 \\ 4x-3y+4z=-1 \end{cases}$

40. $\begin{cases} 4y+3z=-254 \\ 2x-2y-z=100 \\ -x+y-2z=155 \end{cases}$ **41.** $\begin{cases} 2x-y-3z=-10 \\ 2y-z=11 \\ -x+4z=0 \end{cases}$ **42.** $\begin{cases} 3y-4z=15 \\ x+2y-3z=9 \\ -x-y+2z=-5 \end{cases}$

Solve the following sets of systems by the inverse matrix method. See Example 5.

43. **a.** $\begin{cases} x+2y-z=2 \\ 3x+3y-z=-5 \\ 4x+4y-z=1 \end{cases}$ **b.** $\begin{cases} x+2y-z=1 \\ 3x+3y-z=1 \\ 4x+4y-z=1 \end{cases}$ **c.** $\begin{cases} x+2y-z=0 \\ 3x+3y-z=1 \\ 4x+4y-z=1 \end{cases}$

44. **a.** $\begin{cases} -x-y-2z=4 \\ x+3y+3z=0 \\ -3y-2z=9 \end{cases}$ **b.** $\begin{cases} -x-y-2z=1 \\ x+3y+3z=0 \\ -3y-2z=0 \end{cases}$ **c.** $\begin{cases} -x-y-2z=-2 \\ x+3y+3z=-3 \\ -3y-2z=1 \end{cases}$

45. **a.** $\begin{cases} -x+z=6 \\ -x+3y+2z=-11 \\ 2x-4y-3z=13 \end{cases}$ **b.** $\begin{cases} -x+z=-2 \\ -x+3y+2z=2 \\ 2x-4y-3z=-1 \end{cases}$ **c.** $\begin{cases} -x+z=-4 \\ -x+3y+2z=2 \\ 2x-4y-3z=0 \end{cases}$

10.6 Partial Fraction Decomposition

TOPICS

1. The pattern of decompositions
2. Completing the decomposition process

Topic 1: The Pattern of Decompositions

Throughout this text, we have frequently found it useful or necessary to combine fractions. You have by now done this so often, and for such a variety of reasons, that you may not even be consciously aware of the process – the act of finding a common denominator and combining fractions may be second nature. There are occasions, however, when it is helpful to be able to reverse the process. In performing certain operations in Calculus, for instance, the ability to write a single fraction as a sum of simpler fractions comes in very handy. The process of doing so is called **partial fraction decomposition**, and we will find the methods of this chapter useful in the execution of the process.

To be specific, the fractions we will want to decompose are proper rational functions; that is, fractions of the form

$$f(x) = \frac{p(x)}{q(x)}$$

where p and q are polynomials and the degree of p is less than the degree of q (recall that we already know how to perform polynomial division on fractions where the degree of p is greater than or equal to the degree of q). As a consequence of the Fundamental Theorem of Algebra and the Conjugate Roots Theorem (Chapter 4), if $q(x)$ has only real coefficients then it can be written as a product of factors of the form $(ax+b)^m$ and $(ax^2+bx+c)^n$, where m and n are positive integers, a, b, and c are real numbers, and ax^2+bx+c cannot be factored further without resorting to complex coefficients (we say ax^2+bx+c is irreducible). The appearance of such factors tells us how the rational function can be decomposed, as outlined on the next page.

Decomposition Pattern

Given the proper rational function $f(x) = \frac{p(x)}{q(x)}$, assume $q(x)$ has been completely factored as a product of factors of the form $(ax+b)^m$ and $(ax^2+bx+c)^n$, where a, b, and c are real numbers, ax^2+bx+c is irreducible over the real numbers, and m and n are positive integers. Then $f(x)$ can be decomposed as a sum of simpler rational functions, where

1. Each factor of the form $(ax+b)^m$ leads to a sum of the form

$$\frac{A_1}{ax+b} + \frac{A_2}{(ax+b)^2} + \cdots + \frac{A_m}{(ax+b)^m}$$

2. Each factor of the form $(ax^2+bx+c)^n$ leads to a sum of the form

$$\frac{A_1x+B_1}{ax^2+bx+c} + \frac{A_2x+B_2}{(ax^2+bx+c)^2} + \cdots + \frac{A_nx+B_n}{(ax^2+bx+c)^n}$$

example 1

Write the form of the partial fraction decomposition of the rational function

$$f(x) = \frac{p(x)}{x^3+6x^2+12x+8}.$$

Assume that the degree of p is 2 or smaller.

Solution:

The primary task in this problem is to factor the denominator. The Rational Zero Theorem (Section 4.3) tells us that the potential rational zeros of the denominator are $\pm\{1, 2, 4, 8\}$ (remember that the potential rational zeros of a polynomial are the factors of the constant term divided by the factors of the leading coefficient). Synthetic division or long division can then be used to test these potential zeros one by one; the work following uses synthetic division to show that -2 is a zero.

cont'd. on next page ...

$$\begin{array}{r|rrrr} -2 & 1 & 6 & 12 & 8 \\ & & -2 & -8 & -8 \\ \hline & 1 & 4 & 4 & 0 \end{array}$$

From this, we conclude that $x^3+6x^2+12x+8=(x+2)(x^2+4x+4)=(x+2)^3$. Following the guidelines of the decomposition pattern, we now know that $f(x)$ can be written as a sum of rational functions of the form

$$f(x)=\frac{A_1}{x+2}+\frac{A_2}{(x+2)^2}+\frac{A_3}{(x+2)^3}.$$

Of course, we can go no further with the problem above since we were not given a specific polynomial in the numerator. Note, though, that the numerator played no role in our work – any rational function with the denominator $(x+2)^3$ can be decomposed as in Example 1, as long as the degree of the numerator is 2 or smaller.

example 2

Write the form of the partial fraction decomposition of the rational function

$$f(x)=\frac{p(x)}{(x-5)^2(x^2+4)^2}.$$

Assume that the degree of p is 5 or smaller.

Solution:

In this example, the denominator is already appropriately factored; while it is true that x^2+4 can be factored as $(x-2i)(x+2i)$, partial fraction decomposition calls for leaving irreducible quadratics such as x^2+4 in their unfactored state. All that remains is following the decomposition guidelines to write $f(x)$ as:

$$f(x)=\frac{A_1}{x-5}+\frac{A_2}{(x-5)^2}+\frac{B_1x+C_1}{x^2+4}+\frac{B_2x+C_2}{(x^2+4)^2}.$$

Note the choice of letters in the numerators of the decomposition. The names of the unknown coefficients do not matter, of course, but it is important to realize that there are a total of six such coefficients and so six different symbols must be used to denote them.

Topic 2: Completing the Decomposition Process

Assuming the degree of the numerator is smaller than the degree of the denominator, the numerator plays no role in determining the form of the partial fraction decomposition of a given rational function. But of course it must be considered when we need to actually solve for the unknown coefficients appearing in the decomposition. It is at this stage that one of the methods of solving systems of equations may be called for.

example 3

Find the partial fraction decomposition of the rational function

$$f(x)=\frac{-2x+14}{x^2+2x-3}.$$

Solution:

Since $x^2+2x-3=(x+3)(x-1)$, we know

$$\frac{-2x+14}{x^2+2x-3}=\frac{A_1}{x+3}+\frac{A_2}{x-1}.$$

There are many ways we can go about solving for A_1 and A_2, but most begin with the step of eliminating the fractions. Multiplying through by $(x+3)(x-1)$ leads immediately, upon canceling common factors, to the equation

$$-2x+14=A_1(x-1)+A_2(x+3).$$

This particular partial fraction decomposition can be accomplished simply – no advanced methods of solving systems of equations will be necessary. The key observation is that the equation above is true for *all* values of x; thus, we can substitute well-chosen values for x into the equation and quickly solve for the two coefficients. The values of x that are useful are those that make either $x - 1$ or $x + 3$ zero; that is, we will first let $x = 1$ and then let $x = -3$:

$$\text{For } x=1\text{: } -2(1)+14=A_1(1-1)+A_2(1+3)$$
$$12=4A_2$$
$$A_2=3$$

and

$$\text{For } x=-3\text{: } -2(-3)+14=A_1(-3-1)+A_2(-3+3)$$
$$20=-4A_1$$
$$A_1=-5.$$

Thus, $\dfrac{-2x+14}{x^2+2x-3}=\dfrac{-5}{x+3}+\dfrac{3}{x-1}.$

Our answer can be checked by combining these two fractions.

The method shown in Example 3 will not always be sufficient, by itself, to solve for the unknown coefficients. Consider the next example.

example 4

Find the partial fraction decomposition of the rational function

$$f(x)=\frac{3x+1}{(x+2)^3}.$$

Solution:

We already considered rational functions with this particular denominator in Example 1, and we know that

$$\frac{3x+1}{(x+2)^3}=\frac{A_1}{x+2}+\frac{A_2}{(x+2)^2}+\frac{A_3}{(x+2)^3}.$$

Eliminating fractions is again a good first step:

$$3x+1=A_1(x+2)^2+A_2(x+2)+A_3.$$

The obvious convenient value for x is -2, and making this substitution allows us to quickly solve for A_3:

$$\begin{aligned}3(-2)+1&=A_1(-2+2)^2+A_2(-2+2)+A_3\\ -5&=A_3.\end{aligned}$$

However, there are no other values for x that allow us to immediately determine A_1 and A_2. There are still several other ways we can complete the decomposition, though. One is to simply substitute two other values for x and so obtain two equations in two unknowns. Since we now know that $A_3=-5$, we will make the substitutions for x in the equation

$$3x+6=A_1(x+2)^2+A_2(x+2).$$

In the work below, we use $x=0$ and $x=-1$, with these values chosen simply because they make the calculations easy:

$$\begin{aligned}3(0)+6&=A_1(0+2)^2+A_2(0+2)\\ 6&=4A_1+2A_2\end{aligned}$$

and

$$\begin{aligned}3(-1)+6&=A_1(-1+2)^2+A_2(-1+2)\\ 3&=A_1+A_2.\end{aligned}$$

We recognize this as a linear system in the two variables A_1 and A_2, and we can use any of the methods of this chapter to solve it. Below, we have used the method of elimination.

$$\begin{cases} 4A_1 + 2A_2 = 6 \\ A_1 + A_2 = 3 \end{cases} \xrightarrow{-2E_2} \begin{cases} 4A_1 + 2A_2 = 6 \\ -2A_1 - 2A_2 = -6 \end{cases}$$

$$2A_1 \qquad = 0$$

So $A_1 = 0$ and hence $A_2 = 3$. This gives us the partial fraction decomposition

$$\frac{3x+1}{(x+2)^3} = \frac{3}{(x+2)^2} + \frac{-5}{(x+2)^3}.$$

There is another way of thinking about the decomposition process that is very important conceptually. After eliminating fractions, the decomposition in Example 4 has the form

$$3x+1 = A_1(x+2)^2 + A_2(x+2) + A_3.$$

This is an equation relating two polynomials, a fact that is explicitly clear if we multiply out the right-hand side and collect powers of x:

$$3x+1 = A_1x^2 + (4A_1 + A_2)x + (4A_1 + 2A_2 + A_3)$$

As we know, two polynomials are equal only if corresponding coefficients are equal. Equating the coefficients in this example gives us the three equations

$$\begin{cases} 0 = A_1 \\ 3 = 4A_1 + A_2 \\ 1 = 4A_1 + 2A_2 + A_3 \end{cases}.$$

Note that there is no x^2 term on the left.

This is a linear system of three equations in three variables, and again any method from this chapter could be used to solve it. The verification that the solution of this system leads to the answer in Example 4 is left to the reader.

We conclude this section with an example that calls for the application of several skills learned in this text.

example 5

Find the partial fraction decomposition of the rational function

$$f(x) = \frac{3x^5 + 6x^4 + 9x^3 + 7x^2 + 4x - 1}{(x^2 + x + 1)^2}.$$

cont'd. on next page ...

Solution:

First, note that this rational function is improper – the degree of the numerator is greater than the degree of the denominator, so we must perform some polynomial division before the partial fraction decomposition. To do so, we first expand the denominator:

$$\left(x^2+x+1\right)^2=x^4+2x^3+3x^2+2x+1.$$

Polynomial long division (Section 4.2) then gives us the result

$$\frac{3x^5+6x^4+9x^3+7x^2+4x-1}{\left(x^2+x+1\right)^2}=3x+\frac{x^2+x-1}{\left(x^2+x+1\right)^2},$$

and it is the fractional part of this that we will decompose.

Next, note that x^2+x+1 is irreducible. This follows from the use of the quadratic formula to solve the equation $x^2+x+1=0$; since the solutions of this equation are complex numbers, the linear factors of x^2+x+1 contain complex coefficients. The partial fraction decomposition thus has the form

$$\frac{x^2+x-1}{\left(x^2+x+1\right)^2}=\frac{Ax+B}{x^2+x+1}+\frac{Cx+D}{\left(x^2+x+1\right)^2}.$$

(For the sake of variety, and to avoid the necessity for subscripts, we will use different letters to stand for the unknown coefficients.)

Clearing the equation of fractions gives us

$$x^2+x-1=(Ax+B)\left(x^2+x+1\right)+(Cx+D)$$

and hence

$$x^2+x-1=Ax^3+(A+B)x^2+(A+B+C)x+(B+D).$$

From this polynomial equation we derive the system

$$\begin{cases} 0=A \\ 1=A+B \\ 1=A+B+C \\ -1=B+D \end{cases}.$$

This system is easily solved, considering the equations in the order presented, and gives us $A = 0$, $B = 1$, $C = 0$, and $D = -2$. So our final answer is

$$\frac{3x^5+6x^4+9x^3+7x^2+4x-1}{\left(x^2+x+1\right)^2}=3x+\frac{1}{x^2+x+1}+\frac{-2}{\left(x^2+x+1\right)^2}.$$

exercises

Write the form of the partial fraction decomposition of each of the following rational functions. In each case, assume the degree of the numerator is less than the degree of the denominator. See Examples 1 and 2.

1. $f(x)=\frac{p(x)}{x^2-x-6}$

2. $f(x)=\frac{p(x)}{x^2-2x-24}$

3. $f(x)=\frac{p(x)}{x^3+11x^2+40x+48}$

4. $f(x)=\frac{p(x)}{(x+5)^2(x^2+3)^2}$

5. $f(x)=\frac{p(x)}{(x+3)(x^2-4)}$

6. $f(x)=\frac{p(x)}{(x^2+5)(x^2+3x-4)}$

Match each rational expression with the form of its decomposition. The decompositions are labeled a – h.

7. $\frac{2x-1}{(x+2)^3(x-2)}$

8. $\frac{2x-1}{x^2(x^2-4)}$

9. $\frac{2x-1}{x^3-4x^2+4x}$

10. $\frac{2x-1}{x^4-16}$

11. $\frac{2x-1}{x^3(x-2)^2}$

12. $\frac{2x-1}{x^5+2x^4+4x^3+8x^2}$

13. $\frac{2x-1}{x^3(x-2)}$

14. $\frac{2x-1}{x^5+6x^4+12x^3+8x^2}$

a. $\frac{A}{x}+\frac{B}{x^2}+\frac{C}{x^3}+\frac{D}{x-2}+\frac{E}{(x-2)^2}$

b. $\frac{A}{x}+\frac{B}{x^2}+\frac{C}{x+2}+\frac{D}{x-2}$

c. $\frac{A}{x}+\frac{B}{x^2}+\frac{C}{x^3}+\frac{D}{x-2}$

d. $\frac{A}{(x+2)}+\frac{B}{(x+2)^2}+\frac{C}{(x+2)^3}+\frac{D}{x-2}$

e. $\frac{Ax+B}{x^2+4}+\frac{C}{x+2}+\frac{D}{x-2}$

f. $\frac{A}{x}+\frac{B}{x^2}+\frac{C}{x+2}+\frac{Dx+E}{x^2+4}$

g. $\frac{A}{x+2}+\frac{B}{(x+2)^2}+\frac{C}{(x+2)^3}+\frac{D}{x}+\frac{E}{x^2}$

h. $\frac{A}{x}+\frac{B}{x-2}+\frac{C}{(x-2)^2}$

Find the partial fraction decomposition of each of the following rational functions. See Examples 3, 4, and 5.

15. $f(x)=\frac{3x^2+4}{x^3-4x}$

16. $f(x)=\frac{2x}{x^3+7x^2-6x-72}$

17. $f(x)=\frac{4x+2}{(x^3+8x)(x^2+2x-8)}$

18. $f(x)=\frac{5}{x^2+3x-4}$

19. $f(x)=\frac{5x}{x^2-6x+8}$

20. $f(x)=\frac{6x^2-4}{(x^2+3)(x+6)(x+5)}$

21. $f(x)=\frac{6x}{x^3+8x^2+9x-18}$

22. $f(x)=\frac{12x^2+x-1}{x^4+7x^3+5x^2-31x-30}$

23. $f(x)=\frac{1}{x^2-1}$

24. $f(x)=\frac{x+3}{(x^2+3)(x^2+x-6)}$

25. $f(x)=\frac{x^2-4}{(x^4-16)(x^2+2x-8)}$

26. $f(x)=\frac{x+1}{x^3-x}$

27. $f(x)=\frac{x+3}{x^2-4}$

28. $f(x)=\frac{x}{x^3+6x^2+11x+6}$

29. $f(x)=\frac{x}{x^4-16}$

30. $f(x)=\frac{5}{x^2-6x+8}$

31. $f(x)=\frac{2x+3}{(x^2-9)(x^2+4x-12)}$

32. $f(x)=\frac{x}{x^2-7x+12}$

33. $f(x)=\frac{x^2}{x^3+5x^2+3x-9}$

34. $f(x)=\frac{2x}{(x+4)(x^2-2x-3)}$

35. $f(x)=\frac{2x}{x^2-9}$

36. $f(x)=\frac{4x+3}{(x^2-9)(x^2-2x-24)}$

37. $f(x)=\frac{x^2}{x^3+4x^2-12x}$

38. $f(x)=\frac{2}{x^3+7x^2-8x}$

Using your graphing calculator, determine whether the partial fraction decomposition is true or false by graphing the left and right side of the equation on the same x- and y-axis.

39. $\dfrac{x+7}{x^2-x-6}=\dfrac{2}{x-3}-\dfrac{1}{x+2}$

40. $\dfrac{5x^2+20x+6}{x^3+2x^2+x}=\dfrac{6}{x}-\dfrac{1}{x+1}+\dfrac{9}{(x+1)^2}$

41. $\dfrac{4x^2-3x-4}{x^3+x^2-2x}=\dfrac{2}{x}-\dfrac{1}{x-1}+\dfrac{3}{x+2}$

42. $\dfrac{2x+4}{x^3-2x^2}=-\dfrac{2}{x}-\dfrac{2}{x^2}+\dfrac{2}{x-2}$

43. $\dfrac{1}{x^2+x-2}=\dfrac{1}{x-1}+\dfrac{1}{x+2}$

44. $\dfrac{6x^2+2}{x^2(x-3)^3}=\dfrac{4}{x}+\dfrac{2}{x^2}+\dfrac{3x}{x-3}+\dfrac{4}{(x-3)^2}+\dfrac{6x}{(x-3)^3}$

Write the partial fraction decomposition for the rational expression. Then assign a value to the constant, a, and check the result graphically.

45. $\dfrac{1}{x(x+a)}$

46. $\dfrac{1}{x(a-x)}$

47. $\dfrac{1}{a^2-x^2}$

48. $\dfrac{1}{(x+1)(a-x)}$

49. $\dfrac{1}{(x+a)(x+1)}$

10.7 Linear Programming

TOPICS

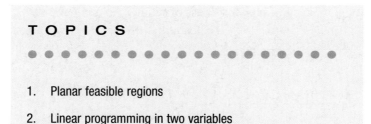

1. Planar feasible regions
2. Linear programming in two variables

Topic 1: Planar Feasible Regions

Many real-world applications of mathematics call for identifying the values that a set of variables can assume. If a particular application imposes limitations on the values that two variables can assume, the collection of all allowable values forms a portion of the plane called the **feasible region** or **region of constraint**. For instance, the constraints $1 \le x \le 5$ and $2 \le y \le 4$ define the rectangular feasible region shown below.

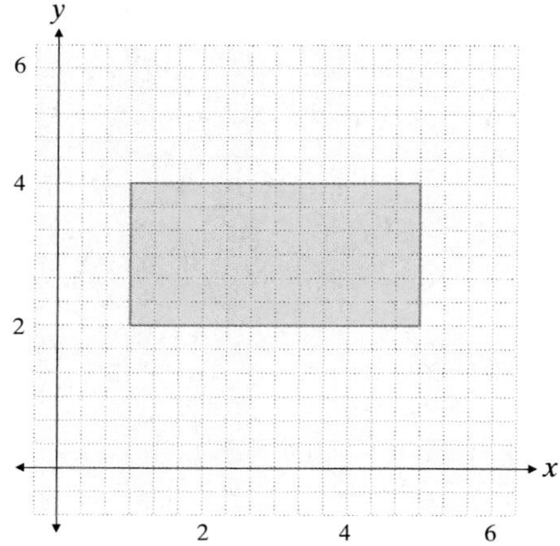

Typically, constraint inequalities involve both variables and the corresponding feasible region is a bit more interesting than a rectangle. The collection of all constraints in a problem comprises a system of inequalities, and the techniques learned in this chapter will be very useful in extracting certain information about the feasible region described by such systems.

Example 1 illustrates how feasible regions may typically be described in an application.

example 1

A family orchard is in the business of selling peaches and nectarines. The owners know that to prevent a certain pest infestation, the number of nectarine trees in the orchard cannot exceed the number of peach trees. Also, because of the space requirements of each type of tree, the number of nectarine trees plus twice the number of peach trees cannot exceed 100 trees. Construct the constraints and graph the feasible region for this situation.

Solution:

If p represents the number of peach trees and n the number of nectarine trees, the paragraph above gives us various constraints on p and n. For instance, the second sentence translates into the inequality $n \le p$ and the third sentence translates into the inequality $n + 2p \le 100$. In addition, we can assume that $n \ge 0$ and $p \ge 0$, as it is not possible to have a negative number of trees. Letting the horizontal axis represent p and the vertical axis n, the portion of the plane satisfying all four of these inequalities appears as shown below (refer to Section 2.5 for a review of graphing inequalities in two variables).

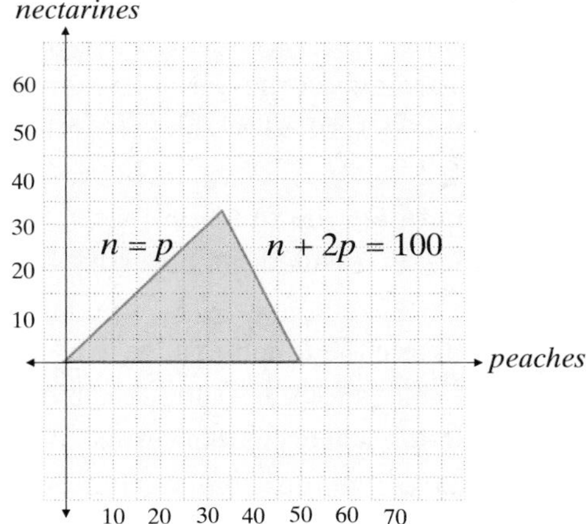

Any point in the shaded feasible region corresponds to a certain number of peach trees and nectarine trees, and all such points meet the stated conditions. The question of how to choose a particular ordered pair then arises; we will soon see how the existence of one more condition resolves this question.

Topic 2: Linear Programming in Two Variables

Real-world applications involving the construction of feasible regions often also possess a quantity that must be either maximized or minimized. Two extremely important examples come from business: profit is a quantity that people usually wish to maximize, and cost is a quantity best kept to a minimum. When the quantity to be optimized (that is, either maximized or minimized) is a linear function of the variables concerned, the collection of mathematical tools known as **linear programming** comes into play.

Linear programming is used in a wide variety of applications, some involving thousands of variables, and a complete study of its techniques is beyond the scope of this text. But the basic concepts are simple and easily applied, especially when restricted to applications calling for just two variables. We begin our study with a description of the steps taken in linear programming, followed by an example that will illustrate why linear programming works.

Linear Programming Method (Two-Variable Case)

1. Identify the variables to be considered in the problem, determine all the constraints on the variables imposed by the problem, and sketch the feasible region described by the constraints.

2. Determine the function that is to be either maximized or minimized. Such a function is called the **objective function**.

3. Evaluate the function at each of the vertices of the feasible region and compare the values found. If the feasible region is *bounded*, the optimum value of the function will occur at a vertex. If the feasible region is *unbounded*, the optimum value of the function may not exist, but if it does it will occur at a vertex.

(A **bounded** feasible region is one in which all the points lie within some fixed distance of the origin. An **unbounded** region has points that are arbitrarily far away from the origin.)

example 2

Find the maximum and minimum values of $f(x,y) = 3x + 2y$ subject to the constraints

$$\begin{cases} x \geq 0 \\ y \geq 0 \\ y \geq 3 - 3x \\ 2x + y \leq 10 \\ x + y \leq 7 \\ y \geq x - 2 \end{cases}.$$

Solution:

The two variables are given to us, as are the constraints; all we must do to complete the first step is sketch the region defined by the constraints. The figure below contains the sketch, with selected boundaries identified by their corresponding equations.

The function to be optimized is also given to us in this problem, so step two is complete. According to the linear programming method, then, the only remaining task is to evaluate the function at the vertices of the feasible region. Why should we expect the maximum and minimum values of f to occur at vertices, and not in the interior of the region?

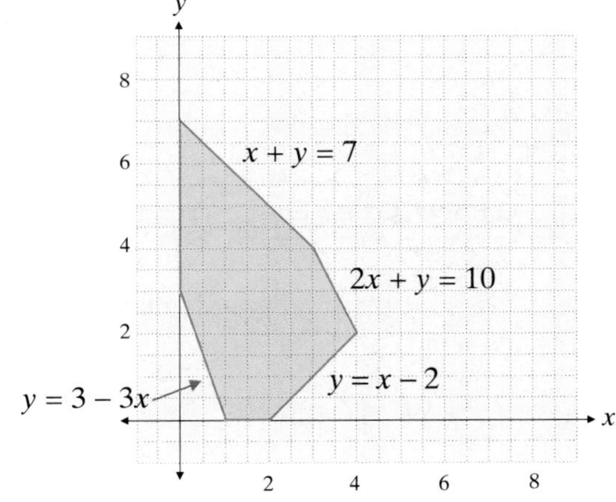

Another diagram will answer this question. The function $f(x,y)$ will, of course, take on a variety of values at different ordered pairs (x, y). But for any particular value, say k, there will be a line of ordered pairs in the plane such that $f(x, y) = k$; such lines, for different values of k, will be parallel, as shown here. Two particular values for k are labeled in the diagram, and these values are the minimum and maximum values of f.

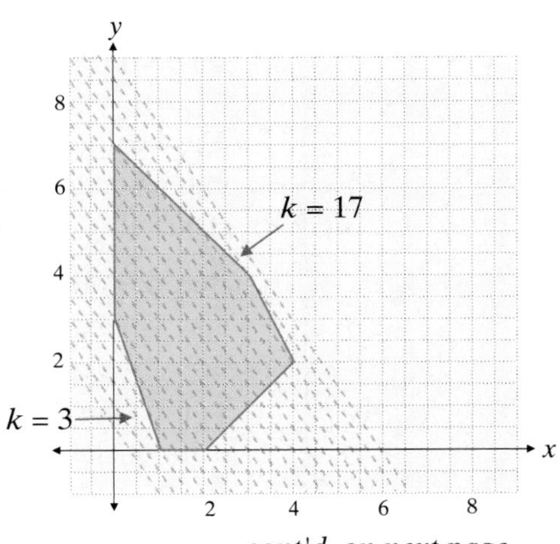

cont'd. on next page ...

The extreme values of a linear function over a planar feasible region bounded by straight lines will thus occur where a line of constant value just touches the region, and this will be at a vertex or one of the edges of the region, if the edge is parallel to the line of constant value. In any event, evaluating the function at the vertices of the feasible region will suffice to identify the extreme values.

It is in the third step that the solution methods of this chapter are useful. Each vertex of a feasible region is the intersection of two edges, so the coordinates of each vertex are found by solving a system of two equations. The six vertices of the region in this problem are determined by the following six systems:

$$\begin{cases} x+y=7 \\ 2x+y=10 \end{cases} \quad \begin{cases} 2x+y=10 \\ y=x-2 \end{cases} \quad \begin{cases} y=x-2 \\ y=0 \end{cases} \quad \begin{cases} y=0 \\ y=3-3x \end{cases} \quad \begin{cases} y=3-3x \\ x=0 \end{cases} \quad \begin{cases} x=0 \\ x+y=7 \end{cases}.$$

Solving each of these systems, we find the vertices to be $(3, 4)$, $(4, 2)$, $(2, 0)$, $(1, 0)$, $(0, 3)$, and $(0, 7)$. Substituting each of these ordered pairs into the objective function we find the maximum value of 17 occurs at the vertex $(3, 4)$, and the minimum value of 3 occurs at the vertex $(1, 0)$. The verification of these facts is left to the reader.

example 3

Find the maximum and minimum values of $f(x, y) = x + 2y$ subject to the constraints

$$\begin{cases} x \geq 0 \\ y \geq 3 - x \\ x - y \leq 1 \\ y \leq 2x + 4 \end{cases}.$$

Solution:

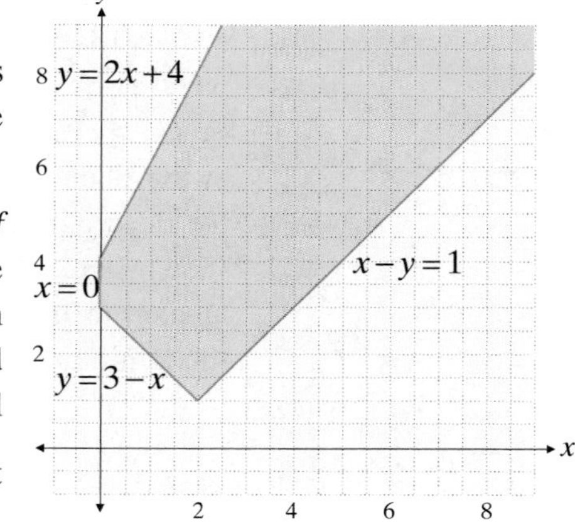

The feasible region defined by the constraints is unbounded; a portion of its sketch appears to the right.

Because the region is unbounded, the function f may not have a maximum or a minimum value over the region. In this case, f has no maximum value. One way to see this is to note that ordered pairs of the form (x, x) lie in the region for all $x \geq \frac{3}{2}$, and $f(x, x) = x + 2x = 3x$ grows without bound as x increases.

On the other hand, f does have a minimum value over the feasible region. The three vertices are $(0, 4)$, $(0, 3)$, and $(2, 1)$. Evaluating the function at these points, we obtain $f(0, 4) = 8$, $f(0, 3) = 6$, and $f(2, 1) = 4$, so the minimum value is 4.

Our last example illustrates how linear programming can be applied in a typical business application.

example 4

A manufacturer makes two models of minidisc players. Model A requires 3 minutes to assemble, 4 minutes to test, and 1 minute to package. Model B requires 4 minutes to assemble, 3 minutes to test, and 6 minutes to package. The manufacturer can allot 7400 minutes total for assembly, 8000 minutes for testing, and 9000 minutes for packaging. Model A generates a profit of \$7.00 and model B generates a profit of \$8.00. How many of each model should be made in order to maximize profit?

Solution:

If we let x denote the number of Model A players made and y the number of Model B players made, the constraint inequalities are

$$\begin{cases} x \geq 0, y \geq 0 \\ 3x + 4y \leq 7400 \\ 4x + 3y \leq 8000 \\ x + 6y \leq 9000 \end{cases}$$

and the sketch of the feasible region appears below.

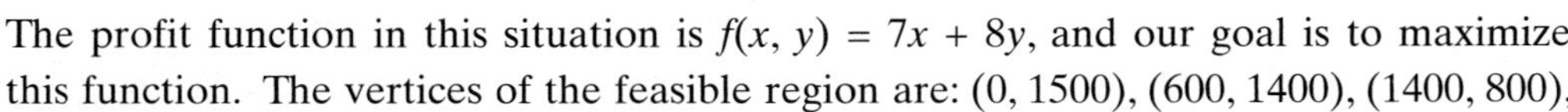

The profit function in this situation is $f(x, y) = 7x + 8y$, and our goal is to maximize this function. The vertices of the feasible region are: $(0, 1500)$, $(600, 1400)$, $(1400, 800)$, $(2000, 0)$, and $(0, 0)$. Of course, manufacturing 0 units of Model A and Model B is not going to be profitable, so we really only need to evaluate f at four vertices:

$$f(0, 1500) = 12{,}000$$
$$f(600, 1400) = 15{,}400$$
$$f(1400, 800) = 16{,}200$$
$$f(2000, 0) = 14{,}000.$$

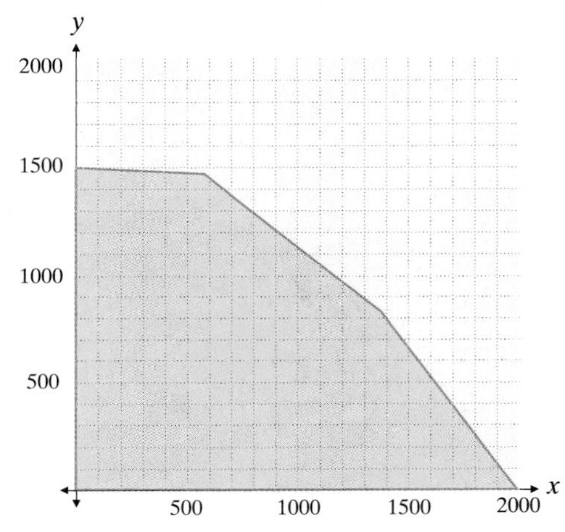

From these calculations, we conclude that the maximum profit is generated from making 1400 units of Model A and 800 units of Model B. The maximum profit would be \$16,200.

exercises

Construct the constraints and graph the feasible regions for the following situations. See Example 1.

1. A plane carrying relief food and water can carry a maximum of 50,000 pounds, and is limited in space to carrying no more than 6000 cubic feet. Each container of water weighs 60 pounds and takes up 1 cubic foot, and each container of food weighs 50 pounds and takes up 10 cubic feet. What is the region of constraint for the number of containers of food and water that the plane can carry?

2. A furniture company makes two kinds of sofas, the Standard model and the Deluxe model. The Standard model requires 40 hours of labor to build, and the Deluxe model requires 60 hours of labor to build. The finish of the Deluxe model, however, uses both teak and fabric, while the Standard uses only fabric, with the result that each Deluxe sofa requires 5 square yards of fabric and each Standard sofa requires 8 square yards of fabric. Given that the company can use 200 hours of labor and 25 square yards of fabric per week building sofas, what is the region of constraint for the number of Deluxe and Standard sofas the company can make per week?

3. Sarah is looking through a clothing catalog and she is willing to spend up to $80 on clothes and $10 for shipping. Shirts cost $12 each plus $2 shipping, and a pair of pants costs $32 plus $3 shipping. What is the region of constraint for the number of shirts and pairs of pants Sarah can buy?

4. Suppose you inherit $75,000 from a previously unknown (and highly eccentric) uncle, and that the inheritance comes with certain stipulations regarding investments. First, the dollar amount invested in bonds must not exceed the dollar amount invested in stocks. Second, a minimum of $10,000 must be invested in stocks and a minimum of $5000 must be invested in bonds. Finally, a maximum of $40,000 can be invested in stocks. What is the region of constraint for the dollar amount that can be invested in the two categories of stocks and bonds?

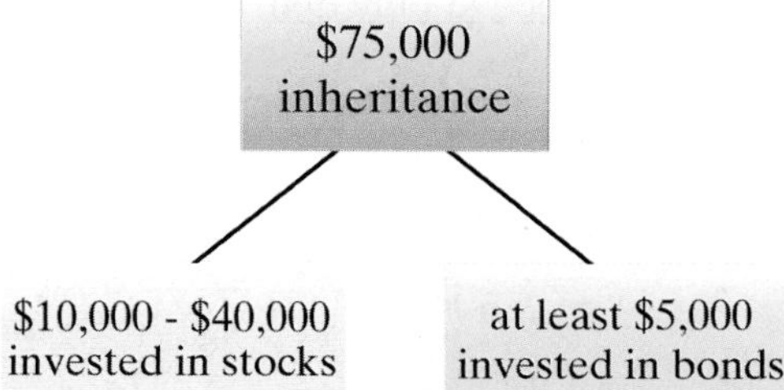

Find the minimum and maximum values of the given functions, subject to the given constraints. See Examples 2 and 3.

5. Objective Function
$f(x, y) = 2x + 3y$
Constraints
$x \geq 0$
$y \geq 0$
$x + y \leq 7$

6. Objective Function
$f(x, y) = 4x + y$
Constraints
$x \geq 0$
$y \geq 0$
$x + y \leq 3$

7. Objective Function
$f(x, y) = 2x + 5y$
Constraints
$x \geq 0$
$y \geq 0$
$2x + y \leq 6$

8. Objective Function
$f(x, y) = 7x + 4y$
Constraints
$x \geq 0$
$y \geq 0$
$3x + y \leq 3$

9. Objective Function
$f(x, y) = 5x + 6y$
Constraints
$0 \leq x \leq 7$
$0 \leq y \leq 10$
$8x + 5y \leq 40$

10. Objective Function
$f(x, y) = 9x + 7y$
Constraints
$0 \leq x \leq 20$
$0 \leq y \leq 10$
$6x + 12y \leq 140$

11. <u>Objective Function</u>
$f(x, y) = 6x + 4y$
<u>Constraints</u>
$0 \leq x \leq 4$
$0 \leq y \leq 5$
$4x + 3y \leq 10$

12. <u>Objective Function</u>
$f(x, y) = 3x + 7y$
<u>Constraints</u>
$0 \leq x \leq 8$
$0 \leq y \leq 6$
$7x + 10y \leq 50$

13. <u>Objective Function</u>
$f(x, y) = 6x + 8y$
<u>Constraints</u>
$x \geq 0; y \geq 0$
$4x + y \leq 16$
$x + 3y \leq 15$

14. <u>Objective Function</u>
$f(x, y) = x + 2y$
<u>Constraints</u>
$x \geq 0; y \geq 0$
$3x + y \leq 45$
$x + 3y \leq 24$

15. <u>Objective Function</u>
$f(x, y) = 6x + y$
<u>Constraints</u>
$x \geq 0; y \geq 0$
$3x + 4y \leq 24$
$3x + 4y \leq 48$

16. <u>Objective Function</u>
$f(x, y) = 15x + 30y$
<u>Constraints</u>
$x \geq 0; y \geq 0$
$5x + 7y \leq 70$
$5x + 7y \leq 140$

17. <u>Objective Function</u>
$f(x, y) = 3x + 10y$
<u>Constraints</u>
$x \geq 0$
$2x + 4y \geq 8$
$5x - y \leq 10$
$x + 3y \leq 40$

18. <u>Objective Function</u>
$f(x, y) = 20x + 30y$
<u>Constraints</u>
$x \geq 0$
$12x + 6y \geq 120$
$9x - 6y \leq 144$
$x + 4y \leq 12$

Use Linear Programming to answer the following questions. See Example 4.

19. A manufacturer produces two models of computers. The times (in hours) required for assembling, testing and packaging each model are listed in the table below.

Process	Model X	Model Y
Assemble	2.5	3
Test	2	1
Package	.75	1.25

The total times available for assembling, testing and packaging are 4000 hours, 2500 hours, and 1500 hours, respectively. The profits per unit are \$50 for Model X and \$52 for Model Y. How many of each type should be produced to maximize profit? What is the maximum profit?

20. A manufacturer produces two types of fans. The time (in minutes) required for assembling, packaging, and shipping each type are listed in the table below.

Process	Type X	Type Y
Assemble	20	25
Package	40	10
Ship	10	7.5

The total times available for assembling, packaging, and shipping are 4000 minutes, 4800 minutes, and 1500 minutes, respectively. The profits per unit are \$4.50 for Type X and \$3.75 for Type Y. How many of each type should be produced to maximize profit? What is the maximum profit?

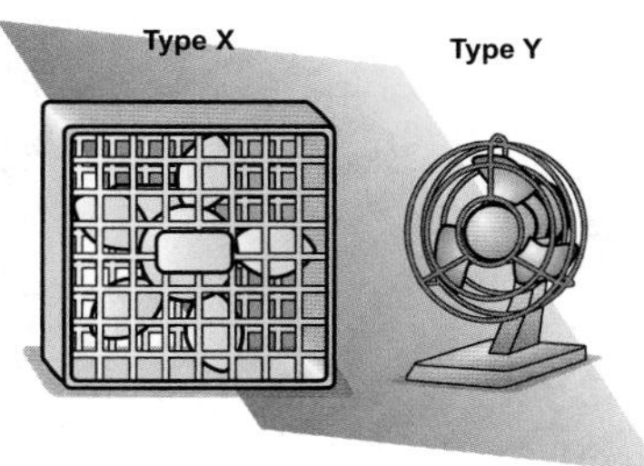

21. Ashley is making a set of patchwork curtains for her apartment. She needs a minimum of 16 yards of the solid material, at least 5 yards of the striped material, and at least 20 yards of the flowered material. She can choose between two sets of precut bundles. The olive-based bundle costs \$10 per bundle and contains 8 yards of the solid material, 1 yard of the striped material, and 2 yards of the flowered material. The cranberry-based bundle costs \$20 per bundle and includes 2 yards of the solid material, 1 yard of the striped material, and 7 yards of the flowered material. How many of each bundle should Ashley buy to minimize her cost and yet buy enough material to complete the curtains? What is her minimum cost?

22. A volunteer has been asked to drop off some supplies at a facility housing victims of a hurricane evacuation. The volunteer would like to bring at least 60 bottles of water, 45 first aid kits, and 30 security blankets on his visit. The relief organization has a standing agreement with two companies that provide victim packages. Company A can provide packages of 5 water bottles, 3 first aid kits, and 4 security blankets at a cost of \$1.50. Company B can provide packages of 2 water bottles, 2 first aid kits, and 1 security blanket at a cost of \$1.00. How many of each package should the volunteer pick up to minimize the cost? What total amount does the relief organization pay?

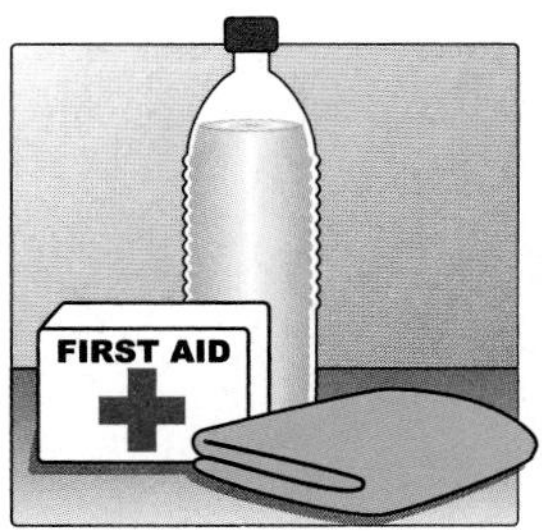

23. On your birthday your grandmother gave you \$25,000, but told you she would like you to invest the money for 10 years before you used any of it. Since you wish to respect your grandmother's wishes, you seek out the advice of a financial adviser. She suggests you invest at least \$15,000 in municipal bonds yielding 6% and no more than \$5,000 in Treasury bills yielding 9%. How much should be placed in each investment so that income is maximized?

24. An independent cell phone manufacturer produces two models: a flip phone and a picture phone. The manufacturer's quota per day is to produce at least 100 flip phones and 80 picture phones. No more than 200 flip phones and 170 picture phones can be produced per day due to limitations on production. A total of at least 200 phones must be shipped every day.
 a. If the production cost is \$5 for a flip phone and \$7 for a picture phone, how many of each model should be produced on a daily basis to minimize cost? What is the minimum cost to produce the minimum amount?
 b. If each flip phone results in a \$2 loss but each picture phone results in a \$5 gain, how many of each model should be manufactured daily to maximize profit? What is the maximum profit if this number of phones were produced?

10.8 Nonlinear Systems of Equations and Inequalities

TOPICS

1. Approximating solutions by graphing
2. Solving nonlinear systems algebraically
3. Solving nonlinear systems of inequalities

Topic 1: Approximating Solutions by Graphing

In this last section of the chapter, we deal with systems of equations in which one or more of the equations is nonlinear. For instance, many of the systems we will solve contain at least one equation with at least one second-degree term.

Because the systems are nonlinear, the matrix-based methods that we have learned do not apply. In fact, the theory of nonlinear systems is much more complex than that of linear systems, and for the most part we will rely on only very general guiding principles to solve them. What this means, in practice, is that you should not be surprised or discouraged if your first approach fails to yield a solution; when this happens, simply try another angle of attack.

The first general guiding principle is to think about the equations that make up the system and to try to get some idea for what real solutions to expect. As we have learned how to graph many classes of equations, we can put our graphing knowledge to good use in applying this principle. Keep in mind, however, that graphing will only help us identify the *real* solutions; as we will see, some nonlinear systems also have solutions containing non-zero imaginary parts.

example 1

Use graphing to guess the real solution(s) of the following system, and then verify that your answer is correct.

$$\begin{cases} x+y=2 \\ x^2+y^2=4 \end{cases}$$

cont'd. on next page ...

Display two graphs simultaneously by entering multiple equations in Y=. When solving for y in this case, enter an equation for both the positive and negative square roots. After selecting Graph, the Trace was turned on to view specific coordinates.

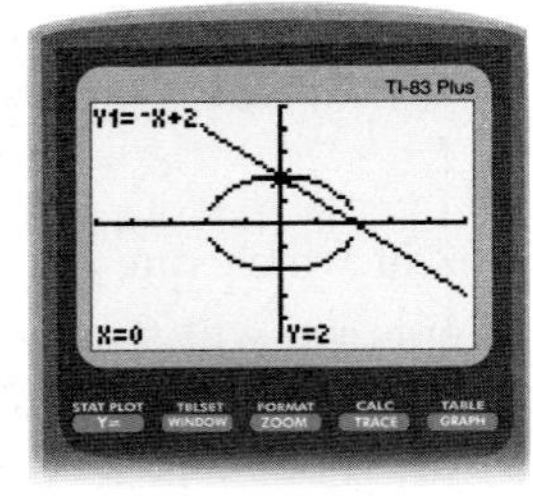

Solution:

The first equation describes a line in the plane with an x-intercept at (2, 0) and a y-intercept of $(0,2)$, and the second describes a circle of radius two centered at the origin. As is always the case when given a system of two equations, we are interested in whether any ordered pairs of numbers satisfy both equations simultaneously. Graphically, we are looking for the points of intersection of the two graphs. The following graph of the system certainly suggests a non-empty solution set.

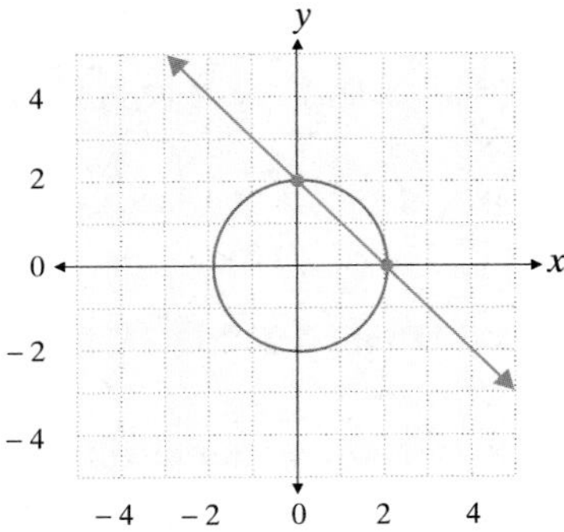

In fact, the graph suggests that $(2,0)$ and $(0,2)$ make up the real solutions of the system. We already know these two points lie on the line, and it can be verified that they also solve the second equation:

$$2^2+0^2=4 \text{ and } 0^2+2^2=4.$$

This simple example points out another very significant difference between linear and nonlinear systems. Recall that the solution set of any linear system is guaranteed to be empty, to contain exactly one element, or to be infinite. As we have now seen, nonlinear systems may have a different number of solutions.

Before we look at another example, verify for yourself (by graphing) that the following systems have, respectively, one real solution and no real solution.

$$\begin{cases} x+y=4 \\ x^2+y^2=8 \end{cases} \quad \begin{cases} x+y=5 \\ x^2+y^2=8 \end{cases}$$

example 2

Use graphing to guess the real solution(s) of the following system, and then verify that your answer is correct.

$$\begin{cases} x+2=y^2 \\ x^2+y^2=4 \end{cases}$$

Solution:

From our work in Section 9.2, we know the first equation represents a horizontally-oriented parabola opening to the right, with vertex at $(-2, 0)$. The second equation is again a circle of radius two centered at the origin. The graph of the system is shown below.

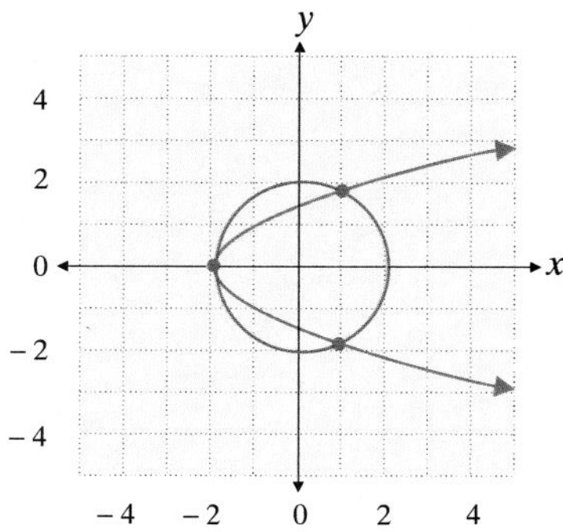

This time, the system appears to have three real solutions. The point $(-2, 0)$ is easily verified to be one of them, but the other two cannot be so readily found just by looking at the graph. However, the x-coordinate of the other two points appears to be 1, and if we replace x with 1 in the two equations we obtain the equation $y^2 = 3$ in *both* cases, meaning that $\left(1, \sqrt{3}\right)$ and $\left(1, -\sqrt{3}\right)$ are the other two solutions of the system. (Checking these solutions verifies that both ordered pairs solve both equations.)

Topic 2: Solving Nonlinear Systems Algebraically

In Example 2, we found it necessary to combine some graphical knowledge with some algebraic work in order to solve the system. The picture corresponding to the system suggested strongly that there were three real solutions, and for that example we can be fairly confident that we found all of them. But graphs can be misleading, especially if they are not drawn accurately. Consider, for example, the following very similar system.

example 3

Solve the following nonlinear system.

$$\begin{cases} x+4=y^2 \\ x^2+y^2=4 \end{cases}$$

cont'd. on next page ...

Solution:

The only difference between this system and the one in Example 2 is that the vertex of the parabola is at $(-4, 0)$, so the picture of the system is as shown.

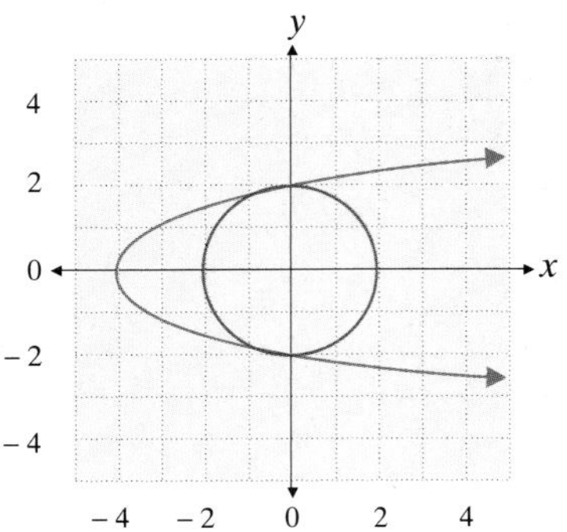

It is not clear, from the graphs of the two equations, exactly how many intersections exist or where those intersections are. A reasonable guess might be the two points $(0, 2)$ and $(0, -2)$, and it can be easily verified that these two points do indeed lie on both equations. But if we were to conclude that these are the only two solutions, we would be wrong.

We need a more reliable method for this example, and algebra comes to the rescue. Although matrix-based solution methods do not apply to nonlinear systems, the simpler methods of substitution and elimination do. For the system

$$\begin{cases} x + 4 = y^2 \\ x^2 + y^2 = 4 \end{cases},$$

the method of substitution works well if we solve both equations for y^2:

$$\begin{cases} x + 4 = y^2 \\ y^2 = 4 - x^2 \end{cases}$$

The two expressions equal to y^2 both contain only the variable x, so if we equate the expressions we will have a single equation in one variable. The resulting equation is solved as shown.

$$\begin{aligned} x + 4 &= 4 - x^2 \\ x &= -x^2 \\ x^2 + x &= 0 \\ x(x+1) &= 0 \\ x &= 0, -1 \end{aligned}$$

We already know that 0 is one possible value for x, as we have verified that (0, 2) and (0, −2) are solutions of the system. But the value of −1 is new, and if we substitute $x = -1$ into the equations, we find that $\left(-1, \sqrt{3}\right)$ and $\left(-1, -\sqrt{3}\right)$ are also solutions. So the system has a total of four solutions, appearing as shown in the following figure.

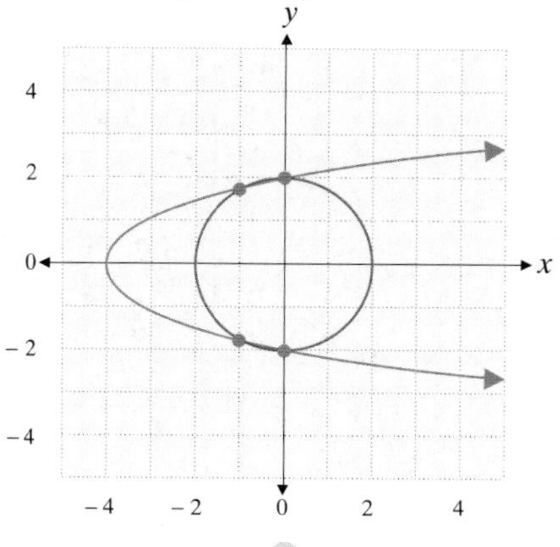

The next example of a nonlinear system is readily solved by the method of elimination.

example 4

Solve the following nonlinear system.

$$\begin{cases} y+1=(x-2)^2 \\ 3-y=(x-1)^2 \end{cases}$$

Solution:

Before solving the system algebraically, you may want to graph these two vertically-oriented parabolas to get some idea of what to expect.

We can eliminate the variable y by adding the two equations, leaving the quadratic equation

$$4=(x-2)^2+(x-1)^2.$$

We know very well how to solve such single-variable equations:

$$\begin{aligned} 4&=(x-2)^2+(x-1)^2 \\ 4&=x^2-4x+4+x^2-2x+1 \\ 0&=2x^2-6x+1 \\ x&=\frac{6\pm\sqrt{36-8}}{4} \\ x&=\frac{3\pm\sqrt{7}}{2} \end{aligned}$$

cont'd. on next page ...

For each of the two values of x, we need to determine the corresponding values of y, and we can use either equation to do so. In the work below, we have used the first equation.

For $x = \frac{3+\sqrt{7}}{2}$

$$y = \left(\frac{3+\sqrt{7}}{2} - 2\right)^2 - 1$$
$$y = \left(\frac{-1+\sqrt{7}}{2}\right)^2 - 1$$
$$y = \frac{8-2\sqrt{7}}{4} - 1$$
$$y = 2 - \frac{\sqrt{7}}{2} - 1$$
$$y = 1 - \frac{\sqrt{7}}{2}$$

For $x = \frac{3-\sqrt{7}}{2}$

$$y = \left(\frac{3-\sqrt{7}}{2} - 2\right)^2 - 1$$
$$y = \left(\frac{-1-\sqrt{7}}{2}\right)^2 - 1$$
$$y = \frac{8+2\sqrt{7}}{4} - 1$$
$$y = 2 + \frac{\sqrt{7}}{2} - 1$$
$$y = 1 + \frac{\sqrt{7}}{2}$$

The two solutions of the system are thus

$$\left(\frac{3+\sqrt{7}}{2}, 1 - \frac{\sqrt{7}}{2}\right) \text{ and } \left(\frac{3-\sqrt{7}}{2}, 1 + \frac{\sqrt{7}}{2}\right).$$

We will conclude with an example that illustrates the last type of complexity we might encounter with nonlinear systems.

example 5

Solve the following nonlinear system.

$$\begin{cases} \frac{(x+1)^2}{9} + y^2 = 1 \\ y^2 = x + 2 \end{cases}$$

Solution:

The first equation describes an ellipse and the second describes a parabola opening to the right, and a graph of the system indicates we should expect two real solutions.

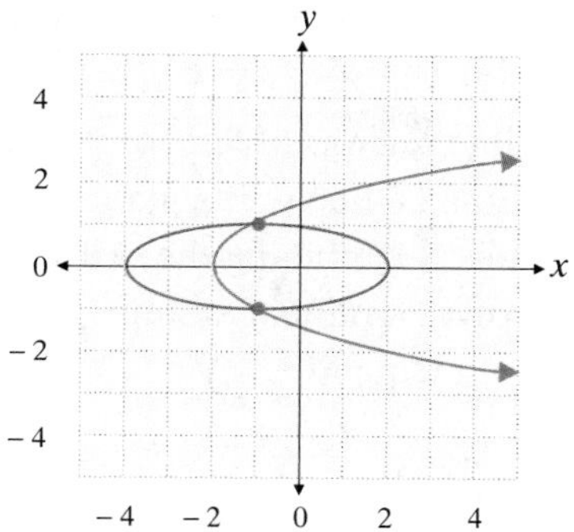

But if we solve this system algebraically, we discover another aspect to the problem. The method of substitution suggests itself, since the second equation is already solved for y^2. When we make the appropriate substitution in the first equation, we obtain

$$\frac{(x+1)^2}{9}+x+2=1.$$

We solve this second-degree equation in x the usual way:

$$\frac{(x+1)^2}{9}+x+2=1$$

$$\frac{(x+1)^2}{9}+x+1=0$$

$$(x+1)^2+9x+9=0$$

$$x^2+11x+10=0$$

$$(x+10)(x+1)=0$$

$$x=-10,\ -1$$

The solution $x=-1$ is not surprising, since our graph certainly suggested this value, but the solution $x=-10$ does not seem to relate to the picture at all. But remember, these are only the x-coordinates of the ordered pairs that solve the system, and we still need to determine the corresponding y-coordinates. We will use the second equation to do this.

For $x=-10$	For $x=-1$
$y^2=-10+2$	$y^2=-1+2$
$y^2=-8$	$y^2=1$
$y=\pm 2i\sqrt{2}$	$y=\pm 1$

Now we know that the two real solutions are $(-1,-1)$ and $(-1,1)$, as was suggested by our graph, and that the two ordered pairs $\left(-10,-2i\sqrt{2}\right)$ and $\left(-10,2i\sqrt{2}\right)$ also solve the system. These last two solutions do not appear on the graph because of their non-real second coordinates.

Topic 3: Solving Nonlinear Systems of Inequalities

The feasible regions of Section 10.7 are portions of the plane that simultaneously satisfy a collection of linear inequalities, and thus represent solutions of linear systems of inequalities. Combining the skills learned there with our newly-acquired knowledge of how to solve nonlinear systems of equations, we can now graph solutions of nonlinear systems of inequalities.

Recall that a linear inequality in two variables divides the Cartesian plane into two regions separated by a line: one region consists of all those points that satisfy the inequality, and the other region consists of those that do not. The boundary line is either a part of the solution region or not, depending on whether the inequality is non-strict or strict. In general, a nonlinear inequality in two variables also separates the plane into regions, with the difference (as we would expect) that the boundary between the regions is not linear. To illustrate, the graphs in Figure 1 below represent the solutions of, respectively, the inequalities $x^2 + y^2 < 1$ and $y \geq x^2 + 1$.

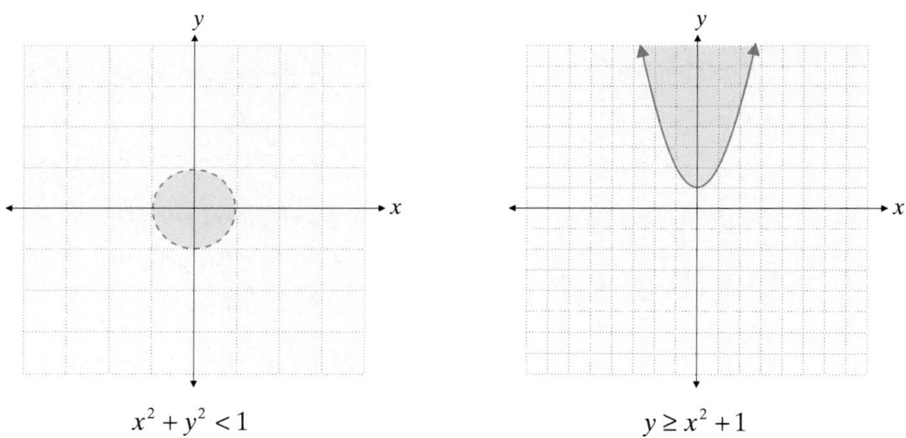

Figure 1: Solutions of Nonlinear Inequalities

As with linear inequalities, the solution regions can often be identified by considering the geometric meaning of the given nonlinear inequality (e.g. the solutions of $x^2 + y^2 < 1$ are those points lying within a radius of 1 of the origin, and the solutions of $y \geq x^2 + 1$ are those points for which the *y*-coordinate lies on or above the parabola $y = x^2 + 1$). Alternatively, a "test point" within a given region can be substituted into the inequality; if the point satisfies the inequality, that region is part of the solution set. All that remains in order to solve nonlinear systems of inequalities is to note that the solution of a system consists of the intersection of the solutions of the individual inequalities. We will illustrate this point with a variation of Example 5.

example 6

Graph the solution of the following nonlinear system of inequalities.

$$\begin{cases} \dfrac{(x+1)^2}{9} + y^2 < 1 \\ y^2 \le x+2 \end{cases}$$

Solution:

The region solving the first inequality is the interior of an ellipse (with the boundary excluded), while the region solving the second inequality consists of those points on and to the right of a rightward-opening parabola. Thus, the solution of the system consists of all those points in the plane lying in both regions; that is, the intersection of the regions. The graph of this intersection appears shaded in the figure below (note the dashed line for the portion of the boundary defined by the ellipse and the solid line for the portion defined by the parabola).

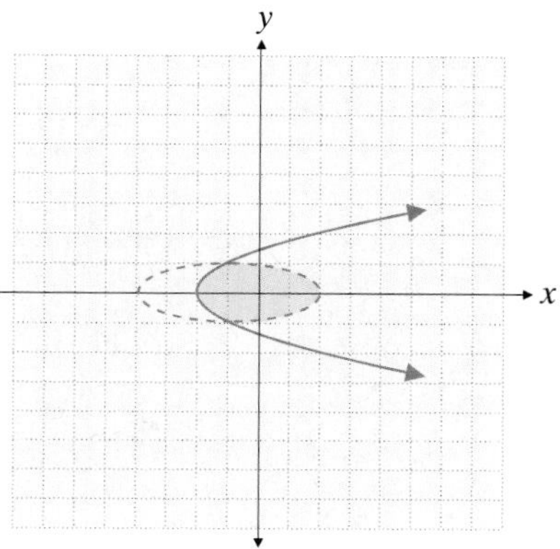

technology note

As we have seen, computer algebra systems and some calculators are capable of solving systems of equations. The command in *Mathematica* for solving nonlinear systems of equations is again **`Solve`**. The illustration below shows the syntax used in *Mathematica* for solving the nonlinear system from Example 5; note how *Mathematica* presents the solution that we found manually.

Section 10.8 Technology Note.nb

```
In[1]:= Solve[{(x+1)^2/9+y^2 == 1, y^2 == x+2}, {x, y}]
Out[1]= {{x → -10, y → -2 ⅈ √2}, {x → -10, y → 2 ⅈ √2}, {x → -1, y → -1}, {x → -1, y → 1}}
```

100%

exercises

Use graphing to guess the real solution(s) of the following systems, and then verify your answer algebraically. See Examples 1 and 2.

1. $\begin{cases} 3x-2y=6 \\ \dfrac{x^2}{4}+\dfrac{y^2}{9}=1 \end{cases}$

2. $\begin{cases} x+2y=2 \\ \dfrac{x^2}{4}+y^2=1 \end{cases}$

3. $\begin{cases} x^2+4y^2=5 \\ x^2+y^2=2 \end{cases}$

4. $\begin{cases} 4x^2+y^2=5 \\ 4(x-2)^2+y^2=5 \end{cases}$

5. $\begin{cases} y=x^2 \\ 2-y=x^2 \end{cases}$

6. $\begin{cases} x^2+(y-1)^2=4 \\ (x-3)^2+(y-1)^2=1 \end{cases}$

7. $\begin{cases} y-x^2=1 \\ y+2=4x^2 \end{cases}$

8. $\begin{cases} x^2+y^2=10 \\ x^2+y=-2 \end{cases}$

9. $\begin{cases} x=y^2-3 \\ x^2+4y^2=4 \end{cases}$

10. $\begin{cases} (x-1)^2+(y-6)^2=9 \\ (x-1)^2+(y+1)^2=16 \end{cases}$

11. $\begin{cases} (x+1)^2+(y-1)^2=4 \\ (x+1)^2+4(y-1)^2=4 \end{cases}$

12. $\begin{cases} x^2+y^2=1 \\ y=x^2-1 \end{cases}$

13. $\begin{cases} (x-2)^2+y^2=4 \\ (x+2)^2+y^2=4 \end{cases}$

14. $\begin{cases} x^2+y^2=9 \\ x^2+y^2-2x-3=1 \end{cases}$

15. $\begin{cases} x^2+y^2=9 \\ \dfrac{x^2}{9}+\dfrac{y^2}{25}=1 \end{cases}$

16. $\begin{cases} x^2+y^2=4 \\ -x^2=2y-1 \end{cases}$

17. $\begin{cases} x^2+\dfrac{y^2}{4}=1 \\ y=0 \end{cases}$

18. $\begin{cases} y=x^2+1 \\ y-1=x^3 \end{cases}$

19. $\begin{cases} 2y^2-3x^2=6 \\ 2y^2+x^2=22 \end{cases}$

20. $\begin{cases} y=x^3 \\ y=\sqrt[3]{x} \end{cases}$

21. $\begin{cases} x=y^2-4 \\ x+13=6y \end{cases}$

22. $\begin{cases} y=2x^2-3 \\ y=-x^2 \end{cases}$

23. $\begin{cases} 2y=x^2-4 \\ x^2+y^2=4 \end{cases}$

24. $\begin{cases} x^2-y^2=5 \\ \dfrac{x^2}{25}+\dfrac{4y^2}{25}=1 \end{cases}$

Solve the following nonlinear systems algebraically. Be sure to check for non-real solutions. See Examples 3, 4, and 5.

25. $\begin{cases} x^2+y^2=30 \\ x^2=y \end{cases}$

26. $\begin{cases} 3x^2+2y^2=12 \\ x^2+2y^2=4 \end{cases}$

27. $\begin{cases} x^2-1=y \\ 4x+y=-5 \end{cases}$

28. $\begin{cases} x^2+y^2=4 \\ 3x^2+4y^2=24 \end{cases}$

29. $\begin{cases} y=\dfrac{4}{x} \\ 2x^2+y^2=18 \end{cases}$

30. $\begin{cases} xy=5 \\ x^2+y^2=10 \end{cases}$

31. $\begin{cases} y-x^2=4 \\ x^2+y^2=16 \end{cases}$

32. $\begin{cases} y-x^2=6x \\ y=4x \end{cases}$

33. $\begin{cases} 2x^2+3y^2=6 \\ x^2+3y^2=3 \end{cases}$

34. $\begin{cases} x^2+y^2=4 \\ \dfrac{x^2}{4}-\dfrac{y^2}{8}=1 \end{cases}$

35. $\begin{cases} 3x^2-y=3 \\ 9x^2+y^2=27 \end{cases}$

36. $\begin{cases} \dfrac{1}{x}+\dfrac{1}{y}=5 \\ \dfrac{1}{x}-\dfrac{1}{y}=-3 \end{cases}$

37. $\begin{cases} x+y^2=2 \\ 2x^2-y^2=1 \end{cases}$

38. $\begin{cases} y-2=(x+3)^2 \\ \dfrac{1}{3}y=(x-1)^2 \end{cases}$

39. $\begin{cases} y-2=(x-2)^2 \\ y+2=(x-1)^2 \end{cases}$

40. $\begin{cases} y^2+2=2x^2 \\ y^2=x^2-6 \end{cases}$

41. $\begin{cases} (x+1)^2+y^2=10 \\ \dfrac{(x-2)^2}{4}+y^2=1 \end{cases}$

42. $\begin{cases} x^2+y^2=10 \\ x^2+y=8 \end{cases}$

43. $\begin{cases} 2x=y-1 \\ \dfrac{x^2}{25}+y^2=1 \end{cases}$

44. $\begin{cases} 2x^2+y^2=4 \\ 2(x-1)^2+y^2=3 \end{cases}$

45. $\begin{cases} x^2+7y^2=14 \\ x^2+y^2=3 \end{cases}$

46. $\begin{cases} x^2+y^2=25 \\ y^2=x-5 \end{cases}$

47. $\begin{cases} y=\dfrac{8}{x^2} \\ x^2+y^2=8 \end{cases}$

48. $\begin{cases} \dfrac{x^2}{25}+\dfrac{y^2}{16}=1 \\ x^2+y^2=16 \end{cases}$

49. $\begin{cases} xy=6 \\ (x-2)^2+(y-2)^2=1 \end{cases}$

50. $\begin{cases} y^2=x+1 \\ \dfrac{x^2}{5}+\dfrac{y^2}{6}=1 \end{cases}$

51. $\begin{cases} y=x^3-1 \\ 3y=2x-3 \end{cases}$

52. $\begin{cases} y+5=(x+1)^2 \\ y-3=(x-3)^2 \end{cases}$

53. $\begin{cases} xy-y=4 \\ (x-1)^2+y^2=10 \end{cases}$

54. $\begin{cases} 2x^2+5y^2=16 \\ 4x^2+3y^2=4 \end{cases}$

55. $\begin{cases} y=\sqrt{x-4}+1 \\ (x-3)^2+(y-1)^2=1 \end{cases}$

56. $\begin{cases} y=\sqrt[3]{x} \\ \sqrt{y}=x \end{cases}$

57. $\begin{cases} y^2 - y - 12 = x - x^2 \\ y - 1 + \dfrac{2x - 12}{y} = 0 \end{cases}$

58. $\begin{cases} y = 7x^2 + 1 \\ x^2 + y^2 = 1 \end{cases}$

59. $\begin{cases} \dfrac{(y+2)^2}{(x+y)} = 1 \\ x = y^2 + 5y + 4 \end{cases}$

60. $\begin{cases} y = x^3 + 8x^2 + 17x + 10 \\ -y = x^3 + 8x^2 + 17x + 10 \end{cases}$

61. $\begin{cases} x = \sqrt{6y+1} \\ y = \sqrt{\dfrac{x^2+7}{2}} \end{cases}$

62. $\begin{cases} \dfrac{-2}{x^2} + \dfrac{1}{y^2} = 8 \\ \dfrac{9}{x^2} - \dfrac{2}{y^2} = 4 \end{cases}$

63. $\begin{cases} x^2 + 3x - 2y^2 = 5 \\ -4x^2 + 6y^2 = 3 \end{cases}$

Draw the graph and determine whether the ordered pairs are solutions to the system of inequalities.

64. $\begin{cases} x \geq 3 \\ y > 4 \end{cases}$ **a.** $(2, 5)$ **b.** $(7, 8)$ **c.** $(5, 0)$ **d.** $(3, 4)$

65. $\begin{cases} y \leq 2x + 1 \\ y < 4 \\ y > x \end{cases}$ **a.** $(1, 2)$ **b.** $(3, 4)$ **c.** $(-1, -1)$ **d.** $(3, 3)$

66. $\begin{cases} y \geq x^2 \\ y < x^3 \\ y \leq 4x \end{cases}$ **a.** $(2, 2)$ **b.** $(2, 4)$ **c.** $(2, 8)$ **d.** $(3, 9)$

67. $\begin{cases} y \geq x^2 - 2 \\ y \leq (x-2)^2 \\ 3y > 2x + 12 \end{cases}$ **a.** $(2, 5)$ **b.** $(7, 8)$ **c.** $(5, 0)$ **d.** $(3, 4)$

68. $\begin{cases} x < 4 \\ y \geq \sqrt{x} \\ 2y > -x \end{cases}$ **a.** $(2, 5)$ **b.** $(7, 8)$ **c.** $(5, 0)$ **d.** $(3, 4)$

Graph the following systems of inequalities. See Example 6.

69. $\begin{cases} y < 2x \\ y > x^2 \end{cases}$

70. $\begin{cases} y \leq 2x + 3 \\ y \geq 0 \\ x \geq 0 \end{cases}$

71. $\begin{cases} y > x^2 \\ -3y \leq x - 9 \end{cases}$

72. $\begin{cases} y \leq x \\ 2y > -x \\ x < 4 \end{cases}$

73. $\begin{cases} y \leq \sqrt{x} \\ 2y > (x-1)^2 - 4 \end{cases}$

74. $\begin{cases} y > x^3 \\ y \leq \sqrt[3]{x} \\ y > 0 \end{cases}$

75. $\begin{cases} y \geq x^3 \\ y \geq -x^3 \\ y < 2(x+1) \end{cases}$

76. $\begin{cases} x^2 + y^2 < 9 \\ -4y \geq x - 12 \end{cases}$

77. $\begin{cases} y \leq \sin x \\ y \geq 0 \end{cases}$

Use the methods of this section to solve the following problems.

78. The area of a certain rectangle is 45 square inches, and its perimeter is 28 inches. Find the dimensions of the rectangle.

79. The product of two positive integers is 88, and their sum is 19. What are the integers?

80. Jack takes half an hour longer than his wife does to make the 210-mile drive between two cities. His wife drives 10 miles an hour faster. How fast do the two drive?

81. To construct the two garden beds shown below, 48.5 meters of fencing are needed. The combined area of the beds is 95 square meters. There are two possibilities for the overall dimensions of the two beds. What are they?

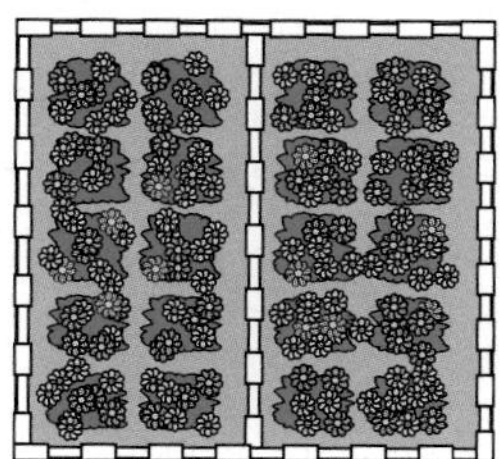

82. The product of two integers is -84, and their sum is -5. What are the integers?

83. Paul and Maria were driving the same 24 mile route, and they departed at the same time. After 20 minutes, Maria was 4 miles ahead of Paul. If it took Paul 10 minutes longer to reach their destination, how fast were they each driving?

84. The surface area of a certain right circular cylinder is $54\pi\text{ cm}^2$ and the volume is $54\pi\text{ cm}^3$. Find the height h and the radius r of this cylinder.

(**Hint:** The formulas for the volume and surface area of a right circular cylinder are as follows: $V = \pi r^2 h$ and $A = 2\pi rh + 2\pi r^2$.)

chapter ten project

Market Share Matrix

Assume you are the sales and marketing director for Joe's Java, a coffee shop located on a crowded city street corner. There are two competing coffee shops on this block – Buck's Café and Tweak's Coffee. The management has asked you to develop a marketing campaign to increase your market share from 25% to at least 35% within 6 months. With the resulting plan to meet this goal, you predict that each month:

a. You will retain 93% of your customers, 4% will go to Buck's Café, and 3% will go to Tweak's Coffee.

b. Buck's Café will retain 91% of their customers, 6% will come to Joe's Java, and 3% will go to Tweak's Coffee.

c. Tweak's Coffee will retain 92% of their customers, 3% will come to Joe's Java and, 5% will go to Buck's Café.

The current percentage of the market is shown in this matrix:

$$x_0 = \begin{bmatrix} 0.25 \\ 0.45 \\ 0.30 \end{bmatrix} \begin{matrix} \textit{Joe's} \\ \textit{Buck's} \\ \textit{Tweak's} \end{matrix}$$

After one month the shares of the coffee shops are:

$$x_1 = Px_0 = \begin{bmatrix} 0.93 & 0.06 & 0.03 \\ 0.04 & 0.91 & 0.05 \\ 0.03 & 0.03 & 0.92 \end{bmatrix} \begin{bmatrix} 0.25 \\ 0.45 \\ 0.30 \end{bmatrix} = \begin{bmatrix} 0.2685 \\ 0.4345 \\ 0.2970 \end{bmatrix}$$

1. Construct a table that lists the market share for all of the coffee shops at the end of each of the first 6 months.

2. Will your campaign be successful based on this model (will you reach 35% market share in 6 months)?

3. What actions do you think Buck's Café and Tweak's Coffee will take as your market share changes?

4. What effect could their actions make on the market?

CHAPTER REVIEW

10.1 Solving Systems by Substitution and Elimination

topics	pages	test exercises
Definition and classification of linear systems of equations • The geometric and algebraic meaning of *linear* (or *simultaneous*) *systems of equations* • The varieties of linear systems of equations, and the meaning of the terms *inconsistent, consistent,* and *dependent*	p. 775 – 776	1 – 3
Solving systems by substitution • The use of the *substitution method* to solve linear systems of equations • Alternatives for describing solutions that consist of more than one point	p. 777 – 779	1 – 3
Solving systems by elimination • The use of the *elimination method* to solve linear systems of equations • The use of elementary row operations to keep track of the steps used in solving a system	p. 779 – 781	1 – 3
Larger systems of equations • The geometric and algebraic meaning of larger linear systems of equations • How the solution process of larger systems relates to that of smaller systems	p. 781 – 784	
Applications of systems of equations • Using systems of equations to solve previously-encountered application problems	p. 784 – 786	

10.2 Matrix Notation and Gaussian Elimination

topics	pages	test exercises
Linear systems, matrices, and augmented matrices • The definition of a *matrix*, the meaning of *rows* and *columns*, and the definition of *matrix order* • Notation commonly used to define and refer to matrices and their elements • The meaning of an *augmented matrix*	p. 791 – 793	4 – 5
Gaussian elimination and row echelon form • The meaning of *Gaussian elimination* as a solution strategy • The definition of *row echelon form*, and how it relates to solving systems of equations • The definition and use of the three *elementary row operations*	p. 794 – 796	6 – 7
Gauss-Jordan elimination and reduced row echelon form • The meaning of *Gauss-Jordan elimination* as a solution strategy • The definition of *reduced row echelon form*	p. 797 – 798	

10.3 Determinants and Cramer's Rule

topics	pages	test exercises
Determinants and their evaluation • The definition of the *determinant* of a matrix • The definition of a matrix element's *minor* and *cofactor*, and the use of minors and cofactors in evaluating determinants • The use of properties of determinants and elementary row and column operations to simplify the computation of determinants	p. 804 – 810	8 – 11
Using Cramer's rule to solve linear systems • The formulas that go by the name Cramer's rule, and how these formulas can be used to solve linear systems of equations	p. 810 – 815	

10.4 The Algebra of Matrices

topics	pages	test exercises
Matrix addition • The definition of *matrix equality* • The meaning of *matrix addition*, and how to perform it	p. 818 – 820	12 – 13
Scalar multiplication • The meaning of *scalar multiplication*, and how to perform it	p. 821 – 822	12 – 13
Matrix multiplication • The meaning of *matrix multiplication*, and how it relates to composition of functions	p. 822 – 827	14 – 15
Interlude: transition matrices • The definition of a *transition matrix* • How to determine the steady state of a system in transition	p. 827 – 829	

10.5 Inverses of Matrices

topics	pages	test exercises
The matrix form of a linear system • How to write a linear system of equations as a single matrix equation	p. 832 – 833	16 – 17
Finding the inverse of a matrix • The meaning of the *identity matrix*, and the effect of multiplying a given square matrix by the appropriate identity • The meaning of the *inverse* of a matrix, and why the inverse of a matrix is useful in solving a given system • The algorithm used to find the inverse of a matrix	p. 834 – 839	18 – 19
Using matrix inverses to solve linear systems • The different interpretations that can be put on solving a system by the inverse matrix method • Why the inverse matrix method is particularly efficient in some cases	p. 839 – 843	

10.6

Partial Fraction Decomposition

topics	pages	test exercises
The pattern of decompositions • The definition of *partial fraction decomposition* • How to write a rational function in the form of a partial fraction decomposition	p. 846 – 848	20 – 23
Completing the decomposition process • Finding the partial fraction decomposition of a rational function	p. 849 – 852	24 – 29

10.7

Linear Programming

topics	pages	test exercises
Planar feasible regions • The definition of *feasible region* or *region of constraints* • How to construct graphs of the feasible region	p. 856 – 857	30 – 31
Linear programming in two variables • How to use the linear programming method	p. 858 – 861	32 – 39

10.8

Nonlinear Systems of Equations and Inequalities

topics	pages	test exercises
Approximating solutions by graphing • Using graphs of systems of nonlinear equations to gain an approximate understanding of the real-number solutions to expect	p. 867 – 869	
Solving nonlinear systems algebraically • Using algebra to determine all the solutions of nonlinear systems of equations	p. 869 – 873	40 – 44
Solving nonlinear systems of inequalities • Using algebra to determine the graphs and solutions of nonlinear systems of inequalities	p. 874 – 875	45 – 48

chapter test

Section 10.1

Use any convenient method to solve the following systems of equations. If a system is dependent, express the solution set in terms of one or more of the variables, as appropriate.

1. $\begin{cases} 3x-y+z=2 \\ -x+y-2z=-4 \\ -6x+2y-2z=-7 \end{cases}$

2. $\begin{cases} 2x-y=13 \\ 5x-2y-z=25 \\ 7x-6z=-2 \end{cases}$

3. $\begin{cases} x+y-z=1 \\ 3x-4y-5z=-1 \\ 6x-3y+z=20 \end{cases}$

Section 10.2

Construct the augmented matrix that corresponds to each of the following systems of equations.

4. $\begin{cases} 2x+(y-z)=3 \\ 2(y-x)+y-2=z \\ 3x-\dfrac{3-z}{2}=4y \end{cases}$

5. $\begin{cases} z-4x=5y \\ 14z+7(x+3y)=21 \\ 8x-y=-2(x-3z) \end{cases}$

Fill in the blanks by performing the indicated elementary row operations.

6. $\left[\begin{array}{cc|c} 3 & 1 & -2 \\ 1 & 2 & 3 \end{array}\right] \xrightarrow{-3R_2+R_1} \underline{?}$

7. $\left[\begin{array}{cc|c} 2 & 3 & 5 \\ -4 & -1 & 2 \end{array}\right] \xrightarrow[2R_1+R_2]{} \underline{?}$

Section 10.3

Evaluate the following determinants.

8. $\begin{vmatrix} x^3 & -x^2 \\ x^2 & x \end{vmatrix}$

9. $\begin{vmatrix} -1 & 3 & 1 \\ 1 & -4 & 0 \\ 0 & 2 & 3 \end{vmatrix}$

10. $\begin{vmatrix} -2 & -1 & -3 & 0 \\ 3 & 3 & 1 & 5 \\ 4 & 0 & 0 & 1 \\ 2 & 0 & 1 & 0 \end{vmatrix}$

11. $\begin{vmatrix} x^4 & x & x & 2x \\ 0 & x & x^3 & x \\ 0 & 0 & x & x \\ 0 & 0 & 0 & x^2 \end{vmatrix}$

Section 10.4

Determine values of the variables that will make the following equations true, if possible.

12. $\begin{bmatrix} w & 5x \\ 2y & z \end{bmatrix} - 3\begin{bmatrix} w & x \\ 2 & -z \end{bmatrix} = \begin{bmatrix} 4 & 2 \\ y-3 & -16 \end{bmatrix}$

13. $\begin{bmatrix} 4x & 2y^2 & z \end{bmatrix} = \begin{bmatrix} 12 \\ 18 \\ -2 \end{bmatrix}$

Evaluate the following matrix products, if possible.

14. $\begin{bmatrix} 7 & 1 & -1 \end{bmatrix}\begin{bmatrix} 1 & 6 \\ 2 & 1 \\ -3 & -3 \end{bmatrix}$

15. $\begin{bmatrix} 4 \\ 5 \\ 6 \end{bmatrix}\begin{bmatrix} -3 & 2 & 3 \end{bmatrix}$

Section 10.5

Write each of the following systems of equations as a single matrix equation.

16. $\begin{cases} \dfrac{x-8y}{3} = -1 \\ -2y-3 = 4x \end{cases}$

17. $\begin{cases} x_1 - x_2 + 2x_3 = -4 \\ 2x_1 - 3x_2 - x_3 = 1 \\ -3x_1 + 6x_3 = 5 \end{cases}$

Find the inverse of each of the following matrices, if possible.

18. $\begin{bmatrix} 4 & -2 \\ 2 & 3 \end{bmatrix}$

19. $\begin{bmatrix} 2 & 2 \\ \frac{1}{2} & 1 \end{bmatrix}$

Section 10.6

Write the form of the partial fraction decomposition of each of the following rational functions. In each case, assume the degree of the numerator is less than the degree of the denominator.

20. $f(x) = \dfrac{p(x)}{x^2 + 4x + 3}$

21. $f(x) = \dfrac{p(x)}{9x^4 - 6x^3 + x^2}$

22. $f(x) = \dfrac{p(x)}{x^3 - 3x^2 + 4x - 12}$

23. $f(x) = \dfrac{p(x)}{x^2 + 3x - 4}$

Find the partial fraction decomposition of each of the following rational functions.

24. $f(x)=\frac{2x}{(x-1)(x+1)}$

25. $f(x)=\frac{x-4}{(2x-5)^2}$

26. $f(x)=\frac{x^2+5}{(x+1)(x^2-2x+3)}$

27. $f(x)=\frac{x+1}{x^3+x}$

28. $f(x)=\frac{2x^2+x+8}{(x^2+4)^2}$

29. $f(x)=\frac{x^2+12x+12}{x^3-4x}$

Section 10.7

Construct the constraints and graph the feasible regions for the following situations.

30. Each bag of nuts contains peanuts and cashews. The total number of nuts in the bag can not exceed 60. There must be at least 20 peanuts and 10 cashews per bag. There can be no more than 40 peanuts or 40 cashews per bag. What is the region of constraint for the number of nuts per bag?

31. You wish to study at least 15 hours (over a 4 day span) for your upcoming Statistics and Biology tests. You need to study a minimum of 6 hours for each test. The maximum you wish to study for Statistics is 10 hours and for Biology is 8 hours. What is the region of constraint for the number of hours you should study for each test?

Find the minimum and maximum values of the given functions, subject to the given constraints.

32. Objective Function
$f(x, y)=6x+10y$
Constraints
$x\geq 0,\ y\geq 0,\ 2x+5y\leq 10$

33. Objective Function
$f(x, y)=2x+y$
Constraints
$x\geq 0,\ x+4y\leq 16,\ 3x-y\leq 9,\ 2x+3y\geq 6$

34. Objective Function
$f(x, y)=5x+2y$
Constraints
$x\geq 0,\ y\geq 0,\ x+y\leq 10,$
$x+2y\geq 10,\ 2x+y\geq 10$

35. Objective Function
$f(x, y)=5x+4y$
Constraints
$x\geq 0,\ y\geq 0,\ 2x+3y\leq 12,$
$3x+y\leq 12,\ x+y\geq 2$

36. Objective Function
$f(x, y) = 70x + 82y$
Constraints
$x \ge 0,\ y \ge 0,\ x \le 10,\ y \le 20,$
$x + y \ge 5,\ x + 2y \le 18$

37. Objective Function
$f(x, y) = 2x + 4y$
Constraints
$x \ge 0,\ y \ge 0,\ 2x + 2y \le 21,$
$x + 4y \le 20,\ x + y \le 18$

Use Linear Programming to answer the following questions.

38. The Krueger's Pottery manufactures two kinds of hand-painted pottery: a vase and a pitcher. There are three processes to create the pottery: throwing (forming the pottery on the potter's wheel), baking, and painting. No more than 90 hours are available per day for throwing, only 120 hours are available per day for baking, and no more than 60 hours per day are available for painting. The vase requires 3 hours for throwing, 6 hours for baking, and 2 hours for painting. The pitcher requires 3 hours for throwing, 4 hours for baking, and 3 hours for painting. The profit for each vase is \$25 and the profit for each pitcher is \$30. How many of each piece of pottery should be produced a day to maximize profit? What would the maximum profit be if Krueger's produced this amount?

39. Pranas produces bionic arms and legs for those that are missing a limb. Pranas can produce at least 20, but no more than 60 arms in a week due to the lab limitations. They can produce at least 15, but no more than 40 legs in a week. To keep their research grant, the company must produce at least 50 limbs. It cost \$450 to produce the bionic arm and \$550 to produce the bionic leg. How many of each should be produced per week to minimize the cost? What would the minimum cost be if Pranas produced this amount?

Section 10.8

Solve the following nonlinear systems algebraically. Be sure to check for non-real solutions.

40. $\begin{cases} 3x^2 + 1 = y \\ -3x + y = 7 \end{cases}$

41. $\begin{cases} x^2 + y^2 = 8 \\ 2x^2 + y^2 = 12 \end{cases}$

42. $\begin{cases} x^2 + 4y^2 = 4 \\ x^2 = y + 4 \end{cases}$

Solve the following problems involving nonlinear systems of equations.

43. The product of two positive integers is 144, and their sum is 25. What are the integers?

44. Kimberly and Scott were driving the same 72 mile route, and they departed at the same time. After 30 minutes, Kimberly was 6 miles ahead of Scott. If it took Scott one more hour than Kimberly to reach their destination, how fast were they each driving?

Draw the graph and determine whether the ordered pairs are solutions to the system of inequalities.

45. $\begin{cases} y^2 \le 9 - x^2 \\ y < |x| \\ y > -|x| \end{cases}$ **a.** $(2,5)$ **b.** $(7,8)$ **c.** $(5,0)$ **d.** $(3,4)$

Graph the following systems of inequalities.

46. $\begin{cases} y \le \sin x \\ y > -\sin x \end{cases}$

47. $\begin{cases} y \le \sqrt{x+1} \\ y > x^2 - 1 \end{cases}$

48. $\begin{cases} x^2 y \le 1 \\ 2y \le x^2 + 2 \\ y < 16x^2 \end{cases}$

chapter ELEVEN

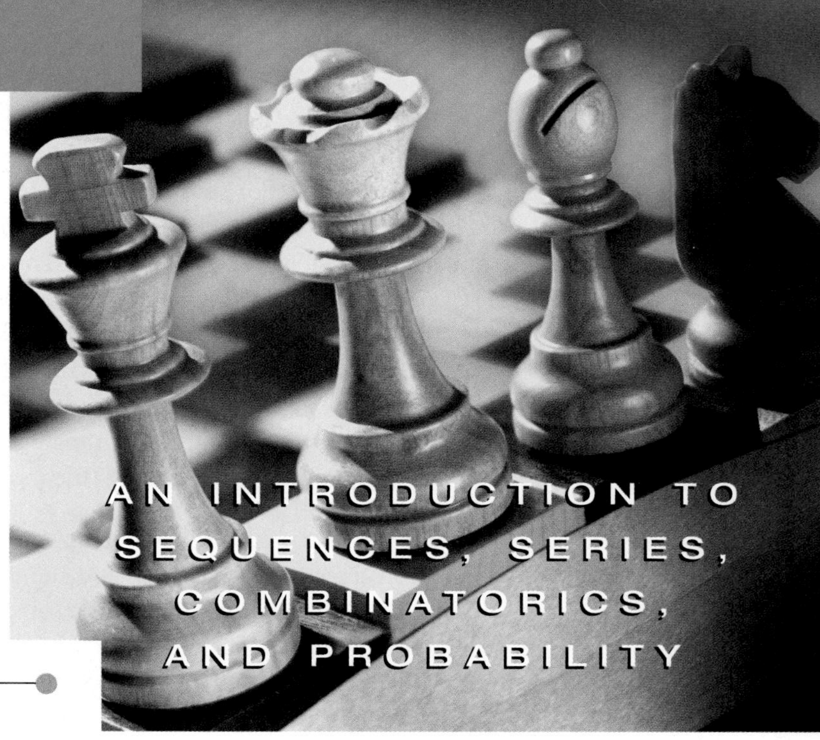

AN INTRODUCTION TO SEQUENCES, SERIES, COMBINATORICS, AND PROBABILITY

OVERVIEW

By the end of this chapter you should be able to articulate both individual terms and general formulas for sequences and series, in addition to finding sums for finite and infinite sequences and series. On page 934 you will find a problem in which the inventor of the game of chess receives a reward for his creation. As his reward, he requests a chess board filled with grains of wheat, with each square on the board having twice the number of grains as the previous square. You will be asked to use tools such as the formula for the *Partial Sum of a Geometric Sequence* on page 924 to find the total number of grains he collects.

Introduction

This last chapter introduces a variety of mathematical topics which, depending on your major, you are likely to encounter and study in greater detail in later courses.

Nearly everyone will, at some point, find it convenient to know the basic facts about sequences and series presented here. (This is not to say that everyone *will* use this material, but those that don't will be forced to go find someone who does when faced with a sequence or series problem.) For instance, sequences and series arise in the financial world in the guise of retirement planning, portfolio analysis, and loan repayment; in sociology in the construction of population models; in the study of computer algorithms; and in medical models of epidemic spread. *Mathematical Induction* is a technique useful for both discovering and proving certain mathematical statements. The counting techniques of *combinatorics*, a branch of mathematics, are immensely useful in a large number of surprisingly diverse situations; the fields of statistics and computer science, to name just two, make heavy use of the methods studied here. Finally, the science of *probability* finds applications ranging from the deadly serious (actuarial mortality tables) to the purely entertaining (the modern "gaming" industry).

Probability, in particular, has a long, convoluted, and fascinating history. People have pondered questions of chance, and the possibility of analyzing or predicting fate, for all of recorded history (and longer, of course). Mathematicians, as a subset of the general population, have been no exception to this rule, but their understanding of probability has had no particularly mathematical aspect until relatively recently. While Italian mathematicians of the Renaissance made some small progress in the area, mathematical probability is usually said to have been born in France in the year 1654. It was in that year that a member of the French nobility, Antoine Gombaud, Chevalier de Mere, posed several questions born from his gambling experience to the mathematician Blaise Pascal (1623 - 1662). Pascal, intrigued by the questions, consulted with Pierre de Fermat (1601 - 1665), another mathematician famous today largely for his so-called "Last Theorem." In a series of letters back and forth, Pascal and Fermat laid the foundation of modern mathematical probability, a field with profound applications within and without the world of gambling.

Pascal

The material in this chapter can be viewed as a bridge between algebra and further study in mathematics. All of the topics introduced in this chapter rely on the mathematical maturity you have gained by studying algebra, but of course there are many more areas of math that are not explored at all in this book. Calculus, in particular, is one such area; many of you will soon be taking at least one semester of calculus, a field of mathematics that extends the static world of algebra to the dynamic world in which we live.

11.1 Sequences and Series

TOPICS

1. Recursively and explicitly defined sequences
2. Summation notation and a few formulas
3. Partial sums and series
4. Interlude: Fibonacci sequences

Topic 1: Recursively and Explicitly Defined Sequences

In attempting to explain – or at least model – many phenomena, it is often useful and very natural to construct a list of associated numbers. For instance, the growth of an isolated population of rabbits over the course of many months might be described with a simple list of the number of rabbits born each month. Or the number of steps necessary for a computer algorithm to sort a certain number of elements might be described with a list of numbers, where the nth number in the list is the number of steps needed to sort n elements.

Such lists are called *sequences*, and they arise often enough that specialized notation is warranted.

Sequences

An **infinite sequence** (or just **sequence**) is a function whose domain is the set of natural numbers $\{1, 2, 3, \ldots\}$. Instead of giving the function a name such as f and referring to the values of the sequence as $f(1), f(2), f(3), \ldots$, we commonly use subscripts and refer to the elements of the sequence as $a_1, a_2, a_3, \ldots$. In other words, a_n stands for the nth term in the sequence. For example, given a sequence

$$1, 2, 4, 8, 16, 24, \ldots$$

we would write

$$a_1 = 1, a_2 = 2, a_3 = 4, a_4 = 8, \text{ and so on.}$$

Finite sequences are defined similarly, but the domain of a finite sequence is a set of the form $\{1, 2, 3, \ldots, k\}$ for some natural number k.

Note that a consequence of sequence notation is that we usually don't have a label, such as f, to attach to a sequence; instead, a given sequence is normally identified by some formula for determining a_n, the nth term of the sequence, and the entire sequence may be referred to as $\{a_n\}$ or $\{a_n\}_{n=1}^{\infty}$. For example, $a_n = 3(n-1)$ could be used to describe the sequence $0, 3, 6, 9, 12, \ldots$ If it is necessary to discuss several sequences at once, we use different letters to denote the different terms. For instance, a discussion may refer to two sequences $\{a_n\}$ and $\{b_n\}$.

example 1

Determine the first five terms of the sequences whose n^{th} terms are as follows:

a. $a_n = 5n - 2$ **b.** $b_n = \dfrac{(-1)^n + 1}{2}$ **c.** $c_n = \dfrac{n}{n+1}$

Solutions:

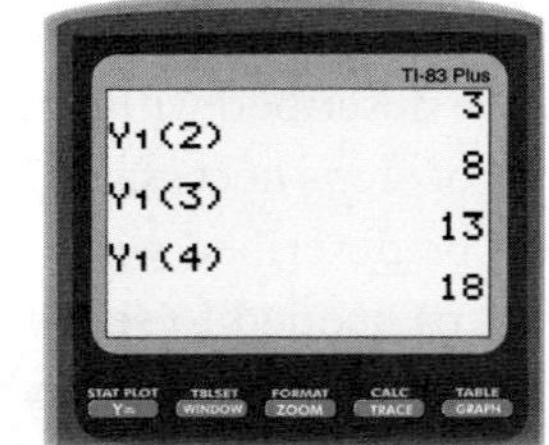

Enter the sequence in Y= as `Y1`. Then select **Vars** and use the right arrow to highlight `Y-Vars`. Press **Enter** and select `Y1`; then use parentheses and enter the desired value for *n*.

a. Replacing n in the formula for the general term with the first five positive integers, we obtain

$$a_1 = 5(1)-2,\quad a_2 = 5(2)-2,\quad a_3 = 5(3)-2,\quad a_4 = 5(4)-2,\quad a_5 = 5(5)-2$$

so the sequence starts out as $3, 8, 13, 18, 23, \ldots$

b. Again, we replace n with the first five positive integers to determine the first five terms of the sequence. Note that $(-1)^n$ is equal to -1 if n is odd and is equal to 1 if n is even. When we add 1 to $(-1)^n$ and divide the result by 2, we obtain either 0 or 1, so the sequence begins $0, 1, 0, 1, 0, \ldots$ This example illustrates the fact that a given value may appear more than once in a sequence.

c. Replacing n in the formula for the general term with the first five positive integers, we obtain

$$c_1 = \frac{1}{1+1},\quad c_2 = \frac{2}{2+1},\quad c_3 = \frac{3}{3+1},\quad c_4 = \frac{4}{4+1},\quad c_5 = \frac{5}{5+1},$$

so the sequence starts out as $\frac{1}{2}, \frac{2}{3}, \frac{3}{4}, \frac{4}{5}, \frac{5}{6}, \ldots$

The three formulas for the general n^{th} term in Example 1 were all examples of **explicit** formulas, so-named because they provide a very direct and easily applied rule for calculating any term in the sequence. In many cases, though, an explicit formula for the general term cannot be found, or at least is not as easily determined as some other sort of formula. In such cases, the terms of a sequence are often defined *recursively*. A **recursive** formula is one that refers to one or more of the terms preceding a_n in the definition for a_n. For instance, if the first term of a sequence is –5, and if it is known that each of the remaining terms of the sequence is 7 more than the term preceding it, the sequence can be defined by the rules $a_1 = -5$ and $a_n = a_{n-1} + 7$ for $n \geq 2$.

example 2

Determine the first five terms of the following recursively defined sequences.

a. $a_1 = 3$ and $a_n = a_{n-1} + 5$ for $n \geq 2$ **b.** $a_1 = 2$ and $a_n = 3a_{n-1} + 1$ for $n \geq 2$

Solutions:

a. We find the first five terms by replacing n with the first five positive integers, just as in Example 1. Note, though, that in using the recursive definition given, we must determine the elements of the sequence in order; that is, to determine a_5, we need to first know a_4. And to determine a_4, we need to first know a_3, and so on back to a_1.

$$a_1 = 3$$
$$a_2 = a_1 + 5 = 3 + 5 = 8$$
$$a_3 = a_2 + 5 = 8 + 5 = 13$$
$$a_4 = a_3 + 5 = 13 + 5 = 18$$
$$a_5 = a_4 + 5 = 18 + 5 = 23$$

Thus, the sequence starts out as 3, 8, 13, 18, 23, ... Note that the sequence defined by this recursive definition appears to be the same as the first sequence in Example 1, defined explicitly by $a_n = 5n - 2$. This is no illusion, and illustrates the fact that a given sequence can often be defined several different ways.

cont'd. on next page ...

b. Using the recursive formula $a_1 = 2$ and $a_n = 3a_{n-1} + 1$ for $n \geq 2$, we obtain

$$a_1 = 2$$
$$a_2 = 3a_1 + 1 = 3(2) + 1 = 7$$
$$a_3 = 3a_2 + 1 = 3(7) + 1 = 22$$
$$a_4 = 3a_3 + 1 = 3(22) + 1 = 67$$
$$a_5 = 3a_4 + 1 = 3(67) + 1 = 202$$

Thus, the sequence starts out as 2, 7, 22, 67, 202, ...

technology note

Computer algebra systems such as *Mathematica* allow sequences to be defined both explicitly and recursively. The illustration below shows how the sequence seen in Examples 1a and 2a can be defined either way; note that `a[5]` has the same value in both cases. *See Appendix A for more guidance on defining functions (explicitly or recursively) in Mathematica.*

Y_1 was defined as $5n - 2$ for this screen.

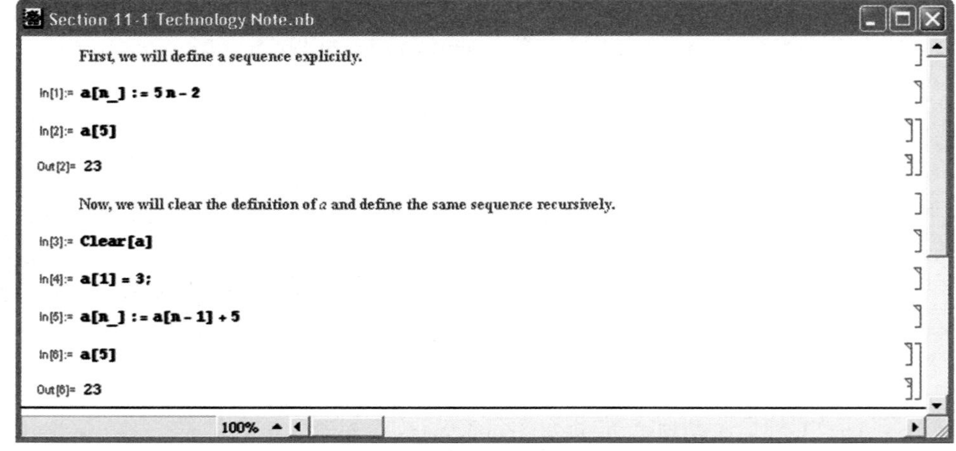

A comparison of Examples 1 and 2 demonstrates that, in general, an explicit formula is far more useful when finding the terms of a sequence than a recursive formula. To convince yourself of this fact, consider the task of calculating a_{100} based on the formula $a_n = 5n - 2$, versus the same task given the formula $a_1 = 3$ and $a_n = a_{n-1} + 5$ for $n \geq 2$.

To a certain extent, the problems in Examples 1 and 2 can be turned around, so that the question becomes one of finding a formula for the general n^{th} term of a sequence given the first few terms of the sequence. In fact, this is often the challenge in modeling situations, where the desire is to extrapolate the behavior of a sequence of numbers by finding a formula that produces the first few terms of the sequence. The catch is that there is always more than one formula that will produce identical terms of a sequence up to a certain point and then differ beyond that. To see why this must be so, consider the following two explicit formulas:

$$a_n = 3n \text{ and } b_n = 3n + (n-1)(n-2)(n-3)(n-4)(n-5)n^2.$$

Do you see why these two formulas will produce identical results for the first five terms, but different results from then on? For this reason, the instructions in Example 3 just ask for possible formulas for the given sequences.

example 3

Find a possible formula for the general n^{th} term of the sequences that begin as follows.

a. $-3, 9, -27, 81, -243, \ldots$

b. $1, 3, 6, 10, 15, \ldots$

Solutions:

a. As a rule, there is no particular method to use to find a formula for the terms of a sequence, other than the very general method of observation. If a formula for a given sequence does not come quickly to mind, it may help to associate each term of the sequence with its place in the sequence, as shown below:

$$\begin{array}{cccccc} 1 & 2 & 3 & 4 & 5 & \ldots \\ \updownarrow & \updownarrow & \updownarrow & \updownarrow & \updownarrow & \updownarrow \\ -3 & 9 & -27 & 81 & -243 & \ldots \end{array}$$

cont'd. on next page ...

If a pattern is still not apparent, try rewriting the terms of the sequence in different ways. For instance, factoring the terms in this case leads to:

$$\begin{array}{cccccc} 1 & 2 & 3 & 4 & 5 & \cdots \\ \updownarrow & \updownarrow & \updownarrow & \updownarrow & \updownarrow & \updownarrow \\ -3 & 3^2 & -3^3 & 3^4 & -3^5 & \cdots \end{array}$$

At this point, it is clear that the nth term of the sequence is the nth power of 3, multiplied by –1 if n is odd. One way of expressing alternating signs in a sequence is to multiply the nth term by $(-1)^n$ (if the odd terms are negative) or by $(-1)^{n+1}$ (if the even terms are negative). In this case, a possible formula for the general n^{th} term is $a_n = (-1)^n (3)^n$, or $a_n = (-3)^n$.

Note that we might also have come up with the recursive formula $a_1 = -3$ and $a_n = -3a_{n-1}$ for $n \geq 2$.

b. A good way to begin would be to again associate each term with its place in the sequence, as follows:

$$\begin{array}{cccccc} 1 & 2 & 3 & 4 & 5 & \ldots \\ \updownarrow & \updownarrow & \updownarrow & \updownarrow & \updownarrow & \updownarrow \\ 1 & 3 & 6 & 10 & 15 & \ldots \end{array}$$

In this case, factoring the terms doesn't seem to help in identifying a pattern, but thinking about the difference between successive terms does:

$$\begin{array}{cccccc} 1 & 2 & 3 & 4 & 5 & \cdots \\ \updownarrow & \updownarrow & \updownarrow & \updownarrow & \updownarrow & \updownarrow \\ 1 & 3=1+2 & 6=3+3 & 10=6+4 & 15=10+5 & \ldots \end{array}$$

This observation leads to the recursive formula $a_1 = 1$ and $a_n = a_{n-1} + n$ for $n \geq 2$. There is an explicit formula that defines the sequence as well, and we will return to this problem soon.

Topic 2: Summation Notation and a Few Formulas

One very common use of sequence notation is to define terms that are to be added together. If the first n terms of a given sequence are to be added, one way of denoting the sum is $a_1 + a_2 + \ldots + a_n$, and in some cases this notation is sufficient. But there are many occasions when the use of ellipsis (which means the "dot dot dot") is too vague. For instance, there is nothing in the expression $a_1 + a_2 + \ldots + a_n$ to indicate what, exactly, the terms in the sequence look like.

Summation notation (also known as *sigma notation*) is the response to this shortcoming. Summation notation borrows the capital Greek letter Σ ("sigma") to denote the operation of summation, as described below.

Summation Notation

The sum $a_1 + a_2 + \ldots + a_n$ is expressed in **summation notation** as $\sum_{i=1}^{n} a_i$.

When this notation is used, the letter i is called the **index of summation**, and a_i often appears as the formula for the ith term of a sequence. In the sum above, all the terms of the sequence beginning with a_1 and ending with a_n are to be added. The notation can be modified to indicate a different first term or last term of the sum.

Note: The index of summation does not have to be i. For example:

$$\sum_{i=1}^{n} a_i = \sum_{j=1}^{n} a_j = \sum_{k=1}^{n} a_k .$$

example 4

Rewrite the following sums in expanded form, and evaluate each one.

a. $\sum_{i=1}^{4} (3i-2)$

b. $\sum_{i=3}^{5} i^2$

cont'd. on next page ...

Solutions:

a. $\sum_{i=1}^{4}(3i-2) = (3\cdot 1-2)+(3\cdot 2-2)+(3\cdot 3-2)+(3\cdot 4-2)$

$= 1+4+7+10$

$= 22$

In order to expand the sum, we replace the index i with the numbers 1 through 4. Note how the formula for the ith term is used to generate the terms of the sum.

b. $\sum_{i=3}^{5} i^2 = 3^2+4^2+5^2$

$= 9+16+25$

$= 50$

This sum differs from the first sum in the formula for the terms and in the beginning and ending point of the sum.

Keep in mind that sigma notation represents nothing more than addition. Thus, the following properties of Σ are nothing more than restatements of familiar properties.

Properties of Sigma

The following properties refer to two sequences $\{a_i\}$ and $\{b_i\}$ and a constant c.

1. $\sum_{i=1}^{n}(a_i+b_i)=\sum_{i=1}^{n}a_i+\sum_{i=1}^{n}b_i$ (terms of a sum can be rearranged.)

2. $\sum_{i=1}^{n}ca_i=c\sum_{i=1}^{n}a_i$ (constants can be factored out of a sum.)

3. $\sum_{i=1}^{n}a_i=\sum_{i=1}^{k}a_i+\sum_{i=k+1}^{n}a_i$ for any $1\le k\le n-1$ (a sum can be broken apart into two smaller sums.)

In addition to the properties listed above, there are formulas for many sums that occur frequently. We will list only a few in this section, and prove only two of them.

Four Summation Formulas

1. $\sum_{i=1}^{n} 1 = n$
2. $\sum_{i=1}^{n} i = \frac{n(n+1)}{2}$
3. $\sum_{i=1}^{n} i^2 = \frac{n(n+1)(2n+1)}{6}$
4. $\sum_{i=1}^{n} i^3 = \frac{n^2(n+1)^2}{4}$

The first formula really requires no proof at all, as it is clearly true if we write the sum in expanded form:

$$\sum_{i=1}^{n} 1 = \underbrace{1+1+\ldots+1}_{n \text{ terms}} = n.$$

The second formula requires a bit more in the way of justification, and several different proofs are possible. One proof begins by letting S stand for the sum; that is,

$$S = \sum_{i=1}^{n} i = 1+2+\ldots+n.$$

Then it is also certainly true, since addition can be performed in any order, that

$$S = n+(n-1)+\ldots+1.$$

Note, now, the result of adding these two equations:

$$\begin{array}{rclllll} S & = & 1 & +2 & +\ldots & +n \\ S & = & n & +(n-1) & +\ldots & +1 \\ \hline 2S & = & (n+1) & +(n+1) & +\ldots & +(n+1) \end{array}$$

Since the term $n + 1$ appears n times on the right-hand side of the last equation, we have $2S = n(n+1)$ or, after dividing both sides by 2,

$$\sum_{i=1}^{n} i = \frac{n(n+1)}{2}.$$

Note that this provides an explicit formula for the nth term of the sequence in Example 3b. Since the nth term is the sum $1 + 2 + \ldots n$, the formula is $a_n = \dfrac{n(n+1)}{2}$.

example 5

Use the above properties and formulas to evaluate the following sums.

a. $\displaystyle\sum_{i=1}^{9}(7i-3)$ **b.** $\displaystyle\sum_{i=4}^{6}3i^2$

Solutions:

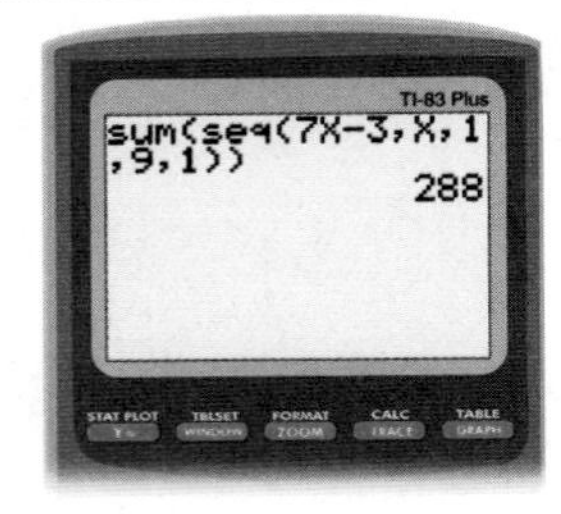

To find the sum of a list, select 2nd Stat (List), use the right arrow to highlight Math, and select 5:sum. To enter the sequence, select 2nd Stat (List), highlight Ops, and select 5:seq. In the parentheses that appear, enter the desired equation, the name of the variable (X), beginning value, ending value, and increment of 1, separated by commas and followed by an end parentheses.

a.
$$\begin{aligned}\sum_{i=1}^{9}(7i-3) &= \sum_{i=1}^{9}7i-\sum_{i=1}^{9}3\\ &= 7\sum_{i=1}^{9}i-3\sum_{i=1}^{9}1\\ &= 7\left(\frac{9\cdot 10}{2}\right)-3\cdot 9\\ &= 315-27\\ &= 288\end{aligned}$$

The first of the three properties of sigma allows us to split the given sum into two sums, and the second property allows us to factor the constants out of each sum. (Note that we have factored a 3 from the second sum in order to apply one of the formulas.)

The first two of the four summation formulas then provide the final answer.

b.
$$\begin{aligned}\sum_{i=4}^{6}3i^2 &= 3\sum_{i=4}^{6}i^2\\ &= 3\sum_{i=1}^{6}i^2-3\sum_{i=1}^{3}i^2\\ &= 3\left(\frac{6\cdot 7\cdot 13}{6}\right)-3\left(\frac{3\cdot 4\cdot 7}{6}\right)\\ &= 273-42\\ &= 231\end{aligned}$$

We can again factor out a constant, but in order to apply the formula for the sum of squares of integers, one further step is necessary. In the second line, we have used the third property of sigma to obtain two sums that begin with $i = 1$, allowing us to use the third summation formula as shown.

Topic 3: Partial Sums and Series

Given a sequence $\{a_i\}$, many applications call for an understanding of sums of the form $a_1 + a_2 + \ldots + a_n$ or for the sum of *all* the terms, which we might express as $a_1 + a_2 + a_3 + \ldots$. The following definition formalizes these two notions.

Partial Sums and Series

Given the infinite sequence $\{a_i\}$, the nth **partial sum** is $S_n = \sum_{i=1}^{n} a_i$. S_n is an example of a **finite series**. The **infinite series** associated with $\{a_i\}$ is the sum $\sum_{i=1}^{\infty} a_i$. Note that the adjective *infinite* refers to the infinite number of terms that appear in the sum.

Many students, when first encountering the definition of an infinite series, have the initial reaction that a sum of an infinite number of terms can't possibly be a finite number, especially if all the terms are positive. But in fact, many examples of infinite series with finite sums exist. Consider, for instance, the fractions of an inch as marked on a typical ruler.

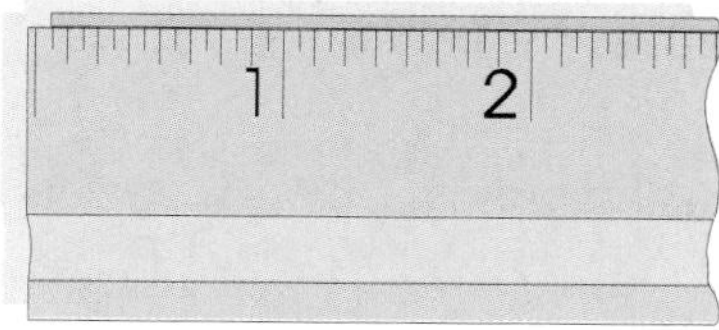

Figure 1: Visualizing an Infinite Series

If half an inch and a quarter of an inch are added, the result is three-quarters of an inch. If an additional eighth of an inch is added, the result is seven-eighths of an inch. If the ruler is of sufficient precision, a sixteenth of an inch can be added, resulting in a total of fifteen-sixteenths of an inch.

In principle, further appropriate fractions can be added, with the result that the sum of the fractions at every stage is closer to, but never exceeds, one inch. In the language of partial sums and series, what we have informally deduced is

$$S_1 = \frac{1}{2} = \frac{1}{2}$$
$$S_2 = \frac{1}{2}+\frac{1}{4} = \frac{3}{4}$$
$$S_3 = \frac{1}{2}+\frac{1}{4}+\frac{1}{8} = \frac{7}{8}$$
$$S_4 = \frac{1}{2}+\frac{1}{4}+\frac{1}{8}+\frac{1}{16} = \frac{15}{16}$$
$$\vdots$$
$$S_n = \frac{1}{2}+\frac{1}{4}+\ldots+\frac{1}{2^n} = \frac{2^n-1}{2^n}$$

and

$$S = \frac{1}{2}+\frac{1}{4}+\frac{1}{8}+\ldots = \sum_{i=1}^{\infty}\frac{1}{2^i} = 1.$$

We will explore such infinite series further when we define and study geometric series.

We say that an infinite series $\sum_{i=1}^{\infty} a_i$ **converges** if the sequence of partial sums $S_n = \sum_{i=1}^{n} a_i$ approaches some fixed real number S, and in this case we write $S = \sum_{i=1}^{\infty} a_i$. If the sequence of partial sums does not approach some fixed real number (either by getting larger and larger in magnitude or by "bouncing around"), we say the series **diverges**. The next example illustrates both possible outcomes.

example 6

Examine the partial sums associated with each infinite series to determine if the series converges or diverges.

a. $\sum_{i=1}^{\infty}\left(\frac{1}{i}-\frac{1}{i+1}\right)$ **b.** $\sum_{i=1}^{\infty}(-1)^i$

Solutions:

a. $S_1 = 1-\frac{1}{2}$

$$S_2 = \left(1-\frac{1}{2}\right)+\left(\frac{1}{2}-\frac{1}{3}\right) = 1-\frac{1}{3}$$

$$S_3 = \left(1-\frac{1}{2}\right)+\left(\frac{1}{2}-\frac{1}{3}\right)+\left(\frac{1}{3}-\frac{1}{4}\right) = 1-\frac{1}{4}$$

$$\vdots$$

$$S_n = 1-\frac{1}{n+1}$$

The partial sums of this series exhibit very convenient behavior that allows us to easily evaluate each one. When most of the terms that appear in a partial sum cancel, as shown at left, the series is called a *telescoping series*.

The partial sums approach 1 as n gets larger, so we say the series converges. In fact, $\sum_{i=1}^{\infty}\left(\frac{1}{i}-\frac{1}{i+1}\right) = 1$.

b. $S_1 = -1$

$S_2 = -1 + 1 = 0$

$S_3 = -1 + 1 - 1 = -1$

$\vdots$

$$S_n = \begin{cases} -1 \text{ if } n \text{ is odd} \\ 0 \text{ if } n \text{ is even} \end{cases}$$

The partial sums of this series oscillate between −1 and 0, and do not approach a fixed number as n gets larger. This series thus diverges.

Topic 4: Interlude: Fibonacci Sequences

One of the most famous sequences in mathematics is one that has a long and colorful history.

Leonardo Fibonacci (meaning "Leonardo, son of Bonaccio") was born in Pisa, *ca.* 1175, and was one of the first of many famous Italian mathematicians of the middle ages. As a child and young adult, he traveled with his father (a merchant) to ports in northern Africa and the Middle East where Arab scholars had collected, preserved, and expanded the mathematics of many different cultures. Among other achievements, Leonardo is known for the book he wrote after returning to Italy, the *Liber abaci*, which exposed European scholars to Hindu-Arabic notation and the mathematics he had acquired, as well as to his own contributions.

One of the problems in the *Liber abaci* gives rise to what we now call the Fibonacci sequence: $1, 1, 2, 3, 5, 8, 13, 21, \ldots$ The problem, loosely translated, asks

How many pairs of rabbits can a single pair produce, if every month each pair of rabbits can beget a new pair and if each pair becomes productive in the second month of existence?

We are to assume that in the first month we begin with a single pair of rabbits. In the second month, the rabbits are too young to reproduce, so the total number of pairs remains 1. But in the third month, the pair produces a new pair of rabbits, and so we have both the original pair and the new pair, for a total of 2 pairs. In the fourth month, we have the 2 pairs that we begin the month with, plus another pair is produced from the 1 pair of rabbits that is old enough to do so, so the total number of pairs is 3. Similarly, in the fifth month we have 3 pairs plus 2 more pairs from those old enough to be productive, for a total of 5 pairs.

If we analyze the pattern that is emerging, we are led to the recursive definition

$$a_1 = 1 \text{ and } a_2 = 1 \text{ and } a_n = a_{n-2} + a_{n-1} \text{ for } n \geq 3.$$

Generalized Fibonacci sequences are defined similarly, with the first two terms given and each successive term defined as the sum of the previous two.

exercises

Determine the first five terms of the sequences whose n^{th} terms are defined as follows. See Examples 1 and 2.

1. $a_n = 7n - 3$

2. $a_n = -3n + 5$

3. $a_n = (-2)^n$

4. $a_n = \dfrac{3n}{n+2}$

5. $a_n = \dfrac{(-1)^n}{n^2}$

6. $a_n = \dfrac{(-1)^{n+1} 2^n}{3^n}$

7. $a_n = \left(-\dfrac{1}{3}\right)^{n-1}$

8. $a_n = \dfrac{n^2}{n+1}$

9. $a_n = \dfrac{(n-1)^2}{(n+1)^2}$

10. $a_n = (-2)^n + n$

11. $a_n = (-n+4)^3 - 1$

12. $a_n = \dfrac{2n^2}{3n-2}$

13. $a_n = (-1)^n \sqrt{n}$

14. $a_n = \dfrac{2^n}{n^2}$

15. $a_n = 4n - 3$

16. $a_n = -5n + 15$

17. $a_n = 2^{n-2}$

18. $a_n = 3^{-n-2}$

19. $a_n = (3n)^n$ **20.** $a_n = \sqrt[2^n]{64}$ **21.** $a_n = \dfrac{5n}{n+3}$

22. $a_n = \dfrac{n^2}{n+2}$ **23.** $a_n = \dfrac{n^2+n}{2}$ **24.** $a_n = (-1)^n n$

25. $a_n = \dfrac{(n+1)^2}{(n-1)^2}$ **26.** $a_n = n^2 + n$ **27.** $a_n = \dfrac{2n-1}{3n}$

28. $a_n = \sqrt{3n} + 1$ **29.** $a_n = -(n-1)^2$

30. $a_n = (n-1)(n+2)(n-3)$ **31.** $a_1 = 2$ and $a_n = (a_{n-1})^2$ for $n \geq 2$

32. $a_1 = -2$ and $a_n = 7a_{n-1} + 3$ for $n \geq 2$ **33.** $a_1 = 1$ and $a_n = na_{n-1}$ for $n \geq 2$

34. $a_1 = -1$ and $a_n = -a_{n-1} - 1$ for $n \geq 2$ **35.** $a_1 = 2$ and $a_n = \sqrt{(a_{n-1})^2 + 1}$ for $n \geq 2$

36. $a_1 = 3$, $a_2 = 1$, and $a_n = (a_{n-2})^{a_{n-1}}$ for $n \geq 3$

Find a possible formula for the general n^{th} term of the sequences that begin as follows. Answers may vary. See Example 3.

37. $5, 12, 19, 26, 33, \ldots$ **38.** $-3, 9, -27, 81, -243, \ldots$ **39.** $-1, 2, -6, 24, -120, \ldots$

40. $\dfrac{1}{3}, \dfrac{2}{4}, \dfrac{3}{5}, \dfrac{4}{6}, \dfrac{5}{7}, \ldots$ **41.** $\dfrac{1}{4}, \dfrac{1}{7}, \dfrac{1}{12}, \dfrac{1}{19}, \dfrac{1}{28}, \ldots$ **42.** $1, \dfrac{1}{2}, \dfrac{1}{6}, \dfrac{1}{24}, \dfrac{1}{120}, \ldots$

43. $-34, -25, -16, -7, 2, \ldots$ **44.** $\dfrac{3}{14}, \dfrac{2}{15}, \dfrac{1}{16}, 0, -\dfrac{1}{18}, \ldots$ **45.** $\dfrac{1}{4}, \dfrac{1}{2}, 1, 2, 4, \ldots$

46. $-1, -6, -11, -16, -21, \ldots$ **47.** $\dfrac{1}{2}, \dfrac{1}{2}, \dfrac{3}{8}, \dfrac{1}{4}, \dfrac{5}{32}, \ldots$ **48.** $1, 4, 15, 64, 325, \ldots$

49. $3, 5, 7, 9, 11, \ldots$ **50.** $2, 5, 10, 17, 26, \ldots$ **51.** $1, \dfrac{1}{4}, \dfrac{1}{9}, \dfrac{1}{16}, \dfrac{1}{25}, \ldots$

52. $-2, 4, -8, 16, -32, \ldots$ **53.** $\dfrac{1}{9}, \dfrac{1}{3}, 1, 3, 9, \ldots$ **54.** $-1, 4, 3, \dfrac{16}{5}, \dfrac{25}{7}, \ldots$

Translate each expanded sum that follows into summation notation, and vice versa. Then use the formulas and properties from this section to evaluate the sums. See Examples 4 and 5.

55. $\sum_{i=1}^{7}(3i-5)$ **56.** $\sum_{i=1}^{5}-3i^2$ **57.** $1+8+27+\ldots+216$

58. $1+4+7+\ldots+22$ **59.** $\sum_{i=3}^{10}5i^2$ **60.** $9+16+25+\ldots+81$

61. $\sum_{i=1}^{6} -3(2)^i$

62. $\sum_{i=6}^{13} (i+3)(i-10)$

63. $9+27+81+\ldots+19683$

64. $\sum_{i=1}^{5} 2i-5$

65. $\sum_{i=1}^{5} (-3)^i$

66. $8+11+14+\ldots+80$

67. $\sum_{i=3}^{8} i^2-3$

68. $\frac{1}{2}+\frac{7}{2}+\frac{17}{2}+\ldots+\frac{199}{2}$

69. $\sum_{i=100}^{103} \frac{2i}{25}$

70. $1-2+4-8+16$

71. $3+\frac{33}{10}+\frac{36}{10}+\ldots+30$

72. $2+4.5+8+\ldots+200$

Find a formula for the nth partial sum S_n of each of the following series. If the series is finite, determine the sum. If the series is infinite, determine if it converges or diverges, and if it converges, determine the sum. See Example 6.

73. $\sum_{i=1}^{100} \left(\frac{1}{i+3}-\frac{1}{i+4}\right)$

74. $\sum_{i=1}^{\infty} \left(\frac{1}{i+3}-\frac{1}{i+4}\right)$

75. $\sum_{i=1}^{\infty} \left(2^i-2^{i-1}\right)$

76. $\sum_{i=1}^{15} \left(2^i-2^{i-1}\right)$

77. $\sum_{i=1}^{49} \left(\frac{1}{2i}-\frac{1}{2i+2}\right)$

78. $\sum_{i=1}^{\infty} \left(\frac{1}{2i}-\frac{1}{2i+2}\right)$

79. $\sum_{i=1}^{100} \ln\left(\frac{i}{i+1}\right)$ (**Hint:** make use of a property of logarithms to rewrite the sum.)

80. $\sum_{i=1}^{\infty} \ln\left(\frac{i}{i+1}\right)$ (**Hint:** make use of a property of logarithms to rewrite the sum.)

Determine the first five terms of the following generalized Fibonacci sequences.

81. $a_1=4,\ a_2=7$, and $a_n=a_{n-2}+a_{n-1}$ for $n\geq 3$

82. $a_1=-9,\ a_2=1$, and $a_n=a_{n-2}+a_{n-1}$ for $n\geq 3$

83. $a_1=10,\ a_2=20$, and $a_n=a_{n-2}+a_{n-1}$ for $n\geq 3$

84. $a_1=-17,\ a_2=13$, and $a_n=a_{n-2}+a_{n-1}$ for $n\geq 3$

85. $a_1=13,\ a_2=-17$, and $a_n=a_{n-2}+a_{n-1}$ for $n\geq 3$

Determine the first five terms of the following recursively defined sequences.

86. $a_1=2,\ a_2=-3$, and $a_n=3a_{n-1}+a_{n-2}$ for $n\geq 3$

87. $a_1=1,\ a_2=-3$, and $a_n=a_{n-1}a_{n-2}$ for $n\geq 3$

11.2 Arithmetic Sequences and Series

TOPICS

1. Characteristics of arithmetic sequences and series
2. The formula for the general term of an arithmetic sequence
3. Evaluating partial sums of arithmetic sequences

Topic 1: Characteristics of Arithmetic Sequences and Series

An arithmetic sequence $\{a_n\}$ is a sequence with the property that the difference between any two consecutive terms is the same fixed constant. Arithmetic sequences arise in many different situations. Suppose, for example, that the parents of a ten-year-old child decide to increase her \$1.00/week allowance by \$0.50/week with the start of each new year. The sequence describing her weekly allowance, beginning with age ten, is then

$$1.00,\ 1.50,\ 2.00,\ 2.50,\ 3.00,\ 3.50, \ldots$$

By the informal definition given above, this is an arithmetic sequence with a difference of 0.50 between consecutive terms. And as a mechanism for increasing a child's allowance, it is not a bad one. As we will see in the next section, though, adult workers would probably not benefit from having yearly raises determined in this manner.

Arithmetic Sequences

A sequence $\{a_n\}$ is an **arithmetic sequence** (sometimes also called an **arithmetic progression**) if there is some constant d such that $a_{n+1} - a_n = d$ for each $n = 1, 2, 3, \ldots$ The constant d is called the **common difference** of the sequence.

As we saw in the last section, every sequence $\{a_n\}$ can be used to determine an associated series $a_1 + a_2 + a_3 + \ldots$, so arithmetic series follow naturally from arithmetic sequences. But the first question that must be asked of any series is whether it converges, and it doesn't take long to realize that any non-trivial arithmetic sequence gives rise to a series that diverges.

Our reasoning goes something like the following. If we denote the first term of the sequence by a_1, the second term is then $a_1 + d$ (where d is the common difference), the third term is $(a_1 + d) + d = a_1 + 2d$, and so on. So if we add up the first n terms of the sequence (that is, if we find the nth partial sum), we have

$$\begin{aligned} S_n &= a_1 + (a_1 + d) + (a_1 + 2d) + \ldots + (a_1 + (n-1)d) \\ &= \sum_{i=1}^{n} (a_1 + (i-1)d) \\ &= \sum_{i=1}^{n} a_1 + \sum_{i=1}^{n} (i-1)d \quad \text{(note the use of a property of } \textstyle\sum \text{)} \end{aligned}$$

For our present purposes, this might already be enough to convince you that the partial sums are getting larger and larger in magnitude as n grows, since S_n consists of a_1 added to itself n times, plus a sum of multiples of d. But we will soon find it useful to know more about S_n, so we continue the analysis:

$$\begin{aligned} S_n &= \sum_{i=1}^{n} a_1 + \sum_{i=1}^{n} (i-1)d \\ &= a_1 \sum_{i=1}^{n} 1 + d \sum_{i=1}^{n} (i-1) \\ &= na_1 + d(0 + 1 + 2 + \ldots + (n-1)) \\ &= na_1 + d \sum_{i=1}^{n-1} i \end{aligned}$$

We have now also used the distributive property of sigma in order to factor out the constants a_1 and d, and we have used the simple formula that tells us adding up n 1's gives us n. We also temporarily expanded the second of the two sums above in order to find a more useful way of writing it, and we can now apply another formula to complete our task.

$$\begin{aligned} S_n &= na_1 + d \sum_{i=1}^{n-1} i \\ &= na_1 + d\left(\frac{(n-1)n}{2}\right) \end{aligned}$$

It is now very clear that the sequence of partial sums is not going to approach some fixed number S, except in the trivial case when $a_1 = 0$ and $d = 0$. That is, except for the series $0 + 0 + 0...$, every arithmetic series diverges. But we will have further use for the above formula for partial sums of arithmetic sequences.

Topic 2: The Formula for the General Term of an Arithmetic Sequence

Another formula we will find useful for arithmetic sequences, as with all sequences, is the explicit formula for the general n^{th} term of the sequence. And the thinking that went into the analysis above has all but given us the formula.

We have already noted that if a_1 is the first term of an arithmetic sequence, and if the common difference is d, then $a_1 + d$ is the second term, $a_1 + 2d$ is the third term, and so on. This pattern is summarized in the box below.

General Term of an Arithmetic Sequence

The explicit formula for the general n^{th} term of an arithmetic sequence is

$$a_n = a_1 + (n-1)d,$$

where d is the common difference for the sequence.

example 1

Find the formula for the general n^{th} term of the following arithmetic sequences.

a. $-3, 2, 7, 12, ...$

b. $9, 1, -7, -15, ...$

Solutions:

a. $d = 2-(-3) = 5$

$a_1 = -3$

$a_n = -3 + 5(n-1)$

To construct the formula, we just need to determine d and the first term. Note that we could have determined d by taking the difference between any two consecutive terms given.

cont'd. on next page ...

b. $d = 1 - 9 = -8$

$a_1 = 9$

$a_n = 9 - 8(n-1)$

This sequence is also arithmetic, though it is an example of a decreasing sequence (the numbers diminish as n gets larger). This means that d is a negative constant.

In modeling situations the mathematical task may not quite take the form seen in Example 1.

example 2

A demographer decides to model the population growth of a small town as an arithmetic progression. He knows that the population in 1992 was 12,790 and that in 1995 the population was 13,150. He wants to treat the population in 1990 as the first term of the arithmetic progression. What is the sought-after formula?

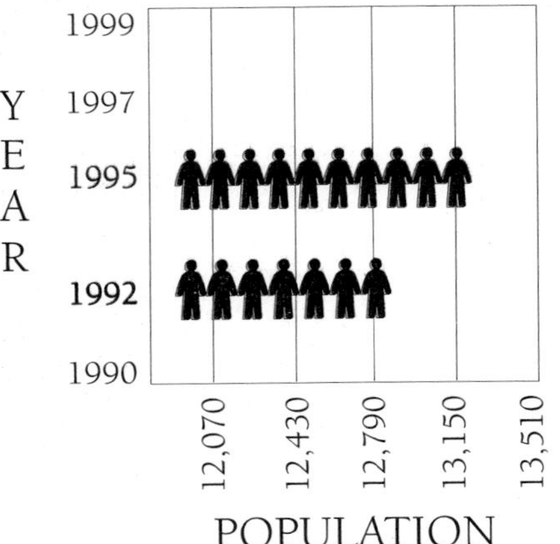

Solution:

Although we do not yet know its value, we know that a_1 represents the population in 1990. Similarly, a_2 represents the population in 1991 and a_3 represents the population in 1992. From what we have been given, we know that $a_3 = 12,790$. We also know that $a_6 = 13,150$ (the population in 1995 is 13,150). If we let d represent the common difference, the population must have grown by $3d$ between 1992 and 1995, so

$$3d = 13{,}150 - 12{,}790 = 360$$

or $d = 120$. Now we can use the fact that $a_3 = 12,790$ to determine a_1, as follows:

$$a_3 = a_1 + 2d$$
$$12{,}790 = a_1 + 2(120)$$
$$12{,}790 - 240 = a_1$$
$$a_1 = 12{,}550$$

Our formula, then, is $a_n = 12{,}550 + 120(n-1)$. Note that to apply the formula, we need to remember $n = 1$ corresponds to 1990, $n = 2$ corresponds to 1991, and so on. We might also find it convenient to simplify the expression and write

$$a_n = 12{,}550 + 120(n-1) = 12{,}430 + 120n.$$

Topic 3: Evaluating Partial Sums of Arithmetic Sequences

To begin our study of partial sums of arithmetic sequences, we will return to the ten-year-old girl whose weekly allowance grows in arithmetic progression.

example 3

Melinda, the ten-year-old in question, knows that her parents plan to increase her weekly allowance by \$0.50/week with the start of each new year, and she currently receives \$1.00/week. Being fiscally minded, she naturally wonders how much money she will have been given by the time she is 18 years old. Assuming that Melinda receives an allowance every week from the current year up through the year she turns 18, what is her total take?

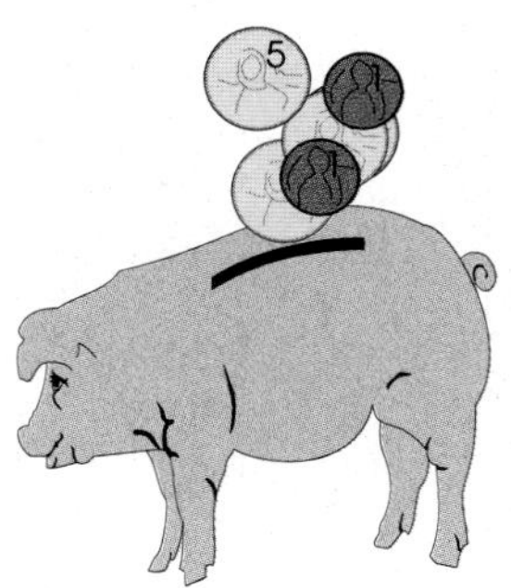

cont'd. on next page ...

Solution:

In the current year, Melinda will receive $(52)(\$1.00) = \$52.00$. Similarly, she will receive $(52)(\$1.50) = \$78.00$ in the next year, so the sum of these two amounts is her total allowance during her 10th and 11th years. We need to extend this reasoning up through Melinda's 18th year, and we can make use of the partial sum formula we have already derived:

$$S_n = na_1 + d\left(\frac{(n-1)n}{2}\right).$$

In order to use this, we note that $a_1 = 52$ (her total allowance in year 1), that $d = 26$ (since her yearly income changes by $(52)(\$0.50) = \$26.00$ from year-to-year), and that S_9 is the sum we are trying to determine (since S_1 corresponds to her income when she is 10, S_9 will be her income up through age 18). So applying the formula, we obtain

$$S_9 = 9 \cdot 52 + (26)\left(\frac{(9-1)\cdot 9}{2}\right) = 468 + 936 = 1404.$$

Melinda will thus receive a total of \$1404 in allowance money from age 10 through age 18.

We can also use the partial sum formula to derive a slightly different formula that is useful in other contexts. Recall that the formula for the general term of an arithmetic sequence is

$$a_n = a_1 + (n-1)d.$$

From this we see that $(n-1)d = a_n - a_1$, and we can use this fact in our partial sum formula to eliminate d, as follows.

$$\begin{aligned} S_n &= na_1 + d\left(\frac{(n-1)n}{2}\right) \\ &= na_1 + \left(\frac{n}{2}\right)(n-1)d \\ &= na_1 + \left(\frac{n}{2}\right)(a_n - a_1) \\ &= \left(\frac{n}{2}\right)(a_1 + a_n) \end{aligned}$$

The box below summarizes these two forms for the partial sums of an arithmetic sequence.

Partial Sums of Arithmetic Sequences

Let $\{a_n\}$ be an arithmetic sequence with common difference d. Then the sum of the first n terms (the nth partial sum) is given by both

$$S_n = na_1 + d\left(\frac{(n-1)n}{2}\right)$$

and

$$S_n = \left(\frac{n}{2}\right)(a_1 + a_n).$$

In practice, the choice of formula is based on whether it is easier to determine d or the last term, a_n, in the sum to be evaluated.

example 4

Determine the sum of the first 100 positive odd integers.

Solution:

First, note that the nth positive odd integer is $2n - 1$, so the 100th positive odd integer is 199. We can use either partial sum formula to evaluate $1 + 3 + ... + 199$. If we choose to use the second formula, we obtain

$$S_{100} = \left(\frac{100}{2}\right)(1+199) = 10{,}000.$$

exercises

Find the explicit formula for the general n^{th} term of the arithmetic sequences described below. See Examples 1 and 2.

1. $-2, 1, 4, 7, 10, \ldots$

2. $5, 7, 9, 11, 13, \ldots$

3. $7, 5, 3, 1, -1, \ldots$

4. $a_2 = 14$ and $a_3 = 19$

5. $a_1 = 5$ and $a_5 = 41$

6. $a_2 = 13$ and $a_4 = 21$

7. $a_3 = -9$ and $d = -6$

8. $a_{12} = 43$ and $d = 3$

9. $a_5 = 100$ and $d = 19$

10. $-37, -20, -3, 14, 31, \ldots$

11. $\frac{7}{2}, \frac{9}{2}, \frac{11}{2}, \frac{13}{2}, \frac{15}{2}, \ldots$

12. $15, 11, 7, 3, -1, \ldots$

13. $a_1 = 12$ and $a_3 = -7$

14. $a_{73} = 224$ and $a_{75} = 230$

15. $a_1 = -1$ and $a_6 = -11$

16. $a_5 = -\frac{5}{2}$ and $d = \frac{3}{2}$

17. $a_4 = 17$ and $d = -4$

18. $a_{34} = -71$ and $d = -2$

Given the initial term and the common difference, find the value of the 7th term of the arithmetic sequences described below. See Examples 1 and 2.

19. $a_1 = 1$ and $d = 2$

20. $a_1 = 4$ and $d = -3$

21. $a_1 = 0$ and $d = \frac{1}{3}$

22. $a_1 = 3$ and $d = \pi$

23. $a_1 = 8$ and $d = -1$

24. $a_1 = \frac{1}{2}$ and $d = 3$

Find the common difference of the given sequence. See Examples 1 and 2.

25. $\{5n - 3\}$

26. $\left\{3n - \frac{1}{2}\right\}$

27. $\{n + 6\}$

28. $\{1 - 4n\}$

29. $\{\sqrt{2} - 2n\}$

30. $\{n\sqrt{3} + 5\}$

Use the given information about each arithmetic sequence to answer the question. See Examples 1 and 2.

31. Given that $a_1 = -3$ and $a_5 = 5$, what is a_{100}?

32. In the sequence 24, 43, 62,..., which term is 955?

33. In the sequence $1, \frac{4}{3}, \frac{5}{3}, \ldots,$ which term is 25?

34. Given that $a_5 = -\frac{5}{3}$ and $a_9 = 1$, what is a_{62}?

35. In the sequence $-16, -9, -2, \ldots,$ what is a_{20}?

36. In the sequence $\frac{1}{4}, \frac{7}{16}, \frac{5}{8}, \ldots,$ which term is $\frac{35}{8}$?

37. In the sequence 2, 5, 8, 11, ..., what is the 9th term?

38. In the sequence 1, 3, 5, 7, ..., what is the 6th term?

39. In the sequence 16, 12, 8, 4, ..., what is the 7th term?

40. In the sequence $\frac{1}{2}, 2, \frac{7}{2}, 5, \ldots,$ what is the 8th term?

41. In the sequence –2, 1, 4, 7, ..., what is the 6th term?

42. In the sequence 9, 6, 3, 0, ..., what is the 10th term?

43. In the sequence 5, 10, 15, 20, ..., what is the 11th term?

44. In the sequence $2\sqrt{2}, 4\sqrt{2}, 6\sqrt{2}, 8\sqrt{2}, \ldots,$ what is the 7th term?

Each of the following sums is a partial sum of an arithmetic sequence; use either formula to determine its value. See Examples 3 and 4.

45. $\sum_{i=1}^{100}(3i-8)$

46. $\sum_{i=1}^{50}(-2i+5)$

47. $\sum_{i=5}^{90}(4i+9)$

48. $3+11+\ldots+795$

49. $25+18+\ldots+(-143)$

50. $-12+2+\ldots+674$

51. $\sum_{i=1}^{37}\left(-\frac{3}{5}i-6\right)$

52. $\sum_{i=100}^{200}(3i+57)$

53. $\sum_{i=2}^{42}(2i-22)$

54. $-90+(-77)+\ldots+92$

55. $7+3+\ldots+(-101)$

56. $4+\frac{81}{20}+\ldots+900$

Each of the following can be answered by finding the partial sum of an arithmetic sequence. See Examples 3 and 4.

57. Melinda borrows $21,000, interest-free, from her parents to help pay for her college education, and promises that upon graduation she will pay back the sum beginning with $1000 the first year and increasing the amount by $1,000 with each successive year. How many years will it take for her to repay the entire $21,000?

58. A certain theater is shaped so that the first row has 30 seats and, moving toward the back, each successive row has two seats more than the previous one. If there are 40 rows, how many seats does the last row contain? How many seats are there altogether?

59. A brick-mason spends a morning moving a pile of bricks from his truck to the work-site by wheelbarrow. Each brick weighs two pounds, and on his first trip he transports 100 pounds. On each successive trip, as he tires, he decides to move one less brick. How many pounds of bricks has he transported after 20 trips?

60. A man decides to lease a car and is told that his payment to the car dealership will be $50 a month for the first month. He is also told that every month after, his payments will increase $25 a month for the next 60 months. How much is his payment after two years? How much has he paid in total after the first two years?

61. The manager of a grocery store decides to create a display of soup cans by placing cans in a row on the floor and then stacking successive rows so that each level of the tower has one less can than the one below it. The manager wants the top row to have 5 cans, and the store has 290 cans that can be used for the display. If all of the cans are used, how many rows will the display have?

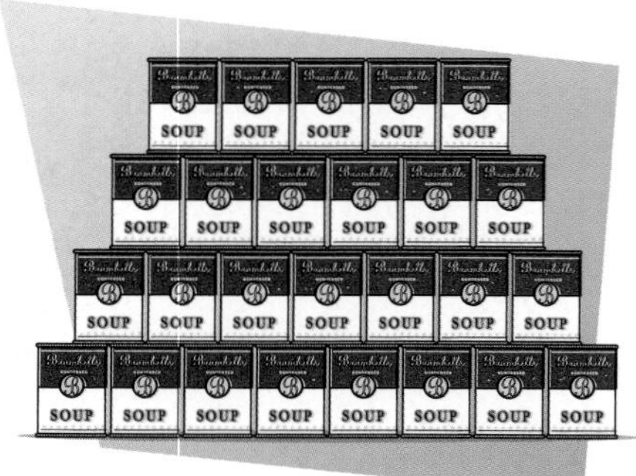

11.3 Geometric Sequences and Series

TOPICS

1. Characteristics of geometric sequences
2. The formula for the general term of a geometric sequence
3. Evaluating partial sums of geometric sequences
4. Evaluating infinite geometric series
5. Interlude: Zeno's paradoxes

Topic 1: Characteristics of Geometric Sequences

We began the last section with a discussion of Melinda, a young girl whose weekly allowance was to increase by a fixed amount (\$0.50/week) with the start of each new year, and we noted that the resulting sequence constituted an arithmetic progression. In this section, we will examine a similar situation as it might apply to Melinda's mother, Marilyn.

Suppose Marilyn is nearing the end of a job-hunt, and has a choice of two positions. Both offer a starting salary of \$40,000/year, but they differ in their projected future salaries. Employer A has a well-defined tier structure for its employees, and offers a yearly increase in grade with a corresponding increase in salary of \$1250/year. Employer B has a flatter hierarchy, and simply offers an increase of 3%/year for all its employees. Assuming the positions are equally desirable in all other respects, which one should Marilyn choose?

As you might expect, real life being what it is, this question is not as easily answered as, say, a problem in a college algebra textbook. One important consideration for Marilyn is the number of years she anticipates working for her next employer. A table comparing the two projected salaries over the next decade or so will help us determine the answer, and the table will be most easily constructed if we can find formulas for the salaries in the nth year.

We know how to find such a formula for Employer A, as the sequence of yearly salaries forms an arithmetic progression with a common difference of \$1250; this means the nth term of the sequence is given by

$$a_n = 40,000 + (n-1)1250.$$

The nth term of the sequence of salaries for Employer B is of a different form. We can begin by noting that since each successive salary is 3% greater than the preceding one, the recursive formula

$$b_1 = 40,000 \text{ and } b_n = b_{n-1} + (0.03)b_{n-1} \text{ for } n \geq 2$$

describes the sequence of salaries. We can then simplify the formula by writing $b_n = (1.03)b_{n-1}$ for $n \geq 2$. With these two formulas in hand, we can now construct the table comparing salaries seen in Figure 1 (all salaries are rounded off to the nearest dollar).

	1	2	3	4	5	6	7	8	9	10
a_n	40,000	41,250	42,500	43,750	**45,000**	46,250	47,500	48,750	50,000	51,250
b_n	40,000	41,200	42,436	43,709	**45,020**	46,371	47,762	49,195	50,671	52,191

Figure 1: Comparison of Salaries of Employers A and B in Year *n*

The first thing we notice is that Employer A offers a higher yearly salary up through the fourth year, but that thereafter Employer B pays more. So if Marilyn anticipates staying in her next job for five years or more, she may want to go with Employer B. This reasoning doesn't consider the *accumulated* salary through year n; for a more accurate comparison of the two jobs, we really need to compare the partial sums of the two sequences and determine when that of Employer B overtakes that of Employer A. We will return to the question of partial sums later in this section.

The salary sequence for Employer B is an example of a *geometric sequence*, and its identifying characteristic is that the *ratio* of consecutive terms in the sequence is a fixed constant. As we noted above, the ratio of any term in the sequence to the preceding term is 1.03. (Remember that for arithmetic sequences, the *difference* between consecutive terms is fixed.)

Geometric Sequences

A sequence $\{a_n\}$ is a **geometric sequence** (sometimes also called a **geometric progression**) if there is some constant $r \neq 0$ so that $\frac{a_{n+1}}{a_n} = r$ for each $n = 1, 2, 3, \ldots$. The constant r is called the **common ratio** of the sequence.

Topic 2: The Formula for the General Term of a Geometric Sequence

Our goal is to develop an explicit formula for the general n^{th} term of a geometric sequence. This can be done if we relate each term to a_1, the first term of the sequence $\{a_n\}$, just as we did for arithmetic sequences.

Since the sequence $\{a_n\}$ is geometric, there is some fixed number r such that $\frac{a_{n+1}}{a_n} = r$ for each $n = 1, 2, 3, \ldots$. In particular,

$$\frac{a_2}{a_1} = r, \text{ or } a_2 = a_1 r.$$

Similarly,

$$\frac{a_3}{a_2} = r, \text{ so } a_3 = a_2 r = (a_1 r)r = a_1 r^2.$$

This pattern persists, leading to the following observation.

General Term of a Geometric Sequence

The explicit formula for the general n^{th} term of a geometric sequence is

$$a_n = a_1 r^{n-1},$$

where r is the common ratio for the sequence.

Given this information, we now know that the yearly salary offered by Employer B above is given by the formula

$$b_n = (40{,}000)(1.03)^{n-1}.$$

example 1

Find the formula for the general term of the following geometric sequences.

a. $\frac{1}{3}, \frac{1}{9}, \frac{1}{27}, \ldots$

b. $3, -6, 12, \ldots$

Solutions:

a. $r = \dfrac{\frac{1}{9}}{\frac{1}{3}} = \frac{3}{9} = \frac{1}{3}$

$a_1 = \frac{1}{3}$

$a_n = \left(\frac{1}{3}\right)\left(\frac{1}{3}\right)^{n-1} = \left(\frac{1}{3}\right)^n$

Since we have been told the sequence is geometric, we could use any two consecutive terms to determine the common ratio *r*. Once we have done this, the formula for the general term follows. Note that in this case we can simplify the formula a bit, as shown.

b. $r = \frac{-6}{3} = -2$

$a_1 = 3$

$a_n = (3)(-2)^{n-1}$

In this sequence, the common ratio is negative. It will always be the case that a geometric sequence consists of terms alternating in sign if and only if the common ratio is negative.

We can also determine the formula for the general term when given other information about a geometric sequence.

example 2

Given that the second term of a geometric sequence is –6 and the fifth term is 162, what is the fourth term?

Solution:

First, the fact that the second term and the fifth term alternate in sign tells us that the common ratio must be negative. (But be careful with this type of reasoning; if, say, the second term and the fourth term have the same sign, the common ratio might be positive or negative.) Given that the sequence is geometric, we know that

$$-6 = a_1 r \text{ and } 162 = a_1 r^4.$$

We can eliminate a_1 from this pair of equations with the following observation:

$$\frac{162}{-6} = \frac{a_1 r^4}{a_1 r},$$

or

$$-27 = r^3.$$

This tells us $r = -3$, a negative number as expected, and we can then determine that $a_1 = 2$. So the general term is given by $a_n = 2(-3)^{n-1}$, and the fourth term is

$$a_4 = 2(-3)^{4-1} = -54.$$

Topic 3: Evaluating Partial Sums of Geometric Sequences

Given a geometric sequence $\{a_n\}$, many applications require us to compute the partial sum $S_n = a_1 + a_2 + \ldots + a_n$ for some fixed n. The ideal situation is to have an explicit formula that tells us what the nth partial sum is for any given geometric sequence.

We can come up with such a formula with a few simple, though not necessarily obvious, steps. The first is to note that if we define $S_n = a_1 + a_2 + \ldots + a_n$, then the formula for the general term of a geometric sequence allows us to write

$$S_n = a_1 + a_1 r + a_1 r^2 + \ldots + a_1 r^{n-1}.$$

If we multiply both sides of the previous equation by r, we obtain

$$rS_n = a_1r + a_1r^2 + a_1r^3 \ldots + a_1r^n.$$

If we now subtract the second equation from the first, most of the terms from the right-hand sides of the equations cancel, as shown below:

$$\begin{aligned} S_n - rS_n &= \left(a_1 + \cancel{a_1r} + \ldots \cancel{a_1r^{n-1}}\right) - \left(\cancel{a_1r} + \cancel{a_1r^2} + \ldots a_1r^n\right) \\ &= a_1 - a_1r^n. \end{aligned}$$

Of course, the goal is to solve for S_n, and we can do this now by factoring the left and right sides of the equation and dividing:

$$\begin{aligned} S_n(1-r) &= a_1(1-r^n) \\ S_n &= \frac{a_1(1-r^n)}{1-r}. \end{aligned}$$

This formula is reproduced in the following box for reference.

Partial Sum of a Geometric Sequence

Let $\{a_n\}$ be a geometric sequence with the common ratio r, and assume r is neither 0 nor 1. Then the nth partial sum of the sequence, $S_n = a_1 + a_1r + a_1r^2 + \ldots + a_1r^{n-1}$, is given by

$$S_n = \frac{a_1(1-r^n)}{1-r}.$$

(Note that the condition $r \neq 1$ is necessary to prevent division by zero in the formula, but it isn't difficult to determine the partial sum without the formula if r happens to be 1. In this case, $S_n = a_1 + a_1 + \ldots + a_1 = na_1$.)

With this formula in hand, along with one of the two partial sum formulas for arithmetic sequences, let us return to the question with which we began this section.

example 3

Recall that Marilyn has the option of taking a job with Employer A, who offers a yearly salary increase of \$1250, or Employer B, who offers a yearly salary increase of 3%. Given that the starting salary is \$40,000 for both, in what year does the accumulated salary paid by Employer B overtake the accumulated salary paid by Employer A?

Solution:

We have already determined that Employer B pays a higher salary beginning with the fifth year of employment, but we need to compare the partial sums of the salaries paid by the two employers up through year n to answer the question above. Recall that the sequence of yearly salaries paid by Employer A is defined by

$$a_n = 40{,}000 + (n-1)1250$$

and that the sequence of yearly salaries paid by Employer B is defined by

$$b_n = 40{,}000(1.03)^{n-1}.$$

Recall also that one partial sum formula for an arithmetic sequence is

$$S_n = na_1 + d\left(\frac{(n-1)n}{2}\right),$$

where d is the common difference for the sequence.

cont'd. on next page ...

In order to keep the partial sums straight, let A_n denote the sum of the salaries paid by Employer A through year n, and let B_n be the same for Employer B. Then

$$A_n = n(40,000) + 1250\left(\frac{(n-1)n}{2}\right) = 40,000n + 625(n^2 - n) = 625n^2 + 39,375n$$

and

$$B_n = \frac{40,000(1-1.03^n)}{1-1.03} = \left(\frac{40,000}{0.03}\right)(1.03^n - 1).$$

We can now use these formulas to compute the accumulated salaries paid by the two employers up through year n, as shown in Figure 2.

	1	2	3	4	5	6	7	8
A_n	40,000	81,250	123,750	167,500	212,500	258,750	306,250	355,000
B_n	40,000	81,200	123,636	167,345	212,365	258,736	306,498	355,693

Figure 2: Accumulated Salaries through Year *n*

Comparing values in the table indicates that the accumulated salary paid by Employer B overtakes that of Employer A in the seventh year.

We can also use the geometric partial sum formula to evaluate certain expressions defined with the sigma notation, as shown in the next example.

example 4

Evaluate $\sum_{i=2}^{7} 5\left(\frac{-1}{2}\right)^i$.

Solution:

The given sum represents a partial sum of a geometric sequence, but as it is written the first term and the common ratio of the sequence are not apparent. So a good way to begin is to write the sum in expanded form:

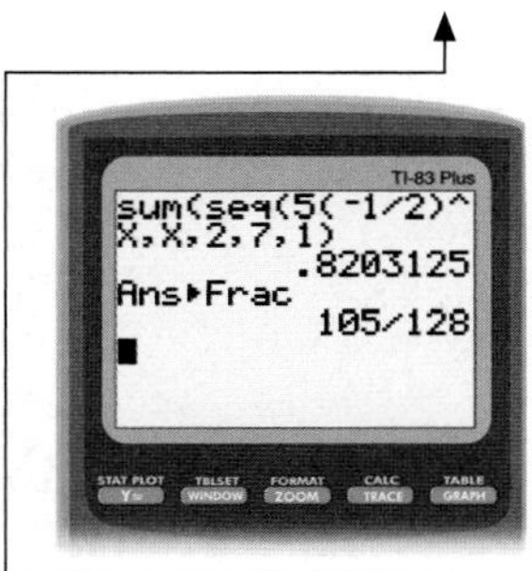

The above screen was performed using the summation and sequence commands as described on page 902. To find the answer in fraction form, select Math and 1.

$$\sum_{i=2}^{7} 5\left(\frac{-1}{2}\right)^i = 5\left(\frac{-1}{2}\right)^2 + 5\left(\frac{-1}{2}\right)^3 + \ldots + 5\left(\frac{-1}{2}\right)^7$$
$$= \frac{5}{4} - \frac{5}{8} + \ldots - \frac{5}{128}$$

From this, we can see that $a_1 = \frac{5}{4}$ and that the common ratio is $r = -\frac{1}{2}$. We also need to observe that there are a total of *six* terms in the sum, so to use the partial sum formula we let $n = 6$. Putting this all together, we have

$$S_6 = \frac{\left(\frac{5}{4}\right)\left(1-\left(-\frac{1}{2}\right)^6\right)}{1-\left(-\frac{1}{2}\right)} = \frac{\left(\frac{5}{4}\right)\left(1-\frac{1}{64}\right)}{\frac{3}{2}} = \frac{\left(\frac{5}{4}\right)\left(\frac{63}{64}\right)}{\frac{3}{2}} = \frac{105}{128}.$$

Topic 4: Evaluating Infinite Geometric Series

In Section 11.1, we used a fairly intuitive approach to think about the result of adding half an inch to a quarter of an inch to an eighth of an inch, and so on. We decided then that if the process were continued indefinitely, the sum of the fractions would be 1. That is,

$$\frac{1}{2} + \frac{1}{4} + \frac{1}{8} + \ldots = 1, \text{ or } \sum_{i=1}^{\infty}\left(\frac{1}{2}\right)^i = 1.$$

Our rationale for this conclusion was that the sequence of partial sums of the sequence

$$\left\{\left(\frac{1}{2}\right)^n\right\}$$

never exceeded 1, yet got closer and closer to 1 as more terms were added.

We now have the machinery to analyze infinite series like the one above more rigorously and more efficiently. For one thing, we know that the sequence

$$\frac{1}{2}, \frac{1}{4}, \frac{1}{8}, \ldots$$

is a geometric sequence, and that both the first term and the common ratio are $\frac{1}{2}$. So we can use the partial sum formula for geometric sequences to determine that

$$S_n = \frac{1}{2}+\frac{1}{4}+\ldots+\frac{1}{2^n} = \frac{\left(\frac{1}{2}\right)\left(1-\left(\frac{1}{2}\right)^n\right)}{\left(1-\frac{1}{2}\right)} = 1-\left(\frac{1}{2}\right)^n$$

where we have made the substitutions

$$a_1 = \frac{1}{2} \text{ and } r = \frac{1}{2}.$$

If we were to write S_n as a single fraction, we would obtain

$$S_n = \frac{2^n - 1}{2^n},$$

the formula we intuitively derived in Section 11.1. In either form, we can see that S_n approaches 1 as $n \to \infty$, but we can now generalize this observation for all convergent geometric series.

Infinite Geometric Series

If $|r|<1$, the infinite geometric series

$$\sum_{n=1}^{\infty} a_1 r^{n-1} = a_1 + a_1 r + a_1 r^2 + a_1 r^3 + \ldots$$

converges, and the sum of the series is given by $S = \frac{a_1}{1-r}$. Alternatively, we can use sigma notation with the index beginning at 0 to write the same series; in this form,

$$\sum_{n=0}^{\infty} a_1 r^n = a_1 + a_1 r + a_1 r^2 + a_1 r^3 + \ldots = \frac{a_1}{1-r}.$$

The proof of this result follows from the fact that if $|r|<1$, then $r^n \to 0$ as $n \to \infty$, so

$$S_n = \frac{a_1(1-r^n)}{1-r} \to \frac{a_1}{1-r}$$

as $n \to \infty$. That is, the sequence of partial sums of such geometric sequences approaches the given formula as n gets larger and larger.

In the case of the series $\frac{1}{2}+\frac{1}{4}+\frac{1}{8}+\ldots$, where $a_1 = \frac{1}{2}$ and $r = \frac{1}{2}$, the sum is

$$S = \frac{\frac{1}{2}}{1-\frac{1}{2}} = \frac{\frac{1}{2}}{\frac{1}{2}} = 1.$$

example 5

Find the sums of the following series.

a. $\sum_{n=1}^{\infty} 5\left(\frac{-1}{2}\right)^{n-1}$ **b.** $\sum_{n=1}^{\infty} 3\left(\frac{1}{10}\right)^{n}$

Solutions:

a. In the case of the first series, it is easy to identify a_1 and r from the way the series is written:

$$a_1 = 5 \text{ and } r = -\frac{1}{2}.$$

Since the common ratio r is less than 1 in magnitude, we can proceed to apply the formula for the sum to determine that

$$\sum_{n=1}^{\infty} 5\left(\frac{-1}{2}\right)^{n-1} = \frac{5}{1-\left(-\frac{1}{2}\right)} = \frac{5}{\frac{3}{2}} = \frac{10}{3}.$$

b. Be careful to note exactly how a series is written. In this case, the form of the series doesn't quite match either of the forms to which the formula $\frac{a_1}{1-r}$ applies. But if we write out the first few terms, we can easily identify the first term of the series and the common ratio:

$$\sum_{n=1}^{\infty} 3\left(\frac{1}{10}\right)^{n} = \frac{3}{10}+\frac{3}{100}+\frac{3}{1000}+\ldots$$

Alternately, we could rewrite the summation as:

$$\sum_{n=1}^{\infty} 3\left(\frac{1}{10}\right)^{n} = \sum_{n=1}^{\infty} 3\left(\frac{1}{10}\right)^{n-1}\left(\frac{1}{10}\right) = \sum_{n=1}^{\infty}\left(\frac{3}{10}\right)\left(\frac{1}{10}\right)^{n-1}.$$

cont'd. on next page ...

From this, we can see that $a_1 = \frac{3}{10}$ and $r = \frac{1}{10}$ (determine r by dividing any term by the preceding term). So

$$\sum_{n=1}^{\infty} 3\left(\frac{1}{10}\right)^n = \frac{\frac{3}{10}}{1-\frac{1}{10}} = \left(\frac{\frac{3}{10}}{1-\frac{1}{10}}\right)\left(\frac{10}{10}\right) = \frac{3}{10-1} = \frac{1}{3}.$$

The last example illustrates an important relation between geometric series and the decimal system we use to write real numbers. Note that what we have really shown is that

$$\sum_{n=1}^{\infty} 3\left(\frac{1}{10}\right)^n = \frac{3}{10} + \frac{3}{100} + \frac{3}{1000} + \ldots = 0.33\bar{3}$$

is the decimal representation of the fraction $\frac{1}{3}$. Similarly,

$$0.\overline{52} = \frac{52}{100} + \frac{52}{10,000} + \frac{52}{1,000,000} + \ldots = \sum_{n=1}^{\infty} 52\left(\frac{1}{100}\right)^n$$

can be written in fractional form by noting that for this series $a_1 = \frac{52}{100}$ and $r = \frac{1}{100}$, so

$$0.\overline{52} = \frac{\frac{52}{100}}{1-\frac{1}{100}} = \frac{52}{99}.$$

Topic 5: Interlude: Zeno's Paradoxes

The fact that the sum of an infinite number of positive quantities may in fact be finite has made people uneasy, at least at first glance, for millennia. The Greek philosopher Zeno (*ca.* 450 BC), a member of a community of thinkers on the island of Elea, used this uneasiness as the basis for a number of paradoxes. His purpose in creating these paradoxes was to point out complexities in the concept of infinity.

One of the most famous of Zeno's paradoxes concerns a race between the Greek hero Achilles and a tortoise. Because Achilles is so much faster, the tortoise is given a head start in the race. The paradox is that, once the race has begun, it will take a certain amount of time for Achilles to reach the tortoise's starting place, and that in this time the tortoise will have been able to move ahead some (small) distance to a new point.

By the time Achilles reaches this new point, the tortoise will have again been able to move ahead, and Achilles will have to repeat the process. Because this argument can be continued indefinitely, Zeno argued, Achilles can never actually catch the tortoise, let alone get ahead of him.

Our knowledge of infinite geometric series now allows us to resolve the paradox. To be specific, let us suppose that Achilles can run 100 feet per second (remember, he's a mythical hero), and that the tortoise begins with a 1000 foot head start. Let us also suppose Achilles can run 10 times as fast as the tortoise (this means the tortoise runs at 10 feet per second, so it's fairly heroic, too). Once the race starts, it takes Achilles 10 seconds to get to the tortoise's starting place, and in this 10 seconds the tortoise has managed to crawl another 100 feet. It takes Achilles an additional 1 second to cover this 100 feet, but during that 1 second the tortoise is able to crawl another 10 feet. If we continue the process, the total amount of time required for Achilles to catch the tortoise is then

$$T = 10 + 1 + \frac{1}{10} + \frac{1}{100} + \ldots$$

This sum is a geometric series with a first term of 10 and a common ratio of $\frac{1}{10}$, so

$$T = \frac{10}{1 - \frac{1}{10}} = \frac{100}{9} = 11.\bar{1} \text{ seconds.}$$

We can also determine that Achilles has run $1{,}111.\bar{1}$ feet in this time, so the tortoise actually only manages to run $111.\bar{1}$ feet before being caught.

exercises

Find the explicit formula for the general term of the geometric sequences described below. See Examples 1 and 2.

1. $-3, -6, -12, -24, -48, \ldots$ **2.** $7, \frac{7}{2}, \frac{7}{4}, \frac{7}{8}, \frac{7}{16}, \ldots$ **3.** $2, -\frac{2}{3}, \frac{2}{9}, -\frac{2}{27}, \frac{2}{81}, \ldots$

4. $a_1 = 5$ and $a_4 = 40$ **5.** $a_2 = -\frac{1}{4}$ and $a_5 = \frac{1}{256}$ **6.** $a_1 = 1$ and $a_4 = -0.001$

7. $a_2 = \frac{1}{7}$ and $r = \frac{1}{7}$ **8.** $a_3 = \frac{9}{16}$ and $r = -\frac{3}{4}$ **9.** $a_3 = 9,\ a_5 = 81$ and $r < 0$

10. $-3, 9, -27, 81, -243, \ldots$ **11.** $3, 2, \frac{4}{3}, \frac{8}{9}, \frac{16}{27}, \ldots$ **12.** $-5, \frac{5}{4}, -\frac{5}{16}, \frac{5}{64}, -\frac{5}{256}, \ldots$

13. $a_3 = 28$ and $a_6 = -224$ **14.** $a_2 = -24$ and $a_5 = -81$ **15.** $a_5 = 1$ and $a_6 = 2$

16. $a_4 = \frac{343}{3}$ and $r = 7$ **17.** $a_2 = \frac{13}{17}$ and $r = \frac{4}{3}$ **18.** $a_4 = 8,\ a_8 = 128$ and $r > 0$

Use the given information about each geometric sequence to answer the question. See Examples 1 and 2.

19. Given that $a_2 = -\frac{5}{2}$ and $a_5 = \frac{5}{16}$, what is a_{15}?

20. Given that $a_1 = 1$ and $a_4 = \frac{8}{27}$, what is the common ratio r?

21. Given that $a_3 = -2$ and $a_4 = -16$, what is a_{13}?

22. Given that $a_2 = 24$ and $a_5 = 375$, what is the common ratio r?

Each of the following sums is a partial sum of a geometric sequence. Use this fact to evaluate the sums. See Examples 3 and 4.

23. $\sum_{i=1}^{10} 3\left(-\frac{1}{2}\right)^i$ **24.** $\sum_{i=5}^{20} 5\left(\frac{3}{2}\right)^i$ **25.** $\sum_{i=10}^{40} 2^i$

26. $1 - \frac{1}{2} + \ldots + \frac{1}{16,384}$ **27.** $2 + 6 + \ldots + 39,366$ **28.** $5 - \frac{5}{3} + \ldots - \frac{5}{19,683}$

29. $1 - 3 + \ldots + 59,049$ **30.** $\sum_{i=4}^{15} 5(-2)^i$ **31.** $1 + \frac{3}{5} + \ldots + \frac{243}{3125}$

Determine if the following infinite geometric series converge. If a given sum converges, find the sum. See Example 5.

32. $\sum_{i=0}^{\infty} -\frac{1}{2}\left(\frac{2}{3}\right)^i$

33. $\sum_{i=1}^{\infty} \left(\frac{4}{5}\right)^i$

34. $\sum_{i=0}^{\infty} \left(-\frac{9}{8}\right)^i$

35. $\sum_{i=0}^{\infty} \left(-\frac{8}{9}\right)^i$

36. $\sum_{i=5}^{\infty} \left(\frac{19}{20}\right)^i$

37. $\sum_{i=0}^{\infty} (-1)^i$

38. $\sum_{i=1}^{\infty} \frac{1}{3}(2)^{i-1}$

39. $\sum_{i=0}^{\infty} 5\left(\frac{6}{11}\right)^i$

40. $\sum_{i=4}^{\infty} \left(\frac{13}{24}\right)^i$

Write each of the following repeating decimal numbers as fractions. See Example 5.

41. $1.\overline{65}$

42. $0.\overline{123}$

43. $-0.\overline{5}$

44. $-3.\overline{8}$

45. $0.\overline{029}$

46. $9.\overline{98}$

47. A rubber ball is dropped from a height of 10 feet, and on each bounce it rebounds up to 80% of its previous height. How far has it traveled vertically at the moment when it hits the ground for the tenth time? If we assume it bounces indefinitely, what is the total vertical distance traveled?

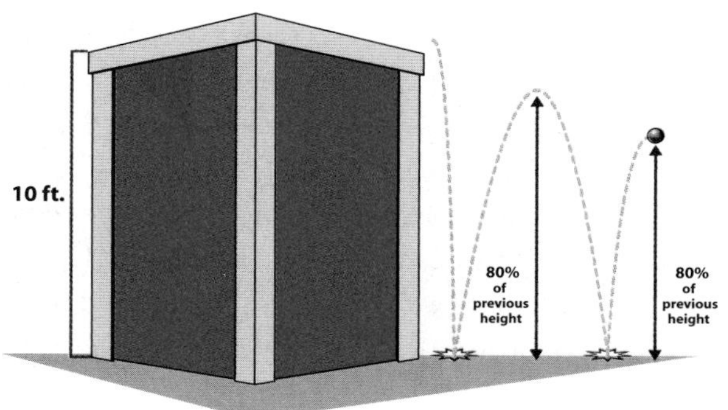

48. If \$10,000 is invested in a simple savings account with an annual interest rate of 4% compounded once a year, what is the value of the account after ten years?

49. If \$10,000 is invested in a simple savings account with an annual interest rate of 4% compounded once a month, what is the value of the account after ten years?

50. An ancient story about the game of chess tells of a king who offered to grant the inventor of the game a wish. The inventor replied "Place a grain of wheat on the first square of the board, 2 grains on the second square, 4 grains on the third, and so on. The wheat will be my reward." How many grains of wheat would the king have had to come up with? (There are 64 squares on a chessboard.)

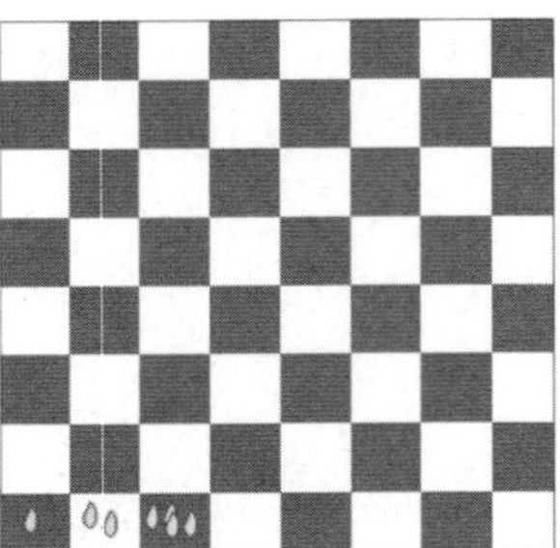

51. An isosceles right triangle is divided into two similar triangles, one of the new triangles is divided into two similar triangles, and this process is continued without end. If the shading pattern seen below is continued indefinitely, what fraction of the original triangle is shaded?

52. Each year, the university admissions committee accepts 3% more students than they accepted in the previous year. If 2130 students were admitted in the first year of this trend, how many total students will have been admitted after 6 years?

11.4 Mathematical Induction

TOPICS

Topic 1: The Role of Induction

The first three sections of this chapter have focused on concepts built on the foundation of the natural numbers: sequences are functions whose domain is the natural numbers, and series are sums based on sequences. This section is similarly focused. While the word "induction" has many meanings in everyday language, the *Principle of Mathematical Induction* has a precise mathematical meaning that depends on properties possessed by the natural numbers.

As a principle, mathematical induction is most often used to prove statements that involve natural numbers in some way. To be precise, suppose that $P(n)$ is a statement about each natural number n. If the statement $P(n)$ is actually true for each n, mathematical induction can be used to prove that fact. As an illustration, consider the statement

$$P(n)\text{: The sum of the first } n \text{ positive integers is } \frac{n(n+1)}{2}.$$

We already know this statement is true for each n, as we proved it using a non-inductive technique in Section 11.1. However, induction provides a more powerful and more general method of proof, and we will see how it can be used to prove similar statements. In particular, we will use induction to prove one of the other summation formulas from Section 11.1.

Before we do that, though, it is important to realize how induction relates to some of the other concepts introduced in this chapter. By now, we have seen several examples of recursive formulas, for instance, $a_1 = 1$ and $a_n = a_{n-1} + n$ for $n \geq 2$ is a recursive formula that generates the sequence of sums $1+2+\cdots+n$. As we know, this recursive formula is inconvenient for calculating, say, a_{100}. Its use for this purpose would require us to first calculate a_{99}, which would in turn require us to calculate a_{98}, and so on back to a_1. In contrast, the explicit formula

$$a_n = \frac{n(n+1)}{2}$$

cont'd. on next page ...

allows us to instantly determine a_{100} or any other such sum. As we will see, the Principle of Mathematical Induction is a very powerful tool for making the transition from a recursive formula to an explicit formula.

Topic 2: Proofs by Mathematical Induction

Consider again the statement

$$P(n)\text{: The sum of the first } n \text{ positive integers is } \frac{n(n+1)}{2}.$$

Suppose we have been able to demonstrate that **if** the statement is true for an integer k, **then** the statement is also true for the next integer, $k + 1$. We can easily show that $P(1)$ is true – that is, the sum of the first 1 positive integers, namely 1, is indeed $\frac{(1)(1+1)}{2}$. So since $P(1)$ is true, we can conclude from our demonstration that $P(2)$ is also true. And since $P(2)$ is true, it follows that $P(3)$ is true. We don't have to literally show that $P(2)$ being true implies that $P(3)$ is true, because we've already demonstrated that, in general, the validity of $P(k)$ implies the validity of $P(k + 1)$. This is the essence of a Proof by Mathematical Induction; the process is summarized below.

Proof by Mathematical Induction

Assume that $P(n)$ is a statement about each natural number n. Suppose that the following two conditions hold:

1. $P(1)$ is true.

2. For each natural number k, if $P(k)$ is true then $P(k + 1)$ is true.

Then the statement $P(n)$ is true for every natural number n.

The construction of an inductive proof thus consists of two steps. The first task, which we will call the **basis step**, is generally very easy; given a statement $P(n)$ simply verify that $P(1)$ is true. The second task, called the **induction step**, normally calls for a bit more work. It is important to realize, though, that the induction step begins with a very powerful assumption: under the *assumption* that $P(k)$ is true, our goal is to prove that $P(k + 1)$ is also true. The remainder of this section will be devoted to illustrating how inductive proofs are constructed.

example 1

Prove that for each natural number n, $\sum_{i=1}^{n} i^2 = \frac{n(n+1)(2n+1)}{6}$.

Solution:

This statement is another of the summation formulas first seen in Section 11.1, but it has not yet been proved. In less formal terms, we wish to prove the statement

$$P(n)\text{: } 1^2 + 2^2 + \cdots + n^2 = \frac{n(n+1)(2n+1)}{6}.$$

Basis Step: Note that $P(1)$ is the statement $1^2 = \frac{(1)(1+1)(2+1)}{6}$ which is clearly true.

Induction Step: We begin this step with the assumption that the statement is true for some integer k, i.e., that $1^2 + 2^2 + \ldots + k^2 = \frac{k(k+1)(2k+1)}{6}$. Our task is to show that $P(k + 1)$ is then also true. To do so, consider the left-hand side of $P(k + 1)$:

$$1^2 + 2^2 + \cdots + k^2 + (k+1)^2.$$

To show that $P(k + 1)$ is true, we need to demonstrate that the sum above is equal to the right-hand side of $P(k + 1)$, namely

$$\frac{(k+1)(k+1+1)(2(k+1)+1)}{6},$$

better written as

$$\frac{(k+1)(k+2)(2k+3)}{6}.$$

We can do this by using the inductive assumption to rewrite all but the last term of the left-hand side; some simple algebra then gives us the desired result:

$$1^2 + 2^2 + \cdots + k^2 + (k+1)^2 = \frac{k(k+1)(2k+1)}{6} + (k+1)^2$$ Use the inductive assumption to rewrite all but the last term.

$$= \frac{k(k+1)(2k+1)}{6} + \frac{6(k+1)^2}{6}$$ Add the two fractions.

cont'd. on next page ...

$$= \frac{(k+1)}{6}\left[k(2k+1)+6(k+1)\right]$$ Factor out the common factors.

$$= \frac{(k+1)}{6}\left[2k^2+k+6k+6\right]$$ Expand the remaining terms.

$$= \frac{(k+1)}{6}\left[2k^2+7k+6\right]$$

$$= \frac{(k+1)}{6}\left[(k+2)(2k+3)\right]$$ Factor the resulting quadratic.

$$= \frac{(k+1)(k+2)(2k+3)}{6}.$$ Rewrite in the desired form.

Every inductive proof is similar in spirit. The single most important step, and the step that is most unfamiliar initially, is the one in which the inductive assumption is used to simplify part of the statement $P(k + 1)$ in order to prove that $P(k + 1)$ is true. Be sure to note how the inductive assumption is used in the following examples as well.

example 2

Prove that for each natural number n, $1+3+5+\cdots+(2n-1)=n^2$.

Solution:

Basis Step: The statement $P(1)$ is the equation $1=1^2$, which is clearly true.

Induction Step: As always, we begin this step with the assumption that $P(k)$ is true; that is, $1+3+5+\cdots+(2k-1)=k^2$. As in Example 1, we then work with the left-hand side of $P(k + 1)$, with the goal of transforming it into the right-hand side:

$$1+3+5+\cdots+(2k-1)+(2(k+1)-1)=k^2+(2(k+1)-1)$$ Use the inductive assumption to rewrite all but the last term.

$$=k^2+(2k+1)$$

$$=(k+1)^2.$$ Factoring gives us the desired result.

Induction can be used to prove statements other than equations, of course. The next example illustrates the proof of an inequality.

example 3

Prove that for each natural number n, $2n \le 2^n$.

Solution:

Basis Step: The statement $P(1)$ is the inequality $2(1) \le 2^1$, or $2 \le 2$. This is clearly true.

Induction Step: Assume that $2k \le 2^k$. Then:

$$\begin{aligned} 2(k+1) &= 2k+2 && \text{Begin by writing the left-hand side of } P(k+1). \\ &\le 2^k+2 && \text{Use the inductive assumption.} \\ &\le 2^k+2^k && \text{With the right-hand side of } P(k+1) \text{ as the goal,} \\ &= 2(2^k) && \text{we use the fact that } k \ge 1 \text{ implies } 2 \le 2^k. \\ &= 2^{k+1}. && \text{Algebra leads to the desired goal.} \end{aligned}$$

We will finish this section with an example that uses the Principle of Mathematical Induction to prove a divisibility fact.

example 4

Prove that for each natural number n, $8^n - 3^n$ is divisible by 5.

Solution:

Basis Step: Clearly, $8^1 - 3^1$ is divisible by 5, so $P(1)$ is true.

Induction Step: Assume that $8^k - 3^k$ is divisible by 5. Then there is some integer p for which $8^k - 3^k = 5p$. Note that

$$\begin{aligned} 8^{k+1} - 3^{k+1} &= 8^k \cdot 8 - 3^k \cdot 3 \\ &= 8^k \cdot (5+3) - 3^k \cdot 3 \\ &= 5 \cdot 8^k + 3 \cdot 8^k - 3 \cdot 3^k \\ &= 5 \cdot 8^k + 3 \cdot (8^k - 3^k) \\ &= 5 \cdot 8^k + 3 \cdot 5p \\ &= 5(8^k + 3p). \end{aligned}$$

This shows that $8^{k+1} - 3^{k+1}$ is a product of 5 and the integer $8^k + 3p$, so we have shown that $8^{k+1} - 3^{k+1}$ is divisible by 5.

exercises

Find S_{k+1} for the given S_k.

1. $S_k = \dfrac{1}{3(k+2)}$

2. $S_k = \dfrac{k^2}{k(k-1)}$

3. $S_k = \dfrac{k(k+1)(2k+1)}{4}$

4. $S_k = \dfrac{1}{(2k-1)(2k+1)}$

Use the Principle of Mathematical Induction to prove the following statements. See Examples 1 through 4.

5. $1+2+3+4+\cdots+n=\dfrac{n(n+1)}{2}$

6. $\dfrac{1}{2}+\dfrac{1}{2^2}+\dfrac{1}{2^3}+\cdots+\dfrac{1}{2^n}=1-\dfrac{1}{2^n}$

7. $2+4+6+8+\cdots+2n=n(n+1)$

8. $\displaystyle\sum_{i=1}^{n}\frac{1}{(2i-1)(2i+1)}=\frac{n}{2n+1}$

9. $4^0+4^1+4^2+\cdots+4^{n-1}=\dfrac{4^n-1}{3}$

10. Prove that $2^n > n^2$ for all $n \geq 5$.

11. $\dfrac{1}{1\cdot 4}+\dfrac{1}{4\cdot 7}+\dfrac{1}{7\cdot 10}+\cdots+\dfrac{1}{(3n-2)(3n+1)}=\dfrac{n}{3n+1}$

12. $5^0 + 5^1 + 5^2 + \cdots + 5^{n-1} = \dfrac{5^n - 1}{4}$

13. $5 + 10 + 15 + \cdots + 5n = \dfrac{5n(n+1)}{2}$

14. Prove that $n^2 \geq 100n$ for all $n \geq 100$.

15. $\left(1+\frac{1}{1}\right)\left(1+\frac{1}{2}\right)\left(1+\frac{1}{3}\right)\cdots\left(1+\frac{1}{n}\right) = n+1$

16. $3 + 5 + 7 + \cdots + (2n+1) = n(n+2)$

17. Prove that $n! > 2^n$ for all $n \geq 4$.

18. $1 + 4 + 7 + 10 + \cdots + (3n-2) = \frac{n}{2}(3n-1)$

19. $-2 - 3 - 4 - \cdots - (n+1) = -\frac{1}{2}\left(n^2 + 3n\right)$

20. Prove that $3^n > 2n+1$ for all $n \geq 2$.

21. Prove that for all natural numbers n, $2^n > n$.

22. $1^3 + 2^3 + 3^3 + 4^3 + \cdots + n^3 = \dfrac{n^2(n+1)^2}{4}$

23. $1 \cdot 2 + 2 \cdot 3 + 3 \cdot 4 + \cdots + n(n+1) = \dfrac{n(n+1)(n+2)}{3}$

24. Prove that if $a > 1$, then $a^n > 1$.

25. Prove that $2^n > 4n$ for all $n \geq 5$.

26. $1^4 + 2^4 + 3^4 + 4^4 + \cdots + n^4 = \dfrac{n(n+1)(2n+1)(3n^2 + 3n - 1)}{30}$

27. $1^5 + 2^5 + 3^5 + 4^5 + \cdots + n^5 = \dfrac{n^2(n+1)^2(2n^2 + 2n - 1)}{12}$

28. $\dfrac{1}{\sqrt{1}} + \dfrac{1}{\sqrt{2}} + \dfrac{1}{\sqrt{3}} + \cdots + \dfrac{1}{\sqrt{n}} > \sqrt{n}$, for all $n \geq 2$

29. $1 + 3 + 5 + 7 + \cdots + (2n - 1) = n^2$

Use the Principle of Mathematical Induction to prove the given properties.

30. $(ab)^n = a^n b^n$, for all positive integers n. (Assume a and b are constant.)

31. $(a^m)^n = a^{mn}$, for all positive integers m and n. (Assume a and m are constant.)

32. If $x_1 > 0$, $x_2 > 0$, ..., $x_n > 0$ then
$\ln(x_1 \cdot x_2 \cdot x_3 \cdot \ldots \cdot x_n) = \ln x_1 + \ln x_2 + \ln x_3 + \cdots + \ln x_n$.

33. 5 is a factor of $(2^{2n-1} + 3^{2n-1})$.

34. 64 is a factor of $(9^n - 8n - 1)$ for all $n \geq 2$.

35. 3 is a factor of $(n^3 + 3n^2 + 2n)$.

36. Prove that for all natural numbers n, $n^3 - n + 3$ is divisible by 3.

37. Prove that for all natural numbers n, $5^n - 1$ is divisible by 4.

38. Prove that for all natural numbers n, $n(n+1)(n+2)$ is divisible by 6.

Use the Principle of Mathematical Induction to prove the following.

39. In the 19th century a mathematical puzzle was published telling of a mythical monastery in Benares, India with three crystal towers holding 64 disks made of gold. The disks are each of a different size and have holes in the middle so that they slide over the towers and sit in a stack with the largest on the bottom and the smallest on the top. The monks of the monastery were instructed to move all of the disks to the third tower following these three rules:

1. Each disk sits over a tower except when it is being moved.
2. No disk may ever rest on a smaller disk.
3. Only one disk at a time may be moved.

According to the puzzle, when the monks complete their task, the world would end! To move n disks requires $H(n) = 2^n - 1$ moves. Prove this is true through mathematical induction.

40. If there are N people in a room, and every person shakes hands with every other person exactly once, then exactly $\frac{n(n-1)}{2}$ handshakes will occur. Prove this is true through mathematical induction.

41. Any monetary value of 4 cents or higher can be composed of twopence (a British two-cent coin) and nickels. Your basis step would be 4 cents = twopence + twopence. Use the fact that $k = 2t + 5n$ where k is the total monetary value, t is the number of twopence, and n is the number of nickels, to prove $k + 1$. (**Hint:** There are 3 induction steps to prove.)

42. What is wrong with this "proof" by induction?

Proposition: All horses are the same color. (In any set of n horses, all horses are the same color.)

Basis Step: If you have only one horse in a group, then all of the horses in that group have the same color.

Induction Step: Assume that in any group of n horses, all horses are the same color. Now take any group of $n + 1$ horses. Remove the first horse from this group and the remaining n horses must be of the same color because of the hypothesis. Now replace the first horse and remove the last horse. Once again, the remaining n horses must be the same color because of the hypothesis. Since the two groups overlap, all $n + 1$ horses must be the same color.

Thus, by induction, any group of n horses are the same color.

11.5 An Introduction to Combinatorics

TOPICS

1. The multiplication principle of counting
2. Permutations
3. Combinations
4. The binomial and multinomial theorems

Topic 1: The Multiplication Principle of Counting

Combinatorics can be informally defined as "the science of counting." Such a simple definition is somewhat misleading, however, and runs the risk of trivializing the topic. A better – though less precise – definition of combinatorics is "the study of techniques used to determine the sizes, or cardinalities, of sets." Even this may sound deceptively simple until a few examples are studied. As we will see, there are many occasions when the size of a perfectly well-defined set may be difficult to determine at first glance. We will also see that in many cases the size of a set is of more importance than the actual elements of the set.

One of the joys of combinatorics is that there are many problems that can be solved with nothing more than a few fairly intuitive ideas, the most basic of which is the *Multiplication Principle of Counting*. Before formally stating the principle, we will use it to solve a problem.

example 1

In the United States, telephone numbers consist of a 3-digit area code followed by a 7-digit local number. Neither the first digit of the area code nor the local number can be 0 or 1. How many such phone numbers are there if:

a. There is a further restriction that the middle digit of the area code must be 0 or 1?
b. There are no further restrictions?

cont'd. on next page ...

Solutions:

a. Notice that this problem, in mathematical terms, concerns finding the cardinality of the set of all phone numbers of a certain form, but the words "set" and "cardinality" don't appear in the statement of the problem. This is very typical. Most of the combinatorics problems we will see use informal language to define a set, and then ask us, with equally informal language, to determine its size.

The method we will use to solve this problem is also very typical. We will count all the possible phone numbers by considering how we could go about constructing them. Every such phone number consists of a string of ten digits, with the restrictions that the first and fourth digits (reading from left to right) can't be either 0 or 1, but the second digit must be either 0 or 1. Since there are ten digits in all (0 through 9), this means there are eight possible ways to choose a digit for the first and fourth "slots", only two possible ways to choose a digit for the second slot, and ten possible ways to choose digits for all the remaining slots. This is illustrated in Figure 1.

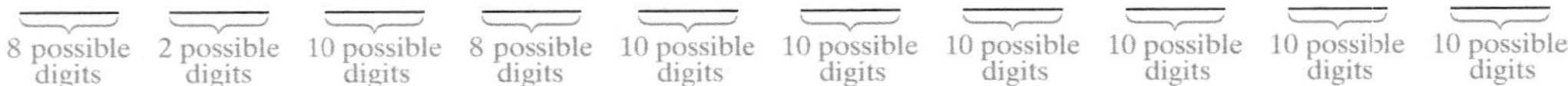

Figure 1: Constructing Allowable Phone Numbers

For the moment, consider how just the first two slots can be filled. Any of the eight allowable digits for the first slot can be paired up with either of the two allowable digits for the second slot, meaning that there are 8×2 ways of filling the first two slots altogether. In fact, all of the sixteen possible choices for the first two slots can be easily listed:

$$\underbrace{\underline{2}\,\underline{0}, \underline{3}\,\underline{0}, \underline{4}\,\underline{0}, \underline{5}\,\underline{0}, \underline{6}\,\underline{0}, \underline{7}\,\underline{0}, \underline{8}\,\underline{0}, \underline{9}\,\underline{0}}_{\text{8 ending in 0}},\ \underbrace{\underline{2}\,\underline{1}, \underline{3}\,\underline{1}, \underline{4}\,\underline{1}, \underline{5}\,\underline{1}, \underline{6}\,\underline{1}, \underline{7}\,\underline{1}, \underline{8}\,\underline{1}, \underline{9}\,\underline{1}}_{\text{8 ending in 1}}$$

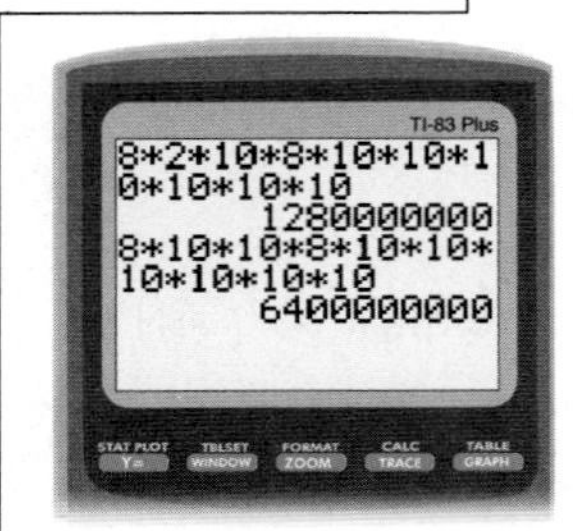

Now, any of these sixteen possible choices for the first two slots can be matched with any of the ten possible digits for the third slot, giving us a total of 160 ways of filling the first three slots. This pattern continues, so that the total number of phone numbers of the required form is

$$8 \times 2 \times 10 \times 8 \times 10 \times 10 \times 10 \times 10 \times 10 \times 10 = 1,280,000,000.$$

b. The same sort of argument applies if there is no restriction on the middle digit of the area code. In this case, the total number of allowable phone numbers is

$$8\times10\times10\times8\times10\times10\times10\times10\times10\times10 = 6,400,000,000.$$

The generalization of the reasoning we used in Example 1 is often stated in terms of a sequence of events.

The Multiplication Principle of Counting

Suppose E_1, E_2, …, E_n is a sequence of events, each of which has a certain number of possible outcomes. Suppose event E_1 has m_1 possible outcomes, and that after event E_1 has occurred, event E_2 has m_2 possible outcomes. Similarly, after event E_2, event E_3 has m_3 possible outcomes, and so on. Then the total number of ways that all n events can occur is $m_1 \cdot m_2 \cdot \ldots \cdot m_n$.

An alternative interpretation of the principle is to think of the events as a sequence of tasks to be completed. In Example 1, for instance, each task consists of selecting a digit for a given slot. There are ten tasks in all, and the product of the number of ways each task can be performed gives us the total number of phone numbers.

example 2

A certain state specifies that all non-personalized license plates consist of two letters followed by four digits, and that the letter O (which could be mistaken for the digit 0) cannot be used. How many such license plates are there?

cont'd. on next page ...

Solution:

Generating such a license plate is a matter of choosing six characters, the first two of which can be any of 25 letters and the last four of which can be any of 10 digits. The total number of such license plates is thus $25 \times 25 \times 10 \times 10 \times 10 \times 10 = 6,250,000$.

Topic 2: Permutations

A *permutation* of a set of objects is a linear arrangement of the objects. In any such arrangement, one of the objects is first, another is second, and so on, so the construction of a permutation of n objects consists of "filling in" n slots. We can thus use the Multiplication Principle of Counting to determine the number of permutations of a given set of objects.

example 3

One brand of combination lock for a door consists of five buttons labeled A, B, C, D, and E, and the installer of the lock can set the combination to be any permutation of the five letters. How many such permutations are there?

Solution:

The difference between this problem and the first two examples is that once a letter has been chosen for a given slot, it can't be reused. So in constructing a combination code, there are five choices for the first letter but only four choices for the second letter, since whatever letter was used first cannot be used again.

Similarly, there are only three choices for the third slot and only two choices for the fourth slot. Finally, whichever of the five letters is left *must* be used for the fifth slot, so there is only one choice. Figure 2 illustrates the slot-filling process.

$\underbrace{\quad}_{\text{5 choices}}\ \underbrace{\quad}_{\text{4 choices}}\ \underbrace{\quad}_{\text{3 choices}}\ \underbrace{\quad}_{\text{2 choices}}\ \underbrace{\quad}_{\text{1 choice}}$

Figure 2: Constructing a Permutation of Five Characters

We can now use the Multiplication Principle of Counting to determine that there are $5\times4\times3\times2\times1=120$ such combinations.

Products of the form $n\times(n-1)\times(n-2)\times\ldots\times2\times1$ occur so frequently that it makes sense to assign some shorthand notation to them, as follows.

Factorial Notation

If n is a positive integer, we use the notation $n!$ (which is read "***n* factorial**") to stand for the product of all the integers from 1 to n. That is,

$$n!=n\times(n-1)\times(n-2)\times\ldots\times2\times1$$

In addition, we define $0!=1$.

(Note that the factorial operation is defined only for positive integers and, as a special case, 0. You may encounter, in a later math class, a related function called the gamma function that extends the factorial idea to all positive real numbers.)

If all permutation problems involved nothing more than a straightforward application of the Multiplication Principle, as in Example 3, there would be little left to say. But typically, the solution of a permutation problem requires counting the number of ways that a linear arrangement of k objects can be made from a collection of n objects, where $k\leq n$. Consider the following problem.

example 4

How many different five-letter combination codes are possible if every letter of the alphabet can be used, but no letter may be repeated?

Solution:

The difference between this problem and the one in Example 3 is that there are now 26 choices for the first letter, 25 choices for the second, and so on. The corresponding "slot diagram" describing the number of ways each slot can be filled appears in Figure 3.

26 choices 25 choices 24 choices 23 choices 22 choices

Figure 3: Five-Letter Permutations Chosen from 26 Characters

The total number of such combination codes is thus $26\times25\times24\times23\times22 = 7{,}893{,}600$.

Products like the one in Example 4 are also very common, and deserve special mention. To motivate the following formula, note that we can state the answer to Example 4 in terms of factorials as follows:

$$26\times25\times24\times23\times22 = \frac{(26\times25\times24\times23\times22)\times21!}{21!} = \frac{26!}{21!}.$$

If we generalize this, we obtain a formula for the number of permutations of length k that can be formed from a collection of n objects, usually expressed as the number of permutations of n objects taken k at a time.

Permutation Formula

The number of permutations of n objects taken k at a time is

$${}_nP_k = \frac{n!}{(n-k)!}.$$

This is a conveniently short way of expressing the product

$${}_nP_k = n\times(n-1)\times\ldots\times(n-k+1).$$

The formula for ${}_nP_k$ is especially useful when the number of factors to be multiplied is large, as in the next example, assuming you have access to a calculator or computer with the factorial function.

example 5

A magician is preparing to demonstrate a card trick that involves 20 cards chosen at random from a standard deck of cards. Once chosen, the 20 cards are arranged in a row, and the order of the cards plays a role in the trick. How many such orderings are possible?

To find the number of permutations, enter the first number, n (52); select Math and use the right arrow to highlight `Prb` (probability). Then select `2:nPr` and enter the second number, r (20).

Solution:

A standard deck of cards contains 52 distinct cards (13 cards in each of 4 suits), and this problem asks for the number of permutations of 52 cards taken 20 at a time. One way to determine this would be to evaluate $52 \times 51 \times \ldots \times 33$ (note that this is a product of 20 numbers), but a far faster way is to use the permutation formula:

$$ {}_{52}P_{20} = \frac{52!}{(52-20)!} = \frac{52!}{32!} \approx \frac{8.07 \times 10^{67}}{2.63 \times 10^{35}} \approx 3.07 \times 10^{32} $$

The two factorials that appear in the formula above have been determined by a calculator. Note that $52!$, $32!$, and the final answer are all very large numbers, so it is convenient to use scientific notation.

Topic 3: Combinations

In contrast to permutations, where the order of the objects under consideration is of great importance, combinations are simply collections of objects. To be specific, a *combination* of n objects taken k at a time is one of the ways of forming a subset of size k from a set of size n, where again $k \leq n$. Combination problems typically ask us to determine the number of different subsets of size k that can be formed.

To emphasize the difference between permutations and combinations, and to begin to understand the combination formula that we will soon derive, consider the following problem.

example 6

Let $S = \{a, b, c, d\}$.

a. How many permutations of size 3 can be formed from the set S?

b. How many combinations of size 3 can be formed from the set S?

Solutions:

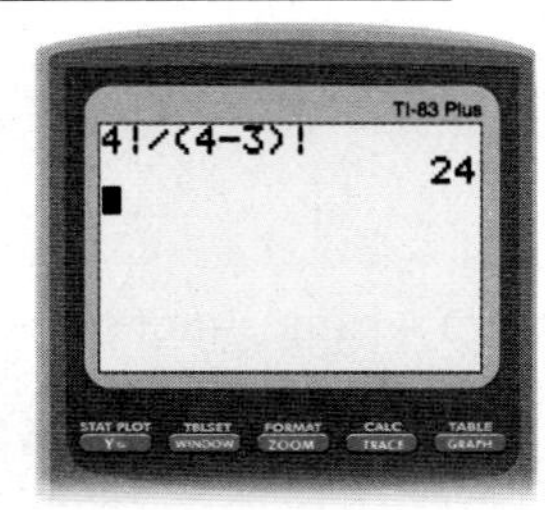

To enter the factorial sign, select Math, highlight `Prb`, and enter `4:!`.

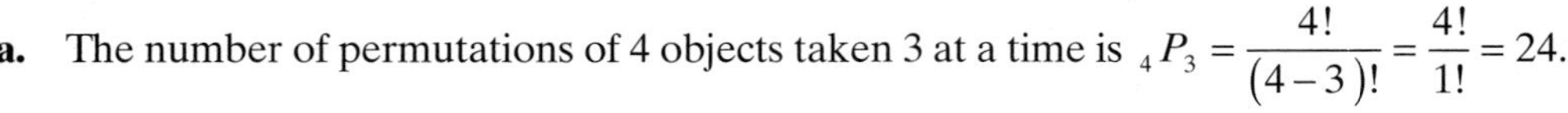

a. The number of permutations of 4 objects taken 3 at a time is ${}_4P_3 = \dfrac{4!}{(4-3)!} = \dfrac{4!}{1!} = 24.$

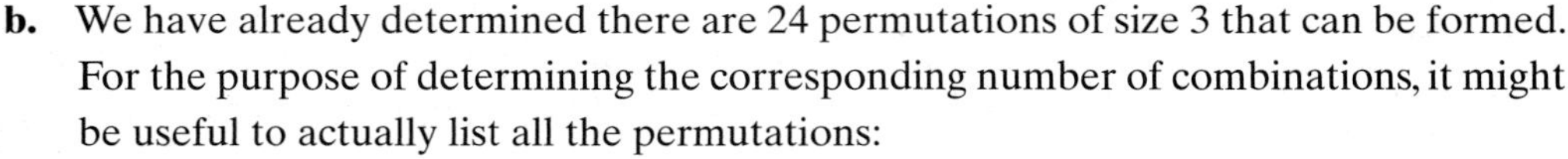

b. We have already determined there are 24 permutations of size 3 that can be formed. For the purpose of determining the corresponding number of combinations, it might be useful to actually list all the permutations:

abc	*acb*	*bac*	*bca*	*cab*	*cba*
abd	*adb*	*bad*	*bda*	*dab*	*dba*
acd	*adc*	*cad*	*cda*	*dac*	*dca*
bcd	*bdc*	*cbd*	*cdb*	*dbc*	*dcb*

If we now view these collections of objects as sets, all six permutations in the first row describe the single set (a, b, c). Similarly, the six permutations in the second row are simply six different ways of describing the set (a, b, d). The third row describes the set (a, c, d) and the fourth row describes the set (b, c, d). In all, there are only four combinations of 4 objects taken 3 at a time.

Even though we don't yet have the formula we seek, let ${}_nC_k$ denote the number of combinations of n objects taken k at a time. Taking a cue from Example 6, we might work toward the desired formula by noting that each of the size k combinations formed from S gives rise to $k!$ permutations of size k, and we know that there are ${}_nP_k$ permutations taken k at a time overall. This means that

$$(k!)({}_nC_k) = {}_nP_k, \text{ or } {}_nC_k = \frac{{}_nP_k}{k!}$$

If we now replace ${}_nP_k$ with $\frac{n!}{(n-k)!}$, we have the following formula.

Combination Formula

The number of combinations of n objects taken k at a time is

$${}_nC_k = \frac{n!}{(k!)(n-k)!}.$$

At this point, we have seen all of the counting techniques we will need. There is much more to the subject of combinatorics, but the three ideas we have discussed – the Multiplication Principle of Counting, the permutation formula, and the combination formula – are sufficient to answer many, many questions. This is especially true when the techniques are combined in various ways, as the next few examples show.

example 7

Suppose a Senate committee consists of 11 Democrats, 10 Republicans, and 1 Independent. The chair of the committee wants to form a sub-committee to be charged with researching a particular issue, and decides to appoint 3 Democrats, 2 Republicans, and the 1 Independent to the sub-committee. How many different sub-committees are possible?

cont'd. on next page ...

To find the number of combinations, enter the first number, n (11); select Math and use the right arrow to highlight Prb (probability). Then select 3:nCr and enter the second number, r (3).

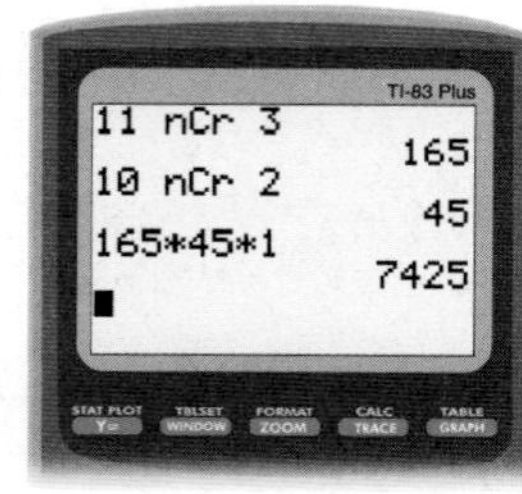

Solution:

This is in large part a combination problem, as the chair needs to form a subset of size 3 from the 11 Democrats, a subset of size 2 from the 10 Republicans, and a subset of size 1 from the 1 Independent. (Note that the order of those chosen is irrelevant, which is why this is a combination problem and not a permutation problem.) The combination formula tells us that these tasks can be done in, respectively, ${}_{11}C_3$, ${}_{10}C_2$, and ${}_1C_1$ ways (of course, there is only 1 way to choose 1 member from a set of 1). These numbers are:

$${}_{11}C_3 = \frac{11!}{3!8!} = \frac{11\times10\times9\times\cancel{8!}}{3!\cancel{8!}} = \frac{990}{6} = 165,$$

$${}_{10}C_2 = \frac{10!}{2!8!} = \frac{10\times9\times\cancel{8!}}{2!\cancel{8!}} = \frac{90}{2} = 45, \text{ and}$$

$${}_1C_1 = \frac{1!}{1!0!} = \frac{1}{1} = 1 \text{ (remember that } 0! = 1\text{)}.$$

Once the appropriate number of people from each party have been chosen, any of the 165 possible groups of 3 Democrats can be matched up with any of the 45 possible groups of 2 Republicans, and then further matched up with the 1 Independent. The Multiplication Principle thus tells us there are $165\times45\times1 = 7425$ ways of forming the desired sub-committee (any one of which, inevitably, is bound to offend one party or another).

example 8

How many different arrangements are there of the letters in the following words?

a. STIPEND **b.** SALAAM **c.** MISSISSIPPI

Solutions:

a. The goal is to count the total number of "words" that can be formed from the letters in the word STIPEND, using each letter once and only once (of course, most of the arrangements will not actually be legitimate English words). Since the order of the letters is important, and since STIPEND contains 7 distinct letters, the answer is

$${}_7P_7 = \frac{7!}{0!} = 7! = 5040.$$

b. The word SALAAM contains 6 letters, but only 4 distinct letters. That is, the 3 A's are indistinguishable, so the answer is not simply ${}_6P_6$. In fact, 6! overcounts the total number of arrangements by a factor of 3!, because any one arrangement of the 6 letters in SALAAM is equivalent to 5 more arrangements. This is because the 3 A's can be permuted in 3! = 6 ways, as is evident if we temporarily color the A's differently:

MAALSA, MAALSA, MAALSA, MAALSA, MAALSA, MAALSA

This means the total number of arrangements is actually

$$\frac{6!}{3!} = \frac{720}{6} = 120.$$

c. MISSISSIPPI contains 11 characters, but 4 of them are S's, 4 of them are I's, 2 of them are P's, and the remaining 1 character is M. If the 11 characters were all distinct, the total number of arrangements of the letters would be $11!$, but we need to divide this number by $4!$ to account for the indistinguishable S's, and then divide again by $4!$ to account for the I's, and then again by $2!$ to account for the P's. This gives us a total of

$$\frac{11!}{4!4!2!} = \frac{39,916,800}{(24)(24)(2)} = 34,650.$$

Topic 4: The Binomial and Multinomial Theorems

Consider the expressions $(x+y)^7$ and $(a+b+c)^3$. The first is a binomial raised to a power, and the second is a trinomial raised to a power. Expressions like these (or even more complicated) can occur in the course of solving an algebra problem, and we often need to expand them in order to work toward the solution of the problem.

We could use only elementary methods to expand such expressions. For example, we could expand $(x+y)^7$ as follows:

$$\begin{aligned}(x+y)^7 &= (x+y)(x+y)(x+y)^5\\ &= \left(x^2+2xy+y^2\right)(x+y)^5\\ &= \left(x^2+2xy+y^2\right)(x+y)(x+y)^4\\ &= \ldots\end{aligned}$$

But such work can be extremely tedious and error-prone if the exponent is larger than, say, 3 or 4. We can use the combinatorics methods we have seen to develop two formulas that greatly simplify the process.

When we expand an expression like $(x+y)^7$, we are really making sure we account for all the possible products that result from taking one term from each factor. For instance, when we multiply the terms that are boxed below, we obtain x^3y^4:

$$(\boxed{x}+y)(x+\boxed{y})(x+\boxed{y})(\boxed{x}+y)(x+\boxed{y})(\boxed{x}+y)(x+\boxed{y})$$

But there are many other choices of terms that also lead to x^3y^4, such as:

$$(x+\boxed{y})(x+\boxed{y})(\boxed{x}+y)(\boxed{x}+y)(\boxed{x}+y)(x+\boxed{y})(x+\boxed{y}).$$

In all, there are 35 different ways of boxing 3 x's and 4 y's, meaning that there is a coefficient of 35 in front of x^3y^4 in the expansion of $(x+y)^7$.

Where did this number of 35 come from? We can relate the expansion of binomials (or, more generally, multinomials) to the "word" problems we studied in Example 8. For instance, the coefficient of x^3y^4 is the number of different arrangements of the letters $xxxyyyy$. Since these 7 letters consist of 3 x's and 4 y's, the total number of such arrangements is

$$\frac{7!}{3!4!}=\frac{5040}{(6)(24)}=35.$$

(Note that the boxed letters above correspond to the "words" $xyyxyxy$ and $yyxxxyy$.)

There is some special notation and language for the type of work we have just done.

Binomial Coefficients

Given non-negative integers n and k, with $k \le n$, we define

$$\binom{n}{k}=\frac{n!}{k!(n-k)!}$$

The expression $\binom{n}{k}$ is called a **binomial coefficient**, as it corresponds to the coefficient of $x^k y^{n-k}$ in the expansion of $(x+y)^n$.

We have already seen the formula for binomial coefficients in another context. Note that

$$\binom{n}{k}={}_nC_k.$$

The last step in expanding a binomial, now that we know how to find the coefficients, is to put all the pieces together. Consider again the expression $(x+y)^7$. When this is expanded, there will be terms containing y^7, xy^6, x^2y^5, x^3y^4, x^4y^3, x^5y^2, x^6y, and x^7 each multiplied by the appropriate binomial coefficient. Note that the sum of the exponents in each term is 7, as this is the total number of x's and y's in each term.

example 9

Expand the expression $(x+y)^7$.

Solution:

Based on the above reasoning, and the binomial coefficient formula,

$$\begin{aligned}&(x+y)^7\\&=\binom{7}{0}x^0y^7+\binom{7}{1}x^1y^6+\binom{7}{2}x^2y^5+\binom{7}{3}x^3y^4+\binom{7}{4}x^4y^3+\binom{7}{5}x^5y^2+\binom{7}{6}x^6y^1+\binom{7}{7}x^7y^0\\&=y^7+7xy^6+21x^2y^5+35x^3y^4+35x^4y^3+21x^5y^2+7x^6y+x^7\end{aligned}$$

The generalization of our work is called the Binomial Theorem.

The Binomial Theorem

Given the positive integer n, and any two expressions A and B,

$$\begin{aligned}(A+B)^n&=\sum_{k=0}^{n}\binom{n}{k}A^kB^{n-k}\\&=\binom{n}{0}A^0B^n+\binom{n}{1}A^1B^{n-1}+\binom{n}{2}A^2B^{n-2}+\ldots+\binom{n}{n-1}A^{n-1}B^1+\binom{n}{n}A^nB^0.\end{aligned}$$

Note that A and B can be more complicated than we have considered so far.

example 10

Expand the expression $(2x-y)^4$.

Solution:

To use the Binomial Theorem, note that $A = 2x$ and $B = -y$. The theorem thus tells us that

$$\begin{aligned}
&(2x-y)^4 \\
&= \binom{4}{0}(2x)^0(-y)^4 + \binom{4}{1}(2x)^1(-y)^3 + \binom{4}{2}(2x)^2(-y)^2 + \binom{4}{3}(2x)^3(-y)^1 + \binom{4}{4}(2x)^4(-y)^0 \\
&= (-y)^4 + 4(2x)(-y)^3 + 6(2x)^2(-y)^2 + 4(2x)^3(-y) + (2x)^4 \\
&= y^4 - 8xy^3 + 24x^2y^2 - 32x^3y + 16x^4.
\end{aligned}$$

We will finish with one last definition and formula. To expand an expression like $(a+b+c)^3$ we need to determine the coefficients of terms of the form $a^{k_1} b^{k_2} c^{k_3}$, where $k_1 + k_2 + k_3 = 3$. Reasoning similar to that preceding the Binomial Theorem leads to the following definition.

Multinomial Coefficients

Let n be a positive integer, and let $k_1, k_2, \ldots, k_r$ be non-negative integers such that $k_1 + k_2 + \ldots + k_r = n$. We define

$$\binom{n}{k_1, k_2, \ldots, k_r} = \frac{n!}{k_1! k_2! \cdots k_r!}.$$

Such an expression is called a multinomial coefficient. It is the coefficient of the term $A_1^{k_1} A_2^{k_2} \cdots A_r^{k_r}$ in the expansion of $(A_1 + A_2 + \ldots + A_r)^n$.

To expand something of the form $(A_1 + A_2 + \ldots + A_r)^n$, we identify all the terms of the form $A_1^{k_1} A_2^{k_2} \cdots A_r^{k_r}$, multiply each one by its corresponding multinomial coefficient, and add them up. This is called the Multinomial Theorem.

The Multinomial Theorem

Given the positive integer n and expressions $A_1, A_2, \ldots, A_r$,

$$(A_1 + A_2 + \ldots + A_r)^n = \sum_{k_1+k_2+\ldots+k_r=n} \binom{n}{k_1, k_2, \ldots, k_r} A_1^{k_1} A_2^{k_2} \cdots A_r^{k_r}.$$

example 11

Expand the expression $(a+b+c)^3$.

Solution:

We will start by noting that the expansion will have terms containing a^3, a^2b, a^2c, ab^2, ac^2, abc, b^3, b^2c, bc^2, and c^3. (Note that these are all the ways that three non-negative integers can add up to 3.) Each of these terms must be multiplied by its corresponding multinomial coefficient. For instance, the coefficient of a^2c is

$$\binom{3}{2,0,1} = \frac{3!}{2!0!1!} = \frac{6}{2} = 3.$$

Similarly, the coefficient of b^3 is

$$\binom{3}{0,3,0} = \frac{3!}{0!3!0!} = \frac{6}{6} = 1.$$

Altogether, we obtain

$$(a+b+c)^3 = a^3 + 3a^2b + 3a^2c + 3ab^2 + 3ac^2 + 6abc + b^3 + 3b^2c + 3bc^2 + c^3.$$

exercises

Use the Multiplication Principle of Counting to answer the following questions. See Examples 1 and 2.

1. Suppose you write down someone's phone number on a piece of paper, but then accidentally wash it along with your laundry. Upon drying the paper, all you can make out of the number is 42? – 3?7?. How many different phone numbers fit this pattern?

2. How many different 7-digit phone numbers contain no odd digits? (Ignore the fact that certain 7-digit sequences are disallowed as phone numbers.)

3. How many different 7-digit phone numbers do not contain the digit 9? (Ignore the fact that certain 7-digit sequences are disallowed as phone numbers.)

4. A certain combination lock allows the buyer to set any combination of five letters, with repetition allowed, but each of the letters must be A, B, C, D, E, or F. How many combinations are possible?

5. In how many different orders can 15 runners finish a race?

6. How many different 4-letter radio-station names can be made, assuming the first letter must be a K or a W? Assume repetition of letters is allowed.

7. How many different 4-letter radio-station names can be made from the call-letters K, N, I, T, assuming the letter K must appear first? Each of the four letters can be used only once.

8. Three men and three women line up in a row for a photograph, and decide men and women should alternate. In how many different ways can this be done? (Don't forget that the left-most person can be a man or a woman.)

9. How many different ways can a 10-question multiple choice test be answered, assuming every question has 5 possible answers?

10. How many different ways can your 12 favorite math books be arranged in a row?

11. How many different 6-character license plates can be formed if all 26 letters and 10 numerical digits can be used with repetition?

12. How many different 6-character license plates can be formed if all 26 letters and 10 numerical digits can be used without repetition?

13. How many different 6-character license plates can be formed if the first 3 places must be letters and the last 3 places must be numerical digits? (Assume repetition is not allowed.)

14. A box of crayons comes with 8 different colored crayons arranged in a single row. How many different ways can the crayons be ordered in the box?

Express the answers to the following permutation problems using permutation notation $\left({}_nP_k\right)$ *and numerically. See Examples 3, 4, and 5.*

15. Suppose you have a collection of 30 cherished math books. How many different ways can you choose 12 of them to arrange in a row?

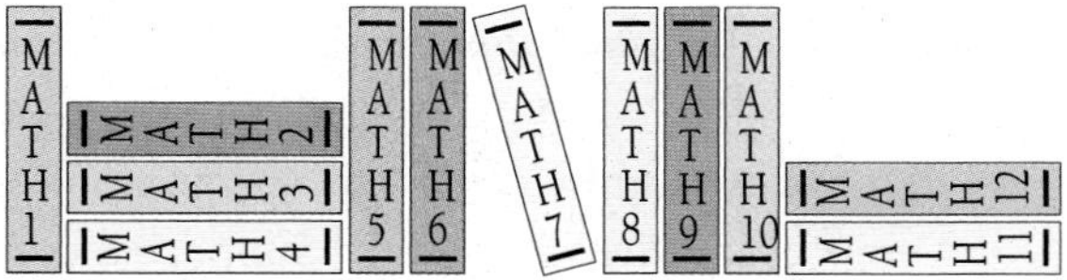

16. In how many different ways can first-place, second-place, and third-place be decided in a 15-person race?

17. Suppose you need to select a user-ID for a computer account, and the system administrator requires that each ID consist of 8 characters with no repetition allowed. The characters you may choose from are the 26 letters of the alphabet (with no distinction between upper-case and lower-case) and the 10 digits. How many choices for a user-ID do you have?

18. How many different 5-letter "words" (they don't have to be actual English words) can be formed from the letters in the word *PLASTIC*?

19. Seven children rush into a room in which six chairs are lined up in a row. How many different ways can six of the seven children choose a chair to sit in? (The seventh remains standing.) How does the answer differ if there are seven chairs in the room?

20. At a meeting of 17 people, a president, vice president, secretary, and treasurer are to be chosen. How many different ways can these positions be filled?

21. Given 26 building blocks, each with a different letter of the alphabet printed on it, how many different 3-letter "words" can be formed?

Solve the following permutation expressions. See Examples 5 and 6.

22. ${}_4P_2$

23. ${}_{15}P_2$

24. ${}_6P_5$

25. ${}_{19}P_{17}$

Solve the following combination expressions. See Example 7.

26. ${}_6C_4$

27. ${}_4C_2$

28. ${}_{12}C_5$

29. ${}_{21}C_{14}$

Express the answer to the following combination problems using combination notation $\left({}_nC_k\right)$ *and numerically. See Examples 6 and 7.*

30. In many countries, it is not uncommon for quite a few political parties to have their representatives in power. Suppose a committee composed of 10 Conservatives, 13 Liberals, 6 Greens, and 4 Socialists decides to form a sub-committee consisting of 3 Conservatives, 4 Liberals, 2 Greens, and 1 Socialist. How many different such sub-committees can be formed?

31. A trade-union asks its members to select 3 people, from a slate of 7, to serve as representatives at a national meeting. How many different sets of 3 can be chosen?

32. Many lottery games are set up so that players select a subset of numbers from a larger set, and the winner is the person whose selection matches that chosen by some random mechanism. The order of the numbers is irrelevant. How many choices of six numbers can be made from the numbers 1 through 49?

33. How many different lines can be drawn through a set of nine points in the plane, assuming that no three of the points are collinear? (Points are said to be *collinear* if a single line containing them can be drawn.)

34. Suppose you are taking a 10-question True-False test, and you are guessing that the professor has arranged it so that five of the answers are True and five are False. How many different ways are there of marking the test with five True answers and five False answers?

35. A caller in a Bingo game draws 5 marked ping pong balls from a basket of 75 and calls the numbers out to the players. How many different combinations are possible assuming that the order is irrelevant?

B	I	N	G	O
10	17	39	49	64
12	21	36	55	62
14	25	FREE!	52	70
7	19	32	56	68
5	24	34	54	71

Use a combination of the techniques seen in this section, if necessary, to answer the following questions. See Examples 1 through 8.

36. How many different "words" can be formed by rearranging the letters in the word *ABYSS*?

37. How many different "words" can be formed by rearranging the letters in the word *BANANA*?

38. How many different ways are there of choosing five cards from a standard 52-card deck and arranging them in a row? How many different five-card hands can be dealt from a standard 52-card deck?

39. Suppose you have 10 Physics texts, 8 Computer Science texts, and 13 Math texts. How many different ways can you select 4 of each to take with you on vacation?

40. Suppose you have 10 Physics texts, 8 Computer Science texts, and 13 Math texts. How many different ways can you select 4 of each and then arrange them in a row on a shelf, so that the books are grouped by discipline?

41. A certain ice cream store has four different kinds of cones and 28 different flavors of ice cream. How many different single-scoop ice cream cones is it possible to order at this ice cream store?

42. If a local pizza shop has three different types of crust, two different kinds of sauce, and five different toppings, how many different one topping pizzas can be ordered?

43. A man has 8 different shirts, 4 different shorts, and 3 different pairs of shoes. How many different outfits can the man choose from?

44. A couple wants to have three children. They want to know the different possible gender outcomes for birth order. How many different birth orders are possible?

45. A student has to make out his schedule for classes next fall. He has to take a math class, a science class, an elective, a history class, and an English class. There are three math classes to choose from, two science classes, four electives, three history classes, and four English classes. How many different schedules could the student have?

46. A basketball team has 12 different people on the team. The team consists of three point guards, two shooting guards, one weak forward, three power forwards, and three centers. How many different starting line-ups are possible? (The starting line-up will consist of one player in each of the 5 positions.)

47. How many 5-letter strings can be formed using the letters V, W, X, Y, and Z, if the same letter cannot be repeated?

48. If at the racetrack nine greyhounds are racing against each other, how many different first, second, and third place finishes are possible?

49. A basketball tryout has four distinct positions available on the team. If 25 people show up for tryouts, how many different ways can the four positions be filled?

50. If a trumpet player is practicing eight different pieces of music, how many different orders can he play the music in?

51. A pizza place has 12 total toppings to choose from. How many different 4-topping pizzas can be ordered?

52. How many different 5-digit numbers can be formed using each of the numbers 6, 8, 1, 9, and 4?

53. If eight cards are chosen randomly from a deck of 52, how many possible groups of eight can be chosen?

54. A baseball team has 15 players on the roster and a batting line-up consists of 9 players. How many different batting line-ups are possible?

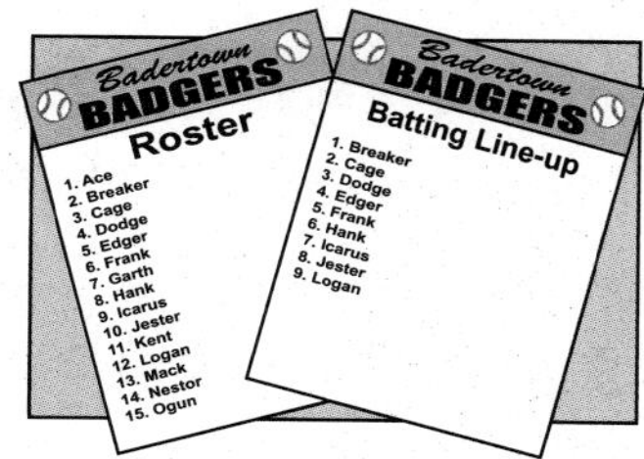

55. How many different ways can two red balls, one orange ball, one black ball, and three yellow balls be arranged?

Use the Binomial and Multinomial Theorems in the following problems. See Examples 9, 10, and 11.

56. Expand the expression $(3x+y)^5$.

57. Expand the expression $(x-2y)^7$.

58. Expand the expression $(x-3)^4$.

59. Expand the expression $\left(x^2-y^3\right)^4$.

60. Expand the expression $\left(6x^2+y\right)^5$.

61. Expand the expression $\left(4x+5y^2\right)^6$.

62. Expand the expression $\left(7x^2+8y^2\right)^4$.

63. Expand the expression $\left(x^3-y^2\right)^5$.

64. Expand the expression $(x+y+z)^2$.

65. Expand the expression $(a-2b+c)^3$.

66. Expand the expression $(2x+5)^6$.

67. Expand the expression $(2x+3y-z)^3$.

68. What is the coefficient of the term containing x^3y in the expansion of $(2x+y)^4$?

69. What is the coefficient of the term containing x^4y^3 in the expansion of $\left(x^2-2y\right)^5$?

70. Find the first four terms in the expansion of $(x+3y)^{16}$.

71. Find the first three terms in the expansion of $(2x+3)^{13}$.

72. Find the first two terms in the expansion of $\left(3x^{\frac{1}{4}}+5y\right)^{17}$.

73. Find the 11th term in the expansion of $(x+2)^{24}$.

74. Find the 17th term in the expansion of $(2x+1)^{21}$.

75. Find the 9th term in the expansion of $(x-6y)^{12}$.

Pascal's triangle is a triangular arrangement of binomial coefficients, the first few rows of which appear as follows:

$$\begin{array}{c} 1 \\ 1 \quad 1 \\ 1 \quad 2 \quad 1 \\ 1 \quad 3 \quad 3 \quad 1 \\ 1 \quad 4 \quad 6 \quad 4 \quad 1 \\ \vdots \end{array}$$

Each number (aside from those on the perimeter of the triangle) is the sum of the two numbers diagonally adjacent to it in the previous row. Pascal's triangle is a useful way of generating binomial coefficients, with the nth row containing the coefficients of a binomial raised to the $(n-1)^{\text{st}}$ power. It can also be used to suggest useful relationships between binomial coefficients.

Prove each of the following such relationships algebraically.

76. $\binom{n}{k}=\binom{n-1}{k-1}+\binom{n-1}{k}$ (Note that this is a restatement of how Pascal's triangle is formed.)

77. $\binom{n}{k}=\binom{n}{n-k}$

78. $\binom{n}{0}=\binom{n}{n}=1$

79. $\binom{n}{0}+\binom{n}{1}+\ldots+\binom{n}{n}=2^n$ (**Hint:** use the Binomial Theorem on $(x+y)^n$ for a convenient choice of x and y.)

11.6 An Introduction to Probability

TOPICS

1. The language of probability
2. Using combinatorics to compute probabilities
3. Unions, intersections, and independent events

Topic 1: The Language of Probability

We all make use of the concept of *probability* nearly every day, and in a wide variety of ways. Most of the time, the use is informal; we might briefly wonder how likely it is we'll run out of gas as we drive to an appointment, or what the probability of rain is during an afternoon game of tennis, or what the chances are of striking it rich by winning the state lottery. In mathematics, probability is much more rigorously defined, and the results of probabilistic analysis are used in many important (and some less important, but entertaining) applications.

Mathematical probability, like combinatorics, is a huge subject, and this section will do no more than introduce the topic. Fittingly, most of the probability questions we will consider can be answered with the combinatorics techniques learned in the last section. To see why, we need to first define the terminology of probability.

An **experiment**, in probability theory, is any activity that results in well-defined **outcomes**. In a given problem, we are usually concerned with finding the probability that one or more of the outcomes will occur, and we use the word **event** to refer to a set of outcomes. (An event may be a set containing just one element; that is, an event may consist of a single outcome of an experiment.) The set of all possible outcomes of a given experiment is called the **sample space** of the experiment. This means that in the language of sets, an event is any subset of the sample space. (Note that the empty set thus meets the definition of an event, as does the entire sample space. We are usually interested in finding the probability of events that fall somewhere between being empty and being everything.)

For example, rolling a standard die qualifies as an experiment, and the outcomes (as in, the number showing on the face-up side when it stops rolling) can be reasonably described with the numbers 1, 2, 3, 4, 5, and 6. So the sample space of this experiment is the set $S = \{1, 2, 3, 4, 5, 6\}$, and the events consist of all the possible subsets of S. We might, for instance, be interested in the event E defined as "the number rolled is even." In terms of sets, $E = \{2, 4, 6\}$. Our intuition tells us that, if the die is a fair one, the probability of E occurring is one-half, as half the numbers are even and half are odd. We denote this by writing $P(E) = \frac{1}{2}$, which is read "the probability of E is one-half."

The intuition we used in this first simple experiment points the way toward the main tool we have for calculating probabilities.

Probabilities when Outcomes are Equally Likely

We say that the outcomes of an experiment are *equally likely* when they all have the same probability of occurring. If E is an event of such an experiment, and if S is the sample space of the experiment, then the probability of E is given by

$$P(E) = \frac{n(E)}{n(S)}$$

where $n(E)$ and $n(S)$ are, respectively, the sizes of the sets E and S.

It is important to note that the formula above has some restrictions. First, it only applies in equally likely situations, so we can't use it to analyze weighted coins, crooked roulette tables, tampered decks of cards, and so on. Secondly, and more subtly, it assumes that the size of the sample space is finite (otherwise, the formula makes no sense), so we can't use it to analyze experiments based on, say, choosing a real number at random.

The formula also tells us, indirectly, that probabilities are always going to be real numbers between 0 and 1, inclusive. The probability of an event E is 0 if E is the empty set, and 1 if E is the entire sample space S. In all other (finite) cases, the size of E is going to be something between 0 and the size of S, so the fraction will yield a number between 0 and 1.

The complement of an event E, denoted E^C, is the set of all outcomes in the sample space that are not in E. Thus the probability of the complement of an event can be derived as:

$$P(E^C) = \frac{n(S) - n(E)}{n(S)} = \frac{n(S)}{n(S)} - \frac{n(E)}{n(S)} = 1 - P(E).$$

example 1

A (fair) die is rolled, and the number on top noted. Determine the probability that the number is:

a. prime **b.** divisible by 5 **c.** 7

Solutions:

a. If we let E denote the event in question, then $E=\{2, 3, 5\}$. Since $n(E)=3$ and the size of the sample space $S=\{1, 2, 3, 4, 5, 6\}$ is 6, $P(E)=\frac{1}{2}$.

b. If we let F denote the event, then $F=\{5\}$, as this is the only integer from 1 to 6 that is divisible by 5. So $P(F)=\frac{1}{6}$.

c. If we let G denote the event, then $G=\{\ \}$. In other words, G is just another name for the empty set, as there is no way for the top face to show a 7. So $P(G)=0$.

Topic 2: Using Combinatorics to Compute Probabilities

Of course, most meaningful probability questions are not as basic as those in Example 1, and what makes a probability problem more interesting usually involves finding the size of an event and/or the sample space. This is where combinatorics comes into play. Consider the following problems.

example 2

A pair of dice is rolled, and the sum of the top faces noted. What is the probability that the sum is:

a. 2 **b.** 5 **c.** 7 or 11

Solutions:

The size of the sample space is the same for all three questions, so it makes sense to determine this first. And, if we want to use the single probability formula that we have, it is important to make sure that all the outcomes of the sample space are equally likely.

cont'd. on next page ...

For this reason, we do *not* want to define the sample space to be all integers between 2 (the smallest possible sum) and 12 (the largest possible sum). The reason for this is that these sums are not all equally likely. For instance, there is only one way for the sum to be 2: both top faces must show a 1. But there are two ways for the sum to be 3: one die (call it die A) shows a 1 and the other (die B) shows a 2, or else die A shows a 2 and die B shows a 1. Similarly, there are *three* ways for the sum of the top faces to be 4. In order to make sense of this situation, and in order to define the sample space properly, it may help to construct a table of ordered pairs. In Figure 1 below, the first number in each pair corresponds to the number showing on die A and the second number corresponds to the number showing on die B.

(A, B)					
(1, 1)	(1, 2)	(1, 3)	(1, 4)	(1, 5)	(1, 6)
(2, 1)	(2, 2)	(2, 3)	(2, 4)	(2, 5)	(2, 6)
(3, 1)	(3, 2)	(3, 3)	(3, 4)	(3, 5)	(3, 6)
(4, 1)	(4, 2)	(4, 3)	(4, 4)	(4, 5)	(4, 6)
(5, 1)	(5, 2)	(5, 3)	(5, 4)	(5, 5)	(5, 6)
(6, 1)	(6, 2)	(6, 3)	(6, 4)	(6, 5)	(6, 6)

Figure 1: Possible Outcomes of a Pair of Dice

Each of these ordered pairs is equally likely to come up, as any of the numbers 1 through 6 are equally likely for the first slot (die A) and similarly for the second slot (die B). And referring to the positions in the ordered pairs as "slots" points to a quick way of determining the size of the sample space. Since there are 6 choices for each slot, the Multiplication Principle of Counting tells us there are 36 possible outcomes of this experiment.

We can now proceed to answer the three specific questions.

a. There is only one ordered pair corresponding to a sum of 2, namely (1, 1), so the probability of this event is $\frac{1}{36}$.

b. There are four ordered pairs corresponding to a sum of 5, namely the set

$$\{(1, 4), (2, 3), (3, 2)(4, 1)\},$$

so the probability of this event is $\frac{4}{36}$, or $\frac{1}{9}$.

c. A sum of 7 or 11 comes from any of the following ordered pairs:

$$\{(1, 6), (2, 5), (3, 4), (4, 3), (5, 2), (6, 1), (5, 6), (6, 5)\}$$

Since there are eight elements in this event, the probability of rolling a 7 or 11 is $\frac{8}{36}$, or $\frac{2}{9}$.

example 3

Suppose you are taking a 10-question True or False test, and that you are woefully unprepared for it. If you decide to guess on each question, what is the probability of getting 8 or more questions right?

Solution:

The sample space for this problem consists of all the possible sequences of 10 answers, each of which is True or False. Combinatorics tells us that there are $2^{10} = 1024$ such sequences (2 choices for each of 10 "slots"), and these 1024 possible sequences are equally likely if your choice of True or False on each question is truly random.

There is only one particular sequence of 10 answers that is completely correct, so your probability of randomly choosing this one sequence out of all 1024 possible sequences is $\frac{1}{1024}$. But there are more ways that you could guess the right answers on exactly 9 of the questions. You might, for instance, miss only question 3, or only question 8, to name two possibilities. This makes it clear, in fact, that there are 10 ways to miss exactly 1 question; stated another way, there are 10 ways to get exactly 9 questions right, so the probability of getting exactly 9 right is $\frac{10}{1024}$.

We now need to figure out the number of ways you could guess the right answer on exactly 8 questions. To do this, let us revisit the question of how you could get exactly 9 answers right. In the last paragraph, we resolved this by realizing that getting 9 right is equivalent to getting 1 wrong, and there are 10 ways of doing this. A more sophisticated way is to realize that there are ${}_{10}C_9$ ways of selecting 9 of the 10 questions (the 9 for which you guess the right answer), and then noting that

$${}_{10}C_9 = \frac{10!}{9!1!} = 10$$

Similarly, there are ${}_{10}C_8$ ways of selecting 8 of the 10 questions, so there are

$${}_{10}C_8 = \frac{10!}{8!2!} = \frac{10 \times 9}{2} = 45$$

ways of getting exactly 8 questions right.

Altogether, there are $45 + 10 + 1 = 56$ ways of getting 8, 9, or 10 questions right, so the probability of this happening is

$$\frac{56}{1024} \approx 0.055$$

In other words, you only have a 5.5% chance of scoring 8 or better by guessing, so studying ahead of time is the way to go (but you knew that already.)

Topic 3: Unions, Intersections, and Independent Events

Since mathematical probability is defined in terms of sets (in the form of sample spaces and events), it is natural to use terms like "union" and "intersection" to describe ways of combining events. The notion of independence of events, on the other hand, is unique to probability. We will look first at probabilities of unions and intersections of events. (For a review of set operations like union and intersection, see Section 1.1.)

If E and F are two subsets of the same sample space S, then $E \cup F$ and $E \cap F$ are also subsets and constitute events in their own right. It makes sense, then, to talk about the probability of the events $E \cup F$ and $E \cap F$, and it's reasonable to suspect that they bear some relation to $P(E)$ and $P(F)$. First, note that if all the outcomes in S are equally likely (as is the case with the problems we study in this section), then

$$P(E \cup F) = \frac{n(E \cup F)}{n(S)}$$

and

$$P(E \cap F) = \frac{n(E \cap F)}{n(S)}$$

so what we are really after are the relations between $n(E \cup F)$, $n(E \cap F)$, $n(E)$, and $n(F)$.

To understand the formula that follows, consider again the specific experiment that consists of rolling a die. Let E be the event "the number rolled is divisible by 2" and let F be the event "the number rolled is divisible by 3." For this small experiment, we can list the elements of each of the four events we are interested in, and then determine each event's cardinality:

$$E = \{2, 4, 6\};\ n(E) = 3$$
$$F = \{3, 6\};\ n(F) = 2$$
$$E \cup F = \{2, 3, 4, 6\};\ n(E \cup F) = 4$$
$$E \cap F = \{6\};\ n(E \cap F) = 1$$

Note that even though union is in some ways the set equivalent of numerical addition, the cardinality of the union of E and F in this example is not the sum of the cardinalities of E and F individually. The reason: both E and F contain the element 6, so if we simply add the sizes of E and F together to get the size of $E \cup F$, we wind up counting 6 twice. This happens in general when we try to determine the cardinality of a union of two sets, and we remedy the situation by subtracting the cardinality of the intersection from the sum of the individual cardinalities. By doing this, we count those elements that lie in both sets just once.

Cardinality of a Union of Sets

Let E and F be two (finite) sets. Then

$$n(E \cup F) = n(E) + n(F) - n(E \cap F).$$

Note that if E and F are *disjoint*, meaning they have no elements in common, then $n(E \cap F) = 0$, so $n(E \cup F) = n(E) + n(F)$.

We can now use this knowledge about cardinalities of unions of sets to find a formula for the probability of the union of two events. Assuming E and F are two subsets of the same sample space S,

$$\begin{aligned} P(E \cup F) &= \frac{n(E \cup F)}{n(S)} \\ &= \frac{n(E) + n(F) - n(E \cap F)}{n(S)} \\ &= \frac{n(E)}{n(S)} + \frac{n(F)}{n(S)} - \frac{n(E \cap F)}{n(S)} \\ &= P(E) + P(F) - P(E \cap F). \end{aligned}$$

(In fact, this statement about probabilities of unions is true in general, not just under the assumptions we have made, but discussion of this fact is best left for a later course in probability.)

Probability of a Union of Two Events

Let E and F be two subsets of the same sample space. Then the probability of the event "E or F", denoted $P(E \cup F)$, is given by the formula

$$P(E \cup F) = P(E) + P(F) - P(E \cap F).$$

The term $P(E \cap F)$ represents the probability of both events E and F happening. If, as sets, events E and F are disjoint (so $E \cap F = \varnothing$), then E and F are said to be *mutually exclusive*. In this case, $E \cap F = \varnothing$ and $P(E \cup F) = P(E) + P(F)$.

example 4

Assume a single die has been rolled and the number showing on top noted. Let E be the event "the number is divisible by 2" and let F be the event "the number is divisible by 3." Find the following probabilities:

a. $P(E \cap F)$ **b.** $P(E \cup F)$

Solutions:

We have already determined the sizes of all the relevant sets in the discussion above, so we are ready to note the answers or apply formulas, as necessary.

a. Since $n(E \cap F) = 1$, and since the sample space has six elements altogether,

$$P(E \cap F) = \frac{1}{6}.$$

b. Since $n(E) = 3$ and $n(F) = 2$,

$$\begin{aligned} P(E \cup F) &= P(E) + P(F) - P(E \cap F) \\ &= \frac{3}{6} + \frac{2}{6} - \frac{1}{6} \\ &= \frac{2}{3}. \end{aligned}$$

The formula for the probability of the union of three or more events can be complicated in general, and will be left for a later course. But if no two events have any elements in common, the formula is less complex. We say that events $E_1, E_2, \ldots, E_n$ are *pairwise disjoint* if $E_i \cap E_j = \varnothing$ whenever $i \neq j$. Another way of saying this is that every possible pair of events in the collection $E_1, E_2, \ldots, E_n$ is a mutually exclusive pair.

Unions of Mutually Exclusive Events

If $E_1, E_2, \ldots, E_n$ are pairwise disjoint, then

$$P(E_1 \cup E_2 \cup \cdots \cup E_n) = P(E_1) + P(E_2) + \ldots + P(E_n).$$

example 5

Suppose Jim has chosen a PIN of 8736 for use at his bank's ATM, and that all PIN's at the bank consist of four digits from 0 to 9. Using the ATM one day, he senses someone looking over his shoulder as he punches in the first two digits, and he decides to cancel the operation and leave. The next day Jim discovers his ATM card is missing, and he's not entirely sure he removed it from the machine the day before. In the worst case scenario, a stranger now has Jim's ATM card and the first two digits of his PIN. As he calls up the bank to cancel the card, he wonders what the chances are the unknown someone can guess the remaining digits in the three tries the ATM allows. What is the probability of this unfortunate event?

Solution:

If the first two digits are indeed known, the stranger has the task of filling in the last two digits, and there are 10 choices for each:

$$\underline{8}\ \ \underline{7}\ \ \underbrace{\_\_\_}_{\text{10 choices}}\ \underbrace{\_\_\_}_{\text{10 choices}}$$

The size of the sample space is thus 100, and any guess at completing the PIN correctly has a probability equal to $\frac{1}{100}$ of being correct. Assuming the stranger tries three different ways of completing the PIN (so that the three events are pairwise disjoint), the probability of Jim's account being broken into is

$$\frac{1}{100}+\frac{1}{100}+\frac{1}{100}=\frac{3}{100}.$$

caution!

A common error often made in finding the answer to Example 5 gives an answer of $\frac{1}{100}+\frac{1}{99}+\frac{1}{98}=\frac{14,701}{485,100}\approx .0303$. This is often due to confusion about an area of statistics called conditional probability. The probability of guessing right on the second guess is not $\frac{1}{99}$, but is actually $\left(\frac{1}{99}\right)\left(\frac{99}{100}\right)$, where $\frac{99}{100}$ is the probability that the first guess is wrong. Of course, $\left(\frac{1}{99}\right)\left(\frac{99}{100}\right)=\frac{1}{100}$.

The last probability idea we will consider is that of *independence*. Most people have an intuitive understanding of what it means for two events to be independent, in terms of probability. Informally, we say that two events are independent if the occurrence of one of them has no effect on the occurrence of the other. Formally, independence of events is related to the probability of their intersection, as follows.

Independent Events

Given two events E_1 and E_2 in the same sample space, we say E_1 and E_2 are **independent** if $P(E_1 \cap E_2) = P(E_1)P(E_2)$. More generally, a collection of events $E_1, E_2, \ldots, E_n$ is **independent** if for any subcollection $E_{n_1}, E_{n_2}, \ldots, E_{n_k}$ of $E_1, E_2, \ldots, E_n$, it is true that

$$P\left(E_{n_1} \cap E_{n_2} \cap \cdots \cap E_{n_k}\right) = P\left(E_{n_1}\right)P\left(E_{n_2}\right)\cdots P\left(E_{n_k}\right).$$

example 6

If a coin is flipped three times, what is the probability of it coming up tails each time?

Solution:

We actually have two good ways of answering this question, one using the notion of independence and one not.

Let E_i be the event "the coin comes up tails on the ith flip." We are interested, then, in the probability of $E_1 \cap E_2 \cap E_3$, the probability that we get tails each time. Since $P(E_i) = \frac{1}{2}$ for each $i = 1, 2, 3$ (remember, the coin is assumed fair), then

$$\begin{aligned} P(E_1 \cap E_2 \cap E_3) &= P(E_1)P(E_2)P(E_3) \\ &= \frac{1}{2} \cdot \frac{1}{2} \cdot \frac{1}{2} \\ &= \frac{1}{8}. \end{aligned}$$

The other way of obtaining the same answer is to consider the sample space made up of all possible three-toss sequences. Since each flip of the coin results in one of two possibilities, the Multiplication Principle of Counting tells us that there are $2^3 = 8$ possible sequences. Only one of these is the sequence consisting of three tails, so the probability of this event is $\frac{1}{8}$.

exercises

Below is the given probability that an event will occur; find the probability that it won't.

1. $P = \frac{2}{5}$

2. $P = 0.72$

3. $P = \frac{4}{13}$

4. $P = 0.15$

5. $P = \frac{2}{3}$

6. $P = 0.49$

Apply the formulas for the probability of intersection or union to the following sets and determine (a) $P(E \cap F)$ and (b) $P(E \cup F)$. $S =$ size of sample space.

7. $S = 8,\ E = \{2,5\},\ F = \{3,7,9\}$

8. $S = 10,\ E = \{1,2,5\},\ F = \{1,2,3,5\}$

9. $S = 5,\ E = \{4,B\},\ F = \{3\}$

10. $S = 8,\ E = \{A\},\ F = \{B,C,D,E\}$

11. $S = 4,\ E = \{1,\beta\},\ F = \{\alpha,2\}$

12. $S = 12,\ E = \{A,C,g,5,n,7,8,t,L\},\ F = \{n,6\}$

13. $S = 16,\ E = \{1,2,A,m,13,Y,8\},\ F = \{1,9,11,m\}$

14. $S = 11,\ E = \{m,7,D,4,\theta\},\ F = \{\phi,D,3,7,m,\Sigma\}$

Determine the sample space of each of the following experiments.

15. A coin is flipped four times, and the result recorded after each flip.

16. A card is drawn at random from the 13 hearts.

17. A coin is flipped, and a card is drawn at random from the 13 hearts.

18. A quadrant of the Cartesian plane is chosen at random.

19. A slot machine lever is pulled; there are 3 slots, each of which can hold 6 different values.

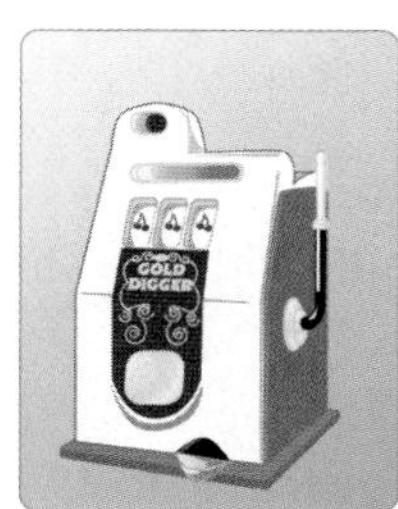

20. An individual die is rolled twice, and each of the two results is recorded.

21. At a casino, a roulette wheel spins until a ball comes to rest in one of the 38 pockets.

22. A lottery drawing consists of 6 randomly drawn numbers from 1 to 20; the order of the numbers matters in this case, and repetition is possible.

Answer the following probability questions. Be careful to properly identify the sample space and the appropriate event in each case. See Examples 1, 2, and 3.

23. An ordinary die is rolled. Find the probability of rolling:

a. A 3 or higher.

b. An even non-prime number.

24. A card is drawn from a standard 52-card deck. Find the probability of drawing:

a. A face card (Jack, Queen, or King) in the suit of hearts.

b. Anything but an Ace.

c. A black (clubs or spades) non-face card.

25. A coin is flipped three times. Find the probability of getting:

a. Heads exactly twice.

b. The sequence Heads, Tails, Heads.

c. Two or more Heads.

26. A state lottery game is won by choosing the same six numbers (without repetition) as those selected by a mechanical device. The numbers are picked from the set $\{1, 2, \ldots, 49\}$, and the order of the numbers chosen is immaterial. What is the probability of winning?

27. What is the probability that a four-digit ATM PIN chosen at random ends in 7, 8, or 9?

28. Assume the probability of a newborn being male is one-half. What is the probability that a family with five children has exactly three boys?

29. What is the probability that a 9-digit driver's license number chosen at random will not have an 8 as a digit?

30. A roulette wheel in a casino has 38 pockets: 18 red, 18 black, and 2 green. Spinning the wheel causes a small ball to randomly drop into one of the pockets. All of the pockets are equally likely. The wheel is spun twice. Find the probability of getting:

a. Green both times.

b. Black at least once.

c. Red exactly once.

31. There is a 25% chance of rain for each of the next 2 days. What is the probability that it will rain on one of the days but not the other?

The following problems all involve unions or intersections of events, or the notion of independent events. See Examples 4, 5, and 6.

32. A pair of dice is rolled. Find the probability that the sum of the top faces is:

a. Seven.

b. Seven or eleven.

c. An even number or a number divisible by three.

d. Ten or higher.

33. A card is drawn from a standard 52-card deck. Find the probability of drawing:

a. A face card or a diamond.

b. A face card but not a diamond.

c. A red face card or a King.

34. A state lottery game is won by choosing the same six numbers (without repetition) as those selected by a mechanical device. The numbers are picked from the set $\{1, 2, \ldots, 49\}$, and the order of the numbers chosen is immaterial. What is the probability of winning if someone buys 1000 tickets? (Of course, no two of the tickets have the same set of six numbers.) How many tickets would have to be bought to raise the probability of winning to one half?

35. Two cards are drawn at random from a standard 52-card deck. What is the probability of them both being Aces if:

a. The first card is drawn, looked at, placed back in the deck, and the deck is then shuffled before the second card is drawn?

b. The two cards are drawn at the same time?

36. What is the probability of being dealt a five-card hand (from a standard 52-card deck) that has four cards of the same rank?

37. The probability of rain today is 75%, and the probability that Bob will forget to put the top up on his convertible is 25%. What is the probability of the inside of his car getting wet?

38. What is the probability of drawing 3 face cards in a row, without replacement, from a 52-card deck?

39. Two dice are rolled, and the difference of the results is calculated by subtracting the smaller value from the larger value. Therefore, the result may range from 0 to 5. Find the probability of each of the following results:

a. 0

b. 1

c. 4

40. A letter is randomly chosen from the word MISSISSIPPI. What is the probability of the letter being an S?

41. Jim works in a company of about 100 employees. If this year five people in the company are going to be randomly laid off, what is the probability that Jim will get laid off?

42. A pack of M&M's contains 10 yellow, 15 green, and 20 red pieces. What is the probability of choosing a green M&M out of the pack?

43. A jar of cookies has 3 sugar cookies, 4 chocolate chip cookies, and 2 peanut butter cookies. What is the probability of randomly choosing a peanut butter cookie out of the jar?

44. A big box of crayons contains 4 different blues, 3 different reds, 5 different greens, and 2 different yellows. What is the probability of randomly choosing a yellow crayon out of the box?

45. A bag of marbles contains 3 blue marbles, 2 red marbles, and 5 orange marbles. What is the probability of randomly picking a blue marble out of the bag?

46. Every week a teacher of a class of 25, randomly chooses a student to wash the blackboards. If there are 15 girls in the class, what is the probability that the student selected will be a boy?

47. If in a raffle 135 tickets are sold, how many tickets must be bought for an individual to have a 20% chance of winning?

48. Bobby is running for student council. At Bobby's school the student council is chosen randomly from all qualified candidates. If there are 42 candidates running, including Bobby, and a total of three positions, what is the probability that Bobby will be selected for the council?

chapter eleven project

Probability

You may be familiar with the casino game of roulette. But have you ever tried to compute the probability of winning on a given bet?

The roulette wheel has 38 total slots. The wheel turns in one direction and a ball is rolled in the opposite direction around the wheel until it comes to rest in one of the 38 slots. The slots are numbered 00, and 0 – 36. Eighteen of the slots between 1 and 36 are colored black and eighteen are colored red. The 0 and 00 slots are colored green and are considered neither even nor odd, and neither red nor black. These slots are the key to the house's advantage.

The following are some common bets in roulette:
A gambler may bet that the ball will land on a particular number, or a red slot, or a black slot, or an odd number, or an even number (not including 0 or 00). He or she could wager instead that the ball will land on a column (one of 12 specific numbers between 1 and 36), or on a street (one of 3 specific numbers between 1 and 36).

The payoffs for winning bets are:
1 to 1 on odd, even, red and black
2 to 1 on a column
11 to 1 on a street
35 to 1 any one number

1. Compute the probability of the ball landing on:
   - **a.** A red slot
   - **b.** An odd number
   - **c.** The number 0
   - **d.** A street (any of 3 specific numbers)
   - **e.** The number 2
2. Based on playing each of the scenarios above (a. – e.) compute the winnings for each bet individually, if $5 is bet each time and all 5 scenarios lead to winnings.
3. If $1 is bet on hitting just one number, what would be the expected payoff? (**Hint:** expected payoff is [*probability of winning* times *payment for a win*] – [*probability of losing* times *payout for a loss*])
4. Given the information in question 3, would you like to play roulette on a regular basis? Why or why not? Why will the casino acquire more money in the long run?

CHAPTER REVIEW

11.1

Sequences and Series

topics	pages	test exercises
Recursively and explicitly defined sequences • The definition of *infinite* and *finite sequences*, and the notation used to define sequences • The meaning of *explicit* and *recursive* formulas, and how to use them • The process of contriving a formula that reproduces a given number of terms of a sequence	p. 893 – 898	1 – 3
Summation notation and a few formulas • The meaning of *summation notation* using the Greek letter sigma • Converting between summation notation and expanded form • Properties of summations • The use of specific summation formulas	p. 899 – 902	4 – 6
Partial sums and series • The meaning of a *series*, and how series are related to sequences • The meaning of a *partial sum* of a series • *Convergence* and *divergence* of series	p. 903 – 905	
Interlude: Fibonacci sequences • Using recursive formulas to define Fibonacci sequences	p. 905 – 906	

11.2

Arithmetic Sequences and Series

topics	pages	test exercises
Characteristics of arithmetic sequences and series • The definition of an *arithmetic sequence*, and the meaning of the *common difference* of an arithmetic sequence	p. 909 – 911	7 – 11
The formula for the general term of an arithmetic sequence • The formula for the general term of an arithmetic sequence, and how to determine it	p. 911 – 913	7 – 9

topics (continued)	pages	test exercises
Evaluating partial sums of arithmetic sequences • The two formulas for partial sums of arithmetic sequences	p. 913 – 915	10 – 11

11.3 Geometric Sequences and Series

topics	pages	test exercises
Characteristics of geometric sequences • The definition of a *geometric sequence*, and the meaning of the *common ratio* of a geometric sequence	p. 919 – 921	12 – 16
The formula for the general term of a geometric sequence • The formula for the general term of a geometric sequence, and how to use it	p. 921 – 923	
Evaluating partial sums of geometric sequences • The formula for the partial sum of a geometric sequence	p. 923 – 927	12 – 13
Evaluating infinite geometric series • The extension of the partial sum formula to the case of infinite geometric series • How to determine convergence or divergence of infinite geometric series • The correspondence between geometric series and decimal notation	p. 927 – 930	14 – 16
Interlude: Zeno's paradoxes • The use of geometric series to resolve apparent paradoxes	p. 930 – 931	

11.4 Mathematical Induction

topics	pages	test exercises
The role of induction • The mathematical meaning of *induction*	p. 935 – 936	17 – 22
Proofs by mathematical induction • Using mathematical induction to prove statements • The conditions for applying proof by mathematical induction	p. 936 – 939	17 – 22

topics (continued)	pages	test exercises
• The two steps – *basis step* and *induction step* – for constructing an inductive proof		

11.5 An Introduction to Combinatorics

topics	pages	test exercises
The multiplication principle of counting • The meaning of the *Multiplication Principle of Counting*, and its use in counting problems	p. 945 – 948	23 – 28
Permutations • The application of the Multiplication Principle of Counting to count the number of *permutations* of a set of objects • The *Permutation Formula* and the meaning and use of the notation ${}_nP_k$	p. 948 – 951	23 – 24, 26
Combinations • The connection between the Permutation Formula and the *Combination Formula* • The meaning and use of the notation ${}_nC_k$	p. 952 – 955	25, 27 – 28
The binomial and multinomial theorems • The formulas for *binomial* and *multinomial coefficients* • The *Binomial* and *Multinomial Theorems*, and their uses	p. 955 – 959	29

11.6 An Introduction to Probability

topics	pages	test exercises
The language of probability • The meaning of *experiment, outcome, event*, and *sample space* in probability theory • The calculation of the probability of an event when all outcomes are *equally likely*	p. 967 – 969	30 – 32
Using combinatorics to compute probabilities • The use of combinatorics in calculating probabilities	p. 969 – 971	32

topics (continued)	pages	test exercises
Unions, intersections, and independent events • The meaning of *unions* and *intersections* of events, and the calculation of probabilities of unions of events • The meaning of *independence* of events	p. 972 – 976	30 – 31

chapter test

Section 11.1

Find a possible formula for the general term of the sequences that begin as follows.

1. $-7, -1, 5, 11, 17, \ldots$ **2.** $\frac{1}{2}, \frac{3}{4}, \frac{9}{8}, \frac{27}{16}, \frac{81}{32}, \ldots$ **3.** $0, 3, 8, 15, 24, 35, \ldots$

Translate each expanded sum that follows into summation notation, and vice versa. Then use the formulas and properties from this section to evaluate the sums.

4. $\sum_{i=3}^{8}(-2i+3)$ **5.** $\sum_{i=2}^{7}(-2)^{i-1}$ **6.** $8 + 27 + 64 + \ldots + 343$

Section 11.2

Find the explicit formula for the general term of the arithmetic sequences described below.

7. $5, 2, -1, -4, -7, \ldots$ **8.** $a_2 = 14$ and $a_4 = 19$ **9.** $a_7 = -43$ and $d = -9$

Determine the value of each of the following partial sums of arithmetic sequences.

10. $\sum_{i=1}^{97}(2i-7)$ **11.** $\sum_{i=1}^{60}(-4i+3)$

Section 11.3

Each of the following sums is a partial sum of a geometric sequence. Use this fact to evaluate the sums.

12. $\sum_{i=3}^{9}3\left(\frac{1}{2}\right)^{i}$ **13.** $5 + 10 + \ldots + 20{,}480$

Determine if the following infinite geometric series converge. If a given sum converges, find the sum.

14. $\sum_{i=0}^{\infty}-3\left(\frac{3}{4}\right)^{i}$ **15.** $\sum_{i=1}^{\infty}\left(-\frac{5}{4}\right)^{i}$ **16.** $\sum_{i=1}^{\infty}\frac{2}{5}\left(\frac{5}{7}\right)^{i}$

Section 11.4

Use the Principle of Mathematical Induction to prove the following statements.

17. $1+4+9+\cdots+n^2=\dfrac{n(n+1)(2n+1)}{6}$

18. $\dfrac{1}{1\cdot 3}+\dfrac{1}{3\cdot 5}+\dfrac{1}{5\cdot 7}+\cdots+\dfrac{1}{(2n-1)(2n+1)}=\dfrac{n}{2n+1}$

19. $5+8+11+\cdots+(3n+2)=\dfrac{n(3n+7)}{2}$

20. $1\cdot 3+2\cdot 4+3\cdot 5+\cdots+n(n+2)=\dfrac{n(n+1)(2n+7)}{6}$

21. Prove that for all natural numbers n, 11^n-7^n is divisible by 4.

22. Prove that for all natural numbers n, 7^n-1 is divisible by 3.

Section 11.5

Use the Multiplication Principle of Counting and the Permutation and Combination formulas to answer the following questions.

23. A license plate must contain 4 numerical digits followed by 3 letters. If the first digit cannot be 0 or 1, how many different license plates can be created?

24. How many different 7-digit phone numbers do not contain the digits 6 or 7?

25. In how many different orders can the letters in the word "aardvark" be arranged?

26. In how many different ways can first place, second place, and third place be awarded in a 10-person shot put competition?

27. At a meeting of 21 people, a president, vice president, secretary, treasurer, and recruitment officer are to be chosen. How many different ways can these positions be filled?

28. A college admissions committee selects 4 out of 12 scholarship finalists to receive merit-based financial aid. How many different sets of 4 recipients can be chosen?

29. Expand the expression $(x+2y+z)^3$.

Section 11.6

Answer the following probability questions. Be careful to properly identify the sample space and the appropriate event in each case.

30. A card is drawn from a standard 52-card deck. Find the probability of drawing:

a. A seven or a club.

b. A face card but not a red Queen.

c. A black three or a spade.

31. What is the probability of being dealt a five-card hand (from a standard 52-card deck) that contains only face cards?

32. There is a 10% chance of rain each individual day for an entire week. What is the probability that it will rain at least once during this seven day period?

Appendix A: Fundamentals of *Mathematica*®

Mathematica is a very powerful and flexible software package with a wide variety of uses. To begin with, *Mathematica* (along with similar products such as *Maple*®, *Matlab*®, and *Derive*™) can be viewed as a sort of super-calculator. It understands the rules of algebra, has a huge number of built-in functions ranging from the trivial to the exotic, and is very good at generating high-quality graphs in one, two, and three dimensions. Beyond that, a package such as *Mathematica* is also a programming environment; it is this aspect of *Mathematica* that allows the user to extend its capabilities to suit specialized needs.

The applications in this text require only a basic understanding of *Mathematica*, and this appendix will serve as a quick guide to its use. It should thus be noted that a complete guide to the use of *Mathematica* can be found within *Mathematica* itself. Once it is installed and running on your computer, you have access to an electronic version of a very large *Mathematica* manual, all available by first clicking on the "Help" button located in the top toolbar (see Figure 1).

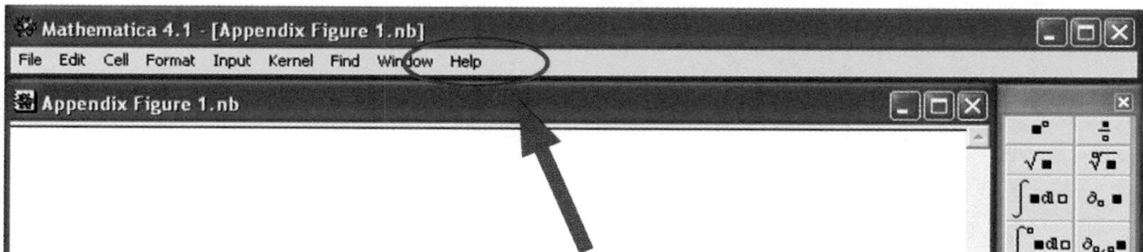

Figure 1: Getting On-Screen Help in *Mathematica*

After clicking on "Help" a drop-down menu appears. The full selection of Help categories appears by clicking on "Help Browser," but a good place to start is with "Getting Started/Demos..." This leads to many useful examples of how Mathematica can be used to solve different sorts of problems.

At first, you will probably be making use of built-in *Mathematica* commands such as **`Solve`** and **`Simplify`** (as opposed to using your own user-defined commands). It is important to realize that *Mathematica* is case-sensitive, and that all built-in commands begin with a capital letter. Once a command has been typed in, you'll need to tell *Mathematica* to execute it. This can be done in one of two ways - either by pressing the "Shift" and "Enter" keys together (known as "Shift-Enter") or, if you are using an extended keyboard, by using the "Enter" key that appears in the numeric keypad area.

Pressing the normal "Enter" key alone will simply move the cursor to the next line and allow you to continue typing, but will not execute any commands.

Each time you press "Shift-Enter" *Mathematica* will execute all the commands contained in a single cell. Different *Mathematica* cells are demarcated by blue bracketing symbols along the right-hand edge of the work area, and you can always start a new cell by positioning the mouse cursor over a blank part of the area and clicking the left mouse button once.

The remainder of this appendix contains examples of a few of the basic *Mathematica* commands used in this text, arranged in the order in which they appear. For instant on-screen help on any command, type the command into *Mathematica* and then press the "F1" key. Doing so will bring up the relevant help pages and, more often than not, provide more examples of how the command is used.

Simplify:

The **Simplify** command is used to simplify mathematical expressions according to the usual rules of algebra. The basic syntax is **Simplify**[*expr*], where *expr* is the expression to be simplified. Note the examples in the two cells shown in Figure 2 below.

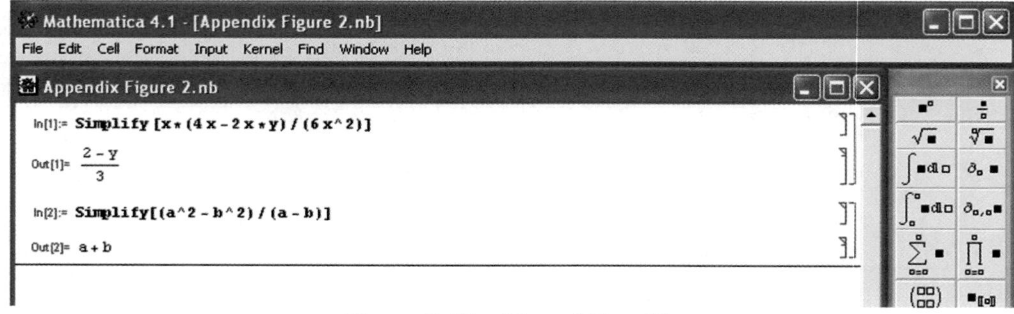

Figure 2: The Use of Simplify

Expand:

This command is used to multiply out factors in an expression. The syntax for the command is **Expand**[*expr*]. Figure 3 shows the use of the command in multiplying out the expression $(x-y)^5$.

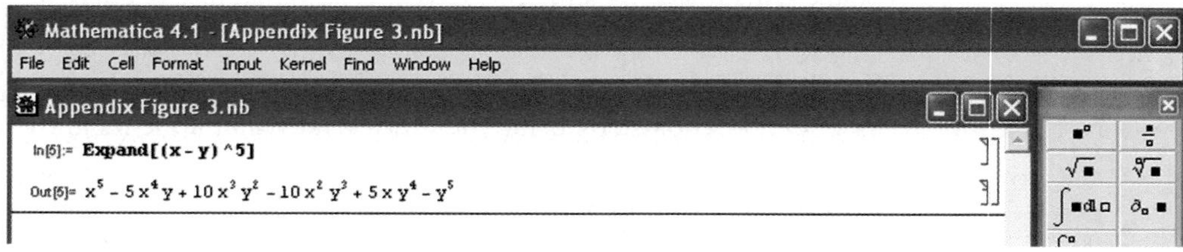

Figure 3: The Use of Expand

Factor:

The **Factor** command is the reverse of the **Expand** command when applied to polynomials. Its basic usage is **Factor**[*poly*], where *poly* is a polynomial expression to be factored.

Solve:

The **Solve** command is very powerful, and is used in several different ways in this text. Its basic usage is **Solve**[*eqns*, *vars*], where *eqns* represents one or more equations and *vars* represents one or more variables. If more than one equation is to be solved, the collection of equations must be enclosed in a set of braces, separated by commas. Similarly, if more than one variable is to be solved for, the variables must be enclosed in a set of braces. Figure 4 shows the use of **Solve** to first solve one equation for one variable, and then to solve a collection of three equations for all three variables. Note how *Mathematica* expresses the solution in each case.

It is important to note that equations in *Mathematica* are expressed with two "=" symbols, as seen in Figure 4. The use of just one "=" is reserved for assigning a permanent value to something. For instance, the expression **x=3** assigns the value of 3 to the symbol x, while the expression **x==3** represents the equation $x = 3$ in *Mathematica*.

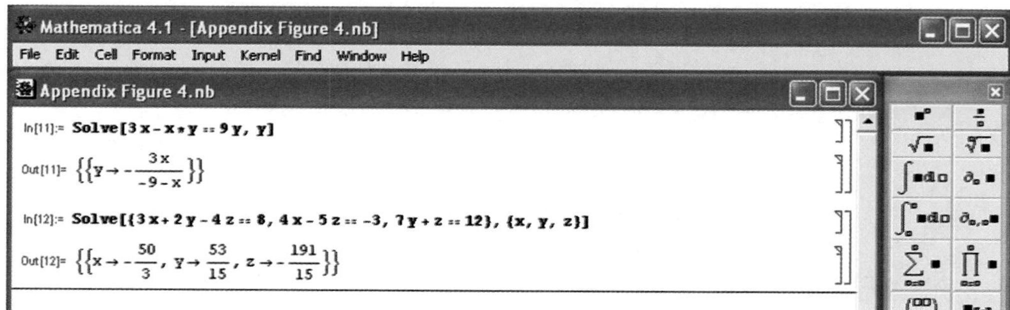

Figure 4: Two Uses of the Solve Command

Together:

The **Together** command is used primarily to combine two or more expressions into one with a common denominator, automatically canceling any common factors that may appear.

Defining functions:

A few rules of syntax must be observed in order to define your own functions in *Mathematica*. The first is that each variable serving as a place-holder in the definition must be followed by the underscore symbol "_" when it appears on the left side of the definition and without the underscore when it appears on the right. The second rule is that ":=" is used in the definition, as opposed to "=" (see the on-screen *Mathematica* help for detailed explanations of these rules). Figure 5 illustrates the definition of the two functions $f(x)=x^2+5$ and $g(x,y)=3x-7y$, followed by an evaluation of each.

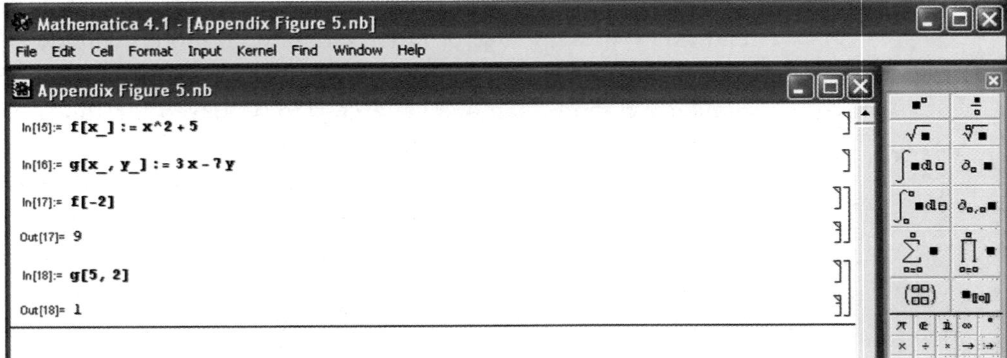

Figure 5: Defining Functions in *Mathematica*

Plot:

The **`Plot`** command is another very versatile command. Its basic usage is **`Plot`**`[`*f*, {*x*, *xmin*, *xmax*}`]`, where *f* is an expression in *x* representing a function to be plotted and *xmin* and *xmax* define the endpoints of the interval on the *x*-axis over which *f* is to be graphed. However, the **`Plot`** command also recognizes many options that modify the details of the resulting picture; these options are best explored via the on-screen help.

ImplicitPlot:

The **`ImplicitPlot`** command (all one word) is used to graph equations in two variables. To use it, you must first execute the command **``<<Graphics`ImplicitPlot` ``**, which extends the collection of built-in commands in *Mathematica* (execute this command only once per *Mathematica* session). Aside from this necessary first step, the usage of **`ImplicitPlot`** is similar to that of **`Plot`**: the syntax is **`ImplicitPlot`**`[`*eqn*, {*x*, *xmin*, *xmax*}`]`, where *eqn* is an equation in *x* and *y*. Figure 6 illustrates its use.

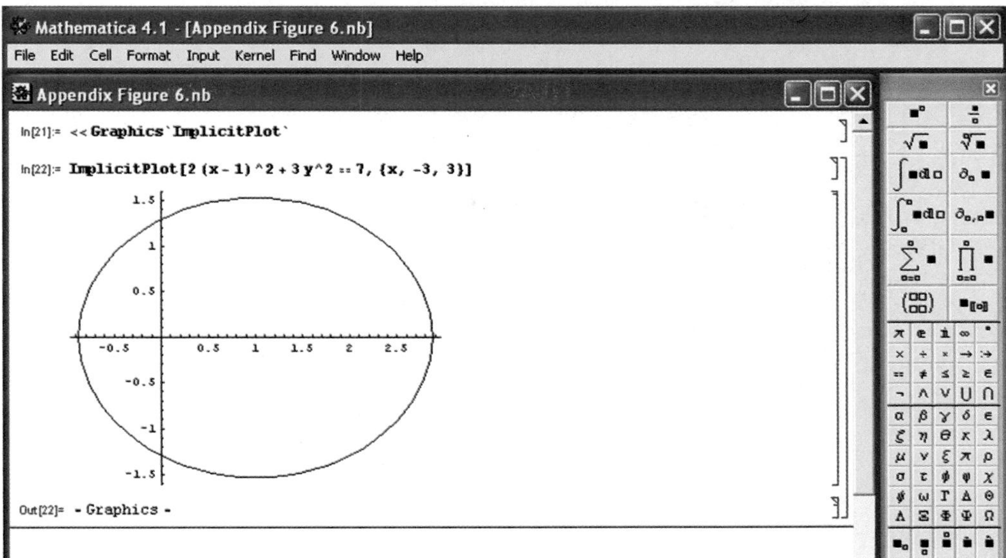

Figure 6: The ImplicitPlot Command

Defining and working with matrices:

A matrix in *Mathematica* is defined as a list consisting of a number of sub-lists. Each sub-list corresponds to a row of the matrix, so an $m \times n$ matrix appears in *Mathematica* as a list of m sub-lists, each of length n. The **MatrixForm** command can be used to view a matrix in a more familiar format. The command **Det** is used to evaluate the determinant of a matrix, and the command **Inverse** finds the inverse of a given matrix. The use of these commands is illustrated in Figure 7.

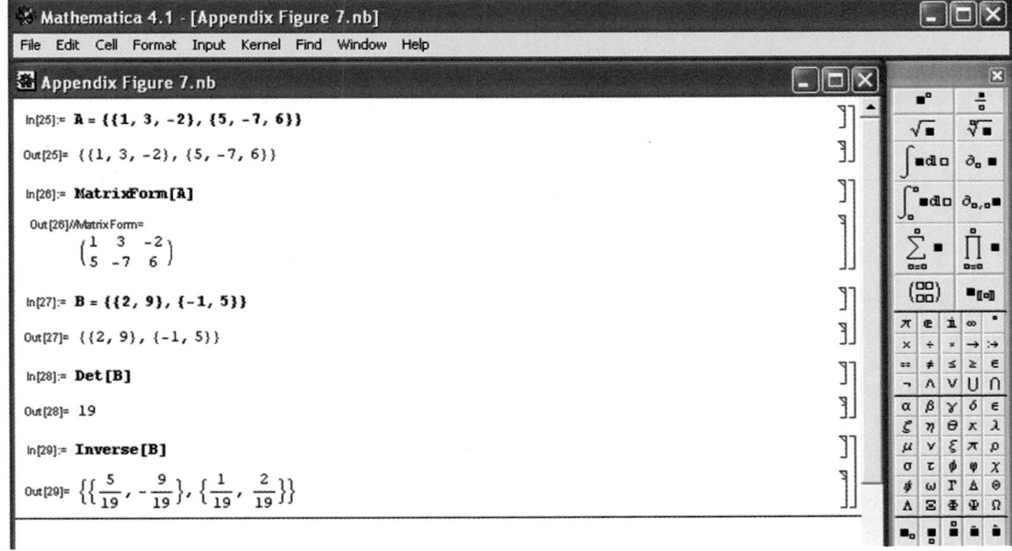

Figure 7: Matrix Commands

chapter one

section 1.1, pages 14-18

1. a. 19 and 2^5 **b.** 19, 2^5, and $\frac{0}{15}$ **c.** 19, 2^5, $\frac{0}{15}$, and -33 **d.** 19, 2^5, $\frac{0}{15}$, -33, and -4.3 **e.** $-\sqrt{3}$

f. all except $\frac{15}{0}$ **g.** $\frac{15}{0}$ **3.** $\frac{7}{3}$ **5.** $\frac{41{,}836}{99{,}999}$ **7.** $-\frac{100}{99}$ **9.** 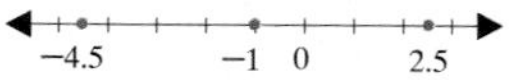

11. 

13. $>, \geq$ **15.** $\leq, \geq$ **17.** $>, \geq$ **19.** $\{3n \mid n \text{ is an integer and } -2 \leq n \leq 3\}$ **21.** $\{n \mid n \text{ is a prime}\}$ **23.** $\{\frac{1}{n} \mid n \text{ is an odd integer}\}$ **25.** $(-5, 4]$ **27.** $[3, 4]$

29. $\mathbb{Z}$ **31.** $[-\pi, 21)$ **33.** $(3, 9]$ **35.** $\mathbb{Z}$ **37.** $[-3, 19)$ **39.** $(0, \infty)$ **41.** $[1, 2]$

43. 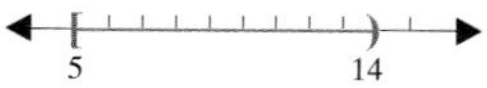

45. 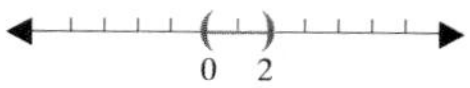

47. 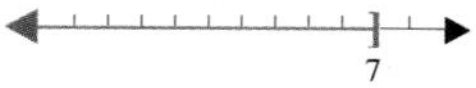

49. -11

51. -5 **53.** 2 **55.** $\sqrt{2}$ **57.** $<, \leq$ **59.** $<, \leq$ **61.** $>, \geq$ **63.** $3x^2y^3$, $-2\sqrt{x+y}$, and $7z$

65. -2 and $\sqrt{x+y}$ **67.** 1, 8.5, and -14 **69.** 8 **71.** $-\frac{\sqrt{2}}{36}+2$ **73.** 4

75. commutative property of addition **77.** associative property of addition

79. associative property of multiplication **81.** distributive **83.** multiplicative cancellation; $\frac{1}{5}$

85. additive cancellation; x **87.** zero-factor property **89.** multiplicative cancellation; 6

91. $-\frac{1}{5}\left(\sqrt{3(3+7)-5}\right)^3$ **93.** $\frac{12{,}167}{54} \approx 225.31$ **95.** $\sqrt{42}-1 \approx 5.48$

section 1.2, pages 36-41

1. 16 **3.** 81 **5.** 64 **7.** 1 **9.** $\frac{1}{7}$ **11.** x^3 **13.** $\frac{3}{t^5}$ **15.** $\frac{x^5}{7}$ **17.** $2n^8$ **19.** $\frac{x^8z^3}{y^3}$ **21.** s^5

23. x^{10} **25.** $-\frac{y^5}{3x^2}$ **27.** $\frac{1}{3zy^2}$ **29.** $27x^2y^4$ **31.** 1 **33.** -0.0000176

35. 2.1×10^{-7} **37.** 5.1×10^3 **39.** 312.12 **41.** 4.6×10^{25} **43.** -11 **45.** 1.5×10^8

47. 1×10^{-8} **49.** Answers will vary **51.** Answers will vary **53.** $\frac{1}{2}bhl$ **55.** 180π in.^{3}

57. $6lw$ ft.^{3} **59.** $14(N+M)$ ft.^{2} **61.** $\frac{1}{15{,}625x^{14}z^8}$ **63.** $\frac{x^4}{4y^2z^6}$ **65.** -3 **67.** not real

69. -2 **71.** $\frac{1}{2}$ **73.** 2 **75.** $3|x|$ **77.** $\frac{x^2|z|}{2}$ **79.** $x^2y^7z^3$ **81.** $\frac{1}{2|xy^3|}$ **83.** $\frac{ab^4}{3c^2}$ **85.** $-\frac{|a|\sqrt{2}}{2}$

87. $2\sqrt{7}+2\sqrt{2}$ **89.** $\sqrt{x}-\sqrt{y}$ **91.** $\frac{2+\sqrt{x}}{4-x}$ **93.** $3x\sqrt[3]{2x}$ **95.** cannot be combined **97.** 0 **99.** $|x^3|$

101. $\frac{1}{8}$ **103.** $(x-z)^{y-4}$ **105.** 4 **107.** $\frac{1}{a}$ **109.** $\sqrt[4]{|y|}$ **111.** x^3

113. Answers will vary **115.** $SA = 5052$ cm^2 (approx. 0.5 m^2)

section 1.3, pages 51-54

1. not a polynomial **3.** degree 11 polynomial of four terms **5.** degree 0 monomial **7.** degree 4 binomial **9.** degree 2 trinomial **11.** degree 5 binomial **13.** $\pi z^5 + 8z^2 - 2z + 1$ **a.** 5 **b.** π **15.** $2s^6 - 10s^5 + 4s^3$ **a.** 6 **b.** 2 **17.** $9y^6 - 3y^5 + y - 2$ **a.** 6 **b.** 9 **19.** $-4x^3y - 6y - x^2z$ **21.** $x^2y + xy^2 + 6x - 6y$ **23.** $-3ab$ **25.** $3a^3b^3 + 21a^3b^2 - 3ab^3 + 2a^2b^2 + 14a^2b - 21ab^2$ **27.** $3a^2 - 2ab - 8b^2$ **29.** $(3a^2b)(1 + a - 3b)$ **31.** $2x(x^5 - 7x^2 + 4)$ **33.** $(a^2 + b)(a - b)$ **35.** $(z^3 + z)(z + 1)$ **37.** $(n-2)(x^2 + y)$ **39.** $(5x^2y - 3)(5x^2y + 3)$ **41.** $(x - 10y)(x^2 + 10xy + 100y^2)$ **43.** $(x+5)(x-3)$ **45.** $(x-1)^2$ **47.** $(x-2)^2$ **49.** $(2x+3)(3x-2)$ **51.** $(5y+1)^2$ **53.** $(2y+1)(3y-8)$ **55.** $2x(2x-1)^{-3/2}$ **57.** $a^{-3}(7a^2 - 2b)$ **59.** $-25x^5 + 15x^4 - 72x^3 - 35x^2 - 9x - 36$ **61.** $x^3 - 3x^3y^2z + x^3z - 2x^2y + 6x^2y^3z - 2x^2yz + xyz - 3xy^3z^2 + xyz^2$ **63.** $(x^2 - y)(3x + y)(x - y^2)$ **65.** $(p - 3q)(2p - q)(p + q)(p + 2q)$

section 1.4, pages 61-62

1. $5i$ **3.** $-3i\sqrt{3}$ **5.** $4i\sqrt{2x}$ **7.** $i\sqrt{29}$ **9.** $1 - 3i$ **11.** $9 + 2i$ **13.** $8 - 6i$ **15.** -9 **17.** $-5 + 6i$ **19.** 10 **21.** 97 **23.** 6 **25.** $-\frac{3}{5} + \frac{4}{5}i$ **27.** $\frac{1}{5} + \frac{2}{5}i$ **29.** $\frac{2}{29} - \frac{5}{29}i$ **31.** i **33.** $\frac{2}{13} + \frac{3}{13}i$ **35.** 1 **37.** $-3\sqrt{2}$ **39.** $-3i$ **41.** $7 + 6i\sqrt{2}$ **43.** $\frac{1}{2} - \frac{5}{2}i$ **45.** $7 - 24i$ **47.** $-56 - 8i$

section 1.5, pages 69-72

1. $\{-5\}$ **3.** $\{-1\}$ **5.** $\{-3\}$ **7.** $\mathbb{R}$ (Identity) **9.** $\varnothing$ (Contradiction) **11.** $\{7\}$ **13.** $\{3.7\}$ **15.** $\{1.05\}$ **17.** $\{-2, 2\}$ **19.** $\left\{\frac{1}{3}\right\}$ **21.** $\{-311, 420\}$ **23.** $\left\{-\frac{4}{5}, 2\right\}$ **25.** $\varnothing$ (Contradiction) **27.** $\{5\}$ **29.** $\left\{-\frac{1}{2}\right\}$ **31.** $\left\{\frac{1}{4}\right\}$ **33.** $T = \frac{PV}{nR}$ **35.** $B = \frac{2A}{h} - b$ **37.** $h = \frac{3V}{s^2}$ **39.** $r = \frac{d}{t_1 + t_2}$ **41.** $\frac{19}{3}$ hours, or 6 hours and 20 minutes **43.** 3.5 hours **45.** 28% **47.** 96.2%, 103.9% **49.** 95, 96, and 97 **51.** \$21.50 **53.** $h = \frac{S}{2\pi r} - r$ **55.** $a = \frac{c}{3-c}$

one

section 1.6, pages 79-80

1. $(-\infty, -3]$  **3.** $(-\infty, 2.25)$  **5.** $\left(-\infty, \frac{3}{2}\right)$

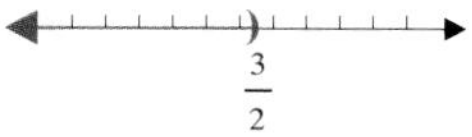

7. $\left(-\infty, -\frac{3}{11}\right]$ 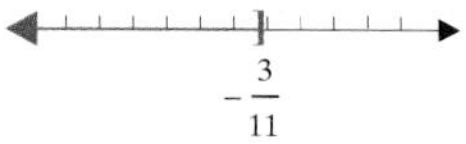 **9.** $(7, \infty)$  **11.** $(35, \infty)$

13. $(1, 5]$  **15.** $[-8, -2)$ **17.** $[4, 8]$

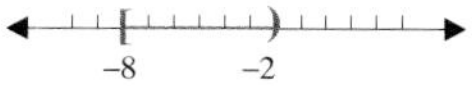

19. $\left(\frac{23}{7}, \frac{25}{7}\right)$ 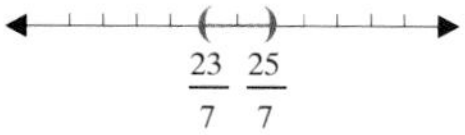 **21.** $\left(-\frac{5}{3}, 1\right]$

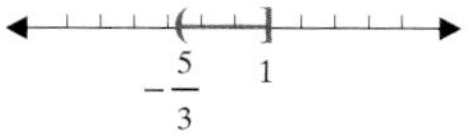

23. $\left(-\infty, \frac{1}{2}\right)\cup\left(\frac{5}{2}, \infty\right)$ 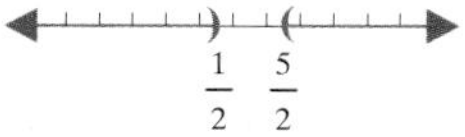 **25.** $(-11, 1)$

27. $\left(-\infty, -\frac{7}{2}\right)\cup\left(\frac{15}{2}, \infty\right)$  **29.** $[-4, 0]$

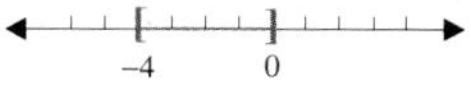

31. $(-\infty, 2)\cup(6, \infty)$  **33.** $\varnothing$  **35.** $(-\infty, \infty)$

37. $[11, \infty)$ **39.** $[0, 5.8)$

section 1.7, pages 93-96

1. $\left\{-1, \frac{3}{2}\right\}$ **3.** $\left\{\frac{-1}{5}, 2\right\}$ **5.** $\{-3, 6\}$ **7.** $\left\{-\frac{2}{5}, \frac{1}{3}\right\}$ **9.** $\{3, 11\}$ **11.** $\left\{0, \frac{7}{3}\right\}$ **13.** $\{0, 6\}$

15. $\{3 \pm 4i\}$ **17.** $\left\{\frac{3}{8}\right\}$ **19.** $\left\{\frac{1}{2} \pm \sqrt{2}\right\}$ **21.** $\{-5, 9\}$ **23.** $\left\{\frac{6}{5}, 6\right\}$ **25.** $\{1, 11\}$ **27.** $\left\{\frac{\pm\sqrt{10}+2}{6}\right\}$

29. $\left\{\frac{\pm\sqrt{15}-5}{2}\right\}$ **31.** $\{-5, -3\}$ **33.** $\left\{-5, \frac{3}{2}\right\}$ **35.** $\{-9, -1\}$ **37.** $\left\{\frac{3 \pm i\sqrt{7}}{8}\right\}$ **39.** $\{-0.78, 2.45\}$

41. $\{-1\}$ **43.** $\left\{\frac{-1 \pm \sqrt{13}}{6}\right\}$ **45.** $\left\{-\frac{3}{2}, 5\right\}$ **47.** 4.5 seconds **49.** 4.8 seconds

51. $(x-3-2i)(x-3+2i)$ **53.** $\left(2x+3-2\sqrt{2}\right)\left(2x+3+2\sqrt{2}\right)$ **55.** $b = -5$ and $c = -24$

57. $\left\{-2, 2, -i\sqrt{3}, i\sqrt{3}\right\}$ **59.** $\left\{1-2i, 1+2i, 1-\sqrt{3}, 1+\sqrt{3}\right\}$ **61.** $\left\{\frac{1}{8}, 27\right\}$ **63.** $\{-2, -1, 2, 3\}$

65. $\left\{-\frac{8}{27}, 1\right\}$ **67.** $\{-4, -3, 3, 4\}$ **69.** $\left\{-\frac{1}{2}, -i, i\right\}$ **71.** $\left\{-2, 1+i\sqrt{3}, 1-i\sqrt{3}\right\}$

73. $\left\{-\frac{3}{2}, \frac{3}{2}, -\frac{3i}{2}, \frac{3i}{2}\right\}$ **75.** $\left\{-\frac{7}{3}, 0, 1\right\}$ **77.** $\left\{-\frac{4}{3}, \frac{2 \pm 2i\sqrt{3}}{3}\right\}$ **79.** $\{-2, 2, -3i, 3i\}$ **81.** $\left\{\frac{5}{2}\right\}$

83. $\{4\}$ **85.** $\{0, 3\}$ **87.** $b = -4$, $c = -12$, and $d = 0$ **89.** $a = 15$, $b = -16$, and $c = -5$

one

91. $\left\{\frac{1}{10}\left(3\pm i\sqrt{331}\right)\right\}$ **93.** $\{-1.796, 1.067\}$ **95.** $\left\{\frac{2}{3}\left(2\pm i\sqrt{2}\right)\right\}$

section 1.8, pages 109-112

1. $\frac{2x+1}{x-5};\ x\neq -3, 5$ **3.** $x(x-1);\ x\neq -3$ **5.** $\frac{x+6}{x+5};\ x\neq -5, 1$ **7.** $\frac{1}{x^2-x+1};\ x\neq -1$

9. $2x+1;\ x\neq -5$ **11.** $2x-3;\ x\neq -7$ **13.** $\frac{x^3+9x^2+11x+19}{(x-3)(x+5)}$ **15.** $\frac{13x}{(x-3)(x+5)}$

17. $\frac{x^3+4x^2-7x+18}{x^2-9}$ **19.** $y-1$ **21.** $(x+2)(2x+3)$ **23.** -6 **25.** $\frac{x^2+9}{6x-3}$ **27.** $\frac{2x^2}{x+1}$

29. $\frac{s-r}{r^2s+s}$ **31.** $\frac{m+n}{mn}$ **33.** $\left\{3\pm\sqrt{10}\right\}$ **35.** $\{-2\}$ **37.** $\varnothing$ **39.** $(-\infty,-3)\cup(-3,3)\cup(3,\infty)$

41. $\left\{\frac{-17\pm\sqrt{145}}{2}\right\}$ **43.** $\frac{35}{12}$ hours, or 2 hours, 55 minutes. **45.** 9.1 hours (to the nearest tenth)

47. 20 weeks **49.** 45 minutes **51.** $\frac{x+10}{(x-5)(x+5)}$ **53.** $\frac{x+1}{(5x-1)}$ **55.** $\frac{x^2+xy+y^2}{x^2y^2}$ **57.** $\{0\}$

59. $\{6\}$ **61.** $\{1\}$ **63.** $\left\{\frac{2}{3}\right\}$ **65.** $\varnothing$ **67.** $\left\{\frac{29}{8}\right\}$ **69.** $\varnothing$ **71.** $\varnothing$

73. $\{-2, 1\}$ **75.** $\varnothing$ **77.** $\left\{\frac{1}{4}, 1\right\}$ **79.** $\{3\}$ **81.** $\{-2, 2\}$ **83.** $\left\{-\frac{8}{27}, \frac{8}{27}\right\}$

85. $a=\sqrt{c^2-b^2}$ **87.** $r=\left(\frac{T^2GM}{4\pi^2}\right)^{1/3}$ **89.** $L=\frac{\sqrt{Z^2-R^2}}{\omega}+\frac{1}{\omega^2C}$

chapter one test, pages 122-126

1. a. $7, 3^3$ **b.** $7, \frac{0}{5}, 3^3$ **c.** $7, \frac{0}{5}, 3^3, -1, -\sqrt{4}$ **d.** $-15.75, 7, \frac{0}{5}, 3^3, -1, -\sqrt{4}$ **e.** none **f.** $-15.75, 7, \frac{0}{5}, 3^3, -1, -\sqrt{4}$ **g.** $\frac{-8}{0}$ **3.** (number line with points at −3, 0, 3, 5) **5.** $<, \leq$ **7.** $\{x \mid x \text{ is an integer and } -2\leq x<4\}$

9. $[4, 11]$ **11.** $[0, 3)$ **13.** $(-7, 9]$ **15.** $\frac{23}{3}$ **17.** -9 **19.** $\sqrt{11}-\sqrt{5}$ **21.** $\frac{4\pi}{3}-36$

23. zero-factor property **25.** $\frac{27z^3}{y^6}$ **27.** 5.224×10^7 **29.** $\frac{|a|}{2}$ **31.** $4|x|$ **33.** $\frac{a^2\sqrt[4]{a}}{3|b|}$

35. third-degree trinomial **37.** $3x^2y+7xy-2z+4xz$ **39.** $20x^3y+10x^2y^2+8x^2y+4xy^2-12x-6y$

41. $(x-2y)(3m+n)$ **43.** $5a^2b\left(2a-3ab^2+b^3\right)$ **45.** $(x+3)(x-4)$ **47.** $5i-5$ **49.** $-\frac{7}{25}+\frac{24}{25}i$

51. $-\frac{5}{17}+\frac{3}{17}i$ **53.** $-8i$ **55.** $y=\frac{14}{9}$; conditional **57.** $x=\frac{19}{3}$; conditional **59.** $\{3, 4\}$

61. $v_0=\frac{h+16t^2}{t}$ **63.** 246.7 miles **65.** $(-1, 7]$ (number line from −1 to 7) **67.** $(-5, -1)$ (number line from −5 to −1)

69. under \$420 **71.** $\left\{3+2\sqrt{3}, 3-2\sqrt{3}\right\}$ **73.** $\{-3,-2\}$ **75.** $\left\{-1, 1, -\sqrt{2}, \sqrt{2}\right\}$ **77.** $\left\{-\sqrt{2}, \sqrt{2}, 4\right\}$

79. $\left\{0, \frac{1\pm\sqrt{13}}{4}\right\}$ **81.** $2z-14;\ z\neq 3$ **83.** $\frac{3a^3+a}{5a+1}$ **85.** $\frac{y^3+6y^2-17y+2}{7y^2+15y+2}$ **87.** $\frac{x+3}{3}$ **89.** $\left\{-\frac{1}{3}\right\}$

91. $\frac{12}{7}$ hrs. ≈ 1 hr. 43 min. **93.** $\{2\}$ **95.** $\{3\}$ **97.** $\{4,5\}$

chapter two

section 2.1, pages 136-138

1.

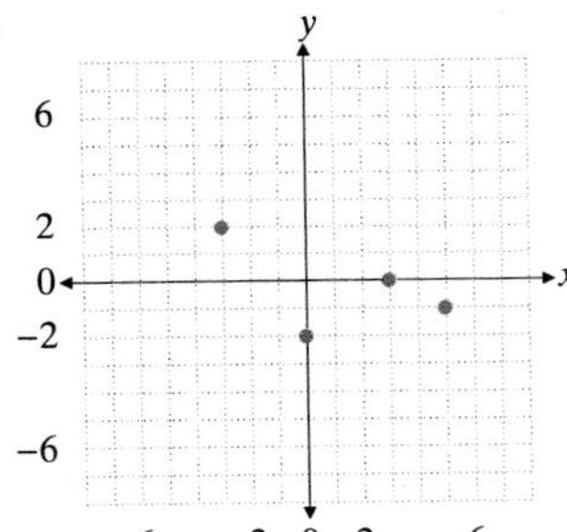

3.

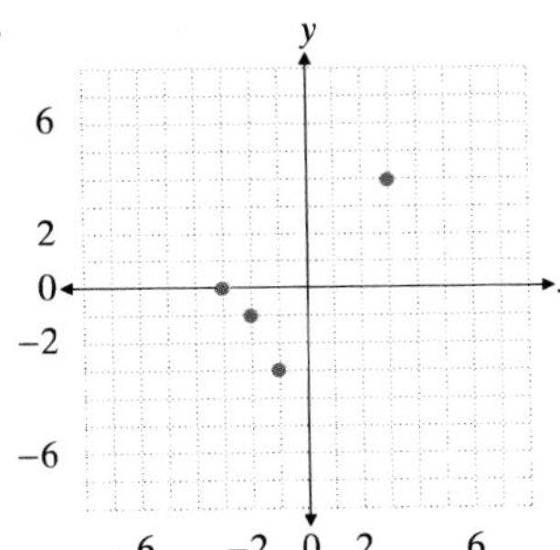

5. III **7.** IV **9.** positive x-axis
11. III **13.** IV **15.** II **17.** IV **19.** I
21. negative y-axis

23. $\left\{(0,-3), (2,0), \left(3, \frac{3}{2}\right), (4,3)\right\}$

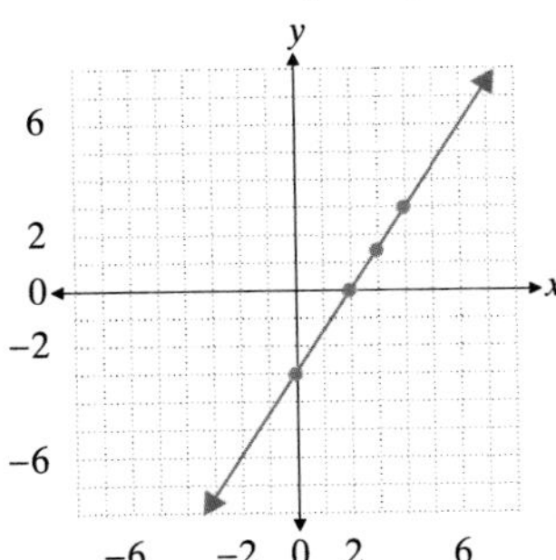

25. $\left\{(0,0), (1,\pm 1), (4,\pm 2), (9,\pm 3), (2,\pm\sqrt{2})\right\}$

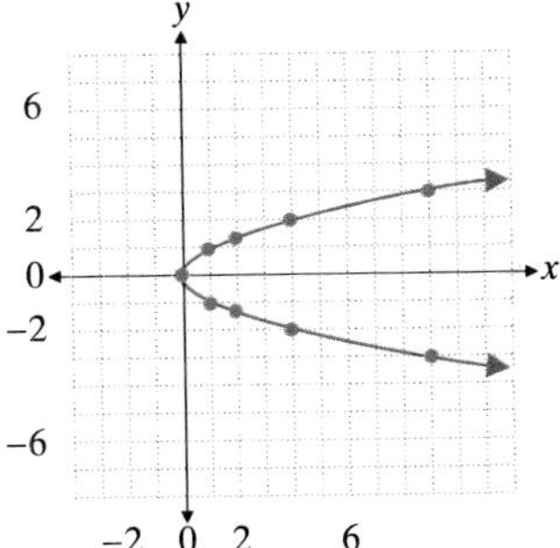

27. $\left\{(0,\pm 3), (\pm 3, 0), (-1, \pm 2\sqrt{2}), (1, \pm 2\sqrt{2}), (\pm\sqrt{5}, 2)\right\}$

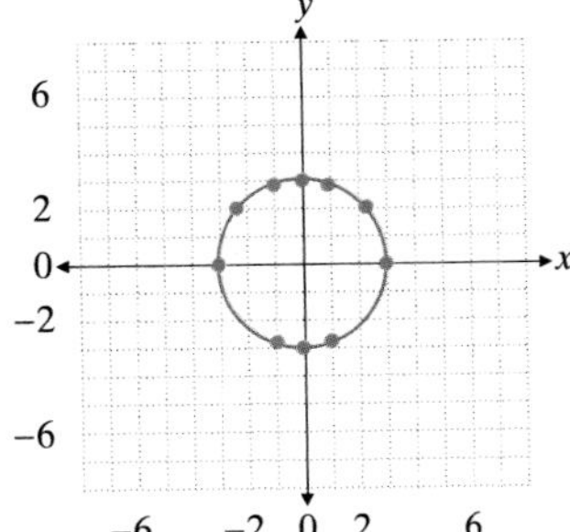

29. a. $\sqrt{34}$ **b.** $\left(\frac{-7}{2}, \frac{1}{2}\right)$ **31. a.** $\sqrt{58}$ **b.** $\left(\frac{3}{2}, \frac{7}{2}\right)$

33. a. $2\sqrt{2}$ **b.** $(-1,-1)$ **35. a.** $4\sqrt{34}$ **b.** $(3,-8)$

37. a. 10 **b.** $(1,-6)$ **39.** 12

41. $2\sqrt{29}+\sqrt{26}+5\sqrt{2}$ **43.** 54

45. area $=\frac{25}{2}$ **47.** 145 miles

section 2.2, pages 144-145

1. Yes **3.** No **5.** No
7. No **9.** Yes **11.** Yes
13. Yes **15.** No **17.** No
19. No **21.** No **23.** Yes

25. x-int: 3, y-int: 4

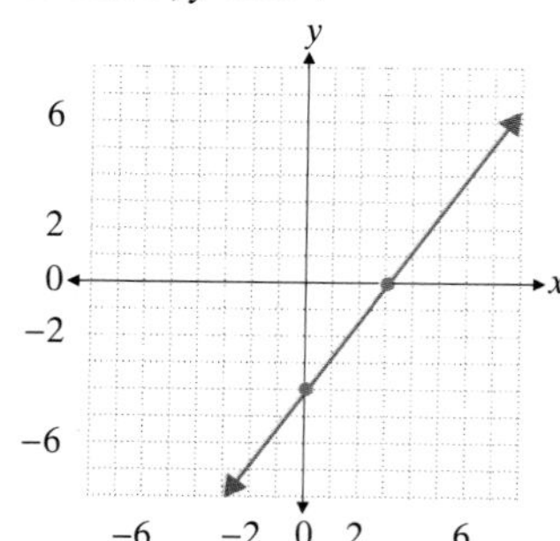

27. x-int: 0.5, y-int: 5

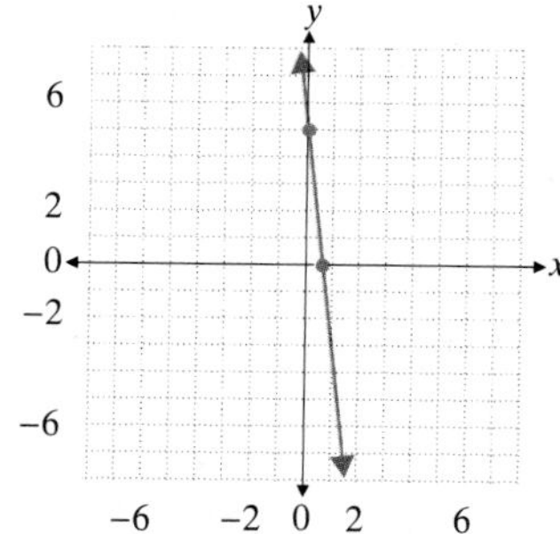

29. x-int: DNE, y-int: 3

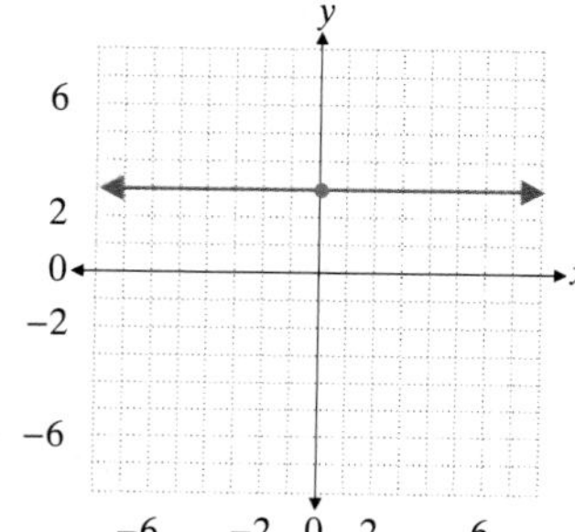

31. x-int: 3.5, y-int: 7

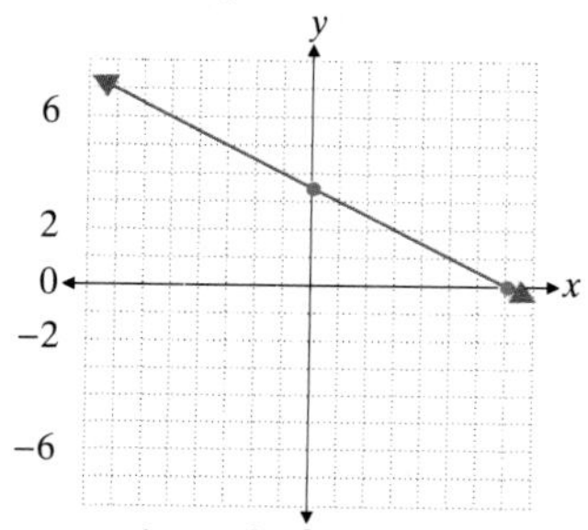

33. x-int: 0, y-int: 0

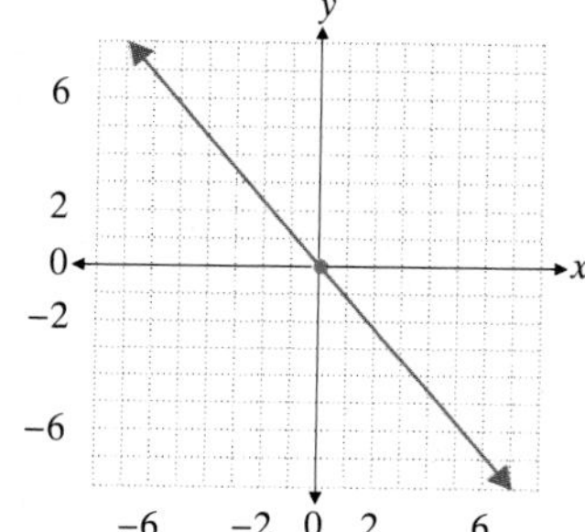

35. x-int: DNE, y-int: 7

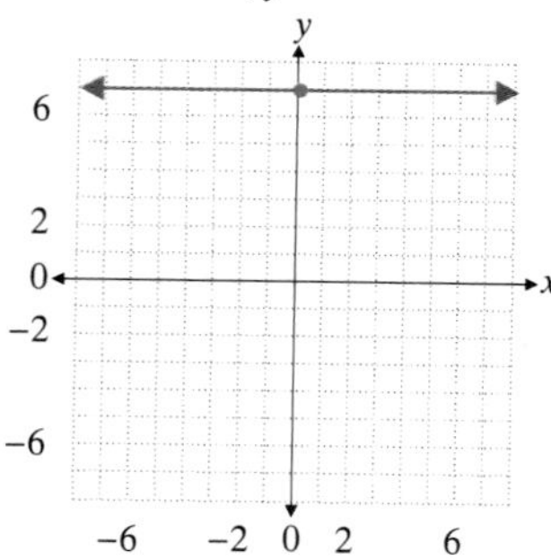

37. x-int: 1, y-int: –1

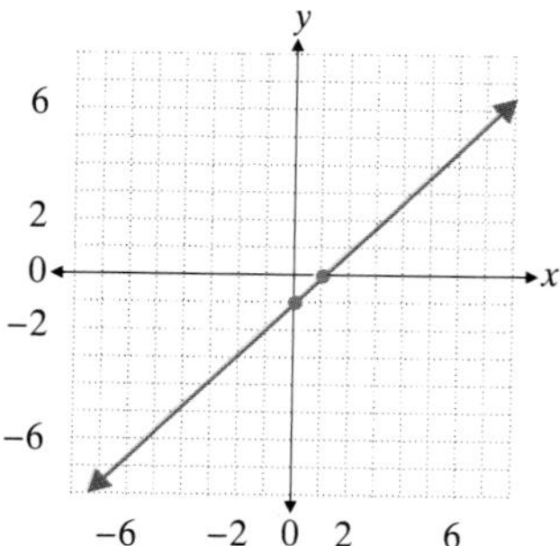

39. x-int: –3, y-int: –2

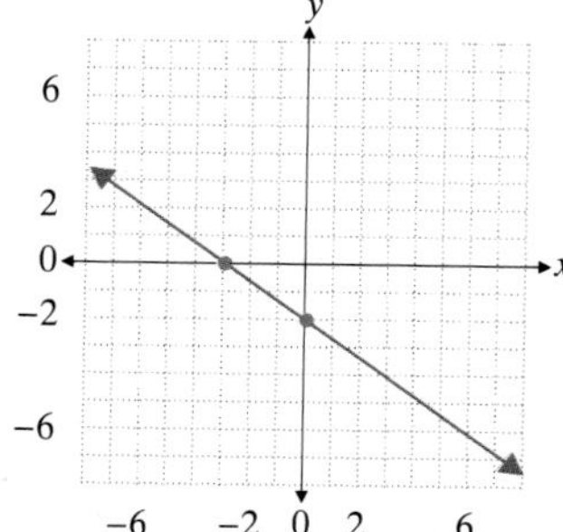

section 2.3, pages 154-156

1. –4 **3.** 0 **5.** undefined **7.** $\frac{2}{3}$ **9.** –7 **11.** –6 **13.** 4 **15.** 0 **17.** undefined

19. $-\frac{5}{6}$ **21.** $\frac{7}{6}$ **23.** $-\frac{5}{2}$

25.
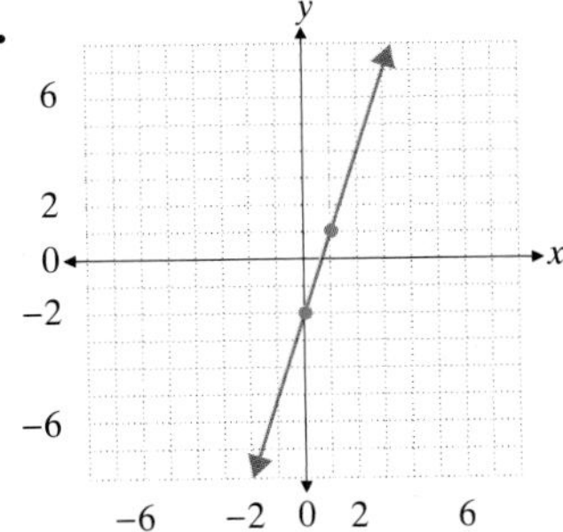

27.
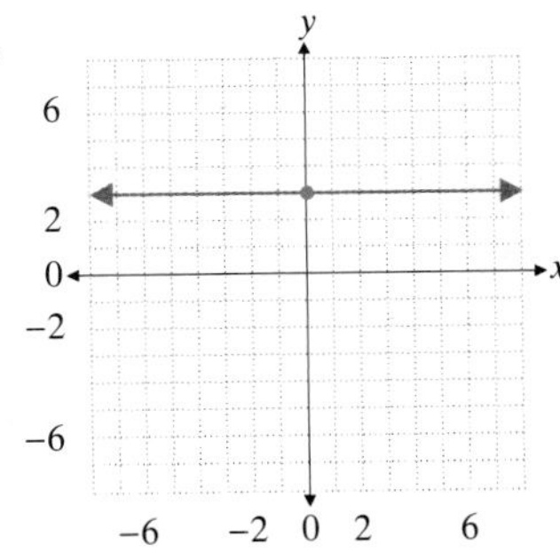

29.
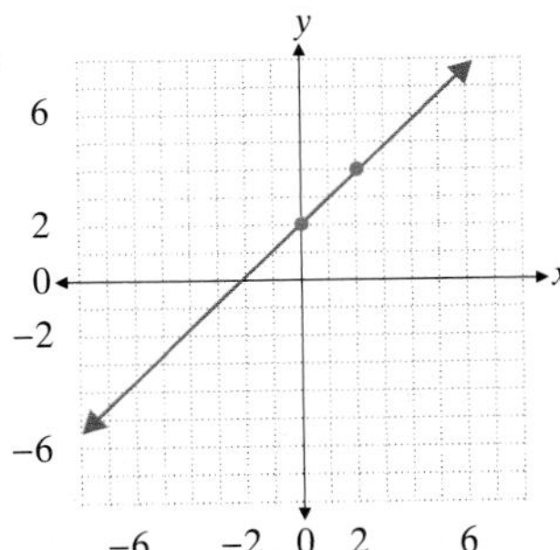

31. $y = \frac{3}{4}x - 3$ **33.** $y = -\frac{5}{2}x - 7$ **35.** $3x - 2y = 3$ **37.** $y = 5$ **39.** $5x - 4y = 30$ **41.** $4x + 3y = 5$

43. $x = 2$ **45.** $y = -1$ **47. a.** \$2225 **b.** \$2100 **c.** \$0.25 **49.** $V = -2766.67t + 51{,}500$; 19 years

51. $m = \frac{31}{99}, b = \frac{20}{11}$ **53.** $m = \frac{-1}{7}, b = \frac{-5}{2}$ **55.** $m = 0, b = 0$

section 2.4, pages 161-163

1. $y = 4x + 9$ **3.** $y = 3x - 2$ **5.** $y = -\frac{1}{10}x - \frac{1}{2}$ **7.** Yes **9.** Yes **11.** No **13.** No **15.** No

17. $y = -\frac{1}{3}x - 1$ **19.** $y = x$ **21.** $y = 3x$ **23.** No **25.** Yes **27.** Yes **29.** No **31.** No **33.** 41.67 ft.

35. $m_1 = \frac{2}{3}, m_2 = \frac{-3}{2}$; Perpendicular **37.** $m_1 = \frac{17}{13}, m_2 = \frac{-17}{13}$; Neither

39. $m_1 = \frac{-13}{18}, m_2 = \frac{18}{13}$; Perpendicular

section 2.5, pages 170-171

1.
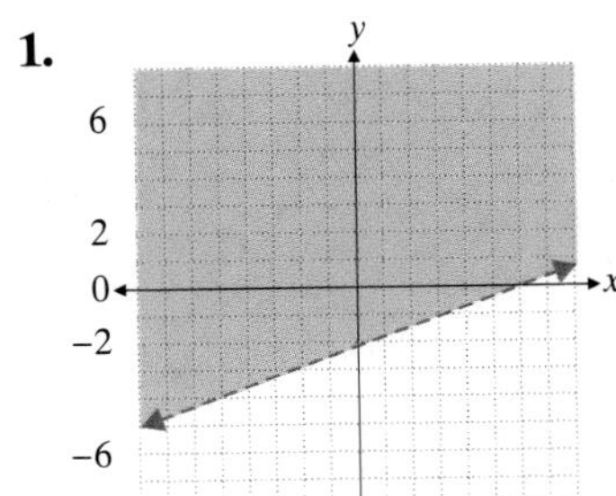

3.
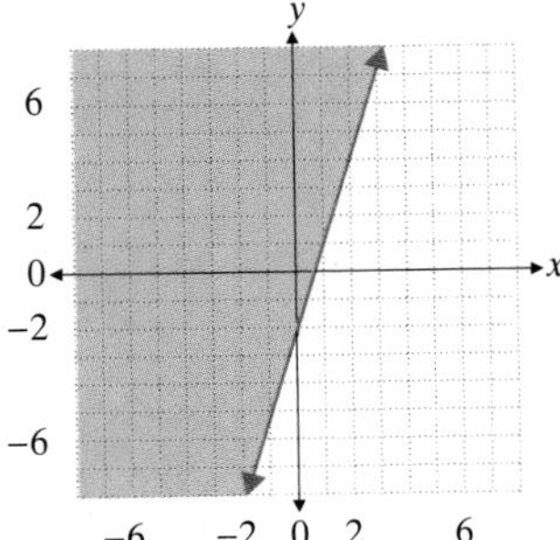

5.
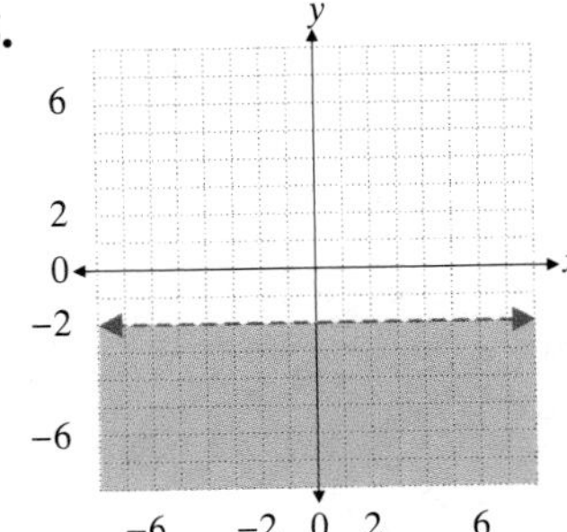

7.

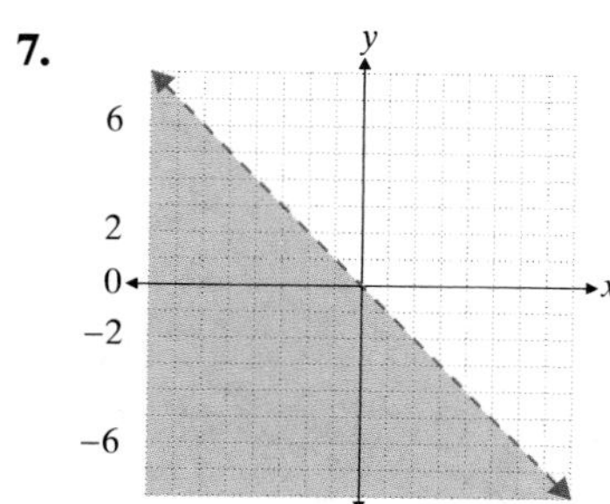

9.

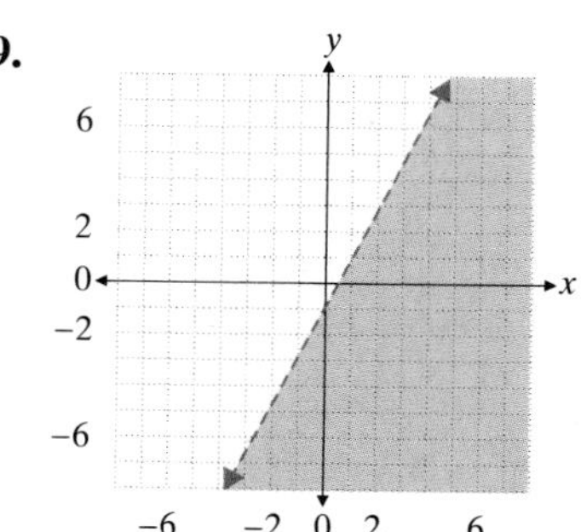

11.

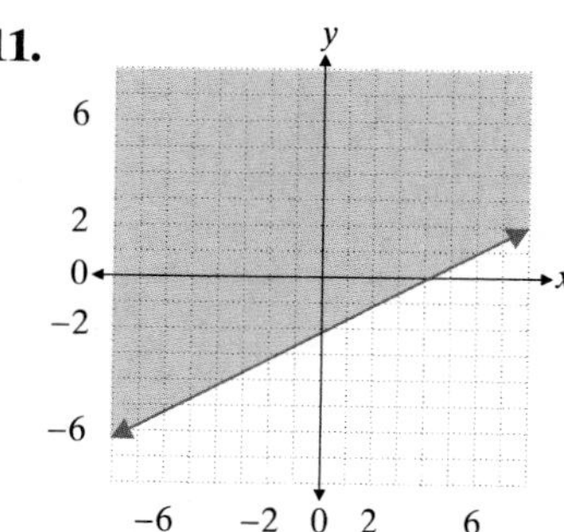

13.

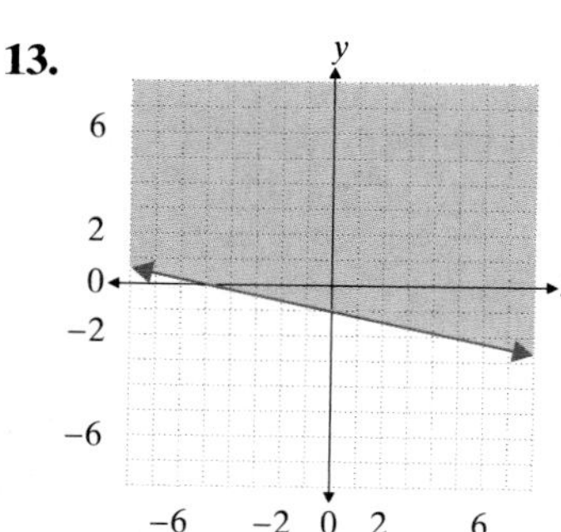

15.

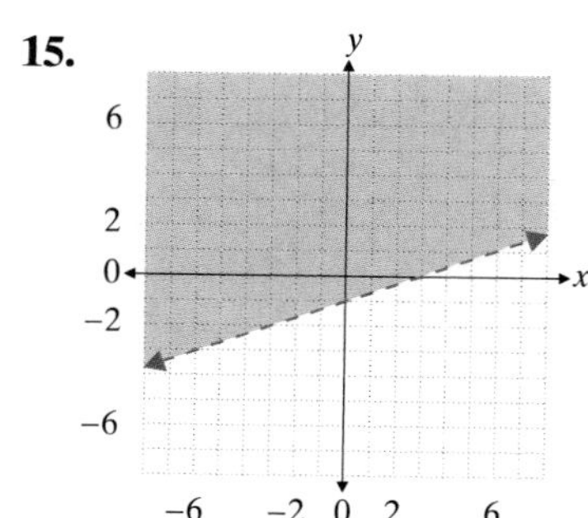

17.

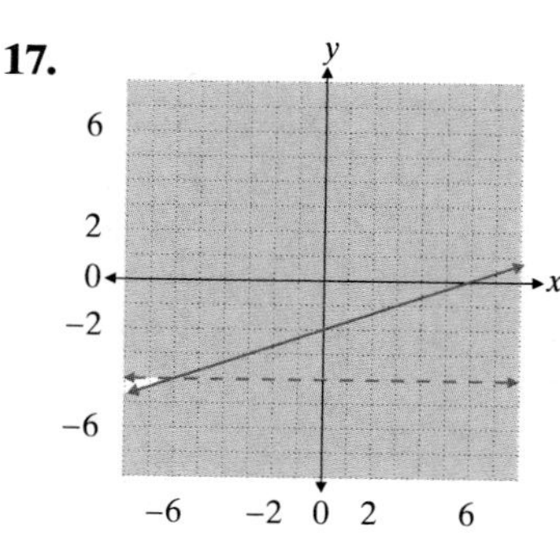

19.

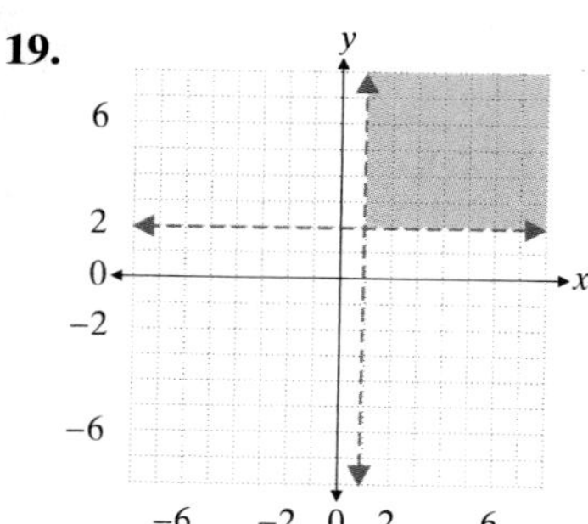

21.

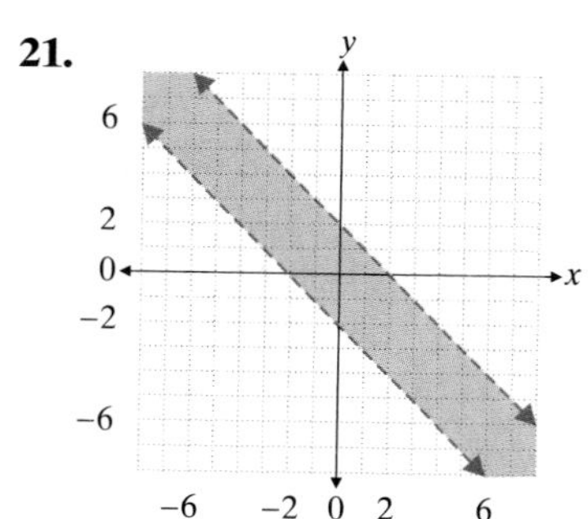

23.

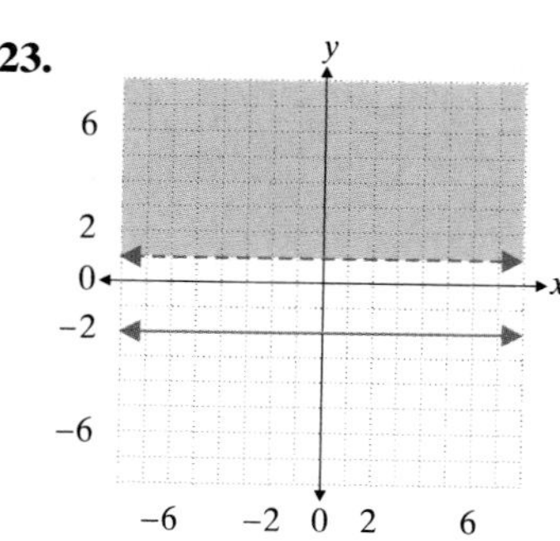

25.

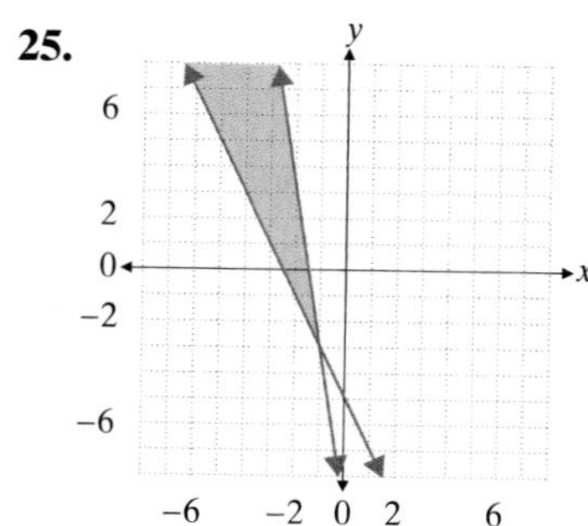

27.

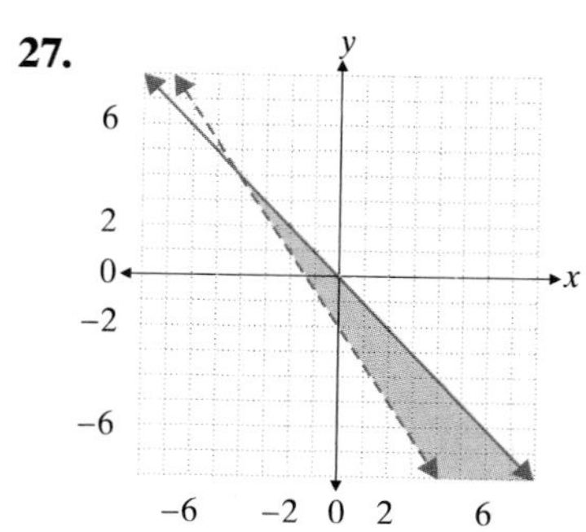

29.

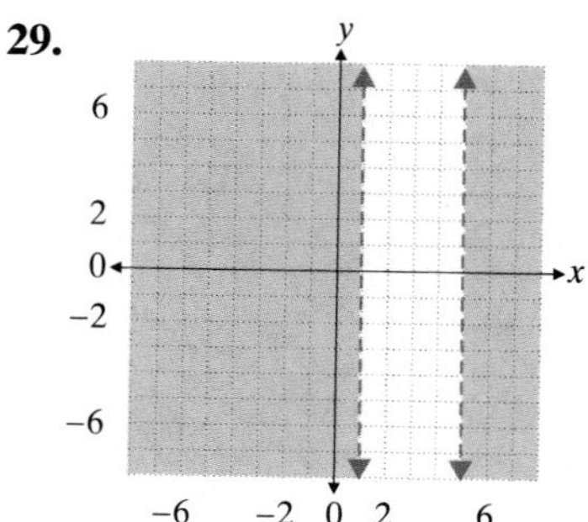

31.
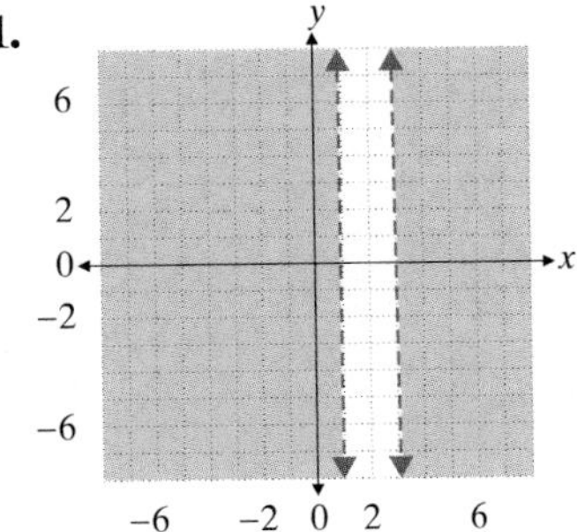

33.
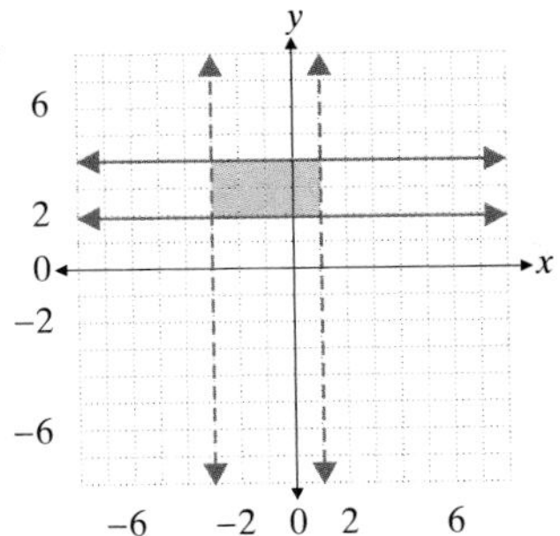

35.
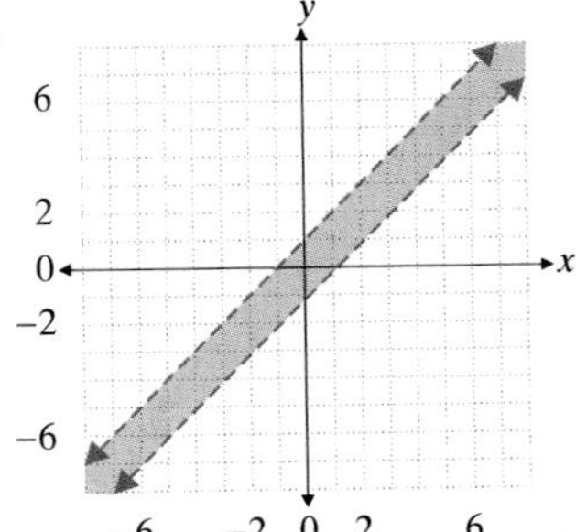

37.
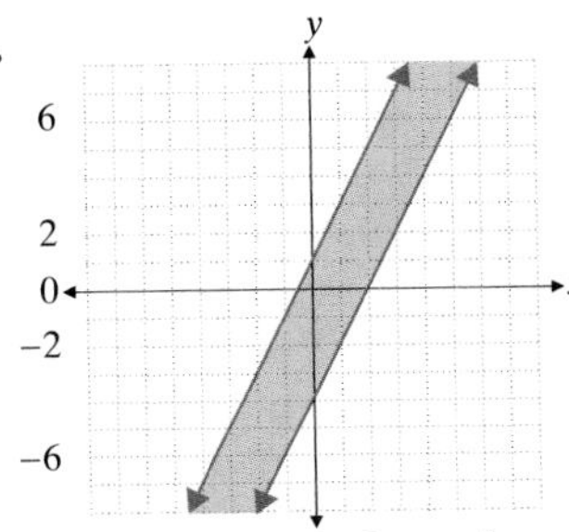

39.
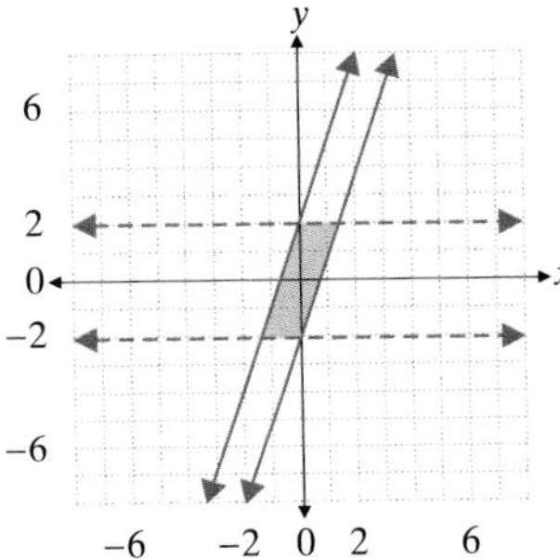

41. $12l + 22o < 150$

43. $2l + 2w \le 300$
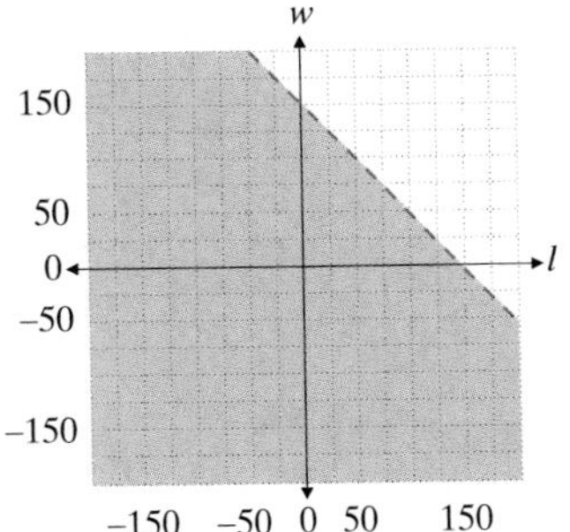

section 2.6, pages 176-178

1. $(x+4)^2 + (y+3)^2 = 25$ **3.** $(x-7)^2 + (y+9)^2 = 9$ **5.** $x^2 + y^2 = 9$ **7.** $\left(x-\sqrt{5}\right)^2 + \left(y-\sqrt{3}\right)^2 = 16$

9. $(x-7)^2 + (y-2)^2 = 4$ **11.** $(x+3)^2 + (y-8)^2 = 2$ **13.** $(x-4)^2 + (y-8)^2 = 10$ **15.** $x^2 + y^2 = 85$

17. $\left(x+\frac{7}{2}\right)^2 + \left(y-\frac{17}{2}\right)^2 = 26.5$ **19.** $(x+6)^2 + \left(y-\frac{3}{2}\right)^2 = 31.25$ **21.** $\left(x+\frac{13}{2}\right)^2 + (y+7)^2 = 91.25$

23. $(x-4)^2 + (y-3)^2 = 25$ **25.** $(x-2)^2 + y^2 = 4$ **27.** $(x-2)^2 + (y-4)^2 = 49$ **29.** $(x+3)^2 + (y+2)^2 = 64$

31.

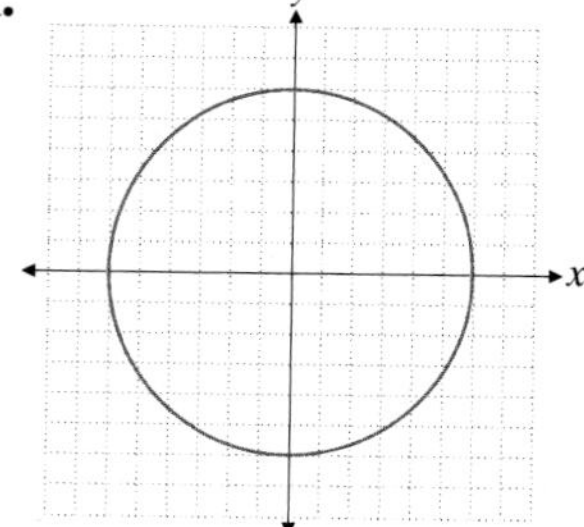

33.

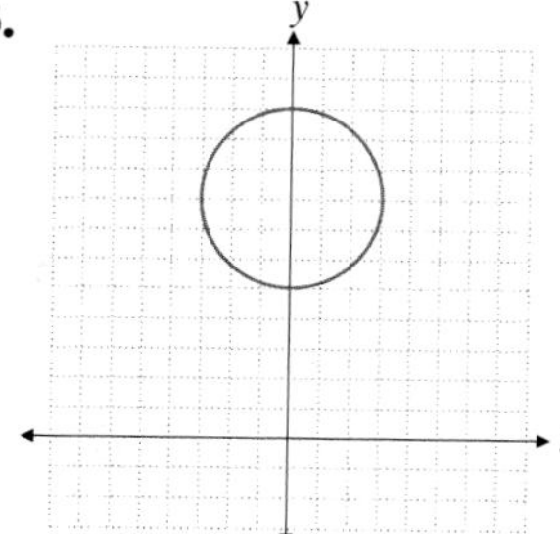

35.

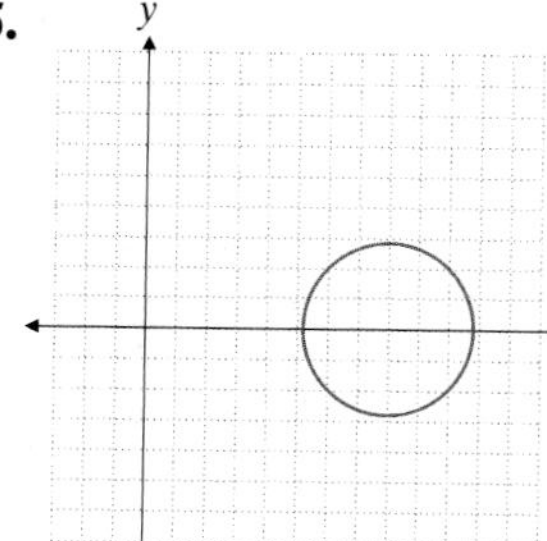

37.

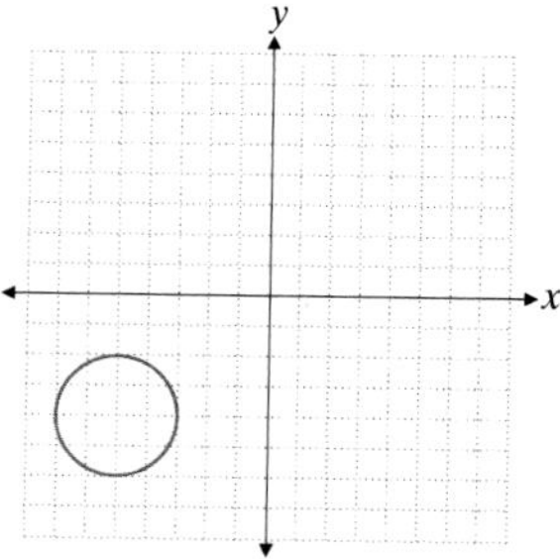

39.

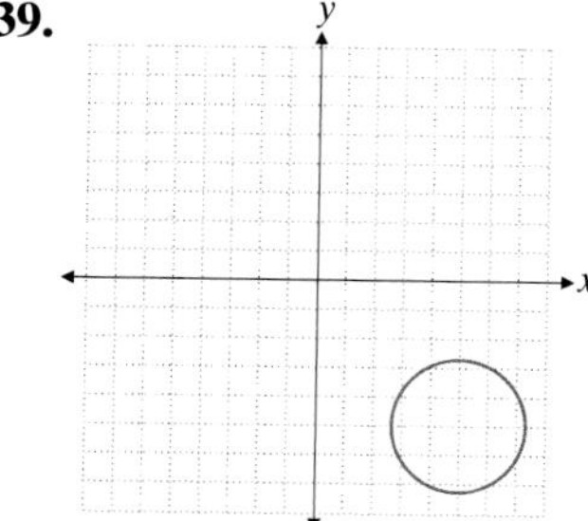

41.

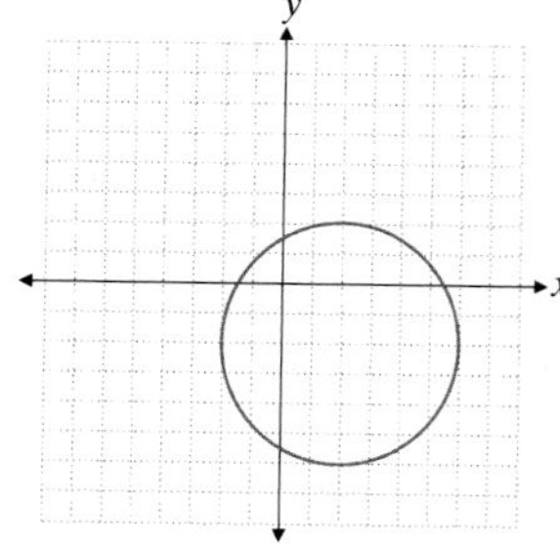

43.

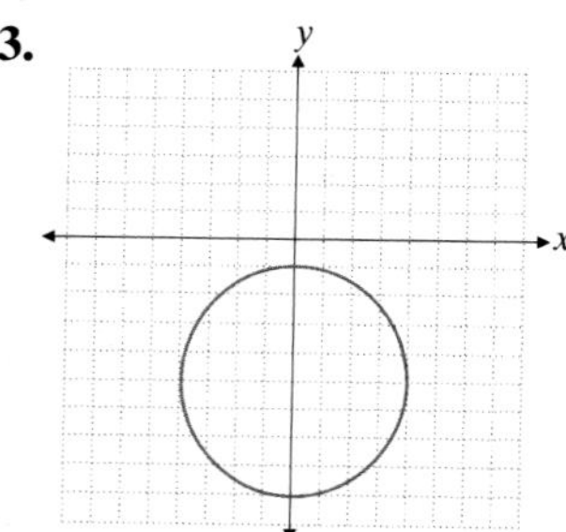

45.

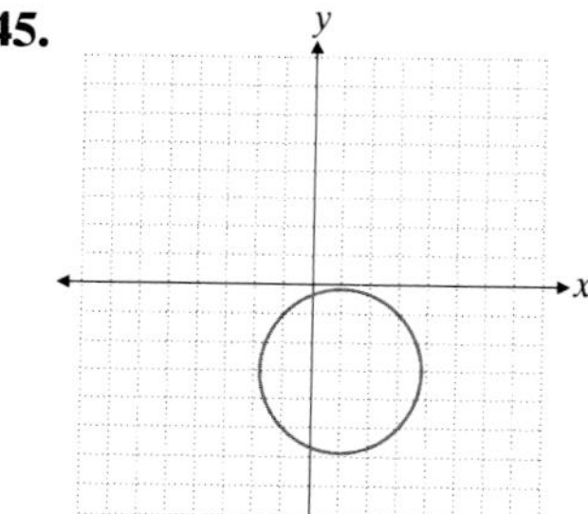

47.

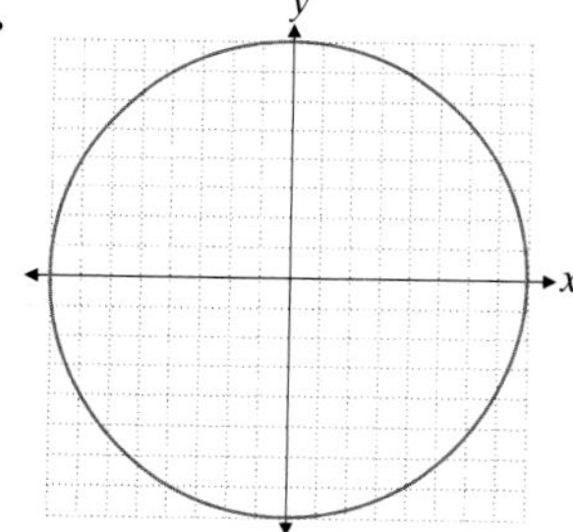

49.

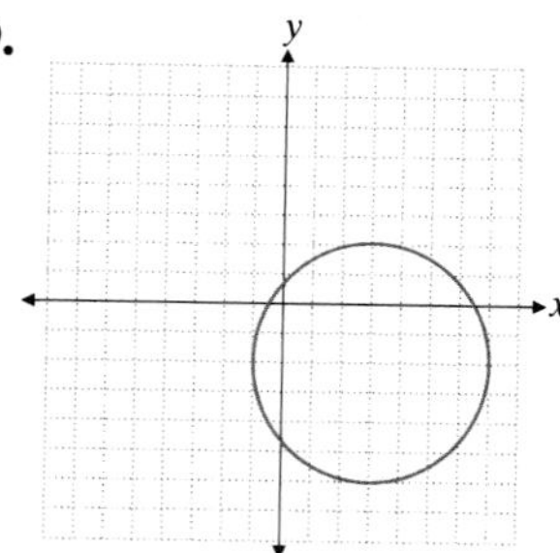

51.

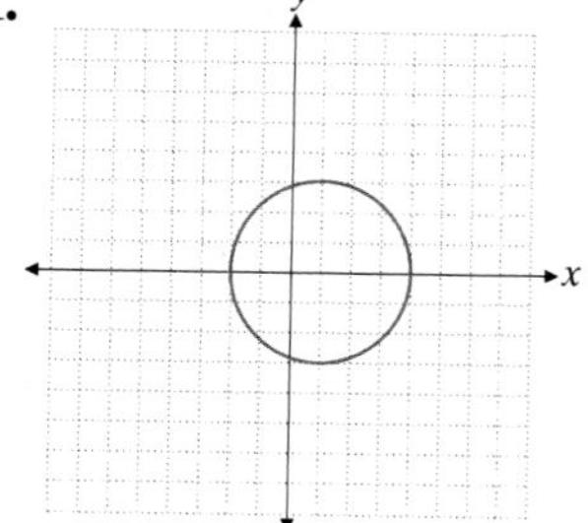

53.

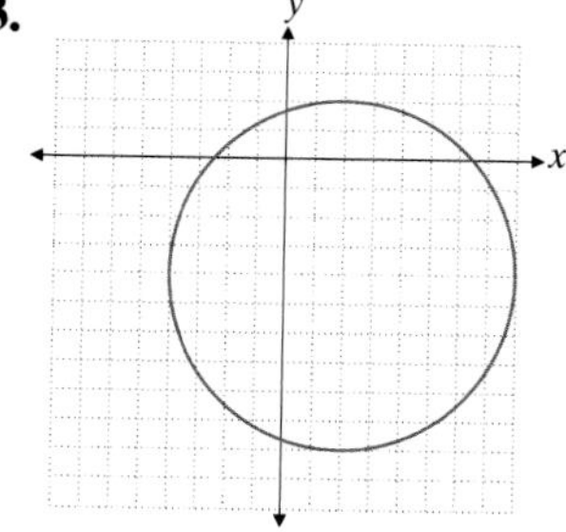

55.

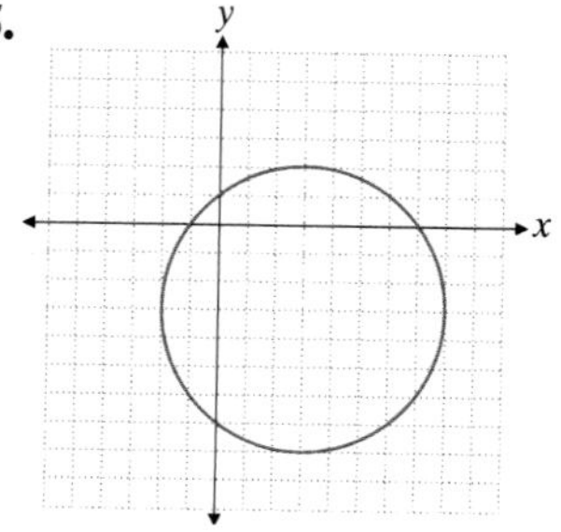

chapter two test, pages 183-185

1. II **3.** III

5.

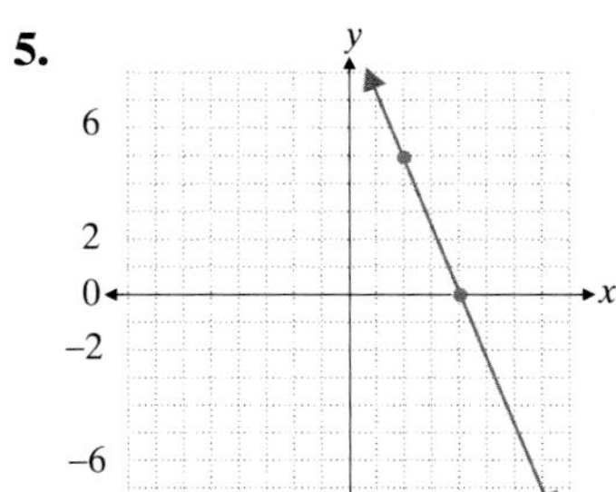

7.

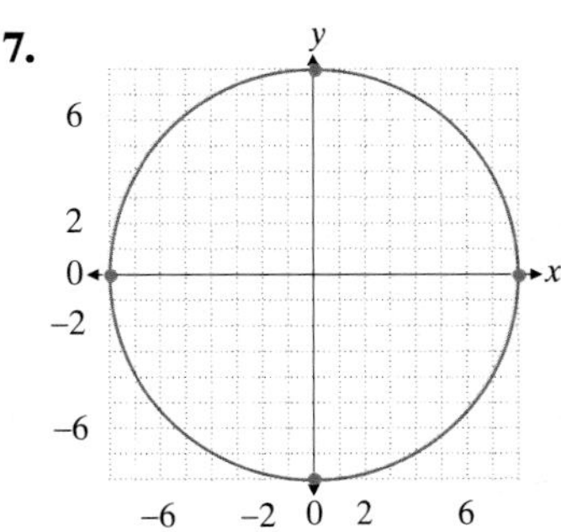

9. a. 10 **b.** (0, –6)

11. $2+3\sqrt{2}+\sqrt{34}$

13. No **15.** Yes

17. x-intercept: $\left(\dfrac{1}{2},0\right)$, y-intercept: (0, 1)

19.

21.

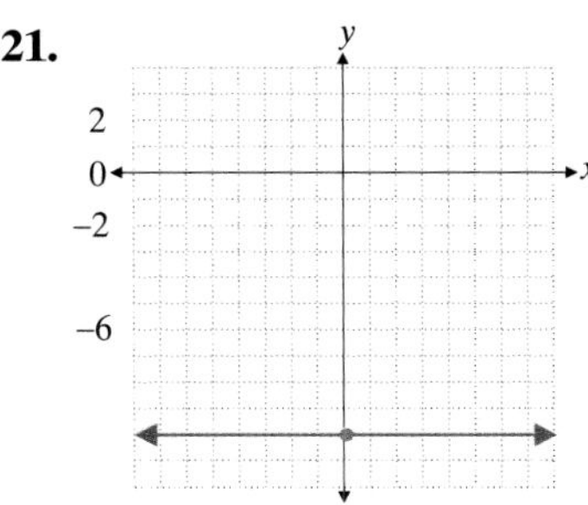

23. –4

25.

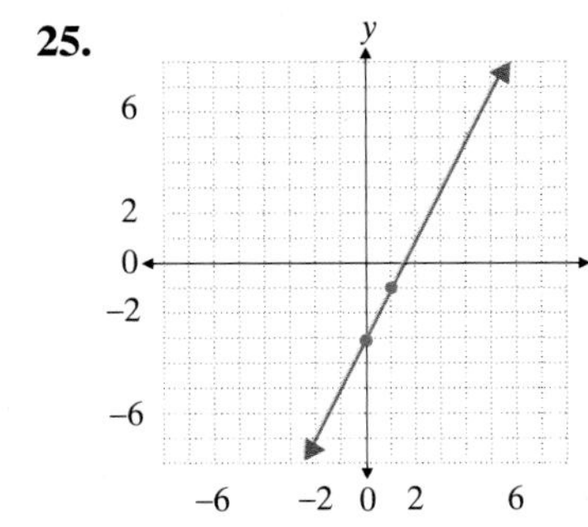

27.

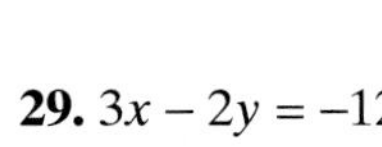

29. $3x-2y=-12$

31. Neither

33. Neither

35. No

37.

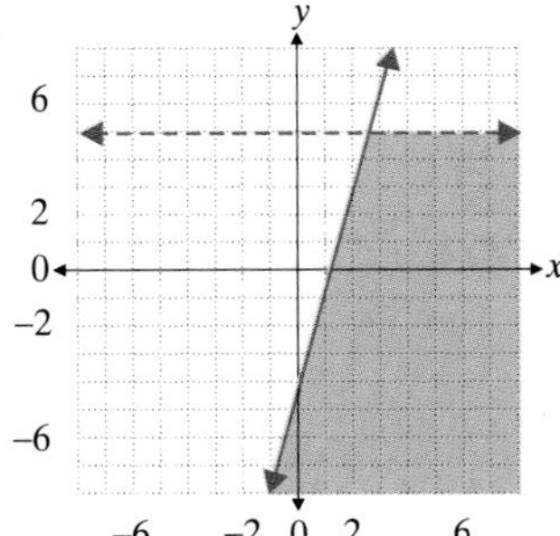

39.

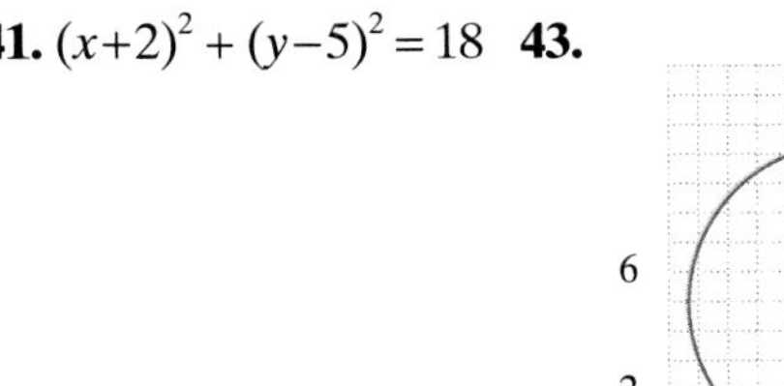

41. $(x+2)^2+(y-5)^2=18$

43.

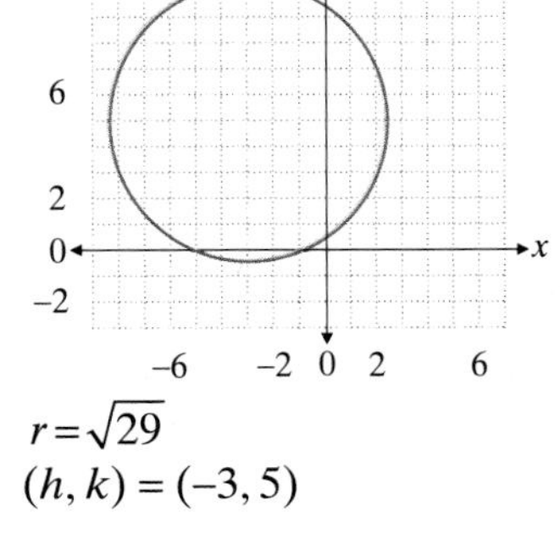

$r=\sqrt{29}$

$(h,k)=(-3,5)$

chapter three

section 3.1, pages 201-205

1. Dom = $\{-2\}$, Ran = $\{-9, 0, 3, 5\}$ **3.** Dom = $\{-2\pi, 1, 3, \pi\}$, Ran = $\{0, 2, 4, 7\}$ **5.** Dom = $\mathbb{Z}$, Ran = even integers **7.** Dom = $\mathbb{Z}$, Ran = $\{\ldots, -2, 1, 4, \ldots\}$ **9.** Dom = Ran = $\mathbb{R}$
11. Dom = $[0, \infty)$, Ran = $\mathbb{R}$ **13.** Dom = $\mathbb{R}$, Ran = $\{-1\}$ **15.** Dom = $\{0\}$, Ran = $\mathbb{R}$
17. Dom = $[-3, 1]$, Ran = $[0, 4]$ **19.** Dom = $[0, 3]$, Ran = $[1, 5]$ **21.** Dom = $[-1, 3]$, Ran = $[-4, 3]$
23. Dom = All males with siblings, Ran = All people who have brothers **25.** Not a function; (–2, 5) and (–2, 3)
27. Function **29.** Not a function; $(1,1)$ and $(1,-3)$ **31.** Function
33. Function; Dom = $(-\infty, 0) \cup (0, \infty)$ **35.** Not a function **37.** Function; Dom = $(-\infty, -2) \cup (-2, \infty)$
39. Function; Dom = $\mathbb{R}$ **41.** Not a function **43.** $f(x) = -6x^2 + 2x,\ f(-1) = -8$
45. $f(x) = \dfrac{-x+10}{3},\ f(-1) = \dfrac{11}{3}$ **47.** $f(x) = -2x - 10,\ f(-1) = -8$ **49. a.** $x^2 + x - 2$ **b.** $2ax + 3a + a^2$
c. $x^4 + 3x^2$ **51. a.** $3x - 1$ **b.** $3a$ **c.** $3x^2 + 2$ **53. a.** $-6x + 16$ **b.** $-6a$ **c.** $-6x^2 + 10$ **55. a.** 5 **b.** 17 **c.** 2
d. $x^2 + 2xh + h^2 - 4x - 4h + 5$ **e.** $2x + h - 4$ **57. a.** –1 **b.** –4 **c.** $\dfrac{7}{2}$ **d.** $\dfrac{3x+3h-2}{2}$ **e.** $\dfrac{3}{2}$
59. a. $2i\sqrt{2}$ **b.** $i\sqrt{10}$ **c.** $i\sqrt{5}$ **d.** $\sqrt{x+h-8}$ **e.** $\dfrac{\sqrt{x+h-8} - \sqrt{x-8}}{h}$ **61. a.** 0 **b.** –8 **c.** 27
d. $(x+h)^3$ **e.** $3x^2 + 3hx + h^2$ **63. a.** 0 **b.** –16 **c.** 9 **d.** $-x^2 + 6x - 2xh + 6h - h^2$ **e.** $-2x + 6 - h$
65. Dom = Cod = $\mathbb{Z}$; Ran = $\{\ldots, -3, 0, 3, \ldots\}$ **67.** Dom = $[0, \infty)$; Cod = $\mathbb{R}$; Ran = $[0, \infty)$
69. Dom = $\mathbb{N}$; Cod = $\mathbb{R}$; Ran = $\left\{\dfrac{1}{2}, 1, \dfrac{3}{2}, 2, \ldots\right\}$ **71.** $\mathbb{R}$ **73.** $[-3, \infty)$ **75.** $(-\infty, 3) \cup (3, \infty)$
77. $g(2) = 3\sqrt{3};\ g(3) = 2\sqrt{7}$ **79.** $g(-19) = 7{,}290{,}099{,}019;\ g(12) = 258{,}474{,}853$
81. $h^5 + 3h^3 + 3h + 6x + 12h^2x + 6h^4x + 18hx^2 + 15h^3x^2 + 12x^3 + 20h^2x^3 + 15hx^4 + 6x^5$
83. $\dfrac{-1}{(x-1)(x+h-1)}$

section 3.2, pages 216-220

1.

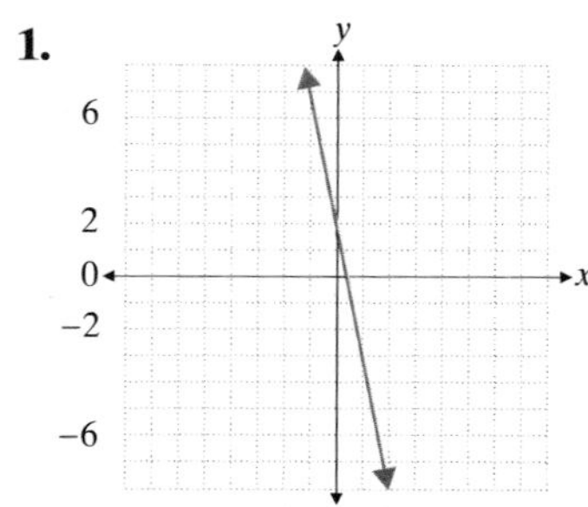

3.

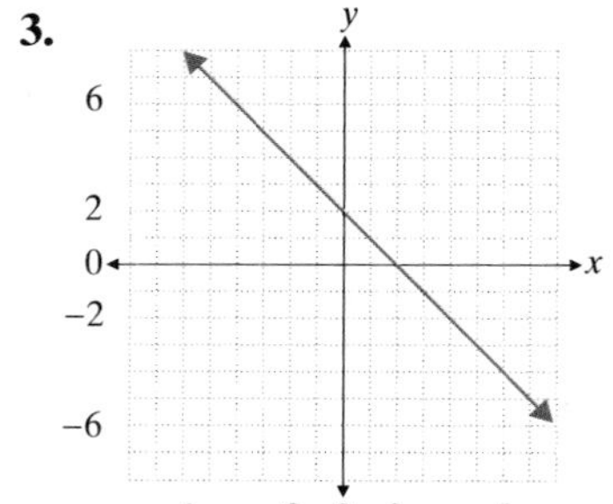

5.

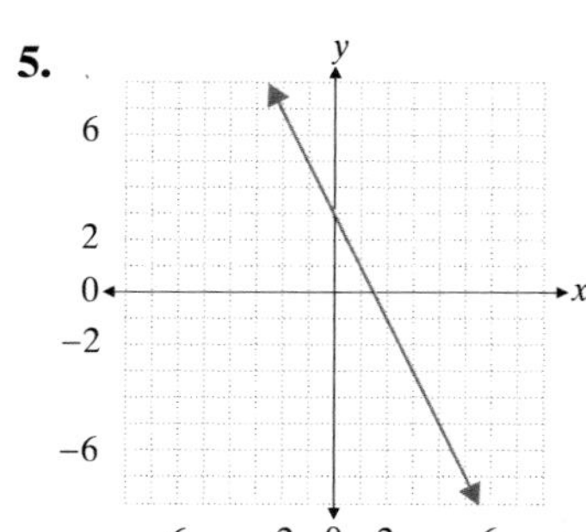

7.

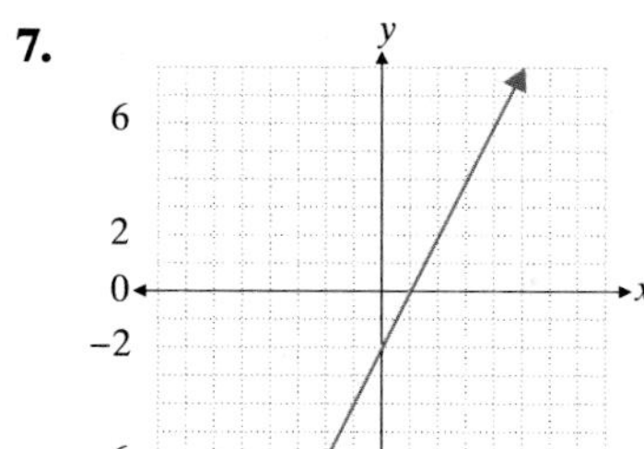

9.

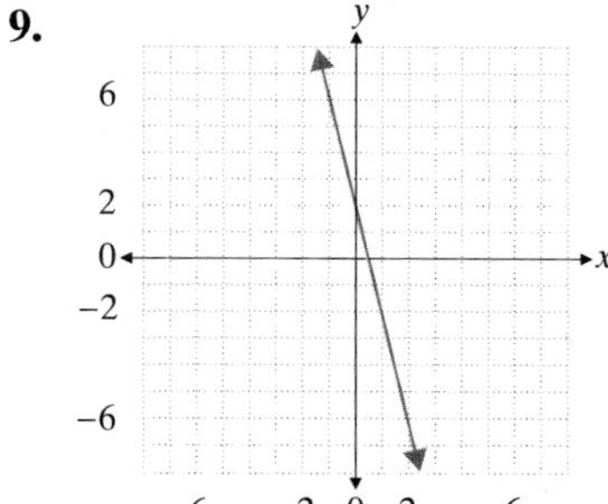

11.

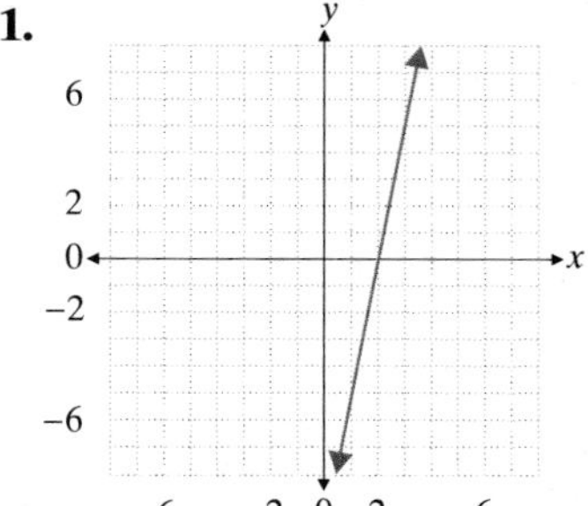

13.

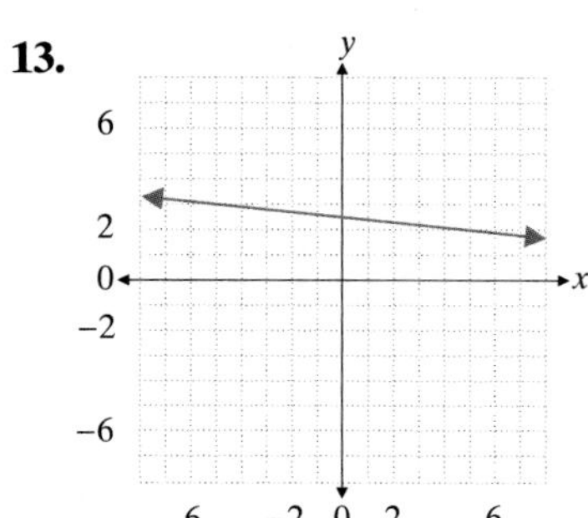

15.

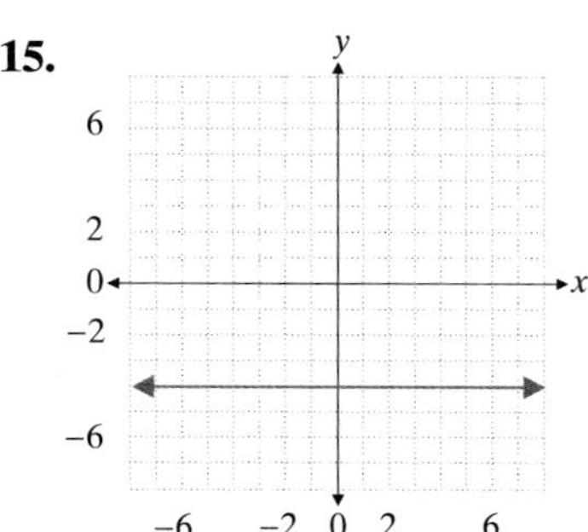

17. vertex: $(-2, -1)$; no x-int.

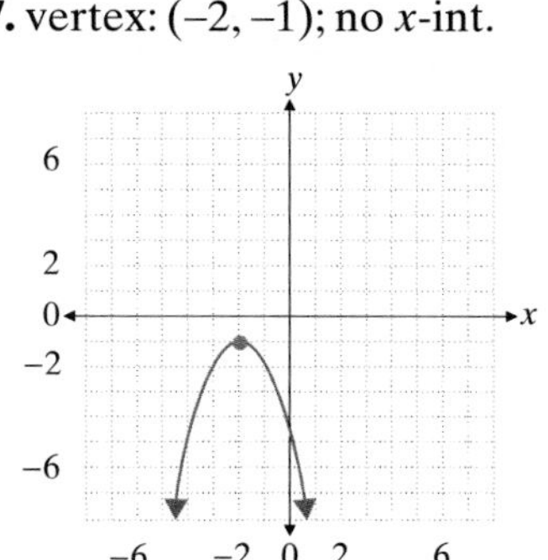

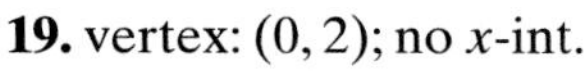

19. vertex: $(0, 2)$; no x-int.

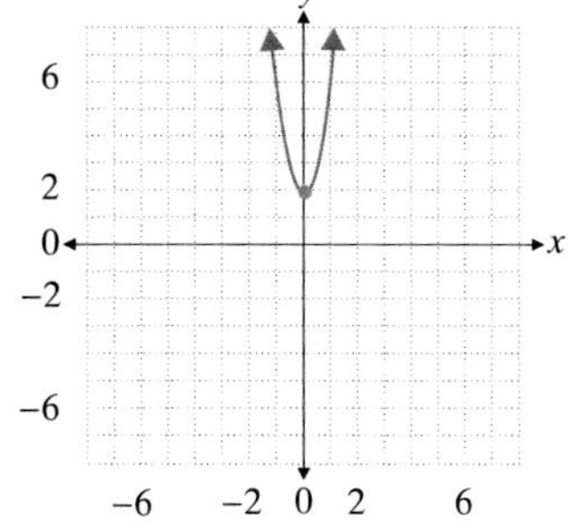

21. vertex: $\left(\frac{1}{2}, \frac{25}{2}\right)$; x-int: $-2, 3$

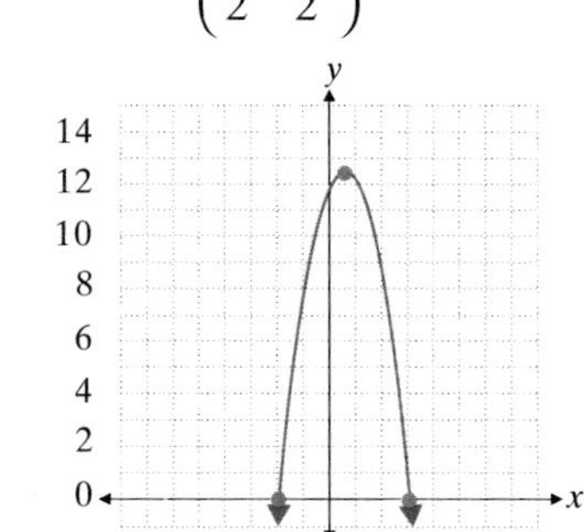

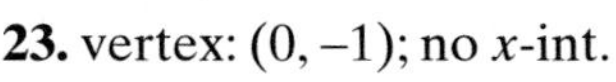

23. vertex: $(0, -1)$; no x-int.

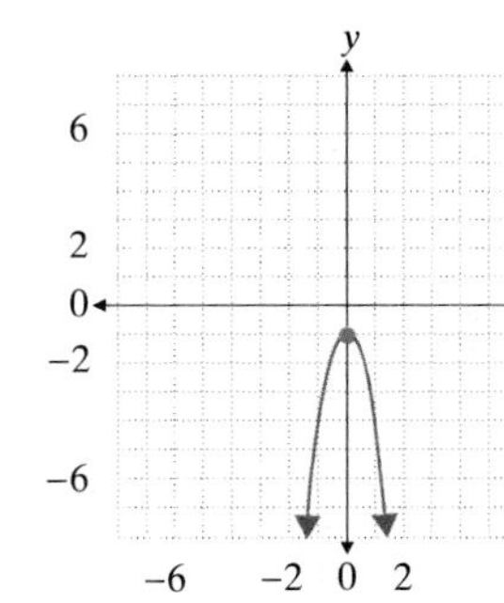

25. vertex: $(-1, 3)$; no x-int.

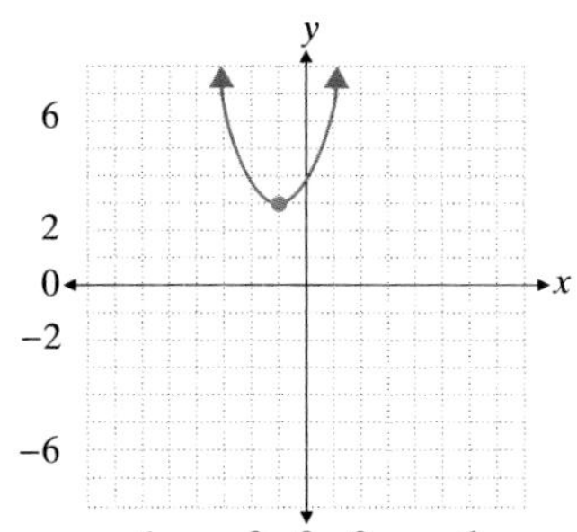

27. vertex: $(1, -4)$; no x-int.

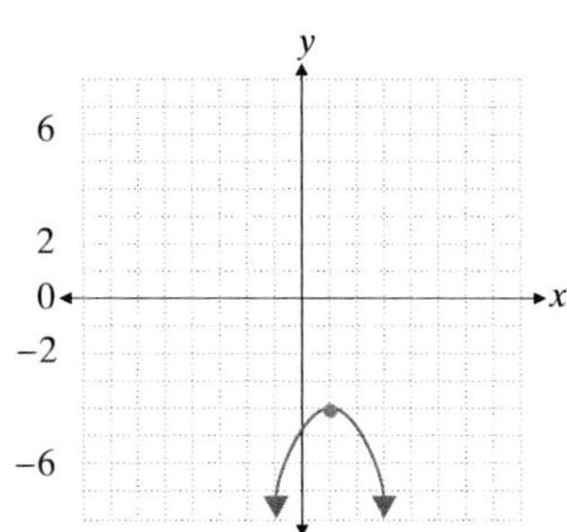

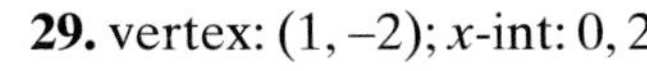

29. vertex: $(1, -2)$; x-int: $0, 2$

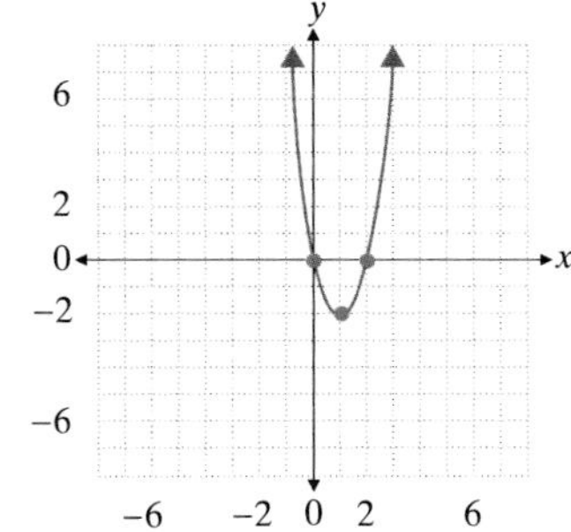

three

31. 50 feet by 100 feet **33.** both width and length are 5 **35.** $(8, 4)$ **37.** 112 feet **39.** 164 feet
41. 11,250 ft.^{2} **43.** 500 rooms **45.** 1500 cars **47.** 6050 ft.^{2} **49.** a **51.** b **53.** f **55.** d

57. vertex: $(-1, 4)$; x-int: $-3, 1$

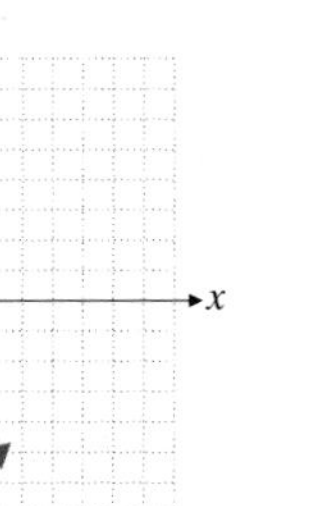

59. vertex: $(2, -4)$; x-int: $0, 4$

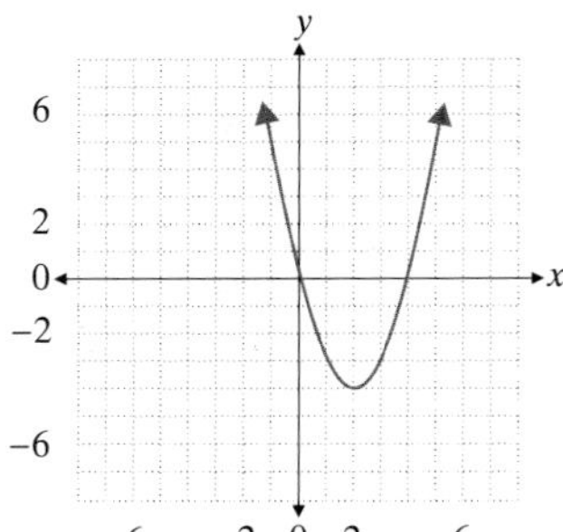

61. vertex: $(-3, -27)$; x-int: $-6, 0$

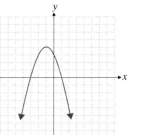

63. vertex: $\left(\frac{4}{3}, -\frac{10}{3}\right)$; x-int: $\frac{4 \pm \sqrt{10}}{3}$

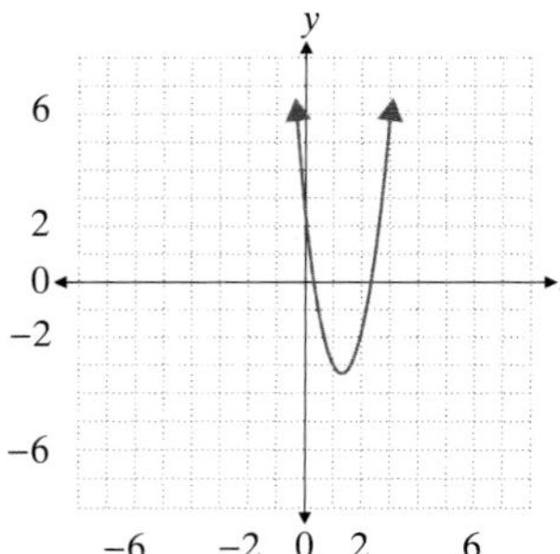

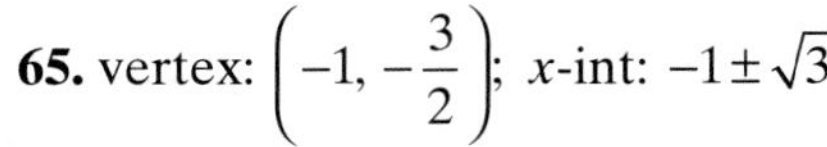

65. vertex: $\left(-1, -\frac{3}{2}\right)$; x-int: $-1 \pm \sqrt{3}$

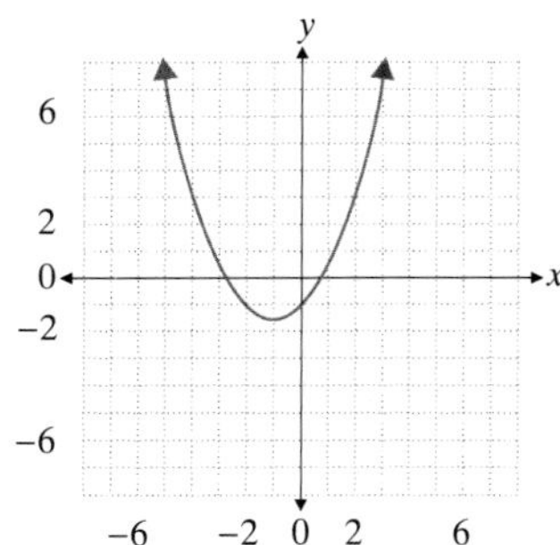

67.

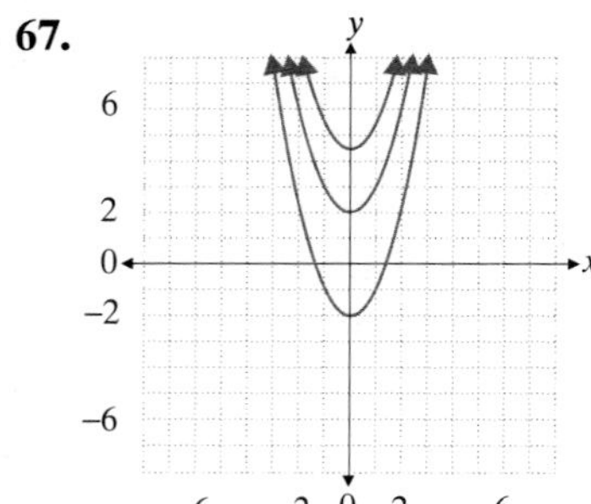

69.

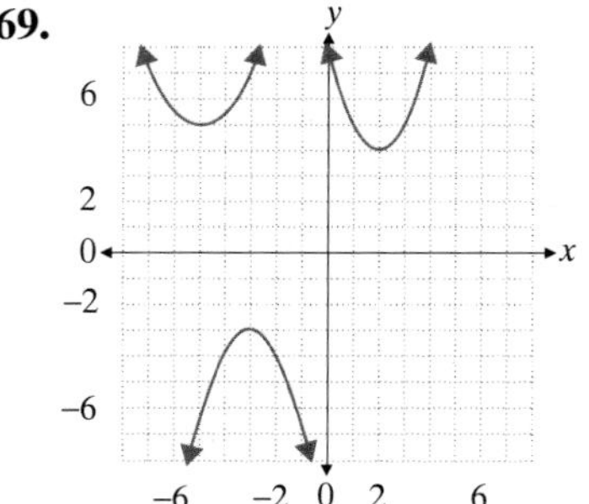

section 3.3, pages 231-232

1.

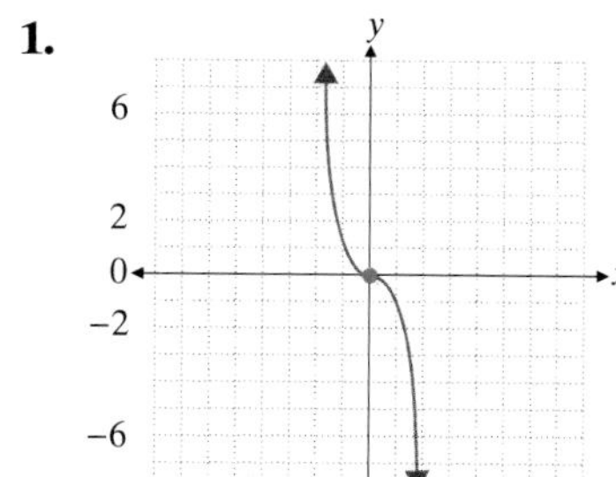

3.

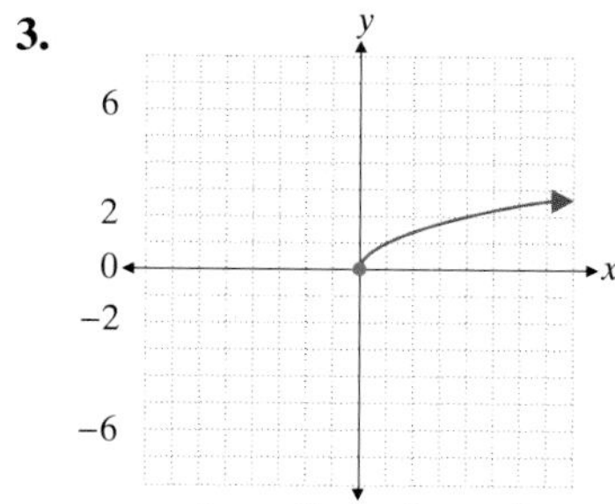

5.

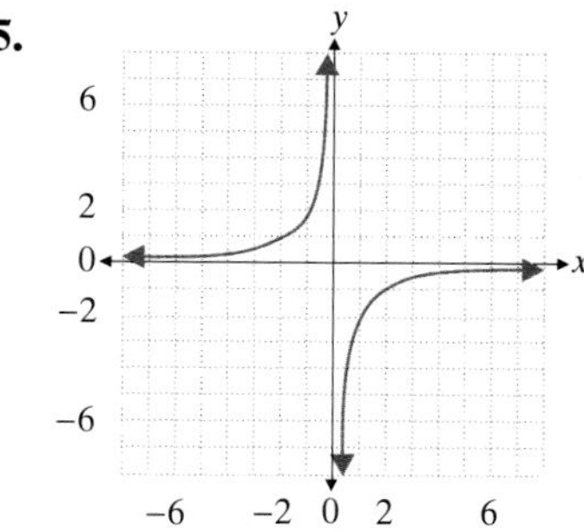

7.

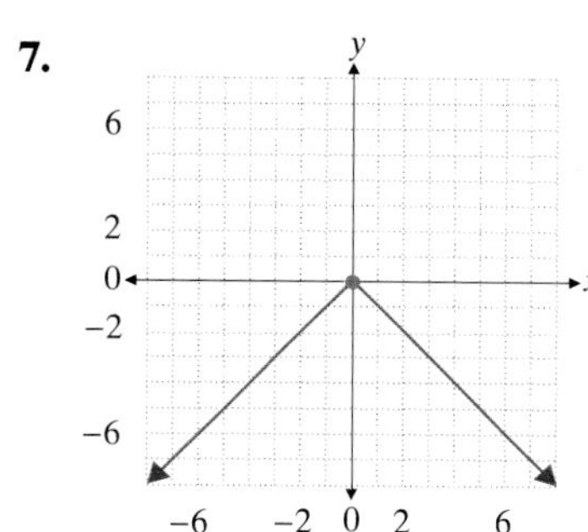

9.

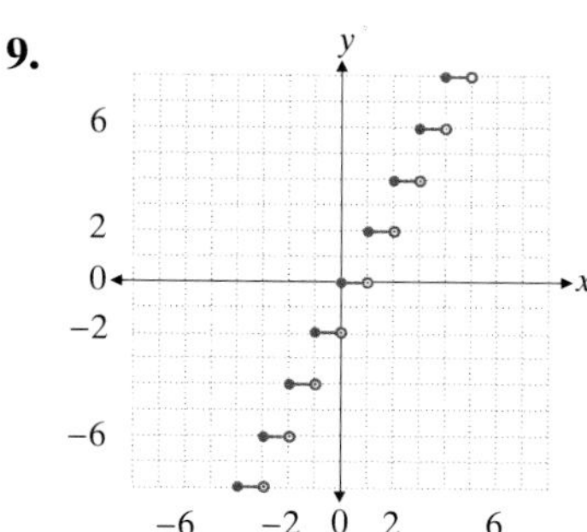

11.

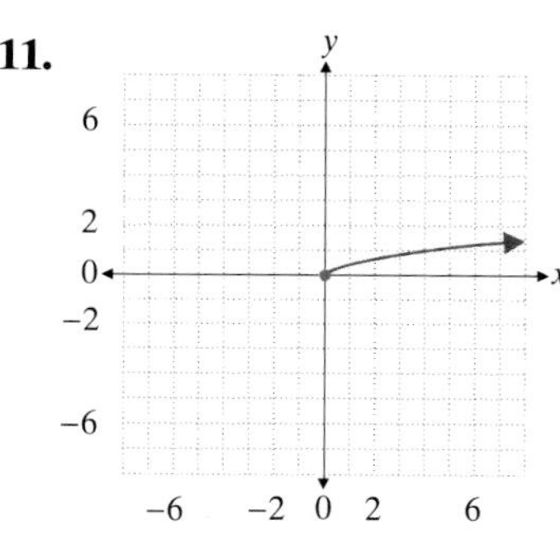

13.

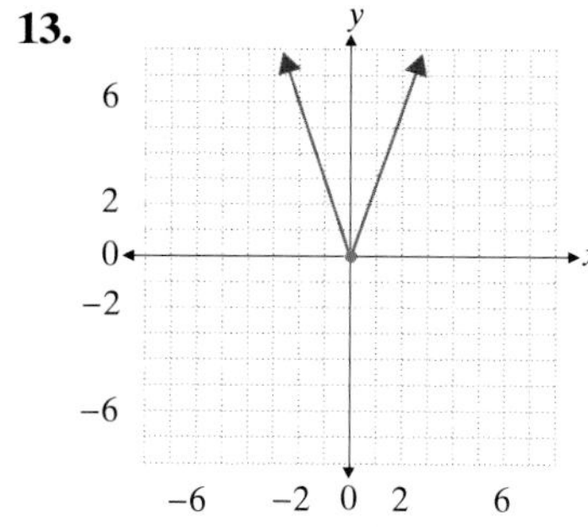

15.

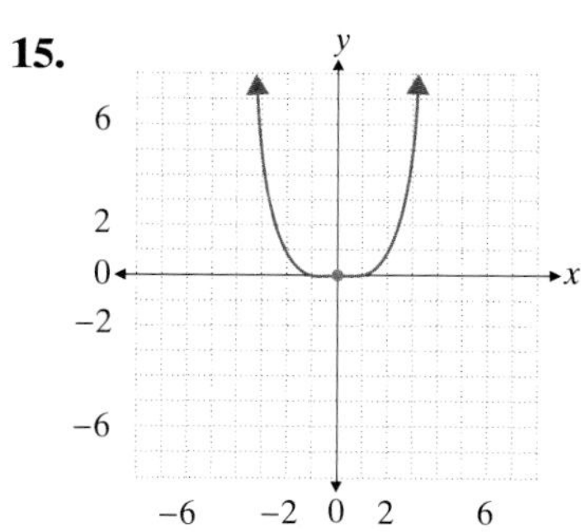

17.

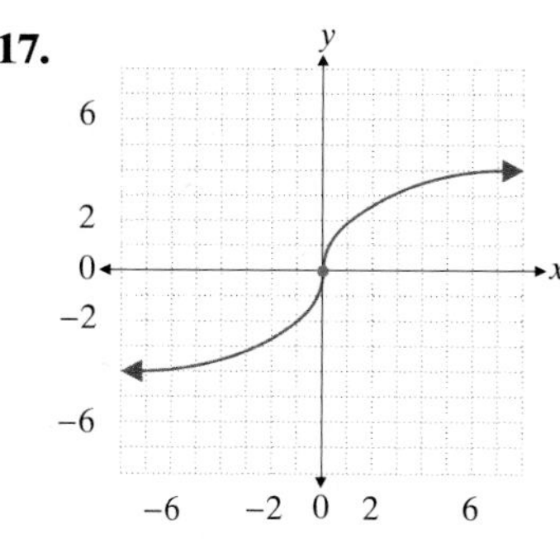

19.

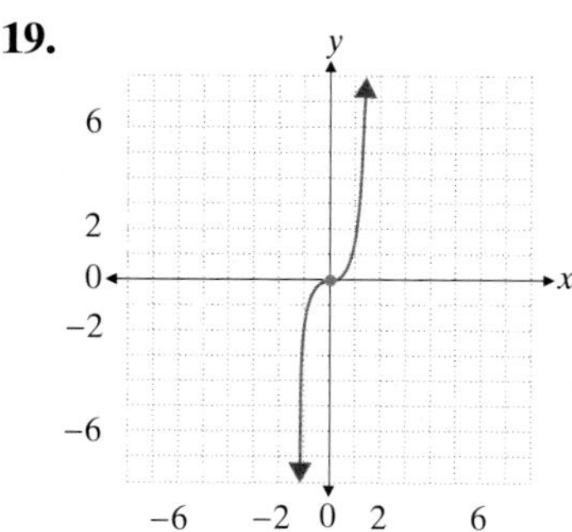

21.

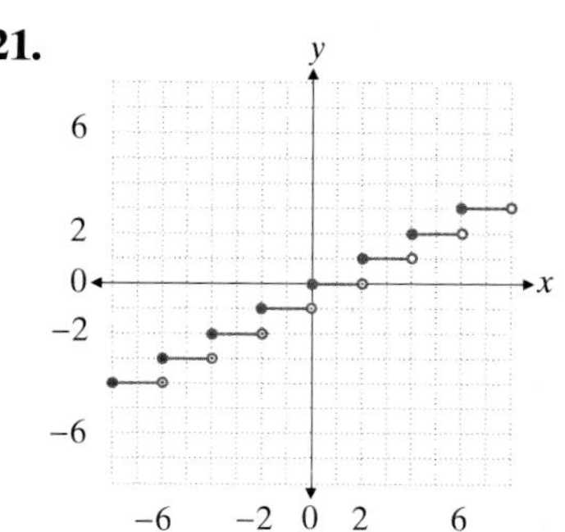

23.

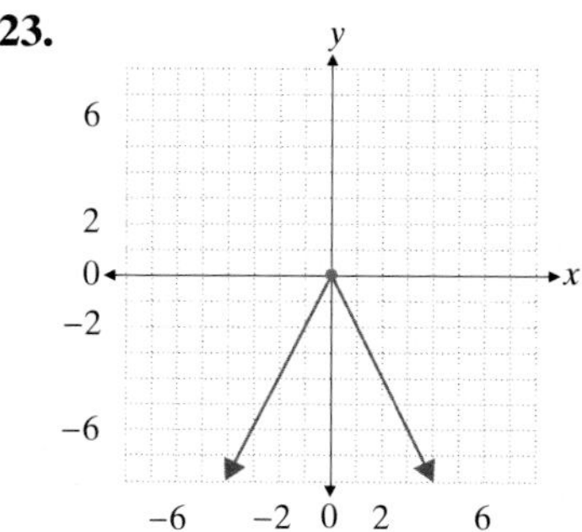

25.
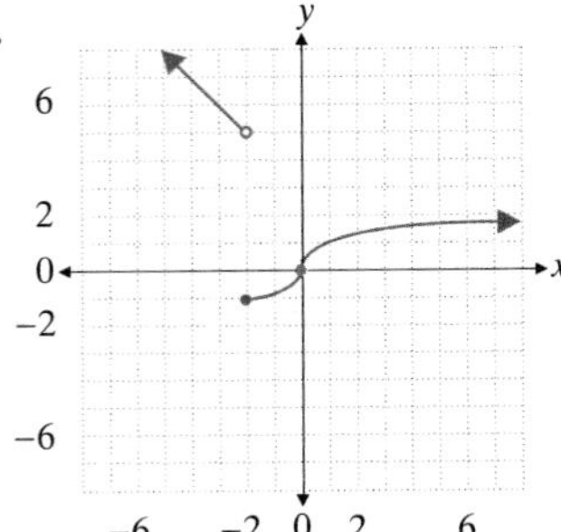

27.
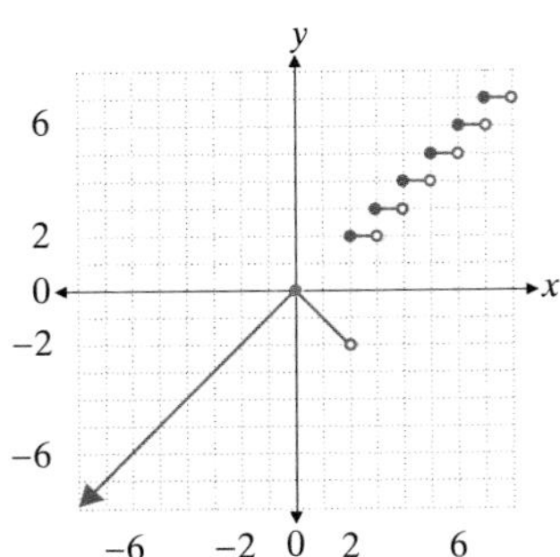

29.
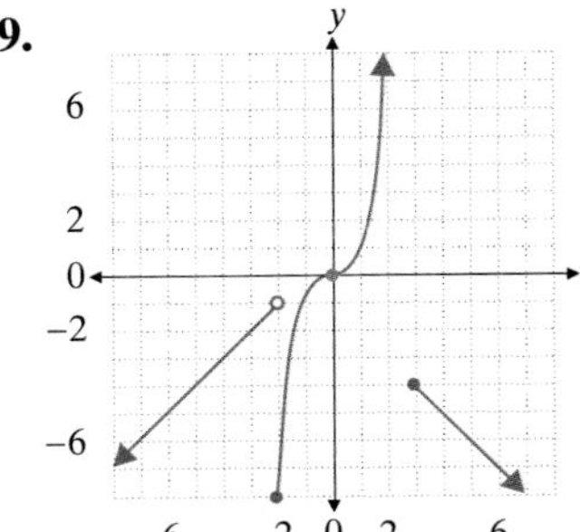

31.
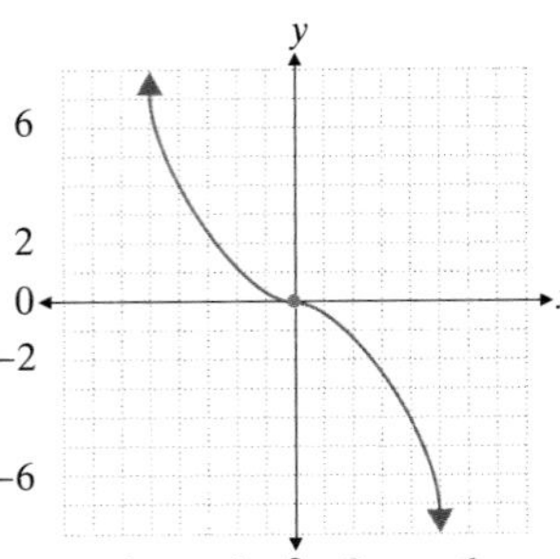

33.
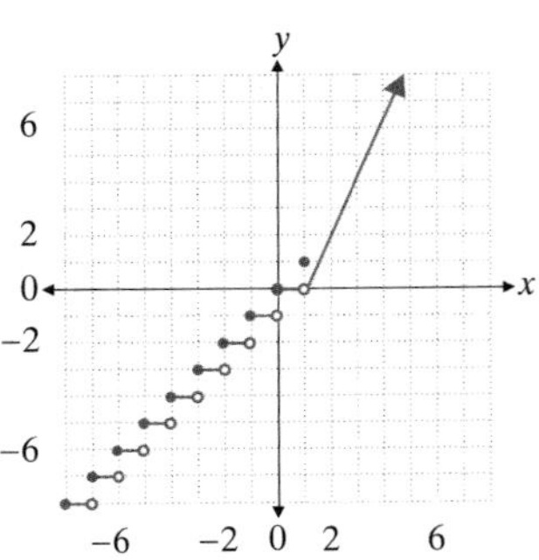

35.
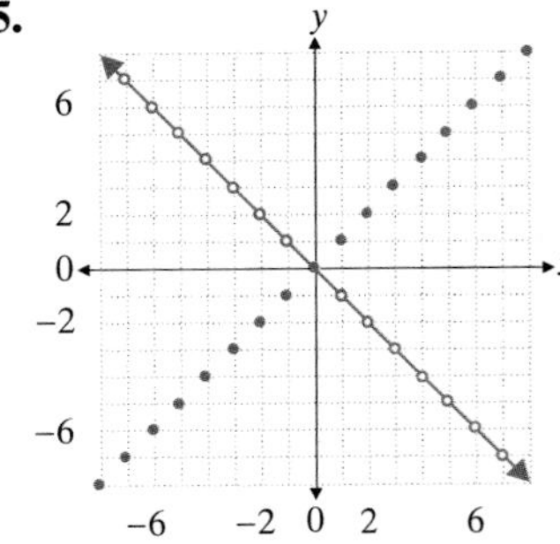

37.
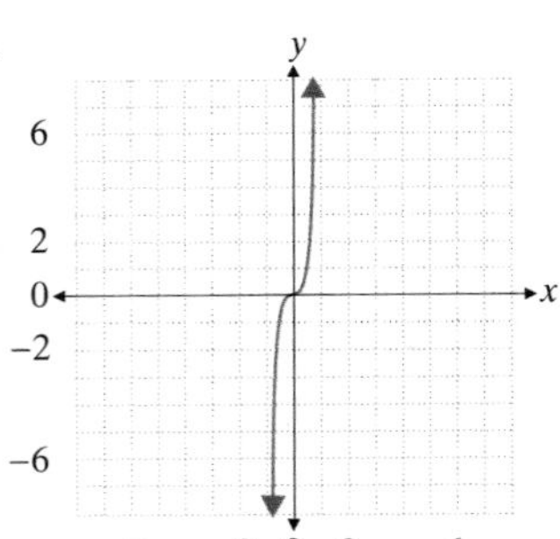

39.
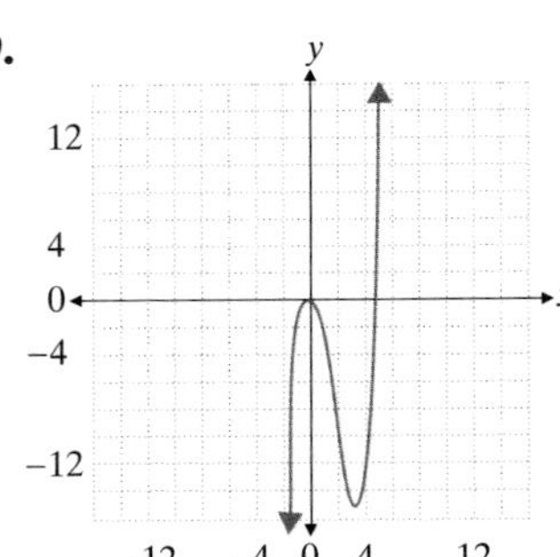

41.
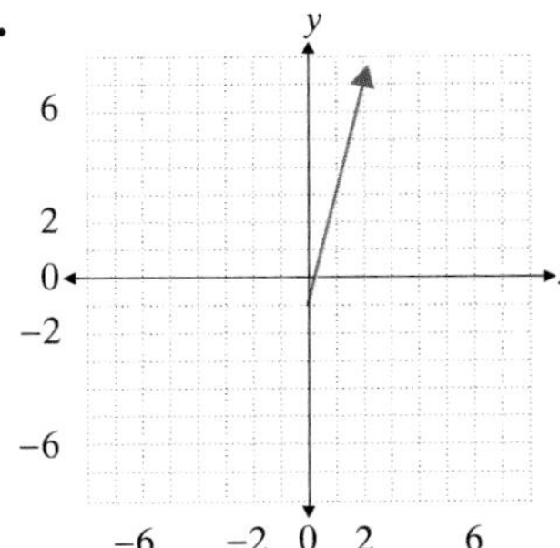

section 3.4, pages 238-243

1. $A = kbh$ **3.** $W = \frac{k}{d^2}$ **5.** $r = \frac{k}{t}$ **7.** $x = ky^3z^2$ **9.** $y = 18\sqrt{5}$ **11.** $y = 60\sqrt[3]{2}$ **13.** $y = .75$

15. $z = 48$ **17.** $z = 112$ **19.** 256 feet **21.** 20.60 **23.** 6.7 meters **25.** 1.25 centimeters

27. 34.54 inches **29.** 164.7872 inches **31.** 9 watts **33.** $P(\sigma, \varepsilon) = \frac{\sigma^2}{2\varepsilon}$ **35.** $10\sqrt{3}$

37. $a = \frac{9b^2}{4}$; 36 **39.** $a = 3bc$; 108 **41.** $P = 2.15g$; \$43 **43.** $F = \frac{15d}{9}$; 12 cm

45. $V = \frac{800}{P}$; 200 cm^3 **47.** $R = \frac{.000009l}{d^2}$; 17.28 ohms

section 3.5, pages 256-260

1. Dom = Ran = $\mathbb{R}$

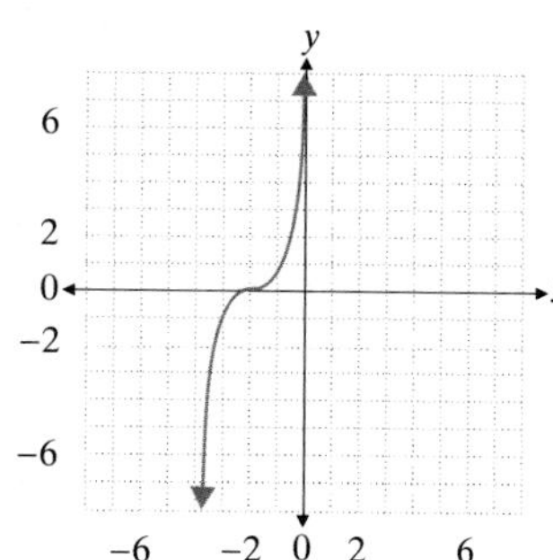

3. Dom = $\mathbb{R}$, Ran = $(-\infty, 2]$

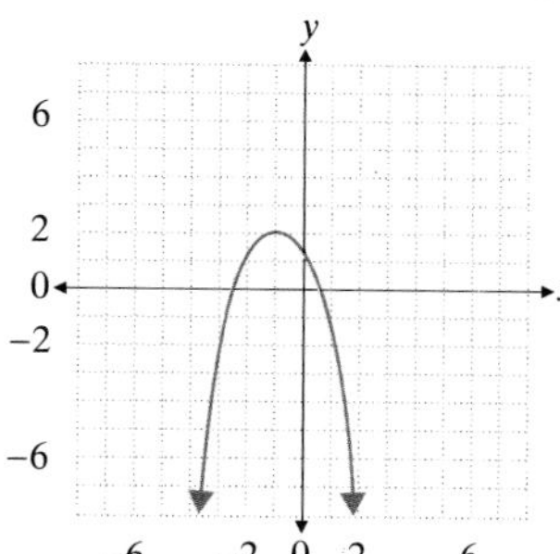

5. Dom = $\mathbb{R}$, Ran = $[0, \infty)$

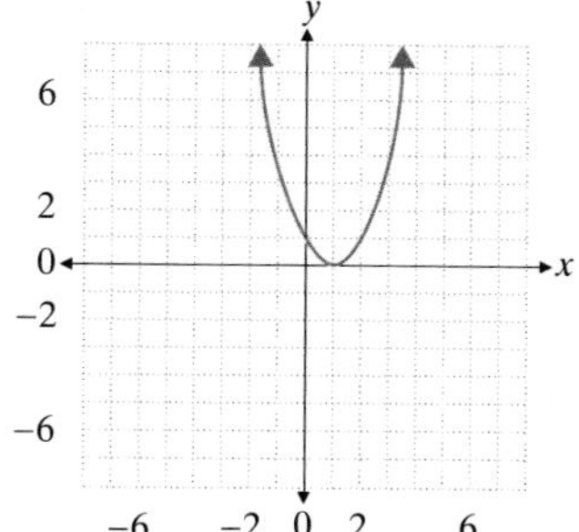

7. Dom = $(-\infty, 2]$, Ran = $[0, \infty)$

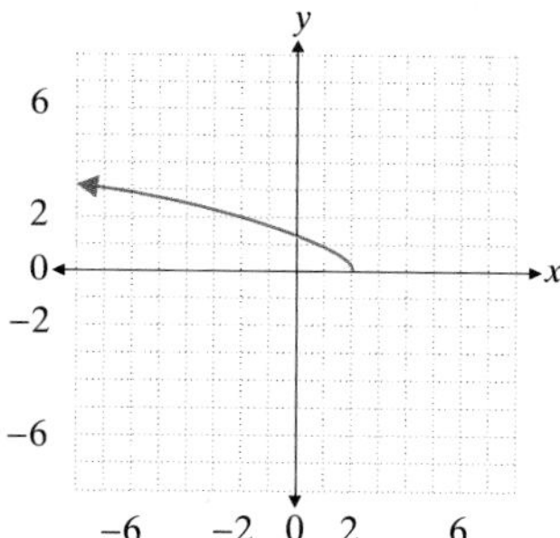

9. Dom = $(-\infty, 3) \cup (3, \infty)$, Ran = $(0, \infty)$

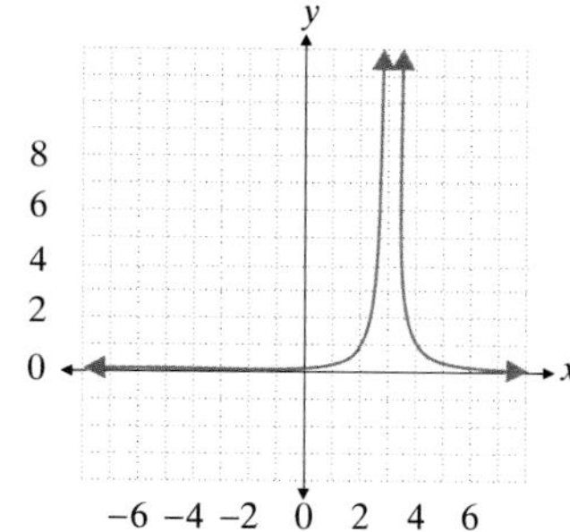

11. Dom = $(-\infty, 2) \cup (2, \infty)$, Ran = $(-\infty, 0) \cup (0, \infty)$

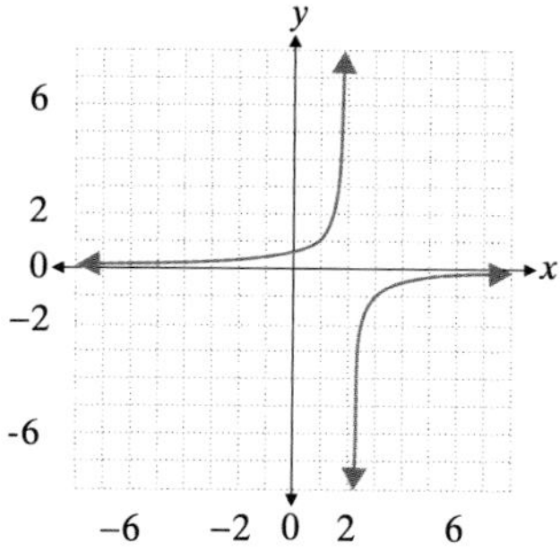

13. Dom = $\mathbb{R}$, Ran = $\mathbb{Z}$

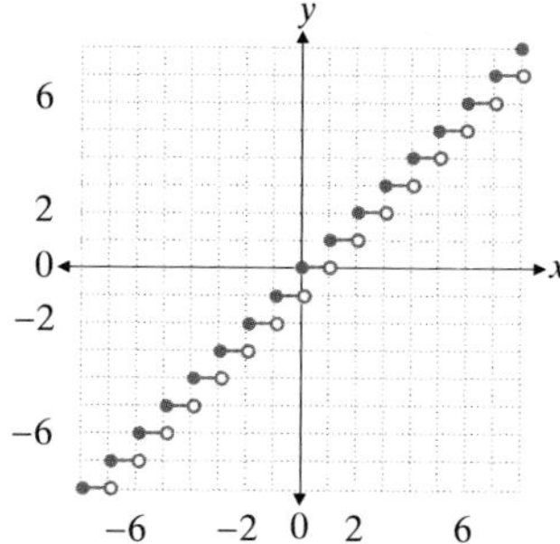

15. Dom = $\mathbb{R}$, Ran = $\mathbb{R}$

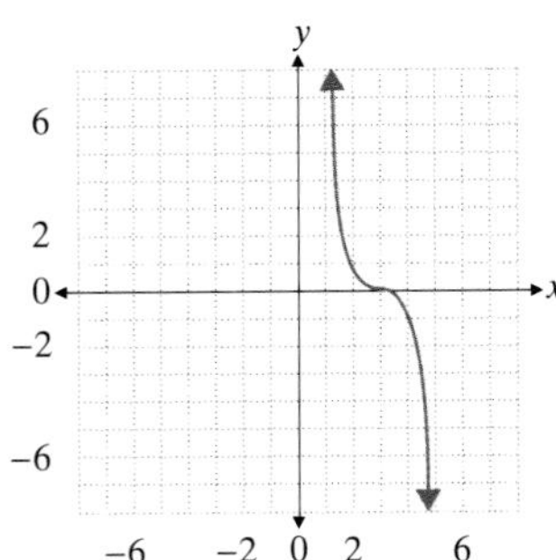

17. Dom = $\mathbb{R}$, Ran = $[-3, \infty)$

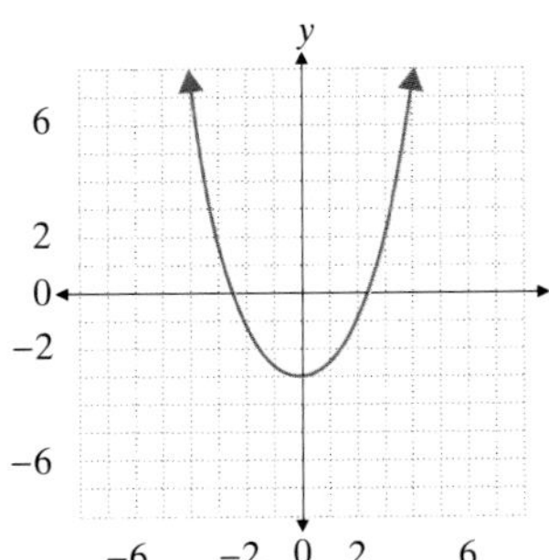

19. Dom = $\mathbb{R}$, Ran = $[0, \infty)$

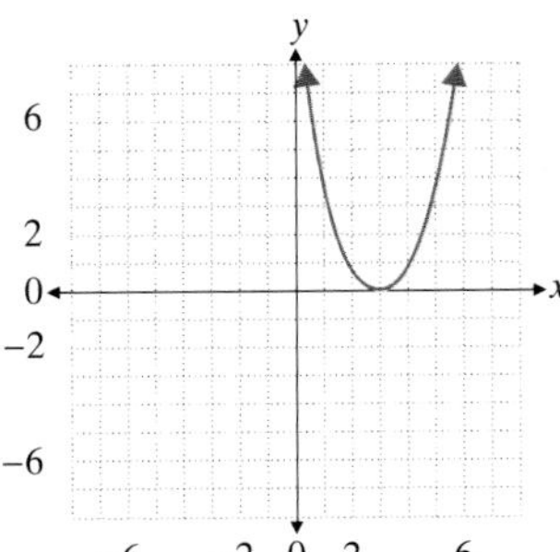

21. Dom = $(-\infty, 1) \cup (1, \infty)$, Ran = $\{-1, 1\}$

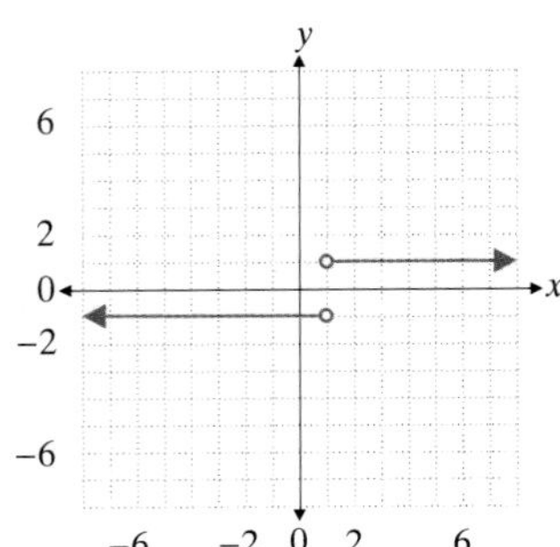

23. Dom = $[1, \infty)$, Ran = $(-\infty, 2]$

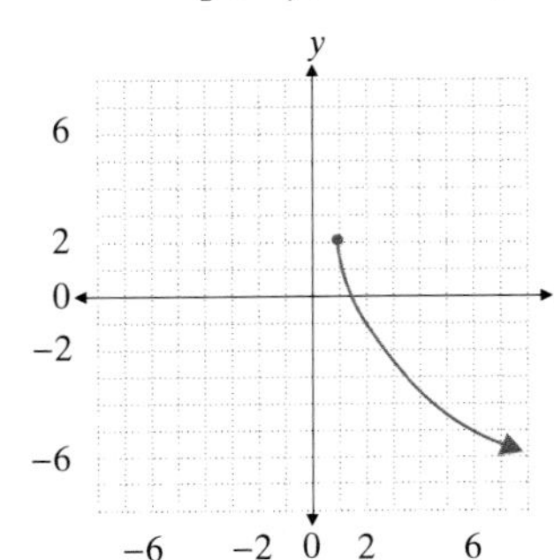

25. $(x-4)^2 + 2$

27. $(-x-2)^2$

29. $(x-10)^3 + 4$

31. $\sqrt{-x} - 3$

33. $-|x-8| + 2$

35. Odd function

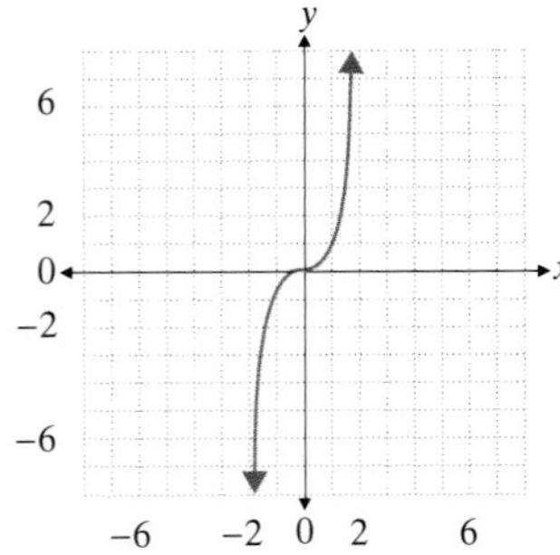

37. Odd function

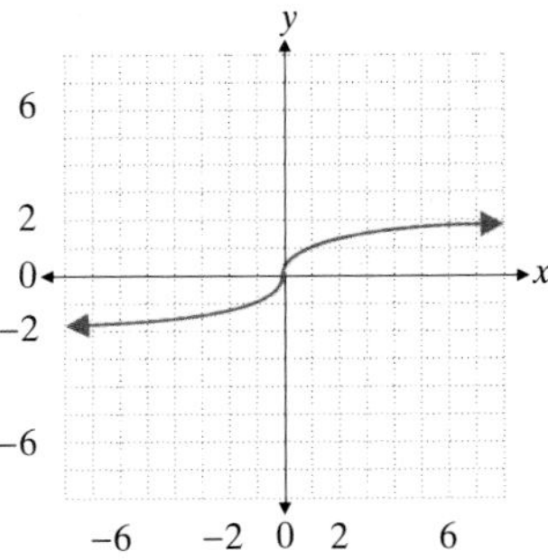

39. None of the above

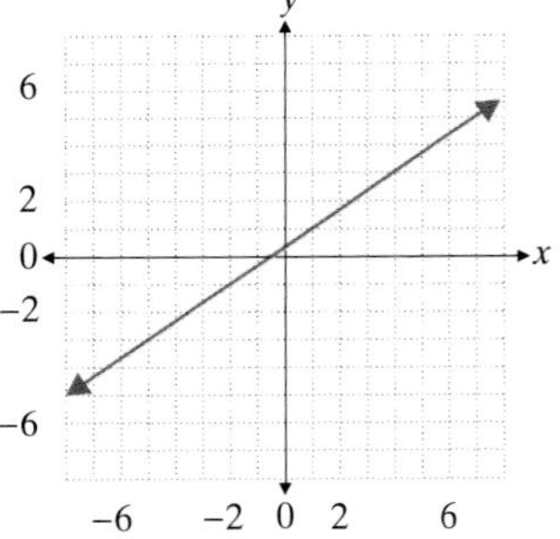

41. Neither

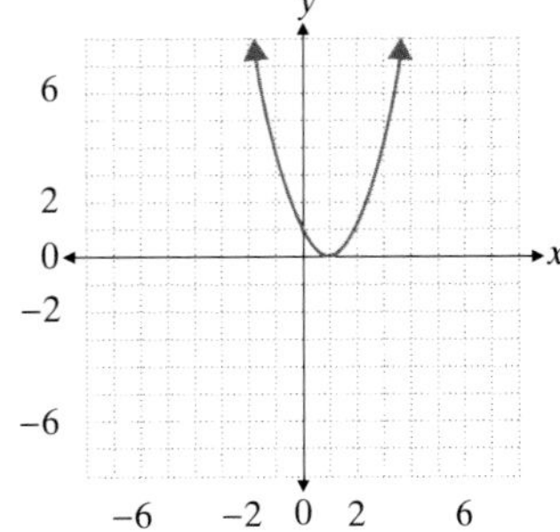

43. x-axis symmetry

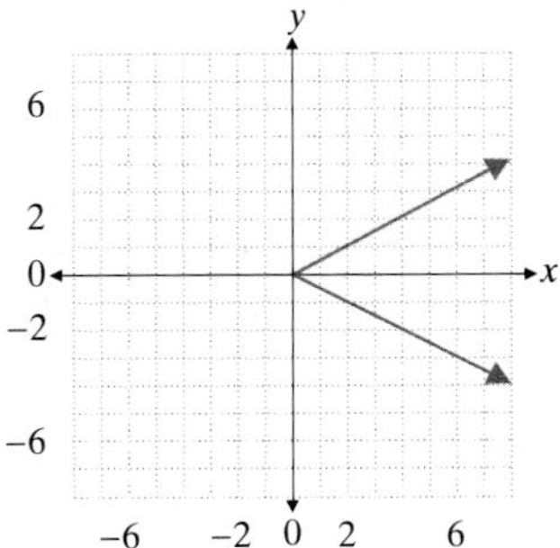

45. Neither

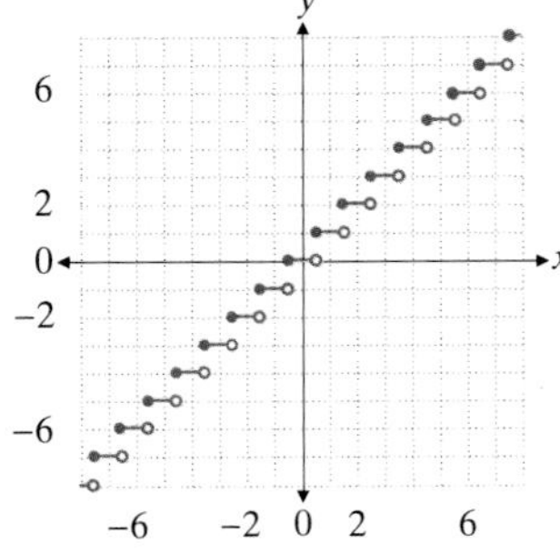

47. Origin symmetry

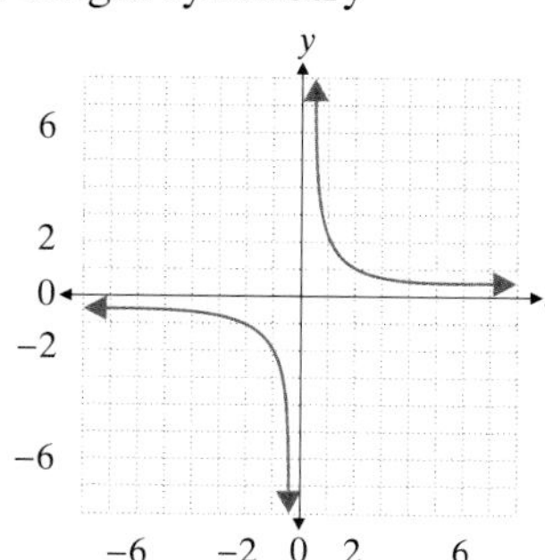

49. Dec. on $(-\infty, -3)$, Inc. on $(-3, \infty)$

51. Dec. on $(-\infty, 1)$, Dec. on $(1, \infty)$

53. Inc. on $(-1, \infty)$

55. Inc. on $(-\infty, 1)$, Dec. on $(1, \infty)$

57. Inc. on $(-\infty, 7)$, Dec. on $(7, \infty)$

59. Dec. on $(-\infty, 1)$, Inc. on $(1,3)$, Dec. on $(3,\infty)$

61.

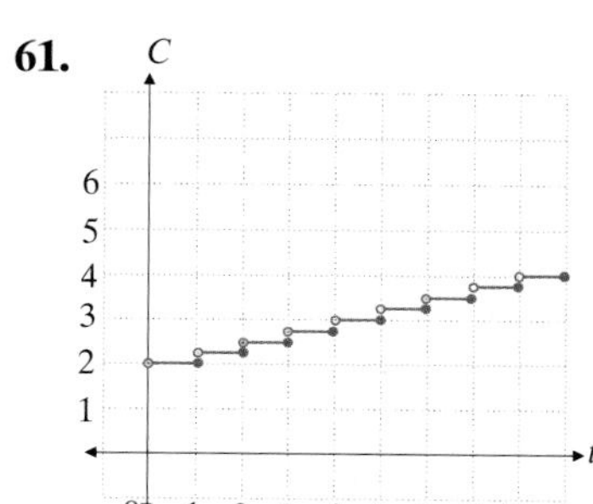

Constant on (0, 1], (1, 2], (2, 3], ...
C(8 min 35 sec) = \$3.84

63.

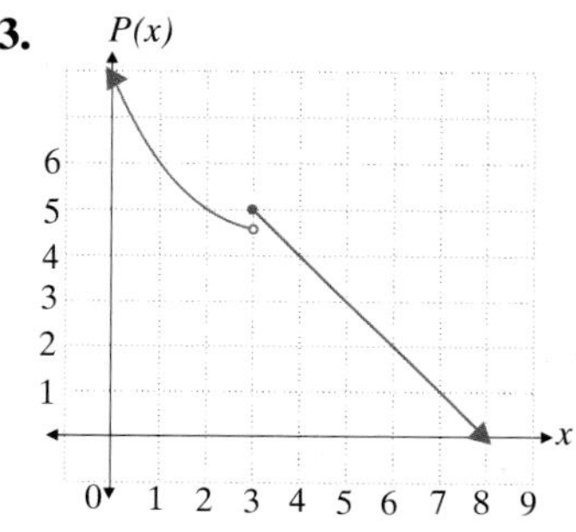

Dec. on (0, 3), Dec. on (3, ∞)

65.

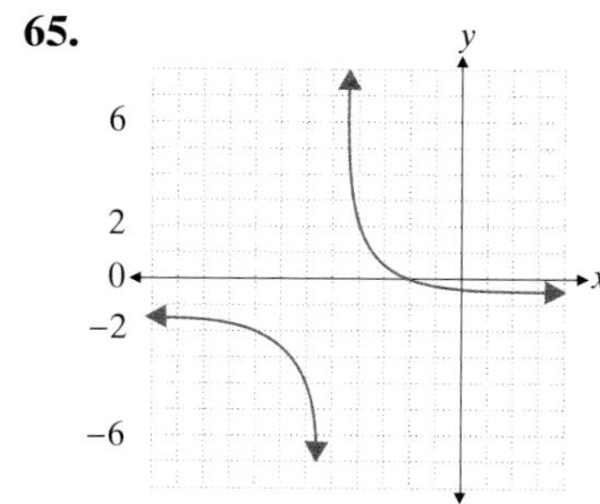

67.

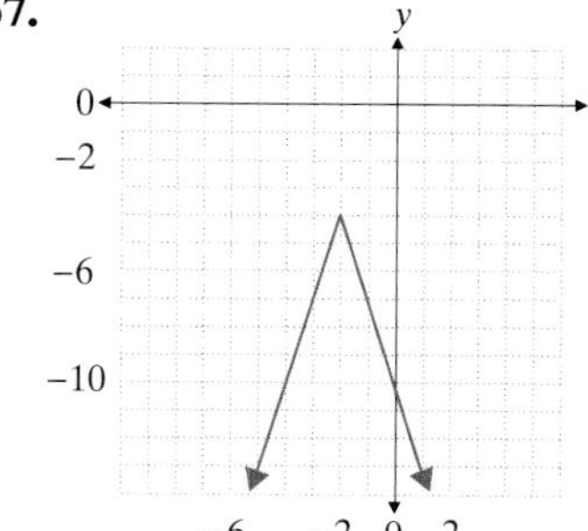

69.

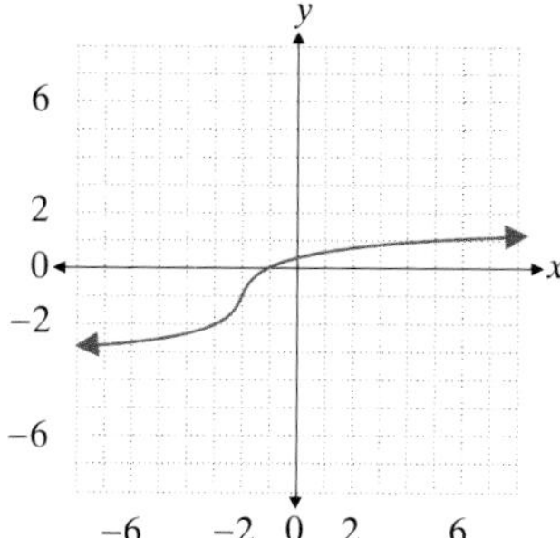

71. $|x+4|-1$ **73.** $-\sqrt{6-x}+2$ **75.** $6-(x-3)^2$

section 3.6, pages 269-273

1. a. 2 **b.** –8 **c.** –15 **d.** $-\frac{3}{5}$ **3. a.** –3 **b.** –1 **c.** 2 **d.** 2 **5. a.** 12 **b.** 18 **c.** –45 **d.** –5

7. a. 3 **b.** 1 **c.** 2 **d.** 2 **9. a.** 6 **b.** 0 **c.** 9 **d.** 1 **11. a.** 5 **b.** –1 **c.** 6 **d.** $\frac{2}{3}$

13. a. $|x|+\sqrt{x}$, Dom = $[0,\infty)$ **b.** $\frac{|x|}{\sqrt{x}}$, Dom = $(0,\infty)$ **15. a.** x^2+x-2, Dom = $\mathbb{R}$ **b.** $\frac{1}{x+1}$, Dom = $(-\infty,-1)\cup(-1,1)\cup(1,\infty)$ **17. a.** x^3+3x-8, Dom = $\mathbb{R}$ **b.** $\frac{3x}{x^3-8}$, Dom = $(-\infty,2)\cup(2,\infty)$

19. a. $-2x^2+[\![x+4]\!]$, Dom = $\mathbb{R}$ **b.** $\dfrac{-2x^2}{[\![x+4]\!]}$, Dom = $(-\infty,-4)\cup[-3,\infty)$ **21.** $-\dfrac{538}{49}$ **23.** $\dfrac{176}{25}$

25. $\dfrac{11}{16}$ **27.** $-\dfrac{1}{4}$ **29.** 5 **31.** 2 **33.** 0 **35.** 8 **37.** 3 **39.** 1

41. a. $\sqrt{x^2-1}$, Dom = $(-\infty,-1]\cup[1,\infty)$ **b.** $x-1$, Dom = $[1,\infty)$ **43. a.** $\dfrac{4-2x}{3x}$, Dom = $(-\infty,0)\cup(0,\infty)$

b. $\dfrac{3}{4x-2}$, Dom = $\left(-\infty,\dfrac{1}{2}\right)\cup\left(\dfrac{1}{2},\infty\right)$ **45. a.** $[\![x^3-2]\!]$, Dom = $\mathbb{R}$ **b.** $[\![x-3]\!]^3+1$, Dom = $\mathbb{R}$

47. a. $9x^4+30x^2+26$, Dom = $\mathbb{R}$ **b.** $3x^4+6x^2+8$, Dom = $\mathbb{R}$

49. a. $\dfrac{x}{2+7x}$, Dom = $\left(-\infty,-\dfrac{2}{7}\right)\cup\left(-\dfrac{2}{7},0\right)\cup(0,\infty)$ **b.** $2x+14$, Dom = $(-\infty,-7)\cup(-7,\infty)$

51. a. $9x^2+6x+1$, Dom = $\mathbb{R}$ **b.** $3x^2+1$, Dom = $\mathbb{R}$ **53. a.** $\sqrt{x^2-2}$, Dom = $\left(-\infty,-\sqrt{2}\right]\cup\left[\sqrt{2},\infty\right)$

b. $x-2$, Dom = $[4,\infty)$ **55.** $g(x)=\sqrt[3]{x}$, $h(x)=3x^2-1$, $g(h(x))$ **57.** $g(x)=|x|+3$, $h(x)=x-2$, $g(h(x))$

59. $g(x)=|x|+7$, $h(x)=x^3-5x$, $g(h(x))$ **61.** $V(r)=3\pi r^3$ **63.** $V(t)=\dfrac{1}{12}\pi r^2t^2$

65. $L(x)=I(x)(M(x)+C(x))$ **67.** $fg(x)=2x^2-1=fg(-x)$ **69.** No **71.** Yes **73.** Yes

75. No **77.** $(f+g)(x)=8+12x+6x^2+x^3+\dfrac{1}{-5+3x}$, $(fg)(x)=\dfrac{8}{-5+3x}+\dfrac{12x}{-5+3x}+\dfrac{6x^2}{-5+3x}+\dfrac{x^3}{-5+3x}$,

$(f\circ g)(x)=\dfrac{1}{-5+3(2+x)^3}$, $(g\circ f)(x)=8+\dfrac{1}{(-5+3x)^3}+\dfrac{6}{(-5+3x)^2}+\dfrac{12}{-5+3x}$

three

section 3.7, pages 282-285

1. Dom = { −2, −1, 2 }, Ran = { −4, 0, 3 }

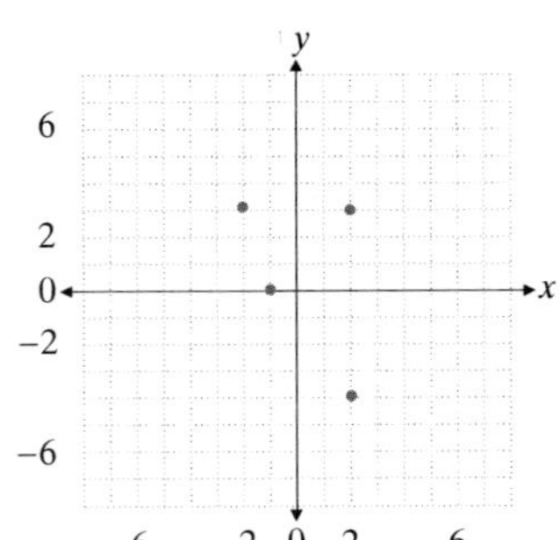

3. Dom = $\mathbb{R}$, Ran = $\mathbb{R}$

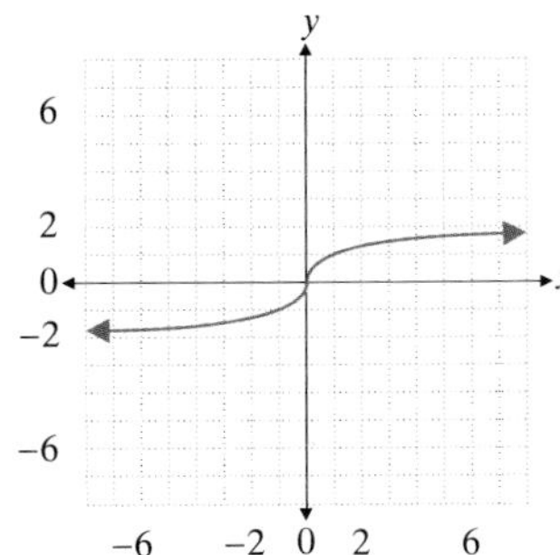

5. Dom = $\mathbb{R}$, Ran = $[0,\infty)$

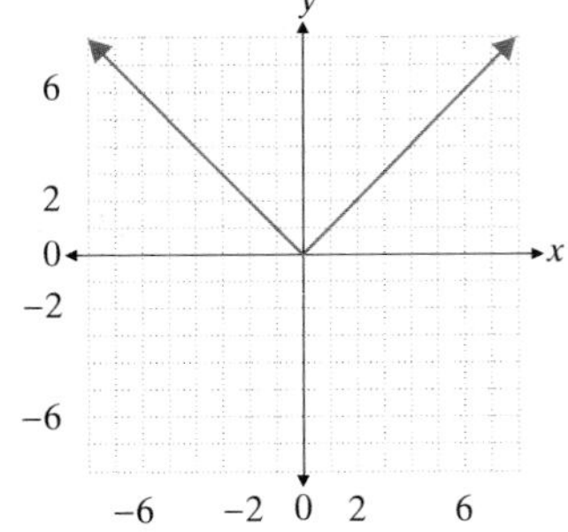

7. Dom = $\mathbb{R}$, Ran = $\mathbb{R}$

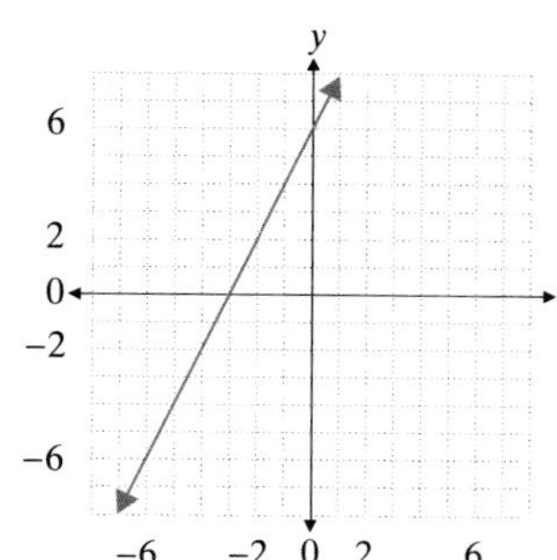

9. Dom = $\mathbb{Z}$, Ran = $\mathbb{R}$

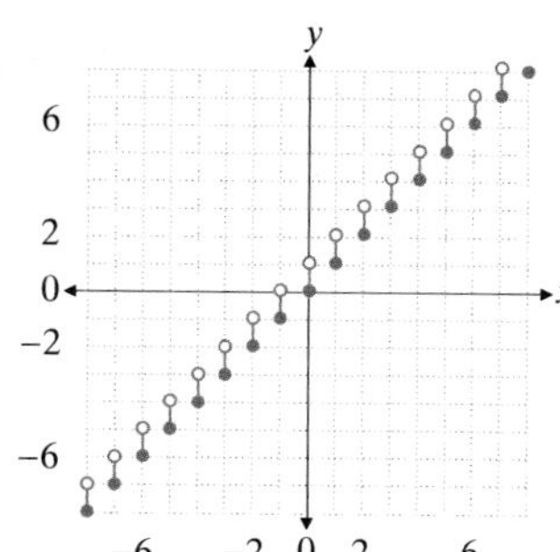

11. Dom = $\mathbb{R}$, Ran = $[-2, \infty)$

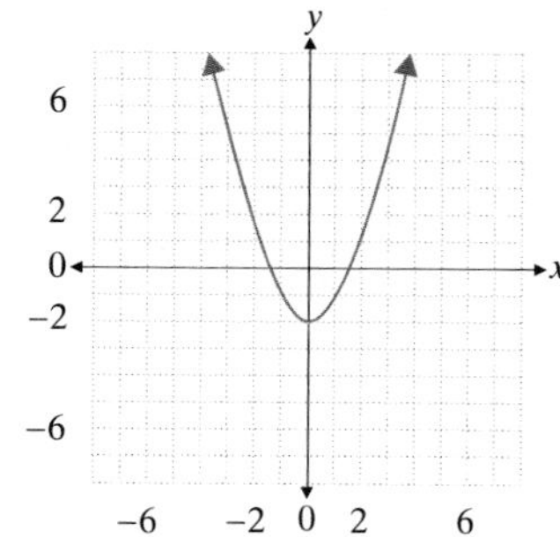

13. Restrict to $[0, \infty)$
15. Inverse exists
17. Inverse exists
19. Inverse exists

21. Restrict to $[2, \infty)$
23. Restrict to $[12, \infty)$
25. $f^{-1}(x) = (x+2)^3$
27. $r^{-1}(x) = \dfrac{-2x-1}{3x-1}$

29. $F^{-1}(x) = (x-2)^{1/3} + 5$
31. $V^{-1}(x) = 2x - 5$
33. $h^{-1}(x) = (x+2)^{5/3}$
35. $J^{-1}(x) = \dfrac{x-2}{3x}$

37. $h^{-1}(x) = (x-6)^{1/7}$
39. $r^{-1}(x) = \dfrac{x^5}{2}$
41. $f^{-1}(x) = \dfrac{x^3}{54}$
43. Answers will vary

45. Answers will vary
47. Answers will vary
49. Answers will vary
51. Answers will vary

53. b
55. e
57. a

59. 73 1 53 13 97 73 29 57 17 73
61. FRISBEE VOLLEYBALL AND HORSESHOES

63. CATCH A WAVE

65. Dom = $[0, \infty)$, Ran = $[-5, \infty)$

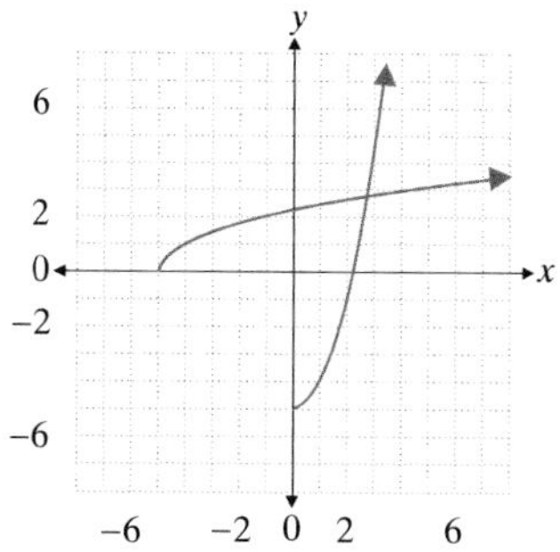

67. Dom = $[3, \infty)$, Ran = $[0, \infty)$

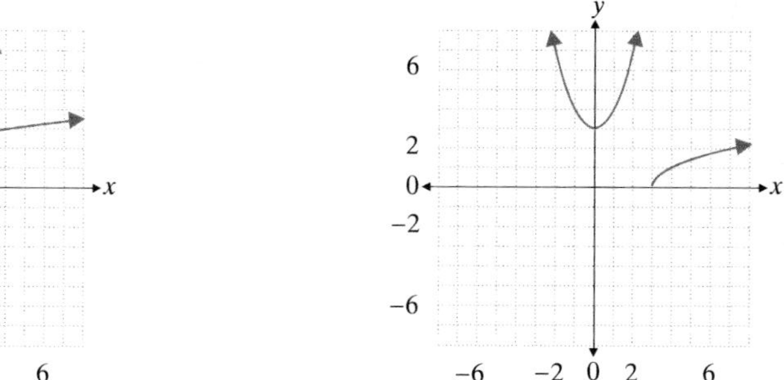

69. Dom = $(-\infty, 2) \cup (2, \infty)$, Ran = $(-\infty, 1) \cup (1, \infty)$

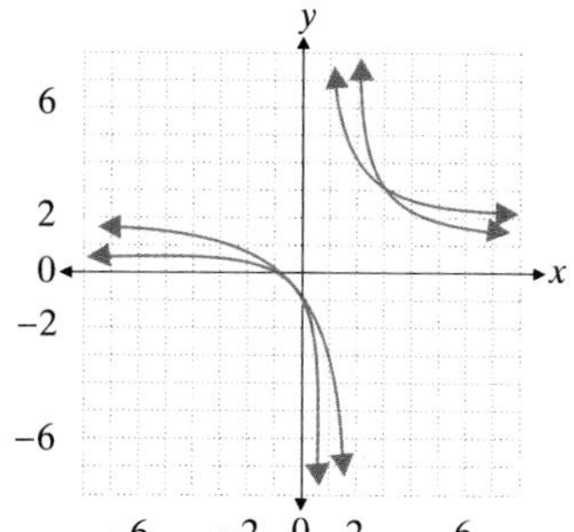

chapter three test, pages 292-295

1. Dom = $\{-2, -3\}$, Ran = $\{-9, -3, 2, 9\}$; No

3. Dom = $[-6, \infty)$, Ran = $\mathbb{R}$; No

5. a. $2x^2 + 6x - 8$ **b.** $4ax + 2a^2 + 10a$ **c.** $2x^4 + 10x^2$

7. Dom = $\mathbb{N}$, Cod = $\mathbb{R}$, Ran = $\left\{\frac{3}{4}, \frac{3}{2}, \frac{9}{4}, \ldots\right\}$

9.

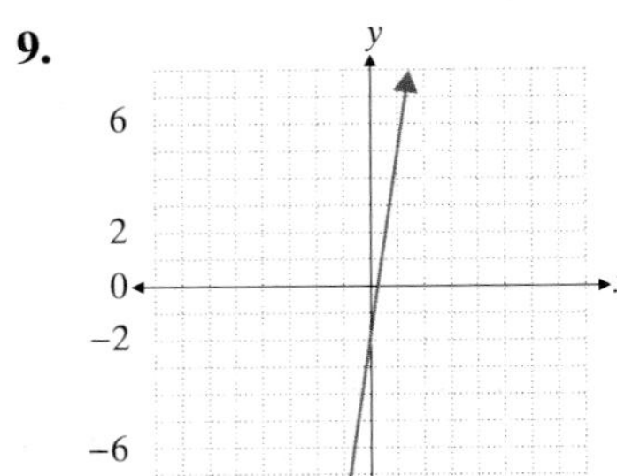

11.

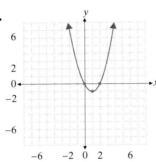

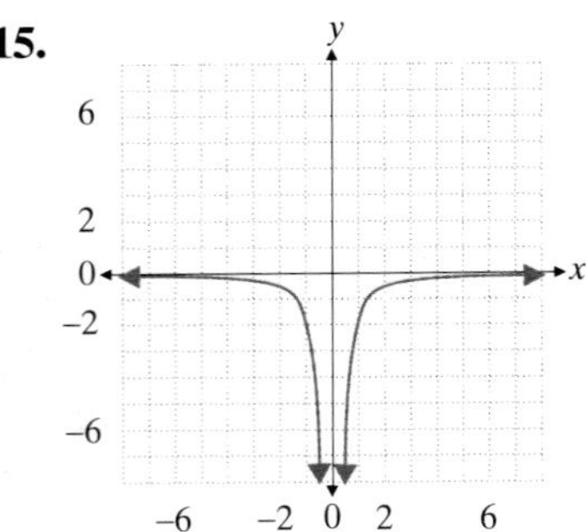

13. $\frac{15}{2}$ and $\frac{15}{2}$

15.

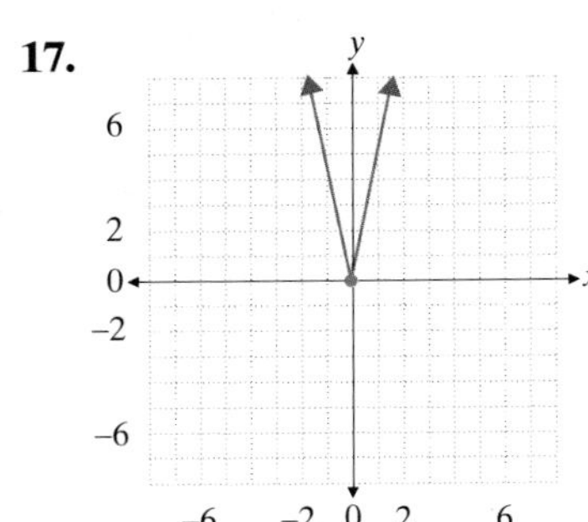

17.

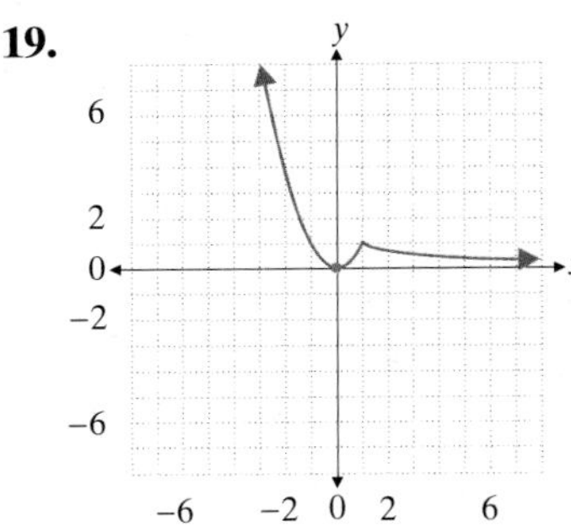

19.

21. $y = 48$

23. about 1226 videos per month

25. Dom = $\mathbb{R}$, Ran = $\mathbb{R}$

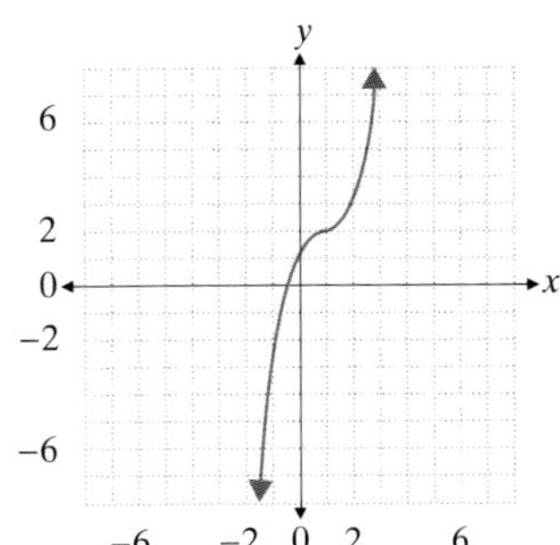

27. odd

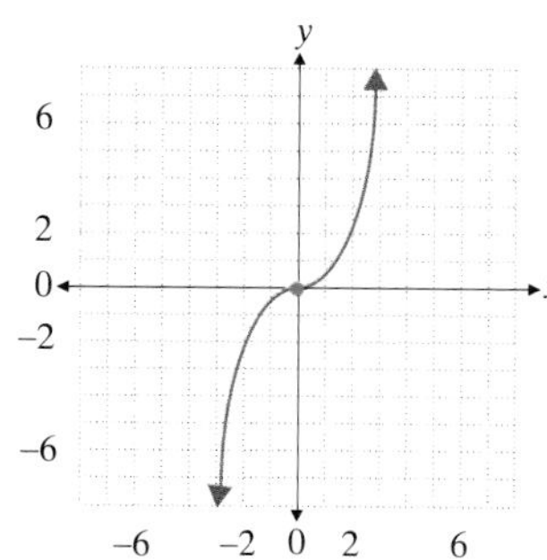

29. y-axis symmetry

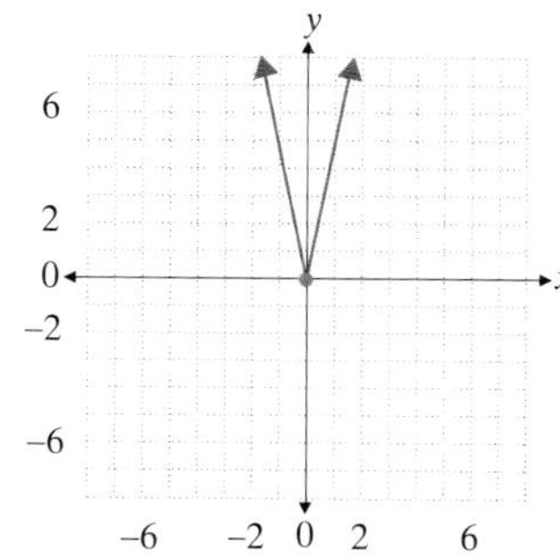

31. Dec. on $(-\infty, 2)$, Inc. on $(2, \infty)$

33. a. $x^2 + \sqrt{x}$, Dom $= [0, \infty)$

b. $x^{3/2}$, Dom $= (0, \infty)$

35. a. $x + 2$ **b.** $x - 2$ **c.** 5

37. $g(x) = \dfrac{\sqrt{x} + 2}{x^2}$,

$h(x) = x + 3$; $g(h(x))$

39. Yes

41. Dom $= \{-5, -2, 1, 5\}$, Ran $= \{-3, -1, 0, 2\}$

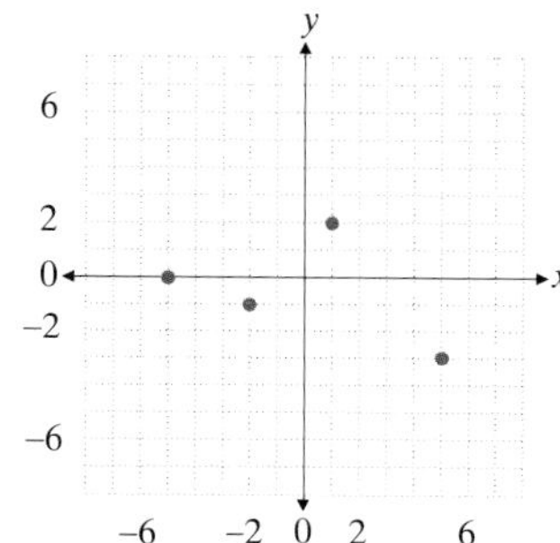

43. $r^{-1}(x) = \dfrac{x + 2}{7x}$

45. $f^{-1}(x) = (x + 6)^5$

four

chapter four

section 4.1, pages 308-310

1. Answers will vary **3.** Answers will vary **5.** Answers will vary **7.** Answers will vary
9. Answers will vary **11.** Answers will vary **13.** Answers will vary **15.** Answers will vary **17.** Yes

19. No **21.** No **23.** $-2, 0, 3$ **25.** $\pm 1, \pm i\sqrt{2}$ **27.** $\dfrac{1}{3}$ **29.** $-8, 0, 9$ **31.** $\dfrac{1}{2}, 5$ **33.** $\pm 2, \pm 3$

35. 4th degree; lead coef $= 2$; $p(x) \to \infty$ as $x \to -\infty$ and ∞ **37.** 4th degree; lead coef $= 24$; $r(x) \to \infty$ as $x \to -\infty$ and ∞

39. 5th degree; lead coef $= -2$; $g(x) \to \infty$ as $x \to -\infty$ and $g(x) \to -\infty$ as $x \to \infty$

41.

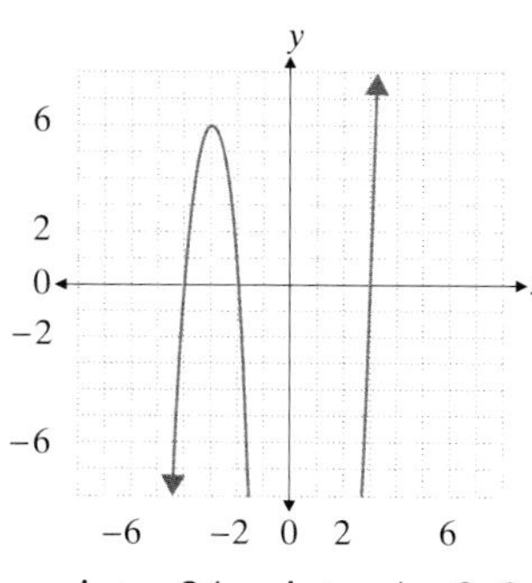

y-int: -24, x-int: $-4, -2, 3$

43.

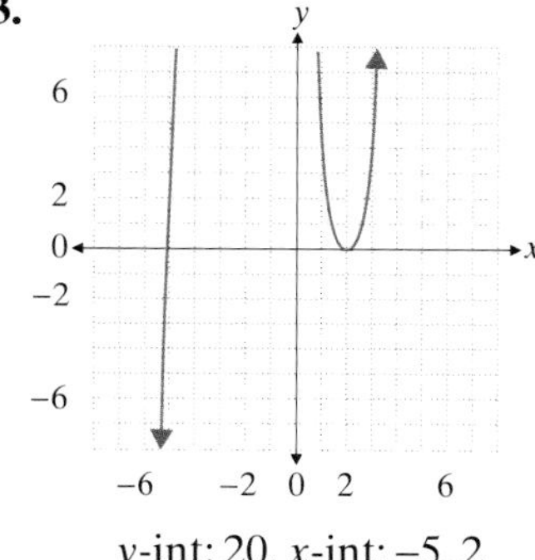

y-int: 20, x-int: $-5, 2$

45.

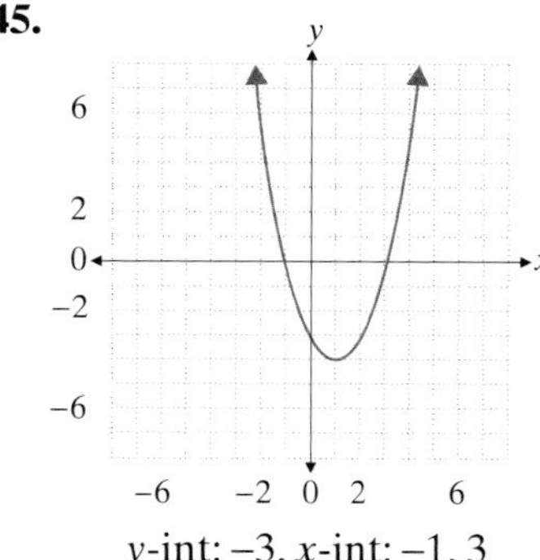

y-int: -3, x-int: $-1, 3$

section 4.5, pages 363-366

1. $x = 1$ **3.** No vertical asymptote **5.** $x = -3, x = 3$ **7.** $x = -1$ **9.** $x = 2$ **11.** $x = 3$
13. $x = 7$ **15.** $x = -2$ **17.** $x = -2,\ x = 2$ **19.** $y = 0$ **21.** $y = x - 2$ **23.** $y = 0$ **25.** $y = 2$
27. $y = 3x + 6$ **29.** $= 0$ **31.** $y = 0$ **33.** $y = x - 11$ **35.** $y = 5x + 4$

37.
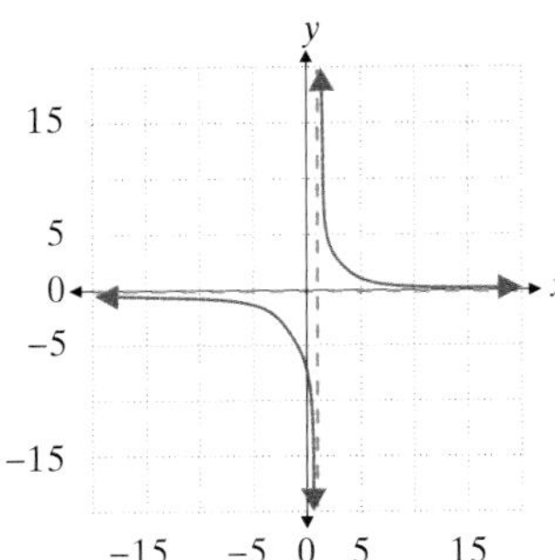

39.
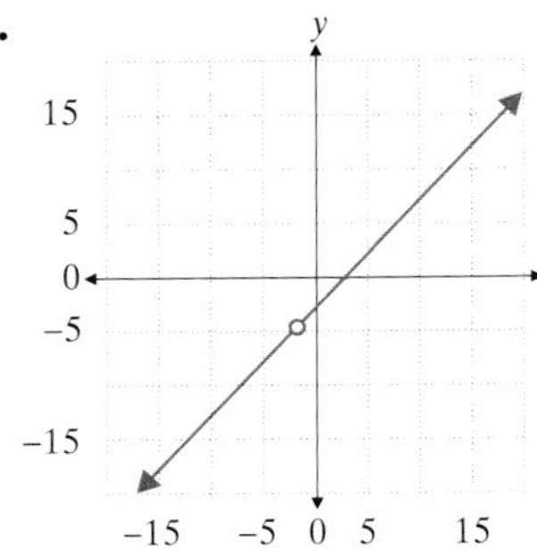

41.
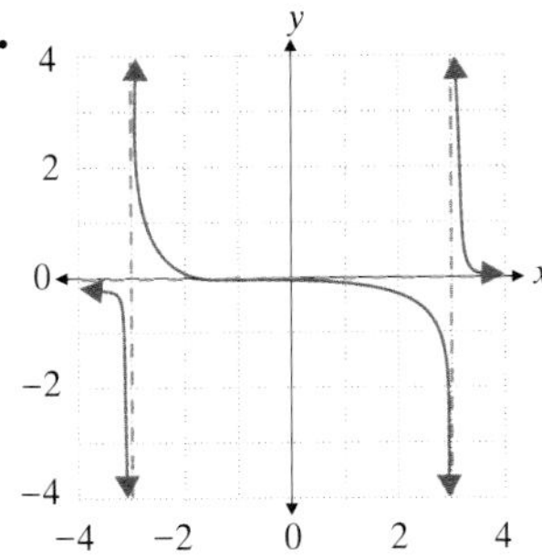

43.
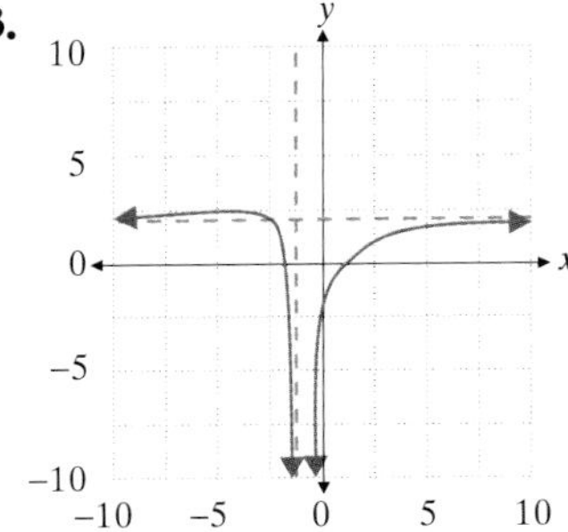

45.
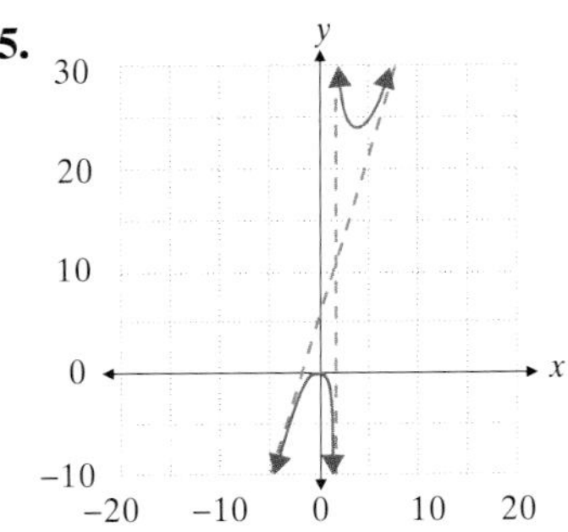

47.
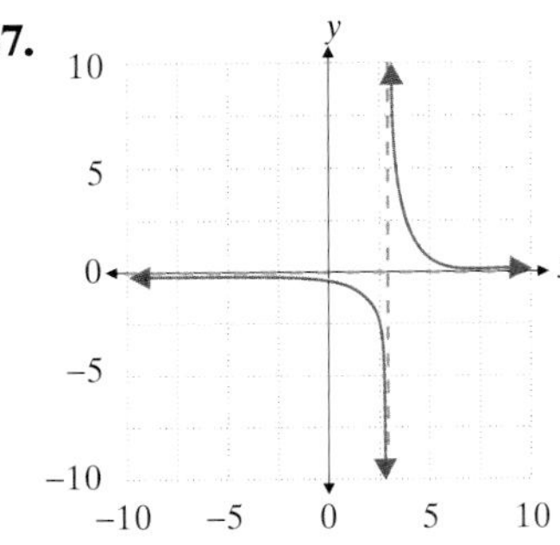

49. $(-\infty, -2)\cup(-1, 1)$ **51.** $(-\infty, -2)\cup(2, \infty)$ **53.** $(-\infty, -2)\cup(-2, 3)$ **55.** $(0, 3)$

57. $(-2, -1)\cup(1, \infty)$ **59.** $(-\infty, -1)\cup\left[-\frac{1}{2}, 0\right)$

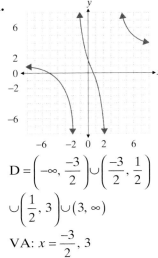

61.
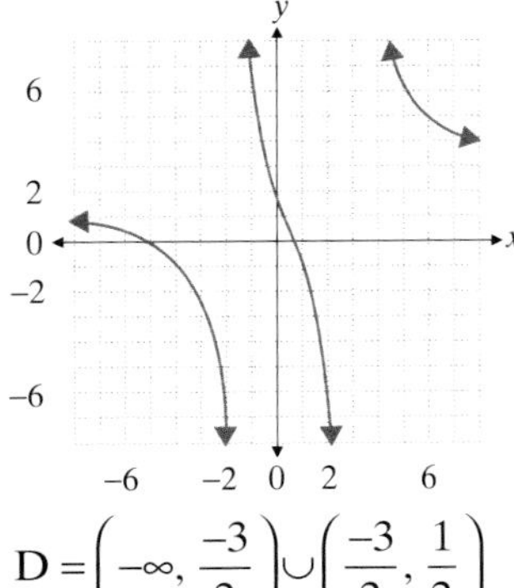

$D = \left(-\infty, \frac{-3}{2}\right)\cup\left(\frac{-3}{2}, \frac{1}{2}\right)\cup\left(\frac{1}{2}, 3\right)\cup(3, \infty)$

VA: $x = \frac{-3}{2}, 3$

63.
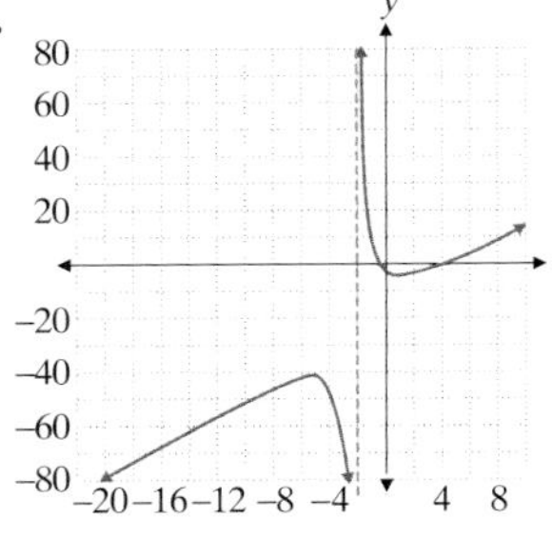

$D = (-\infty, -2)\cup\left(-2, \frac{2}{5}\right)\cup\left(\frac{2}{5}, \infty\right)$

VA: $x = -2$

65.
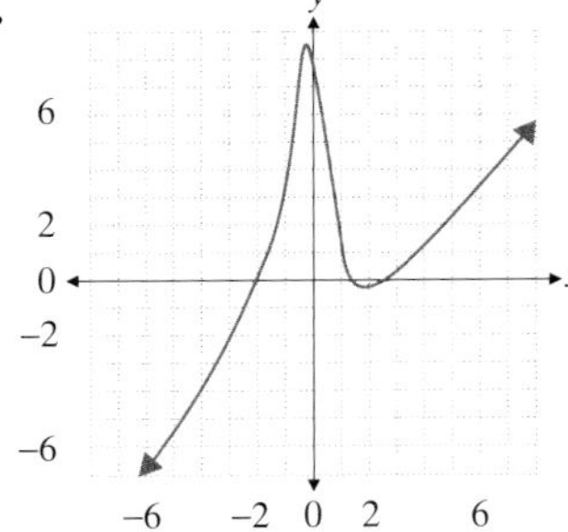

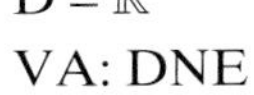

$D = \mathbb{R}$

VA: DNE

69. $c = 8.05, A = 86.21°, B = 58.79°$ **71.** $c = 20.04, B = 44°15', A = 89°38'$ **73.** $c = 19.65, A = 55°42', B = 49°14'$

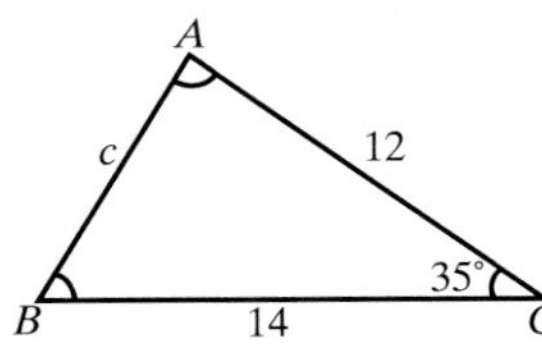

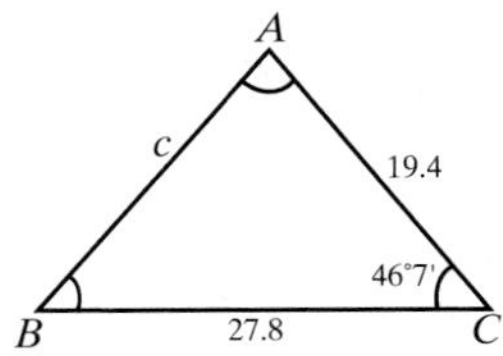

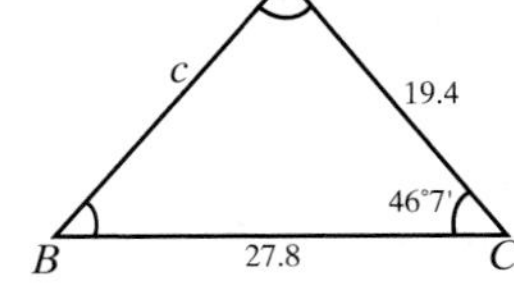

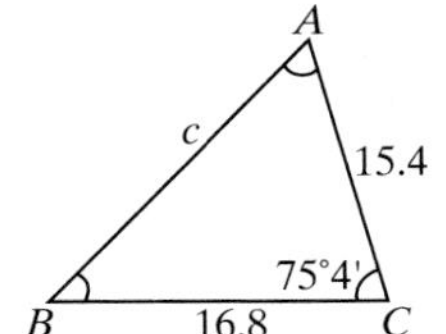

75. $A = 82.62°, B = 80.72°, C = 16.66°$ **77.** $A = 41.93°, B = 38.33°, C = 99.74°$ **79.** $A = 27.66°, B = 111.80°, C = 40.54°$

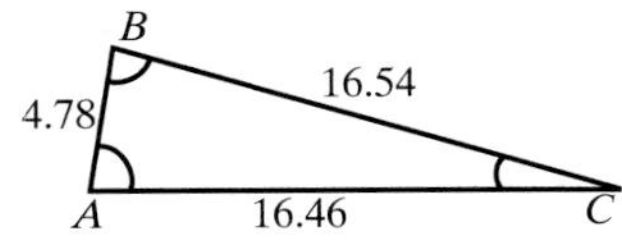

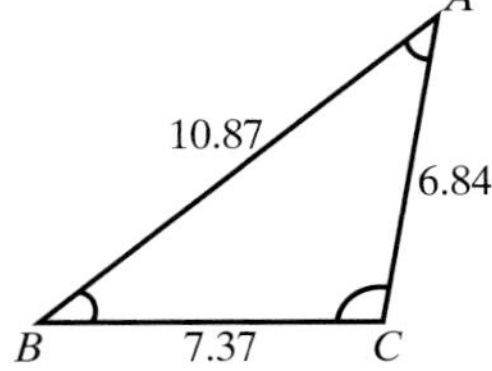

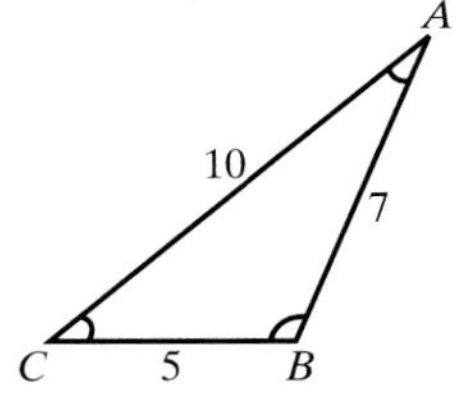

81. 46.82 **83.** 28.32 **85.** 136.67 **87.** 89.29 **89.** 85.5951 ft.^{2} **91.** 178.3882 ft.^{2}

93. a. 61.9372 in.^{2} **b.** 584.2397 in.^{2} **c.** 136.1041 in.^{2}

section 8.2, pages 628-630

1.

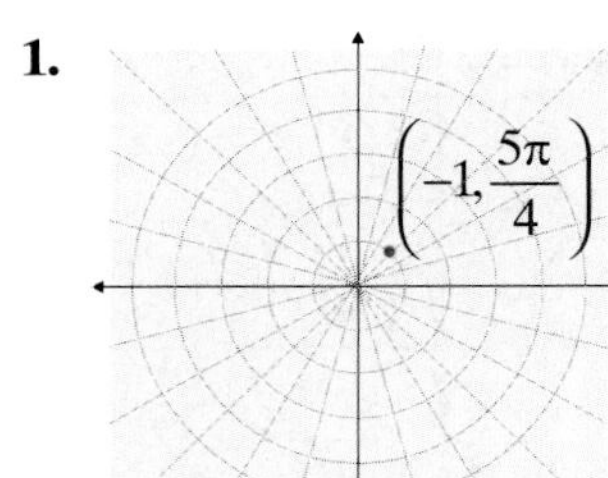

3.

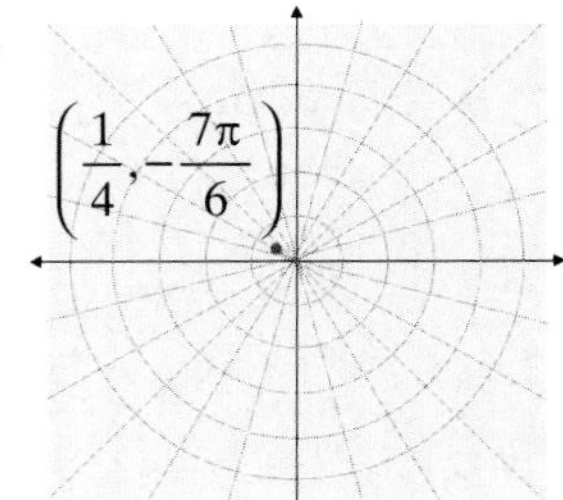

5.

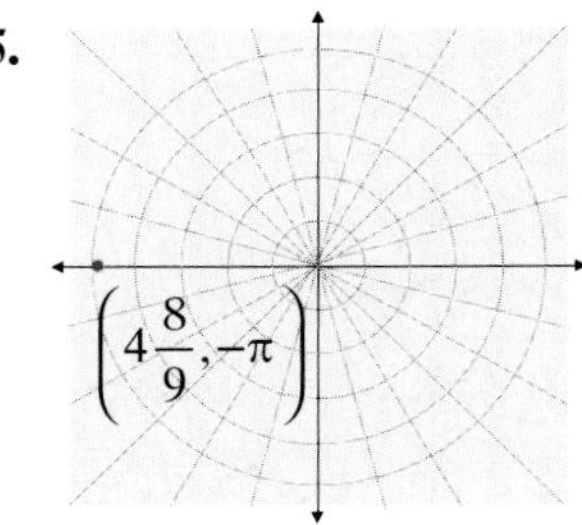

7. $(3.54, -3.54)$ **9.** $(-4.42, -4.42)$ **11.** $(-2.6, -1.5)$

13. $(-3, 0)$ and $(3, \pi)$ **15.** $\left(\sqrt{145}, -0.08\right)$ and $\left(-\sqrt{145}, 3.06\right)$ **17.** $\left(2\sqrt{21}, 1.76\right)$ and $\left(-2\sqrt{21}, -1.38\right)$

19. $r^2 = 25$ **21.** $r\cos\theta = 12$ **23.** $\sin\theta = \cos\theta$

25. $r\cos\theta = 16a$ **27.** $r^2 - 4ar\cos\theta = 0$ **29.** $r^2\sin^2\theta - 4r\cos\theta - 4 = 0$

31. $x^2 + y^2 = 5x$ **33.** $x^2 + y^2 = 49$ **35.** $y = \frac{1}{2}$

37. $x^4 + y^4 + 2x^2y^2 = 2xy$ **39.** $4y + 7x = 12$

41. $x^2 + y^2 = 9$ **43.** $y = -\frac{x}{\sqrt{3}}$ **45.** $x = 7$

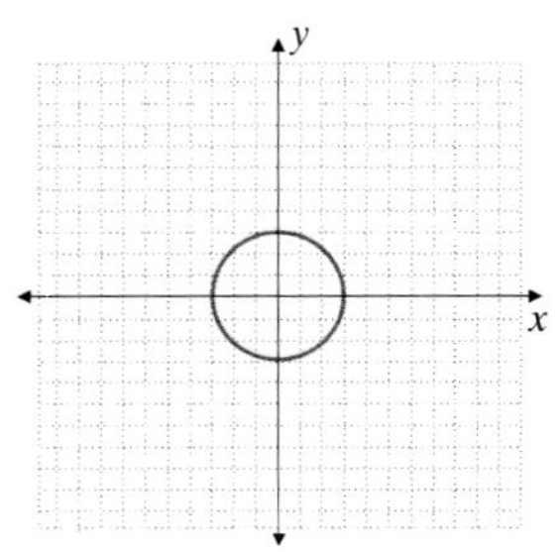

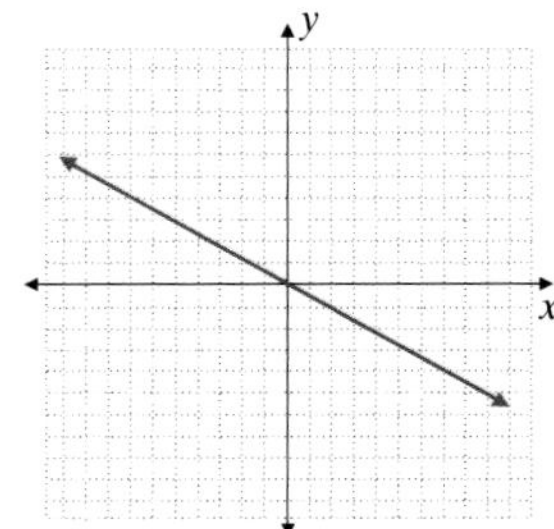

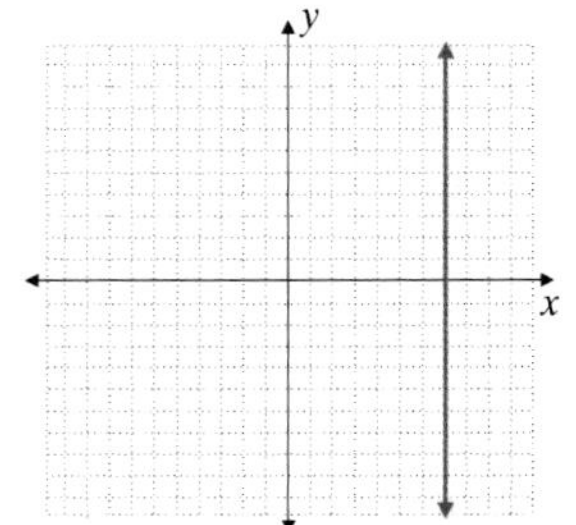

47. 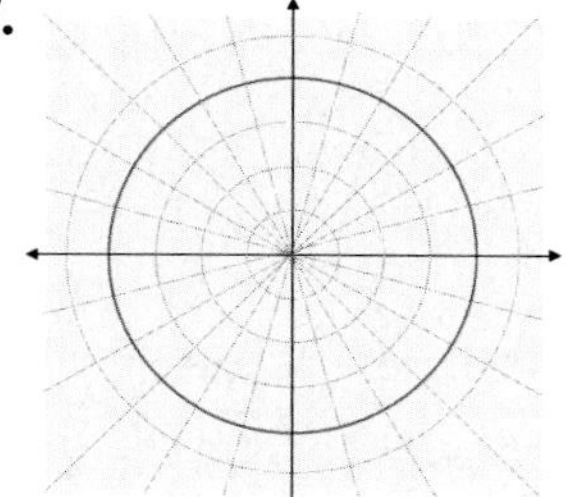

49. 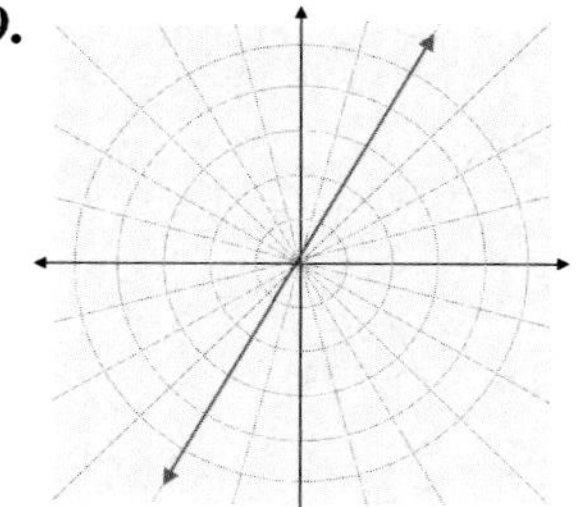

51. 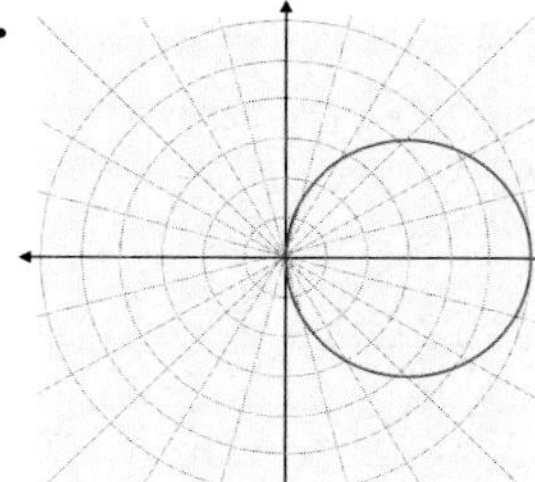

53. 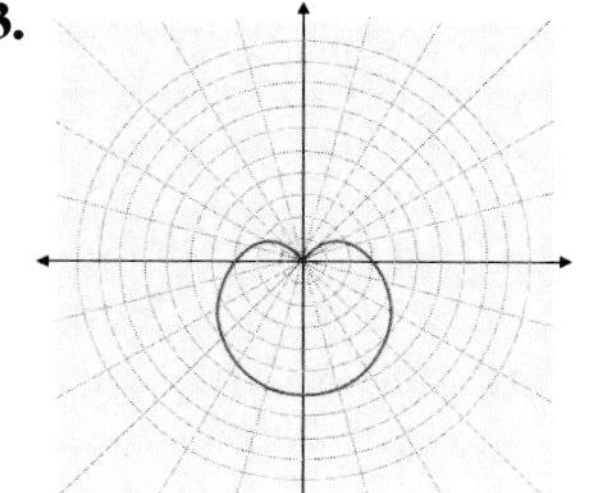

55. 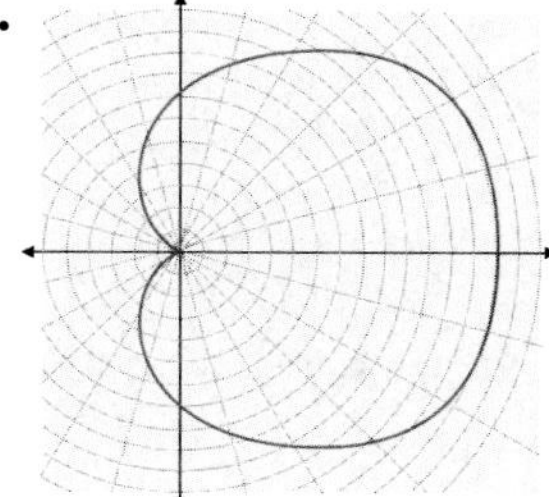

57. 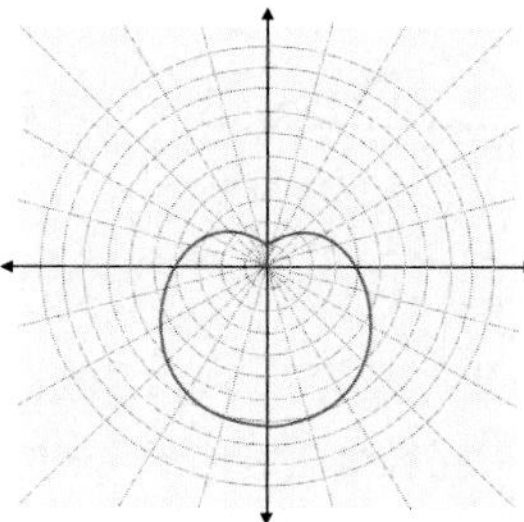

59. 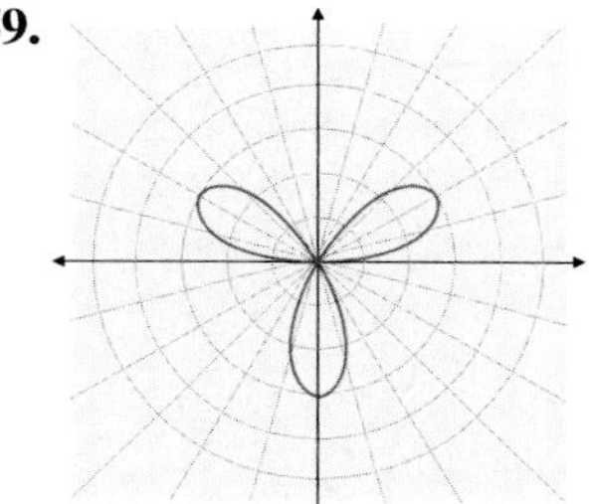

61. 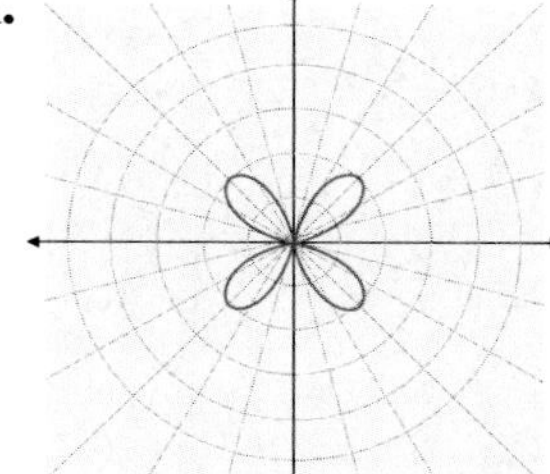

63. 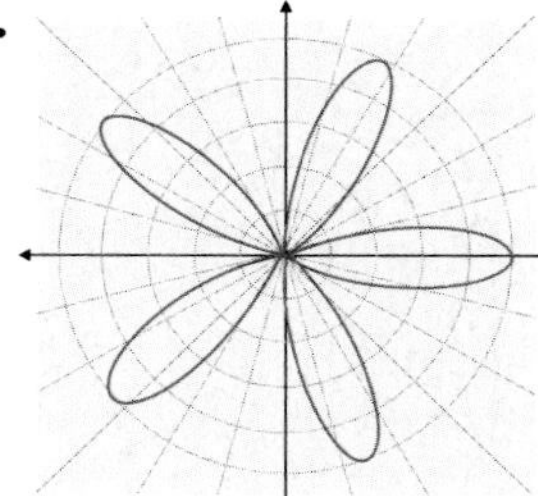

65. 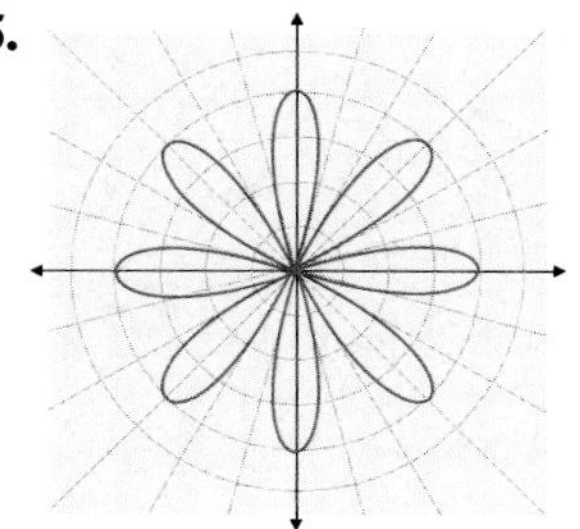

67. 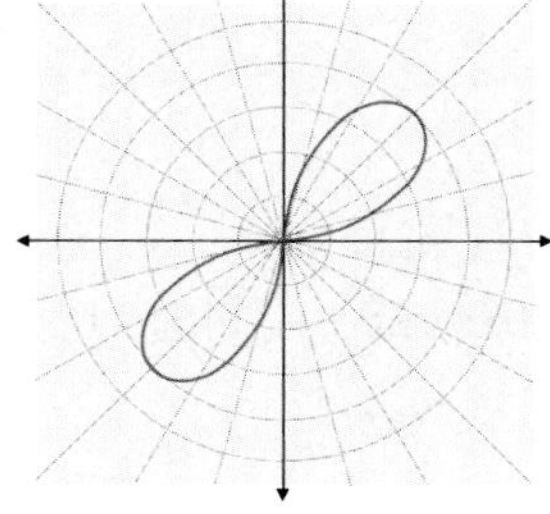

section 8.3, pages 640-643

1.

t	x	y
0	5	0
1	6	-1
2	7	undefined
3	8	$\sqrt{3}$
4	9	1
5	10	$\frac{\sqrt{5}}{3}$
6	11	$\frac{\sqrt{6}}{4}$

3. a. $x = 58.67t,$
$y = -16t^2 + 101.61t + 10$

b. 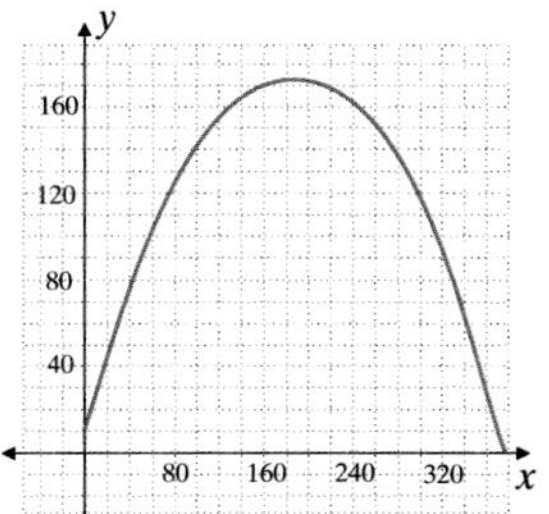

c. 126.42 ft. **d.** 378.42 ft.

e. 6.45 sec **f.** Yes

5. a. $x = 14.05t$, $y = -16t^2 + 15.61t + 7$

b.

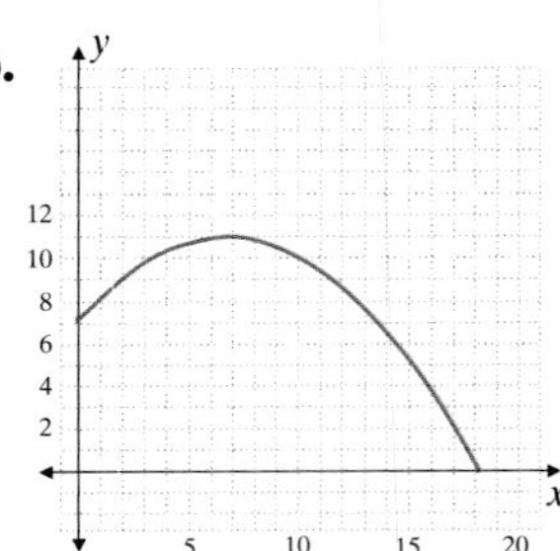

c. No, he won't

7. $y = 3x^2 + 4$ $(x \geq 0)$ as $\sqrt{t-2}$

9. $x = |y+8|$

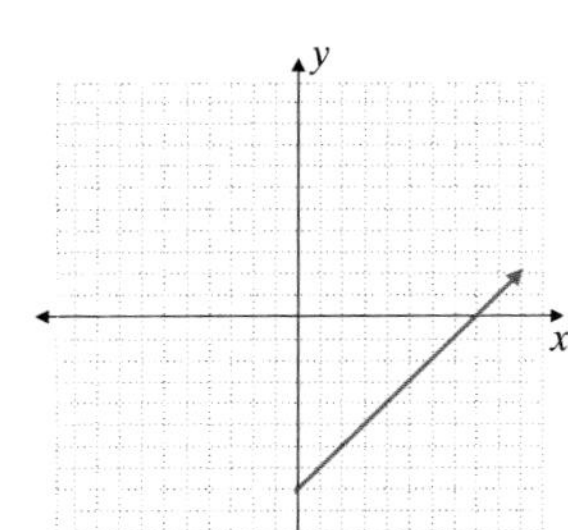

11. $y = \sqrt{\frac{2x}{1-x}}$ $(t \geq 0, \neq -2)$

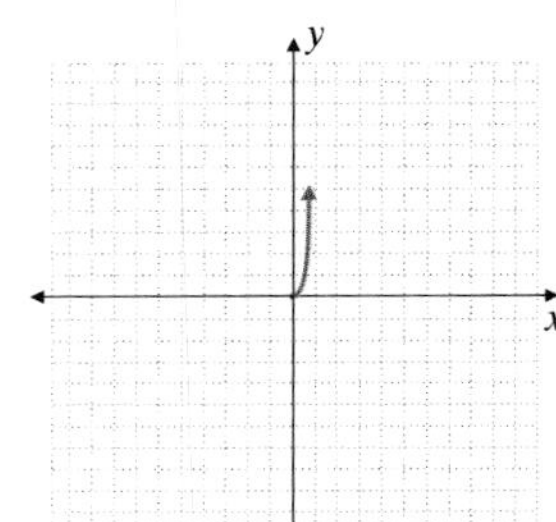

13. $x = \frac{4}{|y-5|}$ $(t \neq 3)$

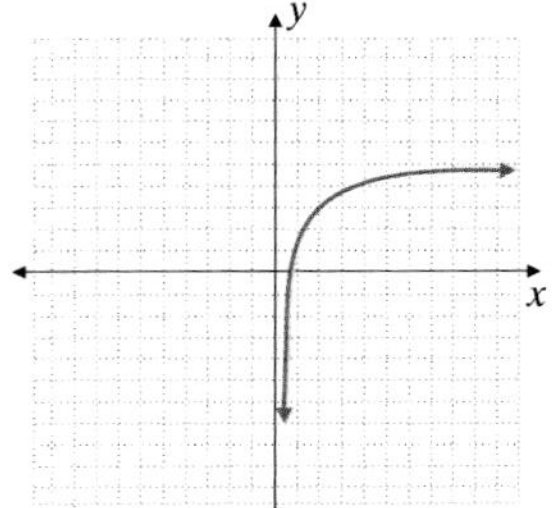

15. $y = \frac{\sqrt{8-2x-x^2}}{6}$

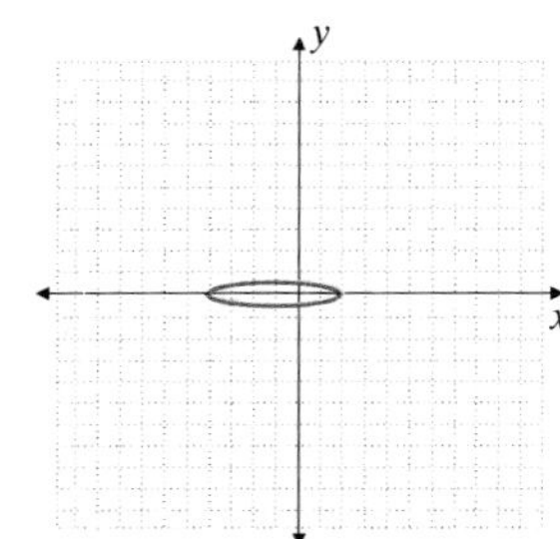

17. $y = \frac{3}{2}x$

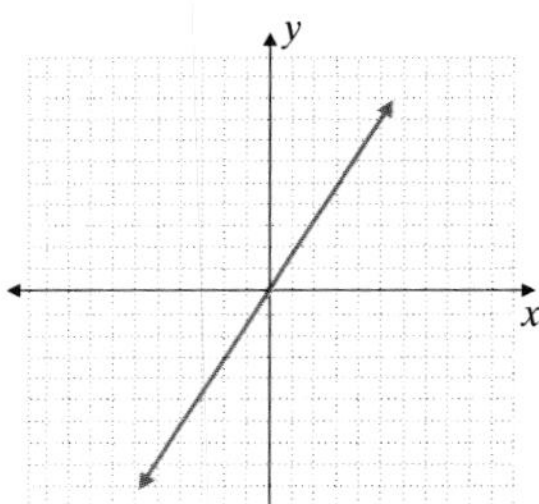

19. $y = 4 - 3\sqrt{1-x^2}$

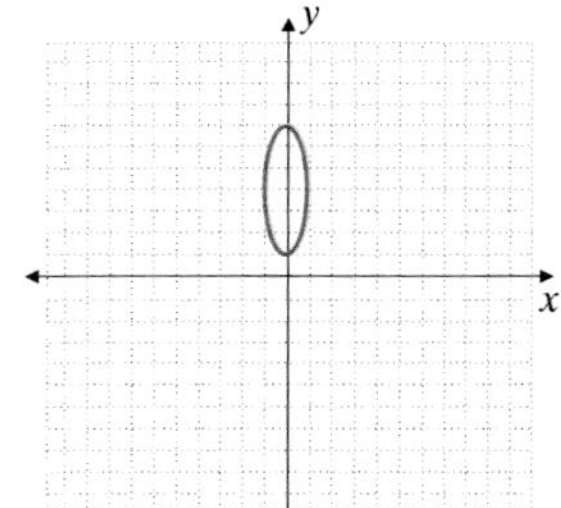

21. $x = t, y = 5t - 2$ **23.** $x = t,\ y = \pm\sqrt{4-4t^2}$ **25.** $x = t,\ y = t^2 + 1$ **27.** $x = t,\ y = \frac{6+t}{4}$

29. $x = t, y = \frac{t+6}{2}$ **31.** $x = t, y = \frac{1}{3t}$ **33.** $x = t,\ y = -2t - 12$ **35.** $x = t,\ y = 3t - 19$

37. $x = t, y = -\frac{3}{2}t + 6$ **39.** $x = \cos\theta,\ y = \sin\theta$ **41.** $x = 7 + 4\cos\theta,\ y = -5 + 4\sin\theta$ **43.** $x = 3\cos\theta,\ y = \sqrt{5}\sin\theta$

45. $x = 5 + 2\sqrt{2}\cos\theta,\ y = -1 + 3\sin\theta$ **47.** $x = 3\sec\theta, y = \sqrt{7}\tan\theta$ **49.** $x = 4\tan\theta,\ y = 3\sec\theta$

51. $x = 12(\theta - \sin\theta),\ y = 12(1 - \cos\theta)$

eight

section 8.4, pages 654-656

1. $\sqrt{34}$

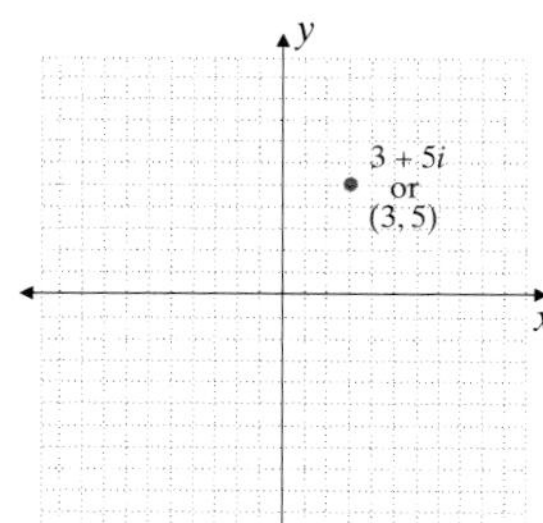

3. $\sqrt{20} = 2\sqrt{5}$

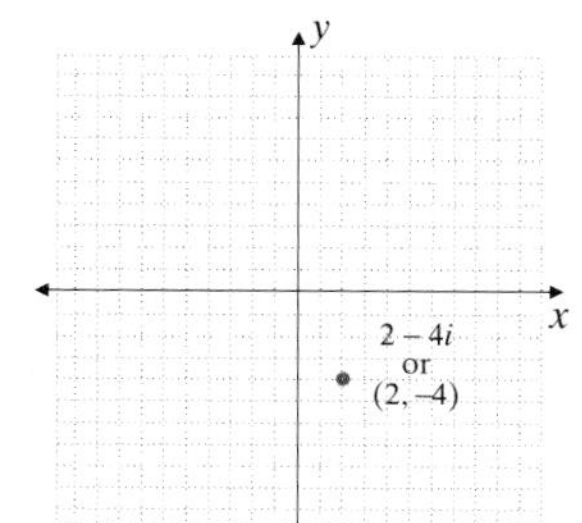

5. $\sqrt{32} = 4\sqrt{2}$

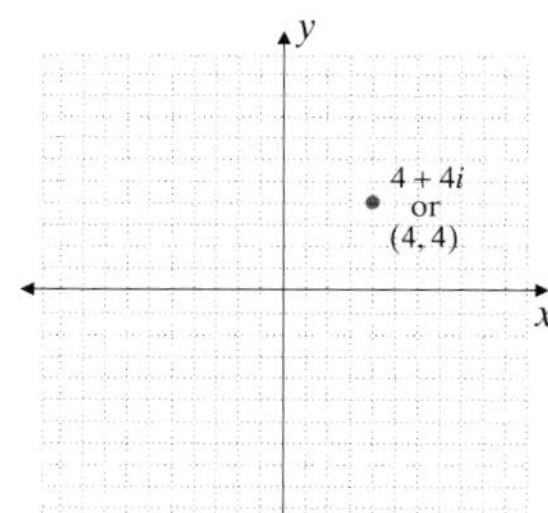

7.

−20 + 17i or (−20, 17); −2 + 3i or (−2, 3); 5 + 5i or (5, 5); 7 + 2i or (7, 2)

9.

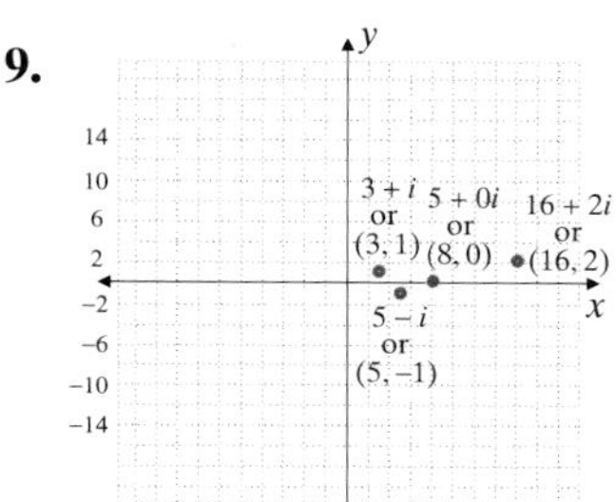

11.

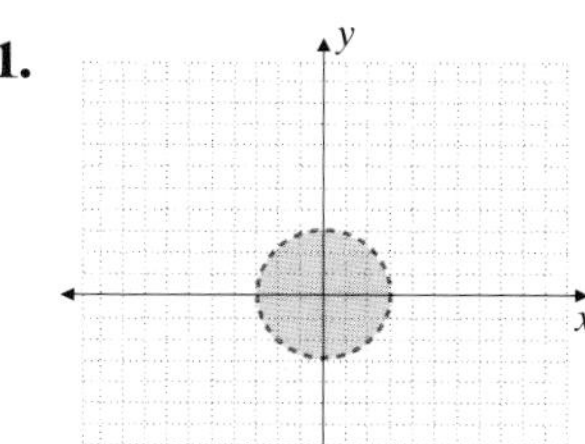

13.

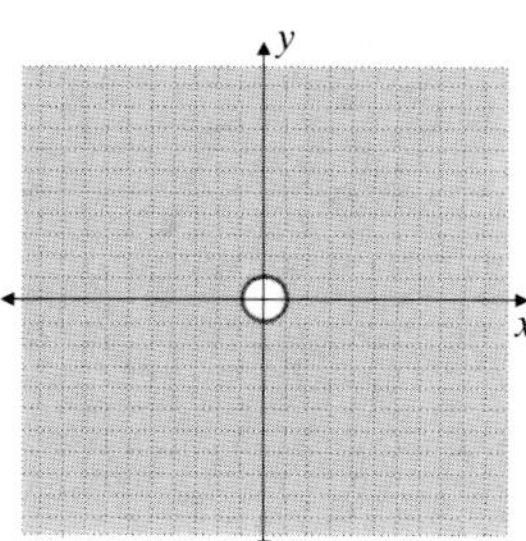

15.

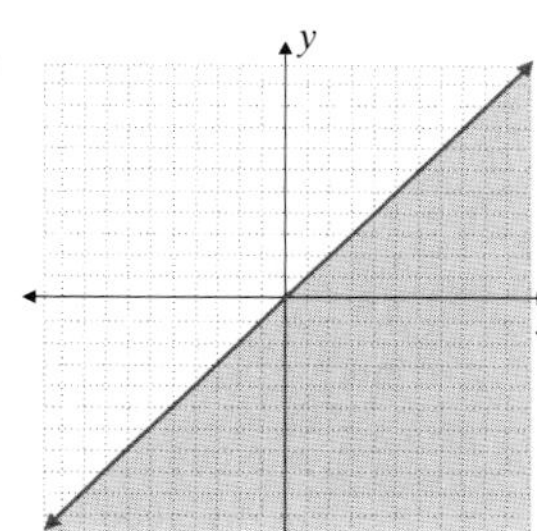

17. $\sqrt{10}\left(\cos(3.46)+i\sin(3.46)\right)$

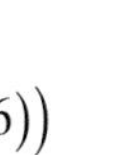

19. $\sqrt{5}\left(\cos(1.11)+i\sin(1.11)\right)$

21. $2\sqrt{5}\left(\cos(0.46)+i\sin(0.46)\right)$

23. $2\left(\cos\left(-\frac{\pi}{4}\right)+i\sin\left(-\frac{\pi}{4}\right)\right)$

25. $5(\cos 0.93+i\sin 0.93)$

27. $8\left(\cos\left(-\frac{\pi}{3}\right)+i\sin\left(-\frac{\pi}{3}\right)\right)$

29. $\frac{-3\sqrt{3}}{2}+\frac{3i}{2}$

31. $-1-i\sqrt{3}$

33. $-\frac{5}{\sqrt{2}}+\frac{5i}{\sqrt{2}}$

35. $\frac{-3\sqrt{3}}{4}+\frac{3i}{4}$

37. $1.01+4.9i$

39. $16\left(\cos 30^\circ+i\sin 30^\circ\right)=8\sqrt{3}+8i$

41. $3\sqrt{6}\left(\cos\frac{17\pi}{12}+i\sin\frac{17\pi}{12}\right)=-1.9-7.1i$

43. $2\sqrt{10}\left(\cos 2.42+i\sin 2.42\right)=\left(-3-\sqrt{3}\right)+i\left(3\sqrt{3}-1\right)$

45. $2\left(\cos 180^\circ+i\sin 180^\circ\right)=-2$

47. $\frac{10}{3}\left(\cos\frac{\pi}{2}+i\sin\frac{\pi}{2}\right)=\frac{10i}{3}$

49. $\frac{1}{\sqrt{2}}\left(\cos\left(-\frac{3\pi}{4}\right)+i\sin\left(-\frac{3\pi}{4}\right)\right)=-\frac{1}{2}-\frac{i}{2}$

51. $2\left(\cos\frac{5\pi}{12}+i\sin\frac{5\pi}{12}\right)=0.52+1.93i$

53.
$z_1 = (-3.46, 2)$
$z_2 = (-2, 0)$
$z_1z_2 = (6.93, -4)$

55.
$z_2 = (1.41, 2)$
$z_1z_2 = (12.83, -3.07)$
$z_1 = (2, -5)$

57.
$z_1 = (1, 1.73)$
$z_2 = (-2.12, -2.12)$
$z_1z_2 = (1.55, -5.8)$

59. $32e^{\frac{\pi}{3}i}$

61. $1.04\times10^{13}e^{2.9\cdot i}$ **63.** $e^{2\pi i}$ **65.** $e^{\frac{\pi i}{4}}, e^{\frac{3\pi i}{4}}, e^{\frac{5\pi i}{4}}, e^{\frac{7\pi i}{4}}$ **67.** $2e^{\frac{\pi i}{12}}, 2e^{\frac{13\pi i}{12}}$ **69.** $4, 4e^{\frac{\pi i}{2}}, 4e^{\pi i}, 4e^{\frac{3\pi i}{2}}$

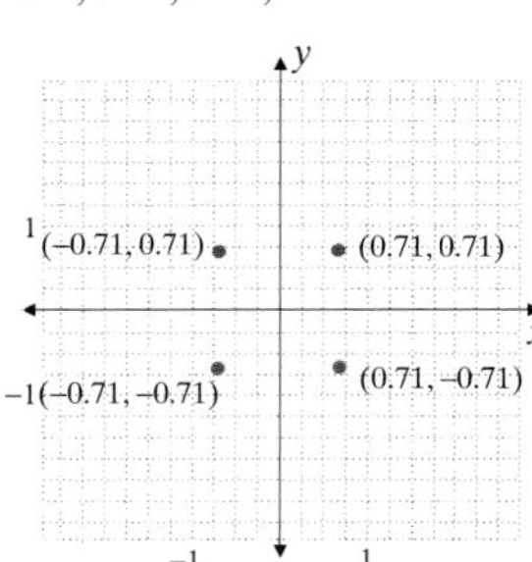

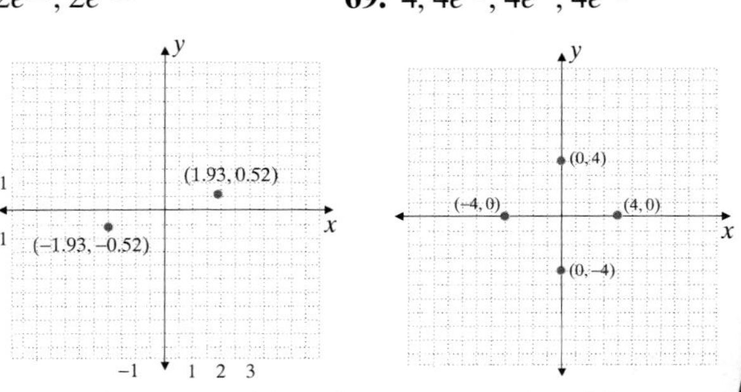

71. $2e^{60^\circ i}, 2e^{240^\circ i}$ **73.** $2\sqrt{2}e^{\frac{\pi i}{12}}, 2\sqrt{2}e^{\frac{13\pi i}{12}}$ **75.** $2e^{\frac{\pi i}{5}}, 2e^{\frac{3\pi i}{5}}, 2e^{\pi i}, 2e^{\frac{7\pi i}{5}}, 2e^{\frac{9\pi i}{5}}$ **77.** $5e^{-\frac{\pi i}{4}}, 5e^{\frac{3\pi i}{4}}$

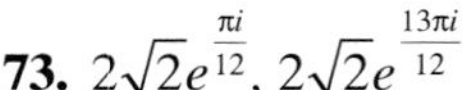

(1, 1.732)
1
−1
(−1, −1.732)
−1 1

section 8.5, pages 666-669

1.
−u

3.

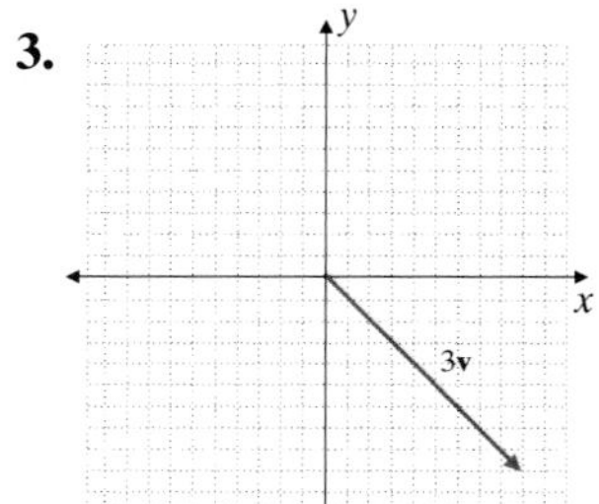

5.
2u − 2v

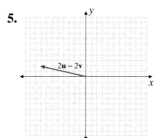

7. $\mathbf{v} = \langle 3, -3\rangle, \|\mathbf{v}\| = 3\sqrt{2}$ **9.** $\mathbf{v} = \langle 5, 3\rangle, \|\mathbf{v}\| = \sqrt{34}$ **11.** $\mathbf{v} = \langle 5, -1\rangle, \|\mathbf{v}\| = \sqrt{26}$ **13.** $\mathbf{v} = \langle -7, 7\rangle, \|\mathbf{v}\| = 7\sqrt{2}$

15. $\mathbf{v} = \langle -4, -6\rangle, \|\mathbf{v}\| = 2\sqrt{13}$ **17. a.** $\langle -2, 8\rangle$ **b.** $\langle 8, -4\rangle$ **c.** $\langle -4, 0\rangle$ **19. a.** $\langle 1, 4\rangle$ **b.** $\langle -11, 12\rangle$ **c.** $\langle 6, -8\rangle$

21. a. $\langle -5, -10\rangle$ **b.** $\langle -8, -2\rangle$ **c.** $\langle 6, 4\rangle$ **23.** $-\mathbf{u} = \langle -1, -1\rangle,\ 2\mathbf{u} - \mathbf{v} = \langle -1, 5\rangle,\ \mathbf{u} + \mathbf{v} = \langle 4, -2\rangle,\ \|\mathbf{u}\| = \sqrt{2},\ \|\mathbf{v}\| = 3\sqrt{2}$

eight

25. $-\mathbf{u} = \langle 4, -4 \rangle$, $2\mathbf{u} - \mathbf{v} = \langle -12, 12 \rangle$, $\mathbf{u} + \mathbf{v} = \langle 0, 0 \rangle$, $\|\mathbf{u}\| = 4\sqrt{2}$, $\|\mathbf{v}\| = 4\sqrt{2}$ **27. a.** $\left\langle \frac{2}{\sqrt{5}}, -\frac{1}{\sqrt{5}} \right\rangle$ **b.** $\mathbf{u} = 6\mathbf{i} - 3\mathbf{j}$
29. a. $\left\langle \frac{-5}{\sqrt{26}}, -\frac{1}{\sqrt{26}} \right\rangle$ **b.** $\mathbf{u} = -5\mathbf{i} - \mathbf{j}$ **31. a.** $\left\langle \frac{2}{\sqrt{13}}, \frac{3}{\sqrt{13}} \right\rangle$ **b.** $\mathbf{u} = 2\mathbf{i} + 3\mathbf{j}$ **33.** $\|\mathbf{v}\| = 5, \theta = 30^\circ$
35. $\|\mathbf{v}\| = 5, \theta = 36.9^\circ$ **37.** $\langle 3\sqrt{3}, 3 \rangle$ **39.** $\langle -9\sqrt{2}, 9\sqrt{2} \rangle$ **41.** $\left\langle -\frac{1}{2}, \frac{\sqrt{3}}{2} \right\rangle$ **43.** $\left\langle \frac{8}{\sqrt{13}}, \frac{12}{\sqrt{13}} \right\rangle$ **45.** $\langle 2\sqrt{3}, 2 \rangle$
47. 38.67 mph, N 77.76° W **49.** 1244.08 lbs.

section 8.6, pages 678-680

1. 17 **3.** 6 **5.** 8 **7.** 1 **9.** −7 **11.** 32 **13.** $\sqrt{13} + 2$ **15.** $\sqrt{37}$ **17.** $\sqrt{53}$ **19.** 123.7° **21.** 14.0°
23. 161.6° **25.** $\frac{\pi}{4}$ **27.** 8°, 69°, 103° **29.** 57.1°, 60.8°, 62.1° **31.** −62.5 **33.** $-32\sqrt{2}$ **35.** $\langle 1, 1 \rangle, \langle 5, 5 \rangle$
37. $\langle 3, 1 \rangle, \langle -6, -2 \rangle$ **39.** neither **41.** orthogonal **43.** $\text{proj}_{\mathbf{v}}\mathbf{u} = \langle 2, 1 \rangle$, $\text{perp}_{\mathbf{v}}\mathbf{u} = \langle -1, 2 \rangle$
45. $\text{proj}_{\mathbf{v}}\mathbf{u} = \left\langle \frac{6}{5}, \frac{2}{5} \right\rangle$, $\text{perp}_{\mathbf{v}}\mathbf{u} = \left\langle \frac{9}{5}, -\frac{27}{5} \right\rangle$ **47.** $\text{proj}_{\mathbf{v}}\mathbf{u} = \left\langle -\frac{60}{17}, -\frac{15}{17} \right\rangle$, $\text{perp}_{\mathbf{v}}\mathbf{u} = \left\langle \frac{9}{17}, -\frac{36}{17} \right\rangle$ **49.** 14 **51.** 3
53. 3479.3 pounds **55.** 109.6 pounds **57.** 579,555.5 ft.-lbs.

chapter eight test, pages 687-691

1. height = 22.5762 ft., angle = 15.8886° **3.** $A = 75.40^\circ$, $B = 47.6^\circ$, $b = 11.45$ or $A = 104.6^\circ$, $B = 18.4^\circ$, $b = 4.89$

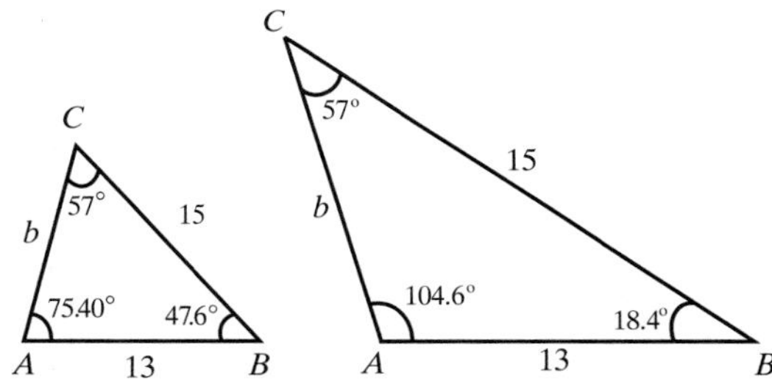

5. $A = 64.99^\circ$, $B = 37.01^\circ$, $a = 12.04$

7. $A = 22^\circ\, 31'$, $C = 63^\circ\, 22'$, $b = 15.62$

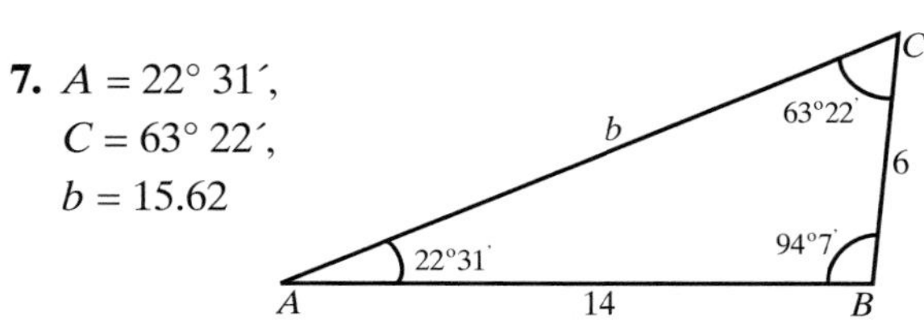

9. $A = 52.04^\circ$, $B = 101.99^\circ$, $C = 25.98^\circ$

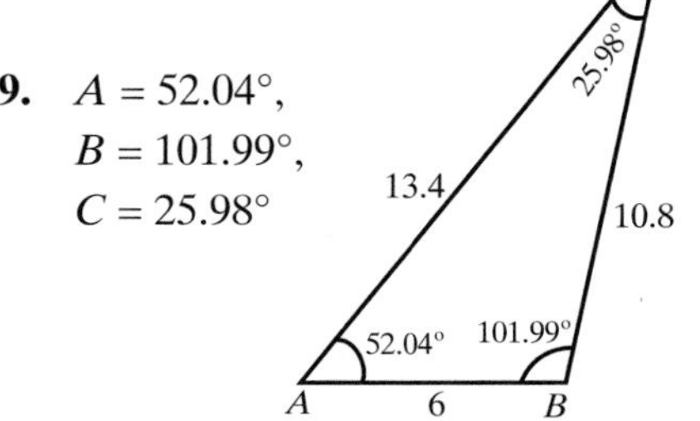

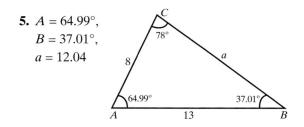

11. (–6.06, –3.5) **13.** (15.62, 0.88) **15.** $r^2 - 9ar\cos\theta = 0$ **17.** $x + y = 4$

19.

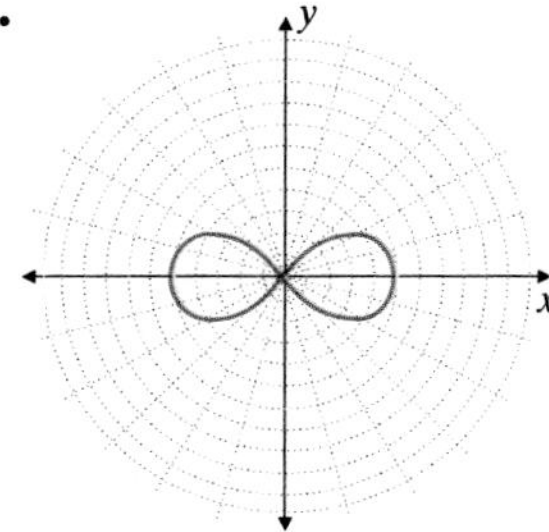

21. $y = |x - 7|$

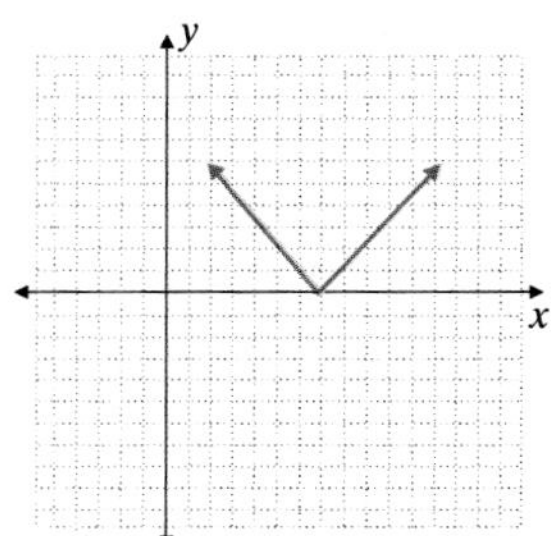

23. $\frac{x^2}{16} + (y-1)^2 = 1$

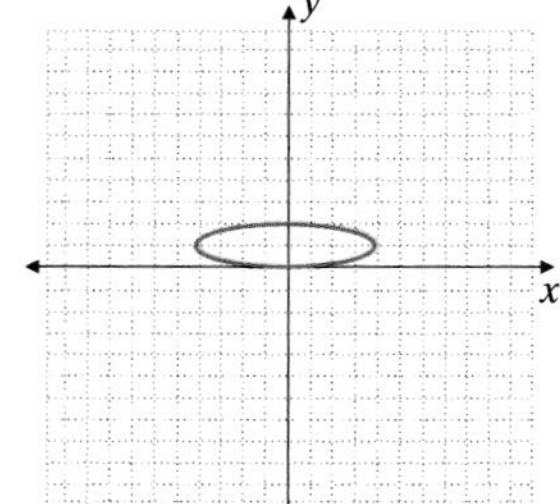

25. $x = t,\ y = 2 - 6t$

27. $x = 1 + \cos\theta,\ y = 1 - \sin\theta$

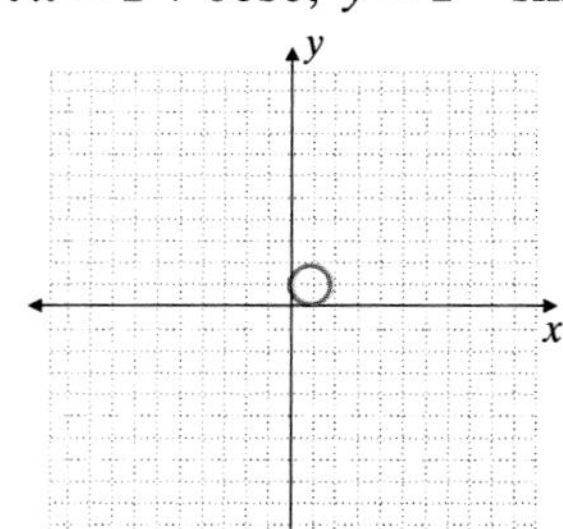

29. $3\sqrt{2}$

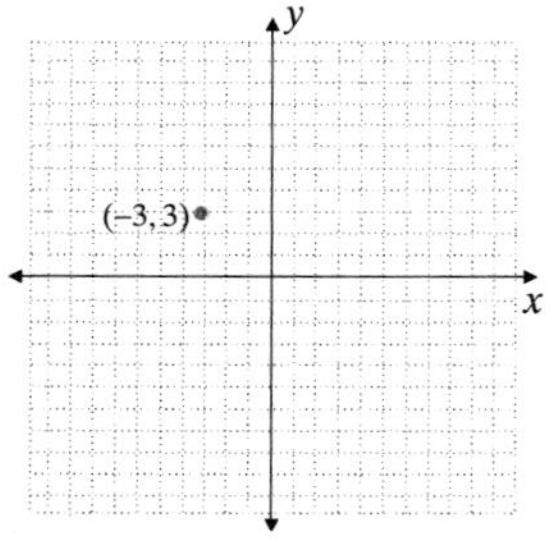

31.

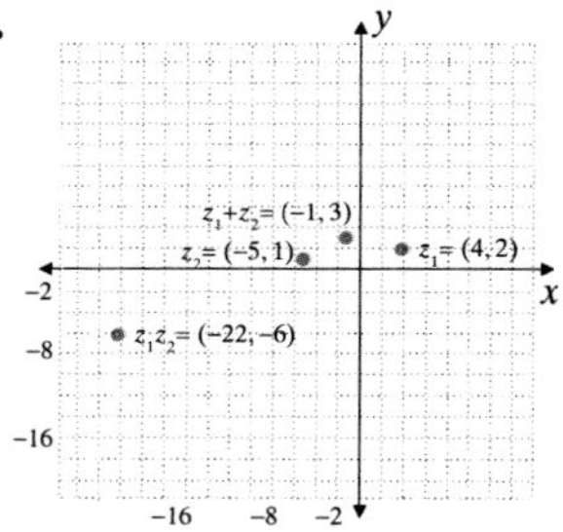

33.

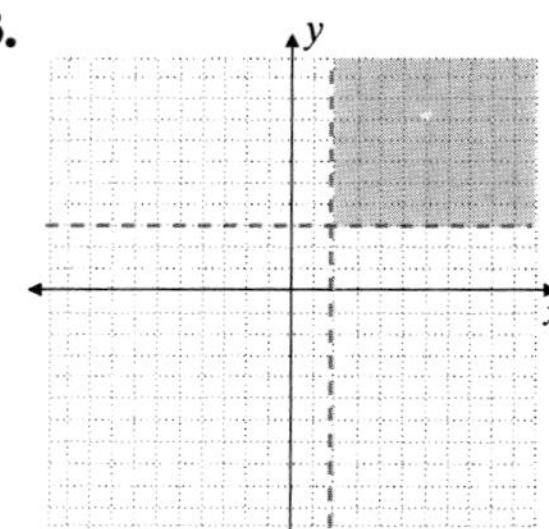

35. $\sqrt{17}\left(\cos(1.33) + i\sin(1.33)\right)$

37. $\frac{3}{2} + \frac{3i\sqrt{3}}{2}$

39. $5(\cos 120° + i\sin 120°), -\frac{5}{2} + \frac{5i\sqrt{3}}{2}$

41. $24(\cos 315° + i\sin 315°), 12\sqrt{2} - 12i\sqrt{2}$

43. $177{,}147(\cos 120° + i\sin 120°)$

45. $5e^{i\left(\frac{7\pi}{12}\right)}, 5e^{i\left(\frac{5\pi}{4}\right)}, 5e^{i\left(\frac{23\pi}{12}\right)}$

47. $2e^{i\left(\frac{\pi}{4}\right)}, 2e^{i\left(\frac{11\pi}{12}\right)}, 2e^{i\left(\frac{19\pi}{12}\right)}$

49. $\mathbf{v} = \langle -10, -6 \rangle;\ \|\mathbf{v}\| = 2\sqrt{34}$

51.

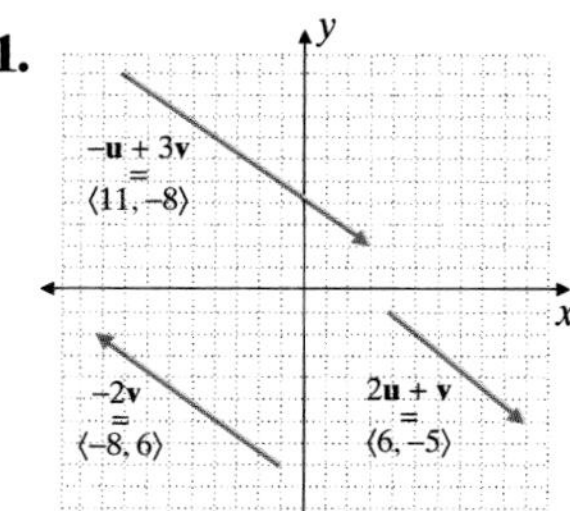

53. $-\mathbf{u} = \langle -5, -1 \rangle;\ 2\mathbf{u} - \mathbf{v} = \langle 7, 1 \rangle;$ $\mathbf{u} + \mathbf{v} = \langle 8, 2 \rangle;\ \|\mathbf{u}\| = \sqrt{26};\ \|\mathbf{v}\| = \sqrt{10}$

55. **a.** $\left\langle \frac{2}{\sqrt{5}}, \frac{1}{\sqrt{5}} \right\rangle$ **b.** $6\mathbf{i} + 3\mathbf{j}$

57. $\|\mathbf{v}\| = \sqrt{26},\ \theta = -11.3°$

59. (3.6, –4.8) **61.** 40.01 mph, W 49.77° N **63.** ⟨–108, –270⟩ **65.** $\sqrt{10}$ **67.** $\frac{5\pi}{12}$ **69.** –36

71. $\text{proj}_{\mathbf{v}}\mathbf{u} = \left\langle \frac{3}{2}, \frac{3}{2} \right\rangle,\ \text{perp}_{\mathbf{v}}\mathbf{u} = \left\langle \frac{5}{2}, \frac{-5}{2} \right\rangle$ **73.** 31 **75.** 417,558.5 ft.-lbs.

chapter nine

section 9.1, pages 705-710

1. Center: $(5, 2)$; Foci: $\left(5, 2 \pm \sqrt{21}\right)$; Vertices: $(5, 7)$, $(5, -3)$ **3.** Center: $(-2, -5)$; Foci: $\left(-2 \pm \sqrt{6}, -5\right)$; Vertices: $(1, -5)$, $(-5, -5)$ **5.** Center: $(-3, 2)$; Foci: $\left(-3 \pm \sqrt{2}, 2\right)$; Vertices: $(-1, 2)$, $(-5, 2)$
7. Center: $(-5, 1)$; Foci: $\left(-5, 1 \pm 2\sqrt{3}\right)$; Vertices: $(-5, 5)$, $(-5, -3)$ **9.** Center: $(-4, 2)$; Foci: $\left(-4 \pm 3\sqrt{2}, 2\right)$; Vertices: $\left(-4 \pm 3\sqrt{3}, 2\right)$ **11.** Center: $(2, 0)$; Foci: $(4, 0)$, $(0, 0)$; Vertices: $\left(2 \pm \sqrt{5}, 0\right)$ **13.** c **15.** b **17.** e **19.** g

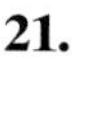

21.

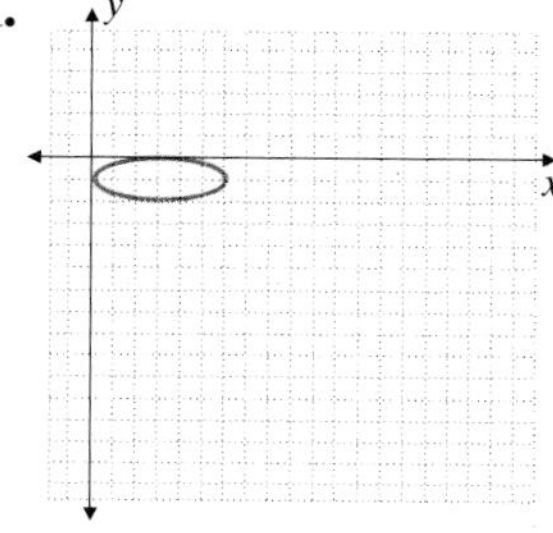

Foci: $\left(3 \pm 2\sqrt{2}, -1\right)$

23.

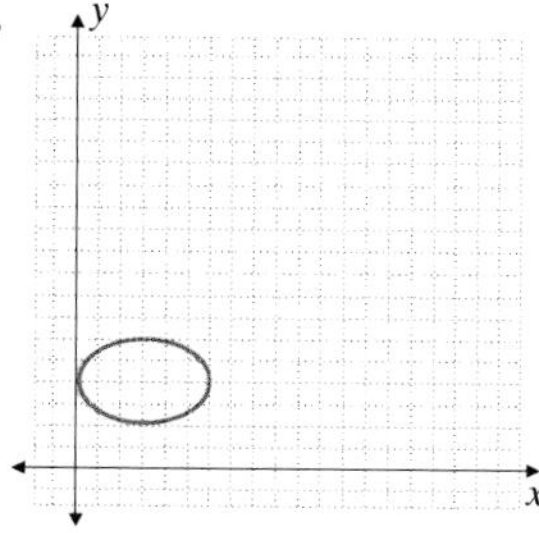

Foci: $\left(3 \pm \sqrt{5}, 4\right)$

25.

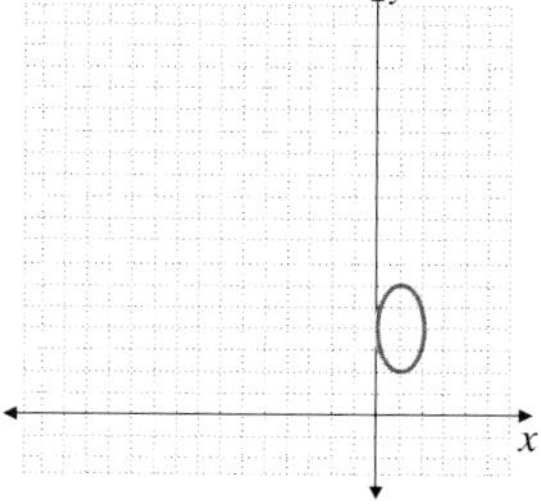

Foci: $\left(1, 4 \pm \sqrt{3}\right)$

27.

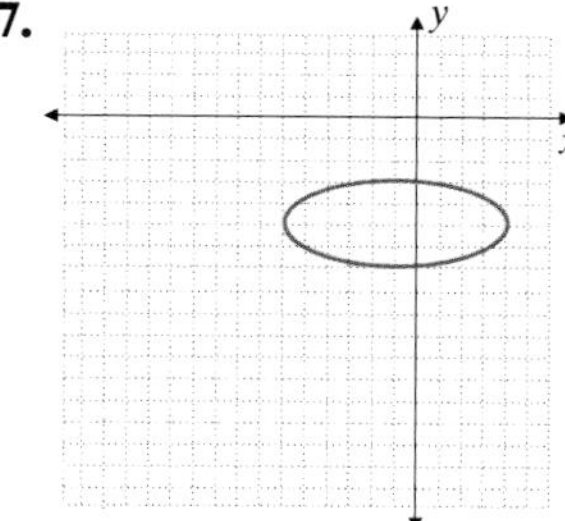

Foci: $\left(-1 \pm \sqrt{21}, -5\right)$

29.

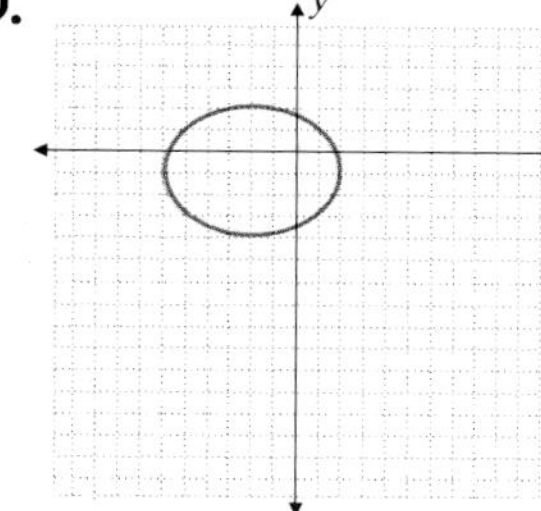

Foci: $\left(-2 \pm \sqrt{7}, -1\right)$

31.

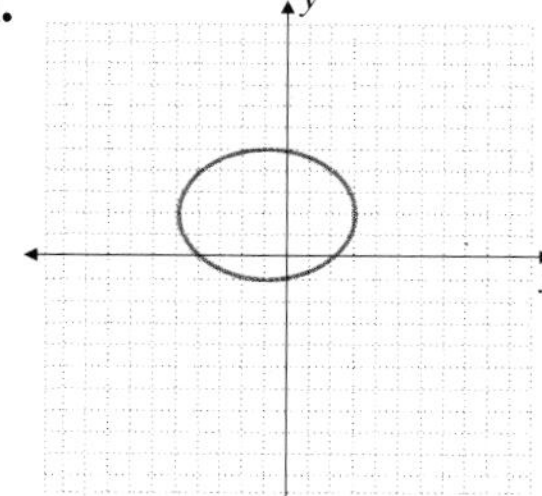

Foci: $\left(-1 \pm \sqrt{7}, 2\right)$

33.

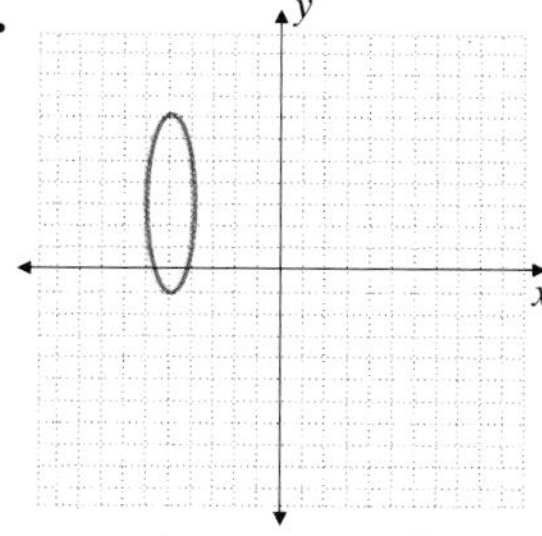

Foci: $\left(-5, 3 \pm \sqrt{15}\right)$

35.

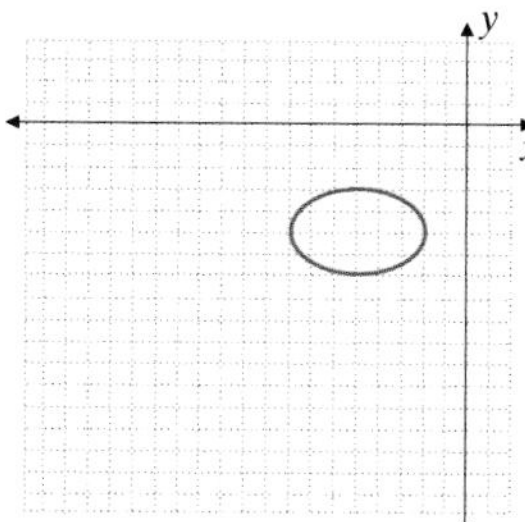

Foci: $\left(-5 \pm \sqrt{5}, -5\right)$

37.

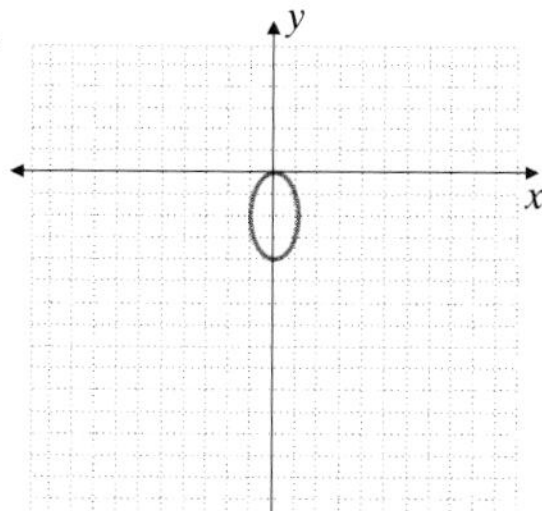

Foci: $\left(0, -2 \pm \sqrt{3}\right)$

39. $\frac{x^2}{16} + \frac{y^2}{25} = 1$ **41.** $(x-1)^2 + \frac{(y-1)^2}{9} = 1$ **43.** $\frac{(x-3)^2}{36} + \frac{y^2}{27} = 1$ **45.** $(x+2)^2 + \frac{(y+3)^2}{4} = 1$

47. $\frac{(x-5)^2}{16} + \frac{(y-3)^2}{15} = 1$ **49.** $\frac{(x-2)^2}{4} + \frac{(y+2)^2}{9} = 1$ **51.** $\frac{(x-1)^2}{9} + \frac{y^2}{16} = 1$

53. $e = \frac{\sqrt{11}}{6}$, major = 24, minor = 20 **55.** $e = \frac{2\sqrt{2}}{3}$, major = 12, minor = 4 **57.** $e = \frac{\sqrt{3}}{2}$, major = 4, minor = 2

59. $e = \frac{\sqrt{2}}{2}$, major = 4, minor = $2\sqrt{2}$ **61.** $e = \frac{\sqrt{42}}{7}$, major = 14, minor = $2\sqrt{7}$ **63.** $e \approx 0.249$

65. 185.93 million miles

67.

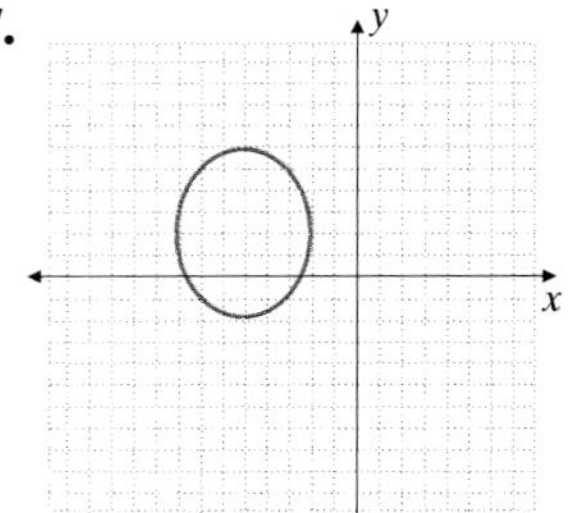

69.

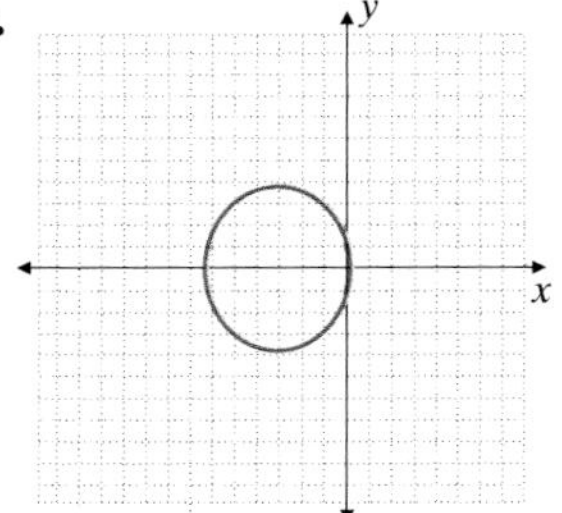

section 9.2, pages 718-721

1.

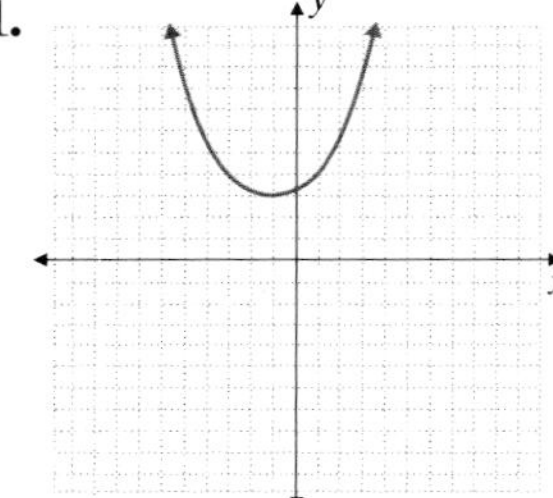

$(-1, 4)$, $y = 2$

3.

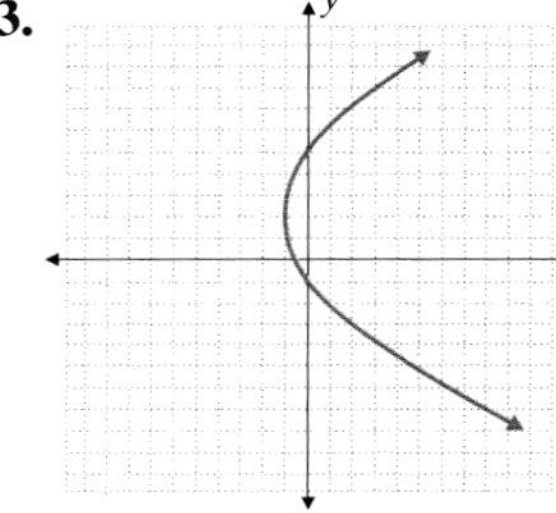

$(1, 2)$, $x = -3$

5.

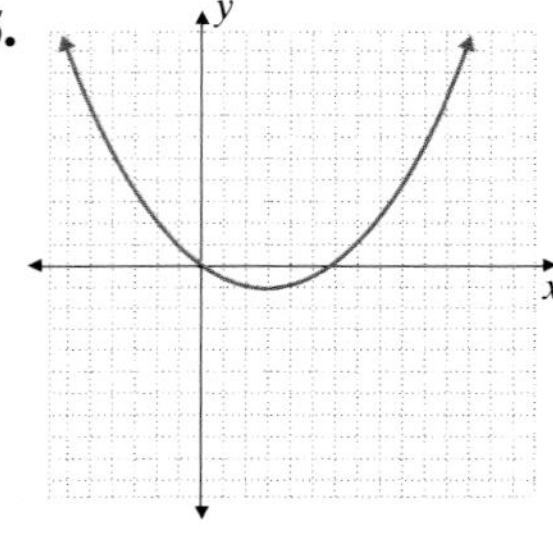

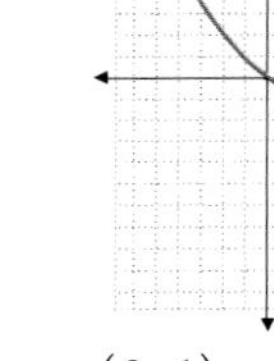

$(3, 1)$, $y = -3$

7.

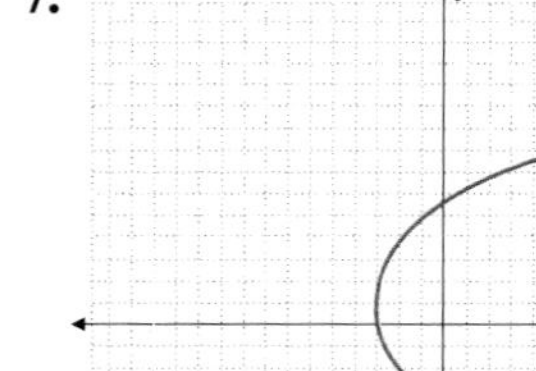

$(-1, 1)$, $x = -5$

9.

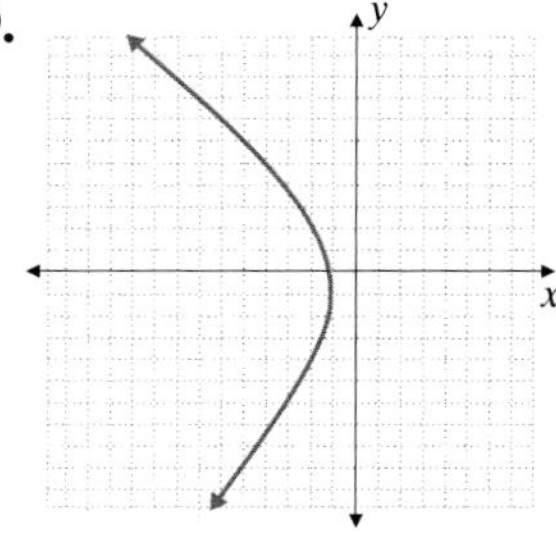

$(-4, -1)$, $x = 2$

11.

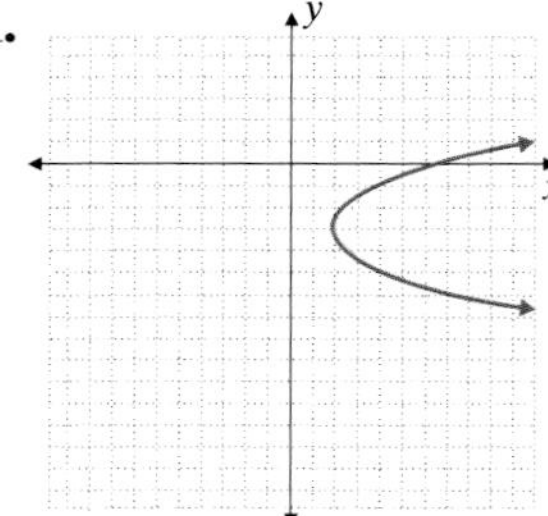

$\left(\frac{5}{2}, -3\right)$, $x = \frac{3}{2}$

13.

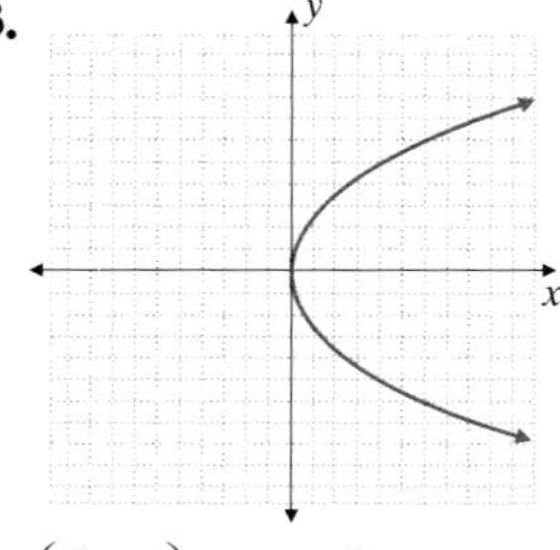

$\left(\frac{3}{2}, 0\right)$, $x = -\frac{3}{2}$

15.

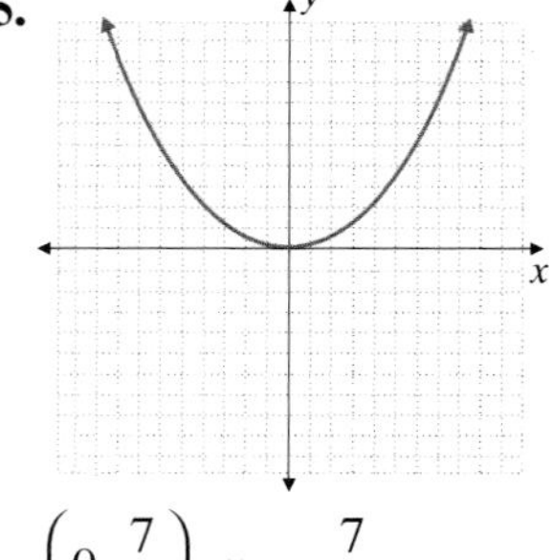

$\left(0, \frac{7}{4}\right)$, $y = -\frac{7}{4}$

17.

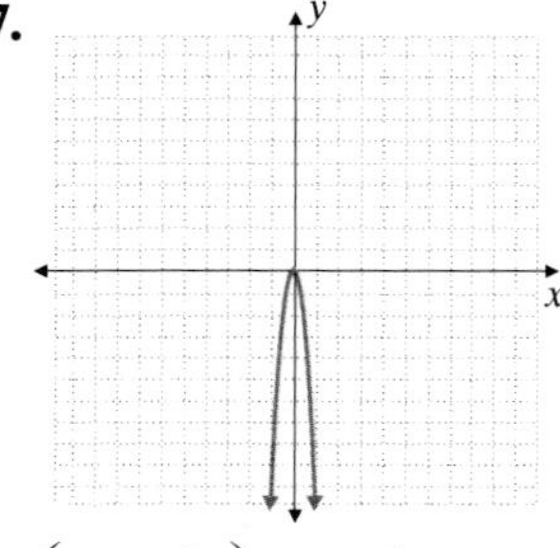

$\left(0, -\frac{1}{48}\right)$, $y = \frac{1}{48}$

19.

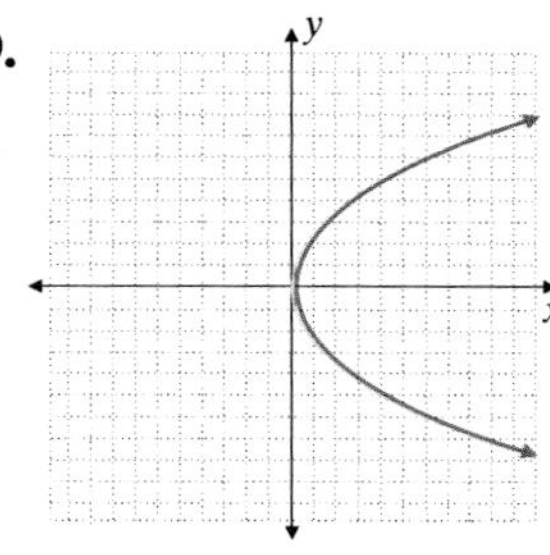

$\left(\frac{3}{2}, 0\right), x = -\frac{3}{2}$

21.

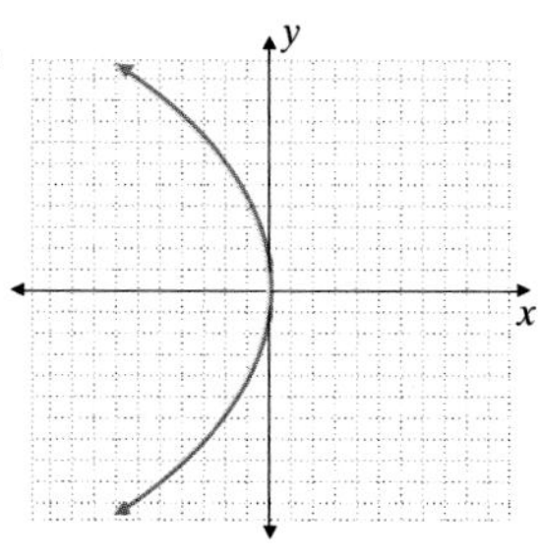

$(-4, 0), x = 4$

23.

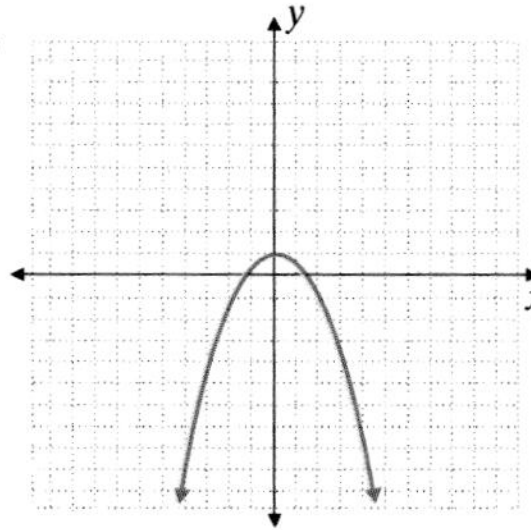

$\left(0, \frac{1}{2}\right), y = \frac{3}{2}$

25. $(y-1)^2 = -4(x+1)$ **27.** $(x-3)^2 = 8(y+1)$ **29.** $(y+2)^2 = 24(x-3)$ **31.** $(x+3)^2 = -2(y+1)$

33. $(y-3)^2 = 10(x+4)$ **35.** $(y+1)^2 = -8(x-2)$ **37.** g **39.** b **41.** e **43.** d

45.

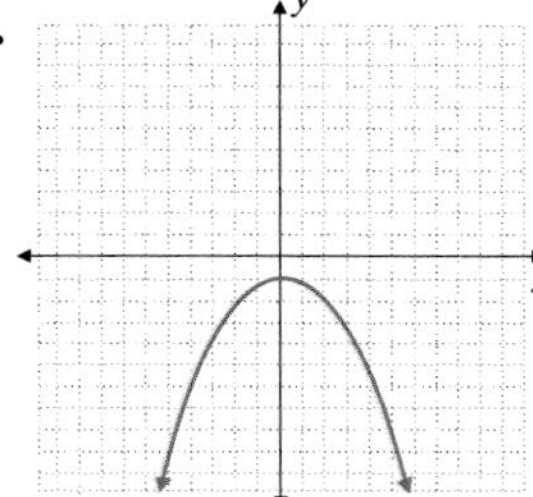

47.

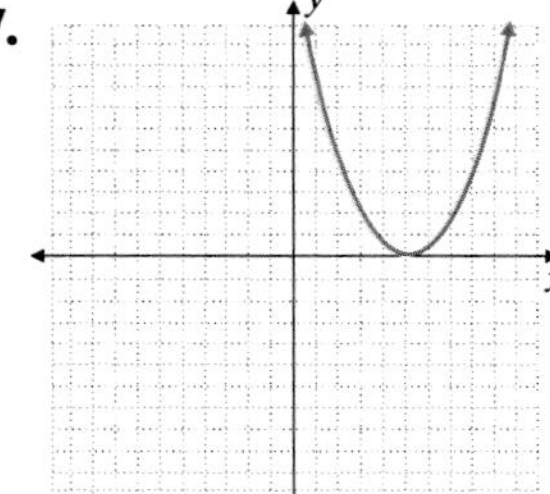

49.

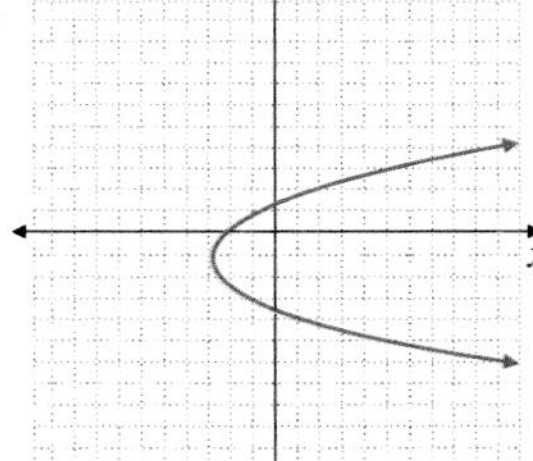

51.

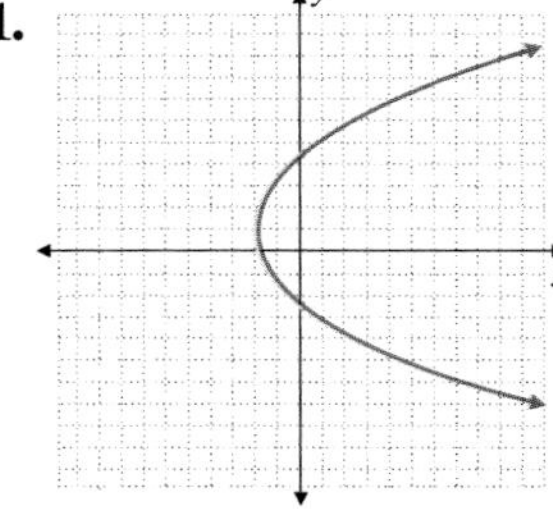

53.

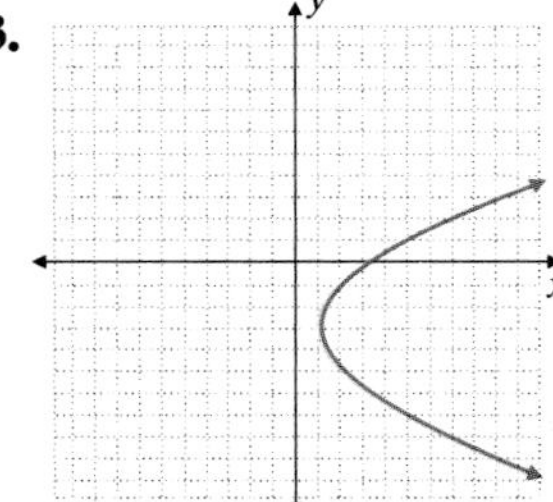

55. 2.25 feet

57.

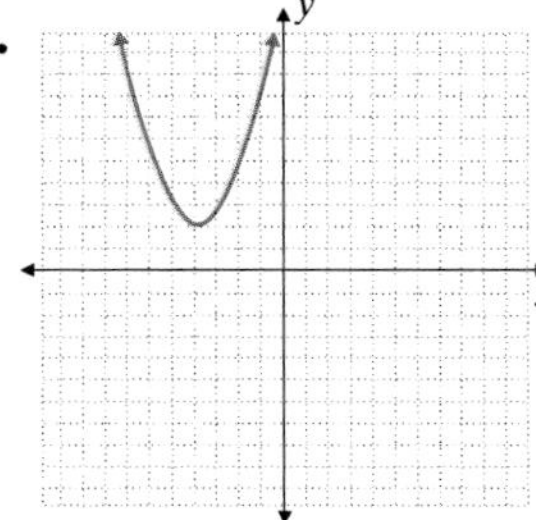

59.

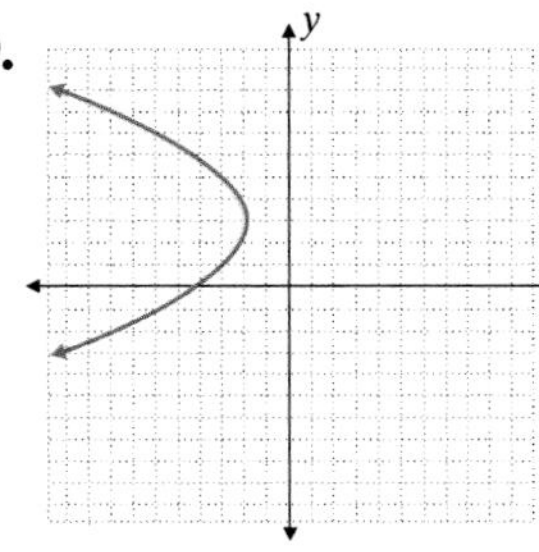

section 9.3, pages 732-736

1.

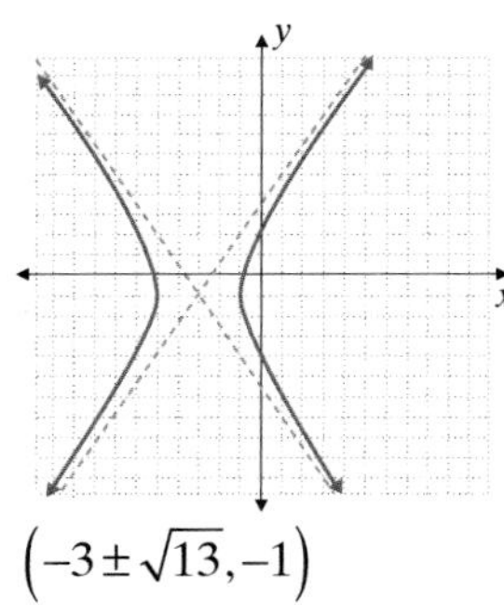

$\left(-3 \pm \sqrt{13}, -1\right)$

3.

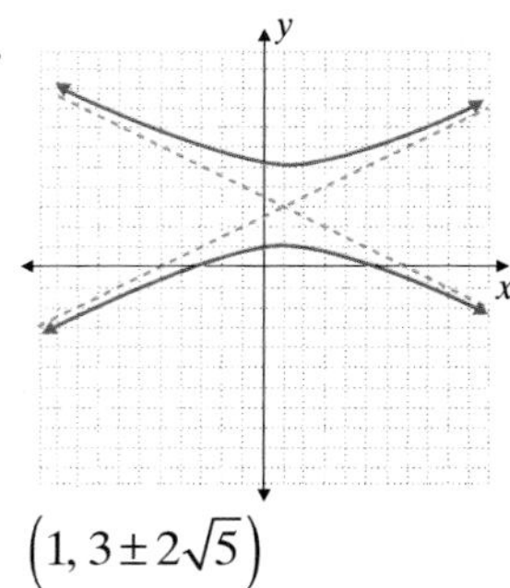

$\left(1, 3 \pm 2\sqrt{5}\right)$

5.

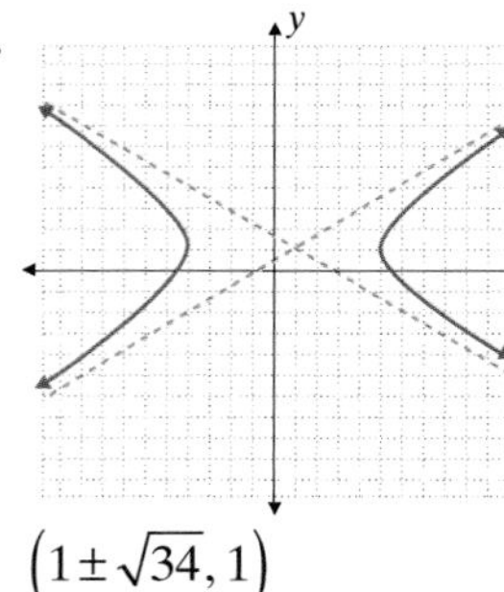

$\left(1 \pm \sqrt{34}, 1\right)$

7.

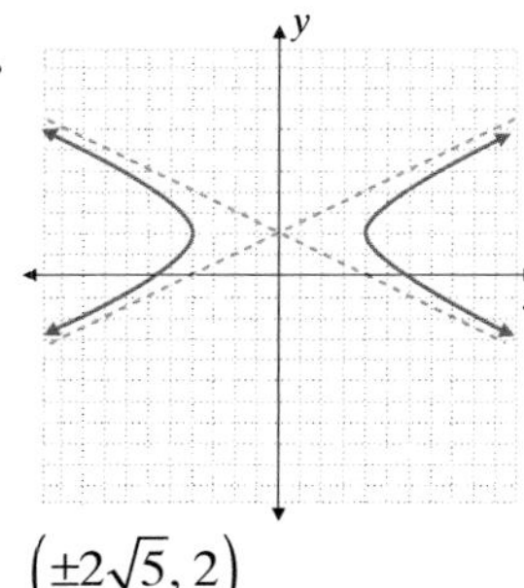

$\left(\pm 2\sqrt{5}, 2\right)$

9.

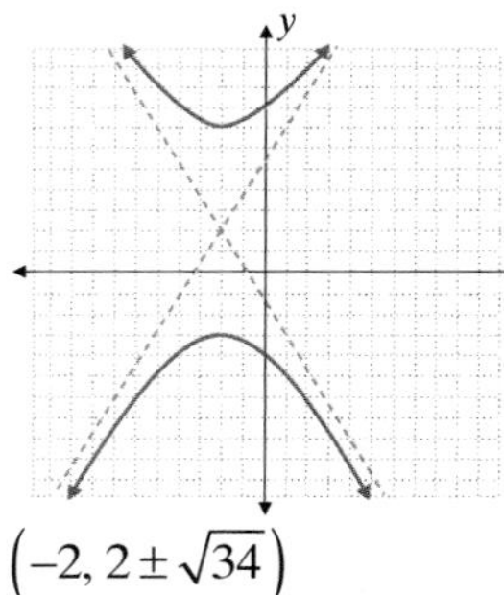

$\left(-2, 2 \pm \sqrt{34}\right)$

11.

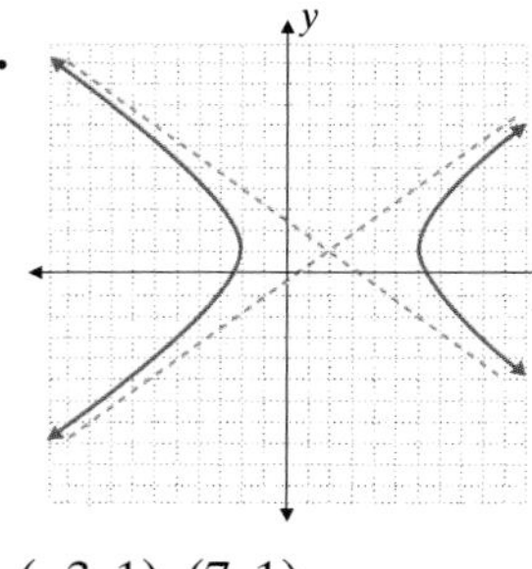

$(-3, 1), (7, 1)$

13. Center: $(-3, 2)$; Foci: $\left(-3 \pm \sqrt{13}, 2\right)$; Vertices: $(-1, 2), (-5, 2)$ **15.** Center: $(1, -4)$; Foci: $\left(1 \pm 2\sqrt{3}, -4\right)$; Vertices: $\left(1 \pm \sqrt{3}, -4\right)$ **17.** Center: $(-2, 1)$; Foci: $\left(-2 \pm \sqrt{30}, 1\right)$; Vertices: $(3, 1), (-7, 1)$

19. Center: $(-3, -1)$; Foci: $\left(-3 \pm 2\sqrt{3}, -1\right)$; Vertices: $(-1, -1), (-5, -1)$ **21.** Center: $(1, 0)$; Foci: $\left(1 \pm \dfrac{\sqrt{5}}{2}, 0\right)$; Vertices: $(2, 0), (0, 0)$ **23.** Center: $(8, 5)$; Foci: $\left(8 \pm 4\sqrt{5}, 5\right)$; Vertices: $(12, 5), (4, 5)$ **25.** $\dfrac{x^2}{4} - \dfrac{y^2}{5} = 1$

27. $y^2 - \dfrac{x^2}{1/4} = 1$ **29.** $\dfrac{x^2}{2/5} - \dfrac{(y-4)^2}{18/5} = 1$ **31.** $\dfrac{(x-6)^2}{9} - \dfrac{(y-5)^2}{7} = 1$ **33.** $\dfrac{(x+4)^2}{4} - \dfrac{(y-3)^2}{16} = 1$

35. $\dfrac{(x-1)^2}{9} - \dfrac{(y+1)^2}{4} = 1$ **37.** $\dfrac{(y+4)^2}{16} - \dfrac{(x-3)^2}{25} = 1$ **39.** a **41.** b **43.** g **45.** h

47.

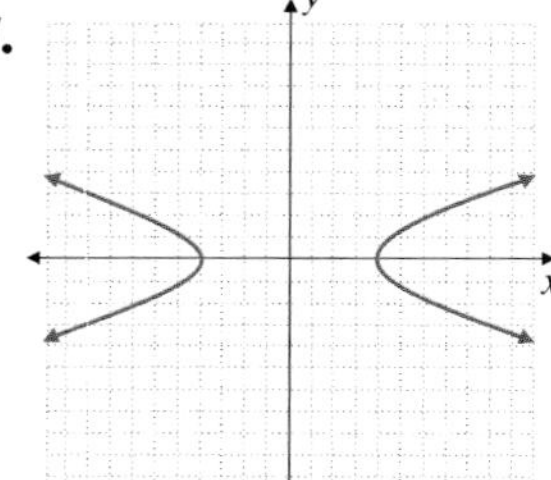

49.

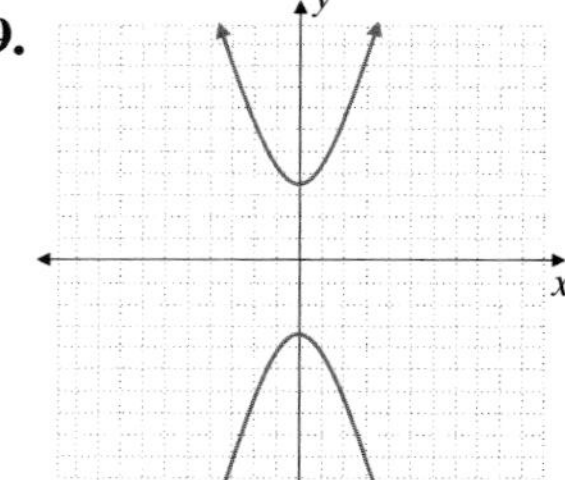

51.

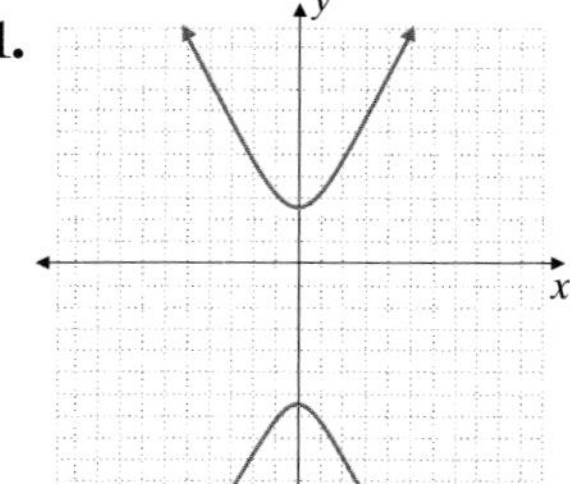

53. $\dfrac{x^2}{\left(6\times10^7\right)^2}-\dfrac{y^2}{\left(7.2\times10^7\right)^2}=1$

55.

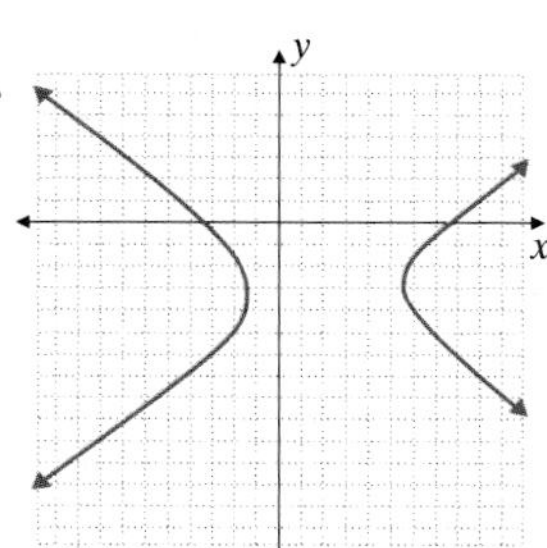

57.

section 9.4, pages 747-750

1. $\left(4\sqrt{3}+3,-4+3\sqrt{3}\right)$ **3.** $\left(\dfrac{-5\sqrt{2}}{16},\dfrac{3\sqrt{2}}{16}\right)$ **5.** $(-1.2097,-13.5476)$ **7.** $(3.3485, 1.5002)$

9. Ellipse **11.** Hyperbola **13.** Hyperbola **15.** Ellipse

17. Hyperbola

$\theta=\dfrac{\pi}{4},\ x'^2-y'^2=4$

19. Parabola

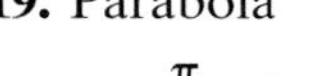

$\theta=\dfrac{\pi}{4},\ y'=-\sqrt{2}x'^2$

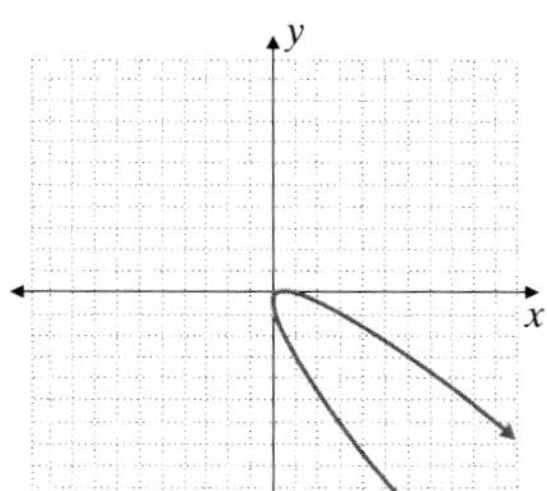

21. Ellipse

$\theta=\dfrac{\pi}{6},\ \dfrac{x'^2}{13}+\dfrac{y'^2}{25}=1$

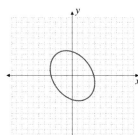

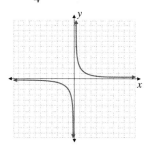

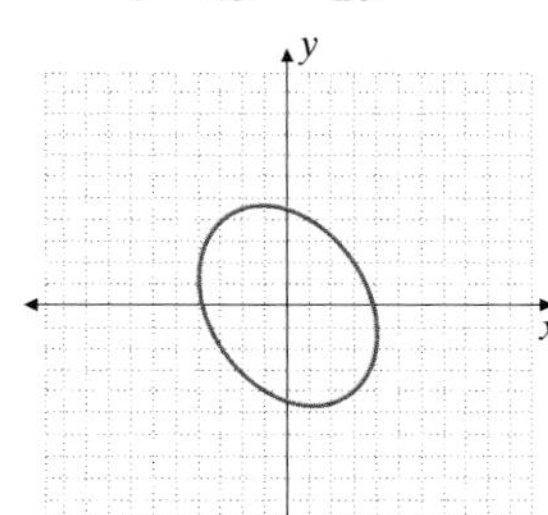

23. Ellipse

$\theta=\dfrac{\pi}{3},\ \dfrac{x'^2}{30}+\dfrac{y'^2}{46}=1$

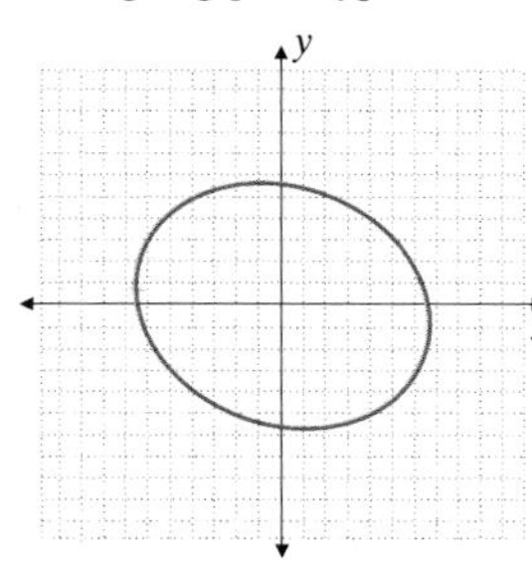

25.

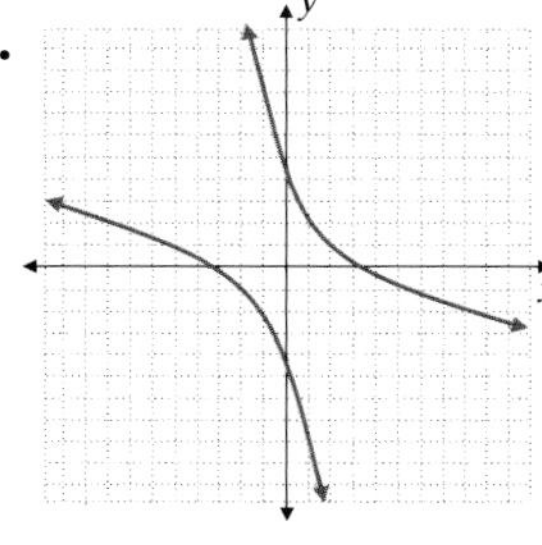

27.

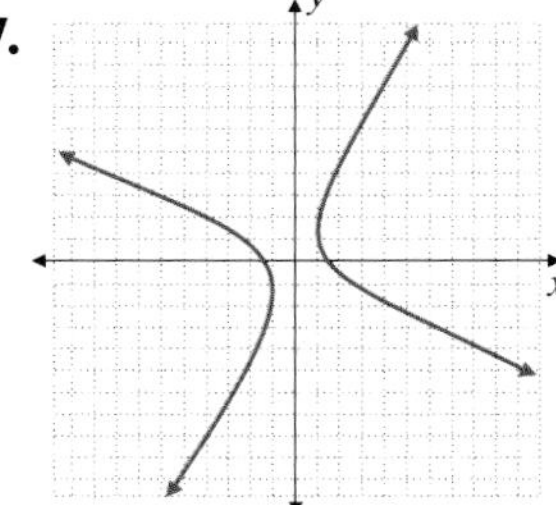

29.

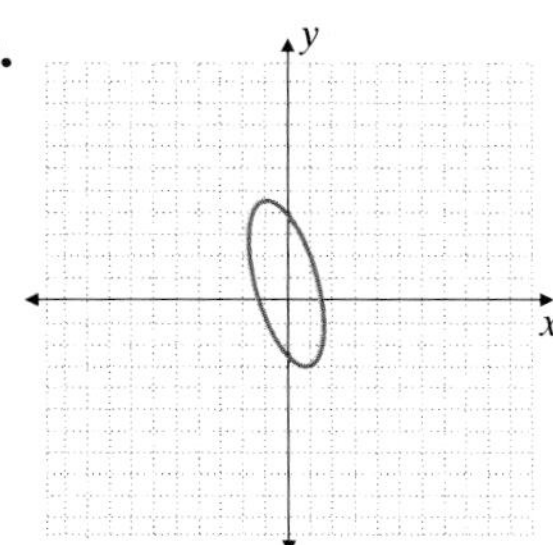

31.

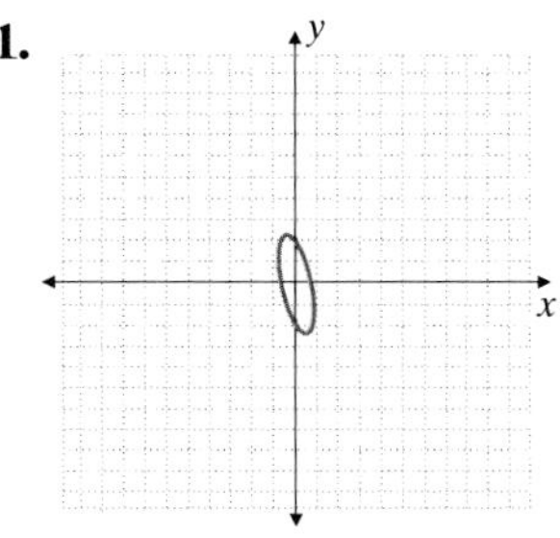

33. c **35.** a **37.** g **39.** e

41. The objective of the rotation of axes is to eliminate the $x'y'$-term. If your final equation contains an $x'y'$-term, you know that a mistake has occurred.

43. a. Use the rotation of axes procedure to obtain the equation $4(x')^2 + 16(y')^2 - 16 = 0$. Now we know $F = -16$ and $F' = -16$. We can plug these values in $F = F'$ and obtain $-16 = -16$, which is true.

b. Use the rotation of axes procedure to obtain the equation $4(x')^2 + 16(y')^2 - 16 = 0$. Now we know $A = 7, C = 13, A' = 4,$ and $C' = 16$. We can plug these values in $A + C = A' + C'$, and obtain $7 + 13 = 4 + 16$, or $20 = 20$, which is true.

c. Use the rotation of axes procedure to obtain the equation $4(x')^2 + 16(y')^2 - 16 = 0$. Now we know $A = 7,\ B = -6\sqrt{3},\ C = 13,\ A' = 4,\ B' = 0,$ and $C' = 16$. We can plug these values in $B^2 - 4AC = (B')^2 - 4A'C'$ and obtain $\left(-6\sqrt{3}\right)^2 - 4(7)(13) = (0)^2 - 4(4)(16)$, or $-256 = -256$, which is true.

section 9.5, pages 760-763

1. c **3.** f **5.** b **7.** Hyperbola, $y = \frac{7}{6}$ **9.** Ellipse, $x = -3$ **11.** Hyperbola, $x = \frac{1}{3}$

13. Ellipse, $x = 5$ **15.** Hyperbola, $x = -\frac{6}{5}$ **17.** Parabola, $y = \frac{3}{2}$ **19.** Hyperbola, $x = -\frac{4}{7}$ **21.** $r = \frac{2}{1 - \cos(\theta)}$

23. $r = \frac{3}{1 - 4\sin(\theta)}$ **25.** $r = \frac{3}{1 + \frac{1}{4}\cos(\theta)}$

27.

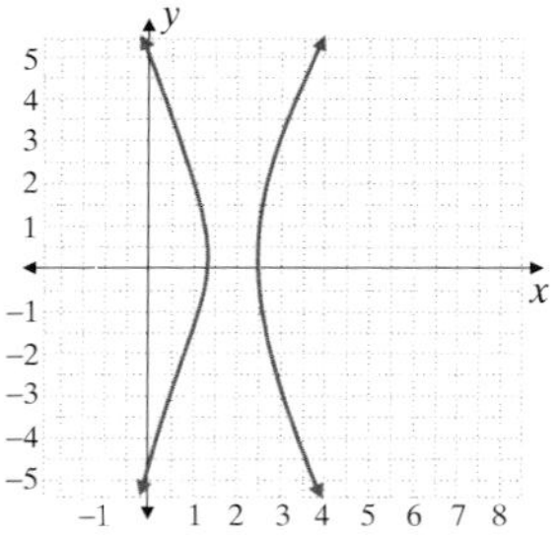

29.

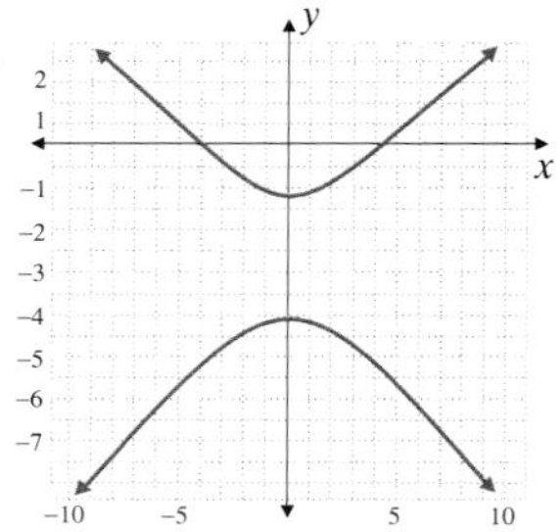

31.

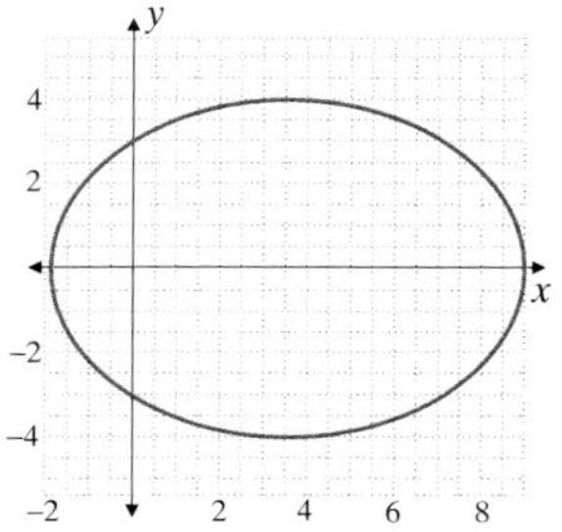

33.

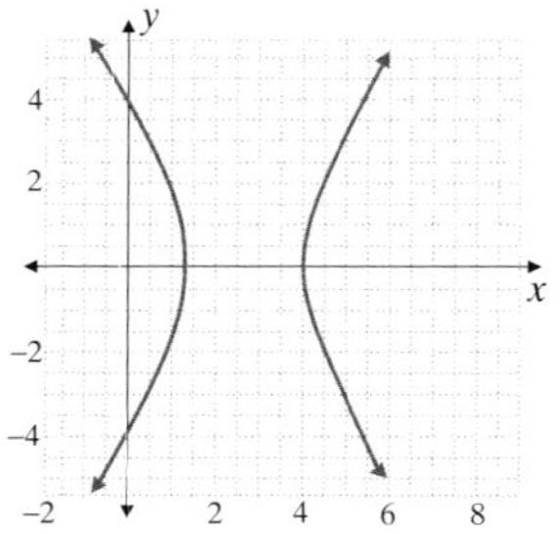

35.

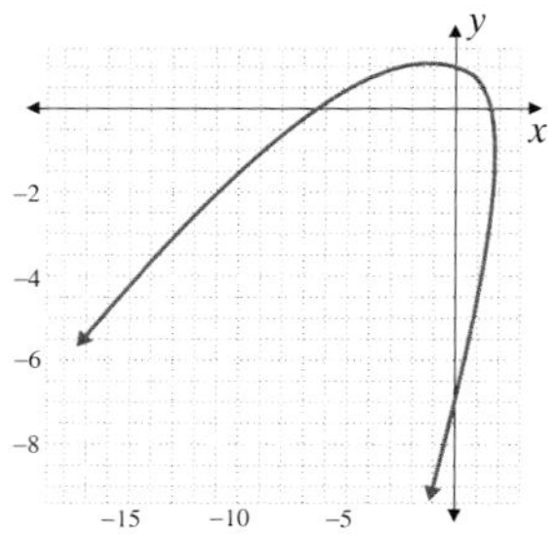

37.

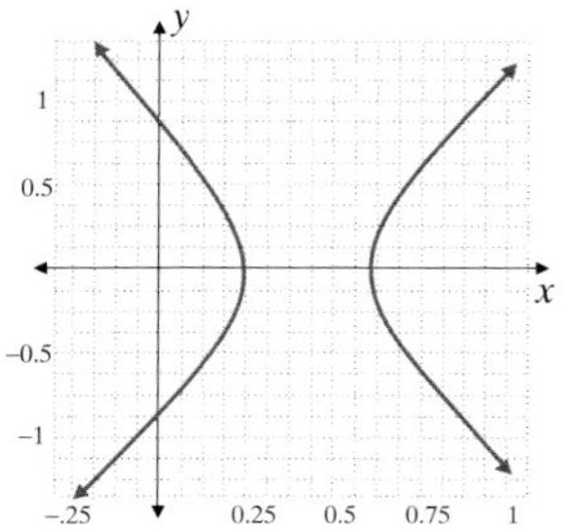

39.

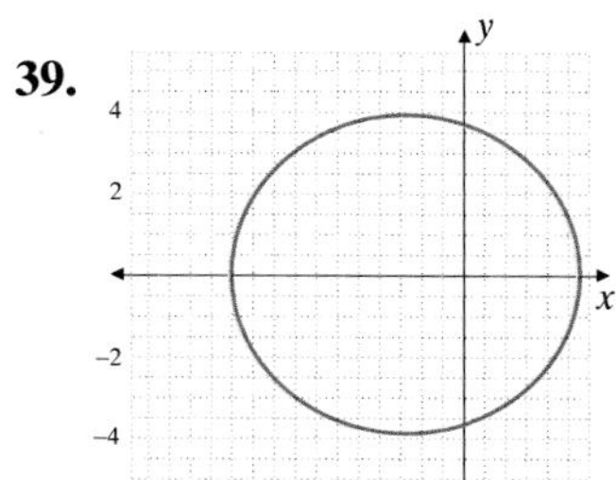

41.

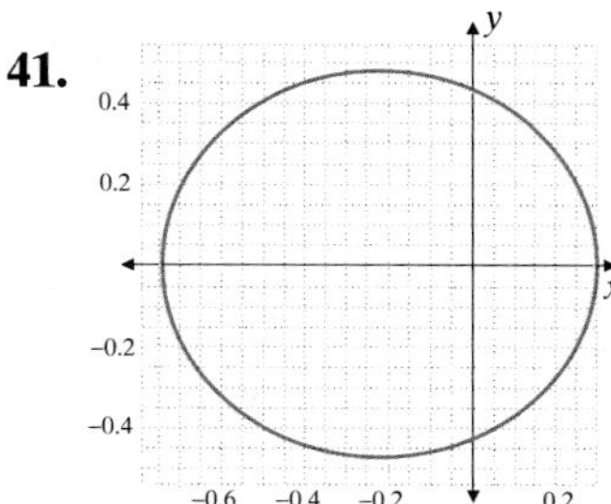

43.

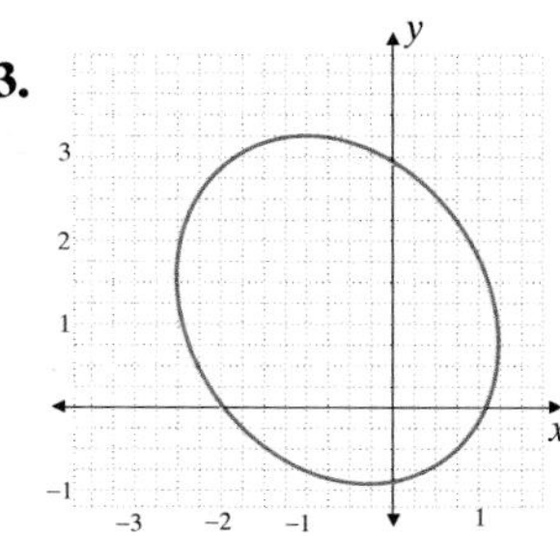

45.

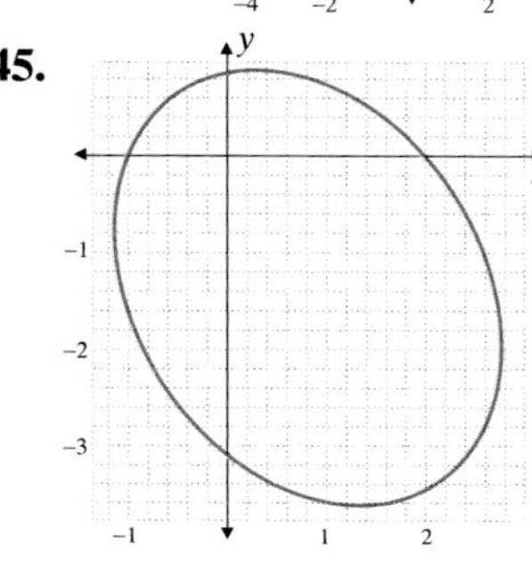

47. Answers will vary

chapter test, pages 769-772

1. $\left(-1, 2 \pm \sqrt{7}\right)$

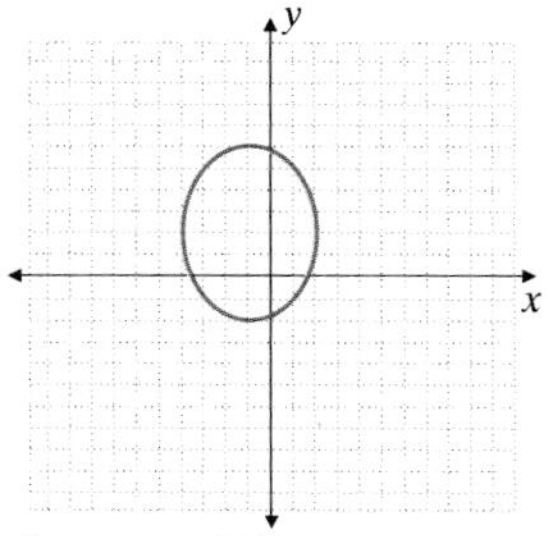

3. $\dfrac{(x+1)^2}{9} + \dfrac{(y-4)^2}{16} = 1$

5. $\dfrac{4(x-2)^2}{9} + \dfrac{4(y+1)^2}{9} = 1$

7. $(-6, -1), x = 0$

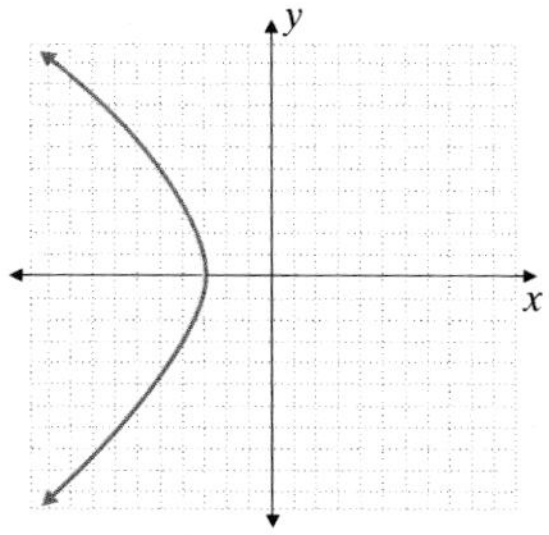

9. $(x+2)^2 = 4(y-3)$

11. $(y+1)^2 = 2\left(x - \dfrac{5}{2}\right)$

13. $(2, -2 \pm 5)$

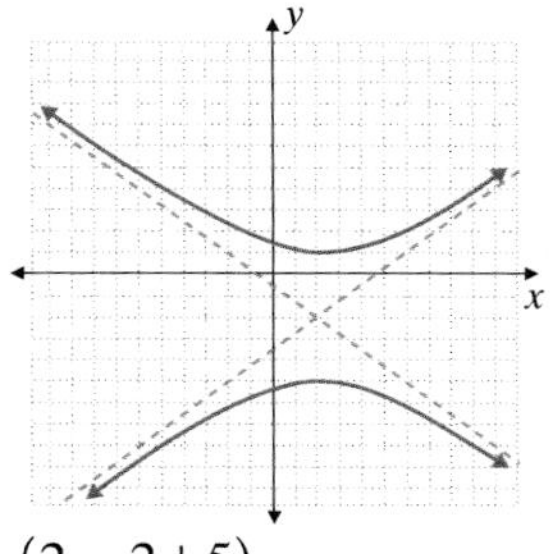

15. $\dfrac{(x-1)^2}{9} - \dfrac{(y+1)^2}{7} = 1$

17. $\dfrac{(y-3)^2}{9} - \dfrac{(x+1)^2}{16} = 1$

19. $\left(\dfrac{-\sqrt{2}}{2}, \dfrac{15\sqrt{2}}{2}\right)$

21. $(-4.3395, -9.0299)$

23. $(3.4823, -32.2999)$

25. Hyperbola, $\theta = \frac{\pi}{4}, x'^2 - y'^2 = 12$

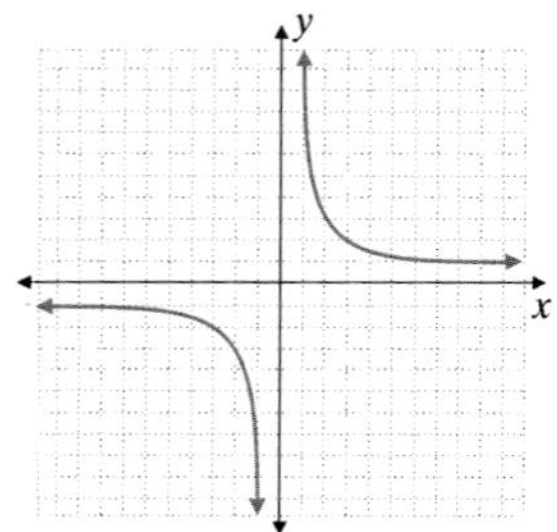

27. Ellipse, $\theta = \frac{\pi}{3}, \frac{\left(x' - \frac{\sqrt{3}}{52}\right)^2}{9} + \frac{\left(y' - \frac{1}{36}\right)^2}{13} = \frac{5}{27{,}378}$

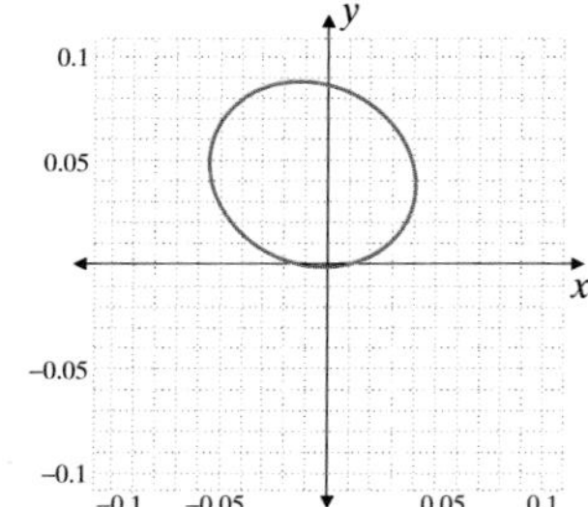

29. Parabola, $\theta = \frac{\pi}{4}, y' = \sqrt{2}x'^2$

31. Hyperbola, $y = 4$

33. Ellipse, $y = \frac{1}{2}$

35. Parabola, $y = \frac{7}{4}$

37. Ellipse, $x = \frac{7}{2}$

39. $r = \frac{12}{1 + 4\sin\theta}$

41. $r = \frac{4}{1 + \frac{1}{4}\cos\theta}$

43. $r = \frac{3}{1 + 9\cos\theta}$

chapter ten

section 10.1, pages 786-790

1. $(-5, 2)$ **3.** $(5, 3)$ **5.** $\varnothing$ **7.** $\left\{\left(\frac{y-3}{2}, y\right) \middle| y \in \mathbb{R}\right\}$ **9.** $(-1, 7)$ **11.** $(3, 11)$

13. $\{(x, 4x+1) \mid x \in \mathbb{R}\}$ **15.** $(2, 19)$ **17.** $(9, 3)$ **19.** $(5, 2)$ **21.** $(2, 0)$

23. $(1, 0.5)$ **25.** $(3, 2)$ **27.** $(8, 13)$ **29.** $(1, -1)$ **31.** $(5, 1)$ **33.** $(14, -16)$

35. $(-4, 1, 3)$ **37.** $(2, 4, 0)$ **39.** $(-7, 9, 4)$ **41.** $(-5, 1)$ **43.** $(5, 6)$ **45.** $\{(-y-2, y) \mid y \in \mathbb{R}\}$

47. $(4, 3)$ **49.** $(2, 3)$ **51.** $\varnothing$ **53.** $(-1, 1)$ **55.** $(1, 3)$ **57.** $(2, 1)$

59. $(3, 5)$ **61.** $(-3, -1)$ **63.** $(3, 0)$ **65.** $(7, -7)$ **67.** $(5, 2, 3)$ **69.** $(1, 2, 3)$

71. $(2, -1, 3)$ **73.** $(-1, 3, 0)$ **75.** $(2, 2, -1)$ **77.** $\left\{\left(\frac{y-z+2}{3}, y, z\right) \middle| y \in \mathbb{R}, z \in \mathbb{R}\right\}$ **79.** $\varnothing$

81. $\{(1, y, 0) \mid y \in \mathbb{R}\}$ **83.** $(9, 1, 1)$ **85.** $(3, 1, -2)$ **87.** $(4, 5, 5)$ **89.** $\left(\frac{49}{3}, -\frac{16}{3}, \frac{5}{4}\right)$

91. $(0, 3, 2)$ **93.** $(5, 2, 3)$ **95.** $(3, 7, -4)$ **97.** $(-2, 3, 4)$ **99.** $(6, 8, 7)$ **101.** $a = -\frac{1}{2}, b = -\frac{3}{2}, c = 5$

103. \$7000 at 8% and \$5000 at 15% **105.** 40 ounces of 12%, 20 ounces of 30%

107. Bob is 51 **109.** \$800 **111.** 50 tickets **113.** \$3000 in the CD, \$1000 in bonds, \$6000 in stocks

section 10.2, pages 799-803

1. a. 3×2 **b.** -1 **c.** none **3. a.** 5×2 **b.** none **c.** 10 **5. a.** 3×4 **b.** none **c.** 286

7. a. 3×2 **b.** 1 **c.** none **9. a.** 2×5 **b.** 5 **c.** 2 **11.** $\left[\begin{array}{ccc|c} \frac{-3}{2} & -1 & 0 & -1 \\ 2 & 2 & 3 & 0 \\ 0 & -1 & 6 & 0 \end{array}\right]$

13. $\left[\begin{array}{ccc|c} \frac{12}{5} & \frac{1}{2} & -\frac{3}{2} & \frac{1}{5} \\ 1 & 0 & 3 & 1 \\ 5 & 2 & 1 & -2 \end{array}\right]$ **15.** $\left[\begin{array}{ccc|c} \frac{3}{2} & 2 & -3 & 6 \\ 3 & -6 & 27 & 0 \\ 2 & 6 & 1 & 3 \end{array}\right]$ **17.** $\left[\begin{array}{ccc|c} -3 & 1 & -2 & -4 \\ \frac{1}{2} & -4 & -1 & 1 \\ 0 & -3 & 3 & 1 \end{array}\right]$

19. $\left[\begin{array}{ccc|c} \frac{1}{2} & -14 & \frac{-1}{4} & -8 \\ \frac{1}{5} & \frac{-7}{6} & \frac{1}{4} & -3 \\ 5 & -5 & \frac{8}{3} & -5 \end{array}\right]$ **21.** $\left[\begin{array}{cc|c} 2 & -5 & 3 \\ 0 & -7 & 5 \end{array}\right]$ **23.** $\left[\begin{array}{cc|c} 1 & 3 & -2 \\ 9 & -2 & 7 \end{array}\right]$ **25.** $\left[\begin{array}{cc|c} 8 & -2 & -4 \\ -6 & 2 & -14 \end{array}\right]$

27. $\left[\begin{array}{cc|c} 4 & 12 & -6 \\ 9 & 9 & 6 \end{array}\right]$ **29.** $\left[\begin{array}{cc|c} 4 & -1 & 5 \\ -6 & 2 & 0 \end{array}\right]$ **31.** $\left[\begin{array}{ccc|c} 18 & -6 & 15 & 42 \\ -7 & 19 & 2 & 3 \\ -4.5 & 5.5 & -2 & 3.5 \end{array}\right]$

33. $\left[\begin{array}{ccc|c} 5 & 18 & 22 & 5 \\ 32 & -9 & -27 & -23 \\ -9 & 21 & 12 & 9 \end{array}\right]$ **35.** $\left[\begin{array}{ccc|c} 0 & 1 & -9 & -3 \\ 1 & 1 & 3 & 4 \\ 0 & 0 & 0 & 0 \end{array}\right]$ **37.** $\left[\begin{array}{cc|c} -1 & 4 & -3 \\ 1 & -6 & \frac{5}{2} \end{array}\right]$ **39.** $\left[\begin{array}{ccc|c} 1 & 5 & -9 & 11 \\ 0 & -1 & 8 & -7 \\ 0 & -17 & 41 & 1 \end{array}\right]$

41. $(3, -1)$ **43.** $(1, 3)$ **45.** $(-7, 3)$ **47.** $\varnothing$ **49.** $(3, 2)$ **51.** $\{(-2y-4, y) \mid y \in \mathbb{R}\}$
53. $\varnothing$ **55.** $(4, 0, 3)$ **57.** $(15, -21, 8)$ **59.** $(3, -5)$ **61.** $\{(x, -3x-2) \mid x \in \mathbb{R}\}$
63. $(-4, 1)$ **65.** $(6, 4)$ **67.** $(-11, -5)$ **69.** $(7, 3, 3)$ **71.** $(2, 2, -1)$
73. $\{(1, y, 0) \mid y \in \mathbb{R}\}$ **75.** $(3, -2, 3)$ **77.** $(2, 3, 4)$ **79.** $(9, -19, 7)$
81. $(1, -2, -1, 3)$

section 10.3, pages 815-817

1. 11 **3.** 15 **5.** $ab - x^2$ **7.** 8 **9.** -10 **11.** -39 **13.** $\{-2, 3\}$ **15.** $\{-5, 1\}$ **17.** $\{-5, -4\}$
19. $\{-6, 4\}$ **21.** $\{2, 5\}$ **23.** 3 **25.** -9 **27.** 2 **29.** -2 **31.** 159 **33.** 78
35. -254 **37.** 404 **39.** 4 **41.** 120 **43.** x^4 **45.** 12 **47.** x^8 **49.** $-.051$
51. 12.595 **53.** 21,334 **55.** $(76, -53)$ **57.** $\{(-y-2, y) \mid y \in \mathbb{R}\}$ **59.** $\varnothing$ **61.** $(1647, 2071)$

63. $\varnothing$ **65.** $\{(-3z-5, -6z-10, z) \mid z \in \mathbb{R}\}$ **67.** $\left\{\left(\frac{-5y-z-5}{2}, \frac{-5y+3z-19}{2}, y, z\right) \middle| y \in \mathbb{R}, z \in \mathbb{R}\right\}$

69. $\{(w, 5w-9, 2w+21, -w+8) \mid w \in \mathbb{R}\}$ **71.** $(-3, -1, 0, -4)$ **73.** $(5, -3, 3)$

ten

section 10.4, pages 829-831

1. $\begin{bmatrix} 5 & -1 \\ 0 & 0 \\ 2 & 13 \end{bmatrix}$ **3.** $\begin{bmatrix} 6 & -3 \\ 18 & 30 \\ -9 & 21 \end{bmatrix}$ **5.** not possible **7.** $\begin{bmatrix} 14 & -14 \\ 8 & 0 \\ -4 & 14 \end{bmatrix}$ **9.** $\begin{bmatrix} -7 & 5 \\ 3 & 10 \\ -3 & -8 \end{bmatrix}$ **11.** not possible

13. $a = 2, b = -2, c = -1$ **15.** not possible **17.** $x = 10, y = 5$ **19.** $x = 3, y = 1$ **21.** not possible

23. $[\,24 \;\; -5\,]$ **25.** $[\,35 \;\; 18\,]$ **27.** not possible **29.** not possible **31.** $[\,-26 \;\; -12\,]$ **33.** not possible

35. $\begin{bmatrix} 21 & -24 \\ 6 & -3 \end{bmatrix}$ **37.** $\begin{bmatrix} 23 & 7 \\ 1 & -4 \end{bmatrix}$ **39.** not possible **41.** In February: 375 for store *A*, 625 for store *B;* In March: 463 for store *A*, 537 for store *B*

43. Answers will vary

section 10.5, pages 843-845

1. $\begin{bmatrix} 14 & -5 \\ 1 & 9 \end{bmatrix}\begin{bmatrix} x \\ y \end{bmatrix} = \begin{bmatrix} 7 \\ 2 \end{bmatrix}$ **3.** $\begin{bmatrix} \frac{3}{5} & -\frac{8}{5} \\ 0 & 1 \end{bmatrix}\begin{bmatrix} x \\ y \end{bmatrix} = \begin{bmatrix} 2 \\ 2 \end{bmatrix}$ **5.** $\begin{bmatrix} 4 & -3 \\ 2 & -4 \end{bmatrix}\begin{bmatrix} x \\ y \end{bmatrix} = \begin{bmatrix} -9 \\ 13 \end{bmatrix}$

7. $\begin{bmatrix} 1 & 2 \\ 9 & -3 \end{bmatrix}\begin{bmatrix} x \\ y \end{bmatrix} = \begin{bmatrix} -6 \\ -14 \end{bmatrix}$ **9.** $\begin{bmatrix} 3 & -7 & 1 \\ 1 & -1 & 0 \\ 0 & 8 & 5 \end{bmatrix}\begin{bmatrix} x_1 \\ x_2 \\ x_3 \end{bmatrix} = \begin{bmatrix} -4 \\ 2 \\ -3 \end{bmatrix}$ **11.** $\begin{bmatrix} 2 & -1 & 3 \\ -1 & 1 & 0 \\ 4 & -5 & 1 \end{bmatrix}\begin{bmatrix} x \\ y \\ z \end{bmatrix} = \begin{bmatrix} 0 \\ 17 \\ -2 \end{bmatrix}$ **13.** $\begin{bmatrix} -\frac{1}{20} & -\frac{1}{5} \\ \frac{1}{4} & 0 \end{bmatrix}$

15. $\begin{bmatrix} -5 & -4 \\ 4 & 3 \end{bmatrix}$ **17.** $\begin{bmatrix} -5 & 0 \\ 2 & 2 \end{bmatrix}$ **19.** not invertible **21.** $\begin{bmatrix} 2 & 1 & -4 \\ -4 & -2 & -3 \\ -1 & -1 & -4 \end{bmatrix}$

23. $\begin{bmatrix} -1 & 2 & -1 \\ 0 & -1 & 1 \\ 0 & -4 & 3 \end{bmatrix}$ **25.** $\begin{bmatrix} -1 & -2 & 1 \\ -2 & 1 & -3 \\ 1 & 2 & 0 \end{bmatrix}$ **27.** $\begin{bmatrix} -2 & 1 & 1 \\ 2 & 0 & -1 \\ -1 & 0 & 1 \end{bmatrix}$ **29.** $\begin{bmatrix} 2 & -1 & 2 \\ 0 & 1 & -1 \\ -3 & -2 & -4 \end{bmatrix}$

31. $\left(-2, -\frac{5}{2}\right)$ **33.** $\left\{\left(\frac{3y-1}{2}, y\right) \middle| y \in \mathbb{R}\right\}$ **35.** $(-2, 0)$ **37.** $(-13, 6, 19)$ **39.** $\varnothing$

41. $(-4, 5, -1)$ **43. a.** $(-13, 19, 23)$, **b.** $(0, 0, -1)$, **c.** $(1, -1, -1)$ **45. a.** $(1, -8, 7)$, **b.** $(3, 1, 1)$, **c.** $(4, 2, 0)$

section 10.6, pages 853-855

1. $\frac{A_1}{x-3} + \frac{A_2}{x+2}$ **3.** $\frac{A_1}{x+3} + \frac{A_2}{x+4} + \frac{A_3}{(x+4)^2}$ **5.** $\frac{A_1}{x+3} + \frac{A_2}{x-2} + \frac{A_3}{x+2}$ **7.** d **9.** h

11. a **13.** c **15.** $\frac{-1}{x} + \frac{2}{x-2} + \frac{2}{x+2}$

17. $\frac{5}{72(x-2)} - \frac{1}{32x} - \frac{7}{288(4+x)} - \frac{17+x}{72(8+x^2)}$ **19.** $\frac{10}{x-4} - \frac{5}{x-2}$ **21.** $\frac{3}{2(x+3)} + \frac{3}{14(x-1)} - \frac{12}{7(x+6)}$

ten

23. $\dfrac{1}{2(x-1)}-\dfrac{1}{2(x+1)}$ **25.** $\dfrac{1}{48(x-2)}-\dfrac{x}{80(x^2+4)}-\dfrac{3}{40(x^2+4)}-\dfrac{1}{120(x+4)}$ **27.** $\dfrac{5}{4(x-2)}-\dfrac{1}{4(x+2)}$

29. $\dfrac{1}{16(x+2)}+\dfrac{1}{16(x-2)}-\dfrac{x}{8(x^2+4)}$ **31.** $\dfrac{1}{24(x+6)}-\dfrac{1}{30(x+3)}-\dfrac{7}{40(x-2)}+\dfrac{1}{6(x-3)}$

33. $\dfrac{15}{16(x+3)}+\dfrac{1}{16(x-1)}-\dfrac{9}{4(x+3)^2}$ **35.** $\dfrac{1}{x+3}+\dfrac{1}{x-3}$ **37.** $\dfrac{1}{4(x-2)}+\dfrac{3}{4(x+6)}$

39. True

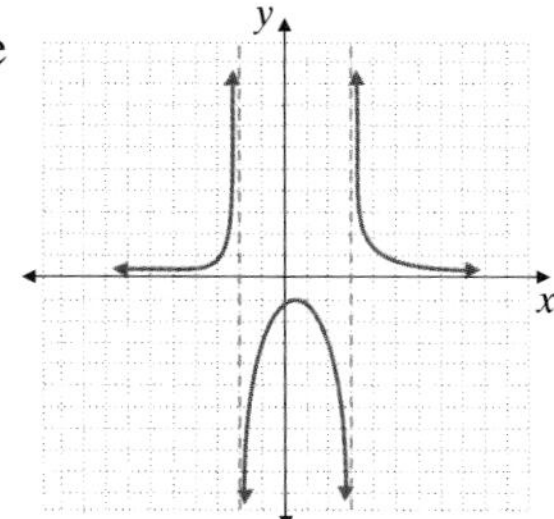

41. True

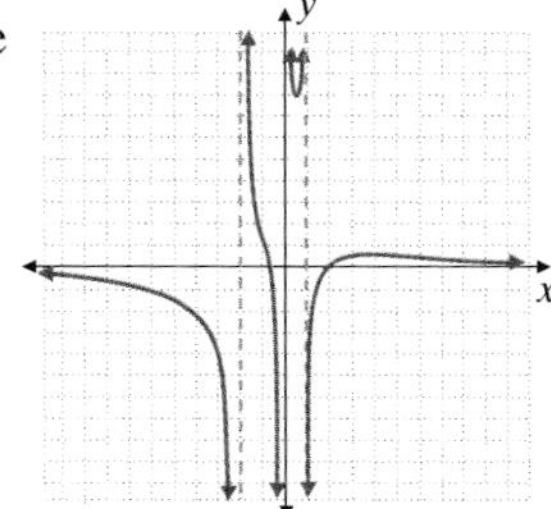

43. False

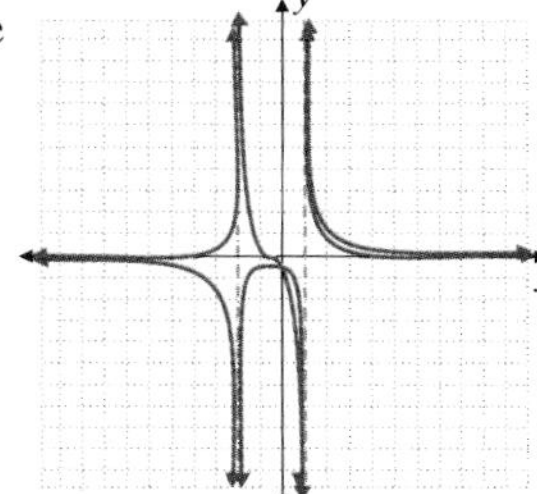

45. $\dfrac{1}{a}\left(\dfrac{1}{x}-\dfrac{1}{x+a}\right)$ **47.** $\dfrac{1}{2a}\left(\dfrac{1}{a+x}+\dfrac{1}{a-x}\right)$

49. $\dfrac{1}{a-1}\left(\dfrac{1}{x+1}-\dfrac{1}{x+a}\right)$

section 10.7, pages 862-866

1.

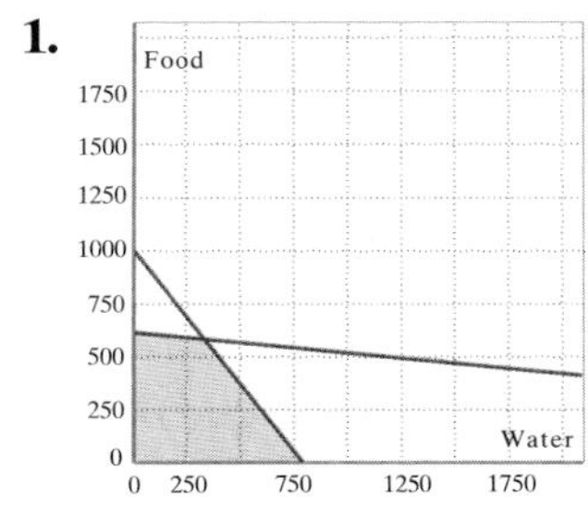

3.

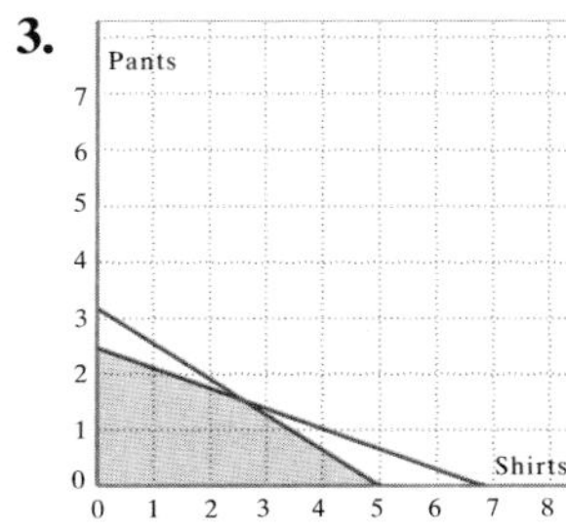

5. Min = 0 at (0, 0), Max = 21 at (0, 7)
7. Min = 0 at (0, 0), Max = 30 at (0, 6)
9. Min = 0 at (0, 0), Max = 48 at (0, 8)
11. Min = 0 at (0, 0), Max = 15 at (2.5, 0)
13. Min = 0 at (0, 0), Max = 50 at (3, 4)
15. Min = 0 at (0, 0), Max = 48 at (8, 0)

17. Min = 15.64 at (2.18, 0.91), Max = 133.33 at (0, 13.33)
19. Model X: 1000 units; Model Y: 500 units; Maximum Profit: $76,000
21. Ashley should buy 3 olive bundles and 2 cranberry bundles. The total cost to make the curtains is $70.
23. $20,000 should be placed in municipal bonds $5000 should be placed in Treasury bills.

section 10.8, pages 876-880

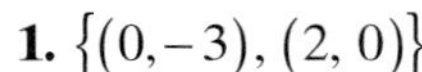

1. $\{(0,-3),(2,0)\}$

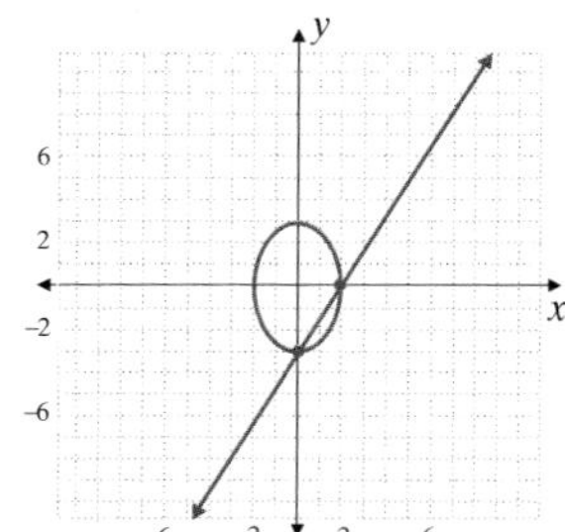

3. $\{(-1,\pm 1),(1,\pm 1)\}$

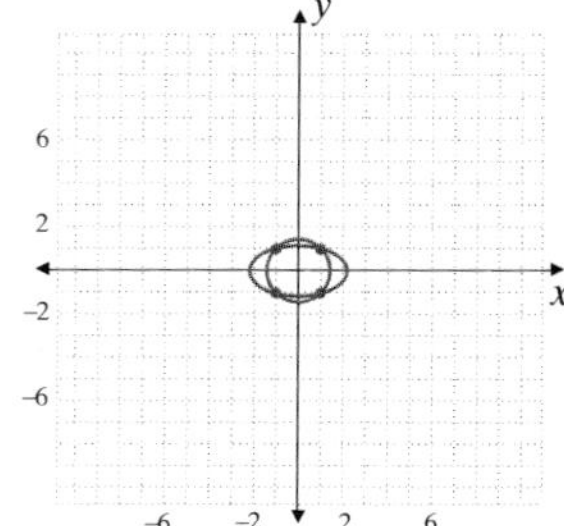

5. $\{(-1,1),(1,1)\}$

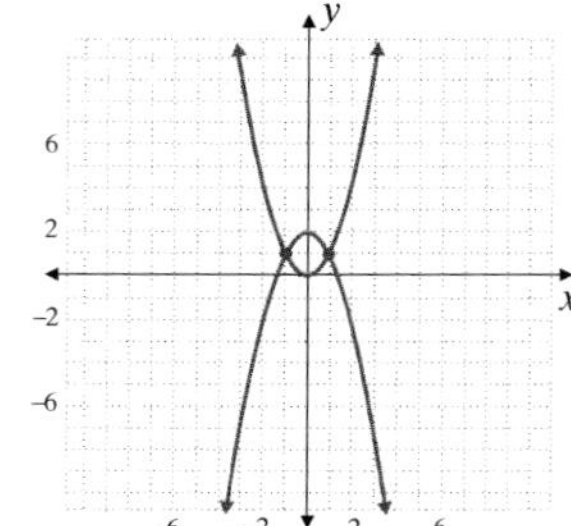

7. $\{(-1,2),(1,2)\}$

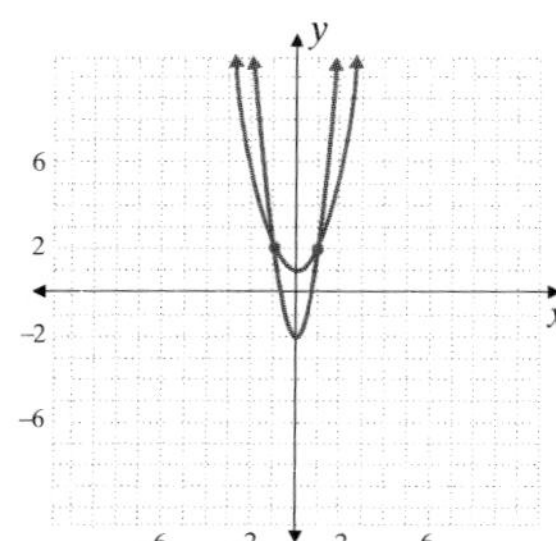

9. no solution

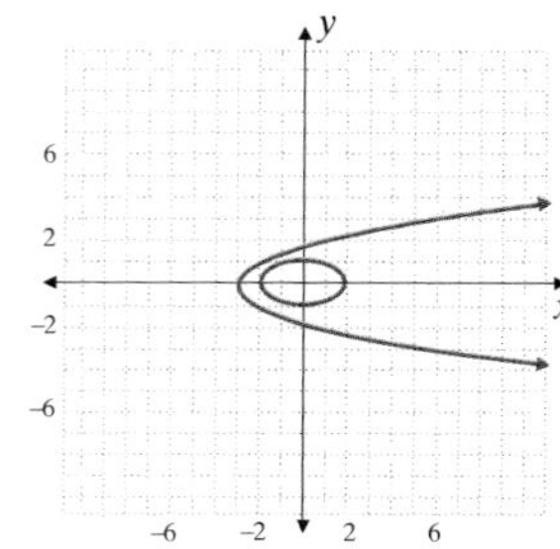

11. $\{(-3,1),(1,1)\}$

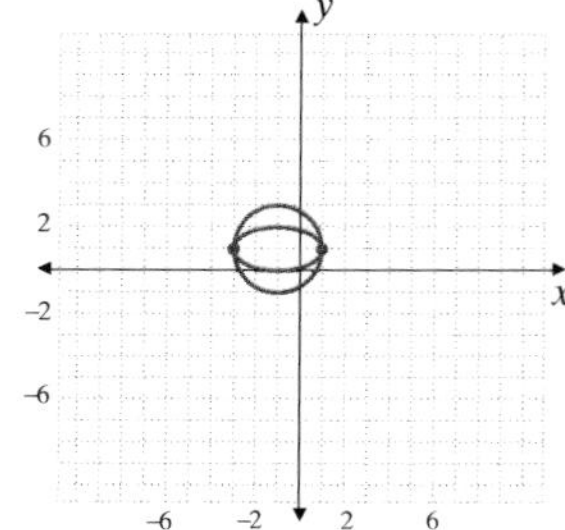

13. $\{(0,0)\}$

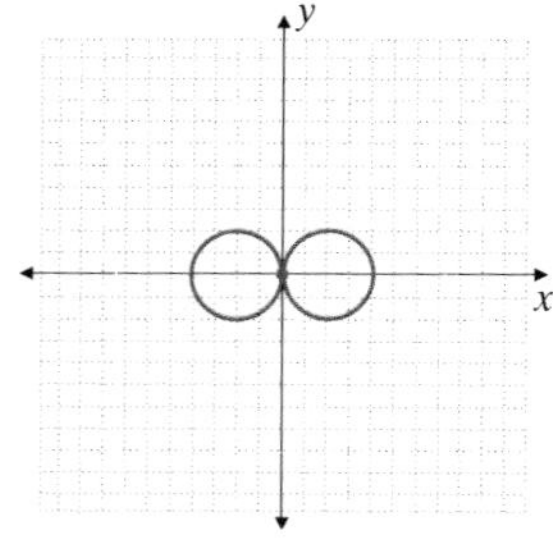

15. $\{(\pm 3,0)\}$

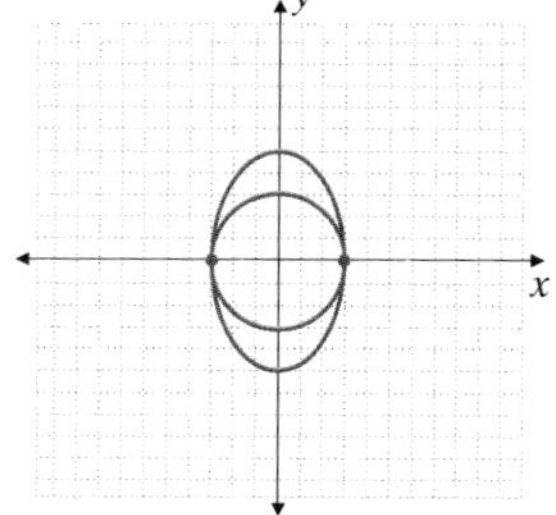

17. $\{(\pm 1,0)\}$

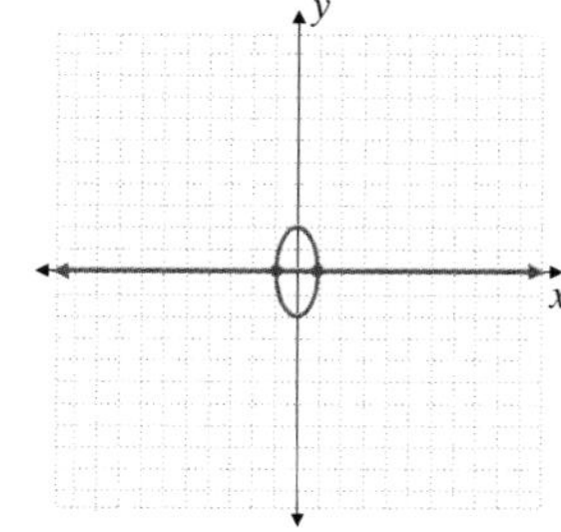

19. $\{(2,\pm 3),(-2,\pm 3)\}$

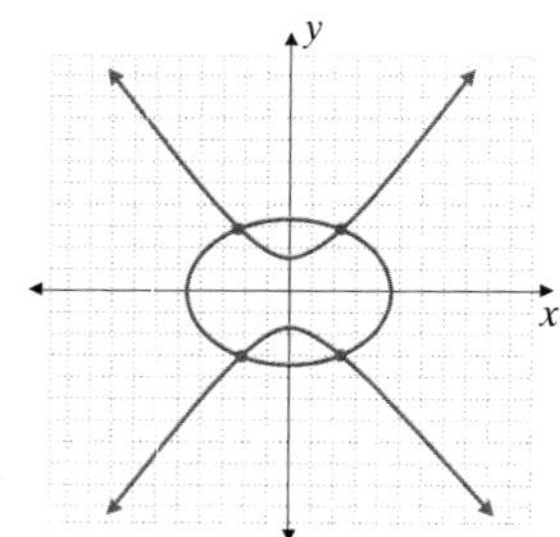

21. $\{(5,3)\}$

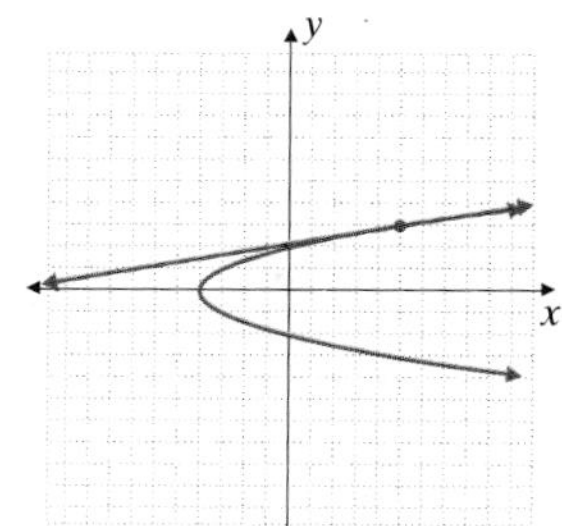

23. $\{(\pm 2,0),(0,-2)\}$

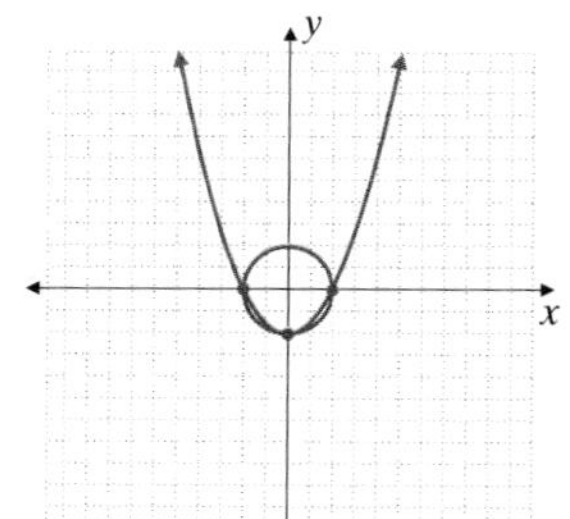

25. $\left\{\left(\pm i\sqrt{6}, -6\right), \left(\pm\sqrt{5}, 5\right)\right\}$ **27.** $\{(-2, 3)\}$ **29.** $\left\{(1, 4), (-1, -4), \left(2\sqrt{2}, \sqrt{2}\right), \left(-2\sqrt{2}, -\sqrt{2}\right)\right\}$ **31.** $\{(\pm 3i, -5), (0, 4)\}$

33. $\left\{\left(-\sqrt{3}, 0\right), \left(\sqrt{3}, 0\right)\right\}$ **35.** $\left\{(\pm i, -6), \left(\pm\sqrt{2}, 3\right)\right\}$ **37.** $\left\{(1, \pm 1), \left(-\frac{3}{2}, \pm\frac{\sqrt{14}}{2}\right)\right\}$ **39.** $\left\{\left(\frac{7}{2}, \frac{17}{4}\right)\right\}$

41. $\left\{\left(-6, \pm i\sqrt{15}\right), (2, \pm 1)\right\}$ **43.** $\left\{(0, 1), \left(-\frac{100}{101}, -\frac{99}{101}\right)\right\}$ **45.** $\left\{\left(-\frac{\sqrt{42}}{6}, \pm\frac{\sqrt{66}}{6}\right), \left(\frac{\sqrt{42}}{6}, \pm\frac{\sqrt{66}}{6}\right)\right\}$

47. $\{(\pm 2, 2), (\pm 2.544, 1.236), (\pm 1.572i, -3.236)\}$ **49.** $\left\{(3, 2), (2, 3), \left(\frac{-1 \pm i\sqrt{23}}{2}, \frac{12}{-1 \pm i\sqrt{23}}\right)\right\}$

51. $\left\{(0, -1), \left(\frac{\sqrt{6}}{3}, \pm\frac{2\sqrt{6}}{9} - 1\right), \left(-\frac{\sqrt{6}}{3}, \pm\frac{2\sqrt{6}}{9} - 1\right)\right\}$

53. $\left\{\left(\sqrt{2}+1, 2\sqrt{2}\right), \left(-\sqrt{2}+1, -2\sqrt{2}\right), \left(2\sqrt{2}+1, \sqrt{2}\right), \left(-2\sqrt{2}+1, -\sqrt{2}\right)\right\}$

55. $\left\{(4, 1), \left(1, 1+i\sqrt{3}\right)\right\}$ **57.** $\{(0, 4), (0, -3), (3, -2), (3, 3)\}$ **59.** $\{(4, 0)\}$

61. $\{(5, 4)\}$ **63.** $\left\{\left(3, \pm\frac{\sqrt{26}}{2}\right), \left(6, \pm\frac{7\sqrt{2}}{2}\right)\right\}$ **65.** a,

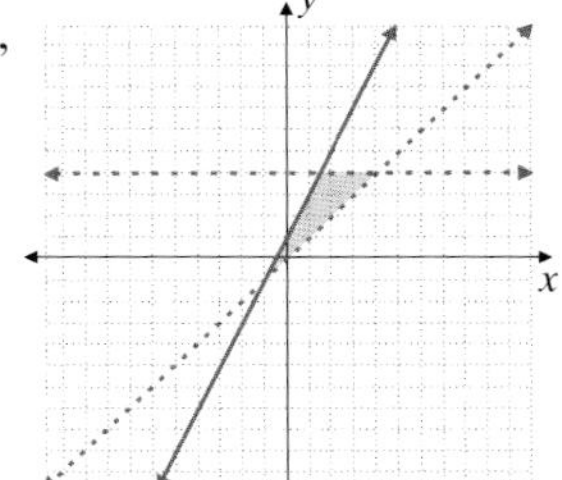

67. none

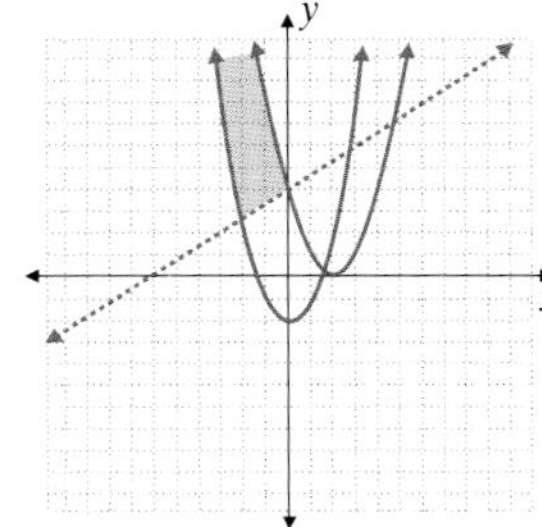

69.

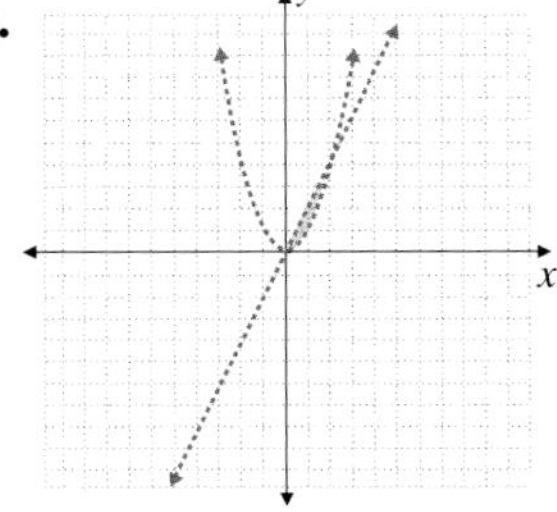

71.

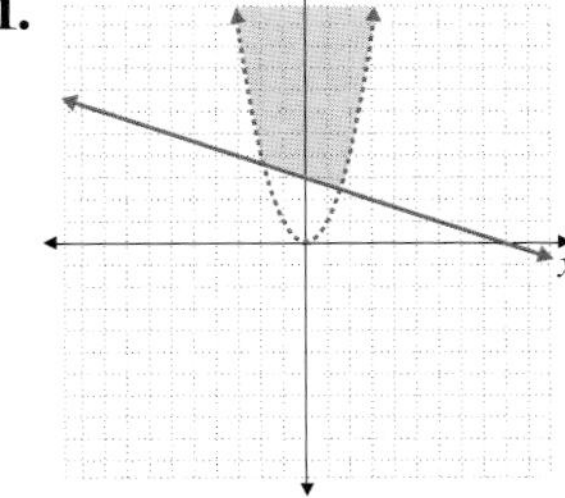

73.

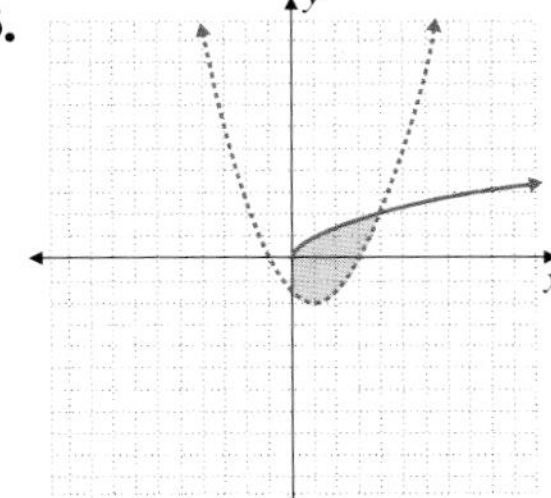

75.

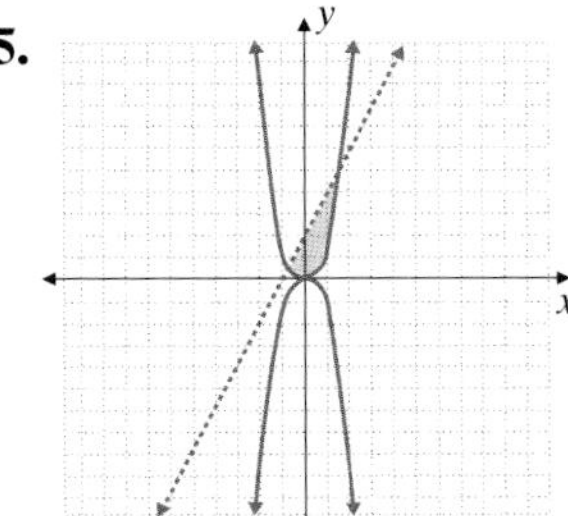

77.

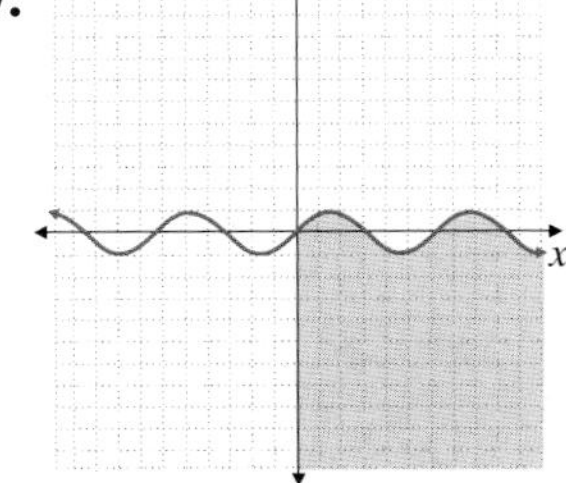

79. 8 and 11 **81.** $6\frac{2}{3} \times 14\frac{1}{4}$ m or $9\frac{1}{2} \times 10$ m **83.** 36 mph and 48 mph

ten

chapter ten test, pages 886-890

1. $\varnothing$ **3.** $(3, 0, 2)$ **5.** $\left[\begin{array}{ccc|c} 4 & 5 & -1 & 0 \\ 1 & 3 & 2 & 3 \\ 10 & -1 & -6 & 0 \end{array}\right]$ **7.** $\left[\begin{array}{cc|c} 2 & 3 & 5 \\ 0 & 5 & 12 \end{array}\right]$ **9.** 5 **11.** x^8 **13.** not possible

15. $\begin{bmatrix} -12 & 8 & 12 \\ -15 & 10 & 15 \\ -18 & 12 & 18 \end{bmatrix}$ **17.** $\begin{bmatrix} 1 & -1 & 2 \\ 2 & -3 & -1 \\ -3 & 0 & 6 \end{bmatrix}\begin{bmatrix} x_1 \\ x_2 \\ x_3 \end{bmatrix} = \begin{bmatrix} -4 \\ 1 \\ 5 \end{bmatrix}$ **19.** $\begin{bmatrix} 1 & -2 \\ -\frac{1}{2} & 2 \end{bmatrix}$ **21.** $\frac{A_1}{x} + \frac{A_2}{x^2} + \frac{A_3}{(3x-1)} + \frac{A_4}{(3x-1)^2}$

23. $\frac{A_1}{x-4} + \frac{A_2}{x-1}$ **25.** $-\frac{3}{2(2x-5)^2} + \frac{1}{2(2x-5)}$ **27.** $\frac{1}{x} + \frac{1-x}{1+x^2}$

29. $\frac{5}{x-2} - \frac{3}{x} - \frac{1}{x+2}$ **31.** $x \ge 6,\ y \ge 6,\ x \le 10,\ y \le 8,\ x + y \ge 15$ **33.** Min = 2 at $(0, 2)$, Max = 11 at $(4, 3)$

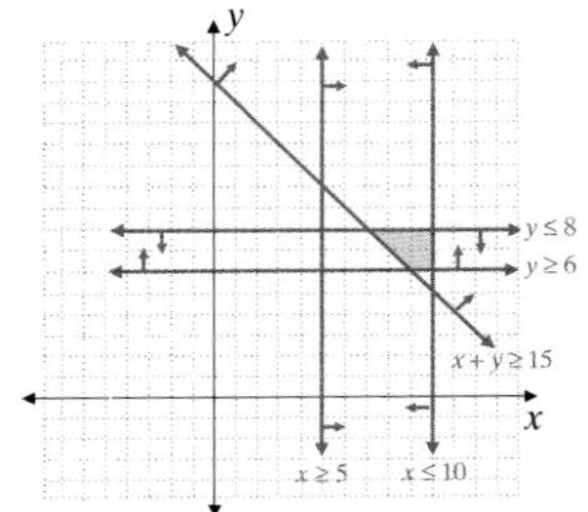

35. Min = 8 at $(0, 2)$, Max = 24 at $(3.43, 1.71)$ **37.** Min = 0 at $(0, 0)$, Max = 27.33 at $\left(\frac{22}{3}, \frac{19}{6}\right)$

39. 35 bionic arms, 15 bionic legs; Minimum cost: \$24,000 **41.** $\{(-2, \pm 2), (2, \pm 2)\}$

43. 9 and 16 **45.** None

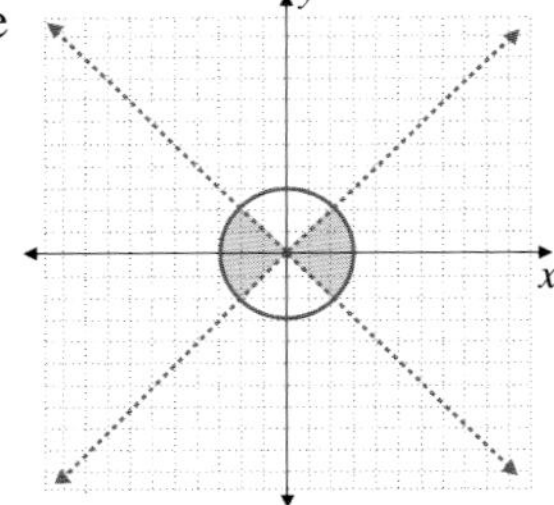

47.

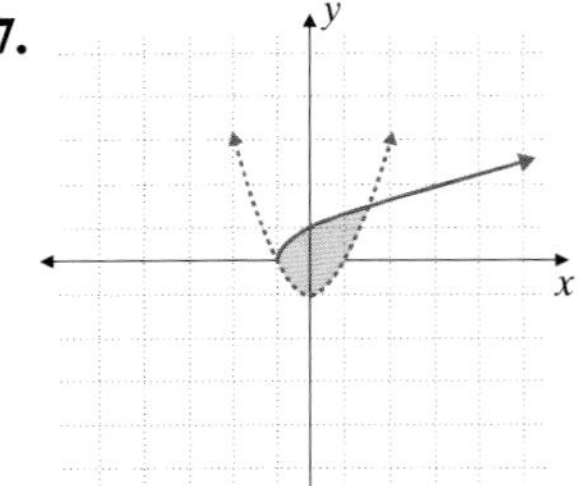

chapter eleven

section 11.1, pages 906-908

1. 4, 11, 18, 25, 32 **3.** –2, 4, –8, 16, –32 **5.** $-1, \frac{1}{4}, \frac{-1}{9}, \frac{1}{16}, \frac{-1}{25}$ **7.** $1, \frac{-1}{3}, \frac{1}{9}, \frac{-1}{27}, \frac{1}{81}$

9. $0, \frac{1}{9}, \frac{1}{4}, \frac{9}{25}, \frac{4}{9}$ **11.** 26, 7, 0, –1, –2 **13.** $-1, \sqrt{2}, -\sqrt{3}, 2, -\sqrt{5}$ **15.** 1, 5, 9, 13, 17 **17.** $\frac{1}{2}$, 1, 2, 4, 8

19. 3, 36, 729, 20736, 759375 **21.** $\frac{5}{4}, 2, \frac{5}{2}, \frac{20}{7}, \frac{25}{8}$ **23.** 1, 3, 6, 10, 15 **25.** undefined, 9, 4, $\frac{25}{9}, \frac{9}{4}$

27. $\frac{1}{3}, \frac{1}{2}, \frac{5}{9}, \frac{7}{12}, \frac{3}{5}$ **29.** 0, –1, –4, –9, –16 **31.** 2, 4, 16, 256, 65536 **33.** 1, 2, 6, 24, 120

35. $2, \sqrt{5}, \sqrt{6}, \sqrt{7}, 2\sqrt{2}$ **37.** $a_n = 7n - 2$ **39.** $a_1 = -1, a_n = -na_{n-1}, n \geq 2$ **41.** $a_n = \frac{1}{n^2+3}$ **43.** $a_n = 9n - 43$

45. $a_n = 2^{n-3}$ **47.** $a_n = \frac{n}{2^n}$ **49.** $a_n = 2n + 1$ **51.** $a_n = n^{-2}$ **53.** $a_n = 3^{n-3}$

55. –2 + 1 + 4 + 7 + 10 + 13 + 16 = 49 **57.** $\sum_{i=1}^{6} i^3 = 441$ **59.** 45 + 80 + 125 + 180 + 245 + 320 + 405 + 500 = 1900

61. –6 – 12 – 24 – 48 – 96 – 192 = –378 **63.** $\sum_{i=2}^{9} 3^i = 29520$ **65.** –3+9–27+81–243 = –183

67. 6 + 13 + 22 + 33 + 46 + 61 = 181 **69.** 8 + 8.08 + 8.16 + 8.24 = 32.48 **71.** $\sum_{i=10}^{100} \frac{3i}{10} = 1501.5$

73. $S_n = \frac{n}{4(n+4)}, S_{100} = \frac{25}{104}$ **75.** $S_n = 2^n - 1$, sum does not converge **77.** $S_n = \frac{n}{2(n+1)}, S_{49} = \frac{49}{100}$

79. $S_n = -\ln(n+1), S_{100} = -\ln 101$ **81.** 4, 7, 11, 18, 29 **83.** 10, 20, 30, 50, 80

85. 13, –17, –4, –21, –25 **87.** 1, –3, –3, 9, –27

section 11.2, pages 916-918

1. $a_n = 3n - 5$ **3.** $a_n = -2n + 9$ **5.** $a_n = 9n - 4$ **7.** $a_n = -6n + 9$ **9.** $a_n = 19n + 5$ **11.** $a_n = n + \frac{5}{2}$

13. $a_n = -\frac{19}{2}n + \frac{43}{2}$ **15.** $a_n = -2n + 1$ **17.** $a_n = -4n + 33$ **19.** 13 **21.** 2 **23.** 2 **25.** 5 **27.** 1

29. –2 **31.** 195 **33.** a_{73} **35.** 117 **37.** 26 **39.** –8 **41.** 13 **43.** 55 **45.** 14,350 **47.** 17,114 **49.** –1475

51. $-\frac{3219}{5}$ **53.** 902 **55.** –1316 **57.** 6 years **59.** 1620 pounds **61.** 20 rows

section 11.3, pages 932-934

1. $a_n = -3(2)^{n-1}$ **3.** $a_n = 2\left(\frac{-1}{3}\right)^{n-1}$ **5.** $a_n = \left(\frac{-1}{4}\right)^{n-1}$ **7.** $a_n = \left(\frac{1}{7}\right)^{n-1}$ **9.** $a_n = (-3)^{n-1}$

11. $a_n = 3\left(\frac{2}{3}\right)^{n-1}$ **13.** $a_n = 7(-2)^{n-1}$ **15.** $a_n = \frac{1}{16}(2)^{n-1}$ **17.** $a_n = \frac{39}{68}\left(\frac{4}{3}\right)^{n-1}$ **19.** $\frac{5}{16,384}$

21. $-2{,}147{,}483{,}648$ **23.** $-\frac{1023}{1024}$ **25.** $2{,}199{,}023{,}254{,}528$ **27.** 59,048 **29.** 44,287

31. $\frac{7448}{3125}=2.38336$ **33.** 4 **35.** $\frac{9}{17}$ **37.** does not converge **39.** 11

41. $\frac{164}{99}$ **43.** $-\frac{5}{9}$ **45.** $\frac{29}{999}$ **47.** 79.26 feet, 90 feet **49.** \$14,908.33 **51.** $\frac{2}{3}$

section 11.4, pages 940-944

1. $S_{k+1}=\frac{1}{3k+9}$ **3.** $S_{k+1}=\frac{(k+1)(k+2)(2k+3)}{4}$

5. Basis Step: $n=1$, $1=1$ and $\frac{1(1+1)}{2}=1$;

Induction Step: If $1+2+3+\cdots+k=\frac{k(k+1)}{2}$, then $(1+2+3+\cdots+k)+(k+1)=\frac{k(k+1)}{2}+(k+1)$

$=\frac{k^2+k+2k+2}{2}=\frac{(k+1)(k+2)}{2}$

7. Basis Step: $n=1$, $2(1)=2$ and $1(1+1)=2$;

Induction Step: If $2+4+6+\cdots+2k=k(k+1)$, then $(2+4+6+\cdots+2k)+2(k+1)=k^2+k+2k+2=(k+1)(k+2)$

9. Basis Step: $n=1$, $4^{1-1}=1$ and $\frac{4^1-1}{3}=1$;

Induction Step: If $4^0+4^1+4^2+\cdots+4^{k-1}=\frac{4^k-1}{3}$, then $4^0+4^1+4^2+\cdots+4^{k-1}+4^{k+1-1}=\frac{4^k-1}{3}+4^k$

$=\frac{4^k-1+3\cdot 4^k}{3}=\frac{4\cdot 4^k-1}{3}=\frac{4^{k+1}-1}{3}$

11. Basis Step: $n=1$, $\frac{1}{(3(1)-2)(3(1)+1)}=\frac{1}{4}$ and $\frac{1}{3(1)+1}=\frac{1}{4}$;

Induction Step: If $\frac{1}{1\cdot 4}+\frac{1}{4\cdot 7}+\frac{1}{7\cdot 10}+\cdots+\frac{1}{(3k-2)(3k+1)}=\frac{k}{3k+1}$, then $\frac{1}{1\cdot 4}+\frac{1}{4\cdot 7}+\frac{1}{7\cdot 10}+\cdots$

$+\frac{1}{(3k-2)(3k+1)}+\frac{1}{(3(k+1)-2)(3(k+1)+1)}=\left[\frac{1}{1\cdot 4}+\frac{1}{4\cdot 7}+\frac{1}{7\cdot 10}+\cdots+\frac{1}{(3k-2)(3k+1)}\right]+\frac{1}{(3k+1)(3k+4)}$

$=\frac{k}{3k+1}+\frac{1}{(3k+1)(3k+4)}=\frac{3k^2+4k+1}{(3k+1)(3k+4)}=\frac{(3k+1)(k+1)}{(3k+1)(3k+4)}=\frac{(k+1)}{(3(k+1)+1)}$

13. Basis Step: $n=1$, $5(1)=5$ and $\frac{5(1)(1+1)}{2}=5$;

Induction Step: If $5+10+15+\cdots+5k=\frac{5k(k+1)}{2}$, then $5+10+15+\cdots+5k+5(k+1)$

$=(5+10+15+\cdots+5k)+5k+5=\frac{5k(k+1)}{2}+5k+5=\frac{5k^2+15k+10}{2}=\frac{5(k+1)(k+2)}{2}=\frac{5(k+1)[(k+1)+1]}{2}$

eleven

15. Basis Step: $n=1$, $1+\frac{1}{1}=2$ and $1+1=2$

Induction Step: If $\left(1+\frac{1}{1}\right)\left(1+\frac{1}{2}\right)\left(1+\frac{1}{3}\right)\cdots\left(1+\frac{1}{k}\right)=k+1$, then $\left(1+\frac{1}{1}\right)\left(1+\frac{1}{2}\right)\left(1+\frac{1}{3}\right)\cdots\left(1+\frac{1}{k}\right)\left(1+\frac{1}{k+1}\right)$
$=(k+1)\left(1+\frac{1}{k+1}\right)=k+1+\frac{k+1}{k+1}=(k+1)+1$

17. Basis Step: $n=4$, $4!=24$ and $2^4=16$, so $4!>2^4$;

Induction Step: If $k!>2^k$, then $(k+1)!=k!(k+1)>2^k(k+1)\geq 2^k(4+1)>2^k\cdot 2=2^{k+1}$

19. Basis Step: $n=1$, $-(1+1)=-2$ and $-\frac{1}{2}\left(1^2+3(1)\right)=-2$;

Induction Step: If $-2-3-4-\cdots-(k+1)=-\frac{1}{2}\left(k^2+3k\right)$, then $-2-3-4-\cdots-(k+1)-((k+1)+1)$
$=-\frac{1}{2}\left(k^2+3k\right)-(k+2)=-\frac{k^2+3k+2k+4}{2}=-\frac{\left(k^2+2k+1\right)+(3k+3)}{2}=-\frac{1}{2}\left((k+1)^2+3(k+1)\right)$

21. Basis Step: $n=1$, since $2^1=2$, $2^1>1$; Induction Step: If $2^k>k$, then $2^{k+1}=2^1\cdot 2^k>2k=k+k\geq k+1$

23. Basis Step: $n=1$, $1(1+1)=2$ and $\frac{1(1+1)(1+2)}{3}=2$;

Induction Step: If $1\cdot 2+2\cdot 3+3\cdot 4+\cdots+k(k+1)=\frac{k(k+1)(k+2)}{3}$, then $\left[1\cdot 2+2\cdot 3+3\cdot 4+\cdots+k(k+1)\right]$
$+(k+1)((k+1)+1)=\frac{k(k+1)(k+2)}{3}+(k+1)(k+2)=\frac{k(k+1)(k+2)+3(k+1)(k+2)}{3}=\frac{(k+1)((k+1)+1)((k+1)+2)}{3}$

25. Basis Step: $n=5$, $2^5=32$ and $4\cdot 5=20$, so $2^5>4\cdot 5$;

Induction Step: If $2^k>4k$, then $2^{k+1}=2\cdot 2^k=2^k+2^k>4k+4k>4k+4=4(k+1)$

27. Basis Step: $n=1$, $1^5=1$ and $\frac{1^2(1+1)^2\left(2(1)^2+2(1)-1\right)}{12}=1$;

Induction Step: If $1^5+2^5+3^5+\cdots+k^5=\frac{k^2(k+1)^2\left(2k^2+2k-1\right)}{12}$, then $\left(1^5+2^5+3^5+\cdots+k^5\right)+(k+1)^5$
$=\frac{k^2(k+1)^2\left(2k^2+2k-1\right)}{12}+(k+1)^5=\frac{(k+1)^2\left[k^2\left(2k^2+2k-1\right)+12(k+1)^3\right]}{12}=\frac{(k+1)^2\left(2k^4+14k^3+34k^2+36k+12\right)}{12}$
$=\frac{(k+1)^2(k+2)^2\left(2k^2+6k+3\right)}{12}=\frac{(k+1)^2(k+2)^2\left(2(k+1)^2+2(k+1)-1\right)}{12}$

29. Basis Step: $n=1$, $2(1)-1=1$ and $1^2=1$;

Induction Step: If $1+3+5+7+\cdots+2k-1=k^2$, then $(1+3+5+7+\cdots+2k-1)+\left(2(k+1)-1\right)=k^2+(2k+1)$
$=(k+1)^2$

31. Basis Step: $n=1$, $\left(a^m\right)^1=a^m$ and $a^{m\cdot 1}=a^m$;

Induction Step: If $\left(a^m\right)^k=a^{mk}$, then $\left(a^m\right)^{k+1}=\left(a^m\right)^k\cdot\left(a^m\right)^1=a^{mk}\cdot a^m=a^{mk+m}\cdot a^{m(k+1)}$

33. Basis Step: $n=1$, $\left(2^{2(1)-1}+3^{2(1)-1}\right)=5$ which is a factor of 5;

Induction Step: If $\left(2^{2(k)-1}+3^{2(k)-1}\right)=5p$ for some integer p, then $\left(2^{2(k+1)-1}+3^{2(k+1)-1}\right)=\left(2^{2k+1}+3^{2k+1}\right)$

$=2^2\cdot 2^{2k-1}+3^2\cdot 3^{2k-1}=4\cdot 2^{2k-1}+9\cdot 3^{2k-1}=4\cdot 2^{2k-1}+(4+5)\cdot 3^{2k-1}=4\cdot 2^{2k-1}+4\cdot 3^{2k-1}+5\cdot 3^{2k-1}$

$=4\left(2^{2k-1}+3^{2k-1}\right)+5\cdot 3^{2k-1}=4(5p)+5\cdot 3^{2k-1}=5\left(4p+3^{2k-1}\right)$

35. Basis Step: $n=1$, $\left(1^3+3(1)^2+2(1)\right)=6=3\cdot 2$, so 3 is a factor;

Induction Step: If $\left(k^3+3k^2+2k\right)=3p$ for some integer p, then $(k+1)^3+3(k+1)^2+2(k+1)$

$=k^3+6k^2+11k+6=\left(k^3+3k^2+2k\right)+\left(3k^2+9k+6\right)=3p+3\left(k^2+3k+2\right)=3\left(p+k^2+3k+2\right)$

37. Basis Step: $n=1$, $5^1-1=4$, which is divisible by 4;

Induction Step: If $\dfrac{5^k-1}{4}=p$ or $5^k-1=4p$ for some integer p, then $5^{k+1}-1=5\cdot 5^k-5+4=5\left(5^k-1\right)+4$

$=5(4p)+4=4(5p+1)$

39. $1+2+4+8+\cdots+2^{n-1}=2^n-1$; Basis Step: $n=1$, $2^{1-1}=1$ and $2^1-1=1$;

Induction Step: If $1+2+4+8+\cdots+2^{k-1}=2^k-1$, then $\left(1+2+4+8+\cdots+2^{k-1}\right)+2^{k+1-1}=\left(2^k-1\right)+2^k$

$=2\cdot 2^k-1=2^{k+1}-1$

41. Induction Steps: If $k=2t+5n$, then...

Case 1 $(t\ge 1, n\ge 1,$ and $k=2t+5n)$: $k+1=2t+5n+1=2t+5n+(6-5)=(2t+6)+(5n-5)=2(t+3)+5(n-1)$

Case 2 $(t=0, n\ge 1,$ and $k=5n)$: $k+1=5n+1=5n+(6-5)=6+(5n-5)=2(3)+5(n-1)$

Case 3 $(t\ge 2, n=0,$ and $k=2t)$: $k+1=2t+1=2t+(5-4)=(2t-4)+5=2(t-2)+5(1)$

section 11.5, pages 960-966

1. $10^3=1000$ **3.** $9^7=4{,}782{,}969$ **5.** $15!\approx 1.308\times 10^{12}$ **7.** $3!=6$ **9.** $5^{10}=9{,}765{,}625$

11. $36^6=2{,}176{,}782{,}336$ **13.** $26\cdot 25\cdot 24\cdot 10\cdot 9\cdot 8=11{,}232{,}000$ **15.** ${}_{30}P_{12}\approx 4.143\times 10^{16}$ **17.** ${}_{36}P_8\approx 1.220\times 10^{12}$ **19.** ${}_7P_6=5040$, ${}_7P_7=5040$ as well (Having a child remain standing is numerically equivalent to putting a seventh chair in the room.) **21.** ${}_{26}P_3=15{,}600$ **23.** 210 **25.** 60,822,550,204,416,000 **27.** 6 **29.** 116,280

31. ${}_7C_3=35$ **33.** ${}_9C_2=36$ **35.** ${}_{75}C_5=17{,}259{,}390$ **37.** $\dfrac{6!}{2!3!}=60$ **39.** ${}_{10}C_4\cdot{}_8C_4\cdot{}_{13}C_4=10{,}510{,}500$

41. 112 cones **43.** 96 outfits **45.** 288 schedules **47.** 120 5-letter strings **49.** 303,600 ways **51.** 495 pizzas **53.** 752,538,150 groups **55.** 420 ways

57. $x^7-14x^6y+84x^5y^2-280x^4y^3+560x^3y^4-672x^2y^5+448xy^6-128y^7$ **59.** $x^8-4x^6y^3+6x^4y^6-4x^2y^9+y^{12}$

61. $4096x^6+30{,}720x^5y^2+96{,}000x^4y^4+160{,}000x^3y^6+150{,}000x^2y^8+75{,}000xy^{10}+15{,}625y^{12}$

63. $x^{15}-5x^{12}y^2+10x^9y^4-10x^6y^6+5x^3y^8-y^{10}$ **65.** $a^3-6a^2b+12ab^2-8b^3+3a^2c-12abc+12b^2c+3ac^2-6bc^2+c^3$

67. $8x^3+36x^2y-12x^2z+54xy^2-36xyz+6xz^2+27y^3-27y^2z+9yz^2-z^3$ **69.** –80

71. $8192x^{13}+159{,}744x^{12}+1{,}437{,}696x^{11}$ **73.** $2{,}008{,}326{,}144x^{14}$ **75.** $831{,}409{,}920x^4y^8$

77. $\dbinom{n}{n-k}=\dfrac{n!}{(n-k)!(n-(n-k))!}=\dfrac{n!}{(n-k)!(n-n+k)!}=\dfrac{n!}{(n-k)!k!}=\dbinom{n}{k}$

eleven

79. $2^n = (1+1)^n = \sum_{k=0}^{n}\binom{n}{k}(1)^k(1)^{n-k} = \sum_{k=0}^{n}\binom{n}{k} = \binom{n}{0}+\binom{n}{1}+\ldots+\binom{n}{n}$

section 11.6, pages 977-981

1. $\frac{3}{5}$ **3.** $\frac{9}{13}$ **5.** $\frac{1}{3}$ **7. a.** 0 **b.** $\frac{5}{8}$ **9. a.** 0 **b.** $\frac{3}{5}$ **11. a.** 0 **b.** 1 **13. a.** $\frac{1}{8}$ **b.** $\frac{9}{16}$

15. The set of all ordered 4-tuples made up of *H*'s and *T*'s. There are 16 such 4-tuples.

17. The set of all ordered pairs that have either an *H* or a *T* in the first slot and one of the 13 hearts in the second slot. There are 26 such ordered pairs.

19. The set of all ordered triples with any of the 6 values in each slot. There are 216 such triples.

21. The set of the 38 pockets. **23. a.** $\frac{2}{3}$ **b.** $\frac{1}{3}$ **25. a.** $\frac{3}{8}$ **b.** $\frac{1}{8}$ **c.** $\frac{1}{2}$ **27.** $\frac{3}{10}$ **29.** 0.3874 **31.** $\frac{3}{8}$

33. a. $\frac{11}{26}$ **b.** $\frac{9}{52}$ **c.** $\frac{2}{13}$ **35. a.** $\frac{1}{169}$ **b.** $\frac{1}{221}$ **37.** 0.1875 **39. a.** $\frac{1}{6}$ **b.** $\frac{5}{18}$ **c.** $\frac{1}{9}$ **41.** $\frac{1}{20}$ **43.** $\frac{2}{9}$

45. $\frac{3}{10}$ **47.** 27 tickets

chapter eleven test, pages 987-989

1. $a_n = 6n-13$ **3.** $a_n = n^2-1$ **5.** $-2+4-8+16-32+64=42$ **7.** $a_n=-3n+8$ **9.** $a_n=-9n+20$

11. –7140 **13.** 40,955 **15.** does not converge

17. Basis Step: $n=1$, $1^2=1$ and $\frac{1(1+1)(2(1)+1)}{6}=1$;

Induction Step: If $1+4+9+\cdots+k^2=\frac{k(k+1)(2k+1)}{6}$, then $(1+4+9+\cdots+k^2)+(k+1)^2$

$=\frac{k(k+1)(2k+1)}{6}+(k+1)^2=\frac{(k+1)\left[2k^2+k+6(k+1)\right]}{6}=\frac{(k+1)\left[2k^2+7k+6\right]}{6}=\frac{(k+1)(k+2)(2k+3)}{6}$

$=\frac{(k+1)((k+1)+1)(2(k+1)+1)}{6}$

19. Basis Step: $n=1$, $(3(1)+2)=5$ and $\frac{1(3(1)+7)}{2}=5$;

Induction Step: If $5+8+11+\cdots+(3k+2)=\frac{k(3k+7)}{2}$, then $5+8+11+\cdots+(3k+2)+(3(k+1)+2)$

$=\frac{k(3k+7)}{2}+(3k+5)=\frac{3k^2+7k+6k+10}{2}=\frac{(k+1)(3k+10)}{2}=\frac{(k+1)(3(k+1)+7)}{2}$

21. Basis Step: $n=1$, $11^1-7^1=4$, which is divisible by 4;

Induction Step: If $\frac{11^k-7^k}{4}=p$ or $11^k-7^k=4p$ for some integer p, then $11^{k+1}-7^{k+1}=11\cdot 11^k-7\cdot 7^k$

$=4\cdot 11^k+7\cdot 11^k-7\cdot 7^k=4\cdot 11^k+7\left(11^k-7^k\right)=4\cdot 11^k+7(4p)=4\left(11^k+7p\right)$

eleven

23. $8 \cdot 10^3 \cdot 26^3 = 140,608,000$ **25.** $\frac{8!}{3!2!} = 3360$ **27.** ${}_{21}P_5 = 2,441,880$

29. $x^3 + 6x^2y + 3x^2z + 12xy^2 + 3xz^2 + 12xyz + 8y^3 + 12y^2z + 6yz^2 + z^3$ **31.** $\frac{33}{108,290}$

applications

Note: **Examples are bold;** *chapter projects are italicized.*

biology and life science

business applications

chemistry

combinatorics

construction

consumer applications

demographics

distance

education

engineering/physics

geometry

interest rate applications

miscellaneous

probability

sound

sports/recreation

time

velocity

weather

technology

Mathmatica Commands

Technology

TI-83 Plus

index

d

e

f

g

h

i

j

k

l

m

n

O

P

q

r

s

t

u

vectors in the cartesian plane

magnitude of a vector

The length of a vector is called its **magnitude** and is denoted $\|\mathbf{u}\|$.

$$\|\mathbf{u}\| = \sqrt{a^2 + b^2}$$

where a is the horizotal displacement and b is the vertical displacement of the vector.

properties of vector addition and scalar multiplication

Assume **u**, **v**, and **w** represent vectors, while a and b represent scalars. Then the following hold:

Vector Addition Properties	Scalar Multiplication Properties
$\mathbf{u}+\mathbf{v}=\mathbf{v}+\mathbf{u}$	$a(\mathbf{u}+\mathbf{v}) = a\mathbf{u} + a\mathbf{v}$
$\mathbf{u}+(\mathbf{v}+\mathbf{w}) = (\mathbf{u}+\mathbf{v})+\mathbf{w}$	$(a+b)\mathbf{u} = a\mathbf{u} + b\mathbf{u}$
$\mathbf{u}+\mathbf{0}=\mathbf{u}$	$(ab)\mathbf{u} = a(b\mathbf{u}) = b(a\mathbf{u})$
$\mathbf{u}+(-\mathbf{u})=\mathbf{0}$	$1\mathbf{u}=\mathbf{u}$, $0\mathbf{u}=\mathbf{0}$, and $a\mathbf{0}=\mathbf{0}$
	$\|a\mathbf{u}\| = \lvert a\rvert\|\mathbf{u}\|$

vector operations using components

Given two vectors $\mathbf{u} = \langle u_1, u_2\rangle$ and $\mathbf{v} = \langle v_1, v_2\rangle$ and a scalar a, $\mathbf{u}+\mathbf{v} = \langle u_1+v_1, u_2+v_2\rangle$, $a\mathbf{u} = \langle au_1, au_2\rangle$, and $\|\mathbf{u}\| = \sqrt{u_1^2 + u_2^2}$.

dot product

Given two vectors $\mathbf{u} = \langle u_1, u_2\rangle$ and $\mathbf{v} = \langle v_1, v_2\rangle$, the **dot product** $\mathbf{u}\cdot\mathbf{v}$ of the two vectors is the scalar defined by

$$\mathbf{u}\cdot\mathbf{v} = u_1v_1 + u_2v_2.$$

elementary properties of the dot product

Given two vectors $\mathbf{u} = \langle u_1, u_2\rangle$ and $\mathbf{v} = \langle v_1, v_2\rangle$ and a scalar a, the following hold:

1. $\mathbf{u}\cdot\mathbf{v} = \mathbf{v}\cdot\mathbf{u}$
2. $0\cdot\mathbf{u} = 0$
3. $\mathbf{u}\cdot(\mathbf{v}+\mathbf{w}) = \mathbf{u}\cdot\mathbf{v} + \mathbf{u}\cdot\mathbf{w}$
4. $a(\mathbf{u}\cdot\mathbf{v}) = (a\mathbf{u})\cdot\mathbf{v} = \mathbf{u}\cdot(a\mathbf{v})$
5. $\mathbf{u}\cdot\mathbf{u} = \|\mathbf{u}\|^2$

the dot product theorem

Let two nonzero vectors **u** and **v** be depicted so that their initial points coincide, and let θ be the smaller of the two angles formed by **u** and **v** (so $0 \le \theta \le \pi$). Then

$$\mathbf{u}\cdot\mathbf{v} = \|\mathbf{u}\|\|\mathbf{v}\|\cos\theta.$$

orthogonal vectors

Two nonzero vectors **u** and **v** are said to be **orthogonal** (or **perpendicular**) if $\mathbf{u}\cdot\mathbf{v} = 0$.

projection of u onto v

Let **u** and **v** be nonzero vectors. The **projection of u onto v** is the vector

$$\text{proj}_{\mathbf{v}}\mathbf{u} = \left(\frac{\mathbf{u}\cdot\mathbf{v}}{\|\mathbf{v}\|^2}\right)\mathbf{v}.$$

amplitude, period and phase shift

Given constants $a, b > 0$, and c, the functions $f(x) = a\sin(bx - c)$ and $g(x) = a\cos(bx - c)$ have **amplitude** $\lvert a\rvert$, **period** $\frac{2\pi}{b}$, and a **phase shift** of $\frac{c}{b}$. The left end-point of one cycle of either function is $\frac{c}{b}$ and the right end-point is $\frac{c}{b} + \frac{2\pi}{b}$.

rotation of axes

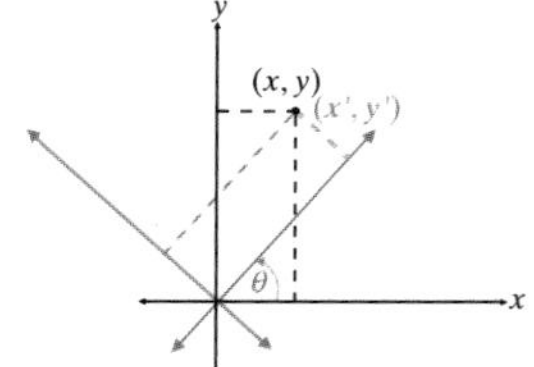

rotation relations

$$x = x'\cos\theta - y'\sin\theta \qquad x' = x\cos\theta + y\sin\theta$$
$$y = x'\sin\theta + y'\cos\theta \qquad y' = -x\sin\theta + y\cos\theta$$

elimination of *xy* term

The graph of the equation $Ax^2 + Bxy + Cy^2 + Dx + Ey + F = 0$ in the xy-plane is the same as the graph of the equation $A'x'^2 + C'y'^2 + D'x' + E'y' + F' = 0$ in the $x'y'$-plane, where the angle of rotation θ between the two coordinate systems satisfies $\cot 2\theta = \dfrac{A-C}{B}$.

polar form of conic sections

A **conic section** consists of all points P in the plane which satisfy the equation

$$\frac{D(P,F)}{D(P,L)} = e,$$

where e is a fixed positive constant. The conic is an ellipse if $0 < e < 1$, a parabola if $e = 1$, and a hyperbola if $e > 1$.

Conic Section Equation	Directrix
$r = \dfrac{ed}{1+e\cos\theta}$	Vertical Directrix $x = d$
$r = \dfrac{ed}{1-e\cos\theta}$	Vertical Directrix $x = -d$
$r = \dfrac{ed}{1+e\sin\theta}$	Horizontal Directrix $y = d$
$r = \dfrac{ed}{1-e\sin\theta}$	Horizontal Directrix $y = -d$

radian and degree measure

$180° = \pi$

$1° = \dfrac{\pi}{180}$ $\qquad$ $\left(\dfrac{180}{\pi}\right)^{\circ} = 1$

$x° = (x)\left(\dfrac{\pi}{180}\right)$ rad $\qquad$ $x \text{ rad} = x\left(\dfrac{180}{\pi}\right)^{\circ}$

$$s = \left(\frac{\theta}{2\pi}\right)(2\pi r)$$
$$= r\theta$$

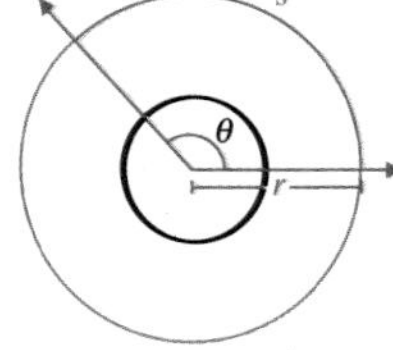

$$A = \left(\frac{\theta}{2\pi}\right)(\pi r^2) = \frac{r^2\theta}{2}$$

$\omega = \dfrac{\theta}{t}$ $\qquad$ $v = \dfrac{s}{t} = \dfrac{r\theta}{t} = r\omega$

trigonometric functions of acute angles

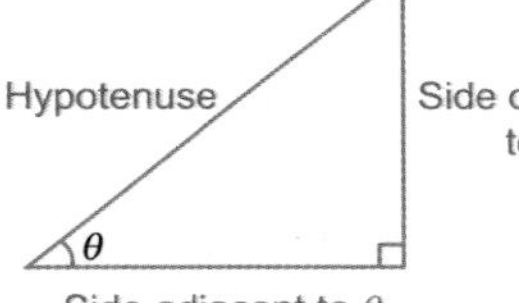

Side opposite to θ

$\sin\theta = \dfrac{\text{opp}}{\text{hyp}}$ $\qquad$ $\csc\theta = \dfrac{1}{\sin\theta} = \dfrac{\text{hyp}}{\text{opp}}$

$\cos\theta = \dfrac{\text{adj}}{\text{hyp}}$ $\qquad$ $\sec\theta = \dfrac{1}{\cos\theta} = \dfrac{\text{hyp}}{\text{adj}}$

$\tan\theta = \dfrac{\text{opp}}{\text{adj}}$ $\qquad$ $\cot\theta = \dfrac{1}{\tan\theta} = \dfrac{\text{adj}}{\text{opp}}$

trigonometric functions of any angle

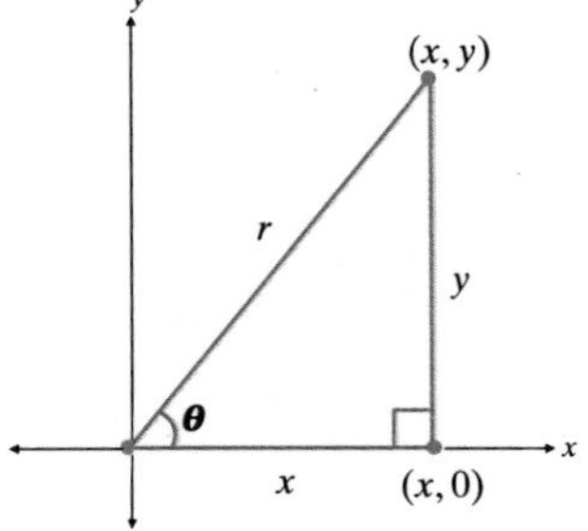

$\sin\theta = \dfrac{y}{r}$ $\qquad$ $\csc\theta = \dfrac{r}{y}$ (for $y \neq 0$)

$\cos\theta = \dfrac{x}{r}$ $\qquad$ $\sec\theta = \dfrac{r}{x}$ (for $x \neq 0$)

$\tan\theta = \dfrac{y}{x}$ (for $x \neq 0$) $\qquad$ $\cot\theta = \dfrac{x}{y}$ (for $y \neq 0$)

cofunction identities

$\sin x = \cos\left(\dfrac{\pi}{2} - x\right)$ $\qquad$ $\csc x = \sec\left(\dfrac{\pi}{2} - x\right)$

$\cos x = \sin\left(\dfrac{\pi}{2} - x\right)$ $\qquad$ $\sec x = \csc\left(\dfrac{\pi}{2} - x\right)$

$\cot x = \tan\left(\dfrac{\pi}{2} - x\right)$ $\qquad$ $\tan x = \cot\left(\dfrac{\pi}{2} - x\right)$

reciprocal identities

$\csc x = \dfrac{1}{\sin x}$ $\qquad$ $\sec x = \dfrac{1}{\cos x}$ $\qquad$ $\cot x = \dfrac{1}{\tan x}$

quotient identities

$\tan x = \dfrac{\sin x}{\cos x}$ $\qquad$ $\cot x = \dfrac{\cos x}{\sin x}$

period identities

$\sin(x + 2\pi) = \sin(x)$ $\qquad$ $\csc(x + 2\pi) = \csc(x)$

$\cos(x + 2\pi) = \cos(x)$ $\qquad$ $\sec(x + 2\pi) = \sec(x)$

$\tan(x + \pi) = \tan(x)$ $\qquad$ $\cot(x + \pi) = \cot(x)$

even/odd identities

$\sin(-x) = -\sin x$ $\qquad$ $\csc(-x) = -\csc x$

$\cos(-x) = \cos x$ $\qquad$ $\sec(-x) = \sec x$

$\tan(-x) = -\tan x$ $\qquad$ $\cot(-x) = -\cot x$

pythagorean identities

$\sin^2 x + \cos^2 x = 1$ $\qquad$ $\tan^2 x + 1 = \sec^2 x$ $\qquad$ $1 + \cot^2 x = \csc^2 x$

commonly encountered angles

θ	Radians	Sin θ	Cos θ	Tan θ
0°	0	0	1	0
30°	$\frac{\pi}{6}$	$\frac{1}{2}$	$\frac{\sqrt{3}}{2}$	$\frac{\sqrt{3}}{3}$
45°	$\frac{\pi}{4}$	$\frac{\sqrt{2}}{2}$	$\frac{\sqrt{2}}{2}$	1
60°	$\frac{\pi}{3}$	$\frac{\sqrt{3}}{2}$	$\frac{1}{2}$	$\sqrt{3}$
90°	$\frac{\pi}{2}$	1	0	—
180°	π	0	–1	0
270°	$\frac{3\pi}{2}$	–1	0	—